NEW ENGLAND INSTITUTE
OF TECHNOLOGY
LEARNING RESOURCES CENTER

# ASM
# MATERIALS
# ENGINEERING
# DICTIONARY

# ASM
# MATERIALS
# ENGINEERING
# DICTIONARY

Edited by

J.R. Davis
Davis & Associates

**The Materials
Information Society**

Library of Congress Catalog Card No.: 92-73858
ISBN: 0-87170-447-1
SAN: 204-7586


Text design and production coordination by
Kathleen Mills Editorial & Production Services

Cover design by Alpha Group


PRINTED IN THE UNITED STATES OF AMERICA

# Contents

# Preface

ASM has a rich tradition in the preparation and publication of definitions related to materials engineering terms. In fact, the growth and changes associated with the Society are mirrored by similar growth in the scope of its various glossaries of engineering terms. For example, the first collection of definitions published by ASM can be found in the 1939 Edition of *Metals Handbook*. At that time, ASM was primarily an organization devoted to heat treating of steels, and the modest 200 terms defined in the 1939 volume were related to ferrous metallurgy. During the 1940s, ASM's technological base expanded to both ferrous and nonferrous materials. When the classic 1948 Edition of *Metals Handbook* was published, the database of metallurgical terms and definitions had grown to approximately 900 entries, and contained all new material on nonferrous metallurgy and powder metallurgy as well as revisions to previously published ferrous terms.

The next significant development came in 1961 with the publication of Volume 1 of the 8th Edition of *Metals Handbook*. The definitions relating to metals and metalworking in this volume numbered some 2500 entries. Continued refinements and additions made in the 1960s, 1970s, and early 1980s culminated in the "Glossary of Terms Related to Metals and Metalworking" published in the *Metals Handbook Desk Edition* in 1985. This glossary included more than 3000 specialized technical terms encountered in metallurgical literature. Included were terms from materials science, physical metallurgy, heat treating, extractive metallurgy, casting, forging, machining, forming, welding and joining, metal cleaning and finishing, electrometallurgy, powder metallurgy, mechanical testing, nondestructive inspection and quality control, metallography, fractography, and failure analysis.

New growth and challenges were posed in the mid-1980s when the American Society for Metals changed its name to ASM International to reflect both its broad-based constituency and its expansion in technical scope to embrace all engineered materials and related processing. To meet the ever-increasing information needs of ASM members, the *Metals Handbook* was expanded to 17 volumes and a significant new book series—the *Engineered Materials Handbook*—was initiated. These concurrent Handbook series resulted in dramatic growth in the database of definitions related to engineered (both metallic and nonmetallic) materials.

The *ASM Materials Engineering Dictionary* resulted from the desire to assemble all the previously published collections of terms into a single format. Although the many glossaries published in the ASM Handbooks may be considered the nucleus for the Dictionary, they are by no means the sole source. In the present volume, much new matter has been introduced and the opportunity has been taken to make revisions and/or additions to existing terms. Every effort has been made to produce a dictionary that is reliable and up-to-date, and it is hoped that the form in which it is presented will prove convenient and handy for reference.

It should be noted, however, that the *ASM Materials Engineering Dictionary* is not a mere collection of words. The approximately 7500 terms are supplemented by more than 700 illustrations and 250 tables. Because ASM is a "materials society," expanded coverage of key material groups has been provided in the form of "Technical Briefs." The 64 Technical Briefs were designed to provide concise overviews of the properties, compositions, and applications of selected materials as well as to direct the reader to more detailed information; the latter goal was accomplished by the inclusion of a recommended reading list in each brief.

In order to produce a volume of such wide coverage, the editor has, of necessity, had to rely upon the help of some very close colleagues. I would like to thank Robert C. Uhl, the Director of Reference Publications at ASM, for his willingness to proceed with the project. My thanks are also extended to my dear friend Kathleen Mills for her creativity in designing the text and for her exacting approach to page layout and other production-related responsibilities. For their perseverance in typing and proofreading the manuscript and for their general editorial assistance, I also say thank you to Heather F. Lampman and Barbara E. Helmrich. Lastly, I must acknowledge the thousands of contributors to the *ASM Handbook* series from whom I drew material. In preparing the Dictionary, I was constantly reminded of how significant their efforts have been to the continued success of ASM International.

*Joseph R. Davis*
Davis & Associates
Chagrin Falls, Ohio

# Guide to the Dictionary

## ALPHABETIZATION

The terms in the *ASM Materials Engineering Dictionary* are alphabetized on a letter-by-letter basis; word spacing, hyphens, commas, and apostrophes in a term are ignored in the sequencing. For example, an ordering of terms would be:

**ABA polymers**
**A-basis**
**aberration**
**ablation**
**ablative plastic**
**ABL bottle**
**abrasion**

## FORMAT

The basic format for an entry provides the term in boldface and the definition in lightface:

**term**. Definition.

A term may be followed by multiple definitions, each introduced by a parenthetical number:

**term**. (1) Definition. (2) Definition. (3) Definition.

A term may have definitions in two or more materials science fields. In these situations, the terms are listed separately with the field in parentheses:

**term** (metals). Definition.
**term** (plastics). Definition.

A simple cross-reference entry appears as:

**term**. See *another term*.

As with the example above, and the example below, italicized words within a definition should be referred to:

**oxyacetylene welding**. An *oxyfuel gas welding* process in which the fuel is acetylene.

In the above example, the user should also refer to the related term "*oxyfuel gas welding*."

A cross-reference may also appear in combination with definitions:

**term**. (1) Definition. (2) Definition. See *another term*.

A term may have a figure(s) or table(s) accompanying it. Figures are called out at the appropriate location within the definition and are placed in parentheses:

**jolt ramming**. Packing sand in a mold by raising and dropping the sand, pattern, and flask on a table. Jolt-type (Fig. 276), jolt squeezers, jarring machines, and jot rammers are machines using this principle.

Tables are normally called out at the end of the definition:

**forgeability**. Definition. Table 22 compares the forgeability of various alloy groups.

## CROSS-REFERENCING

A cross-reference entry directs the user to the defining entry. For example, the user looking up *friction coefficient* finds:

**friction coefficient**. See *coefficient of friction*.

The user then turns to the "C" terms for the definition.

Sometimes a cross-reference is made to a specific definition of a term that has multiple definitions:

**term**. Definition. See also *another term* (2).

In the above example, the second definition accompanying the referenced, italicized term should be referred to.

Cross-references are also made from variant spellings, acronyms, and abbreviations:

**PAN**. See *polyacrylonitrile*.
**PIC**. See *pressure-impregnation-carbonization*.
**P/M**. An acronym for *powder metallurgy*.
**PMMA**. See *polymethyl methacrylate*.

Figures and/or tables are cross-referenced as follows:

**end-quench hardenability test**. Definition. See also the figure accompanying the term *Jominy test*.
**end-relief**. Defined by the figure accompanying the term *single-point tool*.

## TECHNICAL BRIEFS

The 64 Technical Briefs in the dictionary are designed to provide expanded coverage of key material groups. Each

Technical Brief is one or two pages in length, has accompanying figures and/or tables, and a recommended reading list. Technical Briefs are cross-referenced as follows:

> **aluminum and aluminum alloys**. See *Technical Brief 4*.

Generally, Technical Briefs fall within one or two pages from the listed term. The *List of Technical Briefs* on page xxv provides page number assignments for all briefs.

## ALSO KNOWN AS..., etc.

A definition may conclude with a mention of a synonym of the term, a variant spelling, an abbreviation for the term, or other such information, introduced by "Also known as ...," Also spelled ...," "Abbreviated ...," "Symbolized ...," "Compare with ...," and "Contrast with ...." When a term has more than one definition, the positioning of any of these phrases conveys the extent of applicability. For example:

> **term**. (1) Definition. Also known as synonym. (2) Definition. Abbreviated AB.

In the above example, "Also known as ..." applies only to the first definition; "Abbreviated ..." applies only to the second definition.

## CHEMICAL FORMULAS

Chemistry definitions may include either an empirical formula (e.g., for alumina, $Al_2O_3$) or a line formula (e.g., for acrylic acid, $CH_2CHCOOH$), whichever is appropriate. Numbers in metal alloy chemistry definitions, e.g., 70Cu-30Zn, refer to the weight percent of the elements in the alloy.

# List of Figures

An additional 25 figures accompany the various Technical Briefs in this volume (see the *List of Technical Briefs*).

# List of Figures

# List of Figures

# List of Figures

# List of Figures

# List of Figures

# List of Figures

# List of Tables

An additional 71 tables accompany the various Technical Briefs in this volume (see the *List of Technical Briefs*). See also the unnumbered in-text tables accompanying the terms *cast-alloy tool* (p 60), *electromagnetic radiation* (p 139), *karat* (p 241), *Mohs hardness* (p 281), *quartz glass* (p 352), *sieve analysis* (p 416), *tape* (p 473), and *wrought and cast aluminum alloy designations* (p 532).

# List of Technical Briefs

The letters F and T signify that information on the topic will be found in a figure(s) or table(s), respectively.

# List of Technical Briefs

# ASM
# MATERIALS
# ENGINEERING
# DICTIONARY

# A

**A.** The symbol for a repeating unit in a polymer chain.

**$A_{cm}$, $A_1$, $A_3$, $A_4$.** Same as $Ae_{cm}$, $Ae_1$, $Ae_3$, and $Ae_4$.

**ABA copolymers.** Block copolymers with three sequences, but only two domains.

**A-basis.** The "A" mechanical property value is the value above which at least 99% of the population of values is expected to fall, with a confidence of 95%. Also called A-allowable. See also *B-basis*, *S-basis*, and *typical-basis*.

**aberration.** In microscopy, any error that results in image degradation. Such errors may be chromatic, spherical, astigmatic, comatic, distortion, or curvature of field and can result from design or execution, or both.

**abhesive.** A material that resists adhesion. A film or coating applied to surfaces to prevent sticking, heat sealing, and so on, such as a parting agent or mold release agent.

**ablation.** A self-regulating heat and mass transfer process in which incident thermal energy is expended by sacrificial loss of material.

**ablative plastic.** A material that absorbs heat (with a low material loss and char rate) through a decomposition process (pyrolysis) that takes place at or near the surface exposed to the heat. This mechanism essentially provides thermal protection (insulation) of the subsurface materials and components by sacrificing the surface layer. Ablation is an exothermic process.

**ABL bottle.** An internal pressure test vessel about 460 mm (18 in.) in diameter and 610 mm (24 in.) long used to determine the quality and properties of the filament-wound material in the vessel.

**abrasion.** (1) A process in which hard particles or protuberances are forced against and moved along a solid surface. (2) A roughening or scratching of a surface due to *abrasive wear*. (3) The process of grinding or wearing away through the use of abrasives.

**abrasion artifact.** A false structure introduced during an abrasion stage of a surface-preparation sequence.

**abrasion fluid.** A liquid added to an abrasion system. The liquid may act as a lubricant, as a coolant, or as a means of flushing abrasion debris from the abrasion track.

**abrasion process.** An abrasive machining procedure in which the surface of the workpiece is rubbed against a two-dimensional array of abrasive particles under approximately constant load.

**abrasion rate.** The rate at which material is removed from a surface during abrasion. It is usually expressed in terms of the thickness removed per unit of time or distance traversed.

**abrasion resistance.** The ability of a material to resist surface wear.

**abrasion soldering.** A soldering process variation in which the faying surface of the base metal is mechanically abraded during soldering.

**abrasive.** (1) A hard substance used for *grinding*, *honing*, *lapping*, *superfinishing*, *polishing*, pressure blasting, or *barrel finishing*. Abrasives in common use are alumina, silicon carbide, boron carbide, diamond, cubic boron nitride, garnet, and quartz. (2) Hard particles, such as rocks, sand or fragments of certain hard metals, that wear away a surface when they move across it under pressure. See also *superabrasives* (Technical Brief 51).

**abrasive belt.** A coated abrasive product, in the form of a belt, used in production grinding and polishing.

**abrasive blasting.** A process for cleaning or finishing by means of an abrasive directed at high velocity against the workpiece.

**abrasive disk.** (1) A grinding wheel that is mounted on a steel plate, with the exposed flat side being used for grinding. (2) A disk-shaped, coated abrasive product.

**abrasive erosion.** Erosive wear caused by the relative motion of solid particles which are entrained in a fluid, moving nearly parallel to a solid surface. See also *erosion*.

**abrasive flow machining.** Removal of material by a viscous, abrasive media flowing under pressure through or across a workpiece.

**abrasive jet machining.** Material removal from a workpiece, by impingement of fine abrasive particles which are entrained in a focused, high-velocity gas stream.

**abrasive machining.** A machining process in which the points of abrasive particles are used as machining tools. Grinding is a typical abrasive machining process.

**abrasive tumbling.** See *barrel finishing*.

**abrasive wear.** The removal of material from a surface when hard particles slide or roll across the surface under pressure (Fig. 1). The particles may be loose or may be part of another surface in contact with the surface being abraded. Compare with *adhesive wear*.

**abrasive wheel.** A grinding wheel composed of an abrasive grit and a bonding agent.

**abrasivity.** The extent to which a surface, particle, or collection of particles will tend to cause *abrasive wear* when

FIG. 1

*Abrasive wear of the surface of 1020 steel abraded by 220 grit SiC paper, showing characteristic grooves and attached, tiny cutting chips*

forced against a solid surface under relative motion and under prescribed conditions.

**ABS**. See *acrylonitrile-butadiene-styrenes*.

**absolute density**. See *density, absolute*.

**absolute humidity**. The weight of water vapor present in a unit volume of air, such as grams per cubic foot, or grams per cubic meter. The amount of water vapor is also reported in terms of weight per unit weight of dry air, such as grams per pound of dry air, but this value differs from values calculated on a volume basis and should not be referred to as absolute humidity. It is designated as humidity ratio, specific humidity, or moisture content.

**absolute impact velocity**. See *impact velocity*.

**absolute pore size**. The maximum pore opening of a porous material, such as a filter, through which no large particle will pass.

**absolute viscosity**. See *viscosity*.

**absorbance** (*A*). The logarithm to the base 10 of the reciprocal of the transmittance. The preferred term for photography is optical density.

**absorption**. (1) The taking up of a liquid or gas by capillary, osmotic, or solvent action. (2) The capacity of a solid to receive and retain a substance, usually a liquid or gas, with the formation of an apparently homogeneous mixture. (3) Transformation of radiant energy to a different form of energy by interaction with matter. (4) The process by which a liquid is drawn into and tends to fill permeable pores in a porous solid body; also, the increase in mass of a porous solid body resulting from the penetration of a liquid into its permeable pores. See also *adsorption*.

**absorption contrast**. In transmission electron microscopy, image contrast caused by differences in absorption within a sample due to regions of different mass density and thickness.

**absorption edge**. The wavelength or energy corresponding to a discontinuity in the plot of absorption coefficient versus wavelength for a specific medium.

**absorption spectroscopy**. The branch of spectroscopy treating the theory, interpretation, and application of spectra originating in the absorption of electromagnetic radiation by atoms, ions, radicals, and molecules.

**absorptive lens** (eye protection). A filter lens whose physical properties are designed to attenuate the effects of glare and reflected and stray light. See also *filter plate*.

**absorptivity**. A measure of radiant energy from an incident beam as it traverses an absorbing medium, equal to the absorbance of the medium divided by the product of the concentration of the substance and the sample path length. Also known as absorption coefficient.

**Ac$_{cm}$, Ac$_1$, Ac$_3$, Ac$_4$**. Defined under *transformation temperature*.

**AC**. See *acetal (AC) copolymers*, *acetal (AC) homopolymers*, and *acetal (AC) resins*.

**accelerated aging**. A process by which the effects of aging are accelerated under extreme and/or cycling temperature and humidity conditions. The process is meant to duplicate long-time environmental conditions in a relatively short space of time.

**accelerated corrosion test**. Method designed to approximate, in a short time, the deteriorating effect under normal long-term service conditions.

**accelerated-life test**. A method designed to approximate, in a short time, the deteriorating effect obtained under normal long-term service conditions. See also *artificial aging*.

**accelerated testing**. A test performed on materials or assemblies that is meant to produce failures caused by the same failure mechanism as expected in field operation but in significantly shorter time. The failure mechanism is accelerated by changing one or more of the controlling test parameters.

**accelerating potential**. (1) A relatively high voltage applied between the cathode and anode of an electron gun to accelerate electrons. (2) The potential in electron beam welding that imparts the velocity to the electrons, thus giving them energy.

**accelerating voltage**. In various electron beam instruments and x-ray generators, the difference in potential between the filament (cathode) and the anode, causing acceleration of the electrons by 2 to 30 keV. See also *depth of penetration* and *resolution*.

**acceleration period**. In cavitation and liquid impingement erosion, the stage following the incubation period, during which the erosion rate increases from near zero to a maximum value.

**accelerator**. A material that, when mixed with a catalyst or a resin, speeds up the chemical reaction between the catalyst and the resin (usually in the polymerizing of resins or vulcanization of rubbers). Also called promoter.

**acceptable quality level**. (1) The lowest quality level a supplier is permitted to present continually for acceptance. (2) The maximum percentage of defects or number of defective parts considered to be an acceptable average for a given process or technique.

**acceptable weld**. A weld that meets all the requirements and the acceptance criteria prescribed by the welding specifications.

**acceptance test**. A test, or series of tests, conducted by the procuring agency, or an agent thereof, upon receipt, to determine whether an individual lot of materials conforms to the purchase order or contract or to determine the degree of uniformity of the material supplied by the vendor, or both. Compare to *preproduction test* and *qualification test*.

**accepted reference value**. A value that serves as an agreed-on reference for comparison and which is derived as: (1) a theoretical or established value, based on scientific principles, (2) an assigned value, based on experimental work of some national or international standards organization, or (3) a consensus value, based on collaborative experimental work under the auspices of a scientific or engineering group. When the accepted reference value is the theoretical value, it is sometimes referred to as the "true" value.

**access hole**. A hole or series of holes in successive layers of a multilayer printed circuit board that provide access to the surface of the land in one of the layers of the board.

**accessory seal**. On various types of engines, a seal that is employed for sealing an accessory shaft in the gear box, such as a shaft for operating an oil pump, a fuel pump, a generator, a starter, or a de-oiler.

**accordion**. A type of printed circuit connector contact in which the spring is given a Z shape to permit high deflection without causing overstress.

**accumulation period**. See preferred term *acceleration period*.

**accumulator**. An auxiliary cylinder and piston (plunger) mounted on injection molding or blowing machines and used to provide faster molding cycles. In blow molding, the accumulator cylinder is filled (during the time between parison deliveries, or "shots") with melted plastic coming from the main (primary) extruder. The plastic melt is stored, or "accumulated," in this auxiliary cylinder until the next shot or parison is required. At that time the piston in the accumulator cylinder forces the molten plastic into the dies that form the parison.

**accuracy**. (1) The agreement or correspondence between an experimentally determined value and an accepted reference value for the material undergoing testing. The reference value may be established by an accepted standard (such as those established by ASTM), or in some cases the average value obtained by applying the test method to all the sampling units in a lot or batch of the material may be used. (2) The extent to which the result of a calculation or the reading of an instrument approaches the true value of the calculated or measured quantity. Compare with *precision*.

**acetal (AC) copolymers**. A family of highly crystalline thermoplastics prepared by copolymerizing trioxane with small amounts of a comonomer that randomly distributes carbon-carbon bonds in the polymer chain. These bonds, as well as hydroxyethyl terminal units, give the acetal copolymers a high degree of thermal stability and resistance to strong alkaline environments.

**acetal (AC) homopolymers**. Highly crystalline linear polymers formed by polymerizing formaldehyde and capping it with acetate end groups.

**acetal (AC) resins**. Thermoplastics (polyformaldehyde and polyoxymethylene resins) produced by the *addition polymerization* of aldehydes by means of the carbonyl function, yielding unbranched polyoxymethylene chains of great length. The acetal resins, among the strongest and stiffest of all thermoplastics, are also characterized by good fatigue life, resilience, low moisture sensitivity, high solvent and chemical resistance, and good electrical properties. They may be processed by conventional injection molding and extrusion techniques and fabricated by welding methods used for other plastics.

**achromatic**. Free of color. A lens or objective is achromatic when corrected for longitudinal chromatic aberration for two colors. See also *achromatic objective*.

**achromatic lens**. A lens that is corrected for chromatic aberration so that its tendency to refract light differently as a function of wavelength is minimized. See also *achromatic* and *apochromatic lens*.

**achromatic objective**. Objectives are achromatic when corrected chromatically for two colors, generally red and green, and spherically for light of one color, usually in the yellow-green portion of the spectrum.

**acicular alpha** (titanium). A product of nucleation and growth from β to the lower temperature allotrope α phase. It may have a needlelike appearance in a photomicrograph (Fig. 2) and may have needle, lenticular, or flattened bar morphology in three dimensions. See also *alpha*.

**acicular ferrite**. A highly substructured nonequiaxed *ferrite* formed upon continuous cooling by a mixed diffusion and shear mode of transformation that begins at a temperature slightly higher than the transformation temperature range for upper bainite. It is distinguished from *bainite* in that it has a limited amount of carbon available; thus, there is only a small amount of carbide present.

**acicular ferrite steels**. Ultralow carbon (<0.08%) steels having a microstructure consisting of either acicular ferrite (low-carbon bainite) or a mixture of acicular and equiaxed ferrite.

FIG. 2

*Acicular alpha (light) and aged transformed beta in a titanium forging. 500×*

**acicular powder.** A powder composed of needle or sliverlike particles.

**acid.** A chemical substance that yields hydrogen ions ($H^+$) when dissolved in water. Compare with *base*. (2) A term applied to slags, refractories, and minerals containing a high percentage of silica.

**acid-acceptor.** A compound that acts as a stabilizer by chemically combining with acid that may be initially present in minute quantities in a plastic, or that may be formed by the decomposition of the resin.

**acid bottom and lining.** The inner bottom and lining of a melting furnace, consisting of materials like sand, siliceous rock, or silica brick that give an acid reaction at the operating temperature.

**acid copper.** (1) Copper electrodeposited from an acid solution of a copper salt, usually copper sulfate. (2) The solution referred to in (1).

**acid core solder.** See *cored solder*.

**acid embrittlement.** A form of *hydrogen embrittlement* that may be induced in some metals by acid.

**acid extraction.** Removal of phases by dissolution of the matrix metal in an acid. See also *extraction*.

**acid process.** A steelmaking method using an acid refractory-lined furnace. Neither sulfur nor phosphorus is removed.

**acid rain.** Atmospheric precipitation with a pH below 5.6 to 5.7. Burning of fossil fuels for heat and power is the major factor in the generation of oxides of nitrogen and sulfur, which are converted into nitric and sulfuric acids washed down in the rain. See also *atmospheric corrosion*.

**acid refractory.** Siliceous ceramic materials of a high melting temperature, such as silica brick, used for metallurgical furnace linings. Compare with *basic refractories*. *See also refractories* (Technical Brief 37).

**acid steel.** Steel melted in a furnace with an *acid bottom and lining* and under a slag containing an excess of an acid substance such as silica.

**ac noncapacitive arc.** A high-voltage electrical discharge used in spectrochemical analysis to vaporize the sample material. See also *dc intermittent noncapacitive arc*.

**acoustic emission.** A measure of integrity of a material, as determined by sound emission when a material is stressed. Ideally, emissions can be correlated with defects and/or incipient failure.

**acrylics.** See *Technical Brief 1*.

**acrylonitrile.** A monomer with the structure ($CH_2$:CHCN). It is most useful in copolymers. Its copolymer with butadiene is nitrile rubber; acrylonitrile-butadiene copolymers with styrene (SAN) are tougher than polystyrene. Acrylonitrile is also used as a synthetic fiber and as a chemical intermediate.

**acrylonitrile-butadiene-styrenes (ABS).** See *Technical Brief 2*.

**actinic.** Of light, characterized by radiation that causes chemical changes, for example, the effect of light on photographic emulsions. Blue and ultraviolet are the most actinic regions of the spectrum.

**actinide metals.** The group of radioactive elements of atomic numbers 89 through 103 of the periodic system—namely, thorium, protactinium, uranium, neptunium, plutonium, americium, curium, berkelium, californium, einsteinium, fermium, mendelevium, nobelium, and lawrencium.

**activated rosin flux.** A rosin- or resin-base flux containing an additive which increases wetting by the solder.

**activated sintering.** The use of additives, such as chemical additions to the powder or additions to the sintering atmosphere, to improve the densification rate.

**activating.** A treatment that renders nonconductive material receptive to electroless deposition. Also called seeding, catalyzing, and sensitizing, all of which are not preferred terms.

**activation.** (1) The changing of a passive surface of a metal to a chemically active state. Contrast with *passivation*. (2) The (usually) chemical process of making a surface more receptive to bonding with a coating or an encapsulating material.

**activation analysis.** A method of chemical analysis based on the detection of characteristic radionuclides following nuclear bombardment. See also *neutron activation analysis*.

**activation energy.** The energy required for initiating a metallurgical reaction—for example, plastic flow, diffusion, chemical reaction. The activation energy may be calculated from the slope of the line obtained by plotting the natural log of the reaction rate versus the reciprocal of the absolute temperature.

**activator.** The additive used in activated sintering, also called a dopant.

**active.** The negative direction of *electrode potential*. Also used to describe corrosion and its associated potential

## Technical Brief 1: Acrylics

ACRYLIC PLASTICS comprise a broad array of polymers and copolymers in which the major monomeric constituents belong to two families of esters-acrylates and methacrylates. These are used singly or in combination. Hard, clear acrylic sheet is made from methyl methacrylate, whereas molding and extrusion pellets are made from methyl methacrylate copolymerized with small percentages of other acrylates or methacrylates.

The resins produced may be in the form of casting syrups (for cast sheet) or pellets for molding and extrusion. The latter are made either in bulk (continuous solution polymerization), followed by extrusion and pelletizing, or continuously by polymerization in an extruder in which unconverted monomer is removed under reduced pressure and recovered for recycling.

Straight, or unmodified, grades of acrylic plastic are noted for their outstanding optical properties and weatherability. Colorless acrylic plastic is as transparent as the finest plate glass and is capable of giving almost complete transmittance of visible light.

### Nonimpact-modified acrylic grade applications

| Market | Application Injection-molded | Extruded |
|---|---|---|
| Automotive......... | Signal light devices for traffic, aircraft, marine, and bus use; emergency flashers; nameplates and emblems; automotive instrument panels; automotive glazing | ... |
| Medical.......... | Blood cuvettes; medical spikes; urine meters | ... |
| Industrial ........ | Display shelving; video discs; molded letters for signs; lighting diffusers and louvers; HID refractors; instrument panel covers | Lighting; signs; tubing; display shelving and enclosures; mill shapes |
| Consumer........ | Drinking tumblers; faucet knobs; household and bathroom accessories; stationery accessories; camera, projection and viewer lenses; lighting | Furniture components; manometers |
| Miscellaneous ..... | Christmas decorations; drinking tumblers and food service sets | |

Acrylic plastics have outstanding resistance to the effects of sunlight and exposure to the elements over long periods of time. They do not yellow significantly nor do they undergo any significant changes in physical properties. Most of the transparent, translucent, and opaque colors of acrylic have the same outstanding resistance to weathering.

Impact-modified acrylic grades have toughness up to 20 times that of unmodified acrylics. Butadiene-modified grades have the greatest toughness but are not as transparent as acrylic-modified grades. In addition to toughness, the acrylic-modified grades resist changes due to weathering better than do most thermoplastics.

### Impact-modified acrylic grade applications

| Market | Application Injection-molded | Extruded |
|---|---|---|
| Industrial ..................; | Lighting; display shelving; meter covers | Signs; lighting louvers; glazing; display shelving; profiles |
| Medical................... | Bone marrow mixer; medical spikes; chest drainage units; surgical instruments | ... |
| Consumer appliances........ | Refrigerator trays; shelving; microwave oven doors | |
| Consumer miscellaneous .... | Bird feeders; picture frames; cassettes | |
| Consumer recreation........ | ... | Sanitaryware (rigidized and coextrusion); swimming pool panels; spas (rigidized and coextrusion); recreational vehicle components (rigidized) |
| Automotive and marine ..... | Signal light devices; boat windows | |

### Recommended Reading

* R.T. Cassidy, Acrylics, *Engineered Materials Handbook*, Vol 2, ASM International, 1988, p 103-108

range when an electrode potential is more negative than an adjacent depressed corrosion rate (passive) range.

**active area.** In electronic packaging, the internal area of a package bottom, usually a cavity, that is used for substrate attachment. The term is applied preferably to package cases of all-metal construction (as opposed to glass or ceramic).

**active components.** Electronic components, such as transistors, diodes, electron tubes, and thyristors, that can operate on an applied electrical signal in such a way as to change

## Technical Brief 2: Acrylonitrile-Butadiene-Styrene (ABS)

ACRYLONITRILE-BUTADIENE-STYRENE (ABS) resins belong to a very versatile family of engineering thermoplastics, produced by combining three monomers—acrylonitrile, butadiene, and styrene—by a variety of methods involving polymerization, graft polymerization, physical mixing, and combinations thereof. Each monomer is an important component of ABS. Acrylonitrile contributes heat resistance, chemical resistance, and surface hardness to the system. The styrene component contributes processibility, rigidity, and strength. Butadiene contributes toughness and impact resistance.

ABS plastics are two-phase systems. Styrene-acrylonitrile (SAN) forms the continuous matrix phase. The second phase is composed of dispersed polybutadiene particles, which have a layer of SAN grafted onto their surface. The layer of SAN at the interface makes the two phases compatible.

ABS products are grouped into two major divisions: injection molding and extrusion. The primary difference is the melt viscosity, which is significantly lower for injection molding products. Within each division are corresponding product classes. Standard ABS products are grouped by impact strength into medium-, high-, and very-high-impact grades. Versions that provide low surface gloss, high surface gloss, and ultrahigh surface gloss are available. Specialty ABS products include high-heat, plating, clear, flame-retardant, and structural-foam grades. ABS can also be used in alloys and blends. ABS plastics are used in thousands of applications, including appliances, automotive, and construction (pipes).

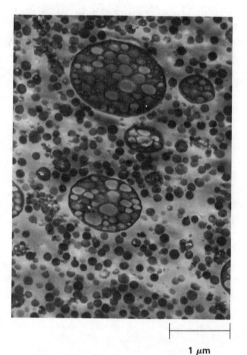

1 μm

*Typical ABS plastic with rubber particles (dark areas) dispersed in the SAN matrix (light background areas)*

## Recommended Reading

- C.A. Johnson and G.B. Hilton, Acrylonitrile-Butadiene-Styrenes, *Engineered Materials Handbook*, Vol 2, ASM International, 1988, p 109-114
- R. Juran, Ed., *Modern Plastics Encyclopedia*, Vol 64 (No. 10A), McGraw-Hill, 1987
- J.M. Margolis, Ed., *Engineering Thermoplastics*, Marcel Decker, 1985

### Typical mechanical and thermal properties of standard ABS grades

| Property | Medium impact | High impact | Very high impact |
|---|---|---|---|
| **Mechanical** | | | |
| Tensile strength at yield per ASTM D 638, MPa (ksi) | 45 (6.5) | 39 (5.6) | 32 (4.7) |
| Tensile modulus per ASTM D 638, GPa ($10^6$ psi) .... | 2.5 (0.36) | 2.2 (0.32) | 1.8 (0.26) |
| Flexural strength at yield per ASTM D 790, MPa (ksi) | 76 (11) | 66 (9.5) | 54 (7.8) |
| Flexural modulus per ASTM D 790, GPa ($10^6$ psi) ... | 2.8 (0.4) | 2.2 (0.32) | 1.8 (0.26) |
| Izod impact strength per ASTM D 256A, 3.2 mm (0.13 in.) bar J/m (ft lbf/in.) | | | |
| At 23 °C (73 °F) ............................ | 160 (3.0) | 270 (5.0) | 400 (7.5) |
| At −40 °C (−40 °F) ......................... | 59 (1.1) | 75 (1.4) | 120 (2.2) |
| Rockwell hardness per ASTM D 785, method A ..... | R108–R118 | R102–R113 | R90–R100 |
| Specific gravity per ASTM D 792 ................. | 1.03–1.07 | 1.01–1.05 | 1.01–1.04 |
| **Thermal** | | | |
| Deflection temperature under load per ASTM D 648, °C (°F) | | | |
| At 0.46 MPa (0.066 ksi) ....................... | 99 (210) | 100 (212) | 93 (200) |
| At 1.8 MPa (0.26 ksi) ......................... | 121 (220) | 106 (222) | 89 (192) |
| Coefficient of linear thermal expansion per ASTM D 696, $10^{-5}$/K ...................... | 9 | 10 | 11 |
| UL temperature index, UL 746C, °C (°F) .......... | 60–75 (140–167) | 60–75 (140–167) | 60 (140) |

its basic characteristics, for example, rectification, amplification, and switching.

**active devices.** Parts of a circuit that are capable of amplification, usually silicon semiconductor devices. Transistors, for instance, are active devices. Components that cannot amplify are passive—for example, resistors and capacitors.

**active metal.** A metal ready to corrode, or being corroded.

**active potential.** The *potential* of a corroding material.

**activity.** A measure of the *chemical potential* of a substance, where the chemical potential is not equal to concentration, that allows mathematical relations equivalent to those for ideal systems to be used to correlate changes in an experimentally measured quantity with changes in chemical potential.

**activity (ion).** The ion concentration corrected for deviations from ideal behavior. Concentration multiplied by activity coefficient.

**activity coefficient.** A characteristic of a quantity expressing the deviation of a solution from ideal thermodynamic behavior; often used in connection with electrolytes.

**actual contact area.** In tribology, the total *area of contact* formed by summing the localized asperity contact areas within the *apparent area of contact*. Also known as *real area of contact*.

**actual slip.** See *macroslip*.

**actual throat.** See *throat of a fillet weld*.

**addition agent.** (1) A substance added to a solution for the purpose of altering or controlling a process. Examples: wetting agents in acid pickles; brighteners or antipitting agents in plating solutions; inhibitors. (2) Any material added to a charge of molten metal in a bath or ladle to bring the alloy to specifications.

**addition polymerization.** A chemical reaction in which simple molecules (monomers) are linked to each other to form long-chain molecules (polymers) by chain reaction.

**additive.** (1) In lubrication, a material added to a lubricant for the purpose of imparting new properties or of enhancing existing properties. Main classes of additives include *anticorrosive*, *antifoam*, *antioxidant*, *antiwear*, *detergent*, *dispersant*, *extreme-pressure*, and *VI improver additives*. (2) In polymer engineering, a substance added to another substance, usually to improve properties, such as plasticizers, initiators, light stabilizers, and flame retardants. See also *filler*.

**additive process.** A process for obtaining conductive patterns by the selective deposition of conductive material on clad or unclad base material. See also *semiadditive process*, *subtractive process*, and *fully additive process*.

**adhere.** To cause two surfaces to be held together by adhesion.

**adherence.** In tribology, the physical attachment of material to a surface (either by adhesion or by other means of attachment) that results from the contact of two solid surfaces undergoing relative motion. Adhesive bonding is not a requirement for adherence because mechanisms such as mechanical interlocking of asperities can also provide a means for adherence. See also *adhesion (adhesive force)* and *adhesion, mechanical*.

**adherend.** A body held to another body by an adhesive. See also *substrate*.

**adherend preparation.** See *surface preparation*.

**adhesion.** (1) In frictional contacts, the attractive force between adjacent surfaces. In physical chemistry, adhesion denotes the attraction between a solid surface and a second (liquid or solid) phase. This definition is based on the assumption of a reversible equilibrium. In mechanical technology, adhesion is generally irreversible. In railway engineering, adhesion often means friction. (2) Force of attraction between the molecules (or atoms) of two different phases. Contrast with *cohesion*. (3) The state in which two surfaces are held together by interfacial forces, which may consist of valence forces, interlocking action, or both. See also *mechanical adhesion* and *specific adhesion*.

**adhesion coefficient.** See *coefficient of adhesion*.

**adhesion promoter.** A coating applied to a substrate, before it is coated with an adhesive, to improve the adhesion of the substrate. Also called primer.

**adhesion promotion.** The chemical process of preparing a surface to provide for a uniform, well-bonded interface.

**adhesive.** A substance capable of holding materials together by surface attachment. Adhesive is a general term and includes, among others, cement, glue, mucilage, and paste. These terms are loosely used interchangeably. Various descriptive adjectives are applied to the term adhesive to indicate certain physical characteristics: *hot-melt adhesives* (Technical Brief 21); *pressure-sensitive adhesives* (Technical Brief 35); *structural adhesives* (Technical Brief 49); *ultraviolet/electron beam cured adhesives* (Technical Brief 58); and *water-based adhesives* (Technical Brief 61). The characteristics of these five groups of adhesives are summarized in Table 1. Table 2 lists the advantages and limitations of the five groups.

**adhesive, anaerobic.** See *anaerobic adhesive*.

**adhesive assembly.** A group of materials or parts, including adhesive, that are placed together for bonding or that have been bonded together. See also *assembly adhesive*.

**adhesive bond.** Attractive forces, generally physical in character, between an adhesive and the base materials. Two principal interactions that contribute to the adhesion are van der Waals bonds and dipole bonds. See also *van der Waals bond*.

**adhesive bonding.** A materials joining process in which an adhesive, placed between the faying surfaces (adherends) solidifies to produce an adhesive bond (Fig. 3).

## Table 1 Characteristics of various types of adhesives

| Structural | Hot melt | Pressure sensitive | Water base | Ultraviolet/electron beam cured |
|---|---|---|---|---|
| Bonds can be stressed to a high proportion of maximum failure load under service environments<br>Most are thermosets.<br>One- or two-component systems<br>Room- or elevated-temperature cures<br>Wide range of costs<br>Various chemical families with varying strengths and flexibilities | 100% solid thermoplastics<br>Melt sharply to a low-viscosity liquid, which is applied to surface<br>Rapid setting, no cure<br>Melt viscosity is an important property.<br>Nonpressure sensitive and pressure sensitive<br>Compounded with additives for tack and wettability | Hold substrates together upon brief application of pressure at room temperature<br>Available as organic-solvent-base, water-base, or hot-melt systems<br>Some require extensive compounding (rubber base) to achieve tackiness, whereas others (polyacrylates) do not.<br>Available supported (most) or unsupported on a substrate<br>Primarily used in tapes and labels | Includes adhesives dissolved or dispersed (latex) in water<br>On porous substrates, water is absorbed or evaporated in order to bond.<br>On nonporous substrates, water must be removed prior to bonding.<br>Some are bonded following reactivation of dried adhesive film under heat and pressure.<br>Many are based on natural (vegetable or animal) adhesives.<br>Nonpressure sensitive (most) or pressure sensitive applications | 100% reactive liquids cured to solids<br>One substrate must be transparent for UV cure, except when dual-curing adhesives are used (see below).<br>Some UV-curable formulations are dual curing; a second cure mechanism introduces heat or moisture or eliminates oxygen (anaerobics).<br>In EB curing, density of material affects penetration.<br>UV/EB-curable formulations have laminating and PSA applications.<br>UV-curable formulations have laminating, PSA, and structural adhesive applications. |

## Table 2 Advantages and limitations of various types of adhesives

| Structural | Hot melt | Pressure sensitive | Water base | Ultraviolet/electron beam cured |
|---|---|---|---|---|
| **Advantages** | | | | |
| High strength<br>Capable of resisting loads<br>Good elevated-temperature resistance (cross-linked)<br>Good solvent resistance<br>Good creep resistance<br>Some available in film form | 100% solids, no solvents<br>Can bond impervious surfaces<br>Rapid bond formation<br>Good gap-filling capability<br>Rigid to flexible bonds<br>Good barrier properties | Labels and tapes have uniform thickness.<br>Permanent tack at room temperature<br>No activation required by heat, water, or solvents<br>Cross-linking of some formulations possible<br>Soft or firm tapes and labels<br>Easy to apply | Low cost, nonflammable, nonhazardous solvent<br>Long shelf life<br>Easy to apply<br>Good solvent resistance<br>Cross-linking of some formulations possible<br>High molecular weight dispersions at high solids content with low viscosity | Fast cure (some in 2 to 60 s)<br>One-component liquid: no mixing, no solvents<br>Heat-sensitive substrates can be bonded; cure is "cool."<br>Many are optically clear.<br>High production rates<br>Good tensile strength |
| **Limitations** | | | | |
| Two-component systems require careful proportioning and mixing.<br>Some have poor peel strength.<br>Some are difficult to remove and repair.<br>Some require heat to cure.<br>Some yield by-products upon cure (condensation polymers). | Thermoplastics have limited elevated-temperature resistance.<br>Poor creep resistance<br>Little penetration due to fast viscosity increase upon cooling<br>Limited toughness at usable viscosities | Many are based on rubbers, requiring compounding.<br>Poor gap fillers<br>Limited heat resistance | Poor water resistance<br>Slow drying<br>Tendency to freeze<br>Low strength under loads<br>Poor creep resistance<br>Limited heat resistance<br>Shrinkage of certain substrates in supported films and tapes | Equipment expensive<br>High material cost<br>UV cures only through transparent materials (or secondary cure required).<br>Difficult curing on parts with complex shapes<br>Many UV cures have poor weatherability because they continue to absorb UV rays. |

FIG. 3

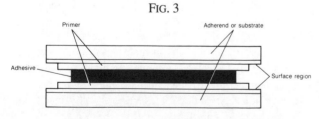

*Idealized adhesively bonded assembly*

**adhesive, cold-setting.** See *cold-setting adhesive.*

**adhesive, contact.** See *contact adhesive.*

**adhesive dispersion.** A two-phase system in which one phase is suspended in a liquid. Compare with *emulsion.*

**adhesive failure.** Rupture of an adhesive bond such that the separation appears to be at the adhesive-adherend interface. Sometimes termed failure in adhesion. Compare with *cohesive failure.*

**adhesive film.** A synthetic resin adhesive, with or without a film carrier fabric, usually of the thermosetting type, in the form of a thin film of resin, used under heat and pressure as an interleaf in the production of bonded structures.

**adhesive, gap-filling.** See *gap-filling adhesive.*

**adhesive, heat-activated.** See *heat-activated adhesive.*

**adhesive, heat-sealing.** See *heat-sealing adhesive.*

**adhesive, hot-melt.** See *hot-melt adhesive* (Technical Brief 21).

**adhesive, hot-setting.** See *hot-setting adhesive.*

**adhesive, intermediate-temperature-setting**. See *intermediate-temperature-setting adhesive*.

**adhesive joint**. Location at which two adherends are held together with a layer of adhesive. See also *bond*.

**adhesive, pressure-sensitive**. See *pressure-sensitive adhesive* (Technical Brief 35).

**adhesive strength**. The strength of the bond between an adhesive and an adherend.

**adhesive, structural**. See *structural adhesive* (Technical Brief 49).

**adhesive system**. An integrated engineering process that analyzes the total environment of a potential bonded assembly to select the most suitable adhesive, application method, and dispensing equipment.

**adhesive wear**. (1) Wear by transference of material from one surface to another during relative motion due to a process of solid-phase welding. Particles that are removed from one surface are either permanently or temporarily attached to the other surface. (2) Wear due to localized bonding between contacting solid surfaces leading to material transfer between the two surfaces or loss from either surface. Compare with *abrasive wear*.

**adiabatic**. Occurring with no addition or loss of heat from the system under consideration.

**adjustable bed**. Bed of a press designed so that the die space height can be varied conveniently.

**admixture**. (1) The addition and homogeneous dispersion of discrete components, before cure of a polymer. (2) A material other than water, aggregates, hydraulic cement, and fiber reinforcement used as an ingredient of concrete or mortar and added to the batch immediately before or during its mixing. (3) Material added to (cement) mortars as a water-repellent or coloring agent or to retard or hasten setting.

**adsorption**. The adhesion of the molecules of gases, dissolved substances, or liquids in more or less concentrated form, to the surfaces of solids or liquids with which they are in contact. The concentration of a substance at a surface or interface of another substance.

**adsorption chromatography**. Chromatography based on differing degrees of adsorption of sample compounds onto a polar stationary phase. See also *liquid-solid chromatography*.

**advanced ceramics**. Ceramic materials that exhibit superior mechanical properties, corrosion/oxidation resistance, or electrical, optical, and/or magnetic properties. This term includes many monolithic ceramics as well as particulate-, whisker-, and fiber-reinforced glass, glass-ceramics, and ceramic-matrix composites. Also known as engineering, fine, or technical ceramics. Contrast with *traditional ceramics*.

**advanced composites**. Composite materials that are reinforced with continuous fibers having a modulus higher than that of fiberglass fibers. The term includes metal-matrix and ceramic-matrix composites, as well as carbon-carbon composites.

**Ae$_{cm}$, Ae$_1$, Ae$_3$, Ae$_4$**. Defined under *transformation temperature*.

**aerate**. To fluff up molding sand to reduce its density.

**aerated bath nitriding**. A type of liquid nitriding in which air is pumped through the molten bath creating agitation and increased chemical activity.

**aeration**. (1) Exposing to the action of air. (2) Causing air to bubble through. (3) Introducing air into a solution by spraying, stirring, or a similar method. (4) Supplying or infusing with air, as in sand or soil.

**aeration cell (oxygen cell)**. See *differential aeration cell*.

**aerodynamic lubrication**. See *gas lubrication*.

**aerostatic lubrication**. See *pressurized gas lubrication*.

**AFS 50-70 test sand**. A rounded quartz sand (Fig. 4) specified for use as an abrasive in the dry sand-rubber wheel abrasive wear test (ASTM G 65). Using the U.S. Sieve Series, none of this sand will be retained on Sieve No. 40, 5% maximum will be retained on Sieve No. 50, 95% minimum will be retained on Sieve No. 70, and none will pass Sieve No. 100. This places all the particle diameters between 425 and 150 µm. See also *sieve analysis*.

FIG. 4

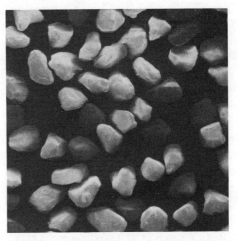

*AFS 50-70 test sand particles used in the ASTM dry sand/rubber wheel abrasion test*

**afterbake**. See *postcure*.

**age hardening**. Hardening by *aging* (heat treatment) usually after rapid cooling or cold working.

**age hardening (of grease)**. The increasing consistency of a lubricating grease with time of storage.

**age softening**. Spontaneous decrease of strength and hardness that takes place at room temperature in certain strain hardened alloys, especially those of aluminum.

**agglomerate**. The clustering together of a few or many particles, whiskers, or fibers, or a combination thereof, into a larger solid mass.

**aggregate**. (1) A dense mass of particles held together by strong intermolecular or atomic cohesive forces. (2) Granular material, such as sand, gravel, crushed stone, or iron blast-furnace slag, used with a cementing medium to form hydraulic-cement concrete or mortar. (3) A hard, coarse material usually of mineral origin used with an epoxy binder (or other resin) in plastic tools. Also used in flooring or as a surface medium.

**aggressive tack**. Synonym for *dry tack*.

**aging**. (1) The effect on materials of exposure to an environment for a prolonged interval of time. (2) The process of exposing materials to an environment for a prolonged interval of time in order to predict in-service lifetime. (3) Generally, the degradation of properties or function with time. In capacitors, the loss of dielectric constant, $K$, by dielectric relaxation. Expressed as a percent change per decade of time.

**aging** (heat treatment). A change in the properties of certain metals and alloys that occurs at ambient or moderately elevated temperatures after hot working or a heat treatment (quench aging in ferrous alloys, natural or artificial aging in ferrous and nonferrous alloys) or after a cold working operation (strain aging). The change in properties is often, but not always, due to a phase change (precipitation), but never involves a change in chemical composition of the metal or alloy. See also *age hardening, artificial aging, interrupted aging, natural aging, overaging, precipitation hardening, precipitation heat treatment, progressive aging, quench aging, step aging,* and *strain aging*.

**agitator**. A device to intensify mixing. Example: a high-speed stirrer or paddle in a blender or drum of a mill.

**air acetylene welding (AAW)**. A fuel gas welding process in which coalescence is produced by heating with a gas flame or flames obtained from the combustion of acetylene with air, without the application of pressure, and with or without the use of filler metal.

**air-assist forming**. A method of thermoforming in which air flow or air pressure is employed to preform plastic sheet partially just before the final pull-down onto the mold using vacuum.

**air bearing**. A bearing using air as a lubricant. See also *gas lubrication* and *pressurized gas lubrication*.

**air bend die**. Angle forming dies in which the metal is formed without striking the bottom of the die. Metal contact is made at only three points in the cross section: the nose of the male die and the two edges of a V-shape die opening.

**air bending**. Bending in an *air bend die*.

**air blasting**. See *blasting or blast cleaning*.

**air-bubble void**. Air entrapment within a molded item or

FIG. 5

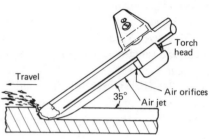

*Cutting action in air carbon arc cutting*

between the plies of reinforcement or within a bond line or encapsulated area; localized, noninterconnected, and spherical in shape.

**air cap** (thermal spraying). A device for forming, shaping, and directing an air pattern for the atomization of wire or ceramic rod.

**air carbon arc cutting (AAC)**. An arc cutting process in which metals to be cut are melted by the heat of a carbon arc and the molten metal is removed by a blast of air (Fig. 5).

**air channel**. A groove or hole that carries the vent from a core to the outside of a mold.

**air classification**. The separation of metal powder into particle-size fractions by means of an air stream of controlled velocity; an application of the principle of *elutriation*.

**air dried**. Refers to the air drying of a casting core or mold without the application of heat.

**air dried strength**. Strength (compressive, shear, or tensile) of a refractory (sand) mixture after being air dried at room temperature.

**air feed**. A *thermal spraying* process variation in which an air stream carries the powdered material to be sprayed through the gun and into the heat source.

**air furnace**. Reverberatory-type furnace in which metal is melted by heat from fuel burning at one end of the hearth, passing over the bath toward the stack at the other end. Heat is also reflected from the roof and sidewalls. See also *reverberatory furnace*.

**air gap**. In extrusion coating, the distance from the die opening to the nip formed by the pressure roll and the chill roll.

**air hammer**. In lubrication, a type of instability, basically a resonance, that occurs in externally pressurized gas bearings.

**air-hardening steel**. A steel containing sufficient carbon and other alloying elements to harden fully during cooling in air or other gaseous media from a temperature above its transformation range. The term should be restricted to steels that are capable of being hardened by cooling in air in fairly large sections, about 2 in. (50 mm) or more in diameter. Same as self-hardening steel.

FIG. 6

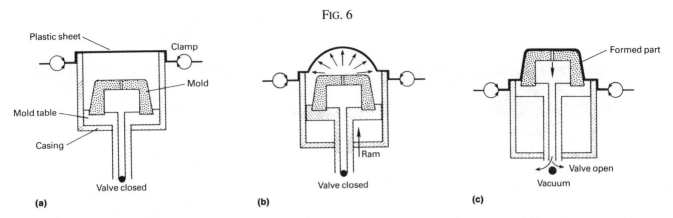

*Air-slip thermoforming. (a) Heated plastic sheet clamped to top of chamber. (b) Pressure buildup from moving mold stretches sheet up. (c) Mold moves to top of chamber. Released pressure allows sheet to drape over mold.*

**air hole.** A hole in a casting caused by air or gas trapped in the metal during solidification.

**air-lift hammer.** A type of gravity-drop hammer in which the ram is raised for each stroke by an air cylinder. Because length of stroke can be controlled, ram velocity and therefore the energy delivered to the workpiece can be varied. See also *drop hammer* and *gravity hammer.*

**air setting.** The characteristic of some materials, such as refractory cements, core pastes, binders, and plastics, to take permanent set at normal air temperatures.

**air-slip forming.** A variation of vacuum snap-back *thermoforming* in which the male mold is enclosed in a box such that when the mold moves forward toward the hot plastic, air is trapped between the mold and the plastic sheet. As the mold advances, the plastic is kept away from it by this air cushion until the full travel of the mold is completed, at which point a vacuum is applied, destroying the cushion and forming the part against the plug (Fig. 6).

**air vent.** A small outlet to prevent entrapment of gases in a molding or tooling fixture.

**alclad.** Composite wrought product comprised of an aluminum alloy core having one or both surfaces a metallurgically bonded aluminum or aluminum alloy coating that is anodic to the core and thus electrochemically protects the core against corrosion.

**alcohols.** Characterized by the hydroxyl (–OH) group they contain, alcohols are valuable starting points for the manufacture of synthetic resins, synthetic rubbers, and plasticizers.

**aldehydes.** Volatile liquids with sharp, penetrating odors that are slightly less soluble in water than are corresponding alcohols.

**algorithm.** A procedure for solving a mathematical problem by a series of operations. In *computed tomography*, the mathematical process used to convert the transmission measurements into a cross-sectional image. Also known as reconstruction algorithm.

**aligning bearing.** A bearing with an external spherical seat surface that provides a compensation for shaft or housing deflection or misalignment. Compare with *self-aligning bearing.*

**alignment.** A mechanical or electrical adjustment of the components of an optical device so that the path of the radiating beam coincides with the optical axis or other predetermined path in the system. See also *mechanical alignment* and *voltage alignment.*

**aliphatic hydrocarbons.** Saturated hydrocarbons having an open-chain structure, for example, gasoline and propane.

**aliquot.** A representative sample of a larger quantity.

**alkali metal.** A metal in group IA of the periodic system—namely, lithium, sodium, potassium, rubidium, cesium, and francium. They form strongly alkaline hydroxides, hence the name.

**alkaline.** (1) Having properties of an alkali. (2) Having a pH greater than 7.

**alkaline cleaner.** A material blended from alkali hydroxides and such alkaline salts as borates, carbonates, phosphates, or silicates. The cleaning action may be enhanced by the addition of surface-active agents and special solvents.

**alkaline earth metal.** A metal in group IIA of the period system—namely, beryllium, magnesium, calcium, strontium, barium, and radium—so called because the oxides or "earths" of calcium, strontium, and barium were found by the early chemists to be alkaline in reaction.

**alkyd.** Resin used in coatings. Reaction products of polyhydric alcohols and polybasic acids.

**alkyd plastic.** Thermoset plastic based on resins composed principally of polymeric esters, in which the recurring ester groups are an integral part of the main polymer chain, and in which ester groups occur in most cross links that may be present between chains.

**alkylation.** (1) A chemical process in which an alkyl radical is introduced into an organic compound by substitution or

FIG. 7

*Alligatoring in a rolled slab*

addition. (2) A refinery process for chemically combining isoparaffin with olefin hydrocarbons.

**alligatoring.** (1) Pronounced wide cracking over the entire surface of a coating having the appearance of alligator hide. (2) The longitudinal splitting of flat slabs in a plane parallel to the rolled surface (Fig. 7). Also called fishmouthing.

**alligator skin.** See *orange peel*.

**all-metal package.** A hybrid circuit package made solely of metal, excluding glass or ceramic. Its main applications are with microwave modules and large plug-ins.

**allophanate.** Reactive product of an isocyanate and the hydrogen atoms in a urethane.

**alloprene.** Chlorinated rubber.

**allotriomorphic crystal.** A crystal whose lattice structure is normal but whose external surfaces are not bounded by regular crystal faces; rather, the external surfaces are impressed by contact with other crystals or another surface such as a mold wall, or are irregularly shaped because of nonuniform growth. Compare with *idiomorphic crystal*.

**allotropy.** (1) A near synonym for *polymorphism*. Allotropy is generally restricted to describing polymorphic behavior in elements, terminal phases, and alloys whose behavior closely parallels that of the predominant constituent element. (2) The existence of a substance, especially an element, in two or more physical states (for example, crystals). See also *graphite*.

**allowance.** (1) The specified difference in limiting sizes (minimum clearance or maximum interference) between mating parts, as computed arithmetically from the specified dimensions and tolerances of each part. (2) In a foundry, the specified clearance. The difference in limiting sizes, such as minimum clearance or maximum interference between mating parts, as computed arithmetically. See also *tolerance*.

**alloy.** (1) A substance having metallic properties and being composed of two or more chemical elements of which at

## Table 3　Chemical compositions for typical alloy steels

| Steel | Composition, wt%(a) | | | | | | | | |
| --- | --- | --- | --- | --- | --- | --- | --- | --- | --- |
| | C | Si | Mn | P | S | Ni | Cr | Mo | Other |
| **Low-carbon quenched and tempered steels** | | | | | | | | | |
| A 514/A 517 grade A | 0.15–0.21 | 0.40–0.80 | 0.80–1.10 | 0.035 | 0.04 | · · · | 0.50–0.80 | 0.18–0.28 | 0.05–0.15 Zr(b) 0.0025 B |
| A 514/A 517 grade F | 0.10–0.20 | 0.15–0.35 | 0.60–1.00 | 0.035 | 0.04 | 0.70–1.00 | 0.40–0.65 | 0.40–0.60 | 0.03–0.08 V 0.15–0.50 Cu 0.0005–0.005 B |
| A 514/A 517 grade R | 0.15–0.20 | 0.20–0.35 | 0.85–1.15 | 0.035 | 0.04 | 0.90–1.10 | 0.35–0.65 | 0.15–0.25 | 0.03–0.08 V |
| A 533 type A | 0.25 | 0.15–0.40 | 1.15–1.50 | 0.035 | 0.04 | · · · | · · · | 0.45–0.60 | · · · |
| A 533 type C | 0.25 | 0.15–0.40 | 1.15–1.50 | 0.035 | 0.04 | 0.70–1.00 | · · · | 0.45–0.60 | · · · |
| HY-80 | 0.12–0.18 | 0.15–0.35 | 0.10–0.40 | 0.025 | 0.025 | 2.00–3.25 | 1.00–1.80 | 0.20–0.60 | 0.25 Cu 0.03 V 0.02 Ti |
| HY-100 | 0.12–0.20 | 0.15–0.35 | 0.10–0.40 | 0.025 | 0.025 | 2.25–3.50 | 1.00–1.80 | 0.20–0.60 | 0.25 Cu 0.03 V 0.02 Ti |
| **Medium-carbon ultrahigh-strength steels** | | | | | | | | | |
| 4130 | 0.28–0.33 | 0.20–0.35 | 0.40–0.60 | · · · | · · · | · · · | 0.80–1.10 | 0.15–0.25 | · · · |
| 4340 | 0.38–0.43 | 0.20–0.35 | 0.60–0.80 | · · · | · · · | 1.65–2.00 | 0.70–0.90 | 0.20–0.30 | · · · |
| 300M | 0.40–0.46 | 1.45–1.80 | 0.65–0.90 | · · · | · · · | 1.65–2.00 | 0.70–0.95 | 0.30–0.45 | 0.05 V min |
| D-6a | 0.42–0.48 | 0.15–0.30 | 0.60–0.90 | · · · | · · · | 0.40–0.70 | 0.90–1.20 | 0.90–1.10 | 0.05–0.10 V |
| **Carburizing bearing steels** | | | | | | | | | |
| 4118 | 0.18–0.23 | 0.15–0.30 | 0.70–0.90 | 0.035 | 0.040 | · · · | 0.40–0.60 | 0.08–0.18 | · · · |
| 5120 | 0.17–0.22 | 0.15–0.30 | 0.70–0.90 | 0.035 | 0.040 | · · · | 0.70–0.90 | · · · | · · · |
| 3310 | 0.08–0.13 | 0.20–0.35 | 0.45–0.60 | 0.025 | 0.025 | 3.25–3.75 | 1.40–1.75 | · · · | · · · |
| **Through-hardened bearing steels** | | | | | | | | | |
| 52100 | 0.98–1.10 | 0.15–0.30 | 0.25–0.45 | 0.025 | 0.025 | · · · | 1.30–1.60 | · · · | · · · |
| A 485 grade 1 | 0.90–1.05 | 0.45–0.75 | 0.95–1.25 | 0.025 | 0.025 | 0.25 | 0.90–1.20 | 0.10 | 0.35 Cu |
| A 485 grade 3 | 0.95–1.10 | 0.15–0.35 | 0.65–0.90 | 0.025 | 0.025 | 0.25 | 1.10–1.50 | 0.20–0.30 | 0.35 Cu |

(a) Single values represent the maximum allowable. (b) Zirconium may be replaced by cerium. When cerium is added, the cerium/sulfur ratio should be approximately 1.5/1, based on heat analysis.

least one is a *metal*. (2) To make or melt an alloy. (3) In plastics, a blend of polymers or copolymers with other polymers or elastomers under selected conditions; for example, styrene-acrylonitrile. Also called polymer blend.

**alloy cast iron**. See *high-alloy cast iron*.

**alloying element**. An element added to and remaining in a metal that changes structure and properties.

**alloy plating**. The codeposition of two or more metallic elements.

**alloy powder, alloyed powder**. A metal powder consisting of at least two constituents that are partially or completely alloyed with each other.

**alloy steel**. Steel containing specified quantities of alloying elements (other than carbon and the commonly accepted amounts of manganese, copper, silicon, sulfur, and phosphorus) within the limits recognized for constructional alloy steels, added to effect changes in mechanical or physical properties. Table 3 lists compositions of typical alloy steels.

**alloy system**. A complete series of compositions produced by mixing in all proportions any group of two or more components, at least one of which is a metal.

**all-position electrode**. In arc welding, a filler-metal electrode for depositing weld metal in the flat, horizontal, overhead, and vertical positions.

**all-weld-metal test specimen**. A test specimen wherein the portion being tested is composed wholly of weld metal.

**allyl plastic**. A thermoset plastic based on resins made by *addition polymerization* of monomers containing allyl groups; for example, diallyl phthalate (DAP).

**allyl resins**. A family of thermoset resins made by *addition polymerization* of compounds containing the group $CH_2$:$CH$–$CH_2$, such as esters of allyl alcohol and dibasic acids. They are available as monomers, partially polymerized prepolymers, or molding compounds. Members of the family are diallyl phthalate (DAP), diallyl isophthalate (DAIP), diallyl maleate (DAM), and diallyl chlorendate (DAC).

**alpha (α)**. The low-temperature allotrope of titanium with a hexagonal close-packed crystal structure that occurs below the β transus.

**alpha-beta structure**. A titanium microstructure containing α and β as the principal phases at a specific temperature (Fig. 8). See also *beta*.

**alpha brass**. A solid-solution phase of one or more alloying elements in copper having the same crystal lattice as copper.

**alpha case**. The oxygen-, nitrogen-, or carbon-enriched α-stabilized, surface in titanium resulting from elevated temperature exposure (Fig. 9). See also *alpha stabilizer*.

**alpha (α) cellulose**. A very pure cellulose prepared by special chemical treatment.

**alpha double prime (α″) (orthorhombic martensite)**. A su-

FIG. 8

*Titanium alloy forging with equiaxed grains of alpha (light) in a matrix of transformed beta (dark) containing fine acicular alpha. 250×*

FIG. 9

*Surface layer of white, oxygen-stabilized alpha (alpha case) in a titanium alloy. 450×*

FIG. 10

*Alpha double prime microstructure in a titanium alloy. 500×*

persaturated, nonequilibrium orthorhombic phase formed by a diffusionless transformation of the β phase in certain titanium alloys (Fig. 10).

**alpha ferrite.** See *ferrite*.

**alpha iron.** The body-centered cubic form of pure iron, stable below 910 °C (1670 °F).

**alpha (α) loss peak.** In dynamic mechanical or dielectric measurement, the first peak in the damping curve below the melt, in order of decreasing temperature or increasing frequency.

**Alpha model 1.** A type of wear-testing machine consisting of a conforming or flat-faced block pressed vertically downward by a deadweight loading arrangement against the circumference of a hardened steel ring that is rotating on a shaft.

**alpha prime (α′) (hexagonal martensite).** A supersaturated, nonequilibrium hexagonal α phase formed by a diffusionless transformation of the β phase in titanium alloys (Fig. 11). It is often difficult to distinguish from acicular α, although the latter is usually less well defined and frequently has curved, instead of straight, sides.

FIG. 11

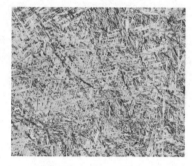

*Alpha prime microstructure in a titanium alloy forging. 100×*

**alpha process.** A *shell molding* and coremaking method in which a thin resin-bonded shell is baked with a less expensive, highly permeable material.

**alpha stabilizer.** An alloying element in titanium that dissolves preferentially in the α phase and raises the α-β transformation temperature.

**alpha transus.** The temperature that designates the phase boundary between the α and α + β fields in titanium alloys.

**alsifer.** A deoxidizer (20 Al, 40 Si, 40 Fe) used for steel.

**alternate immersion test.** A corrosion test in which the specimens are intermittently exposed to a liquid medium at definite time intervals.

**alternate polarity operation.** A resistance welding process variation in which succeeding welds are made with pulses of alternating polarity.

**alternating copolymer.** A copolymer in which each repeat-ing unit is joined to another repeating unit in the polymer chain (–A–B–A–B–).

**alternating current resistance.** The resistance offered by any circuit to the flow of alternating current.

**alternating stress amplitude.** A test parameter of a dynamic fatigue test; one-half the algebraic difference between the maximum and minimum stress in one cycle.

**Alumel.** A nickel-base alloy containing about 2.5 Mn, 2 Al, and 1 Si used chiefly as a component of pyrometric thermocouples.

**alumina.** See *Technical Brief 3*.

**aluminizing.** Forming of an aluminum or aluminum alloy coating on a metal by hot dipping, hot spraying, or diffusion.

**aluminum and aluminum alloys.** See *Technical Brief 4*.

**aluminum bomb.** A bomb-shaped container used in determining the oxygen content in liquid steel.

**aluminum nitride (AlN).** A high-thermal-conductivity ceramic used as an electronic substrate. Also a key component in the production of *sialons*.

**aluminum oxide (Al₂O₃).** See *alumina* (Technical Brief 3).

**amalgam.** A dental alloy produced by combining mercury with alloy particles of silver, tin, copper, and sometimes zinc.

**ambient.** The environment that surrounds and contacts a system or component.

**American wire gage (AWG).** A standard used in the determination of the physical size of a conductor determined by its circular-mil area.

**amine adduct.** Product of the reaction of an amine with a deficiency of a substance containing epoxy groups.

**amino.** Relating to or containing an –NH₂ or –NH group.

**amino resins.** Resins made by the polycondensation of a compound containing amino groups, such as urea or melamine, with an aldehyde, such as formaldehyde, or an aldehyde-yielding material. Melamine-formaldehyde and urea-formaldehyde resins are the most important family members. The resins can be dispersed in water to form colorless syrups. With appropriate catalysts, they can be cured at elevated temperatures.

**ammeter.** An instrument for measuring the magnitude of electric current flow.

**amorphous.** Not having a crystal structure; noncrystalline.

**amorphous plastic.** A plastic that has no crystalline component, no known order or pattern of molecule distribution, and no sharp melting point.

**amorphous powder.** A powder that consists of particles that are substantially noncrystalline in character.

**amorphous solid.** A rigid material whose structure lacks crystalline periodicity; that is, the pattern of its constituent atoms or molecules does not repeat periodically in three dimensions. See also *metallic glass*.

**ampere.** The unit used for measuring the quantity of an

## Technical Brief 3: Alumina

ALUMINA, also referred to as aluminum oxide ($Al_2O_3$), is one of the most commonly used ceramic materials. The natural crystalline mineral is called *corundum*, but the synthetic materials used for abrasive and ceramic materials which are obtained by heating hydrates of alumina, are designated usually as alumina or marketed under trade names. Alumina is widely distributed in nature in combination with silica and other minerals and is an important constituent of the clays used for making porcelain, bricks, pottery, and refractories.

The crushed and graded crystals/particles of alumina when pure are nearly colorless, but the fine powder is white. Off colors are due to impurities. For example, dark alumina is 96% $Al_2O_3$, whereas the white contains 99.5% $Al_2O_3$.

The chief uses for alumina are for the production of abrasives used to grind, clean, or polish materials and structural ceramics, but it is also used for refractories, pigments, catalyst carriers, and in chemicals. The reasons for its wide acceptance as a structural ceramic (it is the most widely used oxide-type ceramic) are

### Properties of various alumina ceramics

| Alumina content, % | Bulk density, g/cm$^3$ | Flexure strength, MPa (ksi) | Fracture toughness, MPa$\sqrt{m}$ (ksi$\sqrt{in.}$) | Hardness, GPa ($10^6$ psi) | Elastic modulus, GPa ($10^6$ psi) | Thermal conductivity, W/m · K (Btu/ft · h · °F) | Linear coefficient of thermal expansion, ppm/°C (ppm/°F) |
|---|---|---|---|---|---|---|---|
| 85 | 3.41 | 317 (46) | 3–4 (2.8–3.7) | 9 (1.3) | 221 (32) | 16.0 (9.24) | 7.2 (4) |
| 90 | 3.60 | 338 (49) | 3–4 (2.8–3.7) | 10 (1.5) | 276 (40) | 16.7 (9.65) | 8.1 (4.5) |
| 94 | 3.70 | 352 (51) | 3–4 (2.8–3.7) | 12 (1.7) | 296 (43) | 22.4 (12.9) | 8.2 (4.6) |
| 96 | 3.72 | 358 (52) | 3–4 (2.8–3.7) | 11 (1.6) | 303 (44) | 24.7 (14.3) | 8.2 (4.6) |
| 99.5 | 3.89 | 379 (55) | 3–4 (2.8–3.7) | 14 (2.0) | 372 (54) | 35.6 (20.6) | 8.0 (4.4) |
| 99.9 | 3.96 | 552 (80) | 3–4 (2.8–3.7) | 15 (2.2) | 386 (56) | 38.9 (22.5) | 8.0 (4.4) |

Source: Coors Ceramic Company

its high hardness, excellent wear and corrosion resistance, and low electrical conductivity. It is also fairly economical to manufacture, involving low-cost alumina powders that can be consolidated by a variety of methods (press and sinter, isostatic pressing, extrusion, etc.). See also *structural ceramics* (Technical Brief 50).

Alumina ceramics actually include a family of materials, typically having alumina contents from 85 to ≥99% $Al_2O_3$, the remainder being a grain-boundary phase. The different varieties of alumina stem from diverse application requirements. For example, 85% alumina ceramics, such as milling media, are used in applications requiring high hardness. Aluminas having purities in the 90 to 97% range are often found in electronic applications as substrate materials, due to their low electrical conductivity. The grain-boundary phase in these materials also allows for a strong bond between the ceramic and the metal conduction paths for integrated circuits. High-purity (>99%) alumina is often used in the production of translucent envelopes for sodium-vapor lamps. Zirconia-toughened aluminas, where alumina is considered the primary or continuous phase (70 to 95%), are also being considered for wear parts such as papermaking machine components.

### Recommended Reading

- W.D. Kingery, H.K. Bowen, and D.R. Uhlmann, *Introduction to Ceramics*, 2nd ed., John Wiley & Sons, 1976
- J.S. Reed, *Introduction to the Principles of Ceramic Processing*, John Wiley & Sons, 1988
- S.S. Schneider, Jr., Ed., *Ceramics and Glasses*, Vol 4, *Engineered Materials Handbook*, ASM International, 1991
- S. Somiya, Ed., *Technical Ceramics*, Academic Press, 1989

electric current flow. One ampere represents a flow of one coulomb per second.

**ampere-turns.** The product of the number of turns in an electromagnetic coil and the current in amperes passing through the coil.

**amperometry.** Chemical analysis by methods that involve measurements of electric currents.

**amphoteric.** A term applied to oxides and hydroxides which can act basic toward strong acids and acidic toward strong alkalis. Substances which can dissociate electrolytically to produce hydrogen or hydroxyl ions according to conditions.

**amplifier.** A negative lens used instead of an eyepiece to project under magnification the image formed by an objective. The amplifier is designed for flatness of field and should be used with an apochromatic objective.

## Technical Brief 4: Aluminum and Aluminum Alloys

ALUMINUM, also called aluminium in England, is a white metal with a bluish tinge obtained chiefly from bauxite. It is the second most abundant metallic element on earth. Aluminum metal is produced by first extracting alumina (aluminum oxide) from bauxite by a chemical process. The alumina is then dissolved in a molten electrolyte, and an electric current is passed through the bath, causing the metallic aluminum to be deposited on the cathode.

The unique combinations of properties provided by aluminum and its alloys make it suitable for a broad range of uses—from soft, highly ductile wrapping foil to the most demanding engineering applications. Its low density and strength-to-weight ratio are its most useful characteristics. It weighs only about 2.7 g/cm$^3$, approximately one-third as much as the same volume of steel, permitting design and construction of strong, lightweight structures—particularly advantageous for aerospace, aircraft, and land- and waterborne vehicles. Aluminum has high resistance to corrosion in atmospheric environments, in fresh and salt water, and in many chemicals. It has no toxic reactions and is highly suitable for processing, handling, storing, and packaging of foods and beverages. The high electrical and thermal conductivities of aluminum account for its use in many applications. It is also nonferromagnetic, a property of importance in the electrical and electronic industries.

Mill products constitute the major share (~80%) of total aluminum product shipments, followed by casting and ingot other than for castings. In decreasing order of current market size, the major application categories are containers and packaging (27.4%), transportation (21.1%), building and construction (17.8%), electrical (9.1%), consumer durables (8.0%), machinery and equipment (5.9%), and others, including exports (<12%).

In addition to being available in wrought and cast form, aluminum alloys are also produced by powder metallurgy processing. Aluminum is also used extensively in *metal-matrix composites*.

## Nominal compositions of selected aluminum alloys
See also *wrought and cast aluminum alloy designations*

| Alloy | Nominal composition, % | Alloy | Nominal composition, % | Alloy | Nominal composition, % |
|---|---|---|---|---|---|
| **Wrought aluminum alloys(a)** | | 7039 | Al-0.27Mn-2.8Mg-0.2Cr-4.0Zn | 380 (380) | Al-9.0Si-3.5Cu |
| | | 7072 | Al-1.0Zn | 384 (384) | Al-12.0Si-3.8Cu |
| 1100 | 0.12Cu-99.00Al (min) | 7075 | Al-1.6Cu-2.5Mg-0.3Cr-5.6Zn | 392 (392) | Al-19.0Si-0.6Cu-0.4Mn-1.0Mg |
| 1230 | 99.30Al (min) | 7079 | Al-0.6Cu-0.2Mn-3.3Mg-0.2Cr-4.3Zn | 413 (13) | Al-12.0Si |
| 2014 | Al-0.8Si-4.4Cu-0.8Mn-0.5Mg | 7178 | Al-2.0Cu-2.7Mg-0.3Cr-6.8Zn | 443 (43) | Al-5.0Si |
| 2024 | Al-4.4Cu-0.6Mn-1.5Mg | | | B443 (43) | Al-5.0Si-0.3Cu max |
| 2025 | Al-0.8Si-4.5Cu-0.8Mn | **Aluminum casting alloys(b)** | | C443 (A43) | Al-5.0Si-2.0Fe max |
| 2117 | Al-2.6Cu-0.35Mg | | | 520 (220) | Al-10.0Mg |
| 2218 | Al-4.0Cu-1.5Mg-2.0Ni | 201 (KO-1) | Al-4.7Cu-0.6Ag-0.3Mg-0.2Ti | D712 (D612, 40E) | Al-0.6Mg-5.3Zn-0.5Cr |
| 2219 | Al-6.3Cu-0.3Mn-0.06Ti-0.1V-0.18Zr | 222 (122) | Al-10.0Cu-0.2Mg | 850 (750) | Al-1.0Cu-1.0Ni-6.5Sn |
| 2618 | Al-2.3Cu-1.6Mg-1.0Ni-1.1Fe-0.07Ti | 224 (···) | Al-5.0Cu-0.4Mn | | |
| 3003 | Al-0.12Cu-1.2Mn | 238 (138) | Al-10.0Cu-4.0Si-0.3Mg | **Aluminum alloy filler metals and brazing alloys** | |
| 5052 | Al-2.5Mg-0.25Cr | A240 (A140) | Al-8.0Cu-0.5Mn-6.0Mg-0.5Ni | | |
| 5083 | Al-0.6Mn-4.45Mg-0.15Cr | 242 (142) | Al-4.0Cu-1.5Mg-2.0Ni | ER2319 | Al-6.2Cu-0.30Mn-0.15Ti |
| 5086 | Al-0.45Mn-4.0Mg-0.15Cr | 295 (195) | Al-4.5Cu-0.8Si | ER4043 | Al-5.2Si |
| 5454 | Al-0.8Mn-2.7Mg-0.12Cr | 308 (A108) | Al-4.5Cu-5.5Si | ER5356 | Al-0.12Mn-5.0Mg-0.12Cr-0.13Ti |
| 5456 | Al-0.8Mn-5.1Mg-0.12Cr | 319 (319) | Al-3.5Cu-6.0Si | 5456 | Al-0.8Mn-5.1Mg-0.12Cr |
| 5457 | Al-0.3Mn-1.0Mg | A332 (A132) | Al-12.0Si-0.8Cu-1.2Mg-2.5Ni | R-SG70A | Al-7Si-0.30Mg |
| 5657 | Al-0.8Mg | 354 (354) | Al-9.0Si-1.8Cu-0.5Mg | 4047 (BAlSi-4) | Al-12Si |
| 6061 | Al-0.6Si-0.27Cu-1.0Mg-0.2Cr | 355 (355) | Al-1.3Cu-5.0Si-0.5Mg | 4245 | Al-10Si-4Cu-10Zn |
| 6063 | Al-0.4Si-0.7Mg | 356 (356) | Al-7.0Si-0.3Mg | 4343 (BAlSi-2) | Al-7.5Si |
| 6151 | Al-0.9Si-0.6Mg-0.25Cr | A356 (A356) | Al-7.0Si-0.3Mg-0.2Fe max | No. 12 brazing sheet | 3003 alloy, 4343 |
| 6351 | Al-1.0Si-0.6Mn-0.6Mg | A357 (A357) | Al-7.0Si-0.5Mg-0.15Ti | | cladding on both sides |
| 7004 | Al-0.45Mn-1.5Mg-4.2Zn-0.15Zr | | | | |

(a) Wrought alloys are identified by Aluminum Association designations. (b) Casting alloys are identified first by Aluminum Association designations (without decimal suffixes) and then, parenthetically, by industry designations.

## Recommended Reading

- R.B.C. Cayless, Alloy and Temper Designation Systems for Aluminum and Aluminum Alloys, *Metals Handbook*, Vol 2, 10th ed., ASM International, 1990, p 15-28
- J.W. Bray, Aluminum Mill and Engineered Wrought Products, *Metals Handbook*, Vol 2, 10th ed., ASM International, 1990, p 29-61
- E.L. Rooy, Cast Aluminum and Aluminum Alloys, *Metals Handbook*, Vol 15, 9th ed., ASM International, 1988, p 743-770
- R.L. Horst *et al.*, Corrosion of Aluminum and Aluminum Alloys, *Metals Handbook*, Vol 13, 9th ed., ASM International, 1987, p 583-609

**Amsler wear machine.** A wear and traction-testing machine consisting of two disk-shaped specimens oriented such that their axes are parallel and whose circumferential, cylindrical surfaces are caused to roll or roll and slide against one another. The rotation rates of each disk may be varied so as to produce varying degrees of sliding and rolling motion.

**amylaceous.** Pertaining to, or of the nature of, starch; starchy.

**anaerobic adhesive.** An adhesive that cures only in the absence of air after being confined between assembled parts.

**analog circuits.** Circuits that provide a continuous (versus discontinuous) relationship between the input and output.

**analog computer.** A computer that processes data that are continuously variable in nature.

**analog-to-digital converter (ADC).** A device that converts a continuously variable electrical signal into discrete signals suitable for analysis by a digital computer.

**analysis.** The ascertainment of the identity or concentration, or both, of the constituents or components of a sample. See also *determination.*

**analyte.** In any analysis, the substance (element, ion, compound, and so on) being identified or determined.

**analytical chemistry.** The science of chemical characterization and measurement. Qualitative analysis is concerned with the description of chemical composition in terms of elements, compounds, or structural units; quantitative analysis is concerned with the precise measurement of amount. A variety of physical measurements are used, including methods based on spectroscopic, electrochemical, radiochemical, chromatographic, and nuclear principles.

**analytical curve.** The graphical representation of a relation between (1) the intensity of the response to measurement (for example, emission, absorbance, and conductivity) and (2) the concentration or mass of the substance being measured. The curve is generated by measuring the responses for standards of known concentration. Also termed standard curve or working curve.

**analytical electron microscopy (AEM).** The technique of materials analysis in the transmission electron microscope equipped to detect and quantify many different signals from the specimen. The technique usually involves a combination of imaging, chemical analysis, and crystallographic analysis by diffraction at high spatial resolution.

**analytical gap.** The region between two electrodes in which the sample is excited in the sources used for emission spectroscopy and spark source mass spectrometry.

**analytical line.** In spectroscopy, the particular spectral line of an element used in the identification or determination of the concentration of that element.

**analytical wavelength.** In spectroscopy, the particular wavelength used for the identification or determination of the concentration of an element or compound.

**analyzer.** An optical device, capable of producing plane polarized light, used for detecting the state of polarization.

**anchorage.** Part of an insert that is molded inside the plastic and held fast by the shrinkage of the plastic.

**anchorite.** A zinc-iron phosphate coating for iron and steel.

**anchor pattern.** A pattern made by blast cleaning abrasives on an adherend surface in preparation for adhesive application prior to bonding. Pattern is examined in profile.

**andalusite.** A mineral of composition $Al_2O_3 \cdot SiO_2$ used in the production of aluminosilicate bricks for use in blast furnaces, steel ladles, and torpedo ladles.

**anelastic deformation.** Any portion of the total deformation of a body that occurs as a function of time when load is applied and which disappears completely after a period of time when the load is removed.

**anelasticity.** The property of solids by virtue of which strain is not a single-value function of stress in the low-stress range where no permanent set occurs.

**angle of attack.** In tribology, the angle between the direction of motion of an impinging liquid or solid particle and the tangent to the surface at the point of impact.

**angle of bend.** The angle between the two legs of the specimen after bending is completed. It is measured before release of the bending force, unless otherwise specified.

**angle of bevel.** See preferred term *bevel angle.*

**angle of bite.** In the rolling of metals, the location where all of the force is transmitted through the rolls; the maximum attainable angle between the roll radius at the first contact and the line of roll centers (Fig. 12). Operating angles less than the angle of bite are termed contact angles or rolling angles.

FIG. 12

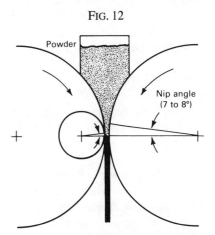

*Angle of bite (nip angle) during rolling*

**angle of contact.** In a ball race, the angle between a diametral plane perpendicular to a ball-bearing axis and a line drawn between points of tangency of the balls to the inner and outer rings (Fig. 13).

**angle of incidence.** (1) In tribology, the angle between the direction of motion of an impinging liquid or solid particle

FIG. 13

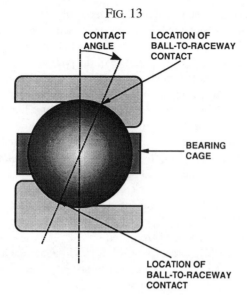

*Angle of contact in an angular-contact bearing*

FIG. 14

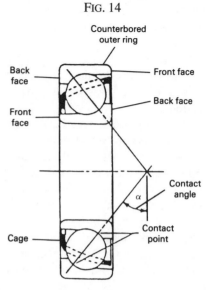

*Angular-contact bearing*

and the normal to the surface at the point of impact. (2) In materials characterization, the angle between an incident radiant beam and a perpendicular to the interface between two media.

**angle of nip.** In rolling, the *angle of bite*. In roll, jaw, or gyratory crushing, the entrance angle formed by the tangents at the two points of contact between the working surfaces and the (assumed) spherical particles to be crushed.

**angle of reflection.** (1) Reflection: the angle between the reflected beam and the normal to the reflecting surface. See also *normal*. (2) Diffraction: the angle between the diffracted beam and the diffracting planes.

**angle of repose.** The angular contour that a powder pile assumes.

**angle-ply laminate.** A laminate having fibers of adjacent plies, oriented at alternating angles.

**angle press.** A hydraulic molding press equipped with horizontal and vertical rams, specially designed for the production of complex plastic moldings having deep undercuts.

**angle wrap.** Tape fabric that is wrapped on a starter dam mandrel at an angle to the centerline.

**angstrom unit (Å).** A unit of linear measure equal to $10^{-10}$ m, or 0.1 nm. Although not an accepted SI unit, it is occasionally used for small distances, such as interatomic distances, and some wavelengths.

**angular aperture.** In optical microscopy, the angle between the most divergent rays that can pass through a lens to form the image of an object. See also *aperture (optical)*.

**angular-contact bearing.** A ball bearing of the grooved type, designed in such a way that when under no load, a line through the outer and inner raceway contacts with the balls

to form an angle with a plane perpendicular to the bearing axis (Fig. 14).

**angularity.** The conformity to, or deviation from, specified angular dimensions in the cross section of a shape or bar.

**anhydride.** A compound from which water has been extracted. An oxide of a metal (basic anhydride) or of a nonmetal (acid anhydride) that forms a base or an acid, respectively, when united with water.

**aniline.** An important organic base ($C_6H_5NH_2$) made by reacting chlorobenzene with aqueous ammonia in the presence of a catalyst. It is used in the production of aniline formaldehyde resins and in the manufacture of certain rubber accelerators and antioxidants.

**aniline-formaldehyde resins.** Members of the aminoplastics family made by the condensation of formaldehyde and aniline in an acid solution. The resins are thermoplastic and are used to a limited extent in the production of molded and laminated insulating materials. Products made from these resins have high dielectric strength and good chemical resistance.

**aniline point.** As applied to a petroleum product, the lowest temperature at which the product is completely miscible with an equal volume of freshly distilled aniline. The aniline point is a guide to the oil composition.

**anion.** A negatively charged ion that migrates through the electrolyte toward the *anode* under the influence of a potential gradient. See also *cation* and *ion*.

**anisotropic.** Exhibiting different properties when tested along axes in different directions. In magnetics, capable of being magnetized more readily in one direction than in a transverse direction.

**anisotropic conductive adhesive.** An adhesive that can be made conductive in the vertical, or *z*, axis while remaining an insulator in the horizontal, or *x* and *y*, axes.

**anisotropic laminate.** One in which the properties are different in different directions along the laminate plane.

**anisotropy.** The characteristic of exhibiting different values of a property in different directions with respect to a fixed reference system in the material.

**anisotropy of laminates.** The difference of the properties along the directions parallel to the length or width of the lamination planes and perpendicular to the lamination.

**anneal** (glass). To prevent or remove objectionable stresses in glassware by controlled cooling from a suitable temperature.

**annealed powder.** A metallic powder that is heat treated to render it soft and compactible.

**annealing** (abrasives). A heating and cooling operation, implying usually a slow cooling used to secure one or more of the following effects: to remove stress, to induce softness, to refine structure, or to alter physical properties.

**annealing carbon.** See *temper carbon.*

**annealing** (glass). A controlled cooling process for glass designed to reduce thermal residual stress to a commercially acceptable level, and, in some cases, modify structure.

**annealing** (metals). A generic term denoting a treatment consisting of heating to and holding at a suitable temperature followed by cooling at a suitable rate, used primarily to soften metallic materials, but also to simultaneously produce desired changes in other properties or in microstructure. The purpose of such changes may be, but is not confined to: improvement of machinability, facilitation of cold work, improvement of mechanical or electrical properties, and/or increase in stability of dimensions. When the term is used unqualifiedly, full annealing is implied. When applied only for the relief of stress, the process is properly called *stress relieving* or stress-relief annealing.

In ferrous alloys, annealing usually is done above the upper critical temperature, but the time-temperature cycles vary widely both in maximum temperature attained and in cooling rate employed, depending on composition, material condition, and results desired. When applicable, the following commercial process names should be used: *black annealing, blue annealing, box annealing, bright annealing, cycle annealing, flame annealing, full annealing, graphitizing,* in-process annealing, *isothermal annealing, malleabilizing,* orientation annealing, *process annealing, quench annealing, spheroidizing, subcritical annealing.*

In nonferrous alloys, annealing cycles are designed to: (a) remove part or all of the effects of cold working (recrystallization may or may not be involved); (b) cause substantially complete coalescence of precipitates from solid

solution in relatively coarse form; or (c) both, depending on composition and material condition. Specific process names in commercial use are *final annealing, full annealing, intermediate annealing, partial annealing, recrystallization annealing,* stress-relief annealing, *anneal to temper.*

**annealing** (plastics). Heating to a temperature at which the molecules have significant mobility, permitting them to reorient to a configuration having less residual stress. In semicrystalline polymers, heating to a temperature at which retarded crystallization or recrystallization can occur.

**annealing point** (glass). That temperature at which internal stresses in a glass are substantially relieved in a matter of minutes.

**annealing range** (glass). The range of glass temperature in which stress in glass can be relieved at a commercially practical rate. For purposes of comparing glasses, the annealing range is assumed to correspond with the temperature between the annealing point and the strain point.

**annealing twin.** A *twin* formed in a crystal during recrystallization.

**annealing twin bands.** See *twin bands.*

**anneal to temper** (metals). A final partial anneal that softens a cold-worked nonferrous alloy to a specified level of hardness or tensile strength.

**annular bearing.** (1) Usually a rolling bearing of short cylindrical form supporting a shaft carrying a radial load. (2) A flat disk-shaped bearing.

**anode.** (1) The electrode of an electrolyte cell at which oxidation occurs. Electrons flow away from the anode in the external circuit. It is usually at the electrode that corrosion occurs and metal ions enter solution. (2) The positive (electron-deficient) electrode in an electrochemical circuit. Contrast with *cathode.*

**anode aperture.** In electron microscopy, the opening in the accelerating voltage anode shield of an electron gun through which the electrons must pass to illuminate or irradiate the specimen.

**anode compartment.** In an electrolytic cell, the enclosure formed by a diaphragm around the anodes.

**anode copper.** Special-shaped copper slabs, resulting from the refinement of *blister copper* in a reverberatory furnace, used as anodes in electrolytic refinement.

**anode corrosion.** The dissolution of a metal acting as an *anode.*

**anode corrosion efficiency.** The ratio of the actual corrosion (weight loss) of an *anode* to the theoretical corrosion (weight loss) calculated by *Faraday's law* from the quantity of electricity that has passed.

**anode effect.** The effect produced by polarization of the *anode* in electrolysis. It is characterized by a sudden increase in voltage and a corresponding decrease in amperage due to the anode becoming virtually separated from the electrolyte by a gas film.

**anode efficiency.** Current efficiency at the *anode*.

**anode film.** (1) The portion of solution in immediate contact with the *anode*, especially if the concentration gradient is steep. (2) The outer layer of the anode itself.

**anode mud.** Deposit of insoluble residue formed from the dissolution of the anode in commercial electrolysis. Sometimes called anode slime.

**anode polarization.** See *polarization*.

**anodic cleaning.** Electrolytic cleaning in which the work is the anode. Also called reverse-current cleaning.

**anodic coating.** A film on a metal surface resulting from an electrolytic treatment at the anode.

**anodic etching.** Development of microstructure by selective dissolution of the polished surface under application of a direct current. Variation with layer formation: anodizing.

**anodic inhibitor.** A chemical substance or mixture that prevents or reduces the rate of the anodic or oxidation reaction. See also *inhibitor*.

**anodic pickling.** *Electrolytic pickling* in which the work is the anode.

**anodic polarization.** The change of the electrode potential in the noble (positive) direction due to current flow. See also *polarization*.

**anodic protection.** (1) A technique to reduce the corrosion rate of a metal by polarizing it into its passive region, where dissolution rates are low. (2) Imposing an external electrical potential to protect a metal from corrosive attack. (Applicable only to metals that show active-passive behavior.) Contrast with *cathodic protection*.

**anodic reaction.** Electrode reaction equivalent to a transfer of positive charge from the electronic to the ionic conductor. An anodic reaction is an oxidation process. An example common in corrosion is $Me \rightarrow Me^{n+} + ne^-$.

**anodizing.** Forming a *conversion coating* on a metal surface by anodic oxidation; most frequently applied to aluminum.

**anolyte.** The electrolyte adjacent to the *anode* in an *electrolytic cell*.

**antechamber.** The entrance vestibule of a continuously operating sintering furnace.

**anticorrosive additive.** A lubricant additive used to reduce corrosion.

**antiextrusion ring.** A ring that is installed on the low-pressure side of a seal or packing, in order to prevent extrusion of the sealing material.

**antiferromagnetic material.** A material wherein interatomic forced hold the elementary atomic magnets (electron spins) of a solid in alignment, a state similar to that of a *ferromagnetic material* but with the difference that equals numbers of elementary magnets (spins) face in opposite directions and are antiparallel, causing the solid

to be weakly magnetic, that is, paramagnetic, instead of ferromagnetic.

**antifoam additive.** An additive used to reduce or prevent foaming. Also known as *foam inhibitor*.

**antifouling.** Intended to prevent fouling of underwater structures, such as the bottom of ships.

**antifriction bearing.** A bearing containing a solid lubricant. See also *roller bearing*, *rolling-element bearing*, and *self-lubricating bearing*.

**antifriction material.** A material that exhibits low-friction or self-lubricating properties.

**antioxidant.** Any additive for the purpose of reducing the rate of oxidation and subsequent deterioration of a material.

**antipitting agent.** An *addition agent* for electroplating solutions to prevent the formation of pits or large pores in the electrodeposit.

**antiscuffing lubricant.** A lubricant that is formulated to avoid scuffing. See also *extreme-pressure lubricant*.

**antiseizure property.** The ability of a bearing material to resist seizure during momentary lubrication failure.

**antistatic agents.** Agents that, when added to a plastic molding material or applied to the surface of a molded object, make it less conductive, thus hindering the fixation of dust or the buildup of electrical charge.

**anti-Stokes Raman line.** A Raman line that has a frequency higher than that of the incident monochromatic radiation.

**antiwear additive.** A lubricant additive used to reduce wear.

**antiweld characteristic.** See *antiseizure property*.

**anvil.** A large, heavy metal block that supports the frame structure and holds the stationary die of a forging hammer. Also, the metal block on which blacksmith forgings are made.

**anvil cap.** Same as *sow block*.

**aperture (electron).** See *anode aperture*, *condenser aperture*, and *physical objective aperture*.

**aperture (optical).** In optical microscopy, the working diameter of a lens or a mirror. See also *angular aperture*.

**aperture size.** The opening of a mesh as in a sieve. See also *sieve analysis*.

**Apiezon oil.** An oil of low vapor pressure used in vacuum technology.

**API gravity (API degree).** A measure of density used in the U.S. petroleum industry where: Degrees API = [141.5/ specific gravity at 60 °F)] – 131.5.

**aplanatic.** Corrected for spherical aberration and coma.

**apochromatic lens.** A lens whose secondary chromatic aberrations have been substantially reduced. See also *achromatic*.

**apochromatic objective.** Objectives corrected chromatically for three colors and spherically for two colors are called apochromats. These corrections are superior to those of the achromatic series of lenses. Because apochromats are not well corrected for lateral color, special eyepieces are used to compensate. See also *achromatic*.

**apparent area of contact.** In tribology, the area of contact between two solid surfaces defined by the boundaries of their macroscopic interface. Contrast with *real area of contact*. See also *Hertzian contact area* and *nominal contact area*.

**apparent density.** (1) The weight per unit volume of a powder, in contrast to the weight per unit volume of the individual particles. (2) The weight per unit volume of a porous solid, where the unit volume is determined from external dimensions of the mass. Apparent density is always less than the true density of the material itself.

**apparent hardness.** The value obtained by testing a sintered object with standard indentation hardness equipment. Because the reading is a composite of pores and solid material, it is usually lower than that of a wrought or cast material of the same composition and condition. Not to be confused with *particle hardness*.

**apparent pore volume.** The total pore volume of a loose powder mass or a green compact. It may be calculated by subtracting the apparent density from the theoretical density of the substance.

**apparent porosity.** The relation of the open pore space to the bulk volume, expressed in percent.

**approach distance.** The linear distance, in the direction of feed, between the point of initial cutter contact and the point of full cutter contact.

**Ar$_{cm}$, Ar$_1$, Ar$_3$, Ar$_4$, Ar′, Ar″.** Defined under *transformation temperature*.

**aramid.** A manufactured organic fiber in which the fiber-forming substance is a long-chain synthetic aromatic polyamide in which at least 85% of the amide linkages are directly attached to two aromatic rings. Aramid fibers, most notably Kevlar fibers, were the first with a high enough tensile modulus and strength to be used as a reinforcement in advanced composites. Key representative properties of aramid (Kevlar) fibers are given in Table 4. See also *Kevlar*.

**arbitration bar.** A test bar, cast with a heat of material, used to determine chemical composition, hardness, tensile strength, and deflection and strength under transverse

loading in order to establish the state of acceptability of the casting.

**arbor.** (1) In machine grinding, the spindle on which the wheel is mounted. (2) In machine cutting, a shaft or bar for holding and driving the cutter. (3) In founding, a metal shape embedded in green sand or dry sand cores to support the sand or the applied load during casting.

**arborescent powder.** See preferred term *dendritic powder*.

**arbor press.** A machine used for forcing arbors or mandrels into drilled or bored parts preparatory to turning or grinding. Also used for forcing bushings, shafts, or pins into or out of holes.

**arbor-type cutter.** A cutter having a hole for mounting on an arbor and usually having a keyway for a driving key.

**arc.** A luminous discharge of electrical current crossing the gap between two electrodes.

**arc blow.** The deflection of an electric arc from its normal path because of magnetic forces.

**arc brazing.** A brazing process in which the heat required is obtained from an electric arc. See *twin carbon arc brazing*.

**arc chamber.** The confined space enclosing the anode and cathode in which the arc is struck.

**arc cutting.** A group of cutting processes which melts the metals to be cut with the heat of an arc between an electrode and the base metal. See *carbon arc cutting*, *metal arc cutting*, *gas metal arc cutting*, *gas tungsten arc cutting*, *plasma arc cutting*, and *air carbon arc cutting*. Compare with *oxygen arc cutting*.

**arc force.** The axial force developed by a plasma.

**arc furnace.** A furnace in which metal is melted either directly by an electric arc between an electrode and the work or indirectly by an arc between two electrodes adjacent to the metal.

**arc gas.** The gas introduced into the arc chamber and ionized by the arc to form a plasma.

**arc gouging.** An arc cutting process variation used to form a bevel or groove.

**arc melting.** Melting metal in an electric arc furnace.

**arc of contact.** The portion of the circumference of a grinding wheel or cutter touching the work being processed.

**arc oxygen cutting.** See preferred term *oxygen arc cutting*.

**arc plasma.** See *plasma*.

## Table 4  Properties of aramid fibers

| Material | Density, g/cm$^3$ | Filament diameter μm | Filament diameter μin. | Tensile modulus(a) GPa | Tensile modulus(a) 10$^6$ psi | Tensile strength(a) GPa | Tensile strength(a) 10$^6$ psi | Tensile elongation, % | Available yarn count, No. filaments |
|---|---|---|---|---|---|---|---|---|---|
| Kevlar 29 (high toughness) . . . . . | 1.44 | 12 | 470 | 83 | 12 | 3.6 2.8(b) | 0.525 0.400(b) | 4.0 | 134-10 000 |
| Kevlar 49 (high modulus) . . . . . . | 1.44 | 12 | 470 | 131 | 19 | 3.6-4.1 | 0.525-0.600 | 2.8 | 134-5000 |
| Kevlar 149 (ultra-high modulus) . . | 1.47 | 12 | 470 | 186 | 27 | 3.4 | 0.500 | 2.0 | 134-1000 |

(a) ASTM D 2343, impregnated strand. (b) ASTM D 885, unimpregnated strand

**arc resistance**. Ability to withstand exposure to an electric voltage. The total time in seconds that an intermittent arc may play across a plastic surface without rendering the surface conductive.

**arc seam weld**. A seam weld made by an arc welding process.

**arc spot weld**. A spot weld made by an arc welding process.

**arc strike**. A discontinuity consisting of any localized remelted metal, heat-affected metal, or change in the surface profile of any part of a weld or base metal resulting from an arc.

**arc time**. The time an arc is maintained in making an arc weld. Also known as *weld time*.

**arc voltage**. The voltage across the welding arc.

**arc welding (AW)**. A group of welding processes which produces coalescence of metals by heating them with an arc, with or without the application of pressure, and with or without the use of filler metal.

**arc welding electrode**. A component of the welding circuit through which current is conducted between the electrode holder and the arc. See *arc welding*.

**arc welding gun**. A device used in semiautomatic, machine, and automatic arc welding to transfer current, guide the consumable electrode, and direct shielding gas when used (Fig. 15).

FIG. 15

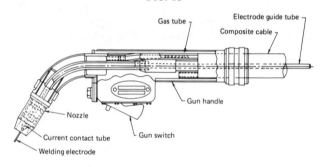

*Typical semiautomatic arc welding gun for gas metal arc welding*

**areal weight**. The weight of a fiber reinforcement per unit area (width × length) of tape or fabric.

**area of contact**. A general term that, without other modifying terminology, is insufficiently specific to be defined precisely. Its use should therefore be avoided. Related terms are *actual area of contact*, *apparent area of contact*, *Hertzian contact area*, *nominal area of contact*, and *real area of contact*.

**argon oxygen decarburization (AOD)**. A secondary refining process for the controlled oxidation of carbon in a steel melt. In the AOD process, oxygen, argon, and nitrogen are injected into a molten metal bath through submerged, side-mounted tuyeres (Fig. 16).

**arm** (resistance welding). A projecting beam extending from

FIG. 16

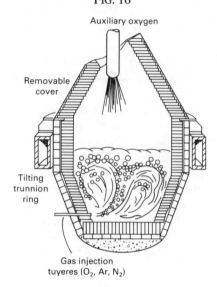

*Schematic of argon oxygen decarburization vessel*

the frame of a resistance welding machine which transmits the electrode force and may conduct the welding current.

**aromatic**. Unsaturated hydrocarbon with one or more benzene ring structures in the molecule.

**aromatic polyester**. A polyester derived from monomers in which all the hydroxyl and carboxyl groups are directly linked to aromatic nuclei.

**arrest line** (ceramics). Rib mark defining the crack front shape of an arrested crack prior to resumption of crack spread under an altered stress configuration (Fig. 17). The duration of rest may be long to infinitesimal. See also *rib mark*.

**arrest lines (marks)**. See *beach marks*.

**arrest mark**. See *dwell mark*.

FIG. 17

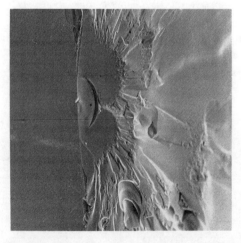

*Arrest line (crescent-shaped flaw) in a fractured glass rod. 65×*

**artifact**. A feature of artificial character, such as a scratch or a piece of dust on a metallographic specimen, that can be erroneously interpreted as a real feature. See also *abrasion artifact*, *mounting artifact*, and *polishing artifact*.

**artificial aging** (heat treatment). Aging above room temperature. See *aging (heat treatment)*. Compare with *natural aging*.

**artificial aging** (plastics). The exposure of a plastic to conditions that accelerate the effects of time. Such conditions include heating, exposure to cold, flexing, application of electric field, exposure to chemicals, ultraviolet light radiation, and so forth. Typically, the conditions chosen for such testing reflect the conditions under which the plastic article will be used. Usually, the length of time the article is exposed to these test conditions is relatively short. Properties such as dimensional stability, mechanical fatigue, chemical resistance, stress cracking resistance, dielectric strength, and so forth, are evaluated in such testing. See also *aging*.

**artificial weathering**. The exposure of plastics to cyclic laboratory conditions, consisting of high and low temperatures, high and low relative humidities, and ultraviolet radiant energy, with or without direct water spray and moving air (wind), in an attempt to produce changes in the properties of the plastics similar to those observed after long-term continuous exposure outdoors. The laboratory exposure conditions are usually more intensified than those encountered in actual outdoor exposure in an attempt to achieve an accelerated effect. Also called accelerated aging.

**as-brazed**. The condition of brazements after brazing, prior to any subsequent thermal, mechanical, or chemical treatments.

**as-cast condition**. Castings as removed from the mold without subsequent heat treatment.

**ash content**. Proportion of the solid residue remaining after a reinforcing substance has been incinerated (charred or intensely heated).

**aspect ratio**. The ratio of length to diameter of a reinforcing fiber.

**asperity**. In tribology, a protuberance in the small-scale topographical irregularities of a solid surface.

**a-spot**. One of many small contact areas through which electrical current can pass when two rough, conductive, solid surfaces are touching.

**assay**. Determination of how much of a sample is the material indicated by the name. For example, for an assay of $FeSO_4$ the analyst would determine both iron and $SO_4^{2-}$ in the sample.

**assembly**. A number of parts or subassemblies or any combination thereof joined together.

**assembly adhesive**. An adhesive that can be used for bonding parts together, such as in the manufacture of a boat, an airplane, furniture, and the like. The term assembly adhesive is commonly used in the wood industry to distinguish such adhesives (formerly called joint glues) from those used in making plywood (sometimes called veneer glues). It describes adhesives used in fabricating finished structures or goods, or subassemblies thereof, as differentiated from adhesives used in the production of sheet materials for sale as such, for example, plywood or laminates.

**assembly line** (adhesives). The time interval between the spreading of the adhesive on the adherend and the application of pressure or heat, or both, to the assembly. For assemblies involving multiple layers or parts, the assembly time begins with the spreading of the adhesive on the first adherend. See also *closed assembly time* and *open assembly time*.

**A-stage**. An early stage in the reaction of certain thermosetting resins in which the material is fusible and still soluble in certain liquids. Synonym for resole. Compare with *B-stage* and *C-stage*.

**astigmatism**. A defect in a lens or optical system that causes rays in one plane parallel to the optical axis to focus at a distance different from those in the plane at right angles to it.

**ASTM grain size number**. See *grain size*.

**ASTM viscosity-temperature equation**. The equation relating kinematic viscosity ($v$) with temperature according to: $\log \log(v + 0.6) = m \log T + C$, where $v$ is the kinematic viscosity in centistokes, $T$ is the absolute temperature, $m$ is the ASTM slope (slope of the temperature-viscosity curve), and $C$ is a constant that depends on the lubricant.

**as-welded**. The condition of weld metal, welded joints, and weldments after welding but prior to any subsequent thermal, mechanical, or chemical treatments.

**atactic stereoisomerism**. A chain of molecules in which the position of the side chains or side atoms is more or less random. See also *isotactic stereoisomerism* and *syndiotactic stereoisomerism*.

**athermal**. Not isothermal. Changing rather than constant temperature conditions.

**athermal transformation**. A reaction that proceeds without benefit of thermal fluctuations—that is, thermal activation is not required. Such reactions are diffusionless and can take place with great speed when the driving force is sufficiently high. For example, many martensitic transformations occur athermally on cooling, even at relatively low temperatures, because of the progressively increasing drive force. In contrast, a reaction that occurs at constant temperature is an *isothermal transformation*; thermal activation is necessary in this case and the reaction proceeds as a function of time.

**atmospheric corrosion**. The gradual degradation or alteration of a material by contact with substances present in the atmosphere, such as oxygen, carbon dioxide, water vapor, and sulfur and chlorine compounds.

**atmospheric riser.** A riser that uses atmospheric pressure to aid feeding. Essentially, a *blind riser* into which a small core or rod protrudes; the function of the core or rod is to provide an open passage so that the molten interior of the riser will not be under a partial vacuum when metal is withdrawn to feed the casting but will always be under atmospheric pressure.

**atom.** The smallest particle of an element that retains the characteristic properties and behavior of the element. See also *atomic structure*, *isotope*, and *nuclear structure*.

**atomic absorption spectrometry.** The measurement of light absorbed at the wavelength of resonance lines by the unexcited atoms of an element.

**atomic fission.** The breakup of the nucleus of an atom in which the combined weight of the fragments is less than that of the original nucleus, the difference being converted to a very large energy release.

**atomic hydrogen welding.** An arc welding process that fuses metals together by heating them with an electric arc maintained between two metal electrodes enveloped in a stream of hydrogen. Shielding is provided by the hydrogen, which also carries heat by molecular dissociation and subsequent recombination. Pressure may or may not be used and filler metal may or may not be used. (This process is now of limited industrial significance.)

**atomic mass unit (amu).** An arbitrarily defined unit expressing the masses of individual atoms. One atomic mass unit is defined as exactly $1/12$ of the mass of an atom of the nuclide $^{12}$C, the predominant isotope of carbon. See also *atomic weight*.

**atomic number (Z).** The number of elementary positive charges (protons) contained within the nucleus of an atom. For an electrically neutral atom, the number of planetary electrons is also given by the atomic number. Atoms with the same $Z$ (isotopes) may contain different numbers of neutrons. Also known as nuclear charge. See also *isotope* and *proton*.

**atomic number contrast.** See *atomic number imaging*.

**atomic number imaging.** In scanning electron microscopy, a technique in which contrast is controlled by atomic number (high atomic number areas appear light, while low atomic number areas appear dark). Usually obtained by imaging based on backscattered electron signal. See also *backscattered electron*.

**atomic percent.** The number of atoms of an element in a total of 100 representative atoms of a substance.

**atomic replica.** A thin replica devoid of structure on the molecular level. It is prepared by the vacuum or hydrolytic deposition of metals or simple compounds of low molecular weight. See also *replica*.

**atomic scattering factor, *f*.** The ratio of the amplitude of the wave scattered by an atom to that scattered by a single electron.

**atomic structure.** The arrangement of the parts of an atom, which consists of a positively charged nucleus surrounded by a cloud of electrons arranged in orbits that can be described in terms of quantum mechanics.

**atomic wear.** Wear between two contacting surfaces in relative motion attributed to migration of individual atoms from one surface to the other.

**atomic weight.** (1) A number assigned to each chemical element that specifies the average mass of its atoms. Because an element may consist of two or more isotopes, each having atoms with well-defined but differing masses, the atomic weight of each element is the average of the masses of its naturally occurring isotopes weighted by the relative proportions of those isotopes. (2) The mean weight of the atom of an element in relation to $^{12}$C = 12.000.

**atomization** (materials characterization). The subdivision of a compound into individual atoms using heat or chemical reactions. This is a necessary step in atomic spectroscopy. See also *atomizer* and *nebulizer*.

**atomization** (powder metallurgy). The disintegration of a molten metal into particles by a rapidly moving gas or liquid stream or by other means (Fig. 18).

FIG. 18

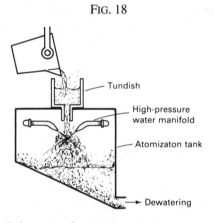

*Schematic of water atomization process*

**atomization** (thermal spraying). The division of molten material at the end of the wire or rod into fine particles.

**atomizer.** A device that atomizes a sample, for example, a burner, plasma, or hydride reaction chamber. See also *atomization* and *nebulizer*.

**atom probe.** An instrument for measuring the mass of a single atom or molecule on a metal surface; it consists of a field ion microscope with a hole in its screen opening into a mass spectrometer; atoms are removed from the specimen by pulsed field evaporation, travel through the hole, and are detected in the mass spectrometer. See also *field-ion microscopy*.

**attack-polishing.** Simultaneous etching and mechanical polishing.

**attenuation.** (1) The fractional decrease of the intensity of an energy flux, including the reduction of intensity resulting from geometrical spreading, absorption, and scattering. (2) The diminution of vibrations or energy over time or distance. The process of making thin and slender, as applied to the formation of fiber from molten glass. (3) The exponential decrease with distance in the amplitude of an electrical signal traveling along a very long, uniform transmission line, due to conductor and dielectric losses.

**attenuation period.** See preferred term *deceleration period*.

**attitude (attitude angle).** In a bearing, the angular position of the line joining the center of the journal to that of the bearing bore, relative to the direction of loading.

**attrition.** Removal of small fragments of surface material during sliding contact.

**attritious wear.** Wear of abrasive grains in grinding such that the sharp edges gradually become rounded. A grinding wheel that has undergone such wear usually has a glazed appearance.

**attritor.** A high-intensity ball mill whose drum is stationary and whose balls are agitated by rotating baffles, paddles, or rods at right angle to the drum axis (Fig. 19).

FIG. 19

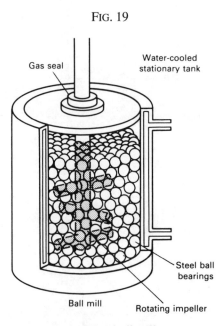

Gas seal

Water-cooled stationary tank

Steel ball bearings

Ball mill

Rotating impeller

*Attrition ball mill*

**attritor grinding.** The intensive grinding or alloying in an attritor. Examples: milling of carbides and binder metal powders and mechanical alloying of hard dispersoid particles with softer metal or alloy powders. See also *mechanical alloying*.

**Auger chemical shift.** The displacement in energy of an Auger electron peak for an element due to a change in chemical bonding relative to a specified element or compound.

**Auger electron.** An electron emitted from an atom with a vacancy in an inner shell. Auger electrons have a characteristic energy detected as peaks in the energy spectra of the secondary electrons generated.

**Auger electron spectroscopy (AES).** A technique for chemical analysis of surface layers that identifies the atoms present in a layer by measuring the characteristic energies of their Auger electrons.

**Auger electron yield.** The probability that an atom with a vacancy in a particular inner shell will relax by an Auger process.

**Auger map.** A two-dimensional image of the specimen surface showing the location of emission of Auger electrons from a particular element. A map is normally produced by rastering the incident electron beam over the specimen surface and simultaneously recording the Auger signal strength for a particular transition as a function of position.

**Auger matrix effects.** Effects that cause changes in the shape of an Auger electron energy distribution or in the Auger signal strength for an element due to the physical environment of the emitting atom and not due to bonding with other elements or changes in concentration.

**Auger process.** The radiationless relaxation of an atom, involving a vacancy in an inner electron shell. An electron (known as an Auger electron) is emitted.

**Auger transition designations.** Transitions are designated by the electron shells involved. The first letter designates the shell containing the initial vacancy; the last two letters designate the shells containing electron vacancies created by Auger emission (for example, KLL and LMN).

**ausforming.** Thermomechanical treatment of steel in the metastable austenitic condition below the recrystallization temperature followed by quenching to obtain martensite and/or bainite.

**austempered ductile iron.** A moderately alloyed *ductile iron* that is austempered for high strength with appreciable ductility. See also *austempering*.

**austempering.** A heat treatment for ferrous alloys in which a part is quenched from the austenitizing temperature at a rate fast enough to avoid formation of ferrite or pearlite and then held at a temperature just above $M_s$ until transformation to bainite is complete. Although designated as bainite in both austempered steel and austempered ductile iron (ADI), austempered steel consists of two phase mixtures containing ferrite and carbide, while austempered ductile iron consists of two phase mixtures containing ferrite and austenite.

**austenite.** A solid solution of one or more elements in face-centered cubic iron (gamma iron). Unless otherwise designated (such as nickel austenite), the solute is generally assumed to be carbon.

**Table 5  Compositions of selected wrought austenitic stainless steels**

| Type | UNS designation | Composition(a), % | | | | | | | |
|---|---|---|---|---|---|---|---|---|---|
| | | C | Mn | Si | Cr | Ni | P | S | Other |
| 201 | S20100 | 0.15 | 5.5-7.5 | 1.00 | 16.0-18.0 | 3.5-5.5 | 0.06 | 0.03 | 0.25 N |
| 205 | S20500 | 0.12-0.25 | 14.0-15.5 | 1.00 | 16.5-18.0 | 1.0-1.75 | 0.06 | 0.03 | 0.32-0.40 N |
| 301 | S30100 | 0.15 | 2.00 | 1.00 | 16.0-18.0 | 6.0-8.0 | 0.045 | 0.03 | ... |
| 302 | S30200 | 0.15 | 2.00 | 1.00 | 17.0-19.0 | 8.0-10.0 | 0.045 | 0.03 | ... |
| 303 | S30300 | 0.15 | 2.00 | 1.00 | 17.0-19.0 | 8.0-10.0 | 0.20 | 0.15 min | 0.6 Mo(b) |
| 304 | S30400 | 0.08 | 2.00 | 1.00 | 18.0-20.0 | 8.0-10.5 | 0.045 | 0.03 | ... |
| 304N | S30451 | 0.08 | 2.00 | 1.00 | 18.0-20.0 | 8.0-10.5 | 0.045 | 0.03 | 0.10-0.16 N |
| 309 | S30900 | 0.20 | 2.00 | 1.00 | 22.0-24.0 | 12.0-15.0 | 0.045 | 0.03 | ... |
| 310 | S31000 | 0.25 | 2.00 | 1.50 | 24.0-26.0 | 19.0-22.0 | 0.045 | 0.03 | ... |
| 316 | S31600 | 0.08 | 2.00 | 1.00 | 16.0-18.0 | 10.0-14.0 | 0.045 | 0.03 | 2.0-3.0 Mo |
| 316LN | S31653 | 0.03 | 2.00 | 1.00 | 16.0-18.0 | 10.0-14.0 | 0.045 | 0.03 | 2.0-3.0 Mo; 0.10-0.16 N |
| 321 | S32100 | 0.08 | 2.00 | 1.00 | 17.0-19.0 | 9.0-12.0 | 0.045 | 0.03 | 5 × %C min Ti |
| 330 | N08330 | 0.08 | 2.00 | 0.75-1.5 | 17.0-20.0 | 34.0-37.0 | 0.04 | 0.03 | ... |
| 347 | S34700 | 0.08 | 2.00 | 1.00 | 17.0-19.0 | 9.0-13.0 | 0.045 | 0.03 | 10 × %C min Nb |
| 348 | S34800 | 0.08 | 2.00 | 1.00 | 17.0-19.0 | 9.0-13.0 | 0.045 | 0.03 | 0.2 Co; 10 × %C min Nb; 0.10 Ta |

(a) Single values are maximum values unless otherwise indicated. (b) Optional

**Table 6  Standard composition ranges for austenitic manganese steel castings**

| ASTM A 128 grade | Composition, % | | | | | | |
|---|---|---|---|---|---|---|---|
| | C | Mn | Cr | Mo | Ni | Si (max) | P (max) |
| A | 1.05–1.35 | 11.0 min | ... | ... | ... | 1.00 | 0.07 |
| B-1 | 0.9–1.05 | 11.5–14.0 | ... | ... | ... | 1.00 | 0.07 |
| B-2 | 1.05–1.2 | 11.5–14.0 | ... | ... | ... | 1.00 | 0.07 |
| B-3 | 1.12–1.28 | 11.5–14.0 | ... | ... | ... | 1.00 | 0.07 |
| B-4 | 1.2–1.35 | 11.5–14.0 | ... | ... | ... | 1.00 | 0.07 |
| C | 1.05–1.35 | 11.5–14.0 | 1.5–2.5 | ... | ... | 1.00 | 0.07 |
| D | 0.7–1.3 | 11.5–14.0 | ... | ... | 3.0–4.0 | 1.00 | 0.07 |
| E-1 | 0.7–1.3 | 11.5–14.0 | ... | 0.9–1.2 | ... | 1.00 | 0.07 |
| E-2 | 1.05–1.45 | 11.5–14.0 | ... | 1.8–2.1 | ... | 1.00 | 0.07 |
| F | 1.05–1.35 | 6.0–8.0 | ... | 0.9–1.2 | ... | 1.00 | 0.07 |

**austenitic corrosion-resistant stainless steel.** A highly alloyed corrosion-resistant stainless steel containing 16% or more chromium, a ferrite-stabilizing element, and sufficient austenite-stabilizing elements such as nickel (up to about 35%), manganese, and nitrogen. Table 5 lists compositions of standard austenitic stainless steels. See also *stainless steels* (Technical Brief 47).

**austenitic grain size.** The size attained by the grains in steel when heated to the austenitic region. This may be revealed by appropriate etching of cross sections after cooling to room temperature.

**austenitic manganese steel.** A cast, wear-resistant material containing about 1.2% C and 12% Mn (Table 6). Used primarily in the fields of earthmoving, mining, quarrying, railroading, ore processing, lumbering, and in the manufacture of cement and clay products. Also known as Hadfield steel.

**austenitic steel.** An alloy steel whose structure is normally austenitic at room temperature (Fig. 20).

FIG. 20

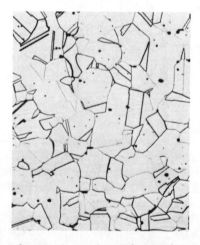

*Equiaxed austenite grains and annealing twins in an austenitic stainless steel. 250×*

**austenitizing**. Forming austenite by heating a ferrous alloy into the transformation range (partial austenitizing) or above the transformation range (complete austenitizing). When used without qualification, the term implies complete austenitizing.

**autoclave**. A closed vessel for conducting and completing either a chemical reaction under pressure and heat or other operation, such as cooling. Widely used for bonding and curing reinforced plastic laminates.

**autoclave molding**. A process in which, after lay-up, winding, or wrapping, an entire assembly is placed in a heated autoclave, usually at 340 to 1380 kPa (50 to 200 psi). Additional pressure permits higher density and improved removal of volatiles from the resin. Lay-up is usually vacuum bagged with a bleeder and release cloth.

**autofrettage**. Prestressing a hollow metal cylinder by the use of momentary internal pressure exceeding the yield strength.

**autogenous weld**. A fusion weld made without the addition of filler metal.

**automated image analysis**. See *image analysis*.

**automatic brazing**. Brazing with equipment which performs the brazing operation without constant observation and adjustment by a brazing operator. The equipment may or may not perform the loading and unloading of the work. See also *machine brazing*.

**automatic gas cutting**. See preferred term *automatic oxygen cutting*.

**automatic mold**. A mold for injection or compression molding of plastics that repeatedly goes through the entire cycle, including ejection, without human assistance.

**automatic oxygen cutting**. Oxygen cutting with equipment which performs the cutting operation without constant observation and adjustment of the controls by an operator. The equipment may or may not perform loading and unloading of the work. See also *machine oxygen cutting*.

**automatic press** (metals). A press in which the work is fed mechanically through the press in synchronism with the press action. An automation press is an automatic press that, in addition, is provided with built-in electrical and pneumatic control equipment.

**automatic press** (plastics). A hydraulic press for compression molding or an injection machine that operates continuously, being controlled mechanically (toggle) or hydraulically, or by a combination of these methods.

**automatic press stop**. A machine-generated signal for stopping the action of a press, usually after a complete cycle, by disengaging the clutch mechanism and engaging the brake mechanism.

**automatic welding**. Welding with equipment which performs the welding operation without adjustment of the controls by a welding operator. The equipment may or may not perform the loading and unloading of the work. See also *machine welding*.

**automation press**. See *automatic press*.

**autoradiography**. An inspection technique in which radiation spontaneously emitted by a material is recorded photographically. The radiation is emitted by radioisotopes that are (a) produced in a metal by bombarding it with neutrons, (b) added to a metal such as by alloying, or (c) contained within a cavity in a metal part. The technique serves to locate the position of the radioactive element or compound.

**auxiliary anode**. In electroplating, a supplementary *anode* positioned so as to raise the current density on a certain area of the *cathode* and thus obtain better distribution of plating.

**auxiliary electrode**. An *electrode* commonly used in polarization studies to pass current to or from a test electrode. It is usually made from a noncorroding material.

**auxiliary magnifier or enlarger** (eye protection). An additional lens or plate, associated with eye protection equipment, used to magnify or enlarge the field of vision.

**average density**. The density measured on an entire body or on a major number of its parts whose measurements are then averaged.

**average erosion rate**. The cumulative erosion divided by the corresponding cumulative exposure duration, that is, the slope of a line from the origin to a specified point on the cumulative erosion-time curve.

**average grain diameter**. The mean diameter of an equiaxed grain section whose size represents all the grain sections in the aggregate being measured. See also *grain size*.

**average linear strain**. See *engineering strain*.

**average molecular weight**. The molecular weight of the most typical chain in a given plastic; it is characteristic of neither the longest nor the shortest chain.

**average particle size**. A single value representing the entire particle size distribution.

**Avogadro's number**. The number of molecules ($6.02 \times 10^{23}$) in a gram-molecular weight of any substance. See also *gram-molecular weight* and *mole*.

**axial**. Longitudinal, or parallel to the axis or centerline of a part. Usually refers to axial compression or axial tension.

**axial load bearing**. See *thrust bearing*.

**axial loading**. The application of pressure on a powder or compact in the direction of the press axis.

**axial rake**. For angular (not helical) flutes, the angle between a plane containing the tooth face and the axial plane through the tooth point. See also *face mill* for definition of nomenclature.

**axial ratio**. The ratio of the length of one axis to that of another, for example, $c/a$, or the continued ratio of three axes, such as $a{:}b{:}c$.

**axial relief**. The relief or clearance behind the end cutting edge of a milling cutter. See also *face mill*.

**axial rolls**. In *ring rolling*, vertically displaceable, taped rolls mounted in a horizontally displaceable frame opposite to, but on the same centerline as, the main roll and rolling mandrel (Fig. 21). The axial rolls control ring height during rolling.

**axial runout**. For any rotating element, the total variation from a true plane of rotation, taken in a direction parallel to the axis of rotation. Compare with *radial runout*.

**axial seal**. See *face seal*.

**axial strain**. The linear strain in a plane parallel to the longitudinal axis of the specimen.

**axial winding**. In filament-wound reinforced plastics, a winding with the filament parallel to, or at a small angle to, the axis (0° helix angle). See also *polar winding*.

**axis** (crystal). The edge of the unit cell of a space lattice. Any one axis of any one lattice is defined in length and direction relative to other axes of that lattice.

**axis** (weld). A line through the length of a weld, perpendicular to and at the geometric center of its cross section.

FIG. 21

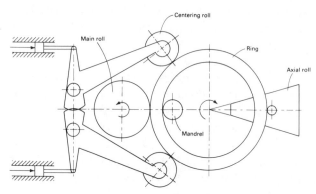

*Axial ring rolling setup*

# B

**B.** The symbol for a repeating unit in a copolymer chain.

**B₁₀-life**. See *rating life*.

**Babbitt metal**. A nonferrous bearing alloy originated by Isaac Babbitt in 1839. Currently, the term includes several tin-base alloys consisting mainly of various amounts of copper, antimony, tin, and lead. Lead-base Babbitt metals are also used. Table 7 lists compositions of tin-base Babbitt bearing alloys.

**back bead**. See preferred term *back weld*.

**back draft**. A reverse taper on a casting pattern or a forging die that prevents the pattern or forged stock from being removed from the cavity.

**back extrusion**. See *backward extrusion*.

**backfill**. Material placed in a drilled hole to fill space around anodes, vent pipe, and buried components of a cathodic protection system (Fig. 22).

**backfire**. The momentary recession of the flame into the welding tip or cutting tip followed by immediate reappearance or complete extinction of the flame. See also *flashback*.

## Table 7  Compositions of tin-base Babbitt alloys

| Designation | Sn(a) | Sb | Pb max(b) | Cu | Fe max | As max | Bi max | Zn max | Al max | Total other max |
|---|---|---|---|---|---|---|---|---|---|---|
| **ASTM B 23 alloys** | | | | | | | | | | |
| Alloy 1 . . . . . . . . . . . . . . . . . . . . . . . . . 91.0 | 91.0 | 4.5 | 0.35 | 4.5 | 0.08 | 0.10 | 0.08 | 0.005 | 0.005 | 0.05 Cd(c) |
| Alloy 2 . . . . . . . . . . . . . . . . . . . . . . . . . 89.0 | 89.0 | 7.5 | 0.35 | 3.5 | 0.08 | 0.10 | 0.08 | 0.005 | 0.005 | 0.05 Cd(c) |
| Alloy 3 . . . . . . . . . . . . . . . . . . . . . . . . . 84.0 | 84.0 | 8.0 | 0.35 | 8.0 | 0.08 | 0.10 | 0.08 | 0.005 | 0.005 | 0.05 Cd(c) |
| Alloy 11 . . . . . . . . . . . . . . . . . . . . . . . . 87.5 | 87.5 | 6.8 | 0.50 | 5.8 | 0.08 | 0.10 | 0.08 | 0.005 | 0.005 | 0.05 Cd(c) |
| **SAE alloys** | | | | | | | | | | |
| SAE 11 . . . . . . . . . . . . . . . . . . . . . . . . . 86.0 | 86.0 | 6.0–7.5 | 0.50 | 5.0–6.5 | 0.08 | 0.10 | 0.08 | 0.005 | 0.005 | 0.20 |
| SAE 12 . . . . . . . . . . . . . . . . . . . . . . . . . 88.0 | 88.0 | 7.0–8.0 | 0.50 | 3.0–4.0 | 0.08 | 0.10 | 0.08 | 0.005 | 0.005 | 0.20 |
| **Intermediate lead-tin alloys** | | | | | | | | | | |
| Lead-tin babbitt . . . . . . . . . . . . . . . . 75 | 75 | 12 | 9.3–10.7 | 3 | 0.08 | 0.15 | . . . | . . . | . . . | . . . |
| ASTM B 102, Alloy PY1815A . . . . . . 65 | 65 | 15 | 17–19 | 2 | 0.08 | 0.15 | . . . | 0.01 | 0.01 | . . . |

(a) Desired minimum in ASTM alloys; specified minimum in SAE alloys. (b) Maximum unless a range is specified. (c) Total named elements, 99.80%

FIG. 22

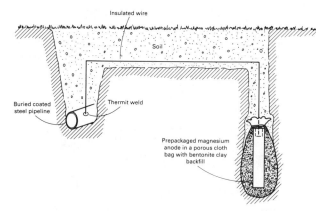

*Use of backfill in cathodic protection of buried pipeline*

**back gouging**. The removal of weld metal and base metal from the other side of a partially welded joint to ensure complete penetration upon subsequent welding from that side.

**background**. Any noise in the signal due to instabilities in the system or to environmental interferences. See also *signal-to-noise ratio*.

**backhand welding**. A welding technique in which the welding torch or gun is directed opposite to the progress of welding (Fig. 23). Sometimes referred to as the "pull gun technique" in gas metal arc welding and flux-cored arc welding. Compare with *forehand welding*. See also *travel angle*, *work angle*, and *drag angle*.

FIG. 23

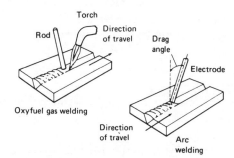

*Backhand welding technique*

**backing**. (1) In grinding, the material (paper, cloth, or fiber) that serves as the base for coated abrasives. (2) In welding, a material placed under or behind a joint to enhance the quality of the weld at the root. It may be a metal backing ring or strip; a pass of weld metal; or a nonmetal such as carbon, granular flux or a protective gas. (3) In plain bearings, that part of the bearing to which the bearing alloy is attached, normally by a metallurgical bond.

**backing bead**. See preferred term *backing weld*.

**backing filler metal**. See *consumable insert*.

**backing film** (metallography). A film used as auxiliary support for the thin replica or specimen-supporting film.

**backing pass**. A pass made to deposit a backing weld.

**backing plate**. In plastic injection molding equipment, a heavy steel plate that is used as a support for the cavity blocks, guide pins, and bushings. In blow molding equipment, it is the steel plate on which the cavities (that is, the bottle molds) are mounted. In casting, a second *bottom board* on which molds are opened.

**backing ring**. Backing in the form of a ring, generally used in the welding of piping.

**backing-split pipe**. See *split pipe backing*.

**backing strap**. See preferred term *backing strip*.

**backing strip**. Backing in the form of a strip.

**backing weld**. Backing in the form of a weld.

**backlash**. Lost motion, play, or movement in moving parts such that the driving element (as a gear) can be reversed for some angle or distance before working contact is again made with a driven element.

**backoff**. A rapid withdrawal of a grinding wheel or cutting tool from contact with a workpiece.

**back pressure**. Resistance of a plastic, because of its viscosity, to continued flow when the mold is closing.

**back-pressure relief port**. An opening in an extrusion die for plastics that allows for the escape of excess material.

**back rake**. The angle on a single-point turning tool corresponding to axial rake in milling. It is the angle measured between the plane of the tool face and the reference plane and lies in a plane perpendicular to the axis of the work material and the base of the tool. See figure accompanying *single-point tool*.

**back reflection**. The diffraction of x-rays at a *Bragg angle* approaching 90°.

**back ring**. A split or multisegment ring in a circumferential seal assembly used for restricting axial leakage flow.

**backscattered electron**. An information signal arising from elastic (electron-nucleus) collisions, wherein the incident electron rebounds from the interaction with a small energy loss. The backscattered electron yield is strongly dependent upon atomic number, qualitatively describes the origin of characteristic rays, and reveals compositional and topographic information about the specimen. See also *atomic number imaging*.

**backstep sequence**. A longitudinal sequence in which the weld bead increments are deposited in the direction opposite to the progress of welding the joint (Fig. 24). See also *block sequence*, *cascade sequence*, *continuous sequence*, *joint building sequence*, and *longitudinal sequence*.

**back taper**. Reverse draft used in a mold to prevent the molded plastic article from drawing freely. See also *undercut*.

**back-to-back ring seal**. An adaptation of the simple ring seal

FIG. 24

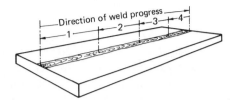

*Backstep welding sequence*

FIG. 25

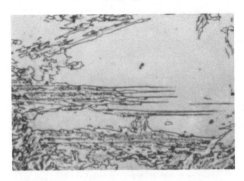

*Upper bainite in a 4360 steel specimen*

FIG. 26

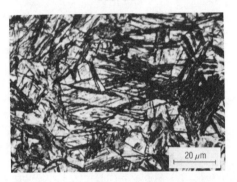

*Lower bainite (dark plates) in a 4150 steel specimen*

that employs two identical elements loaded axially by a spring placed between the rings. The spring forces the elements against mating rings on either side.

**backup** (flash and upset welding). A locator used to transmit all or a portion of the upsetting force to the workpieces or to aid in preventing the workpieces from slipping during upsetting.

**backup coat**. The ceramic slurry of dip coat that is applied in multiple layers to provide a ceramic shell of the desired thickness and strength for use as a casting mold.

**backward extrusion**. Same as *indirect extrusion*. See *extrusion*.

**back weld**. A weld deposited at the back of a single groove weld.

**baffle**. A device used to restrict or divert the passage of fluid through a pipeline or channel.

**bagging**. Applying an impermeable layer of film over an uncured part and sealing the edges so that a vacuum can be drawn.

**baghouse**. A chamber containing bags for filtering solids out of gases.

**bag molding**. A method of molding or bonding plastics or composites involving the application of fluid pressure, usually by means of air, steam, water, or vacuum, to a flexible cover that, sometimes in conjunction with the rigid die, completely encloses the material to be bonded. Also called blanket molding. See also *vacuum bag molding*.

**bag side**. The side of a plastic or composite part that is cured against the vacuum bag.

**bail**. Hoop or arched connection between the core hook and ladle or between crane hook and mold trunnions.

**bainite**. A metastable aggregate of *ferrite* and *cementite* resulting from the transformation of *austenite* at temperatures below the *pearlite* range but above $M_s$, the martensite start temperature. Upper bainite is an aggregate that contains parallel lath-shape units of ferrite, produces the so-called "feathery" appearance in optical microscopy (Fig. 25), and is formed above approximately 350 °C (660 °F). Lower bainite, which has an acicular appearance similar to tempered martensite (Fig. 26), is formed below approximately 350 °C (660 °F).

**bainitic hardening**. Quench-hardening treatment resulting principally in the formation of *bainite*.

**bake**. Heating in an oven to a low controlled temperature to remove gases or to harden a binder.

**baked core**. A casting core that has been heated through sufficient time and temperature to produce the desired physical properties attainable from its oxidizing or thermal-setting binders.

**Bakelite**. A proprietary name for a phenolic thermosetting resin used as a plastic mounting material for metallographic samples.

**baking**. (1) Heating to a low temperature in order to remove gases. (2) Curing or hardening surface coatings such as paints by exposure to heat. (3) Heating to drive off moisture, as in baking of sand cores after molding.

**balance**. (1) (dynamic) Condition existing where the principal inertial axis of a body coincides with its rotational axis. (2) (static) Condition existing where the center of gravity of a body lies on its rotational axis.

**balanced construction**. In woven reinforcements, equal parts of warp and fill fibers. Construction in which reactions to tension and compression loads result in extension or compression deformations only and in which flexural loads produce pure bending of equal magnitude in axial and lateral directions.

**balanced design**. In filament-wound reinforced plastics, a winding pattern so designed that the stresses in all filaments are equal.

**balanced-in-plane contour**. In a filament-wound part, a head contour in which the filaments are oriented within a plane and the radii of curvature are adjusted to balance the stresses along the filaments with the pressure loading.

**balanced laminate**. A laminate in which all laminae at angles other than 0° and 90° occur only in ± pairs (not necessarily adjacent) and are symmetrical around the centerline). See also *symmetrical laminate*.

**balanced twist**. An arrangement of twists in a combination of two or more reinforcing strands that does not kink or twist when the yarn produced is held in the form of an open loop.

**ball bearing**. A rolling-element bearing in which the rolling elements are spherical. See also the figure accompanying the term *rolling-element bearing*.

**ball burnishing**. (1) Same as *ball sizing*. (2) Removing burrs and polishing small stampings and small machined parts by *tumbling* in the presence of metal balls.

**ball clay**. A secondary clay, commonly characterized by the presence of organic matter, high plasticity, high dry strength, long vitrification range, and a light color when fired. Used extensively in traditional ceramics, such as whiteware, wall tile, and china. See also *traditional ceramics* (Technical Brief 57).

**ball complement**. The number of balls contained in a ball bearing.

**ball indented bearing**. A bearing with surface indentations serving as lubricant reservoirs.

**balling up**. The formation of globules of molten brazing filler metal or flux due to lack of wetting of the base metal.

**ball mill**. A machine consisting of a rotating hollow cylinder partly filled with metal balls (usually hardened steel or white cast iron) or sometimes pebbles; used to pulverize crushed ores or other substances such as pigments or ceramics (Fig. 27).

**ball milling**. A method of grinding and mixing material, with or without liquid, in a rotating cylinder or conical mill partially filled with grinding media such as balls or pebbles.

**ball sizing**. Sizing and finishing a hole by forcing a ball of suitable size, finish, and hardness through the hole or by using a burnishing bar or broach consisting of a series of spherical lands of gradually increasing size coaxially arranged. Also called *ball burnishing*, and sometimes ball broaching.

**banbury**. An apparatus for compounding polymeric materials. It is composed of a pair of contrarotating rotors that masticate the materials to form a homogeneous blend. This internal-type mixer produces excellent mixing.

**band density**. In filament winding of composites, the quantity of fiberglass reinforcements per inch of band width, expressed as strands (or filaments) per inch.

**banded structure**. A segregated structure consisting of alternating nearly parallel bands of different composition, typically aligned in the direction of primary hot working (Fig. 28).

**banding**. Inhomogeneous distribution of alloying elements or phases aligned in filaments or plates parallel to the direction of working. See also *banded structure*, *ferrite-pearlite banding*, and *segregation banding*.

**band mark**. An indentation in carbon steel or strip caused by external pressure on the packaging band around cut lengths or coils; it may occur in handling, transit, or storage.

**bands**. (1) Hot-rolled steel strip, usually produced for reroll-

FIG. 27

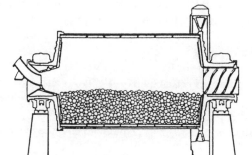

*Conventional ball mill*

FIG. 28

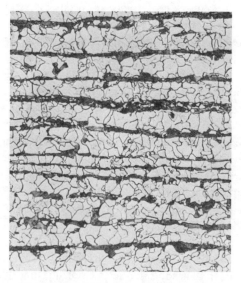

*Hot-rolled 1022 steel showing severe banding. 250×*

ing into thinner sheet or strip. Also known as hot bands or band steel. (2) See *electron bands*.

**band thickness**. In filament winding of composites, the thickness of the reinforcement as it is applied to the mandrel.

**band width**. In filament winding of composites, the width of the reinforcement as it is applied to the mandrel.

**bank sand**. Sedimentary deposits, usually containing less than 5% clay, occurring in banks or pits, used in core-making and in synthetic molding sands. See also *sand*.

**bar**. (1) A section hot rolled from a *billet* to a form, such as round, hexagonal, octagonal, square, or rectangular, with sharp or rounded corners or edges and a cross-sectional area of less than 105 cm$^2$ (16 in.$^2$). (2) A solid section that is long in relationship to its cross-sectional dimensions, having a completely symmetrical cross section and a width or greatest distance between parallel faces of 9.5 mm (³⁄₈) in. or more. (3) An obsolete unit of pressure equal to 100 kPa.

**Barcol hardness**. A hardness value obtained by measuring the resistance to penetration of a sharp steel point under a spring load. The instrument, called the Barcol impressor, gives a direct reading on a 0 to 100 scale. The hardness value is often used as a measure of the degree of cure of a plastic.

**bare electrode**. A filler metal electrode consisting of a single metal or alloy that has been produced into a wire, strip, or bar form and that has had no coating or covering applied to it other than that which was incidental to its manufacture or preservation.

**bare glass**. Glass in the form of yarns, rovings, and fabrics from which the sizing or finish has been removed. Also, such glass before the application of sizing or finish.

**bare metal arc welding (BMAW)**. An arc welding process which produces coalescence of metals by heating them with an electric arc between a bare or lightly coated metal electrode and the work. Neither shielding or pressure is used and filler metal is obtained from the electrode.

**bar folder**. A machine in which a folding bar or wing is used to bend a metal sheet whose edge is clamped between the upper folding leaf and the lower stationary jaw into a narrow, sharp, close, and accurate fold along the edge. It is also capable of making rounded folds such as those used in wiring. A universal folder is more versatile in that it is limited to width only by the dimensions of the sheet.

**barium titanate (BaTiO₃)**. The basic raw material used to make high dielectric constant ceramic capacitors. Used also in high thermal conductivity, thick-film ceramic pastes.

**bark**. The decarburized layer just beneath the scale that results from heating steel in an oxidizing atmosphere.

**Barkhausen effect**. The sequence of abrupt changes in mag-

netic induction occurring when the magnetizing force acting on a ferromagnetic specimen is varied.

**barn**. A unit of area equal to 10$^{-24}$ cm$^2$ used in specifying nuclear cross sections. See also *nuclear cross section*.

**barrel cleaning**. Mechanical or electrolytic cleaning of metal in rotating equipment.

**barrel distortion**. See *negative distortion*.

**barrel finishing**. Improving the surface finish of workpieces by processing them in rotating equipment along with abrasive particles that may be suspended in a liquid. The barrel is normally loaded about 60% full with a mixture of parts, media, compound, and water (Fig. 29).

FIG. 29

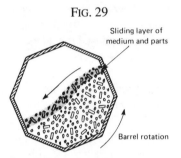

Sliding layer of medium and parts

Barrel rotation

*Action of media and parts during barrel finishing*

**barreling**. Convexity of the surfaces of cylindrical or conical bodies, often produced unintentionally during upsetting or as a natural consequence during compression testing. See also *compression test*.

**barrel plating**. Plating articles in a rotating container, usually a perforated cylinder that operates at least partially submerged in a solution.

**barrier coat**. An exterior coating applied to a composite filament-wound structure to provide protection. In fuel tanks, a coating applied to the inside of the tank to prevent fuel from permeating the side wall.

**barrier film**. The layer of film used during cure to permit removal of air and volatiles from a reinforced plastic or a composite lay-up while minimizing resin loss.

**barrier plastics**. A general term applied to a group of light-weight, transparent, impact-resistant plastics, usually rigid copolymers of high acrylonitrile content. Barrier plastics are generally characterized by gas, aroma, and flavor barrier characteristics approaching those of metal and glass.

**barstock**. Same as *bar*.

**basal plane**. (1) That plane of a hexagonal or tetragonal crystal perpendicular to the axis of highest symmetry. Its Miller indices are (001). (2) A plane perpendicular to the principal axis (*c* axis) in a tetragonal or hexagonal structure.

**base**. (1) A chemical substance that yields hydroxyl ions (OH⁻) when dissolved in water. Compare with *acid*. (2) The surface on which a single-point tool rests when held in a tool post. Also known as heel. See also the figure accompanying the term *single-point tool*. (3) In forging, see *anvil*.

**base bullion**. Crude lead containing recoverable silver, with or without gold.

**base-line technique**. A method for measurement of absorption peaks for quantitative analysis of chemical compounds in which a base line is drawn tangent to the spectrum background; the distance from the base line to the absorption peak is the absorbance due to the sample under study.

**base material**. (1) The material to be welded, brazed, soldered, or cut. See also *base metal* and *substrate*. (2) For a printed circuit board, the insulating material upon which a conductor pattern may be formed. The base material may be rigid or flexible, and it may be a dielectric sheet or insulated metal sheet. See also *dielectrics*.

**base metal**. (1) The metal present in the largest proportion in an alloy; brass, for example, is a copper-base alloy. (2) The metal to be brazed, cut, soldered, or welded. (3) After welding, that part of the metal which was not melted. (4) A metal that readily oxidizes, or that dissolves to form ions. Contrast with *noble metal* (2).

**base metal test specimen**. A test specimen composed wholly of *base metal*.

**basic bottom and lining**. The inner bottom and lining of a melting furnace, consisting of materials such as crushed burned dolomite, magnesite, magnesite bricks, or basic slag that give a basic reaction at the operating temperature.

**basic dynamic load capacity** (of a bearing). The radial load that a rolling-element bearing can support for a rating life of one million revolutions (500 h at $33\frac{1}{3}$ rpm). See also *basic load rating*.

**basic load rating** (*C*). The radial load that a ball bearing can withstand for one million revolutions of the inner ring. The value of the basic load rating depends on bearing type, bearing geometry, accuracy of fabrication, and bearing material. See also *basic dynamic load capacity* and *dynamic load*.

**basic NMR frequency**. The frequency, measured in hertz, of the oscillating magnetic field applied to induce transitions between nuclear magnetic energy levels. See also *magnetic resonance*.

**basic oxygen furnace**. A large tiltable vessel lined with basic refractory material which is the principal type of furnace for modern steelmaking. After the furnace is charged with molten pig iron (which usually comprises 65 to 75% of the charge), scrap steel, and fluxes, a lance is brought down near the surface of the molten metal and a jet of high-velocity oxygen impinges on the metal (Fig. 30). The oxygen reacts with carbon and other impurities in the steel to form liquid compounds that dissolve in the slag and gases that escape from the top of the vessel.

FIG. 30

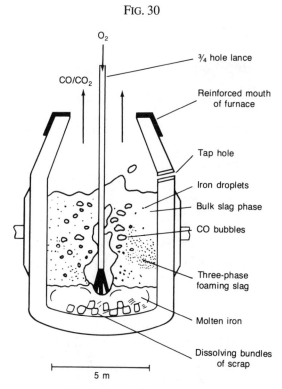

*Basic oxygen furnace vessel*

**basic refractories**. Refractories whose major constituent is lime, magnesia, or both, and which may react chemically with acid refractories, acid slags, or acid fluxes at high temperatures. Basic refractories are used for furnace linings. Compare with *acid refractory*.

**basic static load rating** (of a bearing). A load that, if exceeded on a nonrotating rolling-element bearing, produces a total permanent deformation of rolling element and race at the most heavily stressed contact point of 0.0001 times the ball or roller diameter or greater.

**basic steel**. Steel melted in a furnace with a *basic bottom and lining* and under a slag containing an excess of a basic substance such as magnesia or lime.

**basin**. Same as *pouring basin*.

**basis metal**. The original metal to which one or more coatings are applied.

**basketweave** (titanium). Alpha platelets with or without interleaved β platelets that occur in colonies in a *Widmanstätten structure* (Fig. 31).

**basketweave** (composites). In this type of woven reinforcement, two or more warp threads go over and under two or more filling threads in a repeat pattern (Fig. 32). The basket weave is less stable than the *plain weave* but produces a flatter and stronger fabric. It is also a more pliable fabric than the plain weave and maintains a certain degree of porosity without too much sleaziness, although not as much as the plain weave.

FIG. 31

*Basketweave structure in a hot-worked titanium alloy. 100×*

FIG. 32

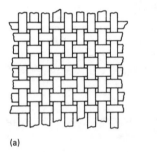

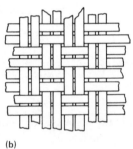

(a)                                                          (b)

*Comparison of plain weave (a) and basketweave (b) fabric construction forms*

**batch**. A quantity of materials formed during the same process or in one continuous process and having identical characteristics throughout. See also *lot*.

**batch furnace**. A furnace used to heat treat a single load at a time. Batch-type furnaces are necessary for large parts such as heavy forgings and are preferred for complex alloy grades requiring long cycles. See also *car furnace* and *horizontal batch furnace*.

**batch sintering**. Presintering or sintering in such a manner that compacts are sintered and removed from the furnace before additional unsintered compacts are placed in the furnace.

**bath**. Molten metal on the hearth of a furnace, in a crucible, or in a ladle.

**bath lubrication**. See *flood lubrication*.

**batt**. Felted fabrics. Structures built by the interlocking action of compressing fibers, without spinning, weaving, or knitting.

**Bauschinger effect**. The phenomenon by which plastic deformation increases yield strength in the direction of plastic flow and decreases it in other directions.

**bauxite**. A whitish to reddish mineral composed largely of hydrates of *alumina* having a composition of $Al_2O_3 \cdot 2H_2O$. It is the most important ore (source) of aluminum, alumina abrasives, and alumina-based refractories.

**Bayer process**. A process for extracting alumina from bauxite ore before the electrolytic reduction. The bauxite is digested in a solution of sodium hydroxide, which converts the alumina to soluble aluminate. After the "red mud" residue has been filtered out, aluminum hydroxide is precipitated, filtered out, and calcined to alumina.

**B-basis**. The "B" mechanical property value is the value above which at least 90% of the population of values is expected to fall, with a confidence of 95%. See also *A-basis*, *S-basis*, and *typical basis*.

**beach marks**. Macroscopic progression marks on a fatigue fracture or stress-corrosion cracking surface that indicate successive positions of the advancing crack front (Fig. 33). The classic appearance is of irregular elliptical or semi-elliptical rings, radiating outward from one or more origins. Beach marks (also known as clamshell marks or arrest marks) are typically found on service fractures where the part is loaded randomly, intermittently, or with periodic variations in mean stress or alternating stress. See also *striation* (metals).

FIG. 33

*Beach marks in a bolt that failed by fatigue*

**bead**. (1) Half-round cavity in a mold, or half-round projection or molding on a casting. (2) A single deposit of weld metal produced by fusion.

**beaded flange**. A flange reinforced by a low ridge, used mostly around a hole.

**beading**. Raising a ridge or projection on sheet metal.

### Table 8  Nominal compositions of high-carbon bearing steels

| Grade | Composition, % | | | | | |
|---|---|---|---|---|---|---|
| | C | Mn | Si | Cr | Ni | Mo |
| AISI 52100 | 1.04 | 0.35 | 0.25 | 1.45 | . . . | . . . |
| ASTM A 485-1 | 0.97 | 1.10 | 0.60 | 1.05 | . . . | . . . |
| ASTM A 485-3 | 1.02 | 0.78 | 0.22 | 1.30 | . . . | 0.25 |
| TBS-9 | 0.95 | 0.65 | 0.22 | 0.50 | 0.25 max | 0.12 |
| SUJ 1(a) | 1.02 | <0.50 | 0.25 | 1.05 | <0.25 | <0.08 |
| 105Cr6(b) | 0.97 | 0.32 | 0.25 | 1.52 | . . . | . . . |
| SHKH15-SHD(c) | 1.00 | 0.40 | 0.28 | 1.48 | <0.30 | . . . |

(a) Japanese grade. (b) German grade. (c) Russian grade

### Table 9  Carburizing bearing steels

| Grade | Composition, % | | | | | |
|---|---|---|---|---|---|---|
| | C | Mn | Si | Cr | Ni | Mo |
| 4118 | 0.20 | 0.80 | 0.22 | 0.50 | . . . | 0.11 |
| 5120 | 0.20 | 0.80 | 0.22 | 0.80 | . . . | . . . |
| 8620 | 0.20 | 0.80 | 0.22 | 0.50 | 0.55 | 0.20 |
| 4620 | 0.20 | 0.55 | 0.22 | . . . | 1.82 | 0.25 |
| 4320 | 0.20 | 0.55 | 0.22 | 0.50 | 1.82 | 0.25 |
| 3310 | 0.10 | 0.52 | 0.22 | 1.57 | 3.50 | . . . |
| SCM420 | 0.20 | 0.72 | 0.25 | 1.05 | . . . | 0.22 |
| 20MnCr5 | 0.20 | 1.25 | 0.27 | 1.15 | . . . | . . . |

**bead weld.** See preferred term *surfacing weld.*

**beam hardening.** The increase in effective energy of a poly-energetic (for example, x-ray) beam with increasing attenuation of the beam. Beam hardening is due to the preferential attenuation of the lower-energy, or soft, radiation.

**bearing.** A support or guide by means of which a moving part is located with respect to other parts of a mechanism.

**bearing area.** (1) The projected bearing or load-carrying area when viewed in the direction of the load. Sometimes used as a synonym for *real area of contact* (this usage is not recommended). (2) The sum of the horizontal intercepts of a surface profile at a given level. (3) The product of the pin (or hole) diameter and the specimen thickness. See also *bearing test.*

**bearing bronzes.** Bronzes used for bearing applications. Two common types of bearing bronzes are copper-base alloys containing 5 to 20 wt% tin and a small amount of phosphorus (*phosphor bronzes*) and copper-base alloys containing up to 10 wt% tin and up to 30 wt% lead (*leaded bronzes*).

**bearing characteristic number.** A dimensionless number that is used to evaluate the operating conditions of plain bearings. See also *capacity number* and *Sommerfeld number.*

**bearing fraction.** The ratio of the bearing area to a reference length.

**bearing steels.** *Alloy steels* used to produce *rolling-element bearings.* Typically, bearings have been manufactured from both high-carbon (1.00%) and low-carbon (0.20%) steels. The high-carbon steels are used in either a through-hardened or a surface induction-hardened condition. Low-carbon bearing steels are carburized to provide the necessary surface hardness while maintaining desirable core properties. Tables 8 and 9 list compositions of typical bearing steels.

**bearing strain.** The ratio of the deformation of the bearing hole, in the direction of the applied force, to the pin diameter. Also, the stretch or deformation strain for a sample under bearing load.

**bearing strength.** The maximum bearing stress that can be sustained. Also, the bearing stress at that point on the stress-strain curve at which the tangent is equal to the bearing stress divided by $n\%$ of the bearing hole diameter.

**bearing stress** (metals). The shear load on a mechanical joint (such as a pinned or riveted joint) divided by the effective bearing area. The effective bearing area of a riveted joint, for example, is the sum of the diameters of all rivets times the thickness of the loaded member.

**bearing stress** (plastics). The applied load in pounds divided by the bearing area. Maximum bearing stress is the maximum load in pounds sustained by the specimen during the test, divided by the original bearing area.

**bearing test.** A method of determining the response to stress (load) of sheet products that are subjected to riveting, bolting, or a similar fastening procedure. The purpose of the test is to determine the *bearing strength* of the material and to measure the *bearing stress* versus the deformation of the hole created by a pin or rod of circular cross section that pierces the sheet perpendicular to the surface (Fig. 34).

**bearing yield strength.** The *bearing stress* at which a material exhibits a specified limiting deviation from the proportionality of *bearing stress* to *bearing strain.*

**bed.** (1) The stationary portion of a press structure that usually

FIG. 34

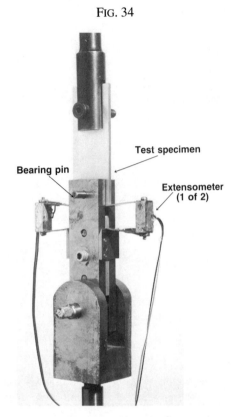

*Typical pin bearing test fixture*

FIG. 35

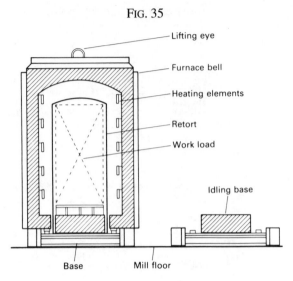

*Bell-type sintering furnace*

rests on the floor or foundation, forming the support for the remaining parts of the press and the pressing load. The *bolster* and sometimes the lower die are mounted on the top surface of the bed. (2) For machine tools, the portion of the main frame that supports the tool, the work, or both. (3) Stationary part of the shear frame that supports the material being sheared and the fixed blade.

**bedding**. Sinking a casting pattern down into the sand to the desired position and ramming the sand around it.

**bedding a core**. Placing an irregularly shaped casting core on a bed of sand for drying.

**Beer's law**. A relationship in which the optical absorbance of a homogeneous sample containing an absorbing substance is directly proportional to the concentration of the absorbing substance. See also *absorptivity*.

**Beilby layer**. A layer of metal disturbed by mechanical working, wear, or mechanical polishing presumed to be without regular crystalline structure (amorphous); originally applied to grain boundaries.

**bell**. A jar-like enclosure for containing a vacuum or a controlled atmosphere in sintering equipment.

**bellows seal**. A type of mechanical seal that utilizes a bellows for providing secondary sealing.

**bell-type furnace**. A furnace for the sintering of large batches of small pieces under a controlled atmosphere (Fig. 35).

**belt furnace**. A continuous-type furnace which uses a mesh-type or cast-link belt to carry parts through the furnace.

**belt grinding**. Grinding with an *abrasive belt*.

**bench molding**. Casting sand molds by hand tamping loose or production patterns at a bench without the assistance of air or hydraulic action.

**bench press**. Any small press that can be mounted on a bench or table.

**bend allowance**. The length of the arc of the neutral axis between the tangent points of a bend.

**bend angle**. The angle through which a bending operation is performed, that is, the supplementary angle to that formed by the two bend tangent lines or planes (Fig. 36).

FIG. 36

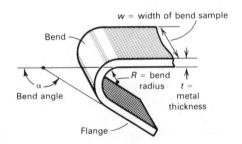

*Terms used in bend testing*

**bender**. Term denoting a die impression, tool, or mechanical device designed to bend forging stock to conform to the general configuration of die impressions to be subsequently used.

**bending**. The straining of material, usually flat sheet or strip metal, by moving it around a straight axis lying in the neutral plane. Metal flow takes place within the plastic range of the metal, so that the bent part retains a *permanent* *set* after removal of the applied stress. The cross section of the bend inward from the neutral plane is in compression; the rest of the bend is in tension. See also *bending stress*.

**bending brake.** A form of open-frame single-action press that is comparatively wide between the housings, with a bed designed for holding long, narrow forming edges or dies. Used for bending and forming strip, plate, and sheet (into boxes, panels, roof decks, and so on). Also known as *press brake*.

**bending dies.** Dies used in presses for bending sheet metal or wire parts into various shapes. The work is done by the punch pushing the stock into cavities or depressions of similar shape in the die or by auxiliary attachments operated by the descending punch.

**bending moment.** The algebraic sum of the couples or the moments of the external forces, or both, to the left or right of any section on a member subjected to bending by couples or transverse forces, or both.

**bending rolls.** Various types of machinery equipped with two or more rolls to form curved sheet and sections (Fig. 37).

FIG. 37

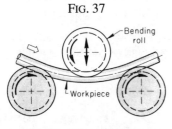

*Three-roll bending setup*

**bending stress.** A stress involving tensile and compressive forces, which are not uniformly distributed. Its maximum value depends on the amount of flexure that a given application can accommodate. Resistance to bending can be termed stiffness.

**bending stress (glass).** A stress system that simultaneously imposes a compressive component at one surface, graduating to an imposed tensile component at the opposite surface of a glass section.

**bending-twisting coupling.** A property of certain classes of laminates that exhibit twisting curvatures when subjected to bending moments.

**bend or twist (defect).** Distortion similar to warpage generally caused during forging or trimming operations. When the distortion is along the length of the part, it is termed bend; when across the width, it is termed twist. When bend or twist exceeds tolerance, it is considered a defect. Corrective action consists of hand straightening, machine straightening, or cold restriking.

**bend radius.** (1) The inside radius of a bend section (Fig. 36). (2) The radius of a tool around which metal is bent during fabrication.

**bend tangent.** A tangent point at which a bending arc ceases or changes.

**bend test.** A test for determining relative ductility of metal that is to be formed (usually sheet, strip, plate, or wire) and for determining soundness and toughness of metal (after welding, for example). The specimen is usually bent over a specified diameter through a specified angle for a specified number of cycles. There are four general types of bend test, named according to the manner in which the forces are applied to the specimen to make the bend: *free bend*, *guided bend*, *semiguided bend*, and *wraparound bend*.

**beneficiation.** Concentration or other preparation of ore for smelting.

**bentonite.** A colloidal claylike substance derived from the decomposition of volcanic ash composed chiefly of the minerals of the montmorillonite family. It is used for bonding molding sand.

**benzene ring.** The six-carbon ring structure found in benzene, $C_6H_6$, and in organic compounds formed from benzene by replacement of one or more hydrogen atoms by other chemical atoms or radicals.

**beryllia.** A colorless to white powder of the composition BeO used in the manufacture of hot-pressed ceramic parts—most notably basic refractories and substrates (heat sinks) in electronics. Also known as beryllium oxide.

**beryllium.** See *Technical Brief 5*.

**beryllium bronze.** See preferred term *beryllium-copper*.

**beryllium-copper.** Copper-base alloys containing not more than 3% Be. Available in both cast and wrought forms, these alloys rank high among copper alloys in attainable strength, while retaining useful levels of electrical and thermal conductivity. Applications for these alloys include electronic components (connector contacts), electrical equipment (switch and relay blades), antifriction bearings, housings for magnetic sensing devices, and resistance welding contacts. Table 10 lists compositions of commercial beryllium-copper alloys.

**beryllium-nickel.** Age-hardenable nickel-base alloys containing up to 2.75% Be. Wrought beryllium nickel alloys are used primarily as mechanical and electrical/electronic components. Cast alloys are used in molds and cores for glass and polymer molding, diamond drill bit matrices, and cast turbine parts. Table 11 lists compositions of commercial beryllium-nickel alloys.

**beryllium oxide (BeO).** See *beryllia*.

**beryllium window.** A very thin (~7.5 μm thick), relatively x-ray transparent window separating the x-ray detector from the vacuum chamber, which serves to protect the detector from damage.

**bessemer process.** A process for making steel by blowing air through molten pig iron contained in a refractory lined vessel so as to remove by oxidation most of the carbon,

## Technical Brief 5: Beryllium

BERYLLIUM is a steel-gray, lightweight, very hard metallic element (symbol Be). The metal is extracted from the minerals beryl ($3BeO \cdot Al_2O_3 \cdot 6SiO_2$) and bertrandite ($4BeO \cdot 2SiO_2 \cdot H_2O$) by chemical reduction.

Almost all of the beryllium in use is a powder metallurgy product. Beryllium powder is consolidated into billets by vacuum hot pressing, hot isostatic pressing, or cold isostatic pressing and sintering. The billets can subsequently be rolled into plate, sheet, and foil or extruded into shapes or tubing at elevated temperatures. Parts with density values in excess of 99.5% of the theoretical value of 1.8477 $g/cm^3$ can be produced.

### Selected physical properties of beryllium

| Property | Amount |
|---|---|
| Elastic modulus, GPa ($10^6$ psi) | 303 (44) |
| Density, $g/cm^3$ ($lb/in.^3$) | 1.8477 (0.067) |
| Thermal conductivity, W/m · K (Btu/h · ft · °F) | 210 (121) |
| Coefficient of thermal expansion, $10^{-6}/°C$ ($10^{-6}/°F$) | 11.5 (6.4) |
| Specific heat at room temperature, kJ/kg · K (Btu/lb · °F) | 2.17 (0.52) |
| Melting point, °C (°F) | 1283 (2341) |
| Mass absorption coefficient (Cu K-alpha), $cm^2/g$ | 1.007 |
| Specific modulus, m (in.)(a) | $16.7 \times 10^6$ ($6.56 \times 10^8$) |

(a) The specific modulus is defined (in inches) from the ratio of the elastic modulus (in psi) and the density (in $lb/in.^3$).

The unusual combination of physical and mechanical properties of unalloyed beryllium make it particularly effective in optical components, precision instruments, and specialized aerospace applications. In each of these applications areas, beryllium is selected because of its combination of low weight, high stiffness, and specific mechanical properties, such as a precise elastic limit. It is also useful because it is transparent to X-rays and other high-energy electromagnetic radiation. In addition, beryllium is used as an alloying element in some copper-base and nickel-base alloys (see the terms *beryllium-copper* and *beryllium-nickel*). Beryllium is also added in small quantities to aluminum and magnesium to achieve grain refinement and oxidation resistance. To date, no beryllium-base alloys have achieved commercial application.

Inhalation of respirable beryllium and its compounds should be avoided. Users should comply with occupational safety and health standards.

### Recommended Reading

- A.J. Stonehouse and J.M. Marder, Beryllium, *Metals Handbook*, Vol 2, 10th ed., ASM International, 1990, p 683-687
- J.M. Marder, Production of Beryllium Powder, *Metals Handbook*, Vol 7, 9th ed., American Society for Metals, 1984, p 169-172
- J.R. Davis, Beryllium P/M Technology, *Metals Handbook*, Vol 7, 9th ed., American Society for Metals, 1984, p 755-764
- J.M. Marder and R. Batich, Metallographic Techniques and Microstructures: Beryllium, *Metals Handbook*, Vol 9, 9th ed., American Society for Metals, 1985, p 389-391

### Chemistry of commercial grades of beryllium

| Beryllium grade | Beryllium components, % | | Maximum impurities, ppm | | | | | |
| --- | --- | --- | --- | --- | --- | --- | --- | --- |
| | Be, min | BeO, max | Al | C | Fe | Mg | Si | Other, each |
| **Structural grades** | | | | | | | | |
| S-65B | 99.0 | 0.7 | 600 | 1000 | 800 | 600 | 600 | 400 |
| S-200F and S-200FH | 98.5 | 1.5 | 1000 | 1500 | 1300 | 800 | 600 | 400 |
| **Instrument grades** | | | | | | | | |
| I-70A | 99.0 | 0.7 | 700 | 700 | 1000 | 700 | 700 | 400 |
| O-50 | 99.0 | 0.5 | 700 | 700 | 1000 | 700 | 700 | 400 |
| I-220B | 98.0 | 2.2 | 1000 | 1500 | 1500 | 800 | 800 | 400 |
| I-400B | 94.0 | 4.25 min | 1600 | 2500 | 2500 | 800 | 800 | 400 |

silicon, and manganese. This process is essentially obsolete in the United States.

**BET** (Brunauer-Emmett-Teller). An instrumental method for determining surface area of a solid sample, measuring monomolecular nitrogen gas adsorption at the temperature of liquid nitrogen (–210 °C, or –345 °F).

**beta (β).** The high-temperature allotrope of titanium with a body-centered cubic crystal structure that occurs above the β transus.

**beta annealing.** Producing a beta phase by heating certain titanium alloys in the temperature range of which this phase forms followed by cooling at an appropriate rate to prevent its decomposition.

**beta eutectoid stabilizer.** An alloying element in titanium

**Table 10  Composition of commercial beryllium-copper alloys**

| UNS number | Be | Co | Ni | Co + Ni | Co + Ni + Fe | Si | Pb | Cu |
|---|---|---|---|---|---|---|---|---|
| **Wrought alloys** | | | | | | | | |
| C17200 ..... | 1.80–2.00 | · · · | · · · | 0.20 min | 0.6 max | · · · | · · · | bal |
| C17300 ..... | 1.80–2.00 | · · · | · · · | 0.20 min | 0.6 max | · · · | 0.20–0.6 | bal |
| C17000 ..... | 1.60–1.79 | · · · | · · · | 0.20 min | 0.6 max | · · · | · · · | bal |
| C17510 ..... | 0.2–0.6 | · · · | 1.4–2.2 | · · · | · · · | · · · | · · · | bal |
| C17500 ..... | 0.4–0.7 | 2.4–2.7 | · · · | · · · | · · · | · · · | · · · | bal |
| C17410 ..... | 0.15–0.50 | 0.35–0.60 | · · · | · · · | · · · | · · · | · · · | bal |
| **Cast alloys** | | | | | | | | |
| C82000 ..... | 0.45–0.80 | · · · | · · · | 2.40–2.70 | · · · | · · · | · · · | bal |
| C82200 ..... | 0.35–0.80 | · · · | 1.0–2.0 | · · · | · · · | · · · | · · · | bal |
| C82400 ..... | 1.60–1.85 | · · · | · · · | 0.20–0.65 | · · · | · · · | · · · | bal |
| C82500 ..... | 1.90–2.25 | · · · | · · · | 0.35–0.70 | · · · | 0.20–0.35 | · · · | bal |
| C82510 ..... | 1.90–2.15 | · · · | · · · | 1.00–1.20 | · · · | 0.20–0.35 | · · · | bal |
| C82600 ..... | 2.25–2.55 | · · · | · · · | 0.35–0.65 | · · · | 0.20–0.35 | · · · | bal |
| C82800 ..... | 2.50–2.85 | · · · | · · · | 0.35–0.70 | · · · | 0.20–0.35 | · · · | bal |

Note: Copper plus additions, 99.5% min

**Table 11  Nominal compositions of commercial beryllium-nickel alloys**

| Product form | Alloy | Be | Cr | Other | Ni |
|---|---|---|---|---|---|
| Wrought | N03360..................... | 1.85–2.05 | · · · | 0.4–0.6 Ti | bal(a) |
| Cast | M220C ......................... | 2.0 | · · · | 0.5 C | bal |
| Cast | 41C ........................... | 2.75 | 0.5 | · · · | bal(b) |
| Cast | 42C ........................... | 2.75 | 12.0 | · · · | bal(b) |
| Cast | 43C ........................... | 2.75 | 6.0 | · · · | bal(b) |
| Cast | 44C ........................... | 2.0 | 0.5 | · · · | bal(b) |
| Cast | 46C ........................... | 2.0 | 12.0 | · · · | bal(b) |
| Cast | Master ...................... | 6 | · · · | · · · | bal(c) |

(a) 99.4 Ni + Be + Ti + Cu min, 0.25 Cu max. (b) 0.1 C max. (c) Master alloys with 10, 25, and 50 wt% Be are also available.

that dissolves preferentially in the β phase, lowers the α-β to β transformation temperature, and results in β decomposition to α plus a compound. This eutectoid reaction can be sluggish for some alloys.

**beta fleck.** Alpha-lean region in the α-β titanium microstructure significantly larger than the primary α width. This β-rich area has a β transus measurably below that of the matrix. Beta flecks have reduced amounts of primary α

FIG. 38

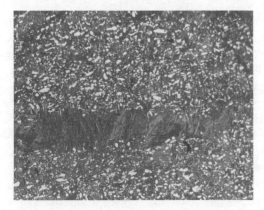

*Dark beta fleck (center of micrograph) in a titanium forging. 75×*

that may exhibit a morphology different from the primary α in the surrounding α-β matrix (Fig. 38).

**beta (β) gage.** A gage consisting of two facing elements, a β-ray-emitting source, and a β-ray detector. Also called beta-ray gage.

**beta isomorphous stabilizer.** An alloying element in titanium that dissolves preferentially in the β phase, lowers the α-β to β transformation temperature without a eutectoid reaction, and forms a continuous series of solid solutions with β-titanium.

**beta (β) loss peak.** In dynamic mechanical or dielectric measurement, the second peak in the damping curve below the melt, in order of decreasing temperature or increasing frequency.

**beta ray.** A ray of electrons emitted during the spontaneous disintegration of certain atomic nuclei.

**beta structure.** A Hume-Rothery designation for structurally analogous body-centered cubic phases (similar to beta brass) or electron compounds that have ratios of three valence electrons to two atoms. Not to be confused with a beta phase on a constitution diagram.

**beta transus.** The minimum temperature above which equilibrium α does not exist in titanium alloys. For β eutectoid additions, the β transus ordinarily is applied to hypoeutectoid compositions or those that lie to the left of the eutectoid composition.

**Betts process**. A process for the electrolytic refining of lead in which the electrolyte contains lead fluosilicate and fluosilicic acid.

**bevel**. See preferred term, *corner angle*, and also the figure accompanying the term *face mill*.

**bevel angle**. The angle formed between the prepared edge of a member and a plane perpendicular to the surface of the member (Fig. 39).

FIG. 39

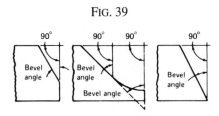

*Examples of bevel angles*

**bevel flanging**. Same as *flaring*.

**bevel groove**. See *groove weld*.

**bias**. A systematic error inherent in a method (such as temperature effects and extraction inefficiencies) or caused by some artifact or idiosyncrasy of the measurement system (such as blanks, contamination, mechanical losses, and calibration errors). Bias may be both positive and negative, and several types can exist concurrently, so that the net bias is all that can be evaluated except under certain conditions.

**bias fabric**. Fabric consisting of warp and fill fibers at an angle to the length of the fabric.

**biaxiality**. In a *biaxial stress* state, the ratio of the smaller to the larger principal stress.

**biaxial load**. A loading condition in which a specimen is stressed in two directions in its plane.

**biaxial stress**. A state of stress in which only one of the *principal stresses* is zero, the other two usually being in tension.

**biaxial winding**. In filament winding, a type of winding in which the helical band is laid in sequence, side by side, with crossover of the fibers eliminated.

**bidirectional laminate**. A reinforced plastic laminate with the fibers oriented in two directions in its plane. A cross laminate. See also *unidirectional laminate*.

**bidirectional seal**. A seal that is designed to seal equally well when the pressure is applied from either direction.

**bifilar eyepiece**. A filar eyepiece with motion in two mutually perpendicular directions.

**bifurcation**. The separation of a material into two sections.

**big-end bearing**. A bearing at the larger (crankshaft) end of a connecting rod in an engine. Also known as *bottom-end bearing*, *crankpin bearing*, and *large-end bearing*. See also *little-end bearing*.

**billet**. (1) A semifinished section that is hot rolled from a metal *ingot*, with a rectangular cross section usually ranging from 105 to 230 cm$^2$ (16 to 36 in.$^2$), the width being less than twice the thickness. Where the cross section exceeds 230 cm$^2$ (36 in.$^2$), the term *bloom* is properly but not universally used. Sizes smaller than 105 cm$^2$ (16 in.$^2$) are usually termed bars. (2) A solid semifinished round or square product that has been hot worked by forging, rolling, or extrusion. See also *bar*.

**billet mill**. A primary rolling mill used for making steel billets.

**bimetal bearing**. A bearing consisting of two layers. Bimetal bearings are usually made with a layer of bearing alloy on a bronze or steel backing.

**bimetal casting**. A casting made of two different metals, usually produced by *centrifugal casting*.

**binary alloy**. An alloy containing only two component elements.

**binary system**. The complete series of compositions produced by mixing a pair of components in all proportions.

**binder** (metals and ceramics). (1) In founding, a material, other than water, added to foundry sand to bind the particles together, sometimes with the use of heat. (2) In powder technology, a cementing medium: either a material added to the powder to increase the green strength of the compact, which is expelled during sintering; or a material (usually of relatively low melting point) added to a powder mixture for the specific purpose of cementing together powder particles that alone would not sinter into a strong body.

**binder** (plastics). (1) The resin or cementing constituent (of a plastic compound) that holds the other components together. The agent applied to fiber mat or preforms to bond the fibers before laminating or molding. (2) A component of an adhesive composition that is primarily responsible for the adhesive forces which hold two bodies together. See also *extender* and *filler*.

**binder metal**. A metal used as a binder. An example would be cobalt in cemented carbides.

**binder phase**. The soft metallic phase that cements the carbide particles in cemented carbides. More generally, a phase in a heterogeneous sintered material that gives solid coherence to the other phase(s) present.

**Bingham solid**. An idealized form of solid that begins to flow appreciably only when a certain stress, called the yield stress or yield point, has been exceeded. The solid subsequently flows at a rate proportional to the difference between the applied stress and this yield stress. Many greases can be regarded as Bingham solids.

**binodal curve**. In a two-dimensional phase diagram, a continuous line consisting of both of the pair of conjugate boundaries of a two-phase equilibrium that join without inflection at a critical point. See also *miscibility gap*.

**biological corrosion**. Deterioration of metals as a result of the metabolic activity of microorganisms. Also known as biofouling.

**bipolar electrode**. An *electrode* in an *electrolytic cell* that is not mechanically connected to the power supply, but is so placed in the electrolyte, between the *anode* and *cathode*, that the part nearer the anode becomes cathodic and the part nearer the cathode becomes anodic. Also called intermediate electrode.

**bipolar field**. A longitudinal magnetic field that creates two magnetic poles within a piece of material. Compare with *circular field*.

**birefringence**. A double-refraction phenomenon in anisotropic materials in which an unpolarized beam of light is divided into two beams with different directions and relative velocities of propagation. The amount of energy transmitted along an optical path through a crystal that exhibits birefringence becomes a function of crystalline orientation.

**birefringent crystal**. A crystalline substance that is anisotropic with respect to the velocity of light.

**birotational seal**. A seal that is designed for applications in which a shaft rotation is in either direction.

**biscuit** (metals). (1) An upset blank for drop forging. (2) A small cake of primary metal (such as uranium made from uranium tetrafluoride and magnesium by bomb reduction). Compare with *derby* and *dingot*.

**biscuit** (plastics). See *cull* and *preform*.

**bismaleimide (BMI)**. A type of polyimide that cures by an addition rather than a condensation reaction, thus avoiding problems with volatiles formation, and which is produced by a vinyl-type polymerization of a prepolymer terminated with two maleimide groups. Intermediate in temperature capability between epoxy and polyimide.

**bit** (soldering). That part of the soldering iron, usually made of copper, which actually transfers heat (and sometimes solder) to the joint.

**bit soldering**. See preferred term *iron soldering*.

**bitumen**. Asphaltlike polymer.

**bituminous coating**. Coal tar or asphalt-based coating.

**bivariant equilibrium**. A stable state among several phases equal to the number of components in a system and in which any two of the external variables of temperature, pressure, or concentration may be varied without necessarily changing the number of phases. Sometimes termed divariant equilibrium.

**black annealing**. Box annealing or pot annealing ferrous alloy sheet, strip, or wire impart a black color to the oxidized surface. See also *box annealing*.

**blackbody**. A hypothetical "body" that completely absorbs all incident radiant energy, independent of wavelength and direction, that is, neither reflects nor transmits any of the incident radiant energy.

**blackheart malleable**. See *malleable cast iron*.

**blacking**. Carbonaceous materials, such as graphite or powdered carbon, usually mixed with a binder and frequently carried in suspension in water or other liquid used as a thin facing applied to surfaces of molds or cores to improve casting finish.

**black light**. Electromagnetic radiation not visible to the human eye. The portion of the spectrum generally used in fluorescent inspection falls in the ultraviolet region between 330 and 400 nm, with the peak at 365 nm.

**black liquor**. The liquid material remaining from pulpwood cooking in the soda or sulfate papermaking process.

**black marking**. Black smudges on the surface of a pultruded plastic product that results from excessive pressures in the die when the pultrusion is rubbing against it or unchromed die surfaces, and that cannot be removed by cleaning or scrubbing or by wiping with solvent.

**black oxide**. A black finish on a metal produced by immersing it in hot oxidizing salts or salt solutions.

**blacksmith welding**. See preferred term *forge welding*.

**bladder**. An elastomeric lining for the containment of hydroproof or hydroburst pressurization medium in filament-wound structures.

**blade-setting angle**. See preferred term *cone angle*.

**blank**. (1) In forming, a piece of sheet metal, produced in cutting dies, that is usually subjected to further press operations. (2) A pressed, presintered, or fully sintered powder metallurgy compact, usually in the unfinished condition and requiring cutting, machining, or some other operation to produce the final shape. (3) A piece of stock from which a forging is made, often called a *slug* or *multiple*. (4) Any article of glass on which subsequent forming or finishing is required.

**blank carburizing**. Simulating the carburizing operation without introducing carbon. This is usually accomplished by using an inert material in place of the carburizing agent, or by applying a suitable protective coating to the ferrous alloy.

**blanket**. Fiber or fabric plies that have been laid up in a complete assembly and placed on or in the mold all at one time (flexible bag process). Also, the type of bag in which the edges are sealed against the mold.

**blankholder**. (1) The part of a drawing or forming die that holds the workpiece against the draw ring to control metal flow. (2) The part of a drawing or forming die that restrains the movement of the workpiece to avoid wrinkling or tearing of the metal.

**blanking**. The operation of punching, cutting, or shearing a piece out of stock to a predetermined shape.

**blank nitriding**. Simulating the nitriding operation without introducing nitrogen. This is usually accomplished by using an inert material in place of the nitriding agent or by applying a suitable protective coating to the ferrous alloy.

FIG. 40

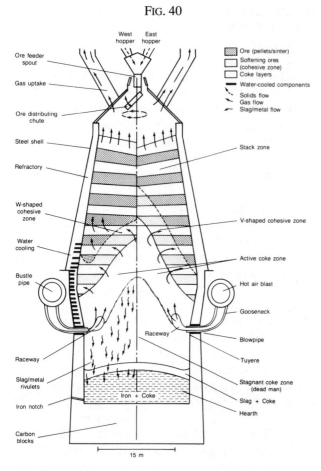

*Principal zones and component parts of an iron blast furnace*

FIG. 41

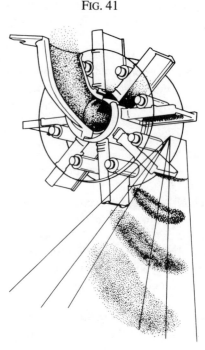

*Centrifugal blast cleaning unit*

**blast furnace**. A shaft furnace in which solid fuel is burned with an air blast to smelt ore in a continuous operation (Fig. 40). Where the temperature must be high, as in the production of pig iron, the air is preheated. Where the temperature can be lower, as in smelting of copper, lead, and tin ores, a smaller furnace is economical, and preheating of the blast is not required.

**blasting or blast cleaning**. A process for cleaning or finishing metal objects with an air blast or centrifugal wheel that throws abrasive particles against the surface of the workpiece (Fig. 41). Small, irregular particles of metal are used as the abrasive in gritblasting; sand, in sandblasting; and steel, in shotblasting.

**bleed**. (1) To give up color when in contact with water or a solvent. Undesirable movement of certain materials in a plastic, such as plasticizers in vinyl, to the surface of the finished article or into an adjacent material; also called migration. (2) Refers to molten metal oozing out of a casting. It is stripped or removed from the mold before complete solidification.

**bleeder cloth**. A woven or nonwoven layer of material used in the manufacture of composite parts to allow the escape of excess gas and resin during cure. The bleeder cloth is removed after the curing process and is not part of the final composite.

**bleeding**. (1) The removal of excess resin from a laminate during cure. The diffusion of color from a plastic part into the surrounding surface or part. (2) Separation of oil (or other fluid) from a grease.

**bleedout**. The excess liquid resin that migrates to the surface of a winding. Primarily pertinent to filament winding.

**bleed-out**. The spread of adhesive away from the bond area.

**blemish**. A nonspecific quality control term designating an imperfection that mars the appearance of a part but does not detract from its ability to perform its intended function.

**blend** (noun). Thoroughly intermingled powders of the same nominal composition.

**blended sand**. A mixture of sands of different grain size and clay content that provides suitable characteristics for foundry use.

**blending**. (1) In powder metallurgy, the thorough intermingling of powders of the same nominal composition (not to be confused with *mixing*). (2) The process of mixing mineral oils to obtain the desired consistency. Blending should be contrasted with compounding, which utilizes additives.

**blind hole**. A hole that is not drilled entirely through.

**blind joint**. A joint, no portion of which is visible.

**blind riser**. A *riser* that does not extend through the top of the mold (Fig. 42).

FIG. 42

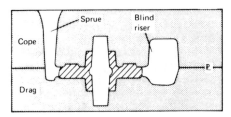

*Example of a blind riser*

**blind sample**. A sample submitted for analysis whose composition is known to the submitter but unknown to the analyst, used to test the efficiency of a measurement process.

**blister** (adhesives). An elevation of the surface of an adherend, the shape of which somewhat resembles a blister on the human skin. Its boundaries may be indefinitely outlined, and it may have burst and become flattened. A blister may be caused by insufficient adhesive; inadequate curing time, temperature, or pressure; or trapped air, water, or solvent vapor.

**blister** (ceramics). A defect consisting of a bubble that forms during fusion and remains when porcelain enamel solidifies.

**blister** (metals). (1) A casting defect, on or near the surface of the metal, resulting from the expansion of gas in a subsurface zone. It is characterized by a smooth bump on the surface of the casting and a hole inside the casting directly below the bump. (2) A raised area, often dome shaped, resulting from loss of adhesion between a coating or deposit and the basis metal.

**blister copper**. An impure intermediate product in the refining of copper, produced by blowing copper *matte* in a converter, the name being derived from the large blisters on the cast surface that result from the liberation of $SO_2$ and other gases.

**blistering**. The development during firing of enclosed or broken macroscopic vesicles or bubbles in a body, or in a glaze or other coating.

**block**. A preliminary forging operation that roughly distributes metal preparatory for *finish*.

**block and finish**. The forging operation in which a part to be forged is blocked and finished in one heat through the use of tooling having both a block impression and a finish impression in the same die block.

**block brazing**. A brazing process in which the heat required is obtained from heated blocks applied to the parts to be joined.

**block copolymer**. An essentially linear copolymer consisting of a small number of repeated sequences of polymeric segments of different chemical structure.

**blocked curing agent**. A curing agent or hardener rendered unreactive, which can be reactivated as desired by physical or chemical means. Compare with *hardener*.

**blocker**. The impression in the dies (often one of a series of impressions in a single die set) that imparts to the forging an intermediate shape, preparatory to forging of the final shape. Also called blocking impression.

**blocker dies**. Forging dies having generous contours, large radii, draft angles of 7° or more, and liberal finish allowances. See also *finish allowance*.

**blocker-type forging**. A forging that approximates the general shape of the final part with relatively generous *finish allowance* and radii. Such forgings are sometimes specified to reduce die costs where only a small number of forgings are described and the cost of machining each part to its final shape is not excessive.

**block, first, second, and finish**. The forging operation in which a part to be forged is passed in progressive order through three tools mounted in one forging machine; only one heat is involved for all three operations.

**block grease**. A grease that is sufficiently hard to retain its shape in block or stick form.

**blocking**. In forging, a preliminary operation performed in closed dies, usually hot, to position metal properly so that in the finish operation the dies will be filled correctly. Blocking can ensure proper working of the material and can increase die life.

**blocking** (adhesives). An undesired adhesion between touching layers of a material, such as occurs under moderate pressure during storage or use.

**blocking** (glass). (1) The process of shaping a gather of glass in a cavity of wood or metal. (2) The process of stirring and fining glass by immersion of a wooden block or other source of bubbles. (3) The process of reprocessing to remove surface imperfections. (4) The mounting of optical glass blanks in a shell for grinding and polishing operations. (5) The process wherein a furnace is idled at reduced temperatures. (6) The process of setting refractory blocks in a furnace.

**blocking impression**. Same as *blocker*.

**block sequence**. A combined longitudinal and buildup sequence for a continuous multiple-pass weld in which separated lengths are completely or partially built up in cross section before intervening lengths are deposited (Fig. 43). See also *backstep sequence* and *longitudinal sequence*.

FIG. 43

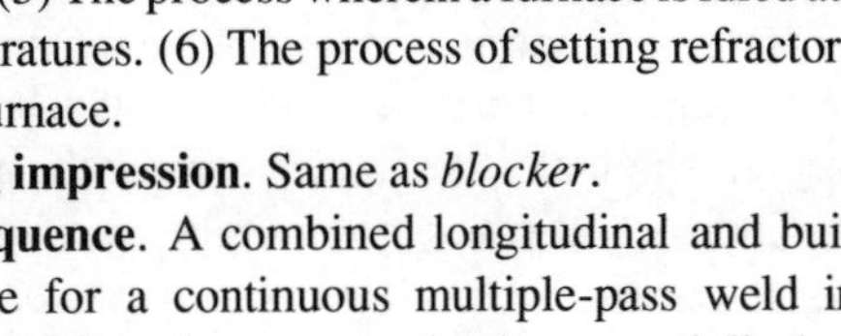

Unwelded spaces filled after deposition of intermittent blocks

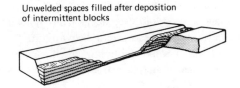

*Block welding sequence to minimize stress*

**blocky alpha**. Alpha phase in titanium alloys that is considerably larger and more polygonal in appearance than the primary α in the sample. It may arise from extended exposure high in the α-β phase field or by slow cooling through the β transus during forging or heat treating. It may be removed by β recrystallization, or all-β working, followed by further α-β work, and may accompany grain-boundary α.

**bloom**. (1) A semifinished hot rolled product, rectangular in cross section, produced on a blooming mill. See also *billet*. For steel, the width of a bloom is not more than twice the thickness, and the cross-sectional area is usually not less than about 230 cm² (36 in.²). Steel blooms are sometimes made by forging. (2) A visible exudation or efflorescence on the surface of an electroplating bath. (3) A bluish fluorescent cast to a painted surface caused by deposition of a thin film of smoke, dust, or oil. (4) A loose, flowerlike corrosion product that forms when certain metals are exposed to a moist environment.

**bloom** (plastics). A noncontinuous surface coating on plastic products that comes from ingredients such as plasticizers, lubricants, antistatic agents, and so on, which are incorporated into the plastic resin, or that occurs by atmospheric contamination. Bloom is the result of ingredients in the plastic coming out of solution and migrating to the surface.

**bloomer**. The mill or other equipment used in reducing steel ingots to blooms.

**blooming mill**. A primary rolling mill used to make blooms.

**blotter**. In grinding, a disk of compressible material, usually blotting-paper stock, used between the grinding wheel and its flanges to avoid concentrated stresses.

**blow**. A term that describes the trapping of gas in castings, causing voids in the metal.

**blow down**. (1) In corrosion prevention, injection of air or water under high pressure through a tube to the anode area for the purpose of purging the annular space and possibly correcting high resistance caused by gas blocking. (2) In connection with boilers or cooling towers, the process of discharging a significant portion of the aqueous solution in order to remove accumulated salts, deposits, and other impurities.

**blowhole**. A hole in a casting or a weld caused by gas entrapped during solidification. See also *porosity*.

**blow hole**. A void produced by the outgassing of trapped air during cure. (2) A void in a solder connection caused by outgassing or a void in a fired dielectric.

**blow holes**. In casting technology, holes in the head plate or blow plate of a core blowing machine through which sand is blown from the reservoir into the *core box*.

**blowing agent**. A compounding ingredient used to produce gas by chemical or thermal action, or both, in the manufacture of hollow or cellular plastic articles.

**blow molding**. A method of fabricating plastics in which a

FIG. 44

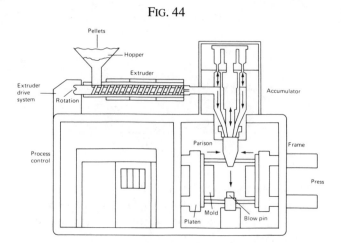

*Blow molding machine for processing thermoplastics*

FIG. 45

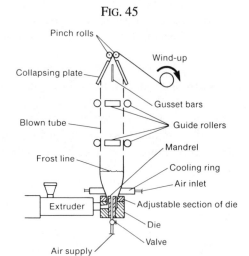

*Typical blown-film extrusion setup*

warm plastic parison (hollow tube) is placed between the two halves of a mold (cavity) and forced to assume the shape of that mold cavity by use of air pressure (Fig. 44). The air pressure is introduced through the inside of the parison and forces the plastic against the surface of the mold, which defines the shape of the product.

**blown-film extrusion**. Technique for making film by extruding the plastic through a circular die, followed by expansion (by the pressure of internal air admitted through the center of the mandrel), cooling, and collapsing of the bubble (Fig. 45).

**blown oil**. Fatty oil that is artificially thickened by blowing air through it.

**blown tubing**. A thermoplastic film that is produced by extruding a tube, applying a slight internal pressure to the tube to expand it while still molten, and subsequently cooling to set the tube. The tube is then flattened through guides and wound up flat on rolls. The size of blown tubing is determined by the flat width in inches as wound rather than by the diameter, as in the case of rigid types of tubing.

**blow pin.** Part of the tooling used to form hollow plastic objects or containers by the blow molding process. It is a tubular tool through which air pressure is introduced into the parison to create the air pressure necessary to form the parison into the shape of the mold. In some blow molding systems, it is a part of, or an extension of, the core pin.

**blowpipe** (brazing and soldering). A device used to obtain a small, accurately directed flame for fine work, such as in the dental and jewelry trades. Any flame may be used, a portion of it being blown to the desired location for the required time by the blowpipe which is usually mouth operated.

**blowpipe** (welding and cutting). See preferred terms *welding torch* and *cutting torch.*

**blow pressure.** The air pressure required to form the parison into the shape of the mold cavity, in a plastic blow molding operation.

**blow rate.** The rate of speed at which air enters, or the time required for air to enter, the parison during the *blow molding* cycle.

**blow-up ratio.** In *blow molding* of plastics, the ratio of the diameter of the product (usually its greatest diameter) to the diameter of the parison from which the product is formed. In blown film extrusion, the ratio between the diameter of the final film tube and the diameter of the die orifice.

**blue annealing.** Heating hot-rolled ferrous sheet in an open furnace to a temperature within the transformation range, then cooling in air to soften the metal. A bluish oxide surface layer forms.

**blue brittleness.** Brittleness exhibited by some steels after being heated to some temperature within the range of about 205 to 370 °C (400 to 700 °F), particularly if the steel is worked at the elevated temperature. Killed steels are virtually free of this kind of brittleness.

**blue dip.** A solution containing a mercury compound, once widely used to deposit mercury on a metal by immersion, usually prior to silver plating.

**blue enamel.** (1) In dry-process porcelain enameling, an area of enamel coating so thin that it appears blue in color. (2) In wet-process enameling, a cover coat applied too thinly to hide the substrate.

**blueing** (plastics). A mold blemish in the form of a blue oxide film on the polished surface of a mold due to abnormally high mold temperatures.

**bluing** (metals). Subjecting the scale-free surface of a ferrous alloy to the action of air, steam, or other agents at a suitable temperature, thus forming a thin blue film of oxide and improving the appearance and resistance to corrosion.

This term is ordinarily applied to sheet, strip, or finished parts. It is used also to denote the heating of springs after fabrication to improve their properties.

**blushing.** (1) Whitening and loss of gloss of a usually organic coating caused by moisture. Also called blooming. (2) The condensation of atmospheric moisture at the adhesive bond line interface.

**BMC.** See *bulk molding compound.*

**BMI.** See *bismaleimide.*

**board hammer.** A type of forging hammer in which the upper die and ram are attached to "boards" that are raised to the striking position by power-driven rollers and let fall by gravity. See also *gravity hammer.*

**boat.** A box or container used to hold the green powder metallurgy compacts during passage through a continuous sintering furnace.

**body.** (1) A loosely used term designating viscosity or consistency. (2) The consistency of an adhesive, which is a function of viscosity, plasticity, and rheological factors.

**body-centered.** Having an atom or group of atoms separated by a translation of $\frac{1}{2}$, $\frac{1}{2}$, $\frac{1}{2}$ from a similar atom or group of atoms. The number of atoms in a body-centered cell must be a multiple of 2. See also the figure accompanying the term *unit cell.*

**body putty.** A pastelike mixture of plastic resin (polyester or epoxy) and talc used in repair of metal surfaces, such as auto bodies.

**bolster.** (1) A plate to which dies may be fastened, the assembly being secured to the top surface of a press bed. In mechanical forging, such a plate is also attached to the ram. (2) Space or filler in a mold for making plastics.

**bolster plate.** A plate to which dies can be fastened; the assembly is secured to the top surface of a press bed. In press forging, such a plate may also be attached to the ram.

**Boltzmann distribution.** In materials characterization, a function giving the probability that a molecule of a gas in thermal equilibrium will have generalized position and momentum coordinates within a given infinitesimal range of values.

**bond.** (1) In grinding wheels and other relatively rigid abrasive products, the material that holds the abrasive grains together. (2) In welding, brazing, or soldering, the junction of joined parts. Where filler metal is used, it is the junction of the fused metal and the heat-affected base metal. (3) In an adhesive bonded or diffusion bonded joint, the line along which the faying surfaces are joined together. (4) In thermal spraying, the junction between the material deposited and the substrate, or its strength. See also *adhesive bond, mechanical bond,* and *metallic bond.*

**bond angle.** The angle formed by the bonds of one atom to other atoms; for example, 109.5° for C–C bonds.

**bond clay.** Any clay suitable for use as a *bonding agent* in molding sand.

**bond coat** (thermal spraying). A preliminary (or prime) coat of material which improves adherence of the subsequent thermal spray deposit.

**bonded film lubricant**. See *bonded solid lubricant*.

**bonded-phase chromatography (BPC)**. Liquid chromatography with a surface-reacted, that is, chemically bonded, organic stationary phase. See also *normal phase chromatography* and *reversed-phase chromatography*.

**bonded solid lubricant**. A solid lubricant dispersed in a continuous matrix of a binder, or attached to a surface by an adhesive material.

**bond face**. The part or surface of an adherend that serves as a substrate for an adhesive.

**bonding**. The joining together of two materials.

**bonding agent**. Any material other than water that, when added to foundry sands, imparts strength either in the green, dry, or fired state.

**bonding force**. The force that holds two atoms together; it results from a decrease in energy as two atoms are brought closer to one another.

**bond length**. The average distance between the centers of two atoms; for example, 0.154 nm (1.54 Å) for C–C bonds.

**bond line**. The cross section of the interface between thermal spray deposits and substrate, or the interface between adhesive and adherend in an adhesive bonded joint.

**bond strength**. (1) The unit load applied to tension, compression, flexure, peel, impact, cleavage, or shear required to break an adhesive assembly with failure occurring in or near the plane of the bond. The term *adherence* is frequently used in place of bond strength. (2) The force required to pull a coating free of a substrate. (3) The degree of cohesiveness that the *bonding agent* exhibits in holding sand grains together.

**bone oil**. A fatty oil obtained by dry distillation of bones.

**book mold**. A split permanent mold hinged like a book.

**borate glass**. A glass in which the essential glass former is boron oxide instead of silica. See also *glass* (Technical Brief 17).

**bore**. A hole or cylindrical cavity produced by a single-point or multipoint tool other than a drill.

**bore seal**. A device in which the outside diameter mates with a bore surface to provide sealing between the two surfaces.

**boriding**. Thermochemical treatment involving the enrichment of the surface layer of an object with borides. This surface-hardening process is performed below the $Ac_1$ temperature. Also referred to as boronizing.

**boring**. Enlarging a hole by removing metal with a single- or occasionally a multiple-point cutting tool moving parallel to the axis of rotation of the work or tool.

**boron carbide**. A black crystalline powder of high hardness, the composition of which is either $B_6C$ or $B_4C$ (the latter being a composite of $B_4C$ and carbon in graphitic form). Applications include loose abrasives (see Table 12), and

**Table 12 Hardnesses of several common abrasives**

| Material | Knoop hardness (100 g), kg/mm$^2$ |
|---|---|
| Sapphire (alumina) | 2000–2050 |
| Tungsten carbide (WC) | 2050–2150 |
| Silicon carbide (SiC) | 2150–2950 |
| Boron carbide (B$_4$C) | 2900–3100 |
| Cubic boron nitride (CBN) | 4500–4600 |
| Diamond (C) | 8000–8500 |

hot pressed shot blast nozzles and other wear-resistant components.

**boron fiber**. A fiber produced by vapor deposition of elemental boron, usually onto a tungsten filament core, to impart strength and stiffness.

**boronizing**. See *boriding*.

**boron nitride (hexagonal)**. A white fluffy powder of composition BN with high chemical and thermal stability and high electrical resistance. Used as a lubricant for high-pressure bearings and in the hot pressed condition for mechanical and electrical parts. See also *cubic boron nitride*.

**borosilicate glass**. Any silicate glass having at least 5% of boron oxide ($B_2O_3$). See also *glass* (Technical Brief 17).

**bort**. (1) Natural diamond of a quality not suitable for gem use. (2) Industrial diamond.

**bosh**. (1) The section of a blast furnace extending upward from the tuyeres to the plane of maximum diameter. (2) A lining of quartz that builds up during the smelting of copper ores and that decreases the diameter of the furnace at the tuyeres. (3) A tank, often with sloping sides, used for washing metal parts or for holding cleaned parts.

**boss**. (1) A relatively short protrusion or projection from the surface of a forging or casting, often cylindrical in shape. Usually intended for drilling and tapping for attaching parts. (2) Projection on a plastic part designed to add strength, to facilitate alignment during assembly, or to provide for a fastening.

**bottle**. See preferred term *cylinder*.

**bottom board**. In casting, a flat base for holding the *flask* in making sand molds.

**bottom blow**. A specific type of *blow molding* technique for plastics that forms hollow arteries by injecting the blowing air into the parison from the bottom of the mold (as opposed to introducing the blowing air at a container opening).

**bottom draft**. Slope or taper in the bottom of a forge depression that tends to assist metal flow toward the sides of depressed areas.

**bottom drill**. A flat-ended twist drill used to convert a cone at the bottom of a drilled hole into a cylinder.

**bottom-end bearing**. See *big-end bearing*.

**bottoming bending**. Press-brake bending process in which

the upper die (punch) enters the lower die and coins or sets the material to eliminate *springback*.

**bottoming tap.** A tap with a *chamfer* of 1 to 1½ threads in length.

**bottom pipe.** An oxide-lined fold or cavity at the butt end of a slab, bloom, or billet; formed by folding the end of an ingot over on itself during primary rolling. Bottom pipe is not *pipe*, in that it is not a shrinkage cavity, and in that sense, the term is a misnomer. Bottom pipe is similar to *extrusion pipe*. It is normally discarded when the slab, bloom, or billet is cropped following primary reduction.

**bottom plate.** In making of plastic parts, the part of the mold that contains the heel radius and the push-up.

**bottom-pour ladle.** A *ladle* from which metal, usually steel, flows through a *nozzle* located at the bottom (Fig. 46).

FIG. 46

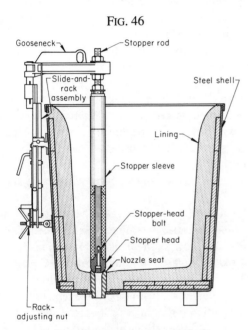

*Schematic of a bottom-pour ladle*

**bottom punch.** In powder metallurgy, the part of the tool assembly that closes the die cavity at the bottom and transfers the pressure to the powder during compaction.

**bottom running or pouring.** Filling of the casting mold cavity from the bottom by means of gates from the runner.

**boundary lubricant.** A lubricant suitable for use in *boundary lubrication* conditions. Fatty acids and soaps are commonly used.

**boundary lubrication.** A condition of lubrication in which the friction and wear between two surfaces in relative motion are determined by the properties of the surfaces and by the properties of the lubricant other than bulk viscosity. See also the figure accompanying the term *lubrication regimes*.

**bow.** (1) A condition of longitudinal curvature in pultruded plastic parts. (2) The tendency of material to curl downward during shearing, particularly when shearing long narrow strips.

**bowing.** Deviation from flatness.

**box annealing.** Annealing a metal or alloy in a sealed container under conditions that minimize oxidation. In box annealing a ferrous alloy, the charge is usually heated slowly to a temperature below the transformation range, but sometimes above or within it, and is then cooled slowly; this process is also called close annealing or pot annealing. See also *black annealing*.

**box furnace.** A furnace used for batch sintering of powder metallurgy parts, normally utilizing a controlled atmosphere-containing sealed retort.

**boxing.** The continuation of a fillet weld around a corner of a member as an extension of the principal weld.

**brackish water.** (1) Water having salinity values ranging from approximately 0.5 to 17 parts per thousand. (2) Water having less salt than seawater, but undrinkable.

**Bragg angle.** The angle between the incident beam and the lattice planes considered.

**Bragg equation.** See *Bragg's law*.

**Bragg's law.** A statement of the conditions under which a crystal will diffract electromagnetic radiation. Bragg's law reads $n\lambda = 2d \sin \theta$, where $n$ is the order of reflection, $\lambda$ is the wavelength of x-rays, $d$ is the distance between lattice planes, and $\theta$ is the Bragg angle, or the angular distance between the incident beam and the lattice planes considered.

**Bragg method.** A method of x-ray diffraction in which a single-crystal is mounted on a spectrometer with a crystal face parallel to the axis of the instrument.

**braiding.** Intertwining two or more systems of yarns in the bias direction to form an integrated structure (Fig. 47).

**brake.** A device for bending sheet metal to a desired angle.

**brale indenter.** A conical 120° diamond indenter with a

FIG. 47

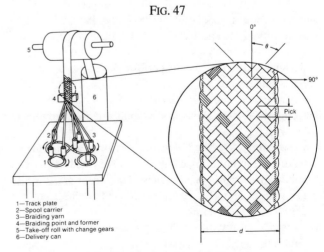

1—Track plate
2—Spool carrier
3—Braiding yarn
4—Braiding point and former
5—Take-off roll with change gears
6—Delivery can

*Flat braider machine and braid*

FIG. 48

*Brale indenter used in Rockwell hardness testing*

conical tip (a 0.2 mm tip radius is typical) used in certain types of Rockwell and scratch hardness tests (Fig. 48).

**branched polymer**. In the molecular structure of polymers, a main chain with attached side chains, in contrast to a linear polymer. Two general types are recognized, short-chain, and long-chain branching.

**branching**. The presence of molecular branches in a polymer. The generation of branch crystals during the crystallization of a polymer.

**brass**. A copper-zinc alloy containing up to 40% Zn, to which smaller amounts of other elements may be added. See also *copper and copper-base alloys* (Technical Brief 13).

**braze**. A weld produced by heating an assembly to suitable temperatures and by using a filler metal having a liquidus above 450 °C (840 °F) and below the solidus of the base metal. The filler metal is distributed between the closely fitted faying surfaces of the joint by capillary action.

**brazeability**. The capacity of a metal to be brazed under the fabrication conditions imposed into a specific suitably designed structure and to perform satisfactorily in the intended service.

**braze interface**. See *weld interface*.

**brazement**. An assembly whose component parts are joined by brazing.

**braze welding**. A method of welding by using a filler metal having a liquidus above 450 °C (840 °F) and below the solidus of the base metals. Unlike *brazing*, in braze welding, the filler metal is not distributed in the joint by capillary attraction.

**brazing**. A group of welding processes that join solid materials together by heating them to a suitable temperature and using a filler metal having a liquidus above 450

°C (840 °F) and below the solidus of the base materials. The filler metal is distributed between the closely fitted surfaces of the joint by capillary attraction.

**brazing alloy**. See preferred term *brazing filler metal*.

**brazing filler metal**. (1) The metal which fills the capillary gap and has a liquidus above 450 °C (840 °F) but below the solidus of the base materials. (2) A nonferrous filler metal used in *brazing* and *braze welding*.

**brazing procedure**. The detailed methods and practices including all joint brazing procedures involved in the production of a brazement. See also *joint brazing procedure*.

**brazing sheet**. Brazing filler metal in sheet form.

**brazing technique**. The details of a brazing operation which, within the limitations of the prescribed brazing procedure, are controlled by the brazer or the brazing operator.

**brazing temperature**. The temperature to which the base metal is heated to enable the filler metal to wet the base metal and form a brazed joint.

**brazing temperature range**. The temperature range within which brazing can be conducted.

**break-away torque**. See *starting torque*.

**breakdown**. (1) An initial rolling or drawing operation, or a series of such operations, for the purpose of reducing a casting or extruded shape prior to the finish reduction to desired size. (2) A preliminary press-forging operation.

**breakdown potential**. The least noble potential where *pitting* or *crevice corrosion*, or both, will initiate and propagate.

**breakdown voltage**. The voltage required, under specific conditions, to cause the failure of an insulating material. See also *dielectric strength* and *arc resistance*.

**breaker plate**. In plastic forming, a perforated plate located at the rear end of an extruder or at the nozzle end of an injector cylinder. It often supports the screens that prevent foreign particles from entering the die, and is used to keep unplasticized material out of the nozzle and to improve distribution of color particles.

**break-in** (noun). See *running-in*.

**break in** (verb). To operate a newly installed bearing, seal, or other tribocomponent in such a manner as to condition its surface(s) for improved functional operation. See also *run in* (verb).

**breaking extension**. The elongation necessary to cause rupture of an adhesively bonded test specimen. The tensile strain at the moment of rupture.

**breaking factor**. The breaking load divided by the original width of an adhesively bonded test specimen, expressed in lb/in.

**breaking length**. A measure of the breaking strength of reinforcing yarn. The length of a specimen the weight of which is equal to the breaking load.

**breaking load**. The maximum load (or force) applied to a test specimen or structural member loaded to rupture.

**breaking stress**. Same as *fracture stress* (1).

**breakout**. Fiber separation or break on surface plies at drilled or machined composite material edges.

**breaks**. Creases or ridges usually in "untempered" or in aged material where the yield point has been exceeded. Depending on the origin of the breaks, they may be termed *cross breaks*, *coil breaks*, edge breaks, or *sticker breaks*.

**breather**. A loosely woven material that serves as a continuous vacuum path over a part but is not in contact with the resin.

**breathing**. The opening and closing of a mold to allow gas to escape early in the plastic molding cycle. Also called degassing; sometimes called bumping, in phenolic molding. When referring to plastic sheeting, the term breathing indicates permeability to air.

**bremsstrahlung**. See *continuum*.

**bridge die**. A two-section extrusion die capable of producing tubing or intricate hollow shapes without the use of a separate mandrel. Metal separates into two streams as it is extruded past a bridge section, which is attached to the main die section and holds a stub mandrel in the die opening; the metal then is rewelded by extrusion pressure before it enters the die opening. Compare with *porthole die*.

**bridging**. (1) Premature solidification of metal across a mold section before the metal below or beyond solidifies. (2) Solidification of slag within a cupola at or just above the tuyeres. (3) Welding or mechanical locking of the charge in a downfeed melting or smelting furnace. (4) In powder metallurgy, the formation of arched cavities in a powder mass. (5) In soldering, an unintended solder connection between two or more conductors, either securely or by mere contact. Also called a crossed joint or solder short.

**bright annealing**. Annealing in a protective medium to prevent discoloration of the bright surface.

**bright dip**. A solution that produces, through chemical action, a bright surface on an immersed metal.

**brightener**. An agent or combination of agents added to an electroplating bath to produce a lustrous deposit.

**bright-field illumination**. For reflected light, the form of illumination that causes specularly reflected surfaces normal to the axis of the microscope to appear bright (Fig. 49). For transmission electron microscopy, the illumination of an object so that it appears on a bright background. Compare with *dark-field illumination*.

**bright finish**. A high-quality finish produced on ground and polished rolls. Suitable for electroplating.

**bright nitriding**. Nitriding in a protective medium to prevent discoloration of the bright surface. Compare with *blank nitriding*.

**bright plate**. An electrodeposit that is lustrous in the as-plated condition.

**bright range**. The range of current densities, other conditions being constant, within which a given electroplating bath produces a bright plate.

**bright stock**. High-viscosity mineral oils that remain after vacuum distillation of crude oil.

**Brillouin zones**. See *electron bands*.

**brine**. Seawater containing a higher concentration of dissolved salt than that of the ordinary ocean.

**Brinell hardness number (HB)**. A number related to the applied load and to the surface area of the permanent impression made by a ball indenter computed from:

$$HB = \frac{2P}{\pi D \left( D - \sqrt{D^2 - d^2} \right)}$$

FIG. 49

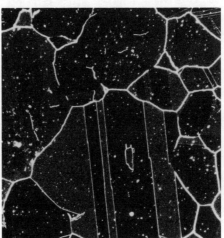

*Comparison of bright-field (left) and dark-field (right) illumination modes*

FIG. 50

Force

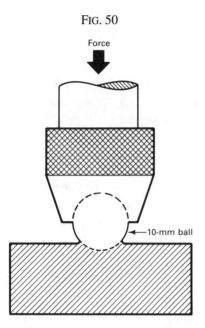

←10-mm ball

*Schematic of Brinell hardness test*

where *P* is applied load, kgf; *D* is diameter of ball, mm; and *d* is mean diameter of the impression, mm.

**Brinell hardness test.** A test for determining the hardness of a material by forcing a hard steel or carbide ball of specified diameter (typically, 10 mm) into it under a specified load (Fig. 50). The result is expressed as the *Brinell hardness number*.

**Brinelling.** (1) Indentation of the surface of a solid body by repeated local impact or impacts, or static overload. Brinelling may occur especially in a rolling-element bearing. (2) Damage to a solid bearing surface characterized by one or more plastically formed indentations brought about by overload. See also *false Brinelling*.

**brine quenching.** A quench in which brine (salt water-chlorides, carbonates, and cyanides) is the quenching medium. The salt addition improves the efficiency of water at the vapor phase or hot stage of the quenching process.

**briquet(te).** A self-sustaining mass of powder of defined shape. See preferred term *compact* (noun).

**brittle.** Permitting little or no plastic (permanent) deformation prior to fracture.

**brittle crack propagation.** A very sudden propagation of a crack with the absorption of no energy except that stored elastically in the body. Microscopic examination may reveal some deformation even though it is not noticeable to the unaided eye. Contrast with *ductile crack propagation*.

**brittle erosion behavior.** Erosion behavior having characteristic properties (e.g., little or no plastic flow, the formation of cracks) that can be associated with *brittle fracture* of the exposed surface. The maximum volume removal occurs at an angle near 90°, in contrast to approximately 25° for *ductile erosion behavior*.

**brittle fracture.** Separation of a solid accompanied by little or no macroscopic plastic deformation. Typically, brittle fracture occurs by rapid crack propagation with less expenditure of energy than for *ductile fracture*. Brittle tensile fractures have a bright, granular appearance and exhibit

FIG. 51

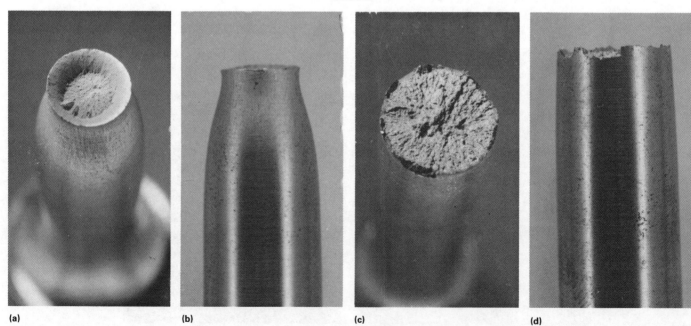

(a)                    (b)                    (c)                    (d)

*Macroscopic appearance of ductile (a and b) and brittle (c and d) tensile fractures*

FIG. 52

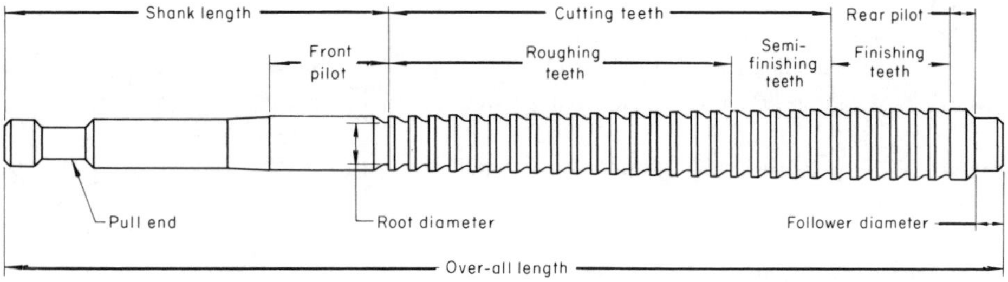

*Essential features and nomenclature of broaches*

little or no necking (Fig. 51). A *chevron pattern* may be present on the fracture surface, pointing toward the origin of the crack, especially in brittle fractures in flat platelike components. Examples of brittle fracture include *transgranular cracking* (*cleavage* and *quasi-cleavage fracture*) and *intergranular cracking* (*decohesive rupture*).

**brittleness**. The tendency of a material to fracture without first undergoing significant *plastic deformation*. Contrast with *ductility*.

**broaching**. Cutting with a tool which consists of a bar having a single edge or a series of cutting edges (i.e., teeth) on its surface (Fig. 52). The cutting edges of multiple-tooth, or successive single-tooth, broaches increase in size and/or change in shape. The broach cuts in a straight line or axial direction when relative motion is produced in relation to the workpiece, which may also be rotating. The entire cut is made in single or multiple passes over the workpiece to shape the required surface contour.

**broad goods**. Fiber woven to form fabric up to 1270 mm (50 in.) wide for reinforcement of plastics. It may or may not be impregnated with resin and is usually furnished in rolls of 25 to 140 kg (50 to 300 lb).

**bronze**. A copper-rich copper-tin alloy with or without small proportions of other elements such as zinc and phosphorus. By extension, certain copper-base alloys containing considerably less tin than other alloying elements, such as manganese bronze (copper-zinc plus manganese, tin, and iron) and leaded tin bronze (copper-lead plus tin and sometimes zinc). Also, certain other essentially binary copper-base alloys containing no tin, such as aluminum bronze (copper-aluminum), silicon bronze (copper-silicon), and beryllium bronze (copper-beryllium). Also, trade designations for certain specific copper-base alloys that are actually brasses, such as architectural bronzes (57 Cu, 40 Zn, 3 Pb) and commercial bronze (90 Cu, 10 Zn). See also *copper and copper-base alloys* (Technical Brief 13).

**bronze welding**. A term erroneously used to denote braze welding. See *braze welding*.

**bronzing**. (1) Applying a chemical finish to copper or cop-

per-alloy surfaces to alter the color. (2) Plating a copper-tin alloy on various materials.

**brush anodizing**. An *anodizing* process similar to *brush plating*.

**brush plating**. Plating with a concentrated solution or gel held in or fed to an absorbing medium, pad, or brush carrying the anode (usually insoluble). The brush is moved back and forth over the area of the cathode to be plated.

**brush polishing (electrolytic)**. A method of *electropolishing* in which the electrolyte is applied with a pad or brush in contact with the part to be polished.

**B-stage**. An intermediate stage in the reaction of certain thermosetting resins in which the material softens when heated and swells when in contact with certain liquids, but may not entirely fuse or dissolve. The resin in an uncured thermosetting adhesive, usually in this stage. Synonym for resitol. Compare with *A-stage* and *C-stage*.

**B-stage resin**. In a thermosetting reaction, a resin that is in an intermediate state of cure when it is sticky, or tacky, and capable of further flow. The cure is normally complete during the laminating cycle. See *C-stage resin* and *prepreg*.

**bubble**. A spherical, internal void or globule of air or other gas trapped within a plastic. See *void*.

**bubbler**. In forming of plastics, a device inserted into a mold force, cavity, or core, that allows water to flow deep inside the hole into which it is inserted and to discharge through the open end of the hole. Uniform cooling of the mold and of isolated mold sections can be achieved in this manner.

**bubbler mold cooling**. In injection molding of plastics, a method of uniformly cooling a mold; a stream of cooling liquid flows continuously into a cooling cavity equipped with a coolant outlet normally positioned at the opposite end.

**bubbly oil**. Oil containing bubbles of gas.

**buckle**. (1) Bulging of a large, flat face of a casting; in investment casting, caused by *dip coat* peeling from the pattern. (2) An indentation in a casting, resulting from expansion of the sand, can be termed the start of an expansion defect. (3) A local waviness in metal bar or sheet, usually transverse to the direction of rolling.

**buckling**. (1) A mode of failure generally characterized by an unstable lateral material deflection due to compressive action on the structural element involved. (2) In metal forming, a bulge, bend, kink, or other wavy condition of the workpiece caused by compressive stresses. See also *compressive stress*.

**buckyball**. An inorganic solid crystal structure which forms from carbon. Buckyballs have a spherical structure with their surfaces made up of from 28 to 450 carbon atoms in hexagonal and pentagonal arrays, with 60 carbon atoms being the most stable (Fig. 53). This soccer ball shaped configuration is reminiscent of R. Buckminster Fuller's geodesic domes, hence the name Buckminster Fullerene or "buckyball."

FIG. 53

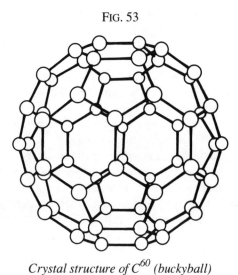

*Crystal structure of $C^{60}$ (buckyball)*

**Bucky diaphragm**. An x-ray scatter-reducing device originally intended for medical radiography but also applicable to industrial radiography in some circumstances. Thin strips of lead, with their widths held parallel to the primary radiation, are used to absorb scattered radiation preferentially; the array of strips is in motion during exposure to prevent formation of a pattern on the film.

**buffer**. (1) A substance which by its addition or presence tends to minimize the physical and chemical effects of one or more of the substances in a mixture. Properties often buffered include pH, oxidation potential, and flame or plasma temperatures. (2) A substance whose purpose is to maintain a constant hydrogen-ion concentration in water solutions, even where acids or alkalis are added. Each buffer has a characteristic limited range of pH over which it is effective.

**buffer gas**. A protective gas curtain at the charge or discharge end of a continuously operating sintering furnace.

**buffing**. Developing a lustrous surface by contacting the work with a rotating *buffing wheel*.

**buffing wheel**. Buff sections assembled to the required face

FIG. 54

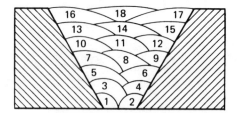

*Buildup weld sequence*

width for use on a rotating shaft between flanges. Sometimes called a buff.

**buff section**. A number of fabric, paper, or leather disks with concentric center holes held together by various types of sewing to provide degrees of flexibility or hardness. These sections are assembled to make wheels for polishing.

**builder**. A material, such as an alkali, a buffer, or a water softener, added to a soap or synthetic surface-active agent to produce a mixture having enhanced detergency. Examples: (1) alkalis—caustic soda, soda ash, and trisodium phosphate. (2) *buffers*—sodium metasilicate and borax; and (3) water softeners—sodium tripolyphosphate, sodium tetraphosphate, sodium hexametaphosphate, and ethylene diamine tetraacetic acid.

**buildup**. Excessive electrodeposition that occurs on high-current-density areas, such as corners or edges.

**buildup sequence**. The order in which the weld beads of a multiple-pass weld are deposited with respect to the cross section of the joint (Fig. 54). See also *block sequence* and *longitudinal sequence*.

**built-up edge**. (1) Chip material adhering to the tool face adjacent to the cutting edge during cutting (Fig. 55).

FIG. 55

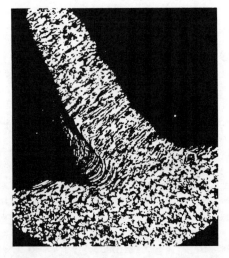

*Continuous machining chip with built-up edge adjacent to tool face*

(2) Material from the workpiece, especially in machining, which is stationary with respect to the tool (Ref 1). See also *wedge formation*.

**built-up laminated wood**. An assembly made by joining layers of lumber with mechanical fastenings so that the grain of all laminations is essentially parallel.

**bulging**. (1) Expanding the walls of a cup, shell, or tube with an internally expanded segmented punch or a punch composed of air, liquids, or semiliquids such as waxes, rubber, and other elastomers. (2) The process of increasing the diameter of a cylindrical shell (usually to a spherical shape) or of expanding the outer walls of any shell or box shape whose walls were previously straight.

**bulk adherend**. With respect to interphase, the adherend, unaltered by the adhesive. Compare with *bulk adhesive*.

**bulk adhesive**. With respect to interphase, the adhesive, unaltered by the adherend. Compare with *bulk adherend*.

**bulk density**. (1) The weight of an object or material divided by its material volume less the volume of its open pores. (2) The density of a plastic molding material in loose form (granular, nodular, and so forth), expressed as a ratio of weight to volume. (3) Metal powder in a container or bin expressed in mass per unit volume.

**bulk factor**. The ratio of the volume of a raw plastic molding compound or powdered plastic to the volume of the finished, solid piece produced from it. The ratio of the density of the solid plastic object to the apparent, or bulk, density of the loose molding powder.

**bulk forming**. Forming processes, such as extrusion, forging, rolling, and drawing, in which the input material is in billet, rod, or slab form and a considerable increase in surface-to-volume ratio in the formed part occurs under the action of largely compressive loading. Compare with *sheet forming*.

**bulk modulus of elasticity (K)**. The measure of resistance to change in volume; the ratio of hydrostatic stress to the corresponding unit change in volume. This elastic constant can be expressed by:

$$K = \frac{\sigma_m}{\Delta} = \frac{-p}{\Delta} = \frac{1}{\beta}$$

where $K$ is the bulk modulus of elasticity, $\sigma_m$ is hydrostatic or mean stress tensor, $p$ is hydrostatic pressure, and $\beta$ is compressibility. Also known as bulk modulus, compression modulus, hydrostatic modulus, and volumetric modulus of elasticity.

**bulk molding compound (BMC)**. Thermosetting resin combined with strand reinforcement, fillers, and so on, into a viscous compound for compression or injection molding.

**bulk sample**. See *gross sample*.

**bulk sampling**. Obtaining a portion of a material that is representative of the entire lot.

**bulk volume**. The volume of the metal powder fill in the die cavity.

**bull block**. A machine with a power-driven revolving drum for cold drawing wire through a drawing die as the wire winds around the drum.

**bulldozer**. Slow-acting horizontal *mechanical press* with a large bed used for bending and straightening. The work is done between dies and can be performed hot or cold. The machine is closely allied to a forging machine.

**bullion**. (1) A semirefined alloy containing sufficient precious metal to make recovery profitable. (2) Refined gold or silver, uncoined.

**bull's-eye structure**. The microstructure of malleable or ductile cast iron when graphite nodules are surrounded by a ferrite layer in a pearlitic matrix (Fig. 56).

FIG. 56

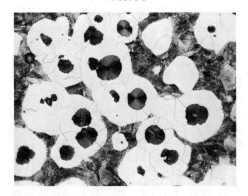

*Ductile iron, with the bull's-eye structure of graphite nodules surrounded by ferrite. 100×*

**bump check**. See *percussion cone*.

**bumper**. A machine used for packing molding sand in a flask by repeated jarring or jolting. See also *jolt ramming*.

**bumping**. (1) Forming a dish in metal by means of many repeated blows. (2) Forming a head. (3) Setting the seams on sheet metal parts. (4) Ramming sand in a flask by repeated jarring and jolting.

**bundle**. A general term for a collection of essentially parallel filaments or fibers used in composite materials.

**buret**. An instrument used to deliver variable and accurately known volumes of a liquid during titration or volumetric analysis. Burets are usually made from uniform-bore glass tubing in capacities of 5 to 100 mL, the most common being 50 mL. See also *titration* and *volumetric analysis*.

**burn**. (1) The heat treatment to which refractory materials are subjected in the firing process. (2) To heat particulate material, grain, aggregate, refractory bricks, or ceramic ware in a furnace or kiln, sufficient to cause desired chemical, structural, or crystalline changes in the material though below the melting point of the major component. To fire, to calcine, or sinter; usually more intense than to dry or to activate.

**burnback time**. See preferred term *meltback time*.

**burned**. Showing evidence of thermal decomposition or

charring through some discoloration, distortion, destruction, or conversion of the surface of the plastic, sometimes to a carbonaceous char.

**burned deposit**. A dull, nodular electrodeposit resulting from excessive plating current density.

**burned-in sand**. A defect consisting of a mixture of sand and metal cohering to the surface of a casting.

**burned-on sand**. A mixture of sand and cast metal adhering to the surface of a casting. In some instances, may resemble *metal penetration*.

**burned plating**. See *burned deposit*.

**burned sand**. Foundry sand in which the binder or bond has been removed or impaired by contact with molten metal.

**burner**. See preferred term *oxygen cutter*.

**burning**. (1) Permanently damaging a metal or alloy by heating to cause either incipient melting or intergranular oxidation. See also *overheating* and *grain-boundary liquidation*. (2) During subcritical annealing, particularly in continuous annealing, production of a severely decarburized and grain-coarsened surface layer that results from excessively prolonged heating to an excessively high temperature. (3) In grinding, getting the work hot enough to cause discoloration or to change the microstructure by tempering or hardening. (4) In sliding contacts, the oxidation of a surface due to local heating in an oxidizing environment. See also *metallurgical burn*.

**burning (firing) of refractories**. The final heat treatment in a kiln to which refractory brick and shapes are subjected in the process of manufacture for the purpose of developing bond and other necessary physical and chemical properties.

**burning in**. See preferred term *flow welding*.

**burning rate**. The tendency of plastic articles to burn at given temperatures.

**burning** (welding). See preferred term *oxygen cutting*.

**burnish**. (1) To alter the original manufactured surface of a sliding or rolling surface to a more polished condition. (2) To apply a substance to a surface by rubbing.

**burnishing**. Finish sizing and smooth finishing of surfaces (previously machined or ground) by displacement, rather than removal, of minute surface irregularities with smooth-point or line-contact fixed or rotating tools.

**burnoff**. (1) Unintentional removal of an autocatalytic deposit from a nonconducting substrate, during subsequent electroplating operations, owing to the application of excessive current or a poor contact area. (2) Removal of volatile lubricants such as metallic stearates from metal powder compacts by heating immediately prior to sintering.

**burnoff rate**. See preferred term *melting rate*.

**burnout**. In casting, firing a mold at a high temperature to remove pattern material residue.

**burn-through**. A term erroneously used to denote excessive melt-through or a hole. See also *melt-through*.

**burn-through weld**. A term erroneously used to denote a seam weld or spot weld.

**burr**. (1) A thin ridge or roughness left on a workpiece (e.g., forgings or sheet metal blanks) resulting from cutting, punching, or grinding. (2) A rotary tool having teeth similar to those on hand files.

**burring**. Same as *deburring*.

**burst strength**. Measure of the ability of a material to withstand internal hydrostatic or gas dynamic pressure without rupture. Hydraulic pressure required to burst a vessel of given thickness.

**bush bearing**. A plain bearing in which the lining is closely fitted into the housing in the form of a bush, usually surfaced with a bearing alloy.

**bushing**. A bearing or guide.

**buster**. A pair of shaped dies used to combine preliminary forging operations, such as edging and blocking, or to loosen scale.

**butadiene**. A gas ($CH_2$:$CH\cdot CH$:$CH_2$), insoluble in water but soluble in alcohol and ether, obtained from the cracking of petroleum, from coal tar benzene, or from acetylene produced from coke and lime. It is widely used in the formation of copolymers with styrene, acrylonitrile, vinyl chloride, and other monomeric substances and imparts flexibility to the resultant moldings.

**butadiene-styrene plastic**. A synthetic resin derived from the copolymerization of butadiene gas and styrene liquids.

**butler finish**. A semilustrous metal finish composed of fine, uniformly distributed parallel lines, usually produced with a soft abrasive buffing wheel; similar in appearance to the traditional hand-rubbed finish on silver.

**butter coat**. A commonly used term to describe a higher than usual amount of surface resin. See also *resin-rich area*.

**butterfly bruise**. See *percussion cone*.

**buttering**. A form of surfacing in which one or more layers of weld metal are deposited on the groove face of one member (for example, a high-alloy weld deposit on steel base metal that is to be welded to a dissimilar base metal). The buttering provides a suitable transition weld deposit for subsequent completion of the butt weld (joint).

**butt fusion**. A method of joining pipe, sheet, or other similar forms of a thermoplastic resin in which the ends of the two pieces to be joined are heated to the molten state and then rapidly pressed together.

**butt joint** (adhesives). A type of edge joint in which the edge faces of the two adherends are at right angles to the other faces of the adherends.

**butt joint** (welding). A joint between two abutting members lying approximately in the same plane (Fig. 57). A welded butt joint may contain a variety of grooves. See also *groove weld*.

**button**. (1) A globule of metal remaining in an assaying crucible or cupel after fusion has been completed. (2) That

FIG. 57

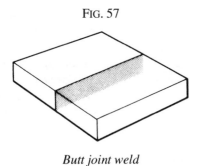

*Butt joint weld*

part of a weld that tears out in destructive testing of a spot, seam, or projection welded specimen.

**butt seam welding**. See *seam welding*.

**butt weld**. An erroneous term for a weld in a butt joint. See *butt joint*.

**butt welding**. Welding a *butt joint*.

**butylene plastics**. Plastics based on resins made by the polymerization of butene or the copolymerization of butene with one or more unsaturated compounds, the butene being the greatest amount by weight.

# C

**cable**. An insulated conductor or group of conductors twisted together for the transmission of electrical energy.

**cage**. In a bearing, a device that partly surrounds the rolling elements and travels with them, the main purpose of which is to space the rolling elements in proper relation to each other (Fig. 58). See also *separator*.

**cake**. (1) A copper or copper alloy casting, rectangular in cross section, used for rolling into sheet or strip. (2) A coalesced mass of unpressed metal powder.

**calcareous coating or deposit**. A layer consisting of a mixture of calcium carbonate and magnesium hydroxide deposited on surfaces being cathodically protected against corrosion, because of the increased pH adjacent to the protected surface.

**calcination**. Heating ores, concentrates, precipitates, or residues to decompose carbonates, hydrates, or other compounds.

**calcine**. (1) A ceramic material or mixture fired to less than fusion for use as a constituent in a ceramic composition. (2) Refractory material, often fireclay, that has been heated to eliminate volatile constituents and to produce desired physical changes.

**calcined gypsum**. A dry powder, primarily calcium sulfate hemihydrate, resulting from calcination of gypsum. Cementitious base for production of most gypsum plasters. Also called plaster of paris or stucco.

**calcium silicon**. An alloy of calcium, silicon, and iron containing 28 to 35% Ca, 60 to 65% Si, and 6% Fe (max), used as a deoxidizer and degasser for steel and cast iron; sometimes called calcium silicide.

**calender**. The passing of plastic sheet material between sets of pressure rollers to produce a smooth finish and a desired thickness.

**calibrate**. To determine, by measurement or comparison with a standard, the correct value of each scale reading on a measuring (test) instrument.

**calibration**. Determination of the values of the significant parameters by comparison with values indicated by a reference instrument or by a set of reference standards.

**calomel electrode**. (1) An *electrode* widely used as a reference electrode of known potential in electrometric measurement of acidity and alkalinity, corrosion studies, voltammetry, and measurement of the potentials of other electrodes. (2) A secondary reference electrode of the

FIG. 58

*Cage in a radial ball bearing (cutaway view)*

composition: Pt/Hg-Hg$_2$Cl$_2$/KCl solution. For 1.0 $N$ KCl solution, its potential versus a hydrogen electrode at 25 °C (77 °F) and one atmosphere is +0.281 V.

**calorimeter.** An instrument capable of making absolute measurements of energy deposition (or absorbed dose) in a material by measuring its change in temperature and imparting a knowledge of the characteristics of its material construction.

**calorizing.** Imparting resistance to oxidation to an iron or steel surface by heating in aluminum powder at 800 to 1000 °C (1470 to 1830 °F).

**camber.** (1) Deviation from edge straightness, usually referring to the greatest deviation of side edge from a straight line. (2) The tendency of material being sheared from sheet to bend away from the sheet in the same plane. (3) Sometimes used to denote crown in rolls where the center diameter has been increased to compensate for deflection caused by the rolling pressure. (4) The planar deflection of a flat cable or flexible laminate from a straight line of specified length. A flat cable or flexible laminate with camber is similar to the curve of an unbanked race track.

**cam press.** A mechanical forming press in which one or more of the slides are operated by cams; usually a double-action press in which the blankholder slide is operated by cams through which the dwell is obtained.

**can.** A sheathing of soft metal that encloses a sintered metal billet for the purpose of hot working (hot isostatic pressing, hot extrusion) without undue oxidation.

**canning.** (1) A dished distortion in a flat or nearly flat sheet metal surface, sometimes referred to as oil canning. (2) Enclosing a highly reactive metal within a relatively inert material for the purpose of hot working without undue oxidation of the active metal.

**capacitance ($C$).** That property of a system of conductors and dielectrics which permits the storage of electrically separated charges when potential differences exist between the conductors. It is the ratio of a quantity, $Q$, of electricity to a potential difference, $V$. A capacitance value is always positive. The units are farads when the charge is expressed in coulombs and the potential in volts: $C = Q/V$.

**capacitor.** A device that can store an electrical charge when voltage is applied. Its impedance is inversely proportional to the frequency of the voltage impressed; that is, it offers little resistance or impedance to high frequencies, but much to low frequencies.

**capacity.** In tensile testing machines, the maximum load and/or displacement for which a machine is designed. Some testing machines have more than one load capacity. These are equipped with accessories that allow the capacity to be modified as desired.

**capacity number ($C_n$).** The product of the *Sommerfeld number* and the square of the length-to-diameter ratio of a journal bearing.

**capillarity.** The ability of a material to conduct liquids through its pore structure by the force of surface tension.

**capillary action.** (1) The phenomenon of intrusion of a liquid into interconnected small voids, pores, and channels in a solid, resulting from surface tension. (2) The force by which liquid, in contact with a solid, is distributed between closely fitted faying surfaces of the joint to be brazed or soldered.

**capillary attraction.** (1) The combined force of adhesion and cohesion that causes liquids, including molten metals, to flow between very closely spaced and solid surfaces, even against gravity. (2) In powder metallurgy, the driving force for the infiltration of the pores of a sintered compact by a liquid.

**capped steel.** A type of steel similar to rimmed steel, usually cast in a bottle-top ingot mold, in which the application of a mechanical or a chemical cap renders the rimming action incomplete by causing the top metal to solidify (Fig. 59). The surface condition of capped steel is much like that of rimmed steel, but certain other characteristics are inter-

FIG. 59

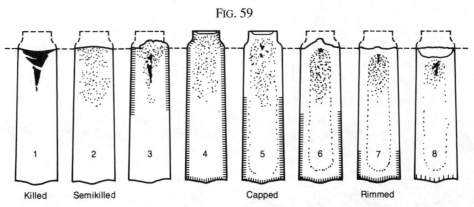

Killed          Semikilled                    Capped              Rimmed

*Eight typical conditions of commercial steel ingots, cast in identical bottle-top molds, in relation to the degree of suppression of gas evolution. The dotted line indicates the height to which the steel originally was poured in each ingot mold.*

mediate between those of *rimmed steel* and those of *semikilled steel*.

**capping** (of abrasive particles). A mechanism of deterioration of abrasive points in which the points become covered by caps of adherent abrasion debris.

**capping** (of powder metallurgy compacts). Partial or complete separation of a powder metallurgy compact into two or more portions by cracks that originate near the edges of the punch faces and that proceed diagonally into the compact.

**caprolactam**. A cyclic amide type of compound containing six carbon atoms. When the ring is opened, caprolactam is polymerizable into a nylon resin known as type 6 nylon or polycaprolactam.

**capture efficiency**. See *collection efficiency*.

**carbanion ion**. Negatively charged organic compound (ion).

**carbide**. A compound of carbon with one or more metallic elements.

**carbide tools**. Cutting or forming tools, usually made from tungsten, titanium, tantalum, or niobium carbides, or a combination of them, in a matrix of cobalt, nickel, or other metals. Carbide tools are characterized by high hardnesses and compressive strengths and may be coated to improve wear resistance (Fig. 60). See also *cemented carbides* (Technical Brief 9).

FIG. 60

*Example of a TiN-coated carbide tool. 1100×*

**carbon**. The element that provides the backbone for all organic polymers. Graphite is a crystalline form of carbon. Diamond is the densest crystalline form of carbon.

**carbonaceous**. A material that contains carbon in any or all of its several allotropic forms.

**carbon arc cutting (CAC)**. An arc cutting process in which metals are severed by melting them with the heat of an arc between a carbon electrode and the base metal.

**carbon arc welding (CAW)**. An arc welding process which produces coalescence of metals by heating them with an

arc between a carbon electrode and the work. No shielding is used. Pressure and filler metal may or may not be used.

**carbon black**. A black pigment produced by the incomplete burning of natural gas or oil. It is widely used as a filler, particularly in the rubber industry. Because it possesses useful ultraviolet protective properties, it is also much used in molding compounds intended for outside weathering applications.

**carbon-carbon composites**. See *Technical Brief 6*.

**carbon dioxide process (sodium silicate/$CO_2$)**. In foundry practice, a process for hardening molds or cores in which carbon dioxide gas is blown through dry clay-free silica sand to precipitate silica in the form of a gel from the sodium silicate binder.

**carbon dioxide welding**. *Gas metal-arc welding* using carbon dioxide as the shielding gas.

**carbon edges**. Carbonaceous deposits in a wavy pattern along the edges of a steel sheet or strip; also known as snaky edges.

**carbon electrode**. A nonfiller material electrode used in arc welding or cutting, consisting of a carbon or graphite rod, which may be coated with copper or other coatings.

**carbon equivalent**. (1) For cast iron, an empirical relationship of the total carbon, silicon, and phosphorus contents expressed by the formula:

$$CE = \%C + 0.3(\%Si) + 0.33(\%P) - 0.027(\%Mn) + 0.4(\%S)$$

(2) For rating of weldability:

$$CE = C + \frac{Mn}{6} + \frac{Ni}{15} + \frac{Cu}{15} + \frac{Cr}{5} + \frac{Mo}{5} + \frac{V}{5}$$

**carbon fiber**. Fiber produced by the pyrolysis of organic precursor fibers, such as rayon, polyacrylonitrile (PAN), and pitch, in an inert environment. The term is often used interchangeably with the term graphite; however, carbon fibers and graphite fibers differ. The basic differences lie in the temperature at which the fibers are made and heat treated, and in the amount of elemental carbon produced. Carbon fibers typically are carbonized in the region of 1315 °C (2400 °F) and assay at 93 to 95% carbon, while graphite fibers are graphitized at 1900 to 2480 °C (3450 to 4500 °F) and assay at more than 99% elemental carbon. See also *pyrolysis*.

**carbon flotation**. In casting, segregation in which free graphite has separated from the molten iron. This defect tends to occur at the upper surfaces of the cope of the castings.

**carbonitriding**. A *case hardening* process in which a suitable ferrous material is heated above the lower transformation temperature in a gaseous atmosphere of such composition as to cause simultaneous absorption of carbon and nitrogen by the surface and, by diffusion, create a concentration gradient. The heat-treating process is completed by cool-

## Technical Brief 6: Carbon-Carbon Composites

CARBON-CARBON COMPOSITES consist of continuous carbon or graphite fibers in a carbon or graphite matrix. These materials have many of the desirable high-temperature properties of conventional carbons and graphites, including high strength, high modulus, high fracture toughness, and low creep. In addition, the high thermal conductivity and low coefficient of thermal expansion,

### High-performance carbon-carbon composite properties

| Reinforcement | Fiber, vol % | Density, g/cm³ | Flexural strength MPa | ksi | Tensile strength MPa | ksi | Flexural modulus GPa | 10⁶ psi | Coefficient of thermal expansion (0-1000 °C, or 32-1830 °F), 10⁻⁶/K |
|---|---|---|---|---|---|---|---|---|---|
| Unidirectional . . . . . . . . 65 | | 1.7 | 827 | 120 | 690 | 100 | 186 | 30 | 1.0 |
| Orthogonal . . . . . . . . . . 55 | | 1.6 | 276 | 40 | ... | ... | 76 | 10 | 1.0 |

coupled with high strength, produce a material with low sensitivity to thermal shock. Carbon-carbon composites can approach the same strengths and moduli as those achieved with resin-matrix composites. Moreover, because their properties are maintained to 2200 °C (4000 °F), they represent a class of material with the highest potential for inert atmosphere or short-duration, high-temperature structural applications.

The methods of manufacturing carbon-carbon composites can be quite varied for flat panels and tubes. In most cases, the fibers are held together through the first steps of the fabrication process with a phenolic resin. The composites are then heat treated to convert the phenolic resin to carbon, usually between 600 and 900 °C (1110 and 1650 °F). After this step, liquid impregnants, such as a resin, a petroleum pitch, or a combination of the two, are typically used to reduce the level of open porosity to 15 to 20%. These impregnants are also converted to carbon at temperatures of 1100 to 2000 °C (2010 to 3630 °F). The final step usually consists of the chemical vapor deposition of carbon from methane-nitrogen gas mixtures at 1100 to 1900 °C (2010 to 3450 °F), which completes the densification of the composite.

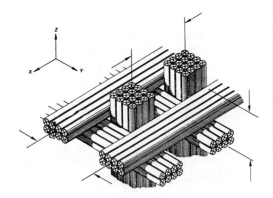

*Three-directional orthogonal preform construction*

Carbon-carbon composites are being developed for a variety of military and aerospace applications. This material is currently being used for reentry thermal protection of the nose cap and wing leading edges of the space shuttle vehicles. Fine-diameter fibers are used to develop turbine rotors and stiff, lightweight space structure components. The frictional characteristics and thermal shock resistance of carbon-carbon composites are exploited in high-performance aircraft brakes. Because they are biocompatible, carbon-carbon composites are also a leading candidate material for prosthetic devices. Further developments in oxidation-resistant coatings should result in additional military and aerospace vehicle applications.

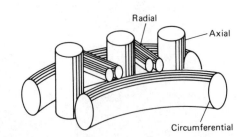

*Geometry of 3-D polar weave preform*

### Recommended Reading

- R.J. Diefendorf, Continuous Carbon Fiber Reinforced Carbon Matrix Composites, *Engineered Materials Handbook*, Vol 1, ASM International, 1987, p 911-914
- L.E. McAllister, Multidirectionally Reinforced Carbon-Graphite Matrix Composites, *Engineered Materials Handbook*, Vol 1, ASM International, 1987, p 915-919
- J.E. Sheehan, Oxidation-Resistant Carbon-Carbon Composites, *Engineered Materials Handbook*, Vol 1, ASM International, 1987, p 920-921
- H.D. Batha and C.R. Rowe, Structurally Reinforced Carbon-Carbon Composites, *Engineered Materials Handbook*, Vol 1, ASM International, 1987, p 922-924

**Table 13  Composition ranges and limits for AISI-SAE standard carbon steels—structural shapes, plate, strip, sheet, and welded tubing**

| AISI-SAE designation | UNS designation | Heat composition ranges and limits(a), % | | AISI-SAE designation | UNS designation | Heat composition ranges and limits(a), % | | AISI-SAE designation | UNS designation | Heat composition ranges and limits(a), % | |
|---|---|---|---|---|---|---|---|---|---|---|---|
| | | C | Mn | | | C | Mn | | | C | Mn |
| 1006 | G10060 | 0.08 max | 0.25-0.45 | 1038 | G10380 | 0.34-0.42 | 0.60-0.90 | 1090 | G10900 | 0.84-0.98 | 0.60-0.90 |
| 1008 | G10080 | 0.10 max | 0.25-0.50 | 1039 | G10390 | 0.36-0.44 | 0.70-1.00 | 1095 | G10950 | 0.90-1.04 | 0.30-0.50 |
| 1009 | G10090 | 0.15 max | 0.60 max | 1040 | G10400 | 0.36-0.44 | 0.60-0.90 | 1524(b) | G15240 | 0.18-0.25 | 1.30-1.65 |
| 1010 | G10100 | 0.08-0.13 | 0.30-0.60 | 1042 | G10420 | 0.39-0.47 | 0.60-0.90 | 1527(b) | G15270 | 0.22-0.29 | 1.20-1.55 |
| 1012 | G10120 | 0.10-0.15 | 0.30-0.60 | 1043 | G10430 | 0.39-0.47 | 0.70-1.00 | 1536(b) | G15360 | 0.30-0.38 | 1.20-1.55 |
| 1015 | G10150 | 0.12-0.18 | 0.30-0.60 | 1045 | G10450 | 0.42-0.50 | 0.60-0.90 | 1541(b) | G15410 | 0.36-0.45 | 1.30-1.65 |
| 1016 | G10160 | 0.12-0.18 | 0.60-0.90 | 1046 | G10460 | 0.42-0.50 | 0.70-1.00 | 1548(b) | G15480 | 0.43-0.52 | 1.05-1.40 |
| 1017 | G10170 | 0.14-0.20 | 0.30-0.60 | 1049 | G10490 | 0.45-0.53 | 0.60-0.90 | 1552(b) | G15520 | 0.46-0.55 | 1.20-1.55 |
| 1018 | G10180 | 0.14-0.20 | 0.60-0.90 | 1050 | G10500 | 0.47-0.55 | 0.60-0.90 | | | | |
| 1019 | G10190 | 0.14-0.20 | 0.70-1.00 | 1055 | G10550 | 0.52-0.60 | 0.60-0.90 | | | | |
| 1020 | G10200 | 0.17-0.23 | 0.30-0.60 | 1060 | G10600 | 0.55-0.66 | 0.60-0.90 | | | | |
| 1021 | G10210 | 0.17-0.23 | 0.60-0.90 | 1064 | G10640 | 0.59-0.70 | 0.50-0.80 | | | | |
| 1022 | G10220 | 0.17-0.23 | 0.70-1.00 | 1065 | G10650 | 0.59-0.70 | 0.60-0.90 | | | | |
| 1023 | G10230 | 0.19-0.25 | 0.30-0.60 | 1070 | G10700 | 0.65-0.76 | 0.60-0.90 | | | | |
| 1025 | G10250 | 0.22-0.28 | 0.30-0.60 | 1074 | G10740 | 0.69-0.80 | 0.50-0.80 | | | | |
| 1026 | G10260 | 0.22-0.28 | 0.60-0.90 | 1078 | G10780 | 0.72-0.86 | 0.30-0.60 | | | | |
| 1030 | G10300 | 0.27-0.34 | 0.60-0.90 | 1080 | G10800 | 0.74-0.88 | 0.60-0.90 | | | | |
| 1033 | G10330 | 0.29-0.36 | 0.70-1.00 | 1084 | G10840 | 0.80-0.94 | 0.60-0.90 | | | | |
| 1035 | G10350 | 0.31-0.38 | 0.60-0.90 | 1085 | G10850 | 0.80-0.94 | 0.70-1.00 | | | | |
| 1037 | G10370 | 0.31-0.38 | 0.70-1.00 | 1086 | G10860 | 0.80-0.94 | 0.30-0.50 | | | | |

(a) Limits on phosphorus and sulfur contents are typically 0.040% maximum phosphorus and 0.050% maximum sulfur. Silicon contents range from 0.080% to approximately 2%. Steels listed in this table can be produced with additions of lead or boron. Leaded steels typically contain 0.15 to 0.35% lead and are identified by inserting the letter "L" in the designation—11L17; boron steels can be expected to contain 0.0005 to 0.003% boron and are identified by inserting the letter "B" in the designation—15B41. (b) Formerly designated 10xx grade

ing at a rate that produces the desired properties in the workpiece.

**carbonium ion.** Positively charged organic compound (ion).

**carbonization.** The conversion of an organic substance into elemental carbon in an inert atmosphere at temperatures ranging from 800 to 1600 °C (1470 to 2910 °F) and higher, but usually at about 1315 °C (2400 °F). Range is influenced by precursor, processing of the individual manufacturer, and properties desired. Should not be confused with car*buri*zation.

**carbonizing flame.** See preferred term *reducing flame*.

**carbon potential.** A measure of the ability of an environment containing active carbon to alter or maintain, under prescribed conditions, the carbon level of a steel. In any particular environment, the carbon level attained will depend on such factors as temperature, time, and steel composition.

**carbon refractory.** A manufactured refractory comprised substantially or entirely of carbon (including graphite).

**carbon restoration.** Replacing the carbon lost in the surface layer from previous processing of a steel by carburizing this layer to substantially the original carbon level. Sometimes called recarburizing.

**carbon steel.** Steel having no specified minimum quantity for any alloying element—other than the commonly accepted amounts of manganese (≤1.65%), silicon (≤0.60%), and copper (≤0.60%)—and containing only an incidental amount of any element other than carbon, silicon, manganese, copper, sulfur, and phosphorus. Low-carbon steels contain up to 0.30% C, medium-carbon steels contain from 0.30 to 0.60% C, and high-carbon steels contain from 0.60 to 1.00% C. Table 13 lists compositions of carbon steels.

**carbon tube furnace.** An electric furnace that has a carbon retort for a resistor element and is especially suitable for the batch or continuous sintering of carbon-insensitive materials such as cemented carbides.

**carbonyl powder.** Metal powders prepared by the thermal decomposition of a metal carbonyl compound such as nickel tetracarbonyl $Ni(CO)_4$ or iron pentacarbonyl $Fe(CO)_5$. See also *thermal decomposition*.

**carburizing.** Absorption and diffusion of carbon into solid ferrous alloys by heating, to a temperature usually above $Ac_3$, in contact with a suitable carbonaceous material. A form of *case hardening* that produces a carbon gradient extending inward from the surface, enabling the surface layer to be hardened either by quenching directly from the carburizing temperature or by cooling to room temperature, then reaustenitizing and quenching.

**carburizing flame.** A gas flame that will introduce carbon into some heated metals, as during a gas welding operation. A carburizing flame is a *reducing flame*, but a reducing flame is not necessarily a carburizing flame.

**car furnace.** A batch-type heat-treating furnace using a car on rails to enter and leave the furnace area. Car furnaces are used for lower stress relieving ranges.

**carrier.** In emission spectrochemical analysis, a material added to a sample to facilitate its controlled vaporization into the analytical gap. See also *analytical gap*.

**carrier gas.** In thermal spraying, the gas used to carry the powdered materials from the powder feeder or hopper to the gun.

FIG. 61

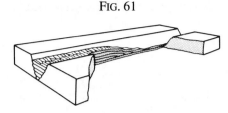

*Cascade welding sequence of reducing stress*

**cascade separator.** A special device to separate metal powder fractions of different particle size or specific gravity.

**cascade sequence.** A welding sequence in which a continuous multiple-pass weld is built up by depositing weld beads in overlapping layers, usually laid in a *backstep sequence* (Fig. 61). Compare with *block sequence*.

**case.** In heat treating, that portion of a ferrous alloy, extending inward from the surface, whose composition has been altered during *case hardening*. Typically considered to be the portion of an alloy (a) whose composition has been measurably altered from the original composition, (b) that appears light when etched (Fig. 62), or (c) that has a higher hardness value than the core. Contrast with *core*.

**case crushing.** A term used to denote longitudinal gouges arising from fracture in case-hardened gears.

**case hardening.** A generic term covering several processes applicable to steel that change the chemical composition of the surface layer by absorption of carbon, nitrogen, or a mixture of the two and, by diffusion, create a concentration gradient. The processes commonly used are *carburizing* and *quench hardening*; *cyaniding*; *nitriding*; and *carbonitriding*. The use of the applicable specific process name is preferred.

**casein.** A protein material precipitated from skimmed milk by the action of either rennet or dilute acid. Rennet casein finds its main application in the manufacture of plastics.

FIG. 62

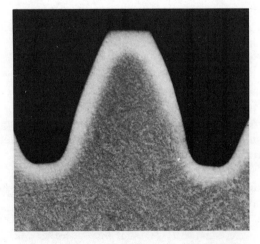

*Uniform case depth in a steel tooth gear. 6×*

Acid casein is a raw material used in a number of industries, including the manufacture of adhesives.

**casein adhesive.** An aqueous colloidal dispersion of *casein* that may be prepared with or without heat; that may contain modifiers, inhibitors, and secondary binders to provide specific adhesive properties; and that includes a subclass, usually identified as casein glue, that ia based on a dry blend of casein, lime, and sodium salts, mixed with water prepared without heat.

**cassette.** A holder used to contain radiographic films during exposure to x-rays or gamma rays, that may or may not contain intensifying or filter screens, or both. A distinction is often made between a cassette, which has a positive means for ensuring contact between screens and film and is usually rigid, and an exposure holder, which is rather flexible.

**CASS test.** Abbreviation for *copper-accelerated salt-spray test*.

**cast (plastics).** To form a "plastic" object by pouring a fluid monomer-polymer solution into an open mold where it finishes polymerizing. Forming plastic film and sheet by pouring the liquid resin onto a moving belt or by precipitation in a chemical bath.

**castable.** In casting, a combination of refractory grain and suitable bonding agent that, after the addition of a proper liquid, is generally poured into place to form a refractory shape or structure which becomes rigid because of chemical action.

**castability.** (1) A complex combination of liquid-metal properties and solidification characteristics that promotes accurate and sound final castings. (2) The relative ease with which a molten metal flows through a mold or casting die.

**cast-alloy tool.** A cutting tool made by casting a cobalt-base alloy and used at machining speeds between those for high-speed steels and cemented carbides. Nominal compositions for two commercially available grades are as follows:

| Element | Tantung G, % | Tantung 144, % |
|---|---|---|
| Cobalt | 42–47 | 40–45 |
| Chromium | 27–32 | 25–30 |
| Tungsten | 14–19 | 16–21 |
| Carbon | 2–4 | 2–4 |
| Tantalum or niobium | 2–7 | 3–8 |
| Manganese | 1–3 | 1–3 |
| Iron | 2–5 | 2–5 |
| Nickel | 7(a) | 7(a) |

(a) Maximum

**cast corrosion-resistant stainless steels.** High chromium-containing (11 to 30% Cr) cast steels specified for liquid corrosion service at temperatures below 650 °C (1200 °F). Corrosion-resistant, or C-type steel castings are classified on the basis of composition using the designation system

## Table 14  Compositions and typical microstructures of corrosion-resistant cast stainless steels

| ACI type | Wrought alloy type(a) | ASTM specifications | Most common end-use microstructure | Composition, %(b) | | | | | |
|---|---|---|---|---|---|---|---|---|---|
| | | | | C | Mn | Si | Cr | Ni | Others(c) |
| **Chromium steels** | | | | | | | | | |
| CA-15............. | 410 | A 743, A 217, A 487 | Martensite | 0.15 | 1.00 | 1.50 | 11.5–14.0 | 1.0 | 0.50Mo(d) |
| CA-15M........... | ... | A 743 | Martensite | 0.15 | 1.00 | 0.65 | 11.5–14.0 | 1.0 | 0.15–1.00Mo |
| CA-40............. | 420 | A 743 | Martensite | 0.40 | 1.00 | 1.50 | 11.5–14.0 | 1.0 | 0.5Mo(d) |
| CA-40F .......... | ... | A 743 | Martensite | 0.2–0.4 | 1.00 | 1.50 | 11.5–14.0 | 1.0 | ... |
| CB-30............431, 442 | | A 743 | Ferrite and carbides | 0.30 | 1.00 | 1.50 | 18.0–22.0 | 2.0 | ... |
| CC-50............. | 446 | A 743 | Ferrite and carbides | 0.30 | 1.00 | 1.50 | 26.0–30.0 | 4.0 | ... |
| **Chromium-nickel steels** | | | | | | | | | |
| CA-6N ........... | ... | A 743 | Martensite | 0.06 | 0.50 | 1.00 | 10.5–12.5 | 6.0–8.0 | ... |
| CA-6NM.......... | ... | A 743, A 487 | Martensite | 0.06 | 1.00 | 1.00 | 11.5–14.0 | 3.5–4.5 | 0.4–1.0Mo |
| CA-28MWV ....... | ... | A 743 | Martensite | 0.20–0.28 | 0.50–1.00 | 1.00 | 11.0–12.5 | 0.50–1.00 | 0.9–1.25Mo; 0.9–1.25W; 0.2–0.3V |
| CB-7Cu-1......... | ... | A 747 | Martensite, age hardenable | 0.07 | 0.70 | 1.00 | 15.5–17.7 | 3.6–4.6 | 2.5–3.2Cu; 0.20–0.35Nb; 0.05N max |
| CB-7Cu-2......... | ... | A 747 | Martensite, age hardenable | 0.07 | 0.70 | 1.00 | 14.0–15.5 | 4.5–5.5 | 2.5–3.2Cu; 0.20–0.35 Nb; 0.05N max |
| CD-4MCu ........ | ... | A 351, A 743, A 744, A 890 | Austenite in ferrite, age hardenable | 0.04 | 1.00 | 1.00 | 25.0–26.5 | 4.75–6.0 | 1.75–2.25Mo; 2.75–3.25Cu |
| CE-30............. | 312 | A 743 | Ferrite in austenite | 0.30 | 1.50 | 2.00 | 26.0–30.0 | 8.0–11.0 | |
| CF-3(e)........... | 304L | A 351, A 743, A 744 | Ferrite in austenite | 0.03 | 1.50 | 2.00 | 17.0–21.0 | 8.0–12.0 | ... |
| CF-3M(e)......... | 316L | A 351, A 743, A 744 | Ferrite in austenite | 0.03 | 1.50 | 2.00 | 17.0–21.0 | 8.0–12.0 | 2.0–3.0Mo |
| CF-3MN ......... | ... | A 743 | Ferrite in austenite | 0.03 | 1.50 | 1.50 | 17.0–21.0 | 9.0–13.0 | 2.0–3.0Mo; 0.10–0.20N |
| CF-8(e)........... | 304 | A 351, A 743, A 744 | Ferrite in austenite | 0.08 | 1.50 | 2.00 | 18.0–21.0 | 8.0–11.0 | ... |
| CF-8C ........... | 347 | A 351, A 743, A 744 | Ferrite in austenite | 0.08 | 1.50 | 2.00 | 18.0–21.0 | 9.0–12.0 | Nb(f) |
| CF-8M ........... | 316 | A 351, A 743, A 744 | Ferrite in austenite | 0.08 | 1.50 | 2.00 | 18.0–21.0 | 9.0–12.0 | 2.0–3.0Mo |
| CF-10............. | ... | A 351 | Ferrite in austenite | 0.04–0.10 | 1.50 | 2.00 | 18.0–21.0 | 8.0–11.0 | ... |
| CF-10M .......... | ... | A 351 | Ferrite in austenite | 0.04–0.10 | 1.50 | 1.50 | 18.0–21.0 | 9.0–12.0 | 2.0–3.0Mo |
| CF-10MC......... | ... | A 351 | Ferrite in austenite | 0.10 | 1.50 | 1.50 | 15.0–18.0 | 13.0–16.0 | 1.75–2.25Mo |
| CF-10SMnN ...... | ... | A 351, A 743 | Ferrite in austenite | 0.10 | 7.00–9.00 | 3.50–4.50 | 16.0–18.0 | 8.0–9.0 | 0.08–0.18N |
| CF-12M .......... | 316 | ... | Ferrite in austenite or austenite | 0.12 | 1.50 | 2.00 | 18.0–21.0 | 9.0–12.0 | 2.0–3.0Mo |
| CF-16F........... | 303 | A 743 | Austenite | 0.16 | 1.50 | 2.00 | 18.0–21.0 | 9.0–12.0 | 1.50Mo max; 0.20–0.35Se |
| CF-20............. | 302 | A 743 | Austenite | 0.20 | 1.50 | 2.00 | 18.0–21.0 | 8.0–11.0 | ... |
| CG-6MMN ....... | ... | A 351, A 743 | Ferrite in austenite | 0.06 | 4.00–6.00 | 1.00 | 20.5–23.5 | 11.5–13.5 | 1.50–3.00Mo; 0.10–0.30Nb; 0.10–30V; 0.20–40N |
| CG-8M .......... | 317 | A 351, A 743, A 744 | Ferrite in austenite | 0.08 | 1.50 | 1.50 | 18.0–21.0 | 9.0–13.0 | 3.0–4.0Mo |
| CG-12............ | ... | A 743 | Ferrite in austenite | 0.12 | 1.50 | 2.00 | 20.0–23.0 | 10.0–13.0 | ... |
| CH-8............. | ... | A 351 | Ferrite in austenite | 0.08 | 1.50 | 1.50 | 22.0–26.0 | 12.0–15.0 | ... |
| CH-10............ | ... | A 351 | Ferrite in austenite | 0.04–0.10 | 1.50 | 2.00 | 22.0–26.0 | 12.0–15.0 | ... |
| CH-20............ | 309 | A 351, A 743 | Austenite | 0.20 | 1.50 | 2.00 | 22.0–26.0 | 12.0–15.0 | ... |
| CK-3MCuN ....... | ... | A 351, A 743, A 744 | Ferrite in austenite | 0.025 | 1.20 | 1.00 | 19.5–20.5 | 17.5–19.5 | 6.0–7.0V; 0.18–0.24N; 0.50–1.00Cu |
| CK-20............ | 310 | A 743 | Austenite | 0.20 | 2.00 | 2.00 | 23.0–27.0 | 19.0–22.0 | |
| **Nickel-chromium steel** | | | | | | | | | |
| CN-3M........... | ... | A 743 | Austenite | 0.03 | 2.00 | 1.00 | 20.0–22.0 | 23.0–27.0 | 4.5–5.5Mo |
| CN-7M........... | ... | A 351, A 743, A 744 | Austenite | 0.07 | 1.50 | 1.50 | 19.0–22.0 | 27.5–30.5 | 2.0–3.0Mo; 3.0–4.0Cu |
| CN-7MS ......... | ... | A 743, A 744 | Austenite | 0.07 | 1.50 | 3.50(g) | 18.0–20.0 | 22.0–25.0 | 2.5–3.0Mo; 1.5–2.0Cu |
| CT-15C .......... | ... | A 351 | Austenite | 0.05–0.15 | 0.15–1.50 | 0.50–1.50 | 19.0–21.0 | 31.0–34.0 | 0.5–1.5V |

(a) Type numbers of wrought alloys are listed only for nominal identification of corresponding wrought and cast grades. Composition ranges of cast alloys are not the same as for corresponding wrought alloys; cast alloy designations should be used for castings only. (b) Maximum unless a range is given. The balance of all compositions is iron. (c) Sulfur content is 0.04% in all grades except: CG-6MMN, 0.030% S (max); CF-10SMnN, 0.03% S (max); CT-15C, 0.03% S (max); CK-3MCuN, 0.010% S (max); CN-3M, 0.030% S (max); CA-6N, 0.020% S (max); CA-28MWV, 0.030% S (max); CA-40F, 0.20–0.40% S; CB-7Cu-1 and -2, 0.03% S (max). Phosphorus content is 0.04% (max) in all grades except: CF-16F, 0.17% P (max); CF-10SMnN, 0.060% P (max); CT-15C, 0.030% P (max); CK-3MCuN, 0.045% P (max); CN-3M, 0.030% P (max); CA-6N, 0.020% P (max); CA-28MWV, 0.030% P (max); CB-7Cu-1 and -2, 0.035% P (max). (d) Molybdenum not intentionally added. (e) CF-3A, CF-3MA, and CF-8A have the same composition ranges as CF-3, CF-3M, and CF-8, respectively, but have balanced compositions so that ferrite contents are at levels that permit higher mechanical property specifications than those for related grades. They are covered by ASTM A 351. (f) Nb, 8 × %C min (1.0% max); or Nb + Ta × %C (1.1% max). (g) For CN-7MS, silicon ranges from 2.50 to 3.50%.

of the High Alloy Product Group of the Steel Founder's Society of America. The first letter of the designation (C) indicates that the alloy is intended for liquid corrosion service. The second letter indicates the nickel content; as the nickel content increases, the second letter of the designation changes from A to Z. Table 14 lists compositions and microstructures of cast C-type steels.

**cast film.** A film made by depositing a layer of liquid plastic onto a surface and stabilizing this form by the evaporation of solvent, by fusing after deposition, or by allowing a melt to cool. Cast films are usually made from solutions or dispersions.

**cast heat-resistant stainless steels.** Iron-chromium, iron-chromium-nickel, and iron-nickel-chromium steel castings specified for service at temperatures 650 °C (1200 °F) to approximately 980 °C (1800 °F). Designated similarly to *cast corrosion-resistant stainless steels*, H-type steels (the "H" denoting high-temperature service) have nickel contents ranging from 0 to 68%. Table 15 lists compositions of cast H-type steels.

## Table 15  Compositions of heat-resistant cast stainless steels

| ACI designation | UNS number | ASTM specifications(a) | C | Cr | Ni | Si (max) |
|---|---|---|---|---|---|---|
| HA | · · · | A 217 | 0.20 max | 8–10 | · · · | 1.00 |
| HC | J92605 | A 297, A 608 | 0.50 max | 26–30 | 4 max | 2.00 |
| HD | J93005 | A 297, A 608 | 0.50 max | 26–30 | 4–7 | 2.00 |
| HE | J93403 | A 297, A 608 | 0.20–0.50 | 26–30 | 8–11 | 2.00 |
| HF | J92603 | A 297, A 608 | 0.20–0.40 | 19–23 | 9–12 | 2.00 |
| HH | J93503 | A 297, A 608, A 447 | 0.20–0.50 | 24–28 | 11–14 | 2.00 |
| HI | J94003 | A 297, A 567, A 608 | 0.20–0.50 | 26–30 | 14–18 | 2.00 |
| HK | J94224 | A 297, A 351, A 567, A 608 | 0.20–0.60 | 24–28 | 18–22 | 2.00 |
| HK30 | · · · | A 351 | 0.25–0.35 | 23.0–27.0 | 19.0–22.0 | 1.75 |
| HK40 | · · · | A 351 | 0.35–0.45 | 23.0–27.0 | 19.0–22.0 | 1.75 |
| HL | J94604 | A 297, A 608 | 0.20–0.60 | 28–32 | 18–22 | 2.00 |
| HN | J94213 | A 297, A 608 | 0.20–0.50 | 19–23 | 23–27 | 2.00 |
| HP | · · · | A 297 | 0.35–0.75 | 24–28 | 33–37 | 2.00 |
| HP-50WZ(c) | · · · | · · · | 0.45–0.55 | 24–28 | 33–37 | 2.50 |
| HT | J94605 | A 297, A 351, A 567, A 608 | 0.35–0.75 | 13–17 | 33–37 | 2.50 |
| HT30 | · · · | A 351 | 0.25–0.35 | 13.0–17.0 | 33.0–37.0 | 2.50 |
| HU | · · · | A 297, A 608 | 0.35–0.75 | 17–21 | 37–41 | 2.50 |
| HW | · · · | A 297, A 608 | 0.35–0.75 | 10–14 | 58–62 | 2.50 |
| HX | · · · | A 297, A 608 | 0.35–0.75 | 15–19 | 64–68 | 2.50 |

(a) ASTM designations are the same as ACI designations. (b) Rem Fe in all compositions. Manganese content: 0.35 to 0.65% for HA, 1% for HC, 1.5% for HD, and 2% for the other alloys. Phosphorus and sulfur contents: 0.04% (max) for all but HP-50WZ. Molybdenum is intentionally added only to HA, which has 0.90 to 1.20% Mo; maximum for other alloys is set at 0.5% Mo. HH also contains 0.2% N (max). (c) Also contains 4 to 6% W, 0.1 to 1.0% Zr, and 0.035% S (max) and P (max)

**casting.** (1) Metal object cast to the required shape by pouring or injecting liquid metal into a mold, as distinct from one shaped by a mechanical process. (2) Pouring molten metal into a mold to produce an object of desired shape. (3) Ceramic forming process in which a body slip is introduced into a porous mold, which absorbs sufficient water from the slip to produce a semirigid article.

**casting copper.** Fire-refined tough pitch copper usually cast from melted secondary metal into ingot bars only, and used for making foundry castings but not wrought products.

**casting defect.** Any imperfection in a casting that does not satisfy one or more of the required design or quality specifications. This term is often used in a limited sense for those flaws formed by improper casting solidification.

**casting section thickness.** The wall thickness of the casting. Because the casting may not have a uniform thickness, the section thickness may be specified at a specific place on the casting. Also, it is sometimes useful to use the average, minimum, or typical wall thickness to describe a casting.

**casting shrinkage.** The amount of dimensional change per unit length of the casting as it solidifies in the mold or die and cools to room temperature after removal from the mold or die. There are three distinct types of casting shrinkage. Liquid shrinkage refers to the reduction in volume of liquid metal as it cools to the liquidus. Solidification shrinkage is the reduction in volume of metal from the beginning to the end of solidification. Solid shrinkage involves the reduction in volume of metal from the solidus to room temperature.

**casting strains.** Strains in a casting caused by *casting stresses* that develop as the casting cools.

**casting stresses.** Residual stresses set up when the shape of a casting impedes contraction of the solidified casting during cooling.

**casting thickness.** See *casting section thickness*.

**casting volume.** The total cubic units (mm$^3$ or in.$^3$) of cast metal in the casting.

**casting yield.** The weight of a casting(s) divided by the total weight of metal poured into the mold, expressed as a percentage.

**cast irons.** See *Technical Brief 7*.

**cast replica.** In *metallography*, a reproduction of a surface in plastic made by the evaporation of the solvent from a solution of the plastic or by polymerization of a monomer on the surface. See also *replica*.

**cast steel.** Steel in the form of a *casting*.

**cast structure.** The metallographic structure of a *casting* evidenced by shape and orientation of grains and by segregation of impurities.

**catalyst.** (1) A substance capable of changing the rate of a reaction without itself undergoing any net change. (2) A substance that markedly speeds up the *cure* of a plastic compound when added in minor quantity, compared to the amounts of primary reactants.

**catastrophic failure.** Sudden failure of a component or assembly that frequently results in extensive secondary damage to adjacent components or assemblies.

**catastrophic period.** In cavitation or liquid impingement erosion, a stage during which the erosion rate increases so dramatically that continued exposure threatens or causes gross disintegration of the exposed surface.

**catastrophic wear.** Sudden surface damage, deterioration, or change of shape caused by wear to such an extent that the life of the part is appreciably shortened or action is impaired.

**catchment efficiency.** See *collection efficiency*.

**catenary.** A measure of the difference in length of reinforcing strands in a specified length of *roving* caused by unequal

## Technical Brief 7: Cast Irons

CAST IRON is a generic term that identifies a large family of cast ferrous alloys that solidify with a eutectic and in which the carbon content exceeds the solubility of carbon in austenite at the eutectic temperature. Cast irons primarily are alloys of iron that contain more than 2% C and from 1 to 3% Si. Wide variations in properties can be achieved by varying the balance between carbon and silicon, by alloying with various metallic or nonmetallic elements, and by varying melting, casting, and heat treating practices.

Cast irons can be classified according to their graphite shape, matrix microstructure (austenitic, ferritic, etc.), or fracture type:

### Range of compositions for typical unalloyed common cast irons

| Type of iron | Composition, % | | | | |
|---|---|---|---|---|---|
| | C | Si | Mn | P | S |
| Gray (FG) | 2.5–4.0 | 1.0–3.0 | 0.2–1.0 | 0.002–1.0 | 0.02–0.25 |
| Compacted graphite (CG) | 2.5–4.0 | 1.0–3.0 | 0.2–1.0 | 0.01–0.1 | 0.01–0.03 |
| Ductile (SG) | 3.0–4.0 | 1.8–2.8 | 0.1–1.0 | 0.01–0.1 | 0.01–0.03 |
| White | 1.8–3.6 | 0.5–1.9 | 0.25–0.8 | 0.06–0.2 | 0.06–0.2 |
| Malleable (TG) | 2.2–2.9 | 0.9–1.9 | 0.15–1.2 | 0.02–0.2 | 0.02–0.2 |

- *White iron* is essentially free of graphite, and most of the carbon content is present as separate grains of hard $Fe_3C$. White iron exhibits a white, crystalline fracture surface, because fracture occurs along the iron carbide plates. It is usually not heat treated, but is stress relieved.

- *Malleable iron* contains compact nodules of graphite flakes called "temper carbon," because they form during an extended annealing of white iron of a suitable composition.

- *Gray iron* contains carbon in the form of graphite flakes. Gray iron exhibits a gray fracture surface, because fracture occurs along the graphite plates (flakes).

- *Mottled iron* falls between gray and white iron, with the fracture showing both gray and white zones.

### Classification of cast iron by commercial designation, microstructure, and fracture

| Commercial designation | Carbon-rich phase | Matrix(a) | Fracture | Final structure after |
|---|---|---|---|---|
| Gray iron | Lamellar graphite | P | Gray | Solidification |
| Ductile iron | Spheroidal graphite | F, P, A | Silver-gray | Solidification or heat treatment |
| Compacted graphite iron | Compacted vermicular graphite | F, P | Gray | Solidification |
| White iron | $Fe_3C$ | P, M | White | Solidification and heat treatment(b) |
| Mottled iron | Lamellar Gr + $Fe_3C$ | P | Mottled | Solidification |
| Malleable iron | Temper graphite | F, P | Silver-gray | Heat treatment |
| Austempered ductile iron | Spheroidal graphite | At | Silver-gray | Heat treatment |

(a) F, ferrite; P, pearlite; A, austenite; M, martensite; At, austempered (bainite). (b) White irons are not usually heat treated, except for stress relief and to continue austenite transformation.

- *Ductile iron*, also known as spheroidal graphite iron, contains spherulitic graphite, in which the graphite flakes form into balls as do cabbage leaves. Ductile iron is so named because in the as-cast form it exhibits measurable ductility.

- *Austempered ductile iron* is a moderately alloyed ductile iron that is austempered for high strength with appreciable ductility. Its microstructure is different from austempered steel, and its heat treatment is a specialty.

- *Compacted graphite iron* contains graphite in the form of thick, stubby flakes. Its mechanical properties are between those of gray and ductile iron.

- *High-alloy iron* contains more than 3% alloy content and is commercially classified separately. High-alloy irons may be a type of white iron, gray iron, or ductile iron. The matrix may be ferritic or austenitic.

### Recommended Reading

- *ASM Handbook*, Vol 4, *Heat Treating*, ASM International, 1991, p 667-708
- *Metals Handbook*, 10th ed., Vol 1, *Properties and Selection: Irons, Steels, and High-Performance Alloys*, ASM International, 1990, p 3-104
- *Metals Handbook*, 9th ed., Vol 15, *Casting*, ASM International, 1988, p 627-710

tension. The tendency of some strands in a taut, horizontal roving to sag more than the others.

**cathode**. The negative *electrode* of an *electrolytic cell* at which reduction is the principal reaction. (Electrons flow toward the cathode in the external circuit.) Typical cathodic processes are cations taking up electrons and being discharged, oxygen being reduced, and the reduction of an element or group of elements from a higher to a lower valence state. Contrast with *anode*.

**cathode compartment**. In an electrolytic cell, the enclosure formed by a diaphragm around the cathode.

**cathode copper**. Copper deposited at the cathode in electrolytic refining.

**cathode efficiency**. Current efficiency at the *cathode*.

**cathode film**. The portion of solution in immediate contact with the *cathode* during *electrolysis*.

**cathode-ray tube (CRT)**. An electronic tube that permits the visual display of electronic signals.

**cathodic cleaning**. *Electrolytic cleaning* in which the work is the *cathode*.

**cathodic corrosion**. Corrosion resulting from a cathodic condition of a structure usually caused by the reaction of an amphoteric metal with the alkaline products of *electrolysis*.

**cathodic disbondment**. The destruction of adhesion between a coating and its substrate by products of a *cathodic reaction*.

**cathodic etching**. See *ion etching*.

**cathodic inhibitor**. In corrosion protection, a chemical substance or mixture that prevents or reduces the rate of the cathodic or reduction reaction.

**cathodic pickling**. Electrolytic pickling in which the work is the *cathode*.

**cathodic polarization**. The change of the *electrode potential* in the active (negative) direction due to current flow. See also *polarization*.

**cathodic protection**. (1) Reduction of corrosion rate by shifting the *corrosion potential* of the electrode toward a less oxidizing potential by applying an external *electromotive force*. (2) Partial or complete protection of a metal from corrosion by making it a *cathode*, using either a galvanic or an impressed current (Fig. 63). Contrast with *anodic protection*.

**cathodic reaction**. Electrode reaction equivalent to a transfer of negative charge from the electronic to the ionic conductor. A cathodic reaction is a reduction process. An example common in corrosion is: $Ox + ne^- \rightarrow Red$.

**cathodoluminescence**. A radiative transition wherein low-energy light photons are released during electron irradiation.

**catholyte**. The *electrolyte* adjacent to the cathode of an electrolytic cell.

**cation**. A positively charged ion that migrates through the

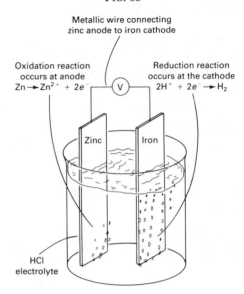

FIG. 63

Metallic wire connecting zinc anode to iron cathode

Oxidation reaction occurs at anode
$Zn \rightarrow Zn^{2+} + 2e^-$

Reduction reaction occurs at the cathode
$2H^+ + 2e^- \rightarrow H_2$

Zinc      Iron

HCl electrolyte

*Cathodic protection of iron by zinc in hydrochloric acid*

electrolyte toward the *cathode* under the influence of a potential gradient. See also *anion* and *ion*.

**cationic detergent**. A detergent in which the *cation* is the active part.

**CAT scanning**. See *computed tomography*.

**caul**. In adhesive bonding, a sheet of material employed singly or in pairs in the hot or cold pressing of assemblies being bonded. A caul is used to protect either the faces of the assembly or the press platens, or both, against marring and staining in order to prevent sticking, facilitate press loading, impart a desired surface texture or finish, and provide uniform pressure distribution. A caul may be made of any suitable material such as aluminum, stainless steel, hardboard, fiberboard, or plastic, the length and width dimensions generally being the same as those of the plates of the press where it is used.

**caulk weld**. See preferred term *seal weld*.

**caul plates**. In fabrication of composites, smooth metal plates, free of surface defects, that are the same size and shape as a composite lay-up, and that contact the lay-up during the curing process in order to transmit normal pressure and temperature, and to provide a smooth surface on the finished laminate.

**caustic**. (1) Burning or corrosive. (2) A hydroxide of a light metal, such as sodium hydroxide or potassium hydroxide.

**caustic cracking**. A form of *stress-corrosion cracking* most frequently encountered in carbon steels or iron-chromium-nickel alloys that are exposed to concentrated hydroxide solutions at temperatures of 200 to 250 °C (400 to 480 °F). Also known as caustic embrittlement.

**caustic dip**. A strongly alkaline solution into which metal is immersed for etching, for neutralizing acid, or for removing organic materials such as greases or paints.

**caustic embrittlement.** An obsolete historical term denoting a form of *stress-corrosion cracking* most frequently encountered in carbon steels or iron-chromium-nickel alloys that are exposed to concentrated hydroxide solutions at temperatures of 200 to 250 °C (400 to 480 °F).

**caustic quenching.** Quenching with aqueous solutions of 5 to 10% sodium hydroxide (NaOH).

**cavitating disk apparatus.** A flow cavitation test device in which cavitating wakes are produced by holes in, or protuberances on, a disk rotating within a liquid-filled chamber. Erosion test specimens are attached flush with the surface of the disk at the location where the bubbles are presumed to collapse.

**cavitation.** The formation and collapse, within a liquid, of cavities or bubbles that contain vapor or gas or both. In general, cavitation originates from a decrease in the static pressure in the liquid. It is distinguished in this way from boiling, which originates from an increase in the liquid temperature. There are certain situations where it may be difficult to make a clear distinction between cavitation and boiling, and the more general definition that is given here is therefore to be preferred. In order to erode a solid surface by cavitation, it is necessary for the cavitation bubbles to collapse on or close to that surface.

**cavitation cloud.** A collection of a large number of cavitation bubbles. The bubbles in a cloud are small, typically less than 1 mm in cross section.

**cavitation corrosion.** A process involving conjoint *corrosion* and *cavitation.*

**cavitation damage.** The degradation of a solid body resulting from its exposure to *cavitation.* This may include loss of material, surface deformation, or changes in properties or appearance.

**cavitation erosion.** Progressive loss of original material from a solid surface due to continuing exposure to *cavitation.*

**cavitation tunnel.** A flow cavitation test facility in which liquid is pumped through a pipe or tunnel, and cavitation is induced in a test section by conducting the flow through a constriction, or around an obstacle, or a combination of these.

**cavity** (metals). The mold or die impression that gives a casting its external shape.

**cavity** (plastics). The space inside a mold into which a resin or molding compound is poured or injected. The female portion of a mold. The portion of the mold that encloses the molded article (often referred to as the die). Depending on the number of such depressions, molds are designated as single cavity or multiple cavity.

**cavity retainer plates.** In forming of plastics, plates in a mold that hold the cavities and forces. These plates are at the mold parting line and usually contain the guide pins and bushings. Also called force retainer plates.

**CCT diagram.** See *continuous cooling transformation diagram.*

**cell.** In honeycomb core, a cell is a single honeycomb unit, usually in a hexagonal shape. See also the figure accompanying the term *honeycomb.*

**cell** (electrochemistry). Electrochemical system consisting of an *anode* and a *cathode* immersed in an *electrolyte.* The anode and cathode may be separate metals or dissimilar areas on the same metal. The cell includes the external circuit, which permits the flow of electrons from the anode toward the cathode. See also *electrochemical cell.*

**cell** (plastics). A single cavity formed by gaseous displacement in a plastic material. See also *cellular plastic.*

**cell feed.** The material supplied to the cell in the electrolytic production of metals.

**cell size.** The diameter of an inscribed circle within a cell of *honeycomb* core.

**cellular adhesive.** Synonym for foamed adhesive.

**cellular plastic.** A plastic with greatly decreased density because of the presence of numerous cells or bubbles dispersed throughout its mass. See also *cell (plastics), foamed plastics,* and *syntactic cellular plastics.*

**cellulose acetate.** An acetic acid ester of cellulose. It is obtained by the action, under rigidly controlled conditions, of acetic acid and acetic anhydride on purified cellulose usually obtained from cotton linters. All three available hydroxyl groups in each glucose unit of the cellulose can be acetylated, but in the material normally used for plastics, it is usual to acetylate fully and then lower the acetyl value (expressed as acetic acid) to 52 to 56% by partial hydrolysis. When compounded with suitable plasticizers, it gives a tough thermoplastic material.

**cellulose acetate butyrate.** An ester of cellulose made by the action of a mixture of acetic and butyric acids and their anhydrides on purified cellulose. It is used in the manufacture of plastics that are similar in general properties to cellulose acetate but are tougher and have better moisture resistance and dimensional stability.

**cellulose ester.** A derivative of cellulose in which the free hydroxyl groups attached to the cellulose chain have been replaced wholly in part by acetic groups; for example, nitrate, acetate, or stearate groups. Esterification is effected by the use of a mixture of an acid with its anhydride in the presence of a catalyst, such as sulfuric acid. Mixed esters of cellulose, such as cellulose acetate butyrate, are prepared by the use of mixed acids and mixed anhydrides. Esters and mixed esters, a wide range of which are known, differ in their compatibility with plasticizers, in molding properties, and in physical characteristics. These esters and mixed esters are used in the manufacture of thermoplastic molding compositions.

**cellulose nitrate**. A nitric acid ester of cellulose manufactured by the action of a mixture of sulfuric acid and nitric acid on cellulose, such as purified cotton linters. The type of cellulose nitrate used for celluloid manufacture usually contains 10.8 to 11.1% nitrogen. The latter figure is the nitrogen content of the dinitrate. Also called *nitrocellulose*.

**cellulose propionate**. An ester of cellulose made by the action of propionic acid and its anhydride on purified cellulose. It is used as the basis of a thermoplastic molding material.

**cellulosic plastics**. Plastics based on cellulose compounds, such as esters (cellulose acetate) and ethers (ethyl cellulose).

**cement**. See *Technical Brief 8*.

**cementation**. The introduction of one or more elements into the outer portion of a metal object by means of diffusion at high temperature.

**cement copper**. Impure copper recovered by *chemical deposition* when iron (most often shredded steel scrap) is brought into prolonged contact with a dilute copper sulfate solution.

**cemented carbides**. See *Technical Brief 9*.

**cementite**. A hard (~800 HV), brittle compound of iron and carbon, known chemically as iron carbide and having the approximate chemical formula $Fe_3C$. It is characterized by an orthorhombic crystal structure. When it occurs as a phase in steel, the chemical composition will be altered by the presence of manganese and other carbide-forming elements. The highest cementite contents are observed in white cast irons (Fig. 64), which are used in applications where high wear resistance is required.

**centane number**. A measure of the ignition quality of a fuel or petroleum product with reference to normal centane high-ignition quality fuel with an arbitrary number of 100.

**center drilling**. Drilling a short, conical hole in the end of a workpiece—a hole to be used to center the workpiece for turning on a lathe.

**center-gated mold**. An injection or transfer mold in which the cavity is filled with plastic molding material, through a sprue or gate, directly into the center of the part.

**centering plug**. A plug fitting both spindle and cutter to ensure concentricity of the cutter mounting.

**centerless grinding**. Grinding the outside or inside diameter of a cylindrical piece which is supported on a work support blade instead of being held between centers and which is rotated by a so-called regulating or feed wheel (Fig. 65).

FIG. 65

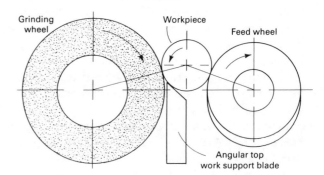

*Basic elements of the centerless grinding operation*

**centerline shrinkage**. Shrinkage or porosity occurring along the central plane or axis of a cast part.

**centrifugal casting** (metals). The process of filling molds by (1) pouring metal into a sand or permanent mold that is revolving about either its horizontal or its vertical axis (Fig. 66) or (2) pouring metal into a mold that is subsequently revolved before solidification of the metal is complete. See also *centrifuge casting*.

**centrifugal casting** (plastics). A method of forming thermoplastic resins in which the granular resin is placed in a rotatable container, heated to a molten condition by the transfer of heat through the walls of the container, and rotated such that the centrifugal force induced will force the molten resin to conform to the configuration of the interior surface of the container. Used to fabricate large-diameter pipes and similar cylindrical items.

**centrifuge casting**. A casting technique in which mold cavities are spaced symmetrically about a vertical axial common downgate. The entire assembly is rotated about that axis during pouring and solidification.

**ceramic** (adjective). (1) Of or pertaining to ceramics, that is, inorganic nonmetallic as opposed to organic or metallic.

FIG. 64

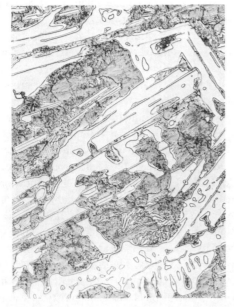

*White cast iron containing massive cementite (white) and pearlite (dark). 500×*

## Technical Brief 8: Cement

CEMENT is a synthetic mineral mixture that, when ground to a powder and mixed with water, forms a stonelike mass. This mass results from a series of chemical reactions whereby the crystalline constituents hydrate, forming a material of high hardness that is extremely resistant to compressive loading. The main uses of cement are in civil engineering, for which, since the late 19th century, it has become indispensable.

The history of cement dates back to the Romans, who found that mixtures of volcanic ash, lime, and clay would harden when wet, and who used it extensively to build structures. In 1757, it was found that burned and ground high calcitic clays would harden when placed in water. In 1824, a patent was granted to a British bricklayer who formulated a new type of cement with improved hardness. Because the color of the material reminded him of the limestone on the Isle of Portland, he named the product portland cement. This cement was made by lightly calcining small batches of lime and clay and grinding the product to fine powder.

The modern manufacturing process is very basic and has not been radically changed since its inception except for the use of computer-controlled equipment, which has greatly improved the consistency of the final product. The four basic cement processing operations are: (1) quarrying and crushing of raw materials, (2) grinding to high fineness and carefully proportioning the mineral constituents, (3) pyroprocessing the raw materials in a rotary calciner, and (4) cooling and grinding the calcined product, or clinker, to obtain a fine powder.

There are four main compounds that compose portland cement: tricalcium silicate, $3CaO \cdot SiO_2$ ($C_3S$); dicalcium silicate, $2CaO \cdot SiO_2$ ($C_2S$); tricalcium aluminate, $3CaO \cdot Al_2O_3$ ($C_4A$); and tetracalcium aluminoferrite, $4CaOAl_2O_3 \cdot Fe_2O_3$ ($C_3AF$). In the United States, portland cements are manufactured to comply with the ASTM Standard Specification for Portland Cement, ASTM C 150. This specification defines five main types of portland cement. Type I, which is made in the greatest quantity, is intended for general-purpose use when the special properties of the other types are not required. The

**Chemical constituents for various types of portland cement**

|  | $3CaO \cdot SiO_2$ | $2CaO \cdot SiO_2$ | $3CaO \cdot Al_2O_3$ | $4CaO \cdot Al_2O_3 \cdot Fe_2O_3$ | $CaSO_4$ | MgO | Free CaO |
|---|---|---|---|---|---|---|---|
| Type I | 45 | 27 | 11 | 8 | 3.1 | 2.9 | 0.5 |
| Type II | 44 | 31 | 5 | 13 | 2.8 | 2.5 | 0.4 |
| Type III | 53 | 19 | 11 | 9 | 4 | 2 | 0.7 |
| Type IV | 28 | 49 | 4 | 12 | 3.2 | 1.8 | 1.9 |
| Type V | 38 | 43 | 4 | 9 | 2.7 | 1.9 | 0.5 |

special properties of the other types when used in concrete are: Type II, moderate sulfate resistance or moderate heat of hydration; Type III, high early strength; Type IV, low heat of hydration; Type V, high sulfate resistance. The chemical and physical differences between the types that produce their special properties lie in the proportions of the cement compounds and in the fineness to which the cement is ground.

**Compressive strength data for typical cements versus time**

| Cement type | 1-day strength | | 3-day strength | | 7-day strength | | 28-day strength | | 91-day strength | |
|---|---|---|---|---|---|---|---|---|---|---|
|  | MPa | ksi | MPa | ksi | MPa | ksi | MPa | ksi | MPa | ksi |
| Type I | 9.3 | 1.3 | 22.5 | 3.3 | 32 | 4.6 | 42 | 6.1 | 50.5 | 7.3 |
| Type II | 14 | 2.0 | 27 | 3.9 | 36.6 | 5.3 | 46.3 | 6.7 | 52.5 | 7.6 |
| Type III | 21 | 3.0 | 37.5 | 5.4 | 44.2 | 6.4 | 52.3 | 7.6 | 56.0 | 8.1 |
| Type IV | ... | ... | 9.6 | 1.4 | 13.9 | 2.0 | 34.3 | 5.0 | ... | ... |
| Type V | ... | ... | 22.1 | 3.2 | 29.5 | 4.3 | 41.3 | 6.0 | ... | ... |
| White portland | ... | ... | 26.5 | 3.8 | 36.2 | 5.3 | 46.5 | 6.7 | ... | ... |
| Portland/blast-furnace | 8.6 | 1.2 | 13.1 | 1.9 | 25.3 | 3.7 | 45.0 | 6.5 | 53.3 | 7.7 |

## Recommended Reading

- G. Frohnsdorff and J.R. Clifton, Portland Cements and Concrete, *Engineered Materials Handbook*, Vol 4, ASM International, 1991, p 918-924
- R.A. Haber and P.A. Smith, Overview of Traditional Ceramics, *Engineered Materials Handbook*, Vol 4, ASM International, 1991, p 3-15

# Technical Brief 9: Cemented Carbides

CEMENTED CARBIDES, also referred to as hard metals, belong to a class of hard, wear-resistant, refractory materials in which the hard carbide particles are bound together, or cemented, by a soft and ductile metal binder. These materials were first developed in Germany in the early 1920s in response to demands for a die material having sufficient wear resistance for drawing tungsten incandescent filament wires to replace the expensive diamond dies then in use. The first cemented carbide produced was tungsten carbide (WC) with a cobalt (Co) binder. Over the years, the basic WC-Co material has been modified to produce a variety of cemented carbides, which are used in a wide range of applications, including metal cutting, mining, construction, rock drilling, metal forming, structural components, and wear parts. Tungsten carbide-based materials with nickel or steel binders have also been produced for specialized applications.

Tungsten carbides are manufactured a powder metallurgy process consisting of (1) processing of the ore and the preparation of the WC powder, (2) preparation (ball milling) of WC powders and grade (alloying) powders, (3) addition of suitable binder material, (4) powder consolidation, and (5) sintering of the compacted part at temperatures between 1300 and 1600 °C (2370 and 2910 °F), most often in vacuum. The sintered product can be directly used or can be ground, polished, and coated to suit a given application.

Approximately 50% of all carbide production is used for machining applications, and a wide variety of compositions are available. "Straight" grades, which consist of WC particles bonded with cobalt, generally contain 3 to 12% Co and carbide grain sizes range from 0.5 to >5 μm. Alloy grades, or steel-cutting grades, contain titanium carbide (TiC), titanium carbonitride (TiCN), titanium nitride (TiN), and/or niobium carbide (NbC). Improved wear resistance of cemented carbide tools is achieved by multilayer hard coatings of TiC, TiCN, TiN, alumina ($Al_2O_3$), and occasionally hafnium carbide (HfC). These coatings are normally applied by *chemical vapor deposition*. See also the figure accompanying *carbide tools*.

Cemented carbides are also being used increasingly for nonmachining applications, such as metal and nonmetallic mining, oil and gas drilling, transportation and construction, metal forming, structural and fluid-handling components, and forestry tools. Straight WC-Co grades are used for the majority of these applications. In general, cobalt contents range from 5 to 30% and WC grain sizes range from <1 to >8 μm.

## Properties of representative cobalt-bonded cemented carbides

| Nominal composition | Grain size | Hardness, HRA | Density g/cm³ | Density oz/in.³ | Transverse strength MPa | Transverse strength ksi | Compressive strength MPa | Compressive strength ksi | Modulus of elasticity GPa | Modulus of elasticity 10⁶ psi | Relative abrasion resistance(a) | Coefficient of thermal expansion, μm/m · K at 200 °C (390 °F) | Coefficient of thermal expansion, μm/m · K at 1000 °C (1830 °F) | Thermal conductivity, W/m · K |
|---|---|---|---|---|---|---|---|---|---|---|---|---|---|---|
| 97WC-3Co | Medium | 92.5–93.2 | 15.3 | 8.85 | 1590 | 230 | 5860 | 850 | 641 | 93 | 100 | 4.0 | · · · | 121 |
| 94WC-6Co | Fine | 92.5–93.1 | 15.0 | 8.67 | 1790 | 260 | 5930 | 860 | 614 | 89 | 100 | 4.3 | 5.9 | · · · |
| | Medium | 91.7–92.2 | 15.0 | 8.67 | 2000 | 290 | 5450 | 790 | 648 | 94 | 58 | 4.3 | 5.4 | 100 |
| | Coarse | 90.5–91.5 | 15.0 | 8.67 | 2210 | 320 | 5170 | 750 | 641 | 93 | 25 | 4.3 | 5.6 | 121 |
| 90WC-10Co | Fine | 90.7–91.3 | 14.6 | 8.44 | 3100 | 450 | 5170 | 750 | 620 | 90 | 22 | · · · | · · · | · · · |
| | Coarse | 87.4–88.2 | 14.5 | 8.38 | 2760 | 400 | 4000 | 580 | 552 | 80 | 7 | 5.2 | · · · | 112 |
| 84WC-16Co | Fine | 89 | 13.9 | 8.04 | 3380 | 490 | 4070 | 590 | 524 | 76 | 5 | · · · | · · · | · · · |
| | Coarse | 86.0–87.5 | 13.9 | 8.04 | 2900 | 420 | 3860 | 560 | 524 | 76 | 5 | 5.8 | 7.0 | 88 |
| 75WC-25Co | Medium | 83–85 | 13.0 | 7.52 | 2550 | 370 | 3100 | 450 | 483 | 70 | 3 | 6.3 | · · · | 71 |
| 71WC-12.5TiC-12TaC-4.5Co | Medium | 92.1–92.8 | 12.0 | 6.94 | 1380 | 200 | 5790 | 840 | 565 | 82 | 11 | 5.2 | 6.5 | 35 |
| 72WC-8TiC-11.5TaC-8.5Co | Medium | 90.7–91.5 | 12.6 | 7.29 | 1720 | 250 | 5170 | 750 | 558 | 81 | 13 | 5.8 | 6.8 | 50 |

(a) Based on a value of 100 for the most abrasion-resistant material

## Recommended Reading

- A.T. Santhanam, P. Tierney, and J.L. Hunt, Cemented Carbides, *Metals Handbook*, 10th ed., Vol 2, ASM International, 1990, p 950-977
- A.T. Santhanam and P. Tierney, Cemented Carbides, *Metals Handbook*, 9th ed., Vol 16, ASM International, 1989, p 71-89
- H.S. Kalish, Corrosion of Cemented Carbides, *Metals Handbook*, 9th ed., Vol 13, ASM International, 1987, p 846-858

Fɪɢ. 66

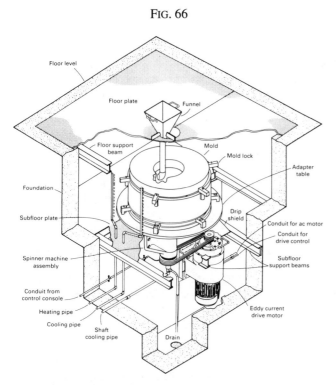

*Typical installation of a vertical centrifugal casting machine*

(2) Pertaining to products manufactured from inorganic nonmetallic substances which are subjected to a high temperature during manufacture or use. (3) Pertaining to the manufacture or use of such articles or materials, such as ceramic process or ceramic science.

**ceramic(s)** (noun). Any of a class of inorganic nonmetallic products which are subjected to a high temperature during manufacture or use (high temperature usually means a temperature above a barely visible red, approximately 540 °C, or 1000 °F). Typically, but not exclusively, a ceramic is a metallic oxide, boride, carbide, or nitride, or a mixture of compound of such materials; that is, it includes anions that play important roles in atomic structures and properties. See also *advanced ceramics, electronic ceramics* (Technical Brief 14), *refractories* (Technical Brief 37), *structural ceramics* (Technical Brief 50), and *traditional ceramics* (Technical Brief 57).

**ceramic color glaze.** An opaque colored glass of satin or gloss finish obtained by spraying the clay body with a compound of metallic oxides, chemicals, and clays. It is fired at high temperatures, fusing the glaze to the body, making them inseparable.

**ceramic glass decorations.** Ceramic glass enamels fused to glassware at temperatures above 245 °C (800 °F) to produce a decoration.

**ceramic glass enamels.** Predominantly colored, silicate glass fluxes used to decorate glassware. Also referred to as ceramic enamels or glass enamels.

**ceramic-matrix composites.** See *Technical Brief 10*.

**ceramic-metal coating.** A mixture of one or more ceramic materials in combination with a metallic phase applied to a metallic substrate which may or may not require heat treatment prior to service. This term may also be used for coatings applied to nonmetallic substrates, for example, graphite.

**ceramic molding.** A precision casting process that employs permanent patterns and fine-grain slurry for making molds. Unlike monolithic investment molds, which are similar in composition, ceramic molds consist of a *cope* and a *drag* or, if the casting shape permits, a drag only.

**ceramic printed board.** A printed board made from ceramic dielectric and cermet materials.

**ceramic process.** The production of articles or coatings from essentially inorganic, nonmetallic materials, the article or coating being made permanent and suitable for utilitarian and decorative purposes by the action of heat at temperatures sufficient to cause sintering, solid-state reactions, bonding, or conversion partially or wholly to the glassy state.

**ceramic rod flame spraying.** A thermal spraying process variation in which the material to be sprayed is in ceramic rod form. See also *flame spraying*.

**ceramic tools.** Cutting tools made from sintered, hot-pressed, or hot isostatically pressed alumina-based or silicon nitride-based ceramic materials. See also *alumina* (Technical Brief 3), *Sialons* (Technical Brief 43), and *silicon nitride* (Technical Brief 45).

**ceramic whiteware.** A fired ware consisting of glazed or unglazed ceramic body which is commonly white and of fine texture. This term designates such product classifications as tile, china, porcelain, semivitreous ware, and earthenware. See also *traditional ceramics* (Technical Brief 57).

**cereal.** An organic *binder*, usually corn flour.

**cermets.** See *Technical Brief 11*.

**C-frame press.** Same as *gap-frame press*.

**CG iron.** Same as *compacted graphite cast iron*.

**C-glass.** A glass with a soda-lime-borosilicate composition that is used for its chemical stability in corrosive environments.

**chafing.** Repeated rubbing between two solid bodies that can result in surface damage and/or wear.

**chafing fatigue.** Fatigue initiated in a surface damaged by rubbing against another body. See also *fretting*.

**chain-intermittent fillet welding.** Depositing a line of intermittent fillet welds on each side of a member at a joint so that the increments on one side are essentially opposite those on the other. Contrast with *staggered-intermittent fillet welding*.

## Technical Brief 10: Ceramic-Matrix Composites

CERAMIC-MATRIX COMPOSITES are candidate materials for high-performance engines and wear-resistant parts. Interest in ceramic composites has been stimulated by the realization that carbon-carbon composites are difficult to protect from oxidation, and that metal-matrix composites have end-use temperature limitations that are below the level needed for engine components.

A wide variety of reinforcing materials, matrices, and corresponding processing methods have been studied. The most successful fiber-reinforced composites have been produced by hot pressing, chemical vapor infiltration, or directed metal oxidation, which is a process that uses accelerated oxidation reactions of molten metals to grow ceramic matrices around pre-

### Typical physical properties of whisker-reinforced ceramics

| Material | Bulk density, g/cm³ | Flexure strength | | Fracture toughness | | Hardness | | Elastic modulus | |
|---|---|---|---|---|---|---|---|---|---|
| | | MPa | ksi | MPa√m | ksi√in. | GPa | 10⁶ psi | GPa | 10⁶ psi |
| $SiC_w$-$Al_2O_3$ | 3.7-3.9 | 600-700 | 87-101 | 5-8 | 4.6-7.3 | 15-16 | 2-2.3 | 380-430 | 55-62 |
| $SiC_w$-$Si_3N_4$ | 3.2-3.3 | 800-1000 | 116-145 | 6-8 | 5.5-7.3 | 15-16 | 2-2.3 | 300-380 | 43-55 |

placed filler or reinforcement material preforms. Much of the work has been on glass and glass-ceramic matrices reinforced with carbon fibers. Because of the low axial coefficient of thermal expansion (CTE) of carbon fibers and the requirements for CTE matching, the more successful composites have been produced with low CTE matrices, such as borosilicate glass (CTE = $3.5 \times 10^{-6}$/K) and lithium aluminosilicate glass-ceramics (CTE =

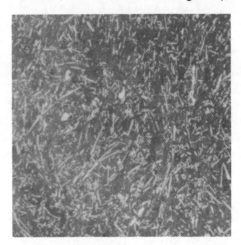

*Typical microstructure of whisker-reinforced alumina ceramic composite containing 20 vol% SiC whiskers*

$1.5 \times 10^{-6}$/K). Other fiber-reinforced ceramic composites include silicon carbide (SiC) fibers in SiC produced by chemical vapor infiltration and deposition, and SiC fiber reinforced alumina ($Al_2O_3$) and zirconium diboride ($ZrB_2$) reinforced zirconium carbide (ZrC) composites produced by directed metal oxidation. Multidirectionally reinforced ceramics, such as fused quartz reinforced silica and $Al_2O_3$ reinforced silica, have also been produced, the latter material being used for large radome structures on ballistic missiles.

Whisker-reinforced ceramic composites are also being studied. Composed of fine equiaxed $Al_2O_3$ grains and needlelike SiC whiskers, $SiC_w$-$Al_2O_3$ composites exhibit promising fracture toughness (6.5 MPa√m, or 5.9 ksi√in.) and strength (600 MPa, or 87 ksi) properties. $Al_2O_3$-$SiC_w$ composites have been used in cutting-tool applications. Composites with whisker loadings higher than 8 vol% must be hot pressed.

The excellent high-temperature strength, oxidation resistance, thermal shock resistance, and fracture toughness of silicon nitride ($Si_3N_4$) has led to the development of $SiC_w$ reinforced $Si_3N_4$. The major phase, $Si_3N_4$, offers many favorable properties, and the SiC whiskers provide significant improvement in the fracture toughness of the composite. Whisker-reinforced $Si_3N_4$ is a leading candidate material for hot-section ceramic-engine components.

### Recommended Reading

- T. Vasilos, Structural Ceramic Composites, *Engineered Materials Handbook*, Vol 1, ASM International, 1987, p 925-932
- J.P. Brazel, Multidirectionally Reinforced Ceramics, *Engineered Materials Handbook*, Vol 1, ASM International, 1987, p 933-940
- P.F. Becher and T.N. Tiegs, Whisker-Reinforced Ceramics, *Engineered Materials Handbook*, Vol 1, ASM International, 1987, p 941-944
- A.W. Urquhart, Directed Metal Oxidation, *Engineered Materials Handbook*, Vol 4, ASM International, 1991, p 232-235

## Technical Brief 11: Cermets

CERMET is an acronym that is used to designate a heterogeneous combination of metal(s) or alloy(s) with one or more ceramic phases, in which the latter constitutes approximately 15 to 85% by volume and in which there is relatively little solubility between metallic and ceramic phases at the preparation temperature. The size of the ceramic component varies, depending on the system and application. It can be as coarse as 50 to 100 μm, as in some types of cermets based on uranium dioxide ($UO_2$) that are used for nuclear reactor fuel elements, or as fine as 1 to 2 μm, as in the micrograin type of cemented carbides (Note: Technically, all metal-bonded tungsten carbide materials should fall into the category of cermets. However, it has been customary in the cutting tool industry to designate all cobalt-bonded tungsten carbide compositions as *cemented carbides*.) If the ceramic component is even finer and is present in small amounts, the material is considered a *dispersion-strengthened material* and therefore falls outside the accepted definition of cermets.

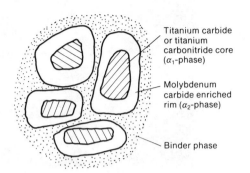

Titanium carbide or titanium carbonitride core ($\alpha_1$-phase)

Molybdenum carbide enriched rim ($\alpha_2$-phase)

Binder phase

*Schematic of cermet microstructure*

Like cemented carbides, cermets contain a metal binder and are produced by powder metallurgy techniques. The metallic binder phase can consist of a variety of elements, alone or in combination, such as nickel, cobalt, iron, chromium, molybdenum, and tungsten; it can also contain other metals, such as stainless steel, superalloys, titanium, zirconium, or some of the lower-melting-point copper or aluminum alloys. The volume fraction of the binder phase depends entirely on the intended properties and end use of the material. It can range anywhere from 15 to 85%, but it is generally kept at the lower half of the scale (for example, 10 to 15%).

Cermets have proven their value in a variety of applications. The most important use of cermets is in cutting tools based on titanium carbide (TiC) or titanium carbonitride (TiC,N). Steel-bonded carbides

### Comparison of high-temperature properties of a TiC cermet and a complex carbonitride cermet

| Composition of cermet | Vickers hardness at 1000 °C (2000 °F), kg/mm$^2$ | Transverse rupture strength at 900 °C (1650 °F) | | Oxidation resistance at 1000 °C (2000 °F) weight gain, mg/cm$^2$ · h | Thermal conductivity at 1000 °C (2000 °F), W/K$^6$ · m |
|---|---|---|---|---|---|
| | | MPa | ksi | | |
| TiC-16.5Ni-9Mo | 500 | 1050 | 152 | 11.8 | 24.7 |
| TiC-20TiN-15WC-10TaC-5.5Ni-11Co-9Mo | 650 | 1360 | 197 | 1.66 | 42.3 |

consisting of 45 vol% TiC are used in wear-resistant parts and in dies and other forming equipment components. Cermets based on $UO_2$, as well as those based on uranium carbide (UC), offer potential for advanced fuel elements. Cermets based on zirconium boride ($ZrB_2$) or silicon carbide (SiC), and others containing alumina ($Al_2O_3$), silicon dioxide ($SiO_2$), boron carbide ($B_4C$), or refractory compounds combined with diamonds, possess unique properties. Several are used commercially in a wide range of applications, including hot-machining tools, shaft seals, valve components and wear parts, ultrahigh-temperature exposed ducts, nozzles, and other rocket engine components, furnace fixtures and hearth elements, grinding wheels, and diamond-containing drill heads and saw teeth.

### Recommended Reading

- C.G. Goetzel and J.L. Ellis, Cermets, *Metals Handbook*, 10th ed., Vol 2, ASM International, 1990, p 978-1007
- W.W. Gruss, Cermets, *Metals Handbook*, 9th ed., Vol 16, ASM International, 1989, p 90-97

**chain length.** In plastics, the length of the stretched linear macromolecule, most often expressed by the number of identical links.

**chain transfer agent.** In plastics, a molecule from which an atom, such as hydrogen, may be readily abstracted by a free radical.

**chalcogenide.** A binary or ternary compound containing a chalcogen (sulfur, selenium, or tellurium) and a more electropositive element. Ternary molybdenum chalcogenides, $M_x - Mo_6X_8$, where M is a cation and $X$ is a chalcogen, are superconducting materials.

**chalking.** (1) Dry, chalklike appearance of deposit on the

FIG. 67

*Electron channeling pattern of vanadium. 20×*

surface of a plastic. (2) The development of loose removable powder at the surface of an organic coating usually caused by weathering.

**chamber furnace**. In powder metallurgy, a batch sintering furnace usually equipped with a retort that can be sealed gastight.

**chamfer**. (1) A beveled surface to eliminate an otherwise sharp corner. (2) A relieved angular cutting edge at a tooth corner.

**chamfer angle**. (1) The angle between a reference surface and the bevel. (2) On a milling contour, the angle between a beveled surface and the axis of the cutter.

**chamfering**. Making a sloping surface on the edge of a member. Also called beveling. See also *bevel angle*.

**channeling**. (1) The tendency of a grease or viscous oil to form air channels in a bearing or gear system, resulting in an incomplete lubricant film. (2) The tendency of a grease to form a channel by working down in a bearing or distribution system, leaving shoulders to act as a reservoir and seal.

**channeling pattern**. A pattern of lines observed in a scanning electron image of a single-crystal surface caused by preferential penetration, or channeling, of the incident beam between rows of atoms at certain orientations (Fig. 67). The pattern provides information on the structure and orientation of the crystal.

**chaplet**. Metal support that holds a core in place within a

casting mold; molten metal solidifies around a chaplet and fuses it into the finished casting (Fig. 68).

**characteristic**. A property of items in a sample or population that when measured, counted, or otherwise observed helps to distinguish between the items.

**characteristic electron energy loss phenomena**. The inelastic scattering of electrons in solids that produces a discrete energy loss determined by the characteristics of the material. The most probable form is due to excitation of valence electrons.

**characteristic radiation**. Electromagnetic radiation of a particular set of wavelengths, produced by and characteristic of a particular element whenever its excitation potential is exceeded. Electromagnetic radiation is emitted as a result of electron transitions between the various energy levels (electron shells) of atoms; the spectrum consists of lines whose wavelengths depend only on the element concerned and the energy levels involved.

**charge**. (1) The materials fed into a furnace. (2) Weights of various liquid and solid materials put into a furnace during one feeding cycle. (3) The weight of plastic material used to load a mold at one time or during one cycle. See also *static charge*.

**charging**. (1) For a lap, impregnating the surface with fine abrasive. (2) Placing materials into a furnace.

**Charpy test**. An impact test in which a V-notched, keyhole-notched, or U-notched specimen, supported at both ends, is struck behind the notch by a striker mounted at the lower end of a bar that can swing as a pendulum (Fig. 69). The energy that is absorbed in fracture is calculated from the height to which the striker would have risen had there been no specimen and the height to which it actually rises after fracture of the specimen. Contrast with *Izod test*.

**charring**. The heating of a reinforced plastic or composite in air to reduce the polymer matrix to ash, allowing the fiber content to be determined by weight.

**chase** (machining). To make a series of cuts each, except for

FIG. 68

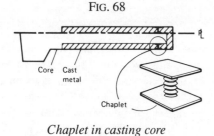

*Chaplet in casting core*

FIG. 69

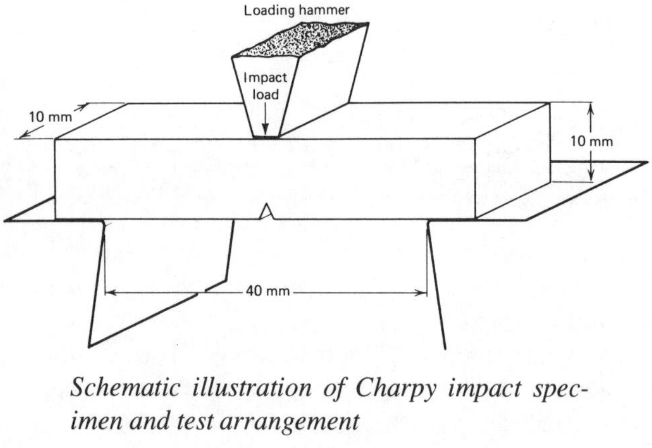

*Schematic illustration of Charpy impact specimen and test arrangement*

the first, following in the path of the cut preceding it, as in chasing a thread.

**chase** (plastic). In plastic part making, an enclosure of any shape, used to shrink-fit parts of a mold cavity in place, prevent spreading or distortion in hobbing, or enclose an assembly of two or more parts in a split cavity block.

**chatter**. In machining or grinding, (1) a vibration of the tool, wheel, or workpiece producing a wavy surface on the work and (2) the finish produced by such vibration. (3) In tribology, elastic vibrations resulting from frictional or other instability.

**chatter marks**. Surface imperfections on the work being ground, usually caused by vibrations transferred from the wheel-work interface during grinding.

**chatter sleek**. See *frictive track*.

**check**. The intermediate section of a flask that is used between the *cope* and the *drag* when molding a shape that requires more than one parting plane.

**checked edges**. Sawtooth edges seen after hot rolling and/or cold rolling.

**checkers**. In a chamber associated with a metallurgical furnace, bricks stacked openly so that heat may be absorbed from the combustion products and later transferred to incoming air when the direction of flow is reversed.

**checking**. The development of slight breaks in a coating that do not penetrate to the underlying surface. See also *checks* (1) and *craze cracking*.

**checks**. (1) Numerous, very fine cracks in a coating or at the surface of a metal part. Checks may appear during processing or during service and are most often associated with thermal treatment or thermal cycling. Also called check marks, *checking*, or *heat checks*. (2) Minute cracks in the surface of a casting caused by unequal expansion or contraction during cooling. (3) Cracks in a die impression corner, generally due to forging strains or pressure, localized at some relatively sharp corner. Die blocks too hard for the depth of the die impression have a tendency to check or develop cracks in impression corners. (4) A series of small cracks resulting from thermal fatigue of hot forging dies.

**chelate**. (1) Five- or six-membered ring formation based on intramolecular attraction of H, O, or N atoms. (2) A molecular structure in which a heterocyclic ring can be formed by the unshared electrons of neighboring atoms. (3) A *coordination compound* in which a heterocyclic ring is formed by a metal bound to two atoms of the associated *ligand*. See also *complexation*.

**chelating agent**. (1) An organic compound in which atoms form more than one coordinate bond with metals in solution. (2) A substance used in metal finishing to control or eliminate certain metallic ions present in undesirable quantities.

**chelation**. A chemical process involving formation of a heterocyclic ring compound that contains at least one metal cation or hydrogen ion in the ring.

**chemical adsorption**. See *chemisorption*.

**chemical blowing agent**. In processing of plastics, an agent that readily decomposes to produce a gas.

**chemical bonding**. The joining together of atoms to form molecules. See also *molecule*.

**chemical conversion coating**. A protective or decorative nonmetallic coating produced *in situ* by chemical reaction of a metal with a chosen environment. It is often used to prepare the surface prior to the application of an organic coating.

**chemical decomposition**. The separating of a compound into its constituents.

**chemical deposition**. The precipitation or plating-out of a metal from solutions of its salts through the introduction of another metal or reagent to the solution.

**chemical etching**. The dissolution of the material of a surface by subjecting it to the corrosive action of an acid or an alkali.

**chemical flux cutting**. An oxygen cutting process in which metals are severed using a chemical flux to facilitate cutting.

**chemically precipitated powder**. A metal powder that is produced as a fine precipitate by chemical displacement.

**chemically strengthened**. Glass that has been ion-exchanged to produce a compressive stress layer at the treated surface.

**chemical machining**. Removing metal stock by controlled selective chemical dissolution.

**chemical metallurgy**. See *process metallurgy*.

**chemical milling**. The machining process in which metal is formed into intricate shapes by masking certain portions and then etching away the unwanted material.

**chemical polishing**. A process that produces a polished surface by the action of a chemical etching solution. The etching solution is compounded so that peaks in the topography of the surface are dissolved preferentially.

**chemical potential**. In a thermodynamic system of several constituents, the rate of change of the Gibbs function of the system with respect to the change in the number of moles of a particular constituent.

**chemical vapor deposited (CVD) carbon**. Carbon deposited on a substrate by pyrolysis of a hydrocarbon, such as methane.

**chemical vapor deposition (CVD)**. (1) A coating process, similar to gas carburizing and carbonitriding, whereby a reactant atmosphere gas is fed into a processing chamber where it decomposes at the surface of the workpiece, liberating one material for either absorption by, or accumulation on, the workpiece. A second material is liberated in gas form and is removed from the processing chamber, along with excess atmosphere gas. (2) Process used in manu-

FIG. 70

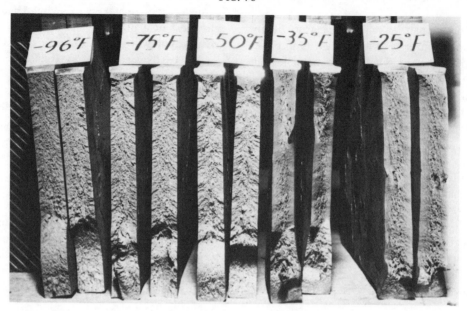

*Chevron patterns in mild steel ship-plate samples*

facture of several composite reinforcements, especially boron and silicon carbide, in which desired reinforcement material is deposited from vapor phase onto a continuous core, for example, boron on tungsten wire (core).

**chemical wear**. See *corrosive wear*.

**chemisorption**. (1) The taking up of a liquid or gas or of a dissolved substance, only one molecular layer in thickness, wherein a new chemical compound or bond is formed between the sorbent surface atoms and those of the sorbate. (2) The binding of an adsorbate to the surface of a solid by forces whose energy levels approximate those of a chemical bond. Contrast with *physisorption*.

**chevron pattern**. A fractographic pattern of radial marks (shear ledges) that look like nested letters "V" (Fig. 70); sometimes called a herringbone pattern. Chevron patterns are typically found on brittle fracture surfaces in parts whose widths are considerably greater than their thicknesses. The points of the chevrons can be traced back to the fracture origin.

**chill**. (1) A metal or graphite insert embedded in the surface of a casting sand mold or core or placed in a mold cavity to increase the cooling rate at that point. (2) White iron occurring on a gray or ductile iron casting, such as the chill in the wedge test. See also *chilled iron*. Compare with *inverse chill*.

**chill coating**. In casting, applying a coating to a *chill* that forms part of the mold cavity so that the metal does not adhere to it, or applying a special coating to the sand surface of the mold that causes the iron to undercool.

**chilled iron**. Cast iron that is poured into a metal mold or against a mold insert so as to cause the rapid solidification that often tends to produce a white iron structure in the casting.

**chill mark**. A wrinkled surface condition on glassware resulting from uneven cooling in the forming process.

**chill ring**. See preferred term *backing ring*.

**chill roll**. A cored roll, usually temperature controlled by circulating water, that cools the web before winding. For chill roll plastic (cast) film, the surface of the roll is highly polished. In extrusion coating, either a polished or matte surface may be used, depending on the surface desired on the finished coating.

**chill roll extrusion**. The extruded plastic film is cooled while being drawn around two or more highly polished chill rolls cored for water cooling for exact temperature control. Also called cast film extrusion.

**chill time**. See preferred term *quench time*.

**chin** (ceramics and glasses). (1) Area along an edge or corner where the material has broken off. (2) An imperfection due to breakage of a small fragment out of an otherwise regular surface.

**china**. A glazed or unglazed vitreous ceramic whiteware used for nontechnical purposes. This term designates such products as dinnerware, sanitary ware, and artware when they are vitreous.

**Chinese-script eutectic**. A configuration of eutectic constituents, found particularly in some cast alloys of aluminum containing iron and silicon and in magnesium alloys containing silicon, that resembles the characters in Chinese script (Fig. 71).

**chip breaker**. (1) Notch or groove in the face of a tool parallel to the cutting edge, designed to break the con-

FIG. 71

*"Chinese script" modification of lamellar eutectic in a magnesium casting*

FIG. 72

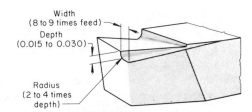

*Typical chip breaker for a turning tool*

tinuity of the chip (Fig. 72). (2) A step formed by an adjustable component clamped to the face of the cutting tool.

**chipping.** (1) Removing seams and other surface imperfections in metals manually with a chisel or gouge, or by a continuous machine, before further processing. (2) Similarly, removing excessive metal.

**chips** (composites). Minor damage to a pultruded surface of a composite material that removes material but does not cause a crack or craze.

**chips** (metals). Pieces of material removed from a workpiece by cutting tools or by an abrasive medium.

**chlorinated hydrocarbon.** An organic compound having chlorine atoms in its chemical structure. Trichloroethylene, methyl chloroform, and methylene chloride are chlorinated hydrocarbon solvents; polyvinyl chloride is a plastic.

**chlorinated lubricant.** A lubricant containing a chlorine compound that reacts with a rubbing surface at elevated temperatures to protect it from sliding damage. See also *extreme-pressure lubricant, sulfochlorinated lubricant,* and *sulfurized lubricant.*

**chlorination.** (1) Roasting ore in contact with chlorine or a

chloride salt to produce chlorides. (2) Removing dissolved gases and entrapped oxides by passing chlorine gas through molten metal such as aluminum and magnesium.

**chlorine extraction.** Removal of phases by formation of a volatile chloride. See also *extraction.*

**chlorofluorocarbon plastics.** Plastics based on polymers made with monomers composed of chlorine, fluorine, and carbon only.

**chlorofluorohydrocarbon plastics.** Plastics based on polymers made with monomers composed of chlorine, fluorine, hydrogen, and carbon only.

**chord modulus.** The slope of the chord drawn between any two specific points on a *stress-strain curve.* See also *modulus of elasticity.*

**chromadizing.** Improving paint adhesion on aluminum or aluminum alloys, mainly aircraft skins, by treatment with a solution of chromic acid. Also called chromidizing or chromatizing. Not to be confused with *chromating* or *chromizing.*

**chromate treatment.** A treatment of metal in a solution of a hexavalent chromium compound to produce a *conversion coating* consisting of trivalent and hexavalent chromium compounds.

**chromatic aberration.** A defect in a lens or optical lens system resulting in different focal lengths for radiation of different wavelengths. The dispersive power of a single positive lens focuses light from the blue end of the spectrum at a shorter distance than from the red end. An image produced by such a lens shows color fringes around the border of the image.

**chromating.** Performing a *chromate treatment.*

**chromatogram.** In materials characterization, the visual display of the progress of a separation achieved by *chromatography.* A chromatogram shows the response of a chromatographic detector as a function of time (Fig. 73).

**chromatography.** The separation, especially of closely related compounds, caused by allowing a solution or mixture to seep through an absorbent (such as clay, gel, or paper),

FIG. 73

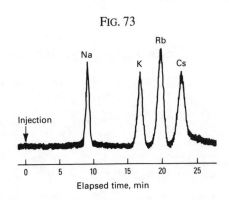

*Ion-exchange chromatogram of radioactive alkali metals*

such that each compound becomes adsorbed in a separate, often colored, layer.

**Chromel.** (1) A 90Ni-10Cr alloy used in thermocouples. (2) A series of nickel-chromium alloys, some with iron, used for heat-resistant applications.

**chrome plating.** (1) Producing a chromate *conversion coating* on magnesium for temporary protection or for a paint base. (2) The solution heat produces the conversion coating.

**chromia.** Formula $Cr_2O_3$, a compound having many properties and derivatives similar to those of *alumina*. Useful either pure or impure (e.g., as chrome ore) in both basic and high-alumina refractories.

**chromium-molybdenum heat-resistant steels.** Alloy steels containing 0.5 to 9% Cr and 0.5 to 1.10% Mo with a carbon content usually below 0.20%. The chromium provides improved oxidation and corrosion resistance, and the molybdenum increases strength at elevated temperatures. Chromium-molybdenum steels are widely used in the oil and gas industries and in fossil fuel and nuclear power plants. Nominal chemical compositions are provided in Table 16.

**chromizing.** A surface treatment at elevated temperature, generally carried out in pack, vapor, or salt baths, in which an alloy is formed by the inward diffusion of chromium into the base metal.

**chuck.** A device for holding work or tools on a machine so that the part can be held or rotated during machining or grinding.

**chucking hog.** A projection forged or cast onto a part to act as a positive means of driving or locating the part during machining.

**chute.** In powder metallurgy, a feeding trough for powder to pass from a fill hopper to the die cavity in an automatic press.

**CIL flow test.** A method of determining the rheology or flow properties of thermoplastic resins. In this test, the amount of the molten resin that is forced through a specified size orifice per unit of time when a specified variable force is applied gives a relative indication of the flow properties of various resins.

**CIP.** The acronym for *cold isostatic pressing.*

**circle grid.** A regular pattern of circles, often 2.5 mm (0.1 in.) in diameter, marked on a sheet metal blank.

**circle-grid analysis.** The analysis of deformed circles to determine the severity with which a sheet metal blank has been deformed.

**circle grinding.** Either *cylindrical grinding* or *internal grinding;* the preferred terms.

**circle shear.** A shearing machine with two rotary disk cutters mounted on parallel shafts driven in unison and equipped with an attachment for cutting circles where the desired piece of material is inside the circle. It cannot be employed to cut circles where the desired material is outside the circle.

**circuit.** (1) In filament winding of composites, one complete traverse of a winding band from one arbitrary point along the winding path to another point on a plane through the starting point and perpendicular to the axis. (2) The interconnection of a number of components in one or more closed paths to perform a desired electrical or electronic function.

**circuit board.** In electronics, a sheet of insulating material laminated to foil that is etched to produce a circuit pattern on one or both sides. Also called printed circuit board or printed wiring board.

**circuit breaker.** A device designed to open and close a circuit by nonautomatic means and to open the circuit automatically on a predetermined overload of current, without injury to itself, when properly applied within its rating.

**circular electrode.** See *resistance welding electrode.*

**circular field.** The magnetic field that (a) surrounds a nonmagnetic conductor of electricity, (b) is completely contained within a magnetic conductor of electricity, or (c) both exists within and surrounds a magnetic conductor. Generally applied to the magnetic field within any magnetic conductor resulting from a current being passed through the part or through a section of the part. Compare with *bipolar field.*

**circular mill.** A measurement used to determine the area of wire. The area of a circle that is one one-thousandth inch in diameter.

## Table 16 Nominal chemical compositions for heat-resistant chromium-molybdenum steels

| Type | UNS designation | Composition, %(a) | | | | | | |
|---|---|---|---|---|---|---|---|---|
| | | C | Mn | S | P | Si | Cr | Mo |
| ½Cr-½Mo | K12122 | 0.10–0.20 | 0.30–0.80 | 0.040 | 0.040 | 0.10–0.60 | 0.50–0.80 | 0.45–0.65 |
| 1Cr-½Mo | K11562 | 0.15 | 0.30–0.60 | 0.045 | 0.045 | 0.50 | 0.80–1.25 | 0.45–0.65 |
| 1¼Cr-½Mo | K11597 | 0.15 | 0.30–0.60 | 0.030 | 0.030 | 0.50–1.00 | 1.00–1.50 | 0.45–0.65 |
| 1¼Cr-½Mo | K11592 | 0.10–0.20 | 0.30–0.80 | 0.040 | 0.040 | 0.50–1.00 | 1.00–1.50 | 0.45–0.65 |
| 2¼Cr-1Mo | K21590 | 0.15 | 0.30–0.60 | 0.040 | 0.040 | 0.50 | 2.00–2.50 | 0.87–1.13 |
| 3Cr-1Mo | K31545 | 0.15 | 0.30–0.60 | 0.030 | 0.030 | 0.50 | 2.65–3.35 | 0.80–1.06 |
| 3Cr-1MoV(b) | K31830 | 0.18 | 0.30–0.60 | 0.020 | 0.020 | 0.10 | 2.75–3.25 | 0.90–1.10 |
| 5Cr-½Mo | K41545 | 0.15 | 0.30–0.60 | 0.030 | 0.030 | 0.50 | 4.00–6.00 | 0.45–0.65 |
| 7Cr-½Mo | K61595 | 0.15 | 0.30–0.60 | 0.030 | 0.030 | 0.50–1.00 | 6.00–8.00 | 0.45–0.65 |
| 9Cr-1Mo | K90941 | 0.15 | 0.30–0.60 | 0.030 | 0.030 | 0.50–1.00 | 8.00–10.00 | 0.90–1.10 |
| 9Cr-1MoV(c) | · · · | 0.08–0.12 | 0.30–0.60 | 0.010 | 0.020 | 0.20–0.50 | 8.00–9.00 | 0.85–1.05 |

(a) Single values are maximums. (b) Also contains 0.02–0.030% V, 0.001–0.003% B, and 0.015–0.035% Ti. (c) Also contains 0.40% Ni, 0.18–0.25% V, 0.06–0.10% Nb, 0.03–0.07% N, and 0.04% Al

**circular resistance seam welding**. See preferred term *transverse resistance seam welding*.

**circular-step bearing**. A flat circular hydrostatic bearing with a central circular recess. See also *step bearing*.

**circumferential ("circ") winding**. In filament-wound reinforced plastics, a winding with the filaments essentially perpendicular to the axis (90° or level winding).

**circumferential resistance seam welding**. See preferred term *transverse resistance seam welding*.

**CIS stereoisomer**. In engineering plastics, a stereoisomer in which side chains or side atoms are arranged on the same side of a double bond present in a chain of atoms.

**clad brazing sheet**. A metal sheet on which one or both sides are clad with brazing filler metal.

**cladding**. (1) A layer of material, usually metallic, that is mechanically or metallurgically bonded to a substrate. Cladding may be bonded to the substrate by any of several processes, such as roll-cladding and explosive forming. (2) A relatively thick layer (>1 mm, or 0.04 in.) of material applied by surfacing for the purpose of improved corrosion resistance or other properties. See also *coating*, *surfacing*, and *hardfacing*.

**clad metal**. A composite metal containing two or more layers that have been bonded together. The bonding may have been accomplished by co-rolling, co-extrusion, welding, diffusion bonding, casting, heavy chemical deposition, or heavy electroplating.

**clamping pressure**. In injection molding and transfer molding of plastics, the pressure that is applied to the mold to keep it closed in opposition to the fluid pressure of the compressed molding material, within the mold cavity (cavities) and the runner system. In blow molding, the pressure exerted on the two mold halves (by the locking mechanism of the blowing table) to keep the mold closed during formation of the container. Normally, this pressure or force is expressed in tons.

**clamshell marks**. Same as *beach marks*.

**classification**. (1) The separation of ores into fractions according to size and specific gravity, generally in accordance with Stokes' law of sedimentation. (2) Separation of a metal powder into fractions according to particle size.

**clay**. A natural mineral aggregate, consisting essentially of hydrous aluminum silicates. It is plastic when sufficiently wetted, rigid when dried en masse, and vitreous when fired to a sufficiently high temperature.

**clean surface**. A surface that is free of foreign material, both visible and invisible.

**cleanup allowance**. See *finish allowance*.

**clearance**. (1) The gap or space between two mating parts. (2) Space provided between the relief of a cutting tool and the surface that has been cut.

**clearance angle**. The angle between a plane containing the flank of the tool and a plane passing through the cutting edge in the direction of relative motion between the cutting edge and the work. See also the figures accompanying the terms *face mill* and *single-point tool*.

**clearance fill**. Any of various classes of fit between mating parts where there is a positive allowance (gap) between the parts, even when they are made to the respective extremes of individual tolerances that ensure the tightest fit between the parts. Contrast with *interference fit*.

**clearance ratio**. In a bearing, the ratio of radial clearance to shaft radius.

**cleavage**. (1) Fracture of a crystal by crack propagation across a crystallographic plane of low index. (2) The tendency to cleave or split along definite crystallographic planes. (3) Breakage of covalent bonds.

**cleavage crack** (crystalline). A crack which proceeds across the grain, that is, a transgranular crack in a single crystal or in a single grain of a polycrystalline material.

**cleavage crack** (glass). Damage produced by the translation of a hard, sharp object across a glass surface. This fracture system typically includes a plastically deformed groove on the damaged surface, together with median and lateral cracks emanating from this groove.

**cleavage fracture**. A fracture, usually of a polycrystalline metal, in which most of the grains have failed by cleavage, resulting in bright reflecting facets (Fig. 74). It is one type of *crystalline fracture* and is associated with low-energy *brittle fracture*. Contrast with *shear fracture*.

**cleavage plane**. A characteristic crystallographic plane or set of planes in a crystal on which *cleavage fracture* occurs easily.

**cleavage strength**. In testing of adhesive bonded assemblies, the tensile load in terms of kgf/mm (lbf/in.) of width required to cause the separation of a test specimen 25 mm (1 in.) in length.

**climb cutting**. Analogous to *climb milling*.

FIG. 74

*Cleavage facets in a fractured Fe-2Si alloy*

FIG. 75

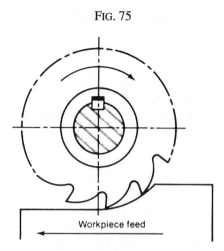

Workpiece feed

*Schematic of climb (down) milling*

**climb milling**. Milling in which the cutter moves in the direction of feed at the point of contact (Fig. 75).

**clinker**. Generally a fused or partly fused by-product of the combustion of coal, but also including lava and portland cement clinker, and partly vitrified slag and brick.

**clip and shave**. In forging, a dual operation in which one cutting surface in the clipping die removes the *flash* and then another shaves and sizes the piece.

**close annealing**. Same as *box annealing*.

**closed assembly time**. The time interval between completion of assembly of the parts for adhesive bonding and the application of pressure or heat, or both, to the assembly.

**closed-cell cellular plastics**. Cellular plastics in which almost all the cells are noninterconnecting.

**closed-die forging**. The shaping of hot metal completely within the walls or cavities of two dies that come together to enclose the workpiece on all sides. The impression for the forging can be entirely either die or divided between the top and bottom dies. Impression-die forgings, often used interchangeably with the term closed-die forging, refers to a closed-die operation in which the dies contain a provision for controlling the flow of excess material, or *flash*, that is generated. By contrast, in flashless forging, the material is deformed in a cavity that allows little or no escape of excess material.

**closed dies**. Forging or forming impression dies designed to restrict the flow of metal to the cavity within the die set, as opposed to open dies, in which there is little or no restriction to lateral flow.

**closed pass**. A pass of metal through rolls where the bottom roll has a groove deeper than the bar being rolled and the top roll has a collar fitting into the groove, thus producing the desired shape free from *flash* or fin.

**closed porosity**. The volume fraction of all pores within a solid mass that are closed off by surrounding dense solid, and hence are inaccessible to each other and to the external surface; they thus are not detectable by gas or liquid penetration.

**close-packed**. A geometric arrangement in which a collection of equally sized spheres (atoms) may be packed together in a minimum total volume.

**close-tolerance forging**. A forging held to unusually close dimensional tolerances so that little or no machining is required after forging. See also *precision forging*.

**closure**. In fabricating of reinforced plastics, the complete coverage of a mandrel with one layer (two plies) of fiber. When the last tape circuit that completes mandrel coverage lays down adjacent to the first without gaps or overlaps, the wind pattern is said to have closed.

**cloth** (composites). See *woven fabric* and *nonwoven fabric*.

**cloth** (powder metallurgy). Metallic or nonmetallic screen or fabric used for screening or classifying powders.

**cloudburst treatment**. A form of *shot peening*.

**cloud point**. The temperature at which a wax cloud first appears on cooling a mineral oil under specified conditions.

**cluster mill**. A rolling mill in which each of the two working rolls of small diameter is supported by two or more backup rolls (Fig. 76).

FIG. 76

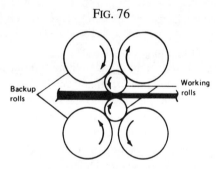

Backup rolls — Working rolls

*Cluster mill*

**coagulation**. Precipitation of a polymer dispersed in a latex.

**coalesced copper**. Massive oxygen-free copper made by briquetting ground, brittle cathode copper, then sintering the briquets in a pressurized reducing atmosphere, followed by hot working.

**coalescence**. (1) The union of particles of a dispersed phase into larger units, usually effected at temperatures below the fusion point. (2) Growth of grains at the expense of the remainder by absorption or the growth of a phase or particle at the expense of the remainder by absorption or reprecipitation.

**coarse fraction**. The large particles in a metal powder spectrum.

**coarse grains**. Grains larger than normal for the particular wrought metal or alloy or of a size that produces a surface roughening known as *orange peel* or alligator skin.

**coarsening**. An increase in grain size, usually, but not necessarily, by *grain growth*.

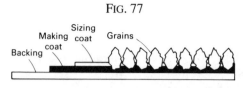

*Typical construction of a coated abrasive product*

**coated abrasive**. An abrasive product (sandpaper, for example) in which a layer of abrasive particles is firmly attached to a paper, cloth, or fiber backing by means of glue or synthetic-resin adhesive (Fig. 77).

**coated electrode**. See preferred term *covered electrode* and *lightly coated electrode*.

**coating**. A relatively thin layer (<1 mm, or 0.04 in.) of material applied by surfacing for the purpose of corrosion prevention, resistance to high-temperature scaling, wear resistance, lubrication, or other purposes.

**coating density**. The ratio of the determined density of a coating to the theoretical density of the material used in the coating process. Usually expressed as percent of theoretical density.

**coating strength**. A measure of the cohesive bond within a coating, as opposed to coating-to-coating substrate bond; the tensile strength of a coating.

**coating stress**. The stresses in a coating resulting from rapid cooling of molten or semimolten particles as they impact the substrate.

**coaxing**. Improvement of the fatigue strength of a specimen by the application of a gradually increasing stress amplitude, usually starting below the fatigue limit.

**cobalt and cobalt alloys**. See *Technical Brief 12*.

**cocoa**. In *fretting wear*, a powdery form of debris, usually consisting of iron oxides, that is expelled from a ferrous metal joint near the location where fretting wear is occurring. Also known as red mud.

**co-curing**. The act of curing a composite laminate and simultaneously bonding it to some other prepared surface, or curing together an inner and outer tube of similar or dissimilar fiber-resin combinations after each has been wound or wrapped separately. See also *secondary bonding*.

**coefficient of adhesion**. (1) The ratio of the normal force required to separate two bodies to the normal load with which they were previously placed together. (2) In railway engineering, sometimes used to signify the coefficient of (limiting) static friction.

**coefficient of compressibility**. See *bulk modulus of elasticity*.

**coefficient of elasticity**. The reciprocal of Young's modulus in a tension test. See also *compliance*.

**coefficient of expansion**. A measure of the change in length or volume of an object, specifically, a change measured by the increase in length or volume of an object per unit length or volume.

**coefficient of friction**. The dimensionless ratio of the friction force ($F$) between two bodies to the normal force ($N$) pressing these bodies together: $\mu$ (or $f$) = ($F/N$)

**coefficient of thermal expansion**. (1) Change in unit of length (or volume) accompanying a unit change of temperature, at a specified temperature. (2) The linear or volume expansion of a given material per degree rise of temperature, expressed at an arbitrary base temperature or as a more complicated equation applicable to a wide range of temperatures.

**coefficient of wear**. See *wear coefficient*.

**coercive force**. The magnetizing force that must be applied in the direction opposite to that of the previous magnetizing force in order to reduce magnetic flux density to zero; thus, a measure of the magnetic retentivity of magnetic materials.

**coextrusion welding**. A solid-state welding process which produces coalescence of the faying surfaces by heating and forcing materials through an extrusion die.

**cogging**. The reducing operation in working an ingot into a billet with a forging hammer or a forging press.

**cogging mill**. A *blooming mill*.

**coherency**. The continuity of lattice of precipitate and parent phase (solvent) maintained by mutual strain and not separated by a phase boundary.

**coherent precipitate**. A crystalline precipitate that forms from solid solution with an orientation that maintains continuity between the crystal lattice of the precipitate and the lattice of the matrix, usually accompanied by some strain in both lattices. Because the lattices fit at the interface between precipitate and matrix, there is no discernible phase boundary.

**coherent radiation**. Radiation in which the phase difference between any two points in the radiation field is constant throughout the duration of the radiation.

**coherent scattering**. In materials characterization, a type of x-ray or electron scattering in which the phase of the scattered beam has a definite (not random) relationship to the phase of the incident beam. Also termed unmodified scattering. See also *incoherent scattering*.

**cohesion**. (1) The state in which the particles of a single substance are held together by primary or secondary valence forces. As used in the adhesive field, the state in which the particles of the adhesive (or adherend) are held together. (2) Force of attraction between the molecules (or atoms) within a single phase. Contrast with *adhesion*.

**cohesive blocking**. The blocking of two similar, potentially adhesive faces.

**cohesive failure**. Failure of an adhesive joint occurring primarily in an adhesive layer.

**cohesive strength**. (1) The hypothetical stress causing tensile

## Technical Brief 12: Cobalt and Cobalt Alloys

COBALT is a tough silver-gray magnetic metal that resembles iron and nickel in appearance and in some properties. Cobalt is useful in applications that utilize its magnetic properties, corrosion resistance, wear resistance, and/or strength at elevated temperatures. Some cobalt-base alloys are biocompatible, which has prompted their use as orthopedic implants.

Much of cobalt today derives from copper and copper-nickel-rich sulfide deposits in Zaire and Zambia in Africa. The ore is subjected to crushing, grinding, and flotation, prior to a magnetic concentrating process. This concentrate is then leached in sulfuric acid and the cobalt and copper extracted by electrolysis.

With an atomic number of 27, cobalt falls between iron and nickel on the periodic table. The density of cobalt is 8.8 g/cm$^3$, similar to that of nickel. At temperatures below 417 °C (783 °F), cobalt exhibits an hcp structure. Between 417 °C (783 °F) and its melting point of 1494 °C (2719 °F), cobalt has an fcc structure.

The single largest application area for cobalt-base alloys is for wear resistance. These alloys are available as castings and weld overlays, with some alloys available in wrought (plate, sheet, and bar) form. In heat-resistant applications, cobalt is more widely used as an alloying element in nickel-base alloys, with cobalt tonnages in excess of those used in cobalt-base heat-resistant alloys. Cobalt is also an important ingredient in:

- Paint pigments
- Nickel-base alloys. See also *superalloys* (Technical Brief 52)
- Cemented carbides. See also *cemented carbides* (Technical Brief 9)
- Tool steels. See also *tool steels* (Technical Brief 56)
- Magnetic materials. See also *permanent magnet materials* (Technical Brief 28) and *soft magnetic materials* (Technical Brief 46)
- Artificial γ-ray sources

### Nominal compositions of various cobalt-base alloys

| Alloy tradename | Co | Cr | W | Mo | C | Fe | Ni | Si | Mn | Others |
|---|---|---|---|---|---|---|---|---|---|---|
| **Cobalt-base wear-resistant alloys** | | | | | | | | | | |
| Stellite 1 | bal | 31 | 12.5 | 1 (max) | 2.4 | 3 (max) | 3 (max) | 2 (max) | 1 (max) | · · · |
| Stellite 6 | bal | 28 | 4.5 | 1 (max) | 1.2 | 3 (max) | 3 (max) | 2 (max) | 1 (max) | · · · |
| Stellite 12 | bal | 30 | 8.3 | 1 (max) | 1.4 | 3 (max) | 3 (max) | 2 (max) | 1 (max) | · · · |
| Stellite 21 | bal | 28 | · · · | 5.5 | 0.25 | 2 (max) | 2.5 | 2 (max) | 1 (max) | · · · |
| Haynes alloy 6B | bal | 30 | 4 | 1 | 1.1 | 3 (max) | 2.5 | 0.7 | 1.5 | · · · |
| Tribaloy T-800 | bal | 17.5 | · · · | 29 | 0.08 (max) | · · · | · · · | 3.5 | · · · | · · · |
| Stellite F | bal | 25 | 12.3 | 1 (max) | 1.75 | 3 (max) | 22 | 2 (max) | 1 (max) | · · · |
| Stellite 4 | bal | 30 | 14.0 | 1 (max) | 0.57 | 3 (max) | 3 (max) | 2 (max) | 1 (max) | · · · |
| Stellite 190 | bal | 26 | 14.5 | 1 (max) | 3.3 | 3 (max) | 3 (max) | 2 (max) | 1 (max) | · · · |
| Stellite 306 | bal | 25 | 2.0 | · · · | 0.4 | · · · | 5 | · · · | · · · | 6 Nb |
| Stellite 6K | bal | 31 | 4.5 | 1.5 (max) | 1.6 | 3 (max) | 3 (max) | 2 (max) | 2 (max) | · · · |
| **Cobalt-base high-temperature alloys** | | | | | | | | | | |
| Haynes alloy 25 (L605) | bal | 20 | 15 | · · · | 0.10 | 3 (max) | 10 | 1 (max) | 1.5 | · · · |
| Haynes alloy 188 | bal | 22 | 14 | · · · | 0.10 | 3 (max) | 22 | 0.35 | 1.25 | 0.05 La |
| MAR-M alloy 509 | bal | 22.5 | 7 | · · · | 0.60 | 1.5 (max) | 10 | 0.4 (max) | 0.1 (max) | 3.5 Ta, 0.2 Ti, 0.5 Zr |
| **Cobalt-base corrosion-resistant alloys** | | | | | | | | | | |
| MP35N, Multiphase alloy | bal | 20 | · · · | 10 | · · · | · · · | 35 | · · · | · · · | · · · |
| Haynes alloy 1233 | bal | 25.5 | 2 | 5 | 0.08 (max) | 3 | 9 | · · · | · · · | 0.1N (max) |

bal, balance

### Recommended Reading

- P. Crook, Cobalt and Cobalt Alloys, *Metals Handbook*, 10th ed., Vol 2, ASM International, 1990, p 446-454
- J.R. Davis, Cast Cobalt Alloys, *Metals Handbook*, 9th ed., Vol 16, ASM International, 1989, p 69-70
- A.I. Asphahani *et al.*, Corrosion of Cobalt-Base Alloys, *Metals Handbook*, 9th ed., Vol 13, ASM International, 1987, p 658-668

FIG. 78

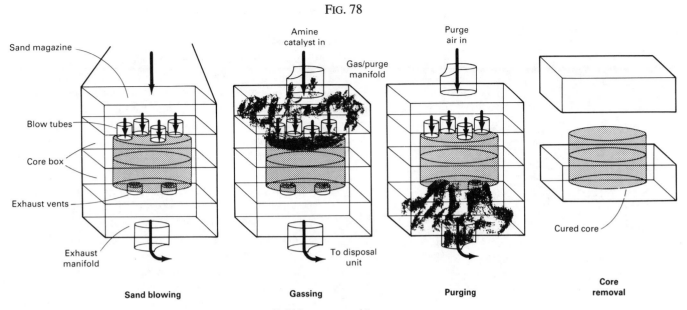

*Cold box coremaking process*

fracture without plastic deformation. (2) The stress corresponding to the forces between atoms. (3) Intrinsic strength of an adhesive. (4) Same as *technical cohesive strength*. (5) Same as *disruptive strength*.

**coil**. (1) An assembly consisting of one or more magnet wire windings. (2) Rolled metal sheet or strip.

**coil breaks**. Creases or ridges in sheet or strip that appear as parallel lines across the direction of rolling and that generally extend the full width of the sheet or strip.

**coil winding**. An electrically continuous length of insulated wire wound on a bobbin, spool, or form.

**coil without support**. A filler metal package type consisting of a continuous length of electrode in coil form without an internal support. It is appropriately bound to maintain its shape.

**coil with support**. A filler metal package type consisting of a continuous length of electrode in coil form wound on an internal support which is a simple cylindrical section without flanges.

**coining**. (1) A closed-die squeezing operation, usually performed cold, in which all surfaces of the work are confined or restrained, resulting in a well-defined imprint of the die upon the work. (2) A *restriking* operation used to sharpen or change an existing radius or profile. (3) The final pressing of a sintered powder metallurgy compact to obtain a definite surface configuration (not to be confused with *re-pressing* or *sizing*).

**coining dies**. Dies in which the coining or sizing operation is performed.

**coin silver**. An alloy containing 90% silver, with copper being the usual alloying element.

**coin straightening**. A combination coining and straightening operation performed in special cavity dies designed to impart a specific amount of working in specified areas of a forging to relieve the stresses developed during heat treatment.

**coke**. A porous, gray, infusible product resulting from the dry distillation of bituminous coal, petroleum, or coal tar pitch that drives off most of the volatile matter. Used as a fuel in cupola melting.

**coke breeze**. In foundry practice, fines from coke screenings, used in blacking mixes after grinding; also briquetted for cupola use.

**coke furnace**. Type of pot or crucible furnace that uses coke as the fuel.

**coke test**. In foundry practice, the first layer of coke placed in the cupola. Also the coke used as the foundation in constructing a large mold in a *flask* or pit.

**cold box process**. In foundry practice, a two-part organic resin binder system mixed in conventional mixers and blown into shell or solid core shapes at room temperature. A vapor mixed with air is blown into the core, permitting instant setting and immediate pouring of metal around it (Fig. 78).

**cold chamber machine**. A die casting machine with an injection system that is charged with liquid metal from a separate furnace (Fig. 79). Compare with *hot chamber machine*.

**cold coined forging**. A forging that has been restruck cold in order to hold closer face distance tolerances, sharpen corners or outlines, reduce section thickness, flatten some particular surface, or in nonheat-treatable alloys, increase hardness.

**cold compacting**. See preferred term *cold pressing*.

FIG. 79

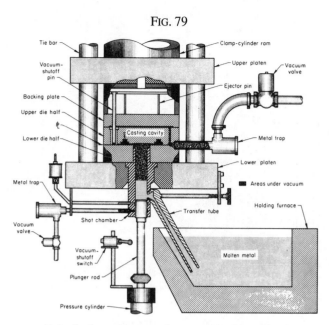

*Principal components of a vertical cold-chamber die casting machine*

**cold cracking.** (1) Cracks in cold or nearly cold cast metal due to excessive internal stress caused by contraction. Often brought about when the mold is too hard or the casting is of unsuitable design. (2) A type of weld cracking that usually occurs below 205 °C (400 °F). Cracking may occur during or after cooling to room temperature, sometimes with a considerable time delay. Three factors combine to produce cold cracks; stress (for example, from thermal expansion and contraction), hydrogen (from hydrogen-containing welding consumables), and a susceptible microstructure (plate martensite is most susceptible to cracking, ferritic and bainitic structures least susceptible). See also *hot cracking*, *lamellar tearing*, and *stress-relief cracking*.

**cold die quenching.** A quench utilizing cold, flat, or shaped dies to extract heat from a part. Cold die quenching is slow, expensive, and is limited to smaller parts with large surface areas.

**cold drawing.** Technique for using standard metalworking equipment and systems for forming thermoplastic sheet at room temperature.

**cold dry die quenching.** Same as *cold die quenching*.

**cold etching.** Development of microstructure at room temperature and below.

**cold extrusion.** See *extrusion*.

**cold finger.** In materials characterization, a liquid-nitrogen-cooled cold trap used to reduce contamination levels in vacuum chambers.

**cold flow.** The distortion that takes place in polymeric materials under continuous load at temperatures within the working range of the material without a phase or chemical change.

**cold forming.** See *cold working*.

**cold form tapping.** Producing internal threads by displacing material rather than removing it as either the tap or the workpiece is rotated. The thread form is produced by a tool, which has neither flutes nor cutting edges, that resembles a simple screw when viewed from the side but the end view shows that both the major and minor diameters have irregular contours for displacing the work material.

**cold heading.** Working metal at room temperature such that the cross-sectional area of a portion or all of the stock is increased. See also *heading* and *upsetting*.

**cold inspection.** A visual (usually final) inspection of forgings for visible imperfections, dimensions, weight, and surface condition at room temperature. The term may also be used to describe certain nondestructive tests such as magnetic-particle, dye-penetrant, and sonic inspection.

**cold isostatic pressing.** Forming technique in which high fluid pressure is applied to a powder (metal or ceramic) part at ambient temperature. Water or oil is used as the pressure medium.

**cold lap.** (1) Wrinkled markings on the surface of an ingot or casting from incipient freezing of the surface and too low a casting temperature. (2) A flaw that results when a workpiece fails to fill the die cavity during the first forging. A seam is formed as subsequent dies force metal over this gap to leave a seam on the workpiece surface. See also *cold shut*.

**cold mill.** A mill for cold rolling of sheet or strip.

**cold molding.** A procedure in which a plastic is shaped at room temperature and subsequently cured by baking.

**cold parison blow molding.** A plastic forming technique in which parisons are extruded or injection molded separately and then stored for subsequent transportation to the blow molding machine for blowing. See also *blow molding*.

**cold pressing** (plastics). A bonding operation in which a plastic assembly is subjected to pressure without the application of heat.

**cold pressing** (powder metallurgy). Forming a powder metallurgy *compact* at a temperature low enough to avoid *sintering*, usually room temperature. Contrast with *hot pressing*.

**cold-press molding.** A plastic molding process in which inexpensive plastic male and female molds are used with room temperature curing resins to produce accurate parts. Limited runs are possible.

**cold rolled sheets.** A metal mill product produced from a hot rolled pickled coil that has been given substantial cold reduction at room temperature. The resulting product usually requires further processing to make it suitable for most common applications. The usual end product is characterized by improved surface, greater uniformity in thickness, and improved mechanical properties compared with hot rolled sheet.

**cold-runner molding.** In plastic part making, a mold in which the sprue-and-runner system (the manifold section) is insulated from the rest of the mold and temperature-controlled to keep the plastic in the manifold fluid. This mold design eliminates scrap loss from sprues and runners.

**cold-setting adhesive.** An adhesive that sets at temperatures below 20 °C (68 °F). See also *hot-setting adhesive, intermediate-temperature-setting adhesive*, and *room-temperature-setting adhesive*.

**cold-setting process.** In foundry practice, any of several systems for bonding mold or core aggregates by means of organic binders, relying on the use of catalysts rather than heat for polymerization (setting).

**cold shortness.** Brittleness that exists in some metals at temperatures below the recrystallization temperature.

**cold shot.** (1) A portion of the surface of an ingot or casting showing premature solidification; caused by splashing of molten metal onto a cold mold wall during pouring. (2) Small globule of metal embedded in, but not entirely fused with, the casting.

**cold shut.** (1) A discontinuity that appears on the surface of cast metal as a result of two streams of liquid meeting and failing to unite. (2) A lap on the surface of a forging or billet that was closed without fusion during deformation. (3) Freezing of the top surface of an ingot before the mold is full.

**cold slug.** The first plastic material to enter an injection mold; so called because in passing through a sprue orifice it is cooled below the effective molding temperature.

**cold-slug well.** In plastic part making, the space provided directly opposite the sprue opening in an injection mold to trap the cold slug.

**cold soldered joint.** A joint with incomplete coalescence caused by insufficient application of heat to the base metal during soldering.

**Coldstream process.** In powder metallurgy, a method of producing cleavage fractures in hard particles through particle impingements in a high-velocity cold gas stream. Also referred to as impact crushing.

**cold stretch.** A pulling operation with little or no heat, usually on extruded filaments, to increase tensile properties of composite materials.

**cold test.** A test in which the pour point of an oil is determined.

**cold treatment.** Exposing steel to suitable subzero temperatures (–85 °C, or –120 °F) for the purpose of obtaining desired conditions or properties such as dimensional or microstructural stability. When the treatment involves the

transformation of retained austenite, it is usually followed by tempering.

**cold trimming.** The removal of flash or excess metal from a forging at room temperature in a trimming press.

**cold welding.** A solid-state welding process in which pressure is used at room temperature to produce coalescence of metals with substantial deformation at the weld. Compare with *hot pressure welding*, *diffusion welding*, and *forge welding*.

**cold work.** Permanent strain in a metal accompanied by strain hardening.

**cold-worked structure.** A microstructure resulting from plastic deformation of a metal or alloy below its recrystallization temperature.

**cold working.** Deforming metal plastically under conditions of temperature and strain rate that induce *strain hardening*. Usually, but not necessarily, conducted at room temperature. Contrast with *hot working*.

**collapse.** Inadvertent densification of cellular plastic material during manufacture resulting from the breakdown of cell structure.

**collapsibility.** The tendency of a sand mixture to break down under the pressures and temperatures developed during casting.

**collapsible tool.** A press tool that can be easily disassembled.

**collar.** The reinforcing metal of a nonpressure thermit weld.

**collaring** (thermal spraying). Adding a shoulder to a shaft or similar component as a protective confining wall for the thermal spray deposit.

**collar oiler.** A collar on a shaft that extends into the oil reservoir and carries oil into a bearing as the shaft rotates. Wipers are usually provided to direct the oil into the bearing.

**collection efficiency.** The cross-sectional area of undisturbed fluid containing particles that will ultimately impinge on a given solid surface, divided by the projected area of the solid surface. Also known as collision efficiency, capture efficiency, catchment efficiency, and impaction ratio.

**collet.** A split sleeve used to hold work or tools during machining or grinding.

**colligative properties.** Properties of plastics based on the number of molecules present. Most important are certain solution properties extensively used in molecular weight characterization.

**collimate.** To make parallel to a certain line or direction.

**collimated.** Rendered parallel.

**collimated roving.** *Roving* for reinforced plastics that has been made using a special process (usually parallel wound), such that the strands are more parallel than in standard roving.

**collimation.** The degree of parallelism of light rays from a given source. A light source with good collimation produces parallel light rays, whereas a poor light source produces divergent, nonparallel light rays.

**collimator**. The x-ray system component that confines the x-ray beam to the required shape. An additional collimator can be located in front of the x-ray detector to further define the portion of the x-ray beam to be measured.

**collision efficiency**. See *collection efficiency*.

**colloid**. A stable (nonsettling) suspension of some material within a fluid host, the dimensions of the former usually being about ≤1 μm. Fogs, smokes, foams, emulsions, sols, and gels are examples.

**colloidal**. A state of suspension in a liquid medium in which extremely small particles are suspended and dispersed but not dissolved.

**collodian replica**. In metallography, a *replica* of a surface cast in nitrocellulose.

**colonies** (titanium). Regions within prior-β grains with α platelets having nearly identical orientations. In commercially pure titanium, colonies often have serrated boundaries. Colonies arise as transformation products during cooling from the β field at cooling rates that induce platelet nucleation and growth.

**colophony**. See *rosin*.

**color buffing**. Producing a final high luster by buffing. Sometimes called *coloring*.

**color center**. In materials characterization, a point lattice defect that produces optical absorption bands in an otherwise transparent crystal.

**color concentrate**. A measured amount of dye or pigment incorporated into a predetermined amount of plastic. The pigmented or colored plastic is then mixed into larger quantities of plastic material to be used for molding. This mixture is added to the bulk of plastic in measured quantity in order to produce a precise, predetermined color of finished articles to be molded.

**color filter**. In metallography, a device that transmits principally a predetermined range of wavelengths. See also *contrast filter* and *filter*.

**colorimeter**. An instrument for measuring the hue, purity, and brightness of a color.

**coloring**. Producing desired colors on metal by a chemical or electrochemical reaction. See also *color buffing*.

**color temperature**. The temperature in degrees Kelvin at which a blackbody must be operated to provide a color equivalent to that of the source in question. See also *blackbody*.

**columnar structure**. A coarse structure of parallel elongated grains formed by unidirectional growth, most often observed in castings (Fig. 80), but sometimes seen in structures resulting from diffusional growth accompanied by a solid-state transformation.

**coma**. In materials characterization, a lens aberration occurring in that part of the image field that is some distance

FIG. 80

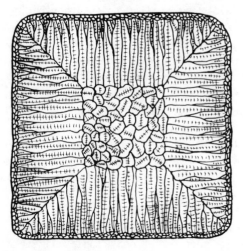

*Schematic cross section of a steel ingot showing typical columnar structure*

from the principal axis of the system. It results from different magnification in the various lens zones. Extraaxial object points appear as short conelike images with the brighter small head toward the center of the field (positive coma) or away from the center (negative coma).

**combination die**. (1) A die-casting die having two or more different cavities for different castings. (2) For forming, see *compound die*.

**combination mill**. An arrangement of a continuous mill for roughing and a *guide mill* or *looping mill* for shaping.

**combination mold**. See *family mold*.

**combined carbon**. Carbon in iron or steel that is combined chemically with other elements; not in the free state as graphite or temper carbon. The difference between the total carbon and the graphite carbon analyses. Contrast with *free carbon*.

**combined cyanide**. The cyanide of a metal-cyanide complex ion.

**combined stresses**. Any state of stress that cannot be represented by a single component of stress; that is, one that is more complicated than simple tension, compression, or shear.

**combing**. Lining up of reinforcing fibers.

**comet tails (on a polished surface)**. A group of comparatively deep unidirectional scratches that form adjacent to a microstructural discontinuity during mechanical polishing. They have the general shape of a comet tail. Comet tails form only when a unidirectional motion is maintained between the surface being polished and the polishing cloth.

**comminution**. (1) Breaking up or grinding an ore into small fragments. (2) Reducing metal to powder by mechanical means. (3) The act or process of reduction of powder particle size, usually but not necessarily by grinding or milling. See also *pulverization.*

**commutator-controlled welding**. Spot or projection welding in which several electrodes, in simultaneous contact with the work, function progressively under the control of an electrical commutating device.

**compact** (noun). The object produced by the compression of metal powder, generally while confined in a die.

**compact** (verb). The operation or process of producing a compact; sometimes called pressing.

**compacted graphite cast iron**. Cast iron having a graphite shape intermediate between the flake form typical of gray cast iron and the spherical form of fully spherulitic ductile cast iron (Fig. 81). An acceptable compacted graphite iron structure is one that contains no flake graphite, <20% spheroidal graphite, and 80% compacted graphite (ASTM A 247, type IV). Also known as CG iron or vermicular iron, compacted graphite cast iron is produced in a manner similar to that for ductile cast iron, but using a technique that inhibits the formation of fully spherulitic graphite nodules. Typical nominal compositions of CG irons contain 3.1 to 4.0% C, 1.7 to 3.0% Si, and 0.1 to 0.6% Mn.

FIG. 81

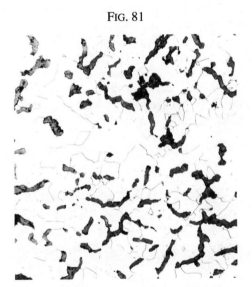

*Microstructure of compacted graphite iron*

**compactibility**. See *compressibility.*

**compacting crack**. A crack in a powder metallurgy compact that is generated during the major phases of the pressing cycle, such as load application, load release, and ejection.

**compacting force**. The force that acts on a powder to be densified expressed in newtons or tons.

**compacting pressure**. In powder metallurgy, the specific compacting force related to the area of contact with the press punch expressed in megapascals, meganewtons per square meter, or tons per square inch.

**compacting tool set**. See *die.*

**compaction**. (1) The act of forcing particulate or granular material together (consolidation) under pressure or impact to yield a relatively dense mass or formed object. Usually followed by drying, curing, or firing in refractory or other ceramic or powder metallurgy processing. (2) In ceramics or powder metallurgy, the preparation of a compact or object produced by the compression of a powder, generally while confined in a die, with or without the inclusion of lubricants, binders, etc., and with or without the concurrent applications of heat. (3) In reinforced plastics and composites, the application of a temporary vacuum bag and vacuum to remove trapped air and compact the lay-up.

**comparison standard**. In metallography, a standard micrograph or a series of micrographs, usually taken at 75 to 100×, used to determine grain size by direct comparison with the image.

**compatibility** (frictional). In tribology, materials that exhibit good sliding behavior, including resistance to adhesive wear, are termed frictionally compatible. Under some conditions materials that are not normally considered compatible in the metallurgical sense (for example, silver and iron) may be very compatible in the frictional sense.

**compatibility** (lubricant). In tribology, a measure of the degree to which lubricants or lubricant components can be mixed without harmful effects such as formation of deposits.

**compatibility** (metallurgical). A measure of the extent to which materials are mutually soluble in the solid state.

**compatibility** (plastics). The ability of two or more substances combined with one another to form a homogeneous composition having useful plastic properties; for example, the suitability of a sizing or finish for use with certain general resin types. Nonreactivity or negligible reactivity between materials in contact.

**compensating eyepiece**. In metallography, an eyepiece designed for use with apochromatic objectives. They are also used to advantage with high-power (oil-immersion) achromatic objectives. Because apochromatic objectives are undercorrected chromatically, these eyepieces are overcorrected. See also *apochromatic objective.*

**complete fusion**. Fusion which has occurred over the entire base material surfaces intended for welding and between all layers and weld beads.

**complete joint fusion**. Joint penetration in which the weld metal completely fills the groove and is fused to the base metal throughout its total thickness.

**complete penetration**. See preferred term *complete joint penetration.*

**complexation**. The formation of complex chemical species by the coordination of groups of atoms termed ligands to a central ion, commonly a metal ion. Generally, the ligand coordinates by providing a pair of electrons that forms an ionic or covalent bond to the central ion. See also *chelate, coordination compound,* and *ligand.*

**complexing agent.** A substance that is an electron donor and that will combine with a metal ion to form a soluble complex ion.

**complex ion.** An ion that may be formed by the addition reaction of two or more other ions.

**complex modulus.** The ratio of stress to strain in which each is a vector that may be represented by a complex number. May be measured in tension or flexure, compression, or shear.

**complex shear modulus.** The vectorial sum of the shear modulus and the loss modulus.

**complex silicate inclusions.** A general term describing silicate inclusions containing visible constituents in addition to the silicate matrix. An example is corundum or spinel crystals occurring in a silicate matrix in steel.

**complex Young's modulus.** The vectorial sum of Young's modulus and the loss modulus.

**compliance.** Tensile compliance is the reciprocal of Young's modulus. Shear compliance is the reciprocal of shear modulus. The term is also used in the evaluation of stiffness and deflection.

**component.** (1) One of the elements or compounds used to define a chemical (or alloy) system, including all phases, in terms of the fewest substances possible. (2) One of the individual parts of a vector as referred to a system of coordinates. (3) An individual functional element in a physically independent body that cannot be further reduced or divided without destroying its stated function, for example, a resistor, capacitor, diode, or transistor.

**component of variance.** A part of a total variance identified with a specified source of variability.

**composite.** See *composite material.*

**composite bearing material.** A solid material composed of a continuous or particulate solid lubricant dispersed throughout a load-bearing matrix to provide continuous replenishment of solid lubricant films as wear occurs, and effective heat transfer from the friction surface.

**composite coating.** A coating on a metal or nonmetal that consists of two or more components, one of which is often particulate in form. Example: a cermet composite coating on a cemented carbide cutting tool (Fig. 82). Also known as multilayer coating.

**composite compact.** A metal powder compact consisting of two or more adhering layers, rings, or other shapes of different metals or alloys with each material retaining its original identity.

**composite electrode.** A welding electrode made from two or more distinct components, at least one of which is filler metal. A composite electrode may exist in any of various physical forms, such as stranded wires, filled tubes, or covered wire.

FIG. 82

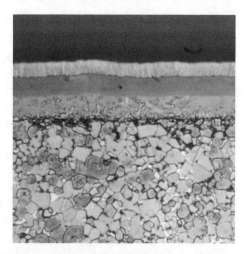

*Composite coating (TiC/TiCN/TiN) on a tungsten carbide substrate. 1500×*

**composite joint.** A joint in which welding is used in conjunction with mechanical joining.

**composite material.** A combination of two or more materials (reinforcing elements, fillers, and composite matrix binder), differing in form or composition on a macroscale. The constituents retain their identities, that is, they do not dissolve or merge completely into one another although they act in concert. Normally, the components can be physically identified and exhibit an interface between one another (Fig. 83). See also *carbon-carbon composites* (Technical Brief 6), *ceramic-matrix composites* (Tech-

FIG. 83

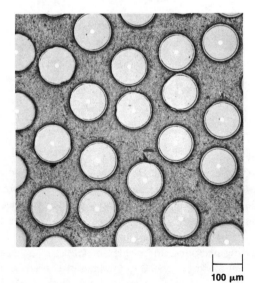

100 μm

*Cross section of a boron-fiber-reinforced aluminum composite material*

FIG. 84

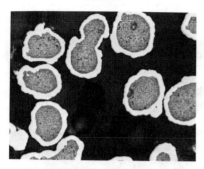

*82Ni-18Al composite powder. 150×*

nical Brief 10), *metal-matrix composites* (Technical Brief 26), and *resin-matrix composites* (Technical Brief 39).

**composite plate.** An electrodeposit consisting of layers of at least two different compositions.

**composite powder.** A powder in which each particle consists of two or more different materials (Fig. 84).

**composite structure.** A structural member (such as a panel, plate, pipe, or other shape) that is built up by bonding together two or more distinct components, each of which may be made of a metal, alloy, nonmetal, or *composite material*. Examples of composite structures include: honeycomb panels, clad plate, electrical contacts, sleeve bearings, carbide-tipped drills or lathe tools, and weldments constructed of two or more different alloys.

**compositional depth profile.** In materials characterization, the atomic concentration measured as a function of the perpendicular distance from the surface.

**compound.** (1) In chemistry, a substance of relatively fixed composition and properties, whose ultimate structural unit (molecule or repeat unit) is comprised of atoms of two or more elements. The number of atoms of each kind in this ultimate unit is determined by natural laws and is part of the identification of the compound. (2) In reinforced plastics and composites, the intimate admixture of a polymer with other ingredients, such as fillers, softeners, plasticizers, reinforcements, catalysts, pigments, or dyes. A thermoset compound usually contains all the ingredients necessary for the finished product, while a thermoplastic compound may require subsequent addition of pigments, blowing agents, and so forth.

**compound compact.** A powder metallurgy *compact* consisting of mixed metals, the particles of which are joined by pressing or sintering, or both, with each metal particle retaining substantially its original composition.

**compound die.** Any die designed to perform more than one operation on a part with one stroke of the press, such as blanking and piercing, in which all functions are performed simultaneously within the confines of the blank size being worked.

**compressibility.** (1) The ability of a powder to be formed into

a compact having well-defined contours and structural stability at a given temperature and pressure; a measure of the plasticity of powder particles. (2) A density ratio determined under definite testing conditions. Also referred to as compactibility.

**compressibility curve.** A plot of the green density of a powder compact with increasing pressure.

**compressibility test.** In powder metallurgy, a test to determine the behavior of a powder under applied pressure. It indicates the degree of densification and cohesiveness of a compact as a function of the magnitude of the pressure.

**compression crack.** See *compacting crack*.

**compression modulus.** See *bulk modulus of elasticity*.

**compression molding.** A technique of thermoset molding in which the plastic molding compound (generally preheated) is placed in the heated open mold cavity, the mold is closed under pressure (usually in a hydraulic press), causing the material to flow and completely fill the cavity, and then pressure is held until the material has cured.

**compression ratio** (plastics). In an extruder screw, the ratio of the volume available in the first flight at the hopper to the volume at the last flight, at the end of the screw.

**compression ratio** (powder metallurgy). The ratio of the volume of the loose powder to the volume of the compact made from it.

**compression set.** In seals, the difference between the thickness of a gasket of static seal before the seal is compressed and after it is released from compression. Compression set is normally expressed as a percentage of the total compression.

**compression test.** A method for assessing the ability of a material to withstand compressive loads. Analyses of structural behavior or metal forming require knowledge of compression stress-strain properties.

**compressive.** Pertaining to forces on a body or part of a body that tend to crush, or compress, the body.

**compressive modulus.** The ratio of compressive stress to compressive strain below the proportional limit. Theoretically equal to Young's modulus determined from tensile experiments.

**compressive strength.** The maximum compressive stress that a material is capable of developing, based on original area of cross section. If a material fails in compression by a shattering fracture, the compressive strength has a very definite value. If a material does not fail in compression by a shattering fracture, the value obtained for compressive strength is an arbitrary value depending upon the degree of distortion that is regarded as indicating complete failure of the material.

**compressive stress.** A stress that causes an elastic body to deform (shorten) in the direction of the applied load. Contrast with *tensile stress*.

**compressometer.** Instrument for measuring change in length

FIG. 85

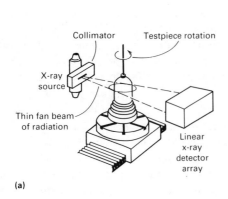

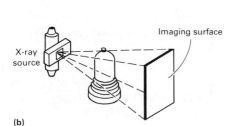

*Comparison of computed tomography (a) and radiography (b)*

over a given gage length caused by application or removal of a force. Commonly used in compression testing of metal specimens.

**Compton scattering**. In materials characterization, the elastic scattering of photons by electrons. Contrast with *Rayleigh scattering*.

**computed tomography (CT)**. The collection of transmission data through an object and the subsequent reconstruction of an image corresponding to a cross section through this object (Fig. 85). Also known as computerized axial tomography, computer-assisted tomography, CAT scanning, or industrial computed tomography.

**concave fillet weld**. A fillet weld having a concave face (Fig. 86).

**concave grating**. In materials characterization, a diffraction grating on a concave mirror surface. See also *diffraction grating* and *plane grating*.

**concave root surface**. A root surface which is concave.

**concavity**. The maximum distance from the face of a concave fillet weld perpendicular to a line joining the toes (Fig. 86).

**concentration**. (1) The mass of a substance contained in a

FIG. 86

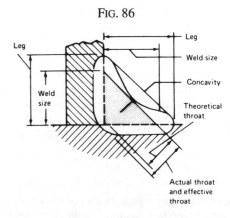

*Concave fillet weld*

unit volume of sample, for example, grams per liter. (2) A process for enrichment of an ore in valuable mineral content by separation and removal of waste material, or *gangue*.

**concentration cell**. An *electrolytic cell*, the *electromotive* *force* of which is caused by a difference in concentration of some component in the electrolyte. This difference leads to the formation of discrete *cathode* and *anode* regions.

**concentration polarization.** That portion of the *polarization* of a cell produced by concentration changes resulting from passage of current through the electrolyte.

**concrete.** (1) A composite material that consists essentially of a binding medium within which are embedded particles or fragments of aggregates (maximum aggregate size >5 mm, or 0.2 in.). In hydraulic-cement concrete, the binder is formed from a mixture of hydraulic cement and water. (2) A homogeneous mixture of portland cement, aggregates, and water and which may contain admixtures. See also *cement* (Technical Brief 8).

**concurrent healing.** The application of supplemental heat to a structure during a welding or cutting operation.

**condensation.** A chemical reaction in which two or more molecules combine, with the resulting separation of water or some other simple substance; the process is called polycondensation if a polymer is formed. See also *polymerization*.

**condensation polymerization.** In plastics technology, a stepwise chemical reaction in which two or more molecules combine, often but not necessarily accompanied by the separation of water or some other simple substance. If a polymer is formed, the process is called polycondensation. See also *polymerization*.

**condensation ratio.** A resin formed by polycondensation, for example, the alkyd, phenolaldehyde, and urea-formaldehyde resins.

**condenser.** A term applied to lenses or mirrors designed to collect, control, and concentrate radiation in an illumination system, such as an optical or scanning electron microscope.

**condenser aperture.** In electron microscopy, an opening in the condenser lens controlling the number of electrons entering the lens and the angular aperture of the illuminating beam.

**condenser lens.** See *condenser*.

**conditioning.** Subjecting a material to a prescribed environmental and/or stress history before testing.

**conditioning heat treatment.** A preliminary heat treatment used to prepare a material for a desired reaction to a subsequent heat treatment. For the term to be meaningful, the exact heat treatment must be specified.

**conditioning time.** See *joint-conditioning time*, *curing time*, and *setting time*.

**conductance** (electrical). A measure of the ability of any material to conduct an electrical charge. Conductance is the ratio of the current flow to the potential difference. The reciprocal of electrical resistance.

**conductance** (thermal). The time rate of heat flow through a unit area of a body, induced by a unit temperature difference between the body surfaces.

**conductivity** (electrical). The reciprocal of volume resistivity. The electrical or thermal conductance of a unit cube of any material (conductivity per unit volume).

**conductivity** (thermal). The time rate of heat flow through unit thickness of an infinite slab of a homogeneous material in a direction perpendicular to the surface, induced by unit temperature difference. Recommended SI units: $W/m \cdot K$.

**conductor.** A wire, cable, or other body capable of carrying an electric current.

**cone.** The conical part of an oxyfuel gas flame next to the orifice of the tip (Fig. 87).

FIG. 87

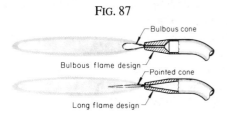

*Inner cones of oxyacetylene welding flames*

**cone angle.** The angle that the cutter axis makes with the direction along which the blades are moved for adjustment, as in adjustable-blade reamers where the base of the blade slides on a conical surface.

**cone resistance value (CVR).** A measure of the yield stress of a grease, obtained by static indentation with a cone. The equilibrium depth of penetration is measured, not the penetration in a given time, which is the penetration value. See also *penetration* (of a grease).

**confidence interval.** That range of values, calculated from estimates of the mean and standard deviation, which is expected to include the population mean with a stated level of confidence. Confidence intervals in the same context also can be calculated for standard deviations, lines, slopes, and points.

**configurations.** In plastics technology, related chemical structures produced by the cleavage and reforming of covalent bonds.

**conformability.** In tribology, that quality of a plain bearing material that allows it to adjust itself to shaft deflections and minor misalignments by deformation or by wearing away of bearing material without producing operating difficulties.

**conformal coating.** A coating that covers and exactly fits the shape of the coated object.

**conformal surfaces.** Surfaces whose centers of curvature are on the same side of the interface. In wear testing, it refers to the case where the curvature of both specimens matches such that the nominal contact area during the testing remains approximately constant. Contrast with *counterformal surfaces.*

**conformations.** Different shapes of polymer molecules resulting from rotation about single covalent bonds in the polymer chain.

**congruent melting.** An isothermal or isobaric melting in which both the solid and liquid phases have the same composition throughout the transformation.

**congruent transformation.** An isothermal or isobaric phase change in metals in which both of the phases concerned have the same composition throughout the process.

**conjugate phases.** In microstructural analysis, those states of matter of unique composition that coexist at equilibrium at a single point in temperature and pressure. For example, the two coexisting phases of a two-phase equilibrium.

**conjugate planes.** Two planes of an optical system such that one is the image of the other.

**connected porosity.** The volume fraction of all pores, voids, and channels within a solid mass that are interconnected with each other and communicate with the external surface, and thus are measurable by gas or liquid penetration. Contrast with *closed porosity.*

**consistency.** (1) In adhesives, that property of a liquid adhesive by virtue of which it tends to resist deformation. Consistency is not a fundamental property but is composed of viscosity, plasticity, and other phenomena. See also *viscosity* and *viscosity coefficient.* (2) An imprecise measure of the degree to which a grease resists deformation under the application of a force.

**consolidation.** In metal-matrix or thermoplastic composites, a processing step in which fiber and matrix are compressed by one of several methods to reduce voids and achieve desired density.

**constantan.** See *resistance alloys* (Technical Brief 40).

**constant life fatigue diagram.** In failure analysis, a plot (usually on rectangular coordinates) of a family of curves, each of which is for a single fatigue life (number of cycles), relating alternating stress, maximum stress, minimum stress, and mean stress. The constant life fatigue diagram is generally derived from a family of *S-N* curves, each of which represents a different stress ratio for a 50% probability of survival. See also *nominal stress, maximum stress, minimum stress, S-N curve, fatigue life,* and *stress ratio.*

**constituent.** (1) One of the ingredients that make up a chemical system. (2) A phase or a combination of phases that occurs in a characteristic configuration in an alloy microstructure. (3) In composites, the principal constituents are the fibers and the matrix.

**constitution diagram.** See *phase diagram.*

**constraint.** Any restriction that limits the transverse contraction normally associated with a longitudinal tension, and that hence causes a secondary tension in the transverse

direction; usually used in connection with welding. Contrast with *restraint.*

**constricted arc** (plasma arc welding and cutting). A plasma arc column that is shaped by a constricting nozzle orifice. See also the figure accompanying the term *plasma arc welding.*

**constricting nozzle** (plasma arc welding and cutting). A water-cooled copper nozzle surrounding the electrode and containing the constricting orifice. See also the figure accompanying the term *cutting torch (plasma arc).*

**constricting orifice** (plasma arc welding and cutting). The hole in the constricting nozzle through which the arc passes. See also the figure accompanying the term *multiport nozzle.*

**constriction resistance.** In electrical contact theory, the resistance that arises from the constriction of current flow lines in order to pass through small areas of contact (a-spots) at the interface of two contacting bodies. See also *a-spot, contact resistance,* and *film resistance.*

**consumable electrode.** A general term for any arc welding electrode made chiefly of filler metal. Use of specific names such as *covered electrode,* bare electrode, flux-cored electrode, and *lightly coated electrode* is preferred.

**consumable-electrode remelting.** A process for refining metals in which an electric current passes between an electrode made of the metal to be refined and an ingot of the refined metal, which is contained in a water-cooled mold (Fig. 88). As a result of the passage of electric current, droplets of molten metal form on the electrode and fall to the ingot. The refining action occurs from contact with the atmosphere, vacuum, or slag through which the

FIG. 88

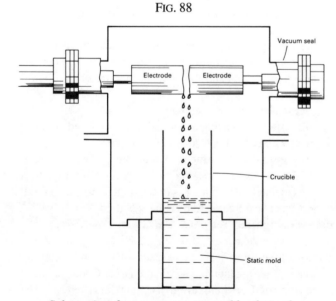

*Schematic of a vacuum consumable-electrode remelting process*

drop falls. See also *electroslag remelting* and *vacuum arc remelting*.

**consumable guide electroslag welding.** An electroslag welding process variation in which filler metal is supplied by an electrode and its guiding member. See also *electroslag welding*.

**consumable insert.** Preplaced filler metal which is completely fused into the root of the joint and becomes part of the weld.

**contact adhesive.** An adhesive that is apparently dry to the touch and that will adhere to itself simultaneously upon contact. An adhesive that, when applied to both adherends and allowed to dry, develops a bond when the adherends are brought together without sustained pressure.

**contact angle.** In lubrication, the angle at which the surface of a liquid drop meets the surface of a solid on which it is placed.

**contact angle** (in a bearing). See *angle of contact*.

**contact area.** The common area between a conductor and a connector across which the flow of electricity takes place.

**contact bond adhesive.** Synonym for contact adhesive.

**contact corrosion.** A term primarily used in Europe to describe *galvanic corrosion* between dissimilar metals.

**contact fatigue.** Cracking and subsequent pitting of a surface subjected to alternating Hertzian stresses such as those produced under rolling contact or combined rolling and sliding. The phenomenon of contact fatigue is encountered most often in rolling-element bearings or in gears, where the surface stresses are high due to the concentrated loads and are repeated many times during normal operation.

**contact infiltration.** In powder metallurgy, the process of infiltration whereby the initially solid infiltrant is placed in direct contact with the compact and the pores are filled with the liquid phase by capillary force after the infiltrant has become molten. See also *infiltrant* and *infiltration*.

**contacting ring seal.** A type of circumferential seal that utilizes a ring-spring loaded radially against a shaft. The ring is either gapped or segmented, in order to have radial flexibility. The seal has overlapping joints for blocking leakage at the gaps. An axial spring load seats the ring against the wall of its containing cartridge.

**contact lubrication.** A little-used term relating to the conditions of lubrication obtained with solid lubricant powders rubbed into a surface. It is recommended that this term not be used.

**contact material.** A metal, composite, or alloy made by the melt-cast method or manufactured by powder metallurgy that is used in devices that make and break electrical circuits or welding electrodes. A majority of contact applications in the electrical industry utilize silver-type contacts, which include the pure metal, alloys, and powder metal combinations. Silver, which has the highest electrical and thermal conductivity of all metals, is also used as a plated, brazed, or mechanically bonded overlay on other contact materials—notably, copper and copper-base materials. Other types of contacts used include the platinum group metals, tungsten, molybdenum, copper, copper alloys, and mercury. See also *electrode* (welding).

**contact molding.** A process for molding reinforced plastics in which reinforcement and resin are placed on a mold. Cure is either at room temperature using a catalyst-promoter system or by heating in an oven, without additional pressure. Also referred to as hand lay-up.

**contact plating.** (1) The plated-on material applied to the basic metal of an electrical contact to provide for required contact resistance and/or specified wear resistance characteristics. (2) A metal plating process wherein the plating current is provided by galvanic action between the work metal and a second metal, without the use of an external source of current.

**contact potential.** In corrosion technology, the potential difference at the junction of two dissimilar substances.

**contact pressure resins.** Liquid resins that thicken or polymerize upon heating, and, when used for bonding laminates, require little or no pressure.

**contact resistance.** (1) The electrical resistance of metallic surfaces at their interface in the *contact area* under specified conditions. (2) The electrical resistance between two contacting bodies, which is the sum of the *constriction resistance* and the *film resistance*.

**contact scanning.** In ultrasonic inspection, a planned systematic movement of the beam relative to the object being inspected, the search unit being in contact with and coupled to this object by a thin film of coupling material.

**contact stress.** Stress that results near the surfaces of two contacting solid bodies when they are placed against one another under a nonzero normal force. This term is not sufficiently precise because it does not indicate the type of stress, although common usage usually implies elastic or Hertz stress.

**contact stress** (glass). The tensile stress component imposed at a glass surface immediately surrounding the contact area between the glass surface and an object generating a locally applied force.

**contact tube.** A device which transfers current to a continuous welding electrode.

**contact weld.** (1) The point of attachment of a contact to its support when accomplished by resistance welding. (2) A contacting failure due to the fusing of contacting surfaces under load conditions to the point that the contacts fail to separate when expected.

**container.** The chamber into which an ingot or billet is inserted prior to extrusion. The container for backward extrusion of cups or cans is sometimes called a die.

**contaminant.** An impurity or foreign substance present in a

FIG. 89

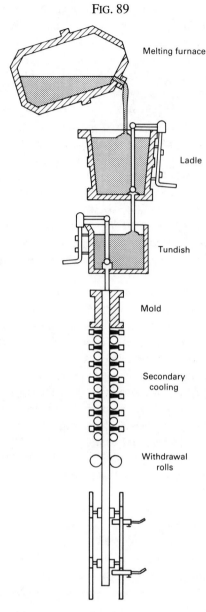

Melting furnace

Ladle

Tundish

Mold

Secondary cooling

Withdrawal rolls

*Main components of a continuous casting strand*

FIG. 90

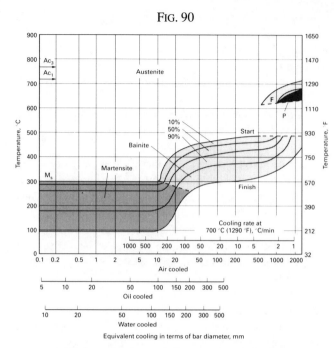

*CCT diagram for an alloy steel with 0.40% C, 1.50% Ni, 1.20% Cr, and 0.30% Mo, plotted as a function of bar diameter*

material or environment that affects one or more properties of the material.

**continuity bond**. In corrosion protection, a metallic connection that provides electrical continuity between metal structures.

**continuous casting**. A casting technique in which a cast shape is continuously withdrawn through the bottom of the mold as it solidifies, so that its length is not determined by mold dimensions (Fig. 89). Used chiefly to produce semi-finished mill products such as billets, blooms, ingots, slabs, strip, and tubes. See also *strand casting*.

**continuous compaction**. In powder metallurgy, the production of relatively long compacts having a uniform cross section, such as sheet, rod, tube, etc., by direct extrusion or rolling of loose powder.

**continuous cooling transformation (CCT) diagram**. Set of curves drawn using logarithmic time and linear temperature as coordinates, which define, for each cooling curve of an alloy, the beginning and end of the transformation of the initial phase (Fig. 90).

**continuous filament yarn**. Yarn formed by twisting two or more continuous filaments into a single, continuous strand.

**continuous furnace** (glass). A glass furnace in which the level of glass remains substantially constant because the feeding of batch continuously replaces the glass withdrawn.

**continuous furnace** (powder metallurgy). A furnace used for the uninterrupted sintering of compacts.

**continuous mill**. A rolling mill consisting of a number of strands of synchronized rolls (in tandem) in which metal undergoes successive reductions as it passes through the various strands.

**continuous phase**. In an alloy or portion of an alloy containing more than one phase, the phase that forms the matrix in which the other phase or phases are dispersed.

**continuous precipitation**. Precipitation from a supersaturated solid solution in which the precipitate particles grow by long-range diffusion without recrystallization of the matrix. Continuous precipitates grow from nuclei distributed more or less uniformly throughout the matrix. They usually are randomly oriented, but may form a

*Widmanstätten structure.* Also called general precipitation. Compare with *discontinuous precipitation* and *localized precipitation.*

**continuous sequence.** A longitudinal welding sequence in which each pass is made continuously from one end of the joint to the other. See also *backstep sequence* and *longitudinal sequence.*

**continuous sintering.** In powder metallurgy, presintering or sintering in such a manner that the objects are advanced through a furnace at a fixed rate by manual or mechanical means; sometimes called stoking.

**continuous spectrum (x-rays).** In materials characterization, the polychromatic radiation emitted by the target of an x-ray tube. It contains all wavelengths above a certain minimum value, known as the short wavelength limit.

**continuous-type furnace.** A furnace used for heat treating materials that progress continuously through the furnace, entering one door and being discharged from another. See also *belt furnace, direct-fired tunnel-type furnace, rotary-retort furnace,* and *shaker-hearth furnace.*

**continuous weld.** A weld extending continuously from one end of a joint to the other or, where the joint is essentially circular, completely around the joint. Contrast with *intermittent weld.*

**continuum.** In materials characterization, the noncharacteristic rays emitted upon irradiation of a specimen and caused by deceleration of the incident electrons by interaction with the electrons and nuclei of the specimen. Also termed bremsstrahlung and white radiation.

**contour forming.** See *roll forming, stretch forming, tangent bending,* and *wiper forming.*

**contour machining.** Machining of irregular surfaces, such as those generated in tracer turning, tracer boring, and *tracer milling.*

**contour milling.** Milling of irregular surfaces. See also *tracer milling.*

**contraction.** The volume change that occurs in metals and alloys upon solidification and cooling to room temperature.

**contrast enhancement (electron optics).** In electron microscopy, an improvement in electron image contrast by the use of an objective aperture diaphragm, shadow casting, or other means. See also *shadowing.*

**contrast filter.** In metallography, a color filter, usually with strong absorption, that uses the special absorption bands of the object to control the contrast of the image by exaggerating or diminishing the brightness difference between differently colored areas.

**contrast perception.** In metallography, the ability to differentiate various components of the object structure by various intensity levels in the image.

**controlled atmosphere.** (1) A specified inert gas or mixture of gases at a predetermined temperature in which selected processes take place. (2) As applied to sintering, to prevent oxidation and destruction of the powder compacts.

**controlled atmosphere chamber.** An inert gas-filled enclosure or cabinet in which plasma spraying or welding can be performed to minimize (or prevent) oxidation of the coating or substrate. The enclosure is usually fitted with viewing ports, glove ports to permit manipulations, and a small separate airlock for introducing or removing components without loss of atmosphere.

**controlled cooling.** Cooling a metal or alloy from an elevated temperature in a predetermined manner to avoid hardening, cracking, or internal damage, or to produce desired microstructure or mechanical properties.

**controlled etching.** Electrolytic etching with selection of suitable etchant and voltage resulting in a balance between current and dissolved metal ions.

**controlled-potential coulometry.** Measurement of the number of coulombs required for an electrochemical reaction occurring under conditions where the working electrode potential is precisely controlled.

**controlled-pressure cycle.** A forming cycle during which the hydraulic pressure in the forming cavity is controlled by an adjustable cam that is coordinated with the punch travel.

**controlled rolling.** A hot-rolling process in which the temperature of the steel is closely controlled, particularly during the final rolling passes, to produce a fine-grain microstructure.

**convection.** The motion resulting in a fluid from the differences in density and the action of gravity. In heat transmission, this meaning has been extended to include both forced and natural motion or circulation.

**conventional forging.** A forging characteristic by design complexity and tolerances that fall within the broad range of general forging practice.

FIG. 91

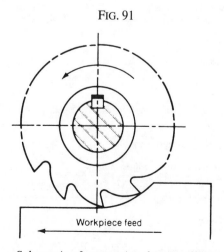

Workpiece feed

*Schematic of conventional (up) milling*

**conventional milling.** Milling in which the cutter moves in the direction opposite to the feed at the point of contact (Fig. 91; see previous page). Contrast with *climb milling*.

**conventional strain.** See *engineering strain* and *strain*.

**conventional stress.** See *engineering stress* and *stress*.

**convergent-beam electron diffraction (CBED).** In *analytical electron microscopy*, a technique of impinging a highly convergent electron beam on a crystal to produce a diffraction pattern composed of disks of intensity. In addition to *d*-spacing and crystal orientation information, the technique can provide information on crystallographic point or space group symmetry.

**conversion coating.** A coating consisting of a compound of the surface metal, produced by chemical or electrochemical treatments of the metal. Examples include chromate coatings on zinc, cadmium, magnesium, and aluminum, and oxide and phosphate coatings on steel. See also *chromate treatment* and *phosphating*.

**converter.** A furnace in which air is blown through a bath of molten metal or matte, oxidizing the impurities and maintaining the temperature through the heat produced by the oxidation reaction. A typical converter is the *argon oxygen decarburization* vessel.

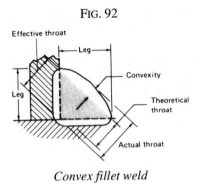

FIG. 92

*Convex fillet weld*

**convex fillet weld.** A fillet weld having a convex face (Fig. 92).

**convexity.** The maximum distance from the face of a convex fillet weld perpendicular to a line joining the toes.

**convex root surface.** A root surface which is convex.

**coolant.** The liquid used to cool the work during grinding and to prevent it from rusting. It also lubricates, washes away chips and grits, and aids in obtaining a finer finish. In metal cutting, the preferred term is cutting fluid.

**cooling channels.** In plastic part making, channels or pas-

---

## Technical Brief 13: Copper and Copper Alloys

COPPER and copper alloys constitute one of the major groups of commercial metals. They are widely used because of their excellent electrical and thermal conductivities, outstanding resistance to corrosion, ease of fabrication, and good strength and fatigue resistance. They are generally nonmagnetic. They can be readily soldered and brazed, and many coppers and copper alloys can be welded by various gas, arc, and resistance methods. For decorative parts, standard alloys having specific colors are readily available. They can be plated, coated with organic substances, or chemically colored to further extend the variety of available finishes.

Pure copper is used extensively for cables and wires, electrical contacts,

### Generic classification of copper alloys

| Generic name | UNS numbers | Composition |
|---|---|---|
| **Wrought alloys** | | |
| Coppers | C10100–C15760 | >99% Cu |
| High-copper alloys | C16200–C19600 | >96% Cu |
| Brasses | C205–C28580 | Cu-Zn |
| Leaded brasses | C31200–C38590 | Cu-Zn-Pb |
| Tin brasses | C40400–C49080 | Cu-Zn-Sn-Pb |
| Phosphor bronzes | C50100–C52400 | Cu-Sn-P |
| Leaded phosphor bronzes | C53200–C54800 | Cu-Sn-Pb-P |
| Copper-phosphorus and copper-silver-phosphorus alloys | C55180–C55284 | Cu-P-Ag |
| Aluminum bronzes | C60600–C64400 | Cu-Al-Ni-Fe-Si-Sn |
| Silicon bronzes | C64700–C66100 | Cu-Si-Sn |
| Other copper-zinc alloys | C66400–C69900 | . . . |
| Copper-nickels | C70000–C79900 | Cu-Ni-Fe |
| Nickel silvers | C73200–C79900 | Cu-Ni-Zn |
| **Cast alloys** | | |
| Coppers | C80100–C81100 | >99% Cu |
| High-copper alloys | C81300–C82800 | >94% Cu |
| Red and leaded red brasses | C83300–C85800 | Cu-Zn-Sn-Pb (75–89% Cu) |
| Yellow and leaded yellow brasses | C85200–C85800 | Cu-Zn-Sn-Pb (57–74% Cu) |
| Manganese and leaded manganese bronzes | C86100–C86800 | Cu-Zn-Mn-Fe-Pb |
| Silicon bronzes, silicon brasses | C87300–C87900 | Cu-Zn-Si |
| Tin bronzes and leaded tin bronzes | C90200–C94500 | Cu-Sn-Zn-Pb |
| Nickel-tin bronzes | C94700–C94900 | Cu-Ni-Sn-Zn-Pb |
| Aluminum bronzes | C95200–C95810 | Cu-Al-Fe-Ni |
| Copper-nickels | C96200–C96800 | Cu-Ni-Fe |
| Nickel silvers | C97300–C97800 | Cu-Ni-Zn-Pb-Sn |
| Leaded coppers | C98200–C98800 | Cu-Pb |
| Miscellaneous alloys | C99300–C99750 | . . . |

---

sageways located within the body of a mold through which a cooling medium can be circulated to control temperature on the mold surface. May also be used for heating a mold by circulating steam, hot oil, or other heated fluid through channels, as in the molding of thermosetting and some thermoplastic materials.

**cooling curve.** A graph showing the relationship between time and temperature during the cooling of a material. It is used to find the temperatures at which phase changes occur. A property or function other than time may occasionally be used—for example, thermal expansion.

**cooling fixture.** In plastic part making, block of metal or wood shaped to hold a molded part to maintain the proper shape or dimensional accuracy of the part after it is removed from the mold until it is cooled enough to retain its shape without further appreciable distortion. Also called a shrink fixture.

**cooling rate.** The average slope of the time-temperature curve taken over a specified time and temperature interval.

**cooling stresses.** Residual stresses in castings resulting from nonuniform distribution of temperature during cooling.

**cooling table.** Same as *hot bed*.

**cool time** (resistance welding). The time interval between successive heat times in multiple-impulse welding or in the making of seam welds.

**coordination catalysis.** Ziegler-type of catalysis for processing plastics. See also *Ziegler-Natta catalysts*.

**coordination compound.** A compound with a central atom or ion bound to a group of ions or molecules surrounding it. Also called coordination complex. See also *chelate*, *complexation*, and *ligand*.

**coordination number.** (1) Number of atoms or radicals coordinated with the central atom in a complex covalent compound. (2) Number of nearest neighboring atoms to a selected atom in crystal structure.

**cope.** In casting, the upper or topmost section of a *flask*, *mold*, or *pattern*.

**copolymer.** A long-chain molecule formed by the reaction of two or more dissimilar monomers. See also *polymer*.

**copolymerization.** See *polymerization*.

**copper-accelerated salt-spray (CASS) test.** An *accelerated corrosion test* for some electrodeposits and for anodic coatings on aluminum.

**copper and copper alloys.** See *Technical Brief 13*.

**copper brazing.** A term improperly used to denote brazing

---

and a wide variety of other parts that are required to pass electrical current. Coppers, and certain *brasses*, *bronzes*, and cupronickels are used extensively for automobile radiators, heat exchangers, home heating systems, panels for absorbing solar energy, and various other applications requiring rapid conduction of heat. Because of their outstanding ability to resist corrosion, coppers, brasses, some bronzes, and cupronickels are used for pipes, valves, and fittings in systems carrying potable water, process water, or other aqueous fluids.

In all classes of copper alloys, certain alloy compositions for wrought products have counterparts among the cast alloys. Most wrought alloys are available in various cold-worked conditions, and the room-temperature strengths and fatigue resistances of these alloys depend on the amount of cold work as well as the alloy content. Typical applications of cold-worked wrought alloys include springs, fasteners, hardware, small gears, cams, and electrical components.

Copper powder metallurgy (P/M) products based on pressed and sintered atomized or hydrometallurgical copper powders are also produced. Applications for copper P/M parts include self-lubricated sintered bearings, structural parts, friction materials, and porous bronze filters. Dispersion-strengthened copper alloys are also produced. See also *dispersion-strengthened materials*.

### Recommended Reading

- D.E. Tyler and W.T. Black, Introduction to Copper and Copper Alloys, *Metals Handbook*, 10th ed., Vol 2, ASM International, 1990, p 217-240
- N.W. Polan *et al.*, Corrosion of Copper and Copper Alloys, *Metals Handbook*, 9th ed., Vol 13, ASM International, 1987, p 610-640
- R.F. Schmidt, D.G. Schmidt, and M. Sahoo, Cast Copper and Copper Alloys, *Metals Handbook*, 9th ed., Vol 15, ASM International, 1988, p 771-785
- E. Klar and D.F. Berry, Copper P/M Products, *Metals Handbook*, 10th ed., Vol 2, ASM International, 1990, p 392-402

with a copper filler metal. See preferred terms *furnace brazing* and *braze welding*.

**copperhead.** A reddish spot in a porcelain enamel coating caused by iron pickup during enameling, iron oxide left on poorly cleaned basis metal, or burrs on iron or steel basis metal that protrude through the coating and are oxidized during firing.

**cordierite.** A talclike mineral of composition $Mg_2Al_4Si_5O_{18}$ used in ceramics to make refractories, filters in diesel engines, and spark plug insulators. Cordierite has an unusually low coefficient of expansion (1.4 to 2.6 $\times$ $10^{-6}$/°C from 25 to 1000 °C) and high thermal shock resistance.

**core.** (1) A specially formed material inserted in a mold to shape the interior or other part of a casting that cannot be shaped as easily by the pattern. (2) In a ferrous alloy prepared for *case hardening*, that portion of the alloy that is not part of the *case*. Typically considered to be the portion that (a) appears dark (with certain etchants) on an etched cross section, (b) has an essentially unaltered chemical composition, or (c) has a hardness, after hardening, less than a specified value. (3) The central member of a *sandwich construction* (honeycomb material, foamed plastic, or solid sheet to which the faces of the sandwich are attached).

**core** (plastics). (1) In plastic part making, a channel in a mold for circulation of heat transfer media. (2) The part of a complex mold that molds undercut parts. Cores are usually withdrawn before the main sections of the mold are opened. Also called core pin.

**core assembly.** In casting, a complex core consisting of a number of sections.

**core binder.** In casting, any material used to hold the grains of core sand together.

**core blow.** A gas pocket in a casting adjacent to a cored cavity and caused by entrapped gases from the core.

**core blower.** A machine for making foundry cores using compressed air to blow and pack the sand into the core box.

**core box.** In casting, a wood, metal, or plastic structure containing a shaped cavity into which sand is packed to make a core.

**core crush.** A collapse, distortion, or compression of the core of a *sandwich construction*.

**core depression.** A localized indentation or gouge in the core of a *sandwich construction*.

**cored bars.** In powder metallurgy, a compact of bar shape heated by its own electrical resistance to a temperature high enough to melt its interior.

**cored mold.** In plastic part making, a mold incorporating passages for electrical heating elements, steam, or water.

**core dryers.** In casting, supports used to hold cores in shape during baking; constructed from metal or sand for conven-

tional baking or from plastic material for use with dielectric core-making equipment.

**cored solder.** A solder wire or bar containing flux as a core.

**core filler.** In casting, a material, such as coke, cinder, and sawdust, used in place of sand in the interiors of large cores; usually added to aid collapsibility.

**core forging.** (1) Displacing metal with a punch to fill a die cavity. (2) The product of such an operation.

**core knockout machine.** In casting, a mechanical device for removing cores from castings.

**coreless induction furnace.** An electric induction furnace for melting or holding molten metals that does not utilize a steel core to direct the magnetic field which stirs the melt (Fig. 93).

FIG. 93

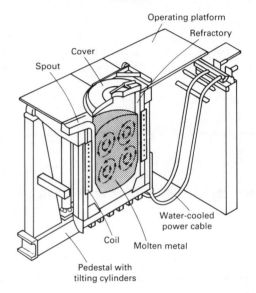

*Cross-sectional view of a coreless-type induction furnace illustrating four-quadrant stirring action*

**core oil.** In casting, a binder for core sand that sets when baked and is destroyed by the heat from the cooling casting.

**core pin.** In plastic part making, a pin used to mold a hole.

**core pin plate.** In plastic part making, a plate holding core pins.

**core plates.** In casting, heat-resistant plates used to support cores during baking; may be metallic or nonmetallic, the latter being a requisite for dielectric core baking.

**core print.** In casting, projections attached to a pattern in order to form recesses in the mold at points where cores are to be supported.

**core rod.** In powder metallurgy, a member of a die assembly used in molding a hole in a compact.

**core sand.** In casting, sand for making cores to which a binding material has been added to obtain good cohesion and permeability after drying; usually low in clays.

**core separation**. In a *sandwich construction*, a partial or complete breaking of the core node bond.

**core shift**. In casting, a variation from the specified dimensions of a cored casting section due to a change in position of the core or misalignment of cores in assembly.

**core splicing**. The joining of segments of a core of a *sandwich construction* by bonding, or by overlapping each segment and then driving them together.

**core vents**. (1) In casting, a wax product, round or oval in form, used to form the vent passage in a core. Also, a metal screen or slotted piece used to form the vent passage in the core box used in a core blowing machine. (2) Holes made in the core for the escape of gas.

**core wash**. In casting, a suspension of a fine refractory applied to cores by brushing, dipping, or spraying to improve the surface of the cored portion of the casting.

**core wires or rods**. In casting, reinforcing wires or rods for fragile cores, often preformed into special shapes.

**coring**. (1) A condition of variable composition between the center and surface of a unit of microstructure (such as a dendrite, grain, carbide particle); results from nonequilibrium solidification, which occurs over a range of temperature (Fig. 94). (2) A central cavity at the butt end of a rod extrusion, sometimes called *extrusion pipe*.

**coring** (plastics). In plastic part making, the removal of excess material from the cross section of a molded part to attain a more uniform wall thickness.

**corner angle**. On face milling cutters, the angle between an angular cutting edge of a cutter tooth and the axis of the cutter, measured by rotation into an axial plane. See the figure accompanying the term *face mill.*

**corner-flange weld**. A flange weld with only one member flanged at the location of welding.

**corner joint**. A joint between two members located ap-

FIG. 94

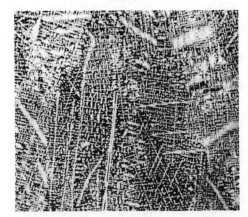

*Dendritic structure in the columnar region of a Cu-30Ni ingot showing coring (variation in solute concentration)*

FIG. 95

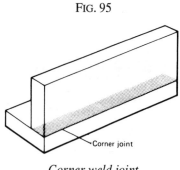

*Corner weld joint*

proximately at right angles to each other in the form of an "L" (Fig. 95).

**corona** (resistance welding). The area sometimes surrounding the nugget of a spot weld at the faying surfaces which provides a degree of solid-state welding.

**correction**. In the case of a testing machine, the difference obtained by subtracting the indicated load from the correct value of the applied load.

**corrodkote test**. An *accelerated corrosion test* for electrodeposits.

**corrosion**. The chemical or electrochemical reaction between a material, usually a metal, and its environment that produces a deterioration of the material and its properties.

**corrosion effect**. A change in any part of the *corrosion system* caused by *corrosion.*

**corrosion embrittlement**. The severe loss of ductility of a metal resulting from corrosive attack, usually *intergranular* and often not visually apparent.

**corrosion-erosion**. See *erosion-corrosion.*

**corrosion fatigue**. The process in which a metal fractures prematurely under conditions of simultaneous corrosion and repeated cyclic loading at lower stress levels or fewer cycles than would be required in the absence of the corrosive environment.

**corrosion fatigue strength**. The maximum repeated stress that can be endured by a metal without failure under definite conditions of corrosion and fatigue and for a specific number of stress cycles and a specified period of time.

**corrosion inhibitor**. See *inhibitor.*

**corrosion potential** ($E_{corr}$). The *potential* of a corroding surface in an electrolyte, relative to a *reference electrode*. Also called rest potential, open-circuit potential, or freely corroding potential.

**corrosion product**. Substance formed as a result of *corrosion.*

**corrosion protection**. Modification of a *corrosion system* so that corrosion damage is mitigated.

**corrosion rate**. *Corrosion effect* on a metal per unit of time. The type of corrosion rate used depends on the technical system and on the type of corrosion effect. Thus, corrosion

**Table 17  Relationships among units commonly used for corrosion rates**

$d$ is metal density in grams per cubic centimeter ($g/cm^3$)

| Unit | Factor for conversion to | | | | | |
|---|---|---|---|---|---|---|
| | mdd | $g/m^2/d$ | $\mu m/yr$ | mm/yr | mils/yr | in./yr |
| Milligrams per square decimeter per day (mdd) | 1 | 0.1 | $36.5/d$ | $0.0365/d$ | $1.144/d$ | $0.00144/d$ |
| Grams per square meter per day ($g/m^2/d$) | 10 | 1 | $365/d$ | $0.365/d$ | $14.4/d$ | $0.0144/d$ |
| Microns per year ($\mu m/yr$) | $0.0274d$ | $0.00274d$ | 1 | 0.001 | 0.0394 | 0.0000394 |
| Millimeters per year (mm/yr) | $27.4d$ | $2.74d$ | 1000 | 1 | 39.4 | 0.0394 |
| Mils per year (mils/yr) | $0.696d$ | $0.0696d$ | 25.4 | 0.0254 | 1 | 0.001 |
| Inches per year (in./yr) | $696d$ | $69.6d$ | 25 400 | 25.4 | 1000 | 1 |

rate may be expressed as an increase in corrosion depth per unit of time (penetration rate, for example, mils/yr) or the mass of metal turned into corrosion products per unit area of surface per unit of time (weight loss, for example, $g/m^2/yr$). The corrosion effect may vary with time and may not be the same at all points of the corroding surface. Therefore, reports of corrosion rates should be accompanied by information on the type, time dependency, and location of the corrosion effect. Table 17 lists relationships among some of the units commonly used for corrosion rates.

**corrosion resistance.** The ability of a material to withstand contact with ambient natural factors or those of a particular, artificially created atmosphere, without degradation or change in properties. For metals, this could be pitting or rusting; for organic materials, it could be crazing.

**corrosion system.** System consisting of one or more metals and all parts of the environment that influence *corrosion.*

**corrosive flux.** A flux with a residue that chemically attacks the base metal. It may be composed of inorganic salts and acids, organic salts and acids, or activated rosins or resins.

**corrosive wear.** *Wear* in which chemical or electrochemical reaction with the environment is significant. See also *oxidative wear.*

**corrosivity.** Tendency of an environment to cause *corrosion* in a given *corrosion system.*

**corrugating.** The forming of sheet metal into a series of straight, parallel alternate ridges and grooves with a rolling mill equipped with matched roller dies or a *press brake* equipped with a specially shaped punch and die.

**corrugations.** In metal forming, transverse ripples caused by a variation in strip shape during hot or cold reduction.

**corundum.** (1) A naturally occurring alumina usually of relatively high purity but containing associated minerals, such as diaspore and various silicates. Commonly coarsely crystalline, but sometimes microcrystalline. (2) Native alumina occurring as rhombohedral crystals and also in masses and variously colored grains. Corundum and its artificial counterparts are abrasives especially suited to the grinding of metals.

**cosmetic pass.** A weld pass made primarily for the purpose of enhancing appearance.

**cottoning.** The formation of weblike filaments of adhesive between the applicator and substrate surface.

**Cottrell process.** Removal of solid particulates from gases with electrostatic precipitation.

**Coulomb friction.** A term used to indicate that the frictional force is proportional to the normal load.

**coulometer.** An electrolytic cell arranged to measure the quantity of electricity by the chemical action produced in accordance with Faraday's law.

**coulometry.** In materials characterization, an electrochemical technique in which the total number of coulombs consumed in electrolysis is used to determine the amount of substance electrolyzed.

**count.** For fabric used in composite fabrication, the number of warp and filling yarns per inch in woven cloth. For yarn, size based on relation of length and weight.

**counterblow equipment.** Forging equipment with two opposed rams that are activated simultaneously to strike repeated blows on the workpiece placed midway between them.

**counterblow forging equipment.** A category of forging equipment in which two opposed rams are activated simultaneously, striking repeated blows on the workpiece at a midway point. Action is vertical (Fig. 96) or horizontal.

**counterblow hammer.** A forging hammer in which both the *ram* and the *anvil* are driven simultaneously toward each other by air or steam pistons (Fig. 96).

**counterboring.** Removal of material to enlarge a hole for part of its depth with a rotary, pilot guided, end cutting tool having two or more cutting lips and usually having straight or helical flutes for the passage of chips and the admission of a cutting fluid.

**counterelectrode.** In emission spectroscopy, the electrode that is used opposite to the self-electrode or supporting electrode and that is not composed of the sample to be analyzed. In voltammetry, the current between the working electrode and counterelectrodes is measured. See also *self-electrode, supporting electrode,* and *auxiliary electrode.*

**counterformal surfaces.** Surfaces whose centers of curvature are on the opposite sides of the interface, as in rolling-element bearings or gear teeth. In wear testing, this term

FIG. 96

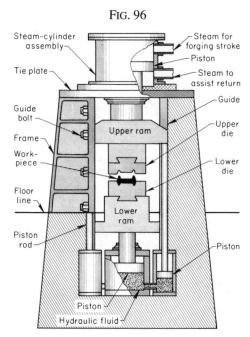

*Principal components of a vertical counterblow hammer with a steam-hydraulic actuating system*

is sometimes used to indicate that the test specimen surfaces are not conformal (for example, with a sphere-on-flat or flat block-on-rotating ring configuration). In such cases, the nominal contact area of at least one of the testpiece surfaces increases as the amount of wear increases. Contrast with *conformal surfaces*.

**counterblock.** A jog in the mating surfaces of dies to prevent lateral die shifting from side thrusts developed in forging irregularly shaped pieces.

**countersinking.** Beveling or tapering the work material around the periphery of a hole creating a concentric surface at an angle less than 90° with the centerline of the hole for the purpose of chamfering holes or recessing screw and rivet heads.

**couple.** See *galvanic corrosion*.

**coupling.** The degree of mutual interaction between two or more elements resulting from mechanical, acoustical, or electrical linkage.

**coupling agent.** In fabricating composites, any chemical designed to react with both the reinforcement and matrix phases of a composite material to form or promote a stronger bond at the interface.

**coupon.** A piece of material from which a test specimen is to be prepared—often an extra piece (as on a casting or forging) or a separate piece made for test purposes (such as a test weldment).

**covalent bond.** A bond in which two atoms share a pair of electrons. Contrast with *ionic bond*.

**cover core.** (1) In casting, a core set in place during the ramming of a mold to cover and complete a cavity partly formed by the withdrawal of a loose part of the pattern. Also used to form part or all of the cope surface of the mold cavity. (2) A core placed over another core to create a flat *parting line*.

**covered electrode.** A composite filler metal electrode consisting of a core of a bare electrode or metal cored electrode to which a covering sufficient to provide a slag layer on the weld metal has been applied. The covering may contain materials providing such functions as shielding from the atmosphere, deoxidation, and arc stabilization and can serve as a source of metallic additions to the weld.

**cover half.** The stationary half of a die-casting die.

**covering power.** (1) The ability of a solution to give satisfactory plating at very low current densities, a condition that exists in recesses and pits. This term suggests an ability to cover, but not necessarily to build up, a uniform coating, whereas *throwing power* suggests the ability to obtain a coating of uniform thickness on an irregularly shaped object. (2) The degree to which a porcelain enamel coating obscures the underlying surface. (3) The ability of a glaze to uniformly and completely cover the surface of the fired ceramic ware.

**cover plate** (welding eye protection). A removable pane of colorless glass, plastic-coated glass, or plastic that covers the filter plate and protects it from weld spatter, pitting, or scratching when used in a helmet, hood, or goggles.

**CO$_2$ process.** See *carbon dioxide process*.

**CO$_2$ welding.** See preferred term *gas metal arc welding*.

**"C" process.** See *Croning process*.

**crack.** (1) A fracture type discontinuity characterized by a sharp tip and high ratio of length and width to opening displacement. (2) A line of fracture without complete separation.

**crack branching.** The separation of a material into two or more segments.

**cracked gas.** A generic term for a gas mixture obtained by thermal decomposition, with or without catalysis, of a gaseous compound. Examples: cracked ammonia (NH$_3$) is a mixture of nitrogen and hydrogen, and cracked natural gas hydrocarbons such as methane (CH$_4$) are a mixture of carbon and hydrogen. Cracked ammonia is also known as *dissociated ammonia*.

**crack extension** (Δ*a*). An increase in crack size. See also *crack length*, *effective crack size*, *original crack size*, and *physical crack size*.

**crack-extension force** (*G*). The elastic energy per unit of new separation area that would be made available at the front of an ideal crack in an elastic solid during a virtual increment of forward crack extension. This definition is useful for either static cracks or running cracks. From past usage, crack extension force is commonly associated with linear-elastic methods of analysis. See also *J-integral*.

**crack-extension resistance ($K_R$).** A measure of the resistance of a material to *crack extension* expressed in terms of the *stress-intensity factor*, the *crack-extension force*, or values of *J* derived using the *J-integral* concept.

**crack growth.** Rate of propagation of a crack through a material due to a static or dynamic applied load.

**cracking.** In lubrication technology, that process of converting unwanted long-chain hydrocarbons to shorter molecules by thermal or catalytic action.

**crack length (depth) (*a*).** In *fatigue* and *stress-corrosion cracking*, the *physical crack size* used to determine the crack growth rate and the *stress-intensity factor*. For a compact-type specimen, crack length is measured from the line connecting the bearing points of load application. For a center-crack tension specimen, crack length is measured from the perpendicular bisector of the central crack. See also *crack size*.

**crack mouth opening displacement (CMOD).** See *crack opening displacement*.

**crack opening displacement (COD).** On a $K_{Ic}$ specimen, the opening displacement of the notch surfaces at the notch and in the direction perpendicular to the plane of the notch and the crack. The displacement at the tip is called the crack tip opening displacement (CTOD); at the mouth, it is called the crack mouth opening displacement (CMOD). See also *stress-intensity factor* for definition of $K_{Ic}$.

**crack plane orientation.** An identification of the plane and direction of a fracture in relation to product geometry. This identification is designated by a hyphenated code, the first letter(s) representing the direction normal to the crack plane and the second letter(s) designating the expected direction of crack propagation.

**crack size (*a*).** A lineal measure of a principal planar dimension of a crack. This measure is commonly used in the calculation of quantities descriptive of the stress and displacement fields. In practice, the value of crack size is obtained from procedures for measurement of *physical crack size*, *original crack size*, or *effective crack size*, as appropriate to the situation under consideration. See also *crack length (depth)*.

**crack tip opening displacement (CTOD).** See *crack opening displacement*.

**crack-tip plane strain.** A stress-strain field near a crack tip that approaches *plane strain* to the degree required by an empirical criterion.

**crank.** Forging shape generally in the form of a "U" with projections at more or less right angles to the upper terminals. Crank shapes are designated by the number of throws (for example, two-throw crank).

**crankpin bearing.** See *big-end bearing*.

FIG. 97

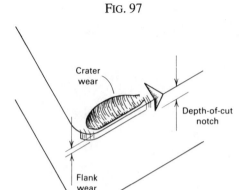

*Crater wear, flank wear, and depth-of-cut notch wear processes produced on a cemented carbide tool insert*

**crank press.** A mechanical press whose slides are actuated by a crankshaft.

**crater.** In arc welding, a depression at the termination of a weld bead or in the molten weld pool.

**crater crack.** A crack in the crater of a weld bead.

**crater fill time.** The time interval following weld time but prior to meltback time during which arc voltage or current reaches a preset value greater or less than welding values. Weld travel may or may not stop at this point.

**crater fill voltage.** The arc voltage value during crater fill time.

**cratering.** Depressions on coated plastic surfaces caused by excess lubricant. Cratering results when paint is too thin and later ruptures, leaving pinholes and other voids. Use of less thinner in the coating can reduce or eliminate cratering, as can less lubricant on the part.

**crater wear.** The wear that occurs on the rake face of a cutting tool due to contact with the material in the chip that is sliding along that face (Fig. 97).

FIG. 98

*Craze cracking on the outer surface of a stainless steel tube. ~4×*

FIG. 99

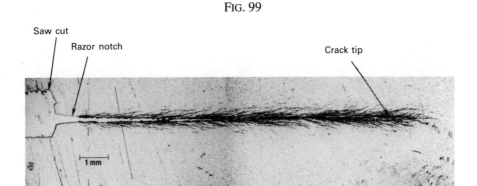

*Crazes surrounding a fatigue fracture in a polypropylene specimen*

**craze cracking**. Irregular surface cracking of a metal associated with thermal cycling (Fig. 98). This term is used more in the United Kingdom than in the United States, where the term checking is used instead.

**crazing** (ceramics). The cracking which occurs in fired glazes or other ceramic coatings due to critical tensile stresses.

**crazing** (plastics). Region of ultrafine cracks, which may extend in a network on or under the surface of a resin or plastic material (Fig. 99). May appear as a white band. Often found in a filament-wound pressure vessel or bottle. In many plastics, craze growth precedes crack growth because crazes are load bearing.

**creel**. In composites fabrication, a spool, along with its supporting structure, that holds the required number of roving balls or supply packages in a desired position for unwinding for the next processing step, that is, weaving, braiding, or filament winding.

**creep**. Time-dependent strain occurring under stress. The creep strain occurring at a diminishing rate is called primary creep; that occurring at a minimum and almost constant rate, secondary creep; and that occurring at an accelerating rate, tertiary creep (Fig. 100).

**creep-feed grinding**. See *grinding*.

**creep limit**. (1) The maximum stress that will cause less than a specified quantity of creep in a given time. (2) The maximum nominal stress under which the creep strain rate decreases continuously with time under constant load and at constant temperature. Sometimes used synonymously with *creep strength*.

**creep rate**. The slope of the creep-time curve at a given time. Deflection with time under a given static load.

**creep recovery**. The time-dependent decrease in strain in a solid, following the removal of force.

**creep-rupture embrittlement**. *Embrittlement* under creep conditions of, for example, aluminum alloys and steels that results in abnormally low rupture ductility. In aluminum alloys, iron in amounts above the solubility limit is known to cause such embrittlement; in steels, the phenomenon is

FIG. 100

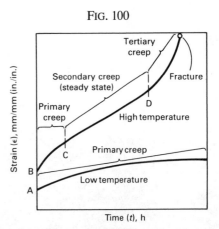

*High-temperature creep curve. A and B denote the elastic strain on loading; C denotes transition from primary (first-stage) to steady-state (second-stage) creep; D denotes transition from steady-state to tertiary (third-stage) creep.*

related to the amount of impurities (for example, phosphorus, sulfur, copper, arsenic, antimony, and tin) present. In either case, failure occurs by *intergranular cracking* of the embrittled material.

**creep-rupture strength**. The stress that causes fracture in a creep test at a given time, in a specified constant environment. This is sometimes referred to as the stress-rupture strength. In glass technology, this is termed the static fatigue strength.

**creep-rupture test**. A test in which progressive specimen deformation and the time for rupture are both measured. In general, deformation is much greater than that developed during a creep test. Also known as stress-rupture test.

**creep strain**. The time-dependent total strain (extension plus initial gage length) produced by applied stress during a creep test.

**creep strength**. The stress that will cause a given *creep strain* in a creep test at a given time in a specified constant environment.

**creep stress**. The constant load divided by the original cross-sectional area of the specimen.

**creep test**. A method of determining the extension of metals under a given load at a given temperature. The determination usually involves the plotting of time-elongation curves under constant load; a single test may extend over many months. The results are often expressed as the elongation (in millimeters or inches) per hour on a given gage length (e.g., 25 mm, or 1 in.).

**crescent crack**. Damage having the appearance of a crescent, produced in a glass surface by the frictive translation of a hard, blunt object across the glass surface. The crescent shape is concave toward the direction of translation on the damaged surface.

**crevice corrosion**. *Localized corrosion* of a metal surface at, or immediately adjacent to, an area that is shielded from full exposure to the environment because of close proximity between the metal and the surface of another material (Fig. 101).

FIG. 101

*Crevice corrosion at tube/tubesheet interface after 3 months of exposure in a natural seawater test*

**crimp**. The waviness of a fiber or fabric used in composites fabrication, which determines the capacity of fibers to cohere under light pressure. Measured by the number of crimps or waves per unit length.

**crimping**. The forming of relatively small *corrugations* in order to set down and lock a seam, to create an arc in a strip of metal, or to reduce an existing arc or diameter. See also *corrugating*.

**critical anodic current density**. The maximum anodic current density observed in the active region for a metal or alloy electrode that exhibits active-passive behavior in an environment.

**critical cooling rate**. The minimum rate of continuous cooling for preventing undesirable transformations. For steel, unless otherwise specified, it is the slowest rate at which austenite can be cooled from above critical temperature to prevent its transformation above the martensite start temperature.

**critical current density**. In an electrolytic process, a current density at which an abrupt change occurs in an operating variable or in the nature of an electrodeposit or electrode film.

**critical curve**. In a binary or higher order *phase diagram*, a line along which the phases of a heterogeneous equilibrium become identical.

**critical damping**. In dynamic mechanical measurement of plastics, that damping required for the borderline condition between oscillatory and nonoscillatory behavior.

**critical diameter**. Diameter of a steel bar that can be fully hardened with 50% martensite at its center.

**critical dimension**. A dimension on a part that must be held within the specified tolerance for the part to function in its application. A noncritical tolerance may be for cost or weight savings or for manufacturing convenience, but is not essential for the products.

**critical flaw size**. The size of a flaw (defect) in a structure that will cause failure at a particular stress level.

**critical humidity**. The *relative humidity* above which the atmospheric corrosion rate of some metals increases sharply.

**critical illumination**. In metallography, the formation of an image of the light source in the object field.

**critical length**. In composites fabrication, the minimum fiber length required for shear loading to its ultimate strength by the matrix.

**critical longitudinal stress**. The longitudinal stress necessary to cause internal slippage and separation of a spun yarn in a fiber-reinforced plastic. The stress necessary to overcome the interfiber friction developed as a result of twist.

**critical micelle concentration**. The concentration of a micelle at which the rate of increase of electrical conductance with increase in concentration levels off or proceeds at a much slower rate. See also *micelle*.

**critical pitting potential** ($E_{cp}$, $E_p$, $E_{pp}$). The lowest value of oxidizing potential at which pits nucleate and grow. It is dependent on the test method used.

**critical point**. (1) The temperature or pressure at which a change in crystal structure, phase, or physical properties occurs. Also termed *transformation temperature*. (2) In an equilibrium diagram, that combination of composition, temperature, and pressure at which the phases of an inhomogeneous system are in equilibrium.

**critical pressure**. That pressure above which the liquid and vapor states are no longer distinguishable.

**critical rake angle**. The rake angle at which the action of a

V-point tool changes from cutting to plowing. See also *rake*.

**critical shear stress**. The shear stress required to cause slip in a designated slip direction on a given slip plane. It is called the critical resolved shear stress if the shear stress is induced by tensile or compressive forces acting on the crystal.

**critical strain**. (1) In mechanical testing, the strain at the *yield point*. (2) The strain just sufficient to cause *recrystallization*; because the strain is small, usually only a few percent, recrystallization takes place from only a few nuclei, which produces a recrystallized structure consisting of very large grains.

**critical stress intensity factor**. See *stress-intensity factor*.

**critical surface**. In a ternary or higher order *phase diagram*, the area upon which the phases in equilibrium become identical.

**critical temperature**. That temperature above which the vapor phase cannot be condensed to liquid by an increase in pressure. Synonymous with *critical point* if pressure is constant.

**critical temperature range**. Synonymous with *transformation ranges*, which is the preferred term.

**Croning process**. In casting, a *shell molding process* that uses a phenolic resin binder. Sometimes referred to as C process or Chronizing.

**crop**. (1) An end portion of an ingot that is cut off as scrap. (2) To shear a bar or billet.

**cross breaks**. Same as *coil breaks*.

**cross-country mill**. A rolling mill in which the mill stands are so arranged that their tables are parallel with a transfer (or crossover) table connecting them. Such a mill is used for rolling structural shapes, rails, and any special form of bar stock not rolled in the ordinary bar mill.

**cross direction**. See *transverse direction*.

**crossed joint**. See *bridging*.

**cross forging**. Preliminary working of forging stock in flat dies to develop mechanical properties, particularly in the center portions of heavy sections.

**cross laminate**. In composites fabrication, a laminate in which some of the layers of material are oriented approximately at right angles to the remaining layers with respect to the grain, or strongest direction in tension. A bidirectional laminate. See also *parallel laminate*.

**cross link**. Intermolecular bonds produced between long-chain molecules in a material to increase molecular size and weight by chemical or electron bombardment, resulting in a change in physical properties in the material, usually improved properties.

**cross linking**. With thermosetting and certain thermoplastic polymers, the setting up of chemical links between the molecular chains. When extensive, as in most thermosetting resins, cross linking makes an infusible super-

molecule of all the chains. In rubbers, the cross linking is just enough to join all molecules into a network.

**cross-linking, degree of**. The fraction of cross-linked polymeric units in the entire system.

**cross-ply laminate**. A *laminate* with plies usually oriented at 0° and 90° only.

**cross rolling**. Rolling of metal or sheet or plate so that the direction of rolling is about 90° from the direction of a previous rolling.

**cross-roll straightener**. A machine having paired rolls of special design for straightening round bars or tubes, the pass being made with the work parallel to the axes of the rolls.

**cross-wire weld**. A weld made at the junction between crossed wires or bars.

**crosswise direction**. In testing of plastics, crosswise refers to the cutting of specimens and to the application of load. For rods and tubes, crosswise is any direction perpendicular to the long axis. For other shapes or materials that are stronger in one direction than in another, crosswise is the direction that is weaker. For materials that are equally strong in both directions, crosswise is an arbitrarily designated direction at right angles to the lengthwise direction.

**crowfoot satin**. In this type of composite fabric weave, there is a three-by-one interlacing; that is, a filling thread floats over three warp threads and then under one (Fig. 102). This type of fabric looks different on one side than the other. Fabrics with this weave are more pliable than either the

FIG. 102

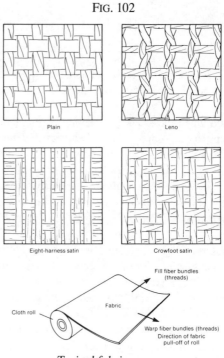

*Typical fabric weaves*

*plain* or *basket weave* and, consequently, are easier to form around curves.

**crown**. (1) The upper part (head) of a forming press frame. On hydraulic presses, the crown usually contains the cylinder; on mechanical presses, the crown contains the drive mechanism. See also *hydraulic press* and *mechanical press*. (2) A shape (crown) ground into a flat roll to ensure flatness of cold (and hot) rolled sheet and strip. (3) A contour on a sheet or roll where the thickness or diameter increases from edge to center.

**crucible furnace**. A melting or holding furnace in which the molten metal is contained in a pot-shaped (hemispherical) shell (Fig. 103). Electric heaters or fuel-fired burners outside the shell generate the heat that passes through the shell (crucible) to the molten metal.

FIG. 103

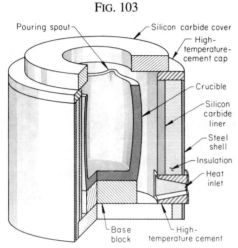

Pouring spout
Silicon carbide cover
High-temperature-cement cap
Crucible
Silicon carbide liner
Steel shell
Insulation
Heat inlet
Base block
High-temperature cement

*Typical lift-out version of a stationary crucible furnace*

**crush**. (1) Buckling or breaking of a section of a casting mold due to incorrect register when the mold is closed. (2) An indentation in the surface of a casting due to displacement of sand when the mold was closed. (3) In a split-journal bearing, the amount by which a bearing half extends above the horizontal split of the bore before it is assembled. Also known as *nip* in the United Kingdom.

**crush forming**. Shaping a grinding wheel by forcing a rotating metal roll into its face so as to reproduce the desired contour.

**crushing**. A process of comminuting large pieces of metal or ore into rough size fractions prior to grinding into powder. A typical machine for this operation is a jaw crusher.

**crushing test**. (1) A radial compressive test applied to tubing, sintered-metal bearings, or other similar products for determining radial crushing strength (maximum load in compression). (2) An axial compressive test for determining quality of tubing, such as soundness of weld in welded tubing.

**crush strip or bead**. In casting, an indentation in the *parting line* of a pattern plate that ensures that *cope* and *drag* will have good contact by producing a ridge of sand that crushes against the other surface of the mold or core.

**cryogenic treatment**. See *cold treatment*.

**cryopump**. A type of vacuum pump that relies on the condensation of gas molecules and atoms on internal surfaces of the pump, which are maintained at extremely low temperatures.

**crystal**. (1) A solid composed of atoms, ions, or molecules arranged in a pattern that is repetitive in three dimensions. (2) That form, or particle, or piece of a substance in which its atoms are distributed in one specific orderly geometrical array, called "lattice," essentially throughout. Crystals exhibit characteristic optical and other properties and growth or cleavage surfaces, in characteristic directions.

**crystal analysis**. A method for determining crystal structure, for example, the size and shape of the unit cell and the location of all atoms within the unit cell.

**crystalline**. That form of a substance which is comprised predominantly of (one or more) crystals, as opposed to glassy or amorphous.

**crystalline fracture**. A pattern of brightly reflecting crystal facets on the fracture surface of a polycrystalline metal, resulting from *cleavage fracture* of many individual crystals. Contrast with *fibrous fracture*, and *silky fracture*; see also *granular fracture*.

**crystalline plastic**. A material having an internal structure in which the atoms are arranged in an orderly three-dimensional configuration. More accurately referred to as a semicrystalline plastic because only a portion of the molecules are in crystalline form.

**crystallinity** (plastics). A regular arrangement of the atoms of a solid in space. In most polymers, including cellulose, this state is usually imperfectly achieved. The crystalline regions (ordered regions) are submicroscopic volumes in which there is some degree of regularity in the arrangement of the component molecules. In these regions there is sufficient geometric order to obtain definite x-ray diffraction patterns.

**crystallite**. A crystalline grain in a metal not bounded by habit planes.

**crystallization**. (1) The separation, usually from a liquid phase on cooling, of a solid crystalline phase. (2) The progressive process in which crystals are first nucleated (started) and then grown in size within a host medium which supplies their atoms. The host may be gas, liquid, or of another crystalline form.

**crystallographic cleavage**. The separation of a crystal along a plane of fixed orientation relative to the three-dimensional crystal structure within which the separation process occurs, with the separation process causing the newly formed surfaces to move away from one another in directions containing major components of motion perpendicular to the fixed plane.

**crystal orientation**. See *orientation*.

**crystal system**. One of seven groups into which all crystals may be divided; triclinic, monoclinic, orthorhombic, hexagonal, rhombohedral, tetragonal, and cubic.

**C-stage**. In processing of plastics, the final stage in the reaction of certain thermosetting resins in which the material is practically insoluble and infusible. Also called resite. The resin in a fully cured thermoset molding is in this stage. See also *A-stage* and *B-stage*.

**CTE mismatch**. The difference in the coefficients of thermal expansion (CTEs) of two materials or components joined together, producing strains and stresses at the joining interfaces or in the attachment structures (solder joints, leads, and so on).

**cube texture**. A texture found in wrought metals in the cubic system in which nearly all the crystal grains have a plane of the type (100) parallel or nearly parallel to the plane of working and a direction of the type [100] parallel or nearly parallel to the direction of elongation.

**cubic**. Having three mutually perpendicular axes of equal length.

**cubic boron nitride (CBN)**. An extremely hard (see the table accompanying the term *boron carbide*) ceramic material synthesized by high-pressure sintering of hexagonal *boron nitride* (Fig. 104). CBN is used in the machining and grinding of ferrous materials such as tool steels, cast irons, hardfacing alloys, and surface-hardened steels. See also *superabrasives* (Technical Brief 51).

**cubic plane**. A plane perpendicular to any one of the three crystallographic axes of the cubic (isometric) system; the *Miller indices* are {100}.

**cull**. Plastic material in a transfer chamber after the mold has been filled. Unless there is a slight excess in the charge, the operator cannot be sure the cavity is filled.

Charge is generally regulated to control thickness of the cull.

**cumulative erosion-time curve**. A plot of cumulative erosion versus cumulative exposure duration, usually obtained by periodic interruption of the erosion test and weighing of the specimen. This is the primary record of an erosion test used to derive other characteristics such as incubation period, maximum erosion rate, terminal erosion rate, and the erosion rate-time curve.

**cup**. (1) A sheet metal part; the product of the first drawing operation. (2) Any cylindrical part or shell closed at one end.

**cupellation**. Oxidation of molten lead containing gold and silver to produce lead oxide, thereby separating the precious metals from the base metal.

**cup fracture (cup-and-cone fracture)**. A mixed-mode fracture, often seen in tensile-test specimens of a ductile material, where the central portion undergoes *plane-strain* fracture and the surrounding region undergoes *plane-stress* fracture. It is called a cup fracture (or cup-and-cone fracture) because one of the mating fracture surfaces looks like a miniature cup—that is, it has a central depressed flat-face region surrounded by a shear lip; the other fracture surface looks like a miniature truncated cone. The figure accompanying the term *brittle fracture* illustrates a cup-and-cone fracture.

**cupping**. (1) The first step in deep drawing. (2) Fracture of severely worked rods or wire where one end has the appearance of a cup and the other that of a cone.

**cupping test**. A mechanical test used to determine the ductility and stretching properties of sheet metal. It consists of measuring the maximum part depth that can be formed before fracture. The test is typically carried out by stretching the test piece clamped at its edges into a circular die using a punch with a hemispherical end. See also *Erichsen test*, *Olsen ductility test*, and *Swift cup test*.

Fɪɢ. 104

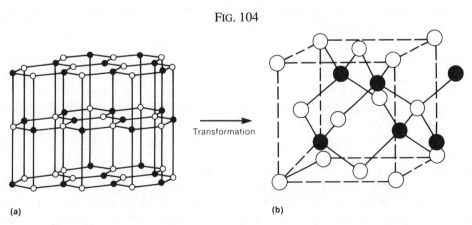

(a)             Transformation             (b)

*Arrangement of boron and nitrogen atoms. (a) The hexagonal arrangement of boron nitride. (b) The cubic arrangement of boron nitride*

FIG. 105

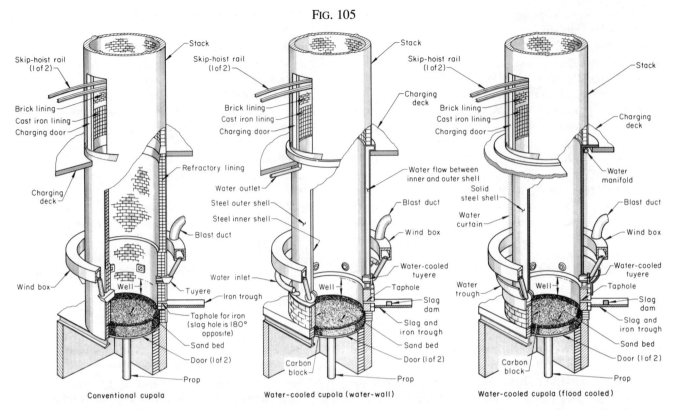

Conventional cupola          Water-cooled cupola (water-wall)          Water-cooled cupola (flood cooled)

*Sectional views of conventional and water-cooled cupolas. The conventional type shown is refractory lined. Water-cooled types incorporate either an enclosed jacket or an open cascade flow.*

**cupola**. A cylindrical vertical furnace for melting metal, especially cast iron, by having the charge come in contact with the hot fuel, usually metallurgical coke (Fig. 105).

**cure**. (1) To change the physical properties of a material (usually from a liquid to a solid) by chemical reaction or by the action of heat and catalysts, alone or in combination, with or without pressure. (2) To irreversibly change, usually at elevated temperatures, the properties of a thermosetting resin by chemical reaction, that is, by condensation, ring closure, or addition. Cure may be accomplished by the addition of curing (cross linking) agents, with or without heat and pressure.

**cure cycle**. The time/temperature/pressure cycle used to cure a thermosetting resin system or prepreg.

**cure monitoring, electrical**. Use of electrical techniques to detect changes in the electrical properties and/or mobility of the resin molecules during cure. A measuring of resin cure.

**cure stress**. A residual internal stress produced during the curing cycle of a composite structure. Normally, these stresses originate when different components of a wet lay-up have different thermal coefficients of expansion.

**Curie temperature**. The temperature marking the transition between ferromagnetism and paramagnetism, or between the ferroelectric phase and the paraelectric phase. Also

known as Curie point. See also *ferromagnetism* and *paramagnetism*.

**curing agent**. A catalytic or reactive agent that causes cross linking of a plastic. Also called a hardener.

**curing temperature**. The temperature to which an adhesive or an assembly is subjected to cure the adhesive. The temperature attained by the adhesive in the process of curing (adhesive curing temperature) may differ from the temperature of the atmosphere surrounding the assembly (assembly curing temperature). See also *drying temperature* and *setting temperature*.

**curing time**. In plastic part making, the time between the instant of cessation of relative movement between the moving parts of a mold and the instant that pressure is released. Also called molding time.

**curing time** (no bake). In foundry practice, the period of time needed before a sand mass reaches maximum hardness.

**curling**. Rounding the edge of sheet metal into a closed or partly closed loop.

**current**. The net transfer of electric charge per unit time. Also called electric current. See also *current density*.

**current-carrying capacity**. The maximum current that can be carried continuously, under specified conditions, by a conductor without causing objectionable degradation of electrical or mechanical properties.

**current decay**. In spot, seam, or projection welding, the controlled reduction of the welding current from its peak amplitude to a lower value to prevent excessively rapid cooling of the weld nugget.

**current density**. The current flowing to or from a unit area of an electrode surface.

**current efficiency**. (1) The ratio of the electrochemical equivalent current density for a specific reaction to the total applied current density. (2) The proportion of current used in a given process to accomplish a desired result; in electroplating, the proportion used in depositing or dissolving metal.

**curtain coating**. A method of coating that may be employed with low-viscosity resins or solutions, suspensions, or emulsions of resins in which the substrate to be coated is passed through and perpendicular to a freely falling liquid curtain (or waterfall). The flow rate of the falling liquid and the linear speed of the substrate passing through the curtain are coordinated in accordance with the thickness of coating desired.

**curvature of field**. In metallography, a property of a lens that causes the image of a plane to be focused into a curved surface instead of a plane.

**cushion**. Same as *die cushion*.

**cut**. In lubricant technology, a product or fraction obtained by distillation within a specified temperature range.

**cut** (foundry practice). (1) To recondition molding sand by mixing on the floor with a shovel or blade-type machine. (2) To form the sprue cavity in a mold. (3) Defect in a casting resulting from erosion of the sand by metal flowing over the mold or cored surface.

**cut-and-carry method**. Stamping method wherein the part remains attached to the strip or is forced back into the strip to be fed through the succeeding stations of a progressive die.

**cut edge**. A mechanically sheared edge obtained by slitting, shearing, or blanking.

**cut layers**. With laminated plastics, a condition of the surface of machined or ground rods and tubes and of sanded sheets in which cut edges of the surface layer or lower laminations are revealed.

**cut-off** (casting). Removing a casting from the sprue by refractory wheel or saw, arc-air torch, or gas torch.

**cut-off** (metal forming). A pair of blades positioned in dies or equipment (or a section of the die milled to produce the same effect as inserted blades) used to separate the forging from the bar after forging operations are completed. Used only when forgings are produced from relatively long bars instead of from individual, precut multiples or blanks. See also *blank* and *multiple*.

**cut-off** (plastics). The line where the two halves of a mold come together. Also called flash groove and pinch-off.

**cutoff wheel**. A thin abrasive wheel for severing or slotting any material or part.

**cutting attachment**. A device for converting an oxyfuel gas welding torch into an oxygen cutting torch.

**cutting down**. Removing roughness or irregularities of a metal surface by abrasive action.

**cutting edge**. The leading edge of a cutting tool (such as a lathe tool, drill or milling cutter) where a line of contact is made with the work during machining. See also the figure accompanying the term *single-point tool*.

**cutting fluid**. A fluid used in metal cutting to improve finish, tool life, or dimensional accuracy. On being flowed over the tool and work, the fluid reduces friction, the heat generated, and tool wear, and prevents galling. It conducts the heat away from the point of generation and also serves to wash the *chips* away.

**cutting head**. The part of a cutting machine or automatic cutting equipment in which a cutting torch or tip is incorporated.

**cutting nozzle**. See preferred term *cutting tip*.

**cutting process**. A process which brings about the severing or removal of metals. See also *arc cutting* and *oxygen cutting*.

**cutting speed**. The linear or peripheral speed of relative motion between the tool and workpiece in the principal direction of cutting.

**cutting tip**. That part of an oxygen cutting torch from which the gases issue.

**cutting torch** (oxyfuel gas). A device used for directing the preheating flame produced by the controlled combustion of fuel gases and to direct and control the cutting oxygen (Fig. 106).

FIG. 106

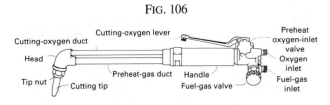

*Typical oxyfuel gas cutting torch*

**cutting torch** (plasma arc). A device used for plasma arc cutting to control the position of the electrode, to transfer current to the arc, and to direct the flow of plasma and shielding gas (Fig. 107).

**cutting wear**. See *abrasive wear*.

**CVD carbon**. See *chemical vapor deposited (CVD) carbon*.

**cyanate resins**. Thermosetting resins that are derived from bisphenols or polyphenols, and are available as monomers, oligomers, blends, and solutions. Also known as cyanate esters, cyanic esters, and triazine esters.

**cyanic copper**. Copper electrodeposited from an alkali-cyanide solution containing a complex ion made up of

FIG. 107

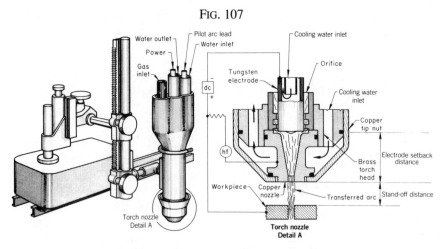

*Components of a plasma arc cutting torch*

univalent copper and the cyanide radical; also the solution itself.

**cyanide slimes**. Finely divided metallic precipitates that are formed when precious metals are extracted from their ores using cyanide solutions.

**cyaniding**. A case-hardening process in which a ferrous material is heated above the lower transformation temperature range in a molten salt containing cyanide to cause simultaneous absorption of carbon and nitrogen at the surface and, by diffusion, create a concentration gradient. *Quench hardening* completes the process.

**cyanoacrylate**. A thermoplastic monomer adhesive characterized by excellent polymerizing and bonding strength.

**cycle**. One complete operation of a plastic molding press from closing time, for example, in one cycle to closing time in the next cycle.

**cycle (N)**. In fatigue, one complete sequence of values of applied load that is repeated periodically. See also *S-N curve*.

**cycle annealing**. An annealing process employing a pre-determined and closely controlled time-temperature cycle to produce specific properties or microstructures.

**cyclic hydrocarbons**. Cyclic or ring compounds; benzene ($C_6H_6$) is a classic example.

**cyclic load**. (1) Repetitive loading, as with regularly recurring stresses on a part, that sometimes leads to fatigue fracture. (2) Loads that change value by following a regular repeating sequence of change.

**cyclic stressing**. See *cyclic load*.

**cyclone**. In powder metallurgy, a collector of fractions of a powder from air, water, or other gases or liquids; the device operates with the aid of centrifugal force that acts on the powder suspension in the fluid.

**cylinder**. A portable container used for transportation and storage of a compressed gas. Used extensively in welding and cutting operations.

**cylinder manifold**. See preferred term *manifold*.

**cylindrical grinding**. Grinding the outer cylindrical surface of a rotating part.

**cylindrical land**. *Land* having zero relief.

# D

**dam**. In composites fabrication, a boundary support or ridge used to prevent excessive edge bleeding or resin runout of a laminate and to prevent crowning of the bag during cure.

**damage tolerance**. (1) A design measure of crack growth rate. Cracks in damage-tolerant designed structures are not permitted to grow to critical size during expected service life. (2) The ability of a part component, such as an aerospace engine, to resist failure due to the presence of flaws, cracks, or other damage for a specified period of usage. The damage tolerance approach is used extensively in the aerospace industry.

**damping**. The loss in energy, as dissipated heat, that results when a material or material system is subjected to an oscillatory load or displacement.

FIG. 108

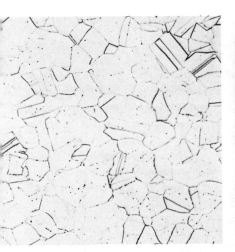 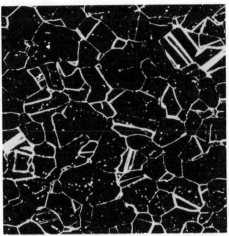 

*Comparison of bright-field (left), dark-field (center), and differential interference contrast (right) illumination. The specimen is an austenitic stainless steel. 400×*

**damping capacity**. The ability of a material to absorb vibration (cyclical stresses) by internal friction, converting the mechanical energy into heat.

**dangler**. The flexible electrode used in *barrel plating* to conduct current to the work.

**Danner process**. A mechanical process for continuously drawing glass cane or tubing from a rotating mandrel.

**dark-field illumination**. The illumination of an object such that it appears bright and the surrounding field dark (Fig. 108). This results from illuminating the object with rays of sufficient obliquity so that none can enter the objective directly. In electron microscopy, the image is formed using only electrons scattered by the object. Contrast with *bright-field illumination*.

**dark reaction**. In adhesives technology, the gelling of material in storage without light initiation.

**dash pot**. A device used in hydraulic systems for damping down vibration. It consists of a piston attached to the part to be damped and fitted into a vessel containing fluid or air. It absorbs shocks by reducing the rate of change in the momentum of moving parts of machinery.

**daubing**. In foundry practice, filling of cracks in molds or cores by specially prepared pastes or coatings to prevent penetration of metal into these cracks during pouring.

**daylight**. The distance, in the open position, between the moving and the fixed tables or the platens of a hydraulic press. In the case of a multiplaten press, daylight is the distance between adjacent platens. Daylight provides space for removal of the molded/formed part from the mold/die.

**dc casting**. Same as *direct chill casting*.

**dc intermittent noncapacitive arc**. A low-voltage electrical discharge used in spectrochemical analysis to vaporize the sample material. Each current pulse has the same polarity as the previous one and lasts for less than 0.1 s. See also *ac noncapacitive arc*.

**dc plasma excitation**. See *plasma-jet excitation*.

**deactivation**. The process of prior removal of the active corrosive constituents, usually oxygen, from a corrosive liquid by controlled corrosion of expendable metal or by other chemical means, thereby making the liquid less corrosive.

**dead-burned**. Term applied to ceramic materials that have been fired to a temperature sufficiently high to render them relatively resistant to moisture and contraction.

**dead center**. (1) A stationary center to hold rotating work. (2) Either of the two points in the path of a moving crank or connecting rod that lie at the ends of its stroke.

**dead roast**. A *roasting* process for complete elimination of sulfur. Also known as sweet roast.

**dead soft**. A *temper* of nonferrous alloys and some ferrous alloys corresponding to the condition of minimum hardness and tensile strength produced by *full annealing*.

**dead time**. The total time during which a spectrometer is processing information and is unavailable to accept input data.

**dealloying**. The selective corrosion of one or more components of a solid solution alloy. Also called parting or *selective leaching*. See also *decarburization, decobaltification, denickelification, dezincification*, and *graphitic-corrosion*.

**debond**. In composites, a deliberate separation of a bonded joint or interface, usually for repair or rework purposes. Also, an unbonded or nonadhered region; a separation at the fiber-matrix interface due to strain incompatibility. In the United Kingdom, the term often refers to accidental damage. See also *disbond* and *delamination*.

**debossed**. An indented or depressed design or lettering that is molded into a plastic article so that it is below the main outside surface of that article.

**debris**. See *wear debris*.

**debulking**. Compacting of a thick composite laminate under moderate heat and pressure and/or vacuum to remove most of the air, to ensure seating on the tool, and to prevent wrinkles.

**deburring**. Removing burrs, sharp edges, or fins from metal parts by filing, grinding, or rolling the work in a barrel containing abrasives suspended in a suitable liquid medium. Sometimes called burring.

**Debye ring**. A continuous circle, concentric about the undeviated beam, produced by monochromatic x-ray diffraction from a randomly oriented crystalline powder. An analogous effect is obtained using electron diffraction.

**Debye-Scherrer method**. A method of x-ray diffraction using monochromatic radiation and a polycrystalline specimen mounted on the axis of a cylindrical strip of film. See also *powder method*.

**decalescence**. A phenomenon, associated with the transformation of alpha iron to gamma iron on the heating (superheating) of iron or steel, revealed by the darkening of the metal surface owing to the sudden decrease in temperature caused by the fast absorption of the latent heat of transformation. Contrast with *recalescence*.

**decarburization**. Loss of carbon from the surface layer of a carbon-containing alloy due to reaction with one or more chemical substances in a medium that contacts the surface (Fig. 109). See also *dealloying*.

**decay constant** ($\lambda$). The constant in the radioactive decay law $dN = -\lambda N dt$, where $N$ is the number of radioactive nuclei present at time $t$. The decay constant is related to half-life $t_{1/2}$ by the expression $t_{1/2} = \ln 2/\lambda$. See also *half-life*.

**deceleration period**. In cavitation or impingement erosion, the stage following the *acceleration period* or the *maximum rate period* (if any), during which the erosion rate has an overall decreasing trend although fluctuations may be superimposed on it.

**decibel (dB)**. Unit that expresses differences of power level. Example: The decibel is ten times the common logarithm of the power ratio. It is used to express power gain in amplifiers or power loss in passive circuits or cables.

**deckle rod**. In forming plastics, a small rod, or similar device, inserted at each end of the extrusion coating die that is used to adjust the length of the die opening.

**decobaltification**. Corrosion in which cobalt is selectively leached from cobalt-base alloys, such as Stellite, or from cemented carbides. See also *dealloying* and *selective leaching*.

**decohesive rupture**. A *brittle fracture* that exhibits little or no bulk plastic deformation and does not occur by dimple rupture, cleavage, or fatigue. This type of fracture is generally the result of a reactive environment or a unique microstructure and is associated almost exclusively with rupture along grain boundaries (Fig. 110).

**decomposition**. Separation of a compound into its chemical elements or components.

**decomposition potential (or voltage)**. The *potential* of a metal surface necessary to decompose the electrolyte of a cell or a component thereof.

FIG. 109

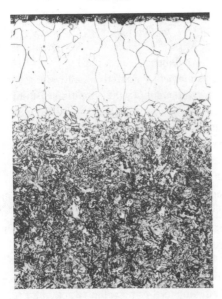

*Decarburization of a spring steel*

FIG. 110

50 µm

*Intergranular decohesive rupture of a nickel-copper specimen that failed in liquid mercury*

FIG. 111

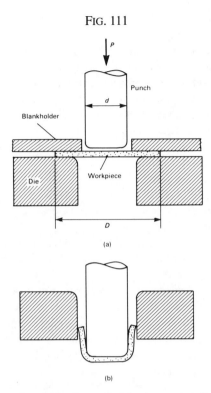

(a)

(b)

*Deep drawing of a cylindrical cup. (a) Before drawing. (b) After drawing*

**decoration** (of dislocations). Segregation of solute atoms to the line of a dislocation in a crystal. In ferrite, the dislocations may be decorated with carbon or nitrogen atoms.

**deep drawing**. Forming deeply recessed parts by forcing sheet metal to undergo plastic flow between dies, usually without substantial thinning of the sheet (Fig. 111).

**deep-draw mold** (plastics). A mold having a core that is appreciably longer than the wall thickness.

**deep etching**. In metallography, *macroetching*, especially for steels, to determine the overall character of the material, that is, the presence of imperfections, such as seams, forging bursts, shrinkage-void remnants, cracks, and coring.

**deep groundbed**. One or more *anodes* installed vertically at a nominal depth of 1.5 m (50 ft) or more below the earth's surface in a drilled hole for the purpose of supplying *cathodic protection* for an underground or submerged metallic structure. See also *groundbed.*

**defect**. (1) A discontinuity whose size, shape, orientation, or location makes it detrimental to the useful service of the part in which it occurs. (2) A discontinuity or discontinuities which by nature or accumulated effect (for example, total crack length) render a part or product unable to meet minimum applicable acceptance standards or specifications. This term designates rejectability. See also *discontinuity* and *flaw.*

**defective**. A quality control term, describing a unit of product or service containing at least one *defect*, or having several lesser imperfections that, in combination, cause the unit not to fulfill its anticipated function. The term defective is not synonymous with *nonconforming* (or rejectable) and should be applied only to those units incapable of performing their anticipated functions.

**defective weld**. A weld containing one or more defects.

**define** (x-rays). To limit a beam of x-rays by passage through apertures to obtain a parallel, divergent, or convergent beam.

**deflashing**. A finishing technique used to remove the flash (excess, unwanted material) on a plastic molding.

**deflection**. In metal forming and forging, the amount of deviation from a straight line or plane when a force is applied to a press member. Generally used to specify the allowable bending of the bed, slide, or frame at rated capacity with a load of predetermined distribution.

**deflection temperature under load (DTUL)**. In testing of plastics, the temperature at which a simple cantilever beam deflects a given amount under load. Formerly called heat distortion temperature.

**deflocculant**. An electrolyte adsorbed on colloidal particles in suspension that charges the particles to create repulsion forces which maintain the particles in a dispersed state, thus reducing the viscosity of the suspension.

**deformability**. See *conformability.*

**deformation**. A change in the form of a body due to stress, thermal change, change in moisture, or other causes. Measured in units of length.

**deformation bands**. Parts of a crystal that have rotated differently during deformation to produce bands of varied orientation without individual grains (Fig. 112).

**deformation curve**. See *stress-strain curve.*

FIG. 112

*Deformation bands on (100) surface of a single crystal of Co-8Fe alloy deformed 44%. 250×*

**deformation limit**. In *drawing*, the limit of deformation is reached when the load required to deform the flange becomes greater than the load-carrying capacity of the cup wall. The deformation limit (limiting drawing ratio, LDR) is defined as the ratio of the maximum blank diameter that can be drawn into a cup without failure, to the diameter of the punch.

**deformation point**. The temperature observed during the measurement of expansivity of glass by the interferometer method at which viscous flow exactly counteracts thermal expansion. The deformation point generally corresponds to a viscosity in the range of 1010 to 1011 Pa · s.

**deformation under load**. The dimensional change of a material under load for a specified time following the instantaneous elastic deformation caused by the initial application of the load. See also *cold flow* and *creep*.

**deformation wear**. Sliding wear involving plastic deformation of the wearing surface. Many forms of wear involve plastic deformation, so this term is imprecise and should not be used.

**degasification**. See *degassing*.

**degasifier**. A substance that can be added to molten metal to remove soluble gases that might otherwise be occluded or entrapped in the metal during solidification.

**degassing** (metals). (1) A chemical reaction resulting from a compound added to molten metal to remove gases from the metal. Inert gases are often used in this operation. (2) A fluxing procedure used for aluminum alloys in which nitrogen, chlorine, chlorine and nitrogen, and chlorine and argon are bubbled up through the metal to remove dissolved hydrogen gases and oxides from the alloy. See also *flux*.

**degassing** (plastics). Opening and closing of a plastics mold to allow gases to escape during molding.

**degradation**. A deleterious change in the chemical structure, physical properties, or appearance of a material.

**degrease**. To remove oil and grease from *adherend* surfaces.

**degreasing**. Removing oil or grease from a surface. See also *vapor degreasing*.

**degree of polymerization**. Number of structural units, or mers, in the average polymer molecule in a sample measure of molecular weight.

**degree of saturation**. The ratio of the weight of water vapor associated with a pound of dry air to the weight of water vapor associated with a pound of dry air saturated at the same temperature.

**degrees of freedom**. In metallography, the number of independent variables, such as temperature, pressure, or concentration, within the phases present that may be adjusted independently without causing a phase change in an alloy system at equilibrium.

**delamination**. The separation of layers in a laminate because

of failure of the adhesive, either in the adhesive itself or at the interface between the adhesive and the adherend.

**delamination wear**. A wear process in which thin layers of material are formed and removed from the wear surface.

**delayed yield**. A phenomenon involving a delay in time between the application of a stress and the occurrence of the corresponding yield-point strain.

**deliquescence**. The absorption of atmospheric water vapor by a crystalline solid until the crystal eventually dissolves into a saturated solution.

**delta ferrite**. See *ferrite*.

**delta iron**. Solid phase of pure iron that is stable from 1400 to 1539 °C (2550 to 2800 °F) and possesses the body-centered cubic lattice.

**delube**. The removal of a lubricant from a powder compact, usually by burnout, or alternatively by treatment with a chemical solvent.

**Demarest process**. A *fluid forming* process in which cylindrical and conical sheet metal parts are formed by a modified rubber bulging punch. The punch, equipped with a hydraulic cell, is placed inside the workpiece, which in turn is placed inside the die (Fig. 113). Hydraulic pressure expands the punch.

FIG. 113

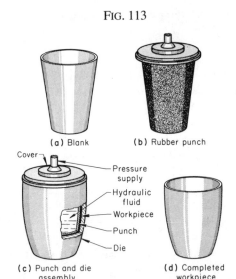

*Forming of a fuel-tank section from a blank using the Demarest process*

**demixing**. (1) The undesirable separation of one or more constituents of a powder mixture. (2) Segregation due to overmixing.

**dendrite**. A crystal that has a treelike branching pattern, being most evident in cast metals slowly cooled through the solidification range (Fig. 114).

**dendritic powder**. Particles usually of electrolytic origin typically having the appearance of a pine tree.

**dendritic segregation**. Inhomogeneous distribution of alloying elements through the arms of dendrites.

FIG. 114

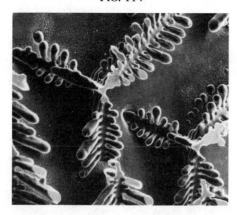

*Scanning electron micrograph of dendrites in a Cu-10Co casting. 150×*

**denickelification.** Corrosion in which nickel is selectively leached from nickel-containing alloys. Most commonly observed in copper-nickel alloys after extended service in fresh water. See also *dealloying* and *selective leaching.*

**denier.** A yarn and filament numbering system in which the yarn number is numerically equal to the weight in grams of 9000 meters (1 denier = $1.111 \times 10^{-7}$ kg/m). Used for continuous filaments. The lower the denier, the finer the yarn.

**densification process.** Consolidation of a loose or bulky material.

**density, absolute.** The mass per unit volume of a solid material, expressed in $g/cm^3$, $kg/m^3$, or $lb/ft^3$.

**density, dry.** In powder metallurgy, the mass per unit volume of an unimpregnated sintered part.

**density ratio.** The ratio of the determined density of a powder compact to the absolute density of metal of the same composition, usually expressed as a percentage. Also referred to as percent theoretical density.

**density, wet.** In powder metallurgy, the mass per unit volume of a sintered part impregnated with oil or other nonmetallic material.

**deoxidation.** Removal of excess oxygen from the molten metal; usually accomplished by adding materials with a high affinity for oxygen.

**deoxidation products.** Those nonmetallic inclusions that form as a result of adding deoxidizing agents to molten metal.

**deoxidized copper.** Copper from which cuprous oxide has been removed by adding a *deoxidizer*, such as phosphorus, to the molten bath.

**deoxidizer.** A substance that can be added to molten metal to remove either free or combined oxygen.

**deoxidizing.** (1) The removal of oxygen from molten metals through the use of a suitable *deoxidizer*. (2) Sometimes refers to the removal of undesirable elements other than oxygen through the introduction of elements or compounds that readily react with them. (3) In metal finishing, the removal of oxide films from metal surfaces by chemical or electrochemical reaction.

**dephosphorization.** The elimination of phosphorus from molten steel.

**depletion.** Selective removal of one component of an alloy, usually from the surface or preferentially from grain-boundary regions. See also *dealloying.*

**depolarization.** A decrease in the *polarization* of an electrode.

**depolarizer.** A substance that produces *depolarization.*

**deposit** (thermal spraying). See preferred term *spray deposit.*

**deposit attack.** See *deposit corrosion.*

**deposit corrosion.** Corrosion occurring under or around a discontinuous deposit on a metallic surface. Also called poultice corrosion.

**deposited metal.** Filler metal that has been added during a welding operation.

**deposition.** The process of applying a material to a base by means of vacuum, electrical, chemical, screening, or vapor methods, often with the assistance of a temperature and pressure container.

**deposition efficiency** (arc welding). The ratio of the weight of deposited metal to the net weight of filler metal consumed, exclusive of stubs.

**deposition efficiency** (thermal spraying). The ratio, usually expressed in percent, of the weight of spray deposit to the weight of the material sprayed.

**deposition rate** (thermal spraying). The weight of material deposited in a unit of time. It is usually expressed as kilograms per hour (kg/h) or pounds per hour (lb/h).

**deposition sequence.** The order in which the increments of weld metal are deposited. See also *longitudinal sequence, buildup sequence,* and *pass sequence.*

**deposit sequence.** See preferred term *deposition sequence.*

**depth dose.** The variation of absorbed dose with distance from the incident surface of a material exposed to radiation. Depth-dose profiles give information about the distribution of absorbed energy in a specific material.

**depth of cut.** The thickness of material removed from a workpiece in a single machining part.

**depth of field.** The depth in the subject over which features can be seen to be acceptably in focus in the final image produced by a microscope.

**depth of fusion.** The distance that fusion extends into the base metal or previous pass from the surface melted during welding.

**depth of penetration** (materials characterization). In various analytical techniques, the distance the probing radiation penetrates beneath the surface of a sample. See also *excitation volume.*

**depth of penetration** (welding). See *joint penetration* and *root penetration*.

**derby**. A massive piece (intermediate in size, extending to more than 45 kg, or 100 lb, and usually cylindrical) of primary metal made by bomb reduction (such as uranium from uranium tetrafluoride reduced with magnesium). Compare with *biscuit* and *dingot*.

**descaling**. (1) Removing the thick layer of oxides formed on some metals at elevated temperatures. (2) A chemical or mechanical process for removing scale or investment material from castings.

**deseaming**. Analogous to *chipping*, the surface imperfections being removed by gas cutting.

**desiccant**. A substance that can be used to dry materials because of its affinity for water.

**design allowables**. Statistically defined (by a test program) material property allowable strengths, usually referring to stress or strain. See also *A-basis*, *B-basis*, *S-basis*, and *typical-basis*.

**desizing**. The process of eliminating sizing, which is generally starch, from *gray* (also *greige*) *goods* before applying special finishes or bleaches (for yarn such as glass or cotton). Also, removing lubricant *size* following the weaving of a cloth.

**desorption**. A process in which an absorbed material is released from another material. See also *absorption*, *adsorption*, and *chemisorption*.

**destaticization**. A treatment of plastic materials that minimizes the effects of static electricity on the surface either by treating the surface with specific materials or by incorporating materials in the molding compound. Minimization of surface static electricity prevents dust and dirt from being attracted to and/or clinging to the surface of the article.

**desulfurizing**. The removal of sulfur from molten metal by reaction with a suitable slag or by the addition of suitable compounds.

**detection limit**. In an analytical method, the lowest mass or concentration of an analyte that can be measured.

**detector, x-ray**. Sensor array used to measure the x-ray intensity. Typical detectors are high-pressure gas ionization, scintillator-photodiode detector arrays, and scintillator-photomultiplier tubes.

**detergent**. A chemical substance, generally used in aqueous solution, that removes *soil*.

**detergent additive**. In lubrication technology, a surface-active additive that helps to keep solid particles suspended in an oil.

**detergent oil**. A heavy-duty oil containing a detergent additive. Detergent oils are used mainly in combustion engines.

**determination**. The ascertainment of the quantity or con-

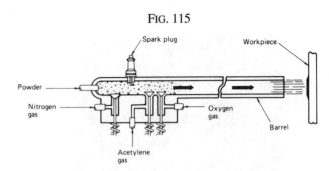

FIG. 115

*Schematic of detonation gun hardfacing process*

centration of a specific substance in a sample. See also *analysis*.

**detonation flame spraying**. A thermal spraying process variation in which the controlled explosion of a mixture of fuel gas, oxygen, and powdered coating material is utilized to melt and propel the material to the workpiece (Fig. 115).

**detritus**. See *wear debris*.

**deuteron**. The nucleus of the atom of heavy hydrogen, deuterium. The deuteron is composed of a proton and a neutron; it is the simplest multinucleon nucleus. Deuterons are used as projectiles in many nuclear bombardment experiments. See also *neutron* and *proton*.

**developed blank**. A sheet metal blank that yields a finished part without trimming or with the least amount of trimming.

**deviation, x-ray**. The angle between the diffracted beam and the transmitted incident beam. It is equal to twice the Bragg angle θ.

**devitrification**. (1) Crystallization in glass. (2) The formation of crystals (seeds) in a glass melt, usually occurring when the melt is too cold. These crystals can appear as defects in glass fibers.

**devitrify**. (1) To convert, partially or completely, from a glassy to a crystalline state; usually by controlled heating. (2) To deprive of glassy luster and transparency or to change from a vitreous to a crystalline condition.

**dewar flask**. A vessel having double walls, the space between being evacuated to prevent the transfer of heat and the surfaces facing the vacuum being heat reflective; used to hold liquid gases and to study low-temperature phenomena.

**dewaxing**. In casting, the process of removing the expendable wax pattern from an investment mold or shell mold; usually accomplished by melting out the application of heat or dissolving the wax with an appropriate solvent.

**dewetting**. (1) The development and formation of a nonwetting condition after wetting has already commenced. (2) A condition that results when molten solder has coated a surface and then receded leaving irregularly shaped mounds of solder separated by areas covered with a thin solder film. The basis metal is not exposed, however.

**dewpoint analyzer**. An atmosphere monitoring device that measures the partial pressure of water vapor in an atmosphere.

**dewpoint temperature**. The temperature at which condensation of water vapor in a space begins for a given state of humidity and pressure as the vapor temperature is reduced; the temperature corresponding to saturation (100% relative humidity) for a given absolute humidity at constant pressure.

**dezincification**. Corrosion in which zinc is selectively leached from zinc-containing alloys leaving a relatively weak layer of copper and copper oxide (Fig. 116). Most commonly found in copper-zinc alloys containing less than 85% copper after extended service in water containing dissolved oxygen. See also *dealloying* and *selective leaching*.

FIG. 116

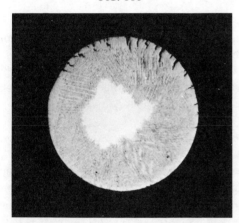

*Dezincification of a silicon brass valve spindle. 4×*

**D-glass**. A high boron content glass made especially for laminates requiring a precisely controlled dielectric constant.

**diadic polyamide**. Polyamide produced by condensation of diamine and a dicarboxylic acid.

**diamagnetic material**. A material whose specific permeability is less than unity and is therefore repelled weakly by a magnet. Compare with *ferromagnetic material* and *paramagnetic material*.

**diametrical strength**. A property that is calculated from the load required to crush a cylindrical sintered test specimen in the direction perpendicular to the axis. See also *radial crushing strength*.

**diamond**. A highly transparent mineral composed entirely of carbon (allotropic form of carbon) having a cubic structure. The hardest material known, it is used as a gemstone and as an abrasive in cutting and grinding applications. Natural diamonds are produced deep within the earth's crust at extremely high pressures and temperatures. Synthetic diamonds are synthesized by subjecting carbon, in

FIG. 117

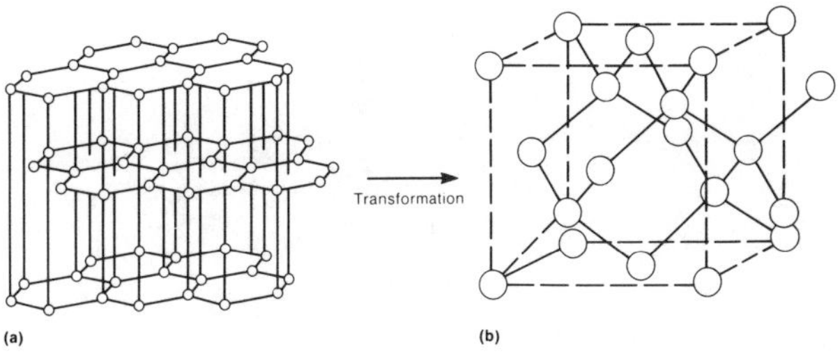

*Arrangement of carbon atoms. (a) The hexagonal arrangement of atoms in graphite. (b) The cubic arrangement of atoms in diamond*

the form of graphite (Fig. 117), to high temperatures and pressures using large special-purpose presses. See also *superabrasives* (Technical Brief 51).

**diamond boring**. Precision boring with a shaped diamond (but not with other tool materials).

**diamond film**. A carbon-composed film, usually deposited by chemical vapor deposition or related process, that has the following three characteristics: (1) a crystalline morphology that can be visually discerned by scanning electron or optical microscopy (Fig. 118), (2) a single-phase crystalline structure identifiable by x-ray and/or electron diffraction, and (3) a Raman spectrum typical for crystalline diamond. See also *diamondlike film*.

FIG. 118

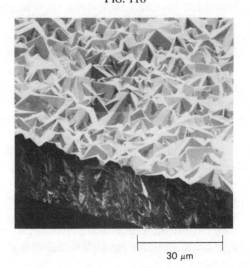

30 μm

*Diamond film grown on silicon*

**diamondlike film**. A hard, noncrystalline carbon film, usually grown by chemical vapor deposition or related techniques, that contains predominantly $sp^2$ carbon-carbon bonds. See also *diamond film*.

**diamond pyramid hardness test**. See *Vickers hardness test*.

**diamond tool**. (1) A diamond, shaped or formed to the con-

FIG. 119

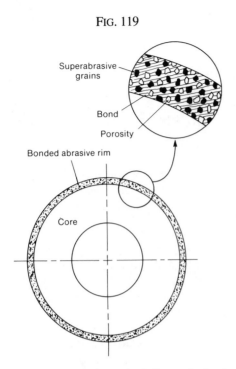

*Construction of a typical diamond wheel*

tour of a single-point cutting tool, for use in precision machining of nonferrous or nonmetallic materials. (2) An insert made from polycrystalline diamond compacts.

**diamond wheels**. A grinding wheel in which crushed and sized industrial diamonds are held in a resinoid, metal, or vitrified bond (Fig. 119).

**diaphragm**. (1) A fixed or adjustable aperture in an optical system. Diaphragms are used to intercept scattered light, to limit field angles, or to limit image-forming bundles or rays. (2) A porous or permeable membrane separating anode and cathode compartments of an electrolytic cell from each other or from an intermediate compartment. (3) Universal die member made of rubber or similar material used to contain hydraulic fluid within the forming cavity and to transmit pressure to the part being formed.

**diaphragm gate**. A gate used in molding annular or tubular plastic articles. The gate forms a solid web across the opening of the part. Also called disc gate.

**dichromate treatment**. A chromate *conversion coating* produced on magnesium alloys in a boiling solution of sodium dichromate.

**didymium**. A natural mixture of the rare-earth elements praseodymium and neodymium, often given the quasi-chemical symbol Di.

**die**. A tool, usually containing a cavity, that imparts shape to solid, molten, or powdered metal primarily because of the shape of the tool itself. Used in many press operations (including blanking, drawing, forging, and forming), in die casting and in forming green powder metallurgy com-

pacts. Die-casting and powder metallurgy dies are sometimes referred to as *molds*. See also *forging dies*.

**die adapter**. In forming plastics, the part of an extrusion die that holds the die block.

**die assembly**. The parts of a die stamp or press that hold the die and locate it for the punches.

**die barrel**. In powder metallurgy presses, a tubular liner for a die cavity.

**die block**. A block, often made of heat-treated steel, into which desired impressions are machined or sunk and from which closed-die forgings or sheet metal stampings are produced using hammers or presses. In forging, die blocks are usually used in pairs, with part of the impression in one of the blocks and the rest of the impression in the other. In sheet metal forming, the female die is used in conjunction with a male punch. See also *closed-die forging*.

**die body**. The stationary or fixed part of a powder pressing die.

**die bolster**. In powder metallurgy presses, the external steel ring that is shrunkfit around the hard parts comprising the die barrel.

**die casting**. (1) A casting made in a die. (2) A casting process in which molten metal is forced under high pressure into the cavity of a metal mold. See also *cold chamber machine* and *hot chamber machine*.

**die cavity**. The machined recess that gives a forging or stamping its shape.

**die check**. A crack in a die impression due to forging and thermal strains at relatively sharp corners. Upon forging, these cracks become filled with metal, producing sharp ragged edges on the part. Usual die wear is the gradual enlarging of the die impression due to erosion of the die material, generally occurring in areas subject to repeated high pressures during forging.

**die clearance**. Clearance between a mated punch and die; commonly expressed as clearance per side. Also called clearance or punch-to-die clearance.

**die cushion**. A press accessory placed beneath (Fig. 120) or within a *bolster plate* or *die block* to provide an additional motion or pressure for stamping or forging operations; actuated by air, oil, rubber, springs, or a combination of these.

**die cutting**. Cutting shapes from metal sheet stock by sharply striking it with a shaped knife-edge known as a steel rule die. Clicking and dinking are other names for die cutting of this kind.

**die fill**. A die cavity filled with powder.

**die forging**. A forging that is formed to the required shape and size through working in machined impressions in specially prepared dies.

**die forming**. The shaping of solid or powdered metal by forcing it into or through the *die cavity*.

FIG. 120

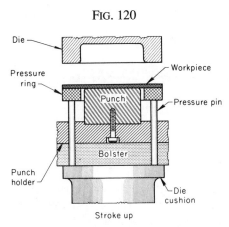

*Setup incorporating die cushion in single-action forming press*

**die height**. In forming, the distance between the fixed and the moving platen when dies are closed.

**die holder**. A plate or block, on which the die block is mounted, having holes or slots for fastening to the *bolster plate* or the *bed* of the press.

**die impression**. The portion of the die surface that shapes a forging or sheet metal part.

**die insert**. A relatively small die that contains part or all of the impression of a forging or sheet metal part and is fastened to the master *die block*.

**die layout**. The transfer of drawing or sketch dimensions to templates or die surfaces for use in sinking dies.

**dielectric**. A nonconductor of electricity. The ability of a material to resist the flow of an electrical current.

**dielectric curing**. The curing of a synthetic thermosetting resin by the passage of an electric charge (produced from a high-frequency generator) through the resin.

**dielectric heating**. The heating of plastic materials by dielectric loss in a high-frequency electrostatic field.

**dielectric loss**. A loss of energy evidenced by the rise in heat of a dielectric placed in an alternating electric field. It is usually observed as a frequency-dependent conductivity.

**dielectric monitoring** (of plastics). Monitoring the cure of thermosets by tracking the changes in their electrical properties during material processing.

**dielectric shield**. In a *cathodic protection* system, an electrically nonconductive material, such as a coating, plastic sheet, or pipe, that is placed between an *anode* and an adjacent *cathode* to avoid current wastage and to improve current distribution, usually on the cathode.

**dielectric strength**. (1) The maximum voltage that a dielectric can withstand, under specified conditions, without resulting in a voltage breakdown (usually expressed as volts per unit dimension). (2) A measure of the ability of a dielectric (insulator) to withstand a potential difference across it without electric discharge.

**dielectrometry**. Use of electrical techniques to measure the changes in loss factor (dissipation) and in capacitance during cure of the resin in a laminate. Also called dielectric spectroscopy.

**die life**. The productive life of a *die impression*, usually expressed as the number of units produced before the impression has worn beyond permitted tolerances.

**die lines**. Lines or markings on formed, drawn, or extruded metal parts caused by imperfections in the surface of the die.

**die lubricant**. (1) A lubricant applied to the working surfaces of dies and punches to facilitate drawing, pressing, stamping, and/or ejection. In powder metallurgy, the die lubricant is sometimes mixed into the powder before pressing into a compact. (2) A compound that is sprayed, swabbed, or otherwise applied on die surfaces or the workpiece during the forging or forming process to reduce friction. Lubricants also facilitate release of the part from the dies and provide thermal insulation. See also *lubricant*.

**die match**. The condition where dies, after having been set up in a press or other equipment, are in proper alignment relative to each other.

**die opening**. (1) In flash or upset welding, the distance between the electrodes, usually measured with the parts in contact before welding has commenced or immediately upon completion of the cycle but before upsetting. (2) In powder metallurgy, the entrance to the die cavity.

**die pad**. In forming, a movable plate or pad in a female die; usually used for part ejection by mechanical means, springs, or fluid cushions.

**die-parting line**. A lengthwise flash or depression on the surface of a pultruded plastic part. The line occurs where separate pieces of the die join together to form the cavity.

**die plate**. The base plate of a press into which the die is sunk.

**die proof**. A casting of a *die impression* made to confirm the accuracy of the impression.

**die pull**. The direction in which the solidified casting must move when it is removed from the die. The die pull direction must be selected such that all points on the surface of the casting move away from the die cavity surfaces.

**die radius**. The radius on the exposed edge of a deep-drawing die, over which the sheet flows in forming drawn shells.

**die scalping**. Removing surface layers from bar, rod, wire, or tube by drawing through a sharp-edged die to eliminate minor surface defects.

**die separation**. The space between the two halves of a die casting die at the parting surface when the dies are closed. The separation may be the result of the internal cavity pressure exceeding the locking force of the machine or warpage of the die due to thermal gradients in the die steel.

**die set**. (1) The assembly of the upper and lower die shoes (punch and die holders), usually including the *guide pins*, *guide pin bushings*, and *heel blocks*. This assembly takes many forms, shapes, and sizes and is frequently purchased as a commercially available unit. (2) Two machined dies used together during the production of a *die forging*.

**die shaft**. The condition that occurs after the dies have been set up in a forging unit in which a portion of the impression of one die is not in perfect alignment with the corresponding portion of the other die. This results in a mismatch in the forging, a condition that must be held within the specified tolerance.

**die shoes**. The upper and lower plates or castings that constitute a *die set* (punch and die holder). Also, a plate or block upon which a *die holder* is mounted, functioning primarily as a base for the complete *die assembly*. This plate or block is bolted or clamped to the *bolster plate* or the face of the press *slide*.

**die sinking**. The machining of the die impressions to produce forgings of required shapes and dimensions.

**die space**. The maximum space (volume), or any part of the maximum space, within a forming or forging press for mounting a die.

**die stamping**. The general term for a sheet metal part that is formed, shaped, or cut by a die in a press in one or more operations.

**die swell ratio**. In forming plastics, the ratio of the outer parison diameter (or parison thickness) to the outer diameter of the die (or die gap). Die swell ratio is influenced by head construction, land length, extrusion speed, and temperature. See also *parison* and *parison swell*.

**die volume**. See preferred term *fill volume*.

**die welding**. See preferred terms *forge welding* and *cold welding*.

**differential aeration cell**. An *electrolytic cell*, the *electro-*

*magnetic force* of which is due to a difference in air (oxygen) concentration at one electrode as compared with that at another electrode of the same material. See also *concentration cell*.

**differential coating**. A coated product having a specified coating on one surface and a significantly lighter coating on the other surface (such as a hot dip galvanized product or electrolytic tin plate).

**differential flotation**. Separating a complex ore into two or more valuable minerals and *gangue* by *flotation*. Also called selective flotation.

**differential heating**. Heating that intentionally produces a temperature gradient within an object such that, after cooling, a desired stress distribution or variation in properties is present within the object.

**differential interference contrast illumination**. A microscopic technique using a beam-splitting double-quartz prism, that is, a modified Wollaston prism placed ahead of the objective together with a polarizer and analyzer in the 90° crossed positions. The two light beams are made to coincide at the focal plane of the objective, revealing height differences as variations in color (Fig. 121). The prism can be moved, shifting the interference image through the range of Newtonian colors.

**differential scanning calorimetry (DSC)**. Measurement of energy absorbed (endotherm) or produced (exotherm). May be applied to melting, crystallization, resin curing, loss of solvents, and other processes involving an energy change. May also be applied to processes involving a change in heat capacity, such as the glass transition.

**differential thermal analysis (DTA)**. An experimental analysis technique in which a specimen and a control are

FIG. 121

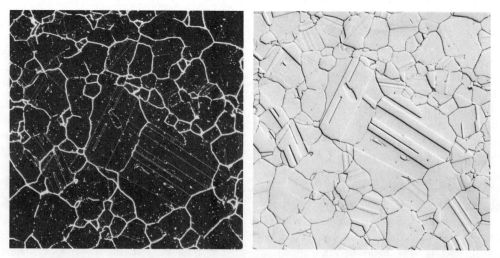

*Comparison of dark-field (left) and differential interference contrast (right) illumination. Note the three-dimensional appearance of the DIC illumination. The specimen is a nickel-base superalloy. 100×*

heated simultaneously and the difference in their temperatures is monitored. The difference in temperatures provides information on relative heat capacities, presence of solvents, changes in structure (that is, phase changes, such as melting of one component in a resin system), and chemical reactions. See also *differential scanning calorimetry*.

**diffraction**. (1) A modification that radiation undergoes, for example, in passing by the edge of opaque bodies or through narrow slits, in which the rays appear to be deflected. (2) Coherent scattering of x-rays by the atoms of a crystal that necessarily results in beams in characteristic directions. Sometimes termed reflection. (3) The scattering of electrons by any crystalline material through discrete angles depending only on the lattice spacings of the material and the velocity of the electrons.

**diffraction contrast**. In electron microscopy, contrast produced by intensity differences in Bragg-diffracted beams from a crystalline material. These differences are caused by regions of varying crystal orientation.

**diffraction grating**. A series of a large number of narrow, close, equally spaced, diffracting slits or grooves capable of dispersing light into its spectrum. See also *concave grating*, and *reflection grating*. Compare with *transmission grating*.

**diffraction pattern**. The spatial arrangement and relative intensities of diffracted beams (Fig. 122).

**diffraction ring**. The diffraction pattern produced by a given

Fig. 122

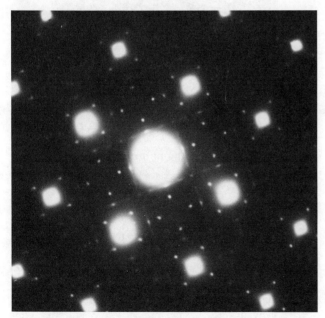

*Diffraction pattern of an ordered tantalum (containing oxygen) structure*

set of planes from randomly oriented crystalline material. See also *Debye ring*.

**diffuse transmittance**. In materials characterization, the transmittance value obtained when the measured radiant energy has experienced appreciable scattering in passing from the source to the receiver. See also *transmittance*.

**diffusion**. (1) Spreading of a constituent in a gas, liquid, or solid, tending to make the composition of all parts uniform. (2) The spontaneous movement of atoms or molecules to new sites within a material. (3) The movement of a material, such as a gas or liquid, in the body of a plastic. If the gas or liquid is absorbed on one side of a piece of plastic and given off on the other side, the phenomenon is called permeability. Diffusion and permeability are not due to holes or pores in the plastic but are caused and controlled by chemical mechanisms.

**diffusion aid**. A solid filler metal sometimes used in *diffusion welding*.

**diffusion bonding**. See preferred terms *diffusion welding* and *diffusion brazing*.

**diffusion brazing**. A brazing process which produces coalescence of metals by heating them to suitable temperatures and by using a filler metal or an *in situ* liquid phase. The filler metal may be distributed by capillary action or may be placed or formed at the faying surfaces. The filler metal is diffused with the base metal to the extent that the joint properties have been changed to approach those of the base metal. Pressure may or may not be applied.

**diffusion coating**. Any process whereby a base metal or alloy is either (1) coated with another metal or alloy and heated to a sufficient temperature in a suitable environment or (2) exposed to a gaseous or liquid medium containing the other metal or alloy, thus causing *diffusion* of the coating or of the other metal or alloy into the base metal with resultant changes in the composition and properties of its surface.

**diffusion coefficient**. A factor of proportionality representing the amount of substance diffusing across a unit area through a unit concentration gradient in unit time.

**diffusion-limited current density**. The *current density*, often referred to as *limiting current density*, that corresponds to the maximum transfer rate that a particular species can sustain because of the limitation of diffusion.

**diffusion porosity**. The porosity that is caused by the diffusion of one metal into another during sintering of a powder metallurgy compact. Also known as Kirkendall porosity.

**diffusion welding**. A solid-state welding process that produces coalescence of the faying surfaces by the application of pressure at elevated temperature. The process does not involve macroscopic deformation, melting, or relative motion of parts. A solid filler metal (*diffusion aid*) may or may not be inserted between the faying surfaces. See also *forge welding*, *hot pressure welding*, and *cold welding*.

**diffusion zone**. The zone of variable composition at the junction between two different materials, such as in welds or between the surface layer and the core of clad materials or sleeve bearings, in which interdiffusion between the various components has taken place.

**digging**. A sudden erratic increase in cutting depth, or in the load on a cutting tool, caused by unstable conditions in the machine setup. Usually the machine is stalled, or either the tool or the workpiece is destroyed.

**digital radiography (DR)**. Radiographic imaging technology in which a two-dimensional set of x-ray transmission measurements is acquired and converted to an array of numbers for display with a computer. *Computed tomography* systems normally have DR imaging capabilities (Fig. 123). Also known as digital radioscopy.

FIG. 123

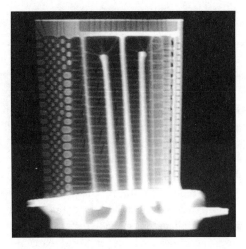

*Digital radiograph of an aircraft engine turbine blade (nickel alloy precision casting)*

**dilatant**. A reversible increase in viscosity with increasing shear stress. Compare with *pseudoplastic behavior*, *rheopectic material*, and *thixotropy*.

**dilatometer**. An instrument for measuring the linear expansion or contraction in a metal resulting from changes in such factors as temperature and allotropy.

**diluent**. In an *organosol*, a liquid component that has little or no solvating action on the resin. Its purpose is to modify the action of the dispersant. The term diluent is commonly used in place of the term plasticizer.

**dilution**. The change in chemical composition of a welding filler metal caused by the admixture of the base metal or previously deposited weld metal in the deposited weld bead. It is normally measured by the percentage of base metal or previously deposited weld metal in the weld bead.

**dilution factor**. When diluting a sample, the ratio of the final

volume or mass after dilution to the volume or mass of the sample before dilution.

**dimensional change**. Object shrinkage or growth resulting from sintering a powder metallurgy compact.

**dimensional stability**. (1) A measure of dimensional change caused by such factors as temperature, humidity, chemical treatment, age, or stress (usually expressed as Δ units per unit). (2) Ability of a plastic part to retain the precise shape in which it was molded, cast, or otherwise fabricated.

**dimer**. A substance (comprising molecules) formed from two molecules of a *monomer*.

**dimerization**. The formation of a *dimer*.

**dimple rupture**. A fractographic term describing *ductile fracture* that occurs through the formation and coalescence of microvoids along the fracture path. The fracture surface of such a ductile fracture appears dimpled when observed at high magnification and usually is most clearly resolved when viewed in a scanning electron microscope (Fig. 124).

FIG. 124

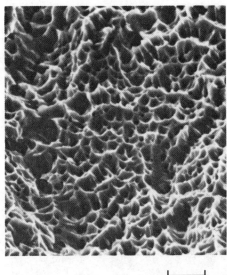

10 μm

*Conical equiaxed dimples in a spring steel specimen*

**dimpling**. (1) The stretching of a relatively small, shallow indentation into sheet metal. (2) In aircraft, the stretching of metal into a conical flange for a countersunk head rivet.

**dingot**. An oversized *derby* (possibly a ton or more) of a metal produced in a bomb reaction (such as uranium from uranium tetrafluoride reduced with magnesium). For these metals, the term "ingot" is reserved for massive units produced in vacuum melting and casting. See also *biscuit* and *derby*.

**dinking**. Cutting of nonmetallic materials or light-gage soft metals by using a hollow punch with a knifelike edge acting against a wooden fiber or resiliently mounted metal plate.

**dip brazing.** A brazing process in which the heat required is furnished by a molten chemical or metal bath. When a molten chemical bath is used, the bath may act as a flux. When a molten metal bath is used, the bath provides the filler metal.

**dip casting.** In forming plastics, the process of submerging a hot mold into a resin. After cooling, the product is removed from the mold.

**dip coat.** (1) In the solid mold technique of investment casting, an extremely fine ceramic precoat applied as a slurry directly to the surface of the pattern to reproduce maximum surface smoothness. This coating is surrounded by coarser, less expensive, and more permeable investment to form the mold. (2) In the shell mold technique of investment casting, an extremely fine ceramic coating called the first coat, applied as a slurry directly to the surface of the pattern to reproduce maximum surface smoothness. The first coat is followed by other dip coats of different viscosity and usually containing different grading of ceramic particles. After each dip, coarser stucco material is applied to the still-wet coating. A buildup of several coats forms an investment shell mold. See also figure accompanying *investment casting*.

**dip coating.** Applying a plastic coating by dipping the article to be coated into a tank of melted resin or plastisol, then chilling the adhering metal.

**diphase cleaning.** Removing *soil* by an emulsion that produces two phases in the cleaning tank: a solvent phase and an aqueous phase. Cleaning is effected by both solvent action and emulsification.

**diphenyl oxide resins.** Thermosetting resins based on diphenyl oxide and possessing excellent handling properties and heat resistance.

**dip plating.** Same as *immersion plating*.

**dip soldering.** A soldering process in which the heat required is furnished by a molten metal bath which provides the solder filler metal.

**direct chill casting.** A continuous method of making ingots for rolling or extrusion by pouring the metal into a short mold. The base of the mold is a platform that is gradually lowered while the metal solidifies, the frozen shell of metal acting as a retainer for the liquid metal below the wall of the mold. The ingot is usually cooled by the impingement of water directly on the mold or on the walls of the solid metal as it is lowered. The length of the ingot is limited by the depth to which the platform can be lowered; therefore, it is often called semicontinuous casting.

**direct current arc furnace.** An electric-arc furnace in which a single electrode positioned at the center of the furnace roof is the *cathode* of the system. Current passes from the electrode through the *charge* or bath to a cathode located at the bottom of the furnace. Current from the bottom of

Fig. 125

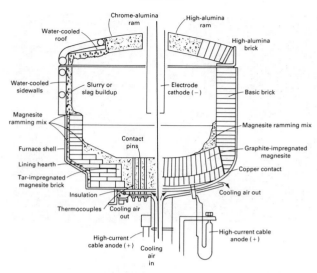

*Cutaway view showing the basic components of a direct current arc furnace*

the furnace then passes through the furnace refractories to a copper base plate to outside cables (Fig. 125). Used in the production of ferroalloys, carbon and alloy steels, and stainless steels. See also *arc furnace*.

**direct-current casting.** Same as *cathodic cleaning*.

**direct current electrode negative (DCEN).** The arrangement of direct current arc welding leads in which the work is the positive pole and the electrode is the negative pole of the welding arc. See also *straight polarity*.

**direct current electrode positive (DCEP).** The arrangement of direct current arc welding leads in which the work is the negative pole and the electrode is the positive pole of the welding arc. See also *reverse polarity*.

**direct current reverse polarity (DCRP).** See *reverse polarity* and *direct current electrode positive*.

**direct current straight polarity (DCSP).** See *straight polarity* and *direct current electrode negative*.

**direct-fired tunnel-type furnace.** A continuous-type furnace where the work is conveyed through a tunnel-type heating zone, and the parts are hung on hooks or fixtures to minimize distortion.

**direct (forward) extrusion.** See *extrusion*.

**direct injection burner.** A burner used in flame emission and atomic absorption spectroscopy in which the fuel and oxidizing gases emerge from separate ports and are mixed in the flame itself. One of the gases, usually the oxidant, is used for nebulizing the sample at the tip of the burner.

**directionally solidified castings.** Investment castings produced by *directional solidification* which exhibit an oriented columnar structure (Fig. 126).

**directional property.** Property whose magnitude varies depending on the relation of the test axis to a specific

FIG. 126

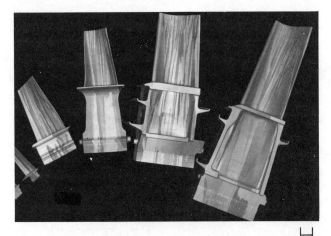

25 mm

*Directionally solidified land-based turbine blades made from investment cast nickel-base superalloys*

direction within the metal. The variation results from preferred orientation or from fibering of constituents or inclusions.

**directional solidification.** Controlled solidification of molten metal in a casting so as to provide feed metal to the solidifying front of the casting (Fig. 127).

**direct quenching.** (1) Quenching carburized parts directly from the carburizing operation. (2) Also used for quenching pearlitic malleable parts directly from the malleabilizing operation.

**direct sintering.** In powder metallurgy, a method whereby the heat needed for sintering is generated in the body itself,

FIG. 127

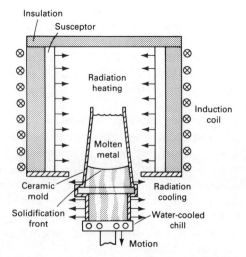

*Schematic showing the directional solidification process*

such as by induction or resistance heating. Contrast with *indirect sintering*.

**dirt content.** A measure of the size and concentration of foreign particles present in a lubricant. Dirt content is usually reported as the number of particles per cubic centimeter, for specified particle sizes.

**disbond.** In adhesive bonded structures, an area within the bonded interface between two adherends in which an adhesion failure or separation has occurred. Also, colloquially, an area of separation between two laminae in the finished laminate (in this case, the term delamination is normally preferred). See also *debond*.

**disbondment.** The destruction of adhesion between a coating and the surface coated.

**discontinuity.** (1) Any interruption in the normal physical structure or configuration of a part, such as cracks, laps, seams, inclusions, or porosity. A discontinuity may or may not affect the utility of the part. (2) An interruption of the typical structure of a weldment, such as a lack of homogeneity in the mechanical, metallurgical, or physical characteristics of the material or weldment. A discontinuity is not necessarily a defect. See also *defect* and *flaw*.

**discontinuous precipitation.** Precipitation from a supersaturated solid solution in which the precipitate particles grow by short-range diffusion, accompanied by recrystallization of the matrix in the region of precipitation. Discontinuous precipitates grow into the matrix from nuclei near grain boundaries, forming cells of alternate lamellae of precipitate and depleted (and recrystallized) matrix. Often referred to as cellular or nodular precipitation. Compare with *continuous precipitation* and *localized precipitation*.

**discontinuous sintering.** Presintering or sintering of the objects in a furnace according to a specified cycle that is tailored to the charge. Examples, batch sintering, bell furnace sintering, box furnace sintering, induction furnace sintering.

**discontinuous yielding.** The nonuniform plastic flow of a metal exhibiting a yield point in which plastic deformation is inhomogeneously distributed along the gage length. Under some circumstances, it may occur in metals not exhibiting a distinct yield point, either at the onset of or during plastic flow.

**dished.** Showing a symmetrical distortion of a flat or curved section of a plastic object so that, as normally viewed, it appears concave, or more concave than intended.

**dishing.** Forming a shallow concave surface, the area being large compared to the depth.

**disk grinding.** Grinding with the flat side of an abrasive disk or segmented wheel (Fig. 128). Also called verticalspindle surface grinding.

**disk machine.** In tribology, a testing machine for rolling or rolling/sliding contact in which two disk-shaped

Fig. 128

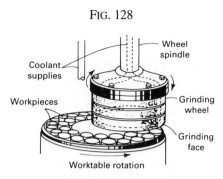

*Disk grinding setup*

rollers, with parallel axes of rotation, make tangential contact on their circumferences as they move relative to one another.

**dislocation**. A linear imperfection in a crystalline array of atoms. Two basic types are recognized: (1) an edge dislocation corresponds to the row of mismatched atoms along the edge formed by an extra, partial plane of atoms within the body of a crystal; (2) a screw dislocation corresponds to the axis of a spiral structure in a crystal, characterized by a distortion that joins normally parallel planes together to form a continuous helical ramp (with a pitch of one interplanar distance) winding about the dislocation. Most prevalent is the so-called mixed dislocation, which is any combination of an edge dislocation and a screw dislocation.

**dislocation etching**. Etching of exit points of dislocations on a surface. Depends on the strain field ranging over a distance of several atoms. Crystal figures (etch pits) are formed at exit points. For example, etch pits for cubic materials are cube faces.

**disordered structure**. The crystal structure of a solid solution in which the atoms of different elements are randomly distributed relative to the available lattice sites. Contrast with *ordered structure*.

**disordering**. Forming a lattice arrangement in which the solute and solvent atoms of a solid solution occupy lattice sites at random. See also *ordering* and *superlattice*.

**dispersant additive**. In lubrication technology, an additive capable of dispersing cold oil sludge.

**dispersant oil**. A heavy-duty oil containing a dispersant additive.

**dispersing agent**. A substance that increases the stability of a suspension of powder particles in a liquid medium by deflocculation of the primary particles.

**dispersion**. Finely divided particles of a material in suspension in another substance.

**dispersion hardening**. See *dispersion strengthening*.

**dispersion-strengthened material**. A metallic material that contains a fine dispersion of nonmetallic phase(s), such as $Al_2O_3$, MgO, $SiO_2$, CdO, $ThO_2$, $Y_2O_3$, or $ZrO_2$ singly or in combination, to increase the hot strength of the metallic matrix. Examples include dispersion-strengthened copper ($Al_2O_3$) used for welding electrodes, silver (CdO) used for electrical contacts, and nickel-chromium ($Y_2O_3$) superalloys used for gas turbine components. See also *mechanical alloying*.

**dispersion strengthening**. The strengthening of a metal or alloy by incorporating chemically stable submicron size particles of a nonmetallic phase that impede dislocation movement at elevated temperature.

**dispersoid**. Finely divided particles of relatively insoluble constituents visible in the microstructure of certain metallic alloys.

**displacement**. The distance that a chosen measurement point on the test specimen displaces normal to the crack plane. See also *crack opening displacement* and *crack plane orientation*.

**displacement angle**. In filament winding, the advancement distance of the winding ribbon on the equator after one complete circuit.

**display resolution**. Number of picture elements (pixels) per unit distance in the object.

**disproportionation**. In the processing of plastics, termination by chain transfer between macroradicals to produce a saturated and an unsaturated polymer molecule.

**disruptive strength**. The stress at which a metal fractures under hydrostatic tension.

**dissipation factor, electrical**. See *quality factor* (2).

**dissociated ammonia**. A frequently used sintering atmosphere. See also *cracked gas*.

**dissociation**. As applied to heterogeneous equilibria, the transformation of one phase into two or more phases of different composition. Compare with *order-disorder transformation*.

**dissociation pressure**. At a designated temperature, the pressure at which a metallic phase will transform into two or more new phases of different composition.

**dissolution etching**. Development of microstructure by surface removal.

**distortion**. Any deviation from an original size, shape, or contour that occurs because of the application of stress or the release of residual stress.

**distortion** (composites). In fabric, the displacement of fill fiber from the 90° angle (right angle) relative to the warp fiber. In a laminate, the displacement of the fibers (especially at radii), relative to their idealized location, due to motion during lay-up and cure.

**distributed impact test**. In impingement erosion testing, an apparatus or method that produces a spatial distribution of impacts by liquid or solid bodies over an exposed surface of a specimen. Examples of such tests are those employing liquid sprays or simulated rainfields. If the impacts are distributed uniformly over the surface, the term uniformly distributed impact test may be used.

FIG. 129

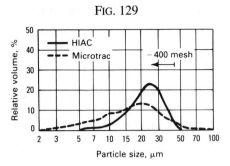

*Particle size distribution curve of –400 mesh copper powder*

FIG. 130

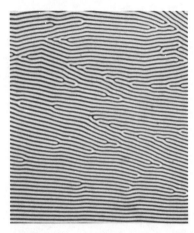

*Magnetic domain structure in garnet. Black and white areas represent domains with opposing magnetic vectors. 100×*

**distribution contour**. The shape of the particle size distribution curve (Fig. 129).

**disturbed metal**. The cold worked metal layer formed at a polished surface during the process of mechanical grinding and polishing.

**divariant equilibrium**. See *bivariant equilibrium*.

**divided cell**. A cell containing a diaphragm or other means for physically separating the *anolyte* from the *catholyte*.

**dividing cone**. A conical powder heap that is divided in quarters for the purpose of sample taking.

**divorced eutectic**. A metallographic appearance in which the two constituents of a eutectic structure appear as massive phases rather than the finely divided mixture characteristic of normal eutectics. Often, one of the constituents of the eutectics is continuous and indistinguishable from an accompanying proeutectic constituent.

**DN value**. The product of bearing bore diameter in millimeters and speed in revolutions per minute.

**doctor blade or bar**. In forming plastics, a straight piece of material used to spread resin, as in application of a thin film of resin for use in hot-melt prepregging or for use as an adhesive film. Also called paste metering blade.

**doctor roll**. In applying adhesives, a roller mechanism that is revolving at a different surface speed, or in an opposite direction, resulting in a wiping action for regulating the adhesive supplied to the spreader roll.

**doily**. In filament winding of composites, the planar reinforcement applied to a local area between windings to provide extra strength in an area where a cutout is to be made, for example, port openings. Usually placed at the knuckle joints of cylinder to dome.

**dolomite**. A type of *limestone* of composition $CaCO_3 \cdot MgCO_3$ used in making *cement* and lime, as a flux in melting iron, as a lining for basic steel furnaces, for filtering, and as a construction stone.

**dolomite brick**. A calcium magnesium carbonate used as a refractory brick that is manufactured substantially or entirely of *dead-burned* dolomite.

**domain** (plastics). A morphological term used in noncrystalline systems, such as block copolymers, in which the chemically different sections of the chain separate, generating two or more amorphous phases.

**domain, magnetic**. A substructure in a ferromagnetic material within which all the elementary magnets (electron spins) are held aligned in one direction by interatomic forces (Fig. 130); if isolated, a domain would be a saturated permanent magnet.

**dome**. In filament winding, the portion of a cylindrical container that forms the spherical or elliptical shell ends of the container.

**domed**. Showing a symmetrical distortion of a flat or curved section of a plastic object so that, as normally viewed, it appears convex, or more convex than intended.

**Donnan exclusion**. In *ion chromatography*, the mechanism by which an ion exchange resin can be made to act like a semipermeable membrane between an interstitial liquid and a liquid occluded inside the resin particles. Highly ionized molecules are excluded from the resin particles by electrostatic forces; weakly ionized or nonionized molecules may pass through the membrane.

**dopant**. (1) An impurity introduced under highly controlled conditions in very small but accurately known quantities into a semiconductor material, such as silicon. Dopants modify the electrical characteristics of the silicon by creating *p* or *n* regions and hence *pn* junctions. (2) A material added to a polymer to change a physical property.

**doped solder**. A solder containing a small amount of an element intentionally added to ensure retention of one or more characteristics of the materials on which it is used.

**doping**. In powder metallurgy, the addition of a small amount of an activator to promote sintering.

**Doppler effect**. The change in the observed frequency of an acoustic or electromagnetic wave due to the relative motion of source and observer. See also *Doppler shift*.

**Doppler shift**. The amount of change in the observed frequency of a wave due to the Doppler effect, usually expressed in hertz. See also *Doppler effect*.

**doré silver**. Crude silver containing a small amount of gold, obtained after removing lead in a cupelling furnace. Same as doré bullion and doré metal.

**dosimeter**. A device for measuring radiation-induced signals that can be related to absorbed dose (or energy deposited) by radiation in materials and is calibrated in terms of the appropriate quantities and units. Also called dose meter.

**dot map**. See *x-ray map*.

**double-acting hammer**. A forging hammer in which the ram is raised by admitting steam or air into a cylinder below the piston, and the blow intensified by admitting steam or air above the piston on the downward stroke.

**double-action die**. A die designed to perform more than one operation in a single stroke of the press.

**double-action forming**. Forming or drawing in which more than one action is achieved in a single stroke of the press.

**double-action mechanical press**. A press having two independent parallel movements by means of two slides, one moving within the other. The inner slide or plunger is usually operated by a crankshaft; the outer or blankholder slide, which dwells during the drawing operation, is usually operated by a toggle mechanism or by cams. See also *slide*.

**double-action press**. A press that provides pressure from two sides, usually opposite each other, such as from top and bottom.

**double aging**. Employment of two different aging treatments to control the type of precipitate formed from a supersaturated matrix in order to obtain the desired properties. The first aging treatment, sometimes referred to as intermediate or stabilizing, is usually carried out at higher temperature than the second.

**double arcing** (plasma arc welding and cutting). A condition in which the main arc does not pass through the constricting orifice but transfers to the inside surface of the nozzle. A secondary arc is simultaneously established between the outside surface of the nozzle and the workpiece. Double arcing usually damages the nozzle.

**double-bevel groove weld**. A groove weld in which the joint edge of one member is beveled from both sides (Fig. 131).

**double cone mixer**. A vessel in the shape of two cones abutting at their base that rotates on an axis through the base, and that provides thorough mixing or blending of a powder by cascading.

**double etching**. In metallography, use of two etching solutions in sequence. The second etchant emphasizes a particular microstructural feature.

**double-J groove weld**. A groove weld in which the joint edge of one member is in the form of two J's, one from either side (Fig. 131).

FIG. 131

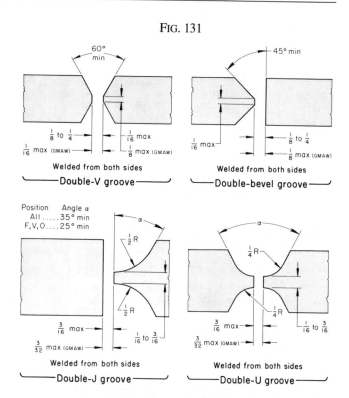

*Double groove weld designs*

**double layer**. The interface between an *electrode* or a suspended particle and an *electrolyte* created by charge-charge interaction leading to an alignment of oppositely charged ions at the surface of the electrode or particle. The simplest model is represented by a parallel plate condensor.

**double pressing**. A method whereby compaction of metal powders is carried out in two steps. It may involve removal of the compact from the die after the first pressing for the purpose of storage, drying, baking, presintering, sintering, or other treatment, before reinserting into a die for the second pressing.

**doubler**. In filament winding of composites, a local area with extra reinforcement, wound integrally with the part, or wound separately and fastened to the part. See also *tabs*.

**double salt**. A compound of two salts that crystallize together in a definite proportion.

**double-shot molding**. In forming plastics, a means of producing two-color parts and/or two different thermoplastic materials by successive molding operations.

**double sintering**. In powder metallurgy, a method consisting of two separate sintering operations with a shape change by machining or coining performed in between.

**double spread**. The application of adhesive to both adherends of a joint.

**double tempering**. A treatment in which a quench-hardened ferrous metal is subjected to two complete tempering cycles, usually at substantially the same temperature, for the purpose of ensuring completion of the tempering reaction and promoting stability of the resulting microstructure.

**double-U groove weld.** A groove weld in which each joint edge is in the form of two J's or two half-U's, one from either side of the member (Fig. 131).

**double-V groove weld.** A groove weld in which each joint edge is beveled from both sides (Fig. 131).

**double-welded joint.** In arc and oxyfuel gas welding, any joint welded from both sides.

**dovetailing** (thermal spraying). A method of surface roughening involving angular undercutting to interlock the spray deposit.

**dowel.** (1) In casting, a wooden or metal pin of various types used in the parting surface of parted patterns and core boxes. (2) In die casting dies, metal pins to ensure correct registry of cover and ejector halves.

**down cutting.** See preferred term *climb cutting.*

**downgate.** Same as *sprue.*

**downhand.** See preferred term *flat position.*

**downhand welding.** See *flat-position welding.*

**down milling.** See preferred term *climb milling.*

**downslope time** (automatic arc welding). The time during which the current is changed continuously from final taper current or welding current to final current.

**downslope time** (resistance welding). The time during which the welding current is continuously decreased.

**downsprue.** Same as *sprue.*

**Dow process.** A process for the production of magnesium by electrolysis of molten magnesium chloride.

**draft.** (1) An angle or taper on the surface of a pattern, core box, punch, or die (or of the parts made with them) that facilitates removal of the parts from a mold or die cavity, or a core from a casting. (2) The change in cross section that occurs during rolling or cold drawing.

**draft angle.** (1) The angle of a taper on a mandrel or mold that facilitates removal of the finished plastic part. (2) The angle of taper, usually 5 to 7°, given to the sides of a forging and the sidewalls of the die impression. See also *draft.*

**drag.** The bottom section of a *flask*, *mold*, or *pattern.*

**drag** (thermal cutting). The offset distance between the actual and the theoretical exit points of the cutting oxygen stream measured on the exit surface of the material (Fig. 132).

**drag angle.** In welding, between the axis of the electrode or torch and a line normal to the plane of the weld when welding is being done with the torch positioned ahead of the weld puddle. See also the figure accompanying the term *backhand welding.*

**drag-in.** Water or solution carried into another solution by the work and its associated handling equipment.

**dragout.** Solution carried out of a bath by the work and its associated handling equipment.

**drag soldering.** A process in which supported, moving printed circuit assemblies or printed wiring assemblies are brought in contact with the surface of a static pool of molten solder.

**drag technique.** A method used in manual arc welding wherein the electrode is in contact with the assembly being welded without being in short circuit. The electrode is usually used without oscillation.

**drainage.** Conduction of electric current from an underground metallic structure by means of a metallic conductor. Forced drainage is that applied to underground metallic structures by means of an applied electromotive force or sacrificial anode. Natural drainage is that from an underground structure to a more negative (anodic) structure, such as the negative bus of a trolley substation.

**drape.** In fabricating composites, the ability of a fabric or prepreg to conform to a contoured surface.

**drape forming.** Method of forming thermoplastic sheet in which the sheet is clamped into a movable frame, heated,

FIG. 133

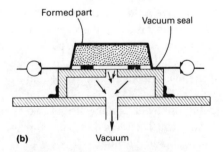

Drape forming. (a) Heated plastic sheet in position over male mold. (b) Mold moves against sheet, or sheet is pulled onto mold.

FIG. 132

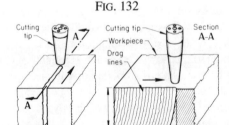

Cross section of work metal during oxyfuel gas cutting showing drag on cutting face

and draped over high points of a male mold. Vacuum is then pulled to complete the forming operation (Fig. 133). Also known as basic male mold forming.

**draw.** A term used to denote the shrinkage that appears on the surface of a casting.

**drawability.** A measure of the *formability* of a sheet metal subject to a drawing process. The term is usually used to indicate the ability of a metal to be deep drawn. See also *drawing* and *deep drawing.*

**draw bead.** (1) An insert or riblike projection on the draw ring or hold-down surfaces that aids in controlling the rate of metal flow during deep draw operations (Fig. 134). Draw beads are especially useful in controlling the rate of metal flow in irregularly shaped stampings.

FIG. 134

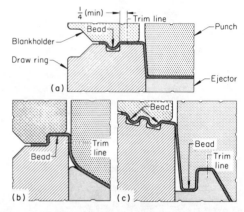

*Use of draw beads. (a) Conventional. (b) Locking. (c) Combined conventional and locking*

**drawbench.** The stand that holds the die and draw head used in drawing of wire, rod, and tubing.

**draw-down ratio.** In forming plastics, the ratio of the thickness of the die opening to the final thickness of the product.

**draw forging.** See *radial forging.*

**draw forming.** A method of curving bars, tubes, or rolled or extruded sections in which the stock is bent around a rotating *form block*. Stock is bent by clamping it to the form block, then rotating the form block while the stock is pressed between the form block and a pressure die held against the periphery of the form block.

**draw head.** Set of rolls or dies mounted on a draw-bench for forming a section from strip, tubing, or solid stock. See also *Turk's-head rolls.*

**drawing.** (1) A term used for a variety of forming operations, such as *deep drawing* a sheet metal blank; *redrawing* a tubular part; and drawing rod, wire, and tube. The usual drawing process with regard to sheet metal working in a press is a method for producing a cuplike form from a sheet metal disk by holding it firmly between blankholding surfaces to prevent the formation of wrinkles while the

punch travel produces the required shape. (2) The process of stretching a thermoplastic to reduce its cross-sectional area, thus creating a more orderly arrangement of polymer chains with respect to each other. (3) A misnomer for tempering (see *temper*).

**drawing** (pattern). In foundry practice, removing a pattern from a mold or a mold from a pattern in production work.

**drawing compound.** (1) A substance applied to prevent *pickup* and *scoring* during drawing or pressing operations by preventing metal-to-metal contact of the work and die. Also known as *die lubricant.* (2) In metalworking, a lubricant having extreme-pressure properties. See also *extreme-pressure lubricant.*

**drawing out.** A stretching operation resulting from forging a series of upsets along the length of the workpiece.

**draw marks.** See *scoring, galling, pickup,* and *die line.*

**drawn fiber.** Fiber for reinforced plastics with a certain amount of orientation imparted by the drawing process by which it is formed.

**drawn shell.** An article formed by drawing sheet metal into a hollow structure having a predetermined geometrical configuration.

**draw plate.** (1) In metal forming, a circular plate with a hole in the center contoured to fit a forming punch; used to support the *blank* during the forming cycle. (2) In casting, a plate attached to a pattern to facilitate drawing of a pattern from the mold.

**draw radius.** The radius at the edge of a die or punch over which sheet metal is drawn.

**draw ring.** A ring-shaped die part (either the die ring itself or a separate ring) over which the inner edge of sheet metal is drawn by the punch.

**draw stock.** The forging operation in which the length of a metal mass (stock) is increased at the expense of its cross section; no *upset* is involved. The operation includes converting ingot to pressed bar using "V," round, or flat dies.

**dresser.** A tool used for *truing* and *dressing* a grinding wheel.

**dressing.** (1) Cutting, breaking down or crushing the surface of a grinding wheel to improve its cutting ability and accuracy. (2) Removing dulled grains from the cutting face of a grinding wheel to restore cutting quality.

**drift.** (1) A flat piece of steel of tapering width used to remove taper shank drills and other tools from their holders. (2) A tapered rod used to force mismated holes into line for riveting or bolting. Sometimes called a drift pin.

**drill.** A rotary end-cutting tool used for making holes; it has one or more cutting lips and an equal number of helical or straight flutes for passage of chips and admission of cutting fluid (Fig. 135).

**drilling.** Hole making with a rotary end-cutting tool having one or more cutting lips and one or more helical or straight flutes or tubes for the ejection of chips and the passage of a cutting fluid.

FIG. 135

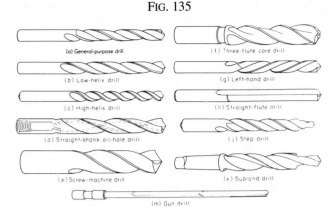

(a) General-purpose drill
(b) Low-helix drill
(c) High-helix drill
(d) Straight-shank oil-hole drill
(e) Screw-machine drill
(f) Three-flute core drill
(g) Left-hand drill
(h) Straight-flute drill
(j) Step drill
(k) Subland drill
(m) Gun drill

*Commonly used types of drills*

1. Center drilling: Drilling a conical hole in the end of a workpiece
2. Core drilling: Enlarging a hole with a chamfer-edged, multiple-flute drill
3. Spade drilling: Drilling wit a flat blade drill tip
4. Step drilling: Using a multiple-diameter drill
5. Gun drilling: Using special straight flute drills with a single lip and cutting fluid at high pressures for deep hole drilling
6. Oil hole or pressurized coolant drilling: Using a drill with one or more continuous holes through its body and shank to permit the passage of a high pressure cutting fluid which emerges at the drill point and ejects chips

**drip feed lubrication.** A system of lubrication in which the lubricant is supplied to the bearing surfaces in the form of drops at regular intervals. Also known as drop feed lubrication.

**drive fit.** A type of *force fit*.

**drop.** A casting imperfection due o a portion of the sand dropping from the cope or other overhanging section of the mold.

**drop etching.** In metallography, placing of a drop of etchant on the polished surface.

**drop forging.** The forging obtained by hammering metal in a pair of closed dies to produce the form in the finishing impression under a *drop hammer*; forging method requiring special dies for each shape.

**drop hammer.** A term generally applied to forging hammers in which energy for forging is provided by gravity, steam, or compressed air (Fig. 136). See also *air-lift hammer*, *board hammer*, and *steam hammer*.

**drop hammer forming.** A process for producing shapes by the progressive deformation of sheet metal in matched dies under the repetitive blows of a gravity-drop or power-drop hammer (Fig. 137). The process is restricted to relatively shallow parts and thin sheet from approximately 0.6 to 1.6 mm (0.024 to 0.064 in.).

**droplet erosion.** Erosive wear caused by the impingement of

FIG. 136

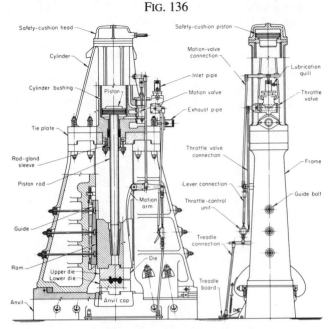

*Principal components of a power-drop hammer with foot control to regulate the force of the blow*

FIG. 137

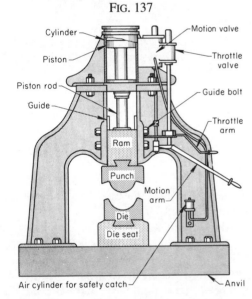

*Schematic of an air-actuated power-drop hammer equipped for drop hammer forming*

liquid droplets on a solid surface. See also *erosion (erosive) wear*.

**drop point.** The temperature at which a drop falls from a grease through a specified orifice. The drop point does not necessarily represent the maximum operating temperature of a grease.

**drop-through.** An undesirable sagging or surface irregularity, usually encountered when brazing or welding near the solidus of the base metal, caused by overheating with rapid diffusion or alloying between the filler metal and the base metal.

**dross.** (1) The scum that forms on the surface of molten metal largely because of oxidation but sometimes because of the rising of impurities to the surface. (2) Oxide and other contaminants that form on the surface of molten solder.

**drum.** A filler metal package type consisting of a continuous length of electrode wound or coiled within an enclosed cylindrical container.

**drum test.** A test of the green strength of powder metallurgy compacts by tumbling them in a drum and examining the sharpness of the edges and corners.

**dry.** To change the physical state of an adhesive on an adherend by the loss of solvent constituents by evaporation or absorption, or both. See also *cure* and *set.*

**dry and baked compression test.** An American Foundrymen's Society test for determining the maximum compressive stress that a baked sand mixture is capable of developing.

**dry blend.** Refers to a plastic molding compound containing all necessary ingredients mixed in a way that produces a dry, free flowing, particulate material. This term is commonly used in connection with polyvinyl chloride molding compounds.

**dry bond adhesive.** See *contact adhesive.*

**dry-bulb.** The temperature of the air as indicated by an accurate thermometer, corrected for radiation if significant.

**dry coloring.** Method commonly used by fabricators for coloring plastics by tumble blending uncolored particles of the plastic material with selected dyes and pigments.

**dry corrosion.** See *gaseous corrosion.*

**dry cyaniding.** (obsolete) Same as *carbonitriding.*

**dry etching.** In metallography, development of microstructure under the influence of gases.

**dry fiber.** In composites fabrication, a condition in which fibers are not fully encapsulated by resin during *pultrusion.*

**dry-film lubrication.** Lubrication that involves the application of a thin film of solid lubricant to the surface or surfaces to be lubricated.

**dry friction.** Friction that occurs between two bodies in the absence of lubrication. This term is inaccurate because it historically implies that there is no intentionally applied lubrication, when in fact solid lubrication conditions can be considered "dry." Therefore, this term should not be used.

**drying.** Removal, by evaporation, of uncombined water or other volatile substance from a ceramic raw material or product, usually expedited by low-temperature heating.

**drying oil.** An oil capable of conversion from a liquid to a solid by slow reaction with oxygen in the air.

**drying temperature.** The temperature to which an adhesive on an adherend, an adhesive in an assembly, or the assembly itself is subjected to dry the adhesive. The temperature attained by the adhesive in the process of drying (adhesive drying temperature) may differ from the temperature of the atmosphere surrounding the assembly (assembly drying temperature). See also *curing temperature* and *setting temperature.*

**drying time.** The period of time during which an adhesive on an adherend or an assembly is allowed to dry with or without the application of heat or pressure, or both. See also *curing time, joint-conditioning time,* and *setting time.*

**dry laminate.** A laminate containing insufficient resin for complete bonding of the reinforcement. See also *resin-starved area.*

**dry lay-up.** Construction of a laminate by the layering of preimpregnated reinforcement (partly cured resin) in a female mold or on a male mold, usually followed by bag molding or autoclave molding. See also *vacuum bag molding.*

**dry lubricant.** See *solid lubricant.*

**dry objective.** Any microscope objective designed for use without liquid between the cover glass and the objective, or, in the case of metallurgical objectives, in the space between objective and specimen.

**dry permeability.** In casting, the property of a molded mass of sand, bonded or unbonded, dried at ~100 to 110 °C (~220 to 230 °F), and cooled to room temperature, that allows the transfer of gases resulting during the pouring of molten metal into a mold.

**dry-running.** In seals, running without liquid present at the seal surface.

**dry sand casting.** The process in which the sand molds are dried at above 100 °C (212 °F) before use.

**dry sand mold.** A casting mold made of sand and then dried at 100 °C (212 °F) or above before being used. Contrast with *green sand mold.*

**dry-sand rubber wheel test.** In wear testing, a term used to describe a standard abrasive wear testing method in which a stream of dry quartz sand is passed between a rotating rubber wheel and a stationary test coupon that is held against it under specified normal force (Fig. 138).

**dry sliding wear.** Sliding wear in which there is no intentional lubricant or moisture introduced into the contact area. See also *unlubricated sliding.*

**dry strength** (adhesives). The strength of an adhesive joint determined immediately after drying under specified conditions or after a period of conditioning in the standard laboratory atmosphere. See also *wet strength.*

**dry strength** (casting). The maximum strength of a molded sand specimen that has been thoroughly dried at ~100 to 110 °C (~220 to 230 °F) and cooled to room temperature. Also known as dry bond strength.

FIG. 138

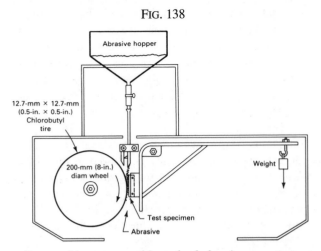

*Dry-sand low-stress rubber wheel abrasive wear tester*

**dry tack**. The property of certain adhesives, particularly nonvulcanizing rubber adhesives, to adhere on contact to themselves at a stage in the evaporation of volatile constituents, even though they seem dry to the touch. Synonym for aggressive tack.

**dry winding**. In composites fabrication, filament winding using preimpregnated roving, as differentiated from wet w inding, in which unimpregnated roving is pulled through a resin bath just before being wound onto a mandrel. See also *wet winding*.

**DSC**. See *differential scanning calorimetry*.

**DTA**. See *differential thermal analysis*.

**DTUL**. See *deflection temperature under load*.

**dual-metal centrifugal casting**. Centrifugal castings produced by pouring a different metal into the rotating mold after the first metal poured has solidified. Also referred to as *bimetal casting*.

**dual-phase steels**. A new class of *high-strength low-alloy steels* characterized by a tensile strength value of approximately 550 MPa (80 ksi) and by a microstructure consisting of about 20% hard martensite particles dispersed in a soft ductile ferrite matrix. The term dual phase refers to the predominance in the microstructure of two phases, ferrite and martensite. However, small amounts of other phases, such as bainite, pearlite, or retained austenite, may also be present. Table 18 lists compositions of typical dual-phase steels.

**ductile crack propagation**. Slow crack propagation that is accompanied by noticeable *plastic deformation* and requires energy to be supplied from outside the body. Contrast with *brittle crack propagation*.

**ductile erosion behavior**. Erosion behavior having characteristic properties (i.e., considerable *plastic deformation*) that can be associated with *ductile fracture* of the exposed solid surface. A characteristic ripple pattern forms on the exposed surface at low values of angle of attack. Contrast with *brittle erosion behavior*.

**ductile fracture**. Fracture characterized by tearing of metal accompanied by appreciable gross plastic deformation and expenditure of considerable energy. Contrast with *brittle fracture* (see also figure accompanying *brittle fracture*).

**ductile iron**. A *cast iron* that has been treated while molten with an element such as magnesium or cerium to induce the formation of free graphite as nodules or spherulites (Fig. 139), which imparts a measurable degree of ductility to the cast metal. Ductile irons typically contain 3.0 to 4.0% C, 1.8 to 2.8% Si, 0.1 to 1.0% Mn, 0.01 to 0.1% P, and 0.01 to 0.03% S. Also known as nodular cast iron, spherulitic graphite cast iron, and spheroidal graphite (SG) iron.

FIG. 139

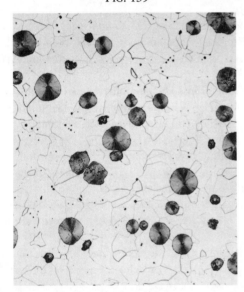

*Microstructure of annealed ductile iron. Note spheroidal shape of graphite nodules. 100×*

**Table 18　Typical dual-phase steel compositions**

| Production method | Composition, wt% | | | | | | |
|---|---|---|---|---|---|---|---|
| | C | Mn | Si | Cr | Mo | V | N |
| Continuous annealing, hot-rolled gage | 0.11 | 1.43 | 0.61 | 0.12 | 0.08 | 0.06 | 0.01 |
| Continuous annealing, cold-rolled gage | 0.11 | 1.20 | 0.40 | . . . | . . . | . . . | . . . |
| Box annealing | 0.12 | 2.10 | 1.40 | . . . | . . . | . . . | . . . |
| As rolled | 0.06 | 0.90 | 1.35 | 0.50 | 0.35 | . . . | . . . |

**ductility.** The ability of a material to deform plastically without fracturing.

**dullness.** A lack of pultruded surface gloss or shine in plastic parts. Can be caused by insufficient cure locally or in large areas, resulting in the dull band created on a pultruded part within the die when the pultrusion process is interrupted briefly.

**dummy block.** In *extrusion*, a thick unattached disk placed between the ram and the billet to prevent overheating of the ram.

**dummy cathode.** (1) A *cathode*, usually corrugated to give variable current densities, that is plated at low current densities to preferentially remove impurities from a plating solution. (2) A substitute cathode that is used during adjustment of operating conditions.

**dummying.** Plating with *dummy cathodes*.

**duoplasmatron.** A type of ion source in which a plasma created by an arc discharge is confined and compressed by a nonuniform magnetic field.

**duplex alloys.** Bearing alloys consisting of two phases, one much softer than the other.

**duplex coating.** See *composite plate*.

**duplex grain size.** The simultaneous presence of two grain sizes in substantial amounts, with one grain size appreciably larger than the others. Also termed mixed grain size.

**duplexing.** Any two-furnace melting or refining process. Also called duplex melting or duplex processing.

**duplex microstructure.** A two-phase structure.

**duplex stainless steels.** Stainless steels having a fine-grained mixed microstructure of ferrite and austenite (Fig. 140) with a composition centered around 26Cr-6.5Ni. The cor-

FIG. 140

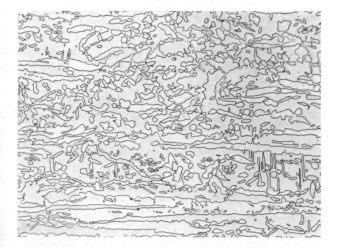

*Photomicrograph of a duplex stainless steel showing elongated austenite islands in the ferrite matrix. 200×*

rosion resistance of duplex stainless steels is like that of austenitic stainless steels. However, duplex stainless steels possess higher tensile and yield strengths and improved resistance to stress-corrosion cracking than their austenitic counterparts.

**duplicate measurement.** A second measurement made on the same (or identical) sample of material to assist in the evaluation of measurement variance.

**duplicate sample.** A second sample randomly selected from a population to assist in the evaluation of sample variance.

**duplicating.** In machining and grinding, reproducing a form from a master with an appropriate type of machine tool, utilizing a suitable tracer or program-controlled mechanism.

**duralumin** (obsolete). A term frequently applied to the class of age-hardenable aluminum-copper alloys containing manganese, magnesium, or silicon.

**durometer reading.** An index that is used for ranking the relative hardness of elastomers. The durometer hardness test involves forcing a 30° tapered indenter into the surface of the specimen using calibrated loading springs. A dial gage indicates the depth of penetration in durometer numbers, which are directly proportional to the load on the spring.

**Durville process.** A casting process that involves rigid attachment of the mold in an inverted position above the crucible. The melt is poured by tilting the entire assembly, causing the metal to flow along a connecting *launder* and down the side of the mold.

**dust.** Specifically, a superfine metal powder having predominantly submicron size particles.

**dusting.** (1) A phenomenon, usually affecting carbon-base electrical motor brushes or other current-carrying contacts, wherein at low relative humidity or high applied current density, a powdery "dust" is produced during operation. (2) Applying a powder, such as sulfur to molten magnesium or graphite to a mold surface.

**duty.** The specification giving the load, ambient temperature, and speed under which surfaces are required to move.

**duty cycle.** For electric welding equipment, the percentage of time that current flows during a specified period. In arc welding, the specified period is 10 min.

**duty parameter.** See *capacity number*.

**dwarf width.** A condition in which the crosswise (of the direction of pultrusion) dimension of a flat surface of a part is less than that which the die would normally yield for a particular plastic or composite. The condition is usually caused by a partial blockage of the pultrusion die cavity caused by buildup, or particles of the composite adhering to the cavity surface. This condition is commonly called a lost edge, when the flat surface has a free edge that is altered by the buildup.

**dwell.** In forming plastics and composites, a pause in the application of pressure or temperature to a mold, made just before it is completely closed, to allow the escape of gas from the molding material. In filament winding, the time that the traverse mechanism is stationary while the mandrel continues to rotate to the appropriate point for the traverse to begin a new pass. In a standard autoclave cure cycle, an intermediate step in which the resin matrix is held at a temperature below the cure temperature for a specified period of time sufficient to produce a desired degree of staging. Used primarily to control resin flow.

**dwell mark**. A fracture surface marking which resembles a pronounced ripple mark, the presence of which indicates that the fracture paused at the location of the dwell mark for some indeterminable length of time. Also known as arrest mark.

**dwell time**. In powder metallurgy, the time period during which maximum pressure is applied to a compact in cold pressing or hot pressing.

**dynamic**. Moving, or having high velocity. Frequently used with high strain rate ($>0.1$ s$^{-1}$) testing of metal specimens. Contrast with *static*.

**dynamic creep**. Creep that occurs under conditions of fluctuating load or fluctuating temperature.

**dynamic electrode force**. The force given in newtons (pounds force) between the electrodes during the actual welding cycle in making spot, seam, or projection welds by resistance welding.

**dynamic friction**. See *kinetic friction*.

**dynamic hot pressing**. In powder metallurgy, a method of applying a vibrational load to the punches or die during hot pressing. See also *hot pressing* and *static hot pressing*.

**dynamic load**. An imposed force that is in motion, that is, one that may vary in magnitude, sense, and direction.

**dynamic mechanical measurement**. A technique in which either the modulus and/or damping of a substance under oscillatory load or displacement is measured as a function of temperature, frequency, or time, or a combination thereof.

**dynamic modulus**. The ratio of stress to strain under cyclic conditions (calculated from data obtained from either free or forced vibration tests, in shear, compression, or tension).

**dynamic seal**. A seal that has rotating, oscillating, or reciprocating motion between its components, as opposed to stationary-type seal such as a gasket.

**dynamic viscosity**. See *viscosity*.

# E

**earing**. The formation of ears or scalloped edges around the top of a drawn shell, resulting from directional differences in the plastic-working properties of rolled metal, with, across, or at angles to the direction of rolling.

**earthenware**. A glazed or unglazed nonvitreous clay-based ceramic *whiteware*.

**eccentric**. The offset portion of the driveshaft that governs the stroke or distance the crosshead moves on a mechanical or manual shear.

**eccentric gear**. A main press-drive gear with an eccentric(s) as an integral part. The unit rotates about a common shaft, with the eccentric transmitting the rotary motion of the gear into the vertical motion of the slide through a connection (Fig. 141).

**eccentricity**. In *journal bearings*, the radial displacement of the journal center from the center of the bearing liner.

**eccentricity ratio**. In a bearing, the ratio of the eccentricity to the radial clearance.

**eccentric press**. A *mechanical press* in which an eccentric, instead of a crankshaft, is used to move the *slide*.

**ECM**. An abbreviation for *electrochemical machining*.

**eddy-current testing**. An electromagnetic nondestructive

FIG. 141

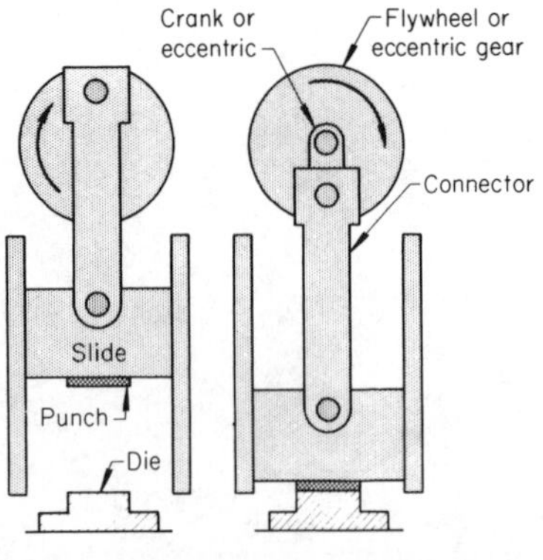

*Operating principle of a mechanical press equipped with eccentric-gear drives*

testing method in which eddy-current flow is induced in the test object. Changes in flow caused by variations in the object are reflected into a nearby coil or coils where they are detected and measured by suitable instrumentation (Fig. 142).

FIG. 142

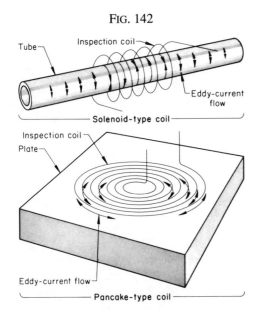

*Two common types of inspection coils and the patterns of eddy-current flow generated by the exciting current in the coils. Solenoid-type coil is applied to cylindrical or tubular parts; pancake-type coil, to a flat surface.*

FIG. 144

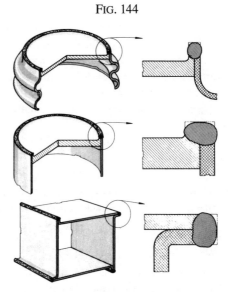

*Typical edge joint designs*

**edge dislocation**. See *dislocation*.

**edge distance**. The distance from the edge of a bearing test specimen to the center of the hole in the direction of applied force. See also *bearing test*.

**edge distance ratio**. In a *bearing test*, the distance from the center of the bearing hole to the edge of a specimen in the direction of the principal stress, divided by the diameter of the hole.

**edge effect** (thermal spraying). Loosening of the bond between the sprayed material and the base material at the edges, due to stresses set up in cooling.

**edge-flange weld**. A flange weld with two members flanged at the location of welding (Fig. 143).

**edge joint** (adhesive bonding). A joint made by bonding the edge faces of two adherends.

**edge joint** (welding). A joint between the edges of two or more parallel or nearly parallel members (Fig. 144).

**edge preparation**. Surface preparation on the edge of a member for welding.

**edger (edging impression)**. The portion of a die impression that distributes metal during forging into areas where it is most needed in order to facilitate filling the cavities of subsequent impressions to be used in the forging sequence. See also *fuller (fullering impression)*.

**edge stability**. An indicator of strength in a green powder metallurgy compact, as may be determined by tumbling in a drum. See also *drum test*.

**edge strain**. Transverse strain lines or Lüders lines ranging from 25 to 300 mm (1 to 12 in.) in from the edges of cold rolled steel sheet or strip. See also *Lüders lines*.

**edge strength**. The resistance of the sharp edges of a powder metallurgy compact against abrasion, as may be determined by tumbling in a drum. See also *drum test*.

**edge-trailing technique**. In metallography, a unidirectional motion perpendicular to and toward one edge of the specimen during abrasion or polishing used to improve edge retention.

**edge weld**. A weld in an *edge joint*.

**edging**. (1) In sheet metal forming, reducing the flange radius by retracting the forming punch a small amount after the

FIG. 143

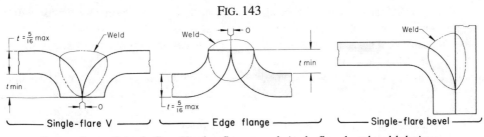

*Comparison of single-flare V, edge flange, and single-flare bevel weld designs*

stroke but before release of the pressure. (2) In rolling, the working of metal in which the axis of the roll is parallel to the thickness dimension. Also called edge rolling. (3) The forging operation of working a bar between contoured dies while turning it 90° between blows to produce a varying rectangular cross section. (4) In a forging, removing flash that is directed upward between dies, usually accomplished using a lathe.

**EDM**. Abbreviation for *electrical discharge machining*.

**effective atomic number**. For an element, the number of protons in the nucleus of an atom. For mixtures, an effective atomic number can be calculated to represent a single element that would have attenuation properties identical to those of the mixture.

**effective crack size ($a_e$)**. The *physical crack size* augmented for the effects of crack-tip plastic deformation. Sometimes the effective crack size is calculated from a measured value of a physical crack size plus a calculated value of a plastic-zone adjustment. A preferred method for calculation of effective crack size compares compliance from the secant of a load-deflection trace with the elastic compliance from a calibration for the type of specimen.

**effective draw**. The maximum limits of forming depth that can be achieved with a multiple-action press; sometimes called maximum draw or maximum depth of draw.

**effective leakage area**. The orifice flow area that will result in the same calculated flow for a given pressure drop as is measured for the seal in question. This concept is useful when comparing the leakage performance of seals of different sizes and designs, and of seals operating under different conditions.

**effective length of weld**. The length of weld throughout which the correctly proportioned cross section exists. In a curved weld, it is measured along the axis of the weld.

**effective rake**. The angle between a plane containing a tooth face and the axial plane through the tooth point as measured in the direction of chip flow through the tooth point. Thus, it is the rake resulting from both cutter configuration and direction of chip flow.

**effective throat**. The minimum distance from the root of a weld to its face. See also *joint penetration*.

**effective yield strength**. An assumed value of uniaxial yield strength that represents the influence of plastic yielding on fracture test parameters.

**E-glass**. A family of glasses with a calcium aluminoborosilicate composition and a maximum alkali content of 2.0%. A general-purpose fiber that is most often used in reinforced plastics, and is suitable for electrical laminates because of its high resistivity. Also called electric glass. See also *glass fiber*.

**eight-harness satin**. A type of fabric weave. The fabric has a seven-by-one weave pattern in which a filling thread floats over seven warp threads and then under one (Fig.

FIG. 145

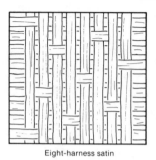 

Eight-harness satin | Crowfoot satin

*Comparison of eight-harness satin and crowfoot satin fabric weaves*

145). Like the crowfoot weave, it looks different on one side than on the other. This weave is more pliable than any of the others and is especially adaptable to forming around compound curves, such as on radomes.

**885-°F (475-°C) embrittlement**. *Embrittlement* of stainless steels upon extended exposure to temperatures between 400 and 510 °C (750 and 950 °F). This type of embrittlement is caused by fine, chromium-rich precipitates that segregate at grain boundaries; time at temperature directly influences the amount of segregation. Grain-boundary segregation of the chromium-rich precipitates increases strength and hardness, decreases ductility and toughness, and changes corrosion resistance. This type of embrittlement can be reversed by heating above the precipitation range.

**ejection**. Removal of a powder metallurgy compact after completion of pressing, whereby the compact is pushed through the die cavity by one of the punches. Also called *knockout*.

**ejection mark**. A surface mark on a plastic part caused by the ejector pin when it pushes the part out of the molded cavity.

**ejector**. A device mounted in such a way that it removes or assists in removing a focused part from a die.

**ejector half**. The movable half of a die-casting die containing the ejector pins.

**ejector punch**. See *knockout punch*.

**ejector rod**. A rod used to push out a formed piece.

**elastic aftereffect**. Time-dependent recovery, toward original dimensions, after the load has been reduced or removed from an elastically or plastically strained body. See also *anelasticity*.

**elastic calibration device**. A device for use in verifying the load readings of a testing machine consisting of an elastic member(s) to which loads may be applied, combined with a mechanism or device for indicating the magnitude (or a quantity proportional to the magnitude) of deformation under load.

**elastic compliance**. A condition under which two bodies in contact, which are subjected to a force, undergo small elastic displacement without slip.

**elastic constants.** The factors of proportionality that relate elastic displacement of a material to applied forces. See also *bulk modulus of elasticity, modulus of elasticity, Poisson's ratio*, and *shear modulus*.

**elastic deformation.** A change in dimensions directly proportional to and in phase with an increase or decrease in applied force.

**elastic electron scatter.** The scatter of electrons by an object without loss of energy, usually an interaction between electrons and atoms.

**elastic energy.** The amount of energy required to deform a material within the elastic range of behavior, neglecting small heat losses due to internal friction. The energy absorbed by a specimen per unit volume of material contained within the gage length being tested. It is determined by measuring the area under the stress-strain curve up to a specified elastic strain. See also *modulus of resilience* and *strain energy*.

**elastic hysteresis.** A misnomer for an anelastic strain that lags a change in applied stress, thereby creating energy loss during cyclic loading. More properly termed *mechanical hysteresis*.

**elasticity.** The property of a material by virtue of which deformation caused by stress disappears upon removal of the stress. A perfectly elastic body completely recovers its original shape and dimensions after release of stress.

**elastic limit.** The maximum stress which a material is capable of sustaining without any permanent strain (deformation) remaining upon complete release of the stress. A material is said to have passed its elastic limit when the load is sufficient to initiate plastic, or nonrecoverable, deformation. See also *proportional limit*.

**elastic modulus.** Same as *modulus of elasticity*.

**elastic ratio.** *Yield point* divided by *tensile strength*.

**elastic recovery.** (1) The fraction of a given deformation that behaves elastically. A perfectly elastic material has an elastic recovery of 1; a perfectly plastic material has an elastic recovery of 0. (2) In hardness testing, the shortening of the original dimensions of the indentation upon release of the applied load.

**elastic resilience.** The amount of energy absorbed in stressing a material up to the elastic limit; or, the amount of energy that can be recovered when stress is released from the elastic limit.

**elastic scattering.** Collisions between particles that are completely described by conservation of energy and momentum. Contrast with *inelastic scattering*.

**elastic strain.** See *elastic deformation*.

**elastic strain energy.** The energy expended by the action of external forces in deforming a body elastically. Essentially all the work performed during elastic deformation is stored as elastic energy, and this energy is recovered upon release of the applied force.

**elastic true strain** ($\varepsilon_e$). Elastic component of the *true strain*.

**elastic waves.** Mechanical vibrations in an elastic medium.

**elastohydrodynamic lubrication.** A condition of lubrication in which the friction and film thickness between two bodies in relative motion are determined by the elastic properties of the bodies, in combination with the viscous properties of the lubricant at the prevailing pressure, temperature, and rate of shear. See also *boundary lubrication*, *plastohydrodynamic lubrication*, and *thin-film lubrication*.

**elastomer.** (1) A material that substantially recovers its original shape and size at room temperature after removal of a deforming force. A material that shows reversible elasticity up to very high strain levels. (2) Any elastic, rubberlike substance, such as natural or synthetic rubber.

**elastomeric tooling.** A tooling system that uses the thermal expansion of rubber materials to form reinforced plastic or composite parts during cure.

**electrical conductivity.** See *conductivity, electrical*.

**electrical discharge grinding.** Grinding by spark discharges between a negative electrode grinding wheel and a positive workpiece separated by a small gap containing a dielectric fluid such as petroleum oil (Fig. 146).

FIG. 146

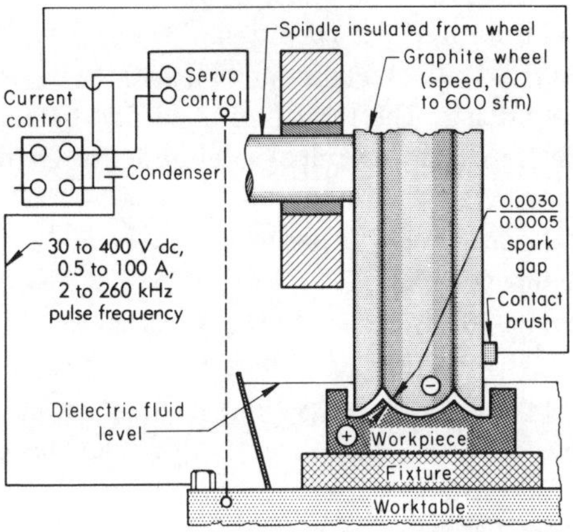

*Setup for electrical discharge grinding. Spark gap given in inches*

**electrical discharge machining (EDM).** Metal removed by a rapid spark discharge between different polarity electrodes, one on the workpiece and the other the tool separated by a gap distance of 0.013 to 0.9 mm (0.0005 to 0.035 in.). The gap is filled with dielectric fluid and metal particles which are melted, in part vaporized and expelled from the gap (Fig. 147).

**electrical discharge wire cutting.** A special form of electrical discharge machining wherein the electrode is a con-

FIG. 147

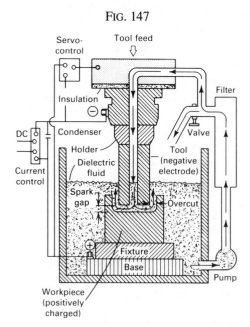

*Typical setup for electrical discharge machining*

FIG. 148

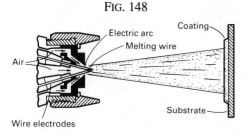

*Schematic of the electric arc spray process*

tinuous moving conductive wire. Also referred to as traveling wire electrical discharge machining.

**electrical disintegration.** Metal removal by an electrical spark acting in air. It is not subject to precise control, the most common application being the removal of broken tools such as taps and drills.

**electrical isolation.** The condition of being electrically separated from other metallic structures or the environment.

**electrical pitting.** The formation of surface cavities by removal of metal as a result of an electrical discharge across an interface.

**electrical porcelain.** Vitrified *whiteware* having an electrical insulating function.

**electrical resistivity.** The electrical resistance offered by a material to the flow of current, times the cross-sectional area of current flow and per unit length of current path; the reciprocal of the conductivity. Also called resistivity or specific resistance.

**electric arc furnace.** See *arc furnace*.

**electric arc spraying.** A *thermal spraying* process using as a heat source an electric arc between two consumable electrodes of a coating material and a compressed gas which is used to atomize and propel the material to the substrate (Fig. 148).

**electric bonding** (thermal spraying). See preferred term *surfacing*.

**electric brazing.** See preferred terms *resistance brazing* and *arc brazing*.

**electric dipole.** The result of a distribution of bound charges, that is, separated charges that are bound to their centers of equilibrium by an elastic force; equal numbers of positive and negative charges must be present in an uncharged medium.

**electric dipole moment.** A quantity characteristic of a distribution of bound charges equal to the vector sum over the charges of the product of the charge and the position vector of the charge.

**electric dipole transition.** A transition of an atom, molecule, or nucleus from one energy state to another, which results from the interaction of electromagnetic radiation with the dipole moment of the molecule, atom, or nucleus.

**electric field effect.** See *Stark effect*.

**electric furnace.** A metal melting or holding furnace that produces heat from electricity. It may operate on the resistance or induction principle. See also *induction furnace*.

**electrochemical admittance.** The inverse of *electrochemical impedance*.

**electrochemical cell.** An electrochemical system consisting of an *anode* and a *cathode* in metallic contact and immersed in an *electrolyte*. The anode and cathode may be different metals or dissimilar areas on the same metal surface. See also figure accompanying *cathodic protection*.

**electrochemical corrosion.** Corrosion that is accompanied by a flow of electrons between cathodic and anodic areas on metallic surfaces.

**electrochemical discharge machining.** Metal removal by a combination of the processes of *electrochemical machining* and *electrical discharge machining*. Most of the metal removal occurs via anodic dissolution (i.e., ECM action). Oxide films which form as a result of electrolytic action through an electrolytic fluid are removed by intermittent spark discharges (i.e., EDM action). Hence, the combination of the two actions.

**electrochemical equivalent.** The weight of an element or group of elements oxidized or reduced at 100% efficiency by the passage of a unit quantity of electricity. Usually expressed as grams per coulomb.

**electrochemical (chemical) etching.** General expression for all developments of microstructure through reduction and oxidation (redox reactions).

**electrochemical grinding.** A process whereby metal is re-

FIG. 149

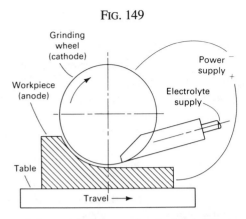

*Schematic of electrochemical grinding*

moved by deplating. The workpiece is the anode; the cathode is a conductive aluminum oxide-copper or metal-bonded diamond grinding wheel with abrasive particles (Fig. 149). Most of the metal is removed by deplating; 0.05 to 10% is removed by abrasive cutting.

**electrochemical impedance.** The frequency-dependent complex-valued proportionality factor ($\Delta E/\Delta i$) between the applied potential or current and the response signal. This factor is the total opposition ($\Omega$ or $\Omega \cdot cm^2$) of an electrochemical system to the passage of charge. The value is related to the *corrosion rate* under certain circumstances.

**electrochemical machining.** Controlled metal removal by anodic dissolution. Direct current passes through flowing film of conductive solution which separates the workpiece from the electrode-tool (Fig. 150). The workpiece is the *anode*, and the tool is the *cathode*.

FIG. 150

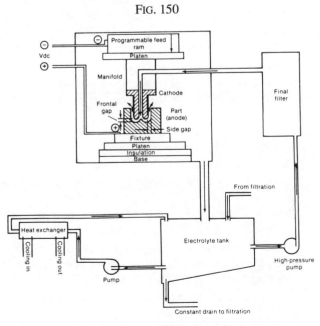

*Schematic of the electrochemical machining system*

**electrochemical potential.** The partial derivative of the total electrochemical free energy of a constituent with respect to the number of moles of this constituent where all factors are kept constant. It is analogous to the *chemical potential* of a constituent except that it includes the electric as well as chemical contributions to the free energy. The *potential* of an electrode in an electrolyte relative to a *reference electrode* measured under open circuit conditions.

**electrochemical reaction.** A reaction caused by passage of an electric current through a medium which contains mobile ions (as in electrolysis); or, a spontaneous reaction made to cause current to flow in a conductor external to this medium (as in a galvanic cell). In either event, electrical connection is made to the external portion of the circuit via a pair of electrodes. See also *electrolyte.*

**electrochemical series.** Same as *electromotive force series.*

**electrocorrosive wear.** Wear of a solid surface which is accelerated by the presence of a corrosion-inducing electrical potential across the contact interface. This process is usually associated with wear in the presence of a liquid electrolyte in the interface. However, moisture from the air can also facilitate this type of wear when a galvanic wear couple exists and the contacting materials are sufficiently reactive.

**electrode.** Compressed graphite or carbon cylinder or rod used to conduct electric current in electric arc furnaces, arc lamps, and so forth.

**electrode (electrochemistry).** One of a pair of conductors introduced into an electrochemical cell, between which the ions in the intervening medium flow in opposite directions and on whose surfaces reactions occur (when appropriate external connection is made). In direct current operation, one electrode or "pole" is positively charged, the other negatively. See also *anode, cathode, electrochemical reaction,* and *electrolyte.*

**electrode (welding).** (1) In arc welding, a current-carrying rod that supports the arc between the rod and work, or between two rods as in twin carbon-arc welding. It may or may not furnish filler metal. See also *bare electrode, covered electrode,* and *lightly coated electrode.* (2) In resistance welding, a part of a resistance welding machine through which current and, in most instances, pressure are applied directly to the work. The electrode may be in the form of a rotating wheel, rotating roll, bar, cylinder, plate, clamp, chuck, or modification thereof. See also *resistance welding electrode.* (3) In arc and plasma spraying, the current-carrying components which support the arc.

**electrode cable.** Same as *electrode lead.*

**electrode deposition.** The weight of weld-metal deposit obtained from a unit length of electrode.

**electrode extension.** The length of unmelted electrode extending beyond the end of the contact tube during welding.

**electrode force.** The force between electrodes in a spot, seam, and projection weld. See also *dynamic electrode force*, *static electrode force*, and *theoretical electrode force*.

**electrode holder**. A device used for mechanically holding the electrode while conducting current to it.

**electrode lead**. The electrical conductor between the source of arc welding current and the electrode holder.

**electrode polarization**. Change of *electrode potential* with respect to a reference value. Often the *free corrosion potential* is used as the reference value. The change may be caused, for example, by the application of an external electrical current or by the addition of an oxidant or reductant.

**electrodeposition**. (1) The deposition of a conductive material from a plating solution by the application of electrical current. (2) The deposition of a substance on an electrode by passing electric current through an electrolyte. *Electrochemical (plating)*, *electroforming*, *electrorefining*, and *electrotwinning* result from electrodeposition.

**electrode potential**. The *potential* of an *electrode* in an *electrolysis* as measured against a *reference electrode*. The electrode potential does not include any resistance losses in potential in either the solution or external circuit. It represents the reversible work to move a unit charge from the electrode surface through the solution to the reference electrode.

**electrode reaction**. Interfacial reaction equivalent to a transfer of charge between electronic and ionic conductors. See also *anodic reaction* and *cathodic reaction*.

**electrode setback**. In plasma arc welding and cutting, the distance the electrode is recessed behind the constricting orifice measured from the outer face of the nozzle.

**electrode skid**. In spot, seam, or projection welding, the sliding of an electrode along the surface of the work.

**electroformed molds**. A mold for forming plastics made by electroplating metal on the reverse pattern of the cavity. Molten steel may then be sprayed on the back of the mold to increase its strength.

**electroforming**. Making parts by electrodeposition on a removable form.

**electrogalvanizing**. The *electroplating* of zinc upon iron or steel.

**electrogas welding (EGW)**. A *vertical position* arc welding process which produces coalescence of metals by heating them with an arc between a continuous filler metal (consumable) electrode and the work. Copper dams (molding shoes) are used to confine the molten weld metal. The electrodes may be either flux cored or solid wire. Shielding may or may not be obtained from an externally supplied gas or mixture. See also *gas metal arc welding* and *flux cored arc welding*.

**electrokinetic potential**. This *potential*, sometimes called zeta potential, is a potential difference in the solution caused by residual, unbalanced charge distribution in the adjoining solution, producing a double layer. The electrokinetic potential is different from the *electrode potential* in that it occurs exclusively in the solution phase; that is, it represents the reversible work necessary to bring a unit charge from infinity in the solution up to the interface in question but not through the interface.

**electroless plating**. (1) A process in which metal ions in a dilute aqueous solution are plated out on a substrate by means of autocatalytic chemical reduction. (2) The deposition of conductive material from an autocatalytic plating solution without the application of electrical current.

**electroluminescence**. The direct conversion of electrical energy into light.

**electrolysis**. (1) Chemical change resulting from the passage of an electric current through an *electrolyte*. (2) The separation of chemical components by the passage of current through an electrolyte.

**electrolyte**. (1) A chemical substance or mixture, usually liquid, containing ions that migrate in an electric field. (2) A chemical compound or mixture of compounds which when molten or in solution will conduct an electric current.

**electrolytic brightening**. Same as *electropolishing*.

**electrolytic cell**. An assembly, consisting of a vessel, electrodes, and an electrolyte, in which *electrolysis* can be carried out.

**electrolytic cleaning**. A process of removing soil, scale, or corrosion products from a metal surface by subjecting it as an *electrode* to an electric current in an electrolytic bath.

**electrolytic copper**. Copper that has been refined by the electrolytic deposition, including cathodes that are the direct product of the refining operation, refinery shapes cast from melted cathodes, and, by extension, fabricators' products made therefrom. Usually when this term is used alone, it refers to electrolytic tough pitch copper without elements other than oxygen being present in significant amounts. See also *tough pitch copper*.

**electrolytic corrosion**. Corrosion by means of electrochemical or mechanical action.

**electrolytic deposition**. Same as *electrodeposition*.

**electrolytic etching**. See *anodic etching*.

**electrolytic extraction**. Removal of phases by using an electrolytic cell containing an electrolyte that preferentially dissolves the metal matrix. See also *extraction*.

**electrolytic grinding**. A combination of grinding and machining wherein a metal-bonded abrasive wheel, usually diamond, is the *cathode* in physical contact with the anodic workpiece, the contact being made beneath the surface of a suitable electrolyte. The abrasive particles produce grinding act as nonconducting spacers permitting simultaneous machining through electrolysis.

**electrolytic machining**. Controlled removal of metal by use

FIG. 151

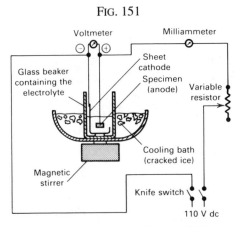

*Basic laboratory setup for electropolishing and electrolytic etching*

of an applied potential and a suitable electrolyte to produce the shapes and dimensions desired.

**electrolytic pickling.** *Pickling* in which electric current is used, the work being one of the electrodes.

**electrolytic polishing.** An electrochemical polishing process in which the metal to be polished is made the *anode* in an electrolytic cell where preferential dissolution at high points in the surface topography produces a specularly reflective surface (Fig. 151). Also referred to as *electropolishing*.

**electrolytic powder.** Powder produced by electrolytic deposition or by pulverizing of an electrodeposit.

**electrolytic protection.** See preferred term *cathodic protection*.

**electrolytic tough pitch.** A term describing the method of raw copper preparation to ensure a good physical- and electrical-grade copper-finished product.

**electromagnetic focusing device.** See *focusing device*.

**electromagnetic forming.** A process for forming metal by the direct application of an intense, transient magnetic field. The workpiece is formed without mechanical con-

tact by the passage of a pulse of electric current through a forming coil (Fig. 152). Also known as magnetic pulse forming.

**electromagnetic interference.** Interference related to accumulated electrostatic charge in a nonconductor.

**electromagnetic lens.** An electromagnet designed to produce a suitably shaped magnetic field for the focusing and deflection of electrons or other charged particles in electron-optical instrumentation.

**electromagnetic radiation.** Energy propagated at the speed of light by an electromagnetic field. The electromagnetic spectrum includes the following approximate wavelength regions:

| Region | Wavelength, Å (metric) |
|---|---|
| Gamma-ray | 0.005 to 1.40 (0.0005 to 0.14 nm) |
| X-ray | 0.1 to 100 (0.01 to 10 nm) |
| Far-ultraviolet | 100 to 2000 (10 to 200 nm) |
| Near-ultraviolet | 2000 to 3800 (200 to 380 nm) |
| Visible | 3800 to 7800 (380 to 780 nm) |
| Near-infrared | 7800 to 30 000 (0.78 to 3 μm) |
| Middle-infrared | $3 \times 10^4$ to $3 \times 10^5$ (3 to 30 μm) |
| Far-infrared | $3 \times 10^5$ to $3 \times 10^6$ (30 to 300 μm) |
| Microwave | $3 \times 10^6$ to $1 \times 10^{10}$ (0.3 mm to 1 m) |

**electromagnetism.** Magnetism caused by the flow of an electric current.

**electromechanical polishing.** An attack-polishing method in which the chemical action of the polishing fluid is enhanced or controlled by the application of an electric current between the specimen and the polishing wheel.

**electrometallurgy.** Industrial recovery or processing of metals and alloys by electric or electrolytic methods.

**electrometric titration.** A family of techniques in which the location of the endpoint of a *titration* involves the measurement of, or observation of changes in, some electrical quantity. Examples of such quantities include potential, current, conductance, frequency, and phase.

**electromotive force.** (1) The force that determines the flow

FIG. 152

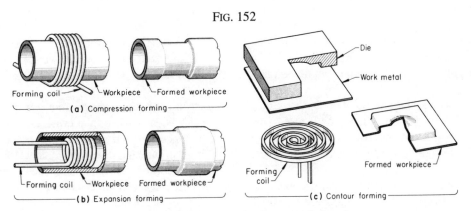

*Three basic methods of electromagnetic forming*

**Table 19　Electromotive force series**
SHE: standard hydrogen electrode

| Electrode reaction | Standard potential at 25 °C (77 °F), volts versus SHE |
|---|---|
| $Au^{3+} + 3e^- \rightarrow Au$ | 1.50 |
| $Pd^{2+} + 2e^- \rightarrow Pd$ | 0.987 |
| $Hg^{2+} + 2e^- \rightarrow Hg$ | 0.854 |
| $Ag^+ + e^- \rightarrow Ag$ | 0.800 |
| $Hg_2^{2+} + 2e^- \rightarrow 2Hg$ | 0.789 |
| $Cu^+ + e^- \rightarrow Cu$ | 0.521 |
| $Cu^{2+} + 2e^- \rightarrow Cu$ | 0.337 |
| $2H^+ + 2e^- \rightarrow H_2$ | (Reference) 0.000 |
| $Pb^{2+} + 2e^- \rightarrow Pb$ | −0.126 |
| $Sn_2 + 2e^- \rightarrow Sn$ | −0.136 |
| $Ni^{2+} + 2e^- \rightarrow Ni$ | −0.250 |
| $Co^{2+} + 2e^- \rightarrow Ni$ | −0.277 |
| $Tl^+ + e^- \rightarrow Tl$ | −0.336 |
| $In^{3+} + 3e^- \rightarrow In$ | −0.342 |
| $Cd^{2+} + 2e^- \rightarrow Cd$ | −0.403 |
| $Fe^{2+} + 2e^- \rightarrow Fe$ | −0.440 |
| $Ga^{3+} + 3e^- \rightarrow Ga$ | −0.53 |
| $Cr^{3+} + 3e^- \rightarrow Cr$ | −0.74 |
| $Cr^{2+} + 2e^- \rightarrow Cr$ | −0.91 |
| $Zn^{2+} + 2e^- \rightarrow Zn$ | −0.763 |
| $Mn^{2+} + 2e^- \rightarrow Mn$ | −1.18 |
| $Zr^{4+} + 4e^- \rightarrow Zr$ | −1.53 |
| $Ti^{2+} + 2e^- \rightarrow Ti$ | −1.63 |
| $Al^{3+} + 3e^- \rightarrow Al$ | −1.66 |
| $Hf^{4+} + 4e^- \rightarrow Hf$ | −1.70 |
| $U^{3+} + 3e^- \rightarrow U$ | −1.80 |
| $Be^{2+} + 2e^- \rightarrow Be$ | −1.85 |
| $Mg^{2+} + 2e^- \rightarrow Mg$ | −2.37 |
| $Na^+ + e^- \rightarrow Na$ | −2.71 |
| $Ca^{2+} + 2e^- \rightarrow Ca$ | −2.87 |
| $K^+ + e^- \rightarrow K$ | −2.93 |
| $Li^+ + e^- \rightarrow Li$ | −3.05 |

FIG. 153

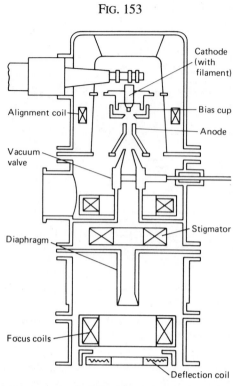

*Electron beam gun*

of electricity; a difference of electric potential. (2) Electrical potential; voltage.

**electromotive force series (emf series).** A series of elements arranged according to their *standard electrode potentials*, with "noble" metals such as gold being positive and "active" metals such as zinc being negative. In corrosion studies, the analogous but more practical *galvanic series* of metals is generally used. The relative positions of a given metal are not necessarily the same in the two series. Table 19 lists the electromotive force series for some common metals.

**electron.** An elementary particle that is the negatively charged constituent of ordinary matter. The electron is the lightest known particle possessing an electric charge. Its rest mass is $m_e \cong 9.1 \times 10^{-28}$ g, approximately $1/1836$ of the mass of the proton or neutron, which are, respectively, the positively charged and neutral constituents of ordinary matter.

**electron bands.** Energy states for the free electrons in a metal, as described by the use of the band theory (zone theory) of electron structure. Also called Brillouin zones.

**electron beam.** A stream of electrons in an electron-optical system.

**electron beam cutting.** A cutting process that uses the heat obtained from a concentrated beam composed primarily of high-velocity electrons, which impinge upon the work-

pieces to be cut; it may or may not use an externally supplied gas.

**electron beam gun.** A device for producing and accelerating electrons. Typical components include the emitter (also called the filament or cathode) which is heated to produce electrons via thermionic emission, a cup (also called the grid or grid cup), and the anode (Fig. 153).

**electron beam gun column.** The electron beam gun plus auxiliary mechanical and electrical components which may include beam alignment, focus, and deflection coils.

**electron beam heat treating.** A selective surface hardening process that rapidly heats a surface by direct bombardment with an accelerated stream of electrons.

**electron beam machining.** Removing material by melting and vaporizing the workpiece at the point of impingement of a focused high-velocity beam of electrons. The machining is done in high vacuum to eliminate scattering of the electrons due to interaction with gas molecules. The most important use of electron beam machining is for hole drilling.

**electron beam microprobe analyzer.** An instrument for selective analysis of a microscopic component or feature in which an electron beam bombards the point of interest in a vacuum at a given energy level. Scanning of a larger area permits determination of the distribution of selected elements. The analysis is made by measuring the wavelengths and intensities of secondary electromagnetic radiation resulting from the bombardment.

**electron beam welding.** A welding process that produces coalescence of metals with the heat obtained from a concentrated beam composed primarily of high-velocity electrons impinging upon the surfaces to be joined. Welding can be carried out at atmospheric pressure (non-vacuum), medium vacuum (approximately $10^{-3}$ to 25 torr), or high vacuum (approximately $10^{-6}$ to $10^{-3}$ torr).

**electron compound.** An intermediate phase on a *phase diagram*, usually a binary phase, that has the same crystal structure and the same ratio of valence electrons to atoms as those of intermediate phases in several other systems. An electron compound is often a solid solution of variable composition and good metallic properties. Occasionally, an ordered arrangement of atoms is characteristic of the compound, in which case the range of composition is usually small. Phase stability depends essentially on electron concentration and crystal structure and has been observed at valence-electron-to-atom ratios of $\frac{3}{2}$, $2\frac{1}{13}$, and $\frac{7}{4}$.

**electron diffraction.** The phenomenon, or the technique of producing diffraction patterns through the incidence of electrons upon matter. See also *diffraction pattern*.

**electron energy loss spectroscopy (EELS).** A spectrographic technique in the electron microscope that analyzes the energy distribution of the electrons transmitted through the specimen. The energy loss spectrum is characteristic of the chemical composition of the region being sampled.

**electron flow.** A movement of electrons in an external circuit connecting an *anode* and *cathode* in a corrosion cell; the current flow is arbitrarily considered to be in an opposite direction to the electron flow.

**electron gun.** A device for producing and accelerating a beam of electrons.

**electronic ceramics.** See *Technical Brief 14*.

**electronic heat control.** A device for adjusting the heating value (rms value) of the current in making a resistance weld by controlling the ignition or firing of the electronic devices in an electronic contactor. The current is initiated each half-cycle at an adjustable time with respect to the zero point on the voltage wave.

**electronic packaging.** The technical discipline of designing a protective enclosure for an electronic circuit so that it will both survive and perform under a plurality of environmental conditions.

**electron image.** A representation of an object formed by a beam of electrons focused by an electron-optical system. See also *image*.

**electron lens.** A device for focusing an electron beam to produce an image of an object.

**electron micrograph.** A reproduction of an image formed by the action of an electron beam on a photographic emulsion.

**electron microscope.** An electron-optical device that produces a magnified image of an object. Detail may be revealed by selective transmission, reflection, or emission of electrons by the object. See also *scanning electron microscope* and *transmission electron microscope*.

**electron microscope column.** The assembly of gun, lenses, specimen, and viewing and plate chambers. See also the figure accompanying the term *scanning electron microscope*.

**electron microscopy.** The study of materials by means of an electron microscope.

**electron microscopy impression.** See *impression*.

**electron multiplier phototube.** See *photomultiplier tube*.

**electron optical axis.** The path of an electron through an electron-optical system, along which it suffers no deflection due to lens fields. This axis does not necessarily coincide with the mechanical axis of the system.

**electron optical system.** A combination of parts capable of producing and controlling a beam of electrons to yield an image of an object.

**electron probe.** A narrow beam of electrons used to scan or illuminate an object or screen.

**electron probe x-ray microanalysis (EPMA).** A technique in analytical chemistry in which a finely focused beam of electrons is used to excite an x-ray spectrum characteristic of the elements in a small region of the sample. Compositional mapping determines elemental location and concentration (Fig. 154).

**electron scattering.** Any change in the direction of propagation or kinetic energy of an electron as a result of a collision.

**electron spin resonance (ESR) spectroscopy.** A form of spectroscopy similar to nuclear magnetic resonance, except that the species studied is an unpaired electron, not a magnetic nucleus.

**electron trajectory.** The path of an electron.

**electron velocity.** The rate of motion of an electron.

**electron wavelength.** The wavelength necessary to account for the deviation of electron rays in crystals by wave-diffraction theory. It is numerically equal to the quotient of *Planck's constant* divided by the electron momentum.

**electrophoresis.** Transport of charged colloidal or macromolecular materials in an electric field.

**electroplate.** The application of a metallic coating on a surface by means of electrolytic action.

**electroplating.** The electrodeposition of an adherent metallic coating on an object serving as a *cathode* for the purpose of securing a surface with properties or dimensions different from those of the basis metal.

**electropolishing.** A technique commonly used to prepare metallographic specimens, in which a high polish is produced making the specimen the *anode* in an *electrolytic cell*, where preferential dissolution at high points smooths the surface. Also referred to as *electrolytic polishing*.

## Technical Brief 14: Electronic Ceramics

THE ELECTRONICS INDUSTRY relies heavily on advanced ceramic materials such as alumina ($Al_2O_3$), alumina-titania ($Al_2O_3$-$TiO_2$), beryllia (BeO), and aluminum nitride (AlN) for substrates; barium titanate ($BaTiO_3$) for capacitors; lead zirconate titanate (PZT), lead magnesium niobate (PMN), and lead magnesium titanium niobate (PMTN) for actuators and transducers; zirconia ($ZrO_2$)-based ceramics for oxygen sensors; zinc oxide (ZnO)-based ma-

### Physical properties of ceramic substrate materials

|  | 96% $Al_2O_3$ | 99.5% $Al_2O_3$ | BeO | AlN | Mullite | Various glass-ceramic materials |
|---|---|---|---|---|---|---|
| Density, g/cm$^3$ | 3.75 | 3.90 | 2.85 | 3.25 | 2.82 | 2.5–2.8 |
| Flexural strength, MPa (ksi) | 400 (58) | 552 (80) | 207 (30) | 345 (50) | 186 (27) | 138 (20) |
| Thermal expansion from 25–500 °C (77–930 °F), $10^{-6}$/°C | 7.4 | 7.5 | 7.5 | 4.4 | 3.7 | 3.0–4.5 |
| Thermal conductivity at 20 °C (68 °F), W/m·°C | 26 | 35 | 260 | 140–220 | 4 | 4–5 |
| Dielectric constant at 1 MHz | 9.5 | 9.9 | 6.7 | 8.8 | 5.4 | 4–8 |
| Dielectric loss at 1 MHz (tan δ) | 0.0004 | 0.0002 | 0.0003 | 0.001–0.0002 | 0.003 | >0.002 |

terials for varistors; nickel oxide (NiO) and iron oxide ($Fe_2O_3$) for temperature sensors; and a variety of glassy materials for a host of devices and packaging. Semiconducting ceramics include materials, such as silicon carbide (SiC), that have properties similar to silicon but that retain their semiconductive properties at elevated properties. Ceramic oxides, which exhibit superconductivity, may soon be used in a number of applications (see also *superconductors* [Technical Brief 53]). In addition, ceramics can be combined with polymers or metals to form composites, or hybrids, having unique electrical and structural properties. Ceramics are important in consumer and military electronics, communications systems, power transmission and distribution systems, automotive and other transportation applications, as well as in computers.

The ceramic properties that are important for electronic applications result from a variety of mechanisms which depend on the bulk material, grain-boundary properties, and surface effects. Important properties include dielectric constant, dielectric strength, electrical conductivity, dielectric loss tangent and power loss, Curie temperature, and piezoelectric constant. The dielectric constant is a measure of the amount of charge a material can withstand before failure. For superconducting ceramics, the critical temperature ($T_c$) is the temperature at which a material becomes superconductive, i.e., has no resistance to the passage of an electrical current. The critical current of a superconductor is the maximum amount of current the superconductor can transmit before reverting to a nonsuperconducting state. For a number of applications, the materials must be able to conduct heat away from electronic circuit elements; hence, the ceramic must also have good thermal conductivity.

### Dielectric constants of ceramic capacitors at 25 °C (75 °F)

| Material | Dielectric constant, κ |
|---|---|
| Teflon | 2.1 |
| Silica glass | 3.8 |
| Polyvinylidene fluoride | 8.4 |
| Alumina | 10 |
| Magnesium oxide | 20 |
| Barium tetratitanate | 40 |
| Titanium dioxide | 100 |
| Calcium titanate | 160 |
| Strontium titanate | 320 |
| Barium titanate | 1000–5000 |
| Barium zirconium titanate | 20,000 |
| Lead magnesium niobate | 20,000 |

### Recommended Reading

- D.C. Cranmer, Overview of Technical, Engineering, and Advanced Ceramics, *Engineered Materials Handbook*, Vol 4, ASM International, 1991, p 16-20
- R.C. Buchanan *et al.*, Electrical/Electronic Applications for Advanced Ceramics, *Engineered Materials Handbook*, Vol 4, 1991, p 1105-1165
- L.L. Hench and D.B. Dove, Ed., *Physics of Electronic Ceramics*, Marcel Dekker, 1972

**electrorefining.** Using electric or electrolytic methods to convert impure metal to purer metal, or to produce an alloy from impure or partly purified raw materials.

**electroslag remelting (ESR).** A *consumable-electrode remelting* process in which heat is generated by the passage

of electric current through a conductive slag (Fig. 155). The droplets of metal are refined by contact with the slag.

**electroslag welding.** A fusion welding process in which the welding heat is provided by passing an electric current through a layer of molten conductive slag (flux) contained

FIG. 154

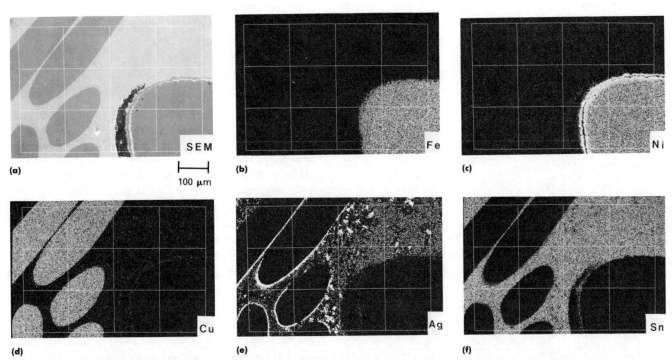

*Photographs of the SEM image (a) and five element EPMA maps taken over an area of a soldered joint. (b) Iron, 55 wt%. (c) Nickel, 27 wt%. (d) Copper, 97 wt%. (e) Silver, 10 wt%. (f) Tin, 84 wt%*

FIG. 155

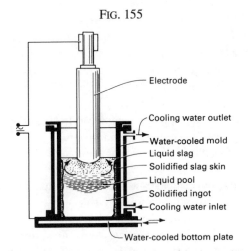

*Schematic of the electroslag remelting process*

FIG. 156

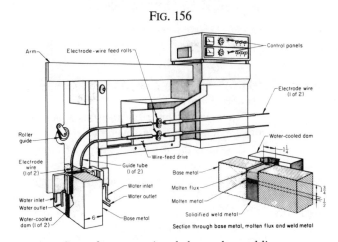

*Setup for conventional electroslag welding*

in a pocket formed by water-cooled dams that bridge the gap between the members being welded (Fig. 156). The resistance heated slag not only melts filler-metal electrodes as they are fed into the slag layer, but also provides shielding for the massive weld puddle characteristic of the process.

**electroslag welding electrode**. A filler metal component of the welding circuit through which current is conducted between the electrode guiding member and the molten slag (flux).

**electrostatic lens**. A lens producing a potential field capable of deflecting electron rays to form an image of an object.

**electrostrictive effect**. The reversible interaction, exhibited by some crystalline materials, between an elastic strain and an electric field. The direction of the strain is independent of the polarity of the field. Compare with *piezo-electric effect*.

**electrotinning**. *Electroplating* tin on an object.

**electrotyping**. The production of printing plates by *electroforming*.

**electrotwinning**. Recovery of a metal from an ore by means of electrochemical processes.

**elliptical bearing**. See *lemon bearing (elliptical bearing)*.

**elongated alpha**. A fibrous structure in titanium alloys brought about by unidirectional metalworking (Fig. 157). It may be enhanced by the prior presence of blocky and/or grain boundary α.

FIG. 157

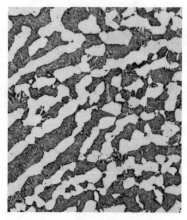

*Elongated alpha (light) in a titanium alloy forging. 250×*

**elongated grain**. A grain with one principal axis slightly longer than either of the other two.

**elongation**. (1) A term used in mechanical testing to describe the amount of extension of a test piece when stressed. (2) In tensile testing, the increase in the gage length, measured after fracture of the specimen within the gage length, usually expressed as a percentage of the original gage length. See also *elongation, percent*.

**elongation at break**. Elongation recorded at the moment of rupture of a specimen, often expressed as a percentage of the original length.

**elongation, percent**. The extension of a uniform section of a specimen expressed as a percentage of the original gage length:

$$\text{Elongation, \%} = \frac{(L_x - L_0)}{L_0} \cdot 100$$

where $L_0$ is the original gage length and $L_x$ is the final gage length.

**elutriation**. A test for particle size in which the speed of a liquid or gas is used to suspend particles of a desired size, with larger sizes settling for removal and weighing, while smaller sizes are removed, collected, and weighed at certain time intervals.

**embeddability**. The ability of a bearing material to embed harmful foreign particles and reduce their tendency to cause scoring or abrasion.

**embedded abrasive**. Fragments of abrasive particles forced into the surface of a workpiece during grinding, abrasion, or polishing.

**embossing**. (1) Technique used to create depressions of a specific pattern in plastic film and sheeting. Such embossing in the form of surface patterns can be achieved on molded parts by the treatment of the mold surface with photoengraving or another process. (2) Raising a design in relief against a surface.

**embossing die**. A die used for producing embossed designs.

**embrittlement**. The severe loss of *ductility* or *toughness* or both, of a material, usually a metal or alloy. Many forms of embrittlement can lead to *brittle fracture*. Many forms can occur during thermal treatment or elevated-temperature service (thermally induced embrittlement). Some of these forms of embrittlement, which affect steels, include *blue brittleness, 885 °F (475 °C) embrittlement, quench-age embrittlement, sigma-phase embrittlement, strain-age embrittlement, temper embrittlement, tempered martensite embrittlement*, and *thermal embrittlement*. In addition, steels and other metals and alloys can be embrittled by environmental conditions (environmentally assisted embrittlement). The forms of environmental embrittlement include *acid embrittlement, caustic embrittlement, corrosion embrittlement, creep-rupture embrittlement, hydrogen embrittlement, liquid metal embrittlement, neutron embrittlement, solder embrittlement, solid metal embrittlement*, and *stress-corrosion cracking*.

**emery**. A naturally occurring, impure form of *alumina*. A less pure form of the oxide than *corundum*.

**emf**. An abbreviation for *electromotive force*.

**emission** (of electromagnetic radiation). The creation of radiant energy in matter, resulting in a corresponding decrease in the energy of the emitting system.

**emission lines**. Spectral lines resulting from emission of electromagnetic radiation by atoms, ions, or molecules during changes from excited states to states of lower energy.

**emission spectrometer**. An instrument that measures percent concentrations of elements in samples of metals and other materials; when the sample is vaporized by an electric spark or arc, the characteristic wavelengths of light emitted by each element are measured with a diffraction grating and an array of photodetectors or photographic plates.

**emission spectroscopy**. The branch of spectroscopy treating the theory, interpretation, and application of spectra originating in the emission of electromagnetic radiation by atoms, ions, radicals, and molecules.

**emission spectrum**. An electromagnetic spectrum produced when radiation from any emitting source, excited by any of various forms of energy, is dispersed.

**emissive electrode**. A filler metal electrode consisting of a core of a bare electrode or a composite electrode to which a very light coating has been applied to produce a stable arc.

**emissivity**. Ratio of the amount of energy or of energetic particles radiated from a unit area of a surface to the amount radiated from a unit area of an ideal emitter under the same conditions.

**emulsion**. (1) A two-phase liquid system in which small droplets of one liquid (the internal phase) are immiscible in, and are dispersed uniformly throughout, a second continuous liquid phase (the external phase). The internal phase is sometimes describe as the dispense phase. (2) A stable dispersion of one liquid in another, generally by means of an emulsifying agent that has affinity for both the continuous and discontinuous phases. The emulsifying agent, discontinuous phase, and continuous phase can together produce another phase that serves as an enveloping (encapsulating) protective phase around the discontinuous phase.

**emulsion calibration curve**. The plot of a function of the relative transmittance of the photographic emulsion versus a function of the exposure. The calibration curve is used in spectrographic analysis to calculate the relative intensity of a radiant source from the density of a photographically recorded image.

**emulsion cleaner**. A cleaner-consisting of organic solvents dispersed in an aqueous medium with the aid of an emulsifying agent.

**emulsion inversion**. An *emulsion* is said to invert when, for example, a water-in-oil emulsion changes to an oil-in-water emulsion.

**emulsion polymerization**. Polymerization of monomers dispersed in an aqueous emulsion.

**enameling iron**. A low-carbon, cold-rolled sheet steel, produced specifically for use as a base metal for porcelain enamel.

**enantiotropy**. The relation of crystal forms of the same substance in which one form is stable above a certain temperature and the other form is stable below that temperature. For example, ferrite and austenite are enantiotropic in ferrous alloys.

**encapsulated adhesive**. An adhesive in which the particles or droplets of one of the reactive components are enclosed in a protective film (microcapsules) to prevent cure until the film is destroyed by suitable means.

**encapsulation**. The enclosure of an item in plastic. Sometimes used specifically in reference to the enclosure of capacitors or circuit board modules.

**end**. In composites fabrication, a strand of *roving* consisting of a given number of filaments gathered together. The group of filaments is considered an end, or strand, before twisting, and a yarn after twist has been applied. An individual warp, yarn, thread, fiber, or roving.

**end-centered**. Having an atom or group of atoms separated by a translation of the type ½, ½, 0 from a similar atom or group of atoms. The number of atoms in an end-centered cell must be a multiple of 2.

**end clearance angle**. See *clearance angle* and also the figures accompanying the terms *face mill* and *single-point tool*.

**end count**. An exact number of ends supplied on a ball of roving. See also *end*.

**end cutting-edge angle**. The angle of concavity between the face cutting edge and the face plane of the cutter. It serves as relief to prevent the face cutting edges from rubbing in the cut. See also the figures accompanying the terms *face mill* and *single-point tool*.

**end mark**. A roll mark caused by the end of a sheet marking the roll during hot or cold rolling.

**end milling**. A method of machining with a rotating cutting tool with cutting edges on both the face end and the periphery. See also *face milling* and *milling*.

**endothermic atmosphere**. A gas mixture produced by the partial combustion of a hydrocarbon gas with air in an endothermic reaction. Also known as endogas.

**endothermic reaction**. Designating or pertaining to a reaction that involves the absorption of heat. See also *exothermic reaction*.

**end-quench hardenability test**. A laboratory procedure for determining the hardenability of a steel or other ferrous alloy; widely referred to as the Jominy test. Hardenability is determined by heating a standard specimen above the upper critical temperature, placing the hot specimen in a fixture so that a stream of cold water impinges on one end, and, after cooling to room temperature is completed, measuring the hardness near the surface of the specimen at regularly spaced intervals along its length. The data are normally plotted as hardness versus distance from the quenched end. See also the figure accompanying the term *Jominy test*.

**end relief**. Defined by the figure accompanying the term *single-point tool*.

**end return**. See preferred term *boxing*.

**endurance**. The capacity of a material to withstand repeated application of stress.

**endurance limit**. The maximum stress that a material can withstand for an infinitely large number of fatigue cycles. See also *fatigue limit* and *fatigue strength*.

**endurance ratio**. The ratio of the *endurance limit* for completely reversed flexural stress to the tensile strength of a given material.

**energy-dispersive spectroscopy (EDS).** A method of x-ray analysis that discriminates by energy levels the characteristic x-rays emitted from the sample. Compare with *wavelength-dispersive spectroscopy.*

**engineering adhesive.** A bonding agent intended to join metal, plastics, wood, glass, ceramics, or rubber. The term differentiates such bonding agents from glues used to join paper and other nondurables.

**engineering ceramics.** Same as *advanced ceramics.*

**engineering plastics.** A general term covering all plastics, with or without fillers or reinforcements, that have mechanical, chemical, and thermal properties suitable for use as construction materials, machine components, and chemical processing equipment components. Included are acrylonitrile-butadiene-styrene, acetal, acrylic, fluorocarbon, nylon, phenoxy, polybutylene, polyaryl ether, polycarbonate, polyether (chlorinated), polyether sulfone, polyphenylene oxide, polysulfone, polyimide, rigid polyvinyl chloride, polyphenylene sulfide, thermoplastic urethane elastomers, and many other reinforced plastics.

**engineering strain (*e*).** A term sometimes used for average linear strain or conventional strain in order to differentiate it from *true strain.* In tension testing it is calculated by dividing the change in the gage length by the original gage length.

**engineering stress (*s*).** A term sometimes used for conventional stress in order to differentiate it from *true stress.* In tension testing, it is calculated by dividing the breaking load applied to the specimen by the original cross-sectional area of the specimen.

**engine oil.** An oil used to lubricate an internal combustion engine.

**Engler viscosity.** A commercial measure of viscosity expressed as the ratio between the time in seconds required for 200 $cm^3$ of a fluid to flow through the orifice of an Engler viscometer at a given temperature under specified conditions and the time required for 200 $cm^3$ of distilled water at 20 °C (70 °F) to flow through the orifice under the same conditions.

**engobe.** A slip coating applied to a ceramic body for imparting color, opacity, or other characteristics, and subsequently covered with a *glaze* (1).

**entraining velocity.** The velocity of a liquid at which bubbles of gas are carried along in the stream.

**entry mark (exit mark).** A slight corrugation caused by the entry or exit rolls of a *roller leveling* unit.

**environment.** The aggregate of all conditions (such as contamination, temperature, humidity, radiation, magnetic and electric fields, shock, and vibration) that externally influence the performance of a material or component.

**environmental cracking.** *Brittle fracture* of a normally ductile material in which the corrosive effect of the environ-

ment is a causative factor. Environmental cracking is a general term that includes *corrosion fatigue, high-temperature hydrogen attack, hydrogen blistering, hydrogen embrittlement, liquid metal embrittlement, solid metal embrittlement, stress-corrosion cracking,* and *sulfide stress cracking.* The following terms have been used in the past in connection with environmental cracking, but are becoming obsolete: caustic embrittlement, delayed fracture, season cracking, static fatigue, stepwise cracking, sulfide corrosion cracking, and sulfide stress-corrosion cracking. See also *embrittlement.*

**environmentally assisted embrittlement.** See *embrittlement.*

**environmental stress cracking.** The susceptibility of a plastic resin to cracking or crazing when in the presence of such chemicals as surface-active agents, or other environments.

**epichlorohydrin.** The basic epoxidizing resin intermediate in the production of epoxy resins. It contains an epoxy group and is highly reactive with polyhydric phenols such as bisphenol A.

**epitaxy.** Growth of an electrodeposit or vapor deposit in which the orientation of the crystals in the deposit are directly related to crystal orientations in the underlying crystalline substrate.

**epoxide.** Compound containing the oxirane structure, a three-member ring containing two carbon atoms and one oxygen atom. The most important members are ethylene oxide and propylene oxide.

**epoxy plastic.** A thermoset polymer containing one or more epoxide groups and curable by reaction with amines, alcohols, phenols, carboxylic acids, acid anhydrides, and mercaptans. An important matrix resin in reinforced composites and in structural adhesives.

**epoxy resin.** A viscous liquid or the brittle, solid-containing epoxide groups that can be cross linked into final form by means of a chemical reaction with a variety of setting agents used with or without heat.

**epoxy smear.** See *resin smear.*

**epsilon (ε).** Designation generally assigned to intermetallic, metal-metalloid, and metal-nonmetallic compounds found in ferrous alloy systems, for example, $Fe_3Mo_2$, FeSi, and $Fe_3P$.

**epsilon carbide.** Carbide with hexagonal close-packed lattice that precipitates during the first stage of tempering of primary martensite. Its composition corresponds to the empirical formula $Fe_{2.4}C$.

**epsilon structure.** Structurally analogous close-packed phases or electron compounds that have ratios of seven valence electrons to four atoms.

**equator.** In *filament winding* of composites, the line in a pressure vessel described by the junction of the cylindrical portion and the end dome. Also called tangent line or point.

FIG. 158

*Equiaxed grains in an unalloyed depleted uranium sample. 100×*

**equiaxed grain structure**. A structure in which the grains have approximately the same dimensions in all directions (Fig. 158).

**equilibrium**. The dynamic condition of physical, chemical, mechanical, or atomic balance which appears to be a condition of rest rather than one of change.

**equilibrium centrifugation**. In resinography, a method for determining the distribution of a molecular weight by spinning a solution of the specimen at a speed such that the molecules of the specimen are not removed from the solvent but are held at a point where the (centrifugal) force tending to remove them is balanced by the dispersive forces caused by thermal agitation.

**equilibrium diagram**. A graph of the temperature, pressure, and composition limits of phase fields in an alloy system as they exist under conditions of thermodynamical equilibrium. In metal systems, pressure is usually considered constant. Compare with *phase diagram*.

**equilibrium (reversible) potential**. The *potential* of an electrode in an electrolytic solution when the forward rate of a given reaction is exactly equal to the reverse rate. The equilibrium potential can only be defined with respect to a specific electrochemical reaction.

**equivalent radial load**. The level of constant radial load on a rolling-element bearing that, when the bearing is stationary with respect to the outer race, will produce the same rating life as a given combination of radial and thrust loads under the same conditions of operation.

**Erichsen test**. A *cupping test* used to assess the ductility of sheet metal. The method consists of forcing a conical or hemispherical-ended plunger into the specimen and measuring the depth of the impression at fracture.

**erosion**. (1) Loss of material from a solid surface due to relative motion in contact with a fluid that contains solid particles. Erosion in which the relative motion of particles is nearly parallel to the solid surface is called *abrasive erosion*. Erosion in which the relative motion of the solid particles is nearly normal to the solid surface is called

FIG. 159

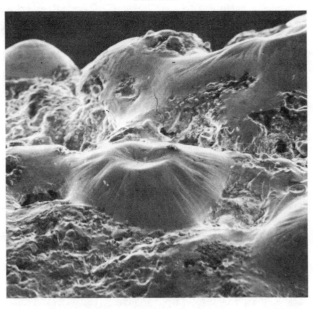

*Erosion of a 70W-30Cu electrical contact due to arcing. 140×*

*impingement erosion* or impact erosion. (2) Progressive loss of original material from a solid surface due to mechanical interaction between that surface and a fluid, a multicomponent fluid, and impinging liquid, or solid particles. (3) Loss of material from the surface of an electrical contact due to an electrical discharge (arcing) (Fig. 159). See also *cavitation erosion*, *electrical pitting*, and *erosion-corrosion*.

**erosion** (brazing). A condition caused by dissolution of the base metal by molten filler metal resulting in a postbraze reduction in the thickness of the base metal.

**erosion-corrosion**. A conjoint action involving *corrosion* and *erosion* in the presence of a moving corrosive fluid, leading to the accelerated loss of material (Fig. 160).

FIG. 160

*The classic appearance of erosion-corrosion in a CN-7M stainless steel pump impeller*

erosion rate. Any determination of the rate of loss of material (erosion) with exposure duration. In certain contexts (for example, ASTM erosion tests), it is given by the slope of the cumulative erosion-time curve.

erosion rate-time curve. A plot of instantaneous erosion rate versus exposure duration, usually obtained by numerical or graphical differentiation of the cumulative erosion-time curve.

erosive wear. See *erosion*.

erosivity. The characteristic of a collection of particles, liquid stream, or a slurry that expresses its tendency to cause erosive wear when forced against a solid surface under relative motion.

error. Deviation from the correct value. In the case of a testing machine, the difference obtained by subtracting the load indicated by the calibration device from the load indicated by the testing machine.

escape peak. An artifact observed in x-ray analysis; manifested as a peak at energy 1.74 keV (the silicon Kα peak) less than the major line detected (Fig. 161). Escape peaks can be avoided by increasing the accelerating voltage.

ester. The reaction product of an alcohol and an acid.

estimate. This particular value, or values, of a parameter computed by an estimation procedure for a given sample.

estimation. A procedure for making a statistical inference about the numerical values of one or more unknown population parameters from the observed values in a sample.

etchant. (1) A chemical solution used to etch a metal to reveal structural details. (2) A solution used to remove, by chemical reaction, the unwanted portion of material from a printed circuit board. (3) Hydrofluoric acid or other agent used to attack the surface of glass for marking or decoration.

etch cleaning. Removing soil by dissolving away some of the underlying metal.

etch cracks. Shallow cracks in hardened steel containing high residual surface stresses, produced by etching in an embrittling acid.

etched metal mask. A mask formed by etching apertures through a solid metal protected by a photoresist.

etch figures. Characteristic markings produced on crystal surfaces by chemical attack, usually having facets parallel to low-index crystallographic planes.

etching. (1) Subjecting the surface of a metal to preferential chemical or electrolytic attack in order to reveal structural details for metallographic examination. (2) Chemically or electrochemically removing tenacious films from a metal surface to condition the surface for a subsequent treatment, such as painting or electroplating. (3) A process by which a printed pattern is formed on a printed circuit board by either chemical or chemical and electrolytic removal of the unwanted portion of conductive material bonded to a base.

etch rinsing. Pouring etchant over a tilted surface until the desired degree of attack is achieved. Used for etchants with severe gas formation.

ethylene plastics. Plastics based on polymers of ethylene or copolymers of ethylene with other monomers, the ethylene being in greatest amount by mass.

ethyl silicate. A strong bonding agent for sand and refractories used in preparing molds in the investment casting process.

Euler angles. In crystallographic texture analysis, three angular parameters that specify the orientation of a body with respect to reference axes.

eutectic. (1) An isothermal reversible reaction in which a liquid solution is converted into two or more intimately mixed solids on cooling, the number of solids formed being the same as the number of components in the system. (2) An alloy having the composition indicated by the

FIG. 161

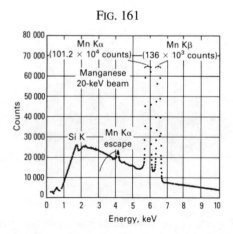

*Examples of escape peaks in energy-dispersive spectra*

FIG. 162

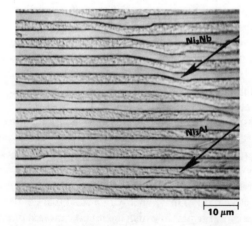

*Lamellar eutectic structure in a Ni₃Al-Ni₃Nb casting that was solidified unidirectionally (left to right)*

eutectic point on a phase diagram. (3) An alloy structure of intermixed solid constituents formed by a eutectic reaction often in the form of regular arrays of lamellae or rods (Fig. 162).

**eutectic arrest.** In a cooling or heating curve, an approximately isothermal segment corresponding to the time interval during which the heat of transformation from the liquid phase to two or more solid phases is evolving.

**eutectic bonding.** The forming of a bond by bringing two solids to their lowest constant melting point at which the molten solids mix and harden upon cooling.

**eutectic carbides.** Carbide formed during freezing as one of the mutually insoluble phases participating in the eutectic reaction of a hypereutectic tool steel. See also *hypereutectic alloy.*

**eutectic-cell etching.** Development of eutectic cells (grains).

**eutectic melting.** Melting of localized microscopic areas whose composition corresponds to that of the eutectic in the system.

**eutectic point.** The composition of a liquid phase in univariant equilibrium with two or more solid phases; the lowest melting alloy of a composition series.

**eutectoid.** (1) An isothermal reversible reaction in which a solid solution is converted into two or more intimately mixed solids on cooling, the number of solids formed being the same as the number of components in the system. (2) An alloy having the composition indicated by the eutectoid point on a phase diagram. (3) An alloy structure of intermixed solid constituents formed by a eutectoid reaction.

**eutectoid point.** The composition of a solid phase that undergoes univariant transformation into two or more other solid phases upon cooling.

**evaporative deposition.** The techniques of condensing a thin film of material on a substrate. The entire process takes place in a high vacuum. The source material may be radioactively heated by bombardment with electrons (electron beam) or may be heated by thermal-conduction techniques.

**evaporation.** The vaporization of a material by heating, usually in a vacuum. In electron microscopy, this process is used for shadowing or to produce thin support films by condensation of the vapors of metals or salts.

**even tension.** In composite reinforcing fibers, the process whereby each end of roving is kept in the same degree of tension as the other ends making up that ball of roving. See also *catenary.*

**Ewald sphere.** In electron diffraction theory, a geometric construction, of radius equal to the reciprocal of the wavelength of the incident radiation, with its surface at the origin of the reciprocal lattice. Any crystal plane will reflect if the corresponding reciprocal lattice point lies on the surface of this sphere.

**exchange current.** When an electrode reaches dynamic equilibrium in a solution, the rate of anodic dissolution balances the rate of cathodic plating. The rate at which either positive or negative charges are entering or leaving the surface at this point is known as the exchange current.

**exchange current density.** The rate of charge transfer per unit area when an electrode reaches dynamic equilibrium (at its reversible potential) in a solution; that is, the rate of anodic charge transfer (oxidation) balances the rate of cathodic charge transfer (reduction).

**excitation index.** In materials characterization, the ratio of the intensities of two selected spectral lines of an element having widely different excitation energies. This ratio serves to indicate the level of excitation energy in the source.

**excitation potential** (x-ray). The applied potential on an x-ray tube required to produce characteristic radiation from the target.

**excitation volume.** The volume within the sample in which data signals originate.

**excrescence.** A term used by some Russian tribologists (in English) describing micro-extrusions on friction surfaces that lead to localized welding. This term is not recommended.

**exfoliation.** Corrosion that proceeds laterally from the sites of initiation along planes parallel to the surface, generally at grain boundaries, forming corrosion products that force metal away from the body of the material, giving rise to a layered appearance (Fig. 163).

**exogenous inclusion.** An *inclusion* that is derived from external causes. Slag, dross, entrapped mold materials, and refractories are examples of inclusions that would be classified as exogenous. In most cases, these inclusions are macroscopic or visible to the naked eye. Compare with *indigenous inclusion.*

**exotherm.** The temperature/time curve of a chemical reaction or a phase change giving off heat, particularly the polymerization of casting resins. The amount of heat given off. The term has not been standardized with respect to sample size, ambient temperature, degree of mixing, and so forth.

**exothermic.** Characterized by the liberation of heat.

**exothermic atmosphere.** A gas mixture produced by the partial combustion of a hydrocarbon gas with air in an exothermic reaction. Also known as exogas.

**exothermic reaction.** A reaction which liberates heat, such as the burning of fuel or when certain plastic resins are cured chemically.

**expandable plastic.** A plastic that can be made cellular by thermal, chemical, or mechanical means. Foam plastics such as expandable polystyrene and foam polyurethane are examples.

FIG. 163

*Exfoliation corrosion of an aluminum test panel*

**expanding**. A process used to increase the diameter of a cup, shell, or tube. See also *bulging*.

**expansion**. (1) An increase in size of a powder metallurgy compact, usually related to an increase in temperature. A decrease in temperature produces an opposite effect. (2) Sometimes used to mean growth.

**expansion film**. An *interference* or *force fit* made by placing a cold (subzero) inside member into a warmer outside member and allowing an equalization of temperature.

**expendable pattern**. A *pattern* that is destroyed in making a casting. It is usually made of wax (*investment casting*) or expanded polystyrene (*lost foam casting*).

**explosion welding**. A solid-state welding process effected by a controlled detonation, which causes the parts to move together at high velocity. The resulting bond zone has a characteristic wavy appearance (Fig. 164).

**explosive compacting**. See *high energy rate compacting*.

**explosive forming**. The shaping of metal parts in which the forming pressure is generated by an explosive charge which takes the place of the punch in conventional forming. A single-element die is used with a blank held over it, and the explosive charge is suspended over the blank at a predetermined distance (standoff distance). The complete assembly is often immersed in a tank of water (Fig. 165). See also *high-energy-rate forming*.

**exposed underlayer**. In composites fabrication, the underlying layer of mat or roving not covered by surface mat in a pultrusion. This condition can be caused by reinforcement shifting, too narrow a surface mat, too wide an

FIG. 164

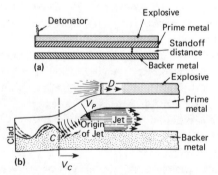

*Schematic of the explosion welding process. $V_C$: collision velocity. $V_P$: plate velocity. D: detonation velocity. C: collision region*

FIG. 165

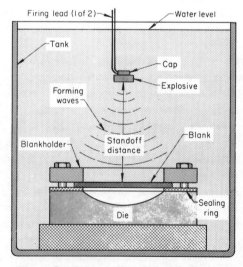

*Unconfined system for explosive forming*

underlying mat, uneven slitting of the surface mat, necking down of the surface mat, or excessive tension in pulling the surface mat off the spindle.

**exposure.** The product of the intensity of a radiant source and the time of irradiation.

**extend.** The addition of fillers or low-cost materials to plastic resins as an economy measure. To add inert materials to improve void-filling characteristics and reduce crazing.

**extended x-ray absorption fine structure (EXAFS).** The weak oscillatory structure extending for several hundred electron volts away from an absorption edge. The oscillations occur because the electromagnetic wave produced by the ionization of the absorbing atom for some energy $E$ has a wavelength $\lambda = 1.225/(E - E_k)^{1/2}$ nm, where $E_k$ is the energy of the absorption edge. For example, a loss of 100 eV above an edge corresponds to a wavelength of 0.12 nm, which is of the order of atomic spacing. Consequently, the wave can be diffracted from neighboring atoms and return to interfere with the outgoing wave. An analysis of EXAFS data reveals important information about atomic arrangements and bonding. Either synchrotron x-radiation or the electron beam in the analytical transmission electron microscope can be used as the excitation source. See also *analytical electron microscopy* and *synchrotron radiation*.

**extenders.** (1) Low-cost materials used to dilute or extend high-cost resins without significant lessening of properties. (2) Substances, generally having some adhesive action, added to an adhesive to reduce the amount of primary binder required per unit area. See also *binder*, *diluent*, *filler*, and *thinner*.

**extensibility.** The ability of a material to extend or elongate upon application of sufficient force, expressed as a percent of the original length.

**extensional-bending composite.** A property of certain classes of laminates that exhibit bending curvatures when subjected to extensional loading.

**extensometer.** An instrument for measuring changes in length over a given gage length caused by application or removal of a force. Commonly used in tension testing (Fig. 166).

**external circuit.** The wires, connectors, measuring devices, current sources, etc., that are used to bring about or measure the desired electrical conditions within the test cell. It is this portion of the cell through which electrons travel.

**externally pressurized seal.** A seal that operates on a thin film at the interface with the mating surface. The film is formed by high-pressure fluid that is brought to the interface at some mid-dam location and that is at a pressure equal to, or higher than, the upstream seal pressure. See also *hydrostatic seal*.

**extinction.** A decrease in the intensity of the diffracted x-ray beam caused by perfection or near perfection of crystal

FIG. 166

*Extensometer attached to a round tensile specimen*

structure. See also *primary extinction* and *secondary extinction*.

**extraction.** A general term denoting chemical methods of isolating phases from a metal matrix.

**extractive metallurgy.** The branch of *process metallurgy* dealing with the winning of metals from their ores. Compare with *refining*.

**extra hard.** A *temper* of nonferrous alloys and some ferrous alloys characterized by values of tensile strength and hardness about one-third of the way from those of *full hard* to those of *extra spring* temper.

**extra spring.** A *temper* of nonferrous alloys and some ferrous alloys corresponding approximately to a cold worked state above *full hard* beyond which further cold work will not measurably increase strength or hardness.

**extreme-pressure lubricant.** A lubricant that imparts increased load-carrying capacity to rubbing surfaces under severe operating conditions. Extreme-pressure lubricants usually contain sulfur, halogens, or phosphorus. The term *antiscuffing lubricant* has been suggested as a replacement for extreme-pressure lubricant.

**extreme-pressure lubrication.** A condition of lubrication in which the friction and wear between two surfaces in relative motion depend upon the reaction of the lubricant with a rubbing surface at elevated temperature.

**extruded hole.** A hole formed by a punch that first cleanly cuts a hole and then is pushed farther through to form a flange with an enlargement of the original hole.

FIG. 167

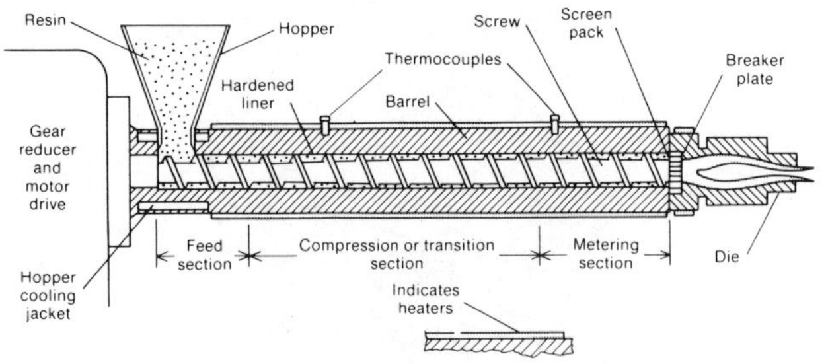

*Typical single-screw extruder for forming plastics*

**extruder** (plastics). A machine that accepts solid particles, such as pellets or powder, or liquid (molten) feed, conveys it through a surrounding barrel by means of a rotating screw, and pumps it, under pressure, through an orifice (Fig. 167).

**extrusion** (ceramics). The process of forcing a mixture of plastic binder and ceramic powder through the opening(s) of a die at relatively high pressure. The material may thus be compacted and emerges in elongated cylindrical or ribbon (or wire, etc.) form having the cross section of the die opening. Ordinarily followed by drying, curing, activating, or firing. (2) The process of forming clay products by forcing the plastic material through a die.

**extrusion** (metals). The conversion of an ingot or billet into lengths of uniform cross section by forcing metal to flow plastically through a die orifice (Fig. 168). In forward (direct) extrusion, the die and ram are at opposite ends of the extrusion stock, and the product and ram travel in the same direction. Also, there is relative motion between the extrusion stock and the die. In backward (indirect) extrusion, the die is at the ram end of the stock and the product travels in the direction opposite that of the ram, either around the ram (as in the impact extrusion of cylinders such as cases for dry cell batteries) or up through the center of a hollow ram. See also *hydrostatic extrusion* and *impact extrusion.*

**extrusion** (plastics). Compacting a plastic material into a powder or granules in a uniform melt and forcing it through an orifice in a more or less continuous fashion to yield a desired shape. While held in the desired shape, the melt must be cooled to a solid state.

**extrusion billet**. A metal slug used as *extrusion stock.*

**extrusion blow molding**. The most common blow molding process, in which a parison is extruded from a plastic melt and is then entrapped between the halves of a mold. The parison is expanded, under air pressure, against the mold cavity to form the part, and is then cooled, removed, and trimmed. See figure accompanying the term *blow molding.*

**extrusion coating**. Using a resin to coat a substrate by extruding a thin film of molten resin and pressing it onto or into the substrate, or both, without the use of an adhesive.

**extrusion defect**. See preferred term *extrusion pipe.*

**extrusion forging**. (1) Forcing metal into or through a die opening by restricting flow in other directions. (2) A part made by the operation.

**extrusion ingot**. A cast metal slug used as *extrusion stock.*

FIG. 168

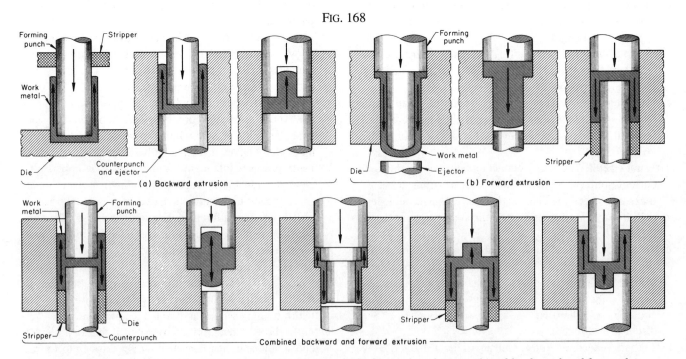

*Displacement of metal in cold extrusion: (a) backward, (b) forward, and (c) combined backward and forward*

**extrusion pipe**. A central oxide-lined discontinuity that occasionally occurs in the last 10 to 20% of an extruded metal bar. It is caused by the oxidized outer surface of the billet flowing around the end of the billet and into the center of the bar during the final stages of extrusion. Also called *coring*.

**extrusion stock**. A rod, bar, or other section used to make extrusions.

**exudation**. The action by which all or a portion of the low melting constituent of a powder metallurgy compact is forced to the surface during sintering; sometimes referred to as bleed out or sweating.

**eyeleting**. The displacing of material about an opening in sheet or plate so that a lip protruding above the surface is formed.

**eyepiece**. A lens or system of lenses for increasing magnification in a microscope by magnifying the image formed by the objective.

# F

**fabrication traces**. Anomalous markings which may appear on crack surfaces of ceramics where the developing crack encounters regions of unusual weakness, density, or elastic modulus, introduced, usually inadvertently, during fabrication.

**fabric fill face**. That side of a woven fabric used in reinforced plastics on which the greatest number of yarns are perpendicular to the *selvage*.

**fabric, nonwoven**. See *nonwoven fabric*.

**fabric prepreg batch**. In composites processing, a *prepreg* containing fabric from one fabric batch and impregnated with one batch of resin in one continuous operation.

**fabric warp face**. That side of a woven fabric used in reinforced plastics on which the greatest number of yarns are parallel to the *selvage*.

**fabric, woven**. See *woven fabric*.

**face** (crystal). An idiomorphic plane surface on a crystal.

**face** (machine tools). In a lathe tool, the surface against which the chips bear as they are formed. See also the figure accompanying the term *single-point tool*.

**face-centered**. Having atoms or groups of atoms separated by translations of $\frac{1}{2}$, $\frac{1}{2}$, 0; $\frac{1}{2}$, 0, $\frac{1}{2}$; and 0, $\frac{1}{2}$, $\frac{1}{2}$ from a similar atom or group of atoms. The number of atoms in a face-centered cell must be a multiple of 4. See also the term accompanying the term *unit cell*.

**face feed**. The application of filler metal to the joint, usually by hand, during brazing and soldering.

**face mill**. See definition of nomenclature in Fig. 169.

**face milling**. Milling a surface that is perpendicular to the cutter axis. See also *milling*.

**face of weld**. The exposed surface of an arc or gas weld on the side from which the welding was done. See also the figure accompanying the term *fillet weld*.

**face pressure**. In seals, the face load divided by the contacting area of the sealing lip. The face load is the sum of the pneumatic or hydraulic force and the spring force. For lip seals and packings, the face load also includes the interference load. See also *face seal*.

Fig. 169

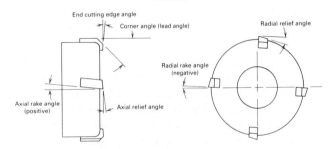

*Face milling nomenclature*

seals and packings, the face load also includes the interference load. See also *face seal*.

**face reinforcement**. Reinforcement of a weld at the side of the joint from which welding was carried out. See also *root reinforcement*.

**face seal**. A device that prevents the leakage of fluids along rotating shafts. Sealing is accomplished by a stationary primary-seal ring bearing against the face of a mating ring mounted on a shaft. Axial pressure maintains the contact between seal ring and mating ring.

**face shield** (welding). A device positioned in front of the eyes and a portion of, or all of, the face, whose predominant function is protection of the eyes and face during welding, brazing, or soldering.

**face-type cutters**. Cutters that can be mounted directly on and driven from the machine spindle nose.

**facing**. (1) In machining, generating a surface on a rotating workpiece by the traverse of a tool perpendicular to the axis of rotation. (2) In foundry practice, any material applied in a wet or dry condition to the face of a mold or core to improve the surface of the casting. See also *mold wash*. (3) For abrasion resistance, see preferred term *hardfacing*.

**fadeometer**. An apparatus for determining the resistance of resins and other materials to fading.

**fagot.** In forging work, a bundle of iron bars that will be heated and then hammered and welded to form a single bar.

**failure.** A general term used to imply that a part in service (a) has become completely inoperable, (b) is still operable but incapable of satisfactorily performing its intended function, or (c) has deteriorated seriously, to the point that is has become unreliable or unsafe for continued use.

**failure criteria.** The limiting conditions relating to the admissibility of the deviation from the characteristic value due to changes after the beginning of stress.

**failure mechanism.** A structural or chemical process, such as corrosion or fatigue, that causes failure.

**fairing.** A secondary structure in airframes and ship hulls, the major function of which is to streamline the airflow or flow of fluid by producing a smooth outline and reducing drag.

**false bottom.** An *insert* put in either member of a die set to increase the strength and improve the life of the die.

**false Brinelling.** (1) Damage to a solid bearing surface characterized by indentations not caused by plastic deformation resulting from overload, but thought to be due to other causes such as *fretting corrosion* (Fig. 170). (2) Local spots appearing when the protective film on a metal is broken continually by repeated impacts, usually in the presence of corrosive agents. The appearance is generally similar to that produced by *Brinelling* but corrosion products are usually visible. It may result from fretting corrosion. This term should be avoided when a more precise description is possible. False Brinelling (race fretting) can be distinguished from true *Brinelling* because in false Brinelling, surface material is removed so that original finishing marks are removed. The borders of a false Brinell mark are sharply defined, whereas a dent caused by a rolling element does not have sharp edges and the finishing marks are visible in the bottom of the dent.

FIG. 170

*Severe damage from false Brinelling on the surface of a shaft that served as the inner raceway for a needle-roller bearing*

**false indication.** In nondestructive inspection, an *indication* that may be interpreted erroneously as an *imperfection*. See also *artifact*.

**false wiring.** Same as *curling*.

**family mold.** A multicavity mold used for forming of plastic in which each of the cavities form one of the component parts of the assembled finished object. The term is often applied to a mold in which parts from different customers are grouped together for economy. Sometimes called combination mold.

**farad (F).** A unit of electric capacity.

**Faraday's law.** (1) The amount of any substance dissolved or deposited in electrolysis is proportional to the total electric charge passed. (2) The amounts of different substances dissolved or deposited by the passage of the same electric charge are proportional to their equivalent weights.

**far-infrared radiation.** Infrared radiation in the wavelength range of 30 to 300 μm ($3 \times 10^5$ to $3 \times 10^6$ Å).

**fat.** An organic ester, the product of a reaction between a fatty acid and glycerol. Fat can be of animal or vegetable materials or can be made synthetically.

**fatigue.** The phenomenon leading to fracture under repeated or fluctuating stresses having a maximum value less than the ultimate tensile strength of the material. *Fatigue failure* generally occurs at loads which applied statically would produce little perceptible effect. Fatigue fractures are progressive, beginning as minute cracks that grow under the action of the fluctuating stress.

**fatigue crack growth rate (*da/dN*).** The rate of crack extension caused by constant-amplitude fatigue loading, expressed in terms of crack extension per cycle of load application, and plotted logarithmically against the *stress intensity factor* range, Δ*K* (Fig. 171).

**fatigue ductility (*D*f).** The ability of a material to deform plastically before fracturing, determined from a constant-strain amplitude, low-cycle fatigue test. Usually expressed in percent in direct analogy with elongation and reduction in area ductility measures.

**fatigue ductility exponent (*c*).** The slope of a log-log plot of the plastic strain range and the fatigue life.

**fatigue failure.** Failure that occurs when a specimen undergoing *fatigue* completely fractures into two parts of has softened or been otherwise significantly reduced in stiffness by thermal heating or cracking.

**fatigue life (*N*).** (1) The number of cycles of stress or strain of a specified character that a given specimen sustains before failure of a specified nature occurs. (2) The number of cycles of deformation required to bring about failure of a test specimen under a given set of oscillating conditions (stresses or strains). See also *S-N curve*.

**fatigue life for *p*% survival.** An estimate of the *fatigue life* that *p*% of the population would attain or exceed at a given

FIG. 171

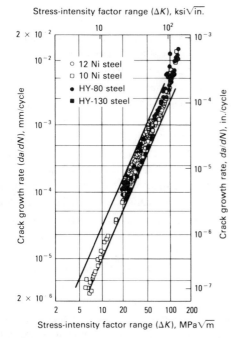

*Fatigue crack growth data for martensitic steels*

stress level. The observed value of the median fatigue life estimates the fatigue life for 50% survival. Fatigue life for *p*% survival values, where *p* is any number, such as 95, 90, etc., may also be estimated from the individual fatigue life values.

**fatigue limit.** The maximum stress that presumably leads to fatigue fracture in a specified number of stress cycles. The value of the *maximum stress* and the *stress ratio* also should be stated. See also *endurance limit*.

**fatigue limit for *p*% survival.** The limiting value of *fatigue strength* for *p*% survival as *N* becomes very large; *p* may be any number, such as 95, 90, etc.

**fatigue notch factor ($K_f$).** The ratio of the *fatigue strength* of an unnotched specimen to the fatigue strength of a notched specimen of the same material and condition; both strengths are determined at the same number of *stress cycles*.

**fatigue notch sensitivity ($q$).** An estimate of the effect of a notch or hole of a given size and shape on the fatigue properties of a material, measured by $q = (K_f - 1)/(K_t - 1)$, where $K_f$ is the *fatigue notch factor* and $K_t$ is the *stress-concentration factor*. A material is said to be fully notch sensitive if $q$ approaches a value of 1.0; it is not notch sensitive if the ratio approaches 0.

**fatigue ratio.** The ratio of fatigue strength to tensile strength. Mean stress and alternating stress must be stated.

**fatigue strength.** The maximum cyclical stress a material can withstand for a given number of cycles before failure occurs.

**fatigue strength at *N* cycles ($S_N$).** A hypothetical value of stress for failure at exactly *N* cycles as determined from an *S-N* diagram. The value of $S_N$ thus determined is subject to the same conditions as those which apply to the *S-N* diagram. The value of $S_N$ that is commonly found in the literature is the hypothetical value of maximum stress, $S_{max}$, minimum stress $S_{min}$, or stress amplitude, $S_a$, at which 50% of the specimens of a given sample could survive *N* stress cycles in which the mean stress $S_m = 0$. This is also known as the *median fatigue strength at N cycles*. See also *S-N curve*.

**fatigue strength for *p*% survival at *N* cycles.** An estimate of the stress level at which *p*% of the population would survive *N* cycles; *p* may be any number, such as 95, 90, etc. The estimates of the fatigue strength for *p*% survival values are derived from particular points of the fatigue life distribution since there is no test procedure by which a frequency distribution of fatigue strength at *N* cycles can be directly observed.

**fatigue-strength reduction factor.** The ratio of the fatigue strength of a member or specimen with no stress concentration to the fatigue strength with stress concentration. This factor has no meaning unless the stress range and the shape, size, and material of the member or specimen are stated.

**fatigue striation** (metals). Parallel lines frequently observed in electron microscope fractographs or fatigue fracture

FIG. 172

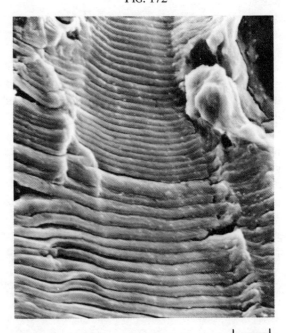

*Increase in fatigue striation spacing in an aluminum alloy test coupon due to higher alternating stress*

surfaces (Fig. 172). The lines are transverse to the direction of local crack propagation; the distance between successive lines represents the advance of the crack front during the one cycle of stress variation.

**fatigue test**. A method for determining the range of alternating (fluctuating) stresses a material can withstand without failing.

**fatigue wear**. (1) Removal of particles detached by fatigue arising from cyclic stress variations. (2) Wear of a solid surface caused by fracture arising from material fatigue. See also *spalling*.

**fatty acid**. An organic acid of aliphatic structure originally derived from fats and fatty oils.

**fatty oil**. A fat (glycerol ester) that is liquid at room temperature.

**faying surface**. The surfaces of materials in contact with each other and joined or about to be joined.

**feather** (ceramics). A striation having the appearance of a feather. See also *striation*.

**feathering**. The tapering of an *adherend* on one side to form a wedge section, as used in an adhesively bonded scarf joint.

**feed**. The rate at which a cutting tool or grinding wheel advances along or into the surface of a workpiece, the direction of advance depending on the type of operation involved.

**feedability**. The ability of a grease to flow to the suction of a pump.

**feeder (feeder head, feedhead)**. In foundry practice, a *riser*.

**feed hopper**. A container used for holding metal powder prior to compacting in a press.

**feeding**. (1) In casting, providing molten metal to a region undergoing solidification, usually at a rate sufficient to fill the mold cavity ahead of the solidification front and to compensate for any shrinkage accompanying solidification. (2) Conveying metal stock or workpieces to a location for use or processing, such as wire to a consumable electrode, strip to a die, or workpieces to an assembler.

**feed lines**. Linear marks on a machined or ground surface that are spaced at intervals equal to the *feed* per revolution or per stroke.

**feed rate** (thermal spraying). The rate at which material passes through the gun in a unit of time. A synonym for *spray rate*.

**feed shoe**. The part of the metal powder feed system that delivers the powder into the die cavity.

**feldspar**. A group of alumina silicate minerals consisting chiefly of microcline ($K_2O \cdot Al_2O_3 \cdot 6SiO_2$), albite ($Na_2O \cdot Al_2O_3 \cdot 6SiO_2$), and/or anorthite ($CaO \cdot Al_2O_3 \cdot 2SiO_2$) used in the production of glass and ceramic whiteware.

**felt**. A fibrous material used in reinforced plastics made up of interlocked fibers by mechanical or chemical action, mois-

ture, or heat. Made from fibers such as asbestos, cotton, glass, and so forth. See also *batt*.

**FEP**. See *fluorinated ethylene propylene*.

**ferrimagnetic material**. (1) A material that macroscopically has properties similar to those of a *ferromagnetic material* but that microscopically also resembles an antiferromagnetic material in that some of the elementary magnetic moments are aligned antiparallel. If the moments are of different magnitudes, the material may still have a large resultant magnetization. (2) A material in which unequal magnetic moments are lined up antiparallel to each other. Permeabilities are of the same order of magnitude as those of ferromagnetic materials, but are lower than they would be if all atomic moments were parallel and in the same direction. Under ordinary conditions the magnetic characteristics of ferrimagnetic materials are quite similar to those of ferromagnetic material.

**ferrite**. (1) A solid solution of one or more elements in body-centered cubic iron. Unless otherwise designated (for instance, as chromium ferrite), the solute is generally assumed to be carbon. On some equilibrium diagrams, there are two ferrite regions separated by an austenite area. The lower area is alpha ferrite; the upper, delta ferrite. If there is no designation, alpha ferrite is assumed. (2) An essentially carbon-free solid solution in which alpha iron is the solvent, and which is characterized by a body-centered cubic crystal structure. Fully ferritic steels are only obtained when the carbon content is quite low. The most obvious microstructural features in such metals are the ferrite grain boundaries (Fig. 173).

FIG. 173

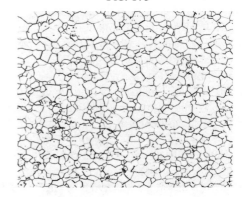

*Low-carbon ferritic sheet steel etched to reveal ferrite grain boundaries. 100×*

**ferrite banding**. Parallel bands of free ferrite aligned in the direction of working. Sometimes referred to as ferrite streaks.

**ferrite number**. An arbitrary, standardized value designating the ferrite content of an austenitic stainless steel weld metal. This value directly replaces percent ferrite or volume percent ferrite and is determined by the magnetic test described in AWS A4.2

**ferrite-pearlite banding.** Inhomogeneous distribution of ferrite and pearlite aligned in filaments or plates parallel to the direction of working. See also the figure accompanying the term *banding*.

**ferrites.** A term referring to magnetic oxides in general, and especially to material having the formula $MOFe_2O_3$, where M is a divalent metal ion or a combination of such ions. Certain ferrites, magnetically "soft" in character, are useful for core applications at radio and higher frequencies because of their advantageous magnetic properties and high volume resistivity. Other ferrites, magnetically "hard" in character, have desirable permanent magnet properties.

**ferrite streaks.** Same as *ferrite banding*.

**ferritic grain size.** The grain size of the ferritic matrix of a steel.

**ferritic malleable.** See *malleable cast iron*.

**ferritic stainless steel.** See *stainless steels* (Technical Brief 47).

**ferritizing anneal.** A treatment given as-cast gray or ductile (nodular) iron to produce an essentially ferritic matrix. For the term to be meaningful, the final microstructure desired or the time-temperature cycle used must be specified.

**ferroalloy.** An alloy of iron that contains a sufficient amount of one or more other chemical elements to be useful as an agent for introducing these elements into molten metal, especially into steel or cast iron.

**ferroelectric.** A crystalline material that exhibits spontaneous electrical polarization, hysteresis, and piezoelectric properties.

**ferroelectric effect.** The phenomena whereby certain crystals may exhibit a spontaneous dipole moment (which is called ferroelectric by analogy with ferromagnetism exhibiting a permanent magnetic moment). Ferroelectric crystals often show several Curie points, domain structures, and hysteresis, much as do ferromagnetic crystals.

**ferrograph.** An instrument used to determine the size distribution of wear particles in lubricating oils of mechanical systems. The technique relies on the debris being capable of being attracted to a magnet.

**ferromagnetic material.** A material that in general exhibits the phenomena of hysteresis and saturation, and whose permeability is dependent on the magnetizing force. Microscopically, the elementary magnets are aligned parallel in volumes called *domains*. The unmagnetized condition of a ferromagnetic material results from the overall neutralization of the magnetization of the domains to produce zero external magnetization.

**ferromagnetic resonance.** Magnetic resonance of a ferromagnetic material. See also *ferromagnetism* and *magnetic resonance*.

**ferromagnetism.** A property exhibited by certain metals, alloys, and compounds of the transition (iron group), rare-earth, and actinide elements in which, below a certain temperature termed the Curie temperature, the atomic magnetic moments tend to line up in a common direction. Ferromagnetism is characterized by the strong attraction of one magnetized body for another. See also *Curie temperature*. Compare with *paramagnetism*.

**ferrous.** Metallic materials in which the principal component is iron.

**fiber** (composites). A general term used to refer to filamentary materials. Often, fiber is used synonymously with filament. It is a general term for a filament with a finite length that is at least 100 times its diameter, which is typically 0.10 to 0.13 mm (0.004 to 0.005 in.). In most cases it is prepared by drawing from a molten bath, spinning, or depositing on a substrate. Fibers can be continuous or specific short lengths (discontinuous), normally no less than 3.2 mm (1/8 in.).

**fiber** (metals). (1) The characteristic of wrought metal that indicates *directional properties* and is revealed by etching of a longitudinal section or is manifested by the fibrous or woody appearance of a fracture. It is caused chiefly by extension of the constituents of the metal, both metallic and nonmetallic, in the direction of working. (2) The pattern of preferred orientation of metal crystals after a given deformation process, usually wiredrawing. See also *fibering* and *preferred orientation*.

**fiber bridging** (composites). Reinforcing fiber material that bridges an inside radius of a pultruded product. This condition is caused by shrinkage stresses around such a radius during cure.

**fiber content.** The amount of fiber present in reinforced plastics and composites, usually expressed as a percentage volume fraction or weight fraction.

**fiber count.** The number of fibers per unit width of ply present in a specified section of a reinforced plastic or composite.

**fiber diameter.** The measurement (expressed in micrometers or microinches) of the diameter of individual filaments used in reinforced plastics or composites.

**fiber direction** (composites). The orientation or alignment of the longitudinal axis of a fiber with respect to a stated reference axis.

**fiber exposure** (composites). A condition in which reinforcing fibers within the base material are exposed in machined, abraded, or chemically attacked areas. See also *weave exposure*.

**fiberglass.** An individual filament made by drawing molten glass. A continuous filament is a glass fiber of great or indefinite length. A staple fiber is a glass fiber of relatively short length, generally less than 430 mm (17 in.), the length depending on the forming or spinning process used.

### Table 20  Compositions of glass fibers

| Glass type | Material, wt% | | | | | | | Total minor oxides |
|---|---|---|---|---|---|---|---|---|
| | Silica | Alumina | Calcium oxide | Magnesia | Boron oxide | Soda | Calcium fluoride | |
| E-glass ........ | 54 | 14 | 20.5 | 0.5 | 8 | 1 | 1 | 1 |
| A-glass ........ | 72 | 1 | 8 | 4 | ... | 14 | ... | 1 |
| ECR-glass ...... | 61 | 11 | 22 | 3 | ... | 0.6 | ... | 2.4 |
| S-glass ........ | 64 | 25 | ... | 10 | ... | 0.3 | ... | 0.7 |

### Table 21  Inherent properties of glass fibers

| | Specific gravity | Tensile strength | | Tensile modulus | | Coefficient of thermal expansion, $10^{-6}$/K | Dielectric constant(a) | Liquidus temperature | |
|---|---|---|---|---|---|---|---|---|---|
| | | MPa | ksi | GPa | $10^6$ psi | | | °C | °F |
| E-glass ......... | 2.58 | 3450 | 500 | 72.5 | 10.5 | 5.0 | 6.3 | 1065 | 1950 |
| A-glass ......... | 2.50 | 3040 | 440 | 69.0 | 10.0 | 8.6 | 6.9 | 996 | 1825 |
| ECR-glass ....... | 2.62 | 3625 | 525 | 72.5 | 10.5 | 5.0 | 6.5 | 1204 | 2200 |
| S-glass ......... | 2.48 | 4590 | 665 | 86.0 | 12.5 | 5.6 | 5.1 | 1454 | 2650 |

(a) At 20 °C (72 °F) and 1 MHz

The four main glasses used for fiberglass are high-alkali glass (A-glass), electrical grade glass (E-glass), a modified E-glass that is chemically resistant (ECR-glass), and high-strength glass (S-glass). The representative chemical compositions of these four glasses are given in Table 20. The inherent properties of these glasses are given in Table 21.

**fiberglass reinforcement.** Major material used to reinforce plastic. Available as mat, roving, fabric, and so forth, it is incorporated into both thermosets and thermoplastics.

**fiber grease.** A type of grease having a pronounced fibrous structure.

**fibering.** Elongation and alignment of internal boundaries, second phases, and inclusions in particular directions corresponding to the direction of metal flow during deformation processing (Fig. 174).

FIG. 174

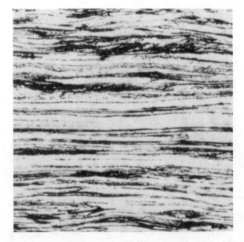

*Fibering in a flat-rolled electrical sheet steel. The structure consists of ferrite grains elongated in the rolling direction. 100×*

**fiber metallurgy.** The technology of producing solid bodies from fibers or chopped filaments, with or without a metal matrix. The fibers may consist of such nonmetals as graphite or aluminum oxide, or of such metals as tungsten or boron. See also *metal-matrix composites* (Technical Brief 26).

**fiber pattern.** Visible fibers on the surface of laminates or molding. The thread size and weave of glass cloth.

**fiber-reinforced composite.** A material consisting of two or more discrete physical phases, in which a fibrous phase is dispersed in a continuous matrix phase. The fibrous phase may be macro-, micro-, or submicroscopic, but it must retain its physical identity so that it could conceivably be removed from the matrix intact (Fig. 175).

**fiber-reinforced plastic (FRP).** A general term for a plastic that is reinforced with cloth, mat, strands, or any other fiber form. See also *resin-matrix composites* (Technical Brief 39).

**fiber show.** Strands or bundles of fibers that are not covered by plastic because they are at or above the surface of a reinforced plastic or composite.

**fiber stress** (metals). Local stress through a small area (a point or line) on a section where the stress is not uniform, as in a beam under a bending load.

**fiber texture** (metals). A texture characterized by having only one preferred crystallographic direction.

**fiber wash** (composites). Splaying out of woven or nonwoven fibers from the general reinforcement direction. Fibers are carried along with bleeding resin during cure.

**fibrillation.** Production of fiber from film.

**fibrous fracture.** A gray and amorphous *fracture* that results when a metal is sufficiently ductile for the crystals to elongate before fracture occurs. When a fibrous fracture is obtained in an impact test, it may be regarded as definite evidence of toughness of the metal. See also *crystalline fracture* and *silky fracture*.

FIG. 175

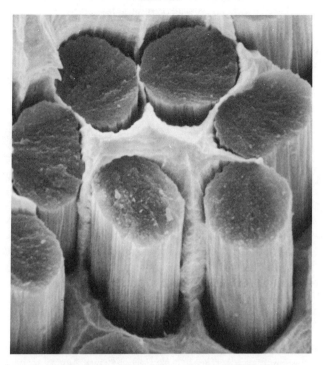

*Fracture surface of a fiber-reinforced composite material (graphite fibers in a 70Ag-30Cu alloy matrix). 3000×*

FIG. 176

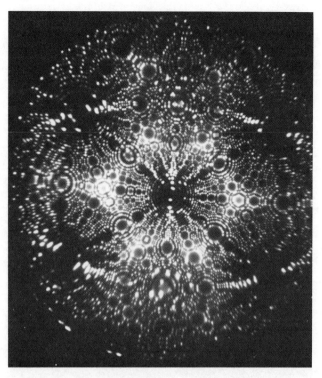

*Typical field ion micrograph of a tungsten sample*

**fibrous structure** (metals). (1) In forgings, a structure revealed as laminations, not necessarily detrimental, on an etched section or as a ropy appearance on a fracture. It is not to be confused with silky or ductile fracture of a clean metal. (2) In wrought iron, a structure consisting of slag fibers embedded in ferrite. (3) In rolled steel plate stock, a uniform, fine-grained structure on a fractured surface, free of laminations or shaletype discontinuities.

**field-emission microscopy**. An image-forming analytical technique in which a strong electrostatic field causes emission of electrons from a sharply rounded point or from a specimen that has been placed on that point. The electrons are accelerated to a phosphorescent screen, or photographic film, producing a visible picture of the variation of emission over the specimen surface.

**field ionization**. The ionization of gaseous atoms and molecules by an intense electric field, often at the surface of a solid.

**field ion microscopy**. An analytical technique in which atoms are ionized by an electric field near a sharp specimen tip; the field then forces the ions to a fluorescent screen which shows an enlarged image of the tip, and individual atoms are made visible (Fig. 176). See also *atom probe*.

**field-of-view**. The maximum diameter of an object that can be imaged by a microscope or other analytic technique.

**filament**. The smallest unit of fibrous material. The basic units formed during drawing and spinning, which are gathered into strands of fiber for use as reinforcements. Filaments usually are of extreme length and very small diameter, usually less than 25 μm (1 mil). Normally, filaments are not used individually. Some textile filaments can function as a reinforcing yarn when they are of sufficient strength and flexibility.

**filamentary shrinkage**. A fine network of shrinkage cavities, occasionally found in steel castings, that produces a radiographic image resembling lace.

**filament winding**. A process for fabricating a reinforced plastic or composite structure in which continuous reinforcements (filament, wire, yarn, tape, and the like), either previously impregnated with a matrix material or impregnated during the winding, are placed over a rotating and removable form or mandrel in a prescribed way to meet certain stress conditions (Fig. 177). Generally the shape is a surface of revolution and may or may not include end closures. When the required number of layers is applied, the wound form is cured and the mandrel is removed. See also *helical winding* and *polar winding*.

**filar eyepiece**. In an optical microscope, an eyepiece having in its focal plane a fiducial line that can be moved using a calibrated micrometer screw. Useful for accurate determination of linear dimensions. Also termed filar micrometer.

FIG. 177

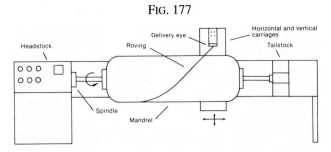

*Typical filament winding machine*

FIG. 178

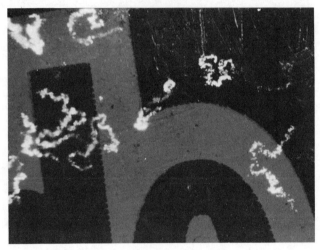

*Filiform corrosion of the aluminum foil vapor barrier on laminated packaging. 10×*

**file hardness**. Hardness as determined by the use of a steel file of standardized hardness on the assumption that a material that cannot be cut with the file is as hard as, or harder than, the file. Files covering a range of hardnesses may be employed; the most common are files heat treated to approximately 67 to 70 HRC.

**filiform corrosion**. Corrosion that occurs under some coatings in the form of randomly distributed threadlike filaments (Fig. 178).

**fill** (composites). Reinforcing yarn oriented at right angles to the warp in a woven fabric.

**fill-and-wipe**. Technique used with plastic parts that are molded with depressed designs; after application of paint, the surplus is wiped off, leaving paint only in the depressed areas. Sometimes called wipe-ins.

**fill density**. See preferred term *apparent density*.

**fill depth**. Synonymous with *fill height*.

**filler**. (1) A relatively inert substance added to a plastic to alter its physical, mechanical, thermal, electrical, or other properties, or to lower cost or density. Sometimes the term is used specifically to mean particulate additives. See also *inert filler* and *reinforced plastics*. (2) A relatively nonadhesive substance added to an adhesive to improve its working properties, permanence, strength, or other qualities. See also *binder* and *extruder*. (3) In lubrication, a substance such as lime, talc, mica, and other powders, added to a grease to increase its consistency or to an oil to increase its viscosity.

**filler metal**. Metal added in making a brazed, soldered, or welded joint.

**filler sheet**. A sheet of deformable or resilient material that, when placed between the assembly to be adhesively bonded and pressure applicator, or when distributed within a stack of assemblies, aids in providing uniform application of pressure over the area to be bonded.

**fillet** (adhesive bonding). A rounded filling or adhesive that fills the corner or angle where two adherends are joined.

**fillet** (metals). (1) Concave corner piece usually used at the intersection of casting sections. Also the radius of metal at such junctions as opposed to an abrupt angular junction. (2) A radius (curvature) imparted to inside meeting surfaces.

**fillet radius**. Bend radius between two abutting walls.

**fillet weld**. A weld, approximately triangular in cross section, joining two surfaces, essentially at right angles to each other in a lap, tee, or corner joint (Fig. 179).

FIG. 179

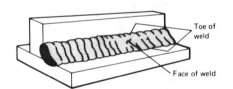

*Example of a fillet weld*

**fillet weld size**. See preferred term *size of weld*.

**fill factor**. In powder metallurgy, the quotient of the fill volume of a powder over the volume of the green compact after ejection from the die. It is the same as the quotient of the powder fill height over the height of the compact. Inverse parameter of *compression ratio*.

**fill height**. In powder metallurgy, the distance between the lower punch face and the top plane of the die body in the fill position of the press tool.

**filling yarn**. The transverse threads or fibers in a woven fabric used in reinforced plastics or composites. Those fibers running perpendicular to the warp. Also called weft.

**fill position**. In powder metallurgy, the position of the press tool which enables the filling of the desired amount of powder into the die cavity.

**fill ratio**. See *compression ratio* (powder metallurgy).

**fill shoe**. See preferred term *feed shoe*.

**fill volume**. The volume that a metal powder fills after flowing loosely into a space that is open at the top, such as a die cavity or a measuring receptacle.

**film adhesive**. A synthetic resin adhesive, usually of thermosetting type, in the form of a thin, dry film of resin with or without a paper or glass carrier.

**film resistance**. The electrical resistance that results from films at contacting surfaces, such as oxides and contaminants, that prevent pure metallic contact.

**film strength**. An imprecise term denoting ability of a surface film to resist rupture by the penetration of asperities during sliding or rolling. A high film strength is primarily inferred from a high load-carrying capacity and is seldom directly measured. It is recommended that this term should not be used.

**film stress**. The compressive or tensile forces appearing in a film, such as internal film stress, which is the intrinsic stress of a film related to its mechanical structure and deposition parameters, or induced film stress, which is the component of film stress related to an external force such as mismatched mechanical properties of the substrate.

**film thickness**. In a dynamic seal, the distance separating the two surfaces that form the primary seal.

**filter**. (1) A porous article or material for separating suspended particulate matter from liquids by passing the liquid through the pores in the filter and sieving out the solids. (2) Any transmission network used in electrical systems for the selective enhancement of a given class of input signals. Also known as electric filter; electric-wave filter. (3) A device employed to reject sound in a particular range of frequencies while passing sound in another range of frequencies. Also known as acoustic filter. (4) A semitransparent optical element capable of absorbing unwanted electromagnetic radiation and transmitting the remainder. A neutral density filter attenuates relatively uniformly from the ultraviolet to the infrared, but in many applications highly wavelength-selective filters are used. See also *neutral filter*.

**filter glass**. See preferred term *filter plate*.

**filter plate** (eye protection). An optical material which protects the eyes against excessive ultraviolet, infrared, and visible radiation.

**fin**. (1) Excess material left on a molded plastic object at those places where the molds or dies mated. Also, the web of material remaining in holes or opening in a molded part, which must be removed in finishing. (2) Metal on a casting caused by an imperfect joint in the mold or die.

**final annealing**. An imprecise term used to denote the last anneal given to a nonferrous alloy prior to shipment.

**final density**. The density of a sintered product.

**final polishing**. A polishing process in which the primary objective is to produce a final surface suitable for microscopic examination.

**fine ceramics**. Same as *advanced ceramics*.

**fine hackle**. See *hackle*.

**fineness**. A measure of the purity of gold or silver expressed in parts per thousand.

**fines** (ceramics, metals, ores). (1) The product that passes through the finest screen in sorting crushed or ground material. (2) Sand grains that are substantially smaller than the predominating size in a batch or lot of foundry sand. (3) The portion of a powder composed of particles smaller than a specified size, usually 44 μm (–325 mesh).

**fines** (plastics). Very small particles (usually under 200 mesh) accompanying larger grains, usually of molding powder.

**fine silver**. Silver with a fineness of 999; equivalent to a minimum content of 99.9% Ag with the remaining content unrestricted.

**finish** (composites). A mixture of materials for treating glass or other fibers. It contains a coupling agent to improve the bond of resin to the fiber and usually includes a lubricant to prevent abrasion, as well as a binder to promote strand integrity. With graphite or other filaments, it may perform any or all of the above functions.

**finish** (metals). (1) Surface condition, quality, or appearance of a metal. (2) Stock on a forging or casting to be removed in finish machining. (3) The forging operation in which the part is forged into its final shape in the finish die. If only one finish operations is scheduled to be performed in the finish die, this operation will be identified simply as finish; first, second, or third finish designations are so termed when one or more finish operations are to be performed in the same finish die.

**finish** (plastics). To complete the secondary work on a molded plastic part so that it is ready for use. Operations such as filling, deflashing, buffing, drilling, tapping, and degating are commonly called finishing operations.

**finish allowance**. (1) The amount of excess metal surrounding the intended final configuration of a formed part; sometimes called forging envelope, machining allowance, or cleanup allowance. (2) Amount of stock left on the surface of a casting for machining.

**finish annealing**. A *subcritical annealing* treatment applied to cold-worked low- or medium-carbon steel. Finish annealing, which is a compromise treatment, lowers residual stresses, thereby minimizing the risk of distortion in machining while retaining most of the benefits to machinability contributed by cold working. Compare with *final annealing*.

**finished steel**. Steel that is ready for the market and has been processed beyond the stages of billets, blooms, sheet bars, slabs, and wire rods.

**finisher (finishing impression)**. The *die impression* that imparts the final shape to a forged part.

**finish grinding**. The final grinding action on a workpiece, of which the objectives are surface finish and dimensional accuracy.

**finishing die**. The die set used in the last forging step.

**finishing temperature**. The temperature at which *hot working* is completed.

**finish machining**. A machining process analogous to *finish grinding*.

**finish trim**. Flash removal from a forging; usually performed by trimming, but sometimes by band sawing or similar techniques.

**firebrick**. A refractory brick, often made from *fireclay*, that is able to withstand high temperature (1500 to 1600 °C, or 2700 to 2900 °F) and is used to line furnaces, ladles, or other molten metal containment components.

**fireclay**. A mineral aggregate that has as its essential constituent the hydrous silicates of aluminum with or without free silica. It is used in commercial refractory products.

**firecracker welding**. A variation of the *shielded metal arc welding* process in which a length of covered electrode is placed along the joint in contact with the parts to be welded; during the welding operation, the stationary electrode is consumed as the arc travels the length of the electrode.

**fired mold**. A shell mold or solid mold that has been heated to a high temperature and is ready for casting.

**fire point**. The temperature at which a material will continue to burn for at least 5 s without the benefit of an outside flame.

**fire-refined copper**. Copper that has been refined by the use of a furnace process only, including refinery shapes and, by extension, fabricators' products made therefrom. Usually, when this term is used alone it refers to fire-refined tough pitch copper without elements other than oxygen being present in significant amounts.

**fire scale**. Intergranular copper oxide remaining below the surface of silver-copper alloys that have been annealed and pickled.

**firing**. The controlled heat treatment of ceramic ware in a kiln or furnace, during the process of manufacture, to develop the desired properties.

**firing range**. (1) The range of fired temperature within which a ceramic composition develops properties which render it commercially useful. (2) The time-temperature interval in which a porcelain enamel or ceramic coating is satisfactorily matured.

**first block, second block, and finish**. The forging operation in which the part to be forged is passed in progressive order through three tools mounted in one forging machine; only one heat is involved for all three operations.

**first-degree blocking**. An adherence between adhesively bonded surfaces under test of such degree that when the upper specimen is lifted, the lower specimen will cling thereto, but may be parted with no evidence of damage to either surface.

**first-order transition**. A change of state associated with crystallization, melting, or a change in crystal structure of a polymer.

**fir-tree crystal**. A type of *dendrite*.

**fisheye** (metals). An area on a steel fracture surface having a characteristic white crystalline appearance.

**fisheye** (plastics). A small, globular mass that has not completely blended into the surrounding pultruded material. This condition is particularly evident in a transparent or translucent material.

**fisheye** (weld defect). A discontinuity found on the fracture surface of a weld in steel that consists of a small pore or inclusion surrounded by an approximately round, bright area (Fig. 180).

FIG. 180

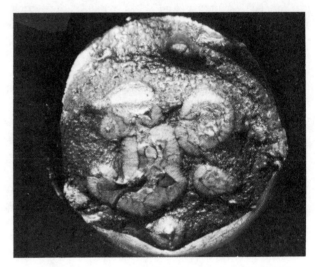

*Fisheyes in as-welded E7018 tensile specimen tested at room temperature. 4×*

**fishmouthing**. See *alligatoring*.

**fishscale**. A scaly appearance in a porcelain enamel coating in which the evolution of hydrogen from the basis metal (iron or steel) causes loss of adhesion between the enamel and the basis metal. Individual scales are usually small, but have been observed in sizes up to 25 mm (1 in.) or more in diameter. The scales are somewhat like blisters that have cracked partway around the perimeter but still remain attached to the coating around the rest of the perimeter.

**fishtail**. (1) In roll forging, the excess trailing end of a forging. It is often used, before being trimmed off, as a tong hold for a subsequent forging operation. (2) In hot rolling or extrusion, the imperfectly shaped trailing end of a bar or special section that must be cut off and discarded as mill scrap.

**fissure**. A small crack-like weld discontinuity with only slight separation (opening displacement) of the fracture surfaces. The prefixes macro or micro indicate relative size.

**fit**. The amount of clearance or interference between mating parts is called actual fit. Fit is the preferable term for the range of clearance or interference that may result from the specified limits on dimensions (limits of size). Refer to ANSI standards.

**fixed-feed grinding.** Grinding in which the wheel is fed into the work, or vice versa, by given increments or at a given rate.

**fixed-land bearing.** See *fixed-pad bearing.*

**fixed-load or fixed-displacement crack extension force curves.** Curves obtained from a fracture mechanics analysis for the test configuration, assuming a fixed applied load or displacement and generating a curve of *crack extension force* versus the *effective crack size* as the independent variable.

**fixed oil.** An imprecise term denoting an oil that is difficult to distill without decomposition.

**fixed-pad bearing.** An axial- or radial-load bearing equipped with fixed pads, the surfaces of which are contoured to promote hydrodynamic lubrication.

**fixed position welding.** Welding in which the work is held in a stationary position.

**fixture.** A device designed to hold parts to be joined in proper relation to each other.

**fixture time.** The shortest time required by an adhesive to develop handling strength such that test specimens can be removed from fixtures, unclamped, or handled without stressing the bond and thereby affecting bond strength. Also referred to as set time.

**fixturing.** The placing of parts to be heat treated in a constraining or semiconstraining apparatus to avoid heat-related distortions. See also *racking.*

**flake (metals).** A short, discontinuous internal crack in ferrous metals attributed to stresses produced by localized transformation and hydrogen-solubility effects during cooling after hot working. In fracture surfaces, flakes appear as bright, silvery areas with a coarse texture (Fig. 181). In deep acid-etched transverse sections, they appear as discontinuities that are usually in the midway to center location of the section. Also termed hairline cracks and shatter cracks.

FIG. 181

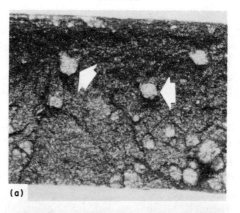

(a)

(b)

*Macroscopic appearance of hydrogen flakes on a fracture surface. (a) Flaking in a 4340 steel. Arrows indicate individual flakes. Actual size. (b) Closeup of fracture surface containing the mating halves of a flake. Note the distinctive shiny appearance of the flake.*

**flake (plastics).** A term used to denote the dry, unplasticized base of cellulosic plastics.

**flake graphite.** Graphitic carbon, in the form of platelets, occurring in the microstructure of *gray iron* (Fig. 182).

**flake powder.** Flat or scalelike particles whose thickness is small compared to the other dimensions.

FIG. 182

| Type A | Type B | Type C | Type D | Type E |
|---|---|---|---|---|

  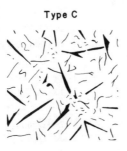  

| Uniform distribution, random orientation | Rosette grouping, random orientation | Superimposed flake size, random orientation | Interdendritic segregation, random orientation | Interdendritic segregation, preferred orientation |
|---|---|---|---|---|

*Types of graphite flakes in gray iron (AFS-ASTM)*

**flaking.** (1) The removal of material from a surface in the form of flakes or scalelike particles. (2) A form of pitting resulting from fatigue. See also *spalling*.

**flame annealing.** Annealing in which the heat is applied directly by a flame.

**flame cleaning.** Cleaning metal surfaces of scale, rust, dirt, and moisture by use of a gas flame.

**flame cutting.** See preferred term *oxygen cutting*.

**flame hardening.** A process for hardening the surfaces of hardenable ferrous alloys in which an intense flame is used to heat the surface layers above the upper transformation temperature, whereupon the workpiece is immediately quenched.

**flame resistance.** Ability of a material to extinguish flame once the source of heat is removed. See also *self-extinguishing resin*.

**flame retardants.** Certain chemicals that are used to reduce or eliminate the tendency of a resin to burn.

**flame spraying.** *Thermal spraying* in which a coating material is fed into an oxyfuel gas flame, where it is melted. Compressed gas may or may not be used to atomize the coating material and propel it onto the substrate. The sprayed material is originally in the form of wire or powder. See the figures accompanying the terms *powder flame spraying* and *wire flame spraying*. The term flame spraying is usually used when referring to a combustion-spraying process, as differentiated from *plasma spraying*.

**flame spraying** (plastics). Method of applying a plastic coating in which finely powdered fragments of the plastic, together with suitable fluxes, are projected through a cone of flame onto a surface.

**flame straightening.** Correcting distortion in metal structures by localized heating with a gas flame.

**flame treating** (plastics). A method of rendering inert thermoplastic objects receptive to inks, lacquers, paints, adhesives, and so forth, in which the object is bathed in an open flame to promote oxidation of the surface of the article.

**flammability.** Measure of the extent to which a material will support combustion.

**flange.** A projecting rim or edge of a part; usually narrow and of approximately constant width for stiffening or fastening.

**flange weld.** A weld made on the edges of two or more members to be joined, usually light gage metal, at least one of the members being flanged. Flange weld, flare-bevel, and flare-V-groove welds may be confused because they have similar geometry before welding. A flange is welded on the edge and a flare is welded in the groove. See the figure accompanying the term *edge-flange weld* which compares flange, flare-bevel, and flare-V-groove welds.

**flank.** The end surface of a tool that is adjacent to the cutting

FIG. 183

*Typical flank wear on a cemented carbide cutting tool insert*

edge and below it when the tool is in a horizontal position, as for turning. See figure accompanying *single-point tool*.

**flank wear.** The loss of relief on the flank of the tool behind the cutting edge due to rubbing contact between the work and the tool during cutting; measured in terms of linear dimension behind the original cutting edge (Fig. 183). See also the figure accompanying the term *crater wear*.

**flapping.** In copper refining, hastening oxidation of molten copper by striking through the slag-covered surface of the melt with a *rabble* just before the bath is poled.

**flare-bevel groove weld.** A weld in a groove formed by a member with a curved surface in contact with a planar member. Compare with *flange weld*. See also the figure accompanying the term *edge-flange weld*.

**flare test.** A test applied to tubing, involving tapered expansion over a cone. Similar to *pin expansion test*.

**flare-V-groove weld.** A weld in a groove formed by two members with curved surfaces. Compare with *flange weld*. See also the figure accompanying the term *edge-flange weld*.

**flaring.** (1) Forming an outward acute-angle flange on a tubular part. (2) Forming a flange by using the head of a hydraulic press.

**flash** (metals). (1) In forging, metal in excess of that required to fill the blocking or finishing forging impression of a set of dies completely. Flash extends out from the body of the forging as a thin plate at the line where the dies meet and is subsequently removed by trimming. Because it cools faster than the body of the component during forging, flash can serve to restrict metal flow at the line where dies meet, thus ensuring complete filling of the impression. See also *closed-die forging*. (2) In casting, a fin of metal that results from leakage between mating mold surfaces. (3) In welding, the material which is expelled or squeezed out of a weld joint and which forms around the weld.

**flash** (plastics). The portion of the charge that flows from or is extruded from the mold cavity during the molding. Extra plastic attached to a molding along the parting line, which must be removed before the part is considered finished.

**flashback**. A recession of the welding or cutting torch flame into or back of the mixing chamber of the torch.

**flashback arrestor**. A device incorporated into an oxygen or oxyfuel welding or cutting torch to limit damage from a *flashback* by preventing propagation of the flame front beyond the point at which the arrester is installed.

**flash butt welding**. See preferred term *flash welding*.

**flash coat**. A thin metallic coating usually less than 0.05 mm (0.002 in.) in thickness.

**flash extension**. That portion of *flash* remaining on a forged part after trimming; usually included in the normal forging tolerances.

**flashing**. In *flash welding*, the heating portion of the cycle, consisting of a series of rapidly recurring localized short circuits followed by molten metal expulsions, during which time the surfaces to be welded are moved one toward the other at a predetermined speed.

**flashing time**. The time during which the flashing action is taking place in *flash welding*.

**flash land**. Configuration in the blocking or finishing impression of forging dies designed to restrict or to encourage the growth of *flash* at the parting line, whichever may be required in a particular case to ensure complete filling of the impression.

**flash line**. The line left on a forging after the flash has been trimmed off.

**flash mold** (plastics). A mold in which the mold faces are perpendicular to the clamping action of the press, so that the greater the clamping force, the tighter the mold seam.

**flash-off time**. See preferred term *flashing time*.

**flash pan**. The machined-out portion of a forging die that permits the flow through of excess metal.

**flash plate**. A very thin final electrodeposited film of metal.

**flash point**. (1) The temperature to which a material must be heated to give off sufficient vapor to form a flammable mixture. (2) The lowest temperature at which the vapor of a lubricant can be ignited under specified conditions.

**flash temperature**. The maximum local temperature generated at some point in a sliding contact. The flash temperature occurs at areas of real contact due to the frictional heat dissipated at these areas. The duration of the flash temperature is often of the order of a microsecond. The term flash temperature may also mean the average temperature over a restricted contact area (for example, between gear teeth).

**flash weld**. A weld made by *flash welding*.

**flash welding**. A resistance welding process that joins metals by first heating abutting surfaces by passage of an electric current across the joint, then forcing the surfaces together by the application of pressure. Flashing and upsetting are

FIG. 184

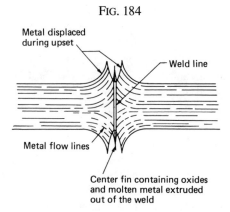

*Typical peaks and flow lines in a flash weld*

FIG. 185

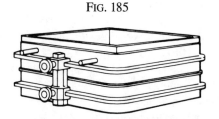

*Flask used to cast metals in sand molds*

accompanied by expulsion of metal from the joint (Fig. 184).

**flask**. A metal or wood frame used for making and holding a sand mold (Fig. 185). The upper part is called the *cope*; the lower, the *drag*. See also the figure accompanying the term *blind riser*.

**flat-die forging**. Forging metal between flat or simple-contour dies by repeated strokes and manipulation of the workpiece. Also known as *open-die forging*, hand forging, or smith forging.

**flat drill**. A rotary end-cutting tool constructed from a flat piece of material, provided with suitable cutting lips at the cutting end. See also figure accompanying the term *drill*.

**flat edge trimmer**. A machine for trimming notched edges on shells. The slide is cam driven so as to obtain a brief dwell at the bottom of the stroke, at which time the die, sometimes called a shimmy die, oscillates to trim the part.

**flat glass**. A general term covering sheet glass, plate glass, and various forms of rolled glass.

**flat-position welding**. Welding from the upper side, the face of the weld being horizontal. See also the figure accompanying the term *welding position*.

**flats**. A longitudinal, flat area on a normally convex surface of a pultruded plastic, caused by shifting of the reinforcement, lack of sufficient reinforcement, or local fouling of the die surface.

**flattening**. (1) A preliminary operation performed on forging stock to position the metal for a subsequent forging operation. (2) The removal of irregularities or distortion in sheets or plates by a method such as *roller leveling* or *stretcher leveling*.

**flattening dies**. Dies used to flatten sheet metal hems; that is, dies that can flatten a bend by closing it. These dies consist of a top and bottom die with a flat surface that can close one section (flange) to another (hem, seam). See also the figure (e) accompanying the term *press-brake forming*.

**flattening test**. A quality test for tubing in which a specimen is flattened to a specified height between parallel plates.

**flat wire**. A roughly rectangular or square mill product, narrower than *strip*, in which all surfaces are rolled or drawn without any previous slitting, shearing, or sawing.

**flaw**. A nonspecific term often used to imply a crack-like discontinuity. See also preferred terms *discontinuity*, *imperfection*, and *defect*.

**flexibility**. The quality or state of a material that allows it to be flexed or bent repeatedly without undergoing rupture. See also *flexure*.

**flexibilizer**. An additive that makes a finished plastic more flexible or tough. See also *plasticizer*.

**flexible cam**. An adjustable pressure-control cam of spring steel strips used to obtain varying pressure during a forming cycle.

**flexible hinge**. See *flexure pivot bearing*.

**flexible molds**. Molds made of rubber or elastomeric plastics, used for casting plastics. They can be stretched to remove cured pieces having undercuts.

**flex roll**. A movable jump roll designed to push up against a metal sheet as it passes through a roller leveler. The flex roll can be adjusted to deflect the sheet any amount up to the roll diameter.

**flex rolling**. Passing metal sheets through a *flex roll* unit to minimize yield-point elongation in order to reduce the tendency for *stretcher strains* to appear during forming.

**flexural failure**. A material failure caused by repeated flexing.

**flexural modulus**. The ratio, within the elastic limit, of the applied stress on a reinforced plastic test specimen in flexure to the corresponding strain in the outermost fibers of the specimen.

**flexural strength**. A property of solid material that indicates its ability to withstand a flexural or transverse load.

**flexural strength** (composites). The maximum stress that can be borne by the surface fibers in a beam in bending. The flexural strength is the unit resistance to the maximum load before failure by bending, usually expressed in force per unit area.

**flexure**. A term used in the study of strength of materials to indicate the property of a body, usually a rod or beam, to bend without fracture. See also *flexibility*.

**flexure pivot bearing**. A type of bearing guiding the moving parts by flexure of an elastic member or members rather

than by rolling or sliding. Only limited movement is possible with a flexure pivot.

**flexure stress** (glass). The tensile component of the bending stress produced on the surface of a glass section opposite to that experiencing a locally impinging force.

**floating bearing**. A bearing designed or mounted to permit axial displacement between shaft and housing.

**floating chase**. In forming of plastics, a mold member, free to move vertically, that fits over a lower plug or cavity, and into which an upper plug telescopes.

**floating die**. (1) In metal forming, a die mounted in a die holder or punch mounted in its holder such that a slight amount of motion compensates for tolerance in the die parts, the work, or the press. (2) A die mounted on heavy springs to allow vertical motion in some trimming, shearing, and forming operations.

**floating die pressing**. The compaction of a metal powder in a floating die, resulting in densification at opposite ends of the compact. Analogous to double action pressing.

**floating plug**. In tube drawing, an unsupported mandrel that locates itself at the die inside the tube, causing a reduction in wall thickness while the die is reducing the outside diameter of the tube (Fig. 186).

FIG. 186

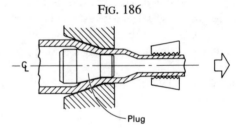

*Drawing with a floating plug*

**floating-ring bearing**. A type of journal bearing that includes a thin ring between the journal and the bearing. The ring floats and rotates at a fraction of the journal rotational speed.

**flocculant**. An electrolyte added to a colloidal suspension to cause the particles to aggregate and settle out as the result of reduction in repulsion between the particles.

**flocculate**. A grouping of primary particles, aggregates, or agglomerates having weaker bonding than either the aggregate or agglomerate structures. Flocculates are usually formed in a gas or liquid suspension and those formed in a liquid can generally be broken up by gentle shaking or stirring.

**flocculation**. Agglomeration of particles in a suspension causing them to settle out.

**flock**. A material obtained by reducing textile fibers to fragments as by cutting, tearing, or grinding, to give various degrees of comminution. Flock can either be fibers in entangled, small masses or beads, usually of irregular broken fibers, or comminuted (powdered) fibers.

**flocking**. A method of coating by spraying finely dispersed textile powders or fibers.

**flock point**. A measure of the tendency of a lubricant to precipitate wax or other solids from solution. Depending on the test used, the flock point is the temperature required for precipitation, or the time required at a given temperature for precipitation.

**flood lubrication**. A system of lubrication in which the lubricant is supplied in a continuous stream at low pressure and subsequently drains away. Also known as *bath lubrication*.

**floor molding**. In foundry practice, making sand molds from loose or production patterns of such size that they cannot be satisfactorily handled on a bench or molding machine, the equipment being located on the floor during the entire operation of making the mold.

**flop forging**. A forging in which the top and bottom die impressions are identical, permitting the forging to be turned upside down during the forging operation.

**floppers**. On metals, lines or ridges that are transverse to the direction of rolling and generally confined to the section midway between the edges of a coil as rolled.

**flospinning**. Forming cylindrical, conical and curvilinear shaped parts by power spinning over a rotating mandrel. See also *spinning*.

**flotation**. The concentration of valuable minerals from ores by agitation of the ground material with water, oil, and flotation chemicals. The valuable minerals are generally wetted by the oil, lifted to the surface by clinging air bubbles, and then floated off.

**flow**. Movement (slipping or sliding) of essentially parallel planes within an element of a material in parallel directions; occurs under the action of *shear stress*. Continuous action in this manner, at constant volume and without disintegration of the material, is termed *yield*, *creep*, or *plastic deformation*.

**flow** (adhesives). Movement of an adhesive during the bonding process, before the adhesive is set.

**flow** (plastics). The movement of resin under pressure, allowing it to fill all parts of a mold.

**flowability**. (1) In casting, a characteristic of a foundry sand mixture that enables it to move under pressure or vibration so that it makes intimate contact with all surfaces of the pattern or core box. (2) In welding, brazing, or soldering, the ability of molten filler metal to flow or spread over a metal surface.

**flow brazing**. Brazing by pouring hot molten nonferrous filler metal over a joint until the brazing temperature is attained. The filler metal is distributed in the joint by capillary action.

**flow brightener**. (1) Melting of an electrodeposit, followed by solidification, especially of tin plate. (2) Fusion (melting) of a chemically or mechanically deposited metallic coating on a substrate, particularly as it pertains to soldering.

**flow cavitation**. Cavitation caused by a decrease in static pressure induced by changes in the velocity of a flowing liquid. Typically this may be caused by flow around an obstacle or through a constriction, or relative to a blade or foil.

**flow factor**. See preferred term *flow rate*.

**flow lines**. (1) Texture showing the direction of metal flow during hot or cold working. Flow lines can often be revealed by etching the surface or a section of a metal part (Fig. 187). (2) In mechanical metallurgy, paths followed by minute volumes of metal during deformation.

FIG. 187

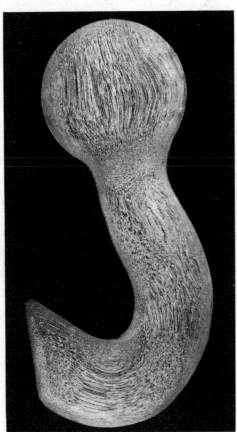

*Hook forged from 4140 steel, showing flow lines in a longitudinal section. 0.5×*

**flow marks**. Wavy surface appearance of an object molded from thermoplastic resins, caused by improper flow of the resin into the mold.

**flow meter**. (1) A device for indicating the rate of gas flow in a system. (2) In powder metallurgy, a metal cylinder whose interior is funnel shaped and whose bottom has a calibrated orifice of standard dimensions to permit passage of a powder and the determination of the *flow rate*.

**flow molding**. The technique of producing leatherlike materials by placing a die-cut plastic blank (solid or expanded vinyl or vinyl-coated substrate) in a mold cavity (usually silicone rubber molds) and applying power via a high-frequency radio frequency generator to melt the plastic such that it flows into the mold to the desired shape and with the desired texture.

**flow rate**. The time required for a metal powder sample of standard weight to flow through an orifice in a standard instrument according to a specified procedure.

**flow soldering**. See *wave soldering*.

**flow stress**. The stress required to produce *plastic deformation* in a solid metal.

**flow test**. A standardized test to measure how readily a metal powder flows. See also *flow rate*.

**flow through**. A forging defect caused by metal flow past the base of a rib with resulting rupture of the grain structure.

**flow welding**. A welding process which produces coalescence of metals by heating them with molten filler metal poured over the surfaces to be welded until the welding temperature is attained and until the required filler metal has been added. The filler metal is not distributed in the joint by capillary action.

**fluid bearing**. See *hydrostatic bearing*.

**fluid-cell process**. A modification of the *Guerin process* for forming sheet metal, the fluid-cell process uses higher pressure and is primarily designed for forming slightly deeper parts, using a rubber pad as either the die or punch. A flexible hydraulic fluid cell forces an auxiliary rubber

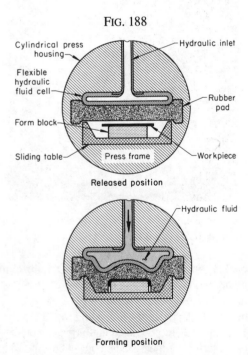

FIG. 188

*Principal components of the fluid-cell forming (Verson-Wheelon) process*

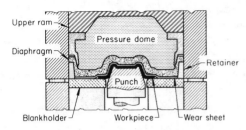

FIG. 189

*Fluid forming in a Hydroform press*

pad to follow the contour of the form block and exert a nearly uniform pressure at all points on the workpiece (Fig. 188). See also *fluid forming* and *rubber-pad forming*.

**fluid erosion**. See *liquid impingement erosion*.

**fluid forming**. A modification of the *Guerin process*, fluid forming differs from the fluid-cell process in that the die cavity, called a pressure dome, is not completely filled with rubber, but with hydraulic fluid retained by cup-shaped rubber diaphragm (Fig. 189). See also *rubber-pad forming*.

**fluid friction**. Frictional resistance due to the viscous or rheological flow of fluids.

**fluidity**. (1) The ability of liquid metal to run into and fill a mold cavity. (2) The reciprocal of *viscosity*.

**fluidized bed**. A contained mass of a finely divided solid that behaves like a fluid when brought into suspension in a moving gas or liquid.

**fluidized-bed coating**. A method of applying a coating of a thermoplastic resin to a heated article that is immersed in a dense-phase fluidized bed of powdered resin and thereafter heated in an oven to provide a smooth, pinhole-free coating (Fig. 190).

**fluidized-bed heating**. Heating carried out in a medium of solid particles suspended in a flow of gas.

**fluidized-bed reduction**. The finely divided solid is a powdered ore or reducible oxide, and the moving gas is reducing; the operation is carried out at elevated temperature in a furnace.

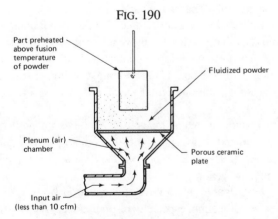

FIG. 190

*Fluidized-bed coating system*

**fluorescence.** (1) Emission of electromagnetic radiation that is caused by the flow of some form of energy into the emitting body and which ceases abruptly when the excitation ceases. (2) Emission of electromagnetic radiation that is caused by the flow of some form of energy into the emitting body and whose decay, when the excitation ceases, is temperature-independent. (3) A type of photoluminescence in which the time interval between the absorption and re-emission of light is very short. Contrast with *phosphorescence.*

**fluorescent magnetic-particle inspection.** Inspection with either dry magnetic particles or those in a liquid suspension, the particles being coated with a fluorescent substance to increase the visibility of the indications (Fig. 191).

FIG. 191

*Wet fluorescent magnetic paint indications of minute grinding cracks in the face of a small sprocket*

**fluorescent penetrant inspection.** Inspection using a fluorescent liquid that will penetrate any surface opening; after the surface has been wiped clean, the location of any surface flaws may be detected by the fluorescence, under ultraviolet light, of back-seepage of the fluid.

**fluorimetry.** See *fluorometric analysis.*

**fluorinated ethylene propylene (FEP).** A member of the fluorocarbon family of plastics that is a copolymer of tetrafluoroethylene and hexafluoropropylene, possessing most of the properties of polytetrafluoroethylene, and having a melt viscosity low enough to permit conventional thermoplastic processing. Available in pellet form for molding and extrusion, and as dispersions for spray or dip coating processes.

**fluorocarbon plastics.** Plastics based on polymers made with monomers composed of fluorine and carbon only.

**fluorocarbons.** The family of plastics including polytetrafluoroethylene, polychlorotrifluoroethylene, polyvinylidene, and fluorinated ethylene propylene. They are characterized by good thermal and chemical resistance, nonadhesiveness, low dissipation factor, and low dielectric constant. They are available in a variety of forms, such as molding materials, extrusion materials, dispersions, film, or tape, depending on the particular fluorocarbon.

**fluorohydrocarbon plastics.** Plastics based on polymers made with monomers composed of fluorine, hydrogen, and carbon only.

**fluorometric analysis.** A method of chemical analysis that measures the fluorescence intensity of the analyte or a reaction product of the analyte and a chemical reagent.

**fluoroplastics.** Plastics based on polymers with monomers containing one or more atoms of fluorine, or copolymers of such monomers with other monomers, with the fluorine-containing monomer(s) being in greatest amount by mass.

**fluoroscopy.** An inspection procedure in which the radiographic image of the subject is viewed on a fluorescent screen, normally limited to low-density materials or thin sections of metals because of the low light output of the fluorescent screen at safe levels of radiation.

**flute.** (1) As applied to drills, reamers, and taps, the channels or grooves formed in the body of the tool to provide cutting edges and to permit passage of cutting fluid and chips. (2) As applied to milling cutters and hobs, the chip space between the back of one tooth and the face of the following tooth.

**fluted bearing.** A sleeve bearing with oil grooves generally in an axial direction.

**fluted core.** An integrally woven reinforcement material consisting of ribs between two skins in a unitized *sandwich construction.*

**flutes.** Elongated grooves or voids that connect widely spaced cleavage planes (Fig. 192).

**fluting.** (1) Forming longitudinal recesses in a cylindrical part, or radial recesses in a conical part. (2) A series of sharp parallel kinks or creases occurring in the arc when sheet metal is roll formed into a cylindrical shape. (3) Grinding the grooves of a twist drill or tap. (4) In bearings, a form of pitting in which the pits occur in a regular pattern so as to form grooves. Ridges may occur with or without burnt craters. The general cause involves vibration together with excessive wear or excessive load. (5) Electric discharge pitting in a rolling-contact bearing subject to vibration. (6) A fracture process whereby *flutes* are produced.

**flux.** (1) In metal refining, a material added to a melt to remove undesirable substances, like sand, ash, or dirt.

FIG. 192

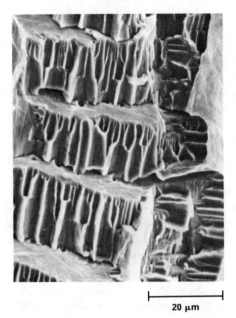

20 μm

*Flutes and cleavage planes in a titanium alloy that failed by stress-corrosion cracking*

Fluxing of the melt facilitates the agglomeration and separation of such undesirable constituents from the melt. It is also used as a protective covering for certain molten metal baths. Lime or limestone is generally used to remove sand, as in iron smelting; sand, to remove iron oxide in copper refining. (2) In brazing, cutting, soldering, or welding, material used to prevent the formation of, or to dissolve and facilitate removal of, oxides and other undesirable substances.

**flux cored arc welding (FCAW).** An arc welding process that joins metal by heating them with an arc between a

FIG. 193

Gas shielded nozzle

Shield produced by flux

Molten slag

Solidified slag

External gas shield

Flux cored electrode

Arc

Weld metal

Weld pool

Metal droplets with slag coating

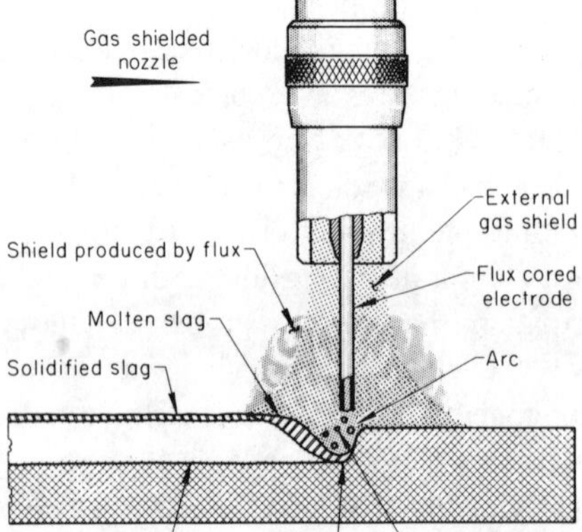

*Schematic of the flux cored arc welding process*

continuous tubular filler-metal electrode and the work (Fig. 193). Shielding is provided by a flux contained within the consumable tubular electrode. Additional shielding may or may not be obtained from an externally supplied gas or gas mixture. See also *electrogas welding*.

**flux cored electrode.** A composite filler metal electrode consisting of a metal tube or other hollow configuration containing ingredients to provide such functions as shielding atmosphere, deoxidation, arc stabilization, and slag formation. Alloying materials may be included in the core. External shielding may or may not be used.

**flux cover.** In metal bath dip brazing and dip soldering, a cover of *flux* over the molten filler metal bath.

**flux density.** In magnetism, the number of *flux lines* per unit area passing through a cross section at right angles. It is given by $B = \mu H$, where $\mu$ and $H$ are permeability and magnetic-field intensity, respectively.

**flux lines.** Imaginary lines used as a means of explaining the behavior of magnetic and other fields. Their concept is based on the pattern of lines produced when magnetic particles are sprinkled over a permanent magnet. Sometimes called magnetic lines of force.

**flux oxygen cutting.** Oxygen cutting with the aid of a flux.

**flux solder connection.** A solder joint characterized by entrapped flux that often causes high electrical resistance in an electronic component.

**fly ash.** A finely divided siliceous material formed during the combustion of coal, coke, or other solid fuels.

**fly cutting.** Cutting with a single-tooth milling cutter.

**flying shear.** A machine for cutting continuous rolled products to length that does not require a halt in rolling, but rather moves along the runout table at the same speed as the product while performing the cutting, and then returns to the starting point in time to cut the next piece.

**foamed plastics.** Resins in sponge form, flexible or rigid, with cells closed or interconnected and density over a range from that of the solid parent resin to $0.030 \, g/cm^3$. Compressive strength of rigid foams is fair, making them useful as core materials for *sandwich constructions*. Also, chemical cellular plastics, the structures of which are produced by gases generated from the chemical interaction of their constituents. See also *expandable plastics*.

**foaming.** In tribology, the production and coalescence of gas bubbles on a liquid lubricant surface.

**foaming agent.** Chemicals added to plastics and rubbers that generate inert gases, such as nitrogen, upon heating, causing the resin to form a cellular structure.

**foam inhibitor.** A surface-active chemical compound used in minute quantities to prevent or reduce *foaming*. Silicone fluids are frequently used as foam inhibitors.

**focal length.** In an optical microscope, the distance from the second principal point to the point on the axis at which parallel rays entering the lens will converge or focus.

**focal point**. See preferred term *focal spot*.

**focal spot**. (1) That area on the target of an x-ray tube that is bombarded by electrons. (2) The area of the x-ray tube from which the x-rays originate. The effective focal-spot size is the apparent size of this area when viewed from the detector. (3) In electron beam and laser beam welding, a spot at which an energy beam has the most concentrated energy level and the smallest cross-sectional area.

**focus**. In microscopy, a point at which rays originating from a point in the object converge or from which they diverge or appear to diverge under the influence of a lens or diffracting system.

**focusing device, electrons**. A device that effectively increases the angular aperture of the electron beam illuminating the object, rendering the focusing more critical.

**focusing, x-rays**. The operation of producing a convergent beam in which all rays meet in a point or line.

**fog quenching**. Quenching in a fine vapor or mist.

**foil**. Metal in sheet form less than 0.15 mm (0.006 in.) thick.

**foil bearing**. A bearing in which the housing is replaced by a flexible foil held under tension against a partition of the journal periphery, lubricant being retained between the journal and the foil.

**foil decorating**. Molding paper, textile, or plastic foils printed with compatible inks directly into a plastic part so that the foil is visible below the surface of the part as integral decoration.

**fold**. (1) A defect in metal, usually on or near the surface, caused by continued fabrication of overlapping surfaces. (2) A forging defect caused by folding metal back onto its own surface during its flow in the die cavity. See also *lap*.

**folded chain**. The conformation of a flexible polymer when present in a crystal. The molecule exits and reenters the same crystal, frequently generating folds.

**follow board**. In foundry practice, a board contoured to a pattern to facilitate the making of a sand mold.

**follow die**. A *progressive die* consisting of two or more parts in a single holder; used with a separate lower die to perform more than one operation (such as piercing and blanking) on a part in two or more stations.

**follower plate**. A plate fitted to the top surface of a grease dispenser.

**foot press**. A small press with a low capacity actuated by foot pressure on a treadle.

**forced-air quench**. A quench utilizing blasts of compressed air against relatively small parts such as a gear.

**force-feed lubrication**. See *pressure lubrication*.

**force fit**. Any of various interference fits between parts assembled under various amounts of force.

**force plug**. The male half of the mold for making plastics that enters the cavity, exerting pressure on the resin and causing it to flow. Also called punch. Sometimes called a core, plunger, or ram.

FIG. 194

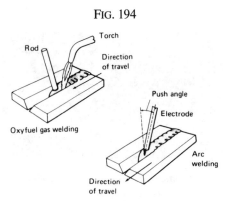

*Forehand welding technique*

**foreign structure**. Any metallic structure that is not intended as part of a *cathodic protection* system of interest.

**forehand welding**. Welding in which the palm of the principal hand (torch or electrode hand) of the welder faces the direction of travel (Fig. 194). It has special significance in oxyfuel gas welding in that the flame is directed ahead of the weld bead, which provides *preheating*. Contrast with *backhand welding*. See also *push angle*, *travel angle*, and *work angle*.

**forgeability**. Term used to describe the relative ability of material to deform without fracture. Also describes the resistance to flow from deformation. Table 22 compares the forgeability of various alloy groups. See also *formability*.

**forge delay time**. In spot, seam, or projection welding, the time between the start of the welding, current or weld interval and the application of forging pressure.

**Table 22  Classification of alloys in order of increasing forging difficulty**

| Alloy group | Approximate forging temperature range | |
| --- | --- | --- |
| | °C | °F |
| **Least difficult** | | |
| Aluminum alloys .......... | 400-550 | 750-1020 |
| Magnesium alloys ......... | 250-350 | 480-660 |
| Copper alloys ............. | 600-900 | 1110-1650 |
| Carbon and low-alloy steels............. | 850-1150 | 1560-2100 |
| Martensitic stainless steels .......... | 1100-1250 | 2010-2280 |
| Maraging steels............ | 1100-1250 | 2010-2280 |
| Austenitic stainless steels .......... | 1100-1250 | 2010-2280 |
| Nickel alloys.............. | 1000-1150 | 1830-2100 |
| Semiaustenitic PH stainless steels ......... | 1100-1250 | 2010-2280 |
| Titanium alloys............ | 700-950 | 1290-1740 |
| Iron-base superalloys ...... | 1050-1180 | 1920-2160 |
| Cobalt-base superalloys .... | 1180-1250 | 2160-2280 |
| Niobium alloys............ | 950-1150 | 1740-2100 |
| Tantalum alloys .......... | 1050-1350 | 1920-2460 |
| Molybdenum alloys........ | 1150-1350 | 2100-2460 |
| Nickel-base superalloys .... | 1050-1200 | 1920-2190 |
| Tungsten alloys .......... | 1200-1300 | 2190-2370 |
| **Most difficult** | | |

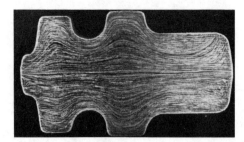

FIG. 195

*Forged structure (flow lines) in a closed-die forging of AISI 4340 steel. ~0.75×*

**forged roll Scleroscope hardness number (HFRSc or HFRSd).** A number related to the height of rebound of a diamond-tipped hammer dropped on a forged steel roll. It is measured on a scale determined by dividing into 100 units the average rebound of a hammer from a forged steel roll of accepted maximum hardness. See also *Scleroscope hardness number* and *Scleroscope hardness test*.

**forged structure.** The macrostructure through a suitable section of a forging that reveals direction of working (Fig. 195).

**forge welding.** Solid-state welding in which metals are heated in a forge (in air) and then welded together by applying pressure or blows sufficient to cause permanent deformation at the interface. The process is most commonly applied to the butt welding of steel.

**forge welding die.** A device used in *forge welding* primarily to form the work while hot and apply the necessary pressure.

**forging.** The process of working metal to a desired shape by impact or pressure in hammers, forging machines (upsetters), presses, rolls, and related forming equipment. Forging hammers, counterblow equipment, and high-energy-rate forging machines apply impact to the workpiece, while most other types of forging equipment apply squeeze pressure in shaping the stock. Some metals can be forged at room temperature, but most are made more plastic for forging by heating. Specific forging processes defined in this book include *closed-die forging, high-energy-rate forging, hot upset forging, isothermal forging, open-die forging, powder forging, precision forging, radial forging, ring rolling, roll forging, rotary forging,* and *rotary swaging*.

**forging billet.** A wrought metal slug used as *forging stock*.

**forging dies.** Forms for making forgings; they generally consist of a top and bottom die. The simplest will form a completed forging in a single impression; the most complex, consisting of several die inserts, may have a number of impressions for the progressive working of complicated shapes. Forging dies are usually in pairs, with part of the impression in one of the blocks and the rest of the impression in the other block.

**forging envelope.** See *finish allowance*.

**forging ingot.** A cast metal slug used as *forging stock*.

**forging machine (upsetter or header).** A type of forging equipment, related to the *mechanical press*, in which the principal forming energy is applied horizontally to the workpiece, which is gripped and held by prior action of the dies. See also *heading, hot upset forging,* and *upsetting*.

**forging plane.** In forging, the plane that includes the principal die face and that is perpendicular to the direction of ram travel. When parting surfaces of the dies are flat, the forging plane coincides with the parting line. Contrast with *parting plane*.

**forging quality.** Term used to describe stock of sufficient quality to make it suitable for commercially satisfactory forgings.

**forging range.** Temperature range in which a metal can be forged successfully. See also Table 22 accompanying the term *forgeability*.

**forging rolls.** Power-driven rolls used in preforming bar or billet stock that have shaped contours and notches for introduction of the work. See also the figure accompanying the term *roll forging*.

**forging stock.** A wrought rod, bar, or other section suitable for subsequent change in cross section by forging.

**forking.** A phenomenon in which a propagating fracture in a ceramic system branches into two or more new fractures, each separated from its immediate neighbor by an acute angle.

**formability.** The ease with which a metal can be shaped through plastic deformation. Evaluation of the formability of a metal involves measurement of strength, ductility, and the amount of deformation required to cause fracture. The term workability is used interchangeably with formability; however, formability refers to the shaping of sheet metal, while workability refers to shaping materials by *bulk forming*. See also *forgeability*.

**form-and-spray.** Technique for thermoforming plastic sheet into an end-product and then backing up the sheet with *spray-up* reinforced plastics.

**form block.** Tooling, usually the male part, used for forming sheet metal contours; generally used in *rubber-pad forming*.

**form cutter.** Any cutter, profile sharpened or cam relieved, shaped to produce a specified form on the work.

**form die.** A die used to change the shape of a sheet metal blank with minimal plastic flow.

**form grinding.** Grinding with a wheel having a contour on its cutting face that is a mating fit to the desired form.

**forming** (ceramics and glasses). (1) The shaping or molding of ceramic ware. (2) The shaping of hot glass.

**forming** (metals). (1) Making a change, with the exception of shearing or blanking, in the shape or contour of a metal part without intentionally altering its thickness. (2) The plastic deformation of a billet or a blanked sheet between tools (dies) to obtain the final configuration. Metalforming processes are typically classified as *bulk forming* and *sheet forming*. Also referred to as metalworking.

**forming** (plastics). A process in which the shape of plastic pieces such as sheets, rods, or tubes is changed to a desired configuration. See also *thermoforming*. The use of the term forming in plastics technology does not include such operations as molding, casting, or extrusion, in which shapes or pieces are made from molding materials or liquids.

**forming limit diagram (FLD)**. A diagram in which the major strains at the onset of necking in sheet metal are plotted vertically and the corresponding minor strains are plotted horizontally (Fig. 196). The onset-of-failure line divides all possible strain combinations into two zones: the safe zone (in which failure during forming is not expected)

and the failure zone (in which failure during forming is expected).

**form-relieved cutter**. A cutter so relieved that by grinding only the tooth face the original form is maintained throughout its life.

**form rolling**. Hot rolling to produce bars having contoured cross sections; not to be confused with *roll forming* of sheet metal or with *roll forging*.

**form tool**. A single-edge, nonrotating cutting tool, circular or flat, that produces its inverse or reverse form counterpart upon a workpiece.

**Formvar**. A plastic material used for the preparation of replicas or for specimen-supporting membranes.

**Formvar replica**. A reproduction of a surface in a plastic Formvar film. See also *replica*.

**forsterite**. A magnesium silicate mineral of composition $2MgO \cdot SiO_2$, which is usually produced synthetically as a ceramic raw material, but it may also be a reaction-produced phase in fired ceramics used in refractory brick.

**forward extrusion**. Same as direct extrusion. See *extrusion*.

**fouling**. An accumulation of deposits. This term includes accumulation and growth of marine organisms on a submerged metal surface (Fig. 197) and also includes the accumulation of deposits (usually inorganic) on heat exchanger tubing. See also *biological corrosion*.

**fouling organisms**. Any aquatic organism with a sessile adult stage that attaches to and fouls underwater structures of ships (Fig. 197).

**foundry**. A commercial establishment or building where metal castings are produced.

FIG. 196

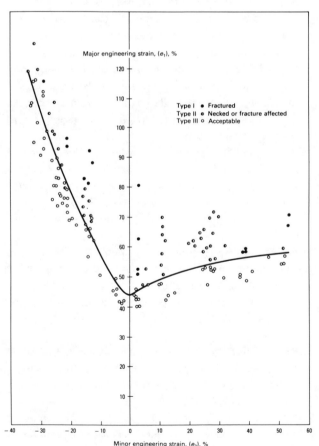

Strain measurements and forming limit diagram for aluminum-killed steel

FIG. 197

Fouling along the periphery of a high-strength steel rudder

**foundry returns**. Metal in the form of gates, sprues, runners, risers, and scrapped castings of known composition returned to the furnace for remelting.

**four-harness satin**. A fabric weave, also called *crowfoot satin* because the weaving pattern, when laid out on cloth design paper, resembles the imprint of crow's foot. In this type of weave there is a three-by-one interlacing. That is, a filling thread floats over the three warp threads and then under one. The two sides of the fabric have different appearances. Fabrics with this weave are more pliable than either the plain or basket weaves and, consequently, are easier to form around curves. See also the figure accompanying the term *crowfoot satin*.

**four-high mill**. A type of rolling mill, commonly used for flat-rolled mill products, in which two large-diameter backup rolls are employed to reinforce two smaller work rolls, which are in contact with the product (Fig. 198). Either the work rolls or the backup rolls may be driven. Compare with *two-high mill* and *cluster mill*.

FIG. 198

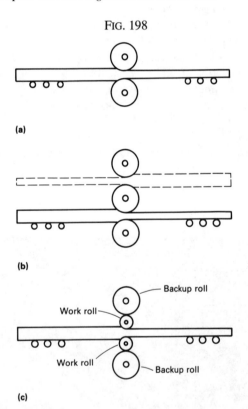

(a)

(b)

Backup roll
Work roll
Work roll
Backup roll

(c)

*The most common types of rolling mills. (a) Two-high. (b) Three-high. (c) Four-high*

**Fourier transform**. (1) A mathematical process for changing the description of a function by giving its value in terms of its sinusoidal spatial (or temporal) frequency components instead of its spatial coordinates (or vice versa). (2) An analytical method used automatically in advanced forms of spectroscopic analysis such as infrared and nuclear magnetic resonance spectroscopy.

**Fourier transform infrared (FT-IR) spectrometry**. A form of infrared spectrometry in which data are obtained as an interferogram, which is then Fourier transformed to obtain an amplitude versus wavenumber (or wavelength) spectrum.

**four-point press**. A press whose slide is actuated by four connections and four cranks, eccentrics, or cylinders, the chief merit being to equalize the pressure at the corners of the slides.

**FP fiber**. A polycrystalline all-alumina fiber (>99% $Al_2O_3$). A ceramic fiber useful for high-temperature (1370 to 1650 °C, or 2500 °F) composites.

**fraction**. In powder metallurgy, the portion of a powder sample that lies between two stated particle sizes.

**fractography**. Descriptive treatment of fracture of materials, with specific reference to photographs of the fracture surface. Macrofractography involves photographs at low magnification (<25×); microfractography, photographs at high magnification (>25×).

**fracture** (composites). The separation of a body. Defined both as rupture of the surface without complete separation of the laminate and as complete separation of a body because of external or internal forces. Fractures in continuous fiber reinforced composites can be divided into three basic fracture types: intralaminar, interlaminar, and translaminar (Fig. 199). Translaminar fractures are those oriented transverse to the laminated plane in which conditions of fiber fracture are generated. Interlaminar fracture, on the other hand, describes failures oriented between plies, whereas intralaminar fractures are those located internally within a *ply*.

**fracture** (metals). The irregular surface produced when a piece of metal is broken. See also *brittle fracture, cleavage fracture, crystalline fracture, decohesive rupture, dimple rupture, ductile fracture, fibrous fracture, granular fracture, intergranular fracture, silky fracture,* and *transgranular fracture*.

**fracture ductility**. The true plastic strain of fracture.

**fracture grain size**. Grain size determined by comparing a fracture of a specimen with a set of standard fractures. For steel, a fully martensitic specimen is generally used, and the depth of hardening and the prior austenitic grain size are determined.

**fracture mechanics**. A quantitative analysis for evaluating structural behavior in terms of applied stress, crack length, and specimen or machine component geometry. See also *linear elastic fracture mechanics*.

**fracture strength**. The normal stress at the beginning of fracture. Calculated from the load at the beginning of fracture during a tension test and the original cross-sectional area of the specimen.

**fracture stress**. The true, normal stress on the minimum

Fig. 199

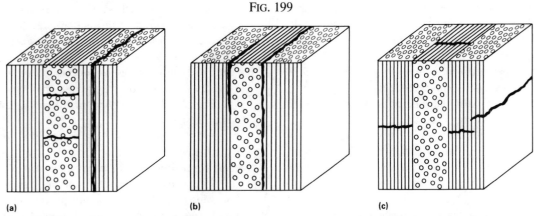

(a)           (b)           (c)

*Fracture modes for fiber-reinforced composites. (a) Intralaminar fracture. (b) Interlaminar fracture. (c) Translaminar fracture*

cross-sectional area at the beginning of fracture. The term usually applies to tension tests of unnotched specimens.

**fracture surface markings.** Fracture surface features that may be used to determine the fracture origin location and the nature of the stress that produced the fracture.

**fracture system.** That family of related fracture surfaces lying within an object, having a common cause and origin.

**fracture test.** Test in which a specimen is broken and its fracture surface is examined with the unaided eye or with a low-power microscope to determine such factors as composition, grain size, case depth, or discontinuities.

**fracture toughness.** A generic term for measures of resistance to extension of a crack. The term is sometimes restricted to results of *fracture mechanics* tests, which are directly applicable in fracture control. However, the term commonly includes results from simple tests of notched or precracked specimens not based on fracture mechanics analysis. Results from tests of the latter type are often useful for fracture control, based on either service experience or empirical correlations with fracture mechanics tests. See also *stress-intensity factor*.

**fragmentation.** The subdivision of a grain into small, discrete crystallite outlined by a heavily deformed network of intersecting slip bands as a result of cold working. These small crystals or fragments differ in orientation and tend to rotate to a stable orientation determined by the slip systems.

**fragmented powder.** A powder obtained by fragmentation and mechanical comminution into fine particles.

**frame.** The main structure of a forming or forging press.

**Frank-Condon principle.** The principle which states that the transition from one energy state to another is so rapid that the nuclei of the atoms involved can be considered stationary during the transition.

**freckling.** A type of segregation revealed as dark spots on a macroetched specimen of a consumable-electrode vacuum-arc-remelted alloy.

**free bend.** The bend obtained by applying forces to the ends of a specimen without the application of force at the point of maximum bending.

**free carbon.** The part of the total carbon in steel or cast iron that is present in elemental form as graphite or *temper carbon*. Contrast with *combined carbon*.

**free corrosion potential.** *Corrosion potential* in the absence of net electrical current flowing to or from the metal surface.

**free-energy diagram.** A graph of the variation with concentration of the Gibbs free energy at constant pressure and temperature.

**free-energy surface.** In a ternary or higher order free-energy diagram, the locus of points representing the Gibbs free energy as a function of concentration, with pressure and temperature constant.

**free ferrite.** (1) Ferrite that is formed directly from the decomposition of hypoeutectoid austenite during cooling, without the simultaneous formation of cementite. (2) Ferrite formed into separate grains and not intimately associated with carbides as in pearlite. Also called *proeutectoid ferrite*.

**free fit.** Any of various clearance fits for assembly by hand and free rotation of parts. See also *running fit*.

**free machining.** Pertains to the machining characteristics of an alloy to which one or more ingredients have been introduced to produce small broken chips, lower power consumption, better surface finish, and longer tool life; among such additions are sulfur or lead to steel, lead to brass, lead and bismuth to aluminum, and sulfur or selenium to stainless steel.

**free radical.** Any molecule or atom that possesses one unpaired electron. In chemical notation, a free radical is symbolized by a single dot (to denote the odd electron) to the right of the chemical symbol.

**free-radical polymerization.** A type of *polymerization* in which the propagating species is a long-chain free radical initiated by the introduction of free radicals from thermal or photochemical decomposition of an initiator molecule.

**free rolling.** Rolling in which no traction is deliberately applied between a rolling element and another surface.

**free rotation.** The rotation of atoms, particularly carbon atoms, about a single bond. Because the energy requirement is only a few kcal, the rotation is said to be free if sufficient thermal energy is available.

**free vibration.** A technique for performing dynamic mechanical measurements in which the sample is deformed, released, and allowed to oscillate freely at the natural resonant frequency of the system. Elastic modulus is calculated from the measured resonant frequency, and damping is calculated from the rate at which the amplitude of the oscillation decays.

**free wall.** The portion of a *honeycomb* cell wall that is not connected to another cell.

**freezing point.** See preferred term *liquidus* and *solidus*. See also *melting point*.

**freezing range.** That temperature range between *liquidus* and *solidus* temperatures in which molten and solid constituents coexist.

**frequency.** The number of cycles per unit time. The recommended unit is the hertz, Hz, which is equal to one cycle per second.

**frequency distribution.** The way in which the frequencies of occurrence of members of a population, or a sample, are distributed according to the values of the variable under consideration.

**Fresnel fringes.** A class of diffraction fringes formed when the source of illumination and the viewing screen are at a finite distance from a diffracting edge. In the electron microscope, these fringes are best seen when the object is slightly out of focus.

**fretting.** A type of wear that occurs between tight-fitting surfaces subjected to cyclic relative motion of extremely small amplitude. Usually, fretting is accompanied by corrosion, especially of the very fine wear debris. Also referred to as *fretting corrosion* and *false Brinelling* (in rolling-element bearings).

**fretting corrosion.** (1) The accelerated deterioration at the interface between contacting surfaces as the result of corrosion and slight oscillatory movement between the two surfaces. (2) A form of *fretting* in which chemical reaction predominates. Fretting corrosion is often characterized by the removal of particles and subsequent formation of oxides (Fig. 200), which are often abrasive and so increase the wear. Fretting corrosion can involve other chemical reaction products, which may not be abrasive.

**fretting fatigue.** (1) Fatigue fracture that initiates at a surface area where fretting has occurred (Fig. 201). The progressive damage to a solid surface that arises from fretting. *Note*: If particles of wear debris are produced, then the term *fretting wear* may be applied.

**fretting wear.** Wear arising as a result of *fretting* (Fig. 202).

**friction.** The resisting force tangential to the common boundary between two bodies when, under the action of an external force, one body moves or tends to move relative to the surface of the other. The term friction is also used, incorrectly, to denote *coefficient of friction*. It is vague and

FIG. 200

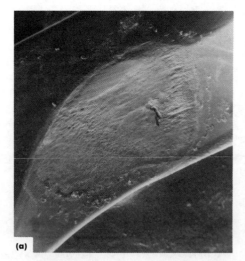
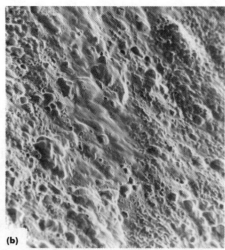

*Fretting corrosion at the contact area between the screw hole of a stainless steel bone plate and the corresponding screw head. (a) Overview of wear on plate hole showing mechanical and pitting corrosion attack. 15×. (b) Higher-magnification view of shallow pitting corrosion attack from periphery of contact area. 355×*

FIG. 201

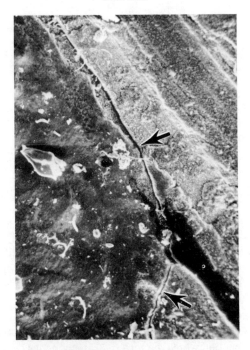

*Fretting scar on fatigued steel specimen showing location of fatigue crack (arrow)*

imprecise unless accompanied by the appropriate modifiers, such as *dry friction* or *kinetic friction*. See also *static coefficient of friction*.

**friction coefficient**. See *coefficient of friction*.

**friction force**. The resisting force tangential to the interface between two bodies when, under the action of an external force, one body moves or tends to move relative to the other.

**friction material**. A sintered material exhibiting a high coefficient of friction designed for use where rubbing or frictional wear is encountered—for example, aircraft brake linings and clutch facings on tractors, heavy trucks, and earth-moving equipment. Friction materials consist of a dispersion of friction-producing ingredients in a metallic matrix. Table 23 lists compositions of typical friction materials.

**friction polymer**. An amorphous organic deposit that is produced when certain metals are rubbed together in the presence of organic liquids or gases. Friction polymer often forms on moving electrical contacts exposed to industrial environments. The varnishlike film will attenuate or modify transmitted signals.

**friction soldering**. See preferred term *abrasion soldering*.

**friction welding** (metals). A solid-state process in which welds are made by holding a nonrotating workpiece in contact with a rotating workpiece under constant or gradually increasing pressure until the interface reaches the welding temperature and rotation can be stopped (Fig. 203).

**friction welding** (plastics). A method of welding thermoplastic materials in which the heat necessary to soften the components is provided by friction.

**frictive track**. In glass, a series of crescent cracks lying along a common axis, paralleling the direction of frictive contact. Also known as a chatter sleek.

**frit**. A glass produced by *fritting*, which contains fluxing material and is employed as a constituent in a glaze, body, or other ceramic composition.

**fritting**. The rapid chilling of the molten glassy material to produce frit.

**frosted area**. See *hackle*.

**frosting**. A form of ball bearing groove damage, appearing as

FIG. 202

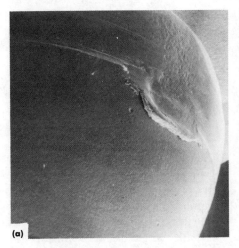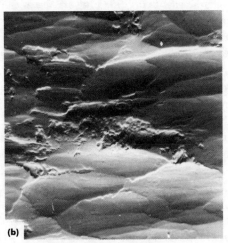

*Fretting wear on head of titanium screw. (a) Material transport and fretting zone. (b) Closeup view of wear structures showing fine wear products. 120×*

**Table 23  Nominal compositions of copper-base and iron-base friction materials**

| Premix | Copper | Iron | Lead | Tin | Zinc | Silicon dioxide | Graphite |
|---|---|---|---|---|---|---|---|
| Copper based ................. | 65-75 | ... | 2-5 | 2-5 | 5-8 | 2-5 | 10-20 |
| Iron based .................. | 10-15 | 50-60 | 2-4 | 2-4 | ... | 8-10 | 10-15 |

FIG. 203

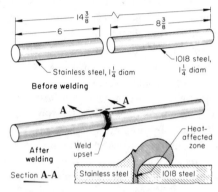

*Pump shaft joined by inertia drive friction welding 1018 steel to stainless steel*

a frosted area, suggestive that surface distress has occurred.

**frost line.** In the extrusion of polyethylene lay-flat film, a ring-shape zone located at the point where the film reaches its final diameter. This zone is characterized by a frosty appearance on the film caused when the film temperature falls below the softening range of the resin.

**frothing.** A technique for applying urethane foam in which blowing agents or tiny air bubbles are introduced under pressure into the liquid mixture of foam ingredients.

**FRP.** See *fiber-reinforced plastic*.

**fuel gases.** Gases usually used with oxygen for heating such as acetylene, natural gas, hydrogen, propane, methylacetylene propadiene stabilized, and other synthetic fuels and hydrocarbons.

**fugitive binder.** An organic substance added to a metal powder to enhance the bond between the particles during compaction and thereby increase the green strength of the compact, and which decomposes during the early stages of the sintering cycle.

**full annealing.** An imprecise term that denotes an annealing cycle to produce minimum strength and hardness. For the term to be meaningful, the composition and starting condition of the material and the time-temperature cycle used must be stated.

**full-automatic plating.** Electroplating in which the work is automatically conveyed through the complete cycle.

**full center.** Mild waviness down the center of a metal sheet or strip.

**full-contour length.** The length of a fully extended polymer chain.

**fuller (fullering impression).** Portion of the die used in hammer forging primarily to reduce the cross section and to lengthen a portion of the forging stock. The fullering impression is often used in conjunction with an *edger (edging impression)*.

**full fillet weld.** A *fillet weld* whose size is equal to the thickness of the thinner member joined.

**full-film lubrication.** A type of lubrication wherein the solid surfaces are separated completely by an elastohydrodynamic fluid film. See also *elastohydrodynamic lubrication* and *lubrication regimes*.

**full hard.** A *temper* of nonferrous alloys and some ferrous alloys corresponding approximately to a cold-worked state beyond which the material can no longer be formed by bending. In specifications, a full hard temper is commonly defined in terms of minimum hardness or minimum tensile strength (or, alternatively, a range of hardness or strength) corresponding to a specific percentage of cold reduction following a full anneal. For aluminum, a full hard temper is equivalent to a reduction of 75% from *dead soft*; for austenitic stainless steels, a reduction of about 50 to 55%.

**full journal bearing.** A journal bearing that surrounds the journal by a full 360°.

**full mold.** A trade name for an expendable pattern casting process in which the polystyrene pattern is vaporized by the molten metal as the mold is poured. See also the figure accompanying the term *lost foam casting*.

**full width at half maximum (FWHM).** A measure of resolution of a spectrum or chromatogram determined by measuring the peak width of a spectral or chromatographic peak at half its maximum height.

**functional group.** A chemical radical or structure that has characteristic properties; examples are hydroxyl and carboxyl groups.

**functionality.** The average number of reaction sites on an individual polymer chain.

**fungus resistance.** The resistance of a material to attack by fungi in conditions promoting their growth.

**furan resins.** Dark-colored thermosetting resins available primarily as liquids ranging from low-viscosity polymers to thick, heavy syrups, which cure to highly cross-linked, brittle substances. Made primarily by polycondensation of furfuryl alcohol in the presence of strong acids, sometimes in combination with formaldehyde or furfuryldehyde.

**furfural resins**. A dark-colored synthetic resin of the thermosetting variety obtained by the condensation of furfural with phenol or its homologs. It is used in the manufacture of molding materials, adhesives, and impregnating varnishes. Properties include high resistance to acids and alkalies.

**furnace brazing**. A mass-production *brazing* process in which the filler metal is preplaced on the joint, then the entire assembly is heated to brazing temperature in a furnace. Usually, a protective furnace atmosphere is required, and wetting of the joint surfaces is accomplished without using a brazing flux.

**furnace soldering**. A soldering process in which the parts to be joined are placed in a furnace heated to a suitable temperature.

**fused coating**. A metallic coating (usually tin or solder alloy) that has been melted and solidified, forming a metallurgical bond to the basis metal.

**fused silica**. A glass made either by flame hydrolysis of silicon tetrachloride or by melting silica, usually in the form of granular quartz.

**fused spray deposit**. A self-fluxing spray deposit which is deposited by conventional *thermal spraying* and subsequently fused using either a heating torch or a furnace. The coatings are usually made of nickel and cobalt alloys to which hard particles, such as tungsten carbide may be added for increased wear resistance.

**fused zone**. See preferred terms *fusion zone*, *nugget*, and *weld interface*.

**fusible alloys**. A group of binary, ternary, quaternary, and quinary alloys containing bismuth, lead, tin, cadmium, and indium. The term "fusible alloy" refers to any of more than 100 alloys that melt at relatively low temperatures, that is, below the melting point of tin-lead solder (183 °C, or 360 °F). The melting points of these alloys range as low as 47 °C (116 °F).

**fusing**. The melting of a metallic coating (usually electrodeposited) by means of a heat-transfer medium, followed by solidification.

**fusion** (plastics). In vinyl dispersions, the heating of a dispersion to produce a homogeneous mixture.

**fusion** (welding). The melting together of filler metal and base metal (substrate), or of base metal only, which results in coalescence. See also *depth of fusion*.

**fusion face**. A surface of the base metal which will be melted during welding.

**fusion spray** (thermal spraying). The process in which the coating is completely fused to the base metal, resulting in a metallurgically bonded, essentially void free coating.

**fusion welding**. Any welding process in which the filler metal and base metal (substrate), or base metal only, are melted together to complete the weld.

**fusion zone**. The area of base metal melted as determined on the cross section of a weld.

**fuzz**. Accumulation of short, broken filaments after passing glass strands, yarns, or rovings over a contact point. Often, weighted and used as an inverse measure of abrasion resistance.

# G

**gag**. A metal spacer inserted so as to render a floating tool or punch inoperative.

**gage**. (1) The thickness of sheet or the diameter of wire. The various standards are arbitrary and differ with regard to ferrous and nonferrous products as well as sheet and wire. (2) An aid for visual inspection that enables an inspector to determine more reliably whether the size or contour of a formed part meets dimensional requirements. (3) An instrument used to measure thickness or length.

**gage length**. The original length of that portion of the specimen over which strain, change of length and other characteristics are measured.

**gagger**. In foundry practice, an irregularly shaped piece of metal used for reinforcement and support in a sand mold.

**gall**. To damage the surface of a powder metallurgy compact or die part, caused by adhesion of powder to the die cavity wall or a punch surface.

**galling**. (1) A condition whereby excessive friction between high spots results in localized welding with subsequent *spalling* and a further roughening of the rubbing surfaces of one or both of two mating parts. (2) A severe form of scuffing associated with gross damage to the surfaces (Fig. 204) or failure. Galling has been used in many ways in tribology; therefore, each time it is encountered its meaning must be ascertained from the specific context of the usage. See also *scoring* and *scuffing*.

**gallium and gallium compounds**. See *Technical Brief 15*.

**galvanic anode**. A metal which, because of its relative position in the galvanic series, provides *sacrificial protection*

FIG. 204

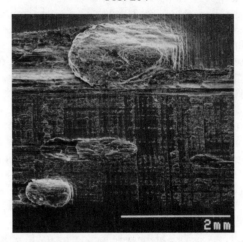

*Galling on the surface of type 316 stainless steel rubbed against a button of the same material at a load of ~1000 N (740 ft · lbf)*

to metals that are more noble in the series, when coupled in an electrolyte. See also *cathodic protection*.

**galvanic cell.** (1) A cell in which chemical change is the source of electrical energy. It usually consists of two dissimilar conductors in contact with each other and with an electrolyte, or of two similar conductors in contact with each other and with dissimilar electrolytes. (2) A cell or system in which a spontaneous oxidation-reduction reaction occurs, the resulting flow of electrons being conducted in an external part of the circuit. See also the figure accompanying the term *cathodic protection*.

**galvanic corrosion.** Corrosion associated with the current of a galvanic cell consisting of two dissimilar conductors in an electrolyte (Fig. 205) or two similar conductors in dissimilar electrolytes. Where the two dissimilar metals are in contact, the resulting reaction is referred to as couple action.

**galvanic couple.** A pair of dissimilar conductors, commonly metals, in electrical contact. See also *galvanic corrosion*.

**galvanic couple potential.** See *mixed potential*.

**galvanic current.** The electric current that flows between metals or conductive nonmetals in a *galvanic couple* (Fig. 205).

FIG. 205

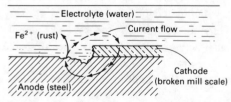

*Schematic showing how breaks in mill scale (Fe₃O₄) can lead to galvanic corrosion of steel*

**Table 24  Galvanic series in seawater at 25 °C (77 °F)**

**Corroded end (anodic, or least noble)**
Magnesium
Magnesium alloys
Zinc
Galvanized steel or galvanized wrought iron
Aluminum alloys
5052, 3004, 3003, 1100, 6053, in this order
Cadmium
Aluminum alloys
2117, 2017, 2024, in this order
Low-carbon steel
Wrought iron
Cast iron
Ni-Resist (high-nickel cast iron)
Type 410 stainless steel (active)
50-50 lead-tin solder
Type 304 stainless steel (active)
Type 316 stainless steel (active)
Lead
Tin
Copper alloy C28000 (Muntz metal, 60% Cu)
Copper alloy C67500 (manganese bronze A)
Copper alloys C46400, C46500, C46600, C46700
   (naval brass)
Nickel 200 (active)
Inconel alloy 600 (active)
Hastelloy alloy B
Chlorimet 2
Copper alloy C27000 (yellow brass, 65% Cu)
Copper alloys C44300, C44400, C44500
   (admiralty brass)
Copper alloys C60800, C61400 (aluminum bronze)
Copper alloy C23000 (red brass, 85% Cu)
Copper C11000 (ETP copper)
Copper alloys C65100, C65500 (silicon bronze)
Copper alloy C71500 (copper nickel, 30% Ni)
Copper alloy C92300, cast (leaded tin bronze G)
Copper alloy C92200, cast (leaded tin bronze M)
Nickel 200 (passive)
Inconel alloy 600 (passive)
Monel alloy 400
Type 410 stainless steel (passive)
Type 304 stainless steel (passive)
Type 316 stainless steel (passive)
Incoloy alloy 825
Inconel alloy 625
Hastelloy alloy C
Chlorimet 3
Silver
Titanium
Graphite
Gold
Platinum

**Protected end (cathodic, or most noble)**

**galvanic series.** A list of metals and alloys arranged according to their relative corrosion potentials in a given environment. Table 24 lists various metals and alloys as they appear in the galvanic series. Compare with *electromotive force series*.

**galvanize.** To coat a metal surface with zinc using any of various processes.

**galvanneal.** To produce a zinc-iron alloy coating on iron or steel by keeping the coating molten after hot-dip galvanizing until the zinc alloys completely with the basis metal.

**galvanometer.** An instrument for indicating or measuring a small electric current by means of a mechanical motion

# Technical Brief 15: Gallium and Gallium Compounds

GALLIUM-BASE COMPONENTS are found in a variety of products, ranging from compact disc players to advanced military electronic warfare systems. Compared with components made of silicon, a material gallium arsenide (GaAs) has replaced in some of these applications, components made of GaAs can emit light, have greater resistance to radiation, and operate at faster speeds and higher temperatures.

Gallium occurs in very low concentrations in the earth's crust and virtually all primary gallium is recovered as a by-product, principally from the processing of bauxite to alumina. Most gallium applications require very high purity levels, and the metal must be refined before use. Commercially available gallium metal ranges in purity from 99.5% to 99.9999+%. The most common impurities are mercury, lead, tin, zinc, and copper. If impurity limits of high-purity gallium are exceeded, optoelectric properties are degraded or destroyed.

The principal use of gallium is in the manufacture of semiconducting compounds. More than 90% of the gallium consumed in the United States is used for optoelectronic devices and integrated circuits. Optoelectronic devices—light-emitting diodes (LEDs), laser diodes, photodiodes, and solar (photovoltaic) cells—take advantage of the ability of GaAs to convert electrical energy into optical energy and vice versa. An LED, which is a semiconductor that emits light when an electric current is passed through it, consists of layers of eptitaxially grown material on a substrate. These epitaxial layers are normally gallium aluminum arsenide (GaAlAs), gallium arsenide phosphide (GaAsP), or indium gallium arsenide phosphide (InGaAsP); the substrate material is either GaAs or gallium phosphide (GaP). Laser diodes operate on the same principle as LEDs, but they convert electrical energy to a coherent light output. Laser diodes principally consist of an eptaxial layer of GaAs, GaAlAs, or InGaAsP on a GaAs substrate. Photodiodes are used to detect a light impulse generated by a source, such as an LED or laser diode, and convert it to an electrical impulse. Photodiodes are fabricated from the same materials as LEDs. Gallium arsenide solar cells have been demonstrated to convert 22% of the available sunlight to electricity, compared with about 16% for silicon solar cells.

## Selected physical properties of GaAs

| Property | Amount |
| --- | --- |
| Molecular weight | 144.6 |
| Melting point, K | 1511 |
| Density, g/cm$^3$ | |
|   At 300 K (solid) | 5.3165 ± 0.0015 |
|   At 1511 K (solid) | 5.2 |
|   At 1511 K (liquid) | 5.7 |
| Lattice constant, nm | 0.5654 |
| Adiabatic bulk modulus, dyne · cm$^{-2}$ | 7.55 × 10$^{11}$ |
| Thermal expansion, K$^{-1}$ | |
|   At 300 K | 6.05 × 10$^{-6}$ |
|   At 1511 K | 7.97 × 10$^{-6}$ |
| Specific heat, J · g$^{-1}$ · K$^{-1}$ | |
|   At 300 K | 0.325 |
|   At 1511 K | 0.42 |
| Thermal diffusivity at 300 K, cm$^2$ · s$^{-1}$ | 0.27 |
| Latent heat, J · cm$^{-3}$ | 3290 |
| Band gap, eV | 1.44 |
| Refractive index at 10 μm | 3.309 |
| Dielectric constant | |
|   Static | 12.85 |
|   Infrared | 10.88 |
| Electron mobility, cm$^2$ · V$^{-1}$ · s$^{-1}$ | |
|   At 77 K | 205 000 |
|   At 300 K | 8500 |
| Hole mobility at 300 K | 400 |
| Intrinsic resistivity at 300 K, Ω · cm | 3.7 × 10$^8$ |

Although ICs currently represent a smaller share of the GaAs market than optoelectronic devices, they are important for military and defense applications. Two types of ICs are produced commercially: analog and digital. Analog ICs are designed to process signals generated by military radar systems, as well as those generated by satellite communications systems. Digital ICs essentially function as memory and logic elements in computers.

Nonsemiconducting applications include the use of gallium oxide for making single-crystal garnets—such as gallium gadolinium garnet (GGG), which is used as the substrate for magnetic domain (bubble) memory devices. Small quantities of metallic gallium are used for low-melting-point alloys, for dental alloys, and as an alloying element in some magnesium, cadmium, and titanium alloys. Gallium is also used in high-temperature thermometers and as a substitute for mercury in switches. Gallium-base superconducting compounds, such as GaV$_3$, have also been developed.

## Recommended Reading

- D.A. Kramer, Gallium and Gallium Compounds, *Metals Handbook*, 10th ed., Vol 1, ASM International, 1990, p 739-749
- M.H. Brodsky, Progress in Gallium Arsenide Semiconductors, *Sci. Am.*, Feb 1990, p 68-75
- K. Zwibel, Photovoltaic Cells, *Chem. Eng. News*, Vol 64 (No. 27), 7 July 1986, p 34-48

derived from electromagnetic or electrodynamic forces produced by the current.

**galvanostatic.** An experimental technique whereby an *electrode* is maintained at a constant current in an *electrolyte*.

**gamma iron.** The face-centered cubic form of pure iron, stable from 910 to 1400 °C (1670 to 2550 °F).

**gamma ray.** A high-energy photon, especially as emitted by a nucleus in a transition between two energy levels. Similar to x-rays but of a nuclear origin; gamma rays have a range of wavelengths from about 0.0005 to 0.14 nm. See also the table accompanying the term *electromagnetic radiation*.

**gamma-ray spectrometry.** See *gamma-ray spectroscopy*.

**gamma-ray spectroscopy.** Determination of the energy distribution of γ-rays emitted by a nucleus. Also known as gamma-ray spectrometry.

**gamma structure.** Structurally analogous phases or electron compounds having ratios of 21 valence electrons to 13 atoms. This is generally a large, complex cubic structure.

**gamma transition.** See *glass transition*.

**gang milling.** Milling with several cutters mounted on the same arbor or with workpieces similarly positioned for cutting either simultaneously or consecutively during a single setup.

**gang slitter.** A machine with a number of pairs of rotary cutters spaced on two parallel shafts, used for *slitting* metal into strips or for trimming the edges of sheets.

**gangue.** The worthless portion of an ore that is separated from the desired part before smelting is commenced.

**gap** (composites). (1) In filament winding, the space between successive windings in which windings are usually intended to lay next to each other. Separations between fibers within a filament winding band. (2) The distance between adjacent plies in a lay-up of unidirectional tape materials.

**gap** (welding). The root opening in a weld joint.

**gap-filling adhesive.** An adhesive subject to reduce shrinkage upon setting, used as a sealant.

**gap-frame press.** A general classification of press in which the uprights or housings are made in the form of a letter C, thus making three sides of the die space accessible.

**garnet.** A generic name for a related group of mineral silicates which have the general chemical formula $A_3B_2(SiO_4)_3$, where $A$ is $Fe^{2+}$, $Mn^{2+}$, Mg, or Ca, and $B$ is Al, $Fe^{3+}$, $Cr^{3+}$, or $Ti^{3+}$. Garnet is used for coating abrasive paper or cloth, for bearing pivots in watches, for electronics, and the finer specimens for gemstones. The hardness of garnet varies from Mohs 6 to 8 (1360 Knoop), the latter being used for abrasive applications.

**gas atomization.** An *atomization* process whereby molten metal is broken up into particles by a rapidly moving inert gas stream (Fig. 206). The resulting particles are nearly spherical with attached satellites (Fig. 207).

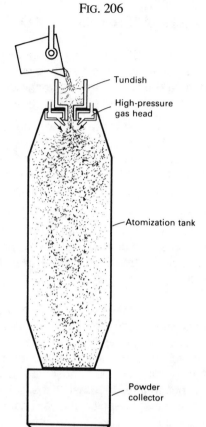

FIG. 206

*Schematic of the gas atomization process*

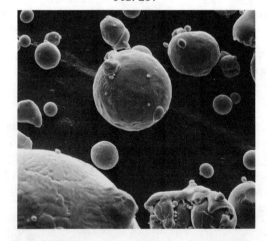

FIG. 207

*Type 316 gas-atomized stainless steel powder with attached satellites. Scanning electron micrograph. 750×*

**gas bearing.** A journal or thrust bearing lubricated with gas.

**gas brazing.** See preferred term *torch brazing*.

**gas carbon arc welding**. A carbon arc welding process variation which produces coalescence of metals by heating them with an electric arc between a single carbon electrode and the work. Shielding is obtained from a gas or gas mixture.

**gas chromatography**. A separation method involving passage of a gaseous mobile phase through a column containing a stationary adsorbent phase; used principally as a quantitative analytical technique for volatile compounds. See also *chromatography, ion chromatography*, and *liquid chromatography*.

**gas classification**. The separation of a powder into its particle size fractions by means of a gas stream of controlled velocity flowing counterstream to the gravity-induced fall of the particles. The method is used to classify submesh-size particles.

**gas classifier**. A device for gas classification; it may be of laboratory size for quality control testing or of industrial capacity to accommodate powder production requirement.

**gas constant**. The constant of proportionality appearing in the equation of state of an ideal gas, equal to the pressure of the gas multiplied by its molar volume divided by its temperature. Also known as universal gas constant.

**gas cutting**. See preferred term *oxygen cutting*.

**gas cyaniding**. A misnomer for *carbonitriding*.

**gaseous corrosion**. Corrosion with gas as the only corrosive agent and without any aqueous phase on the surface of the metal. Also called dry corrosion. See also *hot corrosion* and *sulfidation*.

**gaseous reduction**. (1) The reaction of a metal compound with a reducing gas to produce the metal. (2) The conversion of metal compounds to metallic particles by the use of a reducing gas.

**gas gouging**. See preferred term *oxygen gouging*.

**gas holes**. Holes in castings or welds that are formed by gas escaping from molten metal as it solidifies. Gas holes may occur individually, in clusters, or throughout the solidified metal.

**gasket**. A device, usually made of a deformable material, that is used between two relatively static surfaces to prevent leakage.

**gas lubrication**. A system of lubrication in which the shape and relative motion of the sliding surfaces cause the formation of a gas film having sufficient pressure to separate the surfaces. See also *pressurized gas lubrication*.

**gas mass spectrometry**. An analytical technique that provides quantitative analysis of gas mixtures through the complete range of elemental and molecular gases.

**gas metal arc cutting**. An arc cutting process used to sever metals by melting them with the heat of an arc between a continuous metal (consumable) electrode and the work. Shielding is obtained entirely from an externally supplied gas or gas mixture.

**gas metal arc welding (GMAW)**. An arc welding process which produces coalescence of metals by heating them with an arc between a continuous filler metal (consumable) electrode and the work. Shielding is obtained entirely from an externally supplied gas or gas mixture. Variations of the process include short-circuit arc GMAW, in which the consumable electrode is deposited during repeated short circuits, and pulsed arc GMAW, in which the current is pulsed. See also *globular transfer, short-circuiting transfer*, and *spray transfer*.

**gas plating**. Same as *vapor plating*.

**gas pocket**. A cavity caused by entrapped gas.

**gas porosity**. Fine holes or pores within a metal that are caused by entrapped gas or by the evolution of dissolved gas during solidification.

**gas regulator**. See preferred term *regulator*.

**gas shielded arc welding**. A general term used to describe gas metal arc welding, gas tungsten arc welding, and flux cored arc welding when gas shielding is employed. Typical gases employed include argon, helium, argon-hydrogen mixture, or carbon dioxide.

**gas shielded stud welding**. See *stud arc welding*.

**gassing**. (1) Absorption of gas by a metal. (2) Evolution of gas from a metal during melting operations or upon solidification. (3) Evolution of gas from an electrode during electrolysis.

**gas torch**. See preferred terms *welding torch* and *cutting torch*.

**gas tungsten arc cutting**. An arc-cutting process in which metals are severed by melting them with an arc between a single tungsten (nonconsumable) electrode and the work. Shielding is obtained from a gas or gas mixture.

**gas tungsten arc welding (GTAW)**. An arc welding process which produces coalescence of metals by heating them with an arc between a tungsten (nonconsumable) electrode and the work. Shielding is obtained from a gas or gas mixture. Pressure may or may not be used and filler metal may or may not be used.

**gas welding**. See preferred term *oxyfuel gas welding*.

**gate** (casting). The portion of the runner in a mold through which molten metal enters the mold cavity. The generic term is sometimes applied to the entire network of connecting channels that conduct metal into the mold cavity. See also the figure accompanying the term *gating system*.

**gate** (plastics). In injection and transfer molding of plastics, the orifice through which the melt enters the mold cavity. The gate can have a variety of configurations, depending on product design.

**gated pattern**. In foundry practice, a *pattern* that includes not only the contours of the part to be cast but also the *gates*.

**gate mark**. A surface discontinuity on a molded plastic part caused by the gate through which material enters the cavity.

FIG. 208

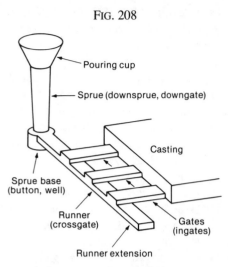

*Basic components of a simple gating system*

FIG. 210

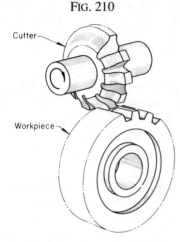

*Relation of cutter and workpiece when milling teeth in a spur gear*

**gathering**. A forging operation that increases the cross section of part of the stock; usually a preliminary operation.

**gathering stock**. Any operation whereby the cross section of a portion of the forging stock is increased beyond its original size.

**gating system**. The complete assembly of sprues, runners, and gates in a mold through which metal flows to enter casting cavity (Fig. 208). The term is also applied to equivalent portions of the *pattern*.

**gear cutting**. Producing tooth profiles of equal spacing on the periphery, internal surface, or face of a workpiece by means of an alternate shear gear-form cutter or a gear generator.

**geared press**. A press whose main crank or eccentric shaft is connected by gears to the driving source.

**gear hobbing**. Gear cutting by use of a tool resembling a worm gear in appearance, having helically spaced cutting

teeth (Fig. 209). In a single-thread hob, the rows of teeth advance exactly one pitch as the hob makes one revolution. With only one hob, it is possible to cut interchangeable gears of a given pitch of any number of teeth within the range of the hobbing machine.

**gear milling**. Gear cutting with a milling cutter that has been formed to the shape of the tooth space to be cut (Fig. 210). The tooth spaces are machined one at a time.

**gear shaping**. Gear cutting with a reciprocating gear-shaped cutter rotating in mesh with the work blank (Fig. 211).

**gear shaving**. A finishing operation performed with a serrated rack or gear-like cutter in mesh with the gear, but with their axis skewed.

**gel**. (1) A colloidal state comprised of interdispersed solid and liquid, in which the solid particles are themselves interconnected or interlaced in three dimensions. (2) A two-

FIG. 209

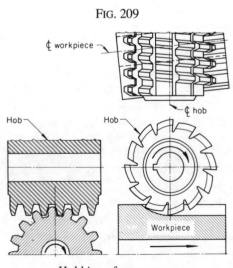

*Hobbing of a spur gear*

FIG. 211

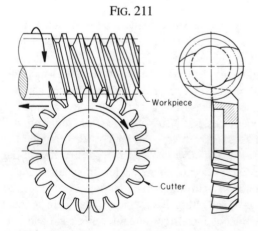

*Relation of cutter and workpiece in generating a worm gear by shaping*

phase colloidal system consisting of a solid and a liquid in more solid form than a *sol*.

**gel** (polymers). The initial jellylike solid phase that develops during the formation of a resin from a liquid. With respect to vinyl plastisols, a state between liquid and solid that occurs in the initial states of heating, or upon prolonged storage. In general, gels have very low strengths and do not flow like a liquid. They are soft, flexible, and may rupture under their own weight unless supported externally. In a cross-linked thermoplastic, gel is the fraction of polymeric material present in the network.

**gelatin replica**. A reproduction of a surface prepared in a film composed of gelatin. See also *replica*.

**gelation**. The point in a resin cure when the resin viscosity has increased to a point such that it barely moves when probed with a sharp instrument.

**gelation time**. (1) That interval of time, in connection with the use of synthetic thermosetting resins, extending from the introduction of a catalyst into a liquid adhesive system until the start of *gel* formation. (2) The time under application of load for a resin to reach a solid state.

**gel coat**. A quick-setting resin applied to the surface of a mold and gelled before *lay-up*. The gel coat becomes an integral part of the finished laminate, and is usually used to improve surface and bonding.

**gelling agent**. See *thickener*.

**gel-permeation chromatography**. See *size-exclusion chromatography*.

**gel point**. (1) The point at which a thermosetting system attains an infinite value of its average molecular weight. (2) The viscosity at which a liquid begins to exhibit pseudoelastic properties. This stage may be conveniently observed from the inflection point on a viscosity-time plot.

**gel time**. The period of time from the initial mixing of the reactants of a liquid material composition to the point in time when *gelation* occurs, as defined by a specific test method.

**general corrosion**. (1) A form of deterioration that is distributed more or less uniformly over a surface. (2) *Corrosion* dominated by uniform thinning that proceeds without appreciable localized attack. See also *uniform corrosion*.

**general precipitate**. A precipitate that is dispersed throughout the metallic matrix.

**geodesic**. The shortest distance between two points on a surface.

**geodesic isotensoid**. Constant stress level in any given *filament* at all points in its path.

**geodesic-isotensoid contour**. In filament-wound reinforced plastic pressure vessels, a dome contour in which the filaments are placed on geodesic paths so that the filaments exhibit uniform tensions throughout their length under pressure loading.

**germanium and germanium compounds**. See *Technical Brief 16*.

**getter**. (1) A special metal or alloy that is placed in a vacuum tube during manufacture and vaporized after the tube has been evacuated; when the vaporized metal condenses, it absorbs residual gases. (2) In powder metallurgy, a substance that is used in a sintering furnace for the purpose of absorbing or chemically binding elements or compounds from the sintering atmosphere that are damaging to the final product.

**gettering box**. In powder metallurgy, a container for the getter substance that is readily accessible to the atmosphere and prevents contamination of the sintered product by direct contact.

**ghost lines**. Lines running parallel to the rolling direction that appear in a sheet metal panel when it is stretched. These lines may not be evident unless the panel has been sanded or painted. Not to be confused with *leveler lines*.

**Gibbs free energy**. The thermodynamic function $\Delta G = \Delta H - T\Delta S$, where $H$ is enthalpy, $T$ is absolute temperature, and $S$ is entropy. Also called free energy, free enthalpy, or Gibbs function.

**Gibbs triangle**. An equilateral triangle used for plotting composition in a ternary system.

**gibs**. Guides or shoes that ensure the proper parallelism, squareness, and sliding fit between metal forming press components such as the slide and the frame. They are usually adjustable to compensate for wear and to establish operating clearance.

**glancing angle**. In materials characterization, the angle (usually small) between an incident x-ray beam and the surface of the specimen.

**glass**. See *Technical Brief 17*.

**glass-ceramics**. See *Technical Brief 18*.

**glass cloth**. Woven glass fiber material. See also *scrim*.

**glass electrode**. A glass membrane *electrode* used to measure pH or hydrogen-ion activity.

**glass fiber**. A fiber spun from an inorganic product of fusion that has cooled to a rigid condition without crystallizing. See also *fiberglass*.

**glass filament**. A form of glass that has been drawn to a small diameter and extreme length. It is standard practice in the fiberglass industry to refer to a specific filament diameter by a specific alphabet designation, as listed in Table 25. Fine fibers, which are used in textile applications, range from D (~6 μm) through G (~10 μm). Conventional plastics reinforcement, however, uses filament diameters that range from G to T (~24 μm).

**glass filament bushing**. The unit through which molten glass is drawn in making glass filaments (Fig. 212).

**glass finish**. A material applied to the surface of a glass reinforcement to improve the bond between the glass and the plastic resin matrix.

## Technical Brief 16: Germanium and Germanium Compounds

GERMANIUM (Ge) is a semiconducting metalloid element found in Group IV A and period 4 of the periodic table. Although it looks like a metal, it is fragile like glass. Its electrical resistivity is about midway between that of metallic conductors and that of good electrical insulators. Its first significant use was in solid-state electronics, and with it the transistor was invented. Germanium is still used in the field of electronics, but its use in the field of infrared optics surpassed its electronic applications in the 1970s.

### Optical properties of germanium

| Wavelength, μm | Refractive index at 300 K (80 °F) | Absorption coefficient, cm$^{-1}$ | Through 1 cm thickness (uncoated) Reflection, % | Absorption, % | Transmission, % |
|---|---|---|---|---|---|
| 1.8 | 4.134 | 7.0 | 37.3 | 62.7 | 0.0 |
| 1.9 | 4.120 | 0.68 | 41.0 | 38.2 | 20.8 |
| 2.0 | 4.108 | 0.010 | 53.6 | 1.0 | 45.4 |
| 4.0 | 4.0255 | 0.0047 | 53.0 | 0.5 | 46.5 |
| 6.0 | 4.0122 | 0.0068 | 52.8 | 0.7 | 46.5 |
| 8.0 | 4.0074 | 0.0150 | 52.4 | 1.5 | 46.1 |
| 10.0 | 4.0052 | 0.0215 | 52.2 | 2.1 | 45.7 |
| 10.6 | 4.0048 | 0.0270 | 52.0 | 2.6 | 45.4 |
| 11.0 | 4.0045 | 0.0295 | 51.8 | 2.9 | 45.3 |
| 11.9 | 4.0040 | 0.200 | 46.9 | 16.4 | 36.7 |
| 12.0 | 4.0039 | 0.170 | 47.6 | 14.4 | 38.0 |
| 13.0 | 4.0035 | 0.160 | 47.8 | 13.7 | 38.5 |
| 14.0 | 4.0032 | 0.149 | 48.2 | 12.8 | 39.0 |
| 15.0 | 4.0029 | 0.385 | 43.3 | 27.1 | 29.6 |
| 16.0 | 4.0026 | 0.530 | 41.4 | 33.4 | 25.2 |
| 18.0 | 4.0022 | 2.00 | 36.3 | 58.2 | 5.5 |
| 20.0 | 4.0018 | 2.15 | 36.2 | 59.0 | 4.8 |

Germanium usually occurs widely dispersed in minerals such as sphalerite; it rarely occurs in concentrated form. Almost all germanium production has been from zinc smelters. Copper smelters are the second largest source. There are only a few actual minerals of germanium, some with germanium concentrations up to about 8%. Most of these have been found in Africa.

For semiconducting applications, single-crystal wafers of germanium are used as substrates for the epitaxial deposition of gallium arsenide (GaAs) or gallium arsenide phosphide (GaAsP) for use as light-emitting diodes or solar cells. These substrates take the place of the more expensive GaAs wafers.

### Table 25  Filament diameter nomenclature

| Alphabet | Filament diameter μm | 10$^{-4}$ in. |
|---|---|---|
| AA | 0.8 - 1.2 | 0.3 - 0.5 |
| A | 1.2 - 2.5 | 0.5 - 1.0 |
| B | 2.5 - 3.8 | 1.0 - 1.5 |
| C | 3.8 - 5.0 | 1.5 - 2.0 |
| D | 5.0 - 6.4 | 2.0 - 2.5 |
| E | 6.4 - 7.6 | 2.5 - 3.0 |
| F | 7.6 - 9.0 | 3.0 - 3.5 |
| G | 9.0 - 10.2 | 3.5 - 4.0 |
| H | 10.2 - 11.4 | 4.0 - 4.5 |
| J | 11.4 - 12.7 | 4.5 - 5.0 |
| K | 12.7 - 14.0 | 5.0 - 5.5 |
| L | 14.0 - 15.2 | 5.5 - 6.0 |
| M | 15.2 - 16.5 | 6.0 - 6.5 |
| N | 16.5 - 17.8 | 6.5 - 7.0 |
| P | 17.8 - 19.0 | 7.0 - 7.5 |
| Q | 19.0 - 20.3 | 7.5 - 8.0 |
| R | 20.3 - 21.6 | 8.0 - 8.5 |
| S | 21.6 - 22.9 | 8.5 - 9.0 |
| T | 22.9 - 24.1 | 9.0 - 9.5 |
| U | 24.1 - 25.4 | 9.5 - 10 |

Fɪɢ. 212

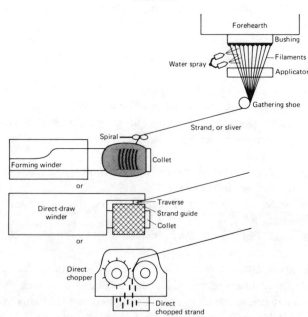

*Fiberglass filament forming process*

The largest use of germanium is in the field of infrared optics. In this application, the transparency of germanium to infrared wavelengths longer than 2 μm and its high refractive index are utilized rather than its electrical properties. Other advantageous properties of germanium for this use are its low price compared with other infrared materials and its good resistance to atmospheric oxidation, moisture, and chemical attack. Infrared devices are principally used for military applications. Infrared systems with germanium usually contain several germanium lenses, a germanium window, and a color-correcting lens made from Ge-Sb-Se glass, a Ge-As-Se glass, or ZnSe. Nonmilitary infrared applications include $CO_2$ lasers, intrusion alarms, and surveillance devices. Germanium is also used as a thin-film coating for infrared materials.

The primary application for germanium dioxide ($GeO_2$) is in the preparation of germanium metal. However, there are several other uses that provide significant markets for the oxide. The largest of these is its use in place of antimony oxide ($Sb_2O_3$) as a catalyst in the reaction of ethylene glycol with terephthalic acid in the production of polyester fibers and polyethylene terephthalate (PET) resins. Germanium oxide produces a polyester fiber that does not yellow with age, which is especially attractive to makers of white shirts and other white fabrics. The PET resins are used in making beverage bottles such as 2- and 3-liter soft drink bottles sold in the Far East. Another significant use of $GeO_2$ is in the production of bismuth germanium oxide crystals ($Bi_4Ge_3O_{12}$), which are used primarily in positron emission tomography scanners. In addition, $GeO_2$ is included in a few special glass formulations, primarily to increase the refractive index of the glass.

Other uses of germanium include the use of germanium tetrachloride in optical fibers, mangesium germanate as a phospor, the addition of lead germanate to barium titanate capacitors, the use of germanium-gold dental alloys, and the use of germanium single crystals as x-ray monochromators for high-energy physics applications.

## Recommended Reading

- J.H. Adams, Germanium and Germanium Compounds, *Metals Handbook*, 10th ed., Vol 1, ASM International, 1990, p 733-737
- F. Glocking, *The Chemistry of Germanium*, Academic Press, 1969

**glass flake**. Thin, irregularly shaped flakes of glass used as a reinforcement in composites.

**glass former**. An oxide that forms a glass easily. Also, one that contributes to the network of silica glass when added to it.

**glass, percent by volume**. The product of the specific gravity of a laminate and the percent glass by weight, divided by the specific gravity of the glass.

**glass stress**. In a filament-wound part, usually a pressure vessel, the stress calculated using the load and the cross-sectional area of the reinforcement only.

**glass transition temperature ($T_g$)**. The temperature at which an amorphous polymer (or the amorphous regions in a partially crystalline polymer) changes from a hard and relatively brittle condition to a viscous or rubbery condition. In this temperature region, many physical properties, such as hardness, brittleness, thermal expansion, and specific heat, undergo significant, rapid changes.

**glassy**. A state or matter that is amorphous or disordered like a liquid in structure, hence capable of continuous composition variation and lacking a true melting point, but softening gradually with increasing temperature. Glasses of commerce are mainly complex silicates in chemical combination with numerous other oxidic substances; made by melting the source materials together, forming in various ways while fluid, and allowing to cool.

**glaze**. (1) A ceramic coating matured to the glassy state on a formed ceramic article, or the material or mixture from which the coating is made. (2) In tribology, a ceramic or other hard, smooth surface film produced by sliding.

**glazing**. Dulling the abrasive grains in the cutting face of a wheel during grinding.

**glide**. (1) Same as *slip*. (2) A noncrystallographic shearing movement, such as of one grain over another.

**globular transfer**. In consumable-electrode arc welding, a type of metal transfer in which molten filler metal passes across the arc as large droplets. Compare with *spray transfer* and *short-circuiting transfer*. See also the figure accompanying the term *short-circuiting transfer*.

**glue**. Originally, a hard gelatin obtained from hides, tendons, cartilage, bones, and so on, of animals and also an adhesive prepared from this substance by heating with water. Through general use, the term is now synonymous with

## Technical Brief 17: Glass

GLASS is an amorphous solid made by fusing silica ($SiO_2$) with a basic oxide. Its characteristic properties are its transparency, its hardness and rigidity at ordinary temperatures, its capacity for plastic working at elevated temperatures, and its resistance to weathering and to most chemicals. The major steps in producing glass products are: (1) melting and refining, (2) forming and shaping, (3) heat treating (annealing and tempering), and (4) finishing (grinding, polishing, and/or decorating).

There are a number of general families of glasses, some of which have many hundreds of variations of compositions. Glasses can be classified according to their product type (flat glass, container glass, etc.) or composition:

- *Soda-lime glasses* are the oldest, lowest in cost, easiest to work, and the most widely used. They account for about 90% of the glass used in the United States. They are composed of silica, sodium oxide or soda ($Na_2O$), and calcium oxide or lime (CaO). These are the glass of ordinary windows, bottles, and drinking glasses.

- *Borosilicate glasses*, which contain boron oxide ($B_2O_3$), are the most versatile of the glasses. They are noted for their excellent chemical durability, for resistance to heat and thermal shock,

### Glass product types and applications

| Glass product type | Application |
|---|---|
| Flat glass | Automotive: cars and trucks<br>Architectural: commercial buildings, storefronts<br>Residential: windows, doors, sunrooms, skylights<br>Patterned glass: shower doors, privacy glass |
| Containers/tableware | Beverage<br>Liquor, beer, wine<br>Food<br>Pharmaceutical, drugs<br>Glasses, plates, cups, bowls, serving dishes |
| Fiberglass | Wool: insulation, filters<br>Textile: plastic or rubber tire reinforcements, fabrics, roof shingle and roll goods reinforcement |
| Specialty glass | Artware, stained glass, lead and lead crystal, lighting, TV picture tubes, ovenware and stovetop, ophthalmic, aviation, tubing, foamed glass, marbles |

### Main categories of silicate glasses

| | Typical composition | Uses | Properties |
|---|---|---|---|
| Soda-lime | 70–75% $SiO_2$<br>12–16% $Na_2O$<br>10–15% CaO | Bottles, glasses, windows | Optically clear, durable |
| Lead ("crystal") glasses | 55–65% $SiO_2$<br>18–38% PbO<br>13–15% $Na_2O$ or $K_2O$ | Decorative items | High refractive index |
| Borosilicate | 70–80% $SiO_2$<br>7–13% $Ba_2O_3$<br>4–8% $Na_2O$ or $K_2O$<br>2–7% $Al_2O_3$ | Chemical apparatus, lamp and tube envelopes | Chemical durability, low thermal expansion |
| Quartz glass | 100% $SiO_2$ | High-temperature uses | High softening temperature, low thermal expansion |
| Aluminosilicate glasses | 52–58% $SiO_2$<br>15–25% $Al_2O_3$<br>4–18% CaO | High-temperature uses, thermometers, combustion tubes, cookware | High softening temperature, low thermal expansion |

the terms bond and adhesive. See also *adhesive, gum, mucilage, paste, resin*, and *sizing*.

**glue-laminated wood.** An assembly made by bonding layers of veneer or lumber with an adhesive so that the grain of all laminations is essentially parallel.

**glue line.** Synonym for *bond line*.

**glue line thickness.** Thickness of layer of cured adhesive.

**gob.** (1) A portion of hot glass delivered by a feeder (Fig. 213).

(2) A portion of hot glass gathered on a punty or pipe.

**gob process.** A process whereby glass is delivered to a forming unit in *gob* form (Fig. 213).

**gold filled.** Covered on one or more surfaces with a layer of gold alloy to form a clad or composite material. Gold-filled dental restorations are an example of such materials.

**goniometer.** In *x-ray spectrometry*, an instrument devised for measuring the angle through which a specimen is rotated or for orienting a sample (for example, a single crystal) in a specific way.

**gooseneck.** In die casting, a spout connecting a molten metal holding pot, or chamber, with a nozzle or sprue hole in the die and containing a passage through which molten metal

and for low coefficients of thermal expansion. Borosilicate glasses are used in such products as sights and gages, lamps, piping, seals for low-expansion alloys, telescope mirrors, electronic tubes, and laboratory glassware.

- *Aluminosilicate glasses* are more costly than the borosilicate types, but are useful at higher temperatures and have greater thermal shock resistance. They are used for high-performance combustion tubes, high-temperature thermometers, and stovetop cookware.

- *Lead glasses*, or lead-alkali glasses, are produced with lead oxide (PbO) ranging from 18 to 38%. They are relatively inexpensive and are noted for high electrical resistivity and a high refractory index. They are used in many optical components, for neon sign tubing, for electric light-bulb stems, and as decorative items.

- *High-silica glasses* (containing 96 to 100% $SiO_2$) include quartz glass and modifications of quartz glass. Quartz glass has extremely high purity (>99.9% $SiO_2$) and high resistance to heat and thermal shock. It is also chemically stable and has high electrical resistance. However, because quartz glass requires high processing temperatures and is high in cost, its use is restricted to such specialty applications as laboratory optical systems and instruments and crucibles for crystal growing. Two types of high-silica glass used for lamps that have many of the favorable properties of quartz glass but require lower processing temperatures are Vycor and doped quartz glass. Vycor is made by melting a borosilicate glass (containing $SiO_2$, 4% $B_2O_3$, and 0.1% $Na_2O$), leaching the $Na_2O \cdot B_2O_3$ phase, and sintering to a high density. Doped quartz is made by the addition of a few hundred ppm of oxides such as BaO, $Al_2O_3$, and $K_2O$ to a pure $SiO_2$ melt.

**Properties of glasses**

| | Density, g/cm³ | Strain point | | Annealing point | | Softening point | | Working point | | CTE at 0–300 °C (32–570 °F), $10^{-7}$ K$^{-1}$ | Young's modulus at 20 °C (70 °F) | | Electrical resistivity at 350 °C (660 °F), $\Omega \cdot cm$ |
|---|---|---|---|---|---|---|---|---|---|---|---|---|---|
| | | °C | °F | °C | °F | °C | °F | °C | °F | | GPa | $10^6$ psi | |
| Soda-lime glass | 2.5 | 490 | 915 | 520 | 970 | 700 | 1290 | 1015 | 1860 | 94 | 72 | 10.4 | $10^5$ |
| Lead glass | 2.8 | 410 | 770 | 445 | 835 | 635 | 1175 | 1000 | 1830 | 93 | 61 | 8.8 | $10^7$ |
| Borosilicate | 2.3 | 520 | 970 | 570 | 1060 | 800 | 1470 | 1200 | 2190 | 40 | 64 | 9.3 | $10^7$ |
| Aluminosilicate | 2.6 | 770 | 1420 | 810 | 1490 | 1025 | 1875 | 1250 | 2280 | 45 | 88 | 12.8 | $10^{11}$ |
| Vycor | 2.2 | 890 | 1635 | 1020 | 1870 | 1530 | 2785 | ... | ... | 7.5 | 72 | 10.4 | $10^8$ |
| Quartz glass | 2.2 | 1070 | 1960 | 1140 | 2085 | 1670 | 3040 | ... | ... | 5.5 | 73 | 10.6 | $10^{10}$ |

### Recommended Reading

- W.R. Prindle *et al.*, Glass Processing, *Engineered Materials Handbook*, Vol 4, ASM International, 1991, p 377-476
- P.S. Danielson *et al.*, Applications for Glasses, *Engineered Materials Handbook*, Vol 4, ASM International, 1991, p 1015-1103
- J.E. Shelby, W.C. Lacourse, and A.G. Clare, Engineering Properties of Oxide Glasses and Other Inorganic Glasses, *Engineered Materials Handbook*, Vol 4, ASM International, 1991, p 845-857

is forced on its way to the die. It is the metal injection mechanism in a *hot chamber machine*.

**gouging.** In welding practice, the forming of a bevel or groove by material removal. See also *back gouging, arc gouging*, and *oxygen gouging*.

**gouging abrasion.** A form of high-stress abrasion in which easily observable grooves or gouges are created on the surface. See also *abrasion*, and *low-stress abrasion*.

**G-P zone.** A *Guinier-Preston zone*.

**gradated coating.** A thermal sprayed deposit composed of mixed materials in successive layers which progressively

change in composition from the constituent material of the substrate to the surface of the sprayed deposit.

**grade.** In powder metallurgy, a specific, nominal chemical analysis powder identified by a code number. Example: cemented carbide manufacturers grade 74 M 60 FWC (74 is usage; M is equipment manufacturer; 60 is nominal HRC; FWC is fine cut tungsten carbide).

**graded abrasive.** An abrasive powder in which the sizes of the individual particles are confined to certain specified limits. See also *grit size*.

**graded coating.** A thermal spray coating consisting of sev-

## Technical Brief 18: Glass-Ceramics

GLASS-CERAMICS are a family of fine-grained crystalline materials made by a process of controlled crystallization from special glass compositions containing nucleating agents. The process for manufacturing

*Microstructure of a potassium fluor-richterite glass-ceramic. The long acicular blades yield a material with unusual strength and toughness. 30,000×*

glass-ceramics consists of melting a glass and forming the desired shape, which is followed by crystallization using a multiple-temperature heat treatment. Glass-ceramics are nonporous and generally either opaque white or transparent. By definition, glass-ceramics are ≥50% crystalline by volume and generally are >90% crystalline. Depending on heat treatment and composition, crystal sizes range from <0.1 m to 10 µm. The range of glass-ceramic compositions is extremely broad, requiring only the ability to form a glass and control its crystallization.

The specific properties of any glass-ceramic are controlled by both the physical properties of the individual crystals and by the textural relationship between the crystals and the residual glass. As a result, glass-ceramics can offer a wide variety of properties not attainable in conventional glasses and ceramics, including durability, strength, and exceptional thermal shock resistance. An example of this phenomenon is Pyroceram tableware, which has a flexural strength of 206 MPa (30 ksi), a density of 2.4 to 2.6 $g/cm^3$, a softening point at 1350 °C (2460 °F), and a coefficient of thermal expansion ranging from 95 up to $115 \times 10^{-7}/°C$.

Glass-ceramic products range from transparent, zero-expansion materials with excellent optical properties and thermal shock resistance to jadelike, highly crystalline materials with excellent strength and toughness. The highest volume is in cookware and tableware consumer items, architectural cladding, and stovetops and stove windows. Reflective telescope mirrors and heat exchangers have also been made from glass-ceramics.

### Base compositions and applications of transparent glass-ceramics

| Material | Composition, wt% | | | | | | | | | | | | | | Commercial applications |
|---|---|---|---|---|---|---|---|---|---|---|---|---|---|---|---|
| | $SiO_2$ | $Al_2O_3$ | MgO | $Na_2O$ | $K_2O$ | ZnO | $Fe_2O_3$ | $Li_2O$ | BaO | $P_2O_5$ | F | $TiO_2$ | $ZrO_2$ | $As_2O_3$ | |
| Vision(a) | 68.8 | 19.2 | 1.8 | 0.2 | 0.1 | 1.0 | 0.1 | 2.7 | 0.8 | ... | ... | 2.7 | 1.8 | 0.8 | Transparent cookware |
| Zerodur(b) | 55.5 | 25.3 | 1.0 | 0.5 | ... | 1.4 | 0.03 | 3.7 | ... | 7.9 | ... | 2.3 | 1.9 | 0.5 | Telescope mirrors |
| Ceran(b) | 63.4 | 22.7 | (d) | 0.7 | (d) | 1.3 | (d) | 3.3 | 2.2 | (d) | (d) | 2.7 | 1.5 | (d) | Black infrared transmission cooktop |
| Narumi(c) | 65.1 | 22.6 | 0.5 | 0.6 | 0.3 | ... | 0.03 | 4.2 | ... | 1.2 | 0.1 | 2.0 | 2.3 | 1.1 | Rangetops; stove windows |

(a) Manufactured by Corning Glass Works. (b) Manufactured by Schott. (c) Manufactured by Nippon Electric. (d) No data available

### Recommended Reading

- L.R. Pinckney, Phase-Separated Glasses and Glass-Ceramics, *Engineered Materials Handbook*, Vol 4, ASM International, 1991, p 433-438
- A.E. McHale, Engineering Properties of Glass-Ceramics, *Engineered Materials Handbook*, Vol 4, ASM International, 1991, p 870-878
- P.W. McMillan, *Glass-Ceramics*, 2nd ed., Academic Press, 1979
- Z. Strnad, Glass-Ceramic Materials, *Glass Science and Technology*, Vol 8, Elsevier, 1986

FIG. 213

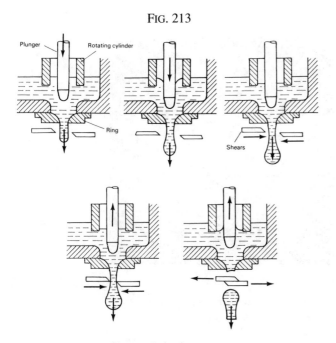

*Glass gob feeder process*

eral successive layers of different materials; for example, starting with 100% metal, followed by one or more layers of metal-ceramic mixtures, and finishing with 100% ceramic.

**gradient coating.** See *gradated coating.*

**gradient elution.** A technique for improving the efficiency of separations achieved by *liquid chromatography.* It refers to a stepwise or continuous change with time in the mobile phase composition.

**graft copolymers.** A chain of one type of polymer to which side chains of a different type are attached or grafted.

**grain.** An individual crystal in a polycrystalline material; it may or may not contain twinned regions and subgrains.

**grain boundary.** A narrow zone in a metal or ceramic corresponding to the transition from one crystallographic orientation to another, thus separating one *grain* from another; the atoms in each grain are arranged in an orderly pattern.

**grain-boundary corrosion.** Same as *intergranular corrosion.* See also *interdendritic corrosion.*

**grain-boundary diffusion.** One of the diffusion mechanisms in *sintering.* It is characterized by a very high diffusion rate because of an abundance of imperfections in the grain boundaries. See also *surface diffusion* and *volume diffusion.*

**grain-boundary etching.** In metallography, the development of intersections of grain faces with the polished surface. Because of severe, localized crystal deformation, grain boundaries have higher dissolution potential than grains themselves. Accumulation of impurities in grain boundaries increases this effect.

**grain-boundary liquation.** An advanced stage of overheating of metals in which material in the region of austenitic grain boundaries melts. Also termed *burning.*

**grain-boundary sulfide precipitation.** An intermediate state of overheating of metals in which sulfide inclusions are redistributed to the austenitic grain boundaries by partial solution at the overheating temperature and reprecipitation during subsequent cooling.

**grain coarsening.** A heat treatment that produces excessively large austenitic grains in metals.

**grain-contrast etching.** In metallography, the development of grain surfaces lying in the polished surface of the microsection. These become visible through differences in reflectivity caused by reaction products on the surface or by differences in roughness.

**grain fineness number.** A system developed by the American Foundrymen's Society (AFS) for rapidly expressing the average grain size of a given sand. It approximates the

### Table 26  Screen scale sieves equivalent for AFS foundry sands

| USA series No. | Tyler screen scale sieves, openings per lineal inch | Sieve opening, mm | Sieve opening, μm | Sieve opening, in., ratio $\sqrt{2}$, or 1.414 | Permissible variation in average opening, ±mm | Wire diameter, mm |
|---|---|---|---|---|---|---|
| 6 | 6 | 3.35 | 3350 | · · · | 0.11 | 1.23 |
| 8(a) | 8(a) | 2.36 | 2360 | 0.0937 | 0.08 | 1.00 |
| 12 | 10 | 1.70 | 1700 | 0.0661 | 0.06 | 0.810 |
| 16(a) | 14(a) | 1.18 | 1180 | 0.0469 | 0.045 | 0.650 |
| 20 | 20 | 0.850 | 850 | 0.0331 | 0.035 | 0.510 |
| 30 | 28 | 0.600 | 600 | 0.0234 | 0.025 | 0.390 |
| 40 | 35 | 0.425 | 425 | 0.0165 | 0.019 | 0.290 |
| 50 | 48 | 0.300 | 300 | 0.0117 | 0.014 | 0.215 |
| 70 | 65 | 0.212 | 212 | 0.0083 | 0.010 | 0.152 |
| 100 | 100 | 0.150 | 150 | 0.0059 | 0.008 | 0.110 |
| 140 | 150 | 0.106 | 106 | 0.0041 | 0.006 | 0.076 |
| 200 | 200 | 0.075 | 75 | 0.0029 | 0.005 | 0.053 |
| 270 | 270 | 0.053 | 53 | 0.0021 | 0.004 | 0.037 |

Note: A fixed ratio exists between the different sizes of the screen scale. This fixed ratio between the different sizes of the screen scale has been taken as 1.414, or the square root of 2 ($\sqrt{2}$). For example, using the USA series equivalent No. 200 as the starting sieve, the width of each successive opening is exactly 1.414 times the opening in the previous sieve. The area or surface of each successive opening in the scale is double that of the next finer sieve or one-half that of the next coarser sieve. That is, the widths of the successive openings have a constant ratio of 1.414, and the areas of the successive openings have a constant ratio of $\sqrt{2}$. This fixed ratio is very convenient; by skipping every other screen, a fixed ratio of width of 2 to 1 exists.
(a) These sieves are not normally used for testing foundry sands.

FIG. 214

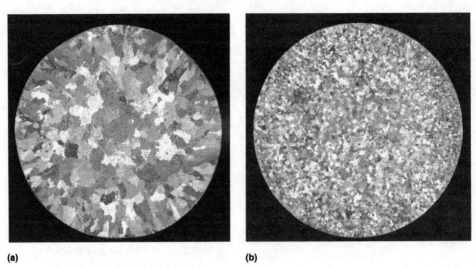

(a)                                                    (b)

*As-cast Al-7Si ingots showing the effects of grain refinement. (a) No grain refiner. (b) Grain refined with Al-5Ti-1B addition. Both 2×*

number of meshes per inch of that sieve that would just pass the sample. Table 26 lists the series of sieves used to run the standard AFS sieve analysis.

**grain flow.** Fiberlike lines on polished and etched sections of forgings caused by orientation of the constituents of the metal in the direction of working during forging. Grain flow produced by proper die design can improve required *mechanical properties* of forgings. See also *flow lines* and *forged structure.*

**grain growth.** (1) An increase in the average size of the grains in polycrystalline material, usually as a result of heating at elevated temperature. (2) In polycrystalline materials, a phenomenon occurring fairly close below the melting point in which the larger grains grow still larger while the smallest ones gradually diminish and disappear. See also *recrystallization.*

**graining.** The process of vigorously stirring or agitating a partially solidified material to develop large grains having a thin oxide coating.

**grain refinement** (metals). The manipulation of the solidification process to cause more (and therefore smaller) grains to be formed and/or to cause the grains to form in specific shapes. The term refinement is usually used to denote a chemical addition to the metal (Fig. 214) but can refer to control of the cooling rate.

**grain refiner.** A material added to a molten metal to induce a finer-than-normal grain size in the final structure (Fig. 214).

**grain size.** (1) For metals, a measure of the areas or volumes of grains in a polycrystalline material, usually expressed as an average when the individual sizes are fairly uniform.

In metals containing two or more phases, grain size refers to that of the matrix unless otherwise specified. Grain size is reported in terms of number of grains per unit area or volume, in terms of average diameter, or as a grain-size number derived from area measurements. (2) For grinding wheels, see preferred term *grit size.*

**grain size distribution.** Measures of the characteristic grain or crystallite dimensions (usually, diameters) in a polycrystalline solid; or of their populations by size increments from minimum to maximum. Usually determined by microscopy.

**gram-equivalent weight.** The mass in grams of a reactant that contains or reacts with Avogadro's number of hydrogen atoms. See also *Avogadro's number.*

**gram-molecular weight.** The mass of a compound in grams equal to its molecular weight.

**granular fracture.** A type of irregular surface produced when metal is broken that is characterized by a rough, grainlike appearance, rather than a smooth or fibrous one. It can be subclassified as *transgranular fracture* or *intergranular fracture.* This type of fracture is frequently called *crystalline fracture*; however, the inference that the metal broke because it "crystallized" is not justified, because all metals are crystalline in the solid state. See also *fibrous fracture* and *silky fracture.*

**granular powder.** A powder having equidimensional but nonspherical particles.

**granular structure.** Nonuniform appearance of finished plastic material due to retention of, or incomplete fusion of, particles of composition, either within the mass or on the surface.

FIG. 215

(a)                                                          (b)

*External surface (a) of a gray cast iron pipe exhibiting severe graphitic corrosion. (b) Closeup view of the graphitically corroded region shown in (a)*

**granulated metal.** Small pellets produced by pouring liquid metal through a screen or by dropping it onto a revolving disk, and, in both instances, chilling with water.

**graphite.** (1) A crystalline allotropic form of carbon. See also the figure accompanying the term *diamond*. (2) Uncombined carbon in cast irons.

**graphite fiber.** A fiber made from a pitch or polyacrylonitrile (PAN) precursor by an oxidation, carbonization, and graphitization process (which provides a graphitic structure). See also *carbon fiber*.

**graphitic carbon.** Free carbon in steel or cast iron.

**graphitic corrosion.** Corrosion of gray iron in which the iron matrix is selectively leached away, leaving a porous mass of graphite behind (Fig. 215); it occurs in relatively mild aqueous solutions and on buried pipe and fittings.

**graphitic steel.** Alloy steel made so that part of the carbon is present as graphite.

**graphitization** (metals). The formation of graphite in iron or steel. Where graphite is formed during solidification, the phenomenon is termed primary graphitization; where formed later by heat treatment, secondary graphitization.

**graphitization** (organic materials). The process of pyrolyzation in an inert atmosphere at temperatures in excess of 1925 °C (3500 °F), usually as high as 2480 °C (4500 °F), and sometimes as high as 5400 °C (9750 °F), converting carbon to its crystalline allotropic form. Temperature depends on precursor and properties desired.

**graphitizing.** Annealing a ferrous alloy such that some or all the carbon precipitates as graphite.

**gravity die casting.** See *permanent mold*.

**gravity hammer.** A class of forging hammer in which energy for forging is obtained by the mass and velocity of a freely falling ram and the attached upper die. Examples are the *board hammer* (Fig. 216) and *air-lift hammer*.

**gravity segregation.** Variable composition of a casting or ingot caused by settling out of heavy constituents, or rising of light constituents, before or during solidification.

**gray body.** A body having the same spectral emittance at all wavelengths.

**gray cast iron.** See *gray iron*.

**gray iron.** A broad class of ferrous casting alloys (*cast irons*) normally characterized by a microstructure of flake graphite in a ferrous matrix. Gray irons usually contain 2.5 to 4% C, 1 to 3% Si, and additions of manganese, depending on the desired microstructure (as low as 0.1% Mn in ferritic gray irons and as high as 1.2% in pearlitics). Sulfur and phosphorus are also present in small amounts as residual impurities. Table 27 lists chemical compositions of typical gray irons. See also the figure accompanying the term *flake graphite*.

**grease.** A lubricant composed of an oil thickened with a soap or other thickener to a semisolid or solid consistency. A lime-base grease is prepared from lubricating oil and calcium soap. Sodium-, barium-, lithium-, and aluminum-base greases are also used. Greases may contain various additives. The liquid phase may also be a synthetic fluid.

**green.** Unsintered (not sintered).

**green ceramic.** An unsintered ceramic.

**green compact.** An unsintered powder metallurgy or ceramic compact.

**green density.** The density of a *green compact*.

**green liquor.** The liquor resulting from dissolving molten smelt from the kraft recovery furnace in water. See also *kraft process* and *smelt*.

**green rot.** A form of high temperature attack on stainless steels, nickel-chromium alloys and nickel-chromium-iron alloys subjected to simultaneous oxidation and carburization. Basically, attack occurs first by precipitation of chromium as chromium carbide, then by oxidation of the carbide particles.

**green sand.** A naturally bonded sand, or a compounded molding sand mixture, that has been "tempered" with water and that is used while still moist.

**green sand core.** (1) A *core* made of *green sand* and used

FIG. 216

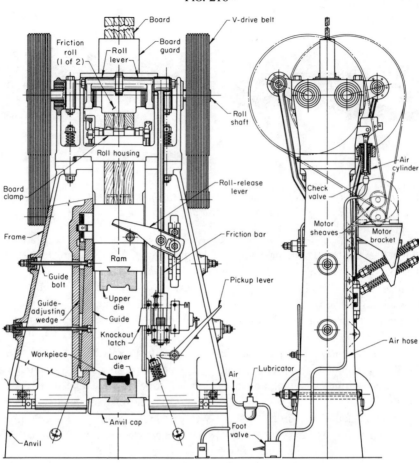

*Principal components of a gravity drop (board) hammer*

as-rammed. (2) A sand core that is used in the unbaked condition.

**green sand mold.** A casting mold composed of moist prepared molding sand. Contrast with *dry sand mold.*

**green strength** (foundry sands). The strength of a tempered sand mixture at room temperature.

**green strength** (plastics). The mechanical strength of material, that, while cure is not complete, allows removal from the mold and handling without tearing or permanent distortion.

**green strength** (powder compacts). (1) The ability of a *green*

*compact* to maintain its size and shape during handling and storage prior to *sintering.* (2) The tensile or compressive strength of a green compact.

**greenware.** A term for formed ceramic articles in the unfired condition.

**greige, gray goods.** Any fabric before finishing, as well as any yarn or fiber before bleaching or dyeing; therefore, fabric with no finish or size.

**grindability.** Relative ease of grinding, analogous to *machinability.*

**grindability index.** A measure of the grindability of a ma-

**Table 27  Typical base compositions of automotive gray cast irons**

| | | Composition, % | | | | |
|---|---|---|---|---|---|---|
| UNS | SAE grade | TC(a) | Mn | Si | P | S |
| F10004 | G1800(b) | 3.40–3.70 | 0.50–0.80 | 2.80–2.30 | 0.15 | 0.15 |
| F10005 | G2500(b) | 3.20–3.50 | 0.60–0.90 | 2.40–2.00 | 0.12 | 0.15 |
| F10006 | G3000(c) | 3.10–3.40 | 0.60–0.90 | 2.30–1.90 | 0.10 | 0.15 |
| F10007 | G3500(c) | 3.00–3.30 | 0.60–0.90 | 2.20–1.80 | 0.08 | 0.15 |
| F10008 | G4000(c) | 3.00–3.30 | 0.70–1.00 | 2.10–1.80 | 0.07 | 0.15 |

(a) TC, total carbon. (b) Ferritic-pearlitic microstructure. (c) Pearlitic microstructure

FIG. 217

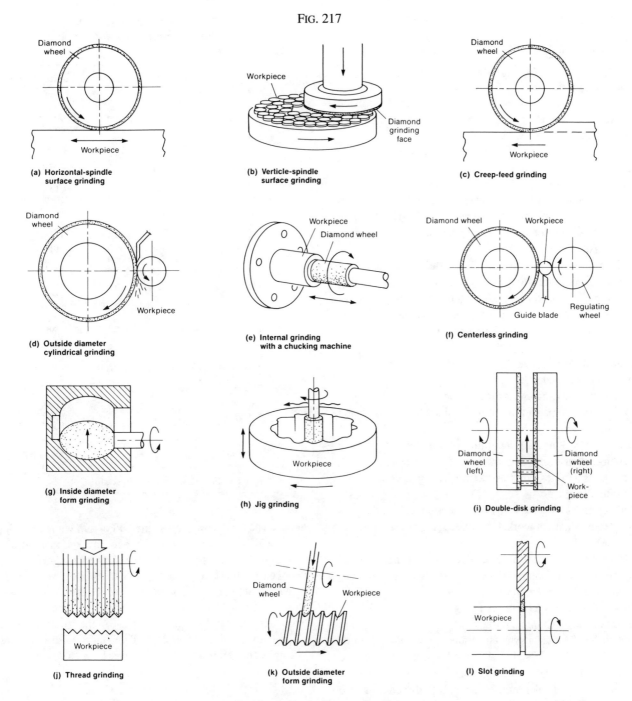

(a) Horizontal-spindle surface grinding

(b) Verticle-spindle surface grinding

(c) Creep-feed grinding

(d) Outside diameter cylindrical grinding

(e) Internal grinding with a chucking machine

(f) Centerless grinding

(g) Inside diameter form grinding

(h) Jig grinding

(i) Double-disk grinding

(j) Thread grinding

(k) Outside diameter form grinding

(l) Slot grinding

*Production grinding applications*

terial under specified grinding conditions, expressed in terms of volume of material removed per unit volume of wheel wear.

**grinding**. Removing material from a workpiece with a grinding wheel or abrasive belt (Fig. 217).

1. Surface grinding: Producing a flat surface with a rotating grinding wheel as the workpiece passes under.

2. Creep-feed grinding: A subset of surface grinding, creep-feed grinding produces deeper (full) depths of cut at slow traverse rates.

3. Cylindrical grinding: Grinding the outside diameters of cylindrical workpieces held between centers.

4. Internal grinding: Grinding the inside of a rotating workpiece by use of a wheel spindle which rotates and reciprocates through the length of depth of the hole being ground.

5. Centerless grinding: Grinding cylindrical surfaces without use of fixed centers to rotate the work. The work is supported and rotates between three fundamental machine components: the grinding wheel, the regulating wheel, and the work guide blade.

6. Gear (form) grinding: Removal of material to obtain correct gear tooth form by grinding. This is one of the more exact methods of finishing gears.

7. Thread grinding: Thread cutting by use of suitably formed grinding wheel.

**grinding burn**. See *burning* (2).

**grinding cracks**. Shallow cracks formed in the surfaces of relatively hard materials because of excessive grinding heat or the high sensitivity of the material. See also *grinding sensitivity*.

**grinding fluid**. An oil- or water-based fluid introduced into grinding operations to (1) reduce and transfer heat during grinding, (2) lubricate during chip formation, (3) wash loose chips or swarf from the grinding belt or wheel, and (4) chemically aid the grinding action or machine maintenance.

**grinding oil**. An oil-type grinding fluid; it may contain additives, but not water.

**grinding relief**. A groove or recess located at the boundary of a surface to permit the corner of the wheel to overhang during grinding.

**grinding sensitivity**. Susceptibility of a material to surface damage such as *grinding cracks*; it can be affected by such factors as hardness, microstructure, hydrogen content, and residual stress.

**grinding stress**. *Residual stress*, generated by grinding, in the surface layer of work. It may be tensile or compressive, or both.

**grinding wheel**. A cutting tool of circular shape made of abrasive grains bonded together. See also the figure accompanying the term *diamond wheel*.

**gripper dies**. The lateral or clamping dies used in a forging machine or mechanical upsetter.

**grit**. Crushed ferrous or synthetic abrasive material in various mesh sizes that is used in abrasive blasting equipment to clean castings. For materials used for grinding belts or grinding wheels, the term *abrasive* is preferred. See also *blasting or blast cleaning*.

**grit blasting**. *Abrasive blasting* with small irregular pieces of steel, malleable cast iron, or hard nonmetallic materials.

**grit size**. Nominal size of abrasive particles in a grinding wheel, corresponding to the number of openings per linear inch in a screen through which the particles can pass. Table 28 lists the mean particle sizes for various grit sizes.

**grizzly**. A set of parallel bars (or grating) used for coarse separation or screening of ores, rock or other material.

**Table 28  Mean particle sizes for grits used in conventional abrasive grinding wheels**

| Grit size | Particle size (mean) | |
|---|---|---|
|  | μm | in. |
| 4 | 6848 | 0.2577 |
| 6 | 5630 | 0.2117 |
| 8 | 4620 | 0.1817 |
| 10 | 3460 | 0.1366 |
| 12 | 2550 | 0.1003 |
| 14 | 2100 | 0.0830 |
| 16 | 1660 | 0.0655 |
| 20 | 1340 | 0.0528 |
| 24 | 1035 | 0.0408 |
| 30 | 930 | 0.0365 |
| 36 | 710 | 0.0280 |
| 46 | 508 | 0.0200 |
| 54 | 430 | 0.0170 |
| 60 | 406 | 0.0160 |
| 70 | 328 | 0.0131 |
| 80 | 266 | 0.0105 |
| 90 | 216 | 0.0085 |
| 100 | 173 | 0.0068 |
| 120 | 142 | 0.0056 |
| 150 | 122 | 0.0048 |
| 180 | 86 | 0.0034 |
| 220 | 66 | 0.0026 |
| 240 | 63 | 0.0024 |
| 280 | 44 | 0.0017 |
| 320 | 32 | 0.0012 |
| 400 | 23 | 0.0009 |
| 500 | 16 | 0.0006 |
| 600 | 8 | 0.0003 |
| 900 | 6 | 0.0002 |
| Levigated alumina | 3 | 0.0001 |

Note: Grit size varies indirectly with particle size.

**groove** (thermal spraying). A method of surface roughening in which grooves are made and the original surface roughened and spread. Also called rotary roughening.

**groove** (welding). An opening or channel in the surface of a part or between two components which provides space to contain a weld.

**groove angle**. The total included angle of the groove between parts to be joined (Fig. 218). Thus, the sum of two bevel angles, either or both of which may be zero degrees.

FIG. 218

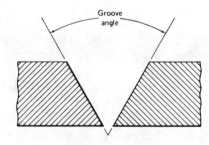

*Weld groove angle*

**groove face.** The portion of a surface or surfaces of a member included in a groove. See also the figure accompanying the term *root of joint*.

**groove radius.** The radius used to form the shape of a J- or U-groove weld joint.

**groove type.** The geometric configuration of a *groove*.

**groove weld.** A weld made in the groove between two members. The standard types are square, single-bevel, single flare-bevel, single flare-V, single-J, single-U, single-V, double-bevel, double flare-bevel, double flare-V, double-J, double-U and double-V. See also the figures accompanying the terms *double-bevel groove weld* and *single-bevel groove weld*.

**Grossmann number (*H*).** A ratio describing the ability of a quenching medium to extract heat from a hot steel workpiece in comparison to still water defined by the following equation:

$$H = \frac{h}{2k}$$

where $h$ is the heat transfer coefficient and $k$ is the conductivity of the metal.

**gross porosity.** In weld metal or in a casting, pores, gas holes or globular voids that are larger and in much greater numbers than those obtained in good practice.

**gross sample.** One or more increments of material taken from a larger quantity (lot) of material for assay or record purposes. Also termed bulk sample or lot sample. See also *increment* and *lot*.

**groundbed.** A buried item, such as junk steel or graphite rods (Fig. 219), that serves as the *anode* for the *cathodic protection* of pipelines or other buried structures. See also *deep groundbed*.

FIG. 219

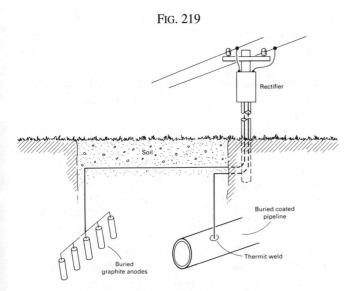

*Impressed-current groundbed cathodic protection of a buried pipeline using graphite anodes*

FIG. 220

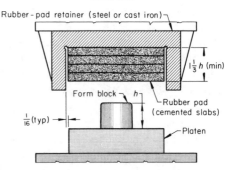

*Tooling and setup for rubber-pad forming by the Guerin process. Dimensions given in inches*

**ground connection.** In arc welding, a device used for attaching the work lead (ground cable) to the work.

**ground-support cable.** A cable construction, usually rugged and heavy, for use in control or power systems.

**group.** The specimens tested at one time, or consecutively, at one stress level. A group may comprise one or more specimens.

**growth (cast iron).** A permanent increase in the dimensions of cast iron resulting from repeated or prolonged heating at temperatures above 480 °C (900 °F) due either to graphitizing of carbides or oxidation.

**guard.** (1) A device, often made of sheet metal or wire screening, that prevents accidental contact with moving parts of machinery. (2) In electroplating, same as *robber*.

**Guerin process.** A *rubber-pad forming* process for forming sheet metal. The principal tools are the rubber pad and form block, or punch (Fig. 220).

**guide.** The parts of a drop hammer or press that guide the up-and-down motion of the ram in a true vertical direction.

**guide bearing.** A bearing used for positioning a slide, or for axial alignment of a long rotating shaft.

**guided bend.** The bend obtained by use of a plunger to force the specimen into a die in order to produce the desired contour of the outside and inside surfaces of the specimen (Fig. 221).

FIG. 221

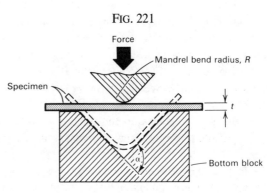

*Guided bend test setup using a closed V-block*

**guided bend test**. A test in which the specimen is bent to a definite shape by means of a punch (mandrel) and a bottom block (Fig. 221).

**guide mill**. A small hand mill with several stands in a train and with guides for the work at the entrance to the rolls.

**guide pin bushings**. Bushings, pressed into a die shoe, that allow the *guide pins* to enter in order to maintain punch-to-die alignment.

**guide pins**. Hardened, ground round pins or posts that maintain alignment between punch and die during die fabrication, set-up, operation, and storage. If the press slide is out of alignment, the guide pins cannot make the necessary correction unless heel plates are engaged before the pins enter the bushings. See also *heel block*.

**Guinier-Preston (G-P) zone**. A small precipitation domain in a supersaturated metallic solid solution. A G-P zone has no well-defined crystalline structure of its own and contains an abnormally high concentration of solute atoms. The formation of G-P zones constitutes the first stage of precipitation and is usually accompanied by a change in properties of the solid solution in which they occur.

**gum** (adhesives). Any of a class of colloidal substances exuded by or prepared from plants, sticky when moist, composed of complex carbohydrates and organic acids, which are soluble or swell in water. The term gum is sometimes used loosely to denote various materials that exhibit gummy characteristics under certain conditions, for example, gum balata, gum benzoin, and gum asphaltum. Gums are included by some in the category of natural resins. See also *adhesive, glue*, and *resin*.

**gum** (lubrication). In lubrication, a rubberlike, sticky deposit, black or dark brown in color, that results from the oxidation and/or polymerization of fuels and lubricating oils. Harder deposits are described as *lacquers* or *varnishes*.

**gun**. See preferred terms *arc welding gun, electron beam gun, resistance welding gun, soldering gun*, and *thermal spraying gun*.

**gun drill**. A drill, usually with one or more flutes and with coolant passages through the drill body, used for deep hole drilling. See also the figure accompanying the term *drill*.

**gun extension** (thermal spraying). An extension tube attached in front of a thermal spraying device to permit spraying within confined areas or deep recesses.

**gutter**. A depression around the periphery of a forging *die impression* outside the *flash pan* that allows space for the excess metal; surrounds the finishing impression and provides room for the excess metal used to ensure a sound forging. A shallow impression outside the parting line.

**gypsum**. The mineral consisting primarily of fully hydrated calcium sulfate, $CaSO_4 \cdot 2H_2O$ or calcium sulfate dihydrate. Most gypsum is ground for wallboard and plasters or used as a flux and fining agent in glass manufacture.

# H

**habit plane**. The plane or system of planes of a crystalline phase along which some phenomenon, such as twinning or transformation, occurs.

**hackle**. (1) A line on a glass crack surface, running parallel to the local direction of cracking, separating parallel but noncoplanar portions of the crack surface. (2) A finely structured fracture surface marking that gives a matte or roughened appearance to the surface, having varying degrees of coarseness. Finely structured hackle is variously known as fine hackle, frosted area, gray area, matte, mist, and stippled area. Coarsely structured hackle is also known as *striation*. See also *mist hackle, shear hackle, twist hackle*, and *wake hackle*.

**hackle marks**. Fine ridges on the fracture surface of the glass, parallel to the direction of propagation of the fracture.

**Hadfield steel**. See *austenitic manganese steel*.

**hair grease**. A grease containing horse hair or wool fiber.

**hairline crack**. See *flake*.

**hairline craze**. Multiple fine surface separation cracks in composites that exceed 6 mm (1/4 in.) in length and do not penetrate in depth the equivalent of a full ply of reinforcement. See also *crazing*.

**half cell**. An *electrode* immersed in a suitable *electrolyte*, designed for measurements of *electrode potential*.

**half hard**. A *temper* of nonferrous alloys and some ferrous alloys characterized by tensile strength about midway between those of *dead soft* and *full hard* tempers.

**half journal bearing**. A journal bearing extending 180° around a journal.

**half-life** ($t_{1/2}$). The time required for one half of an initial (large) number of atoms of a radioactive isotope to decay. Half-life is related to the decay constant $\lambda$ by the expression $t_{1/2} = \ln 2/\lambda$. See also *decay constant*.

**Hall effect**. The development of a transverse electric field in a current-carrying conductor placed in a magnetic field.

**Hall process**. A commercial process for winning aluminum

**Table 29  Capacities of various types of forging hammers**

| Type of hammer | Ram weight | | Maximum blow energy | | Impact speed | | Number of blows per minute |
|---|---|---|---|---|---|---|---|
| | kg | lb | kJ | ft·lb | m/s | ft/s | |
| Board drop ........... | 45-3400 | 100-7500 | 47.5 | 35 000 | 3-4.5 | 10-15 | 45-60 |
| Air or steam lift ....... | 225-7250 | 500-16 000 | 122 | 90 000 | 3.7-4.9 | 12-16 | 60 |
| Electrohydraulic drop .............. | 450-9980 | 1000-22 000 | 108.5 | 80 000 | 3-4.5 | 10-15 | 50-75 |
| Power drop ........... | 680-31 750 | 1500-70 000 | 1153 | 850 000 | 4.5-9 | 15-30 | 60-100 |
| Vertical counterblow ........ | 450-27 215 | 1000-60 000 | 1220 | 900 000 | 4.5-9 | 15-30 | 50-65 |

from alumina by electrolytic reduction of a fused bath of alumina dissolved in cryolite.

**halocarbon plastics.** Plastics based on resins made by the polymerization of monomers composed only of carbon and a halogen or halogens.

**halogen.** Any of the elements of the halogen family, consisting of fluorine, chlorine, bromine, iodine, and astatine.

**hammer.** A machine that applies a sharp blow to the work area through the fall of a ram onto an anvil. The ram can be driven by gravity or power. See also *gravity hammer* and *power-driven hammer*.

**hammer forging.** Forging in which the work is deformed by repeated blows. Table 29 lists the capacities of various types of hammers used for forging. Compare with *press forging*.

**hammering.** The working of metal sheet into a desired shape over a form or on a high-speed hammer and a similar anvil to produce the required dishing or thinning.

**hammer welding.** *Forge welding* by hammering.

**hand brake.** A small manual folding machine designed to bend sheet metal, similar in design and purpose to a *press brake*.

**hand forge (smith forge).** A forging operation in which forming is accomplished on dies that are generally flat. The piece is shaped roughly to the required contour with little or no lateral confinement; operations involving mandrels are included. The term hand forge refers to the operation performed, while hand forging applies to the part produced.

**hand lay-up.** In composites processing, the process of manually placing (and working) successive plies of reinforcing material or resin-impregnated reinforcement in position on a mold.

**handling breaks.** Irregular *breaks* caused by improper handling of metal sheets during processing. These breaks result from bending or sagging of the sheets during handling.

**handling strength.** A low level of strength initially obtained by an adhesive that allows specimens to be handled, moved, or unclamped without causing disruption of the curing process or affecting bond strength.

**hand straightening.** A straightening operation performed on a surface plate to bring a forging within straightness tolerance. A bottom die from a set of finish dies is often used

instead of a surface plate. Hand tools used include mallets, sledges, blocks, jacks, and oil gear presses in addition to regular inspection tools.

**Hansgirg process.** A process for producing magnesium by reduction of magnesium oxide with carbon.

**hard chromium.** Chromium electrodeposited for engineering purposes (such as to increase the wear resistance of sliding metal surfaces) rather than as a decorative coating. It is usually applied directly to basis metal and is customarily thicker (>1.2 μm or 0.05 mils) than a decorative deposit, but not necessarily harder.

**hard drawn.** An imprecise term applied to drawn products, such as wire and tubing, that indicates substantial cold reduction without subsequent annealing. Compare with *light drawn*.

**hard-drawn copper wire.** Copper wire that has been drawn to size and not annealed.

**hardenability.** The relative ability of a ferrous alloy to form martensite when quenched from a temperature above the upper critical temperature. Hardenability is commonly measured as the distance below a quenched surface at which the metal exhibits a specific hardness (50 HRC, for example) or a specific percentage of martensite in the microstructure.

**hardener** (metals). An alloy rich in one or more alloying elements that is added to a melt to permit closer control of composition than is possible by the addition of pure metals, or to introduce refractory elements not readily alloyed with the base metal. Sometimes called *master alloy* or *rich alloy*.

**hardener** (plastics). A substance or mixture added to a plastic composition to promote or control the curing action by taking part in it.

**hardening.** Increasing hardness of metals by suitable treatment, usually involving heating and cooling. When applicable, the following more specific terms should be used: *age hardening, case hardening, flame hardening, induction hardening, precipitation hardening* and *quench hardening*.

**hard face.** A seal facing of high hardness that is applied to a softer material, such as by flame spraying, plasma spraying, electroplating, nitriding, carburizing, or welding.

**hardfacing.** The application of a hard, wear-resistant material to the surface of a component by welding, spraying,

## Table 30  Hardfacing processing

| Process | Heat source | Mode of application | Hardfacing alloy form |
|---|---|---|---|
| Oxyfuel gas welding | Oxyfuel gas | Manual or automatic | Bare cast rods or powder |
| Shielded metal arc welding | Electric arc | Manual | Flux coated rods |
| Open arc welding | Electric arc | Semiautomatic | Flux cored tube wire |
| Gas tungsten arc welding | Inert gas shielded electric arc | Manual or automatic | Bare rods or wire |
| Submerged arc welding | Flux covered electric arc | Semiautomatic | Bare solid or tubular wire |
| Plasma transferred welding | Inert gas shielded plasma arc | Automatic | Powder, hot wire |
| Plasma arc welding | Inert gas shielded plasma arc | Manual or automatic | Same as GTAW |
| Spray and fuse | Oxyfuel gas | Manual | Powder |
| Plasma spray | Plasma arc | Manual or automatic | Powder |
| Detonation gun | Oxyacetylene detonation | Automatic | Powder |

or allied welding processes to reduce wear or loss of material by abrasion, impact, erosion, galling, and cavitation. Table 30 lists various types of hardfacing processes commonly used. See also *surfacing*.

**hardfacing alloys**. Wear-resistant materials available as bare welding rod, flux-coated rod, long-length solid wires, long-length tubular wires, or powders that are deposited by *hardfacing*. Hardfacing materials include a wide variety of alloys, ceramics, and combinations of these materials. Conventional hardfacing alloys are normally classified as steels or low-alloy ferrous materials, chromium white irons, high-alloy ferrous materials, carbides, nickel-base alloys, or cobalt-base alloys. Table 31 lists selection criteria for hardfacing materials.

**hard head**. A hard, brittle, white residue obtained in refining of tin by liquation, containing, among other things, tin, iron, arsenic and copper. Also, a refractory lump of ore only partly smelted.

**hard metal**. A collective term that designates a sintered material with high hardness, strength, and wear resistance, and is characterized by a tough metallic binder phase and particles of carbides, borides, or nitrides of the refractory metals. The term is in general use abroad, while for the carbides the term *cemented carbides* (Technical Brief 9) is preferred in the U.S., and the boride and nitride materials are usually categorized as *cermets* (Technical Brief 11).

**hardness**. A measure of the resistance of a material to surface indentation or abrasion; may be thought of as a function of the stress required to produce some specified type of surface deformation. There is no absolute scale for hardness; therefore, to express hardness quantitatively, each type of test has its own scale of arbitrarily defined hardness. Indentation hardness can be measured by *Brinell, Rockwell, Vickers, Knoop*, and *Scleroscope hardness tests*.

**hardness profile**. Hardness as a function of distance from a fixed reference point (usually from the surface).

**hard solder**. A term erroneously used to denote silver-base brazing filler metals.

## Table 31  Hardfacing alloy selection

| Hardfacing materials | Service conditions |
|---|---|
| Hypoeutectic cobalt-based Alloy No. 1, Laves phase alloys | Metal-to-metal sliding, high-contact stresses |
| Low-alloy hardfacing steels | Metal-to-metal sliding, low-contact stresses |
| Most cobalt-based alloys or nickel-based alloys depending on corrosive environment | Metal-to-metal sliding in combination with corrosion or oxidation |
| High-alloy cast irons | Low-stress abrasion Particle impingement Erosion at low angles |
| Carbides | Low-stress severe abrasion Cutting edge retention |
| Cobalt-based alloys | Cavitation erosion |
| High-alloy manganese steels | Heavy impact |
| Hypoeutectic cobalt-based alloys | Heavy impact with corrosion, oxidation |
| Hypoeutectic cobalt-based Alloys No. 21 and 6, cobalt-based Laves phase alloys | Galling |
| Austenitic manganese steels | Gouging abrasion |
| Cobalt-based alloys and nickel-based carbide-type alloys | Thermal stability Creep resistance at elevated temperatures ($>1000$ °F) |

**hard surfacing**. See preferred terms *surfacing* or *hardfacing*.

**hard temper**. Same as *full hard* temper.

**hard water**. Water that contains certain salts, such as those of calcium or magnesium, which form insoluble deposits in boilers and form precipitates with soap.

**Haring cell**. A four-electrode cell for measurement of electrolyte resistance and electrode polarization during electrolysis.

**harness satin.** A fabric weaving pattern producing a satin appearance. See also *eight-harness satin* and *four-harness satin*.

**Hartmann lines.** See *Lüders lines*.

**haze.** Cloudy appearance under or on the surface of a plastic, not describable by the terms *chalking* or *bloom*.

**H-band steel.** Carbon, carbon-boron, or alloy steel produced to specified limits of hardenability; the chemical composition range may be slightly different from that of the corresponding grade of ordinary carbon or alloy steel.

**HDPE.** See *high-density polyethylene* (Technical Brief 19).

**header.** See *upsetter*.

**heading.** The *upsetting* of wire, rod, or bar stock in dies to form parts that usually contain portions that are greater in cross-sectional area than the original wire, rod, or bar.

**head-to-head.** On a polymer chain, a type of configuration in which the functional groups are on adjacent carbon atoms.

**head-to-tail.** On a polymer chain, a type of configuration in which the functional groups on adjacent polymers are as far apart as possible.

**healed-over scratch.** A scratch in a metallic object that occurred in an earlier mill operation and was partially masked in subsequent rolling. It may open up during forming.

**hearth.** The bottom portions of certain furnaces, such as blast furnaces, air furnaces, and other reverberatory furnaces, that support the charge and sometimes collect and hold molten metal.

**heat.** A stated tonnage of metal obtained from a period of continuous melting in a cupola or furnace, or the melting period required to handle this tonnage.

**heat-activated adhesive.** A dry adhesive that is rendered tacky or fluid by application of heat, or heat and pressure, to the assembly.

**heat-affected zone (HAZ).** That portion of the base metal that was not melted during brazing, cutting, or welding,

FIG. 222

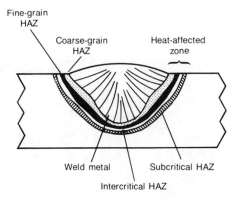

*Various regions of a weld heat-affected zone*

but whose microstructure and mechanical properties were altered by the heat (Fig. 222).

**heat buildup.** In processing of plastics, the rise in temperature in a part resulting from the dissipation of applied strain energy as heat or from applied mold cure heat. See also *hysteresis*.

**heat check.** A pattern of parallel surface cracks that are formed by alternate rapid heating and cooling of the extreme surface metal, sometimes found on forging dies and piercing punches. There may be two sets of parallel cracks, one set perpendicular to the other.

**heat checking.** A process in which fine cracks are formed on the surface of a body in sliding contact due to the buildup of excessive frictional heat.

**heat cleaned.** A condition in which glass or other fibers are exposed to elevated temperatures to remove preliminary sizings or binders not compatible with the resin system to be applied.

**heat-deflection temperature.** The temperature at which a standard plastic test bar deflects a specified amount under a stated load. Now called deflection temperature under load (DTUL).

**heat-disposable pattern.** In foundry practice, a *pattern* formed from a wax- or plastic-base material that is melted from the mold cavity by the application of heat.

**heat distortion.** Distortion or flow of a material or configuration due to the application of heat.

**heat distortion point.** The temperature at which a standard plastic test bar deflects a specified amount under a stated load. Now called deflection temperature.

**heat-fail temperature.** The temperature at which delamination of an adhesively bonded structure occurs under static loading in shear.

**heat forming.** See *thermoforming*.

**heat mark.** Extremely shallow depression or groove in the surface of a plastic visible because of a sharply defined rim or a roughened surface. See also *sinkmark*.

**heat of fusion.** See *latent heat of fusion*.

**heat resistance.** The property or ability of materials to resist the deteriorating effects of elevated temperatures.

**heat-resistant alloy.** An alloy developed for very-high-temperature service where relatively high stresses (tensile, thermal, vibratory, or shock) are encountered and where oxidation resistance is frequently required. Examples include *refractory metals* (Technical Brief 38), chromium-molybdenum pressure vessel steels, *stainless steels* (Technical Brief 47), and *superalloys* (Technical Brief 52).

**heat sealing.** A method of joining plastic films by simultaneous application of heat and pressure to areas in contact.

**heat-sealing adhesive.** A thermoplastic film adhesive that is melted between the adherend surfaces by heat application to one or both of the surfaces.

**heat shock**. A test to determine the stability of a material by sudden exposure to a significantly higher or lower temperature for a short period of time.

**heat sink**. A material that absorbs or transfers heat away from a critical element or part.

**heat time**. In resistance welding, the time that the current flows during any one impulse.

**heat tinting**. Coloration of a metal surface through oxidation by heating to reveal details of the microstructure.

**heat transfer**. Flow of heat by conduction, convection, or radiation.

**heat treatable alloy**. An alloy that can be hardened by heat treatment.

**heat treating film**. A thin coating or film, usually an oxide, formed on the surface of a metal during heat treatment.

**heat treatment**. Heating and cooling a solid metal or alloy in such a way as to obtain desired conditions or properties. Heating for the sole purpose of hot working is excluded from the meaning of this definition.

**heavy-duty oil**. An oil that is stable against oxidation, protects bearings from corrosion, and has detergent and dispersant properties. Heavy-duty oils are suitable for use in gasoline and diesel engines.

**heavy metal**. A sintered tungsten alloy with nickel, copper, and/or iron, the tungsten content being at least 90 wt% and the density being at least 16.8 $g/cm^3$.

**heel**. Synonymous with *base* (1).

**heel block**. A block or plate usually mounted on or attached to a lower die in a forming or forging press that serves to prevent or minimize the deflection of punches or cams.

**helical winding**. In filament-wound items, a winding in which a filament band advances along a helical path, not necessarily at a constant angle, except in the case of a cylinder (Fig. 223).

**hematite**. (1) An iron mineral crystallizing in the rhombohedral system; the most important ore of iron. (2) An iron oxide, $Fe_2O_3$, corresponding to an iron content of approximately 70%. Also known as red hematite, red iron ore, and rhombohedral iron ore.

FIG. 223

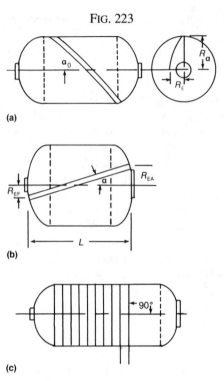

(a)

(b)

(c)

*Basic filament winding modes. (a) Helical. (b) Polar. (c) Hoop*

**hemming**. A bend of 180° made in two steps. First, a sharp-angle bend is made; next the bend is closed using a flat punch and a die (Fig. 224).

**henry**. An electrical unit denoting the inductance of a circuit in which a current varying at the rate of one ampere per second produces an electromotive force of one volt.

**HERF**. A common abbreviation for *high-energy-rate forging* or *high-energy-rate forming*.

**hermetic**. Sealed so that the object is gastight. The test for hermeticity is to fill the object with a test gas, often helium, and observe leak rates when the object is placed in a vacuum. Plastic encapsulation is not hermetic because it allows permeation by gases.

FIG. 224

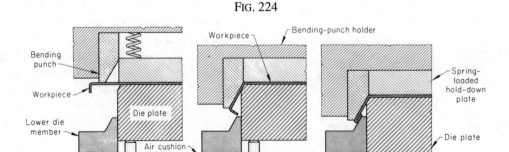

*Compound flanging and hemming die without horizontal motion*

**herringbone bearing**. Any plain, sleeve, or thrust bearing with herringbone-shaped oil grooves.

**herringbone pattern**. Same as *chevron pattern*.

**hertz**. A designation of electrical frequency that denotes cycles per second. Abbreviated Hz.

**hertzian cone crack**. See *percussion cone*.

**Hertzian contact area**. (1) The contact area (also, diameter or radius of contact) between two bodies calculated according to Hertz's equations of elastic deformation. (2) The apparent area of contact between two nonconforming solid bodies pressed against each other, as calculated from Hertz's equations of elastic deformation.

**Hertzian contact pressure**. (1) The pressure at a contact between two solid bodies calculated according to Hertz's equations of elastic deformation. (2) The magnitude of the pressure at any specified location in a Hertzian contact area, as calculated from Hertz's equations of elastic deformation.

**Hertzian stress**. See *contact stress*.

**heterogeneity**. The degree of nonuniformity of composition or properties. Contrast with *homogeneity*.

**heterogeneous**. Of a body of material or matter, comprised of more than one phase (solid, liquid, and gas) separated by boundaries; similarly of a solid, comprised of more than one chemical, crystalline, and/or glassy species, separated by boundaries.

**heterogeneous equilibrium**. In a chemical system, a state of dynamic balance among two or more homogeneous phases capable of stable coexistence in mutual or sequential contact.

**heterogeneous nucleation**. In the crystallization of polymers, the growth of crystals on vessel surfaces, dust, or added nucleating agents.

**Heyn stresses**. Same as *microscopic stresses*.

**hexa**. An abbreviated form of hexamethylenetetramine, a source of reactive methylene for curing *novolacs*.

**hexagonal** (lattices for crystals). Having two equal coplanar axes, $a_1$ and $a_2$, at 120° to each other and a third axis, $c$, at right angles to the other two; $c$ may or may not equal $a_1$ and $a_2$.

**hexagonal close-packed**. (1) A structure containing two atoms per unit cell located at (0, 0, 0) and ($1/3$, $2/3$, $1/2$) or ($2/3$, $1/3$, $1/2$). (2) One of the two ways in which spherical objects can be most closely packed together so that the close-packed planes are alternately staggered in the order A-B-A-B-A-B. See also the figure accompanying the term *unit cell*.

**high aluminum defect**. An α-stabilized region in titanium containing an abnormally large amount of aluminum that may span a large number of β grains (Fig. 225). It contains an inordinate fraction of primary α, but has a microhardness only slightly higher than the adjacent matrix. Also termed type II defects.

**high-conductivity copper**. Copper that, in the annealed condition, has a minimum electrical conductivity of 100% *IACS* as determined by ASTM test methods.

**high-cycle fatigue**. *Fatigue* that occurs at relatively large numbers of cycles. The arbitrary, but commonly accepted, dividing line between high-cycle fatigue and *low-cycle fatigue* is considered to be about $10^4$ to $10^5$ cycles. In practice, this distinction is made by determining whether the dominant component of the *strain* imposed during cyclic loading is elastic (high cycle) or plastic (low cycle), which in turn depends on the properties of the metal and on the magnitude of the nominal *stress*.

**high-density polyethylene (HDPE)**. See *Technical Brief 19*.

**high-energy-rate compacting**. Compacting of a powder at a very rapid rate by the use of explosives in a closed die.

Fig. 225

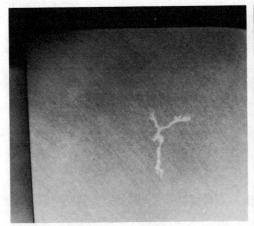 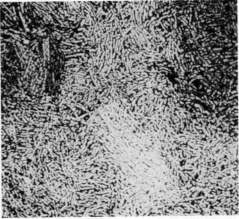

*Titanium alloy billet illustrating (left) the macroscopic (1.25×) appearance of a high aluminum defect and (right) the high volume fraction of elongated α in the area of high aluminum content (50×)*

## Technical Brief 19: High-Density Polyethylenes

HIGH-DENSITY POLYETHYLENES (HDPEs) are thermoplastic materials that are solid in their natural state. Under extrusion conditions of heat, pressure, and mechanical shear, they soften into a highly viscous, molten mass and take the shape of the desired end product. The polymer is characterized by its opacity, chemical inertness, toughness at both low and high temperatures, and moisture barrier and electrical-insulating properties.

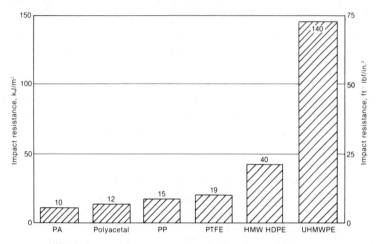

*Comparative impact resistance of HMW HDPEs and UHMWPEs with polyamide (PA), polyacetal, reinforced polypropylene (PP), and polytetrafluoroethylene (PTFE)*

The term HDPE generally includes polyethylene ranging in density from about 0.94 to 0.965 $g/cm^3$. While molecules in low-density polyethylene are branched and linked randomly, those in the higher-density polyethylene are linked in longer chains with fewer side branches, resulting in a more rigid material with greater strength, hardness, and chemical resistance, and higher softening temperature.

The physical properties of HDPE are also affected by the weight-average molecular weight (MW) of the polymer. As the MW increases, the mechanical properties also increase significantly, but the polymer becomes more difficult to process. Polymer grades with MW in the 200,000 to 500,000 range are considered high-performance, high-molecular-weight HDPEs (HMW HDPEs). The combination of high molecular weight and high density provides even higher stiffness and abrasion resistance and extended product service life in critical environmental applications. High-molecular-weight resins also provide excellent environmental stress-corrosion cracking resistance. Polyethylenes with a MW ten times that of HMW HDPE are also available. These materials, referred to as ultrahigh-molecular-weight polyethylenes (UHMWPEs), have the highest abrasion resistance and highest impact strength of any plastic.

### High-molecular-weight pipe applications

| Applications | Typical diameter mm | Typical diameter in. |
|---|---|---|
| Gas distribution | 13–200 | 0.5–8 |
| Oil and gas recovery | 50–910 | 2–36 |
| Domestic water supply | 13–50 | 0.5–2 |
| Sewer and sewer rehab lining | 150–1220 | 6–48 |
| Industrial and mining use | 50–1600 | 2–63 |
| Fiber-optic innerduct | 25–50 | 1–2 |

The four major end-use markets of HMW HDPE are pipe, large-part blow molding, film, and sheet. High-molecular-weight pipe is used in applications that require resistance to environmental cracking and excellent impact resistance at temperatures as low as –50 °C (–60 °F). The major application areas for HMW HDPE blow-molded parts are shipping containers in sizes up to 210 L (55 gal), refuse containers, and bulk storage containers (1040 L, or 275 gal). The largest of the film applications are grocery bags and garbage can liners. Typical applications for HMW sheet include truck bed liners, shipping pallets, and pond liners.

### Recommended Reading

- J.R. Bourgeois and P.W. Blackett, High-Density Polyethylenes (HDPE), *Engineered Materials Handbook*, Vol 2, ASM International, 1988, p 163-166
- H.L. Stein, Ultrahigh Molecular Weight Polyethylenes (UHMWPE), *Engineered Materials Handbook*, Vol 2, ASM International, 1988, p 167-171

FIG. 226

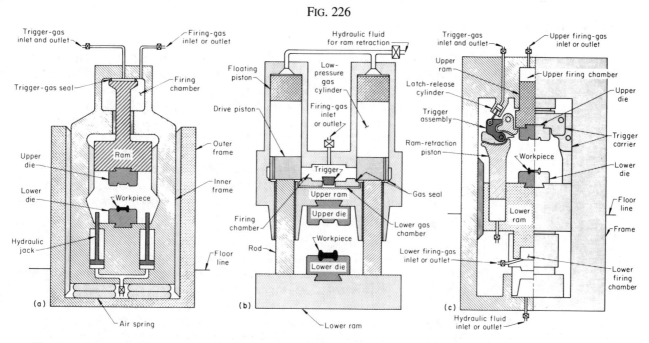

*The three basic machine concepts for high-energy-rate forging. (a) Ram and inner frame machine. (b) Two-ram machine. (c) Controlled-energy-flow machine. Triggering and expansion of the gas in the firing chamber cause the upper and lower rams to move toward eachother at high speed. An outer frame provides guiding surfaces for the rams.*

**high-energy-rate forging (HERF).** A closed-die hot- or cold-forging process in which the stored energy of high-pressure gas is used to accelerate a ram to unusually high velocities in order to effect deformation of the workpiece (Fig. 226). Ideally, the final configuration of the forging is developed in one blow or, at most, a few blows. In high-energy-rate forging, the velocity of the ram, rather than its mass, generates the major forging force. Table 32 compares the die closing speeds of HERF machines with conventional forging methods. Also known as HERF processing, high-velocity forging, and high-speed forging.

**high-energy-rate forming.** A group of forming processes that applies a high rate of strain to the material being formed through the application of high rates of energy transfer. See also *explosive forming, high-energy-rate forging,* and *electromagnetic forming.*

**high-frequency heating.** The heating of materials by dielectric loss in a high-frequency electrostatic field. The material is exposed between electrodes and is heated quickly and uniformly by absorption of energy from the electrical field.

**high-frequency resistance welding.** A resistance welding process that produces coalescence of metals with the heat generated from the resistance of the workpieces to a high-frequency alternating current in the 10 to 500 kHz range and the rapid application of an upsetting force after heating

FIG. 227

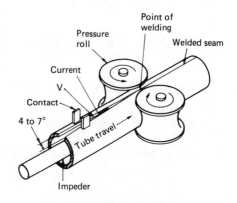

*High-frequency longitudinal butt seam welding of tube*

### Table 32  Forming machines and their die closing speeds

| Press type | Impact speed | |
| --- | --- | --- |
| | m/s | ft/s |
| Hydraulic press | 0.27-0.456 | 0.89-1.50 |
| Crank press | 0.03-1.52 | 0.10-4.99 |
| Toggle press | 0.03-1.52 | 0.10-4.99 |
| Friction screw press | 0.30-1.21 | 0.98-3.97 |
| Drop hammer | 3.65-5.50 | 12.0-18.0 |
| Power hammer | 4.50-9.10 | 14.8-29.9 |
| HERF machine | 5.00-22.0 | 16.0-72.2 |

## Technical Brief 20: High-Impact Polystyrenes

HIGH-IMPACT POLYSTYRENES (HIPS) are thermoplastic resins produced by dissolving polybutadiene rubber in a styrene monomer before polymerizing. Polystyrene (PS) forms the continuous phase, with the rubber phase existing as discrete particles having occlusions of PS. Different production techniques allow this rubber phase

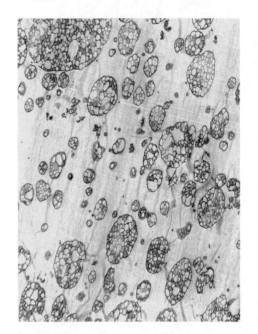

*Morphology of typical commercial HIPS. 5000×*

to be tailored to a wide range of properties. Advances in rubber morphology, control of molecular weight distribution, and additives technology have enabled producers to offer a wide range of standard and specialty HIPS products, which offer good dimensional stability and low-temperature impact properties, and high rigidity. Relative disadvantages of HIPS are their poor high-temperature properties and lower chemical resistance compared with most crystalline polymers.

High-impact polystyrenes are used in myriad applications and industries because of their ease of processing, performance, and low cost. Major industries and markets include packaging and disposables, appliances and consumer electronics, toys and recreation, buildings, and furnishings. In recent years, product development has focused on specialty products. Grades are now available that provide improved resistance to stress cracking from fats, oils, and chlorofluorocarbon blowing agents. Also available are high-gloss/high-toughness, ignition-resistant, glass-filled (up to 40 wt% type E glass), and very-high-impact products.

All of the conventional processing technologies for thermoplastics can be used on HIPS. These include injection molding, structural-foam molding, extrusion, thermoforming, and injection blow molding.

### Recommended Reading

- A. Poloso and M.B. Bradley, High-Impact Polystyrenes (PS, HIPS), *Engineered Materials Handbook*, Vol 2, ASM International, 1988, p 194-199

### Typical properties of specialty HIPS
Representative properties of injection-molded specimens, based on ASTM tests

| Property | Ignition-resistant HIPS | Very-high-impact PS | High-gloss, high-impact PS | Glass-filled HIPS (10% type E glass) |
|---|---|---|---|---|
| **Mechanical** | | | | |
| Melt flow rate, condition G, g/10 min .... | 5.0 | 2.5 | 4.5 | 2.2 |
| Tensile strength at yield, MPa (ksi) ...... | 26.9 (3.9) | 19.3 (2.8) | 23.4 (3.4) | 39.3 (5.7) |
| Tensile strength at break, MPa (ksi)...... | 25.5 (3.7) | 20.7 (3.0) | 24.8 (3.6) | 39.3 (5.7) |
| Tensile modulus, GPa ($10^6$ psi) .......... | 1.79 (0.260) | 1.65 (0.240) | 1.72 (0.250) | 2.14 (0.310) |
| Elongation at break, % ................. | 40 | 50 | 35 | 4.0 |
| Flexural strength, MPa (ksi)............. | 40.0 (5.8) | 33.1 (4.8) | 42.7 (6.2) | 57.9 (8.4) |
| Flexural modulus, GPa ($10^6$ psi) ......... | 2.1 (0.300) | 1.86 (0.270) | 1.93 (0.280) | 3.19 (0.463) |
| Izod impact strength, notched, 23 °C (73 °F), J/m (ft lbf/in.) ................. | 107 (2.0) | 214 (4.0) | 133 (2.5) | 133 (2.5) |
| **Thermal** | | | | |
| Vicat softening temperature, °C (°F) ..... | 100 (212) | 101 (215) | 101 (215) | 111 (230) |
| Heat-deflection temperature under load, at 1.82 MPa (0.264 ksi), °C (°F)....... | 87 (190) | 87 (190) | 85 (185) | 92 (200) |
| Density at 23 °C (73 °F), g/cm³ .......... | 1.15 | 1.0 | ··· | 1.13 |
| Flammability, UL 94 rating ............. | V-0 | ··· | ··· | ··· |
| Gardner gloss at 20 °C (68 °F)........... | ··· | ··· | 85 | ··· |

FIG. 228

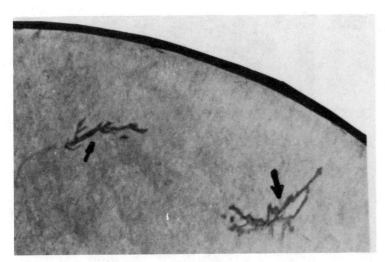 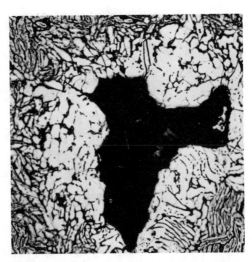

*Titanium alloy billet illustrating (left) macroscopic appearance of high interstitial defects and (right) porosity (black void) normally associated with this type of defect. 25×*

is substantially completed (Fig. 227). The path of the current in the workpiece is controlled by use of the proximity effect (the feed current follows closely the return current conductor).

**high-impact polystyrenes (HIPS).** See *Technical Brief 20*.

**high interstitial defect.** Interstitially stabilized α-phase region in titanium of substantially higher hardness than surrounding material. It arises from very high local nitrogen or oxygen concentrations that increase the β transus and produce the high-hardness, often brittle α phase. Such a defect is often accompanied by a void resulting from thermomechanical working (Fig. 228). Also termed type I or low-density interstitial defects, although they are not necessarily low density.

**highlighting.** Buffing or polishing selected areas of a complex shape to increase the luster or change the color of those areas.

**highly deformed layer.** In tribology, a layer of severely plastically deformed material that results from the shear stresses imposed on that region during sliding contact. See also *Beilby layer* and *white layer*.

**high polymer.** A macromolecular substance that, as indicated by the polymer by which it is identified, consists of molecules that are multiples of the low molecular unit and have a molecular weight of at least 20,000.

**high-pressure laminates.** Laminates molded and cured at pressures not lower than 6.9 MPa (1.0 ksi), and more commonly in the range of 8.3 to 13.8 MPa (1.2 to 2.0 ksi).

**high-pressure molding.** A plastic molding or laminating process in which the pressure used is greater than 1400 kPa (200 psi), but commonly 7000 kPa (1000 psi).

**high-pressure spot.** See *resin-starved area*.

**high pulse current.** In welding, current levels during the high pulse time which produces the high heat level.

**high pulse time.** In welding, the duration of the high current pulse time.

**high residual phosphorus copper.** Deoxidized copper with residual phosphorus present in amounts (usually 0.013 to 0.04%) generally sufficient to decrease appreciably the conductivity of the copper.

**high-speed machining.** High-productivity machining processes which achieve cutting speeds in excess of 600 m/min (2000 sfm) and up to 18,000 m/min (60,000 sfm). Such speeds result in segmented shear-localized chips rather than the continuous chip formation associated with lower-speed machining processes (Fig. 229).

**high-strength low-alloy (HSLA) steels.** Steels designed to provide better mechanical properties and/or greater resistance to atmospheric corrosion than conventional carbon steels. They are not considered to be alloy steels in the normal sense because they are designed to meet specific mechanical properties rather than a chemical composition (HSLA steels have yield strengths greater than 275 MPa, or 40 ksi). The chemical composition of a specific HSLA steel may vary for different product thicknesses to meet mechanical property requirements. The HSLA steels have low carbon contents (0.05 to ~0.25% C) in order to produce adequate formability and weldability, and they have manganese contents up to 2.0%. Small quantities of chromium, nickel, molybdenum, copper, nitrogen, vanadium, niobium, titanium, and zirconium are used in various combinations. The types of HSLA steels commonly used include:

1. Weathering steels, designed to exhibit superior atmospheric corrosion resistance.

FIG. 229

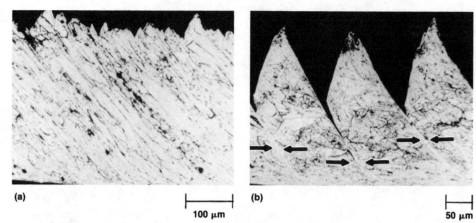

*Examples of (a) continuous chips produced by conventional machining and (b) segmented chip formation associated with high-speed machining. Arrows indicate areas of shear localization.*

2. Control-rolled steels, hot rolled according to a predetermined rolling schedule designed to develop a highly deformed austenite structure that will transform to a very fine equiaxed ferrite structure on cooling.

3. Pearlite-reduced steels, strengthened by very fine-grain ferrite and precipitation hardening but with low carbon content and therefore little or no pearlite in the microstructure.

4. Microalloyed steels, with very small additions (generally <0.10% each) of such elements as niobium, vanadium, and/or titanium for refinement of grain size and/or precipitation hardening.

5. Acicular ferrite steel, very low carbon steels with sufficient hardenability to transform on cooling to a very fine high-strength acicular ferrite (low-carbon bainite) structure rather than the usual polygonal ferrite structure.

6. Dual-phase steels, processed to a microstructure of ferrite containing small, uniformly distributed regions of high-carbon martensite, resulting in a product with low yield strength and a high rate of work hardening, thus providing a high-strength steel of superior formability.

Table 33 lists chemical compositions for various HSLA steels.

**high-stress abrasion.** A form of abrasion in which relatively large cutting forces are imposed on the particles or protuberances causing the abrasion, and that produces significant cutting and deformation of the wearing surface. In metals, high-stress abrasion can result in significant surface strain hardening. This form of abrasion is common in mining and agricultural equipment, and in highly loaded bearings where hard particles are trapped between mating surfaces. See also *low-stress abrasion*.

**high-temperature combustion.** An analytical technique for determining the concentrations of carbon and sulfur in samples. The sample is burned in a graphite crucible in the presence of oxygen, which causes carbon and sulfur to leave the sample as carbon dioxide and sulfur dioxide. These gases are then detected by infrared or thermal conductive means.

**high-temperature hydrogen attack.** A loss of strength and ductility of steel by high-temperature reaction of absorbed hydrogen with carbides in the steel resulting in *decarburization* and internal fissuring.

**hindered contraction.** Contraction where the shape will not permit a metal casting to contract in certain regions in keeping with the coefficient of expansion.

**hinge stress.** The tensile component of the bending stress generated on the same surface of a glass section, but not displaced from the site of a locally impinging force.

**HIP.** See *hot isostatic pressing*.

**histogram.** A plot of frequency of occurrence versus the measured parameter (Fig. 230).

FIG. 230

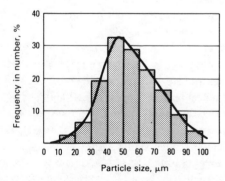

*Histogram and size frequency curve for log-normal powder particle size distribution*

**Table 33  Compositional limits for HSLA steel grades described in ASTM specifications**

| ASTM specification(a) | Type or grade | UNS designation | Heat compositional limits, %(b) | | | | | | | | | |
|---|---|---|---|---|---|---|---|---|---|---|---|---|
| | | | C | Mn | P | S | Si | Cr | Ni | Cu | V | Other |
| A 242 | Type 1 | K11510 | 0.15 | 1.00 | 0.45 | 0.05 | · · · | · · · | · · · | 0.20 min | · · · | · · · |
| A 572 | Grade 42 | · · · | 0.21 | 1.35(c) | 0.04 | 0.05 | 0.30(c) | · · · | · · · | 0.20 min(d) | · · · | (e) |
| | Grade 50 | · · · | 0.23 | 1.35(c) | 0.04 | 0.05 | 0.30(c) | · · · | · · · | 0.20 min(d) | · · · | (e) |
| | Grade 60 | · · · | 0.26 | 1.35(c) | 0.04 | 0.05 | 0.30 | · · · | · · · | 0.20 min(d) | · · · | (e) |
| | Grade 65 | · · · | 0.23(c) | 1.65(c) | 0.04 | 0.05 | 0.30 | · · · | · · · | 0.20 min(d) | · · · | (e) |
| A 588 | Grade A | K11430 | 0.10–0.19 | 0.90–1.25 | 0.04 | 0.05 | 0.15–0.30 | 0.40–0.65 | · · · | 0.25–0.40 | 0.02–0.10 | · · · |
| | Grade B | K12043 | 0.20 | 0.75–1.25 | 0.04 | 0.05 | 0.15–0.30 | 0.40–0.70 | 0.25–0.50 | 0.20–0.40 | 0.01–0.10 | · · · |
| | Grade C | K11538 | 0.15 | 0.80–1.35 | 0.04 | 0.05 | 0.15–0.30 | 0.30–0.50 | 0.25–0.50 | 0.20–0.50 | 0.01–0.10 | · · · |
| | Grade D | K11552 | 0.10–0.20 | 0.75–1.25 | 0.04 | 0.05 | 0.50–0.90 | 0.50–0.90 | · · · | 0.30 | · · · | 0.04 Nb, 0.05–0.15 Zr |
| | Grade K | · · · | 0.17 | 0.5–1.20 | 0.04 | 0.05 | 0.25–0.50 | 0.40–0.70 | 0.40 | 0.30–0.50 | · · · | 0.10 Mo, 0.005–0.05 Nb |
| A 606 | · · · | · · · | 0.22 | 1.25 | · · · | 0.05 | · · · | · · · | · · · | · · · | · · · | · · · |
| A 607 | Grade 45 | · · · | 0.22 | 1.35 | 0.04 | 0.05 | · · · | · · · | · · · | 0.20 min(d) | · · · | (e) |
| | Grade 50 | · · · | 0.23 | 1.35 | 0.04 | 0.05 | · · · | · · · | · · · | 0.20 min(d) | · · · | (e) |
| | Grade 55 | · · · | 0.25 | 1.35 | 0.04 | 0.05 | · · · | · · · | · · · | 0.20 min(d) | · · · | (e) |
| | Grade 60 | · · · | 0.26 | 1.50 | 0.04 | 0.05 | · · · | · · · | · · · | 0.20 min(d) | · · · | (e) |
| | Grade 65 | · · · | 0.26 | 1.50 | 0.04 | 0.05 | · · · | · · · | · · · | 0.20 min(d) | · · · | (e) |
| | Grade 70 | · · · | 0.26 | 1.65 | 0.04 | 0.05 | · · · | · · · | · · · | 0.20 min(d) | · · · | (e) |
| A 618 | Grade Ia | · · · | 0.15 | 1.00 | 0.15 | 0.05 | · · · | · · · | · · · | 0.20 min | · · · | · · · |
| | Grade Ib | · · · | 0.20 | 1.35 | 0.04 | 0.05 | · · · | · · · | · · · | 0.20 min(f) | · · · | · · · |
| | Grade II | K12609 | 0.22 | 0.85–1.25 | 0.04 | 0.05 | 0.30 | · · · | · · · | · · · | 0.02 min | · · · |
| | Grade III | K12700 | 0.23 | 1.35 | 0.04 | 0.05 | 0.30 | · · · | · · · | · · · | 0.02 min | 0.005 Nb min(g) |
| A 633 | Grade A | K01802 | 0.18 | 1.00–1.35 | 0.04 | 0.05 | 0.15–0.30 | · · · | · · · | · · · | · · · | 0.05 Nb |
| | Grade C | K12000 | 0.20 | 1.15–1.50 | 0.04 | 0.05 | 0.15–0.50 | · · · | · · · | · · · | · · · | 0.01–0.05 Nb |
| | Grade D | K02003 | 0.20 | 0.70–1.60(c) | 0.04 | 0.05 | 0.15–0.50 | 0.25 | 0.25 | 0.35 | · · · | 0.08 Mo |
| | Grade E | K12202 | 0.22 | 1.15–1.50 | 0.04 | 0.05 | 0.15–0.50 | · · · | · · · | · · · | 0.04–0.11 | 0.01–0.05 Nb(d), 0.01–0.03 N |
| A 656 | Type 3 | · · · | 0.18 | 1.65 | 0.025 | 0.035 | 0.60 | · · · | · · · | · · · | 0.08 | 0.020 N, 0.005–0.15 Nb |
| | Type 7 | · · · | 0.18 | 1.65 | 0.025 | 0.035 | 0.60 | · · · | · · · | · · · | 0.005–0.15 | 0.020 N, 0.005–0.10 Nb |
| A 690 | · · · | K12249 | 0.22 | 0.60–0.90 | 0.08–0.15 | 0.05 | 0.10 | · · · | 0.40–0.75 | 0.50 min | · · · | · · · |
| A 709 | Grade 50, type 1 | · · · | 0.23 | 1.35 | 0.04 | 0.05 | 0.40 | · · · | · · · | · · · | · · · | 0.005–0.05 Nb |
| | Grade 50, type 2 | · · · | 0.23 | 1.35 | 0.04 | 0.05 | 0.40 | · · · | · · · | · · · | 0.01–0.15 | · · · |
| | Grade 50, type 3 | · · · | 0.23 | 1.35 | 0.04 | 0.05 | 0.40 | · · · | · · · | · · · | (h) | 0.05 Nb max |
| | Grade 50, type 4 | · · · | 0.23 | 1.35 | 0.04 | 0.05 | 0.40 | · · · | · · · | · · · | (i) | 0.015 N max |
| A 715 | · · · | · · · | 0.15 | 1.65 | 0.025 | 0.035 | · · · | · · · | · · · | · · · | V, Ti, Nb added as necessary | |
| A 808 | · · · | · · · | 0.12 | 1.65 | 0.04 | 0.05 max or 0.010 max | 0.15–0.50 | · · · | · · · | · · · | 0.10 | 0.02–0.10 Nb, V+Nb = 0.15 max |
| A 812 | 65 | · · · | 0.23 | 1.40 | 0.035 | 0.04 | 0.15–0.50(j) | · · · | · · · | · · · | V+Nb = 0.02–0.15 | 0.05 Nb max |
| | 80 | · · · | 0.23 | 1.50 | 0.035 | 0.04 | 0.15–0.50 | 0.35 | · · · | · · · | V+Nb = 0.02–0.15 | 0.05 Nb max |
| A 841 | · · · | · · · | 0.20 | (k) | 0.030 | 0.030 | 0.15–0.50 | 0.25 | 0.25 | 0.35 | 0.06 | 0.08 Mo, 0.03 Nb, 0.02 Al total |
| A 871 | · · · | · · · | 0.20 | 1.50 | 0.04 | 0.05 | 0.90 | 0.90 | 1.25 | 1.00 | 0.10 | 0.25 Mo, 0.15 Zr, 0.05 Nb, 0.05 Ti |

(a) For characteristics and intended uses, see Table 10; for mechanical properties, see Table 16. (b) If a single value is shown, it is a maximum unless otherwise stated. (c) Values may vary, or minimum value may exist, depending on product size and mill form. (d) Optional or when specified. (e) May be purchased as type 1 (0.005–0.05 Nb), type 2 (0.01–0.15 V), type 3 (0.05 Nb, max, plus 0.02–0.15 V) or type 4 (0.015 N, max, plus V ≥ 4 N). (f) If chromium and silicon are each 0.50% min, the copper minimum does not apply. (g) May be substituted for all or part of V. (h) Niobium plus vanadium, 0.02 to 0.15%. (i) Nitrogen with vanadium content of 0.015% (max) with a minimum vanadium-to-nitrogen ratio of 4:1. (j) When silicon-killed steel is specified. (k) For plate under 40 mm (1.5 in.), manganese contents are 0.70 to 1.35% or up to 1.60% if carbon equivalents do not exceed 0.47%. For plate over 40 mm (1 to 5 in.), ASTM A 841 specifies manganese contents of 1.00 to 1.60%.

**hob** (machine tool). A rotary cutting tool with its teeth arranged along a helical thread, used for generating gear teeth or other evenly spaced forms on the periphery of a cylindrical workpiece. The hob and the workpiece are rotated in timed relationship to each other while the hob is fed axially or tangentially across or radially into the workpiece. Hobs should not be confused with multiple-thread milling cutters, rack cutters, and similar tools, where the teeth are not arranged along a helical thread. See also the figure accompanying the term *gear hobbing*.

**hob** (plastic molding). A master model used to sink the shape of a mold into a soft steel block.

**hogging.** Machining a part from bar stock, plate, or a simple forging in which much of the original stock is removed.

**Hohman A-6 wear machine.** A widely used type of wear and friction testing machine in which a rotating ring specimen is squeezed between two diametrically opposed rub blocks (Fig. 231). This design is said to eliminate shaft flexure such as that found in other machines whose load application from the rub block to the ring is from one side

FIG. 231

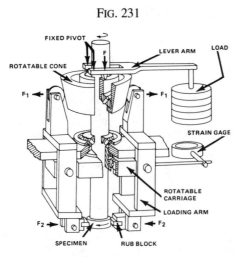

*Hohman A-6 wear machine testing arrangement*

only. Block geometry can be changed from flat to conforming or V-block. This type of machine is designed for use with either lubricated or unlubricated specimens.

**holddown plate (pressure pad).** A pressurized plate designed to hold the workpiece down during a press operation. In practice, this plate often serves as a *stripper* and is also called a stripper plate.

**holding.** In heat treating of metals, that portion of the thermal cycle during which the temperature of the object is maintained constant.

**holding furnace.** A furnace into which molten metal can be transferred to be held at the proper temperature until it can be used to make castings.

**holding temperature.** In heat treating of metals, the constant temperature at which the object is maintained.

**holding time** (heat treating). Time for which the temperature of the heat treated metal object is maintained constant.

**holding time** (joining). In brazing and soldering, the amount of time a joint is held within a specified temperature range.

**hold time** (welding). In resistance welding, the time during which pressure is applied to the work after the current ceases.

**hole expansion test.** A simulative test in which a flat metal sheet specimen with a circular hole in its center is clamped

FIG. 232

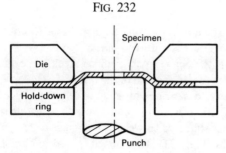

*Schematic of the hole expansion test with a flat-bottomed punch*

between annular die plates and deformed by a punch, which expands and ultimately cracks the edge of the hole (Fig. 232).

**hole flanging.** The forming of an integral collar around the periphery of a previously formed hole in a sheet metal part.

**hole sawing.** The use of a cylindrical saw having end teeth which cut a circular slot through the workpiece leaving a core.

**holidays.** Discontinuities in a coating (such as porosity, cracks, gaps, and similar flaws) that allow areas of basis metal to be exposed to any corrosive environment that contacts the coated surface.

**hollow milling.** Using a special end-cutting mill so designed to leave a core after feeding into or through the workpiece.

**holography.** A technique for recording, and later reconstructing, the amplitude and phase distributions of a wave disturbance; widely used as a method of three-dimensional optical image formation, and also with acoustical and radio waves. In optical image formation, the technique is accomplished by recording on a photographic plate the pattern of interference between coherent light reflected from the object of interest, and light that comes directly from the same source or is reflected from a mirror. In acoustical holography, acoustic beams form an interference pattern of an object and a beam of light interacts with this pattern and is focused to form an optical image (Fig. 233).

**homogeneity.** The degree of uniformity of composition or properties. Contrast with *heterogeneity*.

**homogeneous.** A body of material or matter, alike throughout; hence, comprised of only one chemical composition and phase, without internal boundaries.

**homogeneous carburizing.** Use of a carburizing process to

FIG. 233

*Hologram of a Lincoln U.S. penny, made at 50 MHz using scanning acoustical holography*

FIG. 234

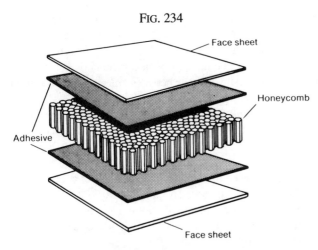

*Example of a bonded honeycomb sandwich assembly*

convert a low-carbon ferrous alloy to one of uniform and higher carbon content throughout the section.

**homogeneous nucleation.** In the crystallization of polymers, the primary nucleated species generated by the polymer molecules.

**homogenizing.** A heat treating practice whereby a metal object is held at high temperature to eliminate or decrease chemical segregation by diffusion.

**homologous.** Belonging to or consisting of a series of organic compounds differentiated by the number of methylene groups ($CH_2$).

**homopolymer.** A polymer resulting from polymerization of a single monomer.

**honeycomb.** Manufactured product of resin-impregnated sheet material (paper, fiberglass, and so on) or metal (aluminum, titanium, and corrosion-resistant alloys) foil, formed into hexagonal-shaped cells (Fig. 234). Used as a core material in composite sandwich constructions. See also *sandwich construction*.

**honing.** A low-speed finishing process used chiefly to produce uniform high dimensional accuracy and fine finish, most often on inside cylindrical surfaces. In honing, very thin layers of stock are removed by simultaneously rotating and reciprocating a bonded abrasive stone or stick that is pressed against the surface being honed with lighter force than is typical of grinding (Fig. 235).

**Hooker process.** Extrusion of a hollow billet or cup through an annulus formed by the die aperture and the mandrel or pilot to form a tube or long cup.

**Hooke's law.** A generalization applicable to all solid material, which states that stress is directly proportional to strain and is expressed as:

$$\frac{\text{Stress}}{\text{Strain}} = \frac{\sigma}{\varepsilon} = \text{constant} = E$$

where $E$ is the modulus of elasticity or Young's modulus. The constant relationship between stress and strain applies only below the proportional limit. See also *modulus of elasticity*.

FIG. 235

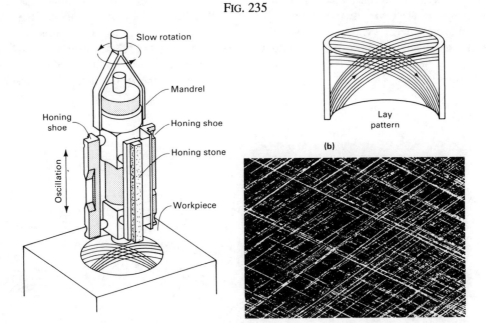

*(a) A honing head containing the abrasive stones, which traverse in a rotary, oscillatory motion. (b) Resulting crosshatched lay pattern. (c) Detail of the inside wall of the hole*

**Hoopes process**. An electrolytic refining process for aluminum, using three liquid layers in the reduction cell.

**hoop stress**. The circumferential stress in a material of cylindrical form subjected to internal or external pressure.

**hopper dryer**. A combination feeding and drying device for extrusion and injection molding of thermoplastics. Hot air flows upward through the hopper containing the feed pellets. See also the figure accompanying the term *extruder*.

**hopper loader**. A curved pipe through which molding plastic powders are pneumatically conveyed from shipping drums to machine hoppers.

**horizontal batch furnace**. A versatile batch-type heat treating furnace that can give light or deep case depths, and because the parts are not exposed to air, horizontal batch furnaces can give surfaces almost entirely free of oxides.

**horizontal-position welding**. (1) Making a fillet weld on the upper side of the intersection of a vertical surface and a horizontal surface. (2) Making a horizontal groove weld on a vertical surface.

**horizontal fixed position** (pipe welding). The position of a pipe joint in which the axis of the pipe is essentially horizontal and the pipe is not rotated during welding (Fig. 236). See also *vertical position* (pipe welding).

**horizontal rolled position** (pipe welding). The position of a pipe joint in which the axis of the pipe is essentially horizontal, and welding is carried out as the pipe is rotated about its axis (Fig. 236). See also *vertical position* (pipe welding).

**horn**. (1) In a resistance welding machine, a cylindrical arm or beam that transmits the electrode pressure and usually conducts the welding current. (2) A cone-shaped member that transmits ultrasonic energy from a transducer to a welding or machining tool. See also *ultrasonic impact grinding* and *ultrasonic welding*.

**horn press**. A mechanical metal forming press equipped with or arranged for a cantilever block or horn that acts as the die or support for the die, used in forming, piercing, setting down, or riveting hollow cylinders and odd-shaped work.

**horn spacing**. The distance between adjacent surfaces of the horns of a resistance welding machine.

**horseshoe thrust bearing**. A tilting-pad thrust bearing in which the top pads are omitted, making an incomplete annulus.

**hot bed**. An area adjacent to the *runout table* where hot rolled metal is placed to cool. Sometimes called a cooling table.

**hot box process**. In foundry practice, resin-base (furan or phenolic) binder process for molding sands similar to shell coremaking; cores produced with it are solid unless mandrelled out.

**hot cathode gun**. See *thermionic cathode gun*.

**hot chamber machine**. A *die casting* machine in which the metal chamber under pressure is immersed in the molten metal in a furnace (Fig. 237). The chamber is sometimes called a gooseneck, and the machine is sometimes called a gooseneck machine.

**hot-cold working**. (1) A high-temperature thermomechanical treatment consisting of deforming a metal above its transformation temperature and cooling fast enough to preserve some or all of the deformed structure. (2) A general term synonymous with *warm working*.

**hot corrosion**. An accelerated corrosion of metal surfaces that results from the combined effect of oxidation and reactions with sulfur compounds and other contaminants, such as chlorides, to form a molten salt on a metal surface that fluxes, destroys, or disrupts the normal protective oxide. See also *gaseous corrosion*.

**hot cracking**. (1) A crack formed in a weldment caused by the segregation at grain boundaries of low-melting constituents in the weld metal. This can result in grain-bound-

FIG. 236

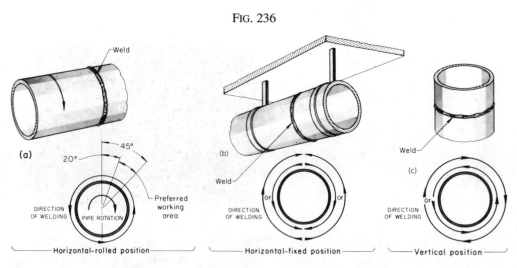

*Positions for horizontal and vertical pipe welding*

FIG. 237

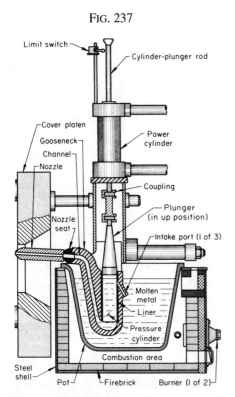

*Principal components of the shot end of a hot-chamber die casting machine*

ary tearing under thermal contraction stresses. Hot cracking can be minimized by the use of low-impurity welding materials and proper joint design. (2) A crack formed in a cast metal because of internal stress developed upon cooling following solidification. A hot crack is less open than a *hot tear* and usually exhibits less oxidation and decarburization along the fracture surface. See also *cold cracking, lamellar tearing*, and *stress-relief cracking*.

**hot densification**. Rapid deformation of a heated metal powder preform in a die assembly for the purpose of reducing porosity. Metal is usually deformed in the direction of the punch travel. See also *hot pressing*.

**hot-die forging**. A hot forging process in which both the dies and the forging stock are heated; typical die temperatures are 110 to 225 °C (200 to 400 °F) lower than the temperature of the stock. Compare with *isothermal forging*.

**hot dip**. Covering a surface by dipping the surface to be coated into a molten bath of the coating material. See also *hot dip coating*.

**hot dip coating**. A metallic coating obtained by dipping the basis metal into a molten metal.

**hot etching**. Development and stabilization of the microstructure at elevated temperature in etchants or gases.

**hot extrusion**. A process whereby a heated *billet* is forced to flow through a shaped die opening. The temperature at which extrusion is performed depends on the material being extruded. Hot extrusion is used to produce long, straight metal products of constant cross section, such as bars, solid and hollow sections, tubes, wires, and strips, from materials that cannot be formed by cold extrusion. Table 34 lists typical billet temperatures for hot extrusion.

**hot forging**. (1) A forging process in which the die and/or forging stock are heated. See also *hot-die forging* and *isothermal forging*. (2) The plastic deformation of a pressed and/or sintered powder compact in at least two directions at temperatures above the recrystallization temperature.

**hot forming**. See *hot working*.

**hot-gas welding**. A technique for joining thermoplastic materials (usually sheet) in which the materials are softened by a jet of hot air from a welding torch and joined together at the softened points. Generally, a thin rod of the same material is used to fill and consolidate the gap.

**hot heated manifold mold**. A thermoplastic injection mold in which the portion of the mold (the manifold) that contains the runner system has its own heating elements, which keep the molding material in a plastic state ready for injection into the cavities, from which the manifold is insulated. See also *thermoplastic injection molding*.

**hot isostatic pressing**. (1) A process for simultaneously heating and forming a compact in which the powder is contained in a sealed flexible sheet metal or glass enclosure and the so-contained powder is subjected to equal pressure from all directions at a temperature high enough to permit plastic deformation and sintering to take place. (2) A process that subjects a component (casting, powder forgings, etc.) to both elevated temperature and isostatic gas pressure in an autoclave (Fig. 238). The most widely used pressurizing gas is argon. When castings are hot isostati-

**Table 34  Typical billet temperatures for hot extrusion**

| Material | Billet temperature | |
|---|---|---|
| | °C | °F |
| Lead alloys | 90–260 | 200–500 |
| Magnesium alloys | 340–430 | 650–800 |
| Aluminum alloys | 340–510 | 650–950 |
| Copper alloys | 650–1100 | 1200–2000 |
| Titanium alloys | 870–1040 | 1600–1900 |
| Nickel alloys | 1100–1260 | 2000–2300 |
| Steels | 1100–1260 | 2000–2300 |

FIG. 238

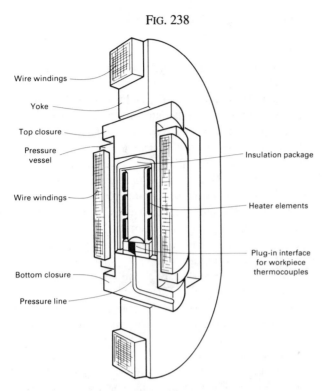

Wire windings
Yoke
Top closure
Pressure vessel
Wire windings
Bottom closure
Pressure line

Insulation package
Heater elements
Plug-in interface for workpiece thermocouples

*Schematic of one type of large, production HIP vessel*

cally pressed, the simultaneous application of heat and pressure virtually eliminates internal voids and microporosity through a combination of plastic deformation, creep, and diffusion.

**hot isostatic pressure welding.** A diffusion welding method that produces coalescence of materials by heating and applying hot inert gas under pressure.

**hot machining.** Machining in which the workpiece shear zone is heated by auxiliary means to reduce the shear strength and increase the machinability of the material.

**hot-melt adhesives.** See *Technical Brief 21*.

**hot mill.** A production line or facility for hot rolling of metals.

**hot press forging.** Plastically deforming metals between dies in presses at temperatures high enough to avoid strain hardening.

**hot pressing.** Simultaneous heating and forming of a powder compact. See also *pressure sintering*.

**hot pressure welding.** A solid-state welding process that produces coalescence of materials with heat and application of pressure sufficient to produce macrodeformation of the base material. Vacuum or other shielding media may be used. See also *forge welding* and *diffusion welding*. Compare with *cold welding*.

**hot quenching.** An imprecise term for various quenching procedures in which a quenching medium is maintained at a prescribed temperature above 70 °C (160 °F).

**hot rod.** Same as *wire rod*.

**hot-runner mold.** A thermoplastic injection mold in which the runners are insulated from the chilled cavities and remain hot so that the center of the runner never cools during the molding cycle. Contrary to usual practice, the runners are not ejected with the molded pieces. Also called insulated-runner mold. See also *thermoplastic injection molding*.

**hot-setting adhesive.** An adhesive that requires a temperature at or above 100 °C (212 °F) to set.

**hot shortness.** A tendency for some alloys to separate along grain boundaries when stressed or deformed at temperatures near the melting point. Hot shortness is caused by a low-melting constituent, often present only in minute amounts, that is segregated at grain boundaries.

**hot stamping.** Engraving operation for plastics in which a design is stamped with heated metal dies onto the face of the plastics.

**hot tear.** A fracture formed in a metal during solidification because of hindered *contraction*. Compare with *hot cracking*.

**hot top.** (1) A reservoir, thermally insulated or heated, that holds molten metal on top of a mold for feeding of the ingot or casting as it contracts on solidifying, thus preventing formation of *pipe* or *voids*. (2) A refractory-lined steel or iron casting that is inserted into the tip of the mold and is supported at various heights to feed the ingot as it solidifies.

**hot trimming.** The removal of *flash* or excess metal from a hot part (such as a forging) in a trimming press.

**hot upset forging.** A *bulk forming* process for enlarging and reshaping some of the cross-sectional area of a bar, tube, or other product form of uniform (usually round) section. It is accomplished by holding the heated forging stock between grooved dies and applying pressure to the end of the stock, in the direction of its axis, by the use of a heading tool, which spreads (upsets) the end by metal displacement (Fig. 239). Also called hot heading or hot upsetting. See also *heading* and *upsetting*.

**hot-wire analyzer.** An electrical atmosphere analysis device that is based on the fact that the electrical resistivity of steel is a linear function of carbon content over a range from 0.05% C to saturation. The device measures the carbon potential of furnace atmospheres (typically). This term is not to be confused with the *hot-wire test* which measures heat extraction rates.

**hot-wire test.** Method used to test heat extraction rates of various quenchants. Faster heat-extracting quenchants will permit more electric current to pass through a standard wire because it is cooled more quickly. Compare with *hot-wire analyzer*.

**hot wire welding.** A variation of arc welding processes in

## Technical Brief 21: Hot-Melt Adhesives

HOT-MELT ADHESIVES are 100% solid thermoplastics that are applied in a molten state and form a bond after cooling to a solid state. In contrast to other adhesives, which achieve the solid state through evaporation of solvents or chemical cure, hot-melt adhesives achieve a solid state and resultant strength by cooling. In general, hot-melt adhesives are solid at temperatures below 79 °C (175 °F). Ideally, as the temperature is increased beyond this point, the material rapidly melts to a low-viscosity fluid that can be easily applied. Upon cooling, the adhesive sets rapidly. Because these adhesives are thermoplastics, the melting-resolidification process is repeatable with the addition and removal of the required amount of heat. Typical application temperatures of hot-melt adhesives are 150 to 290 °C (300 to 550 °F).

Materials that are primarily used as hot-melt adhesives include ethylene and vinyl acetate copolymers (EVA), polyvinyl acetates (PVA), polyethylene (PE), amorphous polypropylene, thermoplastic elastomers such as polyurethane, polyether-amide, and block copolymers (for example, styrene-butadiene-styrene, styrene-isoprene-styrene, and styrene-olefin-styrene), polyamides, and polyesters. Hot-melt adhesives can also be divided into nonpressure-sensitive and pressure-sensitive types. Nonpressure-sensitive adhesives include those for direct bonding and heat sealing. Pressure-sensitive hot-melt adhesives are tacky to the touch and can be bonded by the application of pressure alone at room temperature.

Hot-melt adhesives are used to bond all types of substrates, including metals, glass, plastics, ceramics, rubbers, and wood. Primary areas of application include packaging, book binding, assembly bonding (such as air filters and footwear), and industrial bonding (such as carpet tape and backings).

### Recommended Reading

- M.M. Gauthier, Types of Adhesives, *Engineered Materials Handbook*, Vol 3, ASM International, 1990, p 74-93
- T. Flanagan, in *Adhesives Technology Handbook*, C.V. Cagle, Ed., McGraw-Hill, 1973, chap 8

**Typical properties of hot-melt adhesives**

| Property | EVA/polyolefin homopolymers and copolymers | Polyvinyl acetate | Polyurethane | Polyamides | Polyamide copolymer | Aromatic polyamide |
|---|---|---|---|---|---|---|
| Brookfield viscosity, Pa · s | 1–30 | 1.6–10 | 2 | 0.5–7.5 | 11 | 2.2 |
| Viscosity test temperature, °C (°F) | 204 (400) | 121 (250) | 104 (220) | 204 (400) | 230 (446) | 204 (400) |
| Softening temperature, °C (°F) | 99–139 (211–282) | ... | ... | 93–154 (200–310) | ... | 129–140 (265–285) |
| Application temperature, °C (°F) | ... | 121–177 (250–350) | ... | ... | ... | ... |
| Service temperature range, °C (°F) | −34 to 80 (−30 to 176) | −1 to 120 (30 to 248) | ... | −40 to 185 (−40 to 365) | ... | ... |
| Relative cost(a) | Lowest | Low to medium | Medium to high | High | High | High |
| Bonding substrates | Paper, wood, selected thermoplastics, selected metals, selected glasses | Paper, wood, leather, glass, selected plastics, selected metals | Plastics | Wood, leather, selected plastics, selected metals | Selected metals, selected plastics | Selected metals, selected plastics |
| Applications | Bookbinding, packaging, toys, automotive, furniture, electronics | Tray forming, packaging, binding, sealing cases and cartons, bottle labels, cans, jars | Laminates | Packaging, electronics, furniture, footwear | Packaging, electronics, binding | Electronics, packaging, binding |

(a) Relative to other hot-melt adhesives

which a filler metal wire is resistance heated as it is fed into the molten weld pool.

**hot-worked structure.** The structure of a material worked at a temperature higher than the recrystallization temperature.

**hot working.** (1) The plastic deformation of metal at such a temperature and strain rate that recrystallization takes place simultaneously with the deformation, thus avoiding any *strain hardening*. Also referred to as hot forging and hot forming. (2) Controlled mechanical operations for

FIG. 239

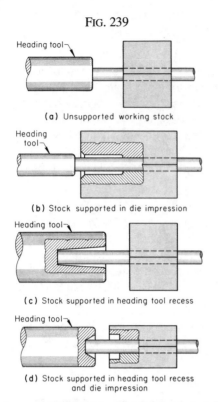

(a) Unsupported working stock

(b) Stock supported in die impression

(c) Stock supported in heading tool recess

(d) Stock supported in heading tool recess and die impression

*Basic type of hot upsetter heading tools and dies showing the extent to which stock is supported*

FIG. 240

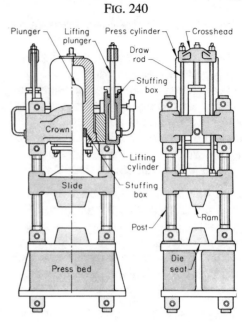

*Principal components of a four-post hydraulic press for closed-die forging*

shaping a product at temperatures above the recrystallization temperature. Contrast with *cold working*.

**hot zone.** The part of a continuous furnace or kiln that is held at maximum temperature. Other zones are the preheat zone and cooling zone.

**hub.** A *boss* that is in the center of a forging and forms a part of the body of the forging.

**hubbing.** The production of forging die cavities by pressing a male master plug, known as a *hub*, into a block of metal.

**Hull cell.** A special electrodeposition cell giving a range of known current densities for test work.

**humidity, absolute.** See *absolute humidity*.

**humidity ratio.** In a mixture of water vapor and air, the mass of water vapor per unit mass of dry air.

**humidity, relative.** See *relative humidity*.

**humidity, specific.** See *specific humidity*.

**humidity test.** A corrosion test involving exposure of specimens at controlled levels of humidity and temperature. Contrast with *salt-fog test*.

**hybrid.** A composite laminate consisting of laminae of two or more composite material systems. A combination of two or more different fibers, such as carbon and glass or carbon and aramid, in a structure. Tapes, fabrics, and other forms may be combined; usually only the fibers differ. See also *interply hybrid* and *intraply hybrid*.

**hydration.** The chemical reaction between cement and water, forming new compounds, most of which have strength-producing properties.

**hydraulic cement.** A cement that sets and hardens by chemical interaction with water and that is capable of doing so under water.

**hydraulic fluid.** A fluid used for transmission of hydraulic pressure or action, not necessarily involving lubricant properties. Hydraulic fluids can be based on oil, water, or synthetic (fire-resistant) liquids.

**hydraulic hammer.** A gravity-drop forging hammer that uses hydraulic pressure to lift the hammer between strokes.

**hydraulic-mechanical press brake.** A mechanical *press brake* that uses hydraulic cylinders attached to mechanical linkages to power the ram through its working stroke.

**hydraulic press** (metal forming). A press in which fluid pressure is used to actuate and control the ram (Fig. 240). Hydraulic presses are used for both open- and closed-die forging.

**hydraulic press** (plastic molding). A press in which the molding force is created by the pressure exerted by a fluid.

**hydraulic press brake.** A *press brake* in which the ram is actuated directly by hydraulic cylinders.

**hydraulic shear.** A shear in which the crosshead is actuated by hydraulic cylinders.

**hydride descaling.** *Descaling* by the action of a hydride in a fused alkali.

**hydride phase.** The phase $TiH_x$ formed in titanium when the hydrogen content exceeds the solubility limit, generally locally due to some special circumstance (Fig. 241).

FIG. 241

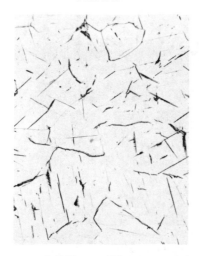

*Hydrogenated (230 ppm H) commercial-purity titanium producing needles of TiH (black). 250×*

**hydride powder**. A powder produced by removal of the hydrogen from a metal hydride.

**hydride process**. In powder metallurgy, the hydrogenation of such reactive metals as titanium and zirconium, followed by comminution of the brittle compound and vacuum treatment to remove the hydrogen from the powder.

**hydrocarbon plastics**. Plastics based on resins made by the polymerization of monomers composed of carbon and hydrogen only.

**hydrodynamic lubrication**. A system of lubrication in which the shape and relative motion of the sliding surfaces causes the formation of a fluid film that has sufficient pressure to separate the surfaces (Fig. 242). See also *elastohydrodynamic lubrication* and *gas lubrication*.

**hydrodynamic machining**. Removal of material by the impingement of a high-velocity fluid against a workpiece. See also *waterjet/abrasive waterjet machining*.

**hydrodynamic seal**. A seal that has special geometric features on one of the mating faces. These features are de-

signed to produce interfacial lift, which arises solely from the relative motion between the stationary and rotating portions of the seal.

**hydrogen-assisted cracking (HAC)**. See *hydrogen embrittlement*.

**hydrogen-assisted stress-corrosion cracking (HSCC)**. See *hydrogen embrittlement*.

**hydrogen blistering**. The formation of blisters on or below a metal surface from excessive internal hydrogen pressure (Fig. 243). Hydrogen may be formed during cleaning, plating, or corrosion.

**hydrogen brazing**. A term sometimes used to denote brazing in a hydrogen-containing atmosphere, usually in a furnace; use of the appropriate process name is preferred.

**hydrogen damage**. A general term for the embrittlement, cracking, blistering, and hydride formation that can occur when hydrogen is present in some metals.

**hydrogen embrittlement**. A process resulting in a decrease of the *toughness* or *ductility* of a metal due to the presence of atomic hydrogen. Hydrogen embrittlement has been recognized classically as being of two types. The first, known as internal hydrogen embrittlement, occurs when the hydrogen enters molten metal which becomes supersaturated with hydrogen immediately after solidification. The second type, environmental hydrogen embrittlement, results from hydrogen being absorbed by solid metals. This can occur during elevated-temperature thermal treatments and in service during electroplating, contact with maintenance chemicals, corrosion reactions, cathodic protection, and operating in high-pressure hydrogen. In the absence of residual stress or external loading, environmental hydrogen embrittlement is manifested in various forms, such as blistering, internal cracking, hydride formation, and reduced ductility. With a tensile stress or stress-intensity factor exceeding a specific threshold, the atomic hydrogen interacts with the metal to induce subcritical crack growth leading to fracture. In the absence of a corrosion reaction (polarized cathodically), the usual term used is hydrogen-assisted cracking (HAC) or hydrogen

FIG. 242

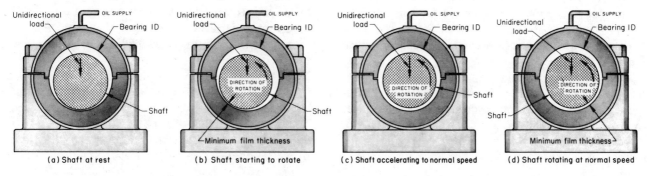

*Step-by-step development of a hydrodynamic lubricating film in a unidirectionally loaded journal bearing*

FIG. 243

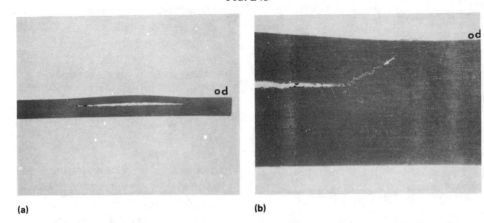

*Hydrogen blister in 19 mm (3/4 in.) steel plate from a spherical tank used to store anhydrous hydrofluoric acid for 13.5 years. (a) Cross section of 150 mm (6 in.) diam blister. (b) Stepwise cracking (arrow) at edge of hydrogen blister shown in (a)*

stress cracking (HSC). In the presence of active corrosion, usually as pits or crevices (polarized anodically), the cracking is generally called *stress-corrosion cracking* (SCC), but should more properly be called hydrogen-assisted stress-corrosion cracking (HSCC). Thus, HSC and electrochemically anodic SCC can operate separately or in combination (HSCC). In some metals, such as high-strength steels, the mechanism is believed to be all, or nearly all, HSC. The participating mechanism of HSC is not always recognized and may be evaluated under the generic heading of SCC.

**hydrogen-induced cracking (HIC).** Same as *hydrogen embrittlement.*

**hydrogen-induced delayed cracking.** A term sometimes used to identify a form of *hydrogen embrittlement* in which a metal appears to fracture spontaneously under a steady stress less than the *yield stress.* There is usually a delay between the application of stress (or exposure of the stressed metal to hydrogen) and the onset of cracking. Also referred to as static fatigue.

**hydrogen loss.** The loss in weight of metal powder or a compact caused by heating a representative sample according to a specified procedure in a purified hydrogen atmosphere. Broadly, a measure of the oxygen content of the sample when applied to materials containing only such oxides as are reducible with hydrogen and no hydride-forming element.

**hydrogen overvoltage.** In electroplating, *overvoltage* associated with the liberation of hydrogen gas.

**hydrogen-reduced powder.** Metal powder produced by hydrogen reduction of a compound.

**hydrogen stress cracking (HSC).** See *hydrogen embrittlement.*

**hydrolysis.** (1) Decomposition or alteration of a chemical

substance by water. (2) In aqueous solutions of electrolytes, the reactions of cations with water to produce a weak base or of anions to produce a weak acid.

**hydrolytic stability.** The ability of an organic or polymeric material to withstand an irreversible change of state when exposed to elevated temperature and humidity.

**hydromatic welding.** See preferred term *pressure-controlled welding.*

**hydromechanical press.** A press in which forces are created partly by a mechanical system and partly by a hydraulic system.

**hydrometallurgy.** Industrial *winning* or *refining* of metals using water or an aqueous solution.

**hydrophilic.** Having an affinity for water; easily wetted by water. Contrast with *hydrophobic.*

**hydrophobic.** Lacking an affinity for, repelling, or failing to absorb water; poorly wetted by water. Contrast with *hydrophilic.*

**hydrostatic bearing.** A bearing in which the solid bodies are separated and supported by a hydrostatic pressure, applied by an external source, to a compressible or incompressible fluid interposed between those bodies.

**hydrostatic compacting.** See *hydrostatic pressing.*

**hydrostatic extrusion.** A method of extruding a *billet* through a die by pressurized fluid instead of the ram used in conventional *extrusion* (Fig. 244).

**hydrostatic lubrication.** A system of lubrication in which the lubricant is supplied under sufficient external pressure to separate the opposing surfaces by a fluid film. See also *pressurized gas lubrication.*

**hydrostatic modulus.** See *bulk modulus of elasticity.*

**hydrostatic mold.** In powder metallurgy, a sealed flexible mold made of rubber, a polymer, or pliable sheet made from a low melting metal such as aluminum.

FIG. 244

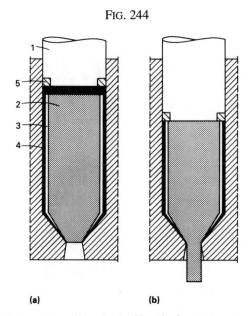

(a)                    (b)

*Schematics of the Hydrafilm (hydrostatic extru-sion) process. (a) Before extrusion. (b) During extrusion. 1, extrusion ram; 2, billet; 3, film of pressurizing liquid; 4, separate film of lubricant; 5, sealing ring*

**hydrostatic pressing.** A special case of isostatic pressing that uses a liquid such as water or oil as a pressure transducing medium and is therefore limited to near room temperature operation.

**hydrostatic seal.** A seal incorporating features that maintain an interfacial film thickness by means of pressure. The pressure is provided either by an external source or by the pressure differential across the seal. The interfacial pressure profile of a seal face is normally speed-dependent; the interfacial pressure profile of the hydrostatic seal is not speed-dependent.

**hydrostatic tension.** Three equal and mutually perpendicular tensile stresses.

**hydroxyl group.** A chemical group consisting of one hydrogen atom and one oxygen atom.

**hygroscopic.** (1) Capable of attracting, absorbing, and retaining atmospheric moisture. (2) Possessing a marked ability to accelerate the condensation of water vapor; applied to condensation nuclei composed of salts that yield aqueous solutions of a very low equilibrium vapor pressure compared with that of pure water at the same temperature. (3) Pertaining to a substance whose physical characteristics are appreciably altered by effects of water vapor. (4) Pertaining to water absorbed by dry soil minerals from the atmosphere; the amounts depend on the physicochemical character of the surfaces, and increase with rising relative humidity.

**hygrothermal effect.** Change in properties of a material

(particularly plastics) due to moisture absorption and temperature change.

**hypereutectic alloy.** In an alloy system exhibiting a *eutectic*, any alloy whose composition has an excess of alloying element compared with the eutectic composition and whose equilibrium microstructure contains some eutectic structure.

**hypereutectoid alloy.** In an alloy system exhibiting a *eutectoid*, any alloy whose composition has an excess of alloying element compared with the eutectoid composition, and whose equilibrium microstructure contains some eutectoid structure.

**hypoeutectic alloy.** In an alloy system exhibiting a *eutectic*, any alloy whose composition has an excess of base metal compared with the eutectic composition, and whose equilibrium microstructure contains some eutectic structure.

**hypoeutectoid alloy.** In an alloy system exhibiting a *eutectoid*, any alloy whose composition has an excess of base metal compared with the eutectoid composition, and whose equilibrium microstructure contains some eutectoid structure.

**hypoid bevel gear.** A gear that transmits rotary motion be-

FIG. 245

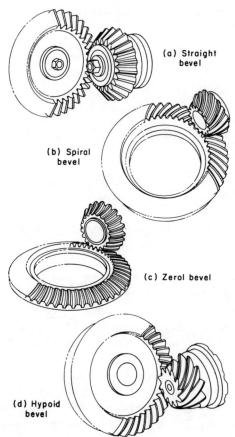

(a) Straight bevel

(b) Spiral bevel

(c) Zerol bevel

(d) Hypoid bevel

*Comparison of a hypoid gear with other types of bevel gears*

FIG. 246

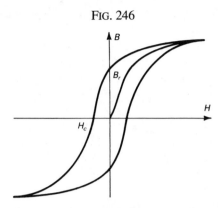

*Schematic of magnetic hysteresis curve for a soft magnetic material. B, magnetic induction; H, magnetizing force; $B_r$, residual magnetic induction; $H_c$, coercive magnetizing force*

tween nonparallel shafts that are usually at 90° to each other. In a hypoid gear set, the axis of the pinion is offset somewhat from the axis of the gear (Fig. 245).

**hypoid gear lubricant (hypoid oil).** A gear lubricant with extreme-pressure characteristics used in hypoid gears.

**hysteresis** (magnetic). The lag of the magnetization of a substance behind any cyclic variation of the applied magnetizing field.

FIG. 247

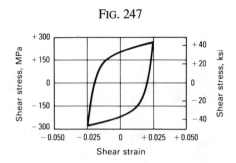

*Torsional hysteresis loop for hot-rolled 1045 steel*

**hysteresis** (mechanical). The phenomenon of permanently absorbed or lost energy that occurs during any cycle of loading or unloading when a material is subjected to repeated loading.

**hysteresis loop** (magnetic). A closed curve that characterizes the magnetization/demagnetization characteristics as a function of the applied magnetic field for magnetic materials (Fig. 246).

**hysteresis loop** (mechanical). In dynamic mechanical measurement, the closed curve representing successive stress-strain status of a material during a deformation cycle (Fig. 247).

**I**

**IACS.** International annealed copper standard; a standard reference used in reporting electrical conductivity. The conductivity of a material, in %IACS, is equal to 1724.1 divided by the electrical resistivity of the material in $n\Omega \cdot m$.

**ideal crack.** A simplified model of a crack used in elastic stress analysis. In a stress-free body, the crack has two smooth surfaces that are coincident and join within the body along a smooth curve called the crack front; in two-dimensional representations, the crack front is called the crack tip.

**ideal-crack-tip stress field.** The singular stress field, infinitesimally close to the crack front, that results from the dominant influence of an ideal crack in an elastic body that is deformed. In a linear-elastic homogeneous body, the significant stress components vary inversely as the square root of the distance from the crack tip. In a linear-elastic body, the crack-tip stress field can be regarded as the superposition of three component stress fields called *modes*.

**ideal critical diameter** ($D_I$). Under an ideal quench condition, the bar diameter that has 50% martensite at the center of the bar when the surface is cooled at an infinitely rapid rate (that is, when $H = \infty$, where $H$ is the quench severity factor or *Grossmann number*).

**identification etching.** Etching to expose particular microconstituents; all others remain unaffected.

**idiomorphic crystal.** An individual crystal that has grown without restraint so that the habit planes are clearly developed. Compare with *allotriomorphic crystal*.

**ignition loss.** (1) The difference in weight before and after burning. (2) The burning off of *binder* or *size*.

**illumination.** The luminous flux density incident on a surface; the ratio of flux to area of illuminated surface.

**image.** A representation of an object produced by radiation, usually with a lens or mirror system.

**image analysis.** Measurement of the size, shape, and distributional parameters of microstructural features by electronic scanning methods, usually automatic or semiautomatic. Image analysis data output can provide individual measurements on each separate feature (feature specific) or field totals for each measured parameter (field specific).

**image contrast.** A measure of the degree of detectable difference in intensity within an image.

**image rotation.** In electron optics, the angular shift of the electron image of an object about the optic axis induced by the tangential component of force exerted on the electrons perpendicular to the direction of motion in the field of a magnetic lens.

**immersed-electrode furnace.** A furnace used for liquid carburizing of parts by heating molten salt baths with the use of electrodes immersed in the liquid (Fig. 248). See also *submerged-electrode furnace*.

Fig. 248

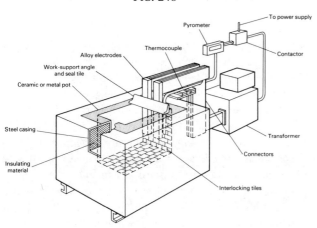

*Internally heated salt bath furnace with immersed electrodes and ceramic tiles*

**immersion cleaning.** Cleaning in which the work is immersed in a liquid solution.

**immersion coating.** A coating produced in a solution by chemical or electrochemical action without the use of external current.

**immersion etching.** Method in which a microsection is dipped face up into etching solution and is moved around during etching. This is the most common etching method.

**immersion lens.** See *immersion objective*.

**immersion objective.** In optical microscopy, an objective in which a medium of high refractive index is used in the object space to increase the numerical aperture and therefore the resolving power of the lens.

**immersion objective** (electron optics). A lens system in which the object space is at a potential or in a medium of index of refraction different from that of the image space.

**immersion plating.** Depositing a metallic coating on a metal immersed in a liquid solution, without the aid of an external electric current. Also called dip plating.

**immiscible.** (1) Of two phases, the inability to dissolve in one another to form a single solution; mutually insoluble. (2) With respect to two or more fluids, not mutually soluble; incapable of attaining homogeneity.

**immunity.** A state of resistance to corrosion or anodic dissolution of a metal caused by thermodynamic stability of the metal.

**impact bruise.** See *percussion cone*.

**impact energy.** The amount of energy, usually given in joules or foot-pound force, required to fracture a material, usually measured by means of an *Izod test* or *Charpy test*. The type of specimen and test conditions affect the values and therefore should be specified.

**impact extrusion.** The process (or resultant product) in which a punch strikes a slug (usually unheated) in a confining die. The metal flow may be either between punch and die or through another opening. The impact extrusion of unheated slugs is often called cold extrusion.

**impaction ratio.** See *collection efficiency*.

**impact line.** A blemish on a drawn sheet metal part caused by a slight change in metal thickness. The mark is called an impact line when it results from the impact of the punch on the blank; it is called a recoil line when it results from transfer of the blank from the die to the punch during forming, or from a reaction to the blank being pulled sharply through the *draw ring*.

**impact load.** An especially severe shock load such as that caused by instantaneous arrest of a falling mass, by shock meeting of two parts (in a mechanical hammer, for example), or by explosive impact, in which there can be an exceptionally rapid buildup of stress.

**impact sintering.** An instantaneous sintering process during *high energy rate compacting* that causes localized heating, welding, or fusion at the particle contacts.

**impact strength.** A measure of the resiliency or toughness of a solid. The maximum force or energy of a blow (given by a fixed procedure) which can be withstood without fracture, as opposed to fracture strength under a steady applied force.

**impact test.** A test for determining the energy absorbed in fracturing a test piece at high velocity, as distinct from static test. The test may be carried out in tension, bending, or torsion, and the test bar may be notched or unnotched. See also *Charpy test, impact energy*, and *Izod test*.

**impact tube.** Same as *Pitot tube*.

**impact value.** The energy absorbed by a specimen of standard design when sheared by a single blow from a testing machine hammer. Expressed in $J/m^2$ or $ft \cdot lbf/in.^2$.

**impact velocity.** The relative velocity between the surface of a solid body and an impacting liquid or solid particle. To describe this velocity completely, it is necessary to specify the direction of motion of the particle relative to the solid surface in addition to the magnitude of the velocity. The following related terms are also in use: (1) absolute impact velocity—the magnitude of the impact velocity, and (2) normal impact velocity—the component of the impact velocity that is perpendicular to the surface of the test solid at the point of impact.

**impact wear**. Wear of a solid surface resulting from repeated collisions between that surface and another solid body. The term *erosion (erosive) wear* is preferred in the case of multiple impacts and when the impacting body or bodies are very small relative to the surface being impacted.

**imperfection**. (1) When referring to the physical condition of a part or metal product, any departure of a quality characteristic from its intended level or state. The existence of an imperfection does not imply nonconformance (see *nonconforming*), nor does it have any implication as to the usability of a product or service. An imperfection must be rated on a scale of severity, in accordance with applicable specifications, to establish whether or not the part or metal product is of acceptable quality. (2) Generally, any departure from an ideal design, state, or condition. (3) In crystallography, any deviation from an ideal space lattice.

**impingement**. A process resulting in a continuing succession of impacts between liquid or solid particles and a solid surface (Fig. 249). In preferred usage, impingement also connotes that the impacting particles are smaller than the solid surface, and that the impacts are distributed over the surface or a portion of the surface. If all impacts are superimposed on the same point or zone, then the term repeated impact is preferred.

**impingement attack**. *Corrosion* associated with turbulent flow of liquid. May be accelerated by entrained gas bub-

bles. See also *erosion-corrosion* and *impingement corrosion*.

**impingement corrosion**. A form of *erosion-corrosion* generally associated with the local impingement of a high-velocity, flowing fluid against a solid surface.

**impingement erosion**. Loss of material from a solid surface due to *liquid impingement* (Fig. 249). See also *erosion*.

**impingement umbrella**. The partial screening of the surface of a solid specimen subjected to solid impingement that sometimes occurs when some of the solid particles rebound from the surface and impede the motion of other impinging particles.

**impregnate**. In reinforcing plastics, to saturate the reinforcement with a resin.

**impregnated fabric**. A fabric impregnated with a synthetic resin. See also *prepreg*.

**impregnation**. (1) Treatment of porous castings with a sealing medium to stop pressure leaks. (2) The process of filling the pores of a sintered compact, usually with a liquid such as a lubricant. (3) The process of mixing particles of a nonmetallic substance in a cemented carbide matrix, as in diamond-impregnated tools (Fig. 250).

**impressed current**. Direct current supplied by a device employing a power source external to the electrode system of a *cathodic protection* installation.

**impression**. (1) In electron microscopy, the reproduction of the surface contours of a specimen formed in a plastic material after the application of pressure, heat, or both. (2) In hardness testing, the imprint or dent made in the specimen by the indenter of a hardness-measuring device. See also *indentation hardness* and *indenter*. (3) A cavity machined into a forging die to produce a desired configuration in the workpiece during forging.

**impression-die forging**. A forging that is formed to the required shape and size by machined impressions in spe-

FIG. 249

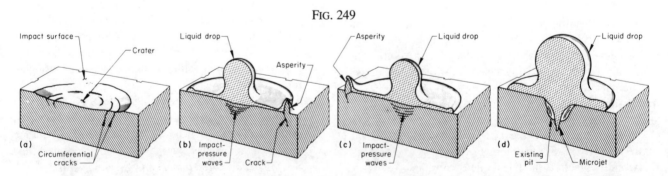

*Processes by which a material is damaged by liquid-impingement erosion. (a) Solid surface showing initial impact of a drop of liquid that produces circumferential cracks in the area of impact or produces shallow craters in very ductile materials. (b) High-velocity radial flow of liquid away from the impact area is arrested by a nearby surface asperity, which cracks at its base. (c) Subsequent impact by another drop of liquid breaks the asperity. (d) Direct hit on a deep pit results in accelerated damage, because shock waves bouncing off the sides of the pit cause the formation of a high-energy microjet within the pit.*

FIG. 250

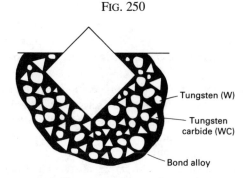

*Schematic of a coarse diamond particle impregnated in a matrix alloy of a diamond drill bit*

FIG. 251

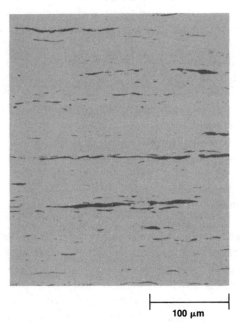

100 μm

*Sulfide inclusions in austenitic stainless steel*

cially prepared dies that exert three-dimensional control on the workpiece.

**impression replica.** A surface replica made by impression. See also *impression* and *replica*.

**impulse** (resistance welding). An impulse of welding current consisting of a single pulse or a series of pulses, separated only by an interpulse time.

**impurities.** (1) Elements or compounds whose presence in a material is undesirable. (2) In a chemical or material, minor constituent(s) or component(s) not included deliberately; usually to some degree or above some level, undesirable.

**inadequate joint penetration.** Weld joint penetration which is less than that specified.

**inclinable press.** A press that can be inclined to facilitate handling of the formed parts. See also *open-back inclinable press*.

**inclined position.** The position of a pipe joint in which the axis of the pipe is at an angle that is approximately 45° to the horizontal and the pipe is not rotated during welding.

**included angle.** See preferred term *groove angle*.

**inclusion.** (1) A physical and mechanical discontinuity occurring within a material or part, usually consisting of solid, encapsulated foreign material. Inclusions are often capable of transmitting some structural stresses and energy fields, but to a noticeably different degree than from the parent material. (2) Particles of foreign material in a metallic matrix. The particles are usually compounds, such as oxides, sulfides (Fig. 251), or silicates, but may be of any substance that is foreign to (and essentially insoluble in) the matrix. See also *exogenous inclusion*, *indigenous inclusion*, and *stringer*.

**inclusion count.** Determination of the number, kind, size, and distribution of nonmetallic inclusions in metals.

**incoherent scattering.** In materials characterization, the deflection of electrons by electrons or atoms that results in a loss of kinetic energy by the incident electron. See also *coherent scattering*.

**incomplete fusion.** In welding, fusion which is less than complete.

**increment.** An individual portion of material collected by a single operation of a sampling device from parts of a lot separated in time or space. Increments may be tested individually or combined (composited) and tested as a unit.

**incubation period.** (1) A period prior to the detection of corrosion while the metal is in contact with a corrodent. (2) In cavitation and impingement erosion, the initial stage of the erosion rate-time pattern during which the erosion rate is zero or negligible compared to later stages. (3) In cavitation and impingement erosion, the exposure duration associated with the initial stage of the erosion rate-time pattern during which the erosion rate is zero or negligible compared to later stages. See also *cavitation erosion* and *impingement erosion*.

**indentation.** In a spot, seam, or projection weld, the depression on the exterior surface of the base metal.

**indentation hardness.** (1) The resistance of a material to indentation. This is the usual type of hardness test, in which a pointed or rounded indenter is pressed into a surface under a substantially static load. (2) Resistance of a solid surface to the penetration of a second, usually harder, body under prescribed conditions. Numerical values used to express indentation hardness are not absolute physical quantities, but depend on the hardness scale used to express hardness. See also *Brinell hardness test*, *Knoop hardness test*, *nanohardness test*, *Rockwell hardness test*, and *Vickers hardness test*.

**indenter**. In hardness testing, a solid body of prescribed geometry, usually chosen for its high hardness, that is used to determine the resistance of a solid surface to penetration.

**index of refraction**. See *refractive index*.

**indication**. In inspection, a response to a nondestructive stimulus that implies the presence of an *imperfection*. The indication must be interpreted to determine if (a) it is a true indication or a *false indication* and (b) whether or not a true indication represents an unacceptable deviation.

**indicator**. A substance that, through some visible change such as color, indicates the condition of a solution or other material as to the presence of free acid, alkali or other substance.

**indices**. See *Miller indices*.

**indigenous inclusion**. An *inclusion* that is native, innate, or inherent in the molten metal treatment. Indigenous inclusions include sulfides, nitrides, and oxides derived from the chemical reaction of the molten metal with the local environment. Such inclusions are small and require microscopic magnification for identification. Compare with *exogenous inclusion*.

**indirect-arc furnace**. An electric-arc furnace in which the metallic charge is not one of the poles of the arc.

**indirect (backward) extrusion**. See *extrusion*.

**indirect sintering**. A process whereby the heat needed for sintering is generated outside the body and transferred to the powder compact by conduction, convection, radiation, etc. Contrast with *direct sintering*.

**individuals**. Conceivable constituent parts of a population. See also *population*.

**induction bonding**. The use of high-frequency (5 to 7 MHz) electromagnetic fields to heat a bonding agent placed between the plastic parts to be joined (Fig. 252). The bonding agent consists of microsized ferromagnetic particles dispersed in a thermoplastic matrix, preferably the parent material of the parts to be bonded. When this binder is exposed to the high-frequency source, the ferromagnetic particles respond and melt the surrounding plastic matrix, which in turn melts the interface surfaces of the parts to be joined.

**induction brazing**. A brazing process in which the surfaces of components to be joined are selectively heated to brazing temperature by electrical energy transmitted to the workpiece by induction, rather than by a direct electrical connection, using an inductor or work coil.

**induction furnace**. An alternating current electric furnace in which the primary conductor is coiled and generates, by electromagnetic induction, a secondary current that develops heat within the metal charge. There are two classifications of induction furnaces: coreless and channel. In a coreless furnace, the refractory-lined crucible is completely surrounded by a water-cooled copper coil, while in the channel furnace the coil surrounds only a small appendage of the unit, called an inductor. The term "channel" refers to the channel that the molten metal forms as a loop within the inductor (Fig. 253). It is this metal loop that forms the secondary of the electrical circuit, with the surrounding copper coil being the primary. In a coreless furnace, the entire metal content of the crucible is the secondary. See also *coreless induction furnace*.

**induction hardening**. A surface-hardening process in which only the surface layer of a suitable ferrous workpiece is

FIG. 253

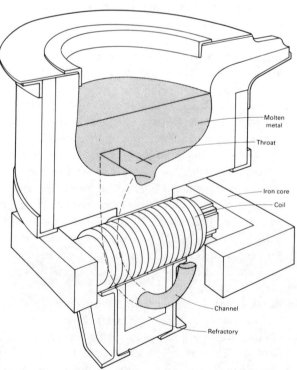

*Cross section of a channel-type induction furnace*

FIG. 252

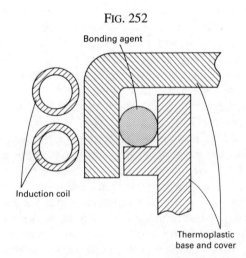

*Induction bonding (welding) process for plastics*

**Table 35  Induction heating applications and some typical products**

| Preheating prior to metalworking | Heat treating | Welding | Melting |
|---|---|---|---|
| **Forging** | **Surface hardening, tempering** | **Seam welding** | **Air melting of steels** |
| Gears | Gears | Oil-country tubular | Ingots |
| Shafts | Shafts | products | Billets |
| Hand tools | Valves | Refrigeration tubing | Castings |
| Ordnance | Machine tools | Line pipe | |
| | Hand tools | | **Vacuum induction melting** |
| **Extrusion** | | | Ingots |
| Structural members | **Through hardening, tempering** | | Billets |
| Shafts | Structural members | | Castings |
| | Spring steel | | "Clean" steels |
| **Heading** | Chain links | | Nickel-base superalloys |
| Bolts | Tubular goods | | Titanium alloys |
| Other fasteners | | | |
| | **Annealing** | | |
| **Rolling** | Aluminum strip | | |
| Slab | Steel strip | | |
| Sheet (can, appliance, and | | | |
| automotive industries) | | | |

heated by electromagnetic induction to above the upper critical temperature and immediately quenched.

**induction heating.** Heating by combined electrical resistance and hysteresis losses induced by subjecting a metal to the varying magnetic field surrounding a coil carrying alternating current. Table 35 lists the various industrial applications for induction heating.

**induction melting.** Melting in an *induction furnace*.

**induction sintering.** Sintering in which the required heat is generated by subjecting the compact to electromagnetic induction. See also *direct sintering*.

**induction soldering.** A soldering process in which the heat required is obtained from the resistance of the work to induced electric current.

**induction tempering.** Tempering of steel using low-frequency electrical *induction heating*.

**induction welding.** Welding in which the required heat is generated by subjecting the workpiece to electromagnetic induction.

**induction work coil.** The *inductor* used when induction heating and melting as well as induction welding, brazing, and soldering.

**inductively coupled plasma (ICP).** An argon plasma excitation source for atomic emission spectroscopy or mass spectroscopy (Fig. 254). It is operated at atmospheric pressure and sustained by inductive coupling to a radio-frequency electromagnetic field. See also *radio frequency*.

**inductor.** A device consisting of one or more associated windings (Fig. 255), with or without a magnetic core, for introducing inductance into an electric circuit.

**industrial atmosphere.** An atmosphere in an area of heavy industry with soot, fly ash, and sulfur compounds as the principal constituents.

**inelastic electron scatter.** See *incoherent scattering*.

**inelastic scattering.** Any collision or interaction that changes the energy of an incident particle. Contrast with *elastic scattering*.

**inert.** Unreactive, stable or indifferent to the presence of other

FIG. 254

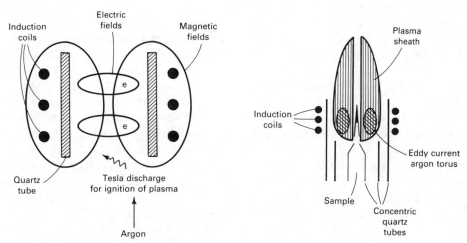

*Electrical and magnetic fields of the inductively coupled plasma (left) and the structure of the ICP torch (right)*

FIG. 255

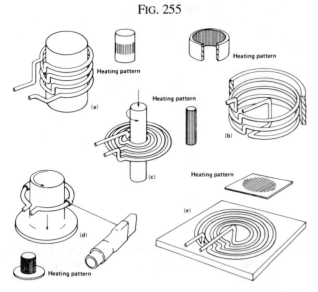

*Typical work coils for high-frequency induction units*

FIG. 256

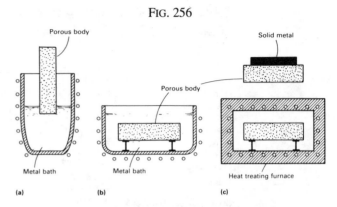

*Capillary infiltration methods. (a) Dip of a small part of a skeleton body into a melt. (b) Complete submersion of a skeleton body into a melt. (c) Positioning of solid infiltrant on top of a skeleton, followed by pore penetration on melting*

materials. A relative term, usually applying under some limited set of circumstances or application.

**inert anode**. An *anode* that is insoluble in the *electrolyte* under the conditions prevailing in the *electrolysis*.

**inert filler**. A material added to a plastic to alter the end-item properties through physical rather than chemical means.

**inert gas**. (1) A gas, such as helium, argon, or nitrogen, which is stable, does not support combustion, and does not form reaction products with other materials. (2) In welding, a gas which does not normally combine chemically with the base metal or filler metal. (3) In processing of plastics, a gas (usually nitrogen) that does not absorb or react with ultraviolet light in a curing chamber.

**inert gas fusion**. An analytical technique for determining the concentrations of oxygen, hydrogen, and nitrogen in a sample. The sample is melted in a graphite crucible in an inert gas atmosphere; individual component concentrations are detected by infrared or thermal conductive methods.

**inert gas metal arc welding**. See preferred term *gas metal arc welding*.

**inert gas tungsten arc welding**. See preferred term *gas tungsten arc welding*.

**infiltrant**. Material used to infiltrate a porous powder compact. The infiltrant as positioned on the compact is called a slug. See also *contact infiltration*.

**infiltration**. The process of filling the pores of a sintered or unsintered compact with a metal or alloy of lower melting temperature (Fig. 256). See also *contact infiltration*.

**inflection point**. Position on a curved line, such as a phase boundary on a *phase diagram*, at which the direction of curvature is reversed.

**infrared (IR)**. Pertaining to that part of the electromagnetic spectrum between the visible light range and the radar range. Radiant heat is in this range, and infrared heaters are frequently used in the thermoforming and curing of plastics and composites. Infrared analysis is used for identification of polymer constituents.

**infrared analyzer**. An atmosphere-monitoring device that measures a gas (usually carbon monoxide, carbon dioxide, and methane) presence based on specific wavelength absorption of infrared energy.

**infrared brazing**. A brazing process in which the heat required is furnished by infrared radiation.

**infrared radiation**. Electromagnetic radiation in the wavelength range of 0.78 to 300 μm (7800 to $3 \times 10^6$ Å). See also *electromagnetic radiation, far-infrared radiation, middle-infrared radiation*, and *near-infrared radiation*.

**infrared soldering**. A soldering process in which the heat required is furnished by infrared radiation.

**infrared spectrometer**. A device used to measure the amplitude of electromagnetic radiation of wavelengths between visible light and microwaves.

**infrared spectroscopy**. The study of the interaction of material systems with electromagnetic radiation in the infrared region of the spectrum. The technique is useful for determining the molecular structure of organic and inorganic compounds by identifying the rotational and vibrational energy levels associated with the various molecules.

**infrared spectrum**. (1) The range of wavelengths of infrared radiation. (2) A display or graph of the intensity of infrared radiation emitted or absorbed by a material as a function of wavelength or some related parameter.

**ingate**. Same as *gate*.

**ingot**. A casting of simple shape, suitable for hot working or remelting.

**ingot iron**. Commercially pure iron.

**inhibitor**. A substance that retards some specific chemical reaction. Picking inhibitors retard the dissolution of metal without hindering the removal of scale from steel. Inhibitors are also used in certain types of monomers and resins to prolong storage life.

**initial current.** The current after starting, but before establishment of welding current.

**initial modulus.** The slope of the initial straight portion of a stress-strain or load-elongation curve. See also *Young's modulus*.

**initial pitting.** Surface fatigue occurring during the early stages of gear operation, associated with the removal of highly stressed local areas and *running-in.*

**initial recovery.** The decrease in strain in a solid during the removal of force before any creep recovery takes place, usually determined at constant temperature. Sometimes referred to as instantaneous recovery.

**initial strain.** The strain produced in a specimen by given loading conditions before creep occurs.

**initial stress.** The stress produced by strain in a specimen before stress relaxation occurs. Also called instantaneous stress.

**initial tangent modulus.** The slope of the stress-strain curve at the beginning of loading. See also *modulus of elasticity.*

**initiation of stable crack growth.** The initiation of slow stable crack advance from the blunted crack tip.

**initiator.** Sources of free radicals, often peroxides or azo compounds. They are used in free-radical polymerizations, for curing thermosetting resins, and as cross-linking agents for elastomers and cross-linked polyethylene.

**injection blow molding.** A *blow molding* process in which an injection molded preform is used instead of an extruded parison.

**injection molding** (ceramics). A process for forming ceramic articles in which a granular ceramic-binder mix is heated until softened, and then forced into a mold cavity, where it cools and resolidifies to produce a part of the desired shape (Fig. 257).

FIG. 257

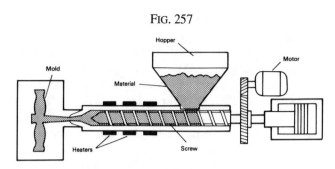

*Cross-sectional side view of a screw-type injection molding machine for ceramics*

**injection molding** (metals). A process similar to plastic injection molding using a plastic coated metal powder of fine particle size (~10 μm).

**injection molding** (plastics). Method of forming a plastic to the desired shape by forcing the heat-softened plastic into a relatively cool cavity under pressure. See also *thermoplastic injection molding* and *thermoset injection molding.*

**inoculant.** Materials that, when added to molten metal, modify the structure and thus change the physical and mechanical properties to a degree not explained on the basis of the change in composition resulting from their use. Ferrosilicon-base alloys are commonly used to inoculate gray irons and ductile irons. Table 36 lists compositions of ferrosilicon inoculants for gray cast iron.

**inoculation.** The addition of a material to molten metal to form nuclei for crystallization. See also *inoculant.*

**inorganic.** Being or composed of matter other than hydrocarbons and their derivatives, or matter that is not of plant or animal origin. Contrast with *organic.*

**inorganic pigments.** Natural or synthetic metallic oxides, sulfides, and other salts that impart heat and light stability, weathering resistance, color, and migration resistance to plastics.

**Table 36  Compositions of ferrosilicon inoculants for gray cast iron**

| Performance category of inoculant | Si | Al | Ca | Ba | Ce | TRE(b) | Ti | Mn | Sr | Others |
|---|---|---|---|---|---|---|---|---|---|---|
| Standard ......... | 46–50 | 0.5–1.25 | 0.60–0.90 | · · · | · · · | · · · | · · · | · · · | · · · | · · · |
| | 74–79 | 1.25 max | 0.50–1.0 | · · · | · · · | · · · | · · · | · · · | · · · | · · · |
| | 74–79 | 0.75–1.5 | 1.0–1.5 | · · · | · · · | · · · | · · · | · · · | · · · | · · · |
| Intermediate...... | 46–50 | 1.25 max | 0.75–1.25 | 0.75–1.25 | · · · | · · · | · · · | 1.25 max | · · · | · · · |
| | 60–65 | 0.8–1.5 | 1.5–3.0 | 4–6 | · · · | · · · | · · · | 7–12 | · · · | · · · |
| | 70–74 | 0.8–1.5 | 0.8–1.5 | 0.7–1.3 0.75–1.25 | · · · | · · · | · · · | · · · | · · · | · · · |
| | 42–44 | · · · | 0.75–1.25 | | · · · | · · · | 9–11 | · · · | · · · | · · · |
| | 50–55 | · · · | 5–7 | · · · | · · · | · · · | 9–11 | · · · | · · · | · · · |
| | 50–55 | · · · | 0.5–1.5 | · · · | · · · | · · · | 9–11 | · · · | · · · | · · · |
| High............. | 36–40 | · · · | · · · | · · · | 9–11 | 10.5–15 | · · · | · · · | · · · | · · · |
| | 46–50 | 0.50 max | 0.10 max | · · · | · · · | · · · | · · · | · · · | 0.60–1.0 | · · · |
| | 73–78 | 0.50 max | 0.10 max | · · · | · · · | · · · | · · · | · · · | 0.60–1.0 | · · · |
| Stabilizing........ | 6–11 | 0.50 max | 0.50 max | · · · | · · · | · · · | · · · | · · · | · · · | 48–52 Cr |

(a) All compositions contain balance of iron. (b) TRE, total rare earths

**inorganic zinc-rich paint**. Coating containing a zinc powder pigment in an *inorganic* vehicle.

**insert** (metals). (1) A part formed from a second material, usually a metal, which is placed in the molds and appears as an integral structural part of the final casting. (2) A removable portion of a die or mold.

**insert** (plastics). An integral part of a plastic molding consisting of metal or other material that may be molded or pressed into position after the molding is completed.

**insert die**. A relatively small die which contains part or all of the impression of a forging, and which is fastened to a master die block.

**inserted-blade cutters**. Cutters having replaceable blades that are either solid or tipped and are usually adjustable.

**instantaneous erosion rate**. The slope of a tangent to the cumulative erosion-time curve at a specified point on that curve.

**instantaneous recovery**. The decrease in strain occurring immediately upon unloading a specimen. A more reproducible value is obtained if the decrease in strain is measured after a given small increment of time (such as 1 min) following unloading. The value is expressed in the same units as strain, that is, the decrease in length divided by the gage length, usually in inches per inch.

**instantaneous strain**. The strain occurring immediately upon loading a creep specimen. Value is measured after loading to obtain a more reproducible value. The value is expressed in the same units as strain, that is, the extension divided by the gage length, usually in inches per inch.

**instrumented impact test**. An impact test in which the load on the specimen is continually recorded as a function of time and/or specimen deflection prior to fracture.

**instrument response time**. The time required for an indicating or detecting device to attain a defined percentage of its steady-state value following an abrupt change in the quantity being measured.

**insulating pads and sleeves**. In foundry practice, insulating material, such as gypsum, diatomaceous earth, and so forth, used to lower the rate of solidification. As sleeves on open risers, they are used to keep the metal liquid, thus increasing the feeding efficiency. Contrast with *chill*.

**insulator**. A material of such low electrical conductivity that the flow of current through it can usually be neglected. Similarly, a material of low thermal conductivity, such as that used to insulate structures.

**integral composite structure**. Composite structure in which several structural elements, which would conventionally be assembled together by bonding or mechanical fasteners after separate fabrication, are instead laid up and cured as a single, complex, continuous structure. The term is sometimes applied more loosely to any composite structure not assembled by mechanical fasteners. All or some parts of the assembly may be co-cured.

**integral skin foam**. Urethane foam with a cellular core structure and a relatively smooth skin.

**integrated circuit (IC)**. An interconnected array of active and passive elements integrated with a single semiconductor substrate or deposited on the substrate by a continuous series of compatible processes, and capable of performing at least one complete electronic circuit function.

**intense quenching**. Quenching in which the quenching medium is cooling the part at a rate at least two and a half times faster than still water. See also *Grossmann number*.

**intensiostatic**. See *galvanostatic*.

**intensity**. The energy per unit area per unit time incident on a surface.

**intensity of scattering**. The energy per unit time per unit area of the general radiation diffracted by matter. Its value depends on the scattering power of the individual atoms of the material, the scattering angle, and the wavelength of the radiation.

**intensity ratio**. The ratio of two (relative) intensities.

**intensity, x-rays**. The energy per unit of time of a beam per unit area perpendicular to the direction of propagation.

**intercept method**. A quantitative metallographic technique in which the desired quantity, such as grain size or inclusion content, is expressed as the number of times per unit length a straight line on a metallographic image crosses particles of the feature being measured.

**intercommunicating porosity**. See preferred term *interconnected porosity*.

**interconnected pore volume**. The volume fraction of pores that are interconnected within the entire pore system of a powder compact or sintered product.

**interconnected porosity**. A network of connecting pores in a sintered object that permits a fluid or gas to pass through the object. Also referred to as interlocking or open porosity.

**intercritical annealing**. Any annealing treatment that involves heating to, and holding at, a temperature between the upper and lower critical temperatures to obtain partial austenitization, followed by either slow cooling or holding at a temperature below the lower critical temperature.

**intercrystalline**. Between the crystals, or grains, of a polycrystalline material.

**intercrystalline corrosion**. See *intergranular corrosion*.

**intercrystalline cracking**. See *intergranular cracking*.

**interdendritic corrosion**. Corrosive attack that progresses preferentially along interdendritic paths. This type of attack results from local differences in composition, such as coring commonly encountered in alloy castings (Fig. 258).

**interdendritic porosity**. Voids occurring between the dendrites in cast metal.

**interface**. The boundary between any two phases. Among the three phases (gas, liquid, and solid), there are five types

FIG. 258

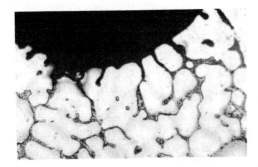

*Interdendritic corrosion in a cast aluminum alloy. 500×*

FIG. 259

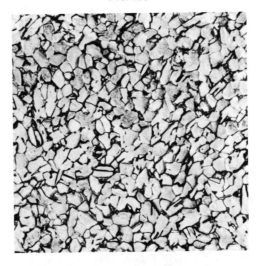

*Equiaxed alpha grains (light) and intergranular beta (dark) in a Ti-6Al-4V bar. 250×*

FIG. 260

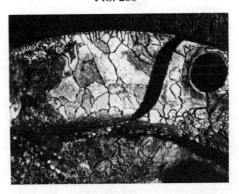

*Intergranular corrosion of a cast stainless steel pump component that contacted HCl-Cl₂ gas fumes*

of interfaces: gas-liquid, gas-solid, liquid-liquid, liquid-solid, and solid-solid.

**interface** (composites). The boundary or surface between two different, physically distinguishable media. With fibers, the contact area between the fibers and the sizing or finish. In a laminate, the contact area between the reinforcement and the laminating resin.

**interface activity**. A measure of the chemical potential between the contacting surfaces of two particles in a compact or two grains in a sintered body.

**interfacial tension**. The contractile force of an interface between two phases.

**interference**. The difference in lateral dimensions at room temperature between two mating components before assembly by expansion, shrinking, or press fitting. Can be expressed in absolute or in relative terms.

**interference filter**. A combination of several thin optical films to form a layered coating for transmitting or reflecting a narrow band of wavelengths by interference effects.

**interference fit**. (1) A joint or mating of two parts in which the male part has an external dimension larger than the internal dimension of the mating female part. Distension of the female by the male creates a stress, which supplies the bonding force for the joint. (2) Any of various classes of fit between mating parts where there is nominally a negative or zero allowance between the parts, and where there is either part interference or no gap when the mating parts are made to the respective extremes of individual tolerances that ensure the tightest fit between the parts. Contrast with *clearance fit.*

**interference of waves**. The process whereby two or more waves of the same frequency or wavelength combine to form a wave whose amplitude is the sum of the amplitudes of the interfering waves.

**interferometer**. An instrument in which the light from a source is split into two or more beams, which are subsequently reunited and interfere after traveling over different paths.

**intergranular**. Between crystals or grains. Also called intercrystalline. Contrast with *transgranular.*

**intergranular beta**. In titanium alloys, β phase situated between α grains. It may be at grain corners, as in the case of equiaxed α-type microstructures in alloys having low β-stabilizer contents (Fig. 259).

**intergranular corrosion**. Corrosion occurring preferentially at grain boundaries, usually with slight or negligible attack on the adjacent grains (Fig. 260). See also *interdendritic corrosion.*

**intergranular cracking**. Cracking or fracturing that occurs between the grains or crystals in a polycrystalline aggregate. Also called intercrystalline cracking. Contrast with *transgranular cracking.*

**intergranular fracture**. *Brittle fracture* of a polycrystalline material in which the fracture is between the grains, or

## Technical Brief 22: Intermetallics

ORDERED INTERMETALLIC compounds constitute a unique class of metallic materials that form long-range ordered crystal structures below a critical temperature, generally referred to as the critical ordering temperature ($T_c$). Ordered intermetallic alloys with relatively low critical ordering temperatures (<700 °C, or 1290 °F) were studied extensively in the 1950s and 1960s, following the discovery of unusual dislocation structures and mechanical behavior associated with ordered lattices. Interest, however, subsided in the latter part of the 1960s because of severe embrittlement problems encountered with the compounds. Most strongly ordered intermetallics are so brittle that they simply cannot be fabricated into useful structural components. Because of the brittleness problem, intermetallics

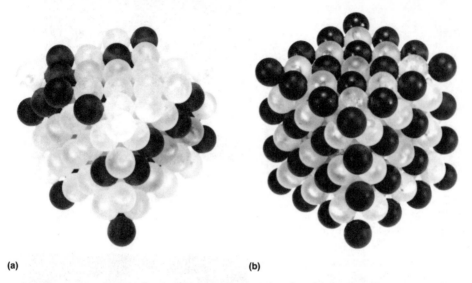

(a)                                                      (b)

*Atomic arrangements of conventional alloys and ordered intermetallic compounds. (a) Disordered crystal structure of a conventional alloy. (b) Long-range ordered crystal structure of an ordered intermetallic compound*

have been used mainly as strengthening constituents in structural materials. For example, high-temperature nickel-base superalloys owe their outstanding strength properties to a fine dispersion of precipitated particles of the ordered $\gamma'$ phase ($Ni_3Al$) embedded in a ductile disordered matrix. See also *superalloys* (Technical Brief 52).

In recent years, alloying and processing have been employed to control the ordered crystal structure, microstructural features, and grain-boundary structure and composition to overcome the brittleness problem of ordered intermetallics. Alloy design work has centered primarily on aluminides of nickel, iron, and titanium.

Fig. 261

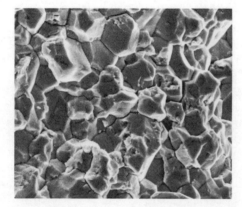

*Intergranular fracture of a maraging steel spring that failed from hydrogen embrittlement. 1000×*

crystals, that form the material (Fig. 261). Also called intercrystalline fracture. Contrast with *transgranular fracture*.

**intergranular penetration**. In welding, the penetration of a filler metal along the grain boundaries of a base metal.

**intergranular stress-corrosion cracking (IGSCC)**. *Stress-corrosion cracking* in which the cracking occurs along grain boundaries.

**interlaminar**. Between two adjacent laminae, for example, an object (such as a flaw, Fig. 262), an event (such as a fracture), or a potential field (such as shear stress). See also *fracture* (composites).

**interlaminar shear**. Shearing force tending to produce a relative displacement between two laminae in a laminate along the plane of their interface.

These materials possess a number of attributes that make them attractive for high-temperature applications. They contain sufficient amounts of aluminum to form, in oxidizing environments, thin films of alumina ($Al_2O_3$) that often are compact and protective. These materials have low densities, relatively high melting points, and good high-temperature strength properties.

The search for new high-temperature structural materials for aerospace applications has stimulated further interest in intermetallics. Titanium aluminides ($Ti_3Al$ and TiAl) with high specific strengths have been developed for jet engine, aircraft, and related structural applications. Nickel aluminides based on $Ni_3Al$ that have a combination of good strength, ductility, and oxidation resistance are currently being developed for use as high-temperature dies, heating elements, and hot components in heat engines and energy conversion systems. It has been found that $Fe_3Al$ and FeAl aluminides possess excellent corrosion resistance in oxidizing, sulfidizing, and molten salt environments.

Nonstructural applications for ordered intermetallics include the use of molybdenum disilicide ($MoSi_2$) for electrical heating elements in high-temperature furnaces, the development of the $Fe_3(Si,Al)$ alloy (Sendust) for magnetic applications, the use of samarium-cobalt materials ($SmCo_5$ and $Sm_2Co_{17}$) for *permanent magnets*, the use of NiTi (Nitinol) as a *shape memory alloy*, and the use of niobium-base materials (such as $Nb_3Sn$) as *superconductors*.

## Recommended Reading

- C.T. Liu, J.O. Stiegler, and F.H. Froes, Ordered Intermetallics, *Metals Handbook*, 10th ed., Vol 2, ASM International, 1990, p 913-942

### Properties of nickel, iron, and titanium aluminides

| Alloy | Crystal structure(a) | Critical ordering temperature ($T_c$) °C | °F | Melting point ($T_m$) °C | °F | Material density g/cm³ | Young's modulus GPa | $10^6$ psi |
|---|---|---|---|---|---|---|---|---|
| $Ni_3Al$ | $L1_2$ (ordered fcc) | 1390 | 2535 | 1390 | 2535 | 7.50 | 179 | 25.9 |
| NiAl | B2 (ordered bcc) | 1640 | 2985 | 1640 | 2985 | 5.86 | 294 | 42.7 |
| $Fe_3Al$ | $D0_3$ (ordered bcc) | 540 | 1000 | 1540 | 2805 | 6.72 | 141 | 20.4 |
|  | B2 (ordered bcc) | 760 | 1400 | 1540 | 2805 | . . . | . . . | . . . |
| FeAl | B2 (ordered bcc) | 1250 | 2280 | 1250 | 2280 | 5.56 | 261 | 37.8 |
| $Ti_3Al$ | $D0_{19}$ (ordered hcp) | 1100 | 2010 | 1600 | 2910 | 4.2 | 145 | 21.0 |
| TiAl | $L1_0$ (ordered tetragonal) | 1460 | 2660 | 1460 | 2660 | 3.91 | 176 | 25.5 |
| $TiAl_3$ | $D0_{22}$ (ordered tetragonal) | 1350 | 2460 | 1350 | 2460 | 3.4 | . . . | . . . |

(a) fcc, face-centered cubic; bcc, body-centered cubic; hcp, hexagonal close packed

**intermediate annealing.** Annealing wrought metals at one or more stages during manufacture and before final treatment.

**intermediate electrode.** Same as *bipolar electrode*.

**intermediate flux.** A soldering flux with a residue that generally does not attack the base metal. The original composition may be corrosive.

**intermediate phase.** In an alloy or a chemical system, a distinguishable homogeneous phase whose composition range does not extend to any of the pure components of the system.

**intermediate temperature setting adhesive.** An adhesive that sets in the temperature range from 30 to 100 °C (87 to 211 °F).

**intermetallic compound.** An intermediate phase in an alloy system, having a narrow range of homogeneity and rela-

tively simple stoichiometric proportions; the nature of the atomic binding can be of various types, ranging from metallic to ionic.

**intermetallic phases.** Compounds, or intermediate solutions, containing two or more metals, which usually have compositions, characteristic properties, and crystal structures different from those of the pure components of the system.

**intermetallics.** See *Technical Brief 22*.

**intermittent weld.** A weld in which the continuity is broken by recurring unwelded spaces.

**internal friction.** The conversion of energy into heat by a material subjected to fluctuating stress.

**internal grinding.** Grinding an internal surface such as that inside a cylinder or hole. See also the figure accompanying the term *grinding*.

FIG. 262

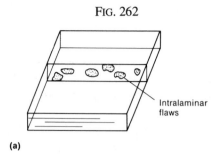

(a)

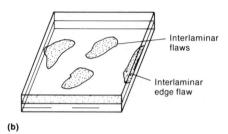

(b)

*Views of (a) effective intralaminar flaws and (b) effective interlaminar flaws*

**internal oxidation.** The formation of isolated particles of corrosion products beneath the metal surface. This occurs as the result of preferential oxidation of certain alloy constituents by inward diffusion of oxygen, nitrogen, sulfur, and so forth. Also called subscale formation.

**internal shrinkage.** A void or network of voids within a casting caused by inadequate feeding of that section during solidification.

**internal shrinkage cracks.** Longitudinal cracks in a composite pultrusion that are found within sections of roving reinforcement. This condition is caused by shrinkage strains during cure that show up in the roving portion of the pultrusion, where transverse strength is low.

**internal standard.** In spectroscopy, a material present in or added to samples that serves as an intensity reference for measurements; used to compensate for variations in sample excitation and photographic processing in emission spectroscopy.

**internal standard line.** In spectroscopy, a spectral line of an *internal standard*, with which the radiant energy of an analytical line is compared.

**internal stress.** See preferred term *residual stress*.

**interpass temperature.** In a multiple-pass weld, the temperature (minimum or maximum as specified) of the deposited weld metal before the next pass is started.

**interphase.** The boundary region between a bulk resin or polymer and an adherend in which the polymer has a high degree of orientation to the adherend on a molecular basis. It plays a major role in the load transfer process between the bulk of the adhesive and the adherend or the fiber and the laminate matrix resin.

**interplanar distance.** The perpendicular distance between adjacent parallel lattice planes.

**interply hybrid.** A reinforced plastic laminate in which adjacent laminae are composed of different materials.

**interpulse time** (resistance welding). The time between successive pulses of current within the same impulse.

**interrupted aging.** Aging at two or more temperatures, by steps, and cooling to room temperature after each step. See also *aging*, and compare with *progressive aging* and *step aging*.

**interrupted-current plating.** Plating in which the flow of current is discontinued for periodic short intervals to decrease anode polarization and elevate the *critical current density*. It is most commonly used in cyanide copper plating.

**interrupted quenching.** A quenching procedure in which the workpiece is removed from the first quench at a temperature substantially higher than that of the quenchant and is then subjected to a second quenching system having a different cooling rate than the first.

**intersection scarp.** A line, of any shape, which is the locus of intersection of two portions of a crack with one another. This is exemplified by intersection of a portion of a slow crack running wet with a portion not wetted (Fig. 263). See also *transition scarp*.

**interstitial solid solution.** A type of solid solution that sometimes forms in alloy systems having two elements of widely different atomic sizes. Elements of small atomic size, such as carbon, hydrogen, and nitrogen, often dissolve in solid metals to form this solid solution. The space lattice is similar to that of the pure metal, and the atoms of carbon, hydrogen, and nitrogen occupy the spaces or interstices between the metal atoms.

**intersystem crossing.** A transition between electronic states that differ in total spin quantum number.

**interval erosion rate.** The slope of a line connecting two specified points on the cumulative erosion-time curve.

FIG. 263

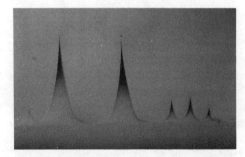

*Intersection scarp on the fracture surface of a glass plate. Direction of fracture was from bottom to top. 60×*

**interval estimate**. The estimate of a parameter given by two statistics, defining the end points of an interval.

**interval test**. Method used to test heat extraction rates of various quenchants. This test measures the increase in temperature of a quenchant when a standard bar of metal is quenched for five seconds. Faster quenchants will exhibit greater temperature increases.

**intracrystalline**. Within or across the crystals or grains of a metal; same as transcrystalline and transgranular.

**intracrystalline cracking**. See *transgranular cracking*.

**intralaminar**. Within a single lamina, for example, an object (such as a void), and event (such as a fracture), or a potential field (such as a temperature gradient). See also the figure accompanying the term *interlaminar*.

**intraply hybrid**. A reinforced plastic laminate in which more than one material is used within a specific layer.

**intrinsic viscosity**. For a polymer, the limiting value of infinite dilution of the ratio of the specific viscosity of the polymer solution to its concentration in moles per liter.

**introfaction**. The change in fluidity and wetting properties of a polymeric impregnating material, produced by the addition of an *introfier*.

**introfier**. A chemical that converts a colloidal solution into a molecular one. See also *introfaction*.

**intumescence**. The swelling or bubbling of a coating usually because of heating (term currently used in space and fire protection applications).

**inverse chill**. The condition in a casting section in which the interior is mottled or white, while the other sections are gray iron. Also known as reverse chill, internal chill, and inverted chill.

**inverse segregation**. A concentration of low-melting constituents in those regions of an alloy in which solidification first occurs (Fig. 264).

FIG. 264

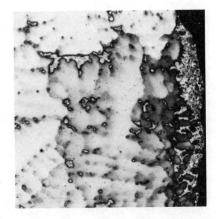

*Horizontal cast strip of tin bronze (5% Sn), showing inverse segregation at the bottom surface of the casting (right). 100×*

**inverted microscope**. A microscope arranged so that the line of sight is directed upward through the objective to the object.

**investing**. In *investment casting*, the process of pouring the investment slurry into a flask surrounding the pattern to form the mold.

**investment**. A flowable mixture, or slurry, of a graded refractory filler, a binder, and a liquid vehicle that, when poured around the patterns, conforms to their shape and subsequently sets hard to form the investment mold (Fig. 265).

**investment casting**. (1) Casting metal into a mold produced by surrounding, or *investing*, an expendable pattern with a refractory slurry coating that sets at room temperature, after which the wax or plastic pattern is removed through the use of heat prior to filling the mold with liquid metal (Fig. 265). Also called *precision casting* or *lost wax process*. (2) A part made by the investment casting process.

**investment compound**. A mixture of a graded refractory filler, a binder and a liquid vehicle, used to make molds for *investment casting*.

**investment precoat**. An extremely fine investment coating applied as a thin slurry directly to the surface of the pattern to reproduce maximum surface smoothness. The coating is surrounded by a coarser, cheaper, and more permeable investment to form the mold. See also *dip coat* and *investment casting*.

**investment shell**. Ceramic mold obtained by alternately dipping a pattern set up in *dip coat* slurry and stuccoing with coarse ceramic particles until the shell of desired thickness is obtained. See also *investment casting*.

**ion**. An atom, or group of atoms, which by loss or gain of one or more electrons has acquired an electric charge. If the ion is formed from an atom of hydrogen or an atom of a metal, it is usually positively charged; if the ion is formed from an atom of a nonmetal or from a group of atoms, it is usually negatively charged. The number of electronic charges carried by an ion is termed its electrovalence. The charges are denoted by superscripts that give their sign and number; for example, a sodium ion, which carries one positive charge, is denoted by $Na^+$; a sulfate ion, which carries two negative charges, by $SO_4^{2-}$. See also *atomic structure* and *chemical bonding*.

**ion beam assisted deposition**. See *ion implantation*.

**ion beam mixing**. See *ion implantation*.

**ion beam sputtering**. See *ion implantation*.

**ion carburizing**. A method of surface hardening in which carbon ions are diffused into a workpiece in a vacuum through the use of high-voltage electrical energy (Fig. 266). Synonymous with plasma carburizing or glow-discharge carburizing.

**ion chromatography**. An area of high-performance liquid chromatography that uses ion exchange resins to separate

FIG. 265

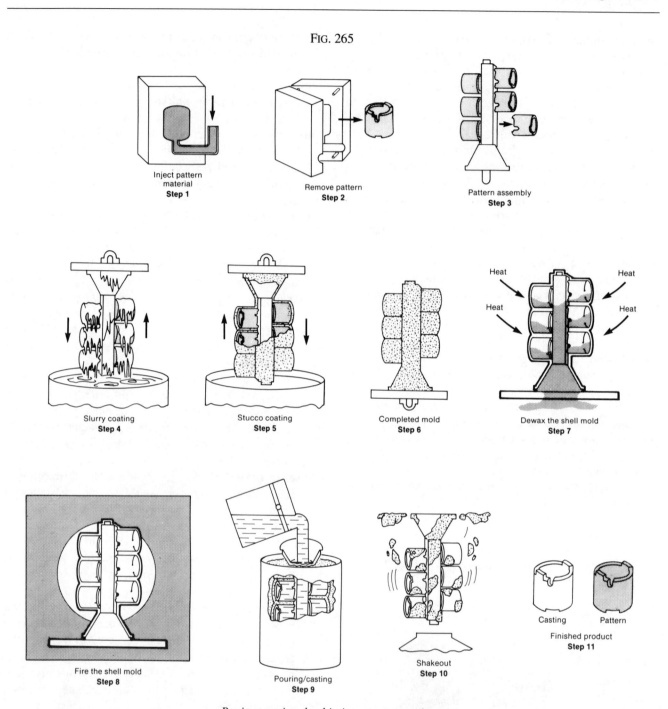

*Basic steps involved in investment casting*

various species of ions in solution and elute them to a suitable detector for analysis. See also *chromatography, gas chromatography,* and *liquid chromatography*.

**ion etching**. Surface removal by bombarding with accelerated ions in vacuum (1 to 10 kV).

**ion exchange**. The reversible interchange of ions between a liquid and solid, with no substantial structural changes in the solid.

**ion exchange** (ceramics). An exchange of ions between two materials in intimate contact, usually accomplished by the application of heat.

**ion-exchange chromatography (IEC)**. Liquid chromatography with a stationary phase that possesses charged functional groups. This technique is applicable to the separation of ionic (charged) compounds. See also *ion chromatography*.

**ion-exchange resins**. Cross-linked polymers that form salts with ions from aqueous solutions.

FIG. 266

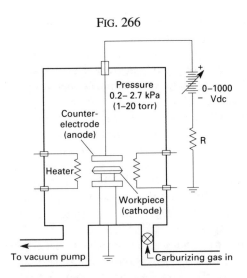

*Schematic of an ion carburizing apparatus*

**ionic bond.** (1) A type of chemical bonding in which one or more electrons are transferred completely from one atom to another, thus converting the neutral atoms into electrically charged ions. These ions are approximately spherical and attract each other because of their opposite charges. (2) A primary bond arising from the electrostatic attraction between two oppositely charged ions. Contrast with *covalent bond*.

**ionic charge.** The positive or negative charge of an ion.

**ionic crystal.** A crystal in which atomic bonds are *ionic bonds*. This type of atomic linkage, also known as (hetero) polar bonding, is characteristic of many compounds (sodium chloride, for instance).

**ion implantation.** The process of modifying the physical or chemical properties of the near surface of a solid (target) by embedding appropriate atoms into it from a beam of ionized particles (Fig. 267). The properties to be modified may be electrical, optical, or mechanical, and they may relate to the semiconducting behavior of the material or its corrosion behavior. The solid may be crystalline, polycrystalline, or amorphous and need not be homogeneous. Related techniques are also used in conjunction with ion implantation to increase the ratio of material introduced

into the substrate per unit area, to provide appropriate mixtures of materials, or to overcome other difficulties involved in surface modification by ion implantation alone. These techniques include:

1. Ion beam sputtering: An ion beam of argon or xenon directed at a target sputters material from the target to a substrate; the sputtered material arrives at the substrate with enough energy to promote good adhesion of the coating to substrate.

2. Ion beam mixing: Deposited layers (electroplating, sputtering) tens or hundreds of nanometers thick are mixed and bonded to the substrate by an argon or xenon ion beam.

3. Plasma ion deposition: Ion beams are used to create coatings having special phases, especially ion beam formed carbon coatings in the diamond phase or ion beam formed boron nitride coatings.

4. Ion beam assisted deposition: Ion beams are combined with physical vapor deposition.

**ionization.** (1) The process in which neutral atoms become charged by gaining or losing an electron. (2) The act of splitting into, or producing, ions.

**ionization chamber.** An enclosure containing two or more electrodes surrounded by a gas capable of conducting an electric current when it is ionized by x-rays or other ionizing rays. It is commonly used for measuring the intensity of such radiation.

**ion migration.** The movement of free ions within a material or across the boundary between two materials under the influence of an applied electric field.

**ion neutralization.** The generic term for a class of charge-exchange processes in which an ion is neutralized by passage through a gas or by interaction with a material surface.

**ion nitriding.** A method of surface hardening in which nitrogen ions are diffused into a workpiece in a vacuum through the use of high-voltage electrical energy (Fig. 268). Sy-

FIG. 267

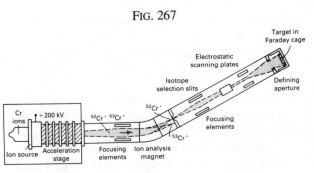

*Schematic of a research-type ion implantation system*

FIG. 268

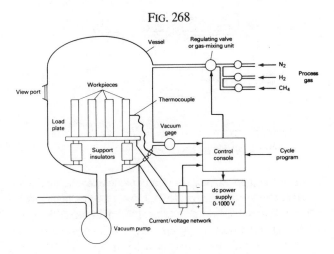

*Typical ion-nitriding vessel*

nonymous with plasma nitriding or glow-discharge nitriding.

**ionomer resins.** A polymer that has ethylene as its major component, but that contains both covalent and ionic bonds. The polymer exhibits very strong interchain ionic forces. The anions hang from the hydrocarbon chain, and the cations are metallic, for example, sodium, potassium, or magnesium. These resins have many of the same features as polyethylene, plus high transparency, tenacity, resilience, and increased resistance to oils, greases, and solvents. Fabrication is accomplished as it is with polyethylene.

**ion-pair chromatography (IPC).** Liquid chromatography with a mobile phase containing an ion that combines with sample ions, creating neutral ion pairs. The ion pairs are typically separated using bonded-phase chromatography. See also *bonded-phase chromatography.*

**ion plating.** A generic term applied to atomistic film deposition processes in which the substrate surface and/or the depositing film is subjected to a flux of high-energy particles (usually gas ions) sufficient to cause changes in the interfacial region or film properties (Fig. 269). Such changes may be in film adhesion to the substrate, film morphology, film density, film stress, or surface coverage by the depositing film material.

FIG. 269

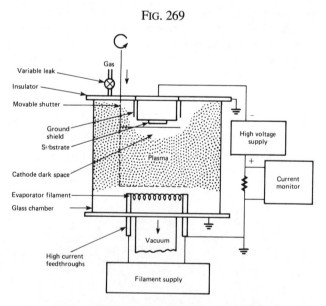

*Ion-plating apparatus using a dc gas discharge and an evaporator filament*

**ion-scattering spectrometry.** A technique to elucidate composition and structure of the outermost atomic layers of a solid material, in which principally mono-energetic, singly charged, low-energy (less than 10 keV) probe ions are scattered from the surface and are subsequently detected and recorded as a function of the energy.

**ion species.** Type and charge of an ion. If an isotope is used, it should be specified.

**IRG transition diagram.** Developed by the International Research Group (IRG) on Wear of Engineering Materials of the Organization of Economic Cooperation and Development (OECD), it is a plot of normal force in newtons (ordinate) versus sliding velocity in meters per second (abscissa) wherein boundaries identify three distinct regions of varying lubricant effectiveness.

**iridescence.** Loss of brilliance in metallized plastics and development of multicolor reflectance. Iridescence is caused by the cold flow of plastic or of coating and by excess heat during vacuum metallizing.

**iron casting.** A part made of *cast iron.*

**ironing.** An operation used to increase the length of a tube or cup through reduction of wall thickness and outside diameter, the inner diameter remaining unchanged (Fig. 270).

FIG. 270

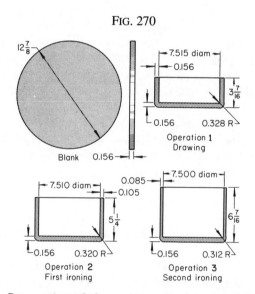

*Progression of shapes in production of a shell with a thick bottom and thin sides in one draw and two ironing operations. Dimensions given in inches.*

**iron-powder electrode.** A welding electrode with a covering containing up to about 50% iron powder, some of which becomes part of the deposit.

**iron rot.** Deterioration of wood in contact with iron-base alloys.

**iron soldering.** A soldering process in which the heat required is obtained from a soldering iron.

**irradiance.** The radiant flux density incident on a surface; the ratio of flux to area of irradiated surface.

**irradiation.** The exposure of a material or object to x-rays, gamma rays, ultraviolet rays, or other ionizing radiation.

**irradiation** (plastics). The bombardment of plastics with a variety of subatomic particles, usually alpha-, beta-, or gamma-rays. Used to initiate the polymerization and copolymerization of plastics and in some cases to bring about changes in the physical properties of a plastic.

**irregular powder**. Particles lacking symmetry.

**irreversible** (plastics). Not capable of redissolving or remelting. Chemical reactions that proceed in a single direction and are not capable of reversal (as applied to thermosetting resins).

**isobar**. In atomic physics, one of two or more atoms that have a common mass number $A$, but differ in atomic number $Z$. Thus, although isobars possess approximately equal masses, they differ in chemical properties; they are atoms of different elements. See also *nuclear structure*.

**isocorrosion diagram**. A graph or chart that shows constant corrosion behavior with changing solution (environment) composition and temperature (Fig. 271).

**isocratic elution**. In liquid chromatography, the use of a mobile phase whose composition is unchanged throughout the course of the separation process.

**isocyanate plastics**. Plastics based on resins made by the condensation of organic isocyanates with other compounds. Generally reacted with polyols on a polyester or polyether backbone molecule, with the reactants being joined by the formation of the urethane linkage. See also *polyurethane* (Technical Brief 32) and *urethane plastics*.

**isomer**. A compound, radical, ion, or nuclide that contains the same number of atoms of the same elements but differs in structural arrangement and properties. See also *stereoisomer*.

**isometric**. A crystal form in which the unit dimension on all three axes is the same.

**isomorphous**. Having the same crystal structure. This usually refers to intermediate phases that form a continuous series of solid solutions.

**isomorphous system**. A complete series of mixtures in all proportions of two or more components in which unlimited mutual solubility exists in the liquid and solid states.

**isostatic mold**. A sealed container of glass or sheet of carbon steel, stainless steel, or a nickel-based alloy. See also *isostatic pressing*.

**isostatic pressing**. A process for forming a powder metallurgy compact by applying pressure equally from all directions to metal powder contained in a sealed flexible mold. See also *cold isostatic pressing* and *hot isostatic pressing*.

**isotactic stereoisomerism**. A type of polymeric molecular structure containing a sequence of regularly spaced asymmetric groups arranged in like configuration in a polymer chain (Fig. 272). Isotactic (and syndiotactic) polymers are crystallizable.

FIG. 271

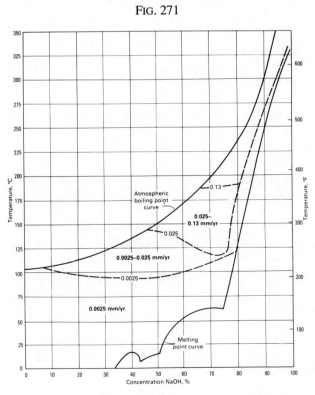

*Isocorrosion diagram for Nickel 200 and Nickel 201 in NaOH*

FIG. 272

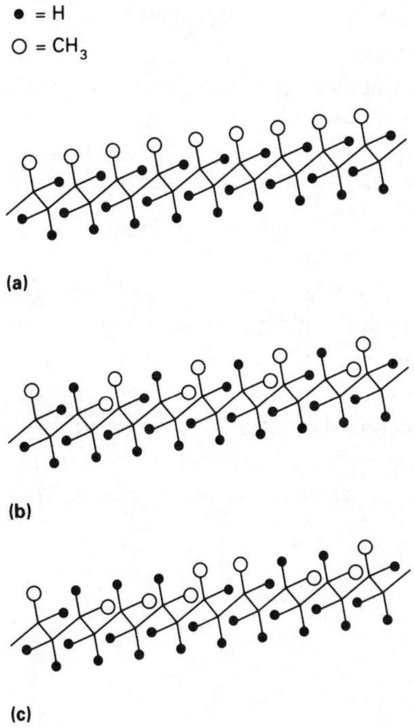

*Stereoisomers of polypropylene. (a) Isotactic. (b) Syndiotactic. (c) Atactic*

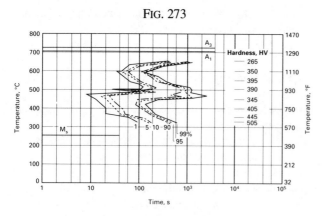

FIG. 273

*Isothermal transformation diagram for a steel with 0.39% C, 0.86% Mn, 0.72% Cr, and 0.97% Ni. The upper C-shape curves describe transformation to pearlite; the lower C-shape curves to bainite. The column on the right side indicates the hardness after completed transformation measured at room temperature.*

**isothermal annealing.** Austenitizing a ferrous alloy, then cooling to and holding at a temperature at which austenite transforms to a relatively soft ferrite-carbide aggregate. See also *austenitizing.*

**isothermal forging.** A hot-forging process in which a constant and uniform temperature is maintained in the workpiece during forging by heating the dies to the same temperature as the workpiece. The process permits the use of extremely slow strain rates, thus taking advantage of the strain rate sensitivity of flow stress for certain alloys (for example, titanium- and nickel-base alloys). The process is capable of producing net shape forgings that are ready to use without machining or near-net shape forgings that require minimal secondary machining.

**isothermal transformation.** A change in phase that takes place at a constant temperature. The time required for transformation to be completed, and in some instances the time delay before transformation begins, depends on the amount of supercooling below (or superheating above) the equilibrium temperature for the same transformation.

**isothermal transformation (IT) diagram.** A diagram that shows the isothermal time required for transformation of austenite to begin and to finish as a function of temperature

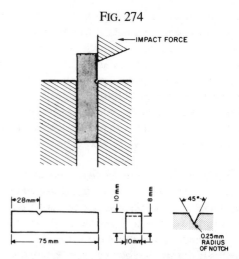

FIG. 274

*Setup and specimen for Izod impact testing*

(Fig. 273). Same as time-temperature-transformation (TTT) diagram or S-curve.

**isotone.** One of two or more atoms that display a constant difference $A - Z$ between their mass number $A$ and their atomic number $Z$. Thus, despite differences in the total number of nuclear constituents, the numbers of neutrons in the nuclei of isotones are the same. See also *nuclear structure.*

**isotope.** One of two or more nuclidic species of an element having an identical number of protons ($Z$) in the nucleus, but a different number of neutrons ($N$). Isotopes differ in mass, but chemically are the same element. See also *nuclear structure.*

**isotropic.** Having uniform properties in all directions. The measured properties of an isotropic material are independent of the axis of testing.

**isotropy.** The condition of having the same values of properties in all directions.

**item.** (1) An object or quantity of material on which a set of observations can be made. (2) An observed value or test result obtained from an object or quantity of material.

**Izod test.** A type of impact test in which a V-notched specimen, mounted vertically, is subjected to a sudden blow delivered by the weight at the end of a pendulum arm (Fig. 274). The energy required to break off the free end is a measure of the impact strength or toughness of the material. Contrast with *Charpy test.*

# J

**jaw crusher**. A machine for the primary disintegration of metal pieces, ores, or agglomerates into coarse powder. See also *crushing*.

**Jeffries' method**. A method for determining grain size based on counting grains in a prescribed area.

**jet molding**. A processing technique for plastics characterized by the fact that most of the heat is applied to the material as it passes through the nozzle or jet, rather than in a heating cylinder, as is done in conventional processes.

**jet pulverizer**. A machine that comminutes metal pieces, ores, or agglomerates by means of pressurized air or steam injected into a chamber.

**jetting** (plastics). The turbulent flow of resin from an undersized gate or thin section into a thicker mold cavity, as opposed to the laminar flow of material progressing radially from a gate to the extremities of the cavity.

**jetting** (welding). The flow process and expulsion of the metal surface that takes place during *explosion welding*.

**jewel bearing**. A bearing made of diamond, sapphire, or a hard substitute metal.

**jig**. A mechanism for holding a part and guiding the tool during machining or assembly operation.

**jig boring**. Boring with a single-point tool where the work is positioned upon a table that can be located so as to bring any desired part of the work under the tool. Thus, holes can be accurately spaced. This type of boring can be done on milling machines or jig borers.

**jig grinding**. Analogous to *jig boring*, where the holes are ground rather than machined.

**J-integral**. A mathematical expression; a line or surface integral that encloses the crack front from one crack surface to the other, used to characterize the *fracture toughness* of a material having appreciable plasticity before fracture. The *J*-integral eliminates the need to describe the behavior of the material near the crack tip by considering the local stress-strain field around the crack front; $J_{Ic}$ is the critical value of the *J*-integral required to initiate growth of a preexisting crack.

**joggle**. An offset in a flat plane consisting of two parallel bends at the same angle but in opposite directions.

**Johnson noise**. See *thermal noise*.

**joint**. The location where two or more members are to be or have been fastened together mechanically or by welding, brazing, soldering, or adhesive bonding.

**joint-aging time**. See *joint-conditioning time*.

**joint buildup sequence**. See *buildup sequence*.

**joint clearance**. The distance between the faying surfaces of a joint. In brazing, this distance is referred to as that which

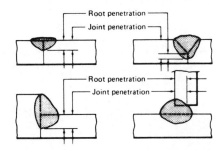

FIG. 275

*Examples of weld joint penetration*

is present before brazing, at the brazing temperature, or after brazing is completed.

**joint-conditioning time**. In adhesive bonding, the time interval between the removal of the joint from the conditions of heat or pressure, or both, used to accomplish bonding and the attainment of approximately maximum bond strength. See also *curing time, drying time*, and *setting time*.

**joint design**. The joint geometry together with the required dimensions of the welded joint.

**joint efficiency**. The strength of a welded joint expressed as a percentage of the strength of the unwelded base metal.

**joint geometry**. The shape and dimensions of a joint in cross section prior to welding.

**joint penetration**. The minimum depth to which a groove or flange weld extends from its face into the joint, exclusive of reinforcement (Fig. 275). Joint penetration may include *root penetration*.

**jolt ramming**. Packing sand in a mold by raising and dropping the sand, pattern, and flask on a table. Jolt-type (Fig.

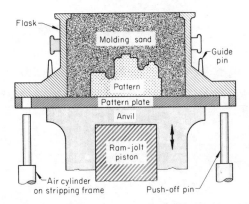

FIG. 276

*Primary components of a jolt-type molding machine*

## FIG. 277

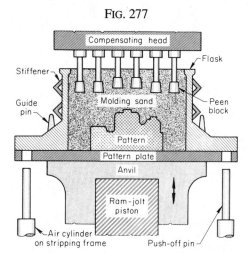

*Jolt squeeze molding machine with compensating heads*

276), jolt squeezers, jarring machines, and jolt rammers are machines using this principle. Also called jar ramming.

**jolt-squeezer machine**. A combination machine that employs a jolt action followed by a squeezing action to compact the sand around the pattern (Fig. 277).

**Jominy test**. A laboratory test for determining the hardenability of steel that involves heating the test specimen to the proper *austenitizing* temperature and then transferring it to a quenching fixture so designed that the specimen is held vertically 12.7 mm (0.5 in.) above an opening through which a column of water can be directed against the bottom face of the specimen (Fig. 278a). While the bottom end is being quenched by the column of water, the opposite end is cooling slowly in air, and intermediate positions along the specimen are cooling at intermediate rates. After the specimen has been quenched, parallel flats 180° apart are ground 0.38 mm (0.015 in.) deep on the cylindrical surface. Rockwell C hardness is measured at intervals of $\frac{1}{16}$ in. (1.6 mm) for alloy steels and $\frac{1}{32}$ in. (0.8 mm) for carbon steels, starting from the water-quenched end. A typical plot of these hardness values and their positions on the test bar indicates the relation between hardness and cooling rate (Fig. 278b), which in effect is the hardenability of the steel. Also referred to as *end-quench hardenability* test.

**Joule-Thomson effect**. A change in temperature in a gas undergoing Joule-Thomson expansion. See also *Joule-Thomson expansion*.

**Joule-Thomson expansion**. The adiabatic, irreversible expansion of a gas flowing through a porous plug or partially open valve. See also *Joule-Thomson effect*.

**journal**. The part of a shaft or axle that rotates or oscillates relative to a radial bearing. A journal is part of a larger unit, for example, a crankshaft or lineshaft, and it is preferred that the term shaft be kept for the whole unit.

## FIG. 278

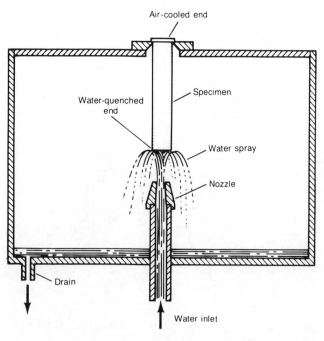

(a)

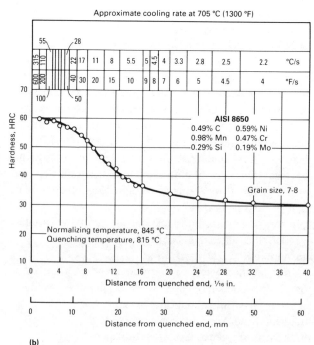

(b)

*Jominy end-quench apparatus (a) and method for presenting end-quench hardenability data (b)*

**journal bearing**. A sliding-type bearing in which a journal rotates or oscillates relative to its housing. A full journal bearing extends 360°, but partial bearings may extend, for example, over 180° or 120°. See also the figure accompanying the term *solid lubrication*.

**jute**. A bast fiber obtained from the stems of several species of the plant *Corchorus* found mainly in India and Pakistan. Used as a filler for plastic molding materials, and as a reinforcement for polyester resins in the fabrication of reinforced plastics.

# K

**karat**. A unit for designating the fineness of gold in an alloy. In this system, 24 karat (24 k) is 1000 fine or pure gold. The most popular jewelry golds are:

| Karat designation | Gold content |
| --- | --- |
| 24 k | 100% Au (99.95% min) |
| 18 k | 18/24ths, or 75% Au |
| 14 k | 14/24ths, or 58.33% Au |
| 10 k | 10/24ths, or 41.67% Au |

**keel block**. A standard test casting, for steel and other high-shrinkage alloys, consisting of a rectangular bar that resembles the keel of a boat, attached to the bottom of a large riser, or shrinkhead. Keel blocks that have only one bar are often called Y-blocks; keel blocks having two bars, double keel blocks. Test specimens are machined from the rectangular bar, and the shrinkhead is discarded.

**kelvin**. A scale of absolute temperatures in which zero is approximately –273.16 °C (–459.69 °F). The color temperature of light is measured in kelvins.

**kerf**. The width of the cut produced during a cutting process.

**Kevlar**. An organic polymer composed of aromatic polyamides (*aramids*) having a para-type orientation (parallel chain with bonds extending from each aromatic nucleus). Often used as a reinforcing fiber.

**keyhole**. A technique of welding in which a concentrated heat source, such as a plasma arc, penetrates completely through a workpiece forming a hole at the leading edge of the molten weld metal. As the heat source progresses, the molten metal fills in behind the hold to form the weld bead (Fig. 279).

FIG. 279

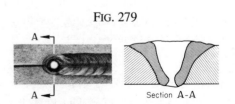

*Keyhole weld in 6 mm (1/4 in.) thick stainless steel plate*

**keyhole specimen**. A type of specimen containing a hole-and-slot notch, shaped like a keyhole, usually used in impact bend tests. See also *Charpy test* and *Izod test*.

***k*-factor**. The ratio between the unknown and standard x-ray intensities used in quantitative analyses.

**Kikuchi lines**. Light and dark lines superimposed on the background of a single-crystal electron-diffraction pattern caused by diffraction of diffusely scattered electrons within the crystal (Fig. 280); the pattern provides structural information on the crystal.

FIG. 280

*Kikuchi lines on a diffraction pattern*

**killed steel**. Steel treated with a strong deoxidizing agent such as silicon or aluminum in order to reduce the oxygen content to such a level that no reaction occurs between carbon and oxygen during solidification. See also the figure accompanying the term *capped steel*.

**kiln**. A large furnace used for baking, drying, or burning firebrick or refractories, or for calcining ores or other substances.

**kiloelectron volt (keV).** A unit of energy usually associated with individual particles. The energy gained by an electron when accelerated across 1000 V.

**kinematic viscosity.** See *viscosity*.

**kinematic wear marks.** In ball bearings, a series of short curved marks on the surface of a bearing race due to the kinematic action of imbedded particles or asperities rolling and spinning at the ball or roller contact points. The length and curvature of these marks depend on the degree of spinning and on the distance from the spinning axis of the rolling element.

**kinetic coefficient of friction.** The *coefficient of friction* under conditions of macroscopic relative motion between two bodies.

**kinetic energy.** The energy that a body possesses because of its motion; in classical mechanics, equal to one half of the body's mass times the square of its speed.

**kinetic friction.** Friction under conditions of macroscopic relative motion between two bodies. This term is sometimes used as a synonym for *kinetic coefficient of friction*; however, it can also be used merely to indicate that the type of friction being indicated is associated with macroscopic motion rather than static conditions.

**Kingsbury bearing.** See *tilting-pad bearing*.

**kink band (deformation).** In polycrystalline materials, a volume of crystal that has rotated physically to accommodate differential deformation between adjoining parts of a grain while the band itself has deformed homogeneously. This occurs by regular bending of the slip lamellae along the boundaries of the band.

**Kirkendall voids.** The formation of voids by diffusion across the interface between two different materials, with the material having the greater diffusion rate diffusing into the other.

**$K_{ISCC}$.** Abbreviation for the critical value of the plane strain *stress-intensity factor* that will produce crack propagation by *stress-corrosion cracking* of a given material in a given environment.

**kish.** Free graphite that forms in molten hypereutectic cast iron as it cools. In castings, the kish may segregate toward the cope surface, where it lodges at or immediately beneath the casting surface.

**klystron.** An evacuated electron-beam tube in which an initial velocity modulation imparted to electrons in the beam subsequently results in density modulation of the beam. This device is used as an amplifier or oscillator in the microwave region.

**knee** (resistance welding). The lower arm supporting structure in a resistance welding machine.

**knife coating.** A method of coating a substrate (usually paper or fabric) in which the substrate, in the form of a continuous moving web, is coated with a material, the thickness of which is controlled by an adjustable knife or bar set at a suitable angle to the substrate.

**knife-line attack.** *Intergranular corrosion* of an alloy, usually stabilized stainless steel, along a line adjoining or in contact with a weld after heating into the sensitization temperature range.

**knitted fabrics.** Fabrics produced by interlooping chains of yarn.

**knock.** In a spark ignition engine, uneven burning of the air/fuel charge that causes violent, explosive combustion and an audible metallic hammering noise. Knock results from premature ignition of the last part of the charge to burn.

**knockout.** (1) Removal of sand cores from a casting. (2) Jarring of an investment casting mold to remove the casting and investment from the flask. (3) A mechanism for freeing formed parts from a die used for stamping, blanking, drawing, forging or heading operations. (4) A partially pierced hole in a sheet metal part, where the slug remains in the hole and can be forced out by hand if a hole is needed.

**knockout mark.** A small protrusion, such as a button or ring of flash, resulting from depression of the *knockout pin* from the forging pressure or the entrance of metal between the knockout pin and the die.

**knockout pin.** A power-operated plunger installed in a die to aid removal of the finished forging.

**knockout punch.** A punch used for ejecting powder compacts.

**Knoop hardness number (HK).** A number related to the applied load and to the projected area of the permanent impression made by a rhombic-based pyramidal diamond indenter having included edge angles of 172° 30′ and 130° 0′ computed from the equation:

$$HK = \frac{P}{0.07028d^2}$$

where $P$ is applied load, kgf; and $d$ is the long diagonal of the impression, mm. In reporting Knoop hardness numbers, the test load is stated.

**Knoop hardness test.** An indentation hardness test using calibrated machines to force a rhombic-based pyramidal diamond indenter having specified edge angles, under specified conditions, into the surface of the material under test and to measure the long diagonal after removal of the load (Fig. 281).

**knuckle area.** In reinforced plastics, the area of transition between sections of different geometry in a filament-wound part, for example, where the skirt joins the cylinder of the pressure vessel. Also called Y-joint.

**knuckle-lever press.** A heavy short-stroke press in which the slide is directly actuated by a single toggle joint that is opened and closed by a connection and crack (Fig. 282). It is used for embossing, coining, sizing, heading, swaging, and extruding.

FIG. 281

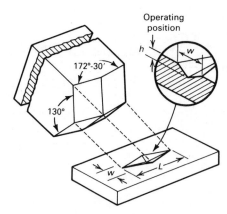

*Pyramidal Knoop indenter and resulting indentation in the workpiece*

FIG. 282

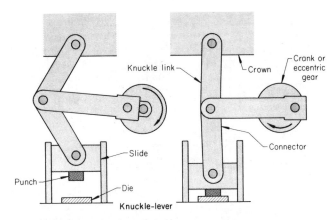

*Knuckle-lever drive system for mechanical presses*

**knurling**. Impressing a design into a metallic surface, usually by means of small, hard rollers that carry the corresponding design on their surfaces.

**K-radiation**. Characteristic x-rays produced by an atom or ion when a vacancy in the K shell is filled by an electron from another shell.

**kraft process**. A wood-pulping process in which sodium sulfate is used in the caustic soda pulp-digestion liquor. Also called kraft pulping or sulfate pulping.

**Kroll process**. A process for the production of metallic titanium sponge by the reduction of titanium tetrachloride with a more active metal, such as magnesium or sodium. The sponge is further processed to granules or powder.

**K-series**. The set of characteristic x-ray wavelengths making up K-radiation for the various elements.

**K shell**. The innermost shell of electrons surrounding the atomic nucleus, having electrons characterized by the principal quantum number 1.

# L

$L_{10}$-**life**. See *rating life*.

**laboratory sample**. A sample, intended for testing or analysis, prepared from a gross sample or otherwise obtained; the laboratory sample must retain the composition of the gross sample. Reduction in particle size is often necessary in the course of reducing the quantity.

**lack of fusion (LOF)**. A condition in a welded joint in which fusion is less than complete.

**lack of penetration (LOP)**. A condition in a welded joint in which joint penetration is less than that specified.

**lack of resin fill-out**. In reinforced plastics, a condition in which an area contains reinforcement not wetted with a sufficient quantity of resin. This condition usually occurs at the edge of a pultrusion.

**lacquer**. (1) A coating formulation based on thermoplastic film-forming material dissolved in organic solvent. The coating dries primarily by evaporation of the solvent.

Typical lacquers include those based on lac, nitrocellulose, other cellulose derivatives, vinyl resins, acrylic resins, and so forth. (2) In lubrication, a deposit resulting from the oxidation and/or polymerization of fuels and lubricants when exposed to high temperatures. Softer deposits are described as *varnishes* or *gums*.

**ladder polymer**. A polymer with two polymer chains cross linked at intervals.

**ladle**. Metal receptacle frequently lined with refractories used for transporting and pouring molten metal. Types include hand, bull, crane, bottom-pour, holding, teapot, shank, and lip-pour (Fig. 283).

**ladle brick**. Refractory brick suitable for lining ladles used to hold molten metal (Fig. 284).

**ladle coating**. The material used to coat metal ladles to prevent iron pickup in aluminum alloys. The material can consist of sodium silicate, iron oxide, and water, applied to the ladle when it is heated.

FIG. 283

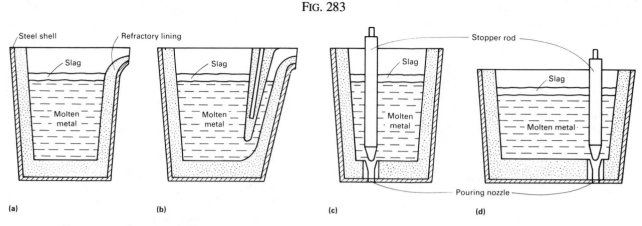

*Four types of pouring ladles. (a) Open lip-pour ladle. (b) Teapot ladle. (c) and (d) Bottom-pour ladles*

FIG. 284

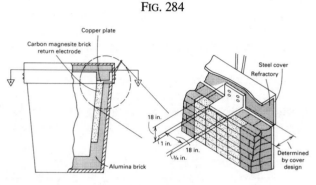

*Schematic of a ladle and corresponding refractory lining*

FIG. 285

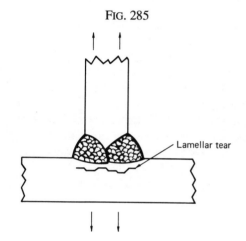

*Lamellar tear caused by thermal contraction strain*

**ladle metallurgy**. Degassing processes for steel carried out in a *ladle*.

**ladle preheating**. The process of heating a ladle prior to the addition of molten metal. This procedure reduces metal heat loss and eliminates moisture-steam safety hazards.

**lamella**. The basic morphological unit of a crystalline polymer, usually ribbonlike or platelike in shape. Generally, about 10 nm thick, 1 μm long, and 0.1 μm wide, if ribbonlike.

**lamellar corrosion**. See *exfoliation corrosion*.

**lamellar tearing**. Occurs in the base metal adjacent to weldments due to high through-thickness strains introduced by weld metal shrinkage in highly restrained joints. Tearing occurs by decohesion and linking along the working direction of the base metal; cracks usually run roughly parallel to the fusion line and are steplike in appearance (Fig. 285). Lamellar tearing can be minimized by designing joints to minimize weld shrinkage stresses and joint restraint. See also *cold cracking*, *hot cracking*, and *stress-relief cracking*.

**lamellar thickness**. A characteristic morphological parameter in plastics, usually estimated from x-ray studies or electron microscopy, that is usually 10 to 50 nm. The average thickness of lamellae in a specimen. See also *lamella*.

**lamina**. A single ply or layer in a laminate, which is made up of a series of layers.

**laminar flow**. The flow of plastic resin in a mold, characterized by solidification of the first layer to contact the mold surface, which then acts as an insulating tube through which material flows to fill the remainder of the cavity. This type of flow is essential to duplication of the mold surface.

**laminate**. To unite laminae with a bonding material, usually with pressure and heat (normally refers to flat sheets, but also includes rods and tubes); a product made by such bonding. Two or more layers of material bonded together

FIG. 286

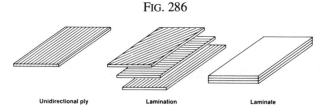

*Example of a laminated structure*

FIG. 287

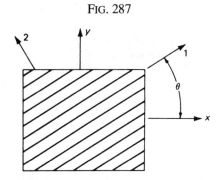

*Laminate coordinate systems: 1, 2, principal material coordinates; x, y, laminate or arbitrary coordinates*

FIG. 288

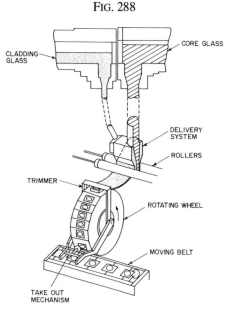

*Process for forming laminated sheet glass*

(Fig. 286). The term can apply to preformed layers joined by adhesives or by heat and pressure. The term also applies to composites of plastics films with other films, or with foil or paper, even though they have been made by spread coating or by extrusion coating. A reinforced laminate usually refers to superimposed layers of resin-impregnated or resin-coated fabrics or fibrous reinforcements that have been bonded, especially by heat and pressure. Products produced with little or no pressure, such as hand lay-ups, filament-wound structures, and spray-ups, are sometimes called contact-pressure laminates. A single resin-impregnated sheet of paper, fabric, or glass mat is not considered a laminate. Such a single-sheet construction may be called a lamina. See also *bidirectional laminate*, *cross laminate*, *parallel laminate*, and *unidirectional laminate*.

**laminate coordinates**. A reference coordinate system used to describe the properties of a laminate, generally in the direction of principal axes, when they exist (Fig. 287).

**laminated glass**. Two or more glasses that are fused together to produce a material with properties that would have been either difficult or impossible to obtain in a single glass. There are many multistep lamination processes. Usually a glass is formed during one process and is subsequently laminated with another glass during a separate process that involves melting the two glasses in separate furnaces and then passing them through a cladding delivery system (Fig. 288).

**laminate orientation**. The geometric configuration of a cross-plied composite laminate with regard to the angles of cross-plying, the number of laminae at each angle, and the exact sequence of laminae lay-up.

**laminate ply**. One fabric-resin or fiber-resin layer (of a product) that is bonded to adjacent layers in the curing process.

**laminate void**. The absence of resin in an area that normally contains resin.

**lamination**. (1) A type of discontinuity with separation or weakness generally aligned parallel to the worked surface of a metal. May be the result of pipe, blisters, seams, inclusions, or segregation elongated and made directional by working. Laminations may also occur in powder metallurgy compacts. (2) In electrical products such as motors, a blanked piece of electrical sheet that is stacked up with several other identical pieces to make a stator or rotor (Fig. 289).

**lampblack**. Fine soot used in the reduction and carburization of tungsten trioxide and titanium dioxide to produce tungsten carbide and titanium carbide powder, respectively.

**lampworking**. Forming glass articles from tubing and cane by heating in a gas flame.

**lancing**. (1) A press operation in which a single-line cut is made in strip stock without producing a detached slug. Chiefly used to free metal for forming, or to cut partial contours for blanked parts, particularly in progressive dies. (2) A misnomer for *oxyfuel gas cutting*.

**land** (metals). (1) For profile-sharpened milling cutters, the relieved portion immediately behind the cutting edge. (2) For reamers, drills, and taps, the solid section between the flutes. (3) On punches, the portion adjacent to the nose that is parallel to the axis and of maximum diameter.

FIG. 289

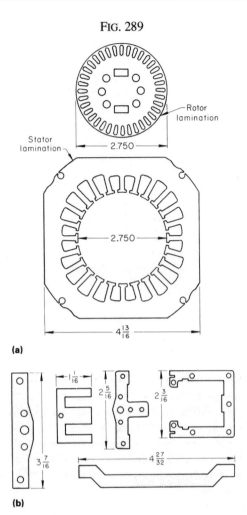

**(a)**

**(b)**

*Typical laminations blanked and pierced from electrical sheet. (a) Laminations for rotating electrical machinery. (b) Laminations for non-rotating machinery. Dimensions given in inches.*

**land** (plastics). (1) The horizontal bearing surface of a semi-positive or flash mold by which excess material escapes. (2) The bearing surface along the top of the flights of a screw in a screw extruder. (3) The surface of an extrusion die parallel to the direction of melt flow. (4) The land region of a nozzle used in injection molding.

**Langelier saturation index**. An index calculated from total dissolved solids, calcium concentration, total alkalinity, pH, and solution temperature that shows the tendency of a water solution to precipitate or dissolve calcium carbonate.

**Lanxide process**. A composite material formation process which involves pressureless metal infiltration into a ceramic preform (Fig. 290). The molten metal is exposed to an oxidizing atmosphere resulting in a matrix material composed of a mixture of the oxidation reaction product and unreacted metal.

FIG. 290

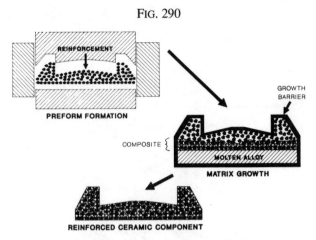

*Schematic of the Lanxide process for making a shaped ceramic-matrix composite component*

FIG. 291

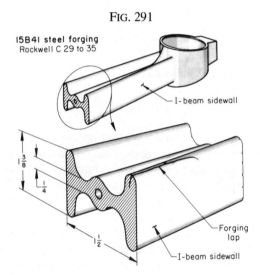

*Forged steel truck connecting rod that failed in service from fatigue initiated at a forging lap. Dimensions given in inches.*

**lap** (composites). (1) In filament winding, the amount of overlay between successive windings, usually intended to minimize gapping. (2) In adhesive bonding, the distance one adherend covers another adherend.

**lap** (metals). A surface imperfection, with the appearance of a seam, caused by hot metal, fins, or sharp corners being folded over and then being rolled or forged into the surface but without being welded (Fig. 291).

**lap joint**. A joint made between two overlapping members (Fig. 292).

**lapping**. A finishing operation using fine abrasive grits loaded into a lapping material such as cast iron. Lapping provides major refinements in the workpiece including extreme accuracy of dimension, correction of minor im-

FIG. 292

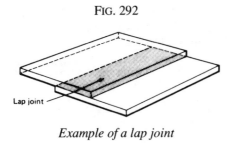

*Example of a lap joint*

perfections of shape, refinement of surface finish, and close fit between mating surfaces.

**large-end bearing**. See *big-end bearing*.

**Larmor frequency**. The classical frequency at which a charged body precesses in a uniform magnetic field. $\omega_L = -eB/2mc$, where $e$ is the electron charge, $B$ is the magnetic field intensity, $m$ is mass, and $c$ is the velocity of light. See also *Larmor period*.

**Larmor period**. The inverse of the Larmor frequency. See also *Larmor frequency*.

**laser**. A device that emits a concentrated beam of electromagnetic radiation (light). Laser beams are used in metalworking to melt, cut, or weld metals; in less concentrated form they are sometimes used to inspect metal parts.

**laser alloying**. See *laser surface processing*.

**laser beam cutting**. A cutting process which severs materials with the heat obtained from the application of a concentrated coherent light beam impinging upon the workpiece to be cut (Fig. 293). The process can be used with (gas-assisted laser beam cutting) or without an externally supplied gas.

**laser beam machining**. Use of a highly focused monofrequency collimated beam of light to melt or sublime material at the point of impingement on a workpiece.

**laser beam welding**. A welding process that joins metal parts using the heat obtained by directing a beam from a *laser* onto the weld joint.

**laser hardening**. A surface-hardening process which uses a laser to quickly heat a surface. Heat conduction into the interior of the part will quickly cool the surface, leaving a shallow martensitic layer.

**laser surface processing**. The use of lasers with continuous outputs of 0.5 to 10 kW to modify the metallurgical structure of a surface and to tailor the surface properties without adversely affecting the bulk properties. The surface modification can take the following three forms. The first is transformation hardening in which a surface is heated so that thermal diffusion and solid-state transformations can take place (Fig. 294). The second is surface melting, which results in a refinement of the structure due to the rapid quenching from the melt. The third is surface (laser) alloying, in which alloying elements are added to the melt pool to change the composition of the surface. The novel structures produced by laser surface melting and alloying can exhibit improved electrochemical and tribological behavior.

**latent curing agent**. A curing agent for plastics that produces long-term stability at room temperature but rapid cure at elevated temperatures.

**latent heat**. Thermal energy absorbed or released when a substance undergoes a phase change.

**latent heat of fusion**. Heat given off by a liquid freezing to a solid, or gained by a solid in melting to a liquid, without a change in temperature.

**latent heat of vaporization**. Heat given off by a vapor condensing to a liquid, or gained by a liquid evaporating to a vapor, without a change in temperature.

**lateral crack**. A crack produced beneath and generally paralleling a glass surface during the unloading phase of mechanical contact with a hard, sharp object. See also *cleavage crack*.

**lateral extrusion**. An operation in which the product is extruded sideways through an orifice in the container wall.

FIG. 293

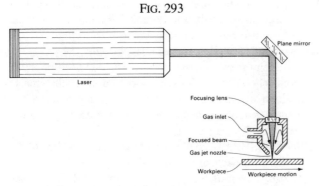

*Schematic of laser beam cutting with a gas jet*

FIG. 294

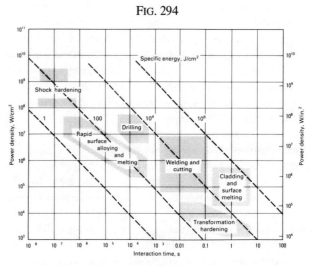

*Interaction times and power densities necessary for various laser surface processes*

FIG. 295

*Lath martensite in a water-quenched 0.20% C steel. 500×*

**Table 37  The seven crystal systems and their corresponding 14 space (Bravais) lattices**

| System | Space lattice | Hermann-Mauguin symbol | Pearson symbol |
|---|---|---|---|
| Triclinic (anorthic) | Primitive | P | aP |
| Monoclinic | Primitive | P | mP |
|  | Base-centered(a) | C | mC |
| Orthorhombic | Primitive | P | oP |
|  | Base-centered(a) | C | oC |
|  | Face-centered | F | oF |
|  | Body-centered | I | oI |
| Tetragonal | Primitive | P | tP |
|  | Body-centered | I | tI |
| Hexagonal | Primitive | R(b) | hP |
| Rhombohedral | Primitive | R | hR |
| Cubic | Primitive | P | cP |
|  | Face-centered | F | cF |
|  | Body-centered | I | cI |

(a) The face that has a lattice point at its center may be chosen as the *c* face (the *xy* plane), denoted by the symbol *C*, or as the *a* or *b* face, denoted by *A* or *B*, because the choice of axes is arbitrary and does not alter the actual translations of the lattice. (b) The symbol *C* may be used for hexagonal crystals, because hexagonal crystals may be regarded as base-centered orthorhombic.

**lateral runout**. Same as *axial runout*.

**latex**. A stable dispersion of polymeric substance (natural or synthetic rubber or plastic) in an essentially aqueous medium.

**lath martensite**. Martensite formed partly in steels containing less than approximately 1.0% C and solely in steels containing less than approximately 0.5% C as parallel arrays of packets of lath-shape units 0.1 to 0.3 μm thick (Fig. 295).

**lattice**. (1) A space lattice is a set of equal and adjoining parallelopipeds formed by dividing space by three sets of parallel planes, the planes in any one set being equally spaced. There are seven ways of so dividing space, corresponding to the seven crystal systems. The unit parallelopiped is usually chosen as the unit cell of the system. See also *crystal system* and *unit cell*. (2) A point lattice is a set of points in space located so that each point has identical surroundings. There are 14 ways of so arranging points in space, corresponding to the 14 Bravais lattices. Table 37 lists the seven crystal systems and their corresponding 14 space (Bravais) lattices.

**lattice constants**. See *lattice parameter*.

**lattice diffusion**. See *volume diffusion*.

**lattice parameter**. The length of any side of a unit cell of a given crystal structure. The term is also used for the fractional coordinates *x*, *y*, and *z* of lattice points when these are variable.

**lattice pattern**. A pattern of a filament winding with a fixed arrangement of open voids.

**Laue method (for crystal analysis)**. A method of x-ray diffraction using a beam of white radiation, a fixed single crystal specimen, and a flat photographic film usually normal to the incident beam. If the film is located on the same side of the specimen as the x-ray source, the method is known as the back reflection Laue method; if on the other side, as the transmission Laue method.

**launder**. (1) A channel for transporting molten metal. (2) A box conduit conveying particles suspended in water.

**lay**. Direction of predominant surface pattern remaining after cutting, grinding, lapping, or other processing (Fig. 296).

**lay (composites)**. The length of twist produced by stranding filaments, such as fibers, wires, or roving. The angle that such filaments make with the axis of the strand during a stranding operation. The length of twist of a filament is usually measured as the distance parallel to the axis of the strand between successive turns of the filament.

**layer**. A stratum of weld metal or surfacing material. The layer may consist of one or more weld beads laid side by side.

**layer bearing**. A bearing constructed in layers. See also *bimetal bearing* and *trimetal bearing*.

**layer-lattice material**. Any material having a layerlike crystal structure, but particularly *solid lubricants* of this type.

**layer level wound**. See preferred term *level wound*.

**layer wound**. See preferred term *level wound*.

**lay-up**. Reinforcing material that is placed in position in the mold. The process of placing the reinforcing material in position in the mold. The resin-impregnated reinforcement. A description of the component materials, geometry, and so forth, of a laminate.

***L*-direction**. The ribbon direction, that is, the direction of the continuous sheets of honeycomb (Fig. 297).

***L/D* ratio**. In bearing technology, the ratio of the axial length of a plain bearing to its diameter.

FIG. 296

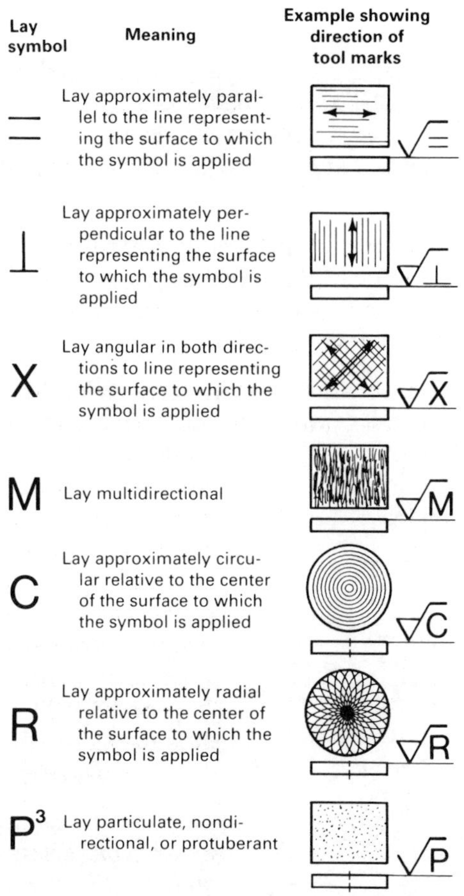

| Lay symbol | Meaning | Example showing direction of tool marks |
|---|---|---|
| — | Lay approximately parallel to the line representing the surface to which the symbol is applied | |
| ⊥ | Lay approximately perpendicular to the line representing the surface to which the symbol is applied | |
| X | Lay angular in both directions to line representing the surface to which the symbol is applied | |
| M | Lay multidirectional | |
| C | Lay approximately circular relative to the center of the surface to which the symbol is applied | |
| R | Lay approximately radial relative to the center of the surface to which the symbol is applied | |
| $P^3$ | Lay particulate, nondirectional, or protuberant | |

*Symbols used to define lay and its direction*

FIG. 297

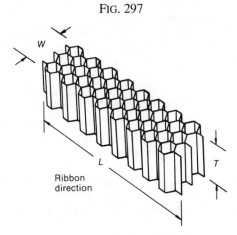

*Honeycomb ribbon direction*

**leaching**. Extracting an element or compound from a solid alloy or mixture by preferential dissolution in a suitable liquid.

**lead**. (1) The axial advance of a helix in one complete turn. (2) The slight bevel at the outer end of a face cutting edge of a face mill.

**lead and lead alloys**. See *Technical Brief 23*.

**lead angle**. In cutting tools, the helix angle of the flutes.

**lead burning**. A misnomer for welding of lead.

**leaded bronzes**. See *bearing bronzes*.

**lead glass**. Glass containing a substantial proportion of lead oxide (PbO).

**lead proof**. See *die proof*.

**leakage field**. The magnetic field that leaves or enters a magnetized part at a magnetic pole.

**leak testing**. A nondestructive test for determining the escape or entry of liquids or gases from pressurized or into evacuated components or systems intended to hold these liquids. Leak testing systems, which employ a variety of gas detectors, are used for locating (detecting and pinpointing) leaks, determining the rate of leakage from one leak or from a system, or monitoring for leakage (Fig. 298).

FIG. 298

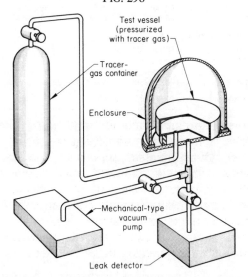

*Schematic of setup for detecting leaks by monitoring an enclosure placed around a vessel pressurized with tracer gas*

**lean mixture**. A gas/air mixture in which too much air is present, so that burning is difficult. The flame is usually noisy, frail, very light blue, short, and inefficient.

**least count**. The smallest value that can be read from an instrument having a graduated scale. Except on instruments provided with a *vernier*, the least count is that fraction of the smallest division which can be conveniently and reliably estimated; this fraction is ordinarily one-fifth or one-tenth, except where the graduations are very closely spaced. Also known as least reading.

## Technical Brief 23: Lead and Lead Alloys

LEAD was one of the first metals known to man, with the oldest lead artifact dating back to about 3000 BC. All civilizations, beginning with the ancient Egyptians, Assyrians, and Babylonians, have used lead for many ornamental and structural purposes. Pipe was one of the earliest applications of lead. The Romans produced 15 standard sizes of water pipe in regular 3 m (10 ft) lengths. Presently, battery applications constitute more than 80% of lead alloy use.

Although there are at least 60 known lead-containing minerals, by far the most important is galena (PbS), which is smelted and refined to produce 99.99% pig lead. Recycling of scrap lead (from batteries, lead sheet, and cable sheathing) is also a major source, providing more than half of the lead used in the United States. Considerable tonnages of scrap solder and bearing materials are also recovered and used again.

The properties of lead that make it useful in a wide variety of applications are density, malleability, lubricity, flexibility, electrical conductivity, and coefficient of thermal expansion, all of which are quite high; and elastic modulus, elastic limit, strength, hardness, and melting point, all of which are quite low. Lead also has good resistance to corrosion under a wide variety of conditions. Lead is easily alloyed with many other metals and casts with little difficulty.

### Compositions of selected lead alloys for battery grids

| UNS designation | Composition, % | | | | | | |
|---|---|---|---|---|---|---|---|
| | As max | Ag max | Ca | Pb | Sb | Sn | Other |
| **Calcium-lead alloys** | | | | | | | |
| L50760 | 0.0005 | 0.001 | 0.06–0.08 | bal | 0.0005 max | 0.0005 max | (a) |
| L50770 | 0.0005 | 0.001 | 0.10 nom | bal | 0.0005 max | 0.0005 max | (a) |
| L50775 | 0.0005 | 0.001 | 0.08–0.11 | bal | 0.0005 max | 0.2–0.4 | (a) |
| L50780 | 0.0005 | 0.001 | 0.08–0.11 | bal | 0.0005 max | 0.4–0.6 | (a) |
| L50790 | 0.0005 | 0.001 | 0.08–0.10 | bal | 0.0005 max | 0.9–1.1 | (a) |
| **Antimony-lead alloys** | | | | | | | |
| L52760 | 0.18 nom | ... | ... | bal | 2.75 nom | 0.2 nom | ... |
| L52765 | 0.3 nom | ... | ... | bal | 2.75 nom | 0.3 nom | ... |
| L52770 | 0.15 nom | ... | ... | bal | 2.9 nom | 0.3 nom | ... |
| L52840 | 0.15 nom | ... | ... | bal | 2.9 nom | 0.3 nom | ... |

(a) 0.005% max Bi and 0.0005% max each for Cu, Zn, Cd, Ni, and Fe

The most significant applications of lead and lead alloys are lead-acid storage batteries (in the grid plates, posts, and connector straps), ammunition, cable sheathing, and building construction materials, such as sheet, pipe, and tin-lead solders (see the table accompanying the term *solder*). Other important applications include counterweights and cast products, such as bearings, ballast, gaskets, type metal, terneplate, and foil. Lead in various forms and combinations is also used as a material for controlling sound and mechanical vibrations and shielding against x-rays and gamma rays. In addition, lead is used as an alloying element in steel and copper alloys to improve machinability, and it is used in fusible alloys. Substantial amounts of lead are also used in the form of lead compounds, including tetraethyl and tetramethyl lead used as antiknock compounds in gasoline engines, litharge (PbO) used in glasses, and various corrosion-inhibiting lead pigments, such as red lead ($Pb_3O_4$).

Because lead presents a health hazard (toxicity), it should not be used to conduct very soft water for drinking, nor should it come into contact with foods. Inhalation of lead dust and fumes should be avoided, and paints containing lead should be removed from structures.

### Recommended Reading

* A.W. Worcester and J.T. O'Reilly, Lead and Lead Alloys, *Metals Handbook*, 10th ed., Vol 2, ASM International, 1990, p 543-556
* J.F. Smith, Corrosion of Lead and Lead Alloys, *Metals Handbook*, 9th ed., Vol 13, ASM International, 1987, p 784-792

**least reading.** See *least count.*

**ledeburite.** The eutectic of the iron-carbon system, the constituents of which are *austenite* and *cementite.* The austenite decomposes into *ferrite* and cementite on cooling below Ar₁, the temperature at which transformation of austenite to ferrite or ferrite plus cementite is completed during cooling.

**left-hand cutting tool.** A cutter all of whose flutes twist away in a counterclockwise direction when viewed from either end.

**legging**. The drawing of filaments or strings when adhesive-bonded substances are separated. Compare with *teeth*. See also *stringiness* and *webbing*.

**leg of fillet weld**. (1) Actual: The distance from the root of the joint to the toe of the fillet weld (Fig. 299). (2) Nominal: The length of a side of the largest right triangle that can be inscribed in the cross section of the weld. See also the figures accompanying the terms *concave fillet weld* and *convex fillet weld*.

FIG. 299

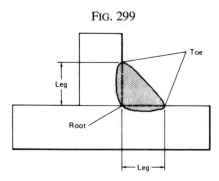

*Leg of a fillet weld*

**Leidenfrost phenomenon**. Slow cooling rates associated with a hot vapor blanket that surrounds a part being quenched in a liquid medium such as water. The gaseous vapor envelope acts as an insulator, thus slowing the cooling rate.

**lemon bearing (elliptical bearing)**. A two-lobed bearing.

**leno weave**. A locking-type weave in which two or more warp threads cross over each other and interlace with one or more filling threads (Fig. 300). It is used primarily to prevent the shifting of fibers in open-weave fabrics.

FIG. 300

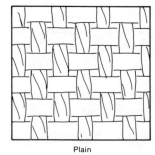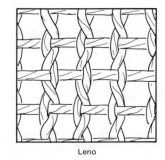

*Comparison of plain and leno weave fabrics*

**lens**. A transparent optical element, so constructed that it serves to change the degree of convergence or divergence of the transmitted rays.

**let-go**. An area in laminated glass over which an initial adhesion between interlayer and glass has been lost.

**leveler lines**. Lines on sheet or strip running transverse to the direction of *roller leveling*. These lines may be seen upon stoning or light sanding after leveling (but before drawing) and can usually be removed by moderate stretching.

**leveling**. Flattening of rolled sheet, strip, or plate by reducing or eliminating distortions. See also *stretcher leveling* and *roller leveling*.

**leveling action**. Action exhibited by a plating solution yielding a plated surface smoother than the basis metal.

**level winding**. See *circumferential winding*.

**level wound**. Spooled or coiled weld filler metal that has been wound in distinct layers such that adjacent turns touch.

**lever rule**. A method that can be applied to any two-phase field of a binary *phase diagram* to determine the amounts of different phases present at a given temperature in a given alloy. A horizontal line, referred to as a tie line, represents the lever, and the alloy composition its fulcrum. The intersection of the tie line with the boundaries of the two-phase field fixes the compositions of the coexisting phases, and the amounts of the phases are proportional to the segments of the tie line between the alloy and the phase compositions.

**levigation**. (1) Separation of fine powder from coarser material by forming a suspension of the fine material in a liquid. (2) A means of classifying a material as to particle size by the rate of settling from a suspension.

**levitation melting**. An *induction melting* process in which the metal being melted is suspended by the electromagnetic field and is not in contact with a container.

**lift beam furnace**. A continuously operating heat treating or sintering furnace. The term in general use by the U.S. furnace industry is *walking beam furnace*.

**liftout**. The mechanism also known as *knockout*.

**lift rod**. Part of the press tooling used for the raising or lifting of one or more punches.

**ligand**. The molecule, ion, or group bound to the central atom in a *chelate* or a *coordination compound*.

**light**. Radiant energy in a spectral range visible to the normal human eye (~380 to 780 nm, or 3800 to 7800 Å). See also *electromagnetic radiation*.

**light drawn**. An imprecise term, applied to drawn products such as wire and tubing, that indicates a lesser amount of cold reduction than for *hard drawn* products.

**light-field illumination**. See *bright-field illumination*.

**light filter**. See *color filter*.

**light fraction**. The first liquid produced during the distillation of a crude oil.

**lightly coated electrode**. A filler-metal electrode used in arc welding, consisting of a metal wire with a light coating, usually of metal oxides and silicates, applied subsequent to the drawing operation primarily for stabilizing the arc. Contrast with *covered electrode*.

**light metal.** One of the low-density metals, such as aluminum (~2.7 g/cm$^3$), magnesium (~1.7 g/cm$^3$), titanium (~4.4 g/cm$^3$), beryllium (~1.8 g/cm$^3$), or their alloys.

**lime.** A general term which includes the various chemical and physical forms of quicklime, hydrated lime, and hydraulic lime. It may be high-calcium, magnesian, or dolomitic. The chemical forms of calcium oxide (CaO), calcium hydroxide (Ca(OH)$_2$, magnesium oxide (MgO), or magnesium hydroxide (Mg(OH)$_2$), alone or in combination may be produced either primarily or as a by-product of materials other than limestone, for example Ca(OH)$_2$ formed by acetylene generation from calcium carbide (CaC$_2$) and water treatment sludges.

**lime glass.** A glass containing a substantial proportion of lime, usually associated with soda and silica. See also *glass* (Technical Brief 17).

**limestone.** An initially sedimentary rock consisting chiefly of calcium carbonate or of the carbonates of calcium and magnesium. Limestone may be high calcium, magnesian, or dolomitic. (1) Dolomitic limestone contains from 35 to 46% magnesium carbonate (MgCO$_3$). (2) Magnesian limestone contains from 5 to 35% MgCO$_3$. (3) High-calcium limestone contains from 0 to 5% MgCO$_3$. Limestone is used extensively in glass formulations.

**limited-coordination specification (or standard).** A specification (or standard) that has not been fully coordinated and accepted by all interested parties. Limited-coordination specifications and standards are issued to cover the need for requirements unique to one particular department. This applies primarily to military agency documents.

**limited solid solution.** A crystalline miscibility series whose composition range does not extend all the way between the components of the system; that is, the system is not *isomorphous*.

**limiting current density.** The maximum current density that can be used to obtain a desired electrode reaction without undue interference such as from *polarization*.

**limiting dome height (LDH) test.** A mechanical test, usually performed unlubricated on sheet metal, that simulates the fracture conditions in a practical press-forming opera-tion (Fig. 301). The results are dependent on the sheet thickness.

**limiting drawing ratio (LDR).** See *deformation limit*.

**limiting static friction.** The resistance to the force tangential to the interface that is just sufficient to initiate relative motion between two bodies under load. The term static friction, which properly describes a tangential resistance called into operation by a force less than this, should not be substituted for limiting static friction.

**lineage structure.** (1) Deviations from perfect alignment of parallel arms of a columnar dendrite as a result of inter-dendritic shrinkage during solidification from a liquid. This type of deviation may vary in orientation from a few minutes to as much as two degrees of arc. (2) A type of substructure consisting of elongated subgrains.

**linear attenuation coefficient.** The fraction of an x-ray beam per unit thickness that a thin object will absorb or scatter (attenuate). A property proportional to the physical density and dependent on the atomic number of the materia and the energy of the x-ray beam.

**linear dispersion.** In spectroscopy, the derivative $dx/d\lambda$, where $x$ is the distance along the spectrum and $\lambda$ is the wavelength. Linear dispersion is usually expressed as mm/Å.

**linear elastic fracture mechanics.** A method of fracture analysis that can determine the stress (or load) required to induce fracture instability in a structure containing a crack-like flaw of known size and shape. See also *fracture mechanics* and *stress-intensity factor*.

**linear expansion.** The increase of a given dimension measured by the expansion of a specimen or component subject to a temperature gradient. See also *coefficient of thermal expansion*.

**linear shrinkage.** The shrinkage in one dimension of a powder compact during sintering. Contrast with *volume shrinkage*.

**linear (tensile or compressive) strain.** The change per unit length due to force in an original linear dimension. An increase in length is considered positive.

**line indices.** The *Miller indices* of the set of planes producing a diffraction line.

**line pair.** In spectroscopy, an analytical line and the internal standard line with which it is compared. See also *internal standard line*.

**liner** (composite). In a filament-wound pressure vessel, the continuous, usually flexible coating on the inside surface of the vessel, used to protect the laminate from chemical attack or to prevent leakage under stress.

**liner** (metals). (1) The slab of coating metal that is placed on the core alloy and is subsequently rolled down to clad sheet as a composite. (2) In extrusion, a removable alloy steel cylindrical chamber, having an outside longitudinal taper firmly positioned in the container or main

FIG. 301

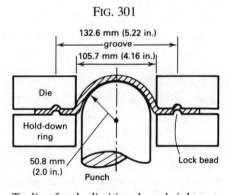

*Tooling for the limiting dome height test*

FIG. 302

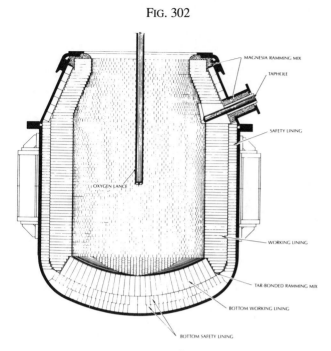

MAGNESIA RAMMING MIX
TAPHOLE
SAFETY LINING
OXYGEN LANCE
WORKING LINING
TAR-BONDED RAMMING MIX
BOTTOM WORKING LINING
BOTTOM SAFETY LINING

*Refractory lining in a basic oxygen furnace*

body of the press, into which the billet is placed for extrusion.

**line reaming.** Simultaneous *reaming* of coaxial holes in various sections of a workpiece with a reamer having cutting faces or piloted surfaces with the desired alignment.

**liners.** Thin strips of metal inserted between the dies and the units into which the dies are fastened.

**lining.** Internal refractory layer of firebrick, clay, sand, or other material in a furnace (Fig. 302) or ladle.

**linishing.** A method of finishing by grinding on a continuous abrasive belt.

**linters.** Short fibers that adhere to the cottonseed after gin-

ning. Used in rayon manufacture as fillers for plastics and as a base for the manufacture of cellulosic plastics.

**lipophilic.** Having an affinity for oil. See also *hydrophilic* and *hydrophobic*.

**lip-pour ladle.** Ladle in which the molten metal is poured over a lip, much as water is poured out of a bucket. See also the figure accompanying the term *ladle*.

**liquation.** (1) The separation of a low melting constituent of an alloy from the remaining constituents, usually apparent in alloys having a wide melting range. (2) Partial melting of an alloy, usually as a result of *coring* or other compositional heterogeneities.

**liquation temperature.** The lowest temperature at which partial melting can occur in an alloy that exhibits the greatest possible degree of segregation.

**liquefied petroleum (LP) gas.** Gases, such as propane and butane, which are usually stored as a liquid under pressure, but are released for use as a gas by a regulator.

**liquid carburizing.** Surface hardening of steel by immersion into a molten bath consisting of cyanides and other salts. Table 38 lists compositions and operating characteristics of liquid carburizing baths.

**liquid chromatography.** A separation method based on the distribution of sample compounds between a stationary phase and a liquid mobile phase. Used extensively in the characterization of organic, inorganic, pharmaceutical, and biochemical compounds. See also *chromatography*, *gas chromatography*, and *ion chromatography*.

**liquid crystal polymer (LCP).** A thermoplastic polymer that contains primarily benzene rings in its backbone (Fig. 303), is melt processable, and develops high orientation during molding with resultant improvement in tensile strength and high-temperature capability. First commercial availability was as an aromatic polyamide. Available with or without fiber reinforcement.

**liquid disintegration.** The process of producing powders by pouring molten metal on a rotating surface.

### Table 38  Operating compositions of liquid carburizing baths

| | Composition of bath, % | |
|---|---|---|
| Constituent | Light case, low temperature 845–900 °C (1550–1650 °F) | Deep case, high temperature 900–955 °C (1650–1750 °F) |
| Sodium cyanide | 10–23 | 6–16 |
| Barium chloride | . . . | 30–55(a) |
| Salts of other alkaline earth metals(b) | 0–10 | 0–10 |
| Potassium chloride | 0–25 | 0–20 |
| Sodium chloride | 20–40 | 0–20 |
| Sodium carbonate | 30 max | 30 max |
| Accelerators other than those involving compounds of alkaline earth metals(c) | 0–5 | 0–2 |
| Sodium cyanate | 1.0 max | 0.5 max |
| Density of molten salt | 1.76 g/cm$^3$ at 900 °C (0.0636 lb/in.$^3$ at 1650 °F) | 2.00 g/cm$^3$ at 925 °C (0.0723 lb/in.$^3$ at 1700 °F) |

(a) Proprietary barium chloride-free deep-case baths are available. (b) Calcium and strontium chlorides have been employed. Calcium chloride is more effective, but its hygroscopic nature has limited its use. (c) Among these accelerators are manganese dioxide, boron oxide, sodium fluoride, and sodium pyrophosphate.

Fig. 303

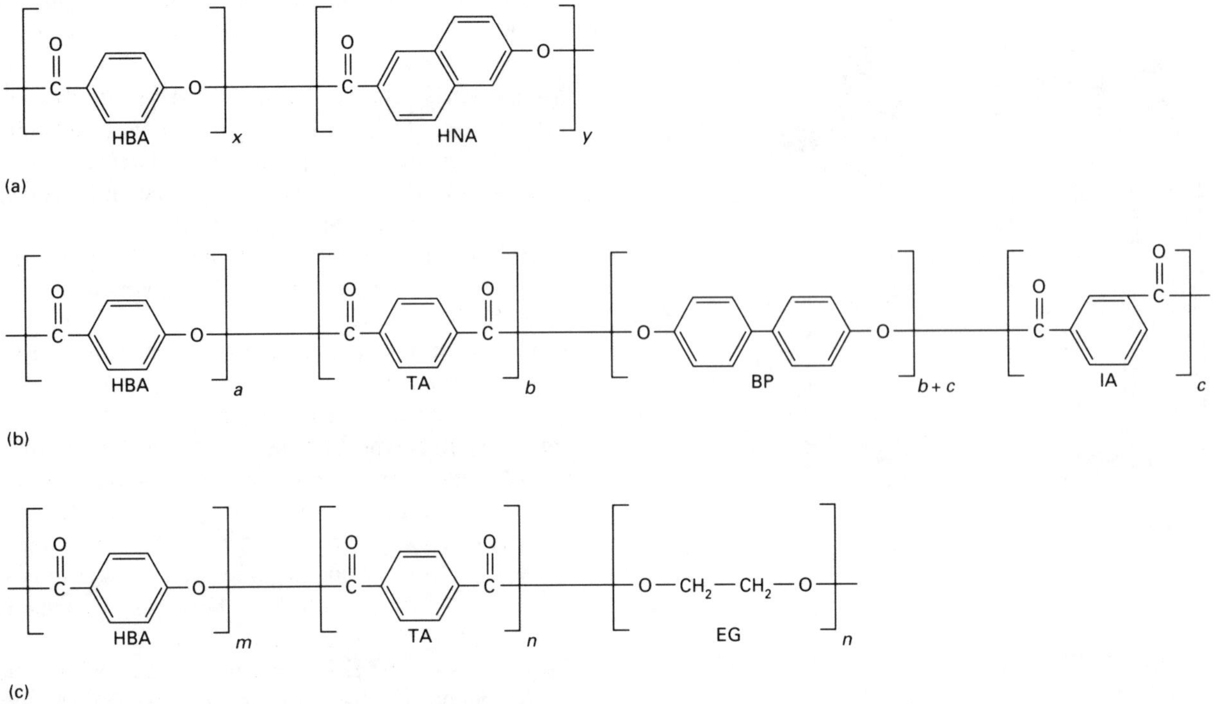

(a)

(b)

(c)

*Principal LCP structures and sources. BP, biphenol; EG, ethylene glycol; HBA, hydroxy napthoic acid; IA, isophthalic acid; TA, terephthalic acid. (a) General-purpose class. (b) A high-temperature variant. (c) Lower-temperature/performance class*

**liquid honing.** Producing a finely polished finish by directing an air-ejected chemical emulsion containing fine abrasives against the surface to be finished.

**liquid impact erosion.** See *erosion (erosive wear)*.

**liquid impingement erosion.** See *erosion (erosive wear)*.

**liquid injection molding (LIM).** A process that involves an integrated system for proportioning, mixing, and dispensing two-component liquid resin formulations and directly injecting the resultant mix into a mold, which is clamped under pressure. Generally used for the encapsulation of electrical and electronic devices. Also, variation on *reaction injection molding*, using mechanical mixing rather than a high-pressure impingement mixer. However, unlike mechanical mixing in other systems, the mixer here does not need to be flushed because a special feed system automatically dilutes the residue in the mixer with part of the polyol needed for the next shot, thereby keeping the ingredients from reacting.

**liquid-liquid chromatography (LLC).** *Liquid chromatography* with a stationary phase composed of a liquid dispersed onto an inert supporting material. Liquid-liquid chromatography has been used in the separation of phenols, aromatic alcohols, organometallic compounds, steroids, drugs, and food products. Also termed liquid-partition chromatography.

**liquid metal embrittlement (LME).** Catastrophic brittle failure of a normally ductile metal when in contact with a liquid metal and subsequently stressed in tension. Table 39 lists a summary of LME couples. See also *solid metal embrittlement*.

**liquid metal infiltration.** Process for immersion of metal fibers in a molten metal bath to achieve a metal-matrix composite; for example, graphite fibers in molten aluminum.

**liquid nitriding.** A method of surface hardening in which molten nitrogen-bearing, fused-salt baths containing both cyanides and cyanates are exposed to parts at subcritical temperatures. A typical commercial bath for liquid nitriding is composed of a mixture of sodium and potassium salts. The sodium salts, which comprise 60 to 70% (by weight) of the total mixture, consist of 96.5% NaCN, 2.5% $Na_2CO_3$, and 0.5% NaCNO. The potassium salts, 30 to 40% (by weight) of the mixture, consist of 96% KCN, 0.6% $K_2CO_3$. 0.75% KCNO, and 0.5% KCl. The operating temperature of this salt bath is 565 °C (1050 °F).

**liquid nitrocarburizing.** A nitrocarburizing process (where both carbon and nitrogen are absorbed into the surface) utilizing molten liquid salt baths below the lower critical temperature. Liquid nitrocarburizing processes are used to

**Table 39　Summary of liquid metal embrittlement couples**

| Solid | | Hg P | Hg A | Cs P | Ga P | Ga A | Na P | In P | In A | Li P | Sn P | Sn A | Bi P | Bi A | Tl P | Cd P | Pb P | Pb A | Zn P | Te P | Sb P | Cu P |
|---|---|---|---|---|---|---|---|---|---|---|---|---|---|---|---|---|---|---|---|---|---|---|
| Sn | P | X | ... | ... | X | ... | ... | ... | ... | ... | ... | ... | ... | ... | ... | ... | ... | ... | ... | ... | ... | ... |
| Bi | P | X | ... | ... | ... | ... | ... | ... | ... | ... | ... | ... | ... | ... | ... | ... | ... | ... | ... | ... | ... | ... |
| Cd | P | ... | ... | X | X | X | ... | ... | ... | ... | X | ... | ... | ... | ... | ... | ... | ... | ... | ... | ... | ... |
| Zn | P | X | X | ... | X | X | ... | X | X | ... | X | ... | ... | ... | ... | ... | ... | X | ... | ... | ... | ... |
|  | LA | ... | X | ... | ... | ... | ... | ... | ... | ... | ... | ... | ... | ... | ... | ... | ... | ... | ... | ... | ... | ... |
| Mg | CA | ... | ... | ... | ... | ... | X | ... | ... | ... | ... | ... | ... | ... | ... | ... | ... | ... | X | ... | ... | ... |
| Al | P | X | X | ... | X | ... | ... | X | ... | ... | ... | X | X | ... | ... | X | X | ... | ... | ... | ... | ... |
|  | CA | X | X | ... | X | ... | X | X | ... | ... | X | ... | X | X | ... | ... | X | ... | X | ... | ... | ... |
| Ge | P | ... | ... | ... | X | ... | ... | X | ... | ... | X | ... | X | ... | X | X | X | ... | ... | ... | X | ... |
| Ag | P | X | X | ... | X | X | ... | ... | ... | X | ... | ... | ... | ... | ... | ... | ... | ... | ... | ... | ... | ... |
|  | LA | ... | ... | ... | ... | ... | ... | ... | ... | X | ... | ... | ... | ... | ... | ... | ... | ... | ... | ... | ... | ... |
| Cu | CP | X | X | ... | ... | ... | X | ... | ... | X | ... | ... | X | X | ... | ... | X | ... | ... | ... | ... | ... |
|  | LA | X | ... | ... | X | ... | ... | ... | ... | ... | ... | ... | ... | ... | ... | ... | ... | ... | ... | ... | ... | ... |
|  | CA | X | X | ... | X | ... | ... | X | X | X | X | (?) | X | ... | ... | ... | X | ... | ... | ... | ... | ... |
| Ni | P | X | ... | ... | ... | ... | ... | ... | ... | X | ... | ... | ... | ... | ... | ... | X | ... | ... | ... | ... | ... |
|  | LA | ... | ... | ... | ... | ... | ... | ... | ... | X | ... | ... | ... | ... | ... | ... | ... | ... | ... | ... | ... | ... |
|  | CA | ... | ... | ... | ... | ... | ... | ... | ... | ... | X | ... | ... | ... | ... | ... | X | ... | ... | ... | ... | ... |
| Fe | P | ... | ... | ... | ... | ... | ... | ... | ... | ... | ... | ... | ... | ... | ... | ... | ... | ... | ... | ... | ... | X |
|  | LA | X | X | ... | X | ... | ... | X | X | X | ... | ... | ... | ... | ... | ... | ... | ... | ... | ... | ... | ... |
|  | CA | ... | ... | ... | ... | ... | ... | X | ... | X | X | ... | ... | ... | ... | X | X | X | X | X | X | X |
| Pd | P | ... | ... | ... | ... | ... | ... | ... | ... | X | ... | ... | ... | ... | ... | ... | ... | ... | ... | ... | ... | ... |
|  | LA | ... | ... | ... | ... | ... | ... | ... | ... | X | ... | ... | ... | ... | ... | ... | ... | ... | ... | ... | ... | ... |
| Ti | CA | ... | X | ... | ... | ... | ... | ... | ... | ... | ... | ... | ... | ... | ... | X | ... | ... | ... | ... | ... | ... |

P, element (nominally pure); A, alloy; C, commercial; L, laboratory

improve wear resistance and fatigue properties of steels and cast irons.

**liquid-partition chromatography (LPC).** See *liquid-liquid chromatography.*

**liquid penetrant inspection.** A type of nondestructive inspection that locates discontinuities that are open to the surface of a metal by first allowing a penetrating dye or fluorescent liquid to infiltrate the discontinuity, removing the excess penetrant, and then applying a developing agent that causes the penetrant to seep back out of the discontinuity and register as an indication (Fig. 304). Liquid penetrant inspection is suitable for both ferrous and nonferrous materials, but is limited to the detection of open surface discontinuities in nonporous solids.

**liquid phase sintering.** Sintering of a compact or loose powder aggregate under conditions where a liquid phase is present during part of the sintering cycle (Fig. 305).

FIG. 305

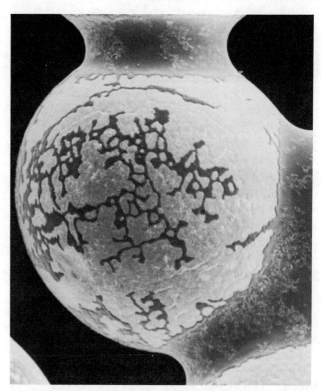

*Wetting of spherical tungsten particles by liquid copper during liquid phase sintering. 500×*

FIG. 304

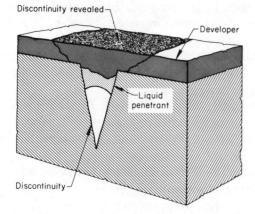

*Sectional view showing result of the migrating action of a liquid penetrant through a developer*

**liquid resin**. An organic, polymeric liquid that becomes a solid when converted to its final state for use.

**liquid shim**. Material used to position components in an assembly where dimensional alignment is critical. For example, epoxy adhesive is introduced into gaps after the assembly is placed in the desired configuration.

**liquid shrinkage**. The reduction in volume of liquid metal as it cools to the liquidus.

**liquid-solid chromatography (LSC)**. Liquid chromatography with silica or alumina as the stationary phase. See also *adsorption chromatography*.

**liquid spray quench**. Same as *spray quenching*.

**liquidus** (glass). The maximum temperature at which equilibrium exists between the molten glass and its primary crystalline phase.

**liquidus** (metals). (1) The lowest temperature at which a metal or an alloy is completely liquid. (2) In a *phase diagram*, the locus of points representing the temperatures at which the various compositions in the system begin to freeze on cooling or finish melting on heating. See also *solidus*.

**liquor finish**. A smooth, bright finish characteristic of wet-drawn wire. Formerly produced by using liquor from fermented grain mash as a drawing lubricant.

**little-end bearing**. A bearing at the smaller (piston) end of a connecting rod in an engine. See also *big-end bearing*.

**live center**. A lathe or grinder center that holds, yet rotates with, the work. It is used in either the headstock or tailstock of a machine to prevent wear and reduce the driving torque.

**load**. (1) In the case of testing machines, a force applied to a testpiece that is measured in units such as pound-force, newton, or kilogram-force. (2) In tribology, the force applied normal to the surface of one body by another contacting body or bodies. The term normal force is more precise and therefore preferred; however, the term normal load is also in use. If applied vertically, the load can be expressed in mass units, but it is preferable to use force units such as newtons (N).

**load-carrying capacity** (of a lubricant). (1) The maximum load that a sliding or rolling system can support without failure. (2) The maximum load or pressure that can be sustained by a lubricant (when used in a given system under specific conditions) without failure of moving bearings or sliding contact surfaces as evidenced by seizure or welding.

**load-deflection curve**. A curve in which the increasing tension, compression, or flexural loads are plotted on the ordinate axis and the deflections caused by those loads are plotted on the abscissa axis.

**loading**. (1) In cutting, building up of a cutting tool back of the cutting edge by undesired adherence of material removed from the work. (2) In grinding, filling the pores of a grinding wheel with material from the work, usually resulting in a decrease in production and quality of finish. (3) In powder metallurgy, filling of the die cavity with powder.

**loading sheet**. In powder metallurgy, the part of a die assembly used as a container for a specific amount of powder to be fed into the die cavity. Sometimes it is part of the feed shoe.

**loading weight**. See preferred term *apparent density*.

**load range (P)**. In fatigue, the algebraic difference between the maximum and minimum loads in a fatigue cycle.

**load ratio (R)**. In fatigue, the algebraic ratio of the minimum to maximum load in a fatigue cycle, that is, $R = P_{min}/P_{max}$. Also known as *stress ratio*.

**loam**. A molding material consisting of sand, silt, and clay, used over brickwork or other structural backup material for making massive castings, usually of iron or steel.

**lobed bearing**. A journal bearing with two or more lobes around its periphery produced by machining or by elastic distortion to increase stability or to provide adjustable clearance.

**local action**. Corrosion due to the action of "local cells," that is, galvanic cells resulting from inhomogeneities between adjacent areas on a metal surface exposed to an *electrolyte*.

**local cell**. A *galvanic cell* resulting from inhomogeneities between areas on a metal surface in an *electrolyte*. The inhomogeneities may be of physical or chemical nature in either the metal or its environment.

**local current density**. Current density at a point or on a small area.

**localized corrosion**. Corrosion at discrete sites, for example, *crevice corrosion*, *pitting*, and *stress-corrosion cracking*.

**localized precipitation**. Precipitation from a supersaturated solid solution similar to *continuous precipitation*, except that the precipitate particles form at preferred locations, such as along slip planes, grain boundaries (Fig. 306), or incoherent twin boundaries.

**local preheating**. Preheating a specific portion of a structure.

**local stress relief heat treatment**. Stress relief heat treatment of a specific portion of a structure.

**locating boss**. A *boss*-shaped feature on a casting to help locate the casting in an assembly or to locate the casting during secondary tooling operations.

**locating ring**. In injection molding machine, a ring that serves to align the nozzle of an injection cylinder with the entrance of the sprue bushing and the mold to the machine platen.

**locational fit**. A clearance or interference *fit* intended for locating mating parts.

**lock**. In forging, a condition in which the flash line is not entirely in one plane. Where two or more plane changes occur, it is called compound lock. Where a lock is placed

FIG. 306

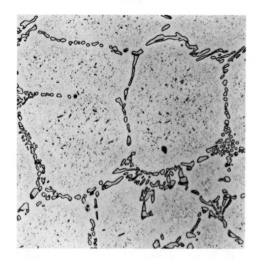

*Localized precipitation of particles at grain boundaries in an Fe-35Ni-16Cr heat-resistant casting. 250×*

FIG. 308

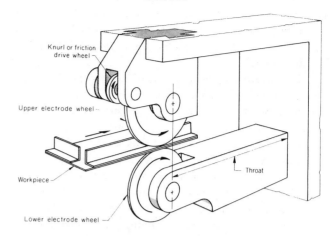

*Longitudinal resistance seam welding machine*

in the die to compensate for die shift caused by a steep lock, it is called a counterlock (Fig. 307).

**locked dies.** Dies with mating faces that lie in more than one plane.

**logarithmic decrement (log decrement).** The natural logarithm of the ratio of successive amplitudes of vibration of a member in free oscillation. It is equal to one-half the specific damping capacity.

**long-chain branching.** A form of molecular branching found in addition polymers as a result of an internal transfer reaction. It primarily influences the melt flow properties.

**longitudinal direction.** That direction parallel to the direction of maximum elongation in a worked material. See also

*normal direction* and the figure accompanying the term *transverse direction*.

**longitudinal field.** A magnetic field that extends within a magnetized part from one or more poles to one or more other poles and that is completed through a path external to the part.

**longitudinal resistance seam welding.** The making of a resistance seam weld in a direction essentially parallel to the throat depth of a resistance welding machine (Fig. 308).

**longitudinal sequence.** The order in which the increments of a continuous weld are deposited with respect to its length. See also *backstep sequence* and *block sequence*.

**long-line current.** Current that flows through the earth from an anodic to a cathodic area of a continuous metallic structure. Usually used only where the areas are separated by considerable distance and where the current results from concentration-cell action.

FIG. 307

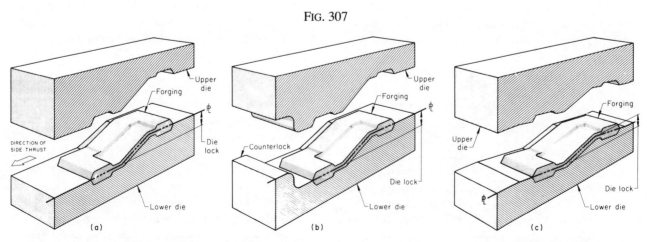

*Locked and counterlocked dies. (a) Locked dies with no means to counteract side thrust. (b) Counterlocked dies. (c) Dies requiring no counterlock because the forging has been rotated to minimize side thrust*

**longos**. Low-angle helical or longitudinal filament windings.

**long period**. A morphological parameter for plastics obtained from small-angle x-ray scattering. It is usually equated to the sum of the *lamellar thickness* and the amorphous thickness.

**long-term etching**. Etching times of a few minutes to hours.

**long transverse**. See *transverse*.

**loop classifier**. A cyclone-type classifier, sometimes connected with a conical ball mill in an airtight system.

**looping mill**. An arrangement of hot rolling stands such that a hot bar, while being discharged from one stand, is fed into a second stand in the opposite direction.

**loop tenacity**. The tenacity or strength value obtained by pulling two loops, such as two links in a chain, against each other to demonstrate the susceptibility of a fibrous material to cutting or crushing; loop strength.

**loose metal**. Refers to an area in a formed panel that is not stiff enough to hold its shape, may be confused with *oil canning*.

**loose powder**. Uncompacted powder.

**loose powder sintering**. Sintering of uncompacted powder using no external pressure.

**loss angle**. See *phase angle*.

**loss factor**. The product of the dissipation factor and the dielectric constant of a dielectric material. See also *tan delta*.

**loss modulus**. A quantitative measure of energy dissipation in polymers, defined as the ratio of stress 90° out of phase with oscillating strain to the magnitude of strain. The loss modulus may be measured in tension or flexure, compression, or shear. See also *complex modulus*.

**loss on ignition (LOI)**. (1) The fractional or percentage weight loss of a material on heating in air from an initial defined state (usually, dried) to a specified temperature, such as 1000 °C (1830 °F), and holding there for a specified period, such as 1 hour. Fixed procedures are designed, usually, such that LOI represents the loss of combined $H_2O$, $CO_2$, certain other volatile inorganics, and combustible organic matter. (2) Weight loss, usually expressed as a percent of the total, after burning off an organic sizing from glass fibers, or an organic resin from a glass fiber laminate.

**lost foam casting**. An *expendable pattern* process in which an expandable polystyrene pattern surrounded by the unbonded sand, is vaporized during pouring of the molten metal (Fig. 309). Also referred to as evaporative pattern casting, evaporative foam casting, the lost pattern process, the cavity-less expanded polystyrene casting process, expanded polystyrene molding, or the full mold process.

**lost wax process**. An *investment casting* process in which a wax pattern is used.

**lot**. (1) A specific amount of material produced at one time using one process and constant conditions of manufacture,

FIG. 309

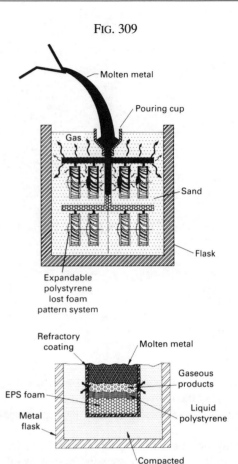

*Pouring of a lost foam casting (top) and the reactions taking place during the pouring operation (bottom)*

and offered for sale as a unit quantity. (2) A quantity of material that is thought to be uniform in one or more stated properties such as isotopic, chemical, or physical characteristics. (3) A quantity of bulk material of similar composition whose properties are under study. (4) A definite quantity of a product or material accumulated under conditions that are considered uniform for *sampling* purposes. Compare with *batch*.

**lot sample**. See *gross sample*.

**low-alloy steels**. A category of ferrous materials that exhibit mechanical properties superior to plain carbon steels as the result of additions of such alloying elements as nickel, chromium, and molybdenum. Total alloy content can range from 2.07% up to levels just below that of stainless steels, which contain a minimum of 10% Cr. For many low-alloy steels, the primary function of the alloying elements is to increase hardenability in order to optimize mechanical properties and toughness after heat treatment. In some cases, however, alloy additions are used to reduce environmental degradation under certain specified service

**Table 40  Compositions of representative AISI/SAE low-alloy steels**

| Steel | C | Mn | P, max | S, max | Si | Ni | Cr | Mo |
|-------|-----|------|--------|--------|----------|-----------|-----------|-----------|
| 1340 | 0.38-0.43 | 1.60-1.90 | 0.035 | 0.040 | 0.20-0.35 | ... | ... | ... |
| 4047 | 0.45-0.50 | 0.70-0.90 | 0.035 | 0.040 | 0.20-0.35 | ... | ... | 0.20-0.30 |
| 4130 | 0.28-0.33 | 0.40-0.60 | 0.035 | 0.040 | 0.20-0.35 | ... | 0.80-1.10 | 0.15-0.25 |
| 4140 | 0.38-0.43 | 0.75-1.00 | 0.035 | 0.040 | 0.20-0.35 | ... | 0.80-1.10 | 0.15-0.25 |
| 4340 | 0.38-0.43 | 0.60-0.80 | 0.035 | 0.040 | 0.20-0.35 | 1.65-2.00 | 0.70-0.90 | 0.20-0.30 |
| 4350 | 0.48-0.53 | 0.60-0.80 | 0.035 | 0.040 | 0.20-0.35 | 1.65-2.00 | 0.70-0.90 | 0.20-0.30 |
| 5046(a) | 0.43-0.48 | 0.75-1.00 | 0.035 | 0.040 | 0.20-0.35 | ... | 0.20-0.35 | ... |
| 5132 | 0.30-0.35 | 0.60-0.80 | 0.035 | 0.040 | 0.20-0.35 | ... | 0.75-1.00 | ... |
| 8645 | 0.43-0.48 | 0.75-1.00 | 0.035 | 0.040 | 0.20-0.35 | 0.40-0.70 | 0.40-0.60 | 0.15-0.25 |
| 8650(a) | 0.48-0.53 | 0.75-1.00 | 0.035 | 0.040 | 0.20-0.35 | 0.40-0.70 | 0.40-0.60 | 0.15-0.25 |
| 8822 | 0.20-0.25 | 0.75-1.00 | 0.035 | 0.040 | 0.15-0.30 | 0.40-0.70 | 0.40-0.60 | 0.30-0.40 |

(a) SAE only

conditions. Table 40 lists chemical compositions of representative low-alloy steels.

**low-cycle fatigue.** *Fatigue* that occurs at relatively small numbers of cycles ($<10^4$ cycles). Low-cycle fatigue may be accompanied by some plastic, or permanent, deformation. Compare with *high-cycle fatigue.*

**low-energy electron diffraction.** A technique for studying the atomic structure of single-crystal surfaces, in which electrons of uniform energy in the approximate range of 5 to 500 eV are scattered from a surface. Those scattered electrons that have lost no energy are selected and accelerated to a fluorescent screen where the diffraction pattern from the surface is observed.

**lower ram.** The part of a pneumatic or hydraulic press that is moving in a lower cylinder and transmits pressure to the lower punch. See also the figure accompanying the term *hydraulic press.*

**low-expansion alloys.** See *Technical Brief 24.*

**low frequency resistance welding cycle.** One positive and one negative pulse of current within the same weld or heat time at a frequency lower than the power supply frequency from which it is obtained.

**low-hydrogen electrode.** A covered arc welding electrode that provides an atmosphere around the arc and molten weld metal that is low in hydrogen.

**low-pressure laminates.** In general, composite laminates molded and cured in the range of pressures from 2760 kPa (400 psi) down to and including pressure obtained by the mere contact of the plies.

**low-pressure molding.** The distribution of relatively uniform low pressure (1400 kPa, or 200 psi, or less) over a resin-bearing fibrous assembly of cellulose, glass, asbestos, or other material, with or without application of heat from an external source, to form a structure possessing specific physical properties.

**low-profile resins.** Special polyester resin systems for reinforced plastics that are combinations of thermoset and thermoplastic resins. Although the terms low-profile and low-shrink are sometimes used interchangeably, there is a difference. Low-shrink resins contain up to 30 wt% thermoplastic polymer, while low-profile resins contain from 30 to 50 wt%. Low shrink offers minimum surface waviness in the molded part (as low as 25 μm per 25 mm, or 1 mil per in., mold shrinkage); low profile offers no surface waviness (from 12.7 to 0 μm per 25 mm, or 0.5 to 0 mil per in., mold shrinkage).

**low-residual-phosphorus copper.** Deoxidized copper with residual phosphorus present in amounts (usually 0.004 to 0.012%) generally too small to decrease appreciably the electrical conductivity of the copper.

**low shaft furnace.** A short shaft-type blast furnace used to produce pig iron and ferroalloys from low-grade ores, using low-grade fuel. The air blast is often enriched with oxygen. Also used for making a variety of other products such as alumina, cementmaking slags, and ammonia synthesis gas.

**low-shrink resins.** See *low-profile resins.*

**low-stress abrasion.** A form of abrasion in which relatively low contact pressures on the abrading particles or protuberances cause only fine scratches and microscopic cutting chips to be produced. See also *high-stress abrasion.*

**L-radiation.** Characteristic x-rays produced by an atom or ion when a vacancy in the L shell is filled by an electron from another shell.

**L-series.** The set of characteristic x-ray wavelengths making up L-radiation for the various elements.

**L shell.** The second shell of electrons surrounding the nucleus of an atom, having electrons with principal quantum number 2.

**lubricant.** (1) Any substance interposed between two surfaces in relative motion for the purpose of reducing the friction or wear between them. This definition implies intentional addition of a substance to an interface; however, species such as oxides and tarnishes on certain metals can also act as lubricants even though they were not added to the system intentionally. (2) A material applied to dies, molds, plungers, or workpieces that promotes the flow of

## Technical Brief 24: Low-Expansion Alloys

LOW-EXPANSION ALLOYS are materials whose dimensions do not change appreciably with temperature. Alloys included in this category are various binary iron-nickel alloys and several ternary alloys of iron combined with nickel-chromium, nickel-cobalt, or cobalt-chromium alloying. Many of the low-expansion alloys are identified by trade names:

- *Invar*, which is a 64Fe-36Ni alloy with the lowest thermal expansion of iron-nickel alloys. Early studies on this alloy demonstrated that the addition of small quantities of manganese, silicon, and carbon amounting to a total of less than 1% produced a material with a coefficient of expansion so low that its length was almost invariable for ordinary changes in temperature. For this reason, the alloy was given the name "Invar."

- *Kovar*, which is a 54Fe-29Ni-17Co alloy with coefficients of expansion closely matching those of standard types of hard (borosilicate) glass

- *Elinvar*, which a 52Fe-36Ni-12Cr alloy with a zero thermoelastic coefficient (that is, an invariable modulus of elasticity over a wide temperature range)

- *Super Invar*, which is a 63Fe-32Ni-5Co alloy with an expansion coefficient smaller than Invar but over a narrower temperature range

- *Stainless Invar*, which is a 37Fe-53Co-10Cr alloy with an exceedingly low, and at times negative, expansion coefficient (over the range from 0 to 100 °C, or 32 to 212 °F) and good corrosion resistance

Besides these common trade names, alloy compositions are also selected to have appropriate expansion characteristics for a par-

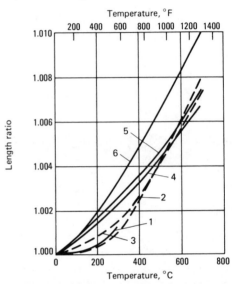

*Thermal expansion of Fe-Ni alloys. Curve 1, 64Fe-31Ni-5Co; curve 2, 64Fe36Ni (Invar); curve 3, 58Fe-42Ni; curve 4, 53Fe-47Ni; curve 5, 48Fe-52Ni; curve 6, carbon steel (0.25% C)*

ticular application. Alloys with coefficients of linear expansion ranging from a small negative value (−0.5 μm/m · K) to a large positive value (20 μm/m · K) have been developed. Low-expansion alloys are used in applications such as rods and tapes for geodetic surveying, compensating pendulums and balance wheels for clocks and watches, moving parts that require control of expansion (such as pistons for some internal-combustion engines), bimetal strip, glass-to-metal seals, thermostatic strip, vessels and piping for storage and transportation of liquefied gas, superconducting systems in power transmissions, integrated-circuit lead frames, components for radios and other electronic devices, and structural components in optical and laser measuring systems.

### Thermal expansion of Fe-Ni alloys between 0 and 38 °C (32 and 100 °F)

| Ni, % | Mean coefficient, μm/m · K |
|---|---|
| 31.4 | 3.395 + 0.00885 $t$ |
| 34.6 | 1.373 + 0.00237 $t$ |
| 35.6 | 0.877 + 0.00127 $t$ |
| 37.3 | 3.457 − 0.00647 $t$ |
| 39.4 | 5.357 − 0.00448 $t$ |
| 43.6 | 7.992 − 0.00273 $t$ |
| 44.4 | 8.508 − 0.00251 $t$ |
| 48.7 | 9.901 − 0.00067 $t$ |
| 50.7 | 9.984 + 0.00243 $t$ |
| 53.2 | 10.045 + 0.00031 $t$ |

### Recommended Reading

- E.L. Frantz, Low-Expansion Alloys, *Metals Handbook*, 10th ed., Vol 2, ASM International, 1990, p 889-896
- K.C. Russel and D. Smith, Ed., *Physical Metallurgy of Controlled Expansion Invar-Type Alloys*, The Metallurgical Society of AIME, 1989

metal, reduces friction and wear, and aids in the release of the finished part.

**lubricant compatibility**. See *compatibility (lubricant)*.

**lubricant residue**. The carbonaceous residue resulting from

lubricant that is burned onto the surface of a hot forged part.

**lubrication**. (1) The reduction of frictional resistance and wear, or other forms of surface deterioration, between two

FIG. 310

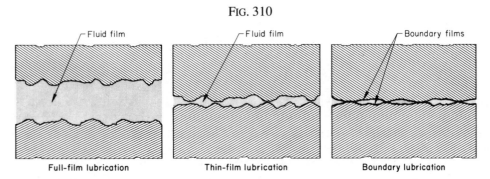

Full-film lubrication          Thin-film lubrication          Boundary lubrication

*Schematic showing three lubrication regimes and the relation of surface roughness to film thickness*

load-bearing surfaces by the application of a *lubricant.* (2) Mixing or incorporating a lubricant with a powder to facilitate compacting and ejecting of the compact from the die cavity; also, applying a lubricant to die walls and/or punch surfaces.

**lubrication regimes.** Ranges of operating conditions for lubricated *tribosystems* that can be distinguished by their frictional characteristics and/or by the manner and amount of separation of the bearing surfaces (Fig. 310). See also *boundary lubrication, elastohydrodynamic lubrication, full-film lubrication, hydrodynamic lubrication, quasi-hydrodynamic lubrication,* and *thin-film lubrication.*

**lubricious (lubricous).** Relating to a substance or surface condition that tends to produce relatively low friction.

**lubricity.** The ability of a lubricant to reduce wear and friction, other than by its purely viscous properties.

**Lüders lines.** Elongated surface markings or depressions in sheet metal, often visible with the unaided eye, caused by discontinuous (inhomogeneous) yielding (Fig. 311). Also known as Lüders bands, Hartmann lines, Piobert lines, or stretcher strains.

**luggin probe.** A small tube or capillary filled with electrolyte, terminating close to the metal surface under study, and used to provide an ionically conducting path without diffusion between an *electrode* under study and a *reference electrode* (Fig. 312).

**luster finish.** A bright as-rolled finish, produced on ground metal rolls; it is suitable for decorative painting or plating, but usually must undergo additional surface preparation after forming.

**lute.** (1) A mixture of fireclay used to seal cracks between a crucible and its cover or between container and cover when heat is to be applied. (2) To seal with clay or other plastic material.

**lyotropic liquid crystal.** A type of liquid crystalline polymer that can be processed only from solution.

FIG. 311

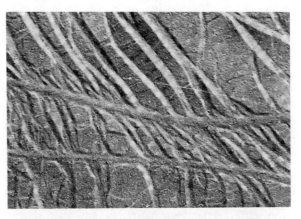

*Rimmed 1008 steel with Lüders lines (stretcher strains) on the surface resulting from the sheet being stretched beyond the yield point during forming. 0.875×*

FIG. 312

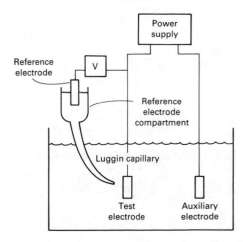

*Potential measurement with a luggin capillary. V, voltmeter*

# M

**macerate.** To chop or shred, as fabric, for use as a filler for a molding resin.

**machinability.** The relative ease of machining a metal.

**machinability index.** A relative measure of the machinability of an engineering material under specified standard conditions. Also known as machinability rating.

**machine forging.** Forging performed in upsetters or horizontal forging machines.

**machine shot capacity.** The maximum weight of thermoplastic resin that can be displaced or injected by the injection (molding) ram in a single stroke.

**machine welding.** Welding with equipment that performs under the continual observation and control of a welding operator. The equipment may or may not load the work. Compare with *automatic welding*.

**machining.** Removing material from a metal part, usually using a cutting tool, and usually using a power-driven machine.

**machining allowance.** See *finish allowance*.

**machining damage** (ceramics and glasses). Atypical or excessively large surface microcracks or damage resulting from the machining process; for example, striations, scratches, impact cracks (Fig. 313). Small surface and subsurface damage is intrinsic to the machining damage.

**machining damage** (metals). See *surface alterations* (metals) and *surface integrity* (metals).

**machining stress.** *Residual stress* caused by machining.

FIG. 313

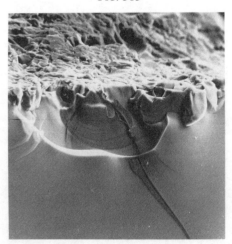

*Machining flaw as fracture origin in glass. Rough surface is the bottom of a groove cut by a diamond saw. Scanning electron micrograph. 200×*

**macro.** In reinforced plastics, the gross properties of a composite as a structural element without consideration of the individual properties or the identity of the constituents.

**macroetching.** Etching a metal surface to accentuate gross structural details, such as grain flow, segregation, porosity, or cracks, for observation by the unaided eye or at magnifications to 25×.

**macrograph.** A graphic representation of the surface of a prepared specimen at a magnification not exceeding 25×. When photographed, the reproduction is known as a photomacrograph.

**macrohardness test.** A term applied to such hardness testing procedures as the Rockwell or Brinell hardness tests to distinguish them from microindentation hardness tests such as the Knoop or Vickers tests. See also *microindentation* and *microindentation hardness number*.

**macropore.** Pores in pressed or sintered powder compacts that are visible with the naked eye.

**macroscopic.** Visible at magnifications at or below 25×.

**macroscopic stress.** Residual stress in a material in a distance comparable to the gage length of strain measurement devices (as opposed to stresses within very small, specific regions, such as individual grains). Compare with *microscopic stress*.

**macroshrinkage.** Isolated, clustered, or interconnected voids in a casting that are detectable macroscopically. Such voids are usually associated with abrupt changes in section size and are caused by feeding that is insufficient to compensate for solidification shrinkage.

**macroslip.** A type of sliding in which all points on one side of the interface are moving relatively to those on the other side in a direction parallel to the interface. See also *microslip*.

**macrostrain.** The mean strain over any finite gage length of measurement large in comparison with interatomic distances. Macrostrain can be measured by several methods, including electrical-resistance strain gages and mechanical or optical extensometers. Elastic macrostrain can be measured by x-ray diffraction.

**macrostress.** Same as *macroscopic stress*.

**macrostructure.** The structure of metals as revealed by macroscopic examination of the etched surface of a polished specimen.

**magnesia.** Magnesium oxide (MgO), used principally in *basic refractories*. Magnesia refractory brick consists mainly of the mineral periclase (MgO) and is available in chemically bonded, pitch-bonded, burned or fired, and burned and pitch-impregnated forms. Historically, natural magnesite ($MgCO_3$) that was calcined provided the raw material for this brick, but with the increased demands for higher temperatures and the introduction of fewer process impurities, more high-purity magnesia from seawater or underground brines has been used. In the seawater and brine processes, the magnesia is obtained by calcining precipitated $Mg(OH)_2$, which provides MgO of purity up to 98%. See also *refractories* (Technical Brief 37).

**magnesium and magnesium alloys.** See *Technical Brief 25*.

**magnetic alignment.** An alignment of the electron-optical axis of the electron microscope so that the image rotates about a point in the center of the viewing screen when the current flowing through a lens is varied. See also *alignment*.

**magnetically hard alloy.** A ferromagnetic alloy capable of being magnetized permanently because of its ability to retain induced magnetization and magnetic poles after removal of externally applied fields; an alloy with high coercive force. The name is based on the fact that the quality of the early permanent magnets was related to their hardness. See also *permanent magnet materials* (Technical Brief 28).

**magnetically soft alloy.** A ferromagnetic alloy that becomes magnetized readily upon application of a field and that returns to practically a nonmagnetic condition when the field is removed; an alloy with the properties of high magnetic permeability, low coercive force, and low magnetic hysteresis loss. See also *soft magnetic materials* (Technical Brief 46).

**magnetic-analysis inspection.** A nondestructive method of inspection to determine the existence of variations in magnetic flux in ferromagnetic materials of constant cross section, such as might be caused by discontinuities and variations in hardness. The variations are usually indicated by a change in pattern on an oscilloscope screen.

**magnetic bearing.** A type of bearing in which the force that separates the relatively moving surfaces is produced by a magnetic field.

**magnetic ceramics.** Inorganic nonmetallic materials having properties associated with the phenomena of magnetism, that is, these materials can produce or conduct magnetic lines of force capable of interacting with electric fields or other magnetic fields. Magnetic ceramics form the basis for numerous devices which rely on soft or hard (permanent) magnets. The soft magnets include materials such as the *ferrites* and *garnets*, while the hard magnets include magnetoplumbites and $\gamma$-$Fe_2O_3$. The applications are diverse, from items such as microwave components to recording tape. They are particularly useful for high-frequency devices and can be found in numerous television and radio applications. In thin films applied to nonmagnetic substrates, these types of ceramics form the basis for magnetic bubble memories for computers.

FIG. 314

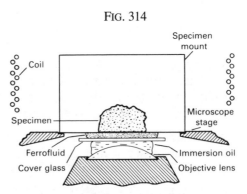

*Components in place for magnetic etching on an inverted microscope*

**magnetic contrast.** In electron microscopy, contrast that arises from the interaction of the electrons in the beam with the magnetic fields of individual magnetic domains in ferromagnetic materials. Special instrumentation is required for this type of work.

**magnetic etching.** The use of magnetized particles in a colloidal suspension (Ferrofluid) to reveal specific features in the microstructure of fully or partially magnetic materials (Fig. 314). Used primarily to observe domain of magnetic materials in order to relate metallographic and domain structures to properties. See also *domain, magnetic*.

**magnetic field.** The region within which a body or current experiences magnetic force.

**magnetic flux.** The rate of flow of magnetic energy across or through a surface.

**magnetic lens.** A device for focusing an electron beam using a magnetic field.

**magnetic-particle inspection.** A nondestructive method of inspection for determining the existence and extent of surface cracks and similar imperfections in ferromagnetic materials (Fig. 315). Finely divided magnetic particles, applied to the magnetized part, are attracted to and outline the pattern of any magnetic-leakage fields created by discontinuities.

**magnetic pole.** The area on a magnetized part at which the magnetic field leaves or enters the part. It is a point of maximum attraction in a magnet.

**magnetic quenchometer test.** Method used to test heat extraction rates of various quenchants. The test works by utilizing the change in magnetic properties of metals at the Curie point—the temperature above which metals lose their magnetism. See also *Curie temperature*.

**magnetic resonance.** A phenomenon in which the magnetic spin systems of certain atoms absorb electromagnetic energy at specific (resonant) natural frequencies of the system.

**magnetic resonance imaging (MRI).** A technique in which an object place in a spatially varying magnetic field is

## Technical Brief 25: Magnesium and Magnesium Alloys

MAGNESIUM is a silvery white metal that is valued chiefly for lightweight components (pure magnesium has a density of approximately 1.7 g/cm$^3$, versus 2.7 g/cm$^3$ for aluminum and 7.8 g/cm$^3$ for steel). Magnesium is produced commercially by the electrolysis of a fused chloride (from brine wells or from seawater) or extracted from mineral ore (most commonly dolomite).

Two major magnesium alloy systems are available. The first includes alloys containing 2 to 10% Al, combined with minor additions of zinc and manganese. The mechanical properties of these alloys are good to 95 to 120 °C (200 to 250 °F). Beyond this, the properties deteriorate rapidly with increasing temperature. The second group consists of magnesium alloyed with various elements (rare earths, zinc, thorium, silver, etc.) except aluminum, all containing a small but effective zirconium content (~0.7%) that imparts a fine grain structure and thus improved mechanical properties. These alloys generally also possess much better elevated-temperature properties.

### Typical magnesium alloy systems and nominal compositions

| Alloy | Element, %(a) | | | | | | | Product form(b) |
|---|---|---|---|---|---|---|---|---|
| | Al | Zn | Mn | Ag | Zr | Th | Re | |
| AM60 | 6 | . . . | 0.2 | . . . | . . . | . . . | . . . | C |
| AZ31 | 3 | 1 | 0.2 | . . . | . . . | . . . | . . . | W |
| AZ61 | 6 | 1 | 0.2 | . . . | . . . | . . . | . . . | W |
| AZ63 | 6 | 3 | 0.2 | . . . | . . . | . . . | . . . | C |
| AZ80 | 8 | 0.5 | 0.2 | . . . | . . . | . . . | . . . | C, W |
| AZ91 | 9 | 1 | 0.2 | . . . | . . . | . . . | . . . | C |
| EZ33 | . . . | 2.5 | . . . | . . . | 0.5 | . . . | 2.5 | C |
| ZM21 | . . . | 2 | 1 | . . . | . . . | . . . | . . . | W |
| HK31 | . . . | 0.1 | . . . | . . . | 0.5 | 3 | . . . | C, W |
| HZ32 | . . . | 2 | . . . | . . . | 0.5 | 3 | . . . | C |
| QE22 | . . . | . . . | . . . | 2.5 | 0.5 | . . . | 2 | C |
| QH21 | . . . | . . . | . . . | 2.5 | 0.5 | 1 | 1 | C |
| ZE41 | . . . | 4.5 | . . . | . . . | 0.5 | . . . | 1.5 | C |
| ZE63 | . . . | 5.5 | . . . | . . . | 0.5 | . . . | 2.5 | C |
| ZK40 | . . . | 4.0 | . . . | . . . | 0.5 | . . . | . . . | C, W |
| ZK60 | . . . | 6.0 | . . . | . . . | 0.5 | . . . | . . . | C, W |

(a) For details, see alloying specifications. (b) C, castings; W, wrought products

Magnesium alloys are produced in both cast and wrought forms. Magnesium alloy castings can be produced by nearly all the conventional casting methods, namely, sand, permanent and semipermanent mold, and shell, investment, and die casting, the latter being the highest in volume. Wrought magnesium alloys are produced as bars, billets, shapes, wire, sheet, plate, and forgings.

Magnesium and magnesium alloys are used in a wide variety of structural and nonstructural applications. Structural applications automotive, industrial, materials handling, commercial, and aerospace equipment. However, it is with nonstructural applications that magnesium finds its greatest use. It is used as an alloying element in alloys of aluminum (the single largest application for magnesium), zinc, lead, and other nonferrous metals. It is used as an oxygen scavenger and desulfurizer in the manufacture on nickel and copper alloys; as a desulfurizer in the iron and steel industry; and as a reducing agent in the production of beryllium, titanium, zirconium, hafnium, and uranium. Magnesium powders are used to manufacture Grignard reagents, which are organometallic halides used in organic synthesis to produce pharmaceuticals, perfumes, and other chemicals. Magnesium powder also finds some use in pyrotechnics, both as pure magnesium and alloyed with 30% or more aluminum. As a galvanic anode, magnesium provides effective corrosion protection for water heaters, underground pipelines, ship hulls, and ballast tanks. Small, lightweight, high-current-output primary batteries also use magnesium as the anode. Gray iron foundries use magnesium and magnesium-containing alloys as ladle addition agents introduced just before the casting is poured. The magnesium makes the graphite particles nodular and greatly improves the toughness and ductility of the cast iron.

### Recommended Reading

- S. Housh, B. Mikucki, and A. Stevenson, Selection and Application of Magnesium and Magnesium Alloys, *Metals Handbook*, 10th ed., Vol 2, ASM International, 1990, p 455-479
- H. Proffitt, Magnesium and Magnesium Alloys Castings, *Metals Handbook*, 9th ed., Vol 15, ASM International, 1988, p 798-810
- J. Hillis *et al.*, Corrosion of Magnesium and Magnesium Alloys, *Metals Handbook*, 9th ed., Vol 13, ASM International, 1987, p 740-754

FIG. 315

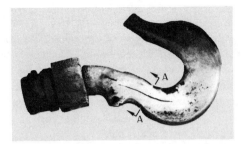

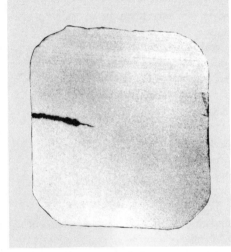

Section **A-A**

*Crane hook showing magnetic-particle indica-
tion of a forging lap and section through hook
showing depth of lap*

subjected to a pulse of radiofrequency radiation, and the
resulting nuclear magnetic resonance spectra are combined to give cross-sectional images.

**magnetic seal.** A seal that uses magnetic material, instead of
springs or bellows, to provide the closing force.

**magnetic separator.** A device used to separate magnetic
from less magnetic or nonmagnetic materials. The crushed
material is conveyed on a belt past a magnet.

**magnetic shielding.** In electron microscopy, shielding for
the purpose of preventing extraneous magnetic fields from
affecting the electron beam in the microscope.

**magnetic writing.** In magnetic-particle inspection, a *false
indication* caused by contact between a magnetized part
and another piece of magnetic material.

**magnetite.** Naturally occurring magnetic oxide or iron
($Fe_3O_4$).

**magnetizing force.** A force field, resulting from the flow of
electric currents or from magnetized bodies, that produces
magnetic induction.

**magnetohydrodynamic lubrication.** Hydrodynamic lubrication in which a significant force contribution arises from

electromagnetic interaction. Magnetohydrodynamic bearings have been proposed for very high-temperature operation, for example, in liquid sodium.

**magnetometer.** An instrument for measuring the magnitude
and sometimes also the direction of a magnetic field, such
as the earth's magnetic field. See also *torque-coil magnetometer.*

**magneton.** A unit of magnetic moment used for atomic,
molecular, or nuclear magnets. The Bohr magneton ($\mu_b$),
which has the value of the classical magnetic moment of
the electron, can theoretically be calculated as:

$$\mu_B = \frac{eh}{2mc} = 9.2731 \times 10^{-20} \text{ erg/G}$$

where *e* and *m* are the electronic charge and mass, respectively; *h* is Planck's constant divided by $2\pi$; and *c* is the
velocity of light. See also *Planck's constant.*

**magnetostriction.** Changes in dimensions of a body resulting
from application of a magnetic field.

**magnetostrictive cavitation test device.** A vibratory cavitation test device driven by a magnetostrictive transducer.

**magnification.** A ratio of the size of an image to its corresponding object. This is usually determined by linear
measurement.

**main bearing.** A bearing supporting the main power-trans-
mitting shaft.

**malleability.** The characteristic of metals that permits *plastic
deformation* in compression without fracture. See also
*ductility.*

**malleable iron.** A cast iron made by prolonged annealing of
*white iron* in which decarburization, graphitization, or
both take place to eliminate some or all of the cementite.
The graphite is in the form of temper carbon. If decarburization is the predominant reaction, the product will
exhibit a light fracture surface; hence whiteheart malleable. Otherwise, the fracture surface will be dark; hence
blackheart malleable. Only the blackheart malleable is
produced in the United States. Ferritic malleable has a
predominantly ferritic matrix (Fig. 316); pearlitic malleable may contain pearlite, spheroidite, or tempered martensite (Fig. 317), depending on heat treatment and desired
hardness. The chemical composition of malleable iron
generally conforms to the ranges given in Table 41. Small
amounts of chromium (0.01 to 0.03%), boron (0.0020%),
copper (~1.0%), nickel (0.5 to 0.8%), and molybdenum
(0.35 to 0.5%) are also sometimes present.

**malleabilizing.** Annealing *white iron* in such a way that some
or all of the combined carbon is transformed into graphite
or, in some cases, so that part of the carbon is removed
completely.

**mandrel** (metals). (1) A blunt-ended tool or rod used to
retain the cavity in a hollow metal product during working. (2) A metal bar around which other metal may be
cast, bent, formed, or shaped. (3) A shaft or bar for holding

FIG. 316

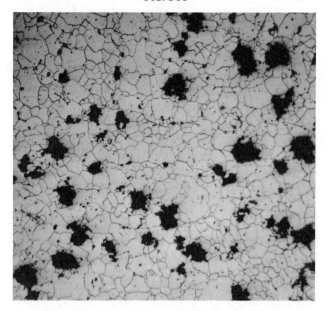

*Structure of annealed ferritic malleable iron showing temper carbon in ferrite. 100×*

FIG. 317

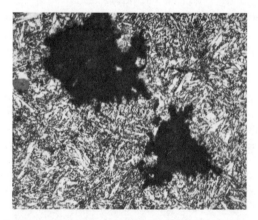

*Pearlitic malleable iron showing graphite nodules (black) in tempered martensite. 500×*

**Table 41  Typical compositions for malleable iron**

| Element | Composition, % Ferritic | Pearlitic |
|---|---|---|
| Total carbon | 2.2–2.9 | 2.0–2.9 |
| Silicon | 0.9–1.9 | 0.9–1.9 |
| Manganese | 0.2–0.6 | 0.2–1.3 |
| Sulfur | 0.02–0.2 | 0.05–0.2 |
| Phosphorus | 0.02–0.2 | 0.02–0.2 |

work to be machined. (4) A form, such as a mold or matrix, used as a cathode in electroforming.

**mandrel** (plastics and reinforced plastics). (1) In blow molding of thermoplastics, part of the tooling that forms the inside of the container neck through which air is forced to form the hot parison to the shape of the mold. (2) In extrusion of thermoplastics, the solid, cylindrical part of the die that forms tubing or pipe. (3) In filament winding of reinforced plastic, the form (usually cylindrical) around which the filaments are wound.

**mandrel forging**. The process of rolling or forging a hollow blank over a mandrel to produce a weldless, seamless ring or tube (Fig. 318). See also *radial forging*.

**manipulator**. A mechanical device for handling an ingot or billet during forging.

**man-made (synthetic) diamond**. A manufactured diamond, darker, blockier, and considered to be more friable than most natural diamonds (Fig. 319).

**Mannesmann process**. A process for piercing tube billets in making seamless tubing. The billet is rotated between two heavy rolls mounted at an angle and is forced over a fixed mandrel.

**manual welding**. Welding wherein the entire welding operation is performed and controlled by hand.

**manufactured unit**. A quantity of finished adhesive or finished adhesively bonded component, processed at one time. The manufactured unit may be a batch or a part thereof.

**maraging**. A precipitation-hardening treatment applied to a special group of iron-base alloys to precipitate one or more intermetallic compounds in a matrix of essentially carbon-free martensite. See also *maraging steels*.

**maraging steels**. A special class of high-strength steels that differ from conventional steels in that they are hardened

FIG. 318

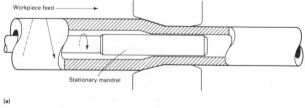

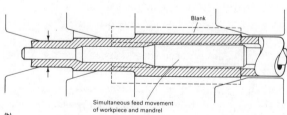

*Forging of tubular parts over a short mandrel (a) and a long mandrel (b) press*

FIG. 319

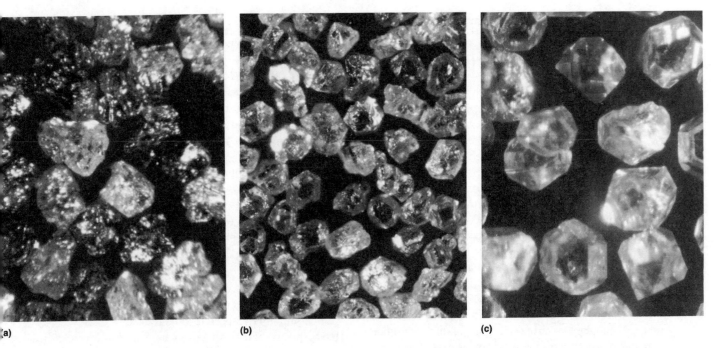

(a)                                          (b)                                          (c)

*Commercially available man-made diamond grains used in various applications. (a) Friable diamond grains especially tailored for resin bond grinding wheels. (b) Diamond grains tailored for use in metal bond grinding wheels. These grains are typically in the 80 to 400 mesh size (350 to 38 μm) range. (c) Diamond grains for use in diamond saw blade applications, such as for the sawing of marble, granite, and concrete. These grains are in the 20 to 60 mesh size (850 to 250 μm) range.*

by a metallurgical reaction that does not involve carbon. Instead, these steels are strengthened by the precipitation of intermetallic compounds at temperatures of about 480 °C (900 °F). The term maraging is derived from martensite age hardening of a low-carbon, iron-nickel lath martensite matrix. Commercial maraging steels are designed to provide specific levels of yield strength from 1030 to 2420 MPa (150 to 350 ksi) with some having yield strengths as high as 3450 MPa (500 ksi). These steels typically have very high nickel, cobalt, and molybdenum contents and

very low carbon contents. Table 42 lists the chemical compositions of common maraging steels.

**Marforming process.** A *rubber-pad forming* process developed to form wrinkle-free shrink flanges and deep-drawn shells. It differs from the *Guerin process* in that the sheet metal blank is clamped between the rubber pad and the blankholder before forming begins (Fig. 320).

**margin.** The cylindrical portion of the *land* of a drill that is not cut away to provide clearance.

**marquenching.** See *martempering.*

## Table 42  Nominal compositions of commercial maraging steels

| Grade | Composition, %(a) | | | | | |
|---|---|---|---|---|---|---|
| | Ni | Mo | Co | Ti | Al | Nb |
| **Standard grades** | | | | | | |
| 18Ni(200) .......................... | 18 | 3.3 | 8.5 | 0.2 | 0.1 | · · · |
| 18Ni(250) .......................... | 18 | 5.0 | 8.5 | 0.4 | 0.1 | · · · |
| 18Ni(300) .......................... | 18 | 5.0 | 9.0 | 0.7 | 0.1 | · · · |
| 18Ni(350) .......................... | 18 | 4.2(b) | 12.5 | 1.6 | 0.1 | · · · |
| 18Ni(Cast)......................... | 17 | 4.6 | 10.0 | 0.3 | 0.1 | · · · |
| 12-5-3(180)(c) ...................... | 12 | 3 | · · · | 0.2 | 0.3 | · · · |
| **Cobalt-free and low-cobalt bearing grades** | | | | | | |
| Cobalt-free 18Ni(200)............... | 18.5 | 3.0 | · · · | 0.7 | 0.1 | · · · |
| Cobalt-free 18Ni(250)............... | 18.5 | 3.0 | · · · | 1.4 | 0.1 | · · · |
| Low-cobalt 18Ni(250)............... | 18.5 | 2.6 | 2.0 | 1.2 | 0.1 | 0.1 |
| Cobalt-free 18Ni(300)............... | 18.5 | 4.0 | · · · | 1.85 | 0.1 | · · · |

(a) All grades contain no more than 0.03% C. (b) Some producers use a combination of 4.8% Mo and 1.4% Ti, nominal. (c) Contains 5% Cr

FIG. 320

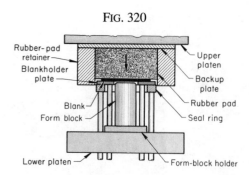

*Tooling and setup for rubber-pad forming by the Marform process*

**martempering**. (1) A hardening procedure in which an austenitized ferrous material is quenched into an appropriate medium at a temperature just above the martensite start temperature of the material, held in the medium until the temperature is uniform throughout, although not long enough for bainite to form, then cooled in air. The treatment is frequently followed by tempering. (2) When the process is applied to carburized material, the controlling martensite start temperature is that of the case. This variation of the process is frequently called marquenching.

**martensite**. A generic term for microstructures formed by diffusionless phase transformation in which the parent and product phases have a specific crystallographic relationship. Martensite is characterized by an acicular pattern in the microstructure in both ferrous and nonferrous alloys. In alloys where the solute atoms occupy interstitial positions in the martensitic lattice (such as carbon in iron), the structure is hard and highly strained; but where the solute atoms occupy substitutional positions (such as nickel in iron), the martensite is soft and ductile. The amount of high-temperature phase that transforms to martensite on cooling depends to a large extent on the lowest temperature attained, there being a rather distinct beginning temperature ($M_s$) and a temperature at which the transformation is essentially complete ($M_f$). See also *lath martensite*, *plate martensite*, and *tempered martensite*.

**martensite range**. The interval between the martensite start ($M_s$) and the martensite finish ($M_f$) temperatures.

**martensitic**. A platelike constituent having an appearance and a mechanism of formation similar to that of martensite. See also *lath martensite* and *plate martensite*.

**martensitic stainless steel**. See *stainless steels* (Technical Brief 47).

**martensitic transformation**. A reaction that takes place in some metals on cooling, with the formation of an acicular structure called *martensite*.

**mash resistance seam welding**. *Resistance seam welding* in which the weld is made in a lap joint, the thickness at the lap being reduced plastically to approximately the thickness of one of the lapped parts.

**mask**. A device for protecting a surface from the effects of blasting and/or coating. Masks are generally either reusable or disposable.

**masking tape**. A tape used as a *resist for stopping-off purposes*.

**masonry cement**. A hydraulic cement for use in mortars for masonry construction, containing one or more of the following materials: portland cement, portland blast furnace slag cement, portland-pozzolan cement, natural cement, slag cement, or hydraulic lime. In addition, masonry cement usually contains one or more materials such as hydrated lime, limestone, chalk, calcareous shell, talc, slag, or clay as prepared for this purpose. See also *cement* (Technical Brief 8).

**mass absorption coefficient**. The linear absorption coefficient divided by the density of the medium.

**mass concentration** (in a slurry). The mass of solid particles per unit mass of mixture, expressed in percent.

**mass spectrometer**. An instrument that is capable of separating ionized molecules of different mass/charge ratios and measuring the respective ion-currents.

**mass spectrometry**. An analytical technique for identification of chemical structures, analysis of mixtures, and quantitative elemental analysis, based on application of the *mass spectrometer*.

**mass spectrum**. A record, graph, or table that shows the relative number of ions of various masses that are produced when a given substance is processed in a *mass spectrometer*.

**master alloy**. An alloy, rich in one or more desired addition elements, that is added to a metal melt to raise the percentage of a desired constituent.

**master alloy powder**. A prealloyed metal powder of high concentration of alloy content, designed to be diluted when mixed with a base powder to produce the desired composition. See also *prealloyed powder*.

**master block**. A forging *die block* used primarily to hold insert dies. See also *die insert*.

**master pattern**. In foundry practice, a pattern embodying a double contraction allowance in its construction, used for making castings to be employed as patterns in production work.

**mat**. A fibrous glass material used as a plastic reinforcement and consisting of randomly oriented chopped filaments, short fibers (with or without a carrier fabric), or swirled filaments loosely held together with a binder. Available in blankets of various widths, weights, and lengths. Also, a sheet formed by filament winding a single-hoop ply of fiber on a mandrel, cutting across its width and laying out a flat sheet.

**match**. A condition in which a point in one metal forming or forging die half is aligned properly with the corresponding point in the opposite die half within specified tolerance.

**matched edges.** Two edges of the die face that are machined exactly at 90° to each other, and from which all dimensions are taken in laying out the die impression and aligning the dies in the forging equipment. Also referred to as match lines.

**matched metal die molding.** A reinforced plastics manufacturing process in which matching male and female metal molds are used (for example, in compression molding) to form the part, with time, pressure, and heat.

**matching draft.** The adjustment of draft angles (usually involving an increase) on parts with asymmetrical ribs and sidewalls to make the surfaces of a forging meet at the parting line.

**match plate.** A plate of metal or other material on which patterns for metal casting are mounted (or formed as an integral part) to facilitate molding (Fig. 321). The pattern is divided along its *parting plane* by the plate.

FIG. 321

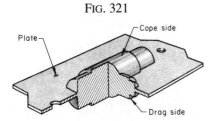

*A typical metal match plate pattern*

**materials characterization.** The use of various analytical methods (spectroscopy, microscopy, chromatography, etc.) to describe those features of composition (both bulk and surface) and structure (including defects) of a material that are significant for a particular preparation, study of properties, or use. Test methods that yield information primarily related to materials properties, such as thermal, electrical, and mechanical properties, are excluded from this definition.

**matrix** (adhesives). The part of an adhesive that surrounds or engulfs embedded filler or reinforcing particles and filaments.

**matrix** (composites). The essentially homogeneous plastic resin in which the fiber reinforcement is embedded. Both thermoplastic and thermoset resins may be used.

**matrix** (metals). The continuous or principal phase in which another constituent is dispersed.

**matrix** (thermal spraying). The major continuous substance of a thermal sprayed coating as opposed to inclusions or particles of materials having dissimilar characteristics.

**matrix isolation.** A technique for maintaining molecules at low temperature for spectroscopic study; this method is particularly well suited for preserving reactive species in a solid, inert environment.

**matrix metal.** The continuous phase of a polyphase alloy, mechanical mixture, or *metal-matrix composite*; the physi-

cally continuous metallic constituent in which separate particles of another constituent are embedded.

**matte.** An intermediate product of *smelting*; an impure metallic sulfide mixture made by melting a roasted sulfide ore, such as an ore of copper, lead, or nickel.

**matte dip.** An etching solution used to produce a dull finish on metal.

**matte finish.** (1) A dull texture produced by rolling sheet or strip between rolls that have been roughened by blasting. (2) A dull finish characteristic of some electrodeposits, such as cadmium or tin.

**maturing temperature.** The temperature, as a function of time and bonding condition, that produces desired characteristics in adhesively bonded components. The term is specific for ceramic adhesives.

**maximum elongation.** The elongation at the time of fracture, including both elastic and plastic deformation of the tensile specimen. Applicable to rubber, plastic, and some metallic materials. Maximum elongation is also called ultimate elongation or break elongation.

**maximum erosion rate.** The maximum instantaneous erosion rate in a test that exhibits such a maximum followed by decreasing erosion rates. Occurrence of such a maximum is typical of many cavitation and liquid impingement tests. In some instances it occurs as an instantaneous maximum; in others it occurs as a steady-state maximum that persists for some time.

**maximum load rate ($P_{max}$).** (1) The load having the highest algebraic value in the load cycle. Tensile loads are considered positive and compressive loads negative. (2) Used to determine the strength of a structural member; the load that can be borne before failure is apparent.

**maximum pore size.** See *absolute pore size*.

**maximum rate period.** In cavitation and liquid impingement erosion, a stage following the acceleration period, during which the erosion rate remains constant (or nearly so) at its maximum value.

**maximum strength.** See *ultimate strength*.

**maximum stress ($S_{max}$).** The stress having the highest algebraic value in the stress cycle, tensile stress being considered positive and compressive stress negative. The *nominal stress* is used most commonly.

**maximum stress intensity factor ($K_{max}$).** The maximum value of the *stress-intensity factor* in a fatigue cycle.

**McQuaid-Ehn grain size.** The austenitic grain size developed in steels by carburizing at 927 °C (1700 °F) followed by slow cooling. Eight standard McQuaid-Ehn grain sizes rate the structure, from No. 8, the finest, to No. 1, the coarsest. The use of standardized ASTM methods for determining grain size is recommended.

**mean stress ($S_m$).** The algebraic average of the maximum and minimum stresses in one cycle, that is, $S_m = (S_{max} + S_{min})/2$. Also referred to as steady component of stress.

FIG. 322

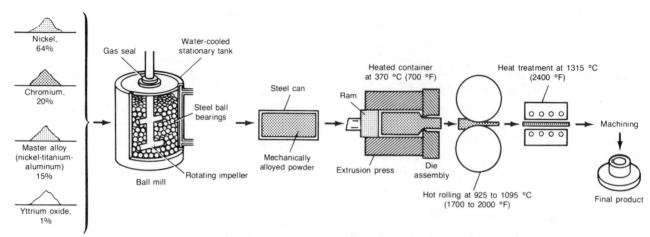

*Schematic showing typical process operations used in the production of mechanically alloyed ODS products*

### Table 43  Nominal compositions of selected mechanically alloyed materials

| Alloy designation | Ni | Fe | Cr | Al | Ti | W | Mo | Ta | Y₂O₃ | C | B | Zr |
|---|---|---|---|---|---|---|---|---|---|---|---|---|
| MA 754 . . . . . . . . . . | bal | · · · | 20 | 0.3 | 0.5 | · · · | · · · | · · · | 0.6 | 0.05 | · · · | · · · |
| MA 758 . . . . . . . . . . | bal | · · · | 30 | 0.3 | 0.5 | · · · | · · · | · · · | 0.6 | 0.05 | · · · | · · · |
| MA 760 . . . . . . . . . . | bal | · · · | 20 | 6.0 | · · · | 3.5 | 2.0 | · · · | 0.95 | 0.05 | 0.01 | 0.15 |
| MA 6000 . . . . . . . . . | bal | · · · | 15 | 4.5 | 2.5 | 4.0 | 2.0 | 2.0 | 1.1 | 0.05 | 0.01 | 0.15 |
| MA 956 . . . . . . . . . . . | · · · | bal | 20 | 4.5 | 0.5 | · · · | · · · | · · · | 0.5 | 0.05 | · · · | · · · |

**mechanical activation.** The acceleration or initiation of a chemical reaction by mechanical exposure of a nascent solid surface. Metal cutting (machining) is an effective method of exposing large areas of fresh surface.

**mechanical adhesion.** Adhesion between surfaces produced solely by the interlocking of protuberances on those surfaces. See also *adherence*.

**mechanical adhesion** (adhesives). Adhesion between surfaces in which the adhesive holds the parts together by interlocking action.

**mechanical alignment.** A method of aligning the geometrical axis of the electron microscope by relative physical movement of the components, usually as a step preceding magnetic or voltage alignment. See also *alignment*.

**mechanical alloying (MA).** An alternate cold welding and shearing of particles of two or more species of greatly differing hardness. The operation is carried out in high-intensity ball mills, such as attritors, and is the preferred method of producing oxide-dispersion-strengthened (ODS) materials (Fig. 322). Mechanical alloying was originally developed for the manufacture of nickel-base superalloys strengthened by both an oxide dispersion and γ′ precipitate. At present, commercial quantities of material are available in the nickel-, iron-, and aluminum-base alloy systems. Table 43 lists compositions of selected mechanically alloyed materials. See also *attritor grinding* and *dispersion-strengthened material*.

**mechanical bond** (thermal spraying). The adherence of a thermal sprayed deposit to a roughened surface by the mechanism of particle interlocking.

**mechanical (cold) crack.** A crack or fracture in a casting resulting from rough handling or from thermal shock, such as may occur at shakeout or during heat treatment.

**mechanical disintegration.** See preferred terms *comminution* and *pulverization*.

**mechanical hysteresis.** Energy absorbed in a complete cycle of loading and unloading within the elastic limit and represented by the closed loop of the stress-strain curves for loading and unloading. Sometimes referred to as elastic, but more properly, mechanical. See also the figure accompanying the term *hysteresis loop* (mechanical).

**mechanically formed plastic.** A cellular plastic having a structure produced by physically incorporated gases.

**mechanical metallurgy.** The science and technology dealing with the behavior of metals when subjected to applied forces; often considered to be restricted to plastic working or shaping of metals.

**mechanical plating.** Plating wherein fine metal powders are peened onto the work by *tumbling* or other means. The process is used primarily to provide ferrous parts with coatings of zinc, cadmium, tin, and alloys of these metals in various combinations.

**mechanical polishing.** A process that yields a specularly reflecting surface entirely by the action of machining

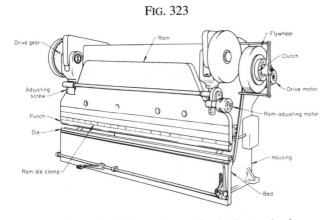

FIG. 323

*Principal components of a mechanical press brake*

tools, which are usually the points of abrasive particles suspended in a liquid among the fibers of a polishing cloth.

**mechanical press.** A press whose slide is operated by a crank, eccentric, cam, toggle links, or other mechanical device. See also the figures accompanying *eccentric gear*, *knuckle-lever press*, and *toggle press*.

**mechanical press brake.** A *press brake* using a mechanical drive consisting of a motor, flywheel, crankshaft, clutch, and eccentric to generate vertical motion (Fig. 323).

**mechanical properties.** The properties of a material that reveal its elastic and inelastic behavior when force is applied, thereby indicating its suitability for mechanical applications; for example, modulus of elasticity, tensile strength, elongation, hardness, and fatigue limit. Compare with *physical properties*.

**mechanical seal.** See *face seal*.

**mechanical stability** (of a grease). Grease shear stability tested in a standard rolling tester.

**mechanical stage.** In microscopy, a device provided for adjusting the position of a specimen, usually by translation in two directions at right angles to each other.

**mechanical testing.** The methods by which the *mechanical properties* of a metal are determined.

**mechanical twin.** A *twin* formed in a crystal by simple shear under external loading.

**mechanical upsetter.** A three-element forging press, with two gripper dies and a forming tool, for flanging or forming relatively deep recesses.

**mechanical wear.** Removal of material due to mechanical processes under conditions of sliding, rolling, or repeated impact. The term mechanical wear includes *adhesive wear*, *abrasive wear*, and *fatigue wear*. Compare with *corrosive wear* and *thermal wear*.

**mechanical working.** The subjecting of metals to pressure exerted by rolls, hammers, or presses in order to change the shape or physical properties of the metal.

**median crack.** Damage produced in glass by the static or translational contact of a hard, sharp object on the glass surface. The crack propagates into the glass perpendicular to the original surface.

**median fatigue life.** The middle value when all of the observed fatigue life values of the individual specimens in a group tested under identical conditions are arranged in order of magnitude. When an even number of specimens are tested, the average of the two middlemost values is used. Use of the sample median rather than the arithmetic mean (that is, the average) is usually preferred.

**median fatigue strength at *N* cycles.** An estimate of the stress level at which 50% of the population would survive *N* cycles. The estimate is derived from a particular point of the fatigue life distribution, since there is no test procedure by which a frequency distribution of fatigue strengths at *N* cycles can be directly observed. Also known as *fatigue strength at N cycles*.

**megaelectron volt (MeV).** A unit of energy usually associated with a particle. The energy gained by an electron accelerated across 1,000,000 V.

**melamine plastics.** Thermosetting plastics made from melamine and formaldehyde resins.

**melt.** (1) To change a solid to a liquid by the application of heat. (2) A charge of molten metal or plastic.

**meltback time.** In arc welding, the time interval at the end of crater fill time to arc outage during which the electrode feed is stopped. Arc voltage and arc length increase and current decreases to zero to prevent the electrode from freezing in the weld deposit.

**melt index.** The amount, in grams, of a thermoplastic resin that can be forced through a 2.0955 mm (0.0825 in.) orifice when subjected to 20.7 N (2160 gf) in 10 min at 190 °C (375 °F).

**melting point** (metals). The temperature at which a pure metal, compound, or eutectic changes from solid to liquid; the temperature at which the liquid and the solid are at equilibrium.

**melting point** (plastics). The term that refers to the first-order transition in crystalline polymers. The fixed point between the solid and liquid phases of a material when approached from the solid phase under a pressure of 101.325 kPa (1 atm).

**melting pressure** (metals). At a stated temperature, the pressure at which the solid phases of an element or congruently melting compound may coexist at equilibrium with liquid of the same composition.

**melting range** (metals). The range of temperatures over which an alloy other than a compound or eutectic changes from solid to liquid; the range of temperatures from *solidus* to *liquidus* at any given composition on a *phase diagram*.

**melting rate.** In electric arc welding, the weight or length of electrode melted in a unit of time. Sometimes called melt-off rate or burn-off rate.

**melting temperature** (glass). The range of furnace temperatures within which melting takes place at a commercially desirable rate and within which the resulting glass generally has a viscosity of $10^{0.5}$ to $10^{1.5}$ Pa · s ($10^{1.5}$ to $10^{2.5}$ P). To compare the melting temperatures of glasses, it is assumed that a glass at its melting temperature has a viscosity of 10 Pa · s ($10^2$ P).

**melting temperature** (metals). See *melting point* (metals).

**melt lubrication**. Lubrication provided by steady melting of a lubricating species. Also known as *phase-change lubrication*.

**melt-off rate**. See *melting rate*.

**melt strength**. The strength of a plastic while in the molten state.

**melt-through**. Complete joint penetration for a joint welded from one side. To prevent melt-through, the welding current and the width of the root opening should be reduced, and travel speed increased.

**membrane**. Any thin sheet or layer.

**Menstruum method**. A method of producing multicarbide powder, such as WC + TiC solid solution, by introducing the individual elements into a molten bath of a noncarbide-forming metal such as cobalt or nickel. The multicarbide is formed above 2100 °C (3800 °F), slowly cooled in the dispersed condition in the Menstruum to room temperature, and finally won by chemical separation.

**mer**. The repeating structural unit of any polymer.

**merchant mill** (obsolete). A mill, consisting of a group of stands of three rolls each arranged in a straight line and driven by one power unit, used to roll rounds, squares or flats of smaller dimensions than would be rolled on a bar mill.

**mesh**. (1) The number of screen openings per linear inch of screen; also called *mesh size*. (2) The screen number on the finest screen of a specified standard screen scale through which almost all of the particles of a powder

sample will pass. See also *sieve analysis* and *sieve classification*.

**mesh-belt conveyor furnace**. A continuously operating furnace that uses a conveyor belt for the transport of the charge. Mesh-belt conveyor furnaces consist of a belt-driven table, a burn-off zone, a slow-cooling zone, a final-cooling zone, and a discharge table (Fig. 324).

**mesh size**. (1) The opening(s) or size of opening(s) in a designated sieve or screen, hence the approximate diameter of particles below which they will pass through and above which they will be retained on the screen. Mesh sizes are given as number of wires per inch of standard screen construction, for example, Tyler or U.S.; these are translated by tables into equivalent particle diameters in inches (in.), millimeters (mm), or micrometers (μm). (2) The width of the aperture in a cloth or wire screen. See also *sieve analysis* and *sieve classification*.

**mesophase**. An intermediate phase in the formation of carbon from a pitch precursor. This is a liquid crystal phase in the form of microspheres, which, upon prolonged heating above 400 °C (750 °F), coalesce, solidify, and form regions of extended order. Heating to above 2000 °C (3630 °F) leads to the formation of graphite structure.

**metal**. (1) An opaque lustrous elemental chemical substance that is a good conductor of heat and electricity and, when polished, a good reflector of light. Most elemental metals are malleable and ductile and are, in general, denser than the other elemental substances. (2) As to structure, metals may be distinguished from nonmetals by their atomic binding and electron availability. Metallic atoms tend to lose electrons from the outer shells, the positive ions thus formed being held together by the electron gas produced by the separation. The ability of these "free electrons" to carry an electric current, and the fact that this ability decreases as temperature increases, establish the prime distinctions of a metallic solid. (3) From a chemical view-

FIG. 324

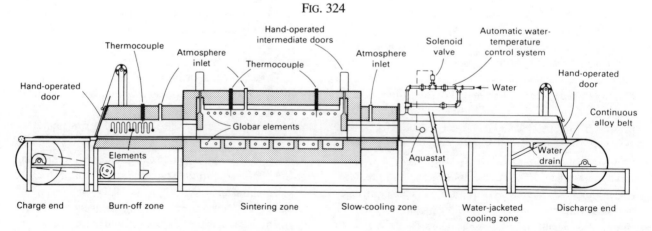

*Longitudinal section of a mesh-belt sintering furnace*

point, an elemental substance whose hydroxide is alkaline. (4) An *alloy*.

**metal-arc cutting.** Any of a group of arc cutting processes which severs metals by melting them with the heat of an arc between a metal electrode and the base metal. See also *shielded metal arc cutting* and *gas metal arc cutting*.

**metal-arc welding.** Any of a group of arc welding processes in which metals are fused together using the heat of an arc between a metal electrode and the work. Use of the specific process name is preferred.

**metal cored electrode.** A composite filler metal welding electrode consisting of a metal tube or other hollow configuration containing alloying ingredients. Minor amounts of ingredients facilitate arc stabilization and fluxing of oxides. External shielding gas may or may not be used.

**metal dusting.** Accelerated deterioration of metals in carbonaceous gases at elevated temperatures to form a dust-like corrosion product.

**metal electrode.** An electrode used in arc welding or cutting which consists of a metal wire or rod that is either bare or covered with a suitable covering or coating.

**metal inert-gas welding.** *Gas metal arc welding* using an inert gas such as argon as the shielding gas.

**metal leaf.** Thin metal sheet, usually thinner than foil, and traditionally produced by beating rather than by rolling.

**metallic bond.** The principal bond between metal atoms, which arises from the increased spatial extension of valence-electron wave functions when an aggregate of metal atoms is brought close together. An example is the bond formed between base metals and filler metals in all welding processes. See also *covalent bond* and *ionic bond*.

**metallic fiber.** Manufactured fiber composed of metal, plastic-coated metal, metal-coated plastic, or a core completely covered by metal.

**metallic glass.** A noncrystalline metal or alloy, commonly produced by drastic supercooling of a molten alloy, by molecular deposition, which involves growth from the vapor phase (e.g., thermal evaporation and sputtering) or from a liquid phase (e.g., electroless deposition and electrodeposition), or by external action techniques (e.g., ion implantation and ion beam mixing). Glassy alloys can be grouped into two major categories. The first group includes the transition metal-metal binary alloy systems, such as Cu-Zr, Ni-Ti, W-Si, and Ni-Nb. The second class consists of transition metal-metalloid alloys. These alloys are usually iron-, nickel-, or cobalt-base systems, may contain film formers (such as chromium and titanium), and normally contain approximately 20 at.% P, B, Si, and/or C as the metalloid component. Also called amorphous alloy or metal. See also *amorphous solid*.

**metallic wear.** Typically, wear due to rubbing or sliding contact between metallic materials that exhibits the characteristics of severe wear, for example, significant plastic

deformation, material transfer, and indications that cold welding of asperities possibly has taken place as part of the wear process. See also *adhesive wear* and *severe wear*.

**metallic whisker.** A fiber composed of a single crystal of metal. See also *whisker*.

**metallization.** A deposited or plated thin metallic film used for its protective or electrical properties.

**metallizing.** (1) Forming a metallic coating by atomized spraying with molten metal or by *vacuum deposition*. Also called spray metallizing. (2) Applying an electrically conductive metallic layer to the surface of a nonconductor. Table 44 lists the various methods for metallizing.

**Table 44  Major metallization techniques**

| Atomistic | Particulate | Bulk |
|---|---|---|
| **Plating** | **Organic medium** | **Organic medium** |
| Electrolytic | Screen printing | Brushing |
| Electroless | Brazing | Roller |
| **Evaporation** | **Powder** | Dipping |
| Vacuum | Flame spraying | Spin |
| Flash | | **Foil** |
| Electron | | Cladding |
| beam | | |
| **Plasma** | | |
| Sputtering | | |
| **Vapor** | | |
| Chemical | | |
| deposition | | |

**metallograph.** An optical instrument designed for visual observation and photomicrography of prepared surfaces of opaque materials at magnifications of 25 to approximately 2000×. The instrument consists of a high-intensity illuminating source, a microscope, and a camera bellows. On some instruments, provisions are made for examination of specimen surfaces using polarized light, phase contrast, oblique illumination, dark-field illumination, and bright-field illumination.

**metallography.** The study of the structure of metals and alloys by various methods, especially by optical and electron microscopy.

**metallurgical bond.** Adherence of a coating to the base material characterized by diffusion, alloying, or intermolecular or intergranular attraction at the interface between the coating and the base material.

**metallurgical burn.** Modification of the microstructure near the contact surface due to frictional temperature rise.

**metallurgical coke.** A coke, usually low in sulfur, having a very high compressive strength at elevated temperatures; used in metallurgical furnaces not only as fuel, but also to support the weight of the charge.

**metallurgy.** The science and technology of metals and alloys. Process metallurgy is concerned with the extraction of metals from their ores and with refining of metals; physical metallurgy, with the physical and mechanical properties of metals as affected by composition, processing, and en-

## Technical Brief 26: Metal-Matrix Composites (MMCs)

METAL-MATRIX COMPOSITES basically consist of a nonmetallic reinforcement incorporated into a metallic matrix. Reinforcements may constitute from 10 to 60 vol% of the composite. Continuous fiber or filament reinforcements include graphite, silicon carbide, boron, $Al_2O_3$, and refractory metals. Fabrication techniques vary from chemical vapor deposition coating of the fibers, liquid-metal infiltration, and diffusion bonding to direct casting to near-net shape. Discontinuous reinforcements consist mainly of SiC in whisker (w) form, particulate (p) types of SiC, $Al_2O_3$, or $TiB_2$, and short or chopped fibers of $Al_2O_3$ or graphite. These MMCs are produced by using modified powder metallurgy techniques.

A metal matrix imparts a metallic nature to the composite in terms of thermal and electrical conductivity, manufacturing options, and interaction with the environment. Metal-matrix composites are capable of providing higher-temperature operating limits than their base metal counterparts, and they can be tailored to impart improved strength, stiffness, creep resistance, abrasion resistance, or dimensional stability.

### Room-temperature properties of unidirectional continuous fiber aluminum-matrix composites

| Property | B/6061 Al | SCS-2/6061 Al | P100 Gr/6061 Al | FP/Al-2Li(a) |
|---|---|---|---|---|
| Fiber content, vol% | 48 | 47 | 43.5 | 55 |
| Longitudinal modulus, GPa ($10^6$ psi) | 214 (31) | 204 (29.6) | 301 (43.6) | 207 (30) |
| Transverse modulus, GPa ($10^6$ psi) | ⋯ | 118 (17.1) | 48 (7.0) | 144 (20.9) |
| Longitudinal strength, MPa (ksi) | 1520 (220) | 1462 (212) | 543 (79) | 552 (80) |
| Transverse strength, MPa (ksi) | ⋯ | 86 (12.5) | 13 (2) | 172 (25) |

(a) FP is the proprietary designation for an alpha alumina ($\alpha$-$Al_2O_3$) fiber developed by E.I. Du Pont de Nemours & Company, Inc.

Unlike resin-matrix composites, they are nonflammable, do not outgas in a vacuum, and suffer minimal attack by organic fluids such as fuels and solvents.

Most commercial work on MMCs has focused on aluminum. The melting point of aluminum is high enough to satisfy many application requirements, yet low enough to render composite processing reasonably convenient. Aluminum can also accommodate a variety of reinforcing agents, including continuous boron, $Al_2O_3$, SiC, and graphite fibers, and various particles, short fibers, and whiskers. Other matrix materiials studied include magnesium, titanium, copper, superalloys, and ordered intermetallic compounds (NiAl and $Ti_3Al$).

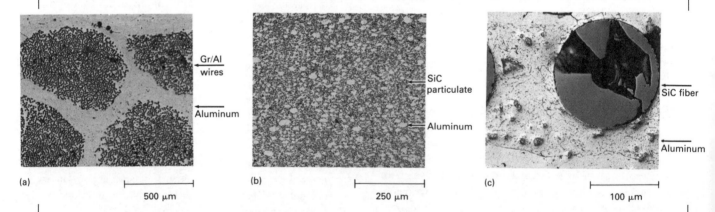

*Cross sections of typical fiber-reinforced MMCs. (a) Continuous-fiber-reinforced Gr/Al composite. (b) Discontinuous SiC(p)/Al composite. (c) Continuous-fiber SiC/Al composite*

### Recommended Reading

- V. Foltz and C.M. Blackmon, Metal-Matrix Composites, *Metals Handbook*, 10th ed., Vol 1, ASM International, 1990, p 903-912
- M. Aylor, Corrosion of Metal-Matrix Composites, *Metals Handbook*, 9th ed., Vol 13, ASM International, 1987, p 859-863
- L. Lachman, S.J. Paprocki, and H.D. Batha, Ed., Metal, Carbon/Graphite, and Ceramic Matrix Composites, *Engineered Materials Handbook*, Vol 1, ASM International, 1987, p 848-944
- P. Rohatgi, Cast Metal-Matrix Composites, *Metals Handbook*, 9th ed., Vol 15, ASM International, 1988, p 840-854

vironmental conditions; and mechanical metallurgy, with the response of metals to applied forces.

**metal-matrix composites**. See *Technical Brief 26*.

**metal penetration**. A surface condition in metal castings in which metal or metal oxides have filled voids between sand grains without displacing them.

**metal powder**. Elemental metals or alloy particles, usually in the size range of 0.1 to 1000 μm.

**metal powder cutting**. A technique that supplements an oxyfuel torch with a stream of iron or blended iron-aluminum powder to facilitate flame cutting of difficult-to-cut materials. The powdered material propagates and accelerates the oxidation reaction, as well as the melting and spalling action of the materials to be cut.

**metal shadowing**. The enhancement of contrast in a microscope by vacuum depositing a dense metal onto the specimen at an angle generally not perpendicular to the surface of the specimen. See also *shadowing*.

**metal spraying**. Coating metal objects by spraying molten metal against their surfaces. See also *thermal spraying* and *flame spraying*.

**metalworking**. See *forming*.

**metastable**. (1) Of a material not truly stable with respect to some transition, conversion, or reaction but stabilized kinetically either by rapid cooling or by some molecular characteristics as, for example, by the extremely high viscosity of polymers. (2) Possessing a state of pseudo-equilibrium that has a free energy higher than that of the true equilibrium state.

**metastable beta**. A β-phase composition in titanium alloys that can be partially or completely transformed to martensite, α, or eutectoid decomposition products with thermal or strain-energy activation during subsequent processing or service exposure.

**methyl methacrylate**. A colorless, volatile liquid derived from acetone, cyanohydrin, methanol, and dilute sulfuric acid and used in the production of *acrylic resins*.

**$M_f$ temperature**. For any alloy system, the temperature at which martensite formation on cooling is essentially finished. See also *transformation temperature* for the definition applicable to ferrous alloys.

**M-glass**. A high beryllia ($BeO_2$) content glass designed especially for high modulus of elasticity.

**mho**. An electrical unit of conductivity that is the conductivity of a body with the resistance of one ohm.

**micelle**. A submicroscopic unit of structure built up from ions or polymeric molecules.

**micro**. In relation to reinforced plastics and composites, the properties of the constituents only, that is, matrix, reinforcements, and interface, and their effects on the properties of the composite.

**microanalysis**. The analysis of samples smaller than 1 mg.

**microbands**. Long, straight bands of highly concentrated *slip* lying on the *slip planes* of individual grains in metals. They are usually 0.1 to 0.2 μm thick, traverse an entire grain, and correspond to the *slip bands* seen on a polished surface.

**microbial corrosion**. See *biological corrosion*.

**microcrack**. A crack of microscopic proportions. Also termed microfissure.

**microcracking**. Cracks formed in composites when thermal stresses locally exceed the strength of the matrix (Fig. 325). Since most microcracks do not penetrate the reinforcing fibers, microcracks in a cross-plied laminate or in a laminate made from cloth prepreg are usually limited to the thickness of a single ply.

FIG. 325

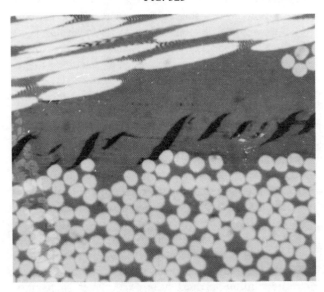

*Inclined microcracks found in short-beam composite shear specimen tested at 132 °C (270 °F). 500×*

**microdynamometer**. An instrument for measuring mechanical force and observing the change in microscopic appearance of a small specimen.

**microelectronics**. The area of electronic technology associated with or applied to the realization of electronic systems from extremely small electronic parts or elements.

**microetching**. Development of microstructure for microscopic examination. The usual magnification exceeds 25× (50× in Europe).

**microfissure**. A crack of microscopic proportions.

**micrograph**. A graphic reproduction of the surface of a specimen at a magnification greater than 25×. If produced by photographic means it is called a photomicrograph (not a microphotograph).

**microhardness**. The hardness of a material as determined by forcing an indenter such as a Vickers or Knoop indenter into the surface of a material under very light load; usually, the indentations are so small that they must be measured with a microscope. Capable of determining hardnesses of different microconstituents within a structure, or of measuring steep hardness gradients such as those encountered in *case hardening*. See also *microhardness test*.

**microhardness number.** A commonly used term for the more technically correct term *microindentation hardness number*.

**microhardness test.** A microindentation hardness test using a calibrated machine to force a diamond indenter of specific geometry, under a test load of 1 to 1000 gram-force, into the surface of the test material and to measure the diagonal or diagonals optically. See also *Knoop hardness test* and *Vickers hardness test*.

**microindentation.** (1) In hardness testing, the small residual impression left in a solid surface when an indenter, typically a pyramidal diamond stylus, is withdrawn after penetrating the surface (Fig. 326). Typically, the dimensions of the microindentations are measured to determine microindentation hardness number, but newer methods measure the displacement of the indenter during the indentation process to use in the hardness calculation. The precise size required to qualify as a microindentation has not been clearly defined; however, typical measurements of the diagonals of such impressions range from approximately 10 to 200 µm, depending on normal force and material. (2) The process of indenting a solid surface, using a hard stylus of prescribed geometry and under a slowly applied normal force, usually for the purpose of determining its microindentation hardness number. See also *Knoop hardness number*, *microindentation hardness number*, and *Vickers hardness number*.

**microindentation hardness number.** A numerical quantity, usually stated in units of pressure ($kg/mm^2$), that expresses the resistance to penetration of a solid surface by a hard indenter of prescribed geometry and under a specified, slowly applied normal force. The prefix "micro" indicates that the indentations produced are typically between 10.0

FIG. 326

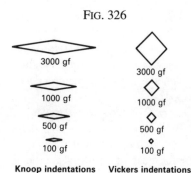

Knoop indentations　　Vickers indentations

*Microindentations made by Knoop and Vickers indenters in the same work metal at the same load*

and 200.0 µm across. See also *Knoop hardness number*, *nanohardness test*, and *Vickers hardness number*.

**micromesh.** A sieve with precisely square openings in the range of 10 to 120 µm produced by electroforming.

**micromesh sizing.** The process of sizing micromesh particles using an air or a liquid suspension process.

**micrometer.** A dimension of 0.001 mm, written with the abbreviation µm. Also referred to as micron (not recommended).

**micropore.** The pores in a sintered product that can only be detected under a microscope.

**microporosity.** Extremely fine *porosity* in castings.

**micropulverizer.** A machine that disintegrates powder agglomerates by strong impacts from small hammers fastened to a solid disk that rotates at very high velocity.

**microradiography.** The technique of passing x-rays through a thin section of a material in contact with a fine-grained photographic film and then viewing the radiograph at 50 to 100× to observe the distribution of constituents and/or defects.

**microscope.** An instrument capable of producing a magnified image of a small object.

**microscopic.** Visible at magnifications above 25×.

**microscopic stress.** Residual stress in a material within a distance comparable to the grain size. See also *macroscopic stress*.

**microscopy.** The science of the interpretive use and applications of microscopes.

**microsegregation.** *Segregation* within a grain, crystal, or small particle. See also *coring*.

**microshrinkage.** A casting imperfection, not detectable microscopically, consisting of interdendritic voids. Microshrinkage results from contraction during solidification where the opportunity to supply filler material is inadequate to compensate for shrinkage. Alloys with wide ranges in solidification temperature are particularly susceptible.

**microshrinkage cavity.** A fine void found microscopically in the low melting metal phase of infiltrated powder metallurgy compacts due to contraction during solidification.

**microslip.** Small relative tangential displacement in a contacting area at an interface, when the remainder of the interface in the contacting area is not relatively displaced tangentially. Microslip can occur in both rolling and stationary contacts. (2) The term microslip is sometimes used to denote the microslip velocity. This usage is not recommended. See also *macroslip* and *slip*.

**microstrain.** The strain over a gage length comparable to interatomic distances. These are the strains being averaged by the *macrostrain* measurement. Microstrain is not measurable by existing techniques. Variance of the microstrain distribution can, however, be measured by x-ray diffraction.

**microstress.** Same as *microscopic stress*.

**microstructure.** The structure of an object, organism, or material as revealed by a microscope at magnifications greater than 25×.

**microwave radiation.** Electromagnetic radiation in the wavelength range of 0.3 mm to 1 m ($3 \times 10^6$ to $10^{10}$ Å). See also *electromagnetic radiation*.

**middle-infrared radiation.** Infrared radiation in the wavelength range of 3 to 30 μm ($3 \times 10^4$ to $3 \times 10^5$ Å). See also *infrared radiation* and *electromagnetic radiation*.

**middling.** A product intermediate between concentrate and tailing and containing enough of a valuable mineral to make retreatment profitable.

**migration.** Movement of entities (such as electrons, ions, atoms, molecules, vacancies, and grain boundaries) from one place to another under the influence of a driving force (such as an electrical potential or a concentration gradient).

**MIG welding.** See preferred term *gas metal arc welding*.

**mil.** An English measure of thickness or diameter equal to 0.0254 mm (0.001 in.). A common designation of wire size, coating thickness, or corrosion loss.

**mild steel.** *Carbon steel* with a maximum of about 0.25% C and containing 0.4 to 0.7% Mn, 0.1 to 0.5% Si, and some residuals of sulfur, phosphorus, and/or other elements.

**mild wear.** A form of wear characterized by the removal of material in very small fragments. Mild wear is an imprecise term, frequently used in research, and contrasted with *severe wear*. In fact, the phenomena studied usually involve the transition from mild to severe wear and the factors that influence this transition. Mild wear may be appreciably greater than can be tolerated in practice. With metallic sliders, mild wear debris usually consists of oxide particles. See also *normal wear* and *severe wear*.

**mill.** (1) A factory in which metals are hot worked, cold worked, or melted and cast into standard shapes suitable for secondary fabrication into commercial products. (2) A production line, usually of four or more *stands*, for hot or cold rolling metal into standard shapes such as bar, rod, plate, sheet, or strip. (3) A single machine for hot rolling, cold rolling, or extruding metal; examples include *blooming mill, cluster mill, four-high mill*, and *Sendzimir mill*. (4) A shop term for a milling cutter. (5) A machine or group of machines for grinding or crushing ores and other minerals. (6) A machine for grinding or mixing material, for example, a ball mill and a paint mill. (7) Grinding or mixing a material, for example, milling a powder metallurgy material.

**milled fiber.** Continuous glass strands hammer milled into very short glass fibers. Useful as inexpensive filler or anticrazing reinforcing fillers for adhesives and engineering plastics.

**mill edge.** The normal edge produced in hot rolling of sheet metal. This edge is customarily removed when hot rolled sheets are further processed into cold rolled sheets.

**Miller indices.** A system for identifying planes and directions in any crystal system by means of sets of integers. The indices of a plane are related to the intercepts of that plane with the axes of a unit cell; the indices of a direction, to the multiples of lattice parameter that represent the coordinates of a point on a line parallel to the direction and passing through the arbitrarily chosen origin of a unit cell.

**Miller number.** A measure of slurry abrasivity as related to the instantaneous mass-loss rate of a standard metal wear block at a specific time on the cumulative abrasion-corrosion time curve. See also *slurry abrasion response number*.

**mill finish.** A nonstandard (and typically nonuniform) surface finish on *mill products* that are delivered without being subjected to a special surface treatment (other than a corrosion-preventive treatment) after the final working or heat-treating step.

**milling** (machining). Using a rotary tool with one or more teeth which engage the workpiece and remove material as the workpiece moves past the rotating cutter.
1. Face milling: Milling a surface perpendicular to the axis of the cutter. Peripheral cutting edges remove the bulk of the material while the face cutting edges provide the finish of the surface being generated.
2. End milling: Milling accomplished with a tool having cutting edges on its cylindrical surfaces as well as on its end. In peripheral end milling, the peripheral cutting edges on the cylindrical surface are used; while in slotting, both end and peripheral cutting edges remove metal.
3. Side and slot milling: Milling of the side or slot of a workpiece using a peripheral cutter.
4. Slab milling: Milling of a surface parallel to the axis of a helical, multiple-toothed cutter mounted on an arbor.
5. Straddle milling: Peripheral milling a workpiece on both sides at once using two cutters spaced as required.

**milling** (powder technology). The mechanical comminution of a material, usually in a ball mill, to alter the size or shape of the individual particles (Fig. 327) to coat one component of a mixture with another, or to create uniform distributions of components.

**milling cutter.** A rotary cutting tool provided with one or more cutting elements, called teeth, which intermittently engage the workpiece and remove material by relative movement of the workpiece and cutter (Fig. 328).

**milling fluid.** An organic liquid, such as hexane, in which *ball milling* is carried out. The liquid serves to reduce the heat of friction and resulting surface oxidation of the particles during grinding, and to provide protection from other surface contamination.

**mill product.** Any commercial product of a *mill*.

FIG. 327

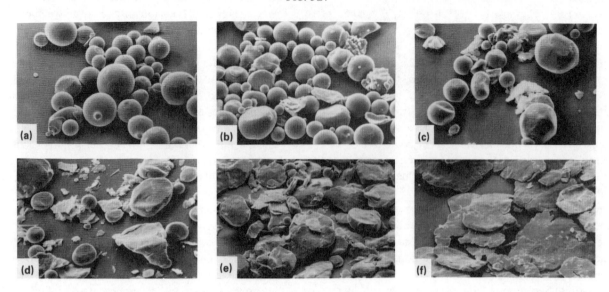

*Effect of milling on the particle shape of Haynes Stellite cobalt-base powder. (a) As-received powder. (b) After 1 h. (c) After 2 h. (d) After 4 h. (e) After 8 h. (f) After 16 h*

FIG. 328

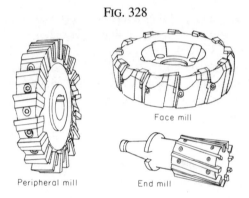

*Three types of milling cutters*

**mill scale.** The heavy oxide layer that forms during the hot fabrication or heat treatment of metals.

**mineral dressing.** Physical and chemical concentration of raw ore into a product from which a metal can be recovered at a profit.

**mineral oil.** A refined hydrocarbon oil without animal or vegetable additives.

**minimized spangle.** A hot dip galvanized coating of very small grain size, which makes the *spangle* less visible when the part is subsequently painted.

**minimum bend radius.** The minimum radius over which a metal product can be bent to a given angle without fracture. See also the figure accompanying the term *bend radius*.

**minimum load ($P_{min}$).** In fatigue, the least algebraic value of applied load in a cycle.

**minimum stress ($S_{min}$).** In fatigue, the stress having the lowest algebraic value in the cycle, tensile stress being considered positive and compressive stress negative.

**minimum stress-intensity factor ($K_{min}$).** In fatigue, the minimum value of the *stress-intensity factor* in a cycle. This value corresponds to the *minimum load* when the *load ratio* > 0 and is taken to be zero when the *load ratio* is ≤ 0.

**minus sieve.** The portion of a powder sample that passes through a standard sieve of a specified number. See also *plus sieve* and *sieve analysis*.

**mirror illumination.** A thin, half-round opaque mirror interposed in a microscope for directing an intense oblique beam of light to the object. The light incident on the object passes through one half of the aperture of the objective, and the light reflected from the object passes through the other half aperture of the objective.

**mirror region.** The comparatively smooth region which symmetrically surrounds a fracture origin in ceramics and glasses (Fig. 329). The mirror region ends in a microscopically irregular manner at the beginning of the mist region. See also *mist hackle*.

**mischmetal.** An natural mixture of rare-earth elements (atomic numbers 57 through 71) in metallic form. It contains about 50% cerium, the remainder being principally lanthanum and neodymium. Mischmetal is used as an alloying additive in ferrous alloys to scavenge sulfur, oxygen, and other impurities and in magnesium alloys to improve high-temperature strength.

**miscible.** Of two phases, the ability of each to dissolve in the

F IG. 329

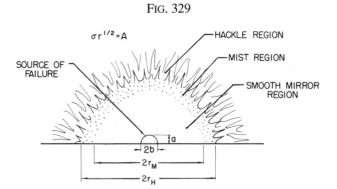

*Schematic of fracture origin showing idealized semielliptical surface flaw and surrounding fracture features known as mirror, mist, and hackle. Crack branching is beyond the hackle.*

other. May occur in a limited range of ratios of the two, or in any ratio.

**miscibility gap**. A region of multiphase equilibrium. It is commonly applied to the specific case in which an otherwise continuous series of liquid or solid solutions is interrupted over a limited temperature range by a two-phase field terminating at a critical point. See also *binodal curve*.

**mismatch**. The misalignment or error in register of a pair of forging dies; also applied to the condition of the resulting forging. The acceptable amount of this displacement is governed by blueprint or specification tolerances. Within tolerances, mismatch is a condition; in excess of tolerance, it is a serious defect. Defective forgings can be salvaged by hot-reforging operations.

**misrun**. Denotes an irregularity on a cast metal surface caused by incomplete filling of the mold due to low pouring temperatures, gas back pressure from inadequate venting of the mold, and inadequate gating.

**mist hackle**. Markings on the surface of a crack in ceramics and glasses accelerating close to the effective terminal velocity, observable first as a mist on the surface and with increasing velocity revealing a fibrous texture elongated in the direction of cracking and coarsening up to the stage at which the crack bifurcates (Fig. 330). Velocity bifurcation or velocity forking is the splitting of a single crack into two mature diverging cracks at or near the effective terminal velocity of about half the transverse speed of sound in the material. See also *bifurcation* and *mirror region*.

**mist lubrication**. Lubrication by an oil mist produced by injecting oil into a gas stream.

**Mitchell bearing**. See *tilting-pad bearing*.

**mixed dislocation**. See *dislocation*.

**mixed grain size**. See *duplex grain size*.

**mixed lubrication**. See *quasi-hydrodynamic lubrication*.

**mixed potential**. The *potential* of a specimen (or specimens

F IG. 330

*Fracture surface of a glass rod broken in bending. Fracture origin is at left; nearly semicircular region is the fracture mirror, bordered by mist and velocity hackle. 50×*

in a *galvanic couple*) when two or more electrochemical reactions are occurring. Also called galvanic couple potential.

**mixing**. In powder metallurgy, the thorough intermingling of powders of two or more different materials (not *blending*).

**mixing chamber**. The part of a torch or furnace burner in which gases are mixed.

**mobile phase**. In chromatography, the gas or liquid that flows through the chromatographic column. A sample compound in the mobile phase moves through the column and is separated from compounds residing in the stationary phase. See also *stationary phase*.

**mock leno weave**. An open fabric weave for composites that resembles a leno and is accomplished by a system of interlacings that draws a group of threads together and leaves a space between that group and the next. The warp threads do not actually cross each other as in a real leno and, therefore, no special attachments are required for the loom. This type of weave is generally used when a high thread count is required for strength and the fabric must remain porous. See also the figure accompanying the term *leno weave*.

**mode**. One of the three classes of crack (surface) displacements adjacent to the crack tip. These displacement modes are associated with stress-strain fields around the crack tip and are designated I, II, and III (Fig. 331). See also *crack-tip plane strain* and *crack opening displacement*.

**modification**. Treatment of molten hypoeutectic (8 to 13% Si) or hypereutectic (13 to 19% Si) aluminum-silicon alloys to improve mechanical properties of the solid alloy by refinement of the size and distribution of the silicon phase. Involves additions of small percentages of sodium,

FIG. 331

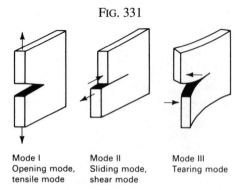

Mode I
Opening mode,
tensile mode

Mode II
Sliding mode,
shear mode

Mode III
Tearing mode

*Fracture loading modes. Arrows show loading direction and relative motion of mating fracture surfaces.*

FIG. 332

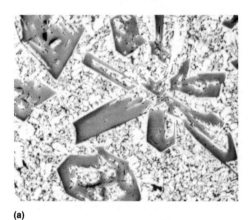

**(a)**

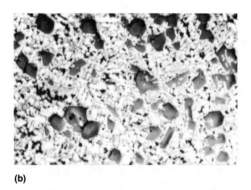

**(b)**

*Effect of phosphorus modification on the microstructure of Al-22Si-1Ni-1Cu alloy. (a) Unrefined. (b) Phosphorus-refined. Both at 100×*

strontium, or calcium (hypoeutectic alloys) or of phosphorus (hypereutectic alloys) (Fig. 332).

**modified acrylic**. A thermoplastic polymer that has been altered to eliminate mixing, curing ovens, and odor and that cures rapidly at room temperature.

**modifier**. Any chemically inert ingredient added to an adhesive formulation to change its properties. Compare with *filler*, *plasticizer*, and *extender*.

**modulus of elasticity ($E$)**. (1) The measure of rigidity or stiffness of a material; the ratio of stress, below the proportional limit, to the corresponding strain. If a tensile stress of 13.8 MPa (2.0 ksi) results in an elongation of 1.0%, the modulus of elasticity is 13.8 MPa (2.0 ksi) divided by 0.01, or 1380 MPa (200 ksi). (2) In terms of the *stress-strain curve*, the modulus of elasticity is the slope of the stress-strain curve in the range of linear proportionality of stress to strain. Also known as *Young's modulus*. For materials that do not conform to Hooke's law throughout the elastic range, the slope of either the tangent to the stress-strain curve at the origin or at low stress, the secant drawn from the origin to any specified point on the stress-strain curve, or the chord connecting any two specific points on the stress-strain curve is usually taken to be the modulus of elasticity. In these cases, the modulus is referred to as the *tangent modulus*, *secant modulus*, or *chord modulus*, respectively.

**modulus of resilience**. The amount of energy stored in a material when loaded to its elastic limit. It is determined by measuring the area under the *stress-strain curve* up to the *elastic limit*. See also *elastic energy*, *resilience*, and *strain energy*.

**modulus of rigidity**. See *shear modulus*.

**modulus of rupture**. Nominal stress at fracture in a bend test or torsion test. In bending, modulus of rupture is the bending moment at fracture divided by the section modulus. In torsion, modulus of rupture is the torque at fracture divided by the polar section modulus. See also *modulus of rupture in bending*, and *modulus of rupture in torsion*.

**modulus of rupture in bending ($S_b$)**. The value of maximum tensile or compressive stress (whichever causes failure) in the extreme fiber of a beam loaded to failure in bending computed from the flexure equation:

$$S_b = \frac{Mc}{I}$$

where $M$ is maximum bending moment, computed from the maximum load and the original moment arm; $c$ is the initial distance from the neutral axis to the extreme fiber where failure occurs; and $I$ is the initial moment of inertia of the cross section about the neutral axis. See also *modulus of rupture*.

**modulus of rupture in torsion ($S_s$)**. The value of maximum shear stress in the extreme fiber of a member of circular cross section loaded to failure in torsion computed from the equation:

$$S_s = \frac{Tr}{J}$$

where *T* is maximum twisting moment, *r* is original outer radius, and *J* is polar moment of inertia of the original cross section. See also *modulus of rupture*.

**modulus of strain hardening**. See preferred term *rate of strain hardening*.

**modulus of toughness**. The amount of work per unit volume of a material required to carry that material to failure under static loading. See also *toughness*.

**Mohs hardness**. The hardness of a body according to a scale proposed by Mohs, based on ten minerals, each of which would scratch the one below it. These minerals, in decreasing order of hardness, are:

| | |
|---|---|
| Diamond | 10 |
| Corundum | 9 |
| Topaz | 8 |
| Quartz | 7 |
| Orthoclase (feldspar) | 6 |
| Apatite | 5 |
| Fluorite | 4 |
| Calcite | 3 |
| Gypsum | 2 |
| Talc | 1 |

**moiety**. A portion of a molecule, generally complex, having a characteristic chemical property.

**moiré pattern**. A pattern developed from interference or light blocking when gratings, screens, or regularly spaced patterns are superimposed on one another.

**moisture absorption**. The pickup of water vapor from air by a polymeric material, in reference to vapor withdrawn from the air only, as distinguished from water absorption, which is the gain in weight due to the absorption of water by immersion.

**moisture content**. The amount of moisture in a polymeric material determined under prescribed conditions and expressed as a percent of the mass of the moist specimen, that is, the mass of the dry substance plus the moisture.

**moisture equilibrium**. The condition reached by a plastic sample when it no longer takes up moisture from, or gives up moisture to, the surrounding environment.

**moisture regain**. The moisture in a polymeric material determined under prescribed conditions and expressed as a percent of the weight of the moisture-free specimens. Moisture regain may result from either sorption or desorption, and differs from moisture content only in the basis used for calculation.

**moisture vapor transmission (MVT)**. A rate at which water vapor passes through a polymeric material at a specified temperature and relative humidity.

**molality**. The number of gram-molecular weights of a compound dissolved in 1 L of solvent. See also *gram-molecular weight*. Compare with *molarity* and *normality*.

**molal solution**. Concentration of a solution expressed in moles of solute divided by 1000 g of solvent.

**molarity**. The number of gram-molecular weights of a compound dissolved in 1 L of solution. See also *gram-molecular weight*. Compare with *molality* and *normality*.

**molar solution**. Aqueous solution that contains 1 mole (gram-molecular weight) of solute in 1 L of the solution.

**mold** (metals). (1) The form, made of sand, metal, or refractory material, that contains the cavity into which molten metal is poured to produce a casting of desired shape. (2) A die.

**mold** (plastics). The cavity into which, or matrix on which, the plastic composition is placed and from which it takes form. To shape plastic parts or finished articles by heat and pressure. The assembly of all the parts that function collectively in the molding process.

**mold cavity** (metals). The space in a mold that is filled with liquid metal to form the casting upon solidification. The channels through which liquid metal enters the mold cavity (sprue, runner, gates) and reservoirs for liquid metal (risers) are not considered part of the mold cavity proper.

**mold coating** (metals). (1) Coating to prevent surface defects on permanent mold castings and die castings. (2) Coating on sand molds to prevent metal penetration and to improve metal finish. Also called mold facing or mold dressing.

**molded edge** (plastics). An edge on a plastic part that is not physically altered after molding for use in final form, particularly one that does not have fiber ends along its length.

**molded net** (plastics). Description of a molded plastic part that requires no additional processing to meet dimensional requirements.

**molding** (plastics). The forming of a polymer or composite into a solid mass of prescribed shape and size by the application of pressure and heat for a given time. The finished part.

**molding compound** (plastics). Plastic material in varying stages of pellet form (Fig. 333) or granulation (powder), consisting of resin, filler, pigments, reinforcements, plasticizers, and other ingredients, ready for use in the molding operation. Also called molding powder.

**molding cycle** (plastics). The period of time required for the complete sequence of operations on a molding press to

FIG. 333

*Phenolic molding pellets*

produce one set of plastic moldings. The operations necessary to produce a set of moldings without reference to the total time.

**molding machine** (metals). A machine for making sand molds by mechanically compacting sand around a pattern. See also the figures accompanying the terms *jolt ramming* and *jolt-squeezer machine*.

**molding press** (metals). A press used to form powder metallurgy *compacts*.

**molding pressure** (plastics). The pressure applied to the ram of an injection machine, compression press, or transfer press to force the softened plastic to fill the mold cavities completely.

**molding sands**. Foundry sands containing over 5% natural clay, usually between 8 and 20%. See also *naturally bonded molding sand*.

**mold jacket** (metals). Wood or metal form that is slipped over a sand mold for support during pouring of a casting.

**mold release agent** (plastics). A lubricant, liquid, or powder (often silicon oils and waxes) used to prevent sticking of molded plastic articles in the cavity.

**mold shift** (metals). A casting defect that results when the parts of the mold do not match at the parting line.

**mold shrinkage** (plastics). (1) The immediate shrinkage that a molded plastic part undergoes when it is removed from a mold and cooled to room temperature. (2) The difference in dimensions, expressed in inches per inch, between a molding and the mold cavity in which it was molded (at normal-temperature measurement). (3) The incremental difference between the dimensions of the molding and the mold from which it was made, expressed as a percentage of the mold dimensions.

**mold surface** (plastics). The side of a laminate that faced the mold (tool) during cure in an autoclave or hydroclave.

**mold wash** (metals). An aqueous or alcoholic emulsion or suspension of various materials used to coat the surface of a casting mold cavity.

**mole**. One mole is the mass numerically equal (in grams) to the relative molecular mass of a substance. It is the amount of substance of a system that contains as many elementary units $(6.02 \times 10^{23})$ as there are atoms of carbon in 0.012 kg of the nuclide $^{12}$C; the elementary unit must be specified and may be an atom, molecule, ion, electron, photon, or even a specified group of such units.

**molecular fluorescence spectroscopy**. An analytical technique that measures the fluorescence emission characteristic of a molecular, as opposed to an atomic, species. The emission results from electronic transitions between molecular states and can be used to detect and/or measure trace amounts of molecular species.

**molecular mass**. The sum of the atomic mass of all atoms in a molecule. In *high polymers*, because the molecular masses of individual molecules vary widely, they must be expressed as averages. The average molecular mass of polymers may be expressed as number-average molecular mass or mass-average molecular map.

**molecular seal**. A seal that is basically of the windback type, but that is used for sealing vapors or gases.

**molecular spectrum**. The spectrum of electromagnetic radiation emitted or absorbed by a collection of molecules as a function of frequency, wave number, or some related quantity.

**molecular structure**. The manner in which electrons and nuclei interact to form a molecule, as elucidated by quantum mechanics and the study of molecular spectra.

**molecular weight**. The sum of the atomic weights of all the atoms in a molecule. Atomic weights (and therefore molecular weights) are relative weights arbitrarily referred to an assigned atomic weight of exactly 12.0000 for the most abundant isotope of carbon, $^{12}$C. See also *atomic weight*.

**molecule**. A molecule may be thought of either as a structure built of atoms bound together by chemical forces or as a structure in which two or more positively charged nuclei are maintained in some definite geometrical configuration by attractive forces from the surrounding cloud of electrons. Besides chemically stable molecules, short-lived molecular fragments termed free radicals can be observed under special circumstances. See also *chemical bonding*, *free radical*, and *molecular structure*.

**molten metal flame spraying**. A thermal spraying process variation in which the metallic material to be sprayed is in the molten condition. See also *flame spraying*.

**molten weld pool**. The liquid state of a weld prior to solidification as weld metal.

**moly-manganese process**. A common method of joining oxide ceramics (most commonly alumina, $Al_2O_3$) whereby a mixture of molybdenum and manganese powders (typically 10 at.% Mn) is applied on the ceramic surface, and then sintered to bond to the ceramic. The sintering temperature is usually above 1400 °C (2500 °F) in a hydrogen ($H_2$) atmosphere with a controlled dew point. Subsequent nickel plating in conjunction with this Mo-Mn metallized layer facilitates the application and adherence of brazing filler metals that would not bond to the original ceramic substrate.

**Mond process**. A process for extracting and purifying nickel. The main features consist of forming nickel carbonyl by reaction of finely divided reduced metal with carbon monoxide, then decomposing the nickel carbonyl to deposit purified nickel on small nickel pellets.

**monochromatic**. Consisting of electromagnetic radiation having a single wavelength or an extremely small range of wavelengths, or particles having a single energy or an extremely small range of energies.

**monochromatic objective**. An *objective* in a microscope, usually of fused quartz, that has been corrected for use with monochromatic light only.

**monochromator**. A device for isolating radiation of one or nearly one wavelength from a beam of many wavelengths.

**monoclinic**. A crystal structure having three axes of any length, with two included angles equal to 90° and one included angle not equal to 90°. See also *unit cell*.

**monofilament**. A single fiber or filament of indefinite length that is strong enough to function as a yarn in commercial textile operation.

**monolayer**. (1) The basic laminate unit from which cross-plied or other laminate types are constructed. (2) A "single" layer of atoms or molecules adsorbed on or applied to a surface.

**monolithic**. An object comprised entirely of one massive piece (although polycrystalline or even heterogeneous) as opposed to being built up of preformed units.

**monolithic refractory**. A refractory which may be installed *in situ* without joints to form an integral structure.

**monomer**. A single molecule that can react with like or unlike molecules to form a polymer. The smallest repeating structure of a polymer (mer). For addition polymers, this represents the original unpolymerized compound.

**monotectic**. An isothermal reversible reaction in a binary system, in which a liquid on cooling decomposes into a second liquid of a different composition and a solid. It differs from a *eutectic* in that only one of the two products of the reaction is below its freezing range.

**monotron hardness test** (obsolete). A method of determining *indentation hardness* by measuring the load required to force a spherical penetrator into a metal to a specified depth.

**monotropism**. The ability of a solid to exist in two or more forms (crystal structures), but in which one form is the stable modification at all temperatures and pressures. *Ferrite* and *martensite* are a monotropic pair below the temperature at which *austenite* begins to form, for example, in steels. Alternate spelling is monotrophism.

**Monte Carlo technique**. Calculation of the trajectory of incident electrons within a given matrix and the pathway of the x-rays generated during interaction (Fig. 334).

**morphology**. The characteristic shape, form, or surface texture or contours of the crystals, grains, or particles of (or in) a material, generally on a microscopic scale.

**morphology** (plastics). The overall physical form of the physical structure of a bulk polymer. Common units are lamellae, spherulites, and domains.

**mortar**. A plastic mixture of cementitious materials, fine aggregates, and water in ratios of about 1:3:0.5 by mass.

**mosaic crystal**. An imperfect single crystal composed of regions that are slightly disoriented relative to each other.

FIG. 334

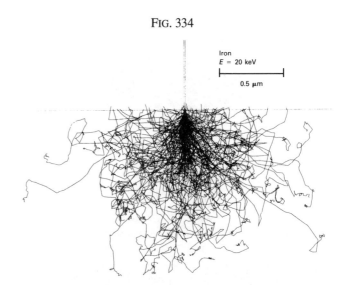

Iron
$E$ = 20 keV

0.5 μm

*Monte Carlo electron trajectory calculation of electron interaction volume in an iron target with a beam energy of 20 keV*

**mosaic structure**. In crystals, a substructure in which adjoining regions have only slightly different orientations.

**Mössbauer effect**. The process in which γ-radiation is emitted or absorbed by nuclei in solid matter without imparting recoil energy to the nucleus and without Doppler broadening of the γ-ray energy.

**Mössbauer spectroscopy**. An analytical technique that measures recoilless absorption of γ-rays that have been emitted from a radioactive source as a function of the relative velocity between the absorber and the source.

**Mössbauer spectrum**. A plot of the relative absorption of γ-rays versus the relative velocity between an absorber and a source of γ-rays.

**mottled cast iron**. Iron that consists of a mixture of variable proportions of gray cast iron and white cast iron; such a material has a mottled fracture appearance.

**mounting**. A means by which a specimen for metallographic examination may be held during preparation of a section surface. The specimen can be embedded in plastic (Fig. 335) or secured mechanically in clamps.

**mounting artifact**. A false structure introduced during the mounting stages of a metallographic surface-preparation sequence.

**mounting resin**. Thermosetting (e.g., Bakelite or diallyl phthalate) or thermoplastic (e.g., methyl methacrylate or polyvinyl chloride) resins used to mount metallographic specimens. Standard plastic mounts usually measure 25 mm (1 in.), 32 mm (1.25 in.), or 38 mm (1.5 in.) in diameter; mount thickness is approximately one-half the mount diameter (Fig. 335).

FIG. 335

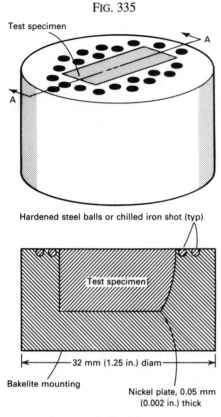

*Metallographic specimen mounted in a thermo-setting resin (Bakelite)*

FIG. 336

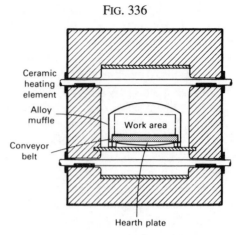

*Muffle-type sintering furnace*

**M shell**. The third layer of electrons surrounding the nucleus of an atom, having electrons characterized by the principal quantum number 3.

**$M_s$ temperature**. For any alloy system, the temperature at which martensite starts to form on cooling. See *transformation temperature* for the definition applicable to ferrous alloys.

**mucilage**. An adhesive prepared from a gum and water, and also in a more general sense, a liquid adhesive that has a low order of bonding strength. See also *adhesive*, *glue*, *paste*, and *sizing*.

**muffle furnace**. A common type of *sintering* furnace for continuous or discontinuous operation. The muffle may be ceramic to support electric heating elements or resistor wire windings, or it may be a gastight metallic retort to retain the furnace atmosphere and support the work trays (Fig. 336); both kinds may be used in the same furnace.

**mulling**. The mixing and kneading of foundry molding sand with moisture and clay to develop suitable properties for molding.

**mullite**. A mineral of theoretical composition $3Al_2O_3 \cdot 2SiO_2$. A relatively stable phase in ceramics produced by the high temperature reaction of alumina and silica or by the ther-mal decomposition of alumina-silica minerals such as kyanite, sillimanite, andalusite, and various clay minerals. Used as a refractory material for firebrick and furnace linings.

**multiaxial stresses**. Any stress state in which two or three principal stresses are not zero.

**multifilament mesh**. Woven material with multiple-strand threads.

**multifilament yarn**. A large number (500 to 2000) of fine, continuous filaments (consisting of 5 to 100 individual filaments), usually with some twist in the yarn to facilitate handling.

**multigrade oil**. An oil having relatively little change in viscosity over a specified temperature range.

**multiple**. A piece of stock for forging that is cut from bar or billet lengths to provide the exact amount of material for a single workpiece.

**multiple die pressing**. The simultaneous compaction of powder into several identical parts with a press tool consisting of a number of components.

**multiple etching**. Sequential etching of a microsection, with specific reagents attacking distinct microconstituents.

**multiple-impulse welding**. Spot, projection, or upset welding with more than one impulse of current during a single machine cycle. Sometimes called pulsation welding.

**multiple-layer adhesive**. A dry-film adhesive, usually supported, with a different adhesive composition on each side; designed to bond dissimilar materials such as the core to face bond of a sandwich composite. See also *honeycomb* and *sandwich construction*.

**multiple-pass weld**. A weld made by depositing filler metal with two or more successive passes (Fig. 337).

**multiple punch press**. A mechanical or hydraulic press that actuates several punches individually and independently of each other.

**multiple-screw extruder**. An extruder machine for processing thermoplastics that has two or four screws (conical or

FIG. 337

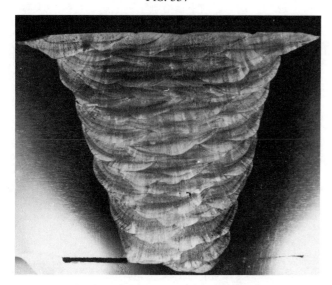

*Multiple-pass weld on type 304 stainless steel plate joined by shielded metal arc welding. 2×*

FIG. 338

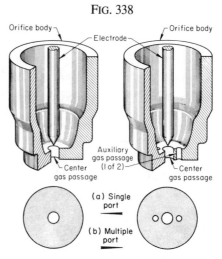

*Single- and multiple-port orifice bodies for a plasma arc torch*

constant depth), in contrast to conventional single-screw extruders. Types include machines with intermeshing counter-rotating screws, intermeshing corotating screws, and nonintermeshing counter-rotating screws.

**multiple-slide press**. A press with individual slides, built into the main slide or connected to individual eccentrics on the main shaft, that can be adjusted to vary the length of stroke and the timing. See also *slide*.

**multiple spot welding**. Spot welding in which several spots are made during one complete cycle of the welding machine.

**multiplier phototube**. See *photomultiplier tube*.

**multiport nozzle** (plasma arc welding and cutting). A constricting nozzle containing two or more orifices located in a configuration to achieve a degree of control over the arc shape (Fig. 338).

***m*-value**. See *strain-rate sensitivity*.

# N

**nanohardness test**. An indentation hardness testing procedure, usually relying on indentation force versus tip displacement data, to make assessments of the resistance of surfaces to penetrations of the order of 10 to 1000 nm deep. The prefix "nano-" normally would imply hardnesses one thousand time smaller than "microhardness"; however, use of this prefix was primarily designed as a means to distinguish this technique from the more traditional microindentation hardness procedures. Most nanohardness testing procedures use three-sided pyramidal diamond indenters.

**napped cloth**. A woven metallurgical polishing cloth in which some fibers are aligned approximately normal to one of its surfaces.

**nascent surface**. A completely uncontaminated surface produced, for example, by cleavage fracture under ideal vacuum conditions.

**native metal**. (1) Any deposit in the earth's crust consisting of uncombined metal. (2) The metal in such a deposit.

**natural aging**. Spontaneous aging of a supersaturated solid solution at room temperature. See also *aging*. Compare with *artificial aging*.

**natural cement**. A *hydraulic cement* produced by calcining a naturally occurring argillaceous limestone at a temperature below the sintering point and then grinding to a fine powder.

**natural diamond**. The densest form of crystallized carbon and the hardest substance known, natural diamond occurs most commonly as well-developed crystals in volcanic pipes or in alluvial deposits. Bort sometimes refers to all diamonds not suitable for gems, or it may refer to off-color flawed or impure diamonds not fit for use for gems or most other industrial applications, but suitable for the preparation of diamond grain and powder for use in lapping or the manufacture of most diamond grinding wheels. This type of bort is also called crushing bort or fragmented bort. See also *diamond* and *man-made diamond*.

**natural draft**. Taper on the sides of a forging, due to its shape or position in the die, that makes added *draft* unnecessary.

**naturally bonded molding sand**. A foundry sand containing sufficient bonding material as mined to be suitable for molding purposes.

**natural strain**. See *true strain*.

**NDE**. See *nondestructive evaluation*.

**NDI**. See *nondestructive inspection*.

**NDT**. See *nondestructive testing*.

**near-infrared radiation**. Infrared radiation in the wavelength range of 0.78 to 3 μm (7800 to 30,000 Å). See also *electromagnetic radiation* and *infrared radiation*.

**near-net shape**. See *net shape*.

**neat oil**. Hydrocarbon oil with or without additives, used

FIG. 339

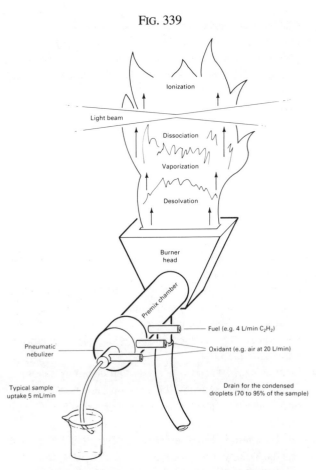

*A pneumatic nebulizer in a flame atomization system for atomic absorption spectrometry*

FIG. 340

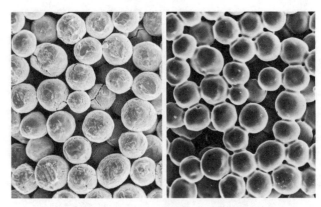

*Neck formation during sintering of copper powder spheres. 150×*

undiluted. This term is used particularly in metal cutting to distinguish these fluids from soluble oils (emulsions).

**neat resin**. Resin to which nothing (additives, reinforcements, and so on) has been added.

**nebulizer**. A device for converting a sample solution into a gas-liquid aerosol for atomic absorption, emission, and fluorescence analysis (Fig. 339). This may be combined with a burner to form a nebulizer burner. See also *atomizer*.

**neck**. The contact area between abutting particles in a powder compact undergoing sintering.

**neck formation**. The growth of interparticle contacts through diffusion processes during sintering (Fig. 340).

**neck-in**. In *extrusion coating* of plastics, the difference between the width of the extruded web as it leaves the die and the width of the coating on the substrate.

**necking**. (1) The reduction of the cross-sectional area of a material in a localized area by uniaxial tension or by stretching (Fig. 341). (2) The reduction of the diameter of a portion of the length of a cylindrical shell or tube.

**necking down**. Localized reduction in area of a specimen during tensile deformation (Fig. 341).

**necking strain**. Same as *uniform strain*.

FIG. 341

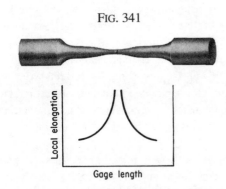

*Variation of local elongation (necking) with position along gage length of tensile specimen*

**needle bearing**. A bearing in which the relatively moving parts are separated by long, thin rollers that have a length-to-diameter ratio exceeding 5.0.

**needle blow**. A specific *blow molding* technique in which the blowing air is injected into the hollow article through a sharpened, hollow needle that pierces the parison (Fig. 342).

**needled mat**. A *mat* for reinforcing plastics formed of strands cut to a short length, then felted together in a needle loom with or without a carrier.

**needles**. Elongated or rodlike powder particles with a high aspect ratio.

**Neel point**. See *Neel temperature*.

**Neel temperature**. The temperature below which spins in an antiferromagnetic material are ordered antiparallel so that there is zero net magnetic moment. Also known as Neel point.

**negative distortion**. The distortion in a microscopic image that occurs when the magnification in the center of the field exceeds that in the edge of the field (Fig. 343). Also termed barrel distortion. Contrast with *positive distortion*.

**negative eyepiece**. An eyepiece in an optical microscope in which the real image of the object forms between the lens elements of the eyepiece.

**negative rake**. Describes a tooth face in rotation whose cutting edge lags the surface of the tooth face. See also the figure accompanying the term *face mill*.

FIG. 342

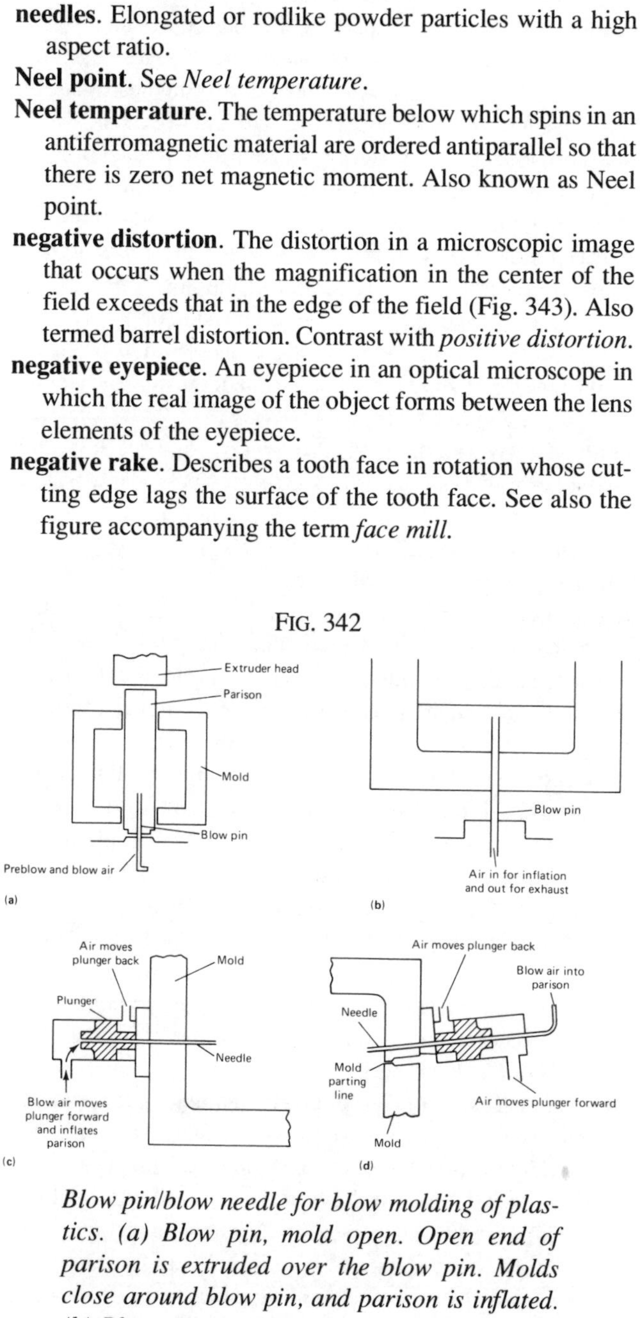

FIG. 343

Normal image            Pincushion image            Barrel image

*Image distortions created in an optical microscope showing normal image, positive (pincushion) distortion, and negative (barrel) distortion*

**negative replica**. A method of reproducing a surface obtained by the direct contact of the replicating material with the specimen. Using this technique, the contour of the replica surface is reversed with respect to that of the original. See also *replica*.

**Nernst equation**. An equation that expresses the exact *electromotive force* of a cell in terms of the activities of products and reactants of the cell. The Nernst equation is given by:

$$E = E^o - \frac{RT}{nF} \ln \frac{(\text{ox})}{(\text{red})}$$

where $E$ is the electrode potential, $E^o$ is the standard electrode potential, $R$ is the gas constant (1.987 cal/K mol), $T$ is the absolute temperature (in degrees Kelvin), $n$ is the number of moles of electrons transferred in the half-cell reaction, $F$ is the Faraday constant ($F = 23,060$ cal/volt equivalent), and (ox) and (red) are the activities of the oxidized and reduced species, respectively.

**Nernst layer, Nernst thickness**. The diffusion layer or the hypothetical thickness of this layer as given by the theory of Nernst. It is defined by:

$$i_d = nFD \left[ \frac{(C_0 - C)}{\delta} \right]$$

where $i_d$ is the diffusion coefficient, $C_0$ is the concentration at the electrode surface, and $\delta$ is the Nernst thickness (0.5 mm in many cases of unstirred aqueous electrolytes).

**nesting**. In reinforced plastics, the placing of plies of fabric such that the yarns of one ply lie in the valleys between the yarns of the adjacent ply. Also called nested cloth.

**net positive suction head**. The difference between total pressure and vapor pressure in a fluid flow, expressed in terms of equivalent height of fluid, or "head," by the following equation:

$$\text{NPSH} = \left( \frac{P_0}{w} \right) + \left( \frac{V^2}{2g} \right) - \left( \frac{P_v}{w} \right)$$

where $P_0$ is the static pressure, $P_v$ is the vapor pressure, $V$ is the flow velocity, $w$ is the specific weight of fluid, and

FIG. 342

Extruder head
Parison
Mold
Blow pin
Preblow and blow air
(a)

Blow pin
Air in for inflation and out for exhaust
(b)

Air moves plunger back
Mold
Plunger
Needle
Blow air moves plunger forward and inflates parison
(c)

Air moves plunger back
Blow air into parison
Needle
Mold parting line
Air moves plunger forward
Mold
(d)

*Blow pin/blow needle for blow molding of plastics. (a) Blow pin, mold open. Open end of parison is extruded over the blow pin. Molds close around blow pin, and parison is inflated. (b) Blow pin, mold closed. (c) Two-stage blow needle through the mold back. (d) Three-stage blow needle close to panel area*

$g$ is the gravitational acceleration. This quantity is used in pump design as a measure of the tendency for cavitation to occur at the pump inlet.

**net shape.** The shape of a powder metallurgy part, casting, or forging that conforms closely to specified dimensions. Such a part requires no secondary machining or finishing. A near-net shape part can be either one in which some but not all of the surfaces are net or one in which the surfaces require only minimal machining or finishing.

**netting analysis.** The analysis of filament-wound structures that assumes that the stresses induced in the structure are carried entirely by the filaments, and the strength of the resin is neglected, and that assumes also that the filaments possess no bending or shearing stiffness and carry only the axial tensile loads.

**network etching.** Formation of networks, especially in *mild steels*, after etching in nitric acid. These networks relate to subgrain boundaries.

**network structure.** A metallic structure in which one constituent occurs primarily at the grain boundaries, partially or completely enveloping the grains of the other constituents.

**Neumann band.** *Mechanical twin* in ferrite.

**neutral filter.** A filter that attenuates the radiant power reaching the detector by the same factor at all wavelengths within a prescribed wavelength region. See also *filter* (4).

**neutral flame.** (1) A gas flame in which there is no excess of either fuel or oxygen in the inner flame. Oxygen from

FIG. 344

(a) Acetylene flame

(b) Carburizing flame

(c) Slightly excess acetylene flame

(d) Neutral flame

(e) Oxidizing flame

(f) Separated flame; separation usually caused by excessive pressure

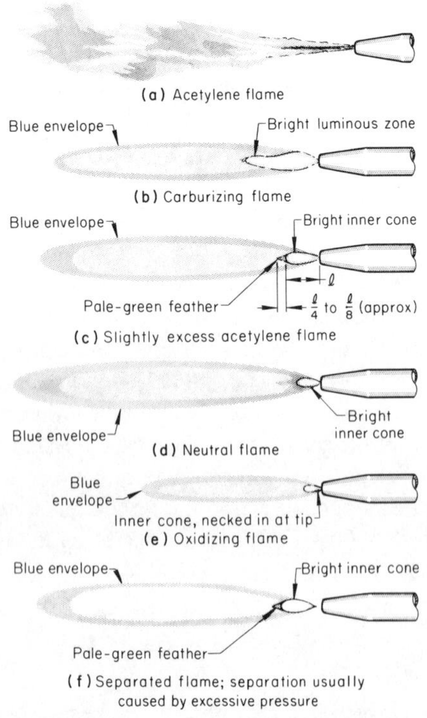

*Flame conditions during oxyfuel thermal cutting*

ambient air is used to complete the combustion of $CO_2$ and $H_2$ produced in the inner flame (Fig. 344). (2) An oxyfuel gas flame in which the portion used is neither oxidizing nor reducing. See also *carburizing flame*, *oxidizing flame*, and *reducing flame*.

**neutralization number.** An ASTM number given to quenching oils that reflects the oil's tendency toward oxidation and sludging. See also *saponification number*.

**neutral oil.** A lubricating oil obtained by distillation, not treated with acid or with alkali.

**neutron.** An elementary particle that has approximately the same mass as the proton, but no electric charge. Rest mass is $1.67495 \times 10^{-27}$ kg. An unbound (extranuclear) neutron is unstable and β-decays with a half-life of 10.6 min.

**neutron absorber.** A material in which a significant number of neutrons entering combine with nuclei and are not re-emitted.

**neutron absorption.** A process in which the collision of a neutron with a nucleus results in the absorption of the neutron into the nucleus with the emission of one or more prompt γ-rays: in certain cases, emission of α-particles, protons, or other neutrons or fission of the nucleus results. Also known as neutron capture.

**neutron activation analysis.** Activation analysis in which the specimen is bombarded with neutrons; identification is made by measuring the resulting radioisotopes. See also *activation analysis*.

**neutron capture.** See *neutron absorption*.

**neutron cross section.** A measure of the probability that an interaction of a given kind will take place between a nucleus and an incident neutron; it is an area such that the number of interactions that occur in a sample exposed to a beam of neutrons is equal to the product of the cross section, the number of nuclei per unit volume in the sample, the thickness of the sample, and the number of neutrons in the beam that would enter the sample if their velocities were perpendicular to it. The usual unit is the barn ($10^{-24}$ cm$^2$). See also *barn*.

**neutron detector.** Any device that detects passing neutrons, for example, by observing the charged particles or γ-rays released in nuclear reactions induced by the neutrons or by observing the recoil of charged particles caused by collisions with neutrons.

**neutron diffraction.** The phenomenon associated with the interference processes that occur when neutrons are scattered by the atoms within solids, liquids, and gases.

**neutron embrittlement.** *Embrittlement* resulting from bombardment with neutrons, usually encountered in metals that have been exposed to a neutron flux in the core of the reactor. In steels, neutron embrittlement is evidenced by a rise in the ductile-to-brittle *transition temperature*.

**neutron flux.** The number of neutrons passing through an area in a unit of time.

FIG. 345

*Thermal neutron radiograph of turbine blades showing residual core material in the upper right corner cooling passage of the blade on the right*

**neutron radiography.** Radiography that uses neutrons generated by a nuclear reactor, accelerator, or by certain radioactive isotopes to form a radiographic image of a test piece (Fig. 345). The neutrons are detected by placing a conventional x-ray film next to a converter screen composed of potentially radioactive materials or next to a transfer screen composed of prompt emission materials which convert the neutron radiation to other types of radiation more easily detected by the film.

**neutron spectrometry.** See *neutron spectroscopy.*

**neutron spectroscopy.** Determination of the energy distribution of neutrons. Scintillation detectors, proportional counters, activation foils, and proton recoil are used.

**neutron spectrum.** The distribution by energy of neutrons impinging on a surface, which can be measured by neutron spectroscopy techniques or sometimes from knowledge of the neutron source. See also *dosimeter.*

**Newtonian fluid.** A fluid exhibiting Newtonian viscosity wherein the shear stress is proportional to the rate of shear. Compare with *dilatant, rheopectic material,* and *thixotropy.*

**nib.** (1) A pressed, preheated, shaped, sintered, hot pressed, rough drilled, or finished compact. (2) A generic term for a piece of cemented carbide intended for use as a wire drawing die.

**nibbling.** Contour cutting of sheet metal by use of a rapidly reciprocating punch that makes numerous small cuts.

**nickel and nickel alloys.** See *Technical Brief 27.*

**Nicol prism.** A prism used in optical microscopes made by cementing together with Canada balsam two pieces of a diagonally cut calcite crystal. In such a prism the ordinary ray is totally reflected at the calcite/cement interface while the orthogonally polarized extraordinary ray is transmitted. The prism can thus be used to polarize light or analyze the polarization of light.

**nip.** In a bearing, the amount by which the outer circumference of a pair of bearing shells exceeds the inner circumference of the housing. Also known as *crush.*

**nip angle.** See *angle of bite.*

**nitride-carbide inclusion types.** A compound with the general formula $M_x(C,N)_y$ observed generally as colored idiomorphic cubic crystals, where $M$ includes titanium, niobium, tantalum, and zirconium.

**nitriding.** Introducing nitrogen into the surface layer of a solid ferrous alloy by holding at a suitable temperature (below $Ac_1$ for ferritic steels) in contact with a nitrogenous material, usually ammonia or molten cyanide of appropriate composition. Quenching is not required to produce a hard case. See also *aerated bath nitriding, bright nitriding,* and *liquid nitriding.*

**nitrocarburizing.** Any of several processes in which both nitrogen and carbon are absorbed into the surface layers of a ferrous material at temperatures below the lower critical temperature and, by diffusion, create a concentration gradient. Nitrocarburizing is performed primarily to provide an antiscuffing surface layer and to improve fatigue resistance. Compare with *carbonitriding.*

**nitrogen blanking.** The use of nitrogen to produce an inert atmosphere.

**NLGI number.** Abbreviation for the National Lubricating Grease Institute number, which is the numerical classification of the consistency of greases, based on the ASTM D 217 test.

**NMR.** See *nuclear magnetic resonance.*

**no-bake binder.** In foundry practice, a synthetic liquid resin sand binder that hardens completely at room temperature, generally not requiring baking; used in a *cold-setting process.*

**noble.** The positive direction of *electrode potential,* thus resembling noble metals such as gold and platinum.

**noble metal.** (1) A metal whose *potential* is highly positive relative to the hydrogen electrode. (2) A metal with marked resistance to chemical reaction, particularly to oxidation and to solution by inorganic acids. The term as often used is synonymous with *precious metal* (Technical Brief 34).

**noble potential.** A *potential* more cathodic (positive) than the standard hydrogen potential.

**node.** The connected portion of adjacent ribbons of honeycomb (Fig. 346).

**no-draft (draftless) forging.** A forging with extremely close tolerances and little or no *draft* that requires minimal machining to produce the final part. Mechanical properties can be enhanced by closer control of grain flow and by retention of surface material in the final component.

**nodular graphite.** (1) Graphite in the nodular form as opposed to flake form (see *flake graphite*). Nodular graphite is characteristic of *malleable iron.* The graphite of nodular

## Technical Brief 27: Nickel and Nickel Alloys

NICKEL AND NICKEL-BASE ALLOYS are vitally important to modern industry because of their ability to withstand a wide variety of severe operating conditions involving corrosive environments, high temperatures, high stresses, and combinations of these factors. There are several reasons for these capabilities. Pure nickel is ductile and tough, because it possesses a face-centered cubic (fcc) structure up to its melting point. Therefore, nickel and its alloys are readily fabricated by conventional methods (wrought, cast, and powder metallurgy products are available), and they offer freedom from the ductile-to-brittle transition behavior of most body-centered cubic (bcc) and noncubic metals. Nickel has good resistance to corrosion in the normal atmosphere, in natural freshwaters, and in deaerated nonoxidizing acids, and it has excellent resistance to corrosion by caustic alkalies. Therefore, nickel offers very useful corrosion resistance itself, and it is an excellent base on which to develop specialized alloys. Its atomic size and nearly complete 3$d$ electron shell enable it to receive large amounts of alloying additions before encountering phase instabilities. This allows a wide variety of alloys to be fashioned in a manner that can adequately capitalize on the unique properties of specific alloying elements. Finally, unique intermetallic phases can form between nickel and some of its alloying elements; this enables the formulation of alloys with very high strengths for both low- and high-temperature services. See also *superalloys* (Technical Brief 52).

Nickel is extracted from sulfide ores, mined principally in Canada, or oxide ores. These ores, which contain about 1 to 3% total nickel, are smelted and refined electrolytically. The single largest use for nickel is as an alloying element in stainless steels. Commercial nickel-base alloys, which account for about 13% of all nickel consumed, are divided into groups or families by their major elemental constituents—for example, nickel-copper, nickel-chromium, nickel-chromium-iron, etc. Other uses for nickel are listed in the table at right.

Nickel and nickel alloys are used for a wide variety of applications, the majority of which involve corrosion resistance and/or heat resistance. Some of these include aircraft gas turbines, steam turbine power plants, turbochargers and valves in reciprocating engines, medical applications (prosthetic devices), heat treating equipment, components used in the chemical and petrochemical

| Use | Amount consumed, % |
|---|---|
| Stainless steel | 57 |
| Alloy steel | 9.5 |
| Nickel-base alloys | 13 |
| Copper-base alloys | 2.3 |
| Plating | 10.4 |
| Foundry | 4.4 |
| Other | 3.3 |

Source: Nickel Development Institute

industries, pollution control equipment, coal gasification and liquefaction systems, and parts used in pulp and paper mills. A number of other applications for nickel alloys involve the unique physical properties of special-purpose alloys, such as *low-expansion alloys* (Technical Brief 24), electrical resistance alloys, soft magnetic alloys, and *shape memory alloys* (Technical Brief 42).

FIG. 346

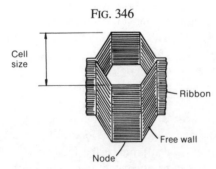

*Honeycomb structure terminology*

or *ductile iron* is spherulitic in form, but called nodular. (2) One of the seven graphite shapes used to classify cast irons (Fig. 347).

**nodular iron.** See preferred term *ductile iron*.

**nodular pearlite.** Pearlite that has grown as a colony with an approximately spherical morphology.

**nodular powder.** Irregularly shaped metal powder having knotted, rounded, or similar shapes.

**noise.** Any undesired signal that tends to interfere with the normal reception or processing of a desired signal.

**NOL ring.** A parallel filament- or tape-wound hoop test specimen developed by the Naval Ordnance Laboratory (NOL) (now the Naval Surface Weapons Laboratory), for measuring various mechanical strength properties of a material, such as tension and compression, by testing the entire ring or segments of it. Also known as parallel fiber reinforced ring.

**nominal area of contact.** The area bounded by the periphery of the region in which macroscopic contact between two

## Nominal chemical compositions of some typical nickel-base alloys

| Common alloy designation | UNS designation | Chemical composition, % ||||||||||| |
|---|---|---|---|---|---|---|---|---|---|---|---|---|
| | | C(a) | Nb | Cr | Cu | Fe | Mo | Ni | Si(a) | Ti | W | Other |
| **Nickel** | | | | | | | | | | | | |
| 200 . . . . . . . . | N02200 | 0.1 | . . . | . . . | 0.25 max | 0.4 max | . . . | 99.2 min | 0.15 | 0.1 max | . . . | . . . |
| 201 . . . . . . . . | N02201 | 0.02 | . . . | . . . | 0.25 max | 0.4 max | . . . | 99.0 min | 0.15 | 0.1 max | . . . | . . . |
| **Nickel-copper** | | | | | | | | | | | | |
| 400 . . . . . . . . | N04400 | 0.15 | . . . | . . . | 31.5 | 1.25 | . . . | bal | 0.5 | . . . | . . . | . . . |
| R-405 . . . . . . . | N04405 | 0.15 | . . . | . . . | 31.5 | 1.25 | . . . | bal | 0.5 | . . . | . . . | 0.0435 |
| **Nickel-molybdenum** | | | | | | | | | | | | |
| B-2 . . . . . . . . | N10665 | 0.01 | . . . | 1.0 max | . . . | 2.0 max | 28 | bal | 0.1 | . . . | . . . | . . . |
| B . . . . . . . . . . | N10001 | 0.05 | . . . | 1.0 max | . . . | 5.0 | 28 | bal | 1.0 | . . . | . . . | . . . |
| **Nickel-chromium-iron** | | | | | | | | | | | | |
| 600 . . . . . . . . | N06600 | 0.08 | . . . | 16.0 | 0.5 max | 8.0 | . . . | bal | 0.5 | 0.3 max | . . . | . . . |
| 601 . . . . . . . . | N06601 | . . . | . . . | 23.0 | . . . | 14.1 | . . . | bal | . . . | . . . | . . . | 1.35Al |
| 800 . . . . . . . . | N08800 | 0.1 | . . . | 21.0 | 0.75 max | 44.0 | . . . | 32.5 | 1.0 | 0.38 | . . . | . . . |
| 800H . . . . . . . | N08810 | 0.08 | . . . | 21.0 | 0.75 max | 44.0 | . . . | 32.5 | 1.0 | 0.38 | . . . | . . . |
| **Nickel-chromium-iron-molybdenum** | | | | | | | | | | | | |
| 825 . . . . . . . . | N08825 | 0.05 | . . . | 21.5 | 2.0 | 29.0 | 3.0 | 42 | 0.5 | 1.0 | . . . | . . . |
| G . . . . . . . . . . | N06007 | 0.05 | 2.0 | 22.0 | 2.0 | 19.5 | 6.5 | 43 | 1.0 | . . . | 1.0 max | . . . |
| G-2/2550 . . . . . | N06975 | 0.03 | . . . | 24.5 | 1.0 | 20.0 | 6.0 | 48 | 1.0 | 1.0 | . . . | . . . |
| G-3 . . . . . . . . | N06985 | 0.015 | 0.8 | 22.0 | 2.0 | 19.5 | 7.0 | 44 | 1.0 | . . . | 1.5 max | . . . |
| H . . . . . . . . . . | | 0.03 | . . . | 22.0 | . . . | 19.0 | 9.0 | 42 | 1.0 | . . . | 2.0 | . . . |
| G-30 . . . . . . . | N06030 | 0.03 | 0.8 | 29.5 | 2.0 | 15.0 | 5.5 | 43 | 1.0 | . . . | 2.5 | . . . |
| **Nickel-chromium-molybdenum-tungsten** | | | | | | | | | | | | |
| N . . . . . . . . . . | N10003 | 0.06 | . . . | 7.0 | 0.35 max | 5.0 max | 16.5 | 71 | 1.0 | 0.5 max | 0.5 max | . . . |
| W . . . . . . . . . . | N10004 | 0.12 | . . . | 5.0 | . . . | 6.0 | 24.0 | 63 | 1.0 | . . . | . . . | . . . |
| 625 . . . . . . . . | N06625 | 0.1 | 4.0 | 21.5 | . . . | 5.0 max | 9.0 | 62 | 0.5 | . . . | . . . | . . . |
| 690 . . . . . . . . | N06690 | 0.02 | . . . | 29.0 | . . . | 10.0 | . . . | 61 | . . . | 0.3 | . . . | . . . |
| C-276 . . . . . . | N10276 | 0.01 | . . . | 15.5 | . . . | 5.5 | 16.0 | 57 | 0.08 | . . . | 4.0 | . . . |
| C-4 . . . . . . . . | N06455 | 0.01 | . . . | 16.0 | . . . | 3.0 max | 15.5 | 65 | 0.08 | . . . | . . . | . . . |
| C-22 . . . . . . . | N06022 | 0.015 | . . . | 22.0 | . . . | 3.0 max | 13.0 | 56 | 0.08 | . . . | 3.0 | . . . |

(a) Maximum

## Recommended Reading

- W.L. Mankins and S. Lamb, Nickel and Nickel Alloys, *Metals Handbook*, 10th ed., Vol 1, ASM International, 1990, p 429-445
- A.I. Asphahani *et al.*, Corrosion of Nickel-Base Alloys, *Metals Handbook*, 9th ed., Vol 13, ASM International, 1987, p 641-657

solid bodies is occurring. This is often taken to mean the area enclosed by the boundaries of a wear scar, even though the real area of contact, in which the solids are touching instantaneously, is usually much smaller. See also *area of contact* and *apparent area of contact*.

**nominal dimension.** The size of the dimension to which the tolerance is applied. For example, if a dimension is 50 mm ± 0.5 mm (2.00 in. ± 0.02 in.), the 50 mm (2.00 in.) is the nominal dimension, and the ±0.5 mm (±0.02 in.) is the tolerance.

**nominal strength.** See *ultimate strength*.

**nominal stress.** The stress at a point calculated on the net cross section without taking into consideration the effect on stress of geometric discontinuities, such as holes, grooves, fillets, and so forth. The calculation is made using simple elastic theory.

**nominal value.** A value assigned for the purpose of a convenient designation. A nominal value exists in name only. In dimensions, it is often an average number with a tolerance in order to fit together with adjacent parts.

**nonconformal surfaces.** (1) Surfaces whose centers of curvature are on the opposite sides of the interface, as in rolling-element bearings or gear teeth. (2) In wear testing, a geometric configuration in which a "point" or "line" of contact is initially established between specimens before the test is started. Examples of nonconformal contacts are ball-on-ring and flat block-on-ring geometries (Fig. 348).

**nonconforming.** A quality control term describing a unit of product or service that does not meet normal acceptance criteria for the specific product or service. A nonconforming unit is not necessarily *defective*.

FIG. 347

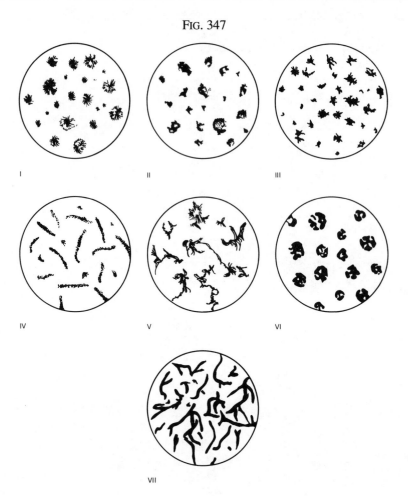

| ASTM type(a) | Equivalent ISO form(b) | Description | ASTM type(a) | Equivalent ISO form(b) | Description |
|---|---|---|---|---|---|
| I | VI | Nodular (spheroidal) graphite | IV | III | Quasi-flake graphite |
| II | VI | Nodular (spheroidal) graphite, imperfectly formed | V | II | Crab-form graphite |
| III | IV | Aggregate, or temper carbon | VI | V | Irregular or open type nodules |
| | | | VII(c) | I | Flake graphite |

(a) As defined in ASTM A 247. (b) As defined in ISO/R 945-1969 (E). (c) Divided into five subtypes: uniform flakes; rosette grouping; superimposed flake size; interdendritic, random orientation; and interdendritic, preferred orientation

*Seven graphite shapes used to classify cast irons*

**noncontact bearing.** A bearing in which no solid contact occurs between relatively moving surfaces. Strictly speaking, a bearing in which *full-film lubrication* is occurring would be considered a noncontact bearing; however, this term is more typically applied to gas bearings and magnetic bearings. See also *gas lubrication* and *magnetic bearing*.

**noncorrosive flux.** A rosin-based soldering flux which in neither its original nor residual form chemically attacks the base metal.

**nondestructive evaluation (NDE).** Broadly considered synonymous with nondestructive inspection (NDI). More specifically, the quantitative analysis of NDI findings to determine whether the material will be acceptable for its function, despite the presence of discontinuities. With NDE, a discontinuity can be classified by its size, shape, type, and location, allowing the investigator to determine whether or not the flaw(s) is acceptable. Damage tolerant design approaches are based on the philosophy of ensuring safe operation in the presence of flaws.

**nondestructive inspection (NDI).** A process or procedure, such as ultrasonic or radiographic inspection, for determining the quality or characteristics of a material, part, or assembly, without permanently altering the subject or its properties. Used to find internal anomalies in a structure without degrading its properties or impairing its serviceability.

FIG. 348

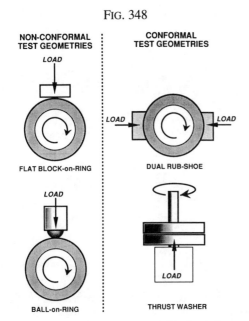

*Nonconformal (left) and conformal (right) test geometries*

FIG. 349

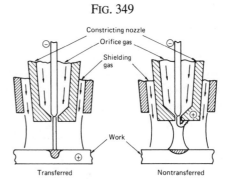

*Transferred and nontransferred plasma arc modes*

**nondestructive testing (NDT).** Broadly considered synonymous with *nondestructive inspection (NDI).*

**nonfill.** A forging condition that occurs when the die impression is not completely filled with metal. Also referred to as underfill.

**nonhygroscopic.** Lacking the property of absorbing and retaining an appreciable quantity of moisture (water vapor) from the air.

**nonmetallic inclusions.** See *inclusions.*

**non-Newtonian viscosity.** The apparent viscosity of a material in which the shear stress is not proportional to the rate of shear.

**nonresonant forced and vibration technique.** A technique for performing dynamic mechanical measurements in which a plastic sample is oscillated mechanically at a fixed frequency. Storage modulus and damping are calculated from the applied strain and the resultant stress and shift in phase angle.

**nonrigid plastic.** For purposes of general classification, a plastic that has a modulus of elasticity either in flexure or in tension of not over 70 MPa (10 ksi) at 23 °C (73 °F) and 50% relative humidity.

**nonsoap grease.** A grease made with a thickener other than soap, such as clay or asbestos.

**nontransferred arc.** In plasma arc welding, cutting, and thermal spraying, an arc established between the electrode and the constricting nozzle (Fig. 349). The workpiece is not in the electrical circuit. Compare with *transferred arc.*

**nonwetting.** (1) The lack of metallurgical *wetting* between molten solder and a metallic surface due to the presence of a physical barrier on the metallic surface. (2) A condition in which a surface has contacted molten solder, but the solder has not adhered to all of the surface; base metal remains exposed.

**nonwoven fabric.** A planar textile structure for reinforced plastics produced by loosely compressing together fibers, yarns, rovings, and so forth, with or without a scrim cloth carrier. Accomplished by mechanical, chemical, thermal, or solvent means, or combinations thereof.

**normal.** An imaginary line forming right angles with a surface or other lines sometimes called the perpendicular. It is used as a basis for determining angles of incidence reflection and refraction.

**normal direction.** That direction perpendicular to the plane of working in a worked material. See also *longitudinal direction* and *transverse direction.*

**normal force.** See *load.*

**normality.** A measure of the number of gram-equivalent weights of a compound per liter of solution. Compare with *molarity.*

**normalized erosion resistance.** The volume loss rate of a specified reference material divided by the volume loss rate of a test material obtained under similar testing and analysis conditions. "Similar testing and analysis conditions" means that the volume loss rates of the two materials are determined at the corresponding portions of the erosion rate-time pattern, for example, the maximum erosion rate or the terminal erosion rate.

**normalizing.** Heating a ferrous alloy to a suitable temperature above the transformation range and then cooling in air to a temperature substantially below the transformation range.

**normal load.** See *load.*

**normal-phase chromatography (NPC).** This refers to liquid-solid chromatography or to bonded-phase chromatography with a polar stationary phase and a nonpolar mobile phase. See also *bonded-phase chromatography* and *liquid-solid chromatography.*

**normal segregation.** Concentration of alloying constituents that have low melting points in those portions of a casting that solidify last. Compare with *inverse segregation*.

**normal solution.** An aqueous solution containing one gram equivalent of the active reagent in 1 L of the solution.

**normal stress.** The stress component that is perpendicular to the plane on which the forces act. Normal stress may be either *tensile* or *compressive*.

**normal wear.** Loss of material within the design limits expected for the specific intended application. The concept of normal wear depends on economic factors, such as the expendability of a worn part.

**nose radius.** The radius of the rounded portion of the cutting edge of a tool. See the figure accompanying the term *single-point tool*.

**nosing.** Closing in the end of a tubular shape to a desired curve contour.

**notch.** See *stress concentration*.

**notch acuity.** Relates to the severity of the *stress concentration* produced by a given notch in a particular structure. If the depth of the notch is very small compared with the width (or diameter) of the narrowest cross section, acuity may be expressed as the ratio of the notch depth to the notch root radius. Otherwise, acuity is defined as the ratio of one-half the width (or diameter) of the narrowest cross section to the notch root radius.

**notch brittleness.** Susceptibility of a material to brittle fracture at points of stress concentration. For example, in a notch tensile test, the material is said to be notch brittle if the *notch strength* is less than the tensile strength of an unnotched specimen. Otherwise, it is said to be notch ductile.

**notch depth.** The distance from the surface of a test specimen to the bottom of the notch. In a cylindrical test specimen, the percentage of the original cross-sectional area removed by machining an annular groove.

**notch ductile.** See *notch brittleness*.

**notch ductility.** The percentage reduction in area after complete separation of the metal in a tensile test of a *notched specimen*.

**notched specimen.** A test specimen that has been deliberately cut or notched, usually in a V-shape, to induce and locate point of failure (Fig. 350).

**notch factor.** Ratio of the resilience determined on a plain specimen to the resilience determined on a notched specimen.

**notching.** Cutting out various shapes from the edge of a strip, blank, or part.

**notching press.** A mechanical press used for notching internal and external circumferences and also for notching along a straight line. These presses are equipped with automatic feeds because only one notch is made per stroke.

**notch rupture strength.** The ratio of applied load to original area of the minimum cross section in a *stress-rupture test* of a *notched specimen*.

FIG. 350

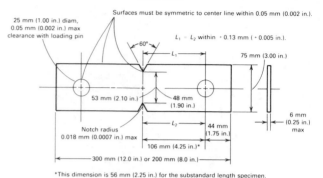

*Standard notched tension specimen*

**notch sensitivity.** The extent to which the sensitivity of a material to fracture is increased by the presence of a *stress concentration*, such as a notch, a sudden change in cross section, a crack, or a scratch. Low notch sensitivity is usually associated with ductile materials, and high notch sensitivity is usually associated with brittle materials.

**notch sharpness.** See *notch acuity*.

**notch strength.** The maximum load on a notched tension-test specimen divided by the minimum cross-sectional area (the area at the root of the notch). Also called notch tensile strength.

**notch tensile strength.** See *notch strength*.

**novolac.** A linear, thermoplastic, two-stage phenolic resin, which, in the presence of methylene or other cross-linking groups, reacts to form a thermoset phenolic. See also *phenolics* (Technical Brief 29).

**nozzle** (metals). (1) Pouring spout of a bottom-pour ladle. (2) On a hot chamber die casting machine, the thick-wall tube that carries the pressurized molten metal from the gooseneck to the die. (3) A device that directs shielding media in a welding or cutting torch. See also the figures accompanying the terms *bottom-pour ladle*, *hot chamber machine*, and *welding torch (arc)*.

**nozzle** (plastics). The hollow-cored metal nose screwed into the injection end of either the heating cylinder of an injection machine or a transfer chamber when this is a separate structure. A nozzle is designed to form under pressure a seal between the heating cylinder or the transfer chamber and the mold. The shape of the front end of a nozzle may be either flat or spherical. See also *reciprocating-screw injection molding*.

**N shell.** The fourth layer of electrons surrounding the nucleus of an atom, having electrons with the principal quantum number 4.

**nuclear cermet fuel.** A sintered fuel rod composed of a fissile carbide or oxide constituent and a metallic matrix.

**nuclear charge**. See *atomic number*.

**nuclear cross section (σ)**. The probability that a nuclear reaction will occur between a nucleus and a particle, expressed in units of area (usually barns). See also *barn*.

**nuclear grade**. Material of a quality adequate for use in nuclear application.

**nuclear magnetic resonance (NMR)**. A phenomenon exhibited by a large number of atomic nuclei that is based on the existence of nuclear magnetic moments associated with quantized nuclear spins. These nuclear moments, when placed in a magnetic field, give rise to distinct nuclear Zeeman energy levels between which spectroscopic transitions can be induced by radio-frequency radiation. Plots of these transition frequencies, termed spectra, furnish important information about molecular structure and sample composition. See also *Zeeman effect*.

**nuclear structure**. The atomic nucleus at the center of the atom, containing more than 99.975% of the total mass of the atom. Its average density is approximately $3 \times 10^{11}$ kg/cm$^3$, its diameter is approximately $10^{-12}$ cm and thus is much smaller than the diameter of the atom, which is approximately $10^{-8}$ cm. The nucleus is composed of protons and neutrons. The number of protons is denoted by $Z$, the number of neutrons by $N$. The total number protons and neutrons in a nucleus is termed the mass number and is denoted by $A = N + Z$. See also *atomic structure*, *electron*, *neutron*, *proton*, *isobar*, *isotone*, and *isotope*.

**nucleating agent** (plastics). A foreign substance, often crystalline, usually added to a crystallizable polymer to increase its rate of solidification during processing.

**nucleation** (metals). The initiation of a phase transformation at discrete sites, with the new phase growing on the nuclei. See also *nucleus* (2).

**nucleus**. (1) The heavy central core of an atom, in which most of the mass and the total positive electric charge are concentrated. (2) The first structurally stable particle capable of initiating recrystallization of a phase or the growth of a new phase and possessing an interface with the parent metallic matrix. The term is also applied to a foreign particle that initiates such action.

**nuclide**. A species of atom distinguished by the constitution of its nucleus. Nuclear constitution is specified by the number of protons, number of neutrons, and energy content or by atomic number, mass number, and atomic mass.

**nugget**. (1) A small mass of metal, such as gold or silver, found free in nature. (2) The weld metal in a spot, seam, or projection weld (Fig. 351).

**nugget size**. The diameter or width of the *nugget* obtained during resistance welding that is measured in the plane of the interface between the pieces joined.

**numerical sphere**. The measure of the light-collecting ability of an objective lens in a light microscope. It is defined as:

$$NA = n \sin \alpha$$

where $n$ is the minimum refraction index of the medium (air or oil) between the specimen and the lens, and $\alpha$ is the half-angle of the most oblique light rays that enter the front lens of the objective.

FIG. 351

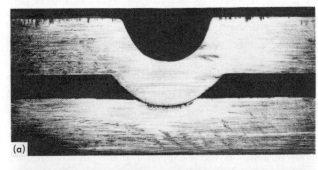
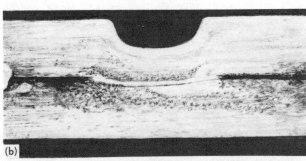
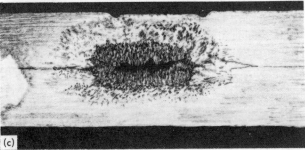
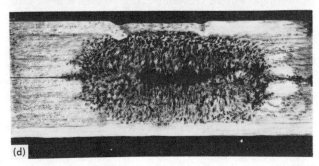

*Development of weld nugget during projection welding*

*n*-value. See *strain-hardening exponent.*

nylon. The generic name, by common usage, for all synthetic *polyamides.*

nylon plastics. Plastics based on a resin composed principally of a long-chain synthetic polymeric amide that has recurring amide groups as an integral part of the main polymer chain. Numerical designations (nylon 6, nylon 6/6, and so on) refer to the monomeric amides from which they are made. Characterized by great toughness and elasticity. See also *polyamides* (Technical Brief 30).

objective. The primary magnifying system of a microscope. A system, generally of lenses, less frequently of mirrors, forming a real, inverted, and magnified image of the object.

objective aperture. See *aperture (electron)* and *aperture (optical).*

oblique evaporation shadowing. The condensation of evaporated material onto a substrate that is inclined to the direct line of the vapor stream to produce shadows. See also *shadowing.*

oblique illumination. Illumination from light inclined at an oblique angle to the optical axis.

observed value. The particular value of a characteristic determined as a result of a test or measurement.

obsidian. A jet-black volcanic natural glass formed by rapid cooling of viscous lava. Although its composition varies, it is highly siliceous.

octahedral plane. In cubic crystals, a plane with equal intercepts on all three axes.

ocular. See *eyepiece.*

Ocvirk number. A dimensionless number used to evaluate the performance of journal bearings, and defined by the following equation:

$$\text{Ocvirk number} = \frac{P}{\eta U}\left(\frac{c}{r}\right)^2\left(\frac{d}{b}\right)^2$$

where $P$ is the load per unit width, $\eta$ is the dynamic viscosity, $U$ is the surface velocity, $c$ is the radial clearance, $r$ is the bearing surface, $b$ is the bearing length, and $d$ is the bearing diameter. This number may be used in its inverted form and is related to the *Sommerfeld number.*

offal. The material trimmed from blanks or formed panels.

offhand grinding. Grinding where the operator manually forces the wheel against the work, or vice versa. It often implies casual manipulation of either grinder or work to achieve the desired result. Dimensions and tolerances frequently are not specified, or are only loosely specified; the operator relies mainly on visual inspection to determine how much grinding should be done. Contrast with *precision grinding.*

offset. The distance along the strain coordinate between the initial portion of a stress-strain curve and a parallel line that intersects the stress-strain curve at a value of stress (commonly 0.2%) that is used as a measure of the *yield strength.* Used for materials that have no obvious *yield point.* See also the figure accompanying the term *stress-strain curve.*

offset yield strength. The stress at which the strain exceeds by a specific amount (the *offset*) an extension of the initial, approximately linear, proportional portion of the stress-strain curve. It is expressed in force per unit area.

off time. In resistance welding, the time that the electrodes are off the work. This term is generally applied where the welding cycle is repetitive.

ohm (Ω). The unit of electrical resistance equal to the resistance through which a current of 1 ampere will flow when there is a potential difference of 1 volt across it.

oil . A liquid of vegetable, animal, mineral, or synthetic origin that feels slippery to the touch.

oil canning. See *canning.*

oil content. The amount of oil which an impregnated part, such as a self-lubricating bearing, retains.

oil cup. A device connected to a bearing that uses a wick, valve, or other means to provide a regulated flow of lubricant.

oil flow rate. The rate at which a specified oil will pass through a porous sintered powder compact under specified test conditions.

oil fog lubrication. See *mist lubrication.*

oil groove. A channel or channels cut in a bearing to improve oil flow through the bearing. A similar groove may be used for grease-filled bearings.

oil hardening. Quench-hardening treatment of steels involving cooling in oil.

oil impregnation. The filling of a sintered skeleton body with oil by capillary attraction or under influence of an external pressure or a vacuum.

oiliness. See *lubricity.*

oilless bearing. See preferred term *self-lubricating bearing.*

oil permeability. A measure of the capacity of a sintered powder metallurgy bearing to allow the flow of an oil through its open pore system.

**Table 45  Typical properties of commercially available quenching and martempering oils**

| Type of quenching oil | No. | API gravity(a) | Flash point(b) °C | Flash point(b) °F | Pour point(c) °C | Pour point(c) °F | Viscosity at 40 °C (100 °F), SUS(d) | Saponification(e) | Ash(f), % | Water(g), % |
|---|---|---|---|---|---|---|---|---|---|---|
| Conventional, no additives | 1 | 33 | 155 | 315 | −12 | 10 | 107 | 0.0 | 0.01 | 0.0 |
| | 2 | 27 | 185 | 365 | −9 | 15 | 111 | 0.0 | 0.03 | 0.0 |
| Fast, with speed improvers | 3 | 33.5 | 190 | 370 | −12 | 10 | 95 | 0.0 | 0.05 | 0.0 |
| | 4 | 35 | 160 | 320 | −4 | 25 | 60 | 0.0 | 0.20 | 0.0 |
| Martempering, without speed improvers | 5 | 31.1 | 235 | 455 | −9 | 15 | 329 | 0.0 | 0.02 | 0.0 |
| | 6 | 28.4 | 245 | 475 | −9 | 15 | 719 | 0.0 | 0.05 | 0.0 |
| | 7 | 26.6 | 300 | 575 | −7 | 20 | 2550 | 0.0 | 0.10 | 0.0 |
| Martempering, with speed improvers | 8 | 28.4 | 230 | 450 | −9 | 15 | 337 | 2.0 | 1.1 | 0.0 |
| | 9 | 27.8 | 245 | 475 | −9 | 15 | 713 | 2.2 | 1.1 | 0.0 |
| | 10 | 25.5 | 300 | 570 | −7 | 20 | 2450 | 2.5 | 1.4 | 0.0 |

ASTM specifications: (a) D 287. (b) D 92. (c) D 97. (d) D 445, D 2161. (e) D 94. (f) D 482. (g) D 95, D 1533

**oil pocket.** A depression designed to retain oil on a sliding surface.

**oil quenching.** Hardening of carbon steel in an oil bath. Oils are categorized as conventional, fast, and martempering. Table 45 lists typical properties of quenching and martempering oils.

**oil ring lubrication.** A system of lubrication for horizontal shafts. A ring of larger diameter rotates with the shaft and collects oils from a container beneath.

**oil starvation.** A condition in which a bearing, or other *tribocomponent*, receives an inadequate supply of lubricant.

**oilstone.** A natural or manufactured abrasive stone, generally impregnated with oil, used for sharpening keen-edged tools.

**oil whirl.** Instability of a rotating shaft associated with instability in the lubricant film. Oil whirl should be distinguished from shaft whirl, which depends only on the stiffness of the shaft.

**olefin.** A group of unsaturated hydrocarbons of the general formula $C_nH_{2n}$ and named after the corresponding paraffins by the addition of -ene or sometimes -ylene to the root.

**olefin plastics.** Plastics based on polymers made by the polymerization of olefins with other monomers, the olefins being at least 50 mass %.

**oligomer.** A polymer consisting of only a few monomer units, for example, a dimer, trimer, tetramer, and so forth, or their mixtures.

**olivine.** A naturally occurring mineral of the composition $(Mg,Fe)_2SiO_4$ that is crushed and used as a foundry molding sand.

**Olsen ductility test.** A *cupping test* in which a piece of sheet metal, restrained except at the center, is deformed by a standard steel ball until fracture occurs (Fig. 352). The height of the cup at the time of fracture is a measure of the ductility.

**omega phase.** A nonequilibrium, submicroscopic phase that forms in titanium alloys as a nucleation growth product; often thought to be a transition phase during the formation of α from β. It occurs in metastable β alloys and can lead to severe embrittlement. It typically occurs during aging

FIG. 352

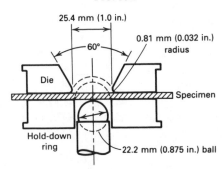

*Schematic of Olsen cup test*

FIG. 353

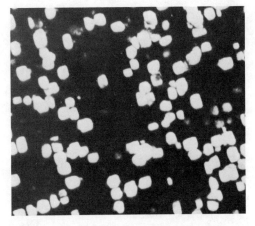

*Cuboidal omega phase (light precipitate) in a titanium-iron alloy. 320,000×*

at low temperature (Fig. 353), but can also be induced by high hydrostatic pressures.

**one-component adhesive.** An adhesive material incorporating a latent hardener or catalyst that is activated by heat.

**one-shot molding.** In the urethane foam field, a system in which the isocyanate, polyol, catalyst, and other additives are mixed together directly and a foam is produced immediately. Compare with *prepolymer molding*.

**open assembly time**. The time interval between the spreading of the adhesive on the adherend and the completion of the assembly of the parts for bonding.

**open-back inclinable press**. A vertical crank press that can be inclined so that the bed will have an inclination generally varying from 0° to 30°. The formed parts slide off through an opening in the back. It is often called an OBI press.

**open-cell cellular plastic**. Foamed or cellular plastic with cells that are generally interconnected.

**open-cell foam**. Foamed or cellular polymeric material with cells that are generally interconnected. Closed cell refers to cells that are not interconnected.

**open-circuit potential**. The *potential* of an electrode measured with respect to a reference electrode or another electrode when no current flows to or from it.

**open-circuit voltage**. The voltage between the output terminals of a welding machine when no current is flowing in the welding current.

**open-die forging**. The hot mechanical forming of metals between flat or shaped dies in which metal flow is not completely restricted. Also known as hand or smith forging (Fig. 354). See also *hand forge (smith forge)*.

FIG. 354

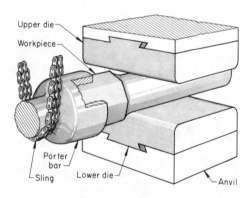

*Open-die forging setup*

**open dies**. Dies with flat surfaces that are used for preforming stock or producing hand forgings.

**open-gap upset welding**. A form of *forge welding* in which the weld interfaces are heated with a fuel gas flame, then forced into intimate contact by the application of force. Not to be confused with *upset welding*, which is a resistance welding process.

**open hearth furnace**. A reverberatory melting furnace with a shallow hearth and a low roof. The flame passes over the charge on the hearth, causing the charge to be heated both by direct flame and by radiation from the roof and sidewalls of the furnace. See also *reverberatory furnace*.

**open pore**. A pore open to the surface of a powder compact. See also *intercommunicating porosity*.

**open porosity**. See *interconnected porosity*.

**open rod press**. A *hydraulic press* in which the slide is guided by vertical, cylindrical rods (usually four) that also serve to hold the crown and bed in position.

**open-sand casting**. Any casting made in a mold that has no *cope* or other covering.

**operating stress**. The stress to which a structural unit is subjected in service.

**optical emission spectroscopy**. Pertaining to *emission spectroscopy* in the near-ultraviolet, visible, or near-infrared wavelength regions of the electromagnetic spectrum. See also *electromagnetic radiation*.

**optical etching**. Development of microstructure under application of special illumination techniques, such as *dark-field illumination, phase contrast illumination, differential interference contrast illumination*, and *polarized light illumination* (see also the figures accompanying these terms).

**optical glass**. Glass of high quality having closely specified optical properties, used in the manufacture of plano or curved windows and refractive and reflective elements for precision instruments and devices. Most products are silicates designed for maximal transmittance in the visible spectrum. Some glasses are prepared from extremely pure raw materials under special conditions to yield high ultraviolet transmission. Among infrared transmitting glasses the chalcide glass $As_2S_3$, fused silica (waterfree, mostly prepared from natural quartz), and calcium-aluminate glasses are most common.

**optical microscope**. An instrument used to obtain an enlarged image of a small object, utilizing visible light. In general it consists of a light source, a condenser, an objective lens, an ocular or eyepiece, and a mechanical stage for focusing and moving the specimen (Fig. 355). Magnification capability of the optical microscope ranges from 1 to 1500×.

**optical pyrometer**. An instrument for measuring the temperature of heated material by comparing the intensity of light emitted with a known intensity of an incandescent lamp filament.

**optoelectronic device**. A device that detects and/or is responsive to electromagnetic radiation (light) in the visible, infrared, and/or ultraviolet spectra regions; emits or modifies noncoherent or coherent electromagnetic radiation in these same regions; or utilizes such electromagnetic radiation for its internal operation.

**orange peel** (ceramics). (1) A surface condition characterized by an irregular waviness of the porcelain enamel resembling an orange skin in texture; sometimes considered a defect. (2) A pitted texture of a fired glaze resembling the surface of rough orange peel.

**orange peel** (metals). A surface roughening in the form of a pebble-grained pattern that occurs when a metal of un-

FIG. 355

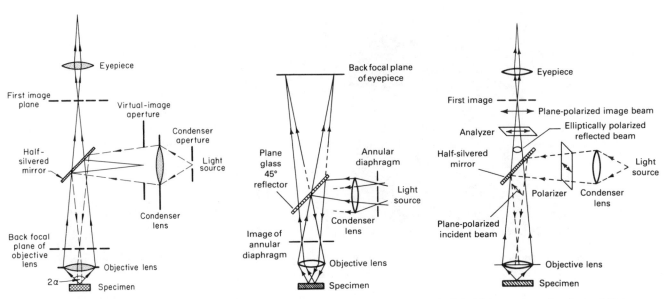

*Principal components of optical microscopes employing bright-field (left), dark-field (center), and polarized-light (right) illumination modes*

FIG. 356

*Two views of a 72% compressed specimen of aluminum alloy 7075-T6 displaying orange peel effect*

usually coarse grain size is stressed beyond its elastic limit (Fig. 356). Also called pebbles and alligator skin.

**orange peel** (plastics). In injection molding of plastics, a part with an undesirable uneven surface somewhat resembling the skin of an orange.

**orbital forging**. See *rotary forging*.

**order** (in x-ray reflection). The factor *n* in *Bragg's law*. In x-ray reflection from a crystal, the order is an integral number that is the path difference measured in wavelengths between reflections from adjacent planes.

**order-disorder transformation**. A phase change among two *solid solutions* having the same crystal structure, but in which the atoms of one phase (disordered) are randomly distributed; in the other, the different kinds of atoms occur in a regular sequence upon the crystal lattice, that is, in an ordered arrangement.

**ordered intermetallics**. See *intermetallics* (Technical Brief 22).

**ordered structure**. The crystal structure of a *solid solution* in which the atoms of different elements seek preferred lattice positions (Fig. 357). Contrast with *disordered structure*.

**order hardening**. A low-temperature *annealing* treatment for metals that permits short-range ordering of solute atoms within a matrix, which greatly impedes dislocation motion.

**ordering**. Forming a *superlattice*.

**ore**. A natural mineral that may be mined and treated for the extraction of any of its components, metallic or otherwise, at a profit.

**ore dressing**. Same as *mineral dressing*.

**organic**. Being or composed of hydrocarbons or their derivatives, or matter of plant or animal origin. Contrast with *inorganic*.

**organic acid**. A chemical compound with one or more carboxyl radicals (COOH) in its structure; examples

FIG. 357

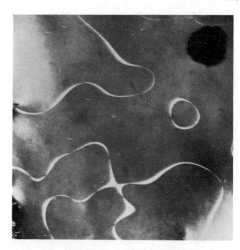

*Ordered structure in an AlFe alloy showing antiphase domain boundaries. Left: Thin-foil electron micrograph. Right: Configuration of atoms with the dashed lines representing phase boundaries*

are butyric acid, $CH_3(CH_2)_2COOH$; maleic acid $HOOCCH:CHCOOH$; and benzoic acid, $C_6H_5COOH$.

**organic fiber**. A fiber derived or composed of matter originating in plant or animal life or composed of chemicals of hydrocarbon origin, either natural or synthetic.

**organic zinc-rich plant**. Coating containing zinc powder pigment and an *organic* resin.

**organosol**. A suspension of a finely divided resin in a volatile, organic liquid. The resin does not dissolve appreciably in the organic liquid at room temperature, but does at elevated temperatures. The liquid evaporates at an elevated temperature, and the residue upon cooling is a homogeneous plastic mass. Plasticizers may be dissolved in the volatile liquid.

**orientation** (crystal). Arrangements in space of the axes of the lattice of a crystal with respect to a chosen reference or coordinate system. See also *preferred orientation*.

**orientation** (plastics). The alignment of the crystalline structure in polymeric materials to produce a highly aligned structure. Orientation can be accomplished by cold drawing or stretching in fabrication. Orientation can generally be divided into two classes, uniaxial and biaxial.

**oriented materials** (plastics). Polymeric materials with molecules and/or macroconstituents that are aligned in a specific way. Oriented materials are anisotropic.

**orifice gas**. In plasma arc welding and cutting, the gas that is directed into the torch to surround the electrode. It becomes ionized in the arc to form the plasma, and issues from the orifice in the torch nozzle as the plasma jet. See also the figures accompanying the terms *nontransferred arc* and *plasma arc welding*.

**original crack size** ($a_0$). The *physical crack size* at the start of testing.

**Orsat analyzer**. A furnace atmosphere analysis device in which gases are absorbed selectively (volumetric basis) by passing them through a series of preselected solvents.

**orthochromatic**. Sensitive to all colors except orange and red.

**orthorhombic**. Having three mutually perpendicular axes of unequal lengths.

**orthotropic**. Having three mutually perpendicular planes of elastic symmetry.

**oscillating die press**. A small high-speed metal forming press in which the die and punch move horizontally with the strip during the working stroke. Through a reciprocating motion, the die and punch return to their original positions to begin the next stroke.

**outgassing**. (1) The release of adsorbed or occluded gases or water vapor, usually by heating, as from a vacuum tube or other vacuum system. (2) Devolatilization of plastics or applied coatings during exposure to vacuum in vacuum metallizing. Resulting parts show voids or thin spots in plating with reduced and spotty brilliance. Additional drying prior to metallizing is helpful, but outgassing is inherent to plastic materials and coating ingredients, including plasticizers and volatile components.

**outlier**. In a set of data, a value so far removed from other values in the distribution that it is probably not a bona fide measurement. There are statistical methods for classifying a data point as an outlier and removing it from a data set.

**out time**. The time a *prepreg* is exposed to ambient temperature, namely, the total amount of time the prepreg is out of the freezer. The primary effects of out time are a decrease in the *drape* and *tack* of the prepreg and the absorption of moisture from the air.

**Table 46 Ovenware compositions and properties**

| Compound/property | Composition, wt%(a) | | | | | |
|---|---|---|---|---|---|---|
| | 1 | 2 | 3 | 4 | 5 | 6 |
| SiO$_2$ | 76.4 | 74.0 | 82.0 | 69.7 | 68.8 | 66.1 |
| Al$_2$O$_3$ | 1.8 | 2.0 | 2.0 | 17.8 | 19.2 | 20.2 |
| B$_2$O$_3$ | 14.7 | 0.5 | 12.0 | ... | ... | ... |
| Li$_2$O | ... | ... | ... | 2.8 | 2.7 | 3.9 |
| Na$_2$O | 5.4 | 13.5 | 4.0 | ... | 0.2 | 0.4 |
| K$_2$O | ... | ... | ... | ... | 0.1 | ... |
| MgO | ... | ... | ... | 2.6 | 1.8 | 1.0 |
| CaO | 0.7 | 10.0 | ... | ... | ... | ... |
| BaO | ... | ... | ... | ... | 0.8 | 1.8 |
| ZnO | ... | ... | ... | 1.0 | 1.0 | 1.0 |
| TiO$_2$ | ... | ... | ... | 4.7 | 2.7 | 3.0 |
| ZrO$_2$ | ... | ... | ... | ... | 1.8 | 1.7 |
| As$_2$O$_3$ | 0.4 | ... | ... | 0.6 | 0.8 | 1.0 |
| Sb$_2$O$_3$ | ... | 0.2 | ... | ... | ... | ... |
| CTE ($\times 10^{-7}$/°C) at 25 to 300 °C | ... | 86 | 37 | 4–20 | 5–7 | 1 |
| Type | Tempered borosilicate | Tempered soda-lime | Tempered borosilicate | β-spodumene glass-ceramic | β-quartz glass-ceramic | β-quartz glass-ceramic |

(a) Numbers refer to the following products/producers: 1, Anchor Hocking; 2, Corning Code 0281; 3, Corning Code 7251; 4, Corning Corning Ware; 5, Corning Visions; 6, Durand Arcoflam Clear-Line

**ovaloid**. A surface of revolution symmetrical about the polar axis that forms the end closure for a filament-wound cylinder.

**oven dry**. The condition of a plastic material that has been heated under prescribed conditions of temperature and humidity until there is no further significant changes in its mass.

**oven soldering**. See preferred term *furnace soldering*.

**ovenware**. Glass and glass-ceramic materials able to withstand thermal downshock of 150°C (270 °F) without breakage that are used for culinary oven use. Table 46 lists ovenware compositions and properties. See also *glass* (Technical Brief 17) and *glass-ceramics* (Technical Brief 18).

**overaging**. *Aging* under conditions of time and temperature greater than those required to obtain maximum change in a certain property, so that the property is altered in the direction of the initial value.

**overbasing**. A technique for increasing the basicity of lubricants. Overbased lubricants are used to assist in neutralizing acidic oxidation products.

**overbending**. Bending metal through a greater arc than that required in the finished part to compensate for springback.

**overdraft**. A condition wherein a metal curves upward on leaving the rolls because of the higher speed of the lower roll.

**overfill**. The fill of a die cavity with an amount of powder in excess of specification.

**overglaze**. A glass coating over another component or element, normally used to give physical or electrical protection.

**overhead-drive press**. A mechanical press with the driving mechanism mounted in or on the crown or upper parts of the uprights. See also the figure accompanying the term *straight-side press*.

**overhead-position welding**. Welding that is performed from the underside of the joint.

**overheating**. Heating a metal or alloy to such a high temperature that its properties are impaired. When the original properties cannot be restored by further heat treating, by mechanical working, or by a combination of working and heat treating, the overheating is known as *burning*.

**overlap**. (1) Pultrusion of weld metal beyond the toe, face, or root of a weld. (2) In resistance seam welding, the area in a given weld remelted by the succeeding weld. See also *face of weld*, *root of weld*, and *toe of weld*.

**overlaying**. See preferred term *surfacing*.

**overlay sheet**. A nonwoven fibrous mat (of glass synthetic fiber, for example) used as the top layer in a cloth or mat *lay-up* to provide a smoother finish, minimize the appearance of the fibrous pattern, or permit machining or grinding to a precise dimension. Also called surfacing mat.

**overmix**. Mixing of a powder longer than necessary to produce adequate distribution of powder particles. Overmixing may cause particle size segregation.

**oversinter**. The sintering of a powder compact at higher temperature or for longer time periods than necessary to obtain the desired microstructure or physical properties. It often leads to swelling due to excessive pore formation.

**oversize powder**. Powder particles larger than the maximum permitted by a particle size specification.

**overspray**. The excess spray material that is not deposited on the part during *thermal spraying*.

**overstressing**. In fatigue testing, cycling at a stress level higher than that used at the end of the test.

**overvoltage.** The difference between the actual electrode potential when appreciable electrolysis begins and the reversible electrode potential.

**oxidation** (carbon fibers). In carbon/graphite fiber processing, the step of reacting the precursor polymer (rayon, polyacrylonitrile (PAN), or pitch) with oxygen, resulting in stabilization of the structure for the hot stretching operation.

**oxidation** (metals). (1) A reaction in which there is an increase in valence resulting from a loss of electrons. Contrast with *reduction*. (2) A corrosion reaction in which the corroded metal forms an oxide (Fig. 358); usually applied to reaction with a gas containing elemental oxygen, such as air. (3) A chemical reaction in which one substance is changed to another by oxygen combining with the substance. Much of the dross from holding and melting furnaces is the result of oxidation of the alloy held in the furnace.

**oxidation grain size.** (1) Grain size determined by holding a metallic specimen at a suitably elevated temperature in a mildly oxidizing atmosphere. The specimen is polished before oxidation and etched afterwards. (2) Refers to the method involving heating a polished steel specimen to a specified temperature, followed by quenching and repolishing. The grain boundaries are sharply defined by the presence of iron oxide.

**oxidation losses.** Reduction in the amount of metal or alloy through *oxidation*. Such losses are usually the largest factor in melting loss.

**oxidative wear.** (1) A *corrosive wear* process in which chemical reaction with oxygen or oxidizing environment predominates. (2) A type of *wear* resulting from the sliding action between two metallic components that generates oxide films on the metal surfaces. These oxide films prevent the formation of a metallic bond between the sliding surfaces, resulting in fine wear debris and low wear rates.

**oxide film replica.** A thin film of an oxide of the specimen to be examined. The replica is prepared by air, oxygen, chemical, or electrochemical oxidation of the parent metal and is subsequently freed mechanically or chemically for examination. See also *replica*.

**oxide-type inclusions.** Oxide compounds occurring as nonmetallic inclusions in metals, usually as a result of deoxidizing additions. In wrought steel products, they may occur as a stinger formation composed of distinct granular or crystalline-appearing particles.

**oxidized steel surface.** Surface having a thin, tightly adhering oxidized skin (from straw to blue in color), extending in from the edge of a coil or sheet.

**oxidizing agent.** A compound that causes *oxidation*, thereby itself being reduced.

**oxidizing atmosphere.** A furnace atmosphere with an oversupply of oxygen that tends to oxidize materials placed in it.

**oxidizing flame.** A gas flame produced with excess oxygen in the inner flame that has an oxidizing effect. See also the figure accompanying the term *neutral flame*.

**oxyacetylene cutting.** An *oxyfuel gas cutting* process in which the fuel gas is acetylene.

**oxyacetylene welding.** An *oxyfuel gas welding* process in which the fuel gas is acetylene.

**oxyfuel gas cutting (OFC).** Any of a group of processes used to sever metals by means of chemical reaction between hot base metal and a fine stream of oxygen. The necessary metal temperature is maintained by gas flames resulting from combustion of a specific fuel gas such as acetylene, hydrogen, natural gas, propane, propylene, or

FIG. 358

 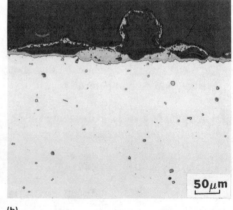

(a)                                          (b)

*Topography (a) and cross section (b) of oxide scale formed on Fe-18Cr alloy at 1100 °C (2010 °F). The bright areas on the alloy surface (a) are areas from which scale has spalled. The buckled scale and locally thickened areas (b) are iron-rich oxide. The thin scale layer adjacent to the alloy is Cr$_2$O$_3$, which controls the oxidation rate.*

**Table 47  Properties of common fuel gases**

|  | Acetylene | Propane | Propylene | Methylacetylene-propadiene (Mapp) | Natural gas |
|---|---|---|---|---|---|
| Chemical formula .............. | $C_2H_2$ | $C_3H_8$ | $C_3H_6$ | $C_3H_4$ (Methylacetylene, propadiene) | $CH_4$ (Methane) |
| Neutral flame temperature | | | | | |
| °F ......................... 5 600 | | 4 580 | 5 200 | 5 200 | 4 600 |
| °C ......................... 3 100 | | 2 520 | 2 870 | 2 870 | 2 540 |
| Primary flame heat emission | | | | | |
| Btu/ft³ ..................... 507 | | 255 | 433 | 517 | 11 |
| MJ/m³ ...................... 19 | | 10 | 16 | 20 | 0.4 |
| Secondary flame heat emission | | | | | |
| Btu/ft³ ..................... 963 | | 2 243 | 1 938 | 1 889 | 989 |
| MJ/m³ ...................... 36 | | 94 | 72 | 70 | 37 |
| Total heat value (after vaporization) | | | | | |
| Btu/ft³ ..................... 1 470 | | 2 498 | 2 371 | 2 406 | 1 000 |
| MJ/m³ ...................... 55 | | 104 | 88 | 90 | 37 |
| Total heat value (after vaporization) | | | | | |
| Btu/lb ..................... 21 500 | | 21 800 | 21 100 | 21 000 | 23 900 |
| kJ/kg ...................... 50 000 | | 51 000 | 49 000 | 49 000 | 56 000 |
| Total oxygen required (neutral flame) | | | | | |
| vol $O_2$/vol fuel ................ 2.5 | | 5.0 | 4.5 | 4.0 | 2.0 |
| Oxygen supplied through torch (neutral flame) | | | | | |
| vol $O_2$/vol fuel ................ 1.1 | | 3.5 | 2.6 | 2.5 | 1.5 |
| ft³oxygen/lb fuel (60 °F) ........ 16.0 | | 30.3 | 23.0 | 22.1 | 35.4 |
| m³oxygen/kg (15.6 °C) ......... 1.0 | | 1.9 | 1.4 | 1.4 | 2.2 |
| Maximum allowable regulator pressure | | | | | |
| psi........................... 15 | | Cylinder | Cylinder | Cylinder | Line |
| kPa .......................... 103 | | | | | |
| Explosive limits in air, % ...... 2.5–80 | | 2.3–9.5 | 2.0–10 | 3.4–10.8 | 5.3–14 |
| Volume-to-weight ratio | | | | | |
| ft³/lb (60 °F) ................. 14.6 | | 8.66 | 8.9 | 8.85 | 23.6 |
| m³/kg (15.6 °C) ............... 0.91 | | 0.54 | 0.55 | 0.55 | 1.4 |
| Specific gravity of gas (60 °F, 15.6 °C) | | | | | |
| Air = 1 .................... 0.906 | | 1.52 | 1.48 | 1.48 | 0.62 |

Source: American Welding Society

Mapp gas (stabilized methylacetylene-propadiene). Table 47 lists the properties of common fuel gases. See also *oxygen cutting* and the figure accompanying the term *cutting torch* (oxyfuel gas).

**oxyfuel gas spraying**. See preferred term *flame spraying*.

**oxyfuel gas welding (OFW)**. Any of a group of processes used to fuse metals together by heating them with gas flames resulting from combustion of a specific fuel gas such as acetylene, hydrogen, natural gas, or propane. The process may be used with or without the application of pressure to the joint, and with or without adding any filler metal. See also the figure accompanying the term *welding torch* (oxyfuel gas).

**oxygas cutting**. See preferred term *oxygen cutting*.

**oxygen arc cutting**. An oxygen cutting process used to sever metals by means of the chemical reaction of oxygen with the base metal at elevated temperatures. The necessary temperature is maintained by an arc between a consumable tubular electrode and the base metal (Fig. 359).

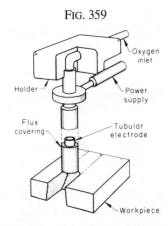

FIG. 359

*Components of an oxygen arc electrode*

**oxygen concentration cell**. See *differential aeration cell*.

**oxygen cutting**. Metal cutting by directing a fine stream of oxygen against a hot metal. The chemical reaction between oxygen and the base metal furnishes heat for localized melting, hence, cutting. In the case of oxidation-resistant metals, the reaction is facilitated by the use of a chemical flux or metal powder. See also *metal powder cutting*.

**oxygen deficiency**. A form of *crevice corrosion* in which galvanic corrosion proceeds because oxygen is prevented from diffusing into the crevice.

**oxygen-free copper**. Electrolytic copper free from cuprous oxide, produced without the use of residual metallic or metalloidal deoxidizers.

**oxygen gouging**. Oxygen cutting in which a bevel or groove is formed.

**oxygen grooving**. See preferred term *oxygen gouging*.

**oxygen lance**. A length of pipe used to convey oxygen either beneath or on top of the melt in a steelmaking furnace, or to the point of cutting in *oxygen lance cutting*. See also the figures accompanying the terms *argon oxygen decarburization* and *basic oxygen furnace*.

**oxygen lance cutting**. An oxygen cutting process used to sever metals with oxygen supplied through a consumable lance; the preheat to start the cutting is obtained by other means.

**oxygen probe**. An atmosphere-monitoring device that elec-tronically measures the difference between the partial pressure of oxygen in a furnace or furnace supply atmosphere and the external air.

**oxyhydrogen cutting**. An *oxyfuel gas cutting* process in which the fuel gas is hydrogen.

**oxyhydrogen welding**. An *oxyfuel gas welding* process in which the fuel gas is hydrogen.

**oxynatural gas cutting**. An *oxyfuel gas cutting* process in which the fuel gas is natural gas.

**oxynatural gas welding**. An *oxyfuel gas welding* process in which the fuel gas is natural gas.

**oxypropane cutting**. An *oxyfuel gas cutting* process in which the fuel gas is propane.

**oxypropane welding**. An *oxyfuel gas welding* process in which the fuel gas is propane.

**ozone**. A powerfully oxidizing allotropic form of the element oxygen. The ozone molecule contains three atoms ($O_3$). Ozone gas is decidedly blue, and both liquid and solid ozone are an opaque blue-black color, similar to that of ink.

**ozone test**. Exposure of material to a high concentration of ozone to give an accelerated indication of degradation expected in normal environments.

# P

**PA**. See *polyamides* (Technical Brief 30).

**package** (composites). Yarn, roving, and so forth for reinforcing plastics in the form of units capable of being unwound and suitable for handling, storing, shipping, and use.

**package** (electronics). In the electronics/microelectronics industry, an enclosure for a single element, an integrated circuit, or a hybrid circuit.

**pack carburizing**. A method of surface hardening of steel in which parts are packed in a steel box with a carburizing compound and heated to elevated temperatures. Common carburizing compounds contain 10 to 10% alkali or alkaline earth metal carbonates (for example, barium carbonate, $BaCO_3$) bound to hardwood charcoal or to coke by oil, tar, or molasses. This process has been largely supplanted by gas and liquid carburizing processes.

**packed density**. See preferred term *tap density*.

**packing density**. See preferred term *apparent density*.

**packing material**. Any material in which powder metallurgy compacts are embedded during the presintering or sintering operation. The material may act as a getter to protect the compacts from contamination. See also *getter*.

**pack nitriding**. A method of surface hardening of steel in which parts are packed in a steel box with a nitriding compound and heated to elevated temperatures.

**pack rolling**. Hot rolling a pack of two or more sheets of metal; scale prevents their being welded together.

**pad**. The general term used for that part of a die which delivers holding pressure to the metal being worked.

**padding**. In foundry practice, the process of adding metal to the cross section of a casting wall, usually extending from a riser, to ensure adequate feed metal to a localized area during solidification where a shrink would occur if the added metal were not present.

**paddle mixer**. A mixer that uses paddles mounted on a rotating shaft or disk to move and mix the metal powder.

**pad lubrication**. A system of lubrication in which the lubricant is delivered to a bearing surface by a pad of felt or similar material.

**PAEK**. See *polyaryletherketone*.

**PAI**. See *polyamide-imide*.

**PAN**. See *polyacrylonitrile*.

**pancake forging**. A rough forged shape, usually flat, that can be obtained quickly with minimal tooling. Considerable machining is usually required to attain the finish size.

**pancake grain structure.** A metallic structure in which the lengths and widths of individual grains are large compared to their thicknesses.

**parabolic reflector.** A reflector for ultraviolet curing of adhesives that projects parallel light beams perpendicular to the assembly, but is not focused. See also *ultraviolet (UV)/electron beam (EB) cured adhesives* (Technical Brief 58).

**parallel laminate.** (1) A composite laminate of woven fabric in which the plies are aligned in the same position as they were on the fabric roll. (2) A series of flat or curved cloth-resin layers stacked uniformly one on top of the other.

**paramagnetic material.** (1) A material whose specific permeability is greater than unity and is practically independent of the magnetizing force. (2) Material with a small positive susceptibility due to the interaction and independent alignment of permanent atomic and electronic magnetic moments with the applied field. Compare with *ferromagnetic material.*

**paramagnetism.** A property exhibited by substances that, when placed in a magnetic field, are magnetized parallel to the field to an extent proportional to the field (except at very low temperatures or in extremely large magnetic fields). Compare with *ferromagnetism.*

**parameter.** In statistics, a constant (usually unknown) defining some property of the frequency distribution of a population, such as a population median or a population standard deviation.

**parameter** (in crystals). See *lattice parameter.*

**parent metal.** See preferred term *base metal.*

**parfocal eyepiece.** Microscope eyepieces, with common focal planes, that are interchangeable without refocusing.

**parison.** The hollow plastic tube from which a plastic component is blow molded. See also the figure accompanying the term *blow molding.*

**parison swell.** In blow molding of plastics, the tendency of the *parison* to enlarge as it emerges from the die. It is expressed as the ratio of the cross-sectional area of the parison to the cross-sectional area of the die opening. See also *blow molding.*

**Parkes process.** A process used to recover precious metals from lead and based on the principle that if 1 to 2% Zn is stirred into the molten lead, a compound of zinc with gold and silver separates out and can be skimmed off.

**partial annealing.** An imprecise term used to denote a treatment given cold-worked metallic material to reduce its strength to a controlled level or to effect stress relief. To be meaningful, the type of material, the degree of cold work, and the time-temperature schedule must be stated.

**partial hydrodynamic lubrication.** See *quasi-hydrodynamic lubrication.*

FIG. 360

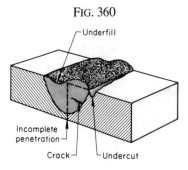

*Partial (incomplete) weld penetration in a plasma arc weld*

**partial joint penetration.** Weld joint penetration which is less than complete (Fig. 360). Compare with *complete joint penetration.*

**partial journal bearing.** A bearing in which the bore extends not more than half the circumference of the journal.

**partially stabilized zirconia.** Zirconia ($ZrO_2$) which contains a mixture of cubic and tetragonal and/or monoclinic phases produced by the addition of small amounts of magnesium oxide (MgO), calcium oxide (CaO), or yttrium oxide ($Y_2O_3$). These materials are used in structural applications where their high strength (600 to 700 MPa, or 87 to 100 ksi) and moderate fracture toughness (11 to 14 MPa$\sqrt{m}$, or 10 to 13 ksi$\sqrt{in.}$) can be exploited. See also *zirconia* (Technical Brief 63).

**particle.** (1) Any small part of matter, such as a molecule, atom, or electron. (2) Any small subdivision of matter, such as a metal or ceramic powder particle.

**particle accelerator.** A device that raises the velocities of charged atomic or subatomic particles to high values.

**particle hardness.** The hardness of an individual ceramic or metal powder particle as measured by a Knoop or Vickers type microhardness indentation test.

**particle-induced x-ray emission (PIXE).** A method of trace elemental analysis in which a beam of ions (usually protons) is directed at a thin foil on which the sample to be analyzed has been deposited; the energy spectrum of the resulting x-rays is measured. See also *particle accelerator* and *proton.*

**particle morphology.** The form and structure of an individual metal or ceramic powder particle.

**particle shape.** The appearance of a metal particle, such as spherical, rounded, angular, acicular, dendritic, irregular, porous, fragmented, blocky, rod, flake, nodular, or plate (Fig. 361).

**particle size.** The controlling lineal dimension of an individual particle as determined by analysis with screens or other suitable instruments. See also *sieve analysis* and *sieve classification.*

**particle size analysis.** See preferred term *sieve analysis.*

FIG. 361

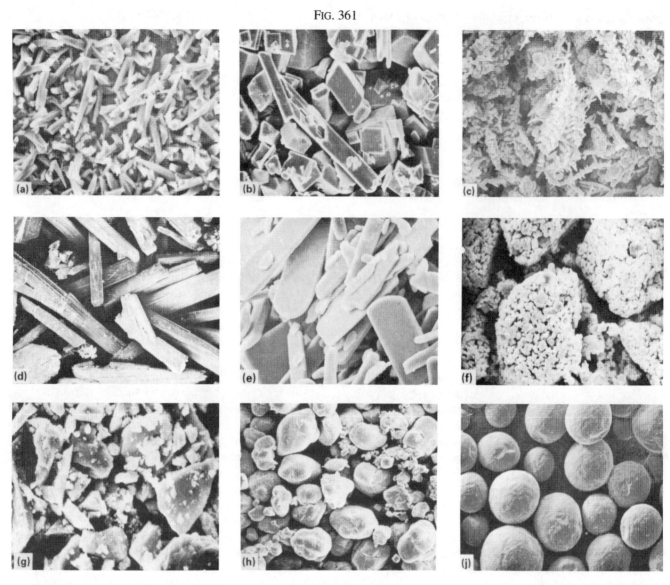

*Common particle shapes in metal powders. (a) Acicular powder particles. (b) Angular powder particles. (c) Dendritic powder particles. (d) Fibrous powder particles. (e) Flaky powder particles. (f) Granular powder particles. (g) Irregular powder particles. (h) Nodular powder particles. (j) Spheroidal powder particles*

**particle size classification.** See preferred term *sieve classification*.

**particle size distribution.** The percentage, by weight or by number, of each fraction into which a powder or sand sample has been classified with respect to sieve number or *particle size*. See the example described with the term *sieve classification*.

**particle size range.** (1) The limits between which a variation in particle size is allowed. (2) Classification of spray powders defined by an upper and lower size limit; for example −200 +300 mesh: a quantity of powder, the largest particles of which will pass through a 200-mesh sieve and the smallest of which will not pass through a 325-mesh sieve. See also *sieve analysis* and *sieve classification*.

**particle sizing.** Segregation of granular material into specified particle size ranges.

**particle spacing.** The distance between the surfaces of two or more adjacent particles in a loose powder or a compact.

**particulate composite.** Material consisting of one or more constituents suspended in a matrix of another material (Fig. 362). These particles are either metallic or nonmetallic.

**parting.** (1) In the recovery of precious metals, the separation of silver from gold. (2) The zone of separation between *cope* and *drag* portions of the mold or flask in sand casting. (3) A composition sometimes used in sand molding to facilitate the removal of the pattern. (4) Cutting simultaneously along two parallel lines or along two lines that

FIG. 362

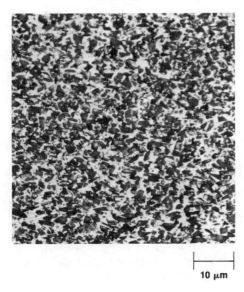

10 μm

*A particulate composite with an aluminum alloy matrix reinforced with 40 vol% SiC particles*

balance each other in side thrust. (5) A shearing operation used to produce two or more parts from a stamping.

**parting agent**. See *mold release agent*.

**parting compound**. A material dusted or sprayed on foundry (casting) patterns to prevent adherence of sand and to promote easy separation of *cope* and *drag* parting surfaces when the cope is lifted from the drag.

**parting line**. (1) The intersection of the parting plane of a casting or plastic mold or the parting plane between forging dies with the mold or die cavity (Fig. 363). (2) A raised line or projection on the surface of a casting, plastic part, or forging that corresponds to said intersection.

FIG. 363

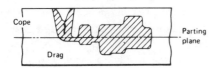

*The parting plane in a casting mold*

**parting sand**. In foundry practice, a fine sand for dusting on sand mold surfaces that are to be separated.

**parts former**. A type of upsetter designed to work on short billets instead of bars and tubes, usually for cold forging.

**parts per billion**. A measure of proportion by weight, equivalent to one unit weight of a material per billion ($10^9$) unit weights of compound.

**parts per million**. A measure of proportion by weight, equivalent to one unit weight of a material per million ($10^6$) unit weights of compound.

**PAS**. See *polyaryl sulfone*.

**pass**. (1) A single transfer of metal through a *stand* of rolls. (2) The open space between two grooved rolls through which metal is processed. (3) The weld metal deposited in one trip along the axis of a weld. See also *weld pass*.

**pass sequence**. See *deposition sequence*.

**passivation**. (1) A reduction of the anodic reaction rate of an electrode involved in corrosion. (2) The process in metal corrosion by which metals become *passive*. (3) The changing of a chemically active surface of a metal to a much less reactive state. Contrast with *activation*. (4) The formation of an insulating layer directly over the semiconductor surface to protect the surface from contaminants, moisture, and so forth.

**passivator**. A type of corrosion *inhibitor* that appreciably changes the potential of a metal to a more noble (positive) value.

**passive**. (1) A metal corroding under the control of a surface reaction product. (2) The state of the metal surface characterized by low corrosion rates in a potential region that is strongly oxidizing for the metal.

**passive-active cell**. A corrosion cell in which the *anode* is a metal in the *active* state and the *cathode* is the same metal in the *passive* state.

**passivity**. A condition in which a piece of metal, because of an impervious covering of oxide or other compound, has a *potential* much more positive than that of the metal in the active state.

**paste**. An adhesive compound having a characteristic plastic-type consistency, that is, a high order of yield value, such as that prepared by heating a mixture of starch and water and subsequently cooling the hydrolyzed product. See also *adhesive, glue, mucilage,* and *sizing*.

**paste brazing filler metal**. A mixture of finely divided brazing filler metal with an organic or inorganic flux or neutral vehicle or carrier. Brazing paste mixtures are usually prepared for *furnace brazing* in a *protective atmosphere*.

**paste extrude**. An extrusion method for plastics in which the extrudable material, a fine powder form mixed with a lubricant, is forced through a die of given size, without heat, as opposed to melt extrude. See also *extruder* (plastics).

**paste soldering filler metal**. A mixture of finely divided metallic solder with an organic or inorganic flux or neutral vehicle or carrier.

**patenting**. In wiremaking, a heat treatment applied to medium-carbon or high-carbon steel before drawing of wire or between drafts. This process consists of heating to a temperature above the transformation range and then cooling to a temperature below $Ae_1$ in air or in a bath of molten lead or salt.

**patent leveling**. Same as *stretcher leveling*.

**patina**. The coating, usually green, that forms on the surface

FIG. 364

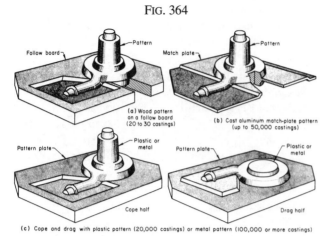

*Three types of patterns for producing a water-pump casting in various quantities*

FIG. 365

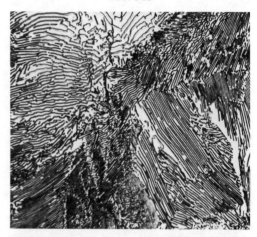

*Pearlite structure in a high-carbon steel (Fe-0.75C). The cementite lamellae are white; the ferrite is dark. 500×*

of metals such as copper and copper alloys exposed to the atmosphere. Also used to describe the appearance of a weathered surface of any metal.

**pattern.** (1) A form of wood, metal, or other material around which molding material is placed to make a mold for casting metals (Fig. 364). (2) A form of wax- or plastic-base material around which refractory material is placed to make a mold for casting metals. (3) A full-scale reproduction of a part used as a guide in cutting.

**pattern draft.** In foundry practice, the paper allowed on the vertical faces of a pattern to permit easy withdrawal of the pattern from the mold or die.

**pattern layout.** A full-size drawing of a foundry (casting) pattern showing its arrangement and structural features.

**patternmaker's shrinkage.** Contraction allowance made on patterns to compensate for the decrease in dimensions as the solidified casting cools in the mold from the freezing temperature of the metal to room temperature. The pattern is made larger by the amount of contraction that is characteristic of the particular metal to be used.

**Pattinson process.** A process for separating silver from lead, in which the molten lead is slowly cooled so that crystals poorer in silver solidify out and are removed, leaving the melt richer in silver.

**PBI.** See *polybenzimidazole.*

**PBT.** See *polybutylene terephthalate.*

**PC.** See *polycarbonates* (Technical Brief 31).

**peak overlap.** In materials characterization techniques such as electron probe x-ray microanalysis or Auger electron spectroscopy, the formation of a single peak when two closely spaced x-ray peaks cannot be resolved; the energy (or wavelength) of the peak is some average of the characteristic energies (or wavelengths) of the original two peaks. See also *full width at half maximum.*

**pearlite.** A metastable lamellar aggregate of *ferrite* and ce-

mentite resulting from the transformation of *austenite* at temperatures above the *bainite* range (Fig. 365).

**pearlitic malleable.** See *malleable cast iron.*

**pearlitic structure.** A microstructure resembling that of the pearlite constituent in steel. Therefore, it is a lamellar structure of varying degrees of coarseness.

**pebbles.** See *orange peel* (metals).

**pedestal bearing.** A bearing that is supported on a column or pedestal rather than on the main body of the machine.

**PEEK.** See *polyether etherketone.*

**peeling.** The detaching of one layer of a coating from another, or from the basis metal, because of poor adherence.

**peel ply.** In composites fabrication, a layer of open-weave material, usually fiberglass or heat set nylon, applied directly to the surface of a prepreg lay-up. The peel ply is removed from the cured laminate immediately before bonding operations, leaving a clean, resin-rich surface that needs no further preparation for bonding, other than the application of a primer if one is required. See also *lay-up* and *prepreg.*

**peel strength** (adhesives). The average load per unit width of bond line required to separate progressively one member from the other over the adhered surfaces at a separation angle of approximately 180° and a separation rate of 152 mm (6 in.) per minute (Fig. 366). It is expressed in force per unit width of specimen (N/mm, or lbf/in.).

**peel test** (adhesives). See *peel strength* (adhesives).

**peel test** (weldments). A destructive method of inspection which mechanically separates a lap joint by peeling.

**peening.** Mechanical working of metal by hammer blows or shot impingement.

**peening wear.** Removal of material from a solid surface caused by repeated impacts on very small areas.

**PEI.** See *polyether-imide.*

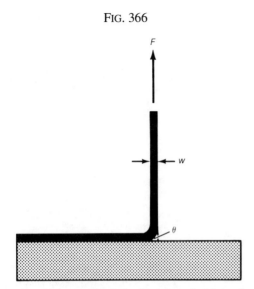

F<sub></sub>

FIG. 366

*The peel test, in which the film is peeled from the substrate at a constant, specified rate by the force F at an angle θ*

**pellet**. In powder metallurgy, a small rounded or spherical solid body that is similar to a shotted particle. See also *shotting*.

**penetrant**. A liquid with low surface tension used in *liquid penetrant inspection* to flow into surface openings of parts being inspected.

**penetrant inspection**. See preferred term *liquid penetrant inspection*.

**penetration** (adhesives). The entering of an adhesive into an adherend. This property of a system is measured by the depth of penetration of the adhesive into the adherend.

**penetration** (metals). (1) In founding, an *imperfection* on a casting surface caused by metal running into voids between sand grains; usually referred to as *metal penetration*. (2) In welding, the distance from the original surface of the base metal to that point at which fusion ceased. See also *joint penetration* and *root penetration*.

**penetration** (of a grease). The depth in $1/10$ mm that a standard cone penetrates the sample in a standard cup under prescribed conditions of weight, time (5 s), and temperature (25 °C, or 77 °F). The result depends on whether or not the grease has been subjected to shear. In unworked penetration, the grease is transferred with as little deformation as possible or is tested in its container. In worked penetration, the grease is subjected to 60 double strokes in a standard device. In prolonged worked penetration, the grease is worked for a specified period before the 60 strokes. The results may be quoted as penetration number or penetration value.

**penetration hardness. Same as** *indentation hardness*.

**penetration hardness number**. Any numerical value ex-

pressing the resistance of a body to the penetration of a second, usually harder, body.

**penetrometer**. In grease technology, an instrument for measuring the consistency of a grease by allowing a cone to penetrate into the grease under controlled conditions. See also *penetration (of a grease)*.

**percent conductivity**. The conductivity of a material expressed as a percentage of that of copper. See also *IACS*.

**percent error**. In the case of a mechanical testing machine, the ratio, expressed as a percentage, of the *error* to the correct value of the applied load.

**percent ferrite**. See preferred term *ferrite number*.

**percent theoretical density, See** *density ratio*.

**percussion cone**. Damage produced by contact stresses generated by mechanical contact of a hard, blunt object with a glass surface. Typically, it has the appearance of a semicircular or circular crack on the damaged surface, propagating into the glass, flaring out with increasing depth into a cone-shaped crack (Fig. 367). Also called impact bruise, butterfly bruise, bump check, and Hertzian cone crack.

FIG. 367

*Percussion cone (Hertzian impact site) in glass. 145×*

**percussion welding**. A resistance welding process which produces coalescence of abutting surfaces using heat from an arc produced by a rapid discharge of electrical energy. Pressure is applied percussively during or immediately following the electrical discharge.

**perforating**. The punching of many holes, usually identical and arranged in a regular pattern, in a sheet, workpiece blank, or previously formed part. The holes are usually round, but may be any shape. The operation is also called multiple punching. See also *piercing*.

**periodic reverse**. Pertains to periodic changes in direction of flow of the current in electrolysis. It applies to the process

FIG. 368

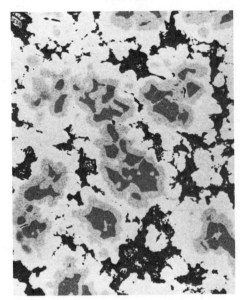

*Peritectic structure of an Sn-17Co alloy. The primary CoSn crystals that form the dark centers are surrounded peritectically by CoSn2 layers (gray), followed by a peritectoid envelope of CoSn3 phase. 200×*

FIG. 369

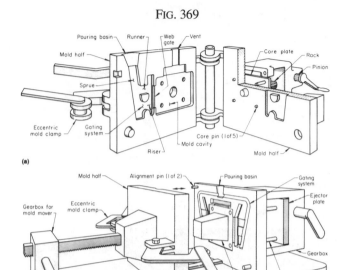

*Two types of manually operated permanent mold casting machines. (a) Simple book-type mold for shallow-cavity castings. (b) Device with straight-line retraction for deep-cavity molds*

and also the machine that controls the time for both directions.

**peripheral clearance angle**. See *clearance angle*.

**peripheral milling**. Milling a surface parallel to the axis of the cutter. See also *milling* and the figure accompanying *milling cutter*.

**peripheral speed**. See preferred term *cutting speed*.

**peritectic**. An isothermal reversible reaction in metals in which a liquid phase reacts with a solid phase to produce a single (and different) solid phase on cooling (Fig. 368).

**peritectic equilibrium**. A reversible univariant transformation in which a solid phase stable only at lower temperature decomposes into a liquid and a solid phase that are conjugate at higher temperature.

**peritectoid**. An isothermal reversible reaction in which a solid phase reacts with a second solid phase to produce a single (and different) solid phase on cooling.

**peritectoid equilibrium**. A reversible univariant transformation in which a solid phase stable only at low temperature decomposes with rising temperature into two or more conjugate solid phases.

**permanence**. (1) The property of a plastic that describes its resistance to appreciable change in characteristics with time and environment. (2) The resistance of an adhesive bond to deteriorating influences.

**permanent magnet materials**. See *Technical Brief 28*.

**permanent mold**. A metal, graphite, or ceramic mold (other than an ingot mold) of two or more parts that is used

repeatedly for the production of many *castings* of the same form (Fig. 369). Liquid metal is usually poured in by gravity.

**permanent set**. The deformation remaining after a specimen has been stressed a prescribed amount in tension, compression, or shear for a specified time period and released for a specified time period. For creep tests, the residual unrecoverable deformation after the load causing the creep has been removed for a substantial and specified period of time. Also, the increase in length, expressed as a percentage of the original length, by which an elastic material fails to return to its original length after being stressed for a standard period of time.

**permeability**. (1) The passage or diffusion (or rate of passage) of a gas, vapor, liquid, or solid through a material (often porous) without physically or chemically affecting it; the measure of fluid flow (gas or liquid) through a material. (2) A general term used to express various relationships between magnetic induction and magnetizing force. These relationships are either "absolute permeability," which is a change in magnetic induction divided by the corresponding change in magnetizing force, or "specific (relative) permeability," the ratio of the absolute permeability to the permeability of free space. (3) In metal casting, the characteristics of molding ma-

## Technical Brief 28: Permanent Magnet Materials

PERMANENT MAGNET is the term used to describe solid materials capable of being magnetized permanently because of their ability to retain induced magnetization and magnetic poles after removal of externally applied fields. Such materials, which are characterized by their high magnetic induction, high resistance to demagnetization, and maximum energy content, include a variety of alloys, intermetallics, and ceramics. Commonly included are certain steels, Alnico (a cast or sintered iron-base alloy containing 7 to 10% Al, 15 to 19% Ni, 13 to 35% Co, 3 to 4% Cu, with an optional 1 to 5% Ti), Cunife (60Cu-20Ni-20Fe), Fe-Co alloys containing vanadium or chromium, platinum-cobalt, hard ferrites ($SrO\text{-}Fe_2O_3$ or $BaO\text{-}6Fe_2O_3$), cobalt rare-earth alloys ($SmCo_5$ or $Sm_2Co_{17}$), and Nd-Fe-B alloys made by powder metallurgy processing. Each type of magnet material possesses unique magnetic and mechanical properties, corrosion resistance, temperature sensitivity, fabrication limitations, and cost. These factors provide designers with a wide range of options in designing magnetic parts.

Permanent magnet materials are based on the cooperation of atomic and molecular moments within a magnet body to produce a high magnetic induction. This induced magnetization is retained because of a strong resistance to demagnetization. These materials are classified as ferromagnetic or ferrimagnetic and do not include diamagnetic or paramagnetic materials. The natural ferromagnetic elements are iron, nickel, and cobalt. Other elements, such as manganese or chromium, can be made ferromagnetic by alloying to induce proper atomic spacing. Ferromagnetic metals combine with other metals or with oxides to form ferrimagnetic substances; ceramic magnets are of this type.

Permanent magnet materials are marketed under a variety of trade names and designations throughout the world. Over the years, the number and range of applications utilizing permanent magnets has increased dramatically. Some of the more predominant applications include aircraft magnetos, alternators, magnetos for lawn mowers, garden tractors, and outboard motors, small and large direct current (dc) motors (including automotive motors), acoustic transducers, magnetic couplings, magnetic resonance imaging, magnetic focusing systems, ammeters and voltmeters, and watt-hour meters.

### Classification of permanent magnetic materials on the basis of application-relevant properties

| Property | Performance High ← | | | | | → Low |
|---|---|---|---|---|---|---|
| Energy (density) | NdFeB | (RE)Co | Alnico, FeCrCo | Ferrites | FeCoVCr | AlNi |
| $(BH)_{max}$, $kJ \cdot m^{-3}$ (MG · Oe) | 320 (40) | ... | ... | ... | ... | 8 (1) |
| Stability against demagnetization | (RE)Co | NdFeB | | Ferrites | Alnico, FeCrCo | FeCoVCr |
| $H_{ci}$, $kA \cdot m^{-1}$ (Oe) | 2000 (25 000) | ... | | ... | ... | 8 (100) |
| $\mu_{rec}$ | 1.0 | ... | | ... | ... | 10 |
| Hysteresis loss | Alnico, FeCoVCr, FeCrCo | | | | | |
| Reversible temperature variation | Alnico, FeCoVCr | FeCrCo | | (RE)Co | NdFeB | Ferrites |
| $\alpha(\phi)$, %/K | −0.01 | ... | ... | ... | ... | −0.2 |
| Curie temperature, | Alnico, FeCrCo | | (RE)Co, FeCoVCr | Ferrites | | NdFeB |
| °C (°F) | 900 (1650) | | ... | ... | | 300 (570) |
| Stability with high-temperature operation | Ferrites | | Alnico, FeCoVCr, FeCrCo | (RE)Co | | NdFeB |
| Possibility for choosing a preferred direction (PD) | Alnico, FeCrCo | | (a) Ferrites; (RE)Co; NdFeB | | | FeCoVCr, FeCrCo, |
| PD cause | Magnetic field with heat treatment | | Magnetic field with shaping by pressing or injection molding | | | Mechanical deformation |
| Elasticity | FeCoVCr, CrVCo | | | | | All others |
| Physical strength (stability with handling) | FeCoVCr, FeCrCo | NdFeB | Alnico, AlNi | Ferrites | | (RE)Co |
| Economy, relative cost per unit of magnetic energy, $ | Ferrites 1 | | Alnico, FeCrCo ... | NdFeB 5 | (RE)Co ... | FeCoVCr 20 |

### Recommended Reading

- J.W. Fiepke, Permanent Magnet Materials, *Metals Handbook*, 10th ed., Vol 2, ASM International, 1990, p 782-803
- M. McCaig and A.E. Clegg, *Permanent Magnets in Theory and Practice*, 2nd ed., Halsted Press, John Wiley & Sons, 1987

terials that permit gases to pass through them. "Permeability number" is determined by a standard test.

**permissible variation**. In mechanical testing machines, the maximum allowable error in the value of the quantity indicated. It is convenient to express permissible variation in terms of the *percent error*. See also *tolerance*.

**perovskites**. A naturally occurring mineral with a structure type that includes no less than 150 compounds; the crystal structure is ideally cubic (Fig. 370). Perovskites are noted for their ferroelectricity, piezoelectricity, pyroelectricity, electrostriction, high permittivity, and optical and electrooptic properties. The hardness of perovskites is approximately 5.5 on the Mohs scale.

Fig. 370

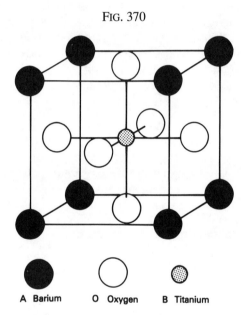

A Barium     O Oxygen     B Titanium

*The ABO₃ perovskite structure typified by BaTiO₃ above its Curie point (cubic symmetry)*

**peroxy compounds**. Polymeric compounds containing O–O linkage.

**perpendicular section**. A section cut perpendicular to a surface of interest in a metallographic specimen. Compare with *taper section*.

**PESV**. See *polyether sulfone*.

**PET**. See *polyether terephthalate*.

**Petroff equation**. An equation describing the viscous power loss in a concentric bearing full of lubricant. The resisting torque on the shaft ($T_0$ = shear stress × shaft radius × bearing area) is given by:

$$T_0 = \frac{\pi^2 \eta N L D^2}{2c}$$

where $\pi = 3.1416$, $\eta$ is the dynamic viscosity, $N$ is the shaft speed, $L$ is the bearing length, $D$ is the shaft diameter, and $2c$ is the diametral clearance.

**petrography**. The study of nonmetallic matter under suitable microscopes to determine structural relationships and to identify the phases or minerals present. With transparent materials, the determination of the optical properties, such as the indices of refraction and the behavior in transmitted polarized light, are means of identification. With opaque materials, the color, hardness, reflectivity, shape, and etching behavior in polished sections are means of identification.

**petroleum oil**. See *mineral oil*.

**pewter**. A tin-base *white metal* containing antimony and copper. Originally, pewter was defined as an alloy of tin and lead, but to avoid toxicity and dullness of finish, lead is excluded from modern pewter. These modern compositions contain 1 to 8% Sb and 0.25 to 3% Cu. Typical pewter products include coffee and tea services, trays, steins, mugs, candy dishes, jewelry, bowls, plates, vases, candlesticks, compotes, decanters, and cordial cups. Table 48 lists chemical compositions of modern pewter alloys.

**pH**. The negative logarithm of the hydrogen-ion activity; it denotes the degree of acidity or basicity of a solution. At 25 °C (77 °F), 7.0 is the neutral value. Decreasing values below 7.0 indicates increasing acidity; increasing values above 7.0, increasing basicity. The pH values range from 0 to 14.

**phase**. A physically homogeneous and distinct portion of a material system.

**phase change**. The transition from one physical state to another, such as gas to liquid, liquid to solid, gas to solid, or vice versa.

**phase-change lubrication**. See *melt lubrication*.

### Table 48 Chemical composition limits for modern pewter

| Specification | Composition, % | | | | | | | |
| --- | --- | --- | --- | --- | --- | --- | --- | --- |
| | Sn | Sb | Cu | Pb max | As max | Fe max | Zn max | Cd max |
| ASTM B 560 | | | | | | | | |
| Type 1(a) | 90–93 | 6–8 | 0.25–2.0 | 0.05 | 0.05 | 0.015 | 0.005 | . . . |
| Type 2(b) | 90–93 | 5–7.5 | 1.5–3.0 | 0.05 | 0.05 | 0.015 | 0.005 | . . . |
| Type 3(c) | 95–98 | 1.0–3.0 | 1.0–2.0 | 0.05 | 0.05 | 0.015 | 0.005 | . . . |
| BS 5140 | bal | 5–7 | 1.0–2.5 | 0.5 | . . . | . . . | . . . | 0.05 |
| | | 3–5 | 1.0–2.5 | 0.5 | . . . | . . . | . . . | 0.05 |
| DIN 17810 | bal | 1–3 | 1–2 | 0.5 | . . . | . . . | . . . | . . . |
| | | 3.1–7.0 | 1–2 | 0.5 | . . . | . . . | . . . | . . . |

(a) Casting alloy, nominal composition 92Sn-7.5Sb-0.5Cu. (b) Sheet alloy, nominal composition 91Sn-7Sb-2Cu. (c) Special-purpose alloy

FIG. 371

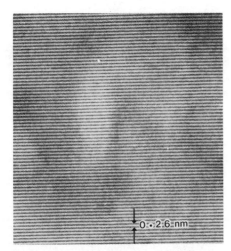

*Lattice image of zinc oxide produced by phase contrast transmission electron microscopy*

**phase contrast** (electron microscopy). Contrast in high-resolution transmission electron microscope images arising from interference effects between the transmitted beam and one or more diffracted beams (Fig. 371).

**phase contrast illumination** (optical microscopy). A special method of controlled illumination ideally suited to observing thin, transparent objects whose structural details vary only slightly in thickness or refractive index. This can also be applied to the examination of opaque materials to determine surface elevation changes.

**phase diagram**. A graphical representation of the temperature and composition limits of phase fields in an alloy or ceramic system as they actually exist under the specific conditions of heating or cooling (Fig. 372). A phase diagram may be an equilibrium diagram, an approximation to an equilibrium diagram, or a representation of metastable conditions or phases. Synonymous with constitution diagram. Compare with *equilibrium diagram*.

**phase rule**. The maximum number of phases (P) that may coexist at equilibrium is two, plus the number of components (C) in the mixture, minus the number of degrees of freedom (F): $P + F = C + 2$.

**phase separation**. The formation of a second liquid portion from a previously homogeneous liquid over time.

**phenolics**. See *Technical Brief 29*.

**phenoxy resin**. A high-molecular-weight thermoplastic polyester resin (polyhydroxy ether) based on bisphenol-A and epichlorohydrin. The material is available in grades suitable for injection molding and extrusion, and for application as coatings and adhesives.

**phenylsilane resins**. Thermosetting copolymers of silicone and phenolic resins. Furnished in solution form.

**phosphate glass**. A glass in which the essential glass former

FIG. 372

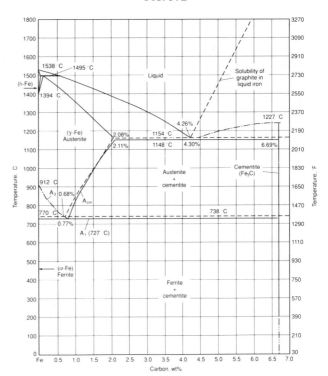

*Iron-carbon phase (equilibrium) diagram up to 6.67 wt% C. Solid lines indicate Fe-Fe₃C diagram; dashed lines indicate iron-graphite diagram*

is phosphorus pentoxide ($P_2O_5$) instead of silica. Also used in *glass-ceramics* and glass tableware.

**phosphating**. Forming an adherent phosphate coating on a metal by immersion in a suitable aqueous phosphate solution. Also called phosphatizing. See also *conversion coating*.

**phosphor**. A substance capable of luminescence, that is, absorbing energy such as x-rays, ultraviolet radiation, electrons, or alpha particles, and emitting a portion of that energy in the visible or near visible region. If emission ceases within $10^{-4}$ s after excitation, the phosphor is fluorescent; if it persists, it is called phosphorescent. See also *fluorescence* and *phosphorescence*.

**phosphor bronzes**. See *bearing bronzes*.

**phosphorescence**. A type of photoluminescence in which the time period between the absorption and re-emission of light is relatively long (of the order of $10^{-4}$ to 10 s or longer). See also *photoluminescence*. Contrast with *fluorescence*.

**phosphorized copper**. General term applied to copper deoxidized with phosphorus. The most commonly used deoxidized copper.

**photoelasticity**. An optical method for evaluating the magnitude and distribution of stresses, using a transparent model

## Technical Brief 29: Phenolics

PHENOLIC MOLDING MATERIALS are high-performance thermosetting plastics. With the heat and pressure of the molding process, thermosetting plastics react to form a cross-linked structure. This structure yields excellent dimensional and thermal stability and superior load-bearing capability at elevated temperatures. Design engineers specify phenolics for close-tolerance precision moldings that must function in hostile environments.

Phenolic resins are products of the condensation reaction of phenol and formaldehyde. Substituted phenols and higher aldehydes may be incorporated to achieve specific properties, such as reactivity and flexibility. A variety of phenolic resins can be produced by adjusting the formaldehyde-to-phenol ratio, the resinification temperature, and the catalyst.

Phenol                Formaldehyde

*Chemical structure of phenolic*

Broadly speaking, two distinct types of resins are produced for use in phenolic molding materials. Single-stage (resole) resins are produced with an alkaline catalyst and a molar excess of formaldehyde. Two-stage (novalac) resins are produced by the acid-catalyzed reaction of phenol and a portion of the required formaldehyde. The resin in the two-stage compound (resole) is capable of further polymerization by the application of heat. In a two-stage compound (novalac), a reactive additive or hardener (formaldehyde supplied as hexamethylenetetramine) causes further polymerization with the application of heat.

Because the unreinforced phenolic polymer is very brittle, a variety of reinforcements and fillers are added to the resin to improve its engineering properties. Usually, a combination of reinforcements and fillers makes up about 45 to 65% of the molding material formulation and imparts specific materials and mold-processing properties. Reinforcement with wood flour or cellulosic fillers yields a molding material with a good balance of properties and cost effectiveness. Glass fiber reinforcement yields substantial improvement in dimensional stability, rigidity, and mechanical properties, including impact strength. Lubricants, colorants, and other modifiers are also used.

of a part, or a thick film of photoelastic material bonded to a real part.

**photoelectric effect.** The liberation of electrons by electromagnetic radiation incident on a substance.

**photoelectric electron-multiplier tube.** See *photomultiplier tube*.

**photoluminescence.** Re-emission of light absorbed by an atom or molecule. The light is emitted in random directions. There are two types of photoluminescence: *fluorescence* and *phosphorescence*.

**photomacrograph.** A *macrograph* produced by photographic means.

**photometer.** A device so designed that it measures the ratio of the radiant power of two electromagnetic beams.

**photomicrograph.** A *micrograph* produced by photographic means.

**photomultiplier tube.** (1) A device in which incident electromagnetic radiation creates electrons by the *photoelectric effect*. These electrons are accelerated by a series of electrodes called dynodes, with secondary emission adding electrons to the stream at each dynode. (2) A light-

sensitive vacuum tube with multiple electrodes providing a highly amplified electrical output.

**photon.** A particle representation of the electromagnetic field. The energy of the photon equals $h\nu$, where $\nu$ is the frequency of the light in hertz, and $h$ is Planck's constant. See also *Planck's constant*.

**photopolymer.** A polymer that changes characteristics when exposed to light of a given frequency.

**photoresist.** (1) A radiation-sensitive material which, when properly applied to a variety of substrates and then properly exposed and developed, masks portions of the substrate with a high degree of integrity. (2) A photosensitive coating that is applied to a laminate and subsequently exposed through a photo tool (film) and developed to create a pattern that can be either plated or etched.

**photosensitive.** Sensitive to light.

**phthalate esters.** The most widely used group of *plasticizers*, produced by the direct action of alcohol on phthalic anhydride. The phthalates are generally characterized by moderate cost, good stability, and good all-around properties.

Industrial applications that have been widely developed for phenolic resins include foundry molds and cores; plywood and particle board; brake and clutch linings; fiberglass, cellulose, and foam insulation; grinding wheels and coated abrasives; adhesives and glues; rubber tackifiers; coatings and varnishes; electrical and decorative laminates; and engineering-gradeplastics. Phenolic-base engineering plastics are used in electrical applications (switchgear, circuit breakers, commutators, etc.), appliance applications (knobs, handles, and heated components for toasters, broilers, and steam irons), and automotive applications (brake caliper pistons, power-assist braking components, water pump housings, etc.).

### Physical, thermal, mechanical, and electrical properties of reinforced phenolic grades

| Property | Cellulose | Mineral | Glass |
|---|---|---|---|
| Specific gravity | 1.35–1.45 | 1.50–1.70 | 1.75–2.10 |
| Shrinkage, mm/mm | 0.006–0.008 | 0.003–0.006 | 0.001–0.003 |
| Water absorption, % | 0.50–0.70 | 0.15–0.40 | 0.05–0.20 |
| UL index, °C (°F) | 150 (300) | 150–170 (300–340) | 150–180 (300–360) |
| Heat-deflection temperature at 1.82 MPa (0.264 ksi), °C (°F) | 165–205 (330–400) | 190–230 (375–450) | 175–260 (350–500) |
| Coefficient of thermal expansion, $10^{-6}$/K | 35–45 | 25–35 | 15–20 |
| Coefficient of thermal conductivity, $W/m^2 \cdot K$ (Btu $\cdot$ in./s $\cdot$ ft$^2$ $\cdot$ °F) | 29–39 (2.0–2.7) | 39–53 (2.7–3.7) | 49–70 (3.4–4.9) |
| Creep modulus at room temperature and 14 MPa (2 ksi), MPa (ksi) | >14 (>2) | >14 (>2) | >17 (>2.5) |
| Rockwell hardness, M | 100–110 | 105–115 | 110–120 |
| Dimensional stability, MIL-M-14, % | 0.50–0.70 | 0.45–0.55 | 0.10–0.25 |
| Compressive strength, MPa (ksi) | 170–205 (25–30) | 170–205 (25–30) | 205–280 (30–40) |
| Flexural strength, MPa (ksi) | 70–90 (10–13) | 70–90 (10–13) | 80–140 (12–20) |
| Elastic modulus, GPa ($10^6$ psi) | 6.9–9.7 (1.0–1.4) | 10.3–17.2 (1.5–2.5) | 17.2–20.7 (2.5–3.0) |
| Izod impact strength, J/m (ft $\cdot$ lbf/in.) | 16–58.7 (0.30–1.10) | 16–24 (0.30–0.45) | 21–800 (0.40–15.0) |
| Dielectric strength, MV/m (V/mil) | | | |
| Short-time | 13.8–15.7 (350–400) | 15.7–16.7 (350–425) | 14.8–16.7 (375–425) |
| Step-by-step | 10.8–13.8 (275–350) | 11.8–15.7 (300–350) | 11.8–14.8 (300–375) |
| Volume resistivity, $\Omega \cdot m$ | $10^{10}$–$10^{11}$ | $10^{10}$–$10^{11}$ | $10^{10}$–$10^{11}$ |

### Recommended Reading

- H.J. Harrington, Phenolics, *Engineered Materials Handbook*, Vol 2, ASM International, 1988, p 242-245
- A. Knop and W. Schieb, *Chemistry and Applications of Phenolic Resins*, Springer-Verlag, 1975

**physical adsorption.** See *physisorption*.

**physical blowing agent.** A gas, such as a fluorocarbon, that is pumped into a mold during the forming of plastics.

**physical crack size ($a_p$).** In fracture mechanics, the distance from a reference plane to the observed crack front. This distance may represent an average of several measurements along the crack front. The reference plane depends on the specimen form, and it is normally taken to be either the boundary or a plane containing either the load line or the centerline of a specimen or plate.

**physical etching.** Development of microstructure through removal of atoms from the surface or lowering the grain-surface potential.

**physical metallurgy.** The science and technology dealing with the properties of metals and alloys, and of the effects of composition, processing, and environment on those properties.

**physical objective aperture.** In electron microscopy, a metal diaphragm centrally pierced with a small hole used to limit the cone of electrons accepted by the objective lens. This improves image contrast, because highly scattered electrons are prevented from arriving at the Gaussian image plane and therefore cannot contribute to background noise.

**physical properties.** Properties of a material that are relatively insensitive to structure and can be measured without the application of force; for example, density, electrical conductivity, coefficient of thermal expansion, magnetic permeability, and lattice parameter. Does not include chemical reactivity. Compare with *mechanical properties*.

**physical testing.** Methods used to determine the entire range of a material's *physical properties*. In addition to density and thermal, electrical, and magnetic properties, physical testing methods may be used to assess simple fundamental physical properties such as color, crystalline form, and melting point.

**physical vapor deposition (PVD).** A coating process whereby the deposition species are transferred and deposited in the form of individual atoms or molecules. The most common PVD methods are sputtering and evaporation. Sputtering, which is the principal PVD process, involves the transport of a material from a source (target) to a

Fig. 373

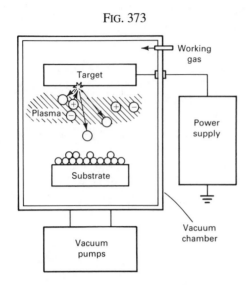

*Schematic of the basic sputtering PVD process*

Fig. 374

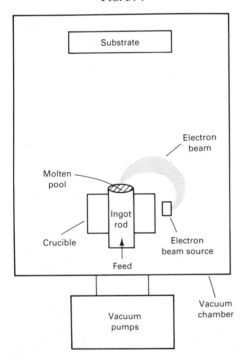

*Schematic of the basic evaporation PVD process*

substrate by means of the bombardment of the target by gas ions that have been accelerated by a high voltage. Atoms from the target are ejected by momentum transfer between the incident ions and the target. These ejected particles move across the vacuum chamber to be deposited on the substrate (Fig. 373). Evaporation, which was the first PVD process used, involves the transfer of material to form a coating by physical means alone, essentially

vaporization. The streaming vapor is generated by melting and evaporating a coating material source bar, by an electron beam in a vacuum chamber (Fig. 374). Because both of these methods are line-of-sight processes, it is necessary to use specially shaped targets or multiple evaporation sources and to rotate or move the substrate uniformly to expose all areas. PVD coatings are used to improve the wear, friction, and hardness properties of cutting tools and as corrosion-resistant coatings.

**physisorption.** The binding of an adsorbate to the surface of a solid by forces whose energy levels approximate those of condensation. Contrast with *chemisorption*.

**PI**. See *polyimide*.

**pi bonding.** Covalent bonding of atoms in which the atomic orbitals overlap along a plane perpendicular to the sigma bond(s) joining the nuclei of two or more atoms. See also *sigma bonding*.

**PIC.** See *pressure-impregnation-carbonization*.

**pick count.** In reinforced plastics, the number of *tows*/mm (in.) of woven fabric in both the *warp* and *fill* directions.

**pickle.** The chemical removal of surface oxides (scale) and other contaminants such as dirt from iron and steel by immersion in an aqueous acid solution (Fig. 375). The most common pickling solutions are sulfuric and hydrochloric acids.

**pickle liquor**. A spent acid-pickling bath.

**pickle patch**. A tightly adhering oxide or scale coating not properly removed during *pickling*.

**pickle stain**. Discoloration of metal due to chemical cleaning without adequate washing and drying.

**pickling**. Removing surface oxides from metals by chemical or electrochemical reaction.

**pickoff**. An automatic device for removing a finished part from the press die after it has been stripped.

**pickup**. (1) Transfer of metal from tools to part or from part to tools during a forming operation. (2) Small particles of oxidized metal adhering to the surface of a *mill product*.

**pick-up roll**. A spreading device where the roll for picking up the adhesive runs in a reservoir of adhesive.

**Pidgeon process**. A process for production of magnesium by reduction of magnesium oxide with ferrosilicon.

**pi electron**. An electron that participates in *pi bonding*.

**piercing**. The general term for cutting (shearing or punching) openings, such as holes and slots, in sheet material, plate, or parts. This operation is similar to *blanking*; the difference is that the slug or pierce produced by piercing is scrap, while the blank produced by blanking is the useful part.

**piezoelectric**. A material or crystal which becomes polarized and its surface becomes charged when a stress is applied to it. Conversely, if the material is subject to an electric field, it will expand in one direction and contract in another.

Fig. 375

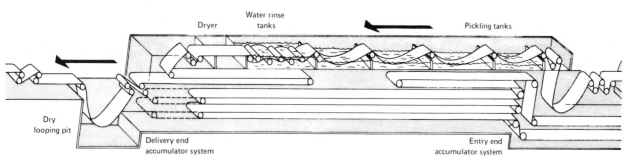

*Schematic of a modern continuous strip pickling line*

**piezoelectric effect**. The reversible interaction, exhibited by some crystalline materials, between an elastic strain and an electric field. The direction of the strain depends on the polarity of the field or vice versa. Compare with *electrostrictive effect*.

**piezoelectric polymers**. Polymers that spontaneously give an electric charge when mechanically stressed or that develop a mechanical response when an electric field is applied. Used as transducers or acoustic sensors.

**pig**. A metal casting used in remelting.

**pig iron**. (1) High-carbon iron made by reduction of iron ore in the blast furnace. (2) Cast iron in the form of *pigs*.

**Pilger tube-reducing process**. See *tube reducing*.

**pilot arc**. A low current continuous arc between the electrode and the constricting nozzle of a plasma arc welding torch used to ionize the gas and facilitate the start of the main welding arc.

**pimple**. An imperfection, such as a small protuberance of varied shape on the surface of a plastic product.

**pin** (for bend testing). The plunger or tool used in making semiguided, guided, or wraparound bend tests to apply the bending force to the inside surface of the bend. In free bends or semiguided bends to an angle of 180°, a shim or block of the proper thickness may be placed between the legs of the specimen as bending is completed. This shim or block is also referred to as a pin or mandrel. See also *mandrel* (metals).

**pinchers**. Surface disturbances on metal sheet or strip that result from rolling processes and that ordinarily appear as fernlike ripples running diagonally to the direction of rolling.

**pinch-off**. In blow molding of plastics, a raised edge around the cavity in the mold that seals off the part and separates the excess material as the mold closes around the *parison*.

**pinch pass**. A pass of sheet metal through rolls to effect a very small reduction in thickness.

**pinch trimming**. The trimming of the edge of a tubular metal part or shell by pushing or pinching the flange or lip over the cutting edge of a stationary punch or over the cutting edge of a draw punch.

**pincushion distortion**. See *positive distortion*.

**pine-tree crystal**. A type of *dendrite*.

**pin expansion test**. A test for determining the ability of a tube to be expanded or for revealing the presence of cracks or other longitudinal weaknesses, made by forcing a tapered pin into the open end of the tube.

**pinhead blister**. See *blister*.

**pinhole eyepiece**. An eyepiece in an *optical microscope*, or a cap to place over an eyepiece, that has a small central aperture instead of an eye lens. It is used in adjusting or aligning microscopes.

**pinhole ocular**. See *pinhole eyepiece*.

**pinhole porosity**. Porosity consisting of numerous small gas holes distributed throughout a metal; found in weld metal, castings, and electrodeposited metal.

**pinholes**. (1) Very small holes that are sometimes found as a type of porosity in a casting because of the microshrinkage or gas evolution during solidification. In wrought products, due to removal of inclusions or microconstituents during macroetching of transverse sections. (2) Small cavities that penetrate the surface of a cured composite or plastic part. (3) In photography, a very small circular aperture.

**pinion**. The smaller of two mating gears.

**pin-on-disk machine**. A tribometer in which one or more

Fig. 376

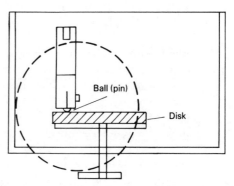

*Pin-on-disk machine test configuration*

relatively moving styli (that is, the "pin" specimen) is loaded against a flat disk specimen surface such that the direction of loading is parallel to the axis of rotation of either the disk or the pin-holding shaft, and a circular wear path is described by the pin motion (Fig. 376). The typical pin-on-disk arrangement resembles that of a traditional phonograph. Either the disk rotates or the pin specimen holder rotates so as to produce a circular path on the disk surface. An arrangement wherein the pin specimen is loaded against the curved circumferential surface of a flat disk is not generally considered to be a pin-on-disk machine.

**Piobert lines**. See *Lüders lines*.

**pipe**. (1) The central cavity formed by contraction in metal, especially ingots (Fig. 377), during solidification. (2) An imperfection in wrought or cast products resulting from such a cavity. (3) A tubular metal product, cast or wrought. See also *extrusion pipe*.

FIG. 377

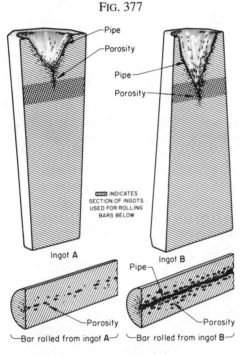

*Longitudinal sections of two types of ingots showing typical pipe and porosity*

**pipet**. A tube, usually made of glass or plastic, used almost exclusively to deliver accurately known volumes of liquids or solutions during titration or volumetric analysis.

**pipe tap**. A *tap* for making internal *pipe threads* within pipe fittings or holes.

**pipe threads**. Internal or external machine threads, usually tapered, of a design intended for making pressure-tight mechanical joints in piping systems.

**Pirani gage**. An instrument used to measure the pressure inside a vacuum chamber. The gage measures electrical resistance in a wire filament which will change in temperature depending on atmospheric pressure.

**piston-pin bearing**. See *little-end bearing*.

**pit**. A small, regular or irregular crater in the surface of a material created by exposure to the environment, for example, corrosion, wear, or thermal cycling (Fig. 378). See also *pitting*.

FIG. 378

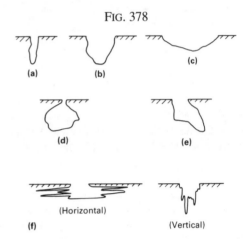

*Variations in the cross-sectional shape of pits. (a) Narrow and deep. (b) Elliptical. (c) Wide and shallow. (d) Subsurface. (e) Undercutting. (f) Shapes determined by microstructural orientation*

**pitch**. A high molecular weight material that is a residue from the destructive distillation of coal and petroleum products. Pitches are used as base materials for the manufacture of certain high-modulus carbon fibers.

**pit molding**. Molding method in which the *drag* is made in a pit or hole in the floor.

**Pitot tube**. An instrument that measures the stagnation pressure of a flowing fluid, consisting of an open tube pointing into the fluid and connected to a pressure-indicating device. Also known as *impact tube*.

**pitting**. (1) Forming small sharp cavities in a surface by corrosion, wear, or other mechanically assisted degradation. (2) *Localized corrosion* of a metal surface, confined to a point or small area, that takes the form of cavities (Fig. 379).

**pitting factor**. Ratio of the depth of the deepest pit resulting from corrosion divided by the average penetration calculated from weight loss (Fig. 380).

**pivot bearing**. An axial-load, radial-load type bearing that supports the end of a rotating shaft or pivot.

**pivoted-pad bearing**. See *tilting-pad bearing*.

**pixel**. Shortened term for picture element. A pixel is the smallest displayable element of a digitized image; a single value of the image matrix. A pixel corresponds to the measurement of a volume element (*voxel*) in the object.

FIG. 379

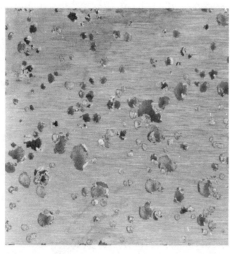

(a)

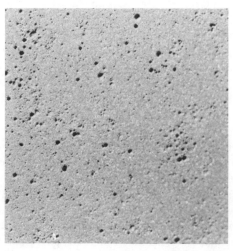

(b)

*Examples of deep pits (a) and shallow pits (b) resulting from pitting corrosion of metallic specimens*

FIG. 380

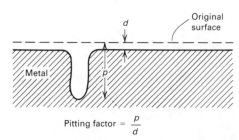

Pitting factor = $\dfrac{p}{d}$

*Schematic illustrating the pitting factor p/d*

**plain bearing**. Any simple sliding-type bearing, as distinct from pad- or rolling-type bearings.

**plain journal bearing**. A plain bearing in which the rela-

FIG. 381

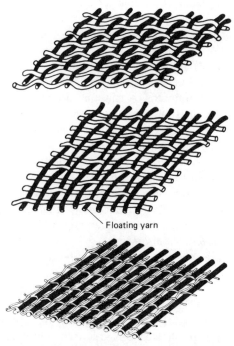

*Comparison of plain fabric weave (top), five-harness satin fabric weave (center), and uni-directional weave (bottom)*

tively sliding surfaces are cylindrical and in which there is relative angular motion. One surface is usually stationary and the force acts perpendicularly to the axis of rotation.

**plain thrust bearing**. A plain bearing of the axial-load type, with or without grooves.

**plain weave**. A weaving pattern in which the warp and fill fibers alternate; that is, the repeat pattern is warp/fill/warp/fill, and so on (Fig. 381). The two sides, or faces, of a plain weave are identical. Properties are significantly reduced relative to a weaving pattern with fewer crossovers. See also the figures accompanying the terms *basketweave* (composites), *crowfoot weave*, and *leno weave*.

**planar**. Lying essentially in a single plane.

**planar anisotropy**. A variation in physical and/or mechanical properties with respect to direction within the plane of material in sheet form. See also *plastic strain ratio*.

**planar helix winding**. In *filament winding* of composites, a winding in which the filament path on each dome lies on a plane that intersects the dome, while a helical path over the cylindrical section is connected to the dome paths. See also the figure accompanying the term *helical winding*.

**planar winding**. In *filament winding* of composites, a winding in which the filament path lies on a plane that intersects the winding surface. See also *polar winding*.

**planchet**. A metal disk with milled edges, ready for coining.

**Planck's constant ($h$)**. A fundamental physical constant, the elementary quantum of action; the ratio of the energy of a photon to its frequency, it is equal to $6.62620 \pm 0.00005 \times 10^{-34}$ J · s.

**plane** (crystal). An idiomorphic face of a crystal. Any atom-containing plane in a crystal.

**plane glass illuminator**. A thin, transparent, flat glass disk interposed in a microscope or a lens imaging system to direct light to the object without reducing the useful aperture of the lens system.

**plane grating**. In materials characterization, an optical component used to disperse light into its component wavelengths by diffraction off a series of finely spaced, equidistant ridges. A plane grating has a flat substrate. See also *concave grating*, *diffraction grating*, *reflection grating*, and *transmission grating*.

**plane strain**. The stress condition in *linear elastic fracture mechanics* in which there is zero strain in a direction normal to both the axis of applied tensile stress and the direction of crack growth (that is, parallel to the crack front); most nearly achieved in loading thick plates along a direction parallel to the plate surface. Under plane-strain conditions, the plane of fracture instability is normal to the axis of the principal tensile stress.

**plane-strain fracture toughness ($K_{Ic}$)**. The crack extension resistance under conditions of *crack-tip plane strain*. See also *stress-intensity factor*.

**plane stress**. The stress condition in *linear elastic fracture mechanics* in which the stress in the thickness direction is zero; most nearly achieved in loading very thin sheet along a direction parallel to the surface of the sheet. Under plane-stress conditions, the plane of fracture instability is inclined 45° to the axis of the principal tensile stress.

**plane-stress fracture toughness ($K_c$)**. In *linear elastic fracture mechanics*, the value of the crack-extension resistance at the instability condition determined from the tangency between the *R-curve* and the critical crack-extension force curve of the specimen. See also *stress-intensity factor*.

**planimetric method**. A method of measuring grain size in which the grains within a definite area are counted. See also *Jeffries' method*.

**planing**. Producing flat surfaces by linear reciprocal motion of work and the table to which it is attached, relative to a stationary single-point cutting tool.

**planishing**. Producing a smooth finish on metal by a rapid succession of blows delivered by highly polished dies or by a hammer designed for the purpose, or by rolling in a planishing mill.

**plasma**. A gas of sufficient energy so that a large fraction of the species present is ionized and thus conducts electricity. Plasmas may be generated by the passage of a current between electrodes, by induction, or by a combination of these methods.

**plasma-arc cutting**. An arc cutting process that severs metals by melting a localized area with heat from a constricted arc and removing the molten metal with a high-velocity jet of hot, ionized gas issuing from the plasma torch. See also the figure accompanying the term *cutting torch* (plasma arc).

**plasma arc welding (PAW)**. An arc welding process that produces coalescence of metals by heating them with a constricted arc between an electrode and the workpiece (transferred arc) or the electrode and the constricting nozzle (nontransferred arc). Shielding is obtained from hot, ionized gas issuing from an orifice surrounding the electrode and may be supplemented by an auxiliary source of shielding gas, which may be an inert gas or a mixture of

FIG. 382

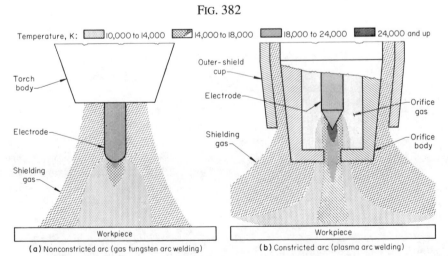

Temperature, K: ☐ 10,000 to 14,000  ▨ 14,000 to 18,000  ▨ 18,000 to 24,000  ▨ 24,000 and up

Torch body

Electrode

Shielding gas

(a) Nonconstricted arc (gas tungsten arc welding)

Outer-shield cup

Electrode

Shielding gas

Orifice gas

Orifice body

Workpiece

(b) Constricted arc (plasma arc welding)

*Comparison of a nonconstricted arc used for GTAW and a constricted arc used for PAW*

gases (Fig. 382). Pressure may or may not be used, and filler metal may or may not be supplied. See also *nontransferred arc*, *transferred arc*, and *welding torch* (arc).

**plasma carburizing**. Same as *ion carburizing*.

**plasma flame** (welding). The zone of intense heat and light emanating from the orifice of the arc chamber resulting from energy liberated as the charged gas particles (ions) recombine with electrons. See also the figure accompanying the term *plasma arc welding*.

**plasma-forming gas**. The gas, in the plasma torch, which is heated to the high-temperature plasma state by the electric arc. See also *welding torch* (arc).

**plasma ion deposition**. See *ion implantation*.

**plasma-jet excitation**. In materials characterization, the use of a high-temperature plasma jet to excite an element in a sample, for example, for inductively coupled plasma atomic emission spectroscopy. Also known as dc plasma excitation.

**plasma metallizing**. See preferred term *plasma spraying*.

**plasma nitriding**. Same as *ion nitriding*.

**plasma spraying**. A *thermal spraying* process in which the coating material is melted with heat from a plasma torch that generates a nontransferred arc; molten powder coating material is propelled against the base metal by the hot, ionized gas issuing from the torch (Fig. 383).

FIG. 383

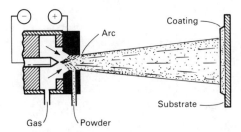

*Schematic of the plasma spray process*

**plasmon**. A quantum of a collective longitudinal wave in the electron gas of a solid.

**plaster molding**. Molding in which a gypsum-bonded aggregate flour in the form of a water slurry is poured over a pattern, permitted to harden, and, after removal of the pattern, thoroughly dried. This technique is used to make smooth nonferrous castings of accurate size.

**plastic**. A material that contains as an essential ingredient an organic polymer of large molecular weight; is solid in its finished state; and, at some stage in its manufacture or its processing into finished articles, can be shaped by flow. Although materials such as rubber, textiles, adhesives, and paint may in some cases meet this definition, they are not considered plastics. The terms plastic, resin, and polymer are somewhat synonymous, but the terms resins and polymers most often denote the basic material as polymerized,

while the term plastic encompasses compounds containing plasticizers, stabilizers, fillers, and other additives.

**plastic deformation**. The permanent (inelastic) distortion of materials under applied stresses that strain the material beyond its *elastic limit*.

**plastic flow** (metals). The phenomenon that takes place when metals are stretched or compressed permanently without rupture.

**plastic flow** (plastics). (1) Deformation of plastics under the action of a sustained hot or cold force. (2) Flow of semisolids in the molding of plastics.

**plastic foam**. See preferred term *cellular plastic*.

**plastic instability**. The stage of deformation in a tensile test where the plastic flow becomes nonuniform and *necking* begins.

**plasticity**. The property of a material which allows it to be repeatedly deformed without rupture when acted upon by a force sufficient to cause deformation and which allows it to retain its shape after the applied force has been removed.

**plasticizer**. (1) A material incorporated in a plastic to increase its workability, flexibility, or distensibility. Normally used in thermoplastics. (2) A material added to a plastic (or polymer) of lower molecular weight to reduce stiffness and brittleness, resulting in a lower glass transition temperature for the polymer. (3) A material added to an adhesive to cause a reduction in melt viscosity, lower the temperature of the second-order transition, or lower the elastic modulus of the solidified adhesive. See also *modifier*.

**plastic memory**. The tendency of a thermoplastic material that has been stretched while hot to return to its unstretched shape upon being reheated.

**plastic replica**. In fractography and metallography, a reproduction in plastic of the surface to be studied. It is prepared by evaporation of the solvent from a solution of plastic, polymerization of a monomer, or solidification of a plastic on the surface.

**plastic strain**. Dimensional change that does not disappear when the initiating stress is removed. Usually accompanied by some *elastic deformation*.

**plastic-strain ratio (*r*-value)**. In formability testing of metals, the ratio of the true width strain to the true thickness strain in a sheet tensile test, $r = \varepsilon_w \varepsilon_t$. A formability parameter that relates to drawing, it is also known as the anisotropy factor. A high *r*-value indicates a material with good drawing properties.

**plastic-zone adjustment (*r*$_Y$)**. In *linear elastic fracture mechanics*, an addition to the *physical crack size* to account for plastic crack-tip deformation enclosed by a linear-elastic stress field. See also *crack-extension resistance (K$_R$)*.

**plastigel**. A *plastisol* exhibiting gel-like flow properties. One having an effective yield value.

**plastisols**. Mixtures of vinyl resins and plasticizers that can be molded, cast, or converted to continuous films by the application of heat. If the mixtures contain volatile thinners as well, they are known as *organosols*.

**plastohydrodynamic lubrication**. A condition of lubrication in which the friction and film thickness between two bodies in relative motion are determined by plastic deformation of the bodies in combination with the viscous properties of the lubricant at the prevailing pressure, temperature, and rate of shear. Compare with *elastohydrodynamic lubrication*.

**plastometer**. An instrument for determining the flow properties of a thermoplastic resin by forcing the molten resin through a die or orifice of specific size at a specified temperature and pressure.

**plate**. A flat-rolled metal product of some minimum thickness and width arbitrarily dependent on the type of metal. Plate thicknesses commonly range from 6 to 300 mm (0.25 to 12 in.); widths from 200 to 2000 mm (8 to 80 in.).

**plate glass**. Flat glass formed by a rolling process, ground and polished on both sides, with surfaces essentially plane and parallel.

**platelet alpha structure**. In titanium alloys, *acicular alpha* of a coarser variety, usually with low aspect ratios (Fig. 384). This microstructure arises from cooling $\alpha$ or $\alpha$-$\beta$ alloys from temperatures at which a significant fraction of $\beta$ phase exists.

**platelets**. Flat particles of metal powder having considerable thickness. The thickness, however, is smaller when compared with the length and widths of the particles. See also *particle shape*.

**plate martensite**. Martensite formed partly in steel containing more than approximately 0.5% C and solely in steel containing more than approximately 1.0% C that appears as lenticular-shape plates (crystals) (Fig. 385).

FIG. 384

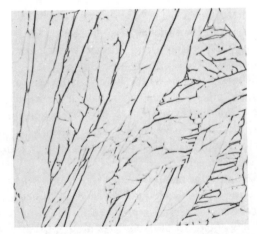

*Hot-worked titanium alloy with coarse, platelet alpha structure. 100×*

FIG. 385

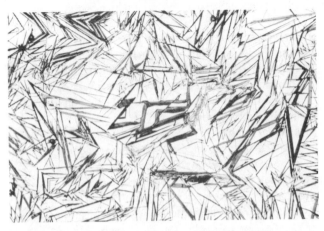

*Plate martensite in an Fe-1.4C alloy. 200×*

**platen**. (1) The sliding member, *slide*, or *ram* of a metal forming press. (2) The mounting plates of a plastic forming press, to which the entire mold assembly is bolted. (3) A part of a resistance welding, mechanical testing, or other machine with a flat surface to which dies, fixtures, backups, or electrode holders are attached and that transmits pressure or force.

**platen force**. The force available at the movable platen in a resistance welding machine that causes upsetting in flash or upset welding.

**plating**. Forming an adherent layer of metal on an object; often used as a shop term for *electroplating*. See also *electrodeposition* and *electroless plating*.

**plating rack**. A fixture used to hold work and conduct current to it during *electroplating*.

**plating range**. The current-density range over which a satisfactory electroplate can be deposited.

**platinum black**. A finely divided form of platinum of a dull black color, usually, but not necessarily produced by reduction of salts in an aqueous solution.

**plenum**. The space between the inside wall of the constricting nozzle and the electrode in a plasma arc torch used for welding, cutting, or thermal spraying (Fig. 386).

**plied yarn**. Yarn for reinforced plastics made by collecting two or more single yarns. Normally, the yarns are twisted together, though sometimes they are collected without twist.

**plowing**. In *tribology*, the formation of grooves by *plastic deformation* of the softer of two surfaces in relative motion.

**plug**. (1) A rod or mandrel over which a pierced tube is forced. (2) A rod or mandrel that fills a tube as it is drawn through a die. (3) A punch or mandrel over which a cup is drawn. (4) A protruding portion of a die impression for forming a corresponding recess in the forging. (5) A false bottom in a die.

FIG. 386

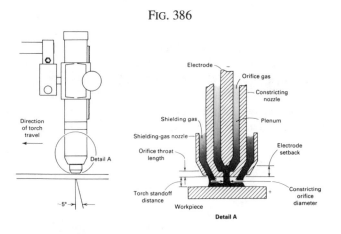

*Components (including the plenum) of a plasma arc cutting torch*

FIG. 387

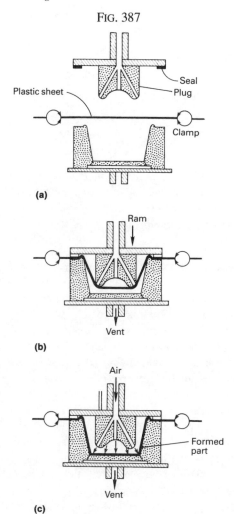

*Plug-assist thermoforming. (a) Heat-softened plastic sheet clamped in position above female mold. (b) Plug stretches sheet, pushing it into female mold. (c) Vacuum pulls sheet off the plug and into the female mold*

FIG. 388

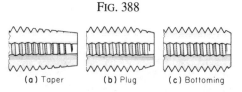

*Chamfers for solid taps*

**plug-assist forming**. A *thermoforming* process in which a plug or male mold is used to partially preform the plastic part before forming is completed using vacuum or pressure (Fig. 387).

**plug tap**. A *tap* with *chamfer* extending from three to five threads (Fig. 388).

**plug weld**. A circular weld made through a hole in one member of a lap or tee joint. Neither a fillet-welded hole nor a spot weld is to be construed as a plug weld. The hole may be partially or completely filled with weld metal.

**plumbage**. A special quality of powdered graphite used to coat molds and, in a mixture of clay, to make crucibles.

**plunge grinding**. *Grinding* wherein the only relative motion of the wheel is radially toward the work.

**plunger**. Ram or piston that forces molten metal into a die in a *die casting* machine. Plunger machines are those having a plunger in continuous contact with molten metal. See also the figures accompanying the terms *cold chamber machine* and *hot chamber machine*.

**plus mesh**. The powder sample retained on a screen of stated size, identified by the retaining mesh number. See also *sieve analysis* and *sieve classification*.

**plus sieve**. The portion of a sample of a granular substance (such as metal powder) retained on a standard sieve of specified number. Contrast with *minus sieve*. See also *sieve analysis* and *sieve classification*.

**ply**. A single layer in a *laminate*. In general, fabrics or felts consisting of one or more layers. Yarn resulting from a twisting operation (for example, three-ply yarn consists of three strands of yarn twisted together). In *filament winding*, a ply is a single pass (two plies forming one layer).

**plymetal**. Sheet consisting of bonded layers of dissimilar metals.

**plywood**. A cross-bonded assembly made of layers of veneer or veneer in combination with a lumber core or plies joined with an adhesive. Two types of plywood are recognized, veneer plywood and lumber core plywood.

**P/M**. The acronym for *powder metallurgy*.

**PMMA**. See *polymethyl methacrylate*.

**PMR polyimides**. A novel class of high temperature resistant polymers. PMR represents *in situ* polymerization of monomer reactants. See also *polyimide (PI)*.

**pneumatic press**. A press that uses air or a gas to deliver the pressure to the upper and lower rams.

**pocket.** In a *rolling-element bearing*, the portion of the case that is shaped to receive the rolling element. Compare with *oil pocket*.

**pocket-thrust bearing.** An externally pressurized thrust bearing having three or more hydrostatic pads with central relieved chambers of pockets supplied with pressurized oil.

**point angle.** In general, the angle at the point of a cutting tool. Most commonly, the included angle at the point of a twist drill, the general-purpose angle being 118° (Fig. 389).

FIG. 389

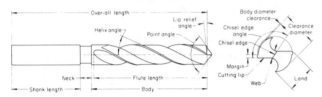

*Design features (including point angle) of a typical straight-shank twist drill*

**point estimate.** The estimate of a *parameter* given by a single statistic.

**pointing.** (1) Reducing the diameter of wire, rod, or tubing over a short length at the end by swaging or hammer forging, turning, or squeezing to facilitate entry into a drawing die and gripping in the drawhead. (2) The operation in automatic machines of chamfering or rounding the threaded end or the head of a bolt.

**poise (P).** The centimeter-gram-second (cgs) unit of dynamic viscosity (1 P = 0.1 Pa · s).

**Poisseuille (Pl).** The meter-kilogram-second (mks) or Système International d'Unités (SI) unit of dynamic viscosity. (1 Pl = 10 P).

**Poisson's ratio (ν).** The absolute value of the ratio of transverse (lateral) strain to the corresponding axial strain resulting from uniformly distributed axial stress below the *proportional limit* of the material.

**poke welding.** Same as *push welding*.

**polar bond.** See *ionic bond*.

**polarimeter.** An instrument used to determine the rotation of the plane of polarization of plane polarized light when it passes through a substance; the light is linearly polarized by a polarizer (such as a *Nicol prism*), passes through the material being analyzed, and then passes through an analyzer.

**polariscope.** Any of several instruments used to determine the effects of substances on polarized light, in which linearly or elliptically polarized light passes through the substance being studied, and then through an analyzer.

**polarity** (welding). See *direct current electrode negative*, *direct current electrode positive*, *straight polarity*, and *reverse polarity*.

**polarization.** (1) The change from the open-circuit electrode potential as the result of the passage of current. (2) A change in the *potential* of an electrode during electrolysis, such that the potential of an *anode* becomes more noble, and that of a *cathode* more active, than their respective reversible potentials. Often accomplished by formation of a film on the electrode surface.

**polarization admittance.** The reciprocal of *polarization resistance (di/dE)*.

**polarization curve.** A plot of *current density* versus *electrode potential* for a specific electrode-electrolyte combination (Fig. 390).

FIG. 390

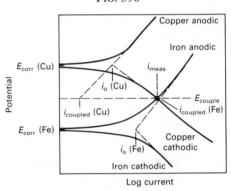

*Polarization curve for the prediction of coupled potential and galvanic current for the Fe-Cu system. i, current; $i_o$, exchange current; $E_{corr}$, corrosion potential*

**polarization resistance.** The slope (*dE/di*) at the *corrosion potential* of a potential (*E*)/current density (*i*) curve (Fig. 390). Also used to describe the method of measuring corrosion rates using this slope.

**polarized light illumination.** A method of illumination in which the incident light is plane polarized before it impinges on the specimen (Fig. 391). See also the figure accompanying the term *optical microscope*.

**polarizer.** In an optical microscope, a *Nicol prism*, polarizing film, or similar device into which normal light passes and from which polarized light emerges (Fig. 392).

**polarizing element.** A general term for a device for producing or analyzing plane-polarized light. It may be a *Nicol prism*, some other form of calcite prism, a reflecting surface, or a polarizing filter. See also *polarizer*.

**polarography.** An electroanalytical technique in which the current between a dropping mercury electrode (DME) and a counterelectrode (both of which are immersed in electrolyte) is measured as a function of the potential difference between the DME and a reference electrode.

**polar winding.** In filament winding of composites, a winding in which the filament path passes tangent to the polar

FIG. 391

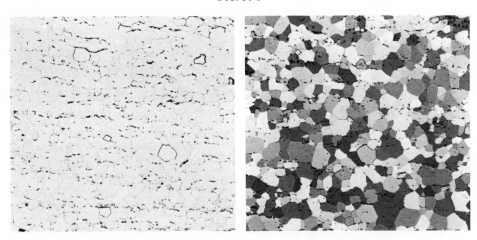

*Use of polarized light for delineation of grain boundaries in polycrystalline zirconium. Left: bright-field illlumination. Right: crossed polarized light. Both at 100×*

FIG. 392

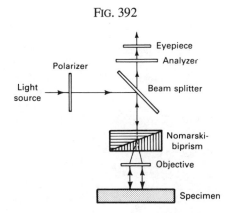

*Placement of the polarizer in an optical microscope with differential interference contrast illumination*

FIG. 393

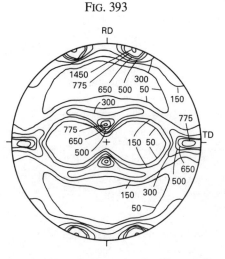

*Measured (111) pole figure for Cu-3Zn. Numbers are arbitrary units denoting x-ray diffraction intensity. RD, rolling direction; TD, transverse direction*

opening at one end of the chamber and tangent to the opposite side of the polar opening at the other end. A one-circuit pattern is inherent in the system. See also the figure accompanying the term *helical winding*.

**pole**. (1) A means of designating the orientation of a crystal plane by stereographically plotting its normal. For example, the north pole defines the equatorial plane. Either of the two regions of a permanent magnet or electromagnet where most of the lines of induction enter or leave.

**pole figure**. A stereoscopic projection of a polycrystalline aggregate showing the distribution of poles, or plane normals, of a specific crystalline plane, using specimen axes as reference axes (Fig. 393). Pole figures are used to characterize preferred orientation in polycrystalline materials.

**polepiece**. In reinforced plastics, the supporting part of the mandrel used in filament winding, usually on one of the axes of rotation.

**poling**. A step in the fire refining of copper to reduce the oxygen content to tolerable limits by covering the bath with coal or coke and thrusting green wood poles below the surface. There is a vigorous evolution of reducing gases, which combine with the oxygen contained in the metal.

**polished surface**. A surface prepared for metallographic inspection that reflects a large proportion of the incident light in a specular manner. See also *polishing* (4).

**polishing.** (1) A surface-finishing process for ceramics and metals utilizing successive grades of abrasive. (2) Smoothing metal surfaces, often to a high luster, by rubbing the surface with a fine abrasive, usually contained in a cloth or other soft lap. Results in microscopic flow of some surface metal together with actual removal of a small amount of surface metal. (3) Removal of material by the action of abrasive grains carried to the work by a flexible support, generally either a wheel or a coated abrasive belt (Fig. 394). (4) A mechanical, chemical, or electrolytic process or combination thereof used to prepare a smooth, reflective surface suitable for microstructural examination that is free of artifacts or damage introduced during prior

FIG. 394

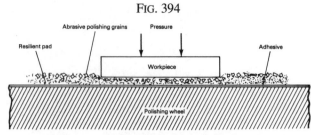

*Schematic showing abrasive grains held in a soft pad for a mechanical polishing application*

sectioning or grinding. See also *electrolytic polishing* and *electropolishing.*

---

## Technical Brief 30: Polyamides

POLYAMIDE (PA) RESINS are thermoplastic polymers that are most commonly regarded as being synonymous with nylons, that is, synthetic polymers that contain an amide group, –CONH–, as a recurring part of the chain.

Nylons are made from diamines and dibasic acids, ω-amino acids, or lactams. Nylons are commonly identified by numbers corresponding to the number of C-atoms in the monomers. Thus, the resins listed above are known as nylon 6/6, nylon 11, and nylon 6, respectively. The molecular weights of nylons range from 11,000 to 34,000. They are usually semicrystalline polymers with melting points in the range of 175 to 300 °C (350 to 570 °F).

Nylon 6/6 and nylon 6 are the most important commercial products. Their melting points are 269 and 228 °C (516 and 440 °F), respectively. Other commercial nylons are 6/9, 6/10, 6/12, 11, 12, and 4/6. The more C-atoms, that is, the lower the concentration of amide groups, the lower the melting point. Nylons are very readily

$$n - H_2N(CH_2)_6NH_2 + n - HOOC(CH_2)_4COOH \rightleftharpoons H[NH(CH_2)_6NHCO(CH_2)_4CO]_nOH + (2n - 1)H_2O$$

hexamethylene      adipic acid      poly(hexamethylene adipamide)
diamine

**(a)**

$$n - H_2N(CH_2)_{10}COOH \rightleftharpoons H[NH(CH_2)_{10}CO]_nOH + (n - 1)H_2O$$

11-aminoundecanoic acid    poly(11-aminoundecanoic acid)

**(b)**

$$n - \begin{array}{c} CH_2 \\ CH_2 \quad CH_2 \\ CH_2 \quad CO \\ CH_2 - NH \end{array} + H_2O \rightleftharpoons H[NH(CH_2)_5CO]_nOH$$

caprolactam      polycaprolactam

**(c)**

*Constituents of nylons. (a) Nylon 6/6 (diamines and dibasic acids). (b) Nylon 11 (ω-amino acids). (c) Nylon 6 (lactams)*

modified by use of monomer mixtures leading to copolymers. These are normally less crystalline, more flexible, and more soluble than the previously mentioned homopolymers.

Additives are used in nylons to improve thermal and photolytic stability, facilitate processing, increase flammability resistance, increase lubricity, and generally improve whatever specific property is required for a specific application. Susceptibility to modification is an important asset. Fiber and mineral reinforcements are widely used. Blending with elastomeric modifiers has yielded nylons with improved toughness (impact strength).

Molded and extruded nylons are used in virtually every industry and market. Transportation represents the

---

**polishing artifact**. A false structure introduced during a polishing stage of a surface-preparation sequence.

**polishing rate**. The rate at which material is removed from a surface during polishing. It is usually expressed in terms of the thickness removed per unit of time or distance traversed.

**polishing wear**. An extremely mild form of wear for which the mechanism has not been clearly identified, but that may involve extremely fine-scale abrasion, plastic smearing of micro-asperities, and/or tribochemical material removal.

**polyacrylate**. A thermoplastic resin made by the polymerization of an acrylic compound.

**polyacrylonitrile (PAN)**. A base material or precursor used in the manufacture of certain carbon fibers. PAN-based carbon fibers have a ribbonlike structure (Fig. 395) and possess high strength (400 GPa, or $10^6$ psi, tensile modulus).

**polyamide-imide (PAI) resins**. A family of polymers based on the combination of trimellitic anhydride with aromatic diamines (Fig. 396). In the uncured form (ortho-amic acid) the polymers are soluble in polar organic solvents. The imide linkage is formed by heating, producing an infusible resin with thermal stability up to 260 °C (500 °F). These resins, which also include graphite- (powder and fiber) and glass fiber-reinforced grades, are used in the automotive

largest single market for nylons (~30% of the market). Applications for unreinforced materials include electrical connectors, wire jackets, and light-duty gears for windshield wipers and speedometers. Toughened nylons are used as stone shields and trim clips. Glass-reinforced nylons are used for engine fans, radiator headers, brake and power-steering fluid reservoirs, as well as other uses. Mineral-reinforced resins are used for mirror housings. Nylons containing both glass and minerals are used as fender extensions. Nylons for electrical and electronic applications constitute about 11% of the market. Applications include plugs, connectors, relays, and antenna-mounting devices. Because of their excellent resistance to fatigue and repeated impact, nylons are widely used in industrial products such as unlubricated gears, bearings, and antifriction parts. Consumer products include toughened nylons for ski boots, ice and roller skate supports, racket sports equipment, kitchen utensils, and toys. Extruded nylon film is widely used for packaging meats and cheeses and for cook-in bags and pouches. Other applications include wire and cable jacketing, tubing used to convey fluids, and nylon filaments for paintbrush bristles, fishing lines, and sewing thread.

## Mechanical property values for nylons 6 and 6/6
Dry, as-molded, approximately 0.2% moisture content

| Property | Nylon 6 | | Nylon 6/6 | | Toughened | |
| | Molding and extrusion compound | Glass fiber reinforced, 30–35% | Molding compound | Glass fiber reinforced, 30–33% | Unreinforced | Glass fiber reinforced, 33% |
|---|---|---|---|---|---|---|
| **Mechanical** | | | | | | |
| Tensile strength at break, MPa (ksi) | ... | 165 (24) | 94.5 (13.7) | 193 (28) | 50 (7.0) | 140 (20.3) |
| Elongation at break, % | 30–100 | 2.2–3.6 | 15–60 | 2.5–34 | 125 | 4–6 |
| Tensile yield strength, MPa (ksi) | 80.7 (11.7) | ... | 55 (8.00) | 170 (25) | ... | ... |
| Compressive strength, rupture or yield, MPa (ksi) | 90–110 (13–16) | 131–165 (19–24) | 86.2–103 (12.5–15.0)(a) | 165–276 (24–40) | ... | 103 (15) |
| Flexural strength, rupture, or yield, MPa (ksi) | 108 (15.7) | 240 (35) | 114–117 (16.5–17.0) | 283 (41) | 59 (8.5) | 206 (29.9) |
| Tensile modulus, GPa ($10^6$ psi) | 2.6 (0.38) | 8.62–10.0 (1.25–1.45) | 1.59–3.79 (0.23–0.55) | 9.0 (1.3) | ... | ... |
| Flexural modulus at 23 °C (73 °F), GPa ($10^6$ psi) | 2.7 (0.39) | 9.65 (1.40) | 2.8–3.1 (0.41–0.45) | 8.96–10.0 (1.30–1.45) | 1.65 (0.240) | 7.58 (1.10) |
| Izod impact, 3.2 mm (⅛ in.) thick specimen, notched, J/m (ft · lbf/in.) | 32–53 (0.6–1.0) | 117–181 (2.2–3.4) | 29–53 (0.55–1.0) | 85–240 (1.6–4.5) | 907 (17.0) | 240 (4.5) |
| Rockwell hardness | R119 | M93–96 | R120 | R101–119 | R100 | R107 |

(a) Yield

## Recommended Reading

- M.I. Kohan, Polyamides (PA), *Engineered Materials Handbook*, Vol 2, ASM International, 1988, p 124-127
- M.I. Kohan, *Nylon Plastics*, Wiley-Interscience, 1973

FIG. 395

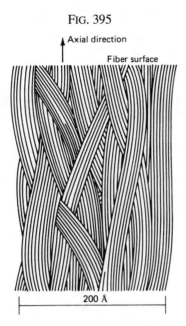

*The undulating ribbon structure of a PAN-based carbon fiber with a 400 GPa (60 × 10⁶ psi) modulus*

FIG. 396

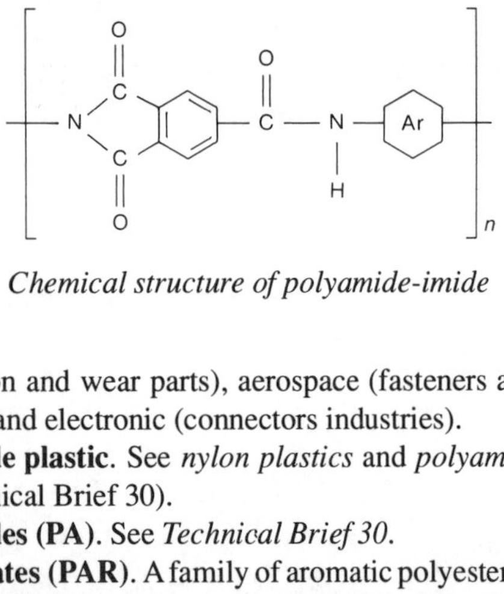

*Chemical structure of polyamide-imide*

(friction and wear parts), aerospace (fasteners and housings), and electronic (connectors industries).

**polyamide plastic.** See *nylon plastics* and *polyamides (PA)* (Technical Brief 30).

**polyamides (PA).** See *Technical Brief 30*.

**polyarylates (PAR).** A family of aromatic polyesters derived from aromatic dicarboxylic acids and diphenols. These thermoplastics have an amorphous molecular structure and are tough, durable, and heat resistant. They have excellent dimensional and ultraviolet stability, electrical properties, flame retardance, and warp resistance.

**polyaryletherketone (PAEK).** A linear aromatic crystalline thermoplastic. A composite with a PAEK matrix may have a continuous-use temperature as high as 250 °C (480 °F). This definition is also germane to polyetheretherketone (PEEK), polyetherketone (PEK), and polyetherketoneketone (PEKK).

**polyaryl sulfone (PAS).** A thermoplastic resin consisting mainly of phenyl and biphenyl groups linked by thermally stable ether and sulfone groups. Its most outstanding property is resistance to high and low temperatures (from –240 to 260 °C, or –400 to 500 °F). It also has good impact resistance, resistance to chemical oils, and most solvents, and good electrical insulating properties. It can be processed by injection molding, extrusion, compression molding, ultrasonic welding, and machining.

**polybenzimidazole (PBI).** A thermoplastic resin that is strong and stable, has a high molecular wight, and contains recurring aromatic units. The resin is produced by the high-temperature, melt condensation reaction of aromatic bis-ortho-diamines and aromatic dicarboxylates (acids, esters, or amides). Although a wide variety of PBIs have been synthesized, the most common is poly (2,2'-(*m*-phenylene)-5-5'-bibenzimidazole (Fig. 397). Parts made from PBI are used in petrochemical, geothermal, chemical process, power generation, aerospace, and transportation markets.

FIG. 397

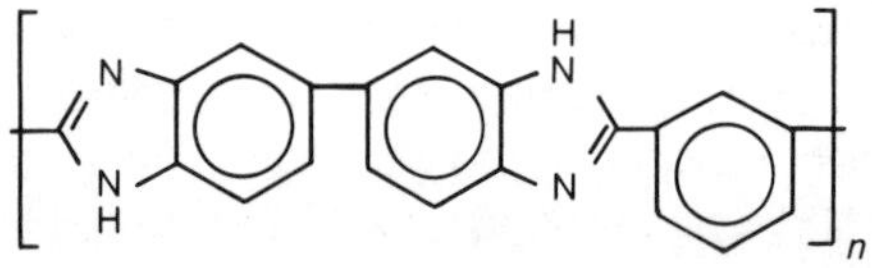

*Chemical structure of polybibenzimidazole*

**polybutylenes.** A group of polymers consisting of isotactic, stereoregular, highly crystalline polymers based on butylene-1. Their properties are similar to those of *polypropylene* and linear polyethylene, with superior toughness, creep resistance, and flexibility.

**polybutylene terephthalate (PBT).** A member of the polyalkyleneterephthalate family that is produced by the transesterification of dimethyl terephthalate with butanediol by means of a catalyzed melt polycondensation (Fig. 398). Properties include high strength, dimensional stability, low moisture absorption, good electrical characteristics, and resistance to heat and chemicals when suitably modified.

FIG. 398

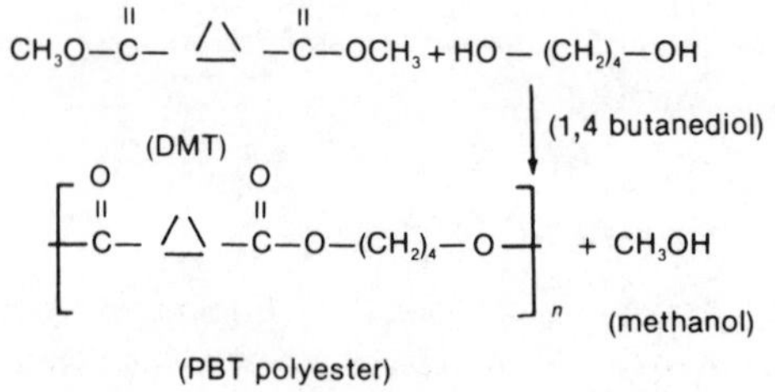

$CH_3O-C-\triangle-C-OCH_3 + HO-(CH_2)_4-OH$

(DMT)

(1,4 butanediol)

$-C-\triangle-C-O-(CH_2)_4-O-$ $+ CH_3OH$

(PBT polyester)

(methanol)

*Chemical structure of polybutylene terephthalate*

## Technical Brief 31: Polycarbonates

POLYCARBONATE (PC) is a high-performance amorphous engineering thermoplastic with exceptionally high impact strength, clarity, heat resistance, and dimensional stability. Polycarbonate is a polyester of carbonic acid. Virtually all general-purpose PC is based on bisphenol A. Additional monomers, such as phthalate, are used commercially to produce copolymers with increased heat resistance. Trifunctional monomers are added for increased melt strength for extrusion blow molding. Polycarbonate is also a versatile blending material and is used as a component in several commercial products. Examples include PC and polyester (polyethylene terephthalate and polybutylene tere-phthalate), PC and acrylonitrile-butadiene-styrene, and PC and styrene-maleic anhydride.

*Chemical structure of polycarbonate*

Polycarbonate has an exceptionally high notched Izod impact strength of 640 to 850 J/m (12 to 16 ft · lbf/in.). Practical impact strength, measured by a falling dart or Dynatup-type tester, is retained to temperatures as low as –50 °C (–60 °F). The heat-deflection temperature of PC, at 1.8 MPa (0.264 ksi), is 125 to 130 °C (260 to 270 °F), with high-heat copolymers of polyphthalate carbonate approaching 160 °C (325 °F). Mold shrinkage is low and very consistent, at 5 to 7 mm/mm. Tolerances can be held to as low as 0.025 mm (0.001 in.), especially with glass-filled grades. Creep resistance is exceptional over a wide temperature range.

Because of their highly aromatic nature, PCs are characterized by a high degree of hydrophobicity, as well as a high glass-transition ($T_g$) temperature and melt strength. A typical PC has a $T_g$ of 150 °C (300 °F) and a melting temperature ($T_m$) of 220 to 230 °C (430 to 445 °F). Although PC can be obtained in a crystalline form (by annealing at 180 °C, or 355 °F, for 24 h), the relatively small difference between high $T_g$ and $T_m$ provides a narrow crystallization window and a lower tendency to crystallize under usual processing conditions, compared with other engineering thermoplastics. The electrical properties of PC are also very good. High corona resistance and insulation properties are achieved, as well as a dielectric constant that is independent of temperature. Other properties include excellent colorability, high gloss, sterilizability, flame retardancy, biocompatibility, and stain resistance.

The aforementioned properties make PCs useful in a broad range of applications. Major markets for PCs include electronic and business equipment (machine housings, computer parts and peripherals, connectors, etc.), appliances (food processors and other electric kitchen components, and refrigerator crisper drawers), transportation (headlights, taillights, seat backs, etc.), sports equipment (helmets, windshields, and sunglass lenses), food service equipment, medical equipment, and sheet products (signs and aircraft interior panels).

### Recommended Reading

- F.W. Liberti, Polycarbonates (PC), *Engineered Materials Handbook*, Vol 2, ASM International, 1988, p 151-152
- S.W. Shalaby and P. Moy, Thermal and Related Properties of Engineering Thermoplastics, *Engineered Materials Handbook*, Vol 2, ASM International, 1988, p 445-459

---

**polycarbonates (PC).** See *Technical Brief 31*.

**polychromator.** A *spectrometer* that has many (typically 20 to 50) detectors for simultaneously measuring light from many spectral lines. Polychromators are commonly used in atomic emission spectroscopy. See also *monochromator*.

**polycondensation.** See *condensation polymerization*.

**polycrystalline.** Pertaining to a solid comprised of many crystals or crystallites, intimately bonded together. May be homogeneous (one substance) or heterogeneous (two or more crystal types or compositions).

**polyester plastics.** Plastics based on resins composed principally of polymeric esters, in which the recurring ester groups are an integral part of the main polymer chain, and in which ester groups occur in most cross links that may be present between chains. See also *thermoplastic polyesters*, *thermosetting polyesters*, and *unsaturated polyesters*.

**polyetheretherketone (PEEK).** A linear aromatic crystalline thermoplastic. A composite with a PEEK matrix may have a continuous-use temperature as high as 250 °C (480 °F). See also *polyaryletherketone (PAEK)*.

**polyether-imide (PEI).** An amorphous high-performance thermoplastic with a chemical structure that consists of repeating aromatic imide and ether units (Fig. 399). PEIs are characterized by high strength and rigidity at room and elevated temperatures, long-term high heat resistance, highly stable dimensional and electrical properties, and broad chemical resistance. Their high glass-transition temperature (215 °C, or 419 °F) allows PEI to be used for short-term use at 200 °C (390 °F).

FIG. 399

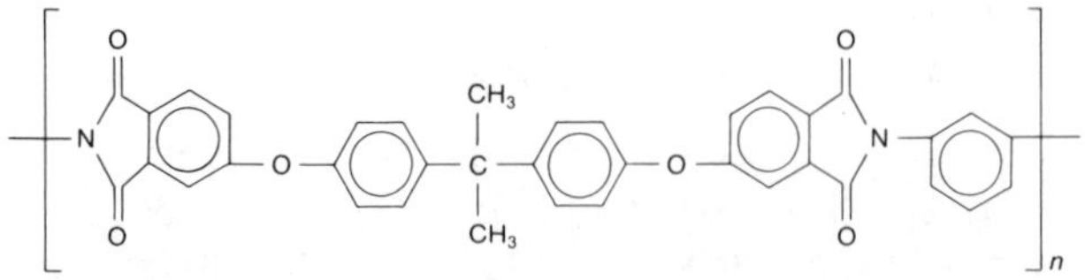

*Chemical structure of polyether-imide*

**polyether sulfone (PES).** A high-temperature engineering thermoplastic consisting of repeating phenyl groups linked by thermally stable ether and sulfone groups (Fig. 400). The material has good transparency and flame resistance, and is one of the lowest smoke-emitting materials available. Both polymer and reinforced grades are available in granular form for extrusion or injection molding.

FIG. 400

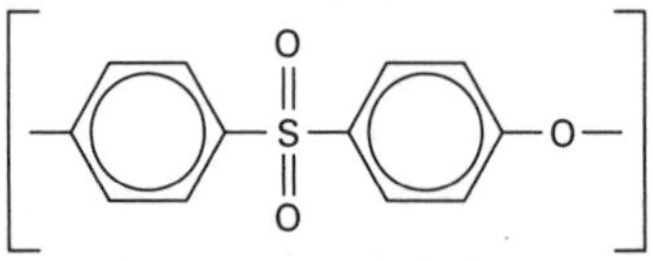

*Chemical structure of polyether sulfone*

**polyethylene (PE) plastics.** Thermoplastic materials composed of ethylene. They are normally translucent, tough, waxy solids that are unaffected by water and by a large range of chemicals. In common usage, these plastics have no less than 85% ethylene and no less than 95% total olefins. See also *high-density polyethylenes* (Technical Brief 19) and *ultrahigh molecular weight polyethylene.*

**polyethylene terephthalate (PET).** A saturated, thermoplastic polyester resin made by condensing ethylene glycol and terephthalic acid and used for fibers, films, and injection-molded parts. It is extremely hard, wear resistant, dimensionally stable, and resistant to chemicals, and it has good dielectric properties. Also known as polyethylene glycol terephthalate.

**polyimide (PI).** A polymer produced by reacting an aromatic dianhydride with an aromatic diamine. It is similar to a polyamide, differing only in the number of hydrogen molecules contained in the groupings. This polymer is suitable for use as a binder or adhesive, and its exceptional thermomechanical properties make it suitable for many high-temperature ($\geq$230 °C, or 450 °F) applications. See also *thermoplastic polyimides (TPI).*

**polymer.** A high molecular weight organic compound, natural or synthetic, with a structure that can be represented by a repeated small unit, the mer. Examples include polyethylene, rubber, and cellulose. Synthetic polymers are formed by addition or condensation polymerization of monomers. Some polymers are *elastomers*, some are *plastics*, and some are *fibers*. When two or more dissimilar monomers are involved, the product is called a copolymer. The chain lengths of commercial thermoplastics vary from ~1000 to >100,000 repeating units. Thermosetting polymers approach infinity after curing, but their resin precursors, often called prepolymers, may be relatively short—6 to 100 repeating units—before curing. The lengths of polymer chains, usually measured by molecular weight, have very significant effects on the performance properties of plastics and profound effects on processability.

**polymerization.** A chemical reaction in which the molecules of a *monomer* are linked together to form large molecules with a molecular weight that is a multiple of the molecular weight of the original substance. When two or more monomers are involved, the process is called copolymerization.

**polymer matrix.** The resin portion of a reinforced or filled plastic. See also *resin-matrix composites* (Technical Brief 39).

**polymethyl methacrylate (PMMA).** A thermoplastic polymer synthesized from methyl methacrylate. It is a transparent solid with exceptional optical properties. Available in the form of sheets, granules, solutions, and emulsions, it is used as facing material in certain composite constructions. See also *acrylic plastic.*

**polymorph.** In crystallography, one crystal form of a polymorphic material. See also *polymorphism.*

**polymorphism.** A general term for the ability of a solid to exist in more than one form. In metals, alloys, and similar substances, this usually means the ability to exist in two or more crystal structures, or in an amorphous state and at least one crystal structure. See also *allotropy, enantiotropy,* and *monotropism.*

**polyol.** An alcohol having many hydroxyl groups. Also known as polyhydric alcohol or polyalcohol. In cellular plastics, usage, the term includes compounds containing alcoholic hydroxyl groups such as polyethers, glycols, polyesters, and castor oil used in urethane foams, and other polyurethanes. See also *alcohols.*

**polyolefins.** Plastics based on a polymer made with an *olefin* as essentially the sole *monomer.*

**polyoxymethylene (POM).** Acetal plastics based on polymers in which oxymethylene is essentially the sole repeating structural unit in the chains. See also *acetal (AC) resins.*

**polyphenylene oxides (PPO).** Thermoplastic, linear, non-crystalline polyethers obtained by the oxidative polycondensation of 2,6-dimethylphenol in the presence of a copper-amine complex catalyst. These resins have a useful temperature range from less than –170 to 190 °C (–275 to 375 °F) with intermittent use up to 205 °C (400 °F) possible, excellent electrical properties, unusual resistance to acids and bases, and processibility on conventional extrusion and injection molding equipment. Also known as polyphenylene ether (PPE).

**polyphenylene sulfides (PPS).** Crystalline polymers having a symmetrical, rigid backbone chain consisting of recurring para-substituted benzene rings and sulfur atoms (Fig. 401). Known for excellent chemical resistance, thermal stability, and fire resistance. Its inertness to organic solvents, inorganic salts, and bases makes it corrosion resistant. Commercial engineering grades are always fiber reinforced.

FIG. 401

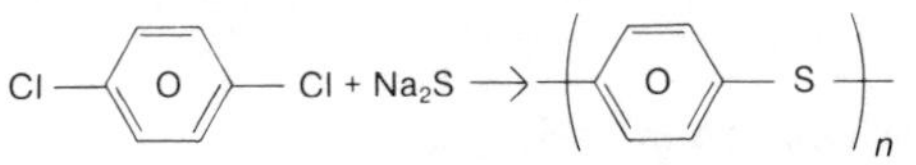

*Chemical structure of polyphenylene sulfide*

**polypropylenes (PP).** Tough, lightweight thermoplastics made by the polymerization of high-purity propylene gas in the presence of an organometallic catalyst at relatively low pressures and temperatures. See also *reinforced polypropylene.*

**polystyrenes (PS).** Water-white homopolymer thermoplastics produced by the polymerization of styrene (vinyl benzene). Has outstanding electrical properties, good thermal and dimensional stability, and staining resistance. Because it is somewhat brittle, it is often copolymerized or blended with other materials to obtain desired properties. See also *high-impact polystyrenes (HIPS)* (Technical Brief 20).

**polysulfide.** A synthetic polymer containing sulfur and carbon linkages, produced from organic dihalides and sodium polysulfide. The material is elastomeric in nature, resistant to light, oil, and solvents, and impermeable to gases.

**polysulfones (PSU).** A family of sulfur-containing thermoplastics made by reacting bisphenol A and 4,4′-dichlorodiphenylsulfone with potassium hydroxide in dimethyl sulfoxide at 130 to 140 °C (265 to 285 °F). The structure of the polymer is benzene rings or phenylene units linked by three different chemical groups, a sulfone group, an ether linkage, and an isopropylidene group (Fig. 402). Polysulfones are characterized by high strength, the highest service temperature of all melt-processible thermoplas-

FIG. 402

*Chemical structure of polysulfone*

tics, low creep, good electrical characteristics, transparency, self-extinguishing properties, and resistance to greases, many solvents, and chemicals. They may be processed by extrusion, injection molding, or blow molding.

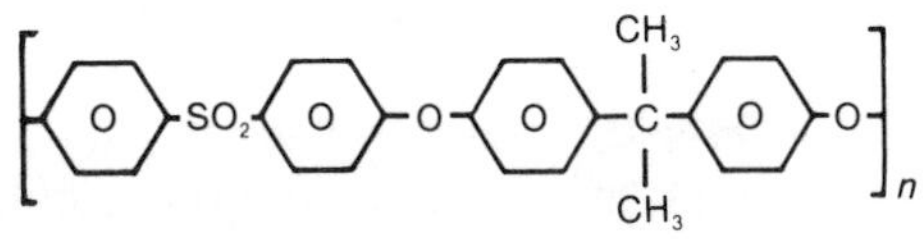

**polyterephthalate.** A thermoplastic polyester in which the terephthalate group is a repeating structural unit in the chain, the terephthalate being greater in amount than other dicarboxylates that may be present.

**polytetrafluoroethylenes (PTFE).** Members of the fluorocarbon family of plastics made by the polymerization of tetrafluoroethylene. PTFE is characterized by its extreme inertness to chemicals, very high thermal stability, and low frictional properties. Among the applications for these materials are bearings, fuel hoses, gaskets and tapes, and coatings for metal and fabric.

**polyurethanes (PUR).** See *Technical Brief 32* and *thermoplastic polyurethanes (TPUR).*

**polyvinyl acetals.** Members of the family of vinyl plastics. Polyvinyl acetal is the general name for resins produced from a condensation of polyvinyl alcohol with an aldehyde. There are three main groups: polyvinyl acetal itself, *polyvinyl butyral*, and *polyvinyl formal*. Polyvinyl acetal resins are thermoplastics that can be processed by casting, extruding, molding, and coating, but their main uses are in adhesives, lacquers, coatings, and films.

**polyvinyl acetate (PVAC).** A thermoplastic material composed of polymers of vinyl acetate in the form of a colorless solid. It is obtainable in the form of granules, solutions, latices, and pastes and is used extensively in adhesives, for paper and fabric coatings, and in bases for inks and lacquers.

**polyvinyl acetate emulsion adhesive.** A latex adhesive in which the polymeric portion comprises polyvinyl acetate, copolymers based mainly on polyvinyl acetate, or a mixture of these and which may contain modifiers and secondary binders to provide specific properties.

**polyvinyl alcohol (PVAL).** A thermoplastic material composed of polymers of the hypothetical vinyl alcohol. Usually a colorless solid, insoluble in most organic solvents and oils, but soluble in water when the content of hydroxy groups in the polymer is sufficiently high. The product is normally granular. It is obtained by the partial hydrolysis or by the complete hydrolysis of polyvinyl esters, usually by the complete hydrolysis of polyvinyl acetate. It is mainly used for adhesives and coatings.

## Technical Brief 32: Polyurethanes

POLYURETHANES (PUR) represent a large family of polymers with widely varying properties and uses. All of these polymers are based on the reaction product of an organic isocyanate with compounds containing a hydroxyl group. Polyurethanes may be thermosetting or thermoplastic, rigid or soft and flexible, cellular or solid. The properties of any of these types may be tailored within wide limits to suit the desired application.

Thermoset PURs are used as foams, coatings, elastomers, adhesives, and sealants. They are reaction products of isocyanates and polyols and can be designed for specific purposes, such as hardness, toughness, elongation, and chemical resistance. The most important commercial polyfunctional isocyanates are toluene diisocyanate (TDI) and polymethylene diphenylene isocyanate (PMDI). TDI is used in the production of flexible PUR foams, while PMDI is primarily used for rigid foam production. Many modified isocyanates based on TDI and PMDI are also commercially available.

*Chemical structures of TDI. (a) 2,4 isomer. (b) 2,6 isomer*

High-performance PUR elastomers are most commonly made from 4,4′-MDI. Aliphatic isocyanates, such as 1,6-hexamethylene diisocyanate, and their derivatives find application in coatings because of superior weatherability.

Many thermoset PUR formulations use mixtures of polyols, cross-linking agents, catalysts, and blowing agents. Flexible foam can be produced by using a 3000 molecular weight triol as the hydroxyl compound, water that produces carbon dioxide upon reaction with isocyanate as the blowing agent, TDI as the isocyanate, a suitable

### Mechanical properties of high-density, semirigid, reaction-injection-molded polyurethane

| Property | High-density integral skin foam | | | | Solid RIM elastomer | | |
|---|---|---|---|---|---|---|---|
| Density, g/cm³ | 0.600 | 0.950 | 1.00 | 1.00 | 1.05 | 1.10 | 1.15 |
| Tensile strength, MPa (ksi) | 5.7 (0.825) | 14 (2.03) | 20 (2.90) | 21 (3.04) | 17 (2.46) | 35 (5.07) | 70 (10.1) |
| Elongation at break, % | 250 | 300 | 180 | 150 | 380 | 300 | ... |
| Shore A hardness | 75 | 92 | ... | ... | ... | ... | ... |
| Shore D hardness | ... | 36 | 56 | 63 | 39 | 64 | 83 |
| Young's modulus, GPa (10⁶ psi) | | | | | | | |
| At −30 °C (−22 °F) | 0.225 (0.033) | 0.475 (0.069) | 0.950 (0.138) | 1.250 (0.181) | 0.300 (0.44)(a) | 1.40 (0.203) | 2.60 (0.290) |
| At 22 °C (71 °F) | 0.027 (0.004) | 0.100 (0.015) | 0.300 (0.044) | 0.500 (0.073) | 0.090 (0.014)(a) | 0.600 (0.087) | 2.35 (0.341) |
| At 65 °C (150 °F) | 0.013 (0.002) | 0.023 (0.003) | 0.120 (0.017) | 0.200 (0.029) | 0.060 (0.009)(a) | 0.360 (0.052) | 1.65 (0.239) |
| Sag, 1 h at 120 °C (250 °F), mm (in.) | ... | ... | 5 (0.20) | 5.5 (0.22) | 7 (0.28) | 5.4 (0.21) | 2.5 (0.10) |
| Tear propagation resistance, kN/m (lbf/in., linear) | 8.3 (45) | 38 (215) | ... | ... | ... | ... | ... |

(a) For glycol-extended systems

**polyvinyl butyral (PVB).** A thermoplastic material derived from a polyvinyl ester in which some or all of the acid groups have been replaced by hydroxyl groups, and some or all of these hydroxyl groups have been replaced by butyral groups by reaction with butyraldehyde. It is a colorless, flexible, tough solid used primarily in interlayers for laminated safety glass. See also *polyvinyl acetals*.

**polyvinyl carbazole.** A thermoplastic resin, brown in color, obtained by reacting acetylene with carbazole. The resin has excellent electrical properties and good heat and chemical resistance. It is used as an impregnant for paper capacitors.

**polyvinyl chloride acetate.** A thermoplastic material composed of copolymers of vinyl chloride and vinyl acetate.

It is a colorless solid with good resistance to water as well as concentrated acids and alkalies. It is obtainable in the form of granules, solutions, and emulsions. Compounded with plasticizers, it yields a flexible material superior to rubber in aging properties. It is widely used for cable and wire coverings, in chemical plants, and in protective garments.

**polyvinyl chlorides (PVC).** See *Technical Brief 33*.

**polyvinyl formal (PVF).** One of the groups of polyvinyl acetal resins made by the condensation of formaldehyde in the presence of polyvinyl alcohol. It is used mainly in combination with cresylic phenolics, for wire coatings, and for impregnations, but can also be molded, extruded, or cast. It is resistant to greases and oils.

**polyvinylidene chloride (PVDC).** A thermoplastic material

catalyst, and silicone. Omitting the blowing agent will produce a solid PUR that may be rigid or elastomeric, depending on the choice of ingredients for the formulation. Significantly reducing the blowing agent will result in a microcellular structure rather than a foam.

Flexible PUR thermoset foams account for about 50% of all the PUR produced in the world. About half is used for cushioning and padding in furniture and beds, and about one-fourth is used by the automotive industry. Rigid thermoset foams are used as the insulation for residential and commercial buildings, as well as by the appliance industry in refrigerators and freezers. The rigid foams have closed cells, and their superior insulating properties result from the fact that the cells entrap the fluorocarbon blowing agent. Integral-skin rigid foams have a cellular core and a higher-density skin. They are used in applications such as bicycle seats and automobile steering wheels. Reinforced PUR thermosets for integral-skin foams provide higher stiffness, lower heat sag, and smaller coefficient of thermal expansion. Typical fillers are glass fibers, glass flakes, mica, and ground minerals. Reaction-injection-molded solid thermoset PURs are being used in the automotive industry for producing fasciae, door panels, and fenders.

Certain combinations of diisocyanates, polyols, and chain extenders generate thermoplastic polyurethanes (TPUR). These materials can also be reinforced with glass or added (alloyed) to other brittle plastics to improve their impact resistance. As with thermosetting PURs, TPURs can be formulated to provide wide-ranging properties suitable for myriad applications. Some primary markets include blown and extruded film and sheet products used as protective tarpaulins and hospital bed covers, sporting equipment (plastic cleats for football and soccer shoes, and cross-country ski poles), reinforced hose and tubing applications, and protective coverings for wire and cabling.

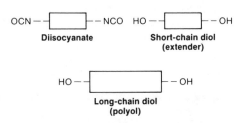

*The three raw material components for TPUR*

## Recommended Reading

- H.F. Hespe, Polyurethanes (PUR), *Engineered Materials Handbook*, Vol 2, ASM International, 1988, p 257-264
- A.A. Sardanopoli, Thermoplastic Polyurethanes (TPUR), *Engineered Materials Handbook*, Vol 2, ASM International, 1988, p 203-208
- K.B. Tator, Organic Coatings and Linings, *Metals Handbook*, 9th ed., Vol 13, ASM International, 1987, p 399-418

composed of polymers of vinylidene chloride (1,1-dichloroethylene). It is a white powder with a softening temperature at 185 to 200 °C (365 to 390 °F). The material is also supplied as a copolymer with acrylonitrile or vinyl chloride, giving products that range from the soft, flexible type to the rigid type. Also known as saran.

**polyvinylidene fluoride.** This recent member of the fluorocarbon family of plastics is a homopolymer of vinylidene fluoride. It is supplied as powders and pellets for molding and extrusion and in solution form for casting. The resin has good tensile and compressive strength and high impact strength.

**POM.** See *polyoxymethylene (POM)*.

**pop-off.** Loss of small portions of a porcelain enamel coating. The usual cause is outgassing of hydrogen or other gases from the basis metal during firing, but pop-off may also occur because of oxide particles or other debris on the surface of the basis metal. Usually, the pits are minute and cone shaped, but when pop-off is the result of severe *fishscale* the pits may be much larger and irregular.

**population.** In statistics, a generic term denoting any finite or infinite collection of individual samples or data points in the broadest concept; an aggregate determined by some property that distinguishes samples that do and do not belong.

**porcelain.** A glazed or unglazed vitreous ceramic whiteware used for technical purposes. This term designates such products as electrical, chemical, mechanical, structural, and thermal wares when they are vitreous. This term is frequently used as a synonym for *china*.

## Technical Brief 33: Polyvinyl Chlorides

POLYVINYL CHLORIDE (PVC) has been used commercially for more than 50 years, since flexible (plasticized) PVC was introduced in the mid-1930s. A rigid engineering grade of PVC became useful in the early 1950s for piping on naval vessels. Applications for rigid PVC have grown steadily to include its use as an engineering thermoplastic; it is now the second most commonly used plastic material, in terms of volume, after polyethylene. PVC is manufactured from sodium chloride (NaCl) and natural gas, with the chloride from the salt comprising 57 wt%. Its chemical structure has the repeating unit of vinyl chloride.

*Chemical structure of PVC*

PVC resin powders must be compounded with additives to make them useful. They cannot be processed neat. Compounded PVC resins are typically used without further additions because they already contain a combination of stabilizers, lubricants, plasticizers, pigments, processing aids, impact modifiers, and fillers (for example, glass). These compounds, available as powders, cubes, or pellets, are ready for further melt processing by extrusion or injection molding.

Alloys or blends of PVC are also available with enhanced properties. Rubbery materials, blended with PVC to improve toughness, are based on rubbers such as acrylonitrile-butadiene-styrene (ABS), methacrylate-butadiene-styrene (MBS), and nitrile rubber; butylacrylate (acrylic and modified acrylic modifiers); and ethylene (chlorinated polyethylene and ethylene/vinyl acetate). Other blending ingredients are used as processing aids to reduce melt fracture during PVC processing or to increase the softening temperature of PVC.

Properties that characterize PVC are its low combustibility, high toughness (notched Izod impact strength of >0.5 J/mm, or >10 ft · lbf/in. at –40 °C, or –40 °F), outstanding weatherability, including good color and impact retention, good tensile and flexural strength retention, and no loss in modulus (stiffness), excellent dimensional control, and low melt viscosity.

Rigid PVC is engineered into a wide range of applications, including water distribution piping; home plumbing; drain, waste, and vent piping; house siding; windows and skylights; chemical tanks and pumps; bathtubs and sinks; business machine housings; appliances; electronics packaging; decorative profiles and trim; outdoor furniture; window roller shutters and vertical blinds; medical apparatus; packaging for foodstuffs; and flooring.

Flexible PVC compounds are used in electrical wire coating and jackets, communication wire coating, seals and gaskets, automotive dash pads and trim, medical tubing, outdoor furniture, wheels, bumpers, shoes, meat wrap, and flooring.

### Thermal and related properties of PVC and other vinyl polymers

| Property | PVC Rigid | PVC Plasticized | PVC 30% glass filled | Chlorinated PVC | PVDC | PVFM | PVB |
|---|---|---|---|---|---|---|---|
| $T_g$, °C (°F) | 75–105 (170–220) | (a) | 75–105(b) (170–220) | 110 (230) | ... | 105 (220) | 49 (120) |
| $T_m$, °C (°F) | (c) | ... | ... | (c) | 210 (410) | (c) | (c) |
| Molding temperature, °C (°F) | | | | | | | |
| Compression | 140–205 (285–400) | 140–195 (285–385) | ... | 170–205 (350–400) | 104–175 (220–350) | 150–175 (300–350) | 140–160 (280–320) |
| Injection | 150–215 (300–415) | 160–195 (320–385) | 130–210 (270–405) | 160–225 (325–440) | 150–205 (300–400) | 150–205 (300–400) | 120–170 (250–340) |
| Heat-deflection temperature, at 1.82 MPa (0.264 ksi), °C (°F) | 140–170 (285–340) | ... | 155 (310) | 202–234 (395–450) | 130–150 (265–300) | 150–170 (300–340) | ... |
| Water absorption, 24 h at 3.2 mm (⅛ in.) thick, % | 0.04–4.0 | 0.15–0.75 | 0.008 | 0.02–0.15 | 0.1 | 0.5–3.0 | 1.0–2.0 |

(a) Variable; can be lower than 75–105 °C (165–220 °F) depending on type and concentration of plasticizer. (b) Irrespective of the filler. (c) Amorphous

### Recommended Reading

- J.W. Summers and E.B. Rabinovitch, Polyvinyl Chlorides (PVC), *Engineered Materials Handbook*, Vol 2, ASM International, 1988, p 209–213

**Table 49  Compositions of frits for ground coat porcelain enamels for sheet steel**

| Constituent | Composition, % | | | |
|---|---|---|---|---|
| | Regular blue-black enamel | Alkali-resistant enamel | Acid-resistant enamel | Water-resistant enamel |
| $SiO_2$ | 33.74 | 36.34 | 56.44 | 48.00 |
| $B_2O_3$ | 20.16 | 19.41 | 14.90 | 12.82 |
| $Na_2O$ | 16.74 | 14.99 | 16.59 | 18.48 |
| $K_2O$ | 0.90 | 1.47 | 0.51 | ... |
| $Li_2O$ | ... | 0.89 | 0.72 | 1.14 |
| CaO | 8.48 | 4.08 | 3.06 | 2.90 |
| BaO | 9.24 | 8.59 | ... | ... |
| ZnO | ... | 2.29 | ... | ... |
| $Al_2O_3$ | 4.11 | 3.69 | 0.27 | ... |
| $ZrO_2$ | ... | 2.29 | ... | 8.52 |
| $TiO_2$ | ... | ... | 3.10 | 3.46 |
| CuO | ... | ... | 0.39 | ... |
| $MnO_2$ | 1.43 | 1.49 | 1.12 | 0.52 |
| NiO | 1.25 | 1.14 | 0.03 | 1.21 |
| $Co_3O_4$ | 0.59 | 1.00 | 1.24 | 0.81 |
| $P_2O_5$ | 1.04 | ... | ... | 0.20 |
| $F_2$ | 2.32 | 2.33 | 1.63 | 1.94 |

**Table 50  Compositions of frits for cover coat porcelain enamels for sheet steel**

| Constituent | Composition, wt% | | |
|---|---|---|---|
| | Titania white enamel | Semi-opaque enamel | Clear enamel |
| $SiO_2$ | 44.67 | 44.92 | 54.26 |
| $B_2O_3$ | 14.28 | 16.40 | 12.38 |
| $Na_2O$ | 8.27 | 8.67 | 6.55 |
| $K_2O$ | 6.99 | 8.12 | 11.32 |
| $Li_2O$ | 0.98 | 0.45 | 1.14 |
| ZnO | ... | 0.74 | ... |
| $ZrO_2$ | 1.98 | 3.34 | 1.40 |
| $Al_2O_3$ | 0.31 | 0.16 | ... |
| $TiO_2$ | 18.49 | 13.05 | 10.04 |
| $P_2O_5$ | 1.32 | 0.88 | ... |
| MgO | 0.5 | ... | ... |
| $F_2$ | 2.21 | 3.27 | 2.91 |

Source: Porcelain Enamel Institute

**porcelain enamel.** A substantially vitreous or glassy, in-organic coating (borosilicate glass) bonded to metal by fusion at a temperature above 425 °C (800 °F). Porcelain enamels are applied primarily to components made of sheet iron or steel, cast iron, aluminum, or aluminum-coated steels. Tables 49 and 50 list the compositions of typical ground coat and cover coat enamels, respectively.

**pore.** (1) A small opening, void, interstice, or channel within a consolidated solid mass or agglomerate, usually larger than atomic or molecular dimensions. (2) A minute cavity in a powder metallurgy compact, sometimes added intentionally. See also *porous P/M parts*. (3) A minute perforation in an electroplated coating.

**pore area.** The effective surface porosity of a sintered powder metallurgy compact to determine the permeability to a test fluid.

**pore channels.** The connections between pores in a sintered body.

**pore formation.** The natural formation of pores during compaction and/or sintering. See also *pore-forming material*.

**pore-forming material.** A substance included in a metal powder mixture that volatilizes during sintering and thereby produces a desired kind of porosity in a finished compact.

**pore size.** Width of a pore in a compacted and/or sintered metal powder or within a particle.

**pore size distribution.** Indicates the volume fractions of different pore size categories in a sintered body, which are determined metallographically.

**pore size range.** The limits between which a variation in pore size in a sintered body is allowed.

**pore structure.** Pattern of pores in a solid body indicating such characteristics as pore shape, *pore size*, and *pore size distribution*.

**porosimeter.** A test apparatus to measure the interconnected porosity in a sintered compact by means of determining its permeability through use of a test fluid such as mercury, which either partially or completely fills the open pores.

**porosity.** (1) Fine holes or pores within a solid; the amount of these pores is expressed as a percentage of the total volume of the solid. (2) Cavity-type discontinuities in weldments formed by gas entrapment during solidification. (3) A characteristic of being porous, with voids or pores resulting from trapped air or shrinkage in a casting. See also *gas porosity* and *pinhole porosity*.

**porous bearing.** A bearing made from porous material such as compressed and sintered metal powder. The pores may act as reservoirs or passages for supplying fluid lubricant, or the bearing may be impregnated with solid lubricant. See also *self-lubricating bearing*.

**porous molds.** Molds for forming plastics that are made up of bonded or fused aggregate (powdered metal, coarse pellets, and so forth) such that the resulting mass contains numerous open interstices of regular or irregular size allowing either air or liquids to pass through the mass of the mold.

**porous P/M parts.** Powder metallurgy components which are characterized by *interconnected porosity*. Primary application areas for porous P/M parts are filters (Fig. 403), damping devices, storage reservoirs for liquids (including *self-lubricating bearing*), and battery elements. Bronzes, stainless steels, nickel-base alloys, titanium, and aluminum are used in P/M porous metal applications.

**porous region** (ceramics). A three-dimensional zone of porosity or microporosity of higher concentration than is normally found in the ceramic matrix.

**porous seam** (ceramics). A two-dimensional area of porosity or microporosity of higher concentration than is normally found in the ceramic matrix.

Fig. 403

*Assorted filters made from porous P/M bronze*

**port**. The opening through which molten metal enters the injection cylinder of a *die casting* plunger machine, or is ladled into the injection cylinder of a cold chamber machine. See also *cold chamber machine* and *plunger*.

**porthole die**. A multiple-section extrusion die capable of producing tubing or intricate hollow shapes without the use of a separate mandrel. Metal is extruded in separate streams through holes in each section and is rewelded by extrusion pressure before it leaves the die. Compare with *bridge die*.

**positioned weld**. A weld made in a joint that has been oriented to facilitate making the weld.

**positive-contact bushing**. A bushing, the inside diameter of which has direct contact with the outside diameter of a shaft or sleeve. Radial or axial clearances are provided in the hole.

**positive-contact seal**. A seal, the primary function of which is achieved by one surface mating with another. Examples include lip, circumferential, and face-type seals. See also *face seal*.

**positive distortion**. In an optical microscope, the distortion in the image that results when the magnification in the center of the field is less than that at the edge of the field. Also termed pincushion distortion. See also the figure accompanying the term *negative distortion*.

**positive eyepiece**. In an optical microscope, an eyepiece in which the real image of the object is formed below the lower lens elements of the eyepiece. See also *eyepiece*.

**positive mold**. A mold for forming plastics designed to trap all the molding material when it closes.

**positive rake**. Describes a tooth face in rotation whose cut-ting edge leads the surface of the tooth face. See also the figure accompanying the term *face mill*.

**positive replica**. A replica whose contours correspond directly to the surface being replicated. Contrast with *negative replica*.

**postcure** (adhesives). A treatment (normally involving heat) applied to an adhesive assembly following the initial cure, to modify specific properties. To expose an adhesive assembly to an additional cure, following the initial cure, for the purpose of modifying specific properties.

**postcure** (plastics). Additional elevated-temperature cure of a plastic, usually without pressure, to improve final properties and/or complete the cure, or decrease the percentage of volatiles in the compound. In certain resins, complete cure and ultimate mechanical properties are attached only by exposure of the cured resin to higher temperatures than those of curing. See also *cure*.

**postforming**. The forming, bending, or shaping of fully cured, C-staged thermoset laminates that have been heated to make them flexible. Upon cooling, the formed laminate retains the contours and shape of the mold over which it has been formed. See also *C-stage*.

**postheat current**. In resistance welding, the current through the welding circuit during *postheat time*.

**postheating**. Heating weldments immediately after welding, for tempering, for stress relieving, or for providing a controlled rate of cooling to prevent formation of a hard or brittle structure. See also *postweld heat treatment*.

**postheat time**. In resistance welding, the time from the end of weld heat time to the end of weld time. See also *postheat current*.

**postweld heat treatment**. Any heat treatment that follows the welding operation.

**pot** (metals). (1) A vessel for holding molten metal. (2) The electrolytic reduction cell used to make such metals as aluminum from a fused electrolyte.

**pot** (plastics). To embed a component or assembly in liquid resin, using a shell, can, or case that remains an integral part of the product after the resin is cured.

**pot annealing**. Same as *box annealing*.

**pot die forming**. Forming products from sheet or plate through the use of a hollow die and internal pressure which causes the preformed workpiece to assume the contour of the die.

**pottery**. A generic term for all fired ceramic wares that contain clay, except for technical, structural, and refractory products. Specifically, however, pottery describes the low-temperature fired porous ware that is usually colored. The term is properly applied to the clay products of primitive peoples, or to decorative art products made of unrefined clays by using unsophisticated methods. See also *traditional ceramics* (Technical Brief 57).

**potential**. Any of various functions from which intensity or velocity at any point in a field may be calculated. The driving influence of an electrochemical reaction. See also *active potential, chemical potential, corrosion potential, critical pitting potential, decomposition potential, electrochemical potential, electrode potential, electrokinetic potential, equilibrium (reversible) potential, free corrosion potential, noble potential, open-circuit potential, protective potential, redox potential,* and *standard electrode potential.*

**potential-pH diagram**. See *Pourbaix (potential-pH) diagram.*

**potentiodynamic (potentiokinetic)**. The technique for varying the *potential* of an electrode in a continuous manner at a preset rate.

**potentiometer**. An instrument that measures electromotive force by balancing against it an equal and opposite electromotive force across a calibrated resistance carrying a definite current.

**potentiometric membrane electrodes**. Electrochemical sensing device that can be used to quantify cationic and anionic substances and gaseous species in aqueous solutions. These devices are also used for analytical titrations. See also *titration.*

**potentiostat**. An instrument for automatically maintaining an electrode in an electrolyte at a constant potential or controlled potentials with respect to a suitable reference electrode.

**potentiostatic**. The technique for maintaining a constant *electrode potential.*

**potentiostatic etching**. The selective corrosion of one or more morphological features of a microstructure that results from

FIG. 404

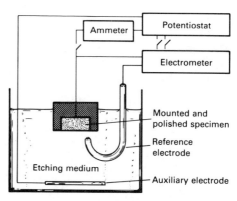

*Arrangement for potentiostatic etching*

FIG. 405

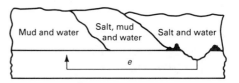

*Schematic showing the mechanism of poultice corrosion. In the example shown, clumps of mud and water have collected, and the varying concentrations of salt and water within the clump encourage corrosion.*

holding the metal to be etched in a suitable etching electrolyte at a controlled potential relative to a reference electrode (Fig. 404). Adjusting the potential makes possible a defined etching of singular phases.

**pot life**. The length of time that a catalyzed thermosetting resin system retains a viscosity low enough to be used in processing. Also called working life.

**poultice corrosion**. A term used in the automotive industry to describe the corrosion of vehicle body parts due to the collection of road salts and debris on ledges and in pockets that are kept moist by weather and washing (Fig. 405). Also called deposit corrosion or attack.

**Pourbaix (potential-pH) diagram**. A plot of the *redox potential* of a corroding system versus the pH of the system, compiled using thermodynamic data and the *Nernst equation.* The diagram shows regions within which the metal itself or some of its compounds are stable (Fig. 406).

**pouring**. The transfer of molten metal from furnace to ladle, ladle to ladle, or ladle into molds.

**pouring basin**. In metal casting, a basin on top of a mold that receives the molten metal before it enters the sprue or downgate. See also the figure accompanying the term *gating system.*

FIG. 406

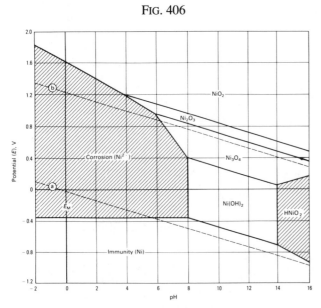

*Potential-pH (Pourbaix) diagram for nickel in an aqueous deaerated acid solution. Shaded areas represent corrosion reactions.*

**pour point**. The lowest temperature at which a lubricant can be observed to flow under specified conditions.

**pour-point depressant**. An additive that lowers the pour point of a lubricant.

**powder**. An aggregate of discrete particles that are usually in the size range of 1 to 1000 µm.

**powder cutting**. See preferred term *chemical flux cutting* and *metal powder cutting*.

**powder feeder**. A mechanical device designed to introduce a controlled flow of powder into the plasma-spray torch. See also *plasma spraying*.

**powder-feed rate**. In thermal spraying, the quantity of powder introduced into the arc per unit time; expressed in pounds/hour or grams/minute.

**powder fill**. In powder metallurgy, the filling of a die cavity with powder prior to compaction.

**powder flame spraying**. A thermal spraying process variation in which the material to be sprayed is in powder form (Fig. 407). See also *flame spraying*.

FIG. 407

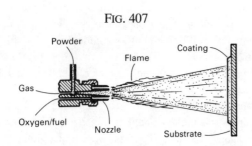

*Schematic of the powder flame spraying process*

FIG. 408

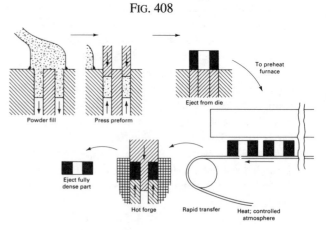

*The powder forging process*

**powder forging**. The plastic deformation of a powder metallurgy *compact* or *preform* into a fully dense finished shape by using compressive force; usually done hot and within closed dies (Fig. 408).

**powder lubricant**. In powder metallurgy, an agent or component incorporated into a mixture to facilitate compacting and ejecting of the compact from its mold.

**powder metallizing**. See preferred term *powder flame spraying*.

**powder metallurgy (P/M)**. The technology and art of producing metal powders and utilizing metal powders for production of massive materials and shaped objects.

**powder metallurgy forging**. See *powder forging*.

**powder metallurgy part**. A shaped object that has been formed from metal powders and sintered by heating below the melting point of the major constituent. A structural or mechanical component made by the powder metallurgy process.

**powder method**. Any method of *x-ray diffraction* involving a polycrystalline and preferably randomly oriented powder specimen and a narrow beam of monochromatic radiation. The powder method is best known as a phase characterization tool because it can routinely differentiate between phases having some chemical composition but different crystal structures (polymorphs). See also *polymorphism*.

**powder molding**. General term used to denote several techniques, such as injection molding, for producing objects of varying size and shape by melting plastic powder, usually against the inside of a mold. The molds are either stationary (for example, in variations of slush molding techniques) or rotating (for example, in variations of rotational molding).

**powder production**. The process by which a metal powder is produced, such as machining, milling, atomization, condensation, reduction, oxide decomposition, carbonyl decomposition, electrolytic deposition, or precipitation from a solution.

**powder rolling**. See preferred term *roll compacting*.

**powder technology**. A broad term encompassing the production and utilization of both metal and nonmetal powders.

**power-driven hammer**. A forging hammer with a steam or air cylinder for raising the ram and augmenting its downward blow. Also known as power drop hammer. See also the figures accompanying the terms *drop hammer* and *gravity hammer*.

**power factor** (electricity). The ratio of the power to the effective values of the electromotive force multiplied by the effective value of current, in volts and amperes, respectively. The cosine of the angle between voltage applied and the resulting current.

**power reel**. A reel that is driven by an electric motor or some other source of power, used to wind or coil strip or wire as it is drawn through a continuous normalizing furnace, through a die, or through rolls.

**PP**. See *polypropylenes*.

**PPE**. Polyphenylene ether. See also *polyphenylene oxides*.

**PPO**. See *polyphenylene oxides*.

**PPS**. See *polyphenylene sulfides*.

**prealloyed powder**. A metallic powder composed of two or more elements that are alloyed in the powder manufacturing process and in which the particles are of the same nominal composition throughout.

**prearc (or prespark) period**. In *emission spectroscopy*, the time interval after the initiation of an arc (or spark) discharge during which the emitted radiation is not recorded for analytical purposes.

**prebond treatment**. Synonym for surface preparation prior to *adhesive bonding*.

**precharge**. In metal forming, the pressure introduced into the cavity prior to forming of the part.

**precious metal**. See *Technical Brief 34*.

**precipitation**. In metals, the separation of a new phase from solid or liquid solution, usually with changing conditions of temperature, pressure, or both.

**precipitation (deposit) etching**. Development of microstructure in a metallographic specimen through formation of reaction products at the surface of the microsection. See also *staining*.

**precipitation hardening**. Hardening in metals caused by the precipitation of a constituent from a supersaturated solid solution. See also *age hardening* and *aging*.

**precipitation heat treatment**. *Artificial aging* of metals in which a constituent precipitates from a supersaturated solid solution.

**precision**. The reproducibility of measurements within a set, that is, the scatter or dispersion of a set of data points about its central axis. Generally expressed as standard deviation or relative standard deviation.

**precision casting**. A metal casting of reproducible, accurate dimensions, regardless of how it is made. Often used interchangeably with *investment casting*.

**precision forging**. A forging produced to closer tolerances than normally considered standard by the industry. With precision forging, a net shape, or at least a near-net shape, can be produced in the as-forged condition. See also *net shape*.

**precision grinding**. Machine grinding to specified dimensions and low *tolerances*.

**precoat**. (1) In investment casting, a special refractory slurry applied to a wax or plastic expendable pattern to form a thin coating that serves as a desirable base for application of the main slurry. See also the figure accompanying the term *investment casting*. (2) To make the thin coating. (3) The thin coating itself.

**precoated metal products**. Mill products that have a metallic, organic, or conversion coating applied to their surfaces before they are fabricated into parts.

**precoating** (joining). Coating the base metal in the joint by dipping, electroplating, or other applicable means prior to brazing or soldering.

**preconditioning**. Any preliminary exposure of a plastic to specified atmospheric conditions for the purpose of favorably approaching equilibrium with a prescribed atmosphere.

**precracked specimen**. A mechanical test specimen that is notched and subjected to alternating stresses until a crack has developed at the root of the notch (Fig. 409).

**precure**. The full or partial setting of a synthetic resin or adhesive in a joint before the clamping operation is complete or before pressure is applied. See also *cure*.

FIG. 409

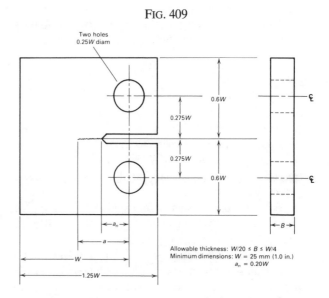

*Standard compact-type precracked specimen for fatigue crack propagation testing*

## Technical Brief 34: Precious Metals

THE EIGHT PRECIOUS METALS, listed in order of their atomic number as found in periods 5 and 6 (groups VIII and Ib) of the periodic table, are ruthenium, rhodium, palladium, silver, osmium, iridium, platinum, and gold. Precious metals, also referred to as noble metals, are of inestimable value to modern civilization. Their functions in coins, jewelry, and bullion, and as catalysts in devices to control auto exhaust emissions, are widely understood. But in certain applications, their functions are not as spectacular and, although vital to the application, are largely unknown except to the users. For example, precious metals are used in dental restorations and dental fillings; thin precious metal films are used to form electronic circuits; and certain organometallic compounds containing platinum are significant drugs for cancer chemotherapy.

Silver is a bright, white metal which, next to gold, is the most easily fabricated metal in the periodic table. It is very soft and ductile in the annealed condition. Silver does not oxidize at room temperature, but it is attacked by sulfur. Nitric, hydrochloric, and sulfuric acids attack silver, but the metal is resistant to many organic acids and to sodium and potassium hydroxide. The primary application for silver (about 50% of the silver demand) is its use for photographic emulsions. The use of silver in photography is based on the ability of exposed silver halide salts to undergo a secondary image amplification process called development. The second largest use is in the electrical and electronic industries for electrical contacts, conductors, and in primary batteries. Other applications include brazing alloys, dental alloys, electroplated ware, sterling ware (see *sterling silver*), and jewelry and coins.

Gold is a bright, yellow, soft, and very ductile metal. Its special properties include corrosion resistance, good reflectance, resistance to sulfidation and oxidation, and high electrical and thermal conductivity. Because gold is easy to fashion, has a bright pleasing color, is nonallergenic, and remains tarnish free indefinitely, it is used extensively in jewelry (about 55% of the gold market). For much the same reasons, it is used in dental alloys and appliances. Gold is also used to a considerable extent in electronic devices, particularly in printed circuit boards, connectors, keyboard contactors, and miniaturized circuitry. Other applications include gold films used as a reflector of infrared radiation in thermal barrier windows for large buildings and space vehicles, fired-on gold organometallic compounds used for decorating glass and china, sliding electrical contacts, and brazing alloys.

The six remaining precious metals are referred to as the platinum-group metals because they are closely related and commonly occur together in nature. Ruthenium, rhodium, and palladium each have a density of approximately 12 g/cm$^3$; osmium, iridium, and platinum each have a density of about 22 g/cm$^3$. The most distinctive trait of the platinum-group metals is their exceptional resistance to corrosion. Of the six metals, platinum has the most outstanding properties and is the most used, the primary application being its use as an automobile exhaust emission catalyst. Second in industrial importance is palladium, which is used primarily in electrical

---

**precursor**. With respect to carbon or graphite fiber, the rayon, polyacrylonitrile (PAN), or pitch fibers from which carbon and graphite fibers are derived. See also *carbon fiber*, *pitch*, and *polyacrylonitrile (PAN)*.

**preferred orientation**. A condition of a polycrystalline aggregate in which the crystal orientations are not random, but rather exhibit a tendency for alignment with a specific direction in the bulk material, commonly related to the direction of working. See also *texture*.

**prefit**. A process for checking the fit of mating detail parts in an assembly prior to *adhesive bonding* to ensure proper bond lines. Mechanically fastened structures are sometimes prefitted to establish shimming requirements.

**preform** (ceramics). A porous ceramic mass in the shape of the desired final part that is infiltrated with metal to form ceramic-metal composite. See also *Lanxide process*.

**preform** (composites and plastics). A preshaped fibrous reinforcement formed by the distribution of chopped fibers or cloth by air, water flotation, or vacuum over the surface of a perforated screen to the approximate contour and thickness desired in the finished part. Also, a preshaped fibrous reinforcement of mat or cloth formed to the desired shape on a mandrel or mock-up before being placed in a mold press. (2) A compressed tablet or biscuit of plastic composition used for efficiency in handling and accuracy in weighing materials. (3) To make plastic molding powder into pellets or tablets.

applications. Rhodium and ruthenium are used as alloying elements in platinum and palladium, while osmium and iridium are used for wear-resistant and heat-resistant applications, respectively.

## Nominal composition and solidification temperatures for silver-base brazing filler metals

| AWS designation(a) | UNS No. | Composition, wt% | | | | | | | | Other elements, total(b) | Solidification temperatures | | | | | |
| | | Ag | Cu | Zn | Cd | Ni | Sn | Li | Mn | | Solidus | | Liquidus | | Brazing temperature range | |
| | | | | | | | | | | | °C | °F | °C | °F | °C | °F |
|---|---|---|---|---|---|---|---|---|---|---|---|---|---|---|---|---|
| BAg-1 | P07450 | 44.0–46.0 | 14.0–16.0 | 14.0–18.0 | 23.0–25.0 | ... | ... | ... | ... | 0.15 | 607 | 1125 | 618 | 1145 | 618–760 | 1145–1400 |
| BAg-1a | P07500 | 49.0–51.0 | 14.5–16.5 | 14.5–18.5 | 17.0–19.0 | ... | ... | ... | ... | 0.15 | 627 | 1160 | 635 | 1175 | 635–760 | 1175–1400 |
| BAg-2 | P07350 | 34.0–36.0 | 25.0–27.0 | 19.0–23.0 | 17.0–19.0 | ... | ... | ... | ... | 0.15 | 607 | 1125 | 702 | 1295 | 702–843 | 1295–1550 |
| BAg-2a | P07300 | 29.0–31.0 | 26.0–28.0 | 21.0–25.0 | 19.0–21.0 | ... | ... | ... | ... | 0.15 | 607 | 1125 | 710 | 1310 | 710–843 | 1310–1550 |
| BAg-3 | P07501 | 49.0–51.0 | 14.5–16.5 | 13.5–17.5 | 15.0–17.0 | 2.5–3.5 | ... | ... | ... | 0.15 | 632 | 1170 | 688 | 1270 | 688–816 | 1270–1500 |
| BAg-4 | P07400 | 39.0–41.0 | 29.0–31.0 | 26.0–30.0 | ... | 1.5–2.5 | ... | ... | ... | 0.15 | 671 | 1240 | 779 | 1435 | 779–899 | 1435–1650 |
| BAg-5 | P07453 | 44.0–46.0 | 29.0–31.0 | 23.0–27.0 | ... | ... | ... | ... | ... | 0.15 | 663 | 1225 | 743 | 1370 | 743–843 | 1370–1550 |
| BAg-6 | P07503 | 49.0–51.0 | 33.0–35.0 | 14.0–18.0 | ... | ... | ... | ... | ... | 0.15 | 688 | 1270 | 774 | 1425 | 774–871 | 1425–1600 |
| BAg-7 | P07563 | 55.0–57.0 | 21.0–23.0 | 15.0–19.0 | ... | ... | 4.5–5.5 | ... | ... | 0.15 | 618 | 1145 | 652 | 1205 | 652–760 | 1205–1400 |
| BAg-8 | P07720 | 71.0–73.0 | bal | ... | ... | ... | ... | ... | ... | 0.15 | 779 | 1435 | 779 | 1435 | 779–899 | 1435–1650 |
| BAg-8a | P07723 | 71.0–73.0 | bal | ... | ... | ... | ... | 0.25–0.50 | ... | 0.15 | 766 | 1410 | 766 | 1410 | 766–871 | 1410–1600 |
| BAg-9 | P07650 | 64.0–66.0 | 19.0–21.0 | 13.0–17.0 | ... | ... | ... | ... | ... | 0.15 | 671 | 1240 | 718 | 1325 | 718–843 | 1325–1550 |
| BAg-10 | P07700 | 69.0–71.0 | 19.0–21.0 | 8.0–12.0 | ... | ... | ... | ... | ... | 0.15 | 691 | 1275 | 738 | 1360 | 738–843 | 1360–1550 |
| BAg-13 | P07540 | 53.0–55.0 | bal | 4.0–6.0 | ... | 0.5–1.5 | ... | ... | ... | 0.15 | 718 | 1325 | 857 | 1575 | 857–968 | 1575–1775 |
| BAg-13a | P07560 | 55.0–57.0 | bal | ... | ... | 1.5–2.5 | ... | ... | ... | 0.15 | 771 | 1420 | 893 | 1640 | 871–982 | 1600–1800 |
| BAg-18 | P07600 | 59.0–61.0 | bal | ... | ... | ... | 9.5–10.5 | ... | ... | 0.15 | 602 | 1115 | 718 | 1325 | 718–843 | 1325–1550 |
| BAg-19 | P07925 | 92.0–93.0 | bal | ... | ... | ... | ... | 0.15–0.30 | ... | 0.15 | 760 | 1400 | 891 | 1635 | 877–982 | 1610–1800 |
| BAg-20 | P07301 | 29.0–31.0 | 37.0–34.0 | 30.0–34.0 | ... | ... | ... | ... | ... | 0.15 | 677 | 1250 | 766 | 1410 | 766–871 | 1410–1600 |
| BAg-21 | P07630 | 62.0–64.0 | 27.5–29.5 | ... | ... | 2.0–3.0 | 5.0–7.0 | ... | ... | 0.15 | 691 | 1275 | 802 | 1475 | 802–899 | 1475–1650 |
| BAg-22 | P07490 | 48.0–50.0 | 15.0–17.0 | 21.0–25.0 | ... | 4.0–5.0 | ... | ... | 7.0–8.0 | 0.15 | 680 | 1260 | 699 | 1290 | 699–830 | 1290–1525 |
| BAg-23 | P07850 | 84.0–86.0 | ... | ... | ... | ... | ... | ... | bal | 0.15 | 960 | 1760 | 970 | 1780 | 970–1038 | 1780–1900 |
| BAg-24 | P07505 | 49.0–51.0 | 19.0–21.0 | 26.0–30.0 | ... | 1.5–2.5 | ... | ... | ... | 0.15 | 660 | 1220 | 705 | 1305 | 705–843 | 1305–1550 |
| BAg-26 | P07250 | 24.0–26.0 | 37.0–39.0 | 31.0–35.0 | ... | 1.5–2.5 | ... | ... | 1.5–2.5 | 0.15 | 705 | 1305 | 800 | 1475 | 800–870 | 1475–1600 |
| BAg-27 | P07251 | 24.0–26.0 | 34.0–36.0 | 24.5–28.5 | 12.5–14.5 | ... | ... | ... | ... | 0.15 | 605 | 1125 | 745 | 1375 | 745–860 | 1375–1575 |
| BAg-28 | P07401 | 39.0–41.0 | 29.0–31.0 | 26.0–30.0 | ... | ... | 1.5–2.5 | ... | ... | 0.15 | 650 | 1200 | 710 | 1310 | 710–843 | 1310–1550 |
| BAg-33 | P07252 | 24.0–26.0 | 29.0–31.0 | 26.5–28.5 | 16.5–18.5 | ... | ... | ... | ... | 0.15 | 607 | 1125 | 682 | 1260 | 682–760 | 1260–1400 |
| BAg-34 | P07380 | 37.0–39.0 | 31.0–33.0 | 26.0–30.0 | ... | ... | 1.5–2.5 | ... | ... | 0.15 | 650 | 1200 | 721 | 1330 | 721–843 | 1330–1550 |

(a) AWS, American Welding Society. (b) The brazing alloy shall be analyzed for the specific elements for which values are shown in this table. If the presence of other elements is indicated in the course of this work, the amount of those elements shall be determined to ensure that their total does not exceed the limit specified for other elements.

## Recommended Reading

- A.R. Robertson, Precious Metals and Their Uses, *Metals Handbook*, 10th ed., Vol 2, ASM International, 1990, p 688-698
- J.R. Davis, Ed., Properties of Precious Metals, *Metals Handbook*, 10th ed., Vol 2, ASM International, 1990, p 699-719
- G.D. Smith and E. Zysk, Corrosion of the Noble Metals, *Metals Handbook*, 9th ed., Vol 13, ASM International, 1987, p 793-807

**preform** (joining). Brazing or soldering filler metal fabricated in a shape or form for a specific application (Fig. 410).

**preform binder**. A resin applied to the chopped strands of a *preform*, usually during its formation, and cured so that the preform will retain its shape and can be handled.

**preformed ceramic core**. In foundry practice, a preformed refractory aggregate inserted in a wax or plastic pattern to shape the interior of that part of a casting which cannot be shaped by the pattern (Fig. 411). The wax is sometimes injected around the preformed core. See also *investment casting*.

**preforming**. (1) The initial pressing of a metal powder to form a compact that is to be subjected to a subsequent pressing operation other than coining or sizing. (2) Preliminary forming operations, especially for impression-die forging.

FIG. 410

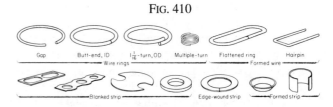

*Typical filler metal preforms used in brazing*

**pregel**. An unintentional, extra layer of cured resin on part of the surface of a reinforced plastic. Not related to gel coat.

FIG. 411

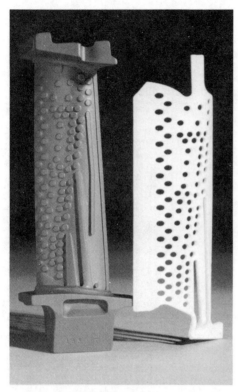

*Investment cast turbine blade with convex wall removed showing complex preformed ceramic core*

**preheat.** See *preheating* and *preheat temperature*.

**preheat current.** In resistance welding, an impulse or series of current impulses which occurs prior to and is separated from welding current.

**preheating** (metals). (1) Heating before some further thermal or mechanical treatment. For tool steel, heating to an intermediate temperature immediately before final austenitizing. For some nonferrous alloys, heating to a high temperature for a long time, in order to homogenize the structure before working. (2) In welding and related processes, heating to an intermediate temperature for a short time immediately before welding, brazing, soldering, cutting, or thermal spraying. (3) In powder metallurgy, an early stage in the sintering procedure when, in a continuous furnace, lubricant or binder burnoff occurs without atmosphere protection prior to actual sintering in the protective atmosphere of the high heat chamber.

**preheating** (plastics). The heating of a polymeric compound before molding or casting to facilitate the operation or reduce the molding cycle.

**preheat temperature.** A specified temperature that the base metal must attain in the welding, brazing, soldering, thermal spraying, or cutting area immediately before these operations are performed.

**preheat time.** In resistance welding, a portion of the preweld interval during which preheat current occurs.

**preimpregnation** (reinforced plastics). The practice of mixing resin and reinforcement and effecting partial cure before use or shipment to the user. See also *prepreg*.

**premix** (metals). (1) A uniform mixture of components prepared by a metal powder producer for direct use in compacting. (2) A term sometimes applied to the preparation of a premix; see preferred term *mixing*.

**premix** (plastics). A plastic molding compound prepared prior to and apart from the molding operations and containing all components required for molding: resin, reinforcement, fillers, catalysts, release agents, and other ingredients.

**premix burner.** A burner used in flame emission and atomic absorption spectroscopy in which the fuel gas is mixed with the oxidizing gas before reaching the combustion zone. See also the figure accompanying the term *nebulizer*.

**premolding.** In composites fabrication, the *lay-up* and partial *cure* at an intermediate cure temperature of a laminated or chopped-fiber detail part to stabilize its configuration for handling and assembly with other parts for final cure.

**preplasticization.** Technique of premelting injection molding plastic powders in a separate chamber, then transferring the melt to the injection chamber. See also *injection molding* (plastics).

**preply.** A composite material *lamina* in the raw-material stage, ready to be fabricated into a finished *laminate*. The lamina is usually combined with other raw laminae before fabrication. A preply includes a fiber system that is placed in position relative to all or part of the required matrix material to constitute the finished lamina. An organic matrix preply is called a *prepreg*.

**prepolymer.** A chemical intermediate with a molecular weight between that of the *monomer* or monomers and the final *polymer* or resin.

**prepolymer molding.** In the urethane foam field, a system in which a portion of the *polyol* is prereacted with the isocyanate to form a liquid *prepolymer* with a viscosity range suitable for pumping or metering. This component is supplied to end-users with a second premixed blend of additional polyol, catalyst, blowing agent, and so forth. When the two components are subsequently mixed, foaming occurs. See also *one-shot molding*.

**prepreg.** In composites fabrication, either ready-to-mold material in sheet form or ready-to-wind material in *roving* form, which may be cloth, mat, unidirectional fiber, or paper, impregnated with resin and stored for use. The resin is partially cured to a *B-stage* and supplied to the fabricator, who lays up the finished shape and completes the cure with heat and pressure. The two distinct types of prepreg available are commercial prepregs, in which the roving is coated with a hot melt or solvent system to produce a specific product to meet specific customer requirements; and wet prepreg, in which the basic resin is installed without solvents or preservatives but has limited room-temperature shelf-life.

**preproduction test.** A test or series of tests conducted by an adhesive manufacturer to determine conformity of an adhesive batch to established production standards, or by a fabricator to determine the quality of an adhesive before parts are produced, or by an adhesive specification custodian to determine conformance of an adhesive to the requirements of a specification not requiring qualification tests. Compare with *acceptance test* and *qualification test*.

**preshadowed replica.** A replica for fractographic or metallographic inspection that is formed by the application of shadowing material to the surface to be replicated. It is formed before the thin replica film is cast or otherwise deposited on the surface. See also *shadowing*.

**presintered blank.** A metal powder compact sintered at a low temperature but at a long enough time to make it sufficiently strong for metal working. See also *presintering*.

**presintered density.** The relative density of a presintered compact. See also *presintering*.

**presintering.** Heating a powder metallurgy compact to a temperature below the final sintering temperature, usually to increase the ease of handling or shaping of a compact or to remove a lubricant or binder (*burnoff*) prior to sintering.

**press.** A machine tool having a stationary bed and a slide or ram that has reciprocating motion at right angles to the bed surface, the slide being guided in the frame of the machine. See also *hydraulic press*, *mechanical press*, *slide*, and *straight-side press*.

**press brake.** An open-frame single-action press used to bend, blank, corrugate, curl, notch, perforate, pierce, or punch sheet metal or plate. See also the figure accompanying the term *mechanical press brake*.

**press-brake forming.** A metal forming process in which the workpiece is placed over an open die and pressed down into the die by a punch that is actuated by the ram portion of a *press brake* (Fig. 412). The process is most widely used for the forming of relatively long, narrow parts that are not adaptable to *press forming* and for applications in which production quantities are too small to warrant the tooling cost for contour *roll forming*.

**press capacity.** The rated force a *press* is designed to exert at a predetermined distance above the bottom of the stroke of the *slide*.

**press clave.** In composites fabrication, a simulated autoclave made by using the platens of a press to seal the ends of an open chamber, providing both the force required to prevent loss of the pressurizing medium and the heat required to cure the laminate inside.

FIG. 412

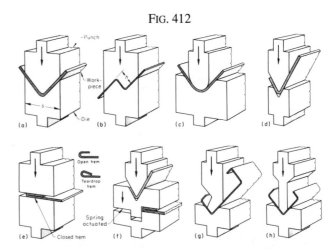

*Dies and punches most commonly used in press-brake forming. (a) 90° V-bending. (b) Offset bending. (c) Radiused 90° bending. (d) Acute-angle bending. (e) Flattening for three types of hems. (f) Combination bending and flattening. (g) Gooseneck punch for multiple bends. (h) Special clearance punch for multiple bends*

**pressed bar.** (1) A powder metallurgy compact in the form of a bar. (2) A green (unsintered) rectangular compact.

**pressed density.** The weight per unit volume of an unsintered compact. Same as green density.

**press fit.** An interference or *force fit* made through the use of a *press*.

**press forging.** The forging of metal between dies by mechanical or hydraulic pressure; usually accompanied with a single work stroke of the press for each die station.

**press forming.** Any sheet metal forming operation performed with tooling by means of a mechanical or hydraulic press (Fig. 413).

**pressing.** (1) In metalworking, the product or process of shallow drawing of sheet or plate. (2) Forming a powder metallurgy part with compressive force. See also *compacting*.

**pressing area.** The clear distance (left to right) between housings, stops, gibs, gibways, or shoulders of strain rods, multiplied by the total distance from front to back on the bed of a metal forming *press*. Sometimes called working area.

**pressing crack.** A rupture in a green powder metallurgy compact that develops during ejection of the compact from the die. Sometimes referred to as a slip crack.

**pressing skin.** The surface of a powder metallurgy compact that is superficially more deformed than the interior due to a preferential alignment of the particles caused by contact with the die wall and punch faces.

**press load.** The amount of force exerted in a given forging or forming operation.

FIG. 413

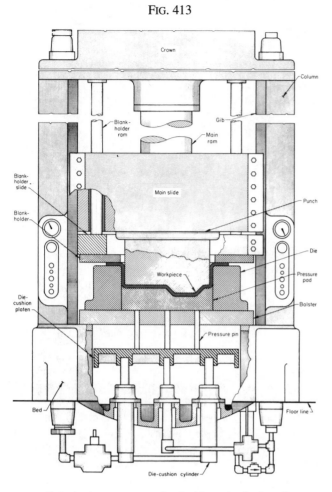

*Principal components of a double-action hydraulic press used for press forming*

**press quenching**. A quench in which hot dies are pressed and aligned with a part before the quenching process begins. Then the part is placed in contact with a quenching medium in a controlled manner. This process avoids part distortion.

**press slide**. See *slide*.

**press tool**. The complete tool assembly used for forming powder metallurgy compacts that consists of the die, a die adaptor, the punches, and, when required, a core rod.

**pressure bag molding**. A process for molding reinforced plastics in which a tailored, flexible bag is placed over the contact lay-up on the mold, sealed, and clamped in place. Fluid pressure, usually provided by compressed air or water, is placed against the bag, and the part is cured. See also *vacuum bag molding*.

**pressure break**. In laminated plastics, a break in one or more outer sheets of the paper, fabric, or other base, which is visible through the surface layer of resin that covers it.

**pressure bubble plug-assist forming**. A *thermoforming* process in which a heated plastic sheet is positioned over

FIG. 414

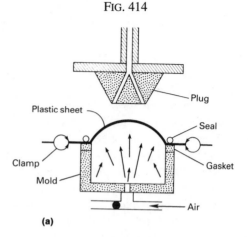

(a)

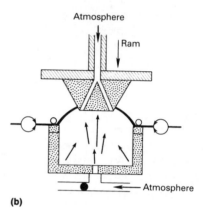

(b)

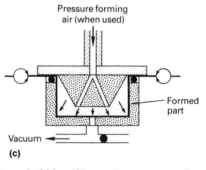

(c)

*Pressure bubble plug-assist vacuum thermoforming. (a) Air blown through bottom of mold cavity stretches the sheet into a bubble shape. (b) Plug pushes sheet into cavity. (c) Vacuum draws sheet against mold cavity surface*

the mold and air is blown up through the base plate channel causing the sheet to billow upward. The plug assist pushes the sheet into the mold cavity and vacuum is applied to transfer the sheet from the plug to the mold surface (Fig. 414).

**pressure casting**. (1) Making castings with pressure on the molten or plastic metal, as in *injection molding*, *die casting*, *centrifugal casting*, cold chamber pressure casting, and *squeeze casting*. (2) A casting made with pressure applied to the molten or plastic metal.

**pressure-controlled welding.** A resistance welding process variation in which a number of spot or projection welds are made with several electrodes functioning progressively under the control of a pressure-sequencing device.

**pressure gas welding.** An oxyfuel gas welding process that produces coalescence simultaneously over the entire area of abutting surfaces by heating them with gas flames obtained from combustion of a fuel gas with oxygen and by application of pressure, without the use of filler metal.

**pressure-impregnation-carbonization (PIC).** A densification process for carbon-carbon composites involving pitch impregnation and carbonization under high temperature and isostatic pressure conditions. This process is carried out in hot isostatic pressing equipment. See also *carbon-carbon composites* (Technical Brief 6).

**pressure intensifier.** In composites fabrication, a layer of flexible material (usually a high-temperature rubber) used to ensure the application of sufficient pressure to a location, such as a radius, in a *lay-up* being cured.

**pressureless sintering.** Sintering of loose powder.

**pressure lubrication.** A system of lubrication in which the lubricant is supplied to a bearing under pressure.

**pressure plate.** A plate located beneath the *bolster plate* in a metal forming *press* that acts against the resistance of a group of cylinders mounted to the pressure plate to provide uniform pressure throughout the press stroke when the press is symmetrically loaded.

**pressure-sensitive adhesives.** See *Technical Brief 35*.

**pressure sintering.** A hot pressing technique that usually employs low loads, high sintering temperatures, continuous or discontinuous sintering, and simple molds to contain the powder. Although the terms pressure sintering and *hot pressing* are used interchangeably, distinct differences exist between the two processes. In pressure sintering, the emphasis is on thermal processing; in hot pressing, applied pressure is the main process variable.

**pressure-viscosity coefficient.** The slope of a graph showing variation in the logarithm of viscosity with pressure. The use of the term pressure-viscosity coefficient assumes a linear relationship.

**pressure welding.** See preferred terms *cold weldng, diffusion welding, forge welding, hot pressure welding, pressure gas welding,* and *solid-state welding*.

**pressurized gas lubrication.** A system of lubrication in which a gaseous lubricant is supplied under sufficient external pressure to separate the opposing surfaces by a gas film.

**pretinning.** See preferred term *precoating*.

**preweld interval.** In resistance welding, the time elapsing between the end of squeeze time and the beginning of welding current in making spot welds and in projection or

FIG. 415

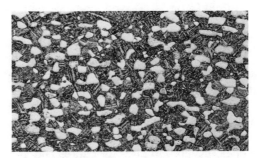

*Primary alpha (light) in a matrix of transformed beta containing acicular alpha in a Ti-6Al-4V alloy. 250×*

upset welding. In flash welding, it is the time during which the material is preheated.

**primary alloy.** An alloy whose major constituent has been refined directly from ore, not recycled scrap metal. Compare with *secondary alloy*.

**primary alpha.** Alpha phase in a titanium alloy crystallographic structure that is retained from the last high-temperature α-β working or heat treatment (Fig. 415). The morphology of α is influenced by the prior thermomechanical history.

**primary creep.** The first, or initial, stage of *creep*, or time-dependent deformation. See also the figure accompanying the term *creep*.

**primary crystals.** The first type of crystals that separate from a melt during solidification.

**primary current distribution.** The current distribution in an *electrolytic cell* that is free of *polarization*.

**primary extinction.** In *x-ray diffraction*, a decrease in intensity of a diffracted x-ray beam caused by perfection of crystal structure extending over such a distance (approximately 1 μm or greater) that interference between multiply reflected beams inside the crystal decreases the intensity of the externally diffracted beam.

**primary gas.** In thermal spraying, the gas constituting the major constituent of the arc gas fed to the gun to produce the plasma.

**primary leakage.** In seals, the leakage of a mechanical seal, with the fluid escaping from the region between the end faces of the primary sealing elements.

**primary metal.** Metal extracted from minerals and free of reclaimed metal scrap. Compare with *secondary metal* and *native metal*.

**primary mill.** A mill for rolling ingots or the rolled products of ingots to blooms, billets, or slabs. This type of mill is often called a *blooming mill* and sometimes called a *cogging mill*.

**primary nucleation.** The mechanism by which crystallization is initiated in plastics, often by an added nucleation agent.

## Technical Brief 35: Pressure-Sensitive Adhesives

PRESSURE-SENSITIVE ADHESIVES (PSAs) are capable of holding substrates together when they are brought into contact under brief pressure at room temperature. PSAs are either unsupported or are supported by various carriers, including paper, cellophane, plastic films, cloth, and metal foil. Both single- and double-sided tapes and films are included. Most of the adhesives are based on rubbers compounded with additives, including tackifiers. PSA materials include, in order of decreasing volume and increasing price, natural rubber, styrene-butadiene rubber, reclaimed rubber, butyl rubber, butadiene-acrylonitrile rubber, thermoplastic elastomers, polyacrylates, polyvinylalkylethers, and silicones.

Materials used as PSAs are usually available as solvent systems or hot melts. Substrates are coated with one of these two types of adhesives, and usually no cure of the material is involved upon its application. Adhesive-coated substrates, in their dry state, are permanently tacky at room temperature and do not require activation by water, solvents, or heat. There are two major classes of PSAs: adhesives that are compounded to form PSAs and adhesives that are inherently pressure sensitive and require little or no compounding. Included in the former category are elastomers, and in the latter category are polyacrylates and polyvinylalkylethers.

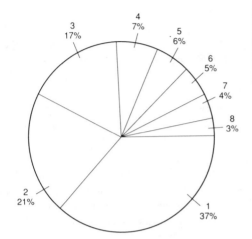

*Pressure-sensitive adhesives market in 1988. (1) Packaging. (2) Hospital/ first aid. (3) Office/graphic arts. (4) Construction. (5) Corrosion protection. (6) Automotive. (7) Electrical. (8) Other*

Tapes, the largest area of application of PSAs, can be classified by construction, function, application, or texture. The construction category includes fabric tapes, paper tapes, foil tapes, film tapes, nonwoven fabric tapes, reinforced tapes, foam tapes, two-faced tapes, and transfer tapes. The backing, rather than the adhesive, is the distinguishing feature of this tape classification. The function category includes masking, holding, sealing, reinforcing, protecting, bundling, stenciling, splicing, identifying, insulating, packaging, and mounting. The application category includes hospital and first aid tapes, office and graphic art tapes, building industry tapes, packaging and surface protection tapes, electrical tapes, automotive industry tapes, shoe industry tapes, and corrosion protection tapes. The texture category includes floor tiles, wall coverings, automobile wood-grained films, and decorative sheets.

### Typical PSA bonding substrates and applications by chemical family

| Property | Natural rubber | Thermoplastic elastomers | Acrylates | Silicones |
|---|---|---|---|---|
| Typical solids content, %(a) | 35–50 | 22–50 | 25–52 | 55–60 |
| Bonding substrates | Wood, metals, leather, paper, plastics, textiles | Wood, metals, leather, paper, glass, plastics, textiles | Glass, metals, plastics, paper | Elastomers, metals, plastics |
| Applications | Tapes General-purpose uses | Apparel, leather, footware, tapes, floor and wall coverings, furniture, fixtures | General industrial uses, tapes, labels, decals, trims, moldings | Tapes, coated films and fabrics, electronics |

(a) Dispersed in solvents

Labels are the second largest application for PSAs. Specific characteristics that are important, and which differentiate labels from tapes, include backing material printability, flatness, ease of die cutting, and release paper properties. The adhesive is applied to the release paper and allowed to dry. This material is then laminated to the label stock.

### Recommended Reading

- M.M. Gauthier, Types of Adhesives, *Engineered Materials Handbook*, Vol 3, ASM International, 1990, p 74-93
- D. Satos, Ed., *Handbook of Pressure Sensitive Adhesive Technology*, Van Nostrand Reinhold, 1982

**primary passive potential (passivation potential).** The potential corresponding to the maximum active current density (critical anodic current density) of an electrode that exhibits active-passive corrosion behavior.

**primary recrystallization.** A process by which nucleation and growth of a new generation of strain-free grains occur in a matrix which has been plastically deformed. See also *secondary recrystallization.*

**primary x-ray.** The emergent beam from the x-ray source.

**primer.** A coating applied to a surface, prior to the application of an adhesive to improve the performance of the bond. The coating can be a low-viscosity fluid that is typically a 10% solution of the adhesive in an organic solvent, which can wet out the adherend surface to leave a coating over which the adhesive can readily flow. See also the figure accompanying the term *adhesive bonding.*

**primes.** Metal products, principally sheet and plate, of the highest quality and free from blemishes or other visible imperfections.

**principal stress (normal).** The maximum or minimum value of the *normal stress* at a point in a plane considered with respect to all possible orientations of the considered plane. On such principal planes the shear stress is zero. There are three principal stresses on three mutually perpendicular planes. The state of stress at a point may be (1) uniaxial, a state of stress in which two of the three principal stresses are zero, (2) biaxial, a state of stress in which only one of the three principal stresses is zero, and (3) triaxial, a state of stress in which none of the principal stresses is zero. Multiaxial stress refers to either biaxial or triaxial stress.

**printed circuit.** An electronic circuit produced by printing an electrically conductive pattern, wiring, or components on a supporting dielectric substrate which may be either rigid or flexible.

**printing.** A metallographic method in which a carrier material is saturated with an etchant and pressed against the surface of the specimen. The etchant reacts with one of the phases, and substances form that react with the carrier material, leaving behind a life-size image. Used for exposing particular elements—for example, sulfur (sulfur prints).

**prism.** A transparent optical element whose entrance and exit apertures are polished plane faces. Using refraction and/or internal reflection, prisms are used to change the direction of propagation of monochromatic light and to disperse polychromatic light into its component wavelengths.

**prismatic plane.** In noncubic crystals, any plane that is parallel to the principal axis (*c* axis).

**probe ion.** In materials characterization, an ionic species intentionally produced by an ion source and directed onto the specimen surface at a known incident angle and a known energy.

FIG. 416

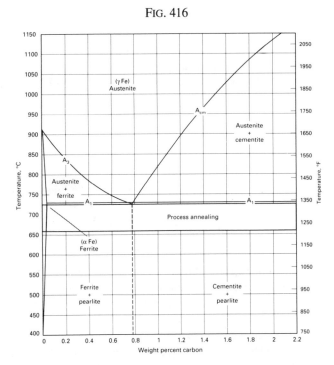

*The iron-carbon binary phase diagram showing region of temperature for process annealing*

**process annealing.** A heat treatment used to soften metal for further cold working. In ferrous sheet and wire industries, heating to a temperature close to but below the lower limit of the transformation range and subsequently cooling for working (Fig. 416). In the nonferrous industries, heating above the recrystallization temperatures at a time and temperature sufficient to permit the desired subsequent cold working.

**processing window.** In forming of plastics, the range of processing conditions, such as stock (melt) temperature, pressure, shear rate, and so on, within which a particular grade of plastic can be fabricated with optimum or acceptable properties by a particular fabricating process, such as extrusion, injection molding, sheet molding, and so forth. The processing window for a particular plastic can vary significantly with design of the part and the mold, with the fabricating machinery used, and with the severity of the end-use stresses.

**process metallurgy.** The science and technology of winning metals from their ores and purifying metals; sometimes referred to as chemical metallurgy. Its two chief branches are *extractive metallurgy* and *refining.*

**process tolerance.** The dimensional variations of a part characteristic of a specific process, once the setup is made.

**proeutectoid carbide.** Primary crystals of *cementite* formed directly in ferrous alloys from the decomposition of *austenite* exclusive of that cementite resulting from the eutectoid reaction. See also *eutectoid.*

Fig. 417

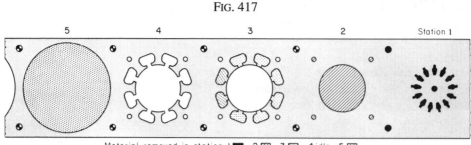

Material removed in station 1 ■, 2 ▨, 3 ▢, 4 idle, 5 ▦

*Blanking and piercing sequence for rotor and stator laminations in a five-station progressive die. Two pilot punches were used at each station. Station 1, pierce pilot holes, rotor slots, and rotor-shaft hole; station 2, pierce stator rivet holes and blank rotor; station 3, pierce stator slots; station 4, idle; station 5, blank stator*

**proeutectoid ferrite.** Primary crystals of *ferrite* formed directly in ferrous alloys from the decomposition of *austenite* exclusive of that ferrite resulting from the eutectoid reaction. See also *eutectoid*.

**proeutectoid phase.** Particles of a phase in ferrous alloys that precipitate during cooling after austenitizing but before the eutectoid transformation takes place. See also *eutectoid*.

**profile (contour) rolling.** In *ring rolling*, a process used to produce seamless rolled rings with a predesigned shape on the outside or the inside diameter, requiring less volume of material and less machining to produce finished parts.

**profiling.** Any operation that produces an irregular contour on a workpiece, for which a tracer or template-controlled duplicating equipment usually is employed.

**progression.** In metal forming equipment, the constant dimension between adjacent stations in a progressive die.

**progressive aging.** Aging by increasing the temperature in steps or continuously during the aging cycle. See also *aging* and compare with *interrupted aging* and *step aging*.

**progressive block sequence.** A welding sequence during which successive blocks are completed progressively along the joint, either from one end to the other or from the center of the joint toward either end. See also the figure accompanying the term *block sequence*.

**progressive die.** A *die* with two or more stations arranged in line for performing two or more operations on a part; one operation is usually performed at each station (Fig. 417).

**progressive forming.** Sequential forming at consecutive stations with a single die or separate dies.

**projected area.** The area of a cavity, or portion of a cavity, in a mold or die casting die measured from the projection on a plane that is normal to the direction of the mold or die opening.

**projection lens.** The final lens in the electron microscope corresponding to an ocular or projector in a compound optical microscope. This lens forms a real image on the viewing screen or photographic film.

**projection welding.** A resistance welding process which produces coalescence of metals with the heat obtained from resistance to electric current through the work parts held together under pressure by electrodes (Fig. 418). The resulting welds are localized at predetermined points by projections, embossments, or intersections. See also the figure accompanying the term *nugget* (2).

**promoter.** A chemical, itself a feeble catalyst, that greatly increases the activity of a given catalyst. Used in formulating plastics. See also *accelerator*.

**proof.** (1) To test a component or system at its peak operating load or pressure. (2) Any reproduction of a *die impression* in any material; often a lead or plaster cast. See also *die proof*.

**proof load.** A predetermined load, generally some multiple of the service load, to which a specimen or structure is submitted before acceptance for use.

**proof pressure.** The test pressure that pressurized components must sustain without detrimental deformation or damage. The proof pressure test is used to give evidence of satisfactory workmanship and material quality.

**proof stress.** (1) A specified stress to be applied to a member or structure to indicate its ability to withstand service loads. (2) The stress that will cause a specified small *permanent set* in a material.

**proportional limit.** The greatest stress a material is capable of developing without a deviation from straight-line proportionality between stress and strain. See also *elastic limit* and *Hooke's law*.

**propylene plastics.** Plastics based on polymers of propylene or copolymers of propylene with other monomers, the propylene being the greatest amount by mass. See also *polypropylenes (PP)*.

**protective atmosphere.** (1) A gas envelope surrounding the part to be brazed, welded, or thermal sprayed, with the gas composition controlled with respect to chemical composition, dew point, pressure, flow rate, and so forth. Examples are inert gases, combusted fuel gases, hydrogen, and vac-

FIG. 418

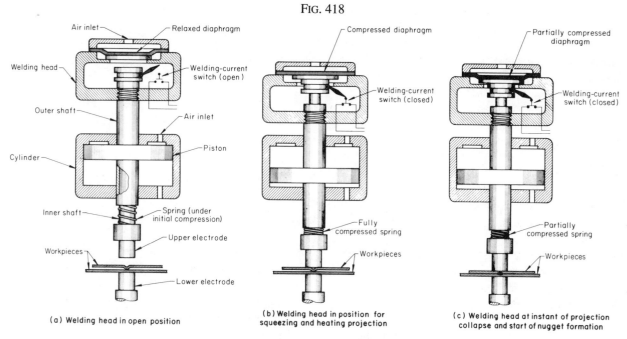

(a) Welding head in open position

(b) Welding head in position for squeezing and heating projection

(c) Welding head at instant of projection collapse and start of nugget formation

*Setup for projection welding*

uum. (2) The atmosphere in a heat treating or sintering furnace designed to protect the parts or compacts from oxidation, nitridation, or other contamination from the environment. Table 51 lists compositions of principal furnace protective atmospheres.

**protective potential.** The threshold value of the *corrosion potential* that has to be reached to enter a *protective potential range*.

**protective potential range.** A range of *corrosion potential* values in which an acceptable corrosion resistance is achieved for a particular purpose.

**prototype.** A model suitable for use in the complete evaluation of form, design, performance, and material processing.

**protuberances.** See *asperities*.

**prow formation.** See *wedge formation*.

**PS.** See *polystyrenes*.

**pseudobinary system.** (1) A three-component or ternary alloy system in which an intermediate phase acts as a component. (2) A vertical section through a ternary diagram.

**pseudocarburizing.** See *blank carburizing*.

**pseudonitriding.** See *blank nitriding*.

**pseudoplastic behavior.** A decrease in viscosity with increasing shear stress. Compare with *dilatant, rheopectic material,* and *thixotropy*.

**PSU.** See *polysulfones*.

**P-T diagram.** A two-dimensional graph of phase relationships in a system of any order by means of the pressure and temperature variables.

**PTFE.** See *polytetrafluoroethylenes*.

**P-T-X diagram.** A three-dimensional graph of phase relationships in a binary system by means of the pressure, temperature, and concentration variables.

## Table 51 Classification and application of principal furnace protective atmospheres

| Class | Description | Common application | Nominal composition, vol% | | | | |
|---|---|---|---|---|---|---|---|
| | | | $N_2$ | CO | $CO_2$ | $H_2$ | $CH_4$ |
| 101 | Lean exothermic | Oxide coating of steel | 86.8 | 1.5 | 10.5 | 1.2 | . . . |
| 102 | Rich exothermic | Bright annealing; copper brazing; sintering | 71.5 | 10.5 | 5.0 | 12.5 | 0.5 |
| 201 | Lean prepared nitrogen | Neutral heating | 97.1 | 1.7 | . . . | 1.2 | . . . |
| 202 | Rich prepared nitrogen | Annealing, brazing stainless steel | 75.3 | 11.0 | . . . | 13.2 | 0.5 |
| 301 | Lean endothermic | Clean hardening | 45.1 | 19.6 | 0.4 | 34.6 | 0.3 |
| 302 | Rich endothermic | Gas carburizing | 39.8 | 20.7 | . . . | 38.7 | 0.8 |
| 402 | Charcoal | Carburizing | 64.1 | 34.7 | . . . | 1.2 | . . . |
| 501 | Lean exothermic-endothermic | Clean hardening | 63.0 | 17.0 | . . . | 20.0 | . . . |
| 502 | Rich exothermic-endothermic | Gas carburizing | 60.0 | 19.0 | . . . | 21.0 | . . . |
| 601 | Dissociated ammonia | Brazing, sintering | 25.0 | . . . | . . . | 75.0 | . . . |
| 621 | Lean combusted ammonia | Neutral heating | 99.0 | . . . | . . . | 1.0 | . . . |
| 622 | Rich combusted ammonia | Sintering stainless powders | 80.0 | . . . | . . . | 20.0 | . . . |

**puckering.** Wrinkling or buckling in a drawn shell in an area originally inside the draw ring.

**puckers.** Local areas on prepreg material where the material has blistered and pulled away from the separator film or release paper. See also *prepreg.*

**puddle.** See preferred term *molten weld pool.*

**pull cracks.** In a casting, cracks that are caused by residual stresses produced during cooling, and that result from the shape of the object.

**pulled compact.** A powder metallurgy compact expanded by internal gas pressure.

**pulled surface.** In laminated plastics, imperfections in the surface, ranging from a slight breaking or lifting in localized areas to pronounced separation of the surface from the body.

**pulp molding.** The process by which a resin-impregnated pulp material is preformed by application of a vacuum and subsequent molding or curing.

**pulsation welding.** Sometimes used as a synonym for *multiple-impulse welding.*

**pulse.** A current of controlled duration through a resistance welding circuit.

**pulsed power welding.** Any arc welding process in which the power is cyclically varied to give short-duration pulses of either voltage or current that are significantly different from the average value.

**pulse time.** The duration of a pulse during resistance welding.

**pultrusion.** A continuous process for manufacturing composites that have a constant cross-sectional shape. The process consists of pulling a fiber-reinforcing material through a resin impregnation bath and then through a shaping die, where the resin is subsequently cured.

**pulverization.** The process of reducing metal powder particle sizes by mechanical means; also called comminution or mechanical disintegration.

**pumping efficiency.** In a bearing, the ratio of actual oil flow to the maximum theoretical flow for a bearing with a 180° oil film operating at an eccentricity ratio of unity.

**punch.** (1) The male part of a die—as distinguished from the female part, which is called the die (Fig. 419). The punch is usually the upper member of the complete die assembly and is mounted on the *slide* or in a *die set* for alignment (except in the inverted die). (2) In double-action draw dies, the punch is the inner portion of the upper die, which is mounted on the plunger (inner slide) and does the drawing. (3) The act of piercing or punching a hole. Also referred to as *punching.* (4) The movable tool that forces material into the die in powder molding and most metal forming operations. (5) The movable die in a trimming press or a forging machine. (6) The tool that forces the stock through the die in rod and tube extrusion and forms the internal surface in can or cup extrusion.

Fig. 419

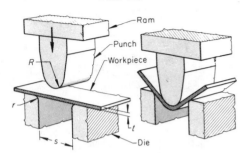

*Punch and die setup for press-brake forming. R, punch radius; r, die radius; s, span width; t, metal thickness*

**punching.** (1) The die shearing of a closed contour in which the sheared out sheet metal part is scrap. (2) Producing a hole by die shearing, in which the shape of the hole is controlled by the shape of the punch and its mating die. Multiple punching of small holes is called *perforating.* See also *piercing.*

**punch press.** (1) In general, any mechanical press. (2) In particular, an endwheel gap-frame press with a fixed bed, used in piercing.

**punch radius.** The radius on the end of the punch that first contacts the work, sometimes called *nose radius.* See also the figure accompanying the term *press.*

**punch-to-die clearance.** See *die clearance.*

**PUR.** See *polyurethanes* (Technical Brief 32).

**purge.** The removal of air from a sintering furnace chamber by replacing it with a vacuum or an inert gas prior to the introduction of the sintering atmosphere.

**purple plague.** A failure mechanism in electronic components which involves the formation of brittle intermetallic compounds at aluminum wire/gold bonding pad intersections (Fig. 420). Both moisture and temperature enhance the formation of such compounds. The term "purple plague" is used because one of the five compounds that can form appears purple to the eye when viewed through a microscope.

**push angle.** The angle between a welding electrode and a line normal to the face of the weld when the electrode is pointing forward along the weld joint. See also the figure accompanying the term *forehand welding.*

**push bench.** Equipment used for drawing moderately heavy-gage tubes by cupping sheet metal and forcing it through a die by pressure exerted against the inside bottom of the cup.

**pusher furnace.** A type of continuous furnace in which parts to be heated are periodically charged into the furnace in containers, which are pushed along the hearth against a line of previously charged containers thus advancing the containers toward the discharge end of the furnace, where

FIG. 420

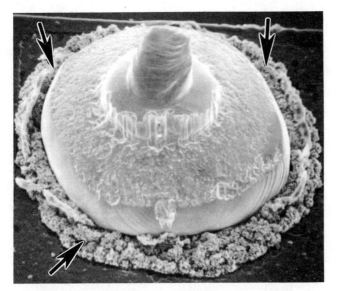

*Scanning electron micrograph of a gold-aluminum intermetallic product (purple plague) around a gold ball bond (arrows). 1000×*

FIG. 421

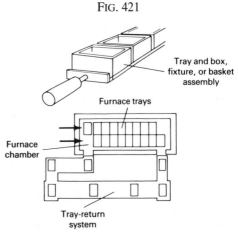

*Schematic of tray movement in a pusher furnace*

they are removed (Fig. 421). Pusher furnaces are widely used for heat treating and sintering of metals.

**pusher-type seal**. A mechanical seal in which the secondary seal is pushed along the shaft or sleeve to compensate for face wear.

**push fit**. A loosely defined fit similar to a *snug fit.*

**push welding**. Spot or projection welding in which the force is applied manually to one electrode, and the work or backing plate takes the place of the other electrode.

**PVAC**. See *polyvinyl acetate.*

**PVAL**. See *polyvinyl alcohol.*

**PVB**. See *polyvinyl butyral.*

**PVC**. See *polyvinyl chloride* (Technical Brief 33).

**PVDC**. See *polyvinylidene chloride.*

**PVDF**. See *polyvinylidene fluoride.*

**PVF**. See *polyvinyl formal.*

*PV* **factor**. The product of bearing pressure and surface velocity traditionally expressed in terms of (lb/in.$^2$) × ft/min); (the ISO equivalent is Pa · m/s).

**pyramidal plane**. In noncubic crystals, any plane that intersects all three axes.

**pyrolysis**. With respect to fibers, the thermal process by which organic precursor fiber materials, such as rayon, polyacrylonitrile (PAN), and pitch, are chemically changed into carbon fiber by the action of heat in an inert

atmosphere. Pyrolysis temperatures can range from 800 to 2800 °C (1470 to 5070 °F), depending on the precursor. Higher processing graphitization temperatures of 1900 to 3000 °C (3450 to 5430 °F) generally lead to higher-modulus carbon fibers, usually referred to as graphite fibers. During the pyrolysis process, molecules containing oxygen, hydrogen, and nitrogen are driven from the precursor fiber, leaving continuous chains of carbon. See also *carbon fiber* and *carbon-carbon composites* (Technical Brief 6).

**pyrometallurgy**. High-temperature *winning* or *refining* of metals.

**pyrometer**. A device for measuring temperatures above the range of liquid thermometers.

**pyrometry**. The measurement of temperatures, for example, by measuring the electrical resistance of wire, the thermoelectric force of a couple, the expansion of solids, liquids, or gases, the specific heat of solids, or the intensity of radiant energy per unit area.

**pyrophoricity**. The property of a substance with a large surface area to self-ignite and burn when exposed to oxygen or air.

**pyrophoric powder**. A powder whose particles self-ignite and burn upon exposure to oxygen or air. Example: fine zirconium powder.

**P-X diagram**. A two-dimensional graph of the isothermal phase relationships in a binary system; the coordinates of the graph are pressure and concentration.

**P-X projection**. A two-dimensional graph of the phase relationships in a binary system produced by making an orthographic projection of the phase boundaries of a *P-T-X diagram* upon a pressure-concentration plane.

# Q

**quadrivariant equilibrium.** In metals, a stable state among several conjugate phases equal to two less than the number of components, that is, having four degrees of freedom.

**qualification test.** A series of tests conducted by the procuring activity, or an agent thereof, to determine the conformance of materials, or materials system, to the requirements of a specification, normally resulting in a qualified products list under the specification. Generally, qualification under a specification requires a conformance to all tests in the specification, or it may be limited to conformance to a specific type or class, or both, under the specification. Compare with *acceptance test* and *preproduction test*.

**qualified products list (QPL).** A list of commercial plastic products that have been pretested and found to meet the requirements of a specification, especially a government specification.

**qualitative analysis.** An analysis in which some or all of the components of a sample are identified. Compare with *quantitative analysis*.

**quality.** (1) The totality of features and characteristics of a product or service that bear on its ability to satisfy a given need (fitness-for-use concept of quality). (2) Degree of excellence of a product or service (comparative concept). Often determined subjectively by comparison against an ideal standard or against similar products or services available from other sources. (3) A quantitative evaluation of the features and characteristics of a product or service (quantitative concept).

**quality characteristics.** Any dimension, mechanical property, physical property, functional characteristic, or appearance characteristic that can be used as a basis for measuring the quality of a unit of product or service.

**quality factor.** (1) In plastics, the ratio of elastic modulus to loss modulus, measured in tension, compression, flexure, or shear. This is a nondimensional term and is the reciprocal of tan delta. (2) The reciprocal of the electrical dissipation factor, which is the ratio of the power loss in a dielectric material to the total power transmitted through it; thus, the imperfection of the dielectric. Equal to the tangent of the loss angle or to the ratio of the dielectric loss to the dielectric constant.

**quantitative analysis.** A measurement in which the amount of one or more components of a sample is determined. Contrast with *qualitative analysis*.

**quantitative metallography.** Determination of specific characteristics of a microstructure by quantitative measurements on micrographs or metallographic images. Quan-

tities so measured include volume concentration of phases, grain size, particle size, mean free path between like particles or secondary phases, and surface area to volume ratio of microconstituents, particles, or grains. See also *image analysis*.

**quantum mechanics.** The modern theory of matter, of electromagnetic radiation, and of the interaction between matter and radiation; also, the mechanics of phenomena to which this theory may be applied. Quantum mechanics, also termed wave mechanics, generalizes and supercedes the older classical mechanics and Maxwell's electromagnetic theory.

**quantum number.** One of the quantities, usually discrete with integer or half-integer values, needed to characterize a quantum state of a physical system.

**quantum theory.** See *quantum mechanics*.

**quarter hard.** A *temper* of nonferrous alloys and some ferrous alloys characterized by tensile strength about midway between that of *dead soft* and *half hard* tempers.

**quartering.** A method of sampling a metal powder by dividing a cone-shaped heap into four parts, selecting one of them randomly, dividing this again into four parts, and repeating the procedure until the sample is small enough for particle size analysis.

**quartz.** One of several crystalline forms of silica ($SiO_2$); others include cristobalite and tridymite. All occur as minerals, also synthetic. Quartz is harder than most minerals, being 7 on the Mohs scale, and the crushed material is used as an abrasive. Fused quartz, or quartz glass is used for lightbulbs, optical glass, crucibles, and for tubes and rods in furnaces. Typical values for the maximum operating temperatures for quartz glass and competing glasses are:

| Glass | Temperature | |
|---|---|---|
| | °C | °F |
| Borosilicate | 500 | 930 |
| Aluminosilicate glass | 650 | 1200 |
| Doped quartz glass | 800 | 1470 |
| Vycor | 900 | 1650 |
| Quartz glass | 1100 | 2010 |

**quartz glass.** See *quartz* and *glass* (Technical Brief 17).

**quasi-binary system.** In a ternary or higher-order system, a linear composition series between two substances each of which exhibits congruent melting, wherein all equilibria, at all temperatures or pressures, involve only phases having compositions occurring in the linear series, so that the series may be represented as a binary on a *phase diagram*.

FIG. 422

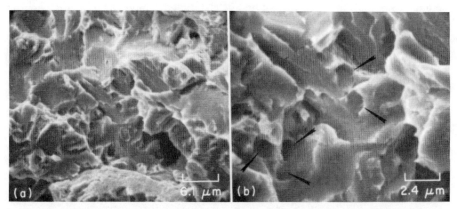

*Quasi-cleavage in the surface of an impact fracture in a specimen of 4340 steel showing small cleavage facets and shallow dimples (marked by arrows). (a) 1650×. (b) 4140×*

**quasi-cleavage fracture**. A fracture mode that combines the characteristics of *cleavage fracture* and *dimple fracture* (Fig. 422). An intermediate type of fracture found in certain high-strength metals.

**quasi-hydrodynamic lubrication**. A loosely defined regime of lubrication, especially in metalworking, where thin-film lubrication predominates.

**quasi-isotropic**. See *isotropic*.

**quasi-isotropic laminate**. A laminate approximating isotropy by orientation of plies in several or more directions (Fig. 423).

**quaternary system**. The complex series of compositions produced by mixing four components in all proportions.

**quench** (plastics). A process of shock-cooling thermoplastic materials from the molten state.

**quench-age embrittlement**. *Embrittlement* of low-carbon steels resulting from precipitation of solute carbon at existing dislocations and from precipitation hardening of the steel caused by differences in the solid solubility of carbon in ferrite at different temperatures. Quench-age embrittlement usually is caused by rapid cooling of the steel from temperatures slightly below $Ac_1$ (the temperature at which austenite begins to form), and can be minimized by quenching from lower temperatures.

**quench aging**. *Aging* induced by rapid cooling after *solution heat treatment*.

**quench annealing**. Annealing an austenitic ferrous alloy by *solution heat treatment* followed by rapid quenching.

**quench cracking**. Fracture of a metal during quenching from elevated temperature. Most frequently observed in hardened carbon steel, alloy steel, or tool steel parts of high hardness and low toughness. Cracks often emanate from fillets, holes, corners, or other stress raisers and result from high stresses due to the volume changes accompanying transformation to martensite.

**quench hardening**. (1) Hardening suitable alpha-beta alloys

FIG. 423

```
0°                    0°
0°                    90°
0°                    45°
0°                    −45°
0°                    −45°
0°                    45°
0°                    90°
0°                    0°
```

**Unidirectional**          **Cross-plied quasi-isotropic**

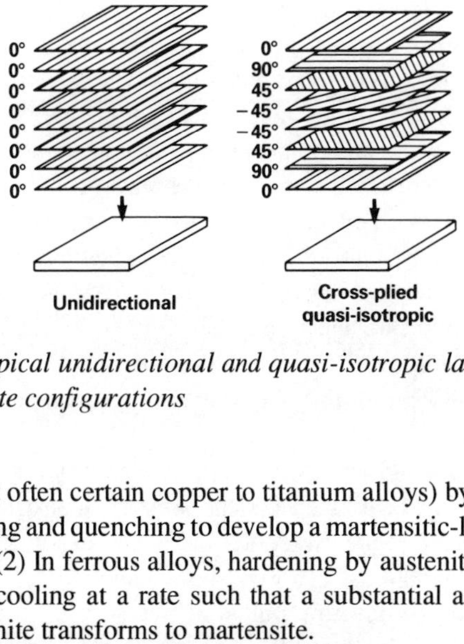

*Typical unidirectional and quasi-isotropic laminate configurations*

(most often certain copper to titanium alloys) by solution treating and quenching to develop a martensitic-like structure. (2) In ferrous alloys, hardening by austenitizing and then cooling at a rate such that a substantial amount of austenite transforms to martensite.

**quenching**. Rapid cooling of metals (often steels) from a suitable elevated temperature. This generally is accomplished by immersion in water, oil, polymer solution, or salt, although forced air is sometimes used. See also *brine quenching, caustic quenching, direct quenching, fog quenching, forced-air quenching, hot quenching, intense quenching, interrupted quenching, oil quenching, press quenching, selective quenching, spray quenching, time quenching,* and *water quenching*.

**quenching crack**. A crack formed in a metal as a result of thermal stresses produced by rapid cooling from a high temperature.

**quenching oil**. Oil used for quenching metals during a heat treating operation. See also the table accompanying the term *oil quenching*.

**quench time**. In resistance welding, the time from the finish of the welding operation to the beginning of tempering. Also called chill time.

**quill**. (1) A hollow or tubular shaft, designed to slide or revolve, carrying a rotating member within itself. (2) Removable spindle projection for supporting a cutting tool or grinding wheel.

**Q value**. A quality factor of a magnetic core material, also called energy factor or coil magnification factor. It represents the ratio of the reactance of a coil to its series resistance. Its specific change with the frequency is a function of the type and composition of the magnetic powder used for the core.

# R

**RA**. See *reduction in area*.

**rabbit**. A hoelike bladed tool or similar device used for stirring molten metal.

**rabbit ear**. Recess in the corner of a metal forming die to allow for wrinkling or folding of the blank.

**race (or raceway)**. The groove or path in which the rolling elements in a rolling-contact bearing operate. See also the figure accompanying the term *rolling-element bearing*.

**racking**. A term used to describe the placing of metal parts to be heat treated on a rack or tray. This is done to keep parts in a proper position to avoid heat-related distortions and to keep the parts separated. See also *fixturing*.

**radial crack**. Damage produced in brittle materials by a hard, sharp, object pressed onto the surface. The resulting crack shape is semi-elliptical and generally perpendicular to the surface.

**radial crushing strength**. The relative capacity of a powder metallurgy bearing specimen to resist fracture induced by a force applied between flat parallel plates in a direction perpendicular to the axis of the specimen.

**radial distribution function analysis**. An *x-ray diffraction* method that gives the distribution of interatomic distances present in a sample along with information concerning the frequency with which the particular distances occur.

**radial draw forming**. The forming of sheet metals by the simultaneous application of tangential stretch and radial compression forces (Fig. 424). The operation is done gradually by tangential contact with the die member. This type of forming is characterized by very close dimensional control.

**radial forging**. A process using two or more moving anvils or dies for producing shafts with constant or varying diameters along their length or tubes with internal or external variations (Fig. 425). Often incorrectly referred to as *rotary forging*. See also the figure accompanying the term *mandrel forging*.

FIG. 424

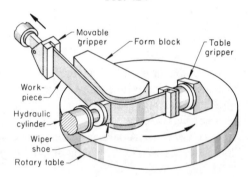

*Radial draw forming setup*

FIG. 425

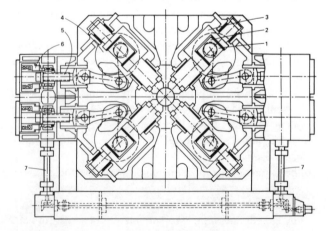

*Cross section through a four-hammer radial forging machine. 1, eccentric shaft; 2, sliding block; 3, connecting rod; 4, adjustment housing; 5, adjusting screw; 6, hydraulic overload protection; 7, hammer adjustment drive shafts*

**radial lip seal**. A radial type of seal that features a flexible sealing member, referred to as a lip. The lip is usually of an elastomeric material. It exerts radial sealing pressure on a mating shaft in order to retain fluids and/or exclude foreign matter.

**radial-load bearing**. A bearing in which the load acts in a radial direction with respect to the axis of rotation.

**radial marks**. Lines on a fracture surface that radiate from the fracture origin and are visible to the unaided eye or at low magnification (Fig. 426). Radial marks result from the intersection and connection of brittle fractures propa-

FIG. 426

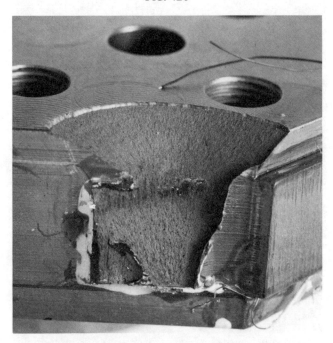

*Radial marks on the fracture surface of a broken AISI 4140 steel load cell. These marks indicate that two crack origins formed at the bottom surface of the cell. Actual size*

FIG. 427

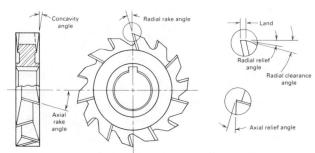

*Radial rake in a slot milling cutter*

gating at different levels. Also known as shear ledges. See also *chevron pattern*.

**radial rake**. The angle between the tooth face and a radial line passing through the cutting edge in a plane perpendicular to the cutter axis (Fig. 427).

**radial roll**. The primary driven roll of the rolling mill for rolling rings in the radial pass. The roll is supported at both ends. Also referred to as main roll. See also the figures accompanying the terms *axial rolls* and *ring rolling*.

**radial rolling force**. In *ring rolling*, the action produced by the horizontal pressing force of the rolling mandrel acting against the ring and the main roll.

**radial runout**. For any rotating element, the total variation from true radial position, taken in a plane perpendicular to the axis of rotation. Compare with *axial runout*.

**radiant energy**. Energy transmitted as *electromagnetic radiation*.

**radiant power (flux)**. The energy emitted by a source or transported in a beam per unit time, expressed, for example, in ergs per second or watts.

**radiant tubes**. Tubes in which fuel is burned for supplying radiant heat to a furnace (Fig. 428). Radiant tubes are made from heat-resistant cast or wrought iron-nickel-chromium alloys or wrought nickel-chromium alloys.

**radiation damage**. A general term for the alteration of properties of a material arising from exposure to ionizing radiation (penetrating radiation), such as x-rays, gamma rays, neutrons, heavy-particle radiation, or fission frag-

FIG. 428

*Radiant tubes for indirect gas-fired furnace heating*

ments in nuclear fuel material. See also *neutron embrittlement*.

**radiation dose.** Accumulated exposure to ionizing radiation during a specified period of time.

**radiation energy.** The energy of a given photon or particle in a beam of radiation, often expressed in electron volts.

**radiation gage.** An instrument for measuring the intensity and quantity of ionizing radiation.

**radiation intensity.** In general, the quantity of radiant energy at a specified location passing perpendicularly through unit area in unit time. It may be given as number of particles or photons per square centimeter per second, or in energy units as $J/m^2 \cdot s$.

**radiation monitoring.** The continuous or periodic measurement of the intensity of radiation received by personnel or present in any particular area.

**radiation quality.** A term describing roughly the spectrum of radiation produced by a radiation source, with respect to its penetrating power or its suitability for a given application.

**radical.** A very reactive chemical intermediate.

**radioactive element.** An element that has at least one isotope that undergoes spontaneous nuclear disintegration to emit positive alpha particles, negative beta particles, or gamma rays.

**radioactivity.** (1) The property of the nuclei of some isotopes to spontaneously decay (lose energy). Usual mechanisms are emission of α, β, or other particles and splitting (fissioning). Gamma rays are frequently, but not always, given off in the process. (2) A particular component from a radioactive source, such as β radioactivity. See also *isotope* and *radioisotope*.

**radioanalysis.** An analytical chemistry technique that uses the radiation properties of a radioactive isotope of an element for its detection and quantitative determination. It can be applied to the detection and measurement of natural and artificial radioactive isotopes. See also *isotope* and *radioisotope*.

**radio frequency.** A frequency at which coherent *electromagnetic radiation* of energy is useful for communication purposes; roughly the range from 10 kHz to 100 GHZ.

**radio frequency (RF) preheating.** A method of preheating plastic molding materials to facilitate the molding operation and/or reduce molding cycle time. The frequencies most commonly used are those between 10 and 100 MHz.

**radio frequency spectrometer.** An instrument that measures the intensity of radiation emitted or absorbed by atoms or molecules as a function of frequency at frequencies from 10 kHz to 100 GHz.

**radio frequency spectroscopy.** The branch of *spectroscopy* concerned with the measurement of the intervals between atomic or molecular energy levels that are separated by frequencies from about $10^5$ to $10^9$ Hz as compared to the frequencies that separate optical energy levels of about 10 × $10^{14}$ Hz.

**radio frequency welding.** A method of welding thermoplastics using a radio frequency field to apply the necessary heat. Also known as high-frequency welding.

**radiograph.** A photographic shadow image resulting from uneven absorption of penetrating radiation in a test object (Fig. 429). See also *radiography*.

FIG. 429

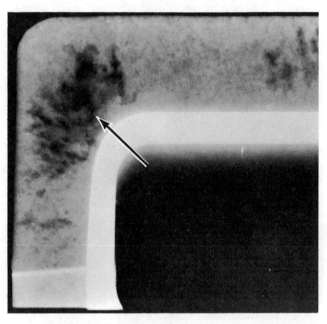

*Radiographic appearance of gross shrinkage porosity (arrow) in an aluminum alloy 319 manifold casting*

**radiography.** A method of nondestructive inspection in which a test object is exposed to a beam of x-rays or gamma rays and the resulting shadow image of the object is recorded on photographic film placed behind the object (Fig. 430), or displayed on a viewing screen or television monitor (real-time radiography). Internal discontinuities are detected by observing and interpreting variations in the image caused by differences in thickness, density, or absorption within the test object. Variations of radiography include *computed tomography*, *fluoroscopy*, and *neutron radiography*. See also *real-time radiography*.

**radio interference.** Undesired conducted or radiated electrical disturbances, including transients, that may interfere with the operation of electrical or electronic communications equipment or other electronic equipment.

**radioisotope.** An isotope that emits ionizing radiation during its spontaneous decay; a radioactive isotope. Also known as radionuclide. See also *isotope* and *radioactivity*.

FIG. 430

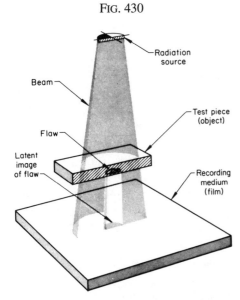

*Diagram of the basic elements of a radiographic system, showing method of detecting and recording an internal flaw in a plate of uniform thickness*

**radiology**. The general term given to material inspection methods that are based on the differential absorption of penetrating radiation—either electromagnetic radiation of very short wavelength or particulate radiation—by the part or testpiece (object) being inspected. Because of differences in density and variations in thickness of the part or differences in absorption characteristics caused by variations in composition, different portions of a testpiece absorb different amounts of penetrating radiation. These variations in the absorption of the penetrating radiation can be monitored by detecting the unabsorbed radiation that passes through the testpiece. See also *radiography*.

**radionuclide**. See *radioisotope*.

**radionuclide tracer element**. A radioactive isotope or an element used to study the movement and behavior of atoms by observing the distribution and intensity of radioactivity.

**radius**. (1) A line segment joining the center of a circle and a point on the circumference or surface of the circle. (2) The length of such a line. (3) To remove the sharp edge or corner of forging stock by means of a radius or form tool.

**radius of bend**. The radius of the cylindrical surface of the pin or mandrel that comes in contact with the inside surface of the bend during bending. In the case of free or semiguided bends to 180° in which a shim or block is used, the radius of bend is one-half the thickness of the shim or block. See also the figure accompanying the term *bend radius*.

**rain erosion**. A form of liquid impingement erosion in which the impinging liquid particles are raindrops. This form of erosion is of particular concern to designers and material selectors for external surfaces of rotary-wing and fixed-wing aircraft. See also *erosion (erosive wear)*.

**rake**. The angular relationship between the tooth face, or a tangent to the tooth face at a given point, and a given reference plane or line. See also the figures accompanying the terms *face mill*, *radial rake*, and *single-point tool*.

**ram**. The moving or falling part of a drop hammer or press to which one of the dies is attached; sometimes applied to the upper flat die of a steam hammer. Also referred to as the *slide*. See also the figures accompanying the terms *drop hammer* and *drop hammer forging*.

**Raman line (band)**. A line (band) that is part of a *Raman spectrum* and corresponds to a characteristic vibrational frequency of the molecule being probed.

**Raman shift**. The displacement in wave number of a *Raman line (band)* from the wave number of the incident monochromatic beam. Raman shifts are usually expressed in units of $cm^{-1}$. They correspond to differences between molecular vibrational, rotational, or electronic energy levels.

**Raman spectroscopy**. Analysis of the intensity of Raman scattering of monochromatic light as a function of frequency of the scattered light.

**Raman spectrum**. The spectrum of the modified frequencies resulting from inelastic scattering when matter is irradiated by a monochromatic beam of radiant energy. Raman spectra normally consist of lines (bands) at frequencies higher and lower than that of the incident monochromatic beam.

**ramming**. (1) Packing foundry sand, refractory, or other material into a compact mass. (2) The compacting of molding (foundry) sand in forming a mold.

**ram travel**. In forming of plastics, the distance the injection ram moves to fill the mold in injection or transfer molding.

**random intermittent welds**. Intermittent welds on one or both sides of a joint in which the weld increments are deposited without regard to spacing.

**random orientation**. A condition of a polycrystalline aggregate in which the orientations of the constituent crystals are completely random relative to each other. Contrast with *preferred orientation*.

**random pattern**. (1) In *filament winding* of composites, a winding with no fixed pattern. If a large number of circuits is required for the pattern to repeat, a random pattern is approached. (2) A winding in which the filaments do not lie in an even pattern.

**random sequence**. A longitudinal welding sequence wherein the weld-bead increments are deposited at random to minimize distortion.

## Technical Brief 36: Rare Earth Metals

THE RARE EARTH METALS include the Group IIIA elements scandium, yttrium, and the lanthanide elements (lanthanum, cerium, praseodymium, neodymium, promethium, samarium, europium, gadolinium, terbium, dysprosium, holmium, erbium, thulium, ytterbium, and lutetium) in the periodic table of the elements. The term "rare" implies that these elements are scarce; in fact, the rare earths are quite abundant. Of the 83 naturally occurring elements, the 16 naturally occurring rare earths as a group lie in the 50th percentile of elemental abundances. Cerium, the most abundant, ranks 28th; thulium, the least abundant, ranks 63rd.

Despite the view by some researchers that the rare earth elements are so chemically similar to one another that collectively they can be considered as one element, a closer examination reveals vast differences in their behaviors and properties. For example, the melting points of the lanthanide elements vary by a factor of almost two between lanthanum (918 °C, or 1684 °F) and lutetium (1663 °C, or 3025 °F). In addition, the modulus of elasticity for these elements varies from as low as ~18 GPa ($2.6 \times 10^6$ psi) for europium to more than 74 GPa ($10.7 \times 10^6$ psi) for thulium.

Rare earth elements are found in nature intimately mixed in varying proportions depending on the ore. Separation into pure component rare earths is done on a large scale by liquid-liquid extraction and by ion exchange on a smaller scale. Because impurities significantly influence the properties of rare earth metals, impurity levels must be kept as low as possible. Research-grade metals are usually ≥99.8 at.% pure, although ≥99.95 at.% metals can be prepared. Commercial-grade rare earth metals are about 98 at.% pure, but occasionally can be as low as 95 at.% pure. Hydrogen and oxygen are the major impurities in both grades.

The primary application for rare earth metals is as an alloying additive. In many of these applications, the rare earths are added in the form of *mischmetal*, which has the approximate rare earth distribution of 50% Ce, 30% La, 15% Nd, and 5% Pr. Additions of pure metals or finely dispersed rare earth-based oxides (primarily $Y_2O_3$) are also used. Materials that rare earths are added to include ductile iron (modify carbon morphology), superalloys (increase operating temperatures), magnesium alloys (improve creep resistance), aluminum alloys (improve tensile strength and corrosion resistance), oxygen-free high-conductivity copper (improve oxidation resistance), and dispersion-strengthened materials (improve high-temperature properties). Other key applications include their use as lighter flints, permanent magnet materials (samarium-cobalt and neodymium-iron-boron magnets), magnetooptical materials, and hydrogen storage batteries.

### Recommended Reading

- K.A. Gschneidner, B.J. Beaudry, and J. Cappellen, Rare Earth Metals, *Metals Handbook*, 10th ed., Vol 2, ASM International, 1990, p 720-732
- K.A. Gschneidner and B.J. Beaudry, Properties of the Rare Earth Metals, *Metals Handbook*, 10th ed., Vol 2, ASM International, 1990, p 1178-1189

**range.** In inspection, the difference between the highest and lowest values of a given *quality characteristic* within a single *sample*.

**range of stress** ($S_r$). The algebraic difference between the maximum and minimum stress in one cycle—that is,

$$S_r = S_{max} - S_{min}$$

**rapid solidification.** The cooling or quenching of liquid (molten) metals at rates that range from 104 to 108 °C/s.

**rare earth metals.** See *Technical Brief 36.*

**ratcheting.** Progressive cyclic inelastic deformation (growth, for example) that occurs when a component or structure is subjected to a cyclic secondary stress superimposed on a sustained primary stress. The process is called thermal ratcheting when cyclic strain is induced by cyclic changes in temperature, and isothermal ratcheting when cyclic strain is mechanical in origin (even though accompanied by cyclic changes in temperature).

**ratchet marks.** Lines or markings on a fatigue fracture surface that results from the intersection and connection of fatigue fractures propagating from multiple origins (Fig. 431). Ratchet marks are parallel to the overall direction of

## Commercial alloys containing rare earth metals

| Designation | Alloy type | Rare earth | Composition, wt% | Remarks |
|---|---|---|---|---|
| AiResist 13 . . . . . . . . . | Co superalloy | Y | 0.1 | High-temperature parts |
| AiResist 213 . . . . . . . . | Co superalloy | Y | 0.1 | Hot corrosion resistance |
| AiResist 215 . . . . . . . . | Co superalloy | Y | 0.17 | Hot corrosion resistance |
| FSX 418 . . . . . . . . . . . | Co superalloy | Y | 0.15 | Oxidation resistance |
| FSX 430 . . . . . . . . . . . | Co superalloy | Y | 0.03–0.1 | Oxidation and hot corrosion resistance |
| Haynes 188 . . . . . . . . | Co superalloy | La | 0.05 | Oxidation resistance, strength |
| Haynes 1002 . . . . . . . | Co superalloy | La | 0.05 | . . . |
| Melco 2 . . . . . . . . . . . | Co superalloy | Y | 0.15 | . . . |
| Melco 9 . . . . . . . . . . . | Co superalloy | Y | 0.13 | . . . |
| Melco 10 . . . . . . . . . . | Co superalloy | Y | 0.10 | . . . |
| Melco 14 . . . . . . . . . . | Co superalloy | Y | 0.18 | . . . |
| C-207 . . . . . . . . . . . . . | Cr | Y | 0.15 | . . . |
| CI-41 . . . . . . . . . . . . . | Cr | Y + La | 0.1 (total) | . . . |
| 253 . . . . . . . . . . . . . . . | Fe superalloy | Ce | 0.055 | . . . |
| GE 1541 . . . . . . . . . . | Fe superalloy | Y | 1.0 | . . . |
| GE 2541 . . . . . . . . . . | Fe superalloy | Y | 1.0 | . . . |
| Haynes 556 . . . . . . . . | Fe superalloy | La | 0.02 | High temperature, up to 1095 °C |
| ICF 42 . . . . . . . . . . . . | High-strength steel | R | . . . | (a) |
| ICF 45 . . . . . . . . . . . . | High-strength steel | R | . . . | (a) |
| ICF 50 . . . . . . . . . . . . | High-strength steel | R | . . . | (a) |
| VAN 50 . . . . . . . . . . . | High-strength steel | Ce | . . . | (a) |
| VAN 60 . . . . . . . . . . . | High-strength steel | Ce | . . . | (a) |
| VAN 70 . . . . . . . . . . . | High-strength steel | Ce | . . . | (a) |
| VAN 80 . . . . . . . . . . . | High-strength steel | Ce | . . . | (a) |
| EK 30A . . . . . . . . . . . | Mg (Zr, Zn) | R | 3.0 | Creep resistance |
| EK 41A . . . . . . . . . . . | Mg (Zr, Zn) | R | 4.0 | Creep resistance |
| EZ 33A . . . . . . . . . . . | Mg (Zr, Zn) | R | 3.0 | Creep resistance |
| QE 22A . . . . . . . . . . . | Mg (Zr, Ag) | R | 1.2–3.0 | Creep resistance |
| QE 222A . . . . . . . . . . | Mg (Zr, Ag) | Dm(b) | 2 | Creep resistance |
| WE 54 . . . . . . . . . . . . | Mg (Zr) | Y + R | 5.25 + 3.5 | High strength, weldability |
| ZE 10A . . . . . . . . . . . | Mg (Zn) | R | 0.17 | Creep resistance |
| ZE 41A . . . . . . . . . . . | Mg (Zr, Zn) | R | 1.2 | . . . |
| ZE 63A . . . . . . . . . . . | Mg (Zr, Zn) | R | 2–3 | Creep resistance |
| ZE 63B . . . . . . . . . . . | Mg (Zr, Zn, Ag) | R | 2–3 | Creep resistance |
| C129Y . . . . . . . . . . . . | Nb | Y | 0.1 | . . . |
| Hastelloy N . . . . . . . . | Ni superalloy | Y | 0.26 | . . . |
| Hastelloy S . . . . . . . . | Ni superalloy | La | 0.05 | High stability |
| Hastelloy T . . . . . . . . | Ni superalloy | La | 0.02 | Low thermal expansion |
| Haynes 214 . . . . . . . . | Ni superalloy | Y | 0.02 | Oxidation resistance |
| Haynes 230 . . . . . . . . | Ni superalloy | La | 0.5 | High-temperature strength |
| Melni 19 . . . . . . . . . . | Ni superalloy | La | 0.17 | . . . |
| Melni 22 . . . . . . . . . . | Ni superalloy | La | 0.16 | . . . |
| René Y . . . . . . . . . . . | Ni superalloy | La | 0.05–0.3 | . . . |
| Udimet 500 + Ce . . . . | Ni superalloy | Ce | . . . | . . . |
| Unimet 700 + Ce . . . . | Ni superalloy | Ce | 0.2–0.5 | . . . |

(a) Rare earth (R) or cerium added for inclusion shape control. (b) Dm, Didymium, alloy of 80Nd-20Pr

crack propagation and are visible to the unaided eye or at low magnification.

**rate of flame propagation.** The speed at which a flame travels through a mixture of gases.

**rate of oil flow.** The rate at which a specified oil will pass through a porous sintered powder metallurgy compact under specific test conditions.

**rate of strain hardening.** Rate of change of *true stress* with respect to *true strain* in the plastic range.

**rating life.** Currently the fatigue life in millions of revolutions or hours at a given operating speed that 90% of a group of substantially identical rolling-element bearings will survive under a given load. The 90% rating life is frequently referred to as $L_{10}$-life or $B_{10}$-life. The rating life in revolutions can be obtained from:

$$L_{10} = \left(\frac{C}{P}\right)^k \times 10^6$$

where $C$ is the basic load rating (in pounds), $P$ is the equivalent radial load (in pounds), and $k$ is a constant (3 for ball bearings and 10/3 for roller bearings). For a rating life in hours use:

FIG. 431

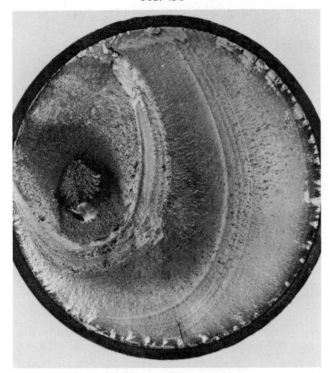

*Ratchet marks (small shiny areas at surface) in a fatigue-fractured AISI 1050 steel shaft*

$$L_{10} = \frac{16,700}{N} \left(\frac{C}{P}\right)^k$$

where $N$ is the rotational speed in rev/min.

**rattail**: A surface imperfection on a casting, occurring as one or more irregular lines, caused by expansion of sand in the mold. Compare with *buckle* (2).

**Rayleigh scattering**. Scattering of *electromagnetic radiation* by independent particles that are smaller than the wavelength of radiation. Contrast with *Compton scattering*.

**Rayleigh step bearing**. A stepped-pad bearing having one step only in each pad.

***R*-curve**. In fracture mechanics, a plot of crack-extension resistance as a function of stable crack extension, which is the difference between either the *physical crack size* or the *effective crack size* and the *original crack size*. *R*-curves normally depend on specimen thickness and, for some materials, on temperature and strain rate.

**RE**. Abbreviation for rare earth. See also *rare earth metals* (Technical Brief 36).

**reactance**. That part of the impedance of an alternating current circuit that is due to capacitance or inductance.

**reaction injection molding (RIM)**. A process for molding polyurethane, epoxy, and other liquid chemical systems. Mixing two to four components in the proper chemical ratio is accomplished by a high-pressure impingement-type mixing head, from which the mixed material is delivered into the mold at low pressure, where it reacts (cures). See also *injection molding* (plastics).

**reaction sintering**. The sintering of a metal powder mixture consisting of at least two components that chemically react during the treatment.

**reaction stress** (welding). The *residual stress* which could not otherwise exist if the members or parts being welded were isolated as free bodies without connection to other parts of the structure.

**reactive diluent**. As used in epoxy formulations, a compound containing one or more epoxy groups that functions mainly to reduce the viscosity of the mixture.

**reactive metal**. A metal that readily combines with oxygen at elevated temperatures to form very stable oxides, for example, titanium, zirconium, and beryllium. Reactive metals may also become embrittled by the interstitial absorption of oxygen, hydrogen, and nitrogen.

**reactive sputtering** (of interference films). A method of revealing the microstructure of metals with the aid of physically deposited interference layers (films) that is based on an optical-contrast mechanism without chemical or morphological alteration of the specimen surface. In interference layer microscopy, light that is incident on the deposited film is reflected at the air/layer and layer/specimen interfaces. Phases with different optical constants appear in various degrees of brightness and colors (Fig. 432). The color of a phase is determined by its optical constants and by the thickness and optical constants of the interference layer. In this process, the film material is

FIG. 432

*Interference-layer micrograph of a cast Sn-18Ag-15Cu alloy. Polished specimen coated with a platinum oxide layer by reactive sputtering. Structure consists of $Ag_3Sn$ (white), Sn (light gray), $Cu_6Sn_5$ (medium gray), and $Cu_3Sn$ (dark gray). 300×*

Fig. 433

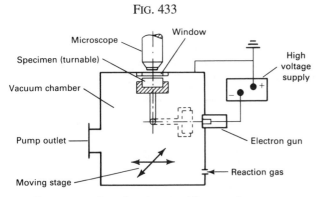

*Gas contrasting chamber used for reactive sputtering and optical examination of interference layers on polished specimens. The results of the reactive sputtering process can be monitored through the viewing window.*

applied by cathode sputtering, transported through a gas chamber containing a reactive gas, and deposited on an anodically connected specimen (Fig. 433). Oxygen is frequently used as reaction gas and interaction with the sputtered material leads to deposition of oxidic interference films. See also *sputtering* and *vacuum deposition* (of interference films).

**reagent**. A substance, chemical, or solution used in the laboratory to detect, measure, or react with other substances, chemicals, or solutions. See also *reagent chemicals*.

**reagent chemicals**. High-purity chemicals used for analytical reactions, for testing of new reactions where the effects of impurities are unknown, and for chemical work where impurities must either be absent or at a known concentration.

**real area of contact**. In *tribology*, the sum of the local areas of contact between two solid surfaces, formed by contacting asperities, that transmit the interfacial force between the two surfaces. Contrast with *apparent area of contact*.

**real-time radiography**. A method of nondestructive inspection in which a two-dimensional radiographic image can be immediately displayed on a viewing screen or television monitor (Fig. 434). This technique does not involve the creation of a latent image; instead, the unabsorbed radiation is converted into an optical or electronic signal, which can be viewed immediately or can be processed in near real time with electronic and video equipment. The principal advantage of real-time radiography over film radiography is the opportunity to manipulate the testpiece during radiographic inspection. This capability allows the inspection of internal mechanisms and enhances the detection of cracks and planar defects by manipulating the part to achieve the proper orientation for flaw detection. See also *radiography*.

**reamed extrusion ingot**. A cast hollow extrusion ingot that has been machined to remove the original inside surface.

**reamer**. A rotary cutting tool with one or more cutting elements called teeth, used for enlarging a hole to desired size and contour (Fig. 435). It is supported principally by the metal around the hole it cuts.

Fig. 434

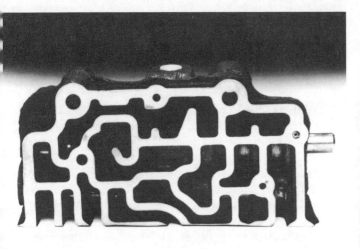

(a)

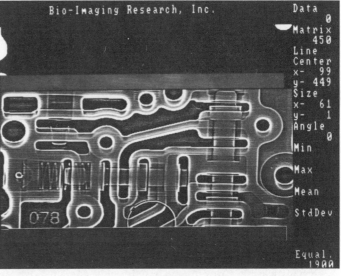

(b)

*Evaluation of cast transmission housing assembly. (a) Photograph of cast part. (b) Real-time radiographic image used to verify the steel spring pin and shuttle valve assembly through material thicknesses ranging from 3 mm (1/8 in.) in the channels to 25 mm (1 in.) in the rib sections of the casting*

F<small>IG</small>. 435

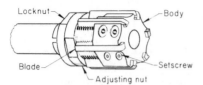

*Inserted-blade adjustable reamer*

F<small>IG</small>. 436

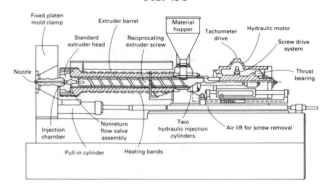

*The injection end of a reciprocating-screw injection molding machine*

**reaming**. An operation in which a previously formed hole is sized and contoured accurately by using a rotary cutting tool (*reamer*) with one or more cutting elements (teeth). The principal support for the reamer during the cutting action is obtained from the workpiece.
1. Form reaming: Reaming to a contour shape.
2. Taper reaming: Using a special reamer for taper pins.
3. Hand reaming: Using a long lead reamer which permits reaming by hand.
4. Pressure coolant reaming (or gun reaming): Using a multiple-lip, end cutting tool through which coolant is forced at high pressure to flush chips ahead of the tool or back through the flutes for finishing of deep holes.

**recalescence**. (1) The increase in temperature that occurs after undercooling, because the rate of liberation of heat during transformation of a material exceeds the rate of dissipation of heat. (2) A phenomenon, associated with the transformation of gamma iron to alpha iron on cooling (supercooling) of iron or steel, that is revealed by the brightening (reglowing) of the metal surface owing to the sudden increase in temperature caused by the fast liberation of the latent heat of transformation. Contrast with *decalescence*.

**recarburize**. (1) To increase the carbon content of molten cast iron or steel by adding carbonaceous material, high-carbon pig iron, or a high-carbon alloy. (2) To carburize a metal part to return surface carbon lost in processing; also known as *carbon restoration*.

**recess**. A groove or depression in a surface.

**reciprocal lattice**. A lattice of points, each representing a set of planes in the crystal lattice, such that a vector from the origin of the reciprocal lattice to any point is normal to the crystal planes represented by that point and has a length that is the reciprocal of the plane spacing.

**reciprocal linear dispersion**. The derivative $d\lambda/dx$, where $\lambda$ is the wavelength and $x$ is the distance along the spectrum. The reciprocal linear dispersion usually is expressed in Å/mm. See also *linear dispersion*.

**reciprocating-screw injection molding**. A combination injection and plasticating unit in which an extrusion device with a reciprocating screw is used to plasticate the material (Fig. 436). The injection of material into a mold can take place by direct extrusion into the mold, by reciprocating

the screw as an injection plunger, or by a combination of the two. When the screw serves as an injection plunger, this unit acts as a holding, measuring, and injection chamber. See also *injection molding* (plastics).

**reclaim rinse**. A nonflowing rinse used to recover *dragout*.

**recoil line**. See *impact line*.

**reconstruction**. The process by which raw digitized detector measurements are transformed into a cross-sectional computed tomography image. See also *computed tomography*.

**recovery**. (1) The time-dependent portion of the decrease in strain following unloading of a specimen at the same constant temperature as the initial test. Recovery is equal to the total decrease in strain minus the instantaneous recovery. The recovery is expressed in the same units as *instantaneous recovery*. (2) Reduction or removal of work-hardening effects in metals without motion of large-angle grain boundaries. (3) The proportion of the desired component obtained by processing an ore, usually expressed as a percentage.

**recrystallization**. (1) The formation of a new, strain-free grain structure from that existing in cold-worked metal, usually accomplished by heating. (2) The change from one crystal structure to another, as occurs on heating or cooling through a critical temperature. (3) A process, usually physical, by which one crystal species is grown at the expense of another or at the expense of others of the same substance but smaller in size. See also *crystallization*.

**recrystallization annealing**. Annealing cold worked metal to produce a new grain structure without phase change.

**recrystallization temperature**. (1) The lowest temperature at which the distorted grain structure of a cold-worked metal is replaced by a new, strain-free grain structure during prolonged heating. Time, purity of the metal, and prior deformation are important factors. (2) The approximate minimum temperature at which complete recrystallization of a cold-worked metal occurs within a specified time.

**recrystallized grain size**. (1) The grain size developed by heating cold-worked metal. The time and temperature are selected so that, although recrystallization is complete, essentially no grain growth occurs. (2) In aluminum and magnesium alloys, the grain size after recrystallization, without regard to grain growth or the recrystallized conditions. See also *recrystallization*.

**recuperator**. (1) Equipment for transferring heat from gaseous products of combustion to incoming air or fuel (Fig. 437). The incoming material passes through pipes surrounded by a chamber through which the outgoing gases pass. (2) A continuous heat exchanger in which heat is conducted from the products of combustion to incoming air through flue walls.

FIG. 437

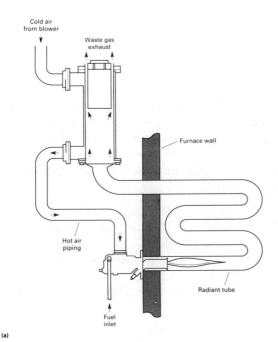

(a)

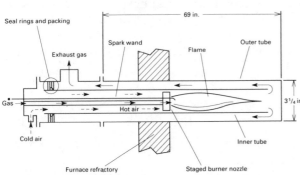

(b)

*Radiant tube recuperator systems. (a) External recuperation with U tubes. (b) Single-ended recuperation with inner and outer tubes made of reaction-bonded silicon carbide*

**red mud**. A residue, containing a high percentage of iron oxide, obtained in purifying bauxite in the production of alumina in the *Bayer process*.

**redox potential**. This *potential* of a reversible oxidation-reduction electrode measured with respect to a *reference electrode*, corrected to the hydrogen electrode, in a given *electrode*.

**redrawing**. The second and successive deep-drawing operations in which cuplike shells are deepened and reduced in cross-sectional dimensions. See also *deep drawing*.

**reduced powder**. Generic term for any metal or nonmetal powder produced by the reduction of an oxide, hydroxide, carbonate, oxalate, or other compound without melting.

**reducing agent**. (1) A compound that causes *reduction*, thereby itself becoming oxidized. (2) A chemical that, at high temperatures, lowers the state of oxidation of other batch chemicals.

**reducing atmosphere**. (1) A furnace atmosphere which tends to remove oxygen from substances or materials placed in the furnace. (2) A chemically active protective atmosphere which at elevated temperature will reduce metal oxides to their metallic state. Reducing atmosphere is a relative term and such an atmosphere may be reducing to one oxide but not to another oxide.

**reducing flame**. (1) A gas flame produced with excess fuel in the inner flame. (2) A gas flame resulting from combustion of a mixture containing too much fuel or too little air. See also the figure accompanying the term *neutral flame*.

**reduction**. (1) In cupping and deep drawing, a measure of the percentage decrease from blank diameter to cup diameter, or of diameter reduction in redrawing. (2) In forging, rolling and drawing, either the ratio of the original to final cross-sectional area or the percentage decrease in cross-sectional area. (3) A reaction in which there is a decrease in valence resulting from a gain in electrons. Contrast with *oxidation*.

**reduction cell**. A pot or tank in which either a water solution of a salt or a fused salt is reduced electrolytically to form free metals or other substances.

**reduction in area (RA)**. The difference between the original cross-sectional area of a tensile specimen and the smallest area at or after fracture as specified for the material undergoing testing. Also known as reduction of area.

**reduction of oxide**. The process of converting a metal oxide to metal by applying sufficient heat in the presence of a solid or gaseous material, such as hydrogen, having a greater attraction for the oxygen than does the metal.

**Redwood viscosity**. A commercial measure of viscosity expressed as the time in seconds required for 50 $cm^3$ of a fluid to flow through a tube of 10 mm length and 1.5 mm diameter at a given temperature. It is recommended that fundamental viscosity units be used.

## Technical Brief 37: Refractories

REFRACTORY MATERIALS are traditionally thought of as nonmetallics (ceramics) that resist degradation by corrosive gases, liquids, or solids at elevated temperatures. These materials must withstand thermal shock caused by rapid heating or cooling, failure attributable to thermal stresses, mechanical fatigue due to other material contacting the refractory itself, and chemical attack activated by the high-temperature environment. Nearly two-thirds of all refractories used by industry are preformed bricks or other fired shapes. The remainder take the form of monolithic materials.

The properties that characterize quality refractory materials depend on the nature of the application. The most important aspect of these materials is referred to as "refractoriness," which refers to the point at which the specimen begins to soften (or melt). Other useful properties include the temperature of failure under load, the dimensional stability (expansion and contraction characteristics), resistance to spalling (fracture, splitting, or flaking of the refractory), porosity and permeability properties, and the capacity to store or transmit thermal energy (thermal conductivity characteristics).

There are a number of ways that refractories can be classified; for example, basic versus acid, or clay versus nonclay (see also *acid refractory* and *basic refractories*). In this brief, refractory materials are classified as basic or high-duty oxides. Silica ($SiO_2$), alumina ($Al_2O_3$), and mullite are used in both categories. Fireclay, magnesia ($MgO$), and chromia ($Cr_2O_3$) are used primarily in basic refractories. Silica brick refractories, which consist almost entirely of $SiO_2$, are subject to devitrification and their use temperature should not exceed 1250 °C (2280 °F). These refractories show a high degree of volume stability and an absence of spalling above 650 °C (1200 °F) compared with fireclay refractories. Fireclay is a basic refractory consisting primarily of hydrated alumino silicates with an $SiO_2$ content of up to 78% and a content of $Al_2O_3$ and other minor constituents remaining below 38%. High-alumina refractories typically vary in $Al_2O_3$ content, from 80 to 99+%. These refractories fall in the basic class, with the exception of pure $Al_2O_3$, which is considered a high-duty refractory oxide. All alumina refractories possess a greater refractoriness and load-bearing ability compared to the fireclay refractories. Mullite is a refractory material grouped with the high-alumina refractories. Composed primarily of the mineral mullite ($3Al_2O_3 \cdot 2SiO_2$) with the remainder of the matrix usually $Al_2O_3$, mullite refractories possess properties similar to those of the high-alumina type.

There are three typical classes of basic refractories whose major raw materials are magnesia and chromia. The first is simply referred to as magnesite (the mineral it originates from); its composition ranges from 80 to 95% $MgO$. Chrome or chromite refractories typically contain 30 to 45% $Cr_2O_3$. Chrome-magnesite refractories are composed of more than 60% $MgO$ with combined $Cr_2O_3$. All three of these basic refractory compositions exhibit high refractoriness, good corrosion resistance, and high thermal conductivity. Other materials used for refractory materials include beryllia, zirconia, silicon carbide, and cordierite.

Refractory materials are most commonly utilized in the iron and steel industries, which account for about 63% of refractory consumption in the United States. Fireclay bricks are used to construct the furnaces to melt and produce steel. If added protection is required, silica, mullite, or high-alumina refractories may be incorporated as needed. The nonferrous metallurgical industries (primarily aluminum and copper) account for up to 10% of refractory consumption. Silica brick is the refractory most commonly used for copper production; high-alumina

---

**reeding.** The operation of forming serrations and corrugations in metals by coining or embossing.

**reel.** (1) A spool or hub for coiling or feeding wire or strip. (2) To straighten and planish a round bar by passing it between contoured rolls.

**reel breaks.** Transverse breaks or ridges on successive inner laps of a coil that results from crimping of the lead end of

the coil into a gripping segmented mandrel. Also called reel kinks.

**reference electrodes.** A nonpolarizable *electrode* with a known and highly reproducible *potential* used for potentiometric and voltammetric analyses. See also *calomel electrode*.

**reference material.** In materials characterization, a material

## Composition and selected properties of basic refractory materials

| Type | Composition | Maximum use temperature in oxygen | | Thermal conductivity, kcal/min · °C | | | Refractoriness under load of 197 kPa (28.5 psi) | |
|------|-------------|------|------|------|------|------|------|------|
| | | °C | °F | At 300 °C (570 °F) | At 800 °C (1470 °F) | At 1200 °C (2190 °F) | °C | °F |
| Silica | 93–96% SiO₃ | 1700 | 3090 | 0.8–1.0 | 1.2–1.4 | 1.6–1.8 | 1650–1700 | 3000–3090 |
| Fireclay | 15–45% Al₂O₃ 55–80% SiO₂ | 1300–1450 | 2370–2640 | 0.8–0.9 | 1.0–1.2 | 2.5–2.8 | 1250–1450 | 2280–2640 |
| Magnesite | 80–95% MgO Fe₂O₃, Al₂O₃ | 1800 | 3270 | 3.8–9.7 | 2.8–4.7 | 2.5–2.8 | 1500–1700 | 2730–3090 |
| Chromite | 30–45% Cr₂O₃ 14–19% MgO 10–17% Fe₂O₃ 15–33% Al₂O₃ | 1700 | 3090 | 1.3 | 1.6 | 1.8 | 1400–1450 | 2550–2640 |
| Chrome magnesite | >60% MgO Fe₂O₃, Al₂O₃ | 1800 | 3270 | 1.9–3.5 | 1.4–2.5 | 1.8 | 1500–1600 | 2730–2910 |

## Composition and selected properties of high-duty refractory oxides

| Type | Composition | Melting point | | Maximum use temperature in oxygen | | Thermal conductivity, kcal/min °C | | | | Refractoriness under load of 196 kPa (28.4 psi) | |
|------|-------------|------|------|------|------|------|------|------|------|------|------|
| | | °C | °F | °C | °F | At 100 °C (212 °F) | At 500 °C (930 °F) | At 1000 °C (1830 °F) | At 1500 °C (2730 °F) | °C | °F |
| Aluminum oxide | 100% Al₂O₃ | 2015 | 3660 | 1950 | 3540 | 26.0 | 9.4 | 5.3 | 5.0 | 2000 | 3630 |
| Beryllium oxide | 100% BeO | 2550 | 4620 | 2400 | 4350 | 189.0 | 56.3 | 17.5 | 13.5 | 2000 | 3630 |
| Magnesium oxide | 100% MgO | 2800 | 5070 | 2400 | 4350 | 31.0 | 12.0 | 6.0 | 5.4 | 2000 | 3630 |
| Silicon dioxide | 100% SiO₂ | ... | ... | 1200 | 2190 | 0.8 | 1.4 | 1.8 | ... | ... | ... |
| Mullite | 72% Al₂O₃ 28% SiO₂ | 1830(a) | 3325(a) | 1850 | 3362 | 5.3 | 3.8 | 3.4 | ... | ... | ... |

(a) Incongruent

refractories are used to produce aluminum. Six percent of all refractories are used by the glassmaking industry. A fused alumina-zirconia-silica (AZS) refractory is most commonly employed.

## Recommended Reading

- Refractories, *Engineered Materials Handbook*, Vol 4, ASM International, 1991, p 895-909
- L.P. Kreitz, Monolithic and Fibrous Refractories, *Engineered Materials Handbook*, Vol 4, ASM International, 1991, p 910-917
- R.A. Haber and P.A. Smith, Overview of Traditional Ceramics, *Engineered Materials Handbook*, Vol 4, ASM International, 1991, p 3-15

of definite composition that closely resembles in chemical and physical nature the material with which an analyst expects to deal; used for calibration or standardization. See also *standard reference material*.

**refining.** The branch of *process metallurgy* dealing with the purification of crude or impure metals. Compare with *extractive metallurgy*.

**reflectance.** The ratio of the radiant power or flux reflected by a medium to the radiant power or flux incident on it; generally expressed as a percentage.

**reflection grating.** An optical component that employs reflection off a series of fine, equidistant ridges, rather than transmission through a pattern of slots, to diffract light into its component wavelengths. The gratings used in optical instrumentation are almost exclusively reflection gratings. See also *concave grating*, *diffraction grating*, *plane grating*, and *transmission grating*.

**reflection method**. The technique of producing a diffraction pattern by x-rays or electrons that have been reflected from a specimen surface.

**reflection (x-ray)**. See *diffraction*.

**reflector sheet**. A clad product consisting of a facing layer of high-purity aluminum capable of taking a high polish, for reflecting heat or light, and a base of commercially pure aluminum or an aluminum-manganese alloy, for strength and formability.

**reflowing**. Melting of an electrodeposit followed by solidification. The surface has the appearance and physical characteristics of a hot dipped surface (especially tin or tin alloy plates). Also called flow brightening.

**reflow soldering**. (1) A process for joining electronic components by tinning the mating surfaces, placing them together, heating until the solder fuses, and allowing to cool in the joined position. (2) A soldering process variation in which preplaced solder is melted to produce a soldered joint or coated surface. The use of this term is not recommended by the American Welding Society.

**reflux**. In materials characterization, heating a substance at the boiling temperature and returning the condensed vapors to the vessel to be reheated.

**refractive index**. The ratio of the phase velocity of monochromatic light in a vacuum to that in a specified medium. Refractive index is generally a function of wavelength and temperature. Also known as index of refraction.

**refractories**. See *Technical Brief 37*.

**refractoriness**. In *refractories*, the capability of maintaining a desired degree of chemical and physical identity at high temperatures and in the environment and conditions of use.

**refractory**. (1) A material (usually an inorganic, nonmetallic, ceramic material) of very high melting point with properties that make it suitable for such uses as furnace linings and kiln construction. (2) The quality of resisting heat. See also *refractories* (Technical Brief 37).

**refractory alloy**. (1) A heat-resistant alloy. (2) An alloy having an extremely high melting point. See also *refractory metals* (Technical Brief 38). (3) An alloy difficult to work at elevated temperatures.

**refractory brick**. A refractory shape, the most widely used size being approximately 23 by 11.3 by 6.4 (or 7.6) cm (9 by 4⁷/₁₆ by 2½ (or 3) in.). Examples of other sizes for brick are 23 by 15.2 by 6.4 (or 7.6) cm (9 by 6 by 2½ (or 3) in.). These rectangular units are often referred to as straight brick to differentiate them from brick shapes having related overall dimensions, such as arch, key, and wedge brick, which have some interfacial angles that depart from 90°.

**refractory metals**. See *Technical Brief 38*.

**regenerator**. Same as *recuperator* except that the gaseous products or combustion heat brick checkerwork in a chamber connected to the exhaust side of the furnace while the incoming air and fuel are being heated by the brick checkerwork in a second chamber, connected to the entrance side. At intervals, the gas flow is reversed so that incoming air and fuel contact hot checkerwork while that in the second chamber is being reheated by exhaust gases.

**regrind**. Waste plastic material (such as sprues, runners, excess parison material, and reject parts from injection molding, blow molding, or extrusion operations) that has been reclaimed by shredding or granulating. Regrind is usually mixed with virgin compound, at a predetermined percentage, for remolding.

**regular reflection**. See *specular reflection*.

**regulator**. A device for controlling the delivery of welding or cutting gas at some substantially constant pressure (Fig. 438).

**regulus**. The impure button, globule, or mass of metal formed beneath the slag in the smelting and reduction of ores. The

FIG. 438

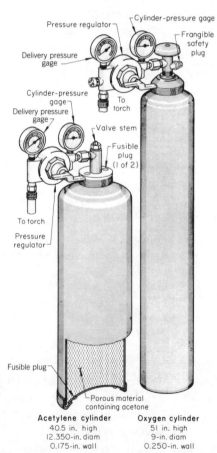

*Gas cylinders and regulators used in oxyfuel gas welding*

## Technical Brief 38: Refractory Metals

THE REFRACTORY METALS include niobium (also known as columbium), tantalum, molybdenum, tungsten, and rhenium. With the exception of the two platinum-group metals, osmium and iridium, they have the highest melting temperatures and lowest vapor pressures of all metals. In fact, tungsten has the highest melting temperature of any element (3140 °C, or 6170 °F). The refractory metals are readily degraded by oxidizing environments at moderately low temperatures, a property that has restricted the applicability of the metals in low-temperature or non-oxidizing high-temperature environments. Protective coating systems have been developed, mostly for niobium alloys, to permit their use in high-temperature oxidizing aerospace applications.

The refractory metals are extracted from ore concentrates, processed into intermediate chemicals, and then reduced to metal, which may be in the form of powder. Refractory metals are almost exclusively produced by powder metallurgy techniques. The pure or alloyed powders are pressed, sintered, and subsequently worked by conventional methods (extrusion, rolling, forging, etc.).

Refractory metals at one time were limited to use in lamp filaments, electron tube grids, heating elements, and electrical contacts; however, they have since found widespread application in the aerospace, electronics, nuclear and high-energy physics, and chemical process industries. Each of the refractory metals, with the exception of rhenium, is consumed in quantities exceeding 900 metric tons (1000 tons) annually on a worldwide basis. Most niobium is consumed as a ferroalloy used in the production of high-strength low-alloy steels and stainless steels; the consumption of niobium-base metals and alloys accounts for about 6% of the total, the majority of which is in aerospace applications. Niobium is also used in low-temperature superconductors. The single largest use for tantalum is as powder and anodes for electronic capacitors, representing about 50% of total consumption. Mill products—sheet and plate, rod and bar, and tubing—constitute nearly 25% of tantalum consumption. Tantalum mill products are used extensively in equipment for the chemical processing industry. The major end use for tungsten is in cemented carbides, which are used for cutting tools and wear-resistant materials. Tungsten carbides make up nearly 60% of tungsten consumption; mill products account for approximately 25%, examples being tungsten wire for lighting, electronic devices, and thermocouples and high-density components such as kinetic energy penetrators. Most molybdenum is used as an alloying addition in steel, irons, and superalloys. Molybdenum-base mill products represent less than 5% of usage, the single most important application being the use of TZM alloy for metalworking dies and high-temperature aerospace and furnace components.

### Compositions of commercially important refractory metal alloys

| Designation | Nominal composition, % |
|---|---|
| **Molybdenum alloys** | |
| Mo-0.5Ti | Mo-0.5Ti-0.02W |
| TZM | Mo-0.5Ti-0.1Zr-0.02W |
| Mo-30W | Mo-30W |
| **Niobium alloys** | |
| Nb-1Zr | Nb-1Zr |
| FS-85 | Nb-27.5Ta-11W-1Zr |
| SCb-291 | Nb-10Ta-10W |
| Cb-752 | Nb-10W-2.5Zr |
| B-66 | Nb-5Mo-5V-1Zr |
| C-103 | Nb-10Hf-1Ti |
| C-129Y | Nb-10W-10Hf-0.15Y |
| **Tantalum alloys** | |
| "63" Metal | Ta-2.5W-0.15Nb |
| Ta-10W | Ta-10W |
| T-111 | Ta-8W-2Hf |
| T-222 | Ta-10W-2.5Hf-0.01C |
| **Tungsten alloys** | |
| W-ThO$_2$ | W-1ThO$_2$; W-2ThO$_2$ |
| W-Mo alloys | Various Mo contents; W-2Mo and W-15 Mo are most common |
| W-Re alloys | Various Re contents up to 26%; W-1.5Re, W-3Re and W-25Re are most common |
| Doped W | 50 ppm Si, 90 ppm K, 15 ppm Al, 35 ppm O |

### Recommended Reading

- J.B. Lambert *et al.*, Refractory Metals and Alloys, *Metals Handbook*, 10th ed., Vol 2, ASM International, 1990, p 557-585
- J.B. Lambert and R.E. Droegkamp, P/M High-Temperature Materials, *Metals Handbook*, 9th ed., Vol 7, ASM International, 1984, p 764-772

name was first applied by alchemists to metallic antimony because it readily alloyed with gold.

**Rehbinder effect.** Modification of the mechanical properties at or near the surface of a solid, attributable to interaction with a *surfactant*.

**reinforced molding compound.** A plastic reinforced with special fillers or fibers (glass, synthetic fibers, minerals, and so forth) to meet specific requirements.

**reinforced plastics.** Molded, formed, filament-wound, tape-wrapped, or shaped plastic parts consisting of resins to which reinforcing fibers, mats, fabrics, and so forth, have been added before the forming operation to provide some

**Table 52  Mechanical properties of reinforced polypropylene**

| Property | 40% talc Homopolymer polypropylene | 40% talc Copolymer polypropylene | 40% mica Homopolymer polypropylene | 40% mica Copolymer polypropylene | 40% calcium carbonate Homopolymer polypropylene | 40% calcium carbonate Copolymer polypropylene | ASTM test method |
|---|---|---|---|---|---|---|---|
| Tensile strength, MPa (ksi) | 33 (4.8) | 21.4 (3.1) | 37.2 (5.4) | 18.6 (2.7) | 23.4 (3.4) | 16.6 (2.4) | D 638 |
| Elongation, % | 8.0 | 10.0 | 2.0 | 3.0 | 15.0 | 9.0 | D 638 |
| Flexural modulus, GPa ($10^6$ psi) | 3.45 (0.500) | 2.35 (0.340) | 4.76 (0.690) | 2.76 (0.400) | 2.48 (0.360) | 1.65 (0.240) | D 790 |
| Notched Izod impact strength, J/m (ft    lbf/in.) | 26.7 (0.5) | 203 (4.0) | 19.5 (0.36) | 90.8 (1.7) | 43 (0.8) | 673 (12.4) | D 256 |
| Deflection temperature under load, °C (°F) | | | | | | | D 648 |
| At 1.82 MPa (0.264 ksi) | 80 (180) | 74 (165) | 99 (210) | 65.5 (150) | 71 (160) | 63 (145) | |
| At 0.45 MPa (0.066 ksi) | 130 (270) | 102 (215) | 143 (290) | 107 (225) | 110 (230) | 94 (201) | |

Source: product data sheets, Quantum Chemical Corporation, USI Division

**Table 53  Comparative properties of reinforcing fibers**

| Fiber | Density, g/cm³ | Average tensile strength(a) GPa | Average tensile strength(a) $10^6$ psi | Modulus of elasticity GPa | Modulus of elasticity $10^6$ psi | Approx cost $/kg | Approx cost $/lb |
|---|---|---|---|---|---|---|---|
| Boron, 100 μm (4000 μin.) | 2.57 | 3.6 | 0.52 | 400 | 60 | 700 | 320 |
| Boron, 140 μm (5600 μin.) | 2.49 | 3.6 | 0.52 | 400 | 60 | 700 | 320 |
| Carbon, AS-4 | 1.75 | 3.1 | 0.45 | 221 | 32.1 | 65 | 30 |
| E-glass | 2.54 | 3.4 | 0.49 | 69 | 10 | 5.5 | 2.5 |
| Aramid | 1.44 | 3.6 | 0.52 | 124 | 18.0 | 45 | 20 |
| SiC | 3.0 | 3.9 | 0.57 | 400 | 60 | ~220(b) | ~100 |

(a) Based on room temperature measurements at 25-mm (1-in.) gage length (b) Projected cost in production quantities

strength properties greatly superior to those of the base resin. See also *resin-matrix composites* (Technical Brief 39).

**reinforced polypropylene.** Polypropylene that is reinforced with mineral fillers, such as talc, mica, and calcium carbonate, as well as glass and carbon fibers. The maximum concentration usually used is 50 wt%, although concentrates with higher levels of filler or reinforcement are available. Table 52 lists properties of reinforced polypropylene.

**reinforced reaction injection molding (RRIM).** A reaction injection molding with a reinforcement added. See also *reaction injection molding.*

**reinforcement.** A strong material bonded into a matrix to improve its mechanical properties. Reinforcements are usually long fibers, chopped fibers, whiskers, particulates, and so forth. The term is not synonymous with *filler*. The most commonly used reinforcement materials are glass (E-glass and S-glass), aramid, silicon carbide, boron, alumina, fused silica, alumina-boria-silica, and carbon/graphite. Table 53 compares properties of various reinforcing fibers. See also *carbon-carbon composites* (Technical Brief 6), *ceramic-matrix composites* (Technical Brief 10), *metal-matrix composites* (Technical Brief 26), and *resin-matrix composites* (Technical Brief 39).

**reinforcement of weld.** Weld metal in excess of the quantity required to fill a joint (Fig. 439). See also *face reinforcement* and *root reinforcement.*

**rejectable.** See preferred term *nonconforming.*

**relative humidity.** (1) The ratio of the molecular fraction of water vapor present in the air to the molecular fraction of

FIG. 439

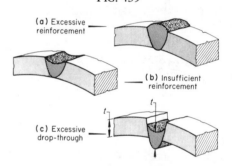

(a) Excessive reinforcement

(b) Insufficient reinforcement

(c) Excessive drop-through

*Three undesirable weld reinforcement conditions encountered in plasma arc welding*

water vapor present in saturated air at the same temperature and barometric pressure. Approximately, it equals the ratio of the partial pressure or density of the water vapor in the air to the saturation pressure or density, respectively, at the same temperature. (2) The ratio, expressed as a percentage, of the amount of water vapor present in a given volume of air at a given temperature to the amount required to saturate the air at that temperature.

**relative rigidity.** In dynamic mechanical measurements of plastics, the ratio of modulus at any temperature, frequency, or time to the modulus at a reference temperature, frequency, or time.

**relative sintering temperature.** In powder metallurgy, the ratio of the sintering temperature to the melting temperature of the substance as expressed on the Kelvin scale.

**relative standard deviation (RSD).** The standard deviation expressed as a percentage of the mean value:

$$\text{RSD} = 100\left(\frac{S}{X}\right)\frac{d^2}{n-1}$$

where $S$ is the standard deviation, $d$ is the difference between the individual results and the average, $n$ is the number of individual results, and $X$ is the average of individual results. Also known as coefficient of variation.

**relative transmittance.** The ratio of the transmittance of the object in question to that of a reference object. For a spectral line on a photographic emulsion, it is the ratio of the transmittance of the photographic image of the spectral line to the transmittance of a clear portion of the photographic emulsion. Relative transmittance may be total, specular, or diffuse. See also *transmittance*.

**relative viscosity.** For a polymer in solution, the ratio of the absolute viscosities of the solution (of stated concentration) and of the pure solvent at the same temperature.

**relaxation curve.** A plot of either the remaining or relaxed stress as a function of time. See also *relaxation rate*.

**relaxation rate.** The absolute value of the slope of a relaxation curve at a given time.

**relaxation time.** The time required for a stress under a sustained constant strain to diminish by a stated fraction of its initial value.

**relaxed stress.** The initial stress minus the remaining stress at a given time during a stress-relaxation test.

**relay.** An electrically controlled device that opens and closes electrical contacts to effect the operation of other devices in the same or another electric circuit.

**release agent.** In forming of plastics, a material that is applied as a thin film to the surface of a mold to keep the resin from bonding to the mold. Also called parting agent. See also *mold release agent*.

**release film.** In forming of plastics, an impermeable layer of film that does not bond to the resin being cured. See also *separator*.

**release paper.** A sheet, serving as a protectant or carrier, or both, for an adhesive film or mass, which is easily removed from the film or mass prior to use.

**reliability.** A quantitative measure of the ability of a product or service to fulfill its intended function for a specified period of time.

**relief.** The result of the removal of tool material behind or adjacent to the cutting edge to provide clearance and prevent rubbing (heel drag). See also *relief angle*.

**relief angle.** The angle formed between a relieved surface and a given plane tangent to a cutting edge or to a point on a cutting edge (Fig. 440). Also known as clearance angle. See also the figure accompanying the term *single-point tool*.

FIG. 440

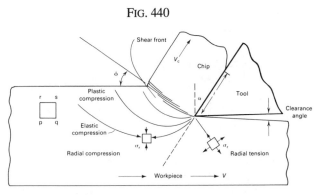

*Relief (clearance) angle during machining with a single-point tool*

**relieving.** Buffing or other abrasive treatment of the high points of an embossed metal surface to produce highlights that contrast with the finish in the recesses.

**remaining stress.** The stress remaining at a given time during a stress-relaxation test. See also *stress relaxation*.

**remanence.** The magnetic induction remaining in a magnetic circuit after removal of the applied magnetizing force. Sometimes called remanent induction.

**repeatability.** A term used to refer to the test result variability associated with a limited set of specifically defined sources of variability within a single laboratory.

**repeated impact.** See *impingement*.

**replica.** A reproduction of a surface in a material. It is usually accomplished by depositing a thin film of suitable material, such as a plastic, onto the specimen surface. This film is subsequently extracted and examined by optical microscopy, scanning electron microscopy, or transmis-

FIG. 441

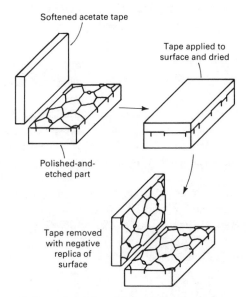

*Schematic of the plastic replica technique*

## Technical Brief 39: Resin-Matrix Composites

RESIN-MATRIX COMPOSITES are advanced engineering materials that contain a reinforcement (such as fibers or particles) supported by an organic (plastic) binder (matrix). Resin-matrix composites were developed in response to demands of the aerospace community for stronger, more lightweight materials. Aluminum alloys,

which provide high strength and fairly high stiffness at low weight, have provided good performance and have been the primary materials used in aircraft structures over the years. However, both corrosion and fatigue in aluminum alloys have produced problems that have been costly to remedy. World War II promoted a need for materials with improved structural properties. In response, fiber-reinforced composites were developed, and by the end of the war, fiberglass-reinforced plastics had been used successfully in filament-wound rocket motors and in various other structural applications. Today resin-matrix composites are seeing significantly expanded levels of use in components where reduced weight is critical. This is primarily a result of their tailorability as well as their high strength- and modulus-to-density ratios. In recent years, these materials have seen applications ranging from mass-produced tennis rackets to complex aerospace structures.

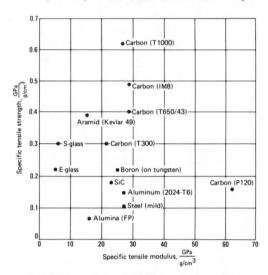

*Specific tensile strength (strength-to-density ratio) versus specific tensile modulus (modulus-to-density ratio) for various commercially available 65 vol% quasi-isotropic epoxy-matrix composites and for steel and aluminum*

Resin-matrix composite materials can be divided into two basic categories, depending upon their type of reinforcement: continuous fiber reinforced and particulate/short fiber reinforced composites. The first of these two materials typically uses a continuous array of oriented fibers, whereas the second category uses randomly dispersed particulates or chopped fibers. Continuous fiber composites, which are the most common type of composite, are typically made up of 3 to 30 μm diam fibers that are oriented and surrounded in a supportive matrix material. Generally, the fibers used in these material systems are several orders of magnitude stiffer and stronger than the surrounding matrix. Glass fiber reinforced organic-matrix composites are the most familiar and widely used, and have had extensive application in industrial, consumer, military, and aerospace markets. Carbon fiber reinforced resin-matrix composites are the most commonly applied advanced (non-fiberglass) composites. They offer extremely high specific properties, high-quality materials that are readily available, reproducible material forms, increasingly favorable cost projections, and comparative ease of manu-

sion electron microscopy, the latter being the most common. Replication techniques can be classified as either surface replication or extraction replication. Surface replicas provide an image of the surface topography of a specimen (Fig. 441), while extraction replicas lift particles from the specimen (Fig. 442). See also *atomic replica, cast replica, collodian replica, Formvar replica, gelatin replica, impression replica, negative replica, oxide film replica, plastic replica, positive replica, preshadowed replica, tape replica method (faxfilm),* and *vapor-deposited replica.*

**replicate.** In electron microscopy, to reproduce using a *replica.*

**repressing.** The application of pressure to a previously pressed and sintered powder metallurgy compact, usually

FIG. 442

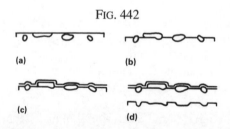

*Production of single-stage extraction replicas for analysis of small particles originally embedded in a matrix. (a) Original material with embedded particles. (b) Particles etched to stand in relief. (c) Specimen coated with a carbon film, usually by vacuum evaporation. (d) Specimen immersed in an aggressive solution to release the particles from the matrix.*

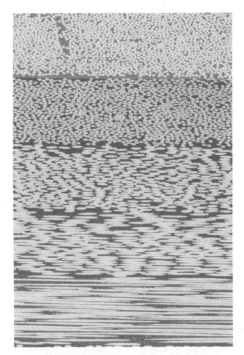

*Cross-sectional optical micrograph illustrating laminated construction typical of continuous-fiber composite materials. 50×*

facture. Composites reinforced with aramid, other organics, boron fibers, and silicon carbide, alumina, and other ceramic fibers are also used. The high stiffness and strength of these fibers control the characteristic engineering properties of the composite.

The continuous matrix phase functions as a supportive element to the fibers. In this capacity, fiber orientation and alignment are maintained, load is transferred between fibers, and strength is provided in nonreinforced directions. A wide variety of matrix materials are available for use with each fiber type. By far the most extensive matrices in current use are those employing either elevated-temperature curing epoxies or room-temperature curing vinyl esters. Other matrix materials include bismaleimide resins, poly-amide resins, and thermoplastic resins such polyether ether-ketone (PEEK), polyphenylene sulfide (PPS), and polyetherimide (PEI).

Although there are a wide variety of processing methods for these materials, continuous fiber reinforced composites are usually fabricated by laminating together and curing multiple fiber plies impregnated with unreacted matrix resin. Such processing facilitates the arranging of plies of material at a variety of angles. This structuring produces a stacked, laminated construction. In the aerospace industry, the angle of plies used in the laminated stack are oriented at fixed angles to the direction of the major load. The most common fixed angles are 0, 45, –45, and 90° to major load axis (see the figure accompanying the term *quasi-isotropic laminate*). By selecting the proper number of plies at each of these angles, composite properties and their anisotropies can be designed for strength, modulus, or even the degree of thermal expansion along each principal direction.

### Recommended Reading

- *Engineered Materials Handbook*, Vol 1, *Composites*, ASM International, 1987
- G. Lubin, Ed., *Handbook of Composite Materials*, Van Nostrand Reinhold, 1982
- M.M. Swartz, *Composite Materials Handbook*, McGraw-Hill, 1984

for the purpose of improving some physical or mechanical property or for dimensional accuracy.

**reprocessed plastic**. A thermoplastic prepared from melt-processed scrap or reject parts by a plastics processor, or from nonstandard or nonuniform virgin material. The term scrap does not necessarily connote feedstock that is less desirable or usable than virgin material. Reprocessed plastic may or may not be reformulated by the addition of fillers, plasticizers, stabilizers, or pigments.

**reproducibility**. A term used to describe test result variability associated with specifically defined components of variance obtained both from within a single laboratory and between laboratories.

**rerolling quality**. Rolled *billets* from which the surface defects have not been completely removed.

**resenes**. The constituents of *rosin* that cannot be saponified with alcoholic alkali, but that contain carbon, hydrogen, and oxygen in the molecule. See also *saponification*.

**reset**. The realigning or adjusting of metal forming dies or tools during a production run; not to be confused with the operation *setup* that occurs before a production run.

**residual elements**. Small quantities of elements unintentionally present in an alloy.

**residual field**. Same as *residual magnetic field*.

**residual gas analysis (RGA)**. The study of residual gases in vacuum systems using mass spectrometry.

**residual magnetic field**. The magnetic field that remains in a part after the magnetizing force has been removed.

**residual method**. Method of *magnetic-particle inspection* in which the particles are applied after the magnetizing force has been removed.

**residual strain.** The strain associated with residual stress.

**residual stress.** (1) The stress existing in a body at rest, in equilibrium, at uniform temperature, and not subjected to external forces. Often caused by the forming or thermal processing curing process. (2) An internal stress not depending on external forces resulting from such factors as cold working, phase changes, or temperature gradients. (3) Stress present in a body that is free of external forces or thermal gradients. (4) Stress remaining in a structure or member as a result of thermal or mechanical treatment or both. Stress arises in fusion welding primarily because the weld metal contracts on cooling from the solidus to room temperature.

**resilience.** (1) The amount of energy per unit volume released on unloading. (2) The capacity of a material, by virtue of high yield strength and low elastic modulus, to exhibit considerable elastic recovery on release of load.

**resin.** A solid, semisolid, or pseudosolid organic material that has an indefinite and often high molecular weight, exhibits a tendency to flow when subjected to stress, usually has a softening or melting range, and usually fracture conchoidally. (2) Liquid resin, that is, an organic polymeric liquid, which, when converted to its final state for sue, becomes a resin. (3) An organic polymer that cross links to form a thermosetting plastic when mixed with a curing agent. (4) In reinforced plastics, the material used to bind together the reinforcement material; the matrix. See also *polymer*, *reinforced plastics*, and *resin-matrix composites* (Technical Brief 39).

**resin content.** The amount of *resin* in a *laminate*, expressed as either a percentage of total weight or total volume.

**resin-matrix composites.** See *Technical Brief 39*.

**resinography.** The science of morphology, structure, and related descriptive characteristics as correlated with composition or conditions and with properties or behavior of resins, polymers, plastics, and their products.

**resinoid.** Any of the class of thermosetting synthetic resins, either in their initial temporarily fusible state or in their final infusible state. See also *novolac* and *thermosetting*.

**resinoid bond.** An organic bond usually of the phenol formaldehyde resin type but sometimes consisting of other synthetic resins.

**resinoid wheel.** A grinding wheel bonded with a synthetic resin. See also the figure accompanying the term *grinding wheels*.

**resin pocket.** In plastics, an apparent accumulation of excess resin in a small, localized section visible on cut edges of molded surfaces, or internal to the structure and nonvisible. See also *resin-rich area*.

**resin-rich area.** A significant thickness of a nonreinforced resin layer of the same composition as that within the base material. In *reinforced plastics*, a localized area filled with resin and lacking reinforcing material. See also *resin pocket*.

**resin-starved area.** In *reinforced plastics*, a localized area of insufficient resin, usually identified by low glass, dry spots, or fiber showing on the surface.

**resin system.** A mixture of *resin* and ingredients such as catalyst, initiator, diluents, and so forth, required for the intended processing method and final product.

**resintering.** (1) A second sintering operation on a powder compact. (2) Sintering a repressed compact. See also *repressing*.

**resin transfer molding (RTM).** A process by which catalyzed resin is transferred or injected into an enclosed mold in which reinforcement has been placed (Fig. 443). The fiberglass reinforcement is usually woven, nonwoven, or knitted fabric.

FIG. 443

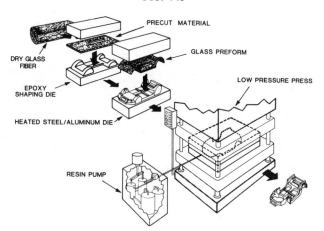

*Schematic of the high-speed resin transfer molding process*

**resist.** (1) Coating material used to mask or protect selected areas of a substrate from the action of an etchant, solder, or plating. (2) A material applied to prevent flow of brazing filler metal into unwanted areas.

**resistance.** The opposition that a device or material offers to the flow of direct current, equal to the voltage drop across the element divided by the current through the element. Also called electrical resistance.

**resistance alloys.** See *Technical Brief 40*.

**resistance brazing.** A resistance joining process in which the workpieces are heated locally and filler metal that is preplaced between the workpieces is melted by the heat obtained from resistance to the flow of electric current through the electrodes and the work (Fig. 444). In the usual application of resistance brazing, the heating current is passed through the joint itself.

FIG. 444

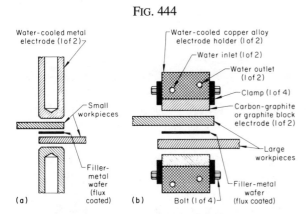

*Arrangements for resistance brazing. (a) For small flat parts or small flat portions of larger components, using opposed water-cooled metal electrodes of the conventional resistance welding type. (b) For large flat parts, typically of a highly conductive metal such as copper, using opposed carbon black electrodes attached to water-cooled copper alloy electrode holders*

**resistance butt welding**. See preferred terms *flash welding* and *upset welding*.

**resistance seam welding**. A resistance welding process which produces coalescence at the faying surfaces by the heat obtained from resistance to electric current through workpieces that are held together under pressure by electrode wheels (Fig. 445). The resulting weld is a series of overlapping resistance spot welds made progressively along a joint by rotating the electrodes. See also the figure accompanying the term *longitudinal resistance seam welding*.

FIG. 445

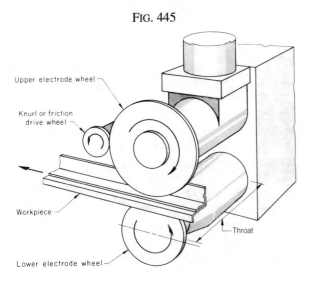

*Position of electrode wheels on a circular resistance seam welding machine*

**resistance soldering**. Soldering in which the joint is heated by electrical resistance. Filler metal is either face fed into the joint or preplaced in the joint.

**resistance spot welding**. A process in which faying surfaces are joined in one or more spots by the heat generated by resistance to the flow of electric current through workpieces that are held together under force by electrodes (Fig. 446). The contacting surfaces in the region of current concentration are heated by a short-time pulse of low-voltage, high-amperage current to form a fused nugget of weld metal. When the flow of current ceases, the electrode force is maintained while the weld metal rapidly cools and solidifies. The electrodes are retracted after each weld, which usually is completed in a fraction of a second.

FIG. 446

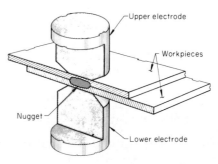

*Resistance spot welding setup*

**resistance welding**. A group of welding processes which produces coalescence of metals with resistance heating and pressure. See also *flash welding*, *projection welding*, *resistance seam welding*, and *resistance spot welding*.

**resistance welding die**. The part of a resistance welding machine, usually shaped to the work contour, in which the parts being welded are held and which conducts the welding current.

**resistance welding electrode**. The part or parts of a resistance welding machine through which the welding current and, in most cases, pressure are applied directly to the workpiece. The electrode may be in the form of a rotating wheel (Fig. 447), rotating roll, bar, cylinder, plate, clamp, chuck, or modification thereof.

**resistance welding gun**. A manipulating device to transfer current and provide electrode force to the weld area (usually in reference to a portable gun).

**resistivity**. See *electrical resistivity*.

**resite**. Synonym for *C-stage*.

**resitol**. Synonym for *B-stage*.

**resol**. Synonym for *A-stage*.

**resole resin**. Linear phenolic resin produced by alkaline condensate of phenol and formaldehyde.

**resolution**. The capacity of an optical or radiation system to separate closely spaced forms or entities; in addition, the degree to which such forms or entities can be discrim-

## Technical Brief 40: Resistance Alloys

ELECTRICAL RESISTANCE ALLOYS include both the types used in instruments and control equipment to measure and regulate electrical characteristics and those used in furnaces and appliances to generate heat. In the former applications, properties near ambient temperature are of primary interest; in the latter, elevated-temperature characteristics are of prime importance. In common commercial terminology, electrical resistance alloys used for control or regulation of electrical properties are called resistance alloys, and those used for generation of heat are referred to as resistance heating alloys.

The primary requirements for resistance alloys are uniform resistivity, stable resistance (no time-dependent aging effects), reproducible temperature coefficient of resistance, and low thermoelectric potential versus copper. Properties of secondary importance are coefficient of expansion, mechanical strength, ductility, corrosion resistance, and ability to be joined to other metals by soldering, brazing, or welding. Alloys must be strong enough to withstand fabrication operations, and it must be easy to procure an alloy that has consistently reproducible properties in order to ensure resistor accuracy.

Resistors for electrical and electronic devices may be divided into two arbitrary classifications: those employed in precision instruments in which overall error is considerably less than 1%, and those employed where less precision is needed. The choice of alloy for a specific resistor application depends on the variation in properties that can be tolerated. Materials for resistors include: copper-nickel (2 to 22% Ni) alloys, generally referred to as radio alloys; copper-manganese-nickel alloys (10 to 13% Mn and 4% Ni), generally referred to as manganins; constantin alloys, whose compositions vary from 50Cu-50Ni to 65Cu-35Ni; nickel-chromium-aluminum alloys that nominally contain 20% Cr, 3% Al, and 2 to 5% of copper, iron, and/or manganese; 80Ni-20Cr alloys; and iron-chromium-aluminum alloys (nominally 73Fe-22Cr-5Al).

Resistance heating elements are used in many varied applications—from small household appliances to large industrial process heating systems and furnaces that may operate continuously at temperatures of 1300 °C (2350 °F) or higher. The primary requirements of materials used for heating elements are high melting point, high electrical resistivity, reproducible temperature coefficient of resistance, good oxidation resistance, absence of volatile components, and resistance to contamination. Other desirable properties are good creep strength, high emissivity, low thermal expansion and low modulus (both of which help minimize thermal fatigue), good resistance to thermal shock, and good strength and ductility at fabricating temperature.

The most commonly used resistance heating alloys are nickel-chromium and nickel-chromium-iron alloys (see

FIG. 447

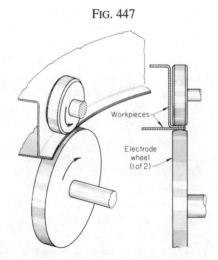

*Electrode placement for resistance welding*

inated. Resolution is usually specified as the minimum distance by which two lines or points in the object must be separated before they can be revealed as separate lines or points in the image. See also *resolving power* and *shape resolution.*

**resolving power.** The ability of a given optical lens system to reveal fine detail in an object. See also *resolution.*

**resonant forced vibration technique.** A technique for performing dynamic mechanical measurements on plastics, in which the sample is oscillated mechanically at the natural resonant frequency of the system. The amplitude of oscillation is maintained constant by the addition of makeup energy. Elastic modulus is calculated from the measured frequency. Damping is calculated from the additional energy required to maintain constant amplitude.

**response curve for *N* cycles.** In fatigue data analysis, a curve fitted to observed values of percentage survival at *N*

## Typical properties of resistance heating materials

| Basic composition | Resistivity(a), $\Omega \cdot$ mm$^2$/m(b) | Average change in resistance(c), %, from 20 °C to: | | | | Thermal expansion, $\mu$m/m · °C, from 20 °C to: | | | Tensile strength | | Density | |
|---|---|---|---|---|---|---|---|---|---|---|---|---|
| | | 260 °C | 540 °C | 815 °C | 1095 °C | 100 °C | 540 °C | 815 °C | MPa | ksi | g/cm$^3$ | lb/in.$^3$ |
| **Nickel-chromium and nickel-chromium-iron alloys** | | | | | | | | | | | | |
| 78.5Ni-20Cr-1.5Si (80-20) ........... | 1.080 | 4.5 | 7.0 | 6.3 | 7.6 | 13.5 | 15.1 | 17.6 | 655–1380 | 95–200 | 8.41 | 0.30 |
| 77.5Ni-20Cr-1.5Si-1Nb ............. | 1.080 | 4.6 | 7.0 | 6.4 | 7.8 | 13.5 | 15.1 | 17.6 | 655–1380 | 95–200 | 8.41 | 0.30 |
| 68.5Ni-30Cr-1.5Si (70-30) ......... | 1.180 | 2.1 | 4.8 | 7.6 | 9.8 | 12.2 | ... | ... | 825–1380 | 120–200 | 8.12 | 0.29 |
| 68Ni-20Cr-8.5Fe-2Si ................ | 1.165 | 3.9 | 6.7 | 6.0 | 7.1 | ... | 12.6 | ... | 895–1240 | 130–180 | 8.33 | 0.30 |
| 60Ni-16Cr-22Fe-1.5Si ............... | 1.120 | 3.6 | 6.5 | 7.6 | 10.2 | 13.5 | 15.1 | 17.6 | 655–1205 | 95–175 | 8.25 | 0.30 |
| 37Ni-21Cr-40Fe-2Si ................. | 1.08 | 7.0 | 15.0 | 20.0 | 23.0 | 14.4 | 16.5 | 18.6 | 585–1135 | 85–165 | 7.96 | 0.288 |
| 35Ni-20Cr-43Fe-1.5Si .............. | 1.00 | 8.0 | 15.4 | 20.6 | 23.5 | 15.7 | 15.7 | ... | 550–1205 | 80–175 | 7.95 | 0.287 |
| 35Ni-20Cr-42.5Fe-1.5Si-1Nb ........ | 1.00 | 8.0 | 15.4 | 20.6 | 23.5 | 15.7 | 15.7 | ... | 550–1205 | 80–175 | 7.95 | 0.287 |
| **Iron-chromium-aluminum alloys** | | | | | | | | | | | | |
| 83.5Fe-13Cr-3.25Al ................ | 1.120 | 7.0 | 15.5 | ... | ... | 10.6 | ... | ... | 620–1035 | 90–150 | 7.30 | 0.26 |
| 81Fe-14.5Cr-4.25Al ................ | 1.25 | 3.0 | 9.7 | 16.5 | ... | 10.8 | 11.5 | 12.2 | 620–1170 | 90–170 | 7.28 | 0.26 |
| 73.5Fe-22Cr-4.5Al ................. | 1.35 | 0.3 | 2.9 | 4.3 | 4.9 | 10.8 | 12.6 | 13.1 | 620–1035 | 90–150 | 7.15 | 0.26 |
| 72.5Fe-22Cr-5.5Al ................. | 1.45 | 0.2 | 1.0 | 2.8 | 4.0 | 11.3 | 12.8 | 14.0 | 620–1035 | 90–150 | 7.10 | 0.26 |
| **Pure metals** | | | | | | | | | | | | |
| Molybdenum...................... | 0.052 | 110 | 238 | 366 | 508 | 4.8 | 5.8 | ... | 690–2160 | 100–313 | 10.2 | 0.369 |
| Platinum......................... | 0.105 | 85 | 175 | 257 | 305 | 9.0 | 9.7 | 10.1 | 345 | 50 | 21.5 | 0.775 |
| Tantalum......................... | 0.125 | 82 | 169 | 243 | 317 | 6.5 | 6.6 | ... | 345–1240 | 50–180 | 16.6 | 0.600 |
| Tungsten ........................ | 0.055 | 91 | 244 | 396 | 550 | 4.3 | 4.6 | 4.6 | 3380–6480 | 490–940 | 19.3 | 0.697 |
| **Nonmetallic heating-element materials** | | | | | | | | | | | | |
| Silicon carbide ..................... | 0.995–1.995 | −33 | −33 | −28 | −13 | 4.7 | ... | ... | 28 | 4 | 3.2 | 0.114 |
| Molybdenum disilicide ............. | 0.370 | 105 | 222 | 375 | 523 | 9.2 | ... | ... | 185 | 27 | 6.24 | 0.225 |
| MoSi$_2$ + 10% ceramic additives...... | 0.270 | 167 | 370 | 597 | 853 | 13.1 | 14.2 | 14.8 | ... | ... | 5.6 | 0.202 |
| Graphite......................... | 9.100 | −16 | −18 | −13 | −8 | 1.3 | ... | ... | 1.8 | 0.26 | 1.6 | 0.057 |

(a)At 20 °C (68 °F). (b) To convert to $\Omega$·circ mil/ft, multiply by 601.53. (c) Changes in resistance may vary somewhat, depending on cooling rate.

table above). Other materials include iron-chromium-aluminum alloys similar in composition to resistor alloys, high-melting-temperature pure metals, and nonmetallic materials, which can be used effectively at temperatures as high as 1900 °C (3450 °F).

### Recommended Reading

- R.A. Watson *et al.*, Electrical Resistance Alloys, *Metals Handbook*, 10th ed., Vol 2, ASM International, 1990, p 822-839

cycles for several stress levels, where $N$ is a preassigned number such as $10^6$, $10^7$, etc. (Fig. 448). It is an estimate of the relationship between applied stress and the percentage of the population that would survive $N$ cycles. See also *S-N curve*.

**rest potential.** See *corrosion potential* and *open-circuit potential*.

**restraint.** Any external mechanical force that prevents a part from moving to accommodate changes in dimension due to thermal expansion or contraction. Often applied to weldments made while clamped in a fixture. Compare with *constraint*.

**restrictor rings.** Rings, usually faced with *white metal*, placed outside a bearing to prevent fluid from being discharged.

**restrike.** Additional compacting of a sintered powder metallurgy compact.

FIG. 448

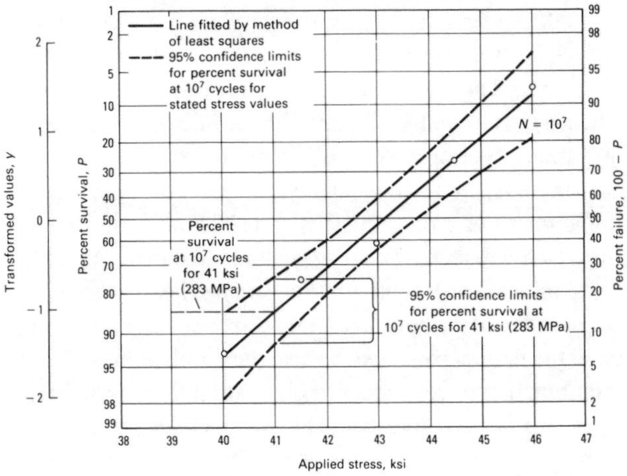

*Typical response curve for N (fatigue) cycles*

**restriking**. (1) The striking of a trimmed but slightly misaligned or otherwise faulty forging with one or more blows to improve alignment, improve surface condition, maintain close tolerances, increase hardness, or effect other improvements. (2) A *sizing* operation in which coining or stretching is used to correct or alter profiles and to counteract distortion. (3) A salvage operation following a primary forging operation in which the parts involved are rehit in the same forging die in which the pieces were last forged.

**resultant field**. The magnetic field which is the result of two or more magnetizing forces impressed on the same area of a magnetizable object. Sometimes called vector field.

**resultant rake**. The angle between the tooth face and an axial plane through the tooth point measured in a plane perpendicular to the cutting edge. The resultant rake of a cutter is a function of three other angles, radial rake, axial rake, and corner angle.

**retainer**. See *cage*.

**retardation plate**. A plate placed in the path of a beam of polarized light in an *optical microscope* for the purpose of introducing a difference in phase. Usually quarter-wave or half-wave plates are used, but if the light passes through them twice, the phase difference is doubled.

**retarder**. Synonym for *inhibitor*.

**retention time ($t_R$)**. In *chromatography*, the amount of time a sample compound spends in the chromatographic column.

**retentivity**. The capacity of a material to retain a portion of the magnetic field set up in it after the magnetizing force has been removed.

**retort**. A vessel used for distillation of volatile materials, as in separation of some metals and in destructive distillation of coal.

**retrogradation**. A change of starch pastes that are used in adhesive formulations from low to high consistency upon aging.

**reverberatory furnace**. A furnace in which the flame used for melting the metal does not impinge on the metal surface itself, but is reflected off the walls of the root of the furnace. The metal is actually melted by the generation of heat from the walls and the roof of the furnace. Although this furnace heating concept is still used in many foundries and smelters, more modern reverberatory furnaces use roof burners or side burners directed toward the metal surface. Such furnaces are designed in either the wet hearth or dry hearth configuration. In a wet hearth furnace, the products of combustion are in direct contact with the top of the molten bath, and the heat transfer is achieved by a combination of convection and radiation (Fig. 449). In a dry hearth furnace, the charge of solid metal is positioned on a sloping hearth above the level of the molten metal so that the charge is completely enveloped by the hot gases

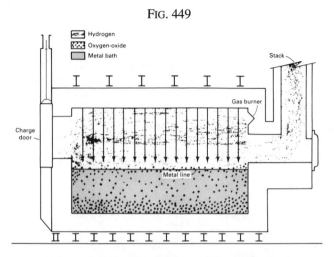

FIG. 449

*Schematic of a wet hearth reverberatory furnace heated by conventional fossil fuel showing the position of the hydrogen and oxygen gases relative to the molten metal bath. Arrows indicate heat radiated from top of furnace chamber.*

FIG. 450

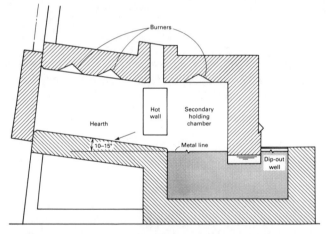

*Schematic of a radiant-fired dry hearth reverberatory furnace illustrating the position of the sloping hearth relative to the molten bath*

(Fig. 450). Heat is rapidly absorbed by the solid charge, which melts and subsequently drains from the sloping hearth into the wet holding basin or chamber.

**reverse-current cleaning**. *Electrolytic cleaning* in which a current is passed between electrodes through a solution, and the part is set up as the anode. Also called *anodic cleaning*.

**reversed-phase chromatography (RPC)**. Bonded-phase chromatography with a nonpolar stationary phase and a polar mobile phase. See also *bonded-phase chromatography*.

**reverse drawing**. *Redrawing* of a sheet metal part in a direction opposite to that of the original drawing.

**reverse flange**. A sheet metal flange made by shrinking, as opposed to one formed by stretching.

**reverse helical winding**. In *filament winding*, as the fiber delivery arm traverses one circuit, a continuous helix is laid down, reversing direction at the polar ends, in contrast to biaxial, compact, or sequential winding. The fibers cross each other at definite equators, the number depending on the helix. The minimum region of crossover is three. See also the figure accompanying the term *helical winding*.

**reverse impact test**. A test in which one side of a sheet of plastic is struck by a pendulum or falling object, and the reverse side is inspected for damage.

**reverse polarity**. Direct-current arc welding circuit arrangement in which the electrode is connected to the positive terminal (Fig. 451). A synonym for *direct current electrode positive (DCEP)*. Contrast with *straight polarity*.

FIG. 452

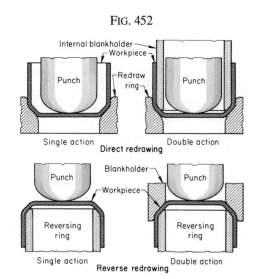

*Direct and reverse redrawing in single-action and double-action dies*

FIG. 451

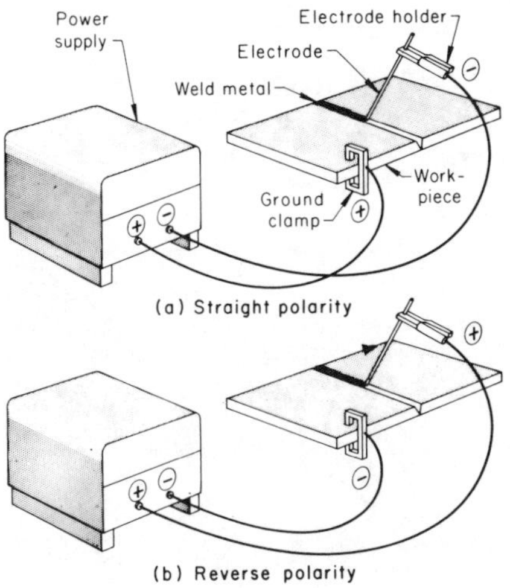

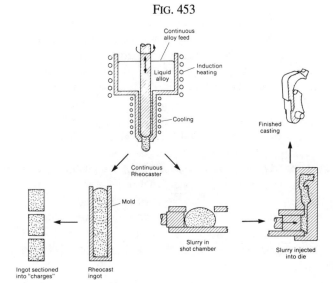

*Connections for straight polarity and reverse polarity welding*

**reverse redrawing**. A second drawing operation in a direction opposite to that of the original drawing (Fig. 452).

**reworked plastic**. A thermoplastic that is reground, pelletized, or solvated after having been previously processed by molding, extrusion, and so forth. In many specifications, the use of reworked material is limited to clean plastic that meets the requirements specified for virgin material and yields a product essentially equal in quality to one made from virgin material only.

**reyn**. The former English unit of dynamic viscosity equal to approximately 14.9 poise.

**Reynold's equation**. A basic equation of *hydrodynamic lubrication*.

**RGA**. See *residual gas analysis*.

FIG. 453

*Schematic of the rheocast process*

**rheocasting**. Casting of a continuously stirred semisolid metal slurry (Fig. 453). The process involves vigorous agitation of the melt during the early stages of solidification to break up solid dendrites into small spherulites. See also *semisolid metal forming*.

**rheodynamic lubrication**. A regime of lubrication in which the rheological (non-Newtonian) properties of the lubricant predominate. This term is especially applied to lubrication with grease.

**rheology**. The science of deformation and the flow of matter.

**rheopectic material**. A material that shows an increase in viscosity with time under a constant shear stress. After removal of the shear stress, the viscosity slowly returns to its original level. Compare with *thixotropy*.

**rheotropic brittleness**. That portion of the brittleness characteristic of non-face-centered cubic metals, where tested in the presence of a stress concentration or at low temperatures or high strain rates, that may be eliminated by prestraining under milder conditions.

**rhombohedral**. Having three equal axes, with the included angles equal to each other, but not equal to 90°. See also the figure accompanying the term *unit cell*.

**rib** (metals). A long V-shaped or radiused indentation used to strengthen large sheet metal panels. (2) A long, usually thin protuberance used to provide flexural strength to a forging (as in a rib-web forging).

**rib** (plastics). A reinforcing member designed into a plastic part to provide lateral, horizontal, hoop, or other structural support.

**rib mark**. A curved line on the crack surface of ceramics or glasses, usually convex in the general direction toward which the crack is running. The term is useful in referring to a mark of this shape until its specific nature is learned.

**riddle**. A sieve used to separate foundry sand or other granular materials into various particle-size grades or to free such a material of undesirable foreign matter.

**ridging wear**. A deep form of scratching in parallel ridges usually caused by plastic flow of the subsurface layer.

**rigging**. The engineering design, layout, and fabrication of pattern equipment for producing castings; including a study of the casting solidification program, feeding and gating, risering, skimmers, and fitting flasks.

**right-hand cutting tool**. A cutter all of whose flutes twist away in a clockwise direction when viewed from either end.

**rigid plastics**. For purposes of general classification, a plastic that has a modulus of elasticity either in flexure or in tension greater than 690 MPa (100 ksi) at 23 °C (73 °F) and 50% relative humidity.

**rigid resin**. A resin having a modulus high enough to be of practical importance, for example, ≥690 MPa (100 ksi).

**RIM**. See *reaction injection molding*.

**rimmed steel**. A low-carbon steel containing sufficient iron oxide to give a continuous evolution of carbon monoxide while the ingot is solidifying, resulting in a case or rim of metal virtually free of voids. Sheet and strip products made from rimmed steel ingots have very good surface quality. See also the figure accompanying the term *capped steel*.

**ring and circle shear**. A cutting or shearing machine with two rotary-disk cutters driven in unison and equipped with a circle attachment for cutting inside circles or rings from sheet metal, where it is impossible to start the cut at the edge of the sheet. One cutter shaft is inclined to the other to provide cutting clearance so that the outside section remains flat and usable. See also *circle shear* and *rotary shear*.

**ringing**. The audible or ultrasonic tone produced in a mechanical part by shock, and having the natural frequency or frequencies of the part. The quality, amplitude, or decay rate of the tone may sometimes be used to indicate the quality or soundness. See also *sonic testing* and *ultrasonic testing*.

**ring riser**. A *riser block* with openings matching those in the metal forming.

**ring rolling**. The process of shaping weldless rings from pierced disks or shaping thick-wall ring-shaped blanks between rolls that control wall thickness, ring diameter, height, and contour (Fig. 454). See also the figure accompanying the term *axial rolls*.

FIG. 454

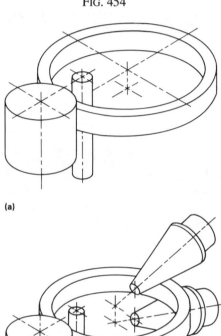

(a)

(b)

*Schematics showing single-pass (radial) rolling (a) and two-pass (radial-axial) ring rolling (b)*

**ring seal**. A piston ring-type seal that assumes its sealing position under the pressure of the fluid to be sealed.

**rinsability**. The relative ease with which a substance can be removed from a metal surface with a liquid such as water.

**ripple formation**. Formation of periodic ridges and valleys transverse to the direction of motion on a solid surface. Also referred to as rippling.

**ripple mark**. See *Wallner lines* (ceramics and glasses).

FIG. 455

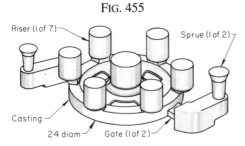

Riser (1 of 7)
Sprue (1 of 2)
Casting
24 diam
Gate (1 of 2)

*Placement of risers in the gating and feeding system used to cast gear blanks*

**riser**. (1) A reservoir of molten metal connected to a casting to provide additional metal to the casting, required as the result of shrinkage before and during solidification (Fig. 455). (2) That section of pipeline extending from the ocean floor up the offshore oil-drilling platform. Also, the vertical tube in a steam generator convection bank that circulates water and steam upward.

**riser blocks**. (1) Plates or pieces inserted between the top of a metal forming press bed or bolster and the die to decrease the height of the die space. (2) Spacers placed between bed and housings to increase *shut height* on a four-piece tie-rod straight-side press.

**rise time**. In urethane foam molding, the time between the pouring of the urethane mix and the completion of foaming.

**river marks** (ceramics and glasses). Cleavage steps on individual grains of a polycrystalline material or on a single crystal. These markings spread out away from the point of origin. These are a special case of *twist hackle*.

**river pattern** (metals). A term used in fractography to describe a characteristic pattern of cleavage steps running parallel to the local direction of crack propagation on the fracture surfaces of grains that have separated by *cleavage* (Fig. 456).

**riveting**. Joining of two or more members of a structure by means of metal rivets, the unheaded end being upset after the rivet is in place.

**roasting**. Heating an ore to effect some chemical change that will facilitate *smelting*.

**robber**. An extra cathode or cathode extension that reduces the current density on what would otherwise be a high-current-density area on work being electroplated.

**Rochelle copper**. (1) A copper electrodeposit obtained from copper cyanide plating solution to which Rochelle salt (sodium potassium tartrate) has been added for grain refinement, better anode corrosion, and cathode efficiency. (2) The solution from which a Rochelle copper electrodeposit is obtained.

**rock candy fracture**. A fracture that exhibits separated-grain facets; most often used to describe an *intergranular fracture* in a large-grained metal (Fig. 457).

FIG. 456

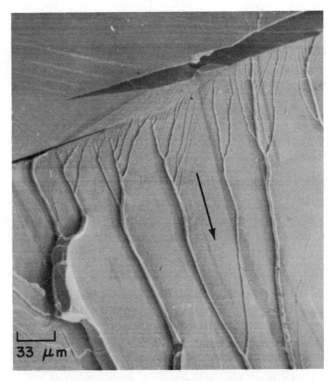

33 $\mu$m

*River patterns in an Fe-0.01C-0.24Mn-0.02Si alloy that was fractured by impact*

FIG. 457

*Surface of a "rock candy" fracture in a bloom of AISI type 302 stainless steel. 0.67×*

**rocking curve**. A method used in *x-ray topography* for determining the degree of imperfection in a crystal by using monochromatic, collimated x-rays reflecting off a "perfect" crystal to probe a second test crystal. A rocking curve is obtained by monitoring the x-ray intensity diffracted by the test cycle as it is slowly rocked or rotated, through the Bragg angle for the reflecting planes.

**rocking shear**. A type of guillotine shear that utilizes a curved blade to shear sheet metal progressively from side to side by a rocker motion.

FIG. 458

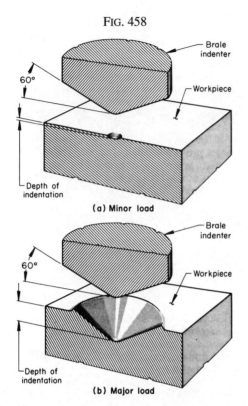

*Indentation in a workpiece made by application of (a) the minor load, and (b) the major load, from a diamond Brale indenter in Rockwell hardness testing*

FIG. 459

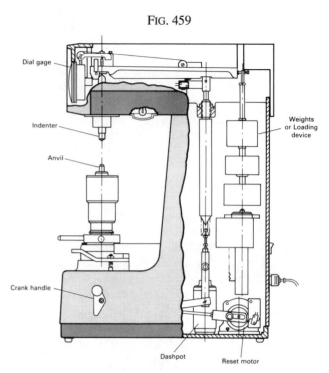

*Schematic of Rockwell testing machine*

FIG. 460

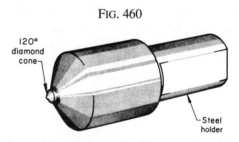

*Diamond-cone Brale indenter used in Rockwell hardness testing (shown at about 2×)*

**Rockrite tube-reducing process**. See *tube reducing*.

**Rockwell hardness number**. A number derived from the net increase in the depth of impression as the load on an indentor is increased from a fixed minor load to a major load and then returned to the minor load (Fig. 458). Various scales of Rockwell hardness numbers have been developed based on the hardness of the materials to be evaluated. The scales are designated by alphabetic suffixes to the hardness designation. For example, 64 HRC represents the Rockwell hardness number of 64 on the Rockwell C scale. Table 54 lists the standard Rockwell hardness scales. See also *Rockwell superficial hardness number*.

**Rockwell hardness test**. An indentation hardness test using a calibrated machine that utilizes the depth of indentation, under constant load, as a measure of hardness (Fig. 459). Either a 120° diamond cone with a slightly rounded point (Fig. 460) or a 1.6 or 3.2 mm (1/16 or 1/8 in.) diam steel ball is used as the indenter. See also *Brale indenter*.

**Rockwell superficial hardness number**. Like the *Rockwell hardness number*, the superficial Rockwell number is expressed by the symbol HR followed by a scale designation. For example, 81 HR30N represents the Rockwell superficial hardness number of 81 on the Rockwell 30N scale.

Table 55 lists the standard Rockwell superficial hardness scales.

**Rockwell superficial hardness test**. The same test as used to determine the Rockwell hardness number except that smaller minor and major loads are used. In Rockwell testing, the minor load is 10 kgf, and the major load is 60, 100, or 150 kgf. In superficial Rockwell testing, the minor load is 3 kgf, and major loads are 15, 30, or 45 kgf. In both tests, the indenter may be either a diamond cone or a steel ball, depending principally on the characteristics of the material being tested.

**rod**. A solid round metal section 9.5 mm (3/8 in.) or greater in diameter, whose length is great in relation to its diameter.

**rod mill**. (1) A *hot mill* for rolling rod. (2) A mill for fine grinding, somewhat similar to a *ball mill*, but employing long steel rods instead of balls to effect grinding.

## Table 54  Standard Rockwell hardness scales

| Scale symbol | Indenter | Major load, kgf | Typical applications |
|---|---|---|---|
| A | Diamond (two scales—carbide and steel) | 60 | Cemented carbides, thin steel, and shallow case-hardened steel |
| B | 1/16-in. (1.588-mm) ball | 100 | Copper alloys, soft steels, aluminum alloys, malleable iron |
| C | Diamond | 150 | Steel, hard cast irons, pearlitic malleable iron, titanium, deep case-hardened steel, and other materials harder than HRB 100 |
| D | Diamond | 100 | Thin steel and medium case-hardened steel and pearlitic malleable iron |
| E | 1/8-in. (3.175-mm) ball | 100 | Cast iron, aluminum and magnesium alloys, bearing metals |
| F | 1/16-in. (1.588-mm) ball | 60 | Annealed copper alloys, thin soft sheet metals |
| G | 1/16-in. (1.588-mm) ball | 150 | Phosphor bronze, beryllium copper, malleable irons. Upper limit HRG 92 to avoid possible flattening of ball |
| H | 1/8-in. (3.175-mm) ball | 60 | Aluminum, zinc, lead |
| K | 1/8-in. (3.175-mm) ball | 150 | Bearing metals and other very soft or thin materials. Use smallest ball and heaviest load that do not produce anvil effect. |
| L | 1/4-in. (6.350-mm) ball | 60 | Bearing metals and other very soft or thin materials. Use smallest ball and heaviest load that do not produce anvil effect. |
| M | 1/4-in. (6.350-mm) ball | 100 | Bearing metals and other very soft or thin materials. Use smallest ball and heaviest load that do not produce anvil effect. |
| P | 1/4-in. (6.350-mm) ball | 150 | Bearing metals and other very soft or thin materials. Use smallest ball and heaviest load that do not produce anvil effect. |
| R | 1/2-in. (12.70-mm) ball | 60 | Bearing metals and other very soft or thin materials. Use smallest ball and heaviest load that do not produce anvil effect. |
| S | 1/2-in. (12.70-mm) ball | 100 | Bearing metals and other very soft or thin materials. Use smallest ball and heaviest load that do not produce anvil effect. |
| V | 1/2-in. (12.70-mm) ball | 150 | Bearing metals and other very soft or thin materials. Use smallest ball and heaviest load that do not produce anvil effect. |

Source: ASTM Standard E 18

## Table 55  Rockwell superficial hardness scales

| Scale symbol | Indenter | Major load, kgf |
|---|---|---|
| 15N | Diamond | 15 |
| 30N | Diamond | 30 |
| 45N | Diamond | 45 |
| 15T | 1/16-in. (1.588-mm) ball | 15 |
| 30T | 1/16-in. (1.588-mm) ball | 30 |
| 45T | 1/16-in. (1.588-mm) ball | 45 |
| 15W | 1/8-in. (3.175-mm) ball | 15 |
| 30W | 1/8-in. (3.175-mm) ball | 30 |
| 45W | 1/8-in. (3.175-mm) ball | 45 |
| 15X | 1/4-in. (6.350-mm) ball | 15 |
| 30X | 1/4-in. (6.350-mm) ball | 30 |
| 45X | 1/4-in. (6.350-mm) ball | 45 |
| 15Y | 1/2-in. (12.70-mm) ball | 15 |
| 30Y | 1/2-in. (12.70-mm) ball | 30 |
| 45Y | 1/2-in. (12.70-mm) ball | 45 |

Note: The Rockwell N scales of a superficial hardness tester are used for materials similar to those tested on the Rockwell C, A, and D scales, but of thinner gage or case depth. The Rockwell T scales are used for materials similar to those tested on the Rockwell B, F, and G scales, but of thinner gage. When minute indentations are required, a superficial hardness tester should be used. The Rockwell W, X, and Y scales are used for very soft materials. The letter N designates the use of the diamond indenter; the letters T, W, X, and Y designate steel ball indenters. Superficial Rockwell hardness values are always expressed by the number suffixed by a number and a letter that show the load and indenter combination. For example, 80 HR30N indicates a reading of 80 on the superficial Rockwell scale using a diamond indenter and a major load of 30 kgf.

FIG. 461

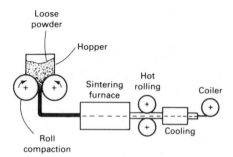

*Roll compacting (powder rolling) process*

FIG. 462

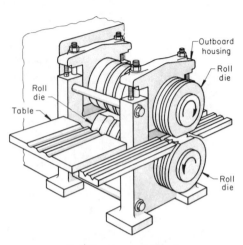

*Roll forging machine*

**roll bending**. Curving sheets, bars, and sections by means of rolls. See also the figure accompanying the term *bending rolls*.

**roll compacting**. Progressive compacting of metal powders by use of a rolling mill (Fig. 461).

**rolled compact**. A compact made by passing metal powder between rollers so as to form a relatively long, sheetlike compact.

**rolled gold**. Same as *gold filled* except that the proportion of gold alloy to the weight of the entire article may be less than 1/20th. Fineness of the gold alloy may not be less than 10K. See also *karat*.

**roller air analyzer**. An air-elutriation apparatus suitable for the particle size determination of metal powders, especially in the subsieve range. See also *elutriation*.

**roller bearing**. A bearing in which the relatively moving parts are separated by rollers. See also the figure accompanying the term *rolling-element bearing*.

**roller hearth furnace**. A modification of the pusher-type continuous furnace that provides for rollers in the hearth or muffle of the furnace whereby friction is greatly reduced and lightweight trays can be used repeatedly without risk of unacceptable distortion and damage to the work. See also *pusher furnace*.

**roller leveler breaks**. Obvious transverse *breaks* usually about 3 to 6 mm (1/8 to 1/4 in.) apart caused by the sheet metal fluting during *roller leveling*. These will not be removed by stretching.

**roller leveler lines**. Same as *leveler lines*.

**roller leveling**. *Leveling* by passing flat sheet metal stock through a machine having a series of small-diameter staggered rolls that are adjusted to produce repeated reverse bending.

**roller stamping die**. An engraved roller used for impressing designs and markings on sheet metal.

**roll flattening**. The flattening of metal sheets that have been rolled in packs by passing them separately through a two-high cold mill with virtually no deformation. Not to be confused with *roller leveling*.

**roll forging**. A process of shaping stock between two driven rolls that rotate in opposite directions and have one or more

matching sets of grooves in the rolls (Fig. 462); used to produce finished parts or preforms for subsequent forging operations.

**roll forming**. Metal forming through the use of power-driven rolls whose contour determines the shape of the product; sometimes used to denote power *spinning*.

**rolling**. The reduction of the cross-sectional area of metal stock, or the general shaping of metal products, through the use of rotating rolls. See also *rolling mills*.

**rolling** (pure rolling with no sliding and no spin). A motion of two relatively moving bodies, of opposite curvature, whose surface velocities in the common contact area are identical with regard to both magnitude and direction. See also *sliding* and *spin*.

**rolling-contact fatigue**. Repeated stressing of a solid surface due to rolling contact between it and another solid surface or surfaces. Continued rolling- contact fatigue of bearing or gear surfaces may result in rolling-contact damage in the form of subsurface fatigue cracks and/or material pitting and spallation.

**rolling-contact wear**. Wear to a solid surface that results from rolling contact between that surface and another solid surface or surfaces.

**rolling direction (in rolled metals)**. See *longitudinal direction*.

**rolling-element bearing**. A bearing in which the relatively moving parts are separated by balls, rollers, or needles (Fig 463).

**rolling mandrel**. In *ring rolling*, a vertical roll of sufficient diameter to accept various sizes of ring blanks and to exert rolling force on an axis parallel to the main roll. See also the figure accompanying the term *axial rolls*.

**rolling mills**. Machines used to decrease the cross-sectional area of metal stock and to produce certain desired shapes as the metal passes between rotating rolls mounted in a framework comprising a basic unit called a *stand*. Cy-

FIG. 463

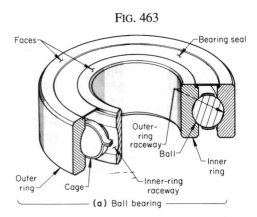

(a) Ball bearing

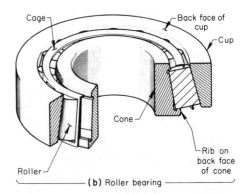

(b) Roller bearing

*Principal components of rolling-element bearings*

lindrical rolls produce flat shapes; grooved rolls produce rounds, squares, and structural shapes. See also *four-high mill*, *Sendzimir mill*, and *two-high mill*.

**rolling velocity**. See *sweep velocity*.

**roll resistance spot welding**. Process for making separated resistance spot welds with one or more rotating circular electrodes. The rotation of the electrodes may or may not be stopped during the making of a weld.

**roll straightening**. The straightening of metal stock of various shapes by passing it through a series of staggered rolls, the rolls usually being in horizontal and vertical planes

FIG. 464

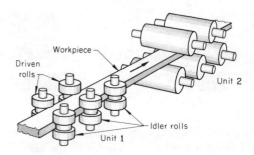

*Arrangement of vertical-shaft and horizontal-shaft rolls in a roll straightener for straightening a rectangular-section bar*

(Fig. 464), or by reeling in two-roll straightening machines.

**roll table**. A conveyor table where rolls furnish the contact surface.

**roll threading**. See preferred term *thread rolling*.

**roll welding**. Solid-state welding in which metals are heated, then welded together by applying pressure, with rolls, sufficient to cause deformation at the faying surfaces. See also *forge welding*.

**room temperature**. A temperature in the range of ~20 to 30 °C (~68 to 85 °F). The term room temperature is usually applied to an atmosphere of unspecified relative humidity.

**room-temperature setting adhesive**. An adhesive that sets (to handling strength) within an hour at temperatures from 20 to 30 °C (68 to 86 °F) and later reaches full strength without heating. Compare with *cold-setting adhesive*, *hot-setting adhesive*, and *intermediate temperature setting adhesive*.

**room-temperature vulcanizing (RTV)**. *Vulcanization* or curing at room temperature by chemical reaction, particularly of silicones and other rubbers.

**root**. See preferred term *root of joint* and *root of weld*.

**root crack**. A crack in either the weld or heat-affected zone at the root of a weld. See also the figure accompanying the term *weld crack*.

**root edge**. A *root face* of zero width.

**root face**. The portion of a weld groove face adjacent to the root of the joint (Fig. 465).

FIG. 465

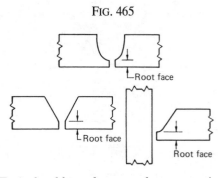

*Typical weld root faces on edge preparations*

**root gap**. See preferred term *root opening*.

**root head**. A weld deposit that extends into or includes part or all of the root of the joint. See also *root of joint*.

**root mean square (rms)**. (1) The square root of the arithmetic mean of the squares of the numbers in a given set of numbers. (2) The effective value of an alternating periodic voltage or current. (3) A term describing the surface roughness of a machined surface. See also *surface roughness*.

**root-mean-square end-to-end distance**. A measure of the average size of a coiled polymer molecule, usually determined by light scattering.

**root of joint**. The portion of a weld joint where the members are closest to each other before welding (Fig. 466). In cross section, this may be a point, a line, or an area.

**root of weld**. The points, as shown in cross section, at which the weld bead intersects the base-metal surfaces either nearest to or coincident with the *root of joint* (Fig. 467).

FIG. 466

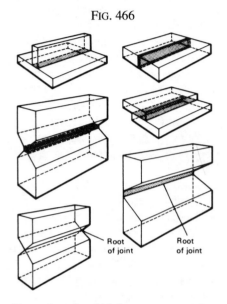

*Examples of weld joint root geometries*

FIG. 467

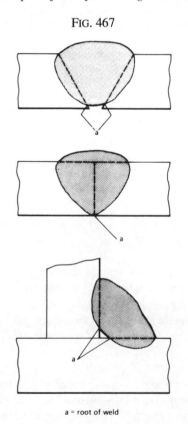

a = root of weld

*Examples of weld root geometries*

FIG. 468

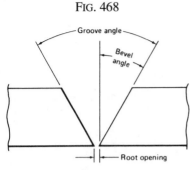

*Root opening in a weld*

**root opening**. In a weldment, the separation between the members at the *root of joint* prior to welding (Fig. 468).

**root pass**. The first bead of a *multiple-pass weld*, laid in the *root of joint*.

**root penetration**. The depth that a weld extends into the *root of joint*, measured on the centerline of the root cross section. See also the figure accompanying the term *joint penetration*.

**root radius**. See preferred term *groove radius*.

**rosebuds**. Concentric rings of distorted coating, giving the effect of an opened rosebud. Noted only on *minimized spangle*.

**rosette**. (1) Rounded configuration of microconstituents in metals arranged in whorls or radiating from a center. (2) Strain gages arranged to indicate at a single position strains in three different directions (Fig. 469). See also *strain gage*.

FIG. 469

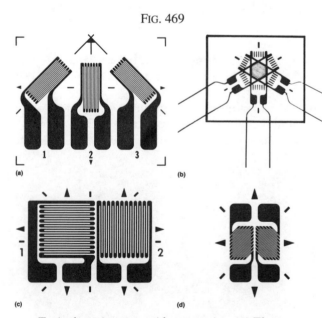

*Typical strain gage grid geometries. (a) Three-element rosette with planar construction. (b) Three-element rosette with stacked construction. (c) Two-element T rosette. (d) Two-element pattern for measuring shear strain*

FIG. 470

*Rosette fracture in a high-strength steel bolt*

**rosette graphite**. Arrangement of graphite flakes in which the flakes extend radially from the center of crystallized areas in gray cast iron. See also the figure accompanying the term *flake graphite*.

**rosette (star) fracture**. A tensile fracture which exhibits a central fibrous zone, an intermediate region of radial shear, and an outer circumferential shear-lip zone (Fig. 470). Rosette fractures are often observed in temper-embrittled steels. See also *shear lip* and *temper embrittlement*.

**rosin**. A resin obtained as a residue in the distillation of crude turpentine from the sap of the pine tree (gum resin) or from an extract of the stumps and other parts of the tree (wood rosin); used in varnishes, lacquers, adhesives, and soldering fluxes. Compare with *resin*.

**rosin flux**. A soldering flux having a rosin base that becomes interactive after being subjected to the soldering temperature. See also *flux*.

**rotary filing and burring**. Machining or smoothing surfaces with contour-fitting rotary tools where only a minimum amount of material is to be removed.

**rotary forging**. A process in which the workpiece is pressed between a flat anvil and a swiveling (rocking) die with a conical working face (Fig. 471); the platens move toward each other during forging. Also called orbital forging. Compare with *radial forging*.

**rotary furnace**. A circular furnace constructed so that the hearth and workpieces rotate around the axis of the furnace during heating (Fig. 472). Also called rotary hearth furnace.

**rotary press**. A machine for forming powder metallurgy parts that is fitted with a rotating table carrying multiple die assemblies in which powder is compacted.

**rotary retort furnace**. A continuous-type furnace in which the work advances by means of an internal spiral, which gives good control of the retention time within the heated chamber (Fig. 473).

**rotary roughening**. A method of surface roughening prior to thermal spraying wherein a revolving roughening tool is

FIG. 471

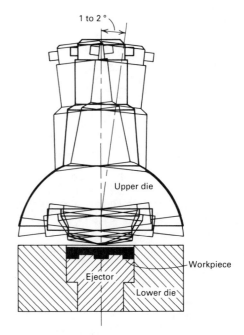

*Schematic of a rotary (rocking-die) forge*

FIG. 472

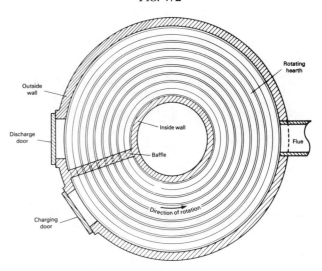

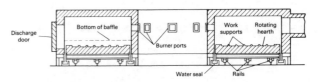

*Schematic arrangement of a relatively small continuous rotary-hearth heating furnace*

pressed against the surface being prepared, while either the work, or the tool, or both, move.

FIG. 473

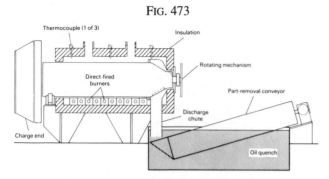

*Rotary retort furnace for continuous carburizing*

**rotary seal**. A mechanical seal that rotates with a shaft and is used with a stationary mating ring.

**rotary shear**. A sheet metal cutting machine with two rotating-disk cutters mounted on parallel shafts driven in unison (Fig. 474).

**rotary swager**. A swaging machine consisting of a power-driven ring that revolves at high speed, causing rollers to engage cam surfaces and force the dies to deliver hammer-like blows on the work at high frequency (Fig. 475). Both straight and tapered sections can be produced.

FIG. 474

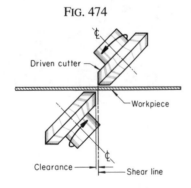

**(a)**

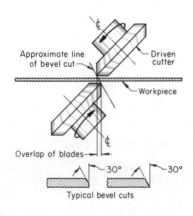

**(b)**

*Two types of rotary shearing. (a) Conventional arrangement of cutters for producing a perpendicular edge. (b) Overlap of cutters for producing a beveled edge*

FIG. 475

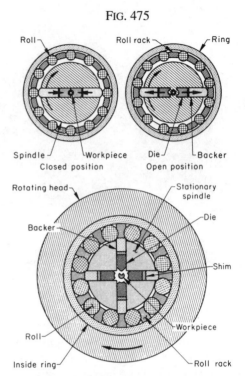

*Principal components and directions of movement in a standard two-die rotary swager (top) and a stationary-spindle rotary swager (bottom)*

**rotary swaging**. A *bulk forming* process for reducing the cross-sectional area or otherwise changing the shape of bars, tubes, or wires by repeated radial blows with one or more pairs of opposed dies.

**rotating electrode powder**. An atomized powder consisting exclusively of solid spherical or near-spherical particles (Fig. 476).

FIG. 476

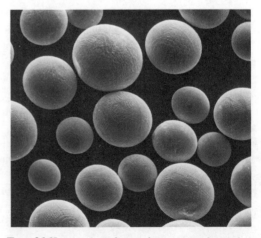

*Type 316L, rotating electrode processed stainless steel powder. Scanning electron micrograph. 190×*

**rotating electrode process.** A method for producing metal powders wherein a consumable metal or alloy electrode is rotated during arc melting at very high speeds, and the molten droplets are propelled by centrifugal force toward the wall of a large collecting chamber while solidifying in flight. If the consumable electrode is melted by an electric arc, the process is called the rotating electrode process, or REP (Fig. 477); if it is melted by a plasma arc, the process is called the plasma rotating electrode process, or PREP (Fig. 478); and if the bar is melted by a laser, the process is called the laser rotating electrode process (LREP). It is also possible to use electron beam melting. The melting technique used depends primarily on the need for cleanness of the powder and on the type of atmosphere required for the alloy.

FIG. 477

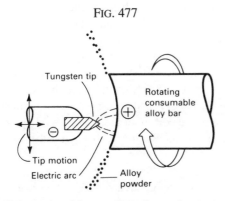

*Schematic of the rotating electrode process*

FIG. 478

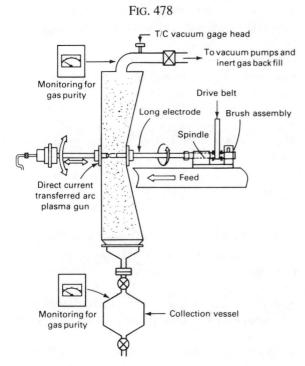

*Schematic of the plasma rotating electrode process*

**rotational casting.** A method used to make hollow articles from thermoplastic materials. The material is charged into a hollow mold capable of being rotated in one or two planes. The hot mold fuses the material into a gel after the rotation has caused it to cover all surfaces. The mold is then chilled, and the product is stripped out.

**rotational molding.** The preferred term for a variation of the *rotational casting* process that uses dry, finely divided (35 mesh, or 500 μ) plastic powders, such as polyethylene, rather than fluid materials. After the powders are heated, they are fused against the mold walls forming a hollow item with uniform wall thickness.

**rouge finish.** A highly reflective finish produced with rouge (finely divided, hydrated iron oxide) or other very fine abrasive, similar in appearance to the bright polish or mirror finish on sterling silver utensils.

**rough blank.** A *blank* for a metal forming or drawing operation, usually of irregular outline, with necessary stock allowance for process metal, which is trimmed after forming or drawing to the desired size.

**rough grinding.** Grinding without regard to finish, usually to be followed by a subsequent operation.

**roughing stand.** The first stand (or several stands) of rolls through which a reheated *billet* passes in front of the finishing stands. See also *rolling mills* and *stand.*

**rough machining.** Machining without regard to finish, usually to be followed by a subsequent operation.

**roughness.** (1) Relatively finely spaced surface irregularities, the heights, widths, and directions of which establish the predominant surface pattern. (2) The microscopic peak-to-valley distances of surface protuberances and depressions (Fig. 479). See also *surface roughness.*

**roughness-width cutoff.** The maximum width of surface irregularities to be included in the measurement of roughness height.

**rough-polishing process.** A polishing process having the primary objective of removing the layer of significant damage produced during earlier machining and abrasion stages of a metallographic preparation sequence. A secondary objective is to produce a finish of such quality that a final polish can be produced easily.

**rough threading.** A method of surface roughening prior to thermal spraying, which consists of cutting threads with the sides and tops of the threads jagged and torn.

**routing.** Cutting out and contouring edges of various shapes in a relatively thin material using a small diameter rotating cutter which is operated at fairly high speeds.

**roving.** A number of yarns, strands, tows, or ends collected into a parallel bundle with little or no twist. Rovings are used to produce fiber-reinforced plastics. See also *woven roving.*

**roving cloth.** A textile fabric, coarse in nature and woven from roving, used in reinforced plastics.

FIG. 479

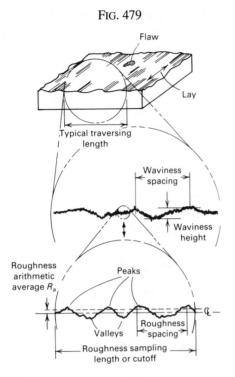

*Schematic of roughness and waviness on a surface with unidirectional lay and one flaw*

**row nucleation**. The mechanism by which stress-induced crystallization is initiated in reinforcing fibers for composites, usually during fiber spinning or hot drawing.

**RRIM**. See *reinforced reaction injection molding*.

**RTM**. See *resin transfer molding*.

**RTV**. See *room-temperature vulcanizing*.

**rubber blanket**. A sheet of rubber or other resilient material used as an auxiliary tool in forming.

**rubber forming**. Forming a sheet metal wherein rubber or another resilient material is used as a functional die part. Processes in which rubber is employed only to contain the hydraulic fluid are not classified as rubber forming.

**rubber-pad forming**. A sheet metal forming operation for shallow parts in which a confined, pliable rubber pad attached to the press slide (ram) is forced by hydraulic pressure to become a mating die for a punch or group of punches placed on the press bed or baseplate. Developed in the aircraft industry for the limited production of a large number of diversified parts, the process is limited to the forming of relatively shallow parts, normally not exceeding 40 mm (1.5 in.) deep. Also known as the Guerin process. Variations of the *Guerin process* include the *Marforming process*, the *fluid-cell process*, and *fluid forming*.

**rubbers**. Cross-linked polymers having glass transition temperatures below room temperature that exhibit highly elastic deformation and have high elongation.

**rubber wheel**. A grinding wheel made with a rubber bond.

**rubbing bearing**. A bearing in which the relatively moving parts slide without deliberate lubrication.

**rub mark**. See *abrasion*.

**rugosities**. See *asperities*.

**run-in** (noun). (1) In *tribology*, an initial transition process occurring in newly established wearing contacts, often accompanied by transients in coefficient of friction, wear rate, or both, that are uncharacteristic of the given tribological system's long-term behavior. (2) In seals, the period of initial operation during which the seal-lip wear rate is greatest and the contact surface is developed.

**run in** (verb). In *tribology*, to apply a specified set of initial operating conditions to a tribological system in order to improve its long-term frictional or wear behavior, or both. The run-in may involve conditions either more severe or less severe than the normal operating conditions of the *tribosystem*, and may also involve the use of special lubricants and/or surface chemical treatments. See also *break in* (verb).

**runner**. (1) A channel through which molten metal flows from one receptacle to another. (2) The portion of the gate assembly of a casting that connects the sprue with the gate(s). (3) Parts of patterns and finished castings corresponding to the portion of the gate assembly described in (2). See also the figure accompanying the term *gating system*.

**runner box**. A distribution box that divides molten metal into several streams before it enters the casting mold cavity.

**runner system**. All the sprues, runners, and gates through which plastic material flows from the nozzle of an injection machine or the pot of a transfer mold to the mold cavity. See also *resin transfer molding* and *transfer molding*.

**running-in**. The process by which machine parts improve in conformity, surface topography, and frictional compatibility during the initial stage of use. Chemical processes, including formation of an oxide skin, and metallurgical processes, such as strain hardening, may contribute.

**runout**. (1) The unintentional escape of molten metal from a mold, crucible, or furnace. (2) An imperfection in a casting caused by the escape of metal from the mold. See also *axial runout* and *radial runout*.

**runout table**. A *roll table* used to receive a rolled or extruded section.

**rupture**. In breaking-strength or creep tests, the point at which a material physically comes apart.

**rupture stress**. The stress at failure. Also known as *breaking stress* or *fracture stress*.

**rust**. A visible corrosion product consisting of hydrated oxides of iron. Applied only to ferrous alloys. See also *white rust*.

**Rutherford backscattering spectroscopy**. A method of determining the concentrations of various elements as a function of depth beneath the surface of a sample, by measuring the energy spectrum of ions which are back-scattered out of a beam directed at the surface.

**Rutherford scattering**. A general term for the classical elastic scattering of energetic ions by the nuclei of a target material.

**rutile**. A particular crystalline form of titanium dioxide ($TiO_2$) used as a pigment and as an additive or component in some ceramic, glass, and glaze manufacture.

***r*-value**. See *plastic strain ratio*.

**RZ powder**. The reduced iron powder made in Germany from the scale of pig iron.

# S

**sacrificial protection**. Reduction of corrosion of a metal in an *electrolyte* by galvanically coupling it to a more anodic metal; a form of *cathodic protection.*

**saddling**. Forming a seamless metal ring by forging a pierced disk over a mandrel (or saddle).

**safety glass**. Glass so constructed, treated, or combined with other materials as to reduce, in comparison with ordinary sheet or plate glass, the likelihood of injury to persons by objects from exterior sources or by these safety glasses when they may be cracked or broken (Fig. 480). Types of safety glass include: (1) laminated safety glass—two or more pieces of glass held together by an intervening layer or layers of plastic materials. It will crack and break under sufficient impact, but the pieces of glass tend to adhere to the plastic and not to fly (Fig. 480b). If a hole is produced, the edges are likely to be less jagged than would be the case with ordinary glass; (2) tempered safety glass—a single piece of specially treated plate, sheet, or float glass. When broken at any point, the entire piece immediately breaks into innumerable small pieces, which may be described as granular, usually with no large jagged edges (Fig. 480c); (3) wire safety glass—a single piece of glass

FIG. 480

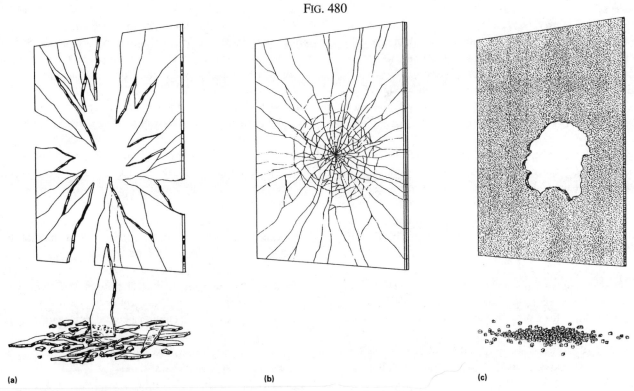

(a)　　　　　　　　　　　　　　　　(b)　　　　　　　　　　　　　　　　(c)

*Break pattern of three states of glass used in commercial and consumer applications. (a) Annealed. (b) Laminated. (c) Tempered*

with a layer of meshed wire completely embedded in the glass, but not necessarily in the center of the sheet. It may or may not be ground and polished on both sides. When the glass is broken, the wire mesh holds the pieces together to a considerable extent.

**sag** (metals). An increase or decrease in the section thickness of a casting caused by insufficient strength of the mold sand of the cope or of the core.

**sag** (plastics). (1) A local extension (often near the die face) of the parison during extrusion by gravitational forces. This causes necking down of the parison. (2) The flow of a molten sheet in a thermoforming operation.

**sagging** (adhesives). Run-off or flow-off of adhesive from an adherend surface due to application of excess or low-viscosity material.

**sagging** (ceramics and glasses). (1) A defect characterized by bending or slumping of a ceramic article fired at excessive temperature. (2) Process of forming glass by reheating until it conforms to the shape of the mold or form on which it rests. (3) A defect characterized by a wavy line or lines appearing on those surfaces of porcelain enamel that have been fired in a vertical position. (4) A defect characterized by irreversible downward bending in a ceramic article insufficiently supported during the firing cycle.

**salt bath heat treatment**. Heat treatment for metals carried out in a bath of molten salt. See also the figures accompanying the terms *immersed-electrode furnaces* and *submerged-electrode furnaces*.

**salt fog test**. An *accelerated corrosion test* in which specimens are exposed to a fine mist of a solution usually containing sodium chloride, but sometimes modified with other chemicals. Also known as salt spray test.

**salting out**. Precipitating a substance in a solution by adding a second substance, usually a salt, without any chemical reaction such as a double decomposition taking place.

**salt spray test**. See *salt fog test*.

**sample**. (1) One or more units of a product (or a relatively small quantity of a bulk material) withdrawn from a *lot* or process stream and then tested or inspected to provide information about the properties, dimensions, or other quality characteristics of the lot or process stream. (2) A portion of a material intended to be representative of the whole.

**sample average**. The sum of all the observed values in a sample divided by the sample size. It is a point estimate of the population mean. Also known as arithmetic mean.

**sample median**. The middle value when all observed values in a sample are arranged in order of magnitude. If an even number of samples are tested, the average of the two middlemost values is used. It is a point estimate of the population median, or 50% point.

**sample percentage**. The percentage of observed values between two stated values of the variable under considera-

FIG. 481

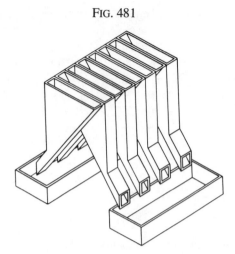

*Sample splitter for metal powders*

tion. It is a point estimate of the percentage of the population between the same two stated values.

**sample splitter**. A device to divide a metal powder pile for sampling (Fig. 481).

**sample standard deviation** (*s*). The square root of the sample variance. It is a point estimate of the population standard deviation, a measure of the "spread" of the frequency distribution of a population. This value of *s* provides a statistic that is used in computing interval estimates and several test statistics. For small sample sizes, *s* underestimates the population standard deviation.

**sample variance** ($s^2$). The sum of the squares of the differences between each observed value and the sample average divided by the sample size minus one. It is a point estimate of the population variance.

**sample thief**. A pointed, hollow, tubular device to withdraw a representative metal powder sample from a shipping drum or other packing unit.

**sampling**. The obtaining of a portion of a material that is adequate for making the required tests or analyses, and that is representative of that portion of the material from which it is taken.

**sand**. A granular material naturally or artificially produced by the disintegration or crushing of rocks or mineral deposits. In casting, the term denotes an aggregate, with an individual particle (grain) size of 0.06 to 2 mm (0.002 to 0.08 in.) in diameter, that is largely free of finer constituents, such as silt and clay, which are often present in natural sand deposits. The most commonly used foundry sand is *silica*; however, *zircon*, *olivine*, aluminum silicates, and other crushed ceramics are used for special applications. See also *grain fineness number*.

**sandblasting**. Abrasive blasting with sand. See also *blasting or blast cleaning* and compare with *shotblasting*.

**sand casting**. Metal castings produced in sand molds.

**sand control**. Testing and regulation of the chemical, physical, and mechanical properties of foundry sand mixtures and their components.

**sand grain distribution**. Variation or uniformity in particle size of a sand aggregate when properly screened by standard screen sizes. See also *grain fineness number*.

**sand hole**. A pit in the surface of a sand casting resulting from a deposit of loose sand on the surface of the mold.

**sand reclamation**. Processing of used foundry sand by thermal, wet, or dry methods so that it can be used in place of new sand without substantially changing the foundry sand practice (Fig. 482).

**sand tempering**. Adding sufficient moisture to molding (foundry) sand to make it workable.

**sandwich construction**. A panel composed of a lightweight core material, such as honeycomb, cellular or foamed plastic, and so forth, to which two relatively thin, dense, high-strength or high-stiffness faces, or skins, are adhered. See also the figure accompanying the term *honeycomb*.

**sandwich testing**. A method of heating a thermoplastic sheet, prior to forming, that consists of heating both sides of the sheet simultaneously.

**sandwich rolling**. Rolling two or more strips of metal in a pack, sometimes to form a roll-welded composite.

**saponification**. The alkaline hydrolysis of fats whereby a soap is formed; more generally, hydrolysis of an ester by an alkali with the formation of an alcohol and a salt of the acid portion.

**saponification number**. (1) A measure of the amount of constituents of petroleum that will easily *saponify* under test conditions, determined by the number of milligrams of sodium hydroxide that is consumed by 1 g of oil under test conditions. Saponification number is a measure of fatty materials compounded in an oil. (2) A number given to quenching oils that reflects the oils amount of compounding with fatty materials, which thereby helps evaluate the condition of these oils in service. See also *neutralization number* and the table accompanying the term *oil quenching*.

**saponify**. (1) To convert into soap. (2) To subject to, or to undergo, saponification.

**satin finish** (metals). A diffusely reflecting surface finish on metals, lustrous but not mirrorlike. One type is a *butler finish*.

**satin finish** (plastics). A type of finish having a satin or velvety appearance, specified for plastics or composites.

**satin weave**. See *four-harness satin*, *eight-harness satin*, and *harness satin*.

**saturated calomel electrode**. A *reference electrode* composed of mercury, mercurous chloride (calomel), and a saturated aqueous chloride solution.

**saturation pressure**. The pressure, for a pure substance at any given temperature, at which vapor and liquid, or vapor and solid, coexist in stable equilibrium.

**Sauter mean diameter**. The diameter of a liquid droplet that has the same ratio of volume-to-surface area as the ratio of total volume-to-total surface area in a distribution of drops, as computed from the equation:

$$SMD = \Sigma_i \, n_i \, d_i^3 / \Sigma_i \, n_i \, d_i^2$$

Fig. 482

(a)                (b)

*Chromite sand before (a) and after (b) sand reclamation. Note the smaller, more rounded grains due to reclamation. Scanning electron micrographs. Both at 50×*

where $i$ is a sampling size interval, $d_i$ is the drop diameter, and $n_i$ is the number of drops in that interval.

**saw burn.** Blackening or carbonization of a cut surface of a pultruded composite material or plastic section. Usually caused by cutting with a dull saw blade, cutting too slowly, or cutting a highly reinforced material with a diamond blade without water.

**saw gumming.** In saw manufacture, grinding away of punch marks or milling marks in the gullets (spaces between the teeth) and, in some cases, simultaneous sharpening of the teeth; in reconditioning of worn saws, restoration of the original gullet size and shape. See also the figure accompanying the term *sawing*.

**sawing.** Using a toothed blade (Fig. 483) or disc to sever parts or cut contours.
1. Circular sawing: Using a circular saw fed into the work by motion of either the workpiece or the blade.
2. Power band sawing: Using a long, multiple-tooth continuous band resulting in a uniform cutting action as the workpiece is fed into the saw.
3. Power hack sawing: Sawing in which a reciprocating saw blade is fed into the workpiece.

FIG. 483

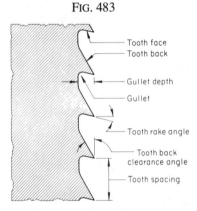

*Standard nomenclature for saw blade teeth*

**Saybolt Universal viscosity.** A commercial measure of viscosity expressed as the time in seconds required for 60 mL of a fluid to flow through the orifice of the standard Saybolt Universal viscometer (calibrated tube) at a given temperature under specified conditions; used for the lighter petroleum products and lubricating oils.

**S-basis.** The S-basis property allowable is the minimum value specified by the appropriate federal, military, Society of Automotive Engineers, American Society for Testing and Materials, or other recognized and approved specifications for the material.

**scab.** A defect on the surface of a casting that appears as a rough, slightly raised surface blemish, crusted over by a thin porous layer of metal, under which is a honeycomb or cavity that usually contains a layer of sand; defect common

to thin-wall portions of the casting or around hot areas of the mold.

**scabbing.** (1) In wear, a loosely used term referring to the formation of bulges in the surface. (2) In fracture mechanics, it is identical with *spalling*.

**scale (glass).** A small particle of foreign material embedded in the surface of molded glass articles.

**scale (metals).** Surface oxidation, consisting of partially adherent layers of corrosion products, left on metals by heating or casting in air or in other oxidizing atmospheres.

**scale (plastics).** A condition in which resin plates or particles are on the surface of a pultrusion. Scales can often be readily removed, sometimes leaving surface voids or depressions.

**scale pit.** (1) A surface depression formed on a forging due to scale remaining in the dies during the forging operation. (2) A pit in the ground in which scale (such as that carried off by cooling water from rolling mills) is allowed to settle out as one step in the treatment of effluent waste water.

**scaling.** (1) Forming a thick layer of oxidation products on metals at high temperature. Scaling should be distinguished from rusting, which involves the formation of hydrated oxides. See also *rust*. (2) Depositing water-insoluble constituents on a metal surface, as in cooling tubes and water boilers.

**scalped extrusion ingot.** A cast, solid, or hollow extrusion *ingot* that has been machined on the outside surface.

**scalping.** Removing surface layers from an *ingot*, *billet*, or slab. See also *die scalping*.

**scanning acoustic microscopy (SAM).** The use of a reflection-type acoustic microscope to generate very high resolution images of surface and near-surface features or defects in a material (Fig. 484). The images are created by mechanically scanning a transducer with an acoustic lens in a raster pattern over the sample. Compared with con-

FIG. 484

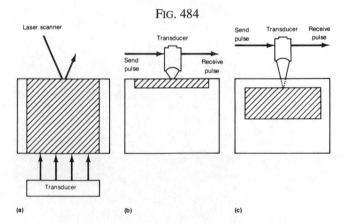

*Simplified comparison of three acoustic microscopy techniques, particularly their zones of application (crosshatched areas) within a sample. (a) SLAM. (b) SAM. (c) C-SAM*

ventional ultrasound imaging techniques, which operate in the 1 to 10 MHz range, SAM is carried out at 100 to 2000 MHz. See also *scanning laser acoustic microscopy (SLAM)*, *ultrasonic C-scan inspection*, and *ultrasonic testing*.

**scanning Auger microscopy (SAM).** An analytical technique that measures the lateral distribution of elements on the surface of a material by recording the intensity of their Auger electrons versus the position of the electron beam.

**scanning electron microscope.** A high-power magnifying and imaging instrument using an accelerated electron beam as an optical device and containing circuitry which causes the beam to traverse or scan an area of sample in the same manner as does an oscilloscope or TV tube (Fig. 485). May utilize reflected (*scanning electron microscopy*) or transmitted (*scanning transmission electron microscopy*) electron optics. The scanning electron microscope provides two outstanding improvements over the *optical microscope*; it extends the resolution limits so that

FIG. 485

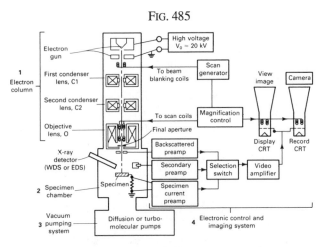

*Basic components of the scanning electron microscope. WDS, wavelength-dispersive spectrometer; EDS, energy-dispersive spectrometer; CRT, cathode-ray tube*

FIG. 486

(a)

(b)

(c)

(d)

*Comparison of light microscope (a and b) and SEM (c and d) fractographs of the test fracture in an alloy X-750 rising-load test specimen. Test was performed in pure water at 95 °C (200 °F). Note the intergranular appearance of the fracture. (a) Bright-field image. (b) Dark-field image. (c) Secondary electron image. (d) Everhart-Thornley backscattered electron image. All 60×*

picture magnifications can be increased from 1000 to 2000× up to 30,000 to 60,000×, and it improves the *depth-of-field* resolution more dramatically, by a factor of approximately 300, thus facilitating its use in fracture studies (Fig. 486).

**scanning electron microscopy (SEM).** An analytical technique in which an image is formed on a cathode-ray tube whose raster is synchronized with the raster of a point beam of electrons scanned over an area of the sample surface. The brightness of the image at any point is proportional to the scattering by or secondary emissions from the point on the sample being struck by the electron beam.

**scanning laser acoustic microscopy (SLAM).** A high-resolution, high-frequency (10 to 500 MHz) ultrasonic inspection technique that produces images of features in a sample throughout its entire thickness. In operation, ultrasound is introduced to the bottom surface of the sample by a piezoelectric transducer, and the transmitted wave is detected on the top side by a rapidly scanning laser beam (Fig. 487). Compare with *scanning acoustic microscopy*.

**scanning transmission electron microscopy (STEM).** An analytical technique in which an image is formed on a cathode-ray tube whose raster is synchronized with the raster of a point beam of electrons scanned over an area of the sample. The brightness of the image at any point is proportional to the number of electrons that are transmitted through the sample at the point where it is struck by the beam.

**scarf.** See preferred term *edge preparation*.

**scarfing.** Cutting surface areas of metal objects, ordinarily by using an oxyfuel gas torch. The operation permits surface

FIG. 487

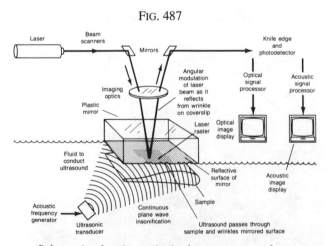

*Schematic showing principal components of a scanning laser acoustic microscope. The unit employs a plane wave piezoelectric transducer to generate the ultrasound and a focused laser beam as a point source detector of the ultrasonic signal. Acoustic images are produced at a rate of 30 images per second.*

imperfections to be cut from ingots, billets, or the edges of plate that are to be beveled for butt welding. See also *chipping*.

**scarf joint** (adhesive bonding). A joint made by cutting away similar angular segments on two adherends and bonding the adherends with the cut areas fitted together. See also *lap joint*.

**scarf joint** (welding). A butt joint in which the plane of the joint is inclined with respect to the main axis of the members (Fig. 488).

FIG. 488

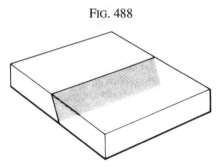

*Example of a scarf butt joint*

**scattering** (of radiant energy). The deviations in the direction of propagation of *radiant energy*.

**scattering of x-rays.** In *radiology*, one of the two ways in which the x-ray beam interacts with matter and the transmitted intensity is diminished in passing through object. The x-rays are scattered in a direction different from the original beam direction. Such scattered x-rays falling on the detector do not add to the radiological image. Because they reduce the contrast in that image, steps are usually taken to eliminate the scattered radiation. Also known as *Compton scattering*.

**schlieren.** Regions of varying refraction in a transparent medium often caused by pressure or temperature differences and detectable especially by photographing the passage of a beam of light.

**Schöniger combustion.** A method of decomposition of organic materials by combusting them in a sealed flask that contains a solution suitable for absorbing the combustion products. The flask is swept with oxygen before ignition.

**scintillator.** A material that produces a rapid flash of visible light when an x-ray photon is absorbed.

**Scleroscope hardness test.** A dynamic indentation hardness test using a calibrated instrument that drops a diamond-tipped hammer from a fixed height onto the surface of the material being tested (Fig. 489). The height of rebound of the hammer is a measure of the hardness of the material.

**Sceleroscope hardness number (HSc or HSd).** A number related to the height of rebound of a diamond-tipped hammer dropped on the material being tested. It is measured on a scale determined by dividing into 100 units the

FIG. 489

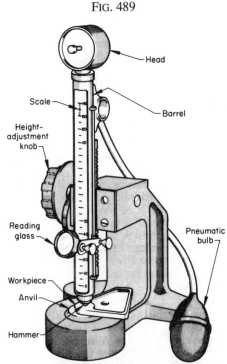

(a) Model C (vertical scale) Scleroscope hardness tester

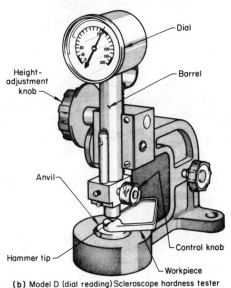

(b) Model D (dial reading) Scleroscope hardness tester

*Principal components of two types of base-mounted Scleroscope hardness testers*

average rebound of the hammer from a quenched (to maximum hardness) and untempered AISI W-5 tool steel test block.

**scorification**. Oxidation, in the presence of fluxes, of molten lead containing precious metals, to partly remove the lead in order to concentrate the precious metals.

**scoring**. (1) The formation of severe scratches in the direction of sliding. Scoring may be due to local solid-phase weld-ing or to abrasion. In the United States, the term *scuffing* is sometimes used as a synonym for scoring. Minor damage should be called *scratching* rather than scoring. (2) In *tribology*, a severe form of wear characterized by the formation of extensive grooves and scratches in the direction of sliding. (3) The act of producing a scratch or narrow groove in a surface by causing a sharp instrument to move along that surface. (4) The marring or scratching of any formed metal part by metal pickup on the punch or die. (5) The reduction in thickness of a material along a line to weaken it intentionally along that line.

**scouring**. (1) A wet or dry cleaning process involving mechanical scrubbing. (2) A wet or dry mechanical finishing operation, using fine abrasive and low pressure, carried out by hand or with a cloth or wire wheel to produce *satin* or *butler*-type finishes.

**scouring abrasion**. Same as *abrasion*.

**scrap**. (1) Products that are discarded because they are defective or otherwise unsuitable for sale. (2) Discarded metallic material, from whatever source, that may be reclaimed through melting and refining.

**scraper**. An exclusion seal that has metallic or other firm lips or scraping elements. It serves to remove foreign material from a reciprocating shaft.

**scratch**. A groove produced in a solid surface by the cutting and/or plowing action of a sharp particle or protuberance moving along that surface.

**scratch hardness test**. A form of hardness test in which a sharp-pointed stylus or corner of a mineral specimen is traversed along a surface so as to determine the resistance of that surface to cutting or abrasion. The *Mohs hardness* test is among the most widely used forms of scratch hardness tests, but is mainly applied to mineralogical specimens or abrasives. The file hardness test was one of the first scratch hardness tests used for evaluating the hardness of metallic materials. The file test, which involves pressing the flat face of a steel file heat treated to approximately 67 to 70 HRC against and slowly across the surface to be tested, is useful in estimating the hardness of steels in the high hardness ranges. It provides information on soft spots and decarburization quickly and easily, and is readily adaptable to odd shapes and sizes that are difficult to test by other methods. Other scratch hardness tests involve using diamond cones, pyramids, and spherical tips, but such scratch hardness tests have not been established and standardized to the extent that macro- and microindentation hardness tests have been.

**scratching**. (1) In *tribology*, the mechanical removal and/or displacement of material from a surface by the action of abrasive particles or protuberances sliding across the surfaces. (2) The formation of fine scratches in the direction of sliding. Scratching may be due to asperities on the harder slider or to hard particles between the surface or embedded in one of them. Scratching is to be considered less damaging than *scoring*. See also *abrasion*, *plowing*, and *ridging wear*.

**scratching abrasion**. See *low-stress abrasion*.

**scratch-resistant coatings**. Coating applied to glass surfaces to reduce the effects of frictive damage. Examples are $SnO_2$ or $TiO_2$ coatings applied to glass containers.

**scratch trace**. In metallographic specimens, a line of etch markings produced on a surface at the site of a preexisting scratch, the physical groove of the scratch having been removed. The scratch trace develops when the deformed material extending beneath the scratch has not been removed with the scratch groove and when the residual deformed material is attacked preferentially during etching.

**screen**. (1) The woven wire or fabric cloth, having square openings, used in a sieve for retaining particles greater than the particular mesh size. U.S. standard, ISO, or Tyler screen sizes are commonly used. (2) One of a set of sieves, designated by the size of the openings, used to classify granular aggregates such as sand, ore, or coke by particle size. (3) A perforated sheet placed in the gating system of a mold to separate impurities from the molten metal.

**screen analysis**. See *sieve analysis*.

**screen classification**. See *sieve classification*.

**screening**. Separation of a powder according to particle size by passing it through a screen having the desired mesh size.

**screw dislocation**. See *dislocation*.

**screw plasticating injection molding**. A technique in which the plastic is heated and converted from pellets to a viscous melt, and then forced into a mold by means of an extruder screw, which is an integral part of the molding machine. Machines are either single stage, in which plastication and injection are done in the same cylinder, or double stage, in which the material is plasticated in one cylinder and then fed to a second for injection into a mold. See also *injection molding* (plastics).

**screw press**. A high-speed press in which the ram is activated by a large screw assembly powered by a drive mechanism (Fig. 490).

**screws**. See *extruder* (plastics).

**screw stock**. Free-machining bar, rod, or wire.

**scrim**. A low-cost reinforcing fabric for composites made from continuous filament yarn in an open-mesh construction. Used in the processing of tape or other *B-stage* material to facilitate handling.

**scruff**. A mixture of tin oxide and iron-tin alloy formed as dross on a tin-coating bath.

**scuffing**. (1) Localized damage caused by the occurrence of solid-phase welding between sliding surfaces, without local surface melting. In the United Kingdom, scuffing implies local solid-phase welding only. In the United

FIG. 490

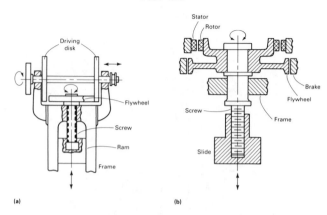

*Two common types of screw press drives. (a) Friction drive. (b) Direct electric drive*

States, scuffing may include abrasive effects and the term *scoring* is sometimes used as a synonym. (2) A mild degree of *galling* that results from the welding of asperities due to frictional heat. The welded asperities break, causing surface degradation. In general, the term scuffing has been used in so many different ways that its use should be avoided whenever possible, and instead replaced with a more precise description of the specific type of *surface damage* being considered.

**sea coal**. Finely ground coal, used as an ingredient in molding sands.

**seal**. (1) In tribology, a device designed to prevent leakage between relatively moving parts. (2) A device designed to prevent the movement of fluid from one chamber to another, or to exclude contaminants.

**sealant**. A material applied to a joint in paste or liquid form that hardens or cures in place, forming a seal against a gas or liquid entry (Fig. 491). The primary application area for sealants is the building construction industry. Sealants for the commercial construction market are designed to act as a barrier to water, air, atmospheric pollution, vibration, insects, dirt, and noise, while compensating for movement that occurs in the structure being sealed. This movement can be caused by many factors, including changes in temperature and moisture content, permanent loading or load transfer, and chemical changes such as sulfate attack and wind loads. Other key sealant markets are the automotive industry (Fig. 492), the aerospace market, and the electrical and electronics market. Major sealant types include oil-base caulks (25 to 30% linseed or soy oil, 6 to 12% fibrous filler, 40 to 60% calcium carbonate filler, 0.05 to 3% pigment, 0 to 2% gelling agents, and 0 to 1% catalyst), latex acrylic polymers, polyvinyl acetate caulks, solvent acrylics, butyl sealants, polysulfides, urethanes, and silicones. Table 56 summarizes properties of such sealant materials.

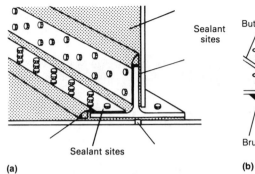

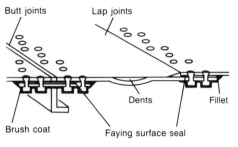

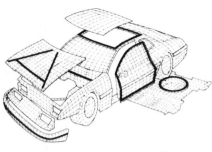

FIG. 491      FIG. 492

Sealant sites. (a) Of integral fuel tank. (b) In structural applications

Automotive body assembly applications (dark areas) for sealants

## Table 56 Summary of sealant properties

| Characteristic | Low performance Oil base, one part | Latex (acrylic), one part | Medium performance Butyl skinning, one part | Acrylic, solvent-release |
|---|---|---|---|---|
| Maximum recommended joint movement, percent of joint width | ±3 | ±5, ±12.5 | ±7.5 | ±10 to 12.5 |
| Life expectancy, years(a) | 2–10 | 2–10 | 5–15 | 5–20 |
| Service temperature range, °C (°F) | −29 to 66 (−20 to 150) | −20 to 82 (−20 to 180) | −40 to 82 (−40 to 180) | −29 to 82 (−20 to 180) |
| Recommended application temperature range, °C (°F)(b) | 4–50 (40–120) | 4–50 (40–120) | 4–50 (40–120) | 4–80 (40–180) |
| Cure time to a tack-free condition, h(c) | 6 | 0.5–1 | 24 | 36 |
| Cure time to specified performance, days(c) | Continues | 5 | Continues | 14 |
| Shrinkage, % | 5 | 20 | 20 | 10–15 |
| Hardness, new (1–6 mo), at 25 °C (75 °F), Shore A scale | ··· | 15–40 | 10–30 | 10–25 |
| Hardness, old (5 years) at 25 °C (75 °F), Shore A scale | ··· | 30–45 | 30–50 | 30–55 |
| Resistance to extension at low temperature | Low to moderate | Moderate to high | Moderate to high | High |
| Primer required for sealant bond to | | | | |
| Masonry | No | Yes | No | No |
| Metal | No | Sometimes | No | No |
| Glass | No | No | No | No |
| Applicable specifications | | | | |
| United States | TT-C-00593b | ASTM C 834, TT-S-00230 | TT-S-001657 | TT-S-00230 |
| Canada | CAN/CGSB 19.6-N87 | | CGSB 19-GP-14M | CGSB 19-GP-5M |

| Characteristic | Polysulfide One part | Two part | High performance Urethane One part | Two part | Silicone One part | Two part |
|---|---|---|---|---|---|---|
| Maximum recommended joint movement, percent of joint width | ±25 | ±25 | ±25 | ±25 | ±25 to +100/−50 | ±12.5 to ±50 |
| Life expectancy, years(a) | 10–20 | 10–20 | 10–20 | 10–20 | 10–50 | 10–50 |
| Service temperature range, °C (°F) | −40 to 82 (−40 to 180) | −51 to 82 (−60 to 180) | −40 to 82 (−40 to 180) | −32 to 82 (−25 to 180) | −54 to 200 (−65 to 400) | −54 to 200 (−65 to 400) |
| Recommended application temperature range, °C (°F)(b) | 4–50 (40–120) | 4–50 (40–120) | 4–50 (40–120) | 4–80 (40–180) | −30 to 70 (−20 to 160) | −30 to 70 (−20 to 160) |
| Cure time to a tack-free condition, h(c) | 24(c) | 36–48(c) | 12–36 | 24 | 1–3 | 0.5–2 |
| Cure time to specified performance, days (c) | 30–45 | 7 | 8–21 | 3–5 | 5–14 | 0.25–3 |
| Shrinkage, % | 8–12 | 0–10 | 0–5 | 0–5 | 0–5 | 0–5 |
| Hardness, new (1–6 mo), at 25 °C (75 °F), Shore A scale | 20–40 | 20–45 | 20–45 | 10–45 | 15–40 | 15–40 |
| Hardness, old (5 years) at 25 °C (75 °F), Shore A scale | 30–55 | 20–55 | 30–55 | 20–60 | 15–40 | 15–50 |
| Resistance to extension at low temperature | Low to high | Low to moderate | Low to high | Low to high | Low | Low |
| Primer required for sealant bond to | | | | | | |
| Masonry | Yes | Yes | Yes | Yes | No | Yes |
| Metal | Yes | Yes | No | No | ... | No |
| Glass | No | No | No | No | No | No |
| Applicable specifications | | | | | | |
| United States | ASTM C-920, TT-C-00230C | ASTM C 920, TT-S-00227E | ASTM C 920, TT-S-00230C | ASTM C 920, TT-S-00227C | ASTM C 920, TT-S-00230C TT-S-001543A | ASTM C 920, TT-S-001543A |
| Canada | CAN 2-19.13M | | CAN 2-19.13M | | CAN/CGSB 19.18 | TT-S-00227E |

Note: Data from manufacturer data sheets; U.S. made sealants are generally considered. (a) Affected by conditions of exposure. (b) Some sealants may require heating in low-temperature environments. (c) Affected by temperature and humidity

**seal coat** (thermal spraying). Material applied to infiltrate the pores of a thermal spray deposit.

**sealing**. (1) Closing pores in anodic coatings to render them less absorbent. (2) Plugging leaks in a casting by introducing thermosetting plastics into porous areas and subsequently setting the plastic with heat.

**sealing face**. The lapped surface of a seal that comes in close proximity to the face of the mating ring of a face seal, thus forming the primary seal. With reference to lip seals, the preferred term is seal contact surface.

**seal nose**. The part of the primary seal ring of a face seal that comes in closest proximity to the mating surface and that, together with the mating surface, forms the primary seal.

**seal weld**. Any weld used primarily to obtain lightness and prevent leakage.

**seam**. (1) On a metal surface, an unwelded fold or lap that appears as a crack, usually resulting from a discontinuity. (2) A surface defect on a casting related to but of lesser degree than a *cold shut*. (3) A ridge on the surface of a casting caused by a crack in the mold face.

**seam weld**. A continuous weld made between or upon overlapping members, in which coalescence may start and occur on the faying surfaces, or may have proceeded from the surface of one member. The continuous weld may consist of a single weld bead or a series of overlapping spot welds. Common seam weld types include (1) lap seam welds joining flat sheets, (2) flange-joint lap seam welds with at least one flange overlapping the mating piece (Fig. 493), and (3) mash seam welds with work metal compressed at the joint to reduce joint thickness.

**seam welding**. (1) Arc or resistance welding in which a series of overlapping spot welds is produced with rotating electrodes or rotating work, or both. (2) Making a longitudinal weld in sheet metal or tubing. See also *resistance seam welding*.

**season cracking**. An obsolete historical term usually applied to *stress-corrosion cracking* of brass.

**secant modulus**. The slope of the secant drawn from the origin to any specified point on the *stress-strain curve*. See also *modulus of elasticity*.

**secondary alloy**. Any alloy whose major constituent is obtained from recycled scrap metal. Compare with *primary alloy*.

**secondary bonding**. The joining together, by the process of adhesive bonding, of two or more already cured composite parts, during which the only chemical or thermal reaction occurring is the curing of the adhesive itself.

**secondary circuit**. That portion of a welding machine which conducts the secondary current between the secondary terminals of the welding transformer and the electrodes, or electrode and work.

**secondary creep**. See *creep*.

**secondary crystallization**. In processing of plastics, the slow crystallization process that occurs after the main solidification process is complete. Often associated with impure molecules.

**secondary electron**. A low-energy electron (0 to 50 eV) emitted from a surface that is struck by particles with higher energies.

**secondary etching**. Development of microstructures deviating from the primary structure of a metal or alloy through transformation and heat treatment in the solid state.

**secondary extinction**. In *x-ray diffraction*, a decrease in the intensity of a diffracted x-ray beam caused by parallelism or near-parallelism of mosaic blocks in a mosaic crystal; the lower blocks are partially screened from the incident radiation by the upper blocks, which have reflected some of it. See also *primary extinction*.

**secondary gas**. In thermal spraying, the gas constituting the minor constituent of the arc gas fed to the gun to produce the plasma. The primary arc gas, usually argon or nitrogen, is supplemented with secondary gases such as nitrogen, helium, and/or hydrogen, in order to increase the temperature of the plasma.

FIG. 493

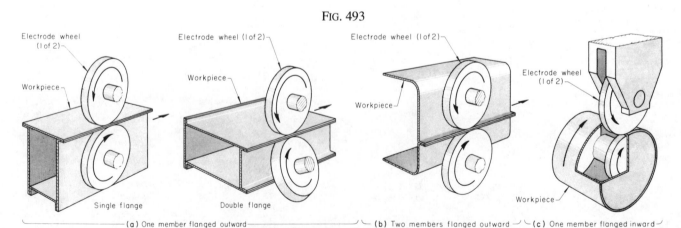

*Typical arrangement of electrode wheels for various flange-joint seam welds*

FIG. 494

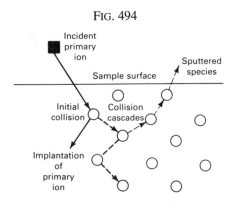

Incident
primary
ion

Sputtered
species

Sample surface

Initial
collision

Collision
cascades

Implantation
of
primary
ion

*The physical effects of primary ion bombard-*
*ment: implantation and sputtering*

**secondary ion.** An ion other than the probe ion that originates from and leaves the specimen surface as a result of bombardment with a beam of primary or probe ions; a sputtered ion (Fig. 494). See also *sputtering*.

**secondary ion mass spectroscopy (SIMS).** An analytical technique that measures the masses of ions emitted from the surface of a material when exposed to a beam of incident ions. The incident ions are usually monoenergetic and are all of the same species, for example, 5-keV $Ne^+$ ions. See also *secondary ion*.

**secondary metal.** Metal recovered from scrap by remelting and refining.

**secondary nucleation.** In processing of plastics, the mechanism by which crystals grow.

**secondary operation.** Any operation performed on a sintered powder metallurgy compact, such as sizing, coining, repressing, impregnation, infiltration, heat or steam treatment, machining, joining, plating, or other surface treatment.

**secondary recrystallization.** The process by which a few large grains are nucleated and grow at the expense of a fine-grained, but essentially strain-free matrix. Also known as abnormal or discontinuous grain growth. See also *primary recrystallization*.

**secondary seal.** A device, such as bellows, piston ring, or O-ring, that allows axial movement of the primary seal of a mechanical face seal without undue leakage.

**secondary-standard dosimetry system.** A system that measures energy deposition indirectly. It requires conversion factors to account for such considerations as geometry, dose rate, relative stopping power, incident energy spectrum, or other effects, in order to interpret the response of the system. Thus, it requires calibration against a primary dosimetry system or by means of a standard radiation source. See also *dosimeter*.

**secondary structure.** In aircraft and aerospace applications, a structure that is not critical to flight safety.

**secondary x-rays.** The x-rays emitted by a specimen following excitation by a primary x-ray beam or an electron beam.

**second-degree blocking.** In adhesive bonding, an adherence of such degree that when surfaces under tests are parted, one surface or the other will be found to be damaged.

**second-phase inhomogeneity.** In ceramics and glasses, a microstructural irregularity related to the nonuniform distribution of a second phase, for example, an atypically large pocket of a second phase or a second-phase zone of composition or crystalline phase structure different than the matrix material.

**sectioning.** The removal of a conveniently sized, representative specimen from a larger sample for metallographic inspection. Sectioning methods include shearing, *sawing* (using hacksaws, band saws, and diamond wire saws, Fig. 495), abrasive cutting, and *electrical discharge machining*.

FIG. 495

0.13 mm
(0.005 in.)

0.25 mm
(0.010 in.)

0.3 mm
(0.012 in.)

0.38 mm
(0.015 in.)

0.08 mm
(0.003 in.)

0.2 mm
(0.008 in.)

| Wire size | | Diamond | Kerf size | |
|---|---|---|---|---|
| mm | in. | size, μm | mm | in. |
| 0.08 | 0.003 | 8 | 0.08 | 0.00325 |
| 0.13 | 0.005 | 20 | 0.14 | 0.0055 |
| 0.2 | 0.008 | 45 | 0.23 | 0.009 |
| 0.25 | 0.010 | 60 | 0.29 | 0.0115 |
| 0.3 | 0.012 | 60 | 0.34 | 0.0135 |
| 0.38 | 0.015 | 60 | 0.42 | 0.0165 |

*Diamond-impregnated wires used for sectioning*

**sedimentation.** The settling of particles suspended and dispersed in a liquid through the influence of an external force, such as gravity or centrifugal force.

**seed.** (1) A small, single crystal of a desired substance added to a solution to induce crystallization. (2) Small particles or agglomerates, crystals, or crystallites introduced in large numbers into a vessel to serve as nuclei or centers for further growth of material on their surfaces. (3) An extremely small gaseous inclusion in glass.

**segment.** In *sampling*, a specifically demarked portion of a *lot*, either actual or hypothetical.

**segment die.** A die made of parts that can be separated for ready removal of the workpiece. Synonymous with *split die*.

**segregation** (metals). (1) Nonuniform distribution of alloying elements, impurities, or microphases in metals and alloys. (2) A casting defect involving a concentration of alloying elements at specific regions, usually as a result of the primary crystallization of one phase with the subsequent concentration of other elements in the remaining liquid. Microsegregation refers to normal segregation on a microscopic scale in which material richer in an alloying element freezes in successive layers on the dendrites (*coring*) and in constituent network. Macrosegregation refers to gross differences in concentration (for example, from one area of a casting to another). See also *inverse segregation* and *normal segregation*.

**segregation** (plastics). A separation of components in a molded article usually denoted by wavy lines and color striations in thermoplastics. In thermoset plastics, usually a separation of resin and filler on the surface.

**segregation banding.** Inhomogeneous distribution of alloying elements aligned in filaments or plates parallel to the direction of working (Fig. 496). See also the figure accompanying the term *banding*.

**segregation (coring) etching.** Development of segregation (*coring*) mainly in macrostructures and microstructures of castings.

**seize.** To prevent a metal part from being ejected from a die as a result of *galling*.

**seizing.** The stopping of a moving part by a mating surface as a result of excessive friction.

**seizure.** The stopping of relative motion as the result of interfacial friction. Seizure may be accompanied by gross surface welding. The term is sometimes used to denote *scuffing*.

**Sejournet process.** See *Ugine-Sejournet process*.

**selected-area diffraction (SAD).** Electron diffraction from a portion of a sample selected by inserting an aperture into the magnification portion of the lens system of a transmission electron microscope. Areas as small as 0.5 μm in diameter can be examined in this way.

**selected area diffraction pattern (SADP).** An electron diffraction pattern obtained from a restricted area of a sample. The sharp spots in the pattern correspond closely to points in the reciprocal lattice of the material being studied. Usually such patterns are taken from a single crystal or a small number of crystals. Any of three types of electron-diffraction patterns can be generated in an SADP: (1) ring patterns from fine-grained polycrystalline materials in which diffraction occurs simultaneously from many grains with different orientations relative to the incident beam (Fig. 497); (2) spot patterns in which diffraction occurs from a single-crystal region of the specimen (Fig. 498); and (3) Kikuchi line patterns in which diffraction occurs from a single-crystal region of the specimen, which is sufficiently thick that the diffracting electrons have undergone simultaneous elastic and inelastic scattering. See also the figures accompanying the terms *diffraction pattern* and *Kikuchi lines*.

**selective block sequence.** A block welding sequence in which successive blocks are completed in a certain order

FIG. 496

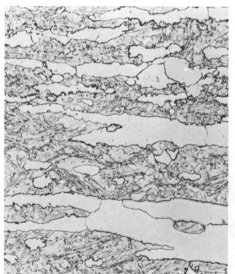 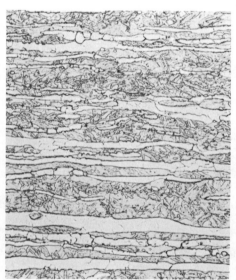 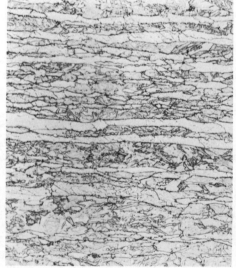

81% reduction                94% reduction                97% reduction

*Type 430 stainless steel hot rolled to various percentages of reduction showing development of a segregated banded structure consisting of alternate layers of ferrite (light) and martensite (dark) as the amount of hot work is increased. 500×*

FIG. 497

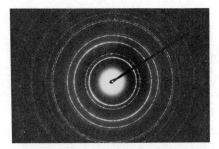

*Ring pattern from polycrystalline aluminum*

FIG. 498

*Spot diffraction pattern from a single crystal of aluminum*

selected to create a predetermined stress pattern. See also the figure accompanying the term *block sequence*.

**selective heating.** Intentionally heating only certain portions of a workpiece.

**selective leaching.** Corrosion in which one element is preferentially removed from an alloy, leaving a residue (often porous) of the elements that are more resistant to the particular environment. Also called *dealloying* or parting. Table 57 lists alloys that are subject to selective leaching. See also *decarburization, decobaltification, denickelification, dezincification,* and *graphitic corrosion*.

**selective quenching.** Quenching only certain portions of an object.

**selective transfer.** A process involving the transfer and attachment of a specific species from one surface to the mating surface during sliding. This term is to be distinguished from the general tribological term *transfer*, which involves the same general process, but which, when used by itself, does not discriminate as to which of the species present in a multiconstituent surface is transferred during sliding.

**selectivity.** In materials characterization, the ability of a method or instrument to respond to a desired substance or constituent and not to others.

**self-absorption.** In *optical emission spectroscopy*, reabsorption of a photon by the same species that emitted it. For example, light emitted by sodium atoms in the center of a flame may be reabsorbed by different sodium atoms near the outer portions of the flame.

**self-acting bearing.** See *gas bearing* and *self-lubricating bearing*.

**self-aligning bearing.** A *rolling-element bearing* with one spherical raceway that automatically provides compensation for shaft or housing deflection or misalignment. Compare with *aligning bearing*.

**self-curing.** See *self-vulcanizing*.

**self-diffusion.** Thermally activated movement of an atom to a new site in a crystal of its own species, as, for example, a copper atom within a crystal of copper.

**self-extinguishing resin.** A resin formulation that will burn

**Table 57  Combinations of alloys and environments subject to selective leaching (dealloying) and elements preferentially removed**

| Alloy | Environment | Element removed |
|---|---|---|
| Brasses | Many waters, especially under stagnant conditions | Zinc (dezincification) |
| Gray iron | Soils, many waters | Iron (graphitic corrosion) |
| Aluminum bronzes | Hydrofluoric acid, acids containing chloride ions | Aluminum (dealuminification) |
| Silicon bronzes | High-temperature steam and acidic species | Silicon (desiliconification) |
| Tin bronzes | Hot brine or steam | Tin (destannification) |
| Copper nickels | High heat flux and low water velocity (in refinery condenser tubes) | Nickel (denickelification) |
| Copper-gold single crystals | Ferric chloride | Copper |
| Monels | Hydrofluoric and other acids | Copper in some acids, and nickel in others |
| Gold alloys with copper or silver | Sulfide solutions, human saliva | Copper, silver |
| High-nickel alloys | Molten salts | Chromium, iron, molybdenum, and tungsten |
| Medium- and high-carbon steels | Oxidizing atmospheres, hydrogen at high temperatures | Carbon (decarburization) |
| Iron-chromium alloys | High-temperature oxidizing atmospheres | Chromium, which forms a protective film |
| Nickel-molybdenum alloys | Oxygen at high temperature | Molybdenum |

in the presence of a flame but will extinguish itself within a specified time after the flame is removed. The term is not universally accepted.

**self-fluxing alloys** (thermal spraying). Certain materials that "wet" the substrate and coalesce when heated to their melting point, without the addition of a fluxing agent.

**self-hardening steel.** See preferred term *air-hardening steel*.

**self-lubricating bearing.** (1) A bearing independent of external lubrication. These bearings may be sealed for life after packing with grease or may contain *self-lubricating material*. (2) A sintered product whose accessible pore volume is filled with a liquid lubricant which automatically produces a lubricating film on the bearing surface during running of the shaft. This is due to a pumping action of the shaft and frictional heat that lowers the viscosity of the oil. After completion of the running cycle, the oil is reabsorbed into the pore system of the bearing by capillary attraction. Porous self-lubricating bearings are divided into three groups: sintered bronze bearings, iron-based sintered bearings, and iron-bronze sintered bearings. The original and most widely used P/M bearing material is 90%Cu-10%Sn bronze, with or without the addition of graphite (1% fine natural graphite is often added to enhance fabrication, as well as to improve bearing properties). The 90%Cu-10%Sn bronze material is superior in bearing performance to the iron-based and iron-bronze compositions, which are lower in cost and used in less severe applications. Table 58 lists compositions of P/M bronze bearings.

**self-lubricating material.** Any solid material that shows low friction without application of a lubricant. Examples are graphite, molybdenum disulfide, and polytetrafluoroethylene. Taken in a broader context, the term can also refer to a composite (powder metallurgy) material into which a lubricous species has been incorporated. See also *self-lubricating bearing* and *solid lubricant*.

**self-reversal.** In optical emission spectroscopy, the extreme case of self-absorption. See also *self-absorption*.

**self-skinning foam.** A urethane foam that produces a tough outer surface over a foam core upon curing.

**self-vulcanizing.** Pertaining to an adhesive that undergoes vulcanization without the application of heat. See also *vulcanization*.

FIG. 499

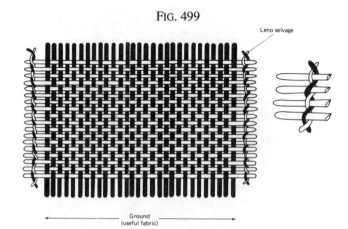

*Full-width plain weave with leno selvage*

**selvage.** The woven-edge portion of a fabric used in reinforced plastics that is parallel to the warp (Fig. 499).

**semiautomatic arc welding.** Arc welding with equipment which controls only the filler metal feed. The advance of the welding is manually controlled.

**semiautomatic brazing.** Brazing with equipment which controls only the brazing filler metal feed. The advance of the brazing is manually controlled.

**semiautomatic plating.** *Plating* in which prepared cathodes are mechanically conveyed through the plating baths, with intervening manual transfers.

**semiblind joint.** A weld joint in which one extremity of the joint is not visible.

**semiconductor.** A solid crystalline material whose electrical conductivity is intermediate between that of a metal and an insulator, ranging from about $10^5$ siemens to $10^{-7}$ siemens per meter, and is usually strongly temperature-dependent.

**semiconductor materials.** See *Technical Brief 41*.

**semicrystalline.** In plastics, materials that exhibit localized crystallinity. See also *crystalline plastic*.

**semifinisher.** An impression in a series of forging dies that only approximates the finish dimensions of the forging. Semifinishers are often used to extend die life or the finishing impression, to ensure proper control of grain flow during forging, and to assist in obtaining desired tolerances.

**Table 58  Chemical composition of self-lubricating sintered bronze bearings**

| | Composition, % | | | |
|---|---|---|---|---|
| | Grade 1 | | Grade 2 | |
| Element | Class A | Class B | Class A | Class B |
|---|---|---|---|---|
| Copper | 87.5-90.5 | 87.5-90.5 | 82.6-88.5 | 82.6-88.5 |
| Tin | 9.5-10.5 | 9.5-10.5 | 9.5-10.5 | 9.5-10.5 |
| Graphite | 0.1 max | 1.75 max | 0.1 max | 1.75 max |
| Lead | (a) | (a) | 2.0-4.0 | 2.0-4.0 |
| Iron | 1.0 max | 1.0 max | 1.0 max | 1.0 max |
| Total other elements by difference | 0.5 max | 0.5 max | 1.0 max | 1.0 max |

(a) Included in other elements
Source: ASTM B 438

FIG. 500

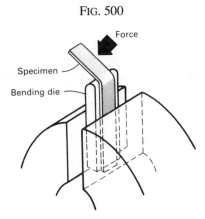

*Semiguided bend test setup*

**semifinishing**. Preliminary operations performed prior to finishing.

**semiguided bend**. The bend obtained by applying a force directly to the specimen in the portion that is to be bent. The specimen is either held at one end and forced around a pin or rounded edge (Fig. 500), or is supported near the ends and bent by a force applied on the side of the specimen opposite the supports and midway between them. In some instances, the bend is started in this manner and finished in the manner of a *free bend*.

**semikilled steel**. Steel that is incompletely deoxidized and contains sufficient dissolved oxygen to react with the carbon to form carbon monoxide and thus offset solidifica-

tion shrinkage. See also the figure accompanying the term *capped steel*.

**semipermanent mold**. A *permanent mold* in which sand cores or plaster are used.

**semipositive mold**. A mold for forming plastics that combines the capabilities of a *flash mold* and *positive mold*. As the two halves of a semipositive mold begin to close, the mold acts much like a flash mold, because the excess material is allowed to escape around the loose-fitting plunger and cavity. As the plunger telescopes further into the cavity, the mold becomes a positive mold with very little clearance, and full pressure is exerted on the material, producing a part of maximum density. This type of mold uses to advantage the free flow of material in a flash mold and the capability of producing dense parts in the positive mold.

**semirigid plastic**. For purposes of general classification, a plastic that has a modulus of elasticity in flexure or in tension between 70 and 690 MPa (10 and 100 ksi) at 23 °C (70 °F) and 50% relative humidity.

**semisolid metal forming**. A two-step casting/forging process in which a billet is cast in a mold equipped with a mixer that continuously stirs the thixotropic melt, thereby breaking up the dendritic structure of the casting into a fine-grained spherical structure (Fig. 501). After cooling, the billet is stored for subsequent use. Later, a slug from the billet is cut, heated to the semisolid state, and forged in a die. Normally the cast billet is forged when 30 to 40% is in the liquid state. See also *rheocasting*.

FIG. 501

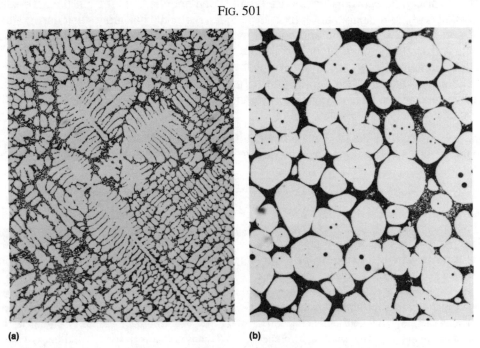

(a)                    (b)

*Comparison of dendritic conventionally cast (a) and nondendritic semisolid formed (b) microstructures of aluminum alloy 357 (Al-7Si-0.5Mg). Both 200×*

## Technical Brief 41: Semiconductor Materials

SEMICONDUCTOR MATERIALS encompass a wide range of metals, metalloids, ceramics, and complex compounds whose electrical resistivity is intermediate between that of a conductor (for example, copper) and an insulator (for example, glass). Useful semiconductors have resistivities ranging from $10^{-3}$ $\Omega \cdot cm$ to $10^{8}$ $\Omega \cdot cm$. Semiconductor materials involve combinations of nearly all the elements in groups IIB, IIIA, IVA, VA, and VIA of the periodic

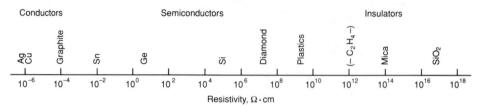

*Resistivity spectrum. Semiconductors have intermediate resistivity that may be altered appreciably by minor structural changes or small additions of foreign materials.*

table. The wide variety of elemental combinations result in a similar wide range in properties (electrical, optical, thermal, and mechanical). Semiconductors are used in rectifiers for changing alternating current to pulsating direct current, and in transistors for amplifying currents. They can also be used for the conversion of heat energy to electric energy, as in solar cells.

The development of semiconducting materials evolved from early experiments on silicon and germanium. Interest in these materials accelerated during World War II with the need to find a material/component that would eliminate the power waste involved in heating the filament of a vacuum tube, and led to the first major breakthrough in semiconductor technology—the development of the transistor made possible by methods to grow high-purity single crystals of silicon and germanium (group IV semiconductors). To date, silicon is still the most important semiconductor material, as it is used in a multitude of applications (see the accompanying table). Polycrystalline semiconductor-grade silicon is produced from pure silane or chlorosilanes by chemical vapor deposition. Total impurity content of the semiconductor-grade silicon is generally less than 0.1 ppm. Single-crystal semiconductor-grade silicon ingots are grown by pulling them from the melt of polycrystalline semiconductor-grade silicon by the floating-zone or Czochralski technique.

Second in value to silicon as semiconductors are the III-V compounds, such as gallium arsenide (GaAs) and gallium phosphide (GaP). The higher carrier mobility of these materials has led to their use in a broad range of optoelectronic and microwave devices, and current research is extending applications to the field of optical communications and ultrahigh-speed integrated circuits. The applications of III-V compounds have led in several cases to the development of new industries. For example, the light-emitting diode revolutionized the field of small alphanumeric displays and played a key role in the evolution of the electronic calculator and digital watch industries. Device applications for GaAs include the heterostructure laser, components for optical-fiber systems, and high-speed digital integrated circuits.

Next in importance are the II-VI and IV-VI semiconductors. The group IIB elements zinc, cadmium, and mercury combine with the group VIA chalcogens oxygen, sulfur, selenium, and tellurium to form 12 compounds. These

**sensitive tint plate.** A gypsum plate used in conjunction with polarizing filters in an *optical microscope* to provide very sensitive detection of birefringence and double refraction.

**sensitivity.** (1) The capability of a method or instrument to discriminate between samples having different concentrations or amounts of the analyte. (2) The smallest difference in values that can be detected reliably with a given measuring instrument.

**sensitization.** In austenitic stainless steels, the precipitation of chromium carbides, usually at grain boundaries, on exposure to temperatures of about 540 to 845 °C (about 1000 to 1550 °F), leaving the grain boundaries depleted of chromium and therefore susceptible to preferential attack by a corroding medium. Welding is the most common cause of sensitization. Weld decay (sensitization) caused by carbide precipitation in the weld heat-affected zone leads to *intergranular corrosion* (Fig. 502).

## Commercial applications of semiconductor materials

| Material | Applications |
|---|---|
| Germanium (Ge) | Tunnel diodes, transistors |
| Silicon (Si) | Diodes, photodiodes, rectifiers, solar cells, SCR, triacs, transistors, integrated circuits |
| Gallium arsenide (GaAs) | Light-emitting diodes, solar cells, transistors, integrated circuits |
| Selenium (Se) | Photoresistors, rectifiers, xerography |
| Silicon carbide (SiC) | Varactors, light-emitting diodes |
| Cadmium sulfide (CdS) | Photoresistors, solar cells |
| Indium antimonide (InSb) | Infrared detectors (photodiodes), Hall effect devices |
| Lead sulfide (PbS) | Infrared detectors (photoresistors) |
| Bismuth telluride ($Bi_2Te_3$) | Thermoelectric elements |
| Lead telluride (PbTe) | Thermoelectric elements |
| Mercury cadmium telluride (HgCdTe) | Infrared detectors (photodiodes) |

## Groupings and crystal structure of various semiconductors

| Material | Group | Structure |
|---|---|---|
| Si | IV | Cubic diamond |
| Ge | IV | Cubic diamond |
| Diamond | IV | Cubic diamond |
| GaAs | III-V | Cubic zincblende |
| GaP | III-V | Cubic zincblende |
| GaSb | III-V | Cubic zincblende |
| InAs | III-V | Cubic zincblende |
| InP | III-V | Cubic zincblende |
| InSb | III-V | Cubic zincblende |
| $\alpha$SiC | IV-IV | Hexagonal |
| $\beta$SiC | IV-IV | Cubic zincblende |
| $\alpha$ZnS | II-VI | Hexagonal |
| $\beta$ZnS | II-VI | Cubic zincblende |
| CdS | II-VI | Hexagonal wurtzite |
| CdSe | II-VI | Hexagonal wurtzite |
| CdTe | II-VI | Cubic zincblende |
| PbS | IV-VI | Cubic rocksalt |
| PbSe | IV-VI | Cubic rocksalt |
| PbTe | IV-VI | Cubic rocksalt |
| $Bi_2Te_3$ | V-VI | Hexagonal |
| Se | VI | Monoclinic hexagonal |

binary chalcogenides, except for CdO and HgO, are the materials generally designated as II-VI compounds. The materials referred to as IV-VI compounds are the nine equiatomic chalcogenides formed by combination of the group IV elements germanium, tin, and lead with sulfur, selenium, and tellurium. The II-VI and IV-VI compounds are employed as media for the generation, transmission, or detection of electromagnetic radiation. One of the principal applications is the use of ZnS and $Zn_{1-x}Cd_xS$ phosphor powders in the screens of cathode ray tubes, such as those used in color television receivers and video-display terminals.

Other semiconductors of commercial importance include silicon carbide (group IV-IV), bismuth telluride (group V-VI), and selenium (group VI).

## Recommended Reading

- W.C. Omara and R. Herring, Ed., *Handbook of Semiconductors*, Noyes Publications, 1990
- R.K. Willardson and A.C. Beer, Ed., *Semiconductors and Semimetals*, Vol 1-18, Academic Press, 1966-present
- G.K. Teal *et al.*, Semiconductor Materials, *Materials and Processes, Part A: Materials*, 3rd ed., Marcel Dekker, 1985, p 219-312

**sensitizing heat treatment.** A heat treatment, whether accidental, intentional, or incidental (as during welding), that causes precipitation of constituents at grain boundaries, often causing the alloy to become susceptible to *intergranular corrosion* or *intergranular stress-corrosion cracking*. See also *sensitization*.

**Sendzimir mill.** A type of *cluster mill* with small-diameter work rolls and larger-diameter backup rolls, backed up by bearings on a shaft mounted eccentrically so that it can be rotated to increase the pressure between the bearing and the backup rolls (Fig. 503). Used to roll precision and very thin sheet and strip.

**separate-application adhesive.** An adhesive consisting of two parts, one part being applied to one adherend and the other part to the other adherend and the two being brought together to form a joint.

**separator.** In rolling-element bearings, the part of a cage that lies between the rolling elements. This term is sometimes

FIG. 502

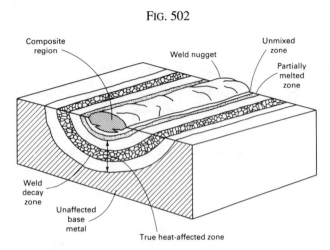

*Schematic of an austenitic stainless steel weldment showing sensitized zone (weld decay zone)*

FIG. 503

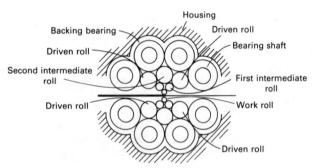

*Sendzimir mill used for precision cold rolling of thin sheet and foil*

used as a synonym for *cage*. See also the figure accompanying the term *rolling-element bearing*.

**separator** (composites). In processing of composites, a permeable layer that allows volatiles and air to escape from the laminate and excess resin to be bled from the laminate into the bleeder plies during cure. Porous Teflon-coated fiberglass is an example. Often placed between lay-up and bleeder to facilitate bleeder systems removal from laminate after cure. See also *bleeder* and *lay-up*.

**sequence timer**. In resistance welding, a device used for controlling the sequence and duration of any or all of the elements of a complete welding cycle except *heat time* or *weld time*.

**sequence weld timer**. Same as *sequence timer* except that either *weld time* or *heat time*, or both, are also controlled.

**sequestering agent**. A material that combines with metallic ions to form water-soluble complex compounds.

**serial sectioning**. A metallographic technique in which an identified area on a section surface is observed repeatedly after successive layers of known thickness have been removed from the surface. It is used to construct a three-

dimensional morphology of structural features. See also *sectioning*.

**series submerged arc welding**. A submerged arc welding process variation in which electric current is established between two (consumable) electrodes which meet just above the surface of the work. The work is not in the electrical circuit. See also *submerged arc welding*.

**series welding**. Resistance welding in which two or more spot, seam, or projection welds are made simultaneously by a single welding transformer with three or more electrodes forming a series circuit (Fig. 504).

FIG. 504

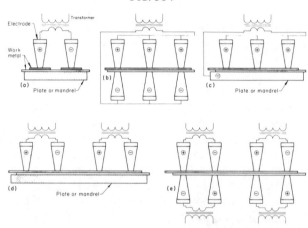

*Comparison of setups of work metal and electrodes for making multiple spot welds using direct (a to c) and series (d and e) welding*

**set**. The shape of the solidifying surface of a metal, especially copper, with respect to concavity or convexity. May also be called pitch.

**set** (polymerization). To convert an adhesive into a fixed or hardened state by chemical or physical action, such as condensation, polymerization, oxidation, vulcanization, gelation, hydration, or evaporation of volatile constituents. See also *cure*.

**set copper**. An intermediate copper product containing about 3.5% cuprous oxide, obtained at the end of the oxidizing portion of the fire-refining cycle.

**set, permanent**. See *permanent set*.

**setting temperature**. The temperature to which an adhesive or an assembly is subjected to set the adhesive. The temperature attained by the adhesive in the process of setting (adhesive setting temperature) may differ from the temperature of the atmosphere surrounding the assembly (assembly setting temperature). See also *curing temperature* and *drying temperature*.

**setting time**. The period of time during which an adhesively bonded assembly is subjected to heat or pressure, or both, to set the adhesive. See also *drying time* and *joint-conditioning time*.

**settling.** (1) Separation of solids from suspension in a fluid of lower density, solely by gravitational effects. (2) A process for removing iron from liquid magnesium alloys by holding the melt at a low temperature after manganese has been added to it.

**set up.** To harden, as in the curing of a polymer resin.

**severe wear.** A form of wear characterized by removal of material in relatively large fragments. Severe wear is an imprecise term, frequently used in research, and contrasted with *mild wear*. In fact, the phenomena studied usually involve the transition from mild to severe wear and the factors that influence that transition. With metals, the fragments are usually predominantly metallic rather than oxidic. Severe wear is frequently associated with heavy loads and/or adhesive contact.

**severity of quench.** Ability of quenching medium to extract heat from a hot steel workpiece; expressed in terms of the *Grossmann number (H)*.

**S-glass.** A magnesium aluminosilicate composition that is especially designed to provide very high tensile strength glass filaments. S-glass and S-2 glass fibers have the same glass composition but different finishes (coatings). S-glass is made to more demanding specifications, and S-2 is considered the commercial grade. See also the tables accompanying the term *fiberglass*.

**shadow angle.** In shadowing of replicas, the angle between the line of motion of the evaporated atoms and the surface being shadowed. See also *replica* and *shadowing*.

**shadow cast replica.** A replica that has been shadowed. See also *replica* and *shadowing*.

**shadowing.** Directional deposition of carbon or a metallic film on a plastic replica so as to highlight features to be analyzed by transmission electron microscopy (Fig. 505). Most often used to provide maximum detail and resolution

FIG. 505

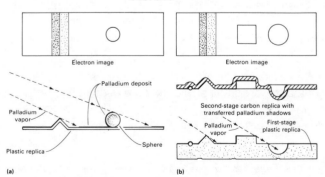

*Shadowing processes for single-stage plastic replicas (a) and two-stage plastic-carbon replicas (b). A sphere of known size can be placed on a replica, as in (a), to calibrate shadow length.*

of the features of fracture surfaces. See also *metal shadowing*, *oblique evaporation shadowing*, and *shadow angle*.

**shadow mask** (thermal spraying). A thermal spraying process variation in which an area is partially shielded during the thermal spraying operation, thus permitting some overspray to produce a feathering at the coating edge.

**shadow microscope.** An electron microscope that forms a shadow image of an object using electrons emanating from a point source located close to the object.

**shaft run-out.** Twice the distance that the center of a shaft is displaced from the axis of rotation; that is, twice the eccentricity.

**shakeout.** Removal of castings from a sand mold. See also *knockout*.

**shaker-hearth furnace.** A continuous type furnace that uses a reciprocating shaker motion to move the parts along the hearth (Fig. 506).

FIG. 506

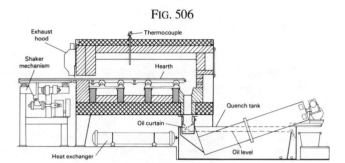

*Schematic of shaker-hearth furnace for continuous carburizing*

**shales.** Abrasive particles of platelike shape. The term is applied particularly to diamond abrasives.

**shank.** (1) The portion of a die or tool by which it is held in position in a forging unit or press. (2) The handle for carrying a small ladle or crucible. (3) The main body of a lathe tool. If the tool is an inserted type, the shank is the portion that supports the insert.

**shank-type cutter.** A cutter having a straight or tapered shank to fit into a machine-tool spindle or adapter.

**shape accuracy.** The shape of a sintered or sized powder metallurgy product as it conforms to or deviates from specified dimensions or tolerances.

**shape factor.** For an elastomeric slab loaded in compression, the ratio of the loaded area to the force-free area.

**shape memory alloys.** See *Technical Brief 42*.

**shaping.** Producing flat surfaces using single-point tools (Fig. 507). The work is held in a vise or fixture, or is clamped directly to the table. The ram supporting the tool is reciprocated in a linear motion past the work.

1. Form shaping: Shaping with a tool ground to provide a specified shape.

## Technical Brief 42: Shape Memory Alloys

SHAPE MEMORY ALLOYS (SMAs) are a group of metallic materials that demonstrate the ability to return to some previously defined shape or size when subjected to the appropriate thermal procedure. Generally, these materials can be plastically deformed at some relatively low temperature, and upon exposure to some higher temperature will return to their shape prior to the deformation. An SMA may be further defined as one that yields a thermoelastic martensite. In this case, the alloy undergoes a martensitic transformation of a type that allows the alloy to be deformed by a twinning mechanism below the shape memory transformation temperature. The deformation is then reversed when the twinned structure reverts upon heating to the parent phase.

Although a relatively wide variety of alloys are known to exhibit the shape memory effect, only those that can recover substantial amounts of strain or that generate significant force upon changing shape are of commercial importance. To date, the alloy systems that have achieved any level of commercial ex-

### Alloys having a shape memory effect

| Alloy | Composition | Transformation-temperature range °C | °F | Transformation hysteresis Δ°C | Δ°F |
|---|---|---|---|---|---|
| Ag-Cd | 44/49 at.% Cd | −190 to −50 | −310 to −60 | ≈15 | ≈25 |
| Au-Cd | 46.5/50 at.% Cd | 30 to 100 | 85 to 212 | ≈15 | ≈25 |
| Cu-Al-Ni | 14/14.5 wt% Al 3/4.5 wt% Ni | −140 to 100 | −220 to 212 | ≈35 | ≈65 |
| Cu-Sn | ≈15 at.% Sn | −120 to 30 | −185 to 85 | | |
| Cu-Zn | 38.5/41.5 wt% Zn | −180 to −10 | −290 to 15 | ≈10 | ≈20 |
| Cu-Zn-X (X = Si, Sn, Al) | a few wt% of X | −180 to 200 | −290 to 390 | ≈10 | ≈20 |
| In-Ti | 18/23 at.% Ti | 60 to 100 | 140 to 212 | ≈4 | ≈7 |
| Ni-Al | 36/38 at.% Al | −180 to 100 | −290 to 212 | ≈10 | ≈20 |
| Ni-Ti | 49/51 at.% Ni | −50 to 110 | −60 to 230 | ≈30 | ≈55 |
| Fe-Pt | ≈25 at.% Pt | ≈−130 | ≈−200 | ≈4 | ≈7 |
| Mn-Cu | 5/35 at.% Cu | −250 to 180 | −420 to 355 | ≈25 | ≈45 |
| Fe-Mn-Si | 32 wt% Mn, 6 wt% Si | −200 to 150 | −330 to 300 | ≈100 | ≈180 |

ploitation are nickel-titanium alloys and copper-base alloys such as Cu-Zn-Al and Cu-Al-Ni. The Ni-Ti alloys have greater shape memory strain (up to 8% versus 4 to 5% for the copper-base alloys), tend to be much more thermally stable, have superior corrosion resistance, and have much higher ductility. The copper-base alloys are much less expensive, can be melted and extruded in air with ease, and have a wider range of potential transformation temperatures.

The Ni-Ti alloys (Nitinols) are equiatomic systems (49 to 51 at.% Ni) that contain a ductile TiNi intermetallic compound. This intermetallic has a moderate solubility range for excess nickel and titanium. This solubility allows alloying with many elements to modify both the mechanical properties and the transformation properties of the system. Excess nickel, in amounts up to about 1%, is the most common alloying addition. Excess nickel strongly depresses the transformation temperature and increases the yield strength of the austenite. Other frequently used elements are iron and chromium (to lower the transformation temperature), and copper (to decrease the hysteresis and lower the deformation stress of the martensite).

Commercial copper-base SMAs are available in ternary Cu-Zn-Al and Cu-Al-Ni alloys, or in their quaternary modifications containing manganese. Elements such as boron, cerium, cobalt, iron, titanium, vanadium, and zirconium are also added for grain refinement. Compositions of Cu-Al-Ni alloys usually fall in the range of 11 to 14.5 wt% Al and 3 to 5 wt% Ni. The Cu-Zn-Al alloys usually contain approximately 40 wt% Zn and less than 6 wt% Al.

There is a wide variety of uses for SMAs. Because the Ni-Ti alloys are extremely corrosion resistant, demonstrate excellent biocompatibility, and can be fabricated to very small sizes, they are used in biomedical devices and as arch wires for orthodontic correction. Copper-base SMAs are used as hydraulic couplings and force actuators. Other applications for SMAs include fluid flow control devices and eyeglass frames.

### Recommended Reading

- D.E. Hodgson, M.H. Wu, and R.J. Biermann, Shape Memory Alloys, *Metals Handbook*, 10th ed., Vol 2, ASM International, 1990, p 897-902
- H. Funakubo, Ed., *Shape Memory Alloys*, Gordon and Breach, 1988
- J. Perkins, Ed., *Shape Memory Effects in Alloys*, Plenum Press, 1975

FIG. 507

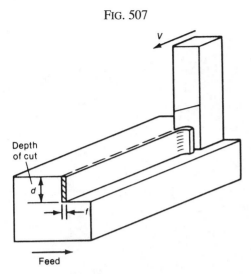

*Schematic of shaping operation*

FIG. 508

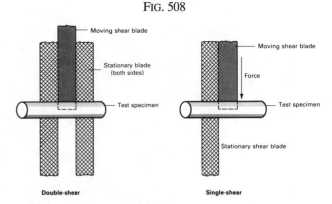

*Schematic of double- and single-shear testing*

2. Contour shaping: Shaping of an irregular surface, usually with the aid of a tracing mechanism.

3. Internal shaping: Shaping of internal forms such as keyways and guides.

**shark's teeth.** In ceramics and glasses, a striation consisting of dagger-like step fractures starting at the scored edge and extending to or nearly to the compression edge.

**sharp-notch strength ($\sigma_s$).** The notch tensile strength measured using specimens with very small notch root radii (approaching the limit for machining capability); values of sharp-notch strength usually depend on notch root radius.

**shatter crack.** See *flake* (metals).

**shaving.** (1) As a finishing operation, the accurate removal of a thin layer of a work surface by straightline motion between a cutter and the surface. (2) Trimming parts such as stampings, forgings, and tubes to remove uneven sheared edges or to improve accuracy.

**Shaw (Osborn-Shaw) Process.** See *ceramic molding.*

**shear.** (1) The type of force that causes or tends to cause two contiguous parts of the same body to slide relative to each other in a direction parallel to their plane of contact (Fig. 508). (2) A machine or tool for cutting metal and other material by the closing motion of two sharp, closely adjoining edges; for example, squaring shear (Fig. 509) and circular shear. (3) An inclination between two cutting edges, such as between two straight knife blades or between the punch cutting edge and the die cutting edge, so that a reduced area will be cut each time. This lessens the necessary force, but increases the required length of the working stroke. This method is referred to as angular shear. (4) The act of cutting by shearing dies or blades, as in shearing lines (Fig. 510).

**shear angle.** The angle that the *shear plane*, in metal cutting, makes with the work surface.

FIG. 509

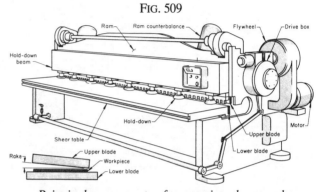

*Principal components of a squaring shear, and detail showing rake angle of blades*

FIG. 510

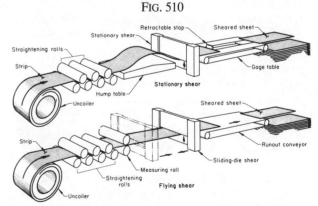

*Two types of shearing lines for cutting coiled strip into flat sheets*

**shear bands.** (1) Bands of very high shear strain that are observed during rolling of sheet metal. During rolling, these form at approximately ±35° to the rolling plane, parallel to the transverse direction. They are independent of grain orientation and at high strain rates traverse the entire thickness of the rolled sheet. (2) Highly localized deformation zones in metals that are observed at very high

FIG. 511

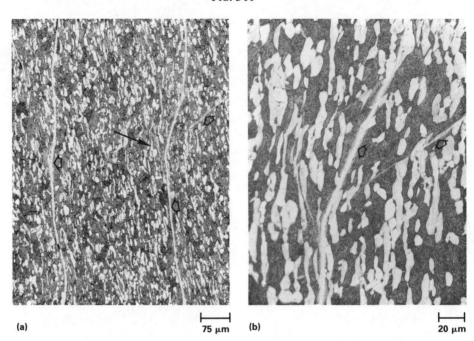

(a)                                                75 μm    (b)                  20 μm

*Appearance of adiabatic shear bands in an explosively ruptured Ti-6Al-4V*
*STA alloy rocket motor. The material exhibits multiple, often intersecting,*
*shear bands (open arrows). Slender arrow points to portion of shear band*
*shown in more detail in (b).*

strain rates, such as those produced by high velocity (100 to 3600 m/s, or 330 to 11,800 ft/s) projectile impacts or explosive rupture. During high-strain-rate shear, also known as adiabatic shear, the bulk of the plastic deformation is concentrated in narrow bands within the relatively undeformed matrix (Fig. 511). These shear bands are believed to occur along slip planes and the local strain rate within the adiabatic shear bands can exceed $10^6 \cdot s^{-1}$.

**shear edge**. The cutoff edge of a mold.

**shear fracture**. A mode of fracture in crystalline materials resulting from translation along slip planes that are preferentially oriented in the direction of the shearing stress (Fig. 512).

**shear hackle**. A *hackle* in ceramics and glasses generated by interaction of a shear component with the principal tension under which the crack is running.

**shear ledges**. See *radial marks*.

**shear lip**. A narrow, slanting ridge along the edge of a fracture surface (Fig. 513). The term sometimes also denotes a narrow, often crescent-shaped, fibrous region at the edge of a fracture that is otherwise of the cleavage type, even though this fibrous region is in the same plane as the rest of the fracture surface.

**shear modulus (*G*)**. The ratio of shear stress to the corresponding shear strain for shear stresses below the proportional limit of the material. Values of shear modulus are

FIG. 512

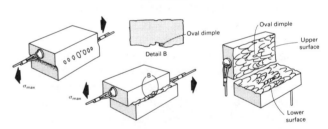

*Ductile fracture due to shear stresses (shear fracture)*

usually determined by torsion testing. Also known as modulus of rigidity.

**shear plane**. A confined zone along which shear takes place in metal cutting. It extends from the cutting edge to the work surface.

**shear rate**. With regard to viscous fluids, the relative rate of flow or movement.

**shear stability**. The ability of a lubricant to withstand shearing without degradation. See also *penetration* (of a grease).

**shear strain**. The tangent of the angular change, caused by a force between two lines originally perpendicular to each other through a point in a body. Also called angular strain.

**shear strength**. The maximum shear stress that a material is capable of sustaining. Shear strength is calculated from the

FIG. 513

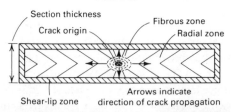

*Appearance of fracture features in steel tensile specimens. Top: Schematic of tensile fracture features in a rectangular specimen. Bottom: A cup-and-cone fracture in AISI 8740 steel with a central fibrous zone containing radial marks between the central fibrous zone and the outer shear lip. 9×*

maximum load during a shear or torsion test and is based on the original cross-sectional area of the specimen.

**shear stress**. (1) The stress component tangential to the plane on which the forces act. (2) A stress that exists when parallel planes in metal crystals slide across each other.

**shear stress** (plastics). The stress developing in a polymer melt when the layers in a cross section are gliding along each other or along the wall of the channel (in laminar flow). Shear stress is equal to force divided by the area sheared, yielding pounds per square inch.

**shearing**. The parting of material that results when one blade forces the material past an opposing blade. See also *shear*.

**shear thickening**. An increase in viscosity of non-Newtonian fluids (for example, polymers and their solutions, slurries, and suspensions) with an increase in shear stress or time. See also *shear thinning*.

**shear thinning**. A decrease in viscosity of non-Newtonian fluids with an increase in shear stress or time. The decrease in viscosity may be temporary or permanent. The latter happens when the shear stress is sufficiently large to rupture a chemical bond, so that the sheared liquid has a lower viscosity than it had prior to shearing. See also *shear thickening*.

**sheath**. (1) The material, usually an extruded plastic or elastomer, applied outermost to a wire or cable. Also called a jacket. (2) A sheet metal or glass covering of a sintered billet to protect it from oxidation or other environmental contamination during hot working. See also *can*.

**sheet**. A flat-rolled metal product of some maximum thickness and minimum width arbitrarily dependent on the type of metal. It has a width-to-thickness ratio greater than about 50. Generally, such flat products under 6.5 mm (¼ in.) thick are called sheets, and those 6.5 mm (¼ in.) thick and over are called plates. Occasionally, the limiting thickness for steel to be designated as sheet steel is No. 10 Manufacturer's Standard Gage for sheet steel, which is 3.42 mm (0.1345 in.) thick. Table 59 lists standard sheet metal gage thicknesses.

**sheet forming**. The plastic deformation of a piece of sheet metal by tensile loads into a three-dimensional shape, often without significant changes in sheet thickness or surface characteristics. Compare with *bulk forming*.

**sheeting**. A form of plastic in which the thickness is very small in proportion to length and width and in which the plastic is present as a continuous phase throughout, with or without filler.

**sheet molding compound (SMC)**. A thermoset composite of fibers, usually an unsaturated polyester resin, and pigments, fillers, and other additives that have been compounded and processed into sheet form to facilitate han-

**Table 59 Sheet metal gage thicknesses**

| Gage | in. | mm | Gage | in. | mm |
|---|---|---|---|---|---|
| 30 | 0.0120 | 0.3048 | 16 | 0.0598 | 1.5189 |
| 29 | 0.0135 | 0.3429 | 15 | 0.0673 | 1.7094 |
| 28 | 0.0149 | 0.3785 | 14 | 0.0747 | 1.8974 |
| 27 | 0.0164 | 0.4166 | 13 | 0.0897 | 2.2784 |
| 26 | 0.0179 | 0.4547 | 12 | 0.1046 | 2.6568 |
| 25 | 0.0109 | 0.5309 | 11 | 0.1196 | 3.0378 |
| 24 | 0.0239 | 0.6071 | 10 | 0.1345 | 3.4163 |
| 23 | 0.0269 | 0.6833 | 9 | 0.1495 | 3.7973 |
| 22 | 0.0299 | 0.7595 | 8 | 0.1644 | 4.1758 |
| 21 | 0.0329 | 0.8357 | 7 | 0.1793 | 4.5542 |
| 20 | 0.0359 | 0.9119 | 6 | 0.1943 | 4.9352 |
| 19 | 0.0418 | 1.0617 | 5 | 0.2092 | 5.3137 |
| 18 | 0.0478 | 1.2141 | 4 | 0.2242 | 5.6947 |
| 17 | 0.0538 | 1.3665 | 3 | 0.2391 | 6.0731 |

FIG. 514

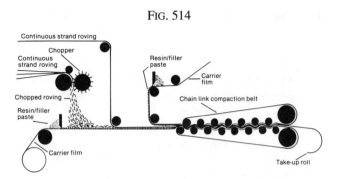

*Sheet molding compound (SMC) manufacture using a configuration that can make chopped-fiber SMC-R; continuous-fiber SMC-C; or continuous, random SMC-C/R material*

dling in the subsequent compression molding operation (Fig. 514). Generally cross linked with styrene.

**sheet separation.** In spot, seam, or projection welding, the gap that exists between faying surfaces surrounding the weld, after the joint has been welded.

**shelf life.** The length of time a material, substance, product, or reagent can be stored under specified environmental conditions and continue to meet all applicable specification requirements and/or remain suitable for its intended function.

**shelf roughness.** Roughness on upward-facing surfaces where undissolved solids have settled on parts during a plating operation.

**shell.** (1) A hollow structure or vessel. (2) An article formed by deep drawing. (3) The metal sleeve remaining when a billet is extruded with a dummy block of somewhat smaller diameter. (4) In shell molding, a hard layer of sand and thermosetting plastic or resin formed over a pattern and used as the mold wall. (5) A tubular casting used in making seamless drawn tube. (6) A pierced forging.

**shell core.** A shell-molded sand core.

**shell hardening.** A surface-hardening process in which a suitable steel workpiece, when heated through and quench hardened, develops a martensite layer or shell that closely follows the contour of the piece and surrounds a core of essentially pearlitic transformation product. This result is accomplished by a proper balance among section size, steel hardenability, and severity of quench.

**shelling.** (1) A term used in railway engineering to describe an advanced phase of *spalling*. (2) A mechanism of deterioration of coated abrasive products in which entire abrasive grains are removed from the cement coating that held the abrasive to the backing layer of the product.

**shell molding.** A foundry process in which a mold is formed from thermosetting resin-bonded sand mixtures brought in contact with preheated (150 to 260 °C, or 300 to 500 °F) metal patterns, resulting in a firm shell with a cavity corresponding to the outline of the pattern. Also called *Croning process.*

**shell tooling.** A mold or bonding fixture for forming plastic parts that consists of a contoured surface shell supported by a substructure to provide dimensional stability.

**shielded carbon arc welding.** A carbon arc welding process variation which produces coalescence of metals by heating them with an electric arc between a carbon electrode and the work. Shielding is obtained from the combustion of a solid material fed into the arc or from a blanket of flux on the work, or both. Pressure may or may not be used, and filler metal may or may not be used.

**shielded metal arc cutting.** A metal arc cutting process in which metals are severed by melting them with the heat of an arc between a covered metal electrode and the base metal.

**shielded metal arc welding (SMAW).** A manual arc welding process in which the heat for welding is generated by an arc established between a flux-covered consumable electrode and a workpiece (Fig. 515). The electrode tip, molten weld pool, arc, and adjacent areas of the workpiece are protected from atmospheric contamination by a gaseous shield obtained from the combustion and decomposition of the electrode covering. Additional shielding is provided for the molten metal in the weld pool by a covering of molten flux or slag. Filler metal is supplied by the core of the consumable electrode and from metal powder mixed with the electrode covering of certain electrodes. Shielded metal arc welding is often referred to as arc welding with stick electrodes, manual metal arc welding, and stick welding.

FIG. 515

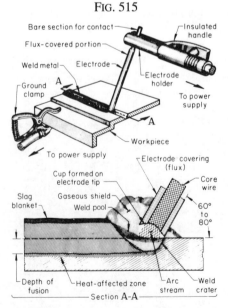

*Setup and fundamentals of operation for shielded metal arc welding*

**shielding.** (1) A material barrier that prevents radiation or a flowing fluid from impinging on an object or a portion of an object. (2) In an electron-optical instrument, the protection of the electron beam from distortion due to extraneous electric and magnetic fields. Because the metallic column of the microscope is at ground potential, it provides electrostatic shielding. See also *magnetic shielding*. (3) Placing an object in an electrolytic bath so as to alter the current distribution on the cathode. A nonconductor is called a shield; a conductor is called a *robber*, a thief, or a guard.

**shielding gas.** (1) Protective gas used to prevent atmospheric contamination during welding. (2) A stream of inert gas directed at the substrate during thermal spraying so as to envelop the plasma flame and substrate; intended to provide a barrier to the atmosphere in order to minimize oxidation.

**shift.** A casting imperfection caused by mismatch of cope and drag or of cores and molds.

**shim.** A thin piece of material used between two surfaces to obtain a proper fit, adjustment, or alignment.

**shimmy die.** See *flat edge trimmer*.

**shock load.** The sudden application of an external force that results in a very rapid build-up of stress—for example, piston loading in internal combustion engines.

**shoe.** (1) A metal block used in a variety of bending operations to form or support the part being processed. (2) An anvil cap or *sow block*. (3) A device for gathering filaments into a strand, in glass fiber forming. See also the figure accompanying the term *glass filament bushing*.

**Shore hardness.** A measure of the resistance of material to indentation by a spring-loaded indenter during Scleroscope hardness testing. The higher the number, the greater the resistance. Normally used for rubber materials. See also *Scleroscope hardness test*.

**short.** An imperfection in a molded plastic part due to an incomplete fill. In reinforced plastics, this may be evident either from an absence of surface film in some areas or as lighter, unfused particles of material showing through a covering surface film, accompanied possibly by thin-skin blisters. In thermoplastics, also called short-shot.

**short beam shear.** A flexural test of a plastic specimen having a low test span-to-thickness ratio (for example, 4:1), such that failure is primarily in shear.

**short circuiting transfer.** In consumable-electrode arc welding, a type of metal transfer similar to globular transfer, but in which the drops are so large that the arc is short circuited momentarily during the transfer of each drop to the weld pool (Fig. 516). See also *gas metal arc welding* and compare with *globular transfer* and *spray transfer*.

**shortness** (adhesives). A qualitative term that describes an adhesive that does not string cotton, or otherwise form filaments or threads during application.

**shortness** (metals). A form of brittleness in metal. It is designated as *cold shortness* or *hot shortness* to indicate the temperature range in which the brittleness occurs.

**shorts.** The product that is retained on a specified screen in the screening of a crushed or ground material. See also *plus sieve*.

**short shot.** Insufficient injection of material into the mold during forming of plastics or composites.

**short-term etching.** In metallographic preparation of specimens, etching times of seconds to a few minutes.

**short transverse.** See *transverse*.

**shot.** (1) Small, spherical particles of metal (Fig. 517). (2) The injection of molten metal into a die casting die. The metal is injected so quickly that it can be compared to the shooting of a gun.

**shotblasting.** Blasting with metal *shot*; usually used to remove deposits or mill scale more rapidly or more effectively than can be done by *sandblasting*.

**shot capacity.** In forming of plastic parts, the maximum weight of material an injection machine can provide from one forward motion of the ram, screw, or plunger.

**shot peening.** A method of cold working metals in which compressive stresses are induced in the exposed surface layers of parts by the impingement of a stream of *shot*,

FIG. 516

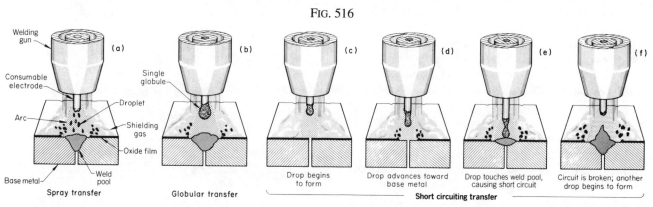

*Modes of metal transfer in gas metal arc welding. (a) Spray transfer. (b) Globular transfer. (c) to (f) Steps in short circuiting transfer*

FIG. 517

*Copper shot*

FIG. 518

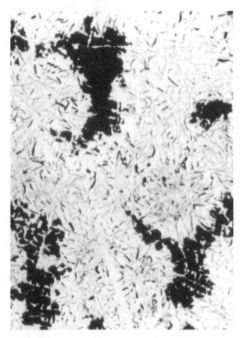

*Typical shrinkage cavities (dark) in gray cast iron. 50×*

directed at the metal surface at high velocity under controlled conditions. It differs from blast cleaning in primary purpose and in the extent to which it is controlled to yield accurate and reproducible results. Although shot peening cleans the surface being peened, this function is incidental. The major purpose of shot peening is to increase fatigue strength. Shot for peening is made of iron, steel, or glass.

**shotting**. The production of *shot* by pouring molten metal in finely divided streams. Solidified spherical particles are formed during descent in a tank of water.

**shoulder**. See preferred term *root face*.

**shrinkage** (ceramics). The fractional reduction in dimensions or volume of a material or object when subjected to drying, calcining, or firing (sintering).

**shrinkage** (metals). (1) The contraction of metal during cooling after hot forging. Die impressions are made oversize according to precise shrinkage scales to allow the forgings to shrink to design dimensions and tolerances. (2) See *casting shrinkage*.

**shrinkage** (plastics). The relative change in dimension from the length measured on the mold when it is cold to the length of the molded plastic object 24 h after it has been taken out of the mold.

**shrinkage cavity**. A void left in cast metal as a result of solidification shrinkage. Shrinkage cavities can appear as either isolated or interconnected irregularly shaped voids (Fig. 518). See also *casting shrinkage*.

**shrinkage cracks**. Cracks that form in metal as a result of the pulling apart of grains by contraction before complete solidification. See also *hot tear*.

**shrinkage rule**. A measuring ruler with graduations expanded to compensate for the change in the dimensions of the solidified casting as it cools in the mold.

**shrinkage stress**. See preferred term *residual stress*.

**shrinkage void**. A cavity type discontinuity in weldments normally formed by shrinkage during solidification.

**shrink fit**. An *interference fit* produced by heating the outside member of mating parts to a practical temperature for easy assembly. Usually the inside member is kept at or near

room temperature. Sometimes the inside member is cooled to increase ease of assembly.

**shrink forming**. Forming of metal wherein the inner fibers of a cross section undergo a reduction in a localized area by the application of heat, cold upset, or mechanically induced pressures.

**shroud**. A protective, refractory-lined metal-delivery system to prevent reoxidation of molten steel when it is poured from ladle to tundish to mold during continuous casting (Fig. 519).

FIG. 519

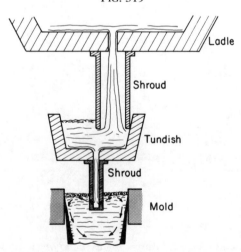

*Continuous casting pouring shrouds*

## Technical Brief 43: Sialons

SIALON is a generic term for a family of compositions produced by reacting silicon nitride ($Si_3N_4$) with aluminum oxide ($Al_2O_3$) and aluminum nitride (AIN) at high temperature. In other words, a Sialon is a $Si_3N_4$-base ceramic in which some of the silicon has been replaced with aluminum and some of the nitrogen with oxygen, resulting in a substituted solid solution referred to as $\beta'$-Sialon with a compositional range of $Si_{6-x}Al_xO_xN_{8-x}$ based on the $\beta$-$Si_6N_8$ unit cell. Two major types have been developed commercially, one consisting of $\beta'$ grains with a residual glass, the other consisting of $\beta'$ grains and a semicontinuous intergranular crystalline phase of yttrium-aluminum-garnet (YAG). Parts made from $\beta'$-Sialon are produced by cold isostatically pressing the powder, followed by sintering at a maximum temperature of 1800 °C (3300 °F) for approximately 1 h; yttrium oxide ($Y_2O_3$) is used as a sintering aid.

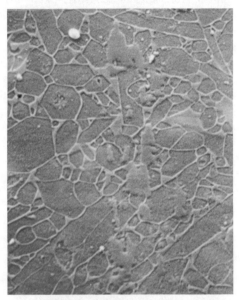

*Microstructure of $\beta'$-Sialon, consisting of $\beta'$ grains cemented by a glassy phase. 10,000×*

A solid solution based on the alpha silicon nitride structure has received less attention, but $\alpha'$-Sialon shows promise commercially because it has greater hardness than $\beta'$-Sialon. In addition, it offers an improved product with less grain-boundary glass as a result of the incorporation of the cation from the sintering additive (for example, $Y_2O_3$) into a solid solution of $M_x(Si,Al)_{12}(O,N)_{16}$ based on the $\alpha$-$Si_{12}N_{16}$ unit cell. Single-phase $\alpha'$-Sialons currently are only available in hot pressed or hot isostatically pressed forms, but composites of $\beta' + \alpha'$ are available produced by pressureless sintering.

Properties of Sialons, which are based on the sintering aids used and the fabrication route followed, are similar to $Si_3N_4$ (see the table accompanying the term *silicon nitride*, Technical Brief 45). The advantages of Sialons are their low coefficient of thermal expansion (2 to $3 \times 10^{-6}$/°C, or 1 to $1.7 \times 10^{-6}$/°F) and good oxidation resistance. The array of applications for Sialons is also similar to that of $Si_3N_4$—namely, automotive applications and machine tool cutting inserts for machining difficult-to-cut materials such as cast irons and wrought nickel-base superalloys.

### Recommended Reading

- S. Hampshire, Engineering Properties of Nitrides, *Engineered Materials Handbook*, Vol 4, ASM International, 1991, p 812-820
- R. Komanduri and S.K. Samanta, Ceramic Cutting Tools, *Metals Handbook*, 9th ed., Vol 16, ASM International, 1989, p 98-104

**shunt.** A device used to divert part of an electric current.

**shut height.** For a metal forming press, the distance from the top of the bed to the bottom of the slide with the stroke down and adjustment up. In general, it is the maximum die height that can be accommodated for normal operation, taking the *bolster plate* into consideration.

**SI** . See *silicones*.

**Sialons.** See *Technical Brief 43*.

**side corings.** In forming of plastics, projections that are used to core a hole in a direction other than the line of closing of a mold, and that must be withdrawn before the part is ejected from the mold. Also called side draw pins.

**side cutting-edge angle.** Defined by the figure accompanying the term *single-point tool*.

**side milling.** Milling with cutters having peripheral and side teeth. They are usually profile sharpened but may be form relieved. See also *milling*.

**side rake.** In a single-point turning tool, the angle between the tool face and a reference plane, corresponding to radial rake in milling. It lies in a plane perpendicular to the tool base and parallel to the rotational axis of the work. See also the figure accompanying the term *single-point tool*.

**side thrust.** The lateral force exerted between the dies by reaction of a forged piece on the die impressions.

**siemens (S).** A unit of electrical conductivity. One siemens of conductance per cubic meter with a potential of 1 V allows the passage of 1 A/m$^2$. See also *conductance* (electrical).

**sieve.** A standard wire mesh or screen used in graded sets to determine the mesh size or particle size distribution of particulate and granular solids. Sieves are stacked in order, with the largest mesh size at the top and a pan at the bottom (Fig. 520). An appropriate sample weight of powder is spread on the top sieve and covered. The stack of sieves is agitated in a prescribed manner (shaking, rotating, or tapping) for a specified period of time. The powder fractions remaining on each sieve and contained in the bottom pan are weighed separately and reported as percentages retained or passed by each sieve. See also *sieve analysis*.

FIG. 520

*Schematic of sieve series stacked in order of size*

**sieve analysis.** A method of determining *particle size distribution*, usually expressed as the weight percentage retained upon each of a series of standard screens of decreasing mesh size. For example, a typical sieve analysis for titanium powder is as follows:

| Mesh size (U.S.) | Particle size, μm | Weight percent retained |
|---|---|---|
| +80 | +177 | 0 |
| −80+100 | −177+149 | 0.1 |
| −100+140 | −149+105 | 11.2 |
| −140+200 | −105+74 | 32.9 |
| −200+230 | −74+64 | 5.0 |
| −230+325 | −64+45 | 23.3 |
| −325 | −45 | 27.5 |

Using the standard sieve designations and screen opening sizes listed in Table 60, the user can determine that −140+299 powder is smaller than approximately 105 μm, yet larger than approximately 74 μm, i.e., ~33% is retained on the 200 mesh screen.

**Table 60  Typical screen (sieve) sizes and related particle sizes**

| Sieve designation Mesh No. | μm | U.S. standard Sieve opening in. | mm | Tyler Standard Mesh No. |
|---|---|---|---|---|
| 80 | 177 | 0.0070 | 0.177 | 80 |
| 100 | 149 | 0.0059 | 0.149 | 100 |
| 120 | 125 | 0.0049 | 0.125 | 115 |
| 140 | 105 | 0.0041 | 0.105 | 150 |
| 170 | 88 | 0.0035 | 0.088 | 170 |
| 200 | 74 | 0.0029 | 0.074 | 200 |
| 230 | 63 | 0.0024 | 0.063 | 250 |
| 270 | 53 | 0.0021 | 0.053 | 270 |
| 325 | 44 | 0.0017 | 0.044 | 325 |
| 400 | 37 | 0.0015 | 0.037 | 400 |

**sieve classification.** The separation of powder into particle size ranges by the use of a series of graded sieves. Also called screen analysis.

**sieve fraction.** That portion of a powder sample that passes through a sieve of specified number and is retained by some finer mesh sieve of specified number. See also *sieve analysis*.

**sieve shaker.** A device for shaking, knocking, or vibrating a single sieve or a stack of sieves. It consists of a frame, a motorized knocker, shaker or vibrator, and fasteners for the sieve(s).

**sieve underside.** The underside of the mesh to which loose powder often adheres. To ensure accuracy of a sieve analysis, this material must be removed and included in the weight determination.

**sigma bonding.** Covalent bonding between atoms in which *s* orbitals or hybrid orbitals between *s* and *p* electrons overlap in cylindrical symmetry along the axis joining the nuclei of the atoms. See also *pi bonding*.

**sigma phase.** A hard, brittle, nonmagnetic intermediate phase with a tetragonal crystal structure, containing 30 atoms per unit cell, space group, *P4/mnm*, occurring in many binary and ternary alloys of the transition elements. The composition of this phase in the various systems is not the same, and the phase usually exhibits a wide range in homogeneity. Alloying with a third transition element usually enlarges the field homogeneity and extends it deep into the ternary section.

**sigma-phase embrittlement.** *Embrittlement* of iron-chromium alloys (most notably austenitic stainless steels) caused by precipitation at grain boundaries of the hard, brittle intermetallic *sigma phase* during long periods of exposure to temperatures between approximately 560 and 980 °C (1050 and 1800 °F). Sigma-phase embrittlement results in severe loss in *toughness* and *ductility*, and can make the embrittled material susceptible to *intergranular corrosion*. See also *sensitization*.

**signal-to-noise ratio.** (1) The ratio of the amplitude of a desired signal at any time to the amplitude of noise signals at the same time. (2) Ratio of the average response to the root-mean-square variation about the average response. Ratio of variances associated with the two parts of the performance measurement. See also *noise*.

**significance level.** The stated probability (risk) that a given test of significance will reject the hypothesis that a specified effect is absent when the hypothesis is true.

**significant.** Statistically significant. An effect of difference between populations is said to be present if the value of a test statistic is significant, that is lies outside the predetermined limits. See also *population*.

**silica (SiO$_2$).** The common oxide of silicon usually found naturally as quartz or in complex combination with other elements such as silicates. Various polymorphs and natural occurrences of silica include cristobalite, tridymite, cryptocrystalline chert, flint, chalcedony, and hydrated opal. Silica is the primary ingredient of sand, refractories, and glass.

**silica flour.** A sand additive, containing about 99.5% silica, commonly produced by pulverizing quartz sand in large ball mills to a mesh size of 80 to 325.

**silica gel.** A precipitated colloidal mass or gel of indefinitely hydrated silica; also the dried or activated product of same. Useful as dessicant, scavenger, and catalyst substrate.

**silicate-type inclusions.** Inclusions composed essentially of silicate glass, normally plastic at forging and hot-rolling temperatures, that appear in steel in the wrought condition as small elongated inclusions usually dark in color under reflected light as normally observed.

**silicon carbide.** See *Technical Brief 44*.

**silicones (SI).** Plastics based on resins in which the main polymer chain consists of alternating silicon and oxygen atoms, with carbon-containing side groups (Fig. 521). Derived from silica (sand) and methyl chlorides and fur-

FIG. 521

(a)

(b)

(c)

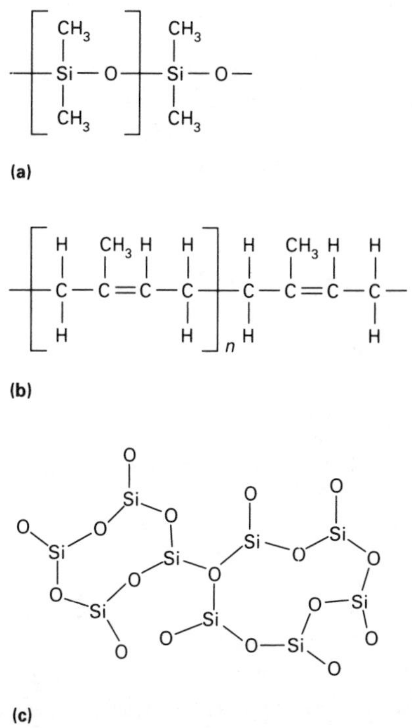

*Chemical structure comparison. (a) Silicone structure. (b) Organic natural rubber structure. (c) Silicon dioxide glass*

nished in different molecular weights, including liquids, solid resins, and elastomers.

**silicon nitride.** See *Technical Brief 45*.

**siliconizing.** Diffusing silicon into solid metal, usually low-carbon steels, at an elevated temperature in order to improve corrosion or wear resistance.

**silky fracture.** A metal fracture in which the broken metal surface has a fine texture, usually dull in appearance. Characteristic of tough and strong metals. Contrast with *crystalline fracture* and *granular fracture*.

**silver soldering.** Nonpreferred term used to denote brazing with a silver-base filler metal. See preferred terms *furnace brazing*, *induction brazing*, and *torch brazing*.

**single-action press.** A metal forming press that provides pressure from one side.

**single-bevel groove weld.** A groove weld in which the joint edge of one member is beveled from one side (Fig. 522).

FIG. 522

Single-bevel groove weld        Single-J groove weld

*Examples of single-bevel and single-J groove welds*

## Technical Brief 44: Silicon Carbide

SILICON CARBIDE (SiC) is a ceramic material that has been in existence for decades, but which has only recently found many applications as an advanced (structural) ceramic. Silicon carbide powder was first produced by E.G. Acheson in the late 1800s. This process, which involves the reaction of a mixture of sand (silica, $SiO_2$) and coke (carbon) in an electric furnace, is still in use today.

In terms of the fabrication of SiC, there are actually three families. The first two are known as direct-sintered SiC and reaction-bonded SiC (also referred to as siliconized SiC). In direct-sintered SiC, submicrometer SiC powder is compacted and sintered at temperatures in excess of 2000 °C (3600 °F), resulting in a high-purity product. Reaction-bonded SiC is processed by forming a porous shape comprised of SiC and carbon powder particles. The shape is then infiltrated with silicon, which acts to bond the SiC particles.

The properties of these two families of SiC are similar in some ways and quite different in others. Both materials have very high hardnesses (27 GPa, or $3.9 \times 10^6$ psi), high thermal conductivities (typically 110 W/m·K), and high strengths (500 MPa, or 73 ksi). However, the fracture toughness of both materials is generally low, on the order of 3 to 4 MPa√m (2.7 to 3.9 ksi√in.). The major differences are found in wear and corrosion resistance. While both are good in each category, direct-sintered SiC has a greater ability to withstand severely corrosive and erosive environments (the limiting factor for reaction-bonded SiC is the silicon metalloid).

The third fabrication family includes SiC fibers that are used to reinforce metal-matrix (most notably aluminum-base and titanium-base alloys) and ceramic-matrix (primarily aluminum oxide, $Al_2O_3$, and silicon nitride, $Si_3N_4$) composites. Two different materials are available. First, SiC filaments of 100 to 150 µm thickness can be prepared by chemical vapor deposition onto tungsten or carbon monofilaments of 40 µm thickness, which act as substrates. Second, fibers with a smaller diameter (10 to 30 µm) and without a central core can be synthesized by melt spinning and heat treatment of organosilicon polymers.

The most common applications for SiC include its use as an abrasive for grinding and cutting operations and as a refractory brick for furnace linings. Other applications include wear parts (rotary seals, components for wire drawing machinery, and mineral processing equipment), heat exchangers (process heat, power generation, and heat recovery), automotive water pump seals, and radiant tubes for metal heat treating furnaces. Silicon carbide is also a semiconducting material with properties similar to silicon, yet it retains its semiconductive properties at elevated temperatures. A number of important electronic ceramics are used in the form of SiC thin films.

### Property comparison of SiC and SiC-reinforced composites

| Product | Density, g/cm³ | Modulus of rupture | | Fracture toughness, MPa√m | Knoop hardness, kg/mm² | Elastic modulus | | Thermal expansion from 25 to 1000 °C (77 to 1830 °F) | | Maximum thermal shock | | Dry erosion resistance test(a); volume lost, cm³/h | Slurry erosion resistance(b) | Sliding abrasion(c), mm³/h |
|---|---|---|---|---|---|---|---|---|---|---|---|---|---|---|
| | | MPa | ksi | | | GPa | psi × 10⁶ | ppm/°C | ppm/°F | Δ°C | Δ°F | | | |
| Silicon nitride-bonded SiC(d) | 2.54 | 48 | 7 | 1.5 | 620 | 152 | 22 | 3.9 | 2.2 | 400 | 720 | 2.48 | 9.0 | 3700 |
| Reaction-bonded SiC(d) | 3.09 | 280 | 40 | 4.93 | 1880 | 380 | 55 | 5.0 | 2.8 | 350 | 630 | 0.10 | 40.0 | 1600 |
| Sintered SiC(d) | 3.10 | 460 | 67 | 4.6 | 2800 | 410 | 59 | 4.0 | 2.2 | 325 | 585 | 0.10 | 91.0 | 1800 |
| SiC ceramic/metal composites(e) | 3.26 | 90 | 13 | 5.5 | 800 | 313 | 45 | 5.4 | 3.0 | 600 | 1080 | 0.20 | 45.0 | 1500 |
| | 3.28 | 140 | 20 | 6.0 | 800 | 313 | 45 | 5.4 | 3.0 | 600 | 1080 | 0.30 | 25.0 | 2300 |

(a) Grit blast testing with a stream of 150-300 µm silica particles at a pressure of 275 KPa (40 psi) at a stationary specimen angle of 30° for 5 min. (b) Values listed are a ratio of volume loss for a reference pin (96% alumina) divided by volume loss of test specimen. Test pins rotated in an aqueous slurry of 40 wt% silica (300-600 µm) at 1750 rev/min for 20 h. (c) Material to be evaluated (wear block) is affixed to a mechanical arm and then immersed in a tray containing a 50% solids silica slurry (200-300 µm). With a 5 lbf load added, the wear block is then slid through the abrasive slurry at 0.16 m/s (0.5 ft/s) via the reciprocating motion of the mechanical arm. The weight loss on the wear block is used to calculate the wear rate in terms of volume loss per unit time. (d) Source: The Carborundum Company. (e) Source: Alanx Products L.P.

### Recommended Reading

- *Engineered Materials Handbook*, Vol 4, *Ceramics and Glasses*, ASM International, 1991
- *Engineered Materials Handbook*, Vol 1, *Composites*, ASM International, 1987

FIG. 523

*Comparison of microstructures in (from left) equiaxed, directionally solidified, and single-crystal blades*

**single-circuit winding**. In forming of plastics and composites, a winding in which the filament path makes a complete traverse of the chamber, after which the following traverse lies immediately adjacent to the previous one. See also *filament winding*.

**single crystal**. A material that is completely composed of a regular array of atoms.

**single-crystal superalloys**. Nickel-base alloys that contain a single crystal, or more accurately, a single grain or primary dendrite (Fig. 523). Because these materials have no grain boundaries, they exhibit improved high-temperature properties and corrosion resistance. Table 61 lists compositions of commercial single-crystal superalloys.

**single-impulse welding**. Spot, projection, or upset welding by a single impulse of current. Where alternating current is used, an impulse may be any fraction or number of cycles.

**single-J groove weld**. A groove weld in which the joint edge of one member is prepared in the form of a J, from one side. See also the figure accompanying the term *single-bevel groove weld*.

**single-lap shear specimen**. In adhesive testing, a specimen made by bonding the overlapped edges of two sheets or strips of material (Fig. 524). In testing, a single-lap specimen is usually loaded in tension at the ends, thereby creating shear stresses at the joint interface.

**single-point tool**. See definition of nomenclature in Fig. 525.

**single-port nozzle** (plasma arc welding and cutting). A constricting nozzle containing one orifice, located below and concentric with the electrode. See also the figure accompanying the term *multiport nozzle*.

**Table 61  Compositions of single-crystal superalloys**

| Alloy | Nominal composition, wt% | | | | | | | | | | | Density, g/cm$^3$ |
| | Cr | Co | Mo | W | Ta | V | Nb | Al | Ti | Hf | Ni | |
|---|---|---|---|---|---|---|---|---|---|---|---|---|
| PWA 1480 | 10 | 5 | ⋯ | 4 | 12 | ⋯ | ⋯ | 5.0 | 1.5 | ⋯ | bal | 8.70 |
| René N-4 | 9 | 8 | 2 | 6 | 4 | ⋯ | 0.5 | 3.7 | 4.2 | ⋯ | bal | 8.56 |
| SRR 99 | 8 | 5 | ⋯ | 10 | 3 | ⋯ | ⋯ | 5.5 | 2.2 | ⋯ | bal | 8.56 |
| RR 2000 | 10 | 15 | 3 | ⋯ | ⋯ | 1 | ⋯ | 5.5 | 4.0 | ⋯ | bal | 7.87 |
| AM1 | 7 | 8 | 2 | 5 | 8 | ⋯ | 1 | 5.0 | 1.8 | ⋯ | bal | 8.59 |
| CMSX-2 | 8 | 5 | 0.6 | 8 | 6 | ⋯ | ⋯ | 5.6 | 1.0 | ⋯ | bal | 8.56 |
| CMSX-3 | 8 | 5 | 0.6 | 8 | 6 | ⋯ | ⋯ | 5.6 | 1.0 | 0.1 | bal | 8.56 |
| CMSX-6 | 10 | 5 | 3 | ⋯ | 2 | ⋯ | ⋯ | 4.8 | 4.7 | 0.1 | bal | 7.98 |

## Technical Brief 45: Silicon Nitride

SILICON NITRIDE ($Si_3N_4$) has been known as a chemical compound for more than a century. The development of a range of ceramic materials based on $Si_3N_4$ began in the 1960s as a result of a search for new materials that combined high-temperature mechanical properties and resistance to oxidation and thermal shock. Interest in $Si_3N_4$ ceramics has intensified in recent years, since the realization that they are suitable substitutes for high-temperature metal alloys in gas-turbine engines. Such engines would have the capacity to operate with higher gas-inlet temperatures than are permitted in conventional metal engines, giving associated increases in efficiency and higher power-to-weight ratios.

Formation of $Si_3N_4$ is normally carried out by the direct reaction of silicon powder with nitrogen at temperatures above ~1200 °C (2190 °F). The resulting $Si_3N_4$ powder can subsequently be densified by various methods to yield a range of products, each one named after the fabrication route employed. The major types are reaction-bonded $Si_3N_4$ (RBSN), hot-pressed $Si_3N_4$ (HPSN), sintered (pressureless) $Si_3N_4$ (SSN), sintered reaction-bonded $Si_3N_4$ (SRBSN), and hot isostatically pressed $Si_3N_4$ (HIPSN). The latter four fabrication methods are used to make dense ceramic products. The properties of $Si_3N_4$-base ceramics must always be considered in relation to the processing route, the densification (sintering) additives, and the resulting microstructure.

RBSN is made by first producing the required shape from silicon powder. The shaped powder form is then nitrided in molecular nitrogen, starting at 1150 °C (2100 °F) and slowly increasing the temperature to over 1400 °C (2550 °F). The resulting product has a density of 12 to 30%. HPSN is formed by the application of both heat and uniaxial pressure in graphite dies heated by induction to temperatures in the range of 1650 to 1850 °C (3000 to 3360 °F) for 1 to 4 h under an applied stress of 15 to 30 MPa (2175 to 4350 psi). Boron nitride is applied as a coating to the graphite die and plungers to prevent reaction of these with the $Si_3N_4$ and to act as a high-temperature solid lubricant to facilitate removal of the hot-pressed material from the die. SSN involves firing the shaped component at 1700 to 1800 °C (3100 to 3275 °F) under a nitrogen atmosphere at 0.1 MPa (14.5 psi). As with hot pressing, additives provide conditions for liquid-phase densification. SRBSN combines the technology of both RBSN and SSN. Additives such as magnesia (MgO) and yttria ($Y_2O_3$) are mixed with the silicon prior to shaping and then nitriding is carried out as for RBSN. A further firing in the range of 1800 to 2000 °C (3275 to 3630 °F) under a nitrogen atmosphere (0.1 to 8 MPa, or 14.5 to 1160 psi), using a protective powder bed to reduce volatilization, allows densification to 98% of theoretical density with only 6% linear

FIG. 524

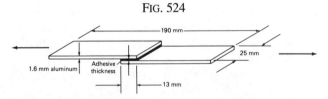

*Standard single-lap shear test for adhesive bonds*

**single relief angle.** Defined by the figure accompanying the term single-point tool.

**single spread.** Application of adhesive to only one adherend of a joint.

**single-stand mill.** A rolling mill designed such that the product contacts only two rolls at a given moment. Contrast with *tandem mill*.

**single-U groove weld.** A groove weld in which each joint edge is prepared in the form of a J or half-U from one side (Fig. 526).

FIG. 525

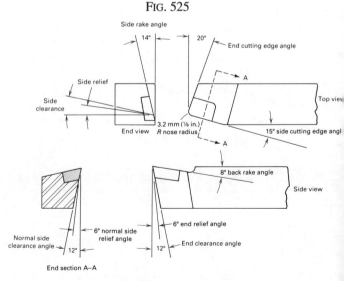

*Single-point tool nomenclature*

shrinkage. The highest $Si_3N_4$ strength levels are obtained by hot isostatic pressing. The components are placed in an autoclave and subjected to high temperature and high pressure using argon or nitrogen as the pressure transmission medium to consolidate either a shaped powder compact or to remove porosity from prefired RBSN, SSN, or SRBSN. In all cases, a small amount of sintering additive is required.

Because of the interest of the automotive industry in $Si_3N_4$, much of the development work during the past few decades has evolved around the use of $Si_3N_4$ for various engine components. The automotive components of interest are turbocharger rotors, pistons, piston liners, and valves. The greatest application of $Si_3N_4$, however, is as cutting-tool materials in metal-machining applications, where machining rates can be dramatically increased by the high-temperature strength of $Si_3N_4$. See also *Sialons* (Technical Brief 43).

### Mechanical properties of silicon nitride ceramics

| Property | Material type | | | | | | | |
|---|---|---|---|---|---|---|---|---|
| | RBSN | HPSN | SSN | SRBSN | HIP-SN | HIP-RBSN | HIP-SSN | Sialon |
| Young's modulus ($E$), GPa ($10^6$ psi) | 120–250 (17.4–36.2) | 310–330 (44.9–47.8) | 260–320 (37.7–46.4) | 280–300 (40.6–43.5) | . . . | 310–330 (44.9–47.8) | . . . | 300 (43.5) |
| Poisson's ratio ($\nu$) | 0.20 | 0.27 | 0.25 | 0.23 | 0.23– | 0.27 | . . . | 0.23 |
| Flexural strength ($\sigma_f$), MPa (ksi) at: | | | | | | | | |
| 25 °C (77 °F) | 150–350 (21.7–50.7) | 450–1000 (65.2–145) | 600–1200 (86.9–173.8) | 500–800 (72.5–115.9) | 600–1200 (86.9–173.8) | 500–800 (72.5–115.9) | 600–1200 (86.9–173.8) | 750–95 (108.7–137.7) |
| 1350 °C (2460 °F) | 140–340 (20.2–49.3) | 250–450 (36.2–65.2) | 340–550 (49.3–79.7) | 350–450 (50.7–65.2) | 350–550 (50.7–79.7) | 250–450 (36.2–65.2) | 300–520 (43.5–75.3) | 300–550 (43.5–79.7) |
| Weibull modulus ($m$) | 19–40 | 15–30 | 10–25 | 10–20 | . . . | 20–30 | . . . | 15 |
| Fracture toughness ($K_{Ic}$), MPa$\sqrt{m}$ (ksi$\sqrt{in.}$) | 1.5–2.8 (1.3–2.5) | 4.2–7.0 (3.8–6.3) | 5.0–8.5 (4.5–7.7) | 5.0–5.5 (4.5–5.0) | 4.2–7.0 (3.8–6.3) | 2.0–5.8 (1.8–5.3) | 4.0–8.0 (3.6–7.2) | 6.0–8.0 (5.4–7.2) |

## Recommended Reading

- S. Hampshire, Engineering Properties of Nitrides, *Engineered Materials Handbook*, Vol 4, ASM International, 1991, p 812-820
- R. Komanduri and S.K. Samanta, Ceramic Cutting Tools, *Metals Handbook*, 9th ed., Vol 16, ASM International, 1989, p 98-104

FIG. 526

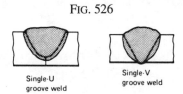

Single-U groove weld

Single-V groove weld

*Examples of single-U and single-V groove welds*

**single-V groove weld.** A groove weld in which each member is beveled from the same side (Fig. 526).

**single welded joint.** In arc and gas welding, any joint welded from one side only.

**sinkhead.** Same as *riser*.

**sinking.** (1) The operation of machining the impression of a desired forging into die blocks. (2) See *tube sinking*.

**sink mark.** A shallow depression on the surface of an injection-molded plastic part due to the collapsing of the surface following local internal shrinkage after the gate seals.

**sinter.** To densify, crystallize, bond together, and/or stabilize a particulate material, agglomerate, or product by heating or firing close to but below the melting point (Fig. 527). Often involves melting of minor components or constituents, and/or chemical reaction. Also, the product of such firing.

**sintered density.** The quotient of the mass (weight) over the volume of the sintered body expressed in grams per cubic centimeter.

**sintered density ratio.** The ratio of the density of the sintered body to the solid, pore-free body of the same composition or theoretical density.

**sintering.** The bonding of adjacent surfaces of particles in a mass of powder or a compact by heating (Fig. 527). Sintering strengthens a powder mass and normally produces densification and, in powdered metals, recrystallization. See also *liquid phase sintering* and *solid-state sintering*.

**sintering atmospheres.** See *protective atmosphere* (2).

FIG. 527

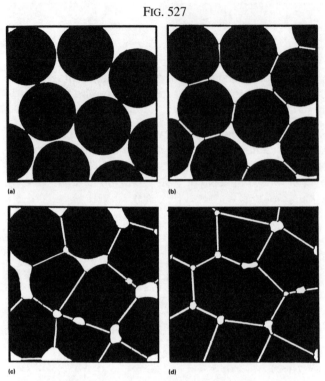

(a)                                     (b)

(c)                                     (d)

*Development of the interparticle bond as a ceramic microstructure is transformed during the sintering process. (a) Loose powder (start of bond growth). (b) Initial stage (the pore volume shrinks). (c) Intermediate stage (grain boundaries form at the contacts). (d) Final stage (pores become smoother)*

FIG. 528

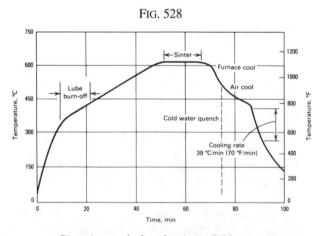

*Sintering cycle for aluminum P/M parts*

**sintering cycle.** A predetermined and closely controlled time-temperature regime for sintering compacts, including the heating and cooling phases (Fig. 528).

**sintering temperature.** The maximum temperature at which a powder compact is sintered. The temperature is either measured directly on the surface of the body by optical pyrometer, or indirectly by thermocouples installed in the furnace chamber.

**sintering time.** The time period during which a powder compact is at sintering temperature.

**sintrate.** In powder metallurgy, controlled heating so that a compact is sintered before the melting point of an infiltrating material is reached. See also *infiltration*.

**size.** In composites manufacturing, a treatment consisting of starch, gelatin, oil, wax, or other suitable ingredients applied to yarn or fibers at the time of formation to protect the surface and aid the process of handling and fabrication or to control the fiber characteristics. The treatment contains ingredients that provide surface lubricity and binding action, but unlike a finish, contains no coupling agent. Before final fabrication into a composite, the size is usually removed by heat cleaning, and a finish is applied.

**size effect.** Effect of the dimensions of a piece of metal on its mechanical and other properties and on manufacturing variables such as forging reduction and heat treatment. In general, the mechanical properties are lower for a larger size.

**size-exclusion chromatography (SEC).** *Liquid chromatography* method that separates molecules on the basis of their physical size. This technique is most often used in the analysis of polymers. Also termed gel permeation chromatography.

**size fraction.** A separated fraction of a powder whose particles lie between specified upper and lower size limits. See also *sieve analysis*.

**size of weld.** (1) The joint penetration in a groove weld (Fig. 529). (2) The lengths of the nominal legs of a fillet weld (Fig. 529). (3) The weld metal thickness measured at the root of a flange weld.

**sizing (adhesives).** The process of applying a material on a surface in order to fill pores and thus reduce the absorption of the subsequently applied adhesive or coating or to otherwise modify the surface properties of the substrate to

FIG. 529

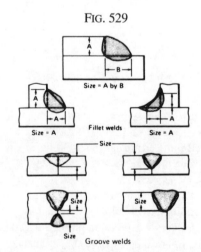

*Size of weld in fillet and groove welds*

improve the adhesion, and also, the material used for this purpose. See also *primer*.

**sizing** (composites). See *size*.

**sizing** (metals). (1) Secondary forming or squeezing operations needed to square up, set down, flatten, or otherwise correct surfaces to produce specified dimensions and tolerances. See also *restriking*. (2) Some burnishing, broaching, drawing, and shaving operations are also called sizing. (3) A finishing operation for correcting ovality in tubing. (4) Final pressing of a sintered powder metallurgy part to obtain a desired dimension.

**sizing content** (composites). The percent of the total strand weight made up of sizing, usually determined by burning off or dissolving the organic sizing. Also known as loss on ignition. See also *size*.

**sizing die**. A die used for the *sizing* of a sintered compact.

**sizing punch**. A punch used for the pressing of a sintered compact during the *sizing operation*.

**sizing knockout**. An ejector punch used for ejecting a sintered compact from one *sizing die*.

**sizing stripper**. A punch used during the *sizing* operation.

**skein**. A continuous filament, strand, yarn, or roving for reinforced plastics, wound up to some measurable length and usually used to measure various physical properties.

**skeleton**. An unsintered or sintered porous powder metallurgy compact with a large proportion of interconnected porosity that makes it suitable for *infiltration*.

**skelp**. The starting stock for making welded pipe or tubing; most often it is strip stock of suitable width, thickness, and edge configuration.

**skidding**. A form of nonuniform relative motion between solid surfaces due to rapid periodic changes in the traction between those surfaces.

**skid-polishing process**. A mechanical polishing process in which the surface of the metallographic specimen to be polished is made to skid across a layer of paste, consisting of the abrasive and the polishing fluid, without contacting the fibers of the polishing cloth.

**skim gate**. In foundry practice, a gating arrangement designed to prevent the passage of slag and other undesirable materials into a casting.

**skimmer**. A tool for measuring scum, slag, and dross from the surface of molten metal.

**skimming**. Removing or holding back dirt or slag from the surface of the molten metal before or during pouring.

**skin** (metals). A thin outside metal layer, not formed by bonding as in cladding or electroplating, that differs in composition, structure, or other characteristics from the main mass of metal.

**skin** (plastics). The relatively dense material that sometimes forms on the surface of a cellular plastic or sandwich construction.

FIG. 530

*Aluminum-killed, hot-rolled 1008 steel, with an open skin lamination that appeared on the surface after rolling. 2×*

**skin drying**. Drying the surface of a foundry mold by direct application of heat.

**skin lamination**. In flat-rolled metals, a surface rupture resulting from the exposure of a subsurface lamination by rolling (Fig. 530).

**skin pass**. See *temper rolling*.

**skip weld**. See preferred term *intermittent weld*.

**skiving**. (1) Removal of a material in thin layers or chips with a high degree of shear or slippage, or both, of the cutting tool. (2) A machining operation in which the cut is made with a form tool with its face so angled that the cutting edge progresses from one end of the work to the other as the tool feeds tangentially past the rotating workpiece.

**skull**. (1) A layer of solidified metal or dross on the walls of a pouring vessel after the metal has been poured. (2) The unmelted residue from a liquated weld filler metal.

**slab**. A flat-shaped semifinished rolled metal ingot with a width not less than 250 mm (10 in.) and a cross-sectional area not less than 105 cm$^2$ (16 in.$^2$).

**slabbing**. The hot working of an *ingot* to a flat rectangular shape.

**slabbing mill**. A primary mill that produces slabs.

**slab milling**. See preferred term *peripheral milling*.

**slack quenching**. The incomplete hardening of steel due to quenching from the austenitizing temperature at a rate slower than the critical cooling rate for the particular steel, resulting in the formation of one or more transformation products in addition to martensite.

**slag**. A nonmetallic product resulting from the mutual dissolution of flux and nonmetallic impurities in smelting, refining, and certain welding operations (see, for example, *electroslag welding*). In steelmaking operations, the slag

FIG. 531

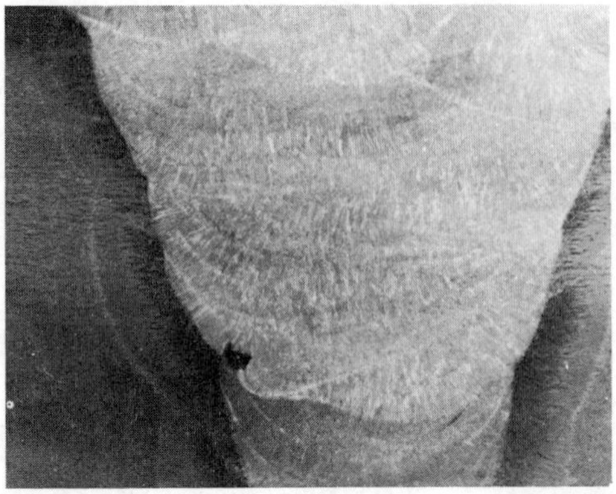

*Slag inclusion (lower left side of weld) in a flux cored arc welded steel plate. 3×*

serves to protect the molten metal from the air and to extract certain impurities.

**slag inclusion**. (1) Slag or dross entrapped in a metal. (2) Nonmetallic solid material entrapped in weld metal or between weld metal and base metal (Fig. 531).

**slant fracture**. A type of fracture in metals, typical of *plane-stress* fractures, in which the plane of separation is inclined at an angle (usually about 45°) to the axis of applied stress.

**sleeve bearing**. A cylindrical plain bearing used to provide radial location for a shaft, which moves axially. Sleeve bearings usually consist of one or more layers of bearing alloy(s), or liner, bonded to a steel backing (Fig. 532). Sleeve bearing is sometimes used to denote *journal bearing*. See also *sliding bearing*.

**slice**. The cross-sectional plane through an object that is scanned to produce the image in *computed tomography*. See also *tomographic plane*.

FIG. 532

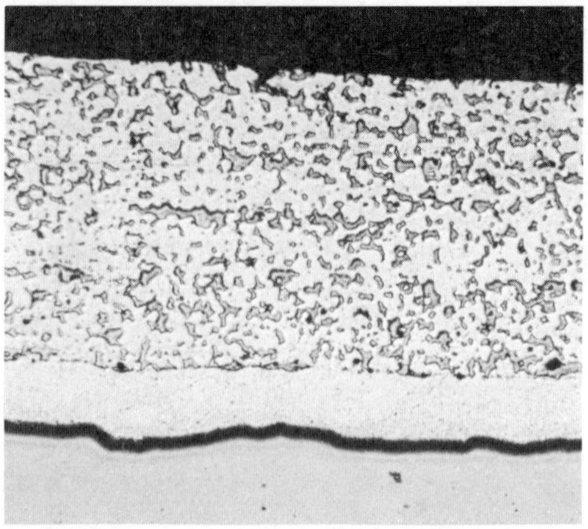

*Tri-metal sleeve bearing consisting of an aluminum-tin alloy strip clad to an unalloyed aluminum bonding layer (center) by warm rolling. The composite strip was subsequently clad to a steel backing strip (bottom) by warm rolling. 100×*

**slices**. Sections of ingots of single-crystal material that have been sawed from the ingot (Fig. 533). Also known as wafers.

**slide**. The main reciprocating member of a metal forming press, guided in the press frame, to which the punch or upper die is fastened; sometimes called the *ram*. The inner slide of a double-action press is called the plunger or punch-holder slide; the outer slide is called the blank-

FIG. 533

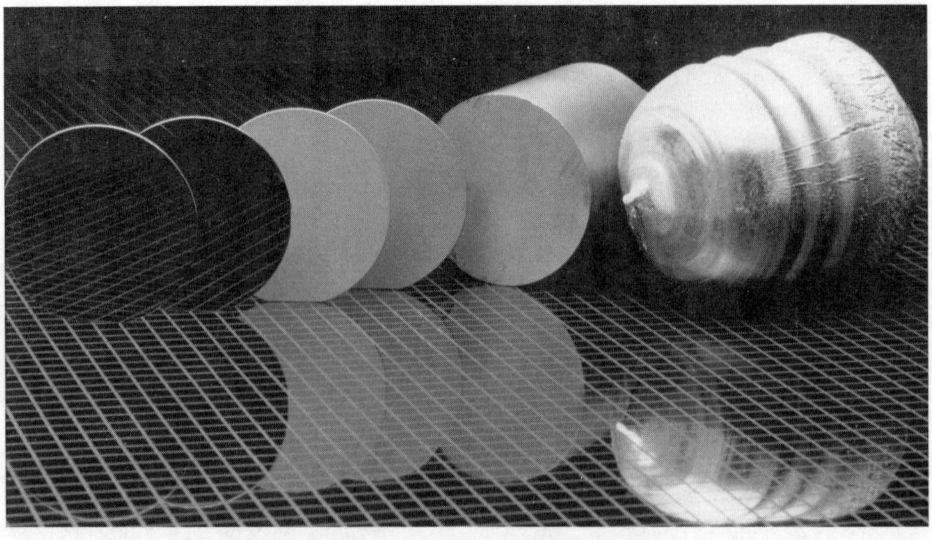

*Slices (wafers) of a single-crystal gallium arsenide ingot*

FIG. 534

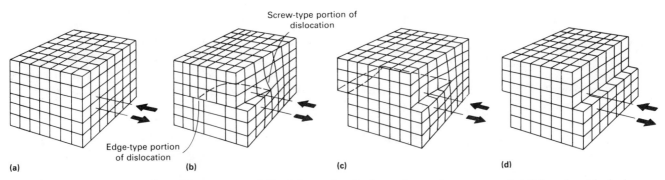

*Schematic representation of four stages of slip deformation by formation and movement of a dislocation (dashed line) through a crystal. (a) Crystal before displacement. (b) Crystal after some displacement. (c) Complete displacement across part of crystal. (d) Complete displacement across entire crystal*

holder slide. The third slide of a triple-action press is called the lower slide, and the slide of a hydraulic press is often called the platen. See also the figures accompanying the terms *press forming* and *straight-side press*.

**slide adjustment.** The distance that a metal forming press slide position can be altered to change the shut height of the die space. The adjustment can be made by hand or by power mechanism.

**slide-roll ratio.** See *slide-sweep ratio*.

**slide-sweep ratio.** The ratio of sliding velocity to sweep velocity, for example, in a pair of gears. In rolling, the slide-sweep ratio is called the slide-roll ratio.

**sliding** (pure sliding with no rolling or spin). A motion of two relatively moving bodies, in which their surface velocities in the common contact area are different with regard to magnitude and/or direction. See also *rolling*, *spin*, and *specific sliding*.

**sliding bearing.** A bearing in which predominantly sliding contact occurs between relatively moving surfaces. Sliding bearings may be either unlubricated, liquid lubricated, grease lubricated, or solid lubricated. See also *sleeve bearing*.

**sliding fit.** A loosely defined fit similar to a *slip fit*.

**sliding velocity.** The difference between the velocities of each of the two surfaces relative to the point of contact.

**slime.** (1) A material of extremely fine particle size encountered in ore treatment. (2) A mixture of metals and some insoluble compounds that forms on the anode in electrolysis.

**slip** (adhesives). In adhesively bonded components or specimens, the relative collinear displacement of the adherends on either side of the adhesive layer in the direction of the applied load.

**slip** (ceramics). (1) A slurry or suspension of fine clay or other ceramic powders in water, having the consistency of cream, that is used in slip casting or as a cement or glaze

preparation. (2) A suspension of colloidal powder in an immiscible liquid (usually water).

**slip** (metals). *Plastic deformation* by the irreversible shear displacement (translation) of one part of a crystal relative to another in a definite crystallographic direction and usually on specific crystallographic plane (Fig. 534). Sometimes called glide.

**slip angle.** The angle at which a tensioned fiber will slide off a filament-wound dome. If the difference between the wind angle and the geodesic angle is less than the slip angle, the fiber will not slide off the dome. Slip angles for different fiber-resin systems vary and must be determined experimentally. See also *filament winding* and *wind angle*.

**slip band.** A group of parallel slip lines so closely spaced as to appear as a single line when observed under an optical microscope (Fig. 535). See also *slip line*.

FIG. 535

*Slip bands on two planes in a single crystal of Co-8Fe alloy that was polished, then plastically deformed. 250×*

FIG. 536

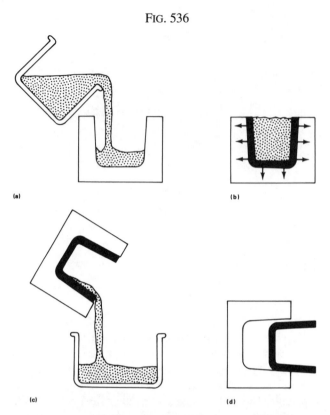

(a)

(b)

(c)

(d)

*Schematic of slip casting operation. (a) Permeable mold is filled with slip. (b) Liquid is extracted from the mold, while forming compacts along mold walls. (c) Excess slip is drained. (d) Casting is removed after partial drying.*

**slip casting**. The ceramic forming process consisting of filling or coating a porous mold with a *slip*, allowing to dry, and removing for subsequent firing (Fig. 536).

**slip coating**. A ceramic material or mixture other than a glaze, applied to a ceramic body and fired to the maturity required to develop specified characteristics.

**slip crack**. See *pressing crack*.

**slip direction**. The crystallographic direction in which the translation of slip takes place. See also the figure accompanying the term *slip* (metals).

**slip fit**. A loosely defined *clearance fit* between parts assembled by hand without force, but implying slipping contact.

**slip flask**. A tapered *flask* that depends on a movable strip of metal to hold foundry sand in position. After closing the mold, the strip is refracted and the flask can be removed and reused. Molds thus made are usually supported by a *mold jacket* during pouring.

**slip forming**. A plastic sheet-forming technique in which some of the sheet material is allowed to slip through the mechanically operated clamping rings during a stretch-forming operation.

**slip-interference theory**. Theory involving the resistance to deformation offered by a hard phase dispersed in a ductile matrix.

**slip line**. Visible traces of slip planes on metal surfaces; the traces are (usually) observable only if the surface has been polished before deformation (Fig. 537). The usual observation on metal crystals (under a light microscope) is of a cluster of slip lines known as a *slip band*.

**slippage**. The movement of adherends with respect to each other during the adhesive bonding process.

**slip plane**. The crystallographic plane in which *slip* occurs in a crystal.

**slitting**. Cutting or shearing along single lines to cut strips from a metal sheet or to cut along lines of a given length or contour in a sheet or workpiece (Fig. 538).

**sliver** (composites). A number of staple or continuous-filament fibers aligned in a continuous strand without twist. See also *continuous filament yarn*, *staple fiber*, and *strand*.

**sliver** (metals). An imperfection consisting of a very thin elongated piece of metal attached by only one end to the parent metal into whose surface it has been worked (Fig. 539).

FIG. 537

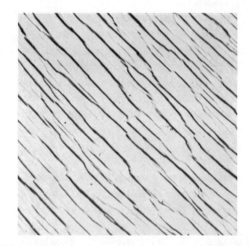

*Wavy slip lines in a single crystal of aluminum that was polished, then plastically deformed. 250×*

FIG. 538

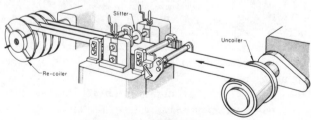

*Typical coil-slitting line*

FIG. 539

*Cold-rolled 1008 steel sheet. Sliver, the result of an ingot scab, is partially welded to the surface. Actual size*

**slope control**. Producing electronically a gradual increase or decrease in the welding current between definite limits and within a selected time interval.

**slot extrusion**. A method of extruding plastic film or sheet in which the molten thermoplastic compound is forced through a slot die (T-shaped or coat hanger shape). Following extrusion, the film or sheet is cooled by passing it through a water bath or over water-cooled rolls (Fig. 540).

**slot furnace**. A common batch furnace for heat treating metals where stock is charged and removed through a slot or opening.

**slotting**. Cutting a narrow aperture or groove with a reciprocating tool in a vertical shaper or with a cutter, broach, or grinding wheel.

**slot weld**. A weld made in an elongated hole in one member of a lap or T-joint joining that member to that portion of the surface of the other member which is exposed through the hole. The hole may be open at one end and may be partially or completely filled with weld metal. A fillet welded slot should not be construed as conforming to this definition.

**slow strain rate technique**. An experimental technique for evaluating susceptibility to *stress-corrosion cracking*. It involves pulling the specimen to failure in uniaxial tension at a controlled slow strain rate while the specimen is in the test environment and examining the specimen for evidence of stress-corrosion cracking (Fig. 541).

**sludge**. A coagulated mass, often containing foreign matter, formed at low temperature in combustion engines from oil oxidation residues, carbon, and water.

**sluffing**. An occurrence during pultrusion of reinforced plastics in which scales peel off or become loose, either partially or entirely, from a pultrusion. Not to be confused with scraping, prying, or physically removing scale from a pultrusion. Sluffing is sometimes spelled sloughing. See also *pultrusion*.

**slug**. (1) A short piece of metal to be placed in a die for forging or extrusion. (2) A small piece of material produced by piercing a hole in sheet material. See also *blank*.

FIG. 541

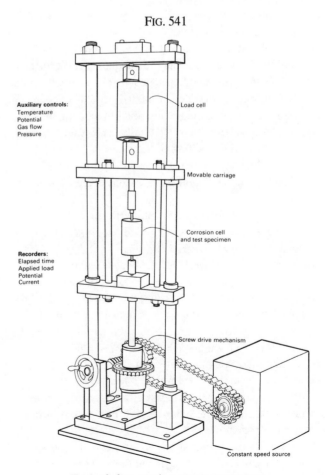

*Typical slow strain rate test apparatus*

FIG. 540

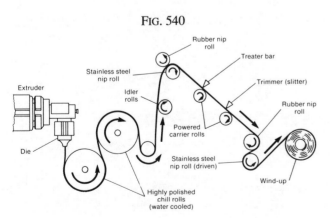

*Slot extrusion method for processing plastic sheet or film featuring chromium-plated chill rollers*

**slugging.** The unsound practice of adding a separate piece of material in a joint before or during welding, resulting in a welded joint in which the weld zone is not entirely built up by adding molten filler metal or by melting and recasting base metal, and which therefore does not comply with design, drawing, or specification requirements.

**slumpability.** The flow of gravity of a grease in a container, allowing it to feed out into a pump or can. Slumpability also influences the leakage of grease from a bearing.

**slurry.** (1) A thick mixture of liquid and solids, the solids being in suspension in the liquid. (2) Any pourable or pumpable suspension of a high content of insoluble particulate solids in a liquid medium, most often water. See also *slip*.

**slurry abrasion response (SAR) number.** A measure of the relative abrasion response of any material in any slurry, as related to the instantaneous rate of mass loss of a specimen at a specific time on the cumulative abrasion-corrosion time curve, converted to volume or thickness loss rate. See also *Miller number*.

**slurry abrasivity.** The relative tendency of a particular moving slurry to produce abrasive and corrosive wear compared with other slurries.

**slurry erosion.** Erosion produced by the movement of a slurry past a solid surface.

**slurry preforming.** Method of preparing reinforced plastic preforms by wet processing techniques similar to those used in the pulp molding industry. For example, glass fibers suspended in water are passed through a screen that passes the water but retains the fibers in the form of a mat.

**slush casting.** A hollow casting usually made of an alloy with a low but wide melting temperature range. After the desired thickness of metal has solidified in the mold, the remaining liquid is poured out. Considered an obsolete practice.

**slushing compound.** An obsolete term for describing oil or grease coatings used to provide temporary protection against *atmospheric corrosion*.

**slushing oil.** A mineral oil containing additives that enable it to protect the parts of a machine against rusting.

**slush molding.** Method for casting thermoplastics, in which the resin in liquid form is poured into a hot mold where a viscous skin forms. The excess slush is drained off, the mold is cooled, and the molding stripped out.

**SMC.** See *sheet molding compound*.

**smearing.** Mechanical removal of material from a surface, usually involving plastic shear deformation, and redeposition of the material as a thin layer on one or both surfaces. See also *transfer*.

**smelting.** Thermal processing wherein chemical reactions take place to produce liquid metal from a beneficiated ore.

**smith forging.** See *hand forge (smith forge)*.

**smut.** A reaction product sometimes left on the surface of a metal after pickling, electroplating, or etching.

**snagging.** (1) Heavy stock removal of superfluous material from a workpiece by using a portable or swing grinder mounted with a coarse grain abrasive wheel. (2) *Offhand grinding* on castings and forgings to remove surplus metal such as gate and riser pads, fins, and parting lines.

**snake.** (1) The product formed by twisting and bending of hot metal rod prior to its next rolling process. (2) Any crooked surface imperfection in a plate, resembling a snake. (3) A flexible mandrel used in the inside of a shape to prevent flattening or collapse during a bending operation.

**snaky edges.** See *carbon edges*.

**snap flask.** A foundry flask hinged on one corner so that it can be opened and removed from the mold for reuse before the metal is poured.

**snap temper.** A precautionary interim stress-relieving treatment applied to high-hardenability steels immediately after quenching to prevent cracking because of delay in tempering them at the prescribed higher temperature.

**S-N curve.** A plot of stress ($S$) against the number of cycles to failure ($N$) (Fig. 542). The stress can be the maximum stress ($S_{max}$) or the alternating stress amplitude ($S_a$). The stress values are usually nominal stress; i.e., there is no adjustment for stress concentration. The diagram indicates the $S$-$N$ relationship for a specified value of the mean stress ($S_m$) or the stress ratio ($A$ or $R$) and a specified probability of survival. For $N$ a log scale is almost always used. For $S$ a linear scale is used most often, but a log scale is sometimes used. Also known as *S-N diagram*.

**S-N curve for 50% survival.** A curve fitted to the median value of fatigue life at each of several stress levels. It is an estimate of the relationship between applied stress and the

FIG. 542

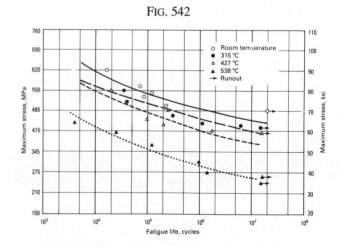

*S-N curves at various temperatures for AISI 4340 steel with an ultimate tensile strength of 1090 MPa (158 ksi). Stress ratio, R, equals –1.0.*

number of cycles-to-failure that 50% of the population would survive.

**S-N curve for p% survival.** A curve fitted to the fatigue life for *p*% survival values at each of several stress levels. It is an estimate of the relationship between applied stress and the number of cycles-to-failure that *p*% of the population would survive. *p* may be any number, such as 95, 90, etc.

**snowflakes.** See *flakes.*

**snug fit.** A loosely defined fit implying the closest clearances that can be assembled manually for firm connection between parts. See also *clearance fit.*

**soak cleaning.** *Immersion cleaning* without electrolysis.

**soaking.** In heat treating of metals, prolonged holding at a selected temperature to effect homogenization of structure or composition. See also *homogenizing.*

**soak time.** The length of time a ceramic material is held at the peak temperature of the firing cycle. See also *firing.*

**soap.** In lubrication, a compound formed by the reaction of a fatty acid with a metal or metal compound. Metallic soaps formed by reaction *in situ* are an important group of boundary lubricants.

**soda ash.** Sodium carbonate ($Na_2CO_3$) obtained from trona, a hydrated sodium carbonate sodium bicarbonate ore ($Na_2CO_3 \cdot NaHCO_3 \cdot 2H_2O$). Soda ash is used in petroleum refining and for soaps and detergents. Its primary use, however, is in glass manufacture. Soda ash is the third major constituent of soda-lime-silica glasses and is the main source of $Na_2O$ in any glass that contains soda. It acts as a flux, reducing the temperature required to melt the silica. See also *Solvay process.*

**softening point.** That temperature at which a glass fiber of uniform diameter elongates at a specific rate under its own weight when measured by standard ASTM test methods. The viscosity at the softening point depends on the density and surface tension. For example, for a glass of density 2.5 $g/cm^2$ and surface tension 300 dynes/cm, the softening point temperature corresponds to a viscosity of 106.6 Pa · s.

**softening range.** The range of temperatures within which a plastic changes from a rigid to a soft state. Actual values depend on the test method. Sometimes erroneously referred to as softening point.

**soft magnetic materials.** See *Technical Brief 46.*

**soft solder.** See preferred term *solder.*

**soft soldering.** See preferred term *soldering.*

**soft temper.** Same as *dead soft* temper.

**soft water.** Water that is free of magnesium or calcium salts.

**soil.** Undesirable material on a surface that is not an integral part of the surface. Oil, grease, and dirt can be soils; a decarburized skin and excess *hard chromium* are not soils. Loose scale is soil; hard scale may be an integral part of the surface and, hence, not soil.

**sol.** A colloidal suspension comprised of discrete or separate solid particles suspended in a liquid. Differs from a solution, though one merges into the other. Compare with *gel.*

**solder.** A filler metal used in soldering which has a liquidus not exceeding 450 °C (840 °F). The most commonly used solders are tin-lead alloys. Table 62 lists nominal compositions and properties, and uses for these alloys. Other solder alloys include tin-antimony, tin-silver, tin-zinc, cadmium-silver, cadmium-zinc, zinc-aluminum, indium-base alloys, bismuth-base alloys (*fusible alloys*), and gold-base solders.

**solderability.** The relative ease and speed with which a surface is wetted by molten solder.

## Table 62  Compositions and properties of tin-lead solders

| Composition, % | | Solidus temperature | | Liquidus temperature | | Pasty range | | Uses |
|---|---|---|---|---|---|---|---|---|
| Tin | Lead | °C | °F | °C | °F | Δ °C | Δ °F | |
| 2 | 98 | 316 | 601 | 322 | 611 | 6 | 10 | Side seams for can manufacturing |
| 5 | 95 | 305 | 581 | 312 | 594 | 7 | 13 | Coating and joining metals |
| 10 | 90 | 268 | 514 | 302 | 576 | 34 | 62 | Sealing cellular automobile radiators, filling seams or dents |
| 15 | 85 | 227 | 440 | 288 | 550 | 61 | 110 | Sealing cellular automobile radiators, filling seams or dents |
| 20 | 80 | 183 | 361 | 277 | 531 | 94 | 170 | Coating and joining metals, or filling dents or seams in automobile bodies |
| 25 | 75 | 183 | 361 | 266 | 511 | 83 | 150 | Machine and torch soldering |
| 30 | 70 | 183 | 361 | 255 | 491 | 72 | 130 | |
| 35 | 65 | 183 | 361 | 247 | 477 | 64 | 116 | General-purpose and wiping solder |
| 40 | 60 | 183 | 361 | 238 | 460 | 55 | 99 | Wiping solder for joining lead pipes and cable sheaths; also for automobile radiator cores and heating units |
| 45 | 55 | 183 | 361 | 227 | 441 | 44 | 80 | Automobile radiator cores and roofing seams |
| 50 | 50 | 183 | 361 | 216 | 421 | 33 | 60 | Most popular general-purpose solder |
| 60 | 40 | 183 | 361 | 190 | 374 | 7 | 13 | Primarily for electronic soldering applications where low soldering temperatures are required |
| 63 | 37 | 183 | 361 | 183 | 361 | 0 | 0 | Lowest-melting (eutectic) solder for electronic applications |

## Technical Brief 46: Soft Magnetic Materials

MAGNETIC MATERIALS are broadly classified into two groups with either hard or soft magnetic characteristics. Hard magnetic materials are characterized by retaining a large amount of residual magnetism after exposure to a strong magnetic field. These materials typically have coercive force, $H_c$, values of several hundred to several thousand oersteds (Oe) and are considered to be permanent magnets (see *permanent magnet materials*, Technical Brief 28). The coercive force is a measure of the magnetizing force required to reduce the magnetic induction to zero after the material has been magnetized. In contrast, soft magnetic materials become magnetized by relatively low-strength magnetic fields, and when the applied field is removed, they return to a state of relatively low residual magnetism. Soft magnetic materials typically exhibit coercive force values of approximately 5 Oe to as low as 0.002 Oe. Soft magnetic behavior is essential in any application involving changing electromagnetic induc-

### Silicon contents, mass densities, and applications of electrical steel sheet and strip

| ASTM specification | AISI type | Nominal (Si + Al) content, % | Assumed density, g/cm$^3$ | Characteristics and applications |
|---|---|---|---|---|
| **Lamination steel** | | | | |
| A 726 or A 840 . . . . . . . . . . . | . . . | 0 | 7.85 | High magnetic saturation; magnetic properties may not be guaranteed; intermittent-duty small motors |
| **Nonoriented electrical steels** | | | | |
| A 677 or A 677M (fully processed) and A 683 or A 683M (semiprocessed) . . . . . . . . | M-47 | 1.05 | 7.80 | Ductile, good stamping properties, good permeability at high inductions; small motors, ballasts, relays |
| | M-45 | 1.85 | 7.75 | Good stamping properties, good permeability at moderate and high inductions, good core loss; small generators, high-efficiency continuous-duty rotating machines, ac and dc |
| | M-43 | 2.35 | 7.70 | |
| | M-36 | 2.65 | 7.70 | Good permeability at low and moderate inductions, low core loss; high reactance cores, generators, stators of high-efficiency rotating machines |
| | M-27 | 2.80 | 7.70 | |
| | M-22(a) | 3.20 | 7.65 | Excellent permeability at low inductions, lowest core loss; small power transformers, high-efficiency rotating machines |
| | M-19(a) | 3.30 | 7.65 | |
| | M-15(a) | 3.50 | 7.65 | |
| **Oriented electrical steels** | | | | |
| A 876 or A 876 M . . . . . . . . | M-6 | 3.15 | 7.65 | Grain-oriented steel has highly directional magnetic properties with lowest core loss and highest permeability when flux path is parallel to rolling direction; heavier thicknesses used in power transformers, thinner thicknesses generally used in distribution transformers. Energy savings improve with lower core loss. |
| | M-5 | 3.15 | 7.65 | |
| | M-4 | 3.15 | 7.65 | |
| | M-3 | 3.15 | 7.65 | |
| **High-permeability oriented steel** | | | | |
| | . . . | 2.9–3.15 | 7.65 | Low core loss at high operating inductions |

(a) ASTM A 677 only

tion, such as solenoids, relays, motors, generators, transformers, magnetic shielding, and so on. Other important characteristics of magnetically soft materials include high permeability, high saturation induction, low hysteresis-energy loss, low eddy-current loss in alternating flux applications, and in specialized cases, constant permeability at low field strengths and/or a minimum or definite change in permeability with temperature.

Cost, availability, strength, corrosion resistance, and ease of processing are among the key factors that influence the final selection of a soft magnetic material. Magnetically soft materials manufactured in large quantities include high-purity irons, low-carbon ($\leq$0.08% C) steels that contain additions of phosphorus (0.03 to 0.15%) and manganese (0.25 to 0.75%) to increase electrical resistivity, silicon (electrical) steels containing 2 to 3.5% Si, iron-nickel alloys with nickel contents ranging from 45 to 79%, iron-cobalt alloys (for example, 49Fe-49Co-2V and Fe-27Co-0.6Cr), ferritic stainless steels, and ferrites (manganese-zinc and nickel-zinc in particular). Soft magnetic amorphous materials are also being produced (see *metallic glass*).

### Recommended Reading

- D.W. Dietrich, Magnetically Soft Materials, *Metals Handbook*, 10th ed., Vol 2, ASM International, 1990, p 761-781
- R. Ball, *Soft Magnetic Materials*, Heyden & Sons, 1979
- C.W. Chen, *Magnetism and Metallurgy of Soft Magnetic Materials*, North-Holland, 1977

**solder embrittlement**. Reduction in mechanical properties of a metal as a result of local penetration of solder along grain boundaries.

**soldering**. A group of processes that join metals by heating them to a suitable temperature below the solidus of the base metals and applying a filler metal having a liquidus not exceeding 450 °C (840 °F). Molten filler metal is distributed between the closely fitted surfaces of the joint by capillary action. See also *solder*.

**soldering flux**. See *flux* (2).

**soldering gun**. An electrical soldering iron with a pistol grip and a quick heating, relatively small bit.

**soldering iron**. A soldering tool having an internally or externally heated metal bit usually made of copper.

**solder short**. See *bridging* (5).

**sol-gel process**. An important ceramic and glass-forming process in which a sol is converted to a gel by partial evaporation of the liquid phase and/or by neutralizing the electric charges on particles which cause them to repel each other. The gel is usually further processed (e.g., formed, dried, and fired).

**solid cutters**. Cutters made of a single piece of material rather than a composite of two or more materials.

**solid density**. See *density, absolute*.

**solid-film lubrication**. Lubrication by application of a *solid lubricant* (Fig. 543).

FIG. 543

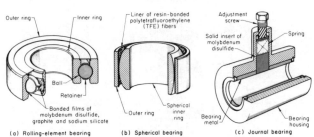

(a) Rolling-element bearing     (b) Spherical bearing     (c) Journal bearing

*Bearing designs using three types of solid lubrication*

**solidification**. The change in state from liquid to solid upon cooling through the melting temperature or melting range.

**solidification range**. The temperature between the liquidus and the solidus.

**solidification shrinkage**. The reduction in volume of metal from beginning to end of solidification. See also *casting shrinkage*.

**solidification shrinkage crack**. A crack that forms, usually at elevated temperature, because of the internal (shrinkage) stresses that develop during solidification of a metal casting. Also termed hot crack.

**solid lubricant**. Any solid used as a powder or thin film on a surface to provide protection from damage during relative movement and to reduce friction and wear. Examples include molybdenum disulfide, graphite, polytetrafluoroethylene (PTFE), and mica.

**solid-metal embrittlement**. The occurrence of *embrittlement* in a material below the melting point of the embrittling species. See also *liquid-metal embrittlement*.

**solid-phase chemical dosimeter**. An apparatus that measures radioactivity by using plastic, dyed plastic, or glass with an optical density, usually in the visible range, that changes when exposed to ionizing radiation. Examples currently in use include dyed polymethyl methacrylate (red perspex), undyed polyvinyl chloride, dyed polyamide (blue dye in a nylon matrix), and dyed polychlorostyrene (green dye in a chlorostyrene matrix). Solid-phase chemical dosimetry is generally considered to be a secondary-standard dosimetry system.

**solid-phase forming**. The use of metalworking technique to form thermoplastics in a solid phase. Procedure begins with a plastic blank that is heated and fabricated (that is, forged) by bulk deformation of the materials in constraining dies by the application of force. Also called solid-state stamping.

**solid shrinkage**. See *casting shrinkage*.

**solid solution**. A single, solid, homogeneous crystalline phase containing two or more chemical species.

**solid state**. Pertaining to circuits and components using semiconductors as substrates.

**solids content**. The percentage by weight of the nonvolatile matter in an adhesive. The actual percentage of the nonvolatile matter in an adhesive will vary according to the analytical procedure that is used. A standard test method must be used to obtain consistent results.

**solid-state sintering**. A sintering procedure for compacts or loose powder aggregates during which no component melts. Contrast with *liquid phase sintering*.

**solid-state welding**. A group of welding processes that join metals at temperatures essentially below the melting points of the base materials, without the addition of a brazing or soldering filler metal. Pressure may or may not be applied to the joint. Examples include *cold welding*, *diffusion welding*, *forge welding*, *hot pressure welding*, and *roll welding*.

**solidus**. (1) The highest temperature at which a metal or alloy is completely solid. (2) In a *phase diagram*, the locus of points representing the temperatures at which various compositions stop freezing upon cooling or begin to melt upon heating. See also *liquidus*.

**soluble oil**. A mineral oil containing additives that enable it to form a stable emulsion with water. Soluble oils are used as cutting or grinding fluids.

**solute**. The component of either a liquid or solid solution that is present to a lesser or minor extent; the component that is dissolved in the *solvent*.

**solution.** In chemistry, a homogeneous dispersion of two or more types of molecular or ionic species. Solutions may be composed of any combination of liquids, solids, or gases, but they always consist of a single phase.

**solution heat treatment.** Heating an alloy to a suitable temperature, holding at that temperature long enough to cause one or more constituents to enter into *solid solution*, and then cooling rapidly enough to hold these constituents in solution.

**solution potential.** *Electrode potential* where half-cell reaction involves only the metal electrode and its ion.

**solvation.** The process of swelling, gelling, or dissolving a resin by a solvent or *plasticizer*.

**Solvay process.** A method for producing *soda ash* that involves the reaction of salt (NaCl) and limestone to form sodium carbonate ($Na_2CO_3$) with calcium chloride ($CaCl_2$) as a by-product.

**solvent.** The component of either a liquid or solid solution that is present to a greater or major extent; the component that dissolves the *solute*.

**solvent-activated adhesive.** A dry-film adhesive that is rendered tacky by the application of a solvent just prior to use.

**solvent adhesive.** An adhesive having a volatile organic liquid as a vehicle. This term excludes water-base adhesives.

**solvent molding.** Process for forming thermoplastic articles by dipping a male mold in a solution or by dispersing the resin and drawing off the solvent, leaving a layer of plastic film adhering to the mold.

**solvus.** In a phase or equilibrium diagram, the locus of points representing the temperature at which solid phases with various compositions coexist with other solid phases, that is, the limits of solid solubility.

**Sommerfeld number.** A dimensionless number that is used to evaluate the performance of journal bearings. It is numerically defined as follows:

$$\frac{P}{\eta U}\left(\frac{c}{r}\right)^2$$

where $P$ is the load per unit width, $\eta$ is the dynamic viscosity, $U$ is the surface velocity, $c$ is the radial clearance, and $r$ is the bearing radius. At lower concentricities it is convenient to use the Sommerfeld number in the form given. Because it tends to infinity as the eccentricity approaches unity, the reciprocal form is frequently used in the case of heavily loaded bearings. The expression:

$$\frac{\eta N}{p}\left(\frac{r}{c}\right)^2$$

in which $N$ is the frequency of rotation and $p$ is the pressure, is sometimes referred to as the Sommerfeld number, particularly in the United States. See also *Ocvirk number*.

**sonic testing.** Any inspection method that uses sound waves (in the audible frequency range, about 20 to 20,000 Hz) to induce a response from a part or test specimen. Sometimes, but inadvisably, used as a synonym for *ultrasonic testing*.

**sorbite** (obsolete). A fine mixture of ferrite and cementite produced either by regulating the rate of cooling of steel or by tempering steel after hardening. The first type is very fine pearlite that is difficult to resolve under the microscope; the second type is tempered martensite.

**source (x-rays).** The area emitting primary x-rays in a diffraction experiment. The actual source is always the focal spot of the x-ray tube, but the virtual source may be a slit or pinhole, depending on the conditions of the experiment.

**sour gas.** A gaseous environment containing hydrogen sulfide and carbon dioxide in hydrocarbon reservoirs. Prolonged exposure to sour gas can lead to *hydrogen damage*, *sulfide-stress cracking*, and/or *stress-corrosion cracking* in ferrous alloys.

**sour water.** Waste waters containing fetid materials, usually sulfur compounds.

**sow block.** A block of heat-treated steel placed between the anvil of the hammer and the forging die to prevent undue wear to the anvil (Fig. 544). Sow blocks are occasionally used to hold insert dies. Also called anvil cap. See also the

FIG. 544

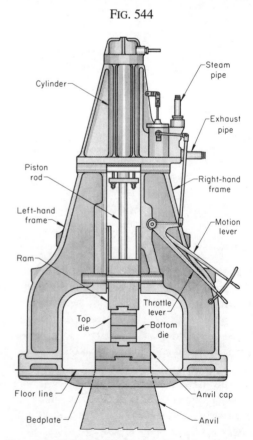

*Location of the sow block (anvil cap) in a double-frame power hammer used for open-die forging*

figures accompanying the terms *drop hammer* and *gravity hammer*.

**space-charge aberration**. In an electron microscope, an aberration resulting from the mutual repulsion of the electrons in a beam. This aberration is most noticeable in low-voltage, high-current beams. This repulsion acts as a negative lens, causing rays, which were originally parallel, to diverge. See also *aberration*.

**space lattice**. A regular, periodic array of points (lattice points) in space that represents the locations of atoms of the same kind in a perfect crystal. The concept may be extended, where appropriate, to crystalline compounds and other substances, in which case the lattice points often represent locations of groups of atoms of identical composition, arrangement, and orientation. See also the table accompanying the term *lattice* and the figure accompanying the term *unit cell*.

**spacer strip**. A metal strip or bar inserted in the root of a joint prepared for groove welding to serve as a backing and to maintain root opening throughout the course of the welding operation (Fig. 545).

FIG. 545

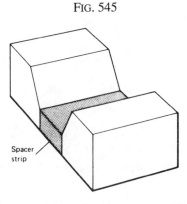

*A spacer bar (strip) for groove welding*

**spacing** (lattice planes). See *interplanar distance*.

**spade drill**. See preferred term *flat drill*.

**spalling** (ceramics). (1) The cracking or rupturing of a refractory unit, which usually results in the detachment of a portion of the unit. (2) A defect characterized by separation of the porcelain enamel from the aluminum base metal without apparent external cause. Spalling can result from the use of improper alloys or enamel formulations, incorrect pretreatment of the base metal, or faulty application and firing procedures.

**spalling** (metals). (1) Separation of particles from a surface in the form of flakes. The term spalling is commonly associated with rolling-element bearings and with gear teeth. Spalling is usually a result of subsurface fatigue and is more extensive than pitting. (2) In *tribology*, the separation of macroscopic particles from a surface in the form of flakes or chips, usually associated with rolling-element

bearings and gear teeth, but also resulting from impact events. (3) The spontaneous chipping, fragmentation, or separation of a surface or surface coating. (4) A chipping or flaking of a surface due to any kind of improper heat treatment or material dissociation.

**spalls**. The primary cause of premature failures of forged hardened steel rolls. Spalls are sections that have broken from the surface of the roll. In nearly all cases, they are observed in the outer hardened zone of the body surface, and they generally exhibit well-defined fatigue beach marks. The most common spalls are the circular spall and the line spall. Circular spalls exhibit subsurface fatigue marks in a circular, semicircular, or elliptical pattern (Fig. 546). They are generally confined to a particular body area. A line spall has a narrow width of subsurface fatigue that extends circumferentially around the body of the roll. Most line spalls originate at or beneath the surface in the outer hardened zone.

FIG. 546

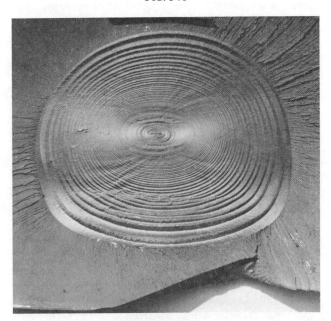

*Circular spall that began at a large subsurface inclusion in a hardened steel roll*

**spangle**. The characteristic crystalline form in which a hot dipped zinc coating solidifies on steel strip.

**spark**. A series of electrical discharges, each of which is oscillatory and has a comparatively high maximum instantaneous current resulting from the breakdown of the analytical gap or the auxiliary gap, or both, by electrical energy stored at high voltage in capacitors. Each discharge is self-initiated and is extinguished when the voltage across the gap, or gaps, is no longer sufficient to maintain it.

**spark erosion**. See *electrical pitting*.

**spark sintering.** In powder metallurgy, a pressure sintering or hot pressing method that provides for the surface activation of the powder particles by electric discharges generated by a high alternating current applied during the early stage of the consolidation process.

**spark source mass spectrometry.** An analytical technique in which a high-voltage spark in a vacuum is used to produce positive ions of a conductive sample material. The ions are injected into a mass spectrometer, and the resulting spectrum is recorded on a photographic plate or measured using an electronic detector. The position of a particular mass spectral signal determines the element and isotope, and the intensity of the signal is proportional to the concentration.

**spark testing.** A method used for the classification of ferrous alloys according to their chemical compositions, by visual examination of the spark pattern or stream that is thrown off when the alloys are held against a grinding wheel rotating at high speed (Fig. 547).

**spatial grain size.** The average size of the three-dimensional grains in polycrystalline materials as opposed to the more conventional grain size determined by a simple average of observations made on a cross section of the material.

**spatial resolution.** A measure of the ability of an imaging system to represent fine detail; the measure of the smallest separation between individually distinguishable structures. See also *resolution*.

**spatter.** The metal particles expelled during arc or gas welding. They do not form part of the weld.

**spatter loss.** The metal lost due to *spatter*.

**specific adhesion.** Adhesion between surfaces that are held together by valence forces of the same type as those that give rise to cohesion.

**specific energy.** In cutting or grinding, the energy expended or work done in removing a unit volume of material.

**specific gravity** (gases). The ratio of the density of a gas to the density of dry air at the same temperature and pressure.

**specific gravity** (solids and liquids). The ratio of the density of a material to the density of some standard material, such as water, at a specified temperature. Also known as relative density.

**specific heat.** (1) The ratio of the amount of heat required to raise a mass of material 1 degree in temperature to the amount required to raise an equal mass of a reference substance, usually water, 1 degree in temperature; both measurements are made at a reference temperature, usually at constant pressure, or constant volume. (2) The quantity of heat required to raise a unit mass of homogeneous material one degree in temperature in a specified way; it is assumed that during the process no phase or chemical change occurs.

**specific humidity.** In a mixture of water vapor and air, the mass of water vapor per unit mass of moist air.

FIG. 547

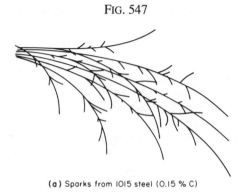

**(a) Sparks from 1015 steel (0.15 % C)**

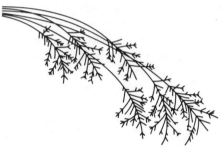

**(b) Sparks from 1045 steel (0.45 % C)**

**(c) Sparks from 1095 steel (1.0 % C)**

*Effect of carbon content of steels on the spark pattern or stream. (a) Sparks from low-carbon steel, showing slight forking effect. (b) Sparks from medium-carbon steel, showing pronounced bursts. (c) Sparks from high-carbon steel, showing the intensity of bursts that is characteristic of steel having high carbon content*

**specific power.** Same as *unit power*.

**specific pressure.** In powder metallurgy, the pressure applied to a green or sintered compact per unit of area of punch cross section.

**specific properties.** Material properties divided by material density.

**specific sliding.** The ratio of the algebraic difference between the surface velocities of two bodies in relative motion to their sum.

**specific surface.** The surface area of a powder expressed in square centimeters per gram of powder or square meters per kilogram of powder.

**specific viscosity**. The relative viscosity of a solution of known concentration of a polymer minus one. It is usually determined for a low concentration of the polymer.

**specific volume**. The volume of a substance per unit mass; the reciprocal of the density.

**specific wear rate**. In journal bearings, the proportionality constant $K$ in the equation:

$$h = Kpvt$$

where $h$ is the radial wear in the bearing, $p$ is the apparent contact pressure, $y$ is the velocity of the journal, and $t$ is the sliding time. The constant $K$ has also been called the *wear factor*, but there are other definitions for the term wear factor that do not necessarily refer to journal bearings or derive their meanings from the above equation.

**specimen**. A test object, often of standard dimensions and/or configuration, that is used for destructive or nondestructive testing. One or more specimens may be cut from each unit of a *sample*.

**specimen chamber** (electron optics). The compartment located in the column of the electron microscope in which the specimen is placed for observation.

**specimen charge** (electron optics). The electrical charge resulting from the impingement of electrons on a nonconducting specimen.

**specimen contamination** (electron optics). The contamination of the specimen caused by the condensation upon it of residual vapors in the microscope under the influence of electron bombardment.

**specimen distortion** (electron optics). A physical change in the specimen caused by desiccation or heating by the electron beam.

**specimen grid**. See *specimen screen*.

**specimen holder** (electron optics). A device that supports the specimen and specimen screen in the correct position in the specimen chamber of the microscope.

**specimen screen** (electron optics). A disk of fine screen, usually 200-mesh stainless steel, copper, or nickel, that supports the replica or specimen support film for observation in the microscope.

**specimen stage**. The part of the microscope that supports the specimen holder and the specimen in the microscope and can be moved in a plane perpendicular to the optic axis from outside the column.

**specimen strain**. A distortion of the specimen resulting from stresses occurring during metallographic preparation or observation. In electron metallography, strain may be caused by stretching during removal of a replica or during subsequent washing or drying.

**spectral background**. In spectroscopy, a signal obtained when no analyte is being introduced into the instrument, or a signal from a species other than that of the analyte.

**spectral distribution curve**. The curve showing the absolute or relative radiant power emitted or absorbed by a substance as a function of wavelength, frequency, or any other directly related variable.

**spectral line**. A wavelength of light with a narrow energy distribution or an image of a slit formed in the focal plane of a spectrometer or photographic plate that has a narrow energy distribution approximately equal to that formed by monochromatic radiation.

**spectral order**. The number of the intensity of a given line from the directly transmitted or specularly reflected light from a *diffraction grating*.

**spectrochemical (spectrographic, spectrometric, spectroscopic) analysis**. The determination of the chemical elements or compounds in a sample qualitatively, semiquantitatively, or quantitatively by measurements of the wavelengths and intensities of spectral lines produced by suitable excitation procedures and dispersed by a suitable optical device.

**spectrogram**. A photographic or graphic record of a spectrum.

**spectrograph**. An optical instrument with an entrance slit and dispersing device that uses photography to record a spectral range. The radiant power passing through the optical system is integrated over time, and the quantity recorded is a function of radiant energy.

**spectrometer**. An instrument with an entrance slit, a dispersing device, and one or more exit slits, with which measurements are made at selected wavelengths within the spectral range, or by scanning over the range. The quantity detected is a function of the radiant power.

**spectrophotometer**. A spectrometer that measures the ratio (or a function of the ratio) of the intensity of two different wavelengths of light. These two beams may be separated in terms of time or space, or both.

**spectrophotometry**. A method for identification of substances and determination of their concentration by measuring light transmittance in different parts of the spectrum.

**spectroscope**. An instrument that disperses radiation into a spectrum for visual observation.

**spectroscopy**. The branch of physical science treating the theory, measurement, and interpretation of spectra.

**spectrum**. The ordered arrangement of electromagnetic radiation according to wavelength, wave number, or frequency.

**specular reflection**. The condition in which all the incident light is reflected at the same angle as the angle of the incident light relative to the normal at the point of incidence. The reflection surface then appears bright, or mirrorlike, when viewed with the naked eye. Sometimes termed regular reflection.

**specular transmittance**. The transmittance value obtained when the measured radiant energy in *emission spectros-*

*copy* has passed from one source to the receiver without appreciable scattering.

**speed of travel**. In welding, the speed with which a weld is made along its longitudinal axis, usually measured in meters per second or inches per minute.

**speiss**. Metallic arsenides and antimonides that result from smelting metal ores such as those of cobalt or lead.

**spelter**. Crude zinc obtained in smelting zinc ores.

**spelter solder**. A brazing filler metal of approximately equal parts of copper and zinc.

**SPF**. See *superplastic forming*.

**spherical aberration**. A lens defect in an optical microscope in which image-forming rays passing through the outer zones of the lens focus at a distance from the principal plane different from that of the rays passing through the center of the lens (Fig. 548). See also *aberration* and *chromatic aberration*.

FIG. 548

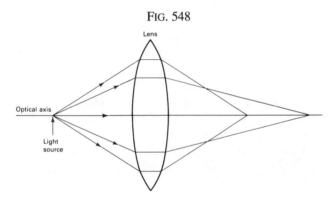

*Spherical aberration. Light rays passing through the outer portion of the lens are more strongly refracted than those passing through the central portion and are focused at a different point along the optical axis. This problem can be minimized by using an aperture to restrict the light path to the central part of the objective.*

**spherical bearing**. A bearing that is self-aligning by virtue of its partially spherical form.

**spherical powder**. A powder consisting of ball-shaped particles (Fig. 549).

**spherical roller bearing**. (1) A spherical bearing containing rollers. (2) A roller bearing containing barrel-shaped or hour glass-shaped rollers riding on spherical (concave or convex) races to provide self-aligning capability.

**spheroidal graphite**. Graphite of spheroidal shape with a polycrystalline radial structure (Fig. 550). This structure can be obtained, for example, by adding cerium or magnesium to the melt. See also *ductile iron* and *nodular graphite*.

**spheroidal powder**. A powder consisting of oval or rounded particles.

FIG. 549

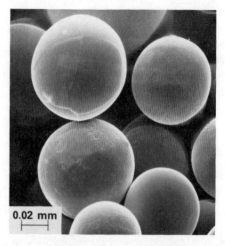

*Spherical titanium alloy powders produced by the plasma rotating electrode process*

FIG. 550

*Spheroidal graphite in a tempered martensite matrix in ductile iron. 500×*

**spheroidite**. An aggregate of iron or alloy carbides of essentially spherical shape dispersed throughout a matrix of *ferrite*.

**spheroidized structure**. A microstructure consisting of a matrix containing spheroidal particles of another constituent (Fig. 551).

**spheroidizing**. Heating and cooling to produce a spheroidal or globular form of carbide in steel. Spheroidizing methods frequently used are:

1. Prolonged holding at a temperature just below $Ae_1$.

2. Heating and cooling alternatively between temperatures that are just above and just below $Ae_1$.

FIG. 551

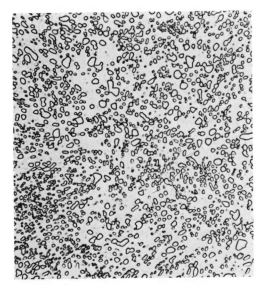

*Spheroidized cementite in AISI W2 carbon-vana-
dium (1.05% C) tool steel. 1000×*

3. Heating to a temperature above $Ae_1$ or $Ae_3$ and then cooling very slowly in the furnace or holding at a temperature just below $Ae_1$.

4. Cooling at a suitable rate from the minimum temperature at which all carbide is dissolved to prevent the re-formation of a carbide network, and then reheating in accordance with method 1 or 2 above. (Applicable to hypereutectoid steel containing a carbide network.)

**spherulite.** A rounded aggregate of radiating lamellar crystals with appearance of a pom-pom in plastics. Spherulites contain amorphous material between the crystals and usually impinge on one another, forming polyhedrons. Spherulites are present in most crystalline plastics and may range in diameter from a few tenths of a micron to several millimeters.

**spherulitic graphite cast iron.** Same as *ductile cast iron.*

**spider.** In a plastic molding press, that part of an ejector mechanism that operates the ejector pins. In extrusion, the membranes supporting a mandrel within the head/die assembly.

**spider die.** Same as *porthole die.*

**spiegeleisen (spiegel).** A pig iron containing 15 to 30% Mn and 4.5 to 6.5% C.

**spiking.** In electron beam welding and laser welding, a condition where the depth of penetration is nonuniform and changes abruptly over the length of the weld.

**spin.** In bearings, rotation of a rolling element about an axis normal to the contact surfaces. See also *rolling* (pure rolling with no sliding and no spin) and *sliding.*

**spindle.** (1) Shaft of a machine tool on which a cutter or grinding wheel may be mounted. (2) Metal shaft to which a mounted wheel is cemented.

**spindle oil.** An oil of low viscosity used to lubricate high-speed light spindles.

**spinel.** Any of a group of compounds of the same crystal type and general formula as magnesium aluminate, $MgAl_2O_4$ or $MgO \cdot Al_2O_3$. That compound itself, which is refractory and chemically near-neutral. Magnesium aluminate spinels are used as an addition to fired magnesia refractory bricks to improve thermal shock resistance. Lithium-based spinels are candidate materials for rechargeable lithium batteries.

**spin glass.** One of a wide variety of materials that contain interacting atomic magnetic moments and also possess some form of disorder, in which the temperature variation of the magnetic susceptibility undergoes an abrupt change in slope at a temperature generally referred to as the freezing temperature.

**spinneret.** A type of extrusion die for plastics that consists of a metal plate with many tiny holes, through which a plastic melt is forced, to make fine fibers and filaments. Filaments may be hardened by cooling in air, water, and so forth, or by chemical action.

**spinning.** The forming of a seamless hollow metal part by forcing a rotating blank to conform to a shaped mandrel that rotates concentrically with the blank (Fig. 552). In the typical application, a flat-rolled metal blank is forced against the mandrel by a blunt, rounded tool; however, other stock (notably, welded or seamless tubing) can be formed. A roller is sometimes used as the working end of the tool.

**spinodal curve.** A graph of the realizable limit of the supersaturation of a solution. See also *spinodal structure.*

**spinodal hardening.** See *aging.*

**spinodal structure.** A fine, homogeneous mixture of two phases that form by the growth of composition waves in a solid solution during suitable heat treatment (Fig. 553). The phases of a spinodal structure differ in composition from each other and from the parent phase, but have the same crystal structure as the parent phase. Spinodal structures are resolvable only at high magnifications such as made possible by *transmission electron microscopy.*

**spin wave.** A sinusoidal variation, propagating through a crystal lattice, of that angular momentum associated with magnetism (mostly spin angular momentum of the electrons). See also *spin glass.*

**spiral-flow test.** A method for determining the flow properties of a thermoplastic resin in which the resin flows along the path of a spiral cavity. The length of the material that flows into the cavity and its weight gives a relative indication of the flow properties of the resin.

**spiral mold cooling.** A method of cooling injection molds or similar molds for forming plastics in which the cooling medium flows through a spiral cavity in the body of the mold. In injection molds, the cooling medium is intro-

FIG. 552

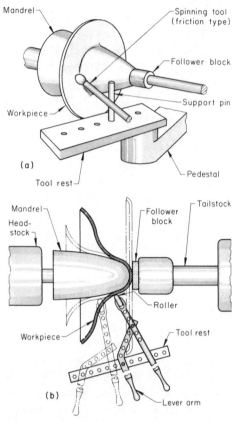

(a)

(b)

*Manual spinning using a lathe. (a) Simple setup using a hand tool applied as a pry bar. (b) Setup using scissorlike levers and roller spinning tool*

FIG. 553

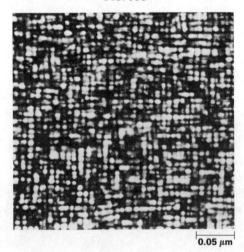

*Spinodal structure in an Fe-25Be (at.%) alloy that was aged 2 h at 400 °C (750 °F). The bright phase is the beryllium-enriched structure; the dark phase is the iron-rich structure. Transmission electron micrograph. 200,000×*

duced at the center of the spiral, near the sprue section, because more heat is localized in this section.

**spit**. See preferred term *flash*.

**splash lubrication**. A system of lubrication in which the lubricant is splashed onto the moving parts.

**splat powder**. A rapidly cooled or quenched powder whose particles have a flat shape and a small thickness compared to other dimensions. Similar to *flake powder*.

**splat quenching**. The process of producing splat powder.

**splay**. The tendency of a rotating drill bit to drill off-center, out-of-round, nonperpendicular holes.

**splay** (plastics). A fanlike surface defect near the gate on a plastic part.

**splay lines**. Lines found in a plastic part after molding, usually due to the flow of material in the mold. Sometimes called silver streaking.

**splice**. The joining of two ends of glass fiber yarn or strand used for reinforcing plastics, usually by means of an air-drying adhesive.

**spline**. Any of a series of longitudinal, straight projections on a shaft that fit into slots on a mating part to transfer rotation to or from the shaft (Fig. 554).

FIG. 554

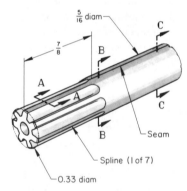

*Spline in a shaft. The sections A–A, B–B, and C–C indicate fracture regions not shown in this figure.*

**split die**. A die made of part that can be separated for ready removal of the workpiece. Also known as segment die.

**split pipe backing**. Backing in the form of a pipe segment used for welding round bars.

**split punch**. A segmented punch or a set of punches in a powder metallurgy forming press that allow(s) a separate positioning for different powder fill heights and compact levels in dual-step and multistep parts. See also *stepped compact*.

**split-ring mold**. A mold for forming plastics in which a split-cavity block is assembled in a chase to permit the forming of undercuts in a molded plastic piece. These parts are ejected from the mold and then separated from the piece.

**split seal.** A seal that has its primary sealing elements split in a plane parallel to the axis of the shaft such that, instead of the rings being continuous, they are essentially two semicircles. Modified designs of lip seals feature units with a single lip separation and with one or more separations of the metallic stiffening mechanisms.

**spodumene.** A mineral of the composition $LiO_2 \cdot Al_2O_3 \cdot 4SiO_2$, which is the principal ore of lithium. It is also used as a melt accelerator in glass manufacture.

**sponge.** A form of metal characterized by a porous condition that is the result of the decomposition or reduction of a compound without fusion. The term is applied to forms of iron, titanium, zirconium, uranium, plutonium, and the platinum-group metals.

**sponge effect.** See *squeeze effect*.

**sponge iron.** A coherent, porous mass of substantially pure iron produced by solid state reduction of iron oxide (mill scale or iron ore).

**sponge iron powder.** Ground and sized sponge iron that may have been purified or annealed or both (Fig. 555).

**sponge titanium powder.** Ground and sized titanium sponge (Fig. 556). See also *Kroll process*.

**spongy.** A porous condition in metal powder particles usually observed in reduced oxides.

**spool.** A type of weld filler metal package consisting of a continuous length of electrode wound on a cylinder (called the barrel) which is flanged at both ends. The flange extends below the inside diameter of the barrel and contains a spindle hole.

**spot drilling.** Making an initial indentation in a work surface, with a drill, to serve as a centering guide in a subsequent machining process.

**spotfacing.** Using a rotary, hole-piloted end-facing tool to produce a flat surface normal to the axis of rotation of the tool on or slightly below the workpiece surface.

**spotting out.** Delayed, uneven staining of metal by entrapment of chemicals during the finishing operation.

**spot weld.** A weld made between or upon overlapping members in which coalescence may start and occur on the faying surfaces or may proceed from the surface of one member. The weld cross section is approximately circular. See also the figure accompanying the term *resistance spot welding*.

**spot welding.** Welding of lapped parts in which fusion is confined to a relatively small circular area. It is generally resistance welding, but may also be gas tungsten-arc, gas metal-arc, or submerged-arc welding. See also the figure accompanying the term *resistance spot welding*.

**spragging.** Intermittent motion arising from design features

FIG. 555

*Scanning electron micrograph of particles of annealed sponge iron powder. 180×*

FIG. 556

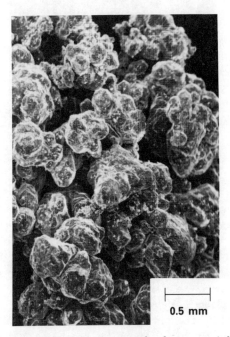

0.5 mm

*Scanning electron micrograph of commercially pure titanium –100 mesh sponge fines*

that allow an increase in tangential force or displacement to produce an increase in normal force.

**spray angle.** In thermal spraying, the angle of particle approach, measured from the surface of the substrate to the axis of the spray nozzle.

**spray deposit.** A coating applied by any of the thermal spray methods. See also *thermal spraying*.

**spray distance**. In thermal spraying, the distance maintained between the gun nozzle and the substrate surface during spraying.

**spray drier**. A large vessel into which a slurry containing metal or ceramic powders is sprayed through orifices in a stationary or revolving head and thrown as droplets into a stream of heated air which dries them. The dried droplets are typically tiny agglomerates, often in hollow bead form, hence free-flowing.

**spray drying**. A powder-producing process in which a slurry of liquids and solids or a solution is atomized into droplets in a chamber through which heated gases, usually air, are passed. The liquids are evaporated from the droplets and the solids are collected continuously from the chamber. The resulting powder consists of free-flowing, spherical agglomerates (Fig. 557).

FIG. 557

*Spray-dried molybdenum powder. 250×*

**sprayed-metal molds**. Molds for forming plastics made by spraying molten metal onto a master until a shell of predetermined thickness is achieved. The shell is then removed and backed with plaster, cement, casting resin, or other suitable material. Used primarily as a mold in the sheet-forming process.

**spraying sequence** (thermal spraying). The order in which different layers of similar or different materials are applied in a planned relationship, such as overlapped, superimposed, or at certain angles.

**spray lay-up**. A wet lay-up for processing of reinforced plastic in which a stream of chopped fibers (usually glass) is fed into a stream of liquid resin in a mold. The direction of the fibers is random, as opposed to the mats or woven fabrics that can be used in hand lay-up. See also *hand lay-up* and *wet lay-up*.

**spray metallizing**. See *metallizing*.

**spray nozzle**. In atomizing of metal powders, an orifice through which a molten metal passes to form a stream that can be further disintegrated by a gas, a liquid, or by mechanical means (Fig. 558).

FIG. 558

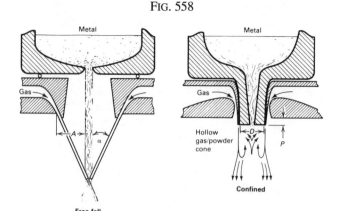

*Atomization nozzle designs. Design characteristics: α, angle formed by free-falling molten metal and impinging gas; A, distance between molten metal and gas nozzle; D, diameter of confined molten metal nozzle; P, protrusion length of metal nozzle*

**spray quenching**. A quenching process using spray nozzles to spray water or other liquids on a part. The quench rate is controlled by the velocity and volume of liquid per unit area per unit of time of impingement.

**spray rate**. Same as *feed rate*.

**spray transfer**. In consumable-electrode arc welding, a type of metal transfer in which the molten filler metal is propelled across the arc as fine droplets. Compare with *globular transfer* and *short-circuiting transfer*. See also the figure accompanying the term *short-circuiting transfer*.

**spread**. The quantity of adhesive per unit joint area applied to an adherend, usually expressed in pounds of adhesive per thousand square feet of joint area.

**spreader**. An axial groove in a plain bearing designed to spread oil along the bearing.

**spreader** (plastics). A streamlined metal block placed in the path of flow of the plastics material in the heating cylinder of extruders and injection molding machines to spread it into thin layers, thus forcing it into intimate contact with the heating areas.

**spreader pockets**. Depressions in a sliding surface designed to distribute lubricant.

**springback**. (1) The elastic recovery of metal after stressing. (2) The extent to which metal tends to return to its original shape or contour after undergoing a forming operation. This is compensated for by overbending or by a secondary operation of *restriking*. (3) In flash, upset, or pressure welding, the deflection in the welding machine caused by the upset pressure.

**spring constant**. The force required to compress a spring or specimen 25 mm (1 in.) in a prescribed test procedure.

**spring temper**. A *temper* of nonferrous alloys and some ferrous alloys characterized by tensile strength and hard-

Fig. 559

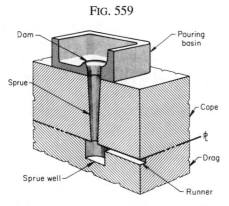

*Section of a typical sand mold, showing pouring basin with a dam, tapered sprue, sprue well, and a portion of the runner*

ness about two-thirds of the way from *full hard* to *extra spring* temper.

**sprue** (metals). (1) The mold channel that connects the *pouring basin* with the runner or, in the absence of a pouring basin, directly into which molten metal is poured (Fig. 559). Sometimes referred to as downsprue or downgate. (2) Sometimes used to mean all gates, risers, runners, and similar scrap that are removed from castings after shake-out.

**sprue** (plastics). A single hole through which thermoset molding compounds are injected directly into the mold cavity.

**spun roving**. A heavy, low-cost glass or aramid fiber strand consisting of filaments that are continuous but doubled back on themselves. See also *roving*.

**sputtering**. The bombardment of a solid surface with a flux of energetic particles (ions) that results in the ejection of atomic species. The ejected material may be used as a source for deposition. See also *physical vapor deposition* and the figure accompanying the term *secondary ion*.

**square drilling**. Making square holes by means of a specially constructed drill made to rotate and also to oscillate so as to follow accurately the periphery of a square guide bushing or template.

**square groove weld**. A groove weld in which the abutting surfaces are square (Fig. 560).

**squaring shear**. A machining tool, used for cutting sheet metal or plate, consisting essentially of a fixed cutting knife (usually mounted on the rear of the bed) and another cutting knife mounted on the front of a reciprocally

Fig. 560

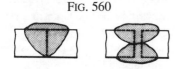

*Examples of square groove welds*

moving crosshead, which is guided vertically in side housings. Corner angles are usually 90°. See also the figure accompanying the term *shear* (2).

**squeeze casting**. A hybrid liquid-metal forging process in which liquid metal is forced into a permanent mold by a hydraulic press (Fig. 561).

**squeeze effect**. (1) The production of lubricant from a porous retainer by application of pressure. Also known as sponge effect. (2) The persistence of a film of fluid between two surfaces that approach each other in the direction of their common normal plane.

**squeeze-out**. Adhesive pressed out at the bond line due to pressure applied on the adherends.

**squeeze time**. In resistance welding, the time between the initial applications of pressure and current.

Fig. 561

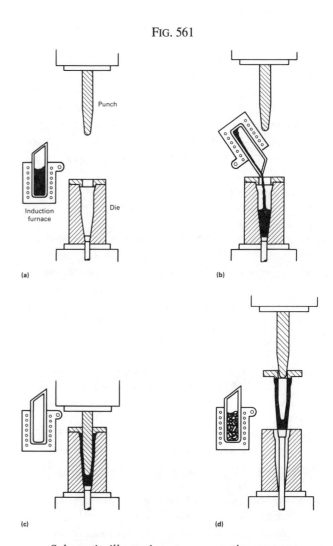

*Schematic illustrating squeeze casting process operations. (a) Melt charge, preheat, and lubricate tooling. (b) Transfer melt into die cavity. (c) Close tooling, solidify melt under pressure. (d) Eject casting, clean dies, charge melt stock.*

**stabilization**. In carbon fiber forming, the process used to render the carbon fiber precursor infusible prior to carbonization.

**stabilizers**. Chemicals used in plastics formulation to help maintain physical and chemical properties during processing and service life. A specific type of stabilizer, known as an ultraviolet stabilizer, is designed to absorb ultraviolet rays and prevent them from attacking the plastic. Heat stabilizers are added to lessen the severity of thermal oxidation processes and their effect on properties.

**stabilizing gas** (plasma spraying). The arc gas, which is ionized to form the plasma. Introduced into the arc chamber tangentially, the relatively cold gas chills the outer surface of the arc stream, tending to constrict the arc, raise its temperature, and force it out of the front anode nozzle in a steady, relatively unfluctuating stream. See also the figures accompanying the terms *plasma spraying* and *transferred arc*.

**stabilizing treatment**. (1) Before finishing to final dimensions, repeatedly heating a ferrous or nonferrous part to or slightly above its normal operating temperature and then cooling to room temperature to ensure dimensional stability in service. (2) Transforming retained austenite in quenched hardenable steels, usually by *cold treatment*. (3) Heating a solution-treated stabilized grade of austenitic stainless steel to 870 to 900 °C (1600 to 1650 °F) to precipitate all carbon as TiC, NbC, or TaC so that *sensitization* is avoided on subsequent exposure to elevated temperature.

**stack cutting**. Thermal cutting of stacked metal plates arranged so that all the plates are severed by a single cut.

**stacking sequence**. A description of a laminate that details the orientations of the plies and their sequence in the laminate. See also the figures accompanying the terms *laminate* and *quasi-isotropic laminate*.

**stack molding** (metals). A foundry practice that makes use of both faces of a mold section, one face acting as the drag and the other as the cope. Sections, when assembled to other similar sections, form several tiers of mold cavities, all castings being poured together through a common sprue.

**stack welding**. *Resistance spot welding* of stacked plates, all being joined simultaneously.

**stage**. A device for holding a specimen in the desired position in the optical path of a microscope.

**staggered-intermittent fillet welding**. Making a line of intermittent fillet welds on each side of a joint so that the increments on one side are not opposite those on the other. Contrast with *chain-intermittent fillet welding*.

**staggered-tooth cutters**. Milling cutters with alternate flutes of oppositely directed helixes.

**staging**. Heating a premixed resin system, such as in a *prepreg*, until the chemical reaction (curing) starts, but stopping the reaction before the gel point is reached. Staging is often used to reduce resin flow in subsequent press molding operations.

**staining**. Precipitation etching that causes contrast by distinctive staining of microconstituents; different interference colors originate from surface layers of varying thickness. Also known as *color etching*.

**stainless steels**. See *Technical Brief 47*.

**staking**. Fastening two parts together permanently by recessing one part within the other and then causing plastic flow at the joint.

**stalagmometer**. An apparatus for determining surface tension. The mass of a drop of liquid is measured by weighing a known number of drops or by counting the number of drops obtained from a given volume of the liquid.

**stamping**. The general term used to denote all sheet metal pressworking. It includes blanking, shearing, hot or cold forming, drawing, bending, or coining.

**stand**. A piece of rolling mill equipment containing one set of work rolls. In the usual sense, any pass of a cold- or hot-rolling mill. See also *rolling mills*.

**standard addition**. In chemical analysis, a method in which small increments of a substance under measurement are added to a sample under test to establish a response function or, by extrapolation, to determine the amount of a constituent originally present in the sample.

**standard deviation**. A measure of the dispersion of observed values or results from the average expressed as the positive square root of the variance.

**standard electrode potential**. The reversible potential for an electrode process when all products and reactions are at unit activity on a scale in which the potential for the standard hydrogen half-cell is zero.

**standard gold**. A gold alloy containing 10% copper; at one time used for legal coinage in the United States.

**standard grain-size micrograph**. A micrograph of a known grain size at a known magnification that is used to determine grain size of metals by direct comparison with another micrograph or with the image of a specimen.

**standardization**. (1) The process of establishing, by common agreement, engineering criteria, terms, principles, practices, materials, items, processes, and equipment parts and components. (2) The adoption of generally accepted uniform procedures, dimensions, materials, or parts that directly affect the design of a product or a facility. (3) In analytical chemistry, the assignment of a compositional value to one standard on the basis of another standard.

**standard reference material**. A reference material, the composition or properties of which are certified by a recognized standardizing agency or group.

**standoff distance**. The distance between a nozzle on a welding or cutting torch and the base metal.

**staple fibers**. Fibers for reinforcing plastics that are of spinnable length manufactured directly or by cutting continuous filaments to short lengths (usually 13 to 50 mm, or ½ to 2 in., long, and 1 to 5 denier). See also *denier*.

**star craze.** Multiple fine surface separation cracks in pultruded reinforced plastics that appear to emanate from a central point and that exceed 6 mm (¼ in.) in length, but do not penetrate the equivalent depth of a full ply of reinforcement. This condition is often caused by impact damage. See also *crazing* (plastics).

**stardusting.** An extremely fine form of roughness on the surface of a metal deposit.

**Stark effect.** A shift in the energy of spectral lines due to an electrical field that is either externally applied or is an internal field caused by the presence of neighboring ions or atoms in a gas, solid, or liquid.

**starting sheet.** A thin sheet of metal used as the cathode in electrolyte refining.

**starting torque.** The torque that is required for initiating rotary motion.

**starved area.** An area in a reinforced plastic part that has an insufficient amount of resin to wet out the reinforcement completely. This condition may be due to improper wetting, impregnation, or resin flow; excessive molding pressure; or incorrect bleeder cloth thickness.

**starved joint.** An adhesive bonded joint that has an insufficient amount of adhesive to produce a satisfactory bond. This condition may result from too thin a spread to fill the gap between the adherends, excessive penetration of the adhesive into the adherend, too short an assembly time, or the use of excessive pressure.

**state of strain.** A complete description of the deformation within a homogeneously deformed volume or at a point. The description requires, in general, the knowledge of the independent components of *strain*.

**state of stress.** A complete description of the stresses within a homogeneously stressed volume or at a point. The description requires, in general, the knowledge of the independent components of *stress*.

**static.** Stationary or very slow. Frequently used in connection with routine testing of metal specimens. Contrast with *dynamic*.

**static coefficient of friction.** The *coefficient of friction* corresponding to the maximum friction force that must be overcome to initiate macroscopic motion between two bodies.

**static electrode force.** The force between the electrodes in making spot, seam, or projection welds by resistance welding under welding conditions, but with no current flowing and no movement in the welding machine.

**static equivalent load ($P_0$).** In rolling-element bearings, the static load which, if applied, would give the same life as that which the bearing will attain under actual conditions of load and rotation. See also *rating life*.

**static fatigue.** A term sometimes used to identify a form of hydrogen embrittlement in which a metal appears to fracture spontaneously under a steady stress less than the yield stress. There almost always is a delay between the application of stress (or exposure of the stressed metal to hydrogen) and the onset of cracking. More properly referred to as *hydrogen-induced delayed cracking*.

**static friction.** See *limiting static friction*.

**static hot pressing.** A method of applying a static load uniaxially during hot pressing of metal or ceramic powders. Contrast with *dynamic hot pressing* and *isostatic hot pressing*. See also *hot pressing*.

**static load rating ($C_0$).** In rolling-element bearings, the static load that corresponds to a permanent deformation of rolling element and race at the most heavily stressed contact of 0.00001 of the rolling-element diameter. See also *rating life*.

**static modulus.** The ratio of stress to strain under static conditions. It is calculated from static stress-strain tests, in shear, compression, or tension. Expressed in force per unit of area.

**static stress.** A stress in which the force is constant or slowly increasing with time, for example, test to failure without shock.

**static viscosity.** See *viscosity*.

**stationary phase.** In *chromatography*, a particulate material packed into the column or a coating on the inner walls of the column. A sample compound in the stationary phase is separated from compounds moving through the column as a result of being in the mobile phase. See also *mobile phase*.

**statistic.** A summary value calculated from the observed values in a sample.

**statistical process control.** The application of statistical techniques for measuring and analyzing the variation in processes.

**statistical quality control.** The application of statistical techniques for measuring and improving the quality of processes and products (includes statistical process control, diagnostic tools, sampling plans, and other statistical techniques).

**stave bearing.** A *sleeve bearing* consisting of several axially held slats or staves on the outer surface of which the bearing material is bonded.

**steadite.** A hard structural constituent of cast iron that consists of a binary eutectic of ferrite (Fig. 562), containing some phosphorus in solution, and iron phosphide ($Fe_3P$). The eutectic consists of 10.2% P and 89.8% Fe. The melting temperature is 1050 °C (1920 °F).

**Stead's brittleness.** A condition of brittleness that causes transcrystalline fracture in the coarse grain structure that results from prolonged annealing of thin sheets of low-carbon steel previously rolled at a temperature below

## Technical Brief 47: Stainless Steels

STAINLESS STEELS are iron-base alloys containing at least 10.5% Cr that achieve their stainless characteristics through the formation of an invisible and adherent chromium-rich oxide surface film. This oxide forms and heals itself in the presence of oxygen. Other elements added to improve particular characteristics include nickel, molybdenum, copper, titanium, aluminum, silicon, niobium, nitrogen, sulfur, and selenium. Carbon is normally present in amounts ranging from less than 0.03% to more than 1.0% in certain grades.

Wrought stainless steels are commonly divided into five groups: martensitic stainless steels, ferritic stainless steels, austenitic stainless steels, duplex (ferritic-austenitic) stainless steels, and precipitation-hardening stainless steels. For a description of cast stainless steels, see *cast corrosion-resistant stainless steels* and *cast heat-resistant stainless steels*.

Martensitic stainless steels are essentially alloys of chromium and carbon that possess a body-centered tetragonal (bct) crystal structure (martensitic) in the hardened condition. They are ferromagnetic, hardenable by heat treatments, and generally resistant to corrosion only in relatively mild environments. Chromium content is generally in the range of 10.5 to 18%, and carbon content may exceed 1.2%. Additions of nitrogen, nickel, and molybdenum in combination with somewhat lower carbon levels produce steels with improved toughness and corrosion resistance. Sulfur or selenium is added to some alloys to improve machinability.

Ferritic stainless steels are essentially iron-chromium (10.5 to 30% Cr) alloys with body-centered cubic (bcc) crystal structures. Some grades may contain molybdenum, silicon, aluminum, titanium, and niobium to confer particular characteristics. Ferritic alloys are ferromagnetic and have good ductility and formability, but their high-temperature strengths are relatively poor compared with the austenitic grades. Ferritic stainless steels are, however, highly resistant to chloride stress-corrosion cracking.

The austenitic stainless steels are the most commonly used stainless steels. These materials have a face-centered cubic (fcc) structure attained through the liberal use of austenitizing elements such as nickel, manganese, and nitrogen. These steels are essentially nonmagnetic in the annealed condition and can be hardened only by cold working. They usually possess excellent cryogenic properties and good high-temperature strength. Chromium content generally varies from 16 to 26%; nickel, up to about 35%; and manganese, up to 15%. Molybdenum, copper, silicon, aluminum, titanium, and niobium may be added to confer certain characteristics, such as enhanced corrosion or oxidation resistance. Sulfur or selenium may be added to improve machinability.

Duplex stainless steels have a mixed structure of bcc ferrite and fcc austenite. The exact amount of each phase is a function of composition and heat treatment. Most alloys are designed to contain about equal amounts of each phase in the annealed condition. The principal alloying elements are chromium (21 to 30%) and nickel (3.5 to 7.5%), but molybdenum (up to 4%), nitrogen, copper, silicon, and tungsten may be added to control structural balance and to impart certain corrosion resistance characteristics.

Precipitation-hardening (PH) stainless steels are chromium-nickel grades that contain precipitation-hardening elements such as copper and aluminum. These grades may have austenitic, semiaustenitic, or martensitic crystal structures. All are hardened by a final aging treatment that precipitates very fine precipitates from a supersaturated solid solution.

The selection of stainless steels is usually based on corrosion resistance and mechanical properties. Stainless steels are firmly established as materials for cooking utensils, fasteners, cutlery, flatware, decorative architectural hardware, and equipment for use in chemical plants, pulp and paper mills, dairy and food- and beverage-processing plants, health and sanitation applications, petroleum and petrochemical plants, textile plants, the pharmaceutical and transportation industries, and the power industry (fossil fuel and nuclear power plants).

## Compositions of representative standard stainless steels

| Type | UNS designation | Composition(a), % | | | | | | | |
|---|---|---|---|---|---|---|---|---|---|
| | | C | Mn | Si | Cr | Ni | P | S | Other |
| **Austenitic types** | | | | | | | | | |
| 201 | S20100 | 0.15 | 5.5-7.5 | 1.00 | 16.0-18.0 | 3.5-5.5 | 0.06 | 0.03 | 0.25 N |
| 205 | S20500 | 0.12-0.25 | 14.0-15.5 | 1.00 | 16.5-18.0 | 1.0-1.75 | 0.06 | 0.03 | 0.32-0.40 N |
| 302 | S30200 | 0.15 | 2.00 | 1.00 | 17.0-19.0 | 8.0-10.0 | 0.045 | 0.03 | ... |
| 304 | S30400 | 0.08 | 2.00 | 1.00 | 18.0-20.0 | 8.0-10.5 | 0.045 | 0.03 | ... |
| 304N | S30451 | 0.08 | 2.00 | 1.00 | 18.0-20.0 | 8.0-10.5 | 0.045 | 0.03 | 0.10-0.16 N |
| 310 | S31000 | 0.25 | 2.00 | 1.50 | 24.0-26.0 | 19.0-22.0 | 0.045 | 0.03 | ... |
| 316 | S31600 | 0.08 | 2.00 | 1.00 | 16.0-18.0 | 10.0-14.0 | 0.045 | 0.03 | 2.0-3.0 Mo |
| 316LN | S31653 | 0.03 | 2.00 | 1.00 | 16.0-18.0 | 10.0-14.0 | 0.045 | 0.03 | 2.0-3.0 Mo; 0.10-0.16 N |
| 321 | S32100 | 0.08 | 2.00 | 1.00 | 17.0-19.0 | 9.0-12.0 | 0.045 | 0.03 | 5 × %C min Ti |
| 347 | S34700 | 0.08 | 2.00 | 1.00 | 17.0-19.0 | 9.0-13.0 | 0.045 | 0.03 | 10 × %C min Nb |
| 348 | S34800 | 0.08 | 2.00 | 1.00 | 17.0-19.0 | 9.0-13.0 | 0.045 | 0.03 | 0.2 Co; 10 × %C min Nb; 0.10 Ta |
| **Ferritic types** | | | | | | | | | |
| 405 | S40500 | 0.08 | 1.00 | 1.00 | 11.5-14.5 | ... | 0.04 | 0.03 | 0.10-0.30 Al |
| 409 | S40900 | 0.08 | 1.00 | 1.00 | 10.5-11.75 | 0.50 | 0.045 | 0.045 | 6 × %C min − 0.75 max Ti |
| 430 | S43000 | 0.12 | 1.00 | 1.00 | 16.0-18.0 | ... | 0.04 | 0.03 | ... |
| 434 | S43400 | 0.12 | 1.00 | 1.00 | 16.0-18.0 | ... | 0.04 | 0.03 | 0.75-1.25 Mo |
| 439 | S43035 | 0.07 | 1.00 | 1.00 | 17.0-19.0 | 0.50 | 0.04 | 0.03 | 0.15 Al; 12 × %C − 1.10 Ti |
| 442 | S44200 | 0.20 | 1.00 | 1.00 | 18.0-23.0 | ... | 0.04 | 0.03 | ... |
| 446 | S44600 | 0.20 | 1.50 | 1.00 | 23.0-27.0 | ... | 0.04 | 0.03 | 0.25 N |
| **Duplex (ferritic-austenitic) types** | | | | | | | | | |
| 329 | S32900 | 0.20 | 1.00 | 0.75 | 23.0-28.0 | 2.50-5.00 | 0.040 | 0.030 | 1.00-2.00 Mo |
| ... | S31803 | 0.03 | 2.00 | 1.00 | 21.0-23.0 | 4.50-6.50 | 0.030 | 0.020 | ... |
| **Martensitic types** | | | | | | | | | |
| 410 | S41000 | 0.15 | 1.00 | 1.00 | 11.5-13.5 | ... | 0.04 | 0.03 | ... |
| 414 | S41400 | 0.15 | 1.00 | 1.00 | 11.5-13.5 | 1.25-2.50 | 0.04 | 0.03 | ... |
| 416Se | S41623 | 0.15 | 1.25 | 1.00 | 12.0-14.0 | ... | 0.06 | 0.06 | 0.15 min Se |
| 420 | S42000 | 0.15 min | 1.00 | 1.00 | 12.0-14.0 | ... | 0.04 | 0.03 | ... |
| 431 | S43100 | 0.20 | 1.00 | 1.00 | 15.0-17.0 | 1.25-2.50 | 0.04 | 0.03 | ... |
| 440A | S44002 | 0.60-0.75 | 1.00 | 1.00 | 16.0-18.0 | ... | 0.04 | 0.03 | 0.75 Mo |
| **Precipitation-hardening types** | | | | | | | | | |
| PH 13-8 Mo | S13800 | 0.05 | 0.20 | 0.10 | 12.25-13.25 | 7.5-8.5 | 0.01 | 0.008 | 2.0-2.5 Mo; 0.90-1.35 Al; 0.01 N |
| 15-5 PH | S15500 | 0.07 | 1.00 | 1.00 | 14.0-15.5 | 3.5-5.5 | 0.04 | 0.03 | 2.5-4.5 Cu; 0.15-0.45 Nb |
| 17-4 PH | S17400 | 0.07 | 1.00 | 1.00 | 15.5-17.5 | 3.0-5.0 | 0.04 | 0.03 | 3.0-5.0 Cu; 0.15-0.45 Nb |

(a) Single values are maximum values unless otherwise indicated. (b) Optional

## Recommended Reading

- S.D. Washko and G. Aggen, Wrought Stainless Steels, *Metals Handbook*, 10th ed., Vol 1, ASM International, 1990, p 841-907
- S. Lampman, Elevated-Temperature Properties of Stainless Steels, *Metals Handbook*, 10th ed., Vol 1, ASM International, 1990, p 930-949
- R.M. Davison, T. DeBold, and M.J. Johnson, Corrosion of Stainless Steels, *Metals Handbook*, 9th ed., Vol 13, ASM International, 1987, p 547-565
- G.F. Vander Voort, Metallography and Microstructures of Wrought Stainless Steels, *Metals Handbook*, 9th ed., Vol 9, ASM International, 1985, p 279-296

Fig. 562

Fig. 562

*Class 30B gray iron, with areas of steadite (eutectic of small, rounded particles in light-colored ferrite) and type A graphite in a pearlite matrix. 1500×*

Fig. 563

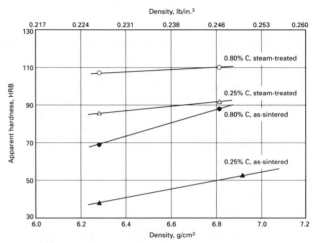

*Effect of steam treating on the hardness of sintered carbon P/M steels as a function of density*

about 705 °C (1300 °F). The fracture usually occurs at about 45° to the direction of rolling.

**steady load**. Loads that do not change in intensity, or change so slowly that they may be regarded as steady.

**steady-rate creep**. See *creep*.

**steadyrest**. In cutting or grinding, a stationary support for a long workpiece.

**steam hammer**. A type of *drop hammer* in which the ram is raised for each stroke by a double-action steam cylinder and the energy delivered to the workpiece is supplied by the velocity and weight of the ram and attached upper die driven downward by steam pressure. The energy delivered during each stroke can be varied.

**steam molding**. A process used to mold plastic parts from preexpanded beads of polystyrene using steam as a source of heat to expand the blowing agent in the material. The steam in most cases is in direct, intimate contact with the beads. It may also be used indirectly, by heating mold surfaces that are in contact with the bead.

**steam treatment**. The treatment of a sintered ferrous part in steam at temperatures between 510 and 595 °C (950 to 1100 °F) in order to produce a layer of black iron oxide (magnetite, or ferrous-ferric oxide, $FeO \cdot Fe_2O_3$) on the exposed surface for the purpose of increasing hardness and wear resistance (Fig. 563).

**steatite**. A compact, massive rock composed principally of *talc* (magnesium silicate). Ground steatite is used in *porcelain enamels* and *ceramic whiteware*.

**Steckel mill**. A cold reducing mill having two working rolls and two backup rolls, none of which is driven. The strip is drawn through the mill by a power reel in one direction as

far as the strip will allow and then reversed by a second power reel, and so on until the desired thickness is attained.

**steels**. See *Technical Brief 48*. For specific types, see *alloy steels, austenitic manganese steels, bearing steels, carbon steels, cast corrosion-resistant stainless steels, cast heat-resistant stainless steels, chromium-molybdenum heat-resistant steels, dual-phase steels, duplex stainless steels, high-strength low-alloy steels, low-alloy steels, stainless steels* (Technical Brief 47), *tool steels* (Technical Brief 56), and *ultrahigh-strength steels.*

**step aging**. Aging of metals at two or more temperatures, by steps, without cooling to room temperature after each step. See also *aging*, and compare with *interrupted aging* and *progressive aging*.

**stepback sequence**. See preferred term *backstep sequence*.

**step bearing**. A plain surface bearing that supports the lower end of a vertical shaft. Other types of bearings may be thus described when they are mounted on a step or bracket. See also *Rayleigh step bearing* and *stepped bearing*.

**step brazing**. The brazing of successive joints on a given part with filler metals of successively lower brazing temperatures so as to accomplish the joining without disturbing the joints previously brazed. A similar result can be achieved at a single brazing temperature if the remelt temperature of prior joints is increased by metallurgical interaction.

**stepdown test**. A test involving the preparation of a series of machined steps progressing inward from the surface of a metal bar (usually steel) for the purpose of detecting by visual inspection the internal laminations caused by inclusion segregates.

**stepped bearing**. A thrust bearing in which the working face consists of one or more shallow steps. A distinction should be drawn between a stepped bearing and a *step bearing*.

FIG. 564

*Stepped P/M parts*

**stepped compact**. A powder metallurgy compact with one (dual step) or more (multistep) abrupt cross-sectional changes, usually obtained by pressing with split punches, each section of which uses a different pressure and a different rate of compaction (Fig. 564). See also *split punch*.

**stepped extrusion**. See *extrusion*.

**step fracture** (glass). See *striation* (glass).

**step fracture** (metals). (1) Cleavage fractures that initiate on many parallel cleavage planes (Fig. 565). (2) Faceted cleavagelike fractures that occur during Stage I fatigue fractures (high-cycle, low-stress fractures) (Fig. 566).

**step soldering**. The soldering of successive joints on a given part with solders of successively lower soldering temperatures so as to accomplish the joining without disturbing the joints previously soldered.

**stereo angle**. One half of the angle through which the specimen is tilted when taking a pair of *stereoscopic micrographs*. The axis of rotation lies in the plane of the specimen.

**stereoisomer**. An isomer in which atoms are linked in the same order but differ in their arrangement. See also *isomer* and *isotactic stereoisomerism*.

FIG. 565

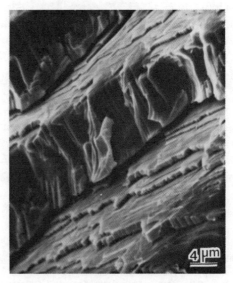

*Cleavage steps in a Cu-25Au (at.%) alloy that failed by transgranular stress-corrosion cracking*

## Technical Brief 48: Steels

STEELS, which are iron-base alloys that contain manganese, usually carbon, and often other alloying elements, constitute the most widely used category of metallic materials, primarily because they can be manufactured relatively inexpensively in large quantities to very precise specifications. They also provide a wide range of mechanical properties, from moderate yield strength levels (200 to 300 MPa, or 30 to 40 ksi) with excellent ductility to yield strengths exceeding 1400 MPa (200 ksi) with fracture toughness levels as high as 110 MPa$\sqrt{m}$ (100 ksi$\sqrt{in}$.).

Steels can be classified on the basis of composition, such as carbon, low-alloy, and high-alloy steel; microstructure, such as ferritic, austenitic, martensitic, and so forth; or product form, such as bar, plate, sheet, strip, tubing, or structural shape. Common use has further subdivided these broad classifications. For example, carbon steels are often classified according to carbon content as low-carbon ($\leq$0.3% C), medium-carbon (0.30 to 0.60% C), or high-carbon (0.60 to 1.00% C) steels. Likewise, alloy steels are classified according to the alloy content as low-alloy steels ($\leq$8% alloying elements) or high-alloy (>8% alloying elements) steels. Alloy steels are also often classified according to the principal alloying element (or elements) present. Thus,

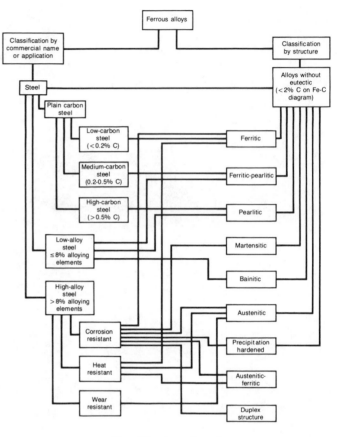

*Classification of steels*

there are nickel steels, chromium steels, chromium-molybdenum steels, and so on. The most widely used system for designating carbon and low-alloy steels is the Society of Automotive Engineers/American Iron and Steel Institute (SAE-AISI) system. This system is applied to semifinished forgings, hot-rolled and cold-finished bars, wire and seamless tubular goods, structural shapes, plates, sheet, strip, and welded tubes. Other systems for classifying carbon and low-alloy steels include the Unified Numbering System (UNS), Aerospace Materials Specifications (AMS), and ASTM (formerly the American Society for Testing and Materials) specifications. Similar classification systems have also been devised for high-alloy steels, such as stainless steels and tool steels.

The performance of steels depends on the properties associated with their microstructures. Because all the phases in steels are crystalline, steel microstructures are made up of various crystals, sometimes as many as three or four different types, which are physically blended by solidification, solid-state phase changes, hot deformation, cold deformation, and heat treatment. Each type of microstructure and product is developed to characteristic property ranges by specific processing routes that control and exploit microstructural changes.

**stereophotogrammetry.** A method of generating topographic maps of fracture surfaces by the use of a stereoscopic microscope interfaced to a microcomputer which calculates the three-dimensional coordinates of the fracture surface and produces the corresponding profile map, contour plot, or carpet plot (Fig. 567).

**stereoradiography.** A technique for producing paired radiographs that may be viewed with a stereoscope to exhibit a shadowgraph in three dimensions with various sections in perspective and spatial relation.

**stereoscopic micrographs.** A pair of micrographs (or fractographs) of the same area, but taken from different angles

## SAE-AISI system of designations for carbon and low-alloy steels

| Numerals and digits | Type of steel and nominal alloy content, % | Numerals and digits | Type of steel and nominal alloy content, % | Numerals and digits | Type of steel and nominal alloy content, % |
|---|---|---|---|---|---|
| **Carbon steels** | | **Nickel-chromium-molybdenum steels** | | **Chromium (bearing) steels** | |
| 10xx(a) ........... | Plain carbon (Mn 1.00 max) | 43xx ............... | Ni 1.82; Cr 0.50 and 0.80; Mo 0.25 | 50xxx ............... | Cr 0.50 |
| 11xx ............... | Resulfurized | 43BVxx ............ | Ni 1.82; Cr 0.50; Mo 0.12 and 0.25; V 0.03 min | 51xxx ............... | Cr 1.02 } C 1.00 min |
| 12xx ............... | Resulfurized and rephosphorized | 47xx ............... | Ni 1.05; Cr 0.45; Mo 0.20 and 0.35 | 52xxx ............... | Cr 1.45 |
| 15xx ............... | Plain carbon (max Mn range: 1.00–1.65) | 81xx ............... | Ni 0.30; Cr 0.40; Mo 0.12 | **Chromium-vanadium steels** | |
| | | 86xx ............... | Ni 0.55; Cr 0.50; Mo 0.20 | 61xx ............... | Cr 0.60, 0.80, and 0.95; V 0.10 and 0.15 min |
| **Manganese steels** | | 87xx ............... | Ni 0.55; Cr 0.50; Mo 0.25 | | |
| 13xx ............... | Mn 1.75 | 88xx ............... | Ni 0.55; Cr 0.50; Mo 0.35 | **Tungsten-chromium steel** | |
| | | 93xx ............... | Ni 3.25; Cr 1.20; Mo 0.12 | 72xx ............... | W 1.75; Cr 0.75 |
| **Nickel steels** | | 94xx ............... | Ni 0.45; Cr 0.40; Mo 0.12 | | |
| 23xx ............... | Ni 3.50 | 97xx ............... | Ni 0.55; Cr 0.20; Mo 0.20 | **Silicon-manganese steels** | |
| 25xx ............... | Ni 5.00 | 98xx ............... | Ni 1.00; Cr 0.80; Mo 0.25 | 92xx ............... | Si 1.40 and 2.00; Mn 0.65, 0.82, and 0.85; Cr 0 and 0.65 |
| **Nickel-chromium steels** | | **Nickel-molybdenum steels** | | | |
| 31xx ............... | Ni 1.25; Cr 0.65 and 0.80 | 46xx ............... | Ni 0.85 and 1.82; Mo 0.20 and 0.25 | **High-strength low-alloy steels** | |
| 32xx ............... | Ni 1.75; Cr 1.07 | 48xx ............... | Ni 3.50; Mo 0.25 | 9xx ............... | Various SAE grades |
| 33xx ............... | Ni 3.50; Cr 1.50 and 1.57 | | | **Boron steels** | |
| 34xx ............... | Ni 3.00; Cr 0.77 | **Chromium steels** | | xxBxx ............... | B denotes boron steel |
| **Molybdenum steels** | | 50xx ............... | Cr 0.27, 0.40, 0.50, and 0.65 | **Leaded steels** | |
| 40xx ............... | Mo 0.20 and 0.25 | 51xx ............... | Cr 0.80, 0.87, 0.92, 0.95, 1.00, and 1.05 | xxLxx ............... | L denotes leaded steel |
| 44xx ............... | Mo 0.40 and 0.52 | | | | |
| **Chromium-molybdenum steels** | | | | | |
| 41xx ............... | Cr 0.50, 0.80, and 0.95; Mo 0.12, 0.20, 0.25, and 0.30 | | | | |

(a) The xx in the last two digits of these designations indicates that the carbon content (in hundredths of a percent) is to be inserted.

For example, sheet steel formability depends on the single-phase ferritic microstructures of low-carbon cold-rolled and annealed steel, while high strength and wear resistance are enhanced by carefully developed microstructures of very fine carbides in fine martensite in fine-grain austenite of high-carbon hardened steels. In addition to the aforementioned wrought processing, cast steels and powder metallurgy steels are also produced.

Steels are used in thousands of applications; leading markets are the automotive industry, rail transportation, construction industry, oil and gas industry, agricultural and industrial equipment, appliances, and containers for packaging and shipping materials. The majority of the steel products produced in the United States are made from carbon steels (~85%), with alloy steel (~10%) and stainless steel (~2 to 5%) products used for applications requiring higher strength or corrosion resistance.

### Recommended Reading

- *Metals Handbook*, 10th ed., Vol 1, *Properties and Selection: Irons, Steels, and High-Performance Alloys*, ASM International, 1990
- *ASM Handbook*, Vol 4, *Heat Treating*, ASM International, 1991
- G. Krauss, *Steels—Heat Treatment and Processing Principles*, ASM International, 1989
- *The Making, Shaping and Treating of Steel*, 10th ed., United States Steel Corp., 1985

so that the two micrographs when properly mounted and viewed reveal the structures of the objects in their three-dimensional relationships (Fig. 568).

**stereoscopic specimen holder.** A specimen holder designed for the purpose of making stereoscopic micrographs that allows the tilting of the specimen through the *stereo angle*.

**stereospecific plastics.** Implies a specific or definite order of arrangement of molecules in space. This ordered regularity of the molecules in contrast to the branched or random arrangement found in other plastics permits close packing of the molecules and leads to high crystallinity (for example, in polypropylene).

FIG. 566

20 µm

*Cleavagelike, crystallographically oriented Stage I fatigue fracture in a cast nickel-chromium alloy*

**sterling silver**. A silver alloy containing at least 92.5% Ag, the remainder being unspecified but usually copper. Sterling silver is used for flat and hollow tableware and for various items of jewelry.

**stern-tube bearing**. The final bearing through which a propeller shaft passes in a boat or ship.

**stick electrode**. A shop term for *covered electrode*.

**stick welding**. See preferred term *shielded metal arc welding*.

FIG. 567

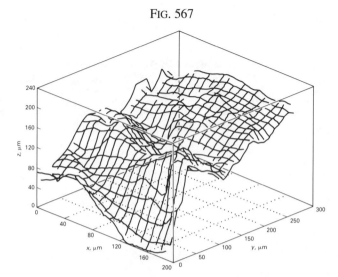

*Fracture surface map (carpet plot) of a Ti-10V-2Fe-3Al specimen by stereophotogrammetry*

**sticker breaks**. Arc-shaped *coil breaks*, usually located near the center of sheet or strip.

**stickout**. See preferred term *electrode extension*.

**stick-slip**. A relaxation oscillation usually associated with a decrease in the *coefficient of friction* as the relative velocity increases. Stick-slip was originally associated with formation and destruction of interfacial junctions on a microscopic scale. This is often the basic cause. The period depends on the velocity and on the elastic characteristics of the system. Stick-slip will not occur if the static friction is equal to or less than the dynamic friction. The motion resulting from stick-slip is sometimes referred to as jerky motion. See also *spragging*.

FIG. 568

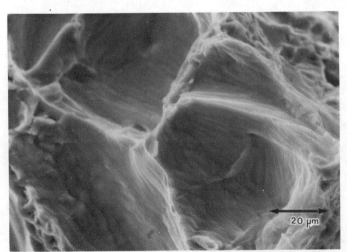
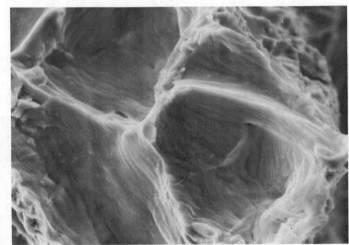

20 µm

*Stereo pair fractographs showing deep dimples in the fracture surface of commercially pure titanium*

**stiffness.** (1) The rate of stress with respect to strain; the greater the stress required to produce a given strain, the stiffer the material is said to be. (2) The ability of a material or shape to resist elastic deflection. For identical shapes, the stiffness is proportional to the modulus of elasticity. For a given material, the stiffness increases with increasing moment of inertia, which is computed from cross-sectional dimensions.

**stippled area.** See *hackle*.

**stitch weld.** See preferred term *intermittent weld*.

**stock.** A general term used to refer to a supply of metal in any form or shape and also to an individual piece of metal that is formed, forged, or machined to make parts.

**stoichiometric.** Having the precise weight relation of the elements in a chemical compound; or (quantities of reacting elements or compounds) being in the same weight relation as the theoretical combining weight of the elements involved.

**stoke (centistoke).** The centimeter-gram-second (cgs) unit of kinematic viscosity.

**Stokes Raman line.** A *Raman line* that has a frequency lower than that of the incident monochromatic beam. See also *Raman spectrum*.

**stoking** (obsolete). Presintering, or sintering, in such a way that powder metallurgy compacts are advanced through the furnace at a fixed rate by manual or mechanical means. See preferred term *continuous sintering*.

**stoneware.** A vitreous or semivitreous ceramic ware of fine texture, made primarily from either nonrefractory fireclay or some combination of clays, fluxes, and silica. Used for cookware, artware, and tableware.

**stop.** A device for positioning stock or parts in a die.

**stopoff.** A material used on the surfaces adjacent to the joint to limit the spread of soldering or brazing filler metal. See also *resist*.

**stopper rod.** A device in a bottom-pour ladle for controlling the flow of metal through the nozzle into a mold. The stopper rod consists of a steel rod, protective refractory sleeves, and a graphite stopper head. See also the figure accompanying the term *bottom-pour ladle*.

**stopping off.** (1) Applying a *resist*. (2) Depositing a metal (copper, for example) in localized areas to prevent carburization, decarburization, or nitriding in those areas. (3) Filling in a portion of a mold cavity to keep out molten metal.

**stops.** Metal pieces inserted between die halves used to control the thickness of a press-molded plastic part. Not a recommended practice, because the resin will receive less pressure, which can result in voids.

**storage life.** The period of time during which a liquid resin, packaged adhesive, or prepreg can be stored under specified temperature conditions and remain suitable for use. Also called shelf life.

**storage modulus.** A quantitative measure of elastic properties in polymers, defined as the ratio of the stress, in phase with the strain, to the magnitude of the strain. The storage modulus may be measured in tension or flexure, compression, or shear.

**storage stability.** A measure of the ability of a lubricant to undergo prolonged periods of storage without showing any adverse conditions due to oxidation, oil separation, contamination, or any type of deterioration.

**stored-energy welding.** Resistance welding with electrical energy accumulated electrostatically, electromagnetically, or electrochemically at a relatively low rate and made available at the higher rate required in welding.

**straddle milling.** Face milling a workpiece on both sides at once using two cutters spaced as required. See also *face milling* and *milling*.

**straightening.** (1) Any bending, twisting, or stretching operation to correct any deviation from straightness in bars, tubes, or similar long parts or shapes (Fig. 569). This deviation can be expressed as either camber (deviation from a straight line) or as total indicator reading (TIR) per unit of length. (2) A finishing operation for correcting misalignment in a forging or between various sections of a forging. See also *roll straightening*.

**straight polarity.** Direct-current arc welding circuit arrangement in which the electrode is connected to the negative terminal. A synonym for *direct current electrode negative (DCEN)*. Contrast with *reverse polarity*.

**straight-side press.** A sheet metal forming press which has a frame made up of a base, or bed; two columns; and a top member, or crown. In most straight-side presses, steel tie rods hold the base and crown against the columns. Straight-side presses have crankshaft, eccentric-shaft, or eccentric-gear drives. In a single-action straight-side press, the slide is equipped with air counterbalances to assist the drive in lifting the weight of the slide and the upper die to the top of the stroke (Fig. 570). Counterbalance cylinders provide a smooth press operation and easy slide adjustment. Die cushions are used in the bed for blankholding and for ejection of the work.

**strain.** The unit of change in the size or shape of a body due to force. Also known as nominal strain. The term is also used in a broader sense to denote a dimensionless number that characterizes the change in dimensions of an object during a deformation or flow process. See also *engineering strain*, *linear strain*, and *true strain*.

**strain-age embrittlement.** A loss in *ductility* accompanied by an increase in hardness and strength that occurs when low-carbon steel (especially rimmed or capped steel) is aged following *plastic deformation*. The degree of *embrittlement* is a function of aging time and temperature, occurring in a matter of minutes at about 200 °C (400 °F), but requiring a few hours to a year at room temperature.

FIG. 569

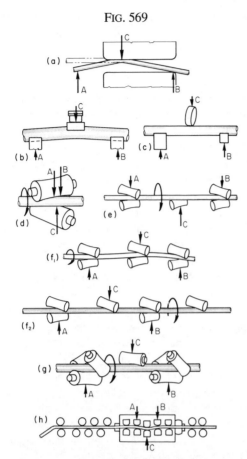

*Principle of straightening by bending. (a) Manual straightening with a grooved block. (b) Straightening in a press. (c) Simplest form of rotary straightening. (d) Two-roll straightening. (e) Five-roll straightening. (f₁ and f₂) Two arrangements of rolls for six-roll straightening. (g) Seven-roll straightening. (h) Wire straightening. In all methods shown, the bar is supported at points A and B, and force at C on the convex side causes straightening.*

FIG. 570

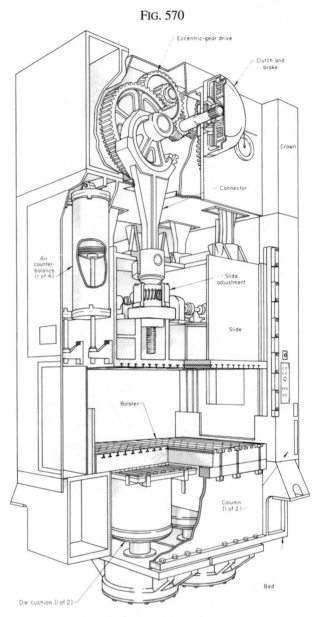

*Principal components of a single-action straight-side mechanical press*

**strain aging**. (1) *Aging* following plastic deformation. (2) The changes in ductility, hardness, yield point, and tensile strength that occur when a metal or alloy that has been cold worked is stored for some time. In steel, strain aging is characterized by a loss of ductility and a corresponding increase in hardness, yield point, and tensile strength.

**strain amplitude**. The ratio of the maximum deformation, measured from the mean deformation to the free length of the unstrained test specimen.

**strain energy**. The potential energy stored in a body by virtue of elastic deformation, equal to the work that must be done to produce this deformation. See also *elastic energy*, *resilience*, and *toughness*.

**strainer core**. In foundry practice, a perforated core in the gating system for preventing slag and other extraneous material from entering the casting cavity.

**strain etching**. Metallographic etching that provides information on deformed and undeformed areas if present side by side. In strained areas, more compounds are precipitated.

**strain gage**. A device for measuring small amounts of strain produced during tensile and similar tests on metal (Fig. 571). A coil of fine wire is mounted on a piece of paper, plastic, or similar carrier matrix (backing material), which is rectangular in shape and usually about 25 mm (1 in.) long. This is glued to a portion of metal under test. As the

FIG. 571

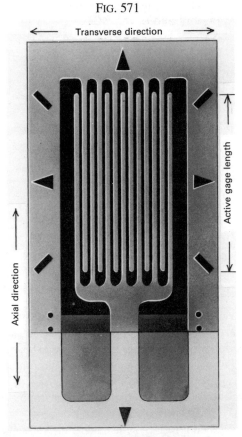

*Modern foil bonded resistance strain gage with a polyimide backing and encapsulation. About 6× actual size*

coil extends with the specimen, its electrical resistance increases in direct proportion. This is known as bonded resistance-strain gage. Other types of gages measure the actual deformation. Mechanical, optical, or electronic devices are sometimes used to magnify the strain for easier reading. See also *rosette*.

**strain hardening**. An increase in hardness and strength of metals caused by plastic deformation at temperatures below the recrystallization range. Also known as work hardening.

**strain-hardening coefficient**. See *strain-hardening exponent*.

**strain-hardening exponent**. The value of *n* in the relationship:

$$\sigma = K\varepsilon^n$$

where $\sigma$ is the *true stress*, $\varepsilon$ is the *true strain*, and *K*, which is called the strength coefficient, is equal to the true stress at a true strain of 1.0. The strain-hardening exponent, also called "*n*-value," is equal to the slope of the true stress/true strain curve up to maximum load, when plotted on log-log coordinates. The *n*-value relates to the ability of a sheet

material to be stretched in metalworking operations. The higher the *n*-value, the better the formability (stretchability).

**strain markings**. Manifestations of prior plastic deformation visible after etching of a metallographic section. These markings may be referred to as slip strain markings, twin strain markings, and so on, to indicate the specific deformation mechanism of which they are a manifestation.

**strain point**. That temperature corresponding to a specific rate of elongation of a glass fiber or a specific rate of midpoint deflection of a glass beam. At the strain point of glass, internal stresses are substantially relieved in a matter of hours.

**strain rate**. The time rate of straining for the usual tensile test. Strain as measured directly on the specimen gage length is used for determining strain rate. Because strain is dimensionless, the units of strain rate are reciprocal time.

**strain-rate sensitivity (*m*-value)**. The increase in stress ($\sigma$) needed to cause a certain increase in plastic strain rate ($\dot{\varepsilon}$) at a given level of plastic strain ($\varepsilon$) and a given temperature (*T*).

$$\text{Strain–rate sensitivity} = m = \left(\frac{\Delta \log \sigma}{\Delta \log \dot{\varepsilon}}\right)_{\varepsilon T}$$

**strain relaxation**. Reduction in internal strain over time.

**strain rods**. (1) Rods sometimes used on gapframe metal forming presses to lessen the frame deflection. (2) Rods used to measure elastic strain, and thus stresses, in frames of metal forming presses.

**strain state**. See *state of strain*.

**strand**. Normally, an untwisted bundle or assembly of continuous filaments used as a unit to reinforce plastics, including slivers, tows, ends, yarn, and so forth. Sometimes a single fiber or filament is called a strand.

**strand casting**. A generic term describing *continuous casting* of one or more elongated shapes such as billets, blooms, or slabs; if two or more shapes are cast simultaneously, they are often of identical cross section.

**strand count**. The number of *strands* in a plied *yarn* or *roving*.

**stranded electrode**. A composite filler metal electrode of stranded wires which may mechanically enclose materials to improve properties, stabilize the arc, or provide shielding.

**strand integrity**. The degree to which the individual filaments making up a *strand* or *end* are held together by the applied sizing. See also *size*.

**strand tensile test**. A tensile test of a single resin-impregnated *strand* of any fiber.

**strata**. In sampling, segments of a lot that may vary with respect to the property under study.

**stray current**. (1) Current flowing through paths other than the intended circuit. (2) Current flowing in electrodeposition by way of an unplanned and undesired bipolar electrode that may be the tank itself or a poorly connected electrode.

**stray-current corrosion**. Corrosion resulting from direct current flow through paths other than the intended circuit. For example, by an extraneous current in the earth.

**strength**. The maximum nominal stress a material can sustain. Always qualified by the type of stress (tensile, compressive, or shear).

**strength coefficient**. See *strain-hardening exponent*.

**stress**. The intensity of the internally distributed forces or components of forces that resist a change in the volume or shape of a material that is or has been subjected to external forces. Stress is expressed in force per unit area. Stress can be normal (tension or compression) or shear. See also *compressive stress, engineering stress, mean stress, nominal stress, normal stress, residual stress, shear stress, tensile stress,* and *true stress.*

**stress** (glass). Any condition of tension or compression existing within the glass, particularly due to incomplete annealing, temperature gradient, or inhomogeneity.

**stress amplitude**. One-half the algebraic difference between the maximum and minimum stresses in one cycle of a repetitively varying stress.

**stress concentration**. On a macromechanical level, the magnification of the level of an applied stress in the region of a notch, void, hole, or inclusion.

**stress concentration factor** ($K_t$). A multiplying factor for applied stress that allows for the presence of a structural discontinuity such as a notch or hole; $K_t$ equals the ratio of the greatest stress in the region of the discontinuity to the nominal stress for the entire section. Also called theoretical stress concentration factor.

**stress corrosion**. Preferential attack of areas under stress in a corrosive environment, where such an environment alone would not have caused corrosion.

**stress-corrosion cracking** (SCC). A cracking process that requires the simultaneous action of a corrodent and sustained tensile stress. This excludes corrosion-reduced sections that fail by fast fracture. It also excludes intercrystalline or transcrystalline corrosion, which can disintegrate an alloy without applied or residual stress. Stress-corrosion cracking may occur in combination with *hydrogen embrittlement*. Table 63 provides a partial listing of some of the more commonly observed alloy-environment combinations that result in SCC.

**stress crack**. External or internal cracks in a plastic caused by tensile stresses less than that of its short-time mechanical strength, frequently accelerated by the environment to which the plastic is exposed. The stresses that cause cracking may be present internally or externally or may be combinations of these stresses. See also *crazing.*

**Table 63 Alloy-environment systems exhibiting SCC**

| Alloy | Environment |
|---|---|
| Carbon steel | Hot nitrate, hydroxide, and carbonate/bicarbonate solutions |
| High-strength steels | Aqueous electrolytes, particularly when containing $H_2S$ |
| Austenitic stainless steels | Hot, concentrated chloride solutions; chloride-contaminated steam |
| High-nickel alloys | High-purity steam |
| α-brass | Ammoniacal solutions |
| Aluminum alloys | Aqueous $Cl^-$, $Br^-$, and $I^-$ solutions |
| Titanium alloys | Aqueous $Cl^-$, $Br^-$, and $I^-$ solutions; organic liquids; $N_2O_4$ |
| Magnesium alloys | Aqueous $Cl^-$ solutions |
| Zirconium alloys | Aqueous $Cl^-$ solutions; organic liquids; $I_2$ at 350 °C (660 °F) |

**stress-cracking failure**. The failure of a plastic by cracking or crazing some time after it has been placed under load. Time-to-failure may range from minutes to years. Causes include molded-in stresses, postfabrication shrinkage or warpage, and hostile environment.

**stress cycle**. The smallest segment of the stress-time function that is repeated periodically.

**stress cycles endured** (*N*). The number of cycles of a specified character (that produce fluctuating stress and strain) that a specimen has endured at any time in its stress history.

**stress equalizing**. A low-temperature heat treatment used to balance stresses in cold-worked material (metals) without an appreciable decrease in the mechanical strength produced by cold working.

**stress fracture**. See *fracture stress.*

**stress-induced crystallization**. The production of crystals in a polymer by the action of stress, usually in the form of an elongation. It occurs in fiber-spinning and rubber elongation and is responsible for enhanced mechanical properties.

**stress-intensity calibration**. A mathematical expression, based on empirical or analytical results, that relates the *stress-intensity factor* to load and crack length for a specific specimen planar geometry. Also known as *K* calibration.

**stress-intensity factor**. A scaling factor, usually denoted by the symbol *K*, used in *linear-elastic fracture mechanics* to describe the intensification of applied stress at the tip of a crack of known size and shape. At the onset of rapid crack propagation in any structure containing a crack, the factor is called the critical stress-intensity factor, or the *fracture toughness*. Various subscripts are used to denote different loading conditions or fracture toughnesses:

$K_c$. Plane-stress fracture toughness. The value of stress

intensity at which crack propagation becomes rapid in sections thinner than those in which plane-strain conditions prevail.

$K_I$. Stress-intensity factor for a loading condition that displaces the crack faces in a direction normal to the crack plane (also known as the opening mode of deformation).

$K_{Ic}$. Plane-strain fracture toughness. The minimum value of $K_c$ for any given material and condition, which is attained when rapid crack propagation in the opening mode is governed by plane-strain conditions.

$K_{Id}$. Dynamic fracture toughness. The fracture toughness determined under dynamic loading conditions; it is used as an approximation of $K_{Ic}$ for very tough materials.

$K_{ISCC}$. Threshold stress intensity factor for stress-corrosion cracking. The critical plane-strain stress intensity at the onset of stress-corrosion cracking under specified conditions.

$K_Q$. Provisional value for plane-strain fracture toughness.

$K_{th}$. Threshold stress intensity for stress-corrosion cracking. The critical stress intensity at the onset of stress-corrosion cracking under specified conditions.

$\Delta K$. The range of the stress-intensity factor during a fatigue cycle. See also *fatigue crack growth rate*.

**stress-intensity factor range (ΔK)**. In fatigue, the variation in the *stress-intensity factor* in a cycle, that is, $K_{max} - K_{min}$. See also the figure accompanying the term *fatigue crack growth rate*.

**stress raisers**. Design features (such as sharp corners) or mechanical defects (such as notches) that act to intensify the stress at these locations.

**stress range**. See *range of stress*.

**stress ratio (*A* or *R*)**. The algebraic ratio of two specified stress values in a stress cycle. Two commonly used stress ratios are: (1) the ratio of the alternating stress amplitude to the mean stress, $A = S_a/S_m$; and (2) the ratio of the minimum stress to the maximum stress. $R = S_{min}/S_{max}$.

**stress relaxation**. The time-dependent decrease in stress in a solid under constant constraint at constant temperature.

**stress-relaxation curve**. A plot of the remaining or relaxed stress as a function of time (Fig. 572). The relaxed stress

equals the initial stress minus the remaining stress. Also known as stress-time curve.

**stress-relief cracking**. Cracking in the *heat-affected zone* or weld metal that occurs during the exposure of weldments to elevated temperatures during postweld heat treatment, in order to reduce residual stresses and improve toughness, or high temperature service. Stress-relief cracking occurs only in metals that can precipitation-harden during such elevated-temperature exposure; it usually occurs as *stress raisers*, is intergranular in nature, and is generally observed in the coarse-grained region of the weld heat-affected zone. Also called postweld heat treatment cracking or stress relief embrittlement.

**stress-relief heat treatment**. Uniform heating of a structure or a portion thereof to a sufficient temperature to relieve the major portion of the residual stresses, followed by uniform cooling.

**stress relieving**. Heating to a suitable temperature, holding long enough to reduce residual stresses, and then cooling slowly enough to minimize the development of new residual stresses.

**stress-rupture strength**. See *creep-rupture strength*.

**stress-rupture test**. See *creep-rupture test*.

**stress state**. See *state of stress*.

**stress-strain curve**. A graph in which corresponding values of stress and strain from a tension, compression, or torsion test are plotted against each other (Fig. 573 to 575). Values of stress are usually plotted vertically (ordinates or *y*-axis) and values of strain horizontally (abscissas or *x*-axis). Also known as deformation curve and stress-strain diagram. See also *engineering strain* and *engineering stress*.

**stretch-bending test**. A simulative test for sheet metal formability in which a strip of sheet metal is clamped at its ends in lock beads and deformed in the center by a punch (Fig.

FIG. 572

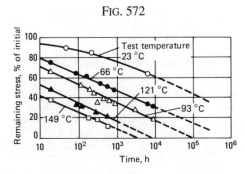

*Stress-relaxation curve for electrolytic tough pitch copper (C11000) alloy*

FIG. 573

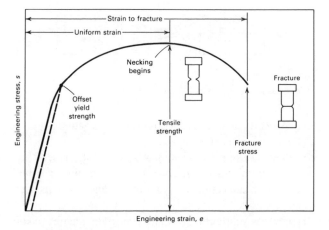

*Engineering stress-strain curve. Intersection of the dashed line with the curve determines the offset yield strength.*

FIG. 574

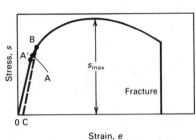

*Typical tension stress-strain curve for ductile metal indicating yielding criteria. Point A, elastic limit; point A', proportional limit; point B, yield strength; line C-B, offset yield strength; 0, intersection of the stress-strain curve with the strain axis*

FIG. 575

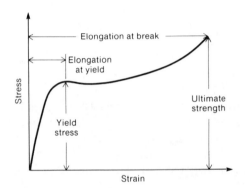

*Generalized tensile stress-strain curve for plastics*

FIG. 576

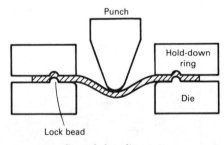

*Stretch-bending test*

576). Test conditions are chosen so that fracture occurs in the region of punch contact.

**stretcher leveling**. The leveling of a piece of sheet metal (that is, removing warp and distortion) by gripping it at both ends and subjecting it to a stress higher than its yield strength.

**stretcher straightening**. A process for straightening rod, tubing, and shapes by the application of tension at the ends of the stock. The products are elongated a definite amount to remove warpage.

**stretcher strains**. Elongated markings that appear on the surface of some sheet materials when deformed just past the yield point. These markings lie approximately parallel to the direction of maximum shear stress and are the result of localized yielding. See also the figure accompanying the term *Lüders lines*.

**stretch former**. (1) A machine used to perform *stretch forming* operations. (2) A device adaptable to a conventional press for accomplishing stretch forming.

**stretch forming**. The shaping of a metal sheet or part, usually of uniform cross section, by first applying suitable tension or stretch and then wrapping it around a die of the desired shape. The four methods of stretch forming are stretch draw forming (Fig. 577), stretch wrapping (Fig. 578), compression forming (Fig. 579), and radial draw forming. See also the figure accompanying the term *radial draw forming*.

FIG. 577

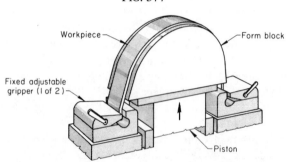

*Stretch draw forming method utilizing a form block*

FIG. 578

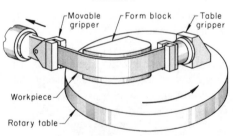

*Stretch wrapping technique*

FIG. 579

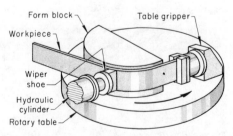

*Compression forming technique*

**stretching**. The extension of the surface of a metal sheet in all directions. In stretching, the flange of the flat blank is securely clamped. Deformation is restricted to the area initially within the die. The stretching limit is the onset of metal failure.

**striation** (glass). A fracture surface marking consisting of a separation of the advancing crack front into separate fracture planes. Also known as coarse hackle, step fracture, or lance. Striations may also be called shark's teeth or whiskers.

**striation** (metals). A fatigue fracture feature, often observed in electron micrographs, that indicates the position of the crack front after each succeeding cycle of stress. The distance between striations indicates the advance of the crack front across that crystal during one stress cycle, and a line normal to the striations indicates the direction of local crack propagation. See also *beach marks* and the figure accompanying the term *fatigue striation* (metals).

**Stribeck curve**. A graph showing the relationship between coefficient of friction and the dimensionless number ($\eta N/P$), where $\eta$ is the dynamic viscosity, $N$ is the speed (revolutions per minute for a journal), and $P$ is the load per unit of projected area (Fig. 580). The symbols $Z$ and $v$ (linear velocity) may be substituted for $\eta$ and $N$, respectively.

FIG. 580

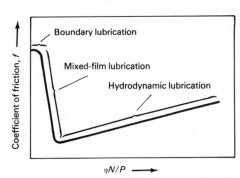

*Stribeck curve: coefficient of friction (and type of lubrication) versus dimensionless variable $\eta N/P$*

**strike**. (1) A thin electrodeposited film of metal to be overlaid with other plated coatings. (2) A plating solution of high covering power and low efficiency designed to electroplate a thin, adherent film of metal.

**striking**. Electrodepositing, under special conditions, a very thin film of metal that will facilitate further plating with another metal or with the same metal under different conditions.

**striking surface**. Those areas on the faces of a set of metal forming dies that are designed to meet when the upper die and lower die are brought together. The striking surface helps protect impressions from impact shock and aids in maintaining longer die life.

FIG. 581

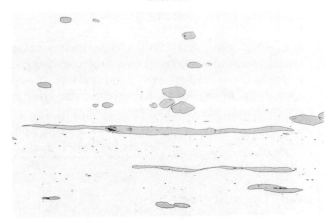

*Manganese sulfide stringers in wrought AISI 1214 free-machining carbon steel. 200×*

**stringer**. In wrought materials, an elongated configuration of microconstituents or foreign material aligned in the direction of working (Fig. 581). The term is commonly associated with elongated oxide or sulfide inclusions in steel.

**stringer bead**. A continuous weld bead made without appreciable transverse oscillation. Contrast with *weave bead*.

**stringiness**. The property of an adhesive that results in the formation of filaments or threads when adhesively bonded surfaces are separated.

**strip**. (1) A flat-rolled metal product of some maximum thickness and width arbitrarily dependent on the type of metal; narrower than *sheet*. (2) A roll-compacted metal powder product. See also *roll compacting* and *rolled compact*. (3) Removal of a powder metallurgy compact from the die. An alternative to ejecting or knockout. See also *ejection*.

**stripped die method**. A specific method of removal of a powder metallurgy compact after pressing, which keeps it in position between the punches while the die is retracted either upward or downward until the compact is fully exposed and freed by an upward withdrawal of the upper punch.

**stripper** (adhesives). A chemical solvent or acid that can remove an adhesive bond.

**stripper** (metals). A plate designed to remove, or strip, sheet metal stock from the punching members during the withdrawal cycle. Strippers are also used to guide small precision punches in close-tolerance dies, to guide scrap away from dies, and to assist in the cutting action. Strippers are made in two types: fixed and movable.

**stripper punch**. A punch that serves as the top or bottom of a metal forming die cavity and later moves farther into the die to eject the part or compact. See also *ejector rod* and *knockout* (3).

## Technical Brief 49: Structural Adhesives

A STRUCTURAL ADHESIVE is defined as a material used to transfer loads between adherends in service environments to which the assembly is typically exposed. Structural adhesives constitute about 35% of the total estimated sales of all adhesives and sealants. Their primary areas of application include automotive, aerospace, appliances, biomedical/dental construction, consumer electronics, fabric, furniture, industrial machines, and marine and sports equipment.

The most common type of structural adhesive is classified as a chemically reactive adhesive. The most widely used materials included in this classification are epoxies, polyurethanes, modified acrylics, cyanoacrylates, and anaerobics. Chemically reactive adhesives can be subdivided into two groups: one-component systems, which include moisture cure and heat-activated cure cate-

### Typical properties of the five most widely used chemically reactive structural adhesives

| Property | Epoxy | Polyurethane | Modified acrylic | Cyanoacrylate | Anaerobic |
|---|---|---|---|---|---|
| Substrates bonded . . . | Most | Most smooth, nonporous | Most smooth, nonporous | Most nonporous metals or plastics | Metals, glass, thermosets |
| Service temperature range, °C (°F) . . . . . | −55 to 121 (−67 to 250) | −157 to 79 (−250 to 175) | −73 to 121 (−100 to 250) | −55 to 79 (−67 to 175) | −55 to 149 (−67 to 300) |
| Impact resistance . . . . | Poor | Excellent | Good | Poor | Fair |
| Tensile shear strength, MPa (ksi). | 15.4 (2.20) | 15.4 (2.20) | 25.9 (3.70) | 18.9 (2.70) | 17.5 (2.50) |
| T-peel strength, N/m (lbf/in.) . . . . . . . . . . | <525 (3) | 14 000 (80) | 5250 (30) | <525 (3) | 1750 (10) |
| Heat cure or mixing required . . . . . . . . . | Yes | Yes | No | No | No |
| Solvent resistance . . . | Excellent | Good | Good | Good | Excellent |
| Moisture resistance . . | Excellent | Fair | Good | Poor | Good |
| Gap limitation, mm (in.). . . . . . . . . . . . . | None | None | 0.762 (0.030) | 0.254 (0.010) | 0.635 (0.025) |
| Odor. . . . . . . . . . . . . . . | Mild | Mild | Strong | Moderate | Mild |
| Toxicity. . . . . . . . . . . | Moderate | Moderate | Moderate | Low | Low |
| Flammability. . . . . . . . | Low | Low | High | Low | Low |

gories, and two-component systems, which are subdivided into mix-in and no-mix systems. One-component formulations that cure by moisture from the surrounding air or by adsorbed moisture from the surface of a substrate include polyurethanes, cyanoacrylates, and silicones. A one-component heat-activated system usually consists of two components that are premixed. Chemical families in this group include epoxies and epoxy-nylons, polyurethanes, polyimides, polybenzimidazoles, and phenolics.

**stripping.** (1) Removing a coating from a metal surface. (2) Removing a foundry pattern from the mold or the core box from the core.

**structural adhesives.** See *Technical Brief 49*.

**structural bond.** An adhesive bond that joins basic load-bearing parts of an assembly. The load may be either static or dynamic.

**structural ceramics.** See *Technical Brief 50*.

**structural foams.** Expanded plastic materials having integral solid skins and porous cores that exhibit outstanding rigidity. Structural foams involve a variety of thermoplas-

tic resins as well as urethanes. Table 64 lists mechanical properties of rigid integral skin structural foams. See also *polyurethanes* (Technical Brief 32).

**structural glass.** (1) Flat glass, usually colored or opaque, and frequently ground and polished, used for structural purposes. (2) Glass block, usually hollow, used for structural purposes.

**structural reaction injection molding (SRIM).** A molding process that is similar in practice to resin transfer molding. SRIM derives its name from the RIM process from which the resin chemistry and injection techniques have been

### Table 64 Mechanical properties of rigid integral skin polyurethane foams

| Property | System I | System II |
|---|---|---|
| Density, g/cm³ . . . . . . . . . . . . . . . . . . . . . . . . . . . . . . . . . . | 0.400 | 0.600 |
| Flexural strength, MPa (ksi). . . . . . . . . . . . . . . . . . . . . . . | 25 (3.63) | 45 (6.53) |
| Flexural stress at 10 mm (0.40 in.) deflection, MPa (ksi). . . . . . . . . . . . . . . . . . . . . . . . . . . . . . . . . . . . | 25 (3.63) | 44 (6.38) |
| Tensile strength, MPa (ksi). . . . . . . . . . . . . . . . . . . . . . . | 8 (1.16) | 18 (2.61) |
| Elongation at break, % . . . . . . . . . . . . . . . . . . . . . . . . . . | 7 | 7 |
| Flexural modulus, GPa (10⁶ psi) . . . . . . . . . . . . . . . . . . . | 0.60 (0.087) | 1.05 (0.162) |
| Tensile modulus, GPa (10⁶ psi) . . . . . . . . . . . . . . . . . . . | 0.35 (0.051) | 0.60 (0.087) |

Two-component mix-in systems consist of two separate components that must be metered in the proper ratio, mixed, and then dispersed. Chemical families in this group include epoxies, modified acrylics, polyurethanes, silicones, and phenolics. Two-component no-mix systems consist of two separate components that do not require careful metering because no mixing is involved. Adhesive is applied to one surface, while an accelerator is applied to a second surface. The surfaces are then joined. Modified acrylics are included in this group.

Other types of structural adhesives include evaporation or diffusion, hot-melt, delayed-tack, film, pressure-sensitive, and conductive adhesives. Evaporation or diffusion adhesives include materials based on organic solvents or water. In solvent-base systems, which include rubbers, phenolics, and polyurethanes, the adhesive solution is coated on the porous substrates. Following solvent evaporation and/or absorption into the substrates, the surfaces are joined. Water-base adhesives comprise materials that are totally soluble or dispersive in water (see *water-base adhesives*, Technical Brief 61). Hot-melt adhesives are 100% solid thermoplastics that are very loosely classified as structural adhesives because most will not withstand elevated-temperature loads without creep (see *hot-melt adhesives*, Technical Brief 21). Delayed-tack adhesives remain tacky following heat activation and cooling. Tack time ranges from minutes to days over a wide temperature range. Tack adhesives include styrene-butadiene copolymers, polyvinyl acetates, and polystyrene. Film adhesives are two-sided and one-sided tapes and films that are applied quickly and easily. Examples are nylon-epoxies, elastomer-epoxies, epoxy-phenolics, and high-temperature-resistant polyimides. Pressure-sensitive adhesives are capable of holding substrates together when they are brought into contact under brief pressure at room temperature (see *pressure-sensitive adhesives*, Technical Brief 35). Conductive adhesives include both electrically and thermally conductive materials that are added as fillers to the adhesive (usually epoxies). The most commonly used electrically conductive filler is silver in powder or flake form. Gold, copper, and aluminum are also used. Thermally conductive fillers include alumina (the most common), beryllia, boron nitride, and silica.

## Recommended Reading

- M.M. Gauthier, Types of Adhesives, *Engineered Materials Handbook*, Vol 3, ASM International, 1990, p 74-93
- A.H. Landrock, *Adhesives Technology Handbook*, Noyes Publications, 1985

adapted. The term structural is added to indicate the reinforced nature of the composite components manufactured by this process. In the SRIM process a preformed reinforcement is placed in a closed mold, and a reactive resin mixture is mixed under high pressure in a specially designed mix head (Fig. 582). Upon mixing, the reacting liquids flow at low pressure through a runner system and fill the mold cavity, impregnating the reinforcement material in the process. Once the mold cavity has filled, the resin quickly completes its reaction. A completed component can often be removed from the mold in as little as 1 min.

**structural shape**. A piece of metal of any of several designs accepted as standard by the structural branch of the iron and steel industries.

**structure**. As applied to a crystal, the shape and size of the unit cell and the location of all atoms within the unit cell. As applied to microstructure, the size, shape, and arrangement of phases. See also *unit cell*.

**structure factor**. A mathematically formulated term that relates the positions and identities of atoms in a crystalline material to the intensities of x-ray or electron beams diffracted from particular crystallographic planes.

**stud welding**. An arc welding process in which the contact surfaces of a stud, or similar fastener, and a workpiece are heated and melted by an arc drawn between them. The stud is then plunged rapidly onto the workpiece to form a weld (Fig. 583). Partial shielding may be obtained by the use of a ceramic ferrule surrounding the stud. Shielding gas or flux may or may not be used. The two basic methods of stud welding are known as stud arc welding, which produces a large amount of weld metal around the stud base and a relatively deep penetration into the base metal, and capacitor discharge stud welding, which produces a very small amount of weld metal around the stud base and shallow penetration into the base metal.

**styrene-acrylonitrile (SAN)**. A copolymer of about 70% styrene and 30% acrylonitrile, with higher strength, rigidity, and chemical resistance than can be attained with polystyrene alone. These copolymers may be blended

## Technical Brief 50: Structural Ceramics

ADVANCED (STRUCTURAL) CERAMIC materials are being used increasingly for load-bearing applications. Such applications require materials that have high strength at room temperature and/ or retain high strength at elevated temperatures, resist deformation (slow crack growth or creep), are damage tolerant, and are resistant to corrosion and oxidation and/or to abrasion and friction. Ceramics appropriate for such use offer a significant weight savings over metals. Applications include heat exchangers, automotive engine components such as turbocharger rotors and roller cam followers, power generation components,

### Industry, use, properties, and applications for structural ceramics

| Industry | Use | Property | Application |
|---|---|---|---|
| Fluid handling | Transport and control of aggressive fluids | Resistance to corrosion, mechanical erosion, and abrasion | Mechanical seal faces, meter bearings, faucet valve plates, spray nozzles, micro-filtration membranes |
| Mineral processing power generation | Handling ores, slurries, pulverized coal, cement clinker, and flue gas neutralizing compounds | Hardness, corrosion resistance, and electrical insulation | Pipe linings, cyclone linings, grinding media, pump components, electrostatic precipitator insulators |
| Wire manufacturing | Wear applications and surface finish | Hardness, toughness | Capstans and draw blocks, pulleys and sheaves, guides, rolls, dies |
| Pulp and paper | High-speed paper manufacturing | Abrasion and corrosion resistance | Slitting and sizing knives, stock-preparation equipment |
| Machine tool and process tooling | Machine components and process tooling | Hardness, high stiffness-to-weight ratio, low inertial mass, and low thermal expansion | Bearings and bushings, close tolerance fittings, extrusion and forming dies, spindles, metal-forming rolls and tools, coordinate-measuring machine structures |
| Thermal processing | Heat recovery, hot-gas cleanup, general thermal processing | Thermal stress resistance, corrosion resistance, and dimensional stability at extreme temperatures | Compact heat exchangers, heat exchanger tubes, radiant tubes, furnace components, insulators, thermocouple protection tubes, kiln furniture |
| Internal combustion engine components | Engine components | High-temperature resistance, wear resistance, and corrosion resistance | Exhaust port liners, valve guides, head faceplates, wear surface inserts, piston caps, bearings, bushings, intake manifold liners |
| Medical and scientific products | Medical devices | Inertness in aggressive environments | Blood centrifuge, pacemaker components, surgical instruments, implant components, lab ware |

cutting tools, biomedical implants, and processing equipment used for fabricating a variety of polymer, metal, and ceramic parts. The materials can be either monolithics or composites. A major obstacle to be overcome before these materials see more widespread use is their cost. Many of the processes used for fabrication

with butadiene as a terpolymer or grafted onto the butadiene to make acrylonitrile-butadiene-styrene resins. They are transparent and have high heat-deflection properties, excellent gloss, chemical resistance, hardness, rigidity, dimensional stability, and load-bearing capability.

**styrene-maleic anhydride (SMA).** Copolymers made by the copolymerization of styrene and maleic anhydride, with higher heat resistance than the parent styrenic and acrylonitrile-butadiene-styrene families. For structurally demanding applications, SMAs are reinforced with glass fibers. Most reinforced grades contain 20 wt% reinforcement.

**styrene-rubber plastics.** Plastics based on styrene polymers and rubbers, the styrene polymers being the greatest amount by mass.

**styrofoam pattern.** An expendable pattern of foamed plastic, especially expanded polystyrene, used in manufacturing castings by the lost foam process. See also the figure accompanying the term *lost foam casting*.

**sub-boundary structure (subgrain structure).** A network of low-angle boundaries, usually with misorientations less than 1° within the main grains of a microstructure.

**subcritical annealing.** An annealing treatment in which a steel is heated to a temperature below the $A_1$ temperature, then cooled slowly to room temperature. See also *transformation temperature*.

**subgrain.** A portion of a crystal or *grain*, with an orientation slightly different from the orientation of neighboring portions of the same crystal.

**submerged arc welding.** Arc welding in which the arc, between a bare metal electrode and the work, is shielded by a blanket of granular, fusible material overlying the joint (Fig. 584). Pressure is not applied to the joint, and filler

are labor intensive or have high rejection rates, resulting in unacceptably high costs for the final products.

The most important of the bulk monolithic materials for high-temperature structural applications are silicon nitride ($Si_3N_4$), silicon aluminum oxynitride (Sialon), silicon carbide (SiC), partially stabilized or transformation-toughened zirconia ($ZrO_2$), and alumina ($Al_2O_3$). These materials, each of which is covered in separate Technical Briefs in this volume, can exhibit high strengths (>500 MPa, or 70 ksi), moderate to high fracture toughness (4 to 14 MPa$\sqrt{m}$, or 3.6 to 12.7 ksi$\sqrt{in}$.), and low creep rates (<$10^{-9} \cdot s^{-1}$) at 1300 °C (2370 °F). Other candidates for structural applications are aluminum titanate ($Al_2TiO_5$), which has received much attention because of its good thermal shock resistance, and boron carbide ($B_4C$), the chief advantages of which are its high hardness (29 GPa, or $4.2 \times 10^6$ psi) and low density (2.50 g/cm$^3$).

Two-phase structural ceramics include siliconized silicon carbide (Si/SiC) and ceramic-matrix composites. The former is fabricated by first making a green body, or preform, of SiC, which can be infiltrated with liquid silicon to fill the open space in the preform. Silicon contents range from about 15 to 50%. Ceramic composites are designed to have improved damage tolerance or increased toughness through the addition of second-phase reinforcements in the form of particulates, whiskers, or fibers. Particulate-reinforced ceramics include transformation-toughened zirconia, in which $ZrO_2$ particulates act to deflect cracks from the main propagation path. Several whisker-reinforced compositions are available, the most successful to date being $SiC_w$-$Al_2O_3$ used as cutting tools. A number of continuous fibers are available for reinforcement of polymer, metal, glass, glass-ceramic, and ceramic matrices. These include compositions of carbon and graphite, silicon carbide, silicon carboxynitride, silicon nitride, alumina, mullite, and a variety of glasses. Single-crystal fibers of sapphire are also available.

The favorable thermal, chemical, and tribological properties of some of the structural ceramics can also be achieved by the use of ceramic coatings on other materials such as metals. A number of these ceramic coatings are in use, including $ZrO_2$, titanium nitride (TiN), titanium carbide (TiC), SiC, and diamond.

## Recommended Reading

- Structural Applications for Technical, Engineering, and Advanced Ceramics, M.K. Ferber and V.J. Tennery, Ed., *Engineered Materials Handbook*, Vol 4, ASM International, 1991, p 959-1013
- G.L. DePoorter, T.K. Brog, and M.J. Readey, Structural Ceramics, *Metals Handbook*, 10th ed., Vol 2, ASM International, 1990, p 1019-1024

metal is obtained from the consumable electrode (and sometimes from a supplementary welding rod).

**submerged-electrode furnace.** A furnace used for liquid carburizing of parts by heating molten salt baths with the use of electrodes submerged in the ceramic lining (Fig. 585). See also *immersed-electrode furnace*.

**submicron powder.** Any powder whose particles are smaller than ~1 μm.

**submicroscopic.** Below the resolution of a microscope.

**subsample.** A portion taken from a *sample*. A laboratory sample may be a subsample of a gross sample; similarly, a test portion may be a subsample of a laboratory sample.

**subsieve analysis.** Size distribution of particles that will pass through a standard 325-mesh sieve having 44-μm openings. See also the table accompanying the term *sieve analysis*.

**subsieve fraction.** Particles that will pass through a 44-μm

(325-mesh) screen. See also the table accompanying the term *sieve analysis*.

**subsieve size.** See preferred term *subsieve fraction*.

**sub-sow block (die holder).** A block used as an adapter in order to permit the use of forging dies that otherwise would not have sufficient height to be used in the particular unit or to permit the use of dies in a unit with different *shank* sizes. See also *sow block*.

**substitutional element.** An alloying element with an atomic size and other features similar to the solvent that can replace or substitute for the solvent atoms in the lattice and form a significant region of solid solution in the *phase diagram*.

**substitutional solid solution.** A *solid solution* in which the solvent and solute atoms are located randomly at the atom sites in the crystal structure of the solution. See also *interstitial solid solution*.

FIG. 582

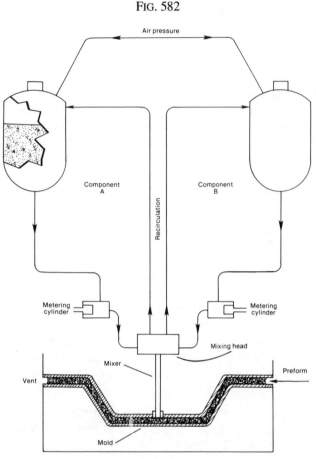

*Schematic of the structural reaction injection molding process*

FIG. 584

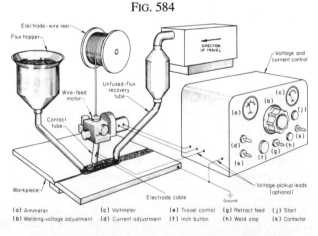

(a) Ammeter      (c) Voltmeter      (e) Travel control      (g) Retract feed      (j) Start
(b) Welding-voltage adjustment      (d) Current adjustment      (f) Inch button      (h) Weld stop      (k) Contactor

*Typical automatic submerged arc welding unit*

**substrate**. (1) The material, workpiece, or substance on which the coating is deposited. (2) A material upon the surface of which an adhesive-containing substance is spread for any purpose, such as bonding or coating. A broader term than *adherend*. (3) In electronic devices, a

FIG. 583

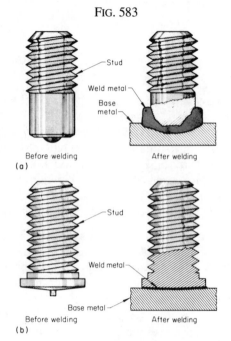

Before welding          After welding
(a)

Before welding          After welding
(b)

*Sections through studs prepared by (a) stud arc welding and (b) capacitor discharge stud welding*

FIG. 585

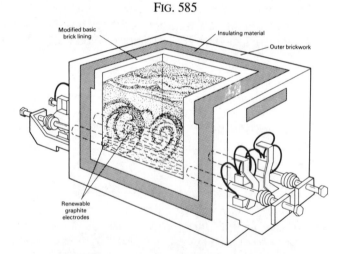

*Internally heated salt bath furnace with submerged electrodes. This furnace has a modified brick lining for use with carburizing salts.*

body, board, or layer of material, on which some other active or useful material(s) or component(s) may be deposited or laid, for example, electronic circuitry laid on an alumina ceramic board. (4) In catalysts, the formed, porous, high-surface area carrier on which the catalytic agent is widely and thinly distributed for reasons of performance and economy.

**substrate preparation**. The set of operations, including cleaning, degreasing, and finishing applied to the base material prior to applying a coating; intended to ensure an adequate bond to the coating.

**substrate temperature.** In thermal spraying, the temperature attained by the base material as the coating is applied. Proper control of the substrate temperature by intermittent spraying or by the application of external cooling will minimize stresses caused by substrate and coating thermal expansion differences.

**substructure.** Same as *sub-boundary structure*.

**subsurface corrosion.** Formation of isolated particles of corrosion products beneath a metal surface. This results from the preferential reactions of certain alloy constituents to inward diffusion of oxygen, nitrogen, or sulfur.

**subzero machining.** Using refrigerant or other means for cooling the workpiece during, or before, machining.

**suck-back.** See preferred term *concave root surface*.

**sulfidation.** The reaction of a metal or alloy with a sulfur-containing species to produce a sulfur compound that forms on or beneath the surface on the metal or alloy (Fig. 586).

**sulfide spheroidization.** A stage of overheating ferrous metals in which sulfide inclusions are partly or completely spheroidized.

**sulfide stress cracking (SSC).** Brittle fracture by cracking under the combined action of *tensile stress* and *corrosion* in the presence of water and hydrogen sulfide. See also *environmental cracking*.

**sulfide-type inclusions.** In steels, nonmetallic inclusions composed essentially of manganese iron sulfide solid solutions (Fe,Mn)S. They are characterized by plasticity at hot-rolling and forging temperatures and, in the hot-worked product, appear as dove-gray elongated inclusions varying from a threadlike to oval outline. See also the figure accompanying the term *stringer*.

**sulfochlorinated lubricant.** A lubricant containing chlorine and sulfur compounds, which react with a rubbing surface at elevated temperatures to form a protective film. There may be a synergistic effect, producing a faster reaction than with sulfur or chlorine additives alone.

**sulfur dome.** An inverted container, holding a high concentration of sulfur dioxide gas, used in die casting to cover a pot of molten magnesium to prevent burning.

**sulfurized lubricant.** A lubricant containing sulfur or a sulfur compound that reacts with a rubbing surface at elevated temperatures to form a protective film. The shear strength of the sulfide film formed on ferrous materials is lower than that of the metal but greater than that of the film formed by reaction with a chlorinated lubricant.

**sulfur print.** A macrographic method of examining for distribution of sulfide inclusions by placing a sheet of wet acidified photographic paper in contact with the polished sheet surface to be examined.

**sum peak.** An artifact encountered in x-ray analysis during pulse pileup where two x-rays simultaneously entering the

FIG. 586

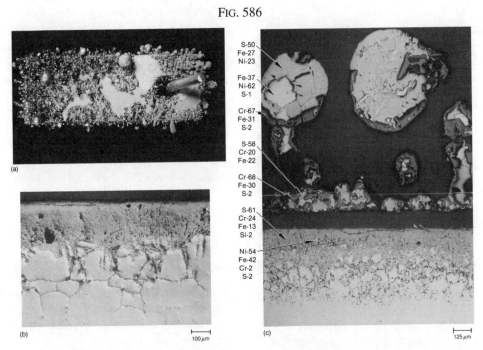

*Sulfidation attack of Alloy 800 test coupons exposed to a coal gasification environment. (a) and (b) Macrograph and cross-sectional micrograph, respectively, of test coupon. (c) Higher-magnification micrograph showing external sulfides, sulfide scale, and intergranular sulfidation*

detector are counted as one x-ray, the energy of which is equal to the sum of both x-rays. See also *escape peak*.

**superabrasives**. See *Technical Brief 51*.

**superalloys**. See *Technical Brief 52*.

**superconductivity**. A property of many metals, alloys, compounds, oxides, and organic materials at temperatures near absolute zero by virtue of which their electrical resistivity vanishes and they become strongly diamagnetic. See also *superconductors* (Technical Brief 53).

**superconductors**. See *Technical Brief 53*.

**supercooling**. Cooling of a substance below the temperature at which a change of state would ordinarily take place without such a change of state occurring, for example, the cooling of a liquid below its freezing point without freezing taking place; this results in a *metastable* state.

**superficial hardness test**. See *Rockwell superficial hardness test*.

**superfines**. The portion of a metal powder that is composed of particles smaller than a specified size, usually 10 μm.

**superfinishing**. An abrasive process utilizing either a curved bonded honing stick (stone) for a cylindrical workpiece (Fig. 587) or a cup wheel for flat and spherical work. A large contact area, 30% approximately, exists between workpiece and abrasive. The object of superfinishing is to remove surface fragmentation and to correct inequalities in geometry, such as grinding feed marks and chatter marks. Also known as microhoning. See also *honing*.

FIG. 587

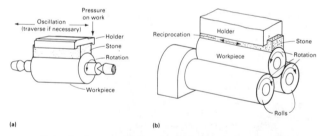

(a)                                              (b)

*Principal components and their motion in (a) cylindrical and (b) centerless superfinishing (microhoning)*

**superheating**. (1) Heating of a substance above the temperature at which a change of state would ordinarily take place without a change of state occurring, for example, the heating of a liquid above its boiling point without boiling taking place; this results in a metastable state. (2) Any increment of temperature above the melting point of a metal; sometimes construed to be any increment of temperature above normal casting temperatures introduced for the purpose of refining, alloying, or improving fluidity.

**superlattice**. See *ordered structure*.

**superplastic forming (SPF)**. A strain rate sensitive sheet metal forming process that uses characteristics of ma-

FIG. 588

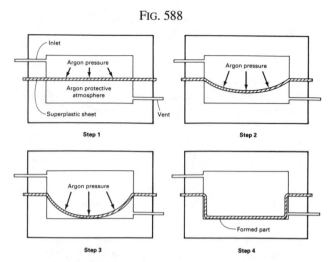

*Schematic of the blow forming technique for superplastic forming*

terials exhibiting high tensile elongation. Superplastic forming methods include: blow molding, in which gas pressure is imposed on a superplastic diaphragm, causing the material to form into the die configuration (Fig. 588); vacuum forming, a process similar to blow molding except that the forming pressure is limited to atmospheric pressure (100 kPa, or 15 psi) versus the maximum pressure of 700 to 3400 kPa (100 to 500 psi) for blow molding; thermoforming methods adopted from plastics technology (see *thermoforming*), which involve a moving or adjustable die member in conjunction with gas pressure or vacuum; and superplastic forming/diffusion bonding (SPF/DB), which combines blow molding and solid-state bonding. See also *diffusion bonding* and *superplasticity*.

**superplasticity**. The ability of certain metals (most notably aluminum- and titanium-base alloys) to develop extremely high tensile elongations at elevated temperatures and under controlled rates of deformation. Table 65 lists properties of superplastic alloys.

**supersaturated**. A metastable solution in which the dissolved material exceeds the amount the solvent can hold in normal equilibrium at the temperature and other conditions that prevail.

**supersonic**. Pertains to phenomena in which the speed is higher than that of sound. Not synonymous with ultrasonic; see also *ultrasonic frequency*.

**supplemental operation**. See *secondary operation*.

**support pins**. Rods or pins of precise length used to support the overhang of irregularly shaped punches in metal forming presses.

**support plate**. A plate that supports a draw ring or draw plate in a sheet metal forming press. It also serves as a spacer. See also *draw plate* and *draw ring*.

**supporting electrode**. An electrode, other than a self-elec-

## Technical Brief 51: Superabrasives

SUPERABRASIVES collectively refer to diamond and cubic boron nitride (CBN), both of which are produced by synthetic means. Diamond is the hardest material known, and CBN is the second hardest. Because of their high hardness, abrasion resistance, and other unique properties, these materials find extensive use in a wide variety of abrasive or cutting applications.

The primary objective in the synthesis of diamond and CBN is to transform a crystal structure from a soft hexagonal form to a hard cubic form. In the case of carbon, for example, hexagonal carbon (graphite) would be transformed into cubic

### Typical properties of abrasives

| | Abrasive | | | |
|---|---|---|---|---|
| | Superabrasive | | Conventional | |
| | Diamond | Cubic boron nitride | Aluminum oxide | Silicon carbide |
| Chemical composition .................... | Carbon | BN | $Al_2O_3$ | SiC |
| Density, $g/cm^3$ ........................ | 3.52 | 3.48 | 3.92 | 3.21 |
| Knoop hardness, HK (GPa) ............ | 60–110 | 40–70 | 21 | 24 |
| Relative thermal conductivity ........... | 100–350 | 35–120 | 1 | 10 |
| Coefficient of thermal expansion, $\times 10^{-6}$ mm/mm/°C........................ | 4.8 | 5.6 | 8.6 | 4.5 |
| Threshold temperature for degradation (ambient conditions), °C (°F)......... | 800 (1470) | 1400 (2550) | 1750 (3180) | 1500 (2730) |

carbon (diamond) (see the figure accompanying the term *diamond*). The synthesis of CBN or diamond grit is normally achieved by subjecting hexagonal carbon or boron nitride to high temperatures and high pressures with large special-purpose presses. By the simultaneous application of heat and pressure, hexagonal carbon or boron nitride can be transformed into a hard cubic form. Synthetic CBN and diamond grains can be used as loose abrasives, as bonded abrasives in grinding wheels and hones, and as bonded abrasives in single-point cutting tools. When used for the latter application, diamond and CBN are referred to as *ultrahard tool materials*.

The synthesis of diamond was first demonstrated by the General Electric Company in the 1950s. This invention subsequently led to the rapid growth of diamond for industrial applications. In addition to its high hardness, diamond is an excellent heat conductor and has an extremely low coefficient of friction. However, at temperatures above 800 °C (1475 °F), diamond tends to graphitize, thus losing its value as a wear-resistant abrasive. Diamond suffers rapid wear and chemical dissolution (erosion) when abraded against iron. Therefore, it is not normally used as an abrasive against ferrous materials. Diamond is used extensively to grind stone, concrete, carbides, glass, ceramics, plastics, and composites. Polycrystalline diamond tool (machining) applications include nonferrous materials (aluminum-silicon alloys, copper alloys, and tungsten carbides), fiberboard, composites (graphite-epoxy, carbon-carbon, and fiberglass-reinforced plastic), and ceramics.

Like diamond, CBN also has a cubic structure (see the figure accompanying the term *cubic boron nitride*). In its hexagonal close-packed structure, hexagonal boron nitride is similar to graphite and is used as a solid lubricant. Borazon, the General Electric trade name for CBN, was first synthesized in 1959. Unlike diamond, CBN is not very reactive with iron and therefore is used for grinding and machining of ferrous materials such as hardened steels and cast irons. Cubic boron nitride is also used for precision grinding and machining of nickel- and cobalt-base superalloys and hardfacing materials.

### Recommended Reading

- T.J. Clark and R.C. DeVries, Superabrasives and Ultrahard Tool Materials, *Metals Handbook*, 10th ed., Vol 2, ASM International, 1990, p 1008-1018
- K. Subramanian, Superabrasives, *Metals Handbook*, 9th ed., Vol 16, ASM International, 1989, p 453-471

trode, on which the sample is supported during spectrochemical analysis.

**surface alterations** (metals). Irregularities or changes on the surface of a material due to machining or grinding operations. The types of surface alterations associated with metal removal practices include mechanical (for example, plastic deformation, hardness variations, cracks, etc.), metallurgical (for example, phase transformations, twinning, recrystallization, and untempered or overtempered martensite), chemical (for example, intergranular attack, embrittlement, and pitting), thermal (heat-affected zone, recast, or redeposited metal, and resolidified material), and electrical surface alterations (conductivity change or resistive heating).

## Technical Brief 52: Superalloys

SUPERALLOYS are heat-resistant alloys based on nickel (Ni), iron-nickel (Fe-Ni), or cobalt (Co) that exhibit a combination of mechanical strength and resistance to surface degradation that is unmatched by other metallic alloys. Superalloys are primarily used in gas turbines, coal conversion plants, and chemical process industries, and for other specialized applications requiring high heat and corrosion resistance.

Superalloys consist of a face-centered cubic (fcc) austenitic gamma ($\gamma$) phase matrix plus a variety of secondary phases. The principal secondary phases are carbides (MC, $M_{23}C_6$, $M_6C$, and $M_7C_3$) in all superalloy types and gamma prime ($\gamma'$) fcc ordered $Ni_3(Al,Ti)$ intermetallic compound in Ni- and Fe-Ni-base alloys. In alloys containing niobium and tantalum, the primary strengthening phase is gamma double prime ($\gamma''$), a body-centered tetragonal phase. Superalloys derive their strength from solid-solution hardeners and precipitating phases. Carbides may provide limited strengthening directly (e.g., through dispersion hardening) or, more commonly, indirectly (e.g., by stabilizing grain boundaries against excessive shear). In addition to those elements that produce solid-solution hardening and promote carbide and $\gamma$ formation, other elements (e.g., boron, zirconium, hafnium, and cerium) are added to enhance mechanical and/or chemical properties.

The three types of superalloys (Fe-Ni, Ni-, and Co-base) are further subdivided into wrought, cast, and powder metallurgy alloys. Cast alloys can be further broken down into polycrystalline, directionally solidified, and single-crystal superalloys (see also *directionally solidified castings* and *single-crystal superalloys*). The most important class of Fe-Ni-base superalloys includes those alloys which are strengthened by intermetallic-compound precipitation in an fcc matrix. The most common precipitate is $\gamma'$, typified by alloys A-286 and Incoloy 901, but some alloys precipitate $\gamma''$, typified by Inconel 718. Another class of cast Fe-Ni-base superalloys is hardened by carbides, nitrides, and carbonitrides; some tungsten and molybdenum may be added to produce solid-solution hardening. Other Fe-Ni-base alloys are modified stainless steels primarily strengthened by solid-solution hardening.

The most important class of Ni-base superalloys is strengthened by intermetallic-compound precipitation in an fcc matrix. The strengthening precipitate is $\gamma'$, typified by Waspaloy and Udimet 700. Another class is represented by Hastelloy X, which is essentially solid-solution strengthened, but which also derives some strengthening from carbide precipitation produced through a working-plus-aging schedule. A third class includes oxide-dispersion-strengthened (ODS) alloys, which are strengthened by dispersions of inert particles such as yttria (see also *dispersion-strengthened material* and *mechanical alloying*).

Cobalt-base superalloys are strengthened by solid-solution alloying and carbide precipitation. Unlike the Fe-Ni

**surface area.** (1) The area, per unit weight of a granular or powdered solid, of all external and internal surfaces that ar accessible to a penetrating gas or liquid. Surface area is given as square meters per kilogram ($m^2$/kg) or square centimeters per gram ($cm^2$/g). (2) The actual area of the surface of a casting or cavity. The surface area is always greater than the *projected area*.

**surface checking.** Same as *checks*.

**surface contact points.** In powder technology, the points at which abutting particles make contact during contacting and which grow into necks during sintering. See also *neck* and *neck formation*.

**surface damage.** In tribology, damage to a solid surface resulting from mechanical contact with another substance, surface, or surfaces moving relatively to it and involving the displacement or removal of material. In certain contexts, *wear* is a form of surface damage in which material is progressively removed. In another context, surface damage involves a deterioration of function of a solid surface even though there is no material loss from that surface. Surface damage may therefore precede wear.

**surface diffusion.** One of the primary diffusion mechanisms during sintering. It is predominant for smaller particles and lower sintering temperatures as compared to other diffusion mechanisms, such as lattice or volume diffusion, which are prevalent for larger particles and higher temperatures. See also *volume diffusion*.

**surface distress.** In bearings and gears, damage to the contacting surfaces that occurs through intermittent solid contact involving some degree of sliding and/or surface fatigue. Surface distress can occur in numerous forms depending on the conditions under which the bearing or gear was operated and on the nature of the interaction between the contacting surfaces.

## Compositions of selected superalloys

| Alloy | Cr | Ni | Co | Mo | W | Nb | Ti | Al | Fe | C | Other |
|---|---|---|---|---|---|---|---|---|---|---|---|
| **Fe-Ni-base** | | | | | | | | | | | |
| 19-9DL | 19.0 | 9.0 | ... | 1.25 | 1.25 | 0.4 | 0.3 | ... | 66.8 | 0.30 | 1.10 Mn; 0.6 Si |
| Incoloy 800 | 21.0 | 32.5 | ... | ... | ... | ... | 0.38 | 0.38 | 45.7 | 0.05 | 0.8 Mn; 0.5 Si |
| A-286 | 15.0 | 26.0 | ... | 1.25 | ... | ... | 2.0 | 0.2 | 55.2 | 0.04 | 0.005 B; 0.3 V |
| V-57 | 14.8 | 27.0 | ... | 1.25 | ... | ... | 3.0 | 0.25 | 48.6 | 0.08 max | 0.01 B; 0.5 max V |
| Incoloy 901 | 12.5 | 42.5 | ... | 6.0 | ... | ... | 2.7 | ... | 36.2 | 0.10 max | ... |
| Inconel 718 | 19.0 | 52.5 | ... | 3.0 | ... | 5.1 | 0.9 | 0.5 | 18.5 | 0.08 max | 0.15 max Cu |
| Hastelloy X | 22.0 | 49.0 | 1.5 max | 9.0 | 0.6 | ... | ... | 2.0 | 15.8 | 0.15 | ... |
| **Ni-base** | | | | | | | | | | | |
| Waspaloy | 19.5 | 57.0 | 13.5 | 4.3 | ... | ... | 3.0 | 1.4 | 2.0 max | 0.07 | 0.006 B; 0.09 Zr |
| M252 | 19.0 | 56.5 | 10.0 | 10.0 | ... | ... | 2.6 | 1.0 | <0.75 | 0.15 | 0.005 B |
| Udimet 500 | 19.0 | 48.0 | 19.0 | 4.0 | ... | ... | 3.0 | 3.0 | 4.0 max | 0.08 | 0.005 B |
| Udimet 700 | 15.0 | 53.0 | 18.5 | 5.0 | ... | ... | 3.4 | 4.3 | <1.0 | 0.07 | 0.03 B |
| Astroloy | 15.0 | 56.5 | 15.0 | 5.25 | ... | ... | 3.5 | 4.4 | <0.3 | 0.06 | 0.03 B; 0.06 Zr |
| René 80 | 14.0 | 60.0 | 9.5 | 4.0 | 4.0 | ... | 5.0 | 3.0 | ... | 0.17 | 0.015 B; 0.03 Zr |
| IN-100 | 10.0 | 60.0 | 15.0 | 3.0 | ... | ... | 4.7 | 5.5 | <0.6 | 0.15 | 1.0 V; 0.06 Zr; 0.015 B |
| René 95 | 14.0 | 61.0 | 8.0 | 3.5 | 3.5 | 3.5 | 2.5 | 3.5 | <0.3 | 0.16 | 0.01 B; 0.05 Zr |
| MAR-M 247 | 8.25 | 59.0 | 10.0 | 0.7 | 10.0 | ... | 1.0 | 5.5 | <0.5 | 0.15 | 0.015 B; 0.05 Zr; 1.5 Hf; 3.0 Ta |
| IN MA-754 | 20.0 | 78.5 | ... | ... | ... | ... | 0.5 | 0.3 | ... | ... | 0.6 $Y_2O_3$ |
| IN MA-6000E | 15.0 | 68.5 | ... | 2.0 | 4.0 | ... | 2.5 | 4.5 | ... | 0.05 | 1.1 $Y_2O_3$; 2.0 Ta; 0.01 B; 0.15 Zr |
| **Co-base** | | | | | | | | | | | |
| Haynes 25(L-605) | 20.0 | 10.0 | 50.0 | ... | 15.0 | ... | ... | ... | 3.0 | 0.10 | 1.5 Mn |
| Haynes 188 | 22.0 | 22.0 | 37.0 | ... | 14.5 | ... | ... | ... | 3.0 max | 0.10 | 0.90 La |
| S-816 | 20.0 | 20.0 | 42.0 | 4.0 | 4.0 | 4.0 | ... | ... | 4.0 | 0.38 | ... |
| X-40 | 22.0 | 10.0 | 57.5 | ... | 7.5 | ... | ... | ... | 1.5 | 0.50 | 0.5 Mn; 0.5 Si |
| WI-52 | 21.0 | ... | 63.5 | ... | 11.0 | ... | ... | ... | 2.0 | 0.45 | 2.0 Nb + Ta |
| MAR-M 302 | 21.5 | ... | 58.0 | ... | 10.0 | ... | ... | ... | 0.5 | 0.85 | 9.0 Ta; 0.005 B; 0.2 Zr |
| MAR-M 509 | 23.5 | 10.0 | 54.5 | ... | 7.0 | ... | 0.2 | ... | ... | 0.6 | 0.5 Zr; 3.5 Ta |
| J-1570 | 20.0 | 28.0 | 46.0 | ... | ... | ... | 4.0 | ... | 2.0 | 0.2 | ... |

and Ni-base alloys, no intermetallic phase has been found that will strengthen Co-base alloys to the same degree that $\gamma'$ or $\gamma''$ strengthens the other superalloys.

## Recommended Reading

- N.S. Stoloff, Wrought and P/M Superalloys, *Metals Handbook*, 10th ed., Vol 1, ASM International, 1990, p 950-980
- G.L. Erickson, Polycrystalline Cast Superalloys, *Metals Handbook*, 10th ed., Vol 1, ASM International, 1990, p 981-994
- K. Harris, G.L. Erickson, and R.E. Schwer, Directionally Solidified and Single-Crystal Superalloys, *Metals Handbook*, 10th ed., Vol 1, 1990, p 995-1006
- G.F. Vander Voort and H.M. James, Metallography and Microstructures of Wrought Heat-Resistant Alloys, *Metals Handbook*, 9th ed., Vol 9, ASM International, 1985, p 305-379

**surface film.** Any continuous contamination on the surface of a powder particle.

**surface finish.** (1) The geometric irregularities in the surface of a solid material. Measurement of surface finish shall not include inherent structural irregularities unless these are the characteristics being measured. (2) Condition of a surface as a result of a final treatment. See also *roughness*.

**surface grinding.** Producing a plane surface by grinding.

**surface hardening.** A generic term covering several processes applicable to a suitable ferrous alloy that produces, by quench hardening only, a surface layer that is harder or more wear resistant than the core. There is no significant alteration of the chemical composition of the surface layer. The processes commonly used are *carbonitriding*, *carburizing*, *induction hardening*, *flame hardening*, *nitriding*, and *nitrocarburizing*. Use of the applicable specific process name is preferred.

**surface integrity** (metals). A technology that involves the specification and manufacture of unimpaired or enhanced surfaces through the control of the many possible alterations produced in a surface layer during manufacture. Surface integrity is achieved by the proper selection and control of manufacturing processes and the ability to estimate their effects on the significant engineering properties of work materials. See also *surface alterations* (metals).

**surface modification.** The alteration of surface composition or structure by the use of energy or particle beams. Elements may be added to influence the surface characteristics of the substrate by the formation of alloys, metastable alloys or phases, or amorphous layers. Surface-modified layers are distinguished from conversion or coating layers by their greater similarity to metallurgical alloying versus chemically reacted, adhered, or physically bonded layers. However, surface structures are produced

## Technical Brief 53: Superconductors

SUPERCONDUCTORS are materials that exhibit a complete disappearance of electrical resistance on lowering the temperature below a critical temperature ($T_c$). For all superconductors presently known, the critical temperatures are well below room temperature, and they are usually attained by cooling with liquefied gas (helium or nitrogen), either at or below atmospheric pressure. A superconducting material must also exhibit perfect diamagnetism, that is, complete exclusion of an applied magnetic field from the bulk of the superconductor. Superconductivity permits electric power generators and transmission lines to have capacities many times greater than recently possible. It also allows the development of levitated transit systems capable of high speeds and provides an economically feasible way of producing the large magnetic fields required for the confinement of ionized gases in controlled thermonuclear fusion.

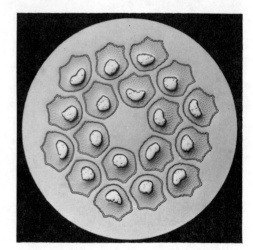

*Cross section of a multifilamentary $Nb_3Sn$ superconductor wire. 75×*

Superconductivity is observed in a broad range of materials. These include more than half of the metallic elements and a wide range of compounds and alloys. To date, however, the materials that have received the most attention are niobium-titanium superconductors (the most widely used superconductor), A15 compounds (in which class the important intermetallic $Nb_3Sn$ lies), ternary molybdenum chalcogenides, and high-temperature ceramic superconductors. The chalcogenides and ceramics, however, are only in the research stage.

Niobium-titanium superconductors are actually composite wires that consist of Nb-Ti filaments (<10 μm in diameter) embedded in an oxygen-free, high-purity (99.99%) copper matrix. Commercially pure aluminum (alloy 1100, 99.0% Al) and copper-nickel alloys (typically in concentrations of 90:10 or 70:30) matrices have also been utilized. The filament alloy most widely used is Nb-46.5Ti. Binary Nb-Ti compositions in the range of 45 to 50% Ti exhibit $T_c$ values of 9.0 to 9.3 K. Composite conductors containing as few as one to as many as 25,000 filaments have been processed by advanced extrusion and wire-drawing techniques. The primary applications for Nb-Ti superconductors are magnets for use in magnetic resonance imaging (MRI) devices used in hospitals and high-energy physics pulsed accelerator-magnet applications. An example of the latter application is the proposed Superconducting Supercollider for studying the elementary particles of which all matter is composed and the forces through which matter interacts.

that differ significantly from those obtained by conventional metallurgical processes. This latter characteristic further distinguishes surface modification from other conventional processes, such as amalgamation or thermal diffusion. Two types of surface modification methods commonly employed are *ion implantation* and *laser surface processing*.

**surface preparation.** The operations necessary to produce a desired or specified surface condition.

**surface preparation** (adhesives). Physical and/or chemical preparation of an adherend to make it suitable for adhesive bonding.

**surface roughness.** Fine irregularities in the *surface texture* of a material, usually including those resulting from the

inherent action of the production process (Fig. 589). Surface roughness is usually reported as the arithmetic roughness average, $R_a$, and is given in micrometers or microinches. See also the figure accompanying the term *roughness*.

**surface tension.** (1) The force acting on the surface of a liquid, tending to minimize the area of the surface. (2) The force existing in a liquid-vapor phase interface that tends to diminish the area of the interface. This force acts at each point on the interface in the plane tangent to that point.

**surface texture.** The roughness, waviness, lay, and flaws associated with a surface as defined in the figure accompanying the term *roughness*. See also *lay*.

**surface treatment.** In composites fabrication, a material

A15 superconductors are brittle intermetallic $A_3B$ compounds with a body-centered cubic crystal structure. Of the 76 known A15 compounds, 46 are known to be superconducting. Because of its ease of fabrication, $Nb_3Sn$ is the most commercially important A15 compound. Like Nb-Ti superconductors, $Nb_3Sn$ is also assembled into multifilamentary wires. Applications for $Nb_3Sn$-base superconductors include large commercial magnets, power generators and power transmission lines, and devices for magnetically confining high-energy plasma for thermonuclear fusion.

The ternary molybdenum chalcogenides represent a vast class of materials whose general formula is $M_xMo_6X_8$, where M is a cation and X a chalcogen (sulfur, selenium, or tellurium). Most of the research on these materials has centered around $PbMo_6S_8$ and $SnMo_6S_8$, the former having a $T_c$ of 14 to 15 K.

High-temperature superconductors ($T_c$ values exceeding 90 K) are ceramic oxides in wire, tape, or thin-film form. The systems being studied include Y-Ba-Cu-O (most notably $YBa_2Cu_3O_7$), Bi-Sr-Ca-Cu-O, and Tl-Ba-Ca-Cu-O.

## Approximate superconducting properties of selected superconducting materials

| Material | Type | Critical temperature, $T_c$ at 0 T | Thermodynamic critical field, T, at $\mu_o H_c$ | $\mu_o H_{c1}$ | $\mu_o H_{c2}$ | Magnetic penetration depth ($\lambda$), nm | Coherence length ($\xi$), nm | Critical current density ($J_c$), kA · mm$^{-2}$ |
|---|---|---|---|---|---|---|---|---|
| Pb | I | 7.3 | 0.0803(a) | · · · | · · · | 40 | 83 | · · · |
| Nb | II | 9.3 | 0.37 | 0.25 | 0.41 | 30 | 40 | · · · |
| Nb45-50-Ti | II | 8.9–9.3 | 0.16 | 0.009 | 10.5–11.0 | 500 | 10 | 3 (at 5 T) |
| Nb$_3$Sn | II | 18 | 0.46 | 0.034 | 19–25 | 200 | 6 | 10 (at 5 T) |
| Nb$_3$Ge | II | 23 | 0.16 | 0.004 | 36–41 | 650 | 4 | 10 (at 5 T) |
| NbN | II | 16–18 | 0.16 | 0.004 | 20–35 | 600 | 5 | 10 (at 0 T) |
| PbMo$_6$S$_8$ | II | 14–15 | 0.4 | 0.005 | 40–55 | 240 | 4 | 0.8 (at 5 T) |
| YBa$_2$Cu$_3$O$_7$ | II | 92 | 0.5 | 0.05(b) | 60(b) | 150(b) | 15(b) | 1 (at 77 K, 0 T)(d) |
|  |  |  | 0.03 | 0.01(c) | >200(c) | 1000(c) | 2–3(c) |  |

(a) Thermodynamic critical field at 0 K. (b) Measured with field parallel to the $c$-axis. (c) Measured with field parallel to the $a$-$b$ plane. (d) Epitaxial thin film, current in the $a$-$b$ plane

## Recommended Reading

- Superconducting Materials, T.S. Kreilick, Ed., *Metals Handbook*, 10th ed., Vol 2, ASM International, 1990, p 1027-1089
- R.B. Poeppel *et al.*, High-Temperature Superconductors, *Engineered Materials Handbook*, Vol 4, ASM International, 1991, p 1156-1160

(size or finish) applied to fibrous material during the forming operation or in subsequent processes. For carbon fiber surface treatment, the process used to enhance bonding capability of fiber to resin. See also *size*.

**surface void.** A void which is located at the surface of a material and is a consequence of processing, that is, a surface reaction layer, as distinguished from a volume distributed flaw such as a *pore* or *inclusion*.

**surfacing.** The deposition of filler metal (material) on a base metal (substrate) to obtain desired properties or dimensions. See also *buttering*, *cladding*, *coating*, and *hardfacing*.

**surfacing mat.** A very thin mat, usually 180 to 510 μm (7 to 10 mils) thick, of highly filamentized fiberglass, used primarily to produce a smooth surface on a reinforced plastic laminate, or for precise machining or grinding.

**surfacing weld.** A type of weld composed of one or more stringer or weave beads deposited on an unbroken surface to obtain desired properties or dimensions. See also *stringer bead* and *weave bead*.

**surfactant.** (1) A chemical substance characterized by a strong tendency to form adsorbed interfacial films when in solution, emulsion, or suspension, thus producing effects such as low surface tension, penetration, boundary lubrication, wetting, and dispersing. (2) A compound that affects interfacial tensions between two liquids. It usually reduces surface tension. See also *Rehbinder effect*.

**swabbing.** Wiping of the surface of a metallographic spec-

**Table 65  Superplastic properties of several aluminum and titanium alloys**

| Alloy | Test temperature °C | Test temperature °F | Strain rate, s$^{-1}$ | Strain rate sensitivity, $m$ | Elongation, % |
|---|---|---|---|---|---|
| **Aluminum** | | | | | |
| Statically recrystallized | | | | | |
| Al-33Cu | 400–500 | 752–930 | $8 \times 10^{-4}$ | 0.8 | 400–1000 |
| Al-4.5Zn-4.5Ca | 550 | 1020 | $8 \times 10^{-3}$ | 0.5 | 600 |
| Al-6 to 10Zn-1.5Mg-0.2Zr | 550 | 1020 | $10^{-3}$ | 0.9 | 1500 |
| Al-5.6Zn-2Mg-1.5Cu-0.2Cr | 516 | 961 | $2 \times 10^{-4}$ | 0.8–0.9 | 800–1200 |
| Dynamically recrystallized | | | | | |
| Al-6Cu-0.5Zr (Supral 100) | 450 | 840 | $10^{-3}$ | 0.3 | 1000 |
| Al-6Cu-0.35Mg-0.14Si (Supral 220) | 450 | 840 | $10^{-3}$ | 0.3 | 900 |
| Al-4Cu-3Li-0.5Zr | 450 | 840 | $5 \times 10^{-3}$ | 0.5 | 900 |
| Al-3Cu-2Li-1Mg-0.2Zr | 500 | 930 | $1.3 \times 10^{-3}$ | 0.4 | 878 |
| **Titanium** | | | | | |
| α/β | | | | | |
| Ti-6Al-4V | 840–870 | 1545–1600 | $1.3 \times 10^{-4}$ to $10^{-3}$ | 0.75 | 750–1170 |
| Ti-6Al-5V | 850 | 1560 | $8 \times 10^{-4}$ | 0.70 | 700–1100 |
| Ti-6Al-2Sn-4Zr-2Mo | 900 | 1650 | $2 \times 10^{-4}$ | 0.67 | 538 |
| Ti-4.5Al-5Mo-1.5Cr | 871 | 1600 | $2 \times 10^{-4}$ | 0.63–0.81 | >510 |
| Ti-6Al-4V-2Ni | 815 | 1499 | $2 \times 10^{-4}$ | 0.85 | 720 |
| Ti-6Al-4V-2Co | 815 | 1499 | $2 \times 10^{-4}$ | 0.53 | 670 |
| Ti-6Al-4V-2Fe | 815 | 1499 | $2 \times 10^{-4}$ | 0.54 | 650 |
| Ti-5Al-2.5Sn | 1000 | 1830 | $2 \times 10^{-4}$ | 0.49 | 420 |
| β and near β | | | | | |
| Ti-15V-3Cr-3Sn-3Al | 815 | 1499 | $2 \times 10^{-4}$ | 0.5 | 229 |
| Ti-13Cr-11V-3Al | 800 | 1470 | ... | ... | <150 |
| Ti-8Mn | 750 | 1380 | ... | 0.43 | 150 |
| Ti-15Mo | 800 | 1470 | ... | 0.60 | 100 |
| α | | | | | |
| CP Ti | 850 | 1560 | $1.7 \times 10^{-4}$ | ... | 115 |

FIG. 589

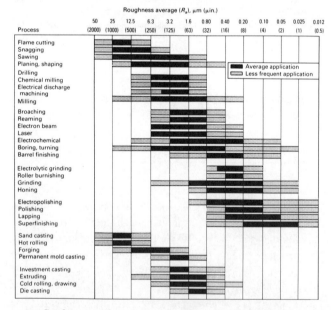

*Surface roughness produced by common production methods*

imen with a cotton ball saturated with etchant to remove reaction products simultaneously.

**swage.** (1) The operation of reducing or changing the cross-section area of stock by the fast impact of revolving dies.

(2) The tapering of bar, rod, wire, or tubing by forging, hammering, or squeezing; reducing a section by progressively tapering lengthwise until the entire section attains the smaller dimension of the taper. See also the figure accompanying the term *rotary swager*.

**swaging.** Tapering bar, rod, wire, or tubing by forging, hammering, or squeezing; reducing a section by progressively tapering lengthwise until the entire section attains the smaller dimension of the taper. See also *rotary swaging*.

**swarf.** Intimate mixture of grinding chips and fine particles of abrasive and bond resulting from a grinding operation.

**sweat.** Exudation of a low-melting phase during solidification of metals. Also known as sweatback. For tin bronzes, it is called tin sweat. See also *exudation*.

**sweating.** (1) Exudation of bearing material or lubricant due to high temperature. (2) A soldering technique in which two or more parts are precoated (tinned), the reheated and joined without adding more solder. Also called sweat soldering.

**sweating out.** Bringing small globules of one of the low-melting constituents of an alloy to the surface during heat treatment, such as lead out of bronze.

**sweat soldering.** See *sweating* (2).

**sweep.** A type of foundry pattern that is a template cut to the profile of the desired mold shape that, when revolved around a stake or spindle, produces that shape in the mold.

**sweeps.** Floor and table sweepings containing precious metal particles.

**sweep velocity.** The mean of the surface velocities of two bodies at the area of contact. Occasionally the sum of the velocities is quoted instead of the mean. In rolling, the sweep velocity is also called the rolling velocity.

**sweet roast.** Same as *dead roast*.

**Swift cup test.** A simulative test for determining formability of sheet metal in which circular blanks of various diameters are clamped in a die ring and deep drawn into a cup by a flat-bottomed cylindrical punch (Fig. 590). The ratio of the largest blank diameter that can be drawn successfully to the cup diameter is known as the *limiting drawing ratio (LDR)* or *deformation limit*.

FIG. 590

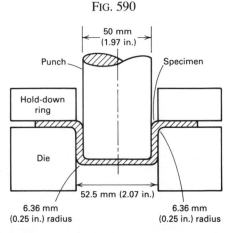

*Standard tooling for the Swift flat-bottomed cup test*

**swing forging machine.** Equipment for continuously hot reducing ingots, blooms, or billets to square flats, rounds, or rectangles by the crank-driven oscillating action of paired dies.

**swing frame grinder.** A grinding machine suspended by a chain at the center point so that it may be turned and swung in any direction for grinding of billets, large castings, or other heavy work. Principal use is removing surface imperfections and roughness.

**symmetrical laminate.** A *laminate* in which the stacking sequence of plies below its midplane is a mirror image of the stacking sequence above the midplane.

**synchronous initiation.** In resistance welding, the initiation

and termination of each half-cycle of welding transformer primary current so that all half-cycles of such current are identical in making spot and seam welds or in making projection welds.

**synchronous timing.** See preferred term *synchronous initiation*.

**synchrotron.** A device for accelerating charged particles by directing them along a roughly circular path in a magnetic guide field. As the particles pass through accelerating cavities placed along their orbit, their kinetic energy is increased repetitively, multiplying their initial energy by factors of hundreds or thousands. See also *synchrotron radiation*.

**synchrotron radiation.** Electromagnetic radiation emitted by charged particles in circular motion at relativistic energies.

**syndiotactic stereoisomerism.** A polymer molecule in which side atoms or side groups alternate regularly on opposite sides of the chain. See also the figure accompanying the term *isotactic stereoisomerism*.

**syneresis** (of a grease). See *bleeding* (2).

**syneresis.** Spontaneous separation of a liquid from a gel due to contraction of the gel.

**syntactic cellular plastics.** Reinforced plastics made by mixing hollow microspheres of glass, epoxy, phenolic, and so forth, into fluid resins (with additives and curing agents) to form a moldable, curable, lightweight, fluid mass; as opposed to foamed plastic, in which the cells are formed by gas bubbles released in the liquid plastic by either chemical or mechanical action. Also known as syntactic foams.

**syntectic.** An isothermal reversible reaction in which a solid phase, on absorption of heat, is converted to two conjugate liquid phases.

**syntectic equilibrium.** A reversible univariant transformation in which a solid phase that is stable only at lower temperature decomposes into two conjugate liquid phases that remain stable at higher temperature.

**synthetic cold rolled sheet.** A hot rolled pickled sheet given a sufficient final temper pass to impart a surface approximating that of cold rolled steel.

**synthetic oil.** Oil produced from chemical synthesis rather than from petroleum. Examples are esters, ethers, silicons, silanes, and halogenated hydrocarbons.

# T

**$T_g$** . See *glass transition temperature*.

**tabs**. Extra lengths of reinforced plastic or other material at the ends of a tensile specimen to promote failure away from the grips. Also called doublers.

**tack**. The property of an adhesive that enables it to form a bond of measurable strength immediately after adhesive and adherend are brought into contact under low pressure. See also *dry tack*, *tack range*, and *tacky-dry*.

**tacking**. Making *tack welds*.

**tack range**. The period of time in which an adhesive will remain in the *tacky-dry* condition after application to the adherend, under specified conditions of temperature and humidity.

**tack welds**. (1) Small, scattered welds made to hold parts of a weldment in proper alignment while the final welds are being made. (2) Intermittent welds to secure weld backing bars (Fig. 591). See also *backing* (2).

FIG. 591

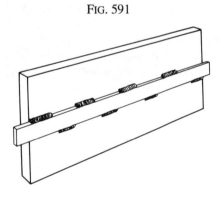

*Use of short, intermittent tack welds to secure backing bars*

**tacky-dry**. The condition of an adhesive when the volatile constituents have evaporated or been absorbed sufficiently to leave it in a desired tacky state.

**taconite**. A siliceous iron formation from which certain iron ores of the Lake Superior region are derived; consists chiefly of fine-grain silica mixed with magnetite and hematite.

**Tafel line, Tafel slope, Tafel diagram**. When an electrode is polarized, it frequently will yield a current/potential relationship over a region that can be approximated by: $\eta = \pm \beta \log (i/i_0)$, where $\eta$ is the change in open-circuit potential, $i$ is the current density, and $\beta$ and $i_0$ are constants. The constant $\beta$ is also known as the Tafel slope. If this behavior is observed, a plot on semilogarithmic coordinates is

FIG. 592

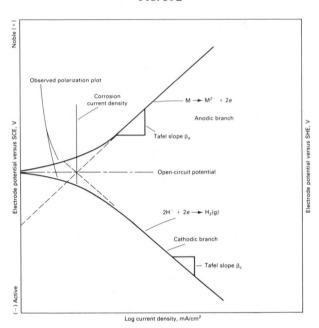

*Experimentally measured Tafel polarization plot*

known as the Tafel line and the overall diagram is termed a Tafel diagram (Fig. 592).

**tailings**. The discarded portion of a crushed ore, separated during concentration.

**talc**. A whitish, greenish, or grayish hydrated magnesium silicate, $Mg_3Si_4O_{10}(OH)_2$, mineral which is extremely soft (hardness is 1 on the Mohs scale) and has a characteristic soapy or greasy feel. Used as an ingredient in *ceramic whiteware* and ceramic wall and floor tiles.

**tamp**. To form or compact a ceramic mixture or a particulate solid by tamping or repeated impact, usually performed manually.

**tan delta (tan δ)**. (1) The ratio of the loss modulus to the *storage modulus*, measured in compression, $K$; tension or flexure, $E$; or shear, $G$. See also *complex modulus* and *loss modulus*. (2) The ratio of the out-of-phase components of the dielectric constant (that is, the loss) to the in-phase component of the dielectric constant (that is, the permittivity). See also *loss factor*.

**tandem die**. Same as *follow die*.

**tandem mill**. A rolling mill consisting of two or more stands arranged so that the metal being processed travels in a straight line from stand to stand. In continuous rolling, the various stands are synchronized so that the strip can be

FIG. 593

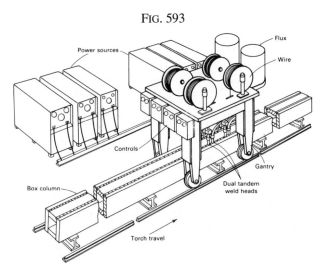

*Tandem arc welding equipment suitable for high-current groove or fillet joints*

| Parameter | Typical range |
|---|---|
| Thickness, mm (in. $\times 10^{-3}$) . . . . . . . | 0.08-0.25 (3-10) |
| Resin content, % . . . . . . . . . . . . . . | 28-45 nominal $\pm$ 2 |
| Dry fiber areal weight, g/m² (oz/ft²) . . | 30-300 (0.10-1.0) |
| Width, mm (in.). . . . . . . . . . . . . . . | 25-1525 (1-60) |
| Package size, kg (lb) . . . . . . . . . . . | 4.5-225 (10-500) |

rolled in all stands simultaneously. Contrast with *single-stand mill*. See also *rolling mills*.

**tandem seal**. A multiple-seal arrangement consisting of two seals mounted one after the other, with the faces of the seal heads oriented in the same direction.

**tandem welding**. Arc welding in which two or more electrodes are in a plane parallel to the line of travel (Fig. 593).

**tangent bending**. The forming of one or more identical bends having parallel axes by wiping sheet metal around one or more radius dies in a single operation. The sheet, which may have side flanges, is clamped against the radius die and then made to conform to the radius die by pressure from a rocker-plate die that moves along the periphery of the radius die. See also *wiper forming (wiping)*.

**tangential stress**. See *shear stress*.

**tangent modulus**. The slope of the *stress-strain curve* at any specified stress or strain. See also *modulus of elasticity*.

**tank voltage**. The total voltage between the anode and cathode of a plating bath or electrolytic cell during electrolysis. It is equal to the sum of: (a) the equilibrium reaction potential, (b) the *IR* drop, and (c) the electrode potentials.

**tap**. A cylindrical or conical thread-cutting tool with one or more cutting elements having threads of a desired form on the periphery (Fig. 594). By a combination of rotary and axial motions, the leading end cuts an internal thread, the tool deriving its principal support from the thread being produced.

**tap density**. The apparent density of a powder, obtained when the volume receptacle is tapped or vibrated during loading under specified conditions.

**tape**. In composites manufacture, a unidirectional *prepreg* that consists of a thin sheet of fiber reinforced uncured resin usually wound on a cardboard core. Typical tape dimensions are as follows:

**tapered land bearing**. A *thrust bearing* containing pads of fixed taper.

**tapered roller bearing**. A *rolling-element bearing* containing tapered rollers.

**tape replica method (faxfilm)**. A method of producing a *replica* by pressing the softened surface of tape or plastic sheet material onto the surface to be replicated.

**taper section**. A section cut obliquely (acute angle) through a surface and prepared metallographically (Fig. 595). The angle is often chosen to increase the vertical magnification of surface features by a factor of 5 or 10. Taper sectioning is usually carried out for microstructural examination of coated metal specimens.

**tape wrapped**. In composites fabrication, wrapping of heated fabric tape onto a rotating mandrel, which is subsequently cooled to firm the surface for the next tape layer application.

**tapping**. (1) Producing internal threads with a cylindrical cutting tool having two or more peripheral cutting elements shaped to cut threads of the desired size and form. By a combination of rotary and axial motion, the leading end of the tap cuts the thread while the tap is supported mainly by the thread it produces. See also the figure accompanying the term *tap*. (2) Opening the outlet of a melting furnace to remove molten metal. (3) Removing molten metal from a furnace.

**target**. That part of an x-ray tube in an x-ray spectrometer which the electrons strike and from which x-rays are emitted.

FIG. 594

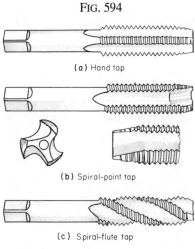

(a) Hand tap

(b) Spiral-point tap

(c) Spiral-flute tap

*Three basic styles of solid taps*

FIG. 595

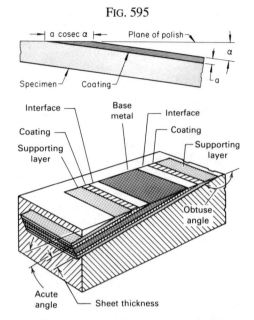

*Schematic of taper sectioning (top), as applied to a coated specimen. Taper magnification equals cosecant of taper angle α. Bottom: Components of a taper section of tinplate*

**tarnish.** Surface discoloration of a metal caused by formation of a thin film of corrosion product.

**Taylor process.** A process for making extremely fine metal wire by inserting a piece of larger-diameter wire into a glass tube and stretching the two together at high temperature.

**Taylor vortices.** In a *journal bearing*, vortices formed in a liquid occupying the annular space between two concentric cylinders.

**teapot ladle.** A ladle in which, by means of an external spout, metal is removed from the bottom rather than the top of the ladle (Fig. 596). See also the figure accompanying the term *ladle*.

**technical ceramics.** Same as *advanced ceramics*.

FIG. 596

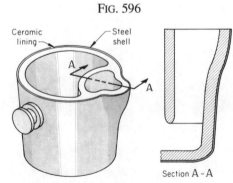

*Typical teapot ladle used for pouring small- to medium-size steel castings*

FIG. 597

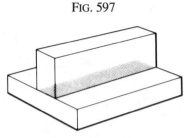

*Example of a tee weld joint*

**technical glass.** A term that usually refers to glasses designed with some specific property essential for a mechanical, industrial, or scientific device.

**tee joint.** A joint in which the members are oriented in the form of a T (Fig. 597).

**teeming.** Pouring molten metal from a ladle into ingot molds. The term applies particularly to the specific operation of pouring either iron or steel into ingot molds.

**teeth.** The resultant surface irregularities or projections formed by the breaking of filaments or strings which may form when adhesive-bonded substrates are separated. See also *legging, stringiness*, and *webbing*.

**telegraphing.** In a laminate or other type of composite construction, a condition in which irregularities, imperfections, or patterns of an inner layer are visibly transmitted to the surface. Telegraphing is occasionally referred to as photographing.

**telomer.** A polymer composed of molecules having terminal groups incapable of reacting with additional monomers, under the conditions of the synthesis, to form larger polymer molecules of the same chemical type. See also *monomer* and *polymer*.

**temper** (ceramics). To moisten and mix clay, plaster, or mortar to proper consistency.

**temper** (glass). (1) The degree of residual stress in annealed glass measured polarimetrically or by polariscopic comparison with a reference standard. (2) Term sometimes employed in referring to *tempered glass*.

**temper** (metals). (1) In heat treatment, reheating hardened steel or hardened cast iron to some temperature below the eutectoid temperature for the purpose of decreasing hardness and increasing toughness. The process also is sometimes applied to normalized steel. (2) In tool steels, temper is sometimes used, but inadvisedly, to denote the carbon content. (3) In nonferrous alloys and in some ferrous alloys (steels that cannot be hardened by heat treatment), the hardness and strength produced by mechanical or thermal treatment, or both, and characterized by a certain structure, mechanical properties, or reduction in area during cold working. (4) To moisten *green sand* for casting molds with water.

**temperature coefficient of resistance.** The amount of resistance change of a material per degree of temperature rise.

**temperature stress.** The maximum stress that can be applied to a material at a given temperature without physical deformation.

**temper brittleness.** See *temper embrittlement.*

**temper carbon.** Clusters of finely divided graphite, such as that found in malleable iron, that are formed as a result of decomposition of cementite, for example, by heating white cast iron above the ferrite-austenite transformation temperature and holding at these temperatures for a considerable period of time. Also known as annealing carbon. See also the figures accompanying the term *malleable iron.*

**temper color.** A thin, tightly adhering oxide skin (only a few molecules thick) that forms when steel is tempered at a low temperature, or for a short time, in air or a mildly oxidizing atmosphere. The color, which ranges from straw to blue depending on the thickness of the oxide skin, varies with both tempering time and temperature (Fig. 598).

**tempered glass.** Glass that has been subjected to a thermal treatment characterized by rapid cooling to produce a compressively stressed surface layer. See also the figure accompanying the term *safety glass.*

FIG. 598

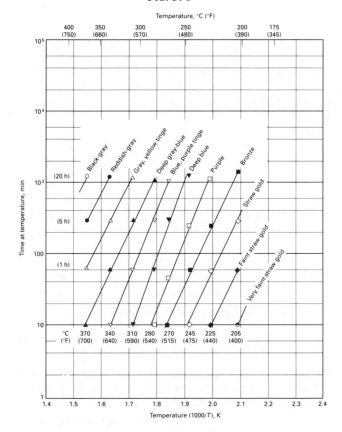

*Temper colors after heating 1035 steel in circulating air (atmospheric pressure)*

**tempered layer.** A surface or subsurface layer in a steel specimen that has been tempered by heating during some stage of the metallographic preparation sequence (usually grinding). When observed in a section after etching, the layer appears darker than the base material.

**tempered martensite.** The decomposition products that result from heating martensite below the ferrite-austenite transformation temperature. Under the optical microscope, darkening of the martensite needles is observed in the initial stages of tempering. Prolonged tempering at high temperatures produces spheroidized carbides in a matrix of ferrite. At the higher resolution of the electron microscope, the initial stage of tempering is observed to result in a structure containing a precipitate of fine iron carbide particles. At approximately 260 °C (500 °F), a transition occurs to a structure of larger and elongated cementite particles in a ferrite matrix. With further tempering at higher temperatures, the cementite particles become spheroidal, decreased in number, and increased in size. Table 66 lists the various reactions that develop during tempering.

**tempered martensite embrittlement.** *Embrittlement* of high-strength alloy steels caused by tempering in the temperature range of 205 to 370 °C (400 to 700 °F); also called 350 °C or 500 °F embrittlement. Tempered martensite embrittlement is thought to result from the combined effects of cementite precipitation on prior-austenite grain boundaries or interlath boundaries and the segregation of impurities at prior-austenite grain boundaries. It differs from *temper embrittlement* in the strength of the material and the temperature exposure range. In temper embrittlement, the steel is usually tempered at a relatively high temperature, producing lower strength and hardness, and embrittlement occurs upon slow cooling after tempering and during service at temperatures within the embrittlement range. In tempered martensite embrittlement, the steel is tempered within the embrittlement range, and service exposure is usually at room temperature.

**temper embrittlement.** *Embrittlement* of low-alloy steels caused by holding within or cooling slowly through a temperature range (generally 300 to 600 °C, or 570 to 1110 °F) just below the transformation range. Embrittlement is the result of the segregation at grain boundaries of impurities such as arsenic, antimony, phosphorus, and tin; it is usually manifested as an upward shift in ductile-to-brittle transition temperature. Temper embrittlement can be reversed by retempering above the critical temperature range, then cooling rapidly. Compare with *tempered martensite embrittlement.*

**tempering (glass).** The process of rapidly cooling glass from near its softening point to induce compressive stresses on the surface balanced by interior tension, thereby imparting increased strength.

**Table 66  Tempering reactions in steel**

| Temperature range | | Reaction and symbol (if designated) | Comments |
|---|---|---|---|
| °C | °F | | |
| −40 to 100 | −40 to 212 | Clustering of two to four carbon atoms on octahedral sites of martensite (A1); segregation of carbon atoms to dislocations and boundaries | Clustering is associated with diffuse spikes around fundamental electron diffraction spots of martensite |
| 20 to 100 | 70 to 212 | Modulated clusters of carbon atoms on (102) martensite planes (A2) | Identified by satellite spots around electron diffraction spots of martensite |
| 60 to 80 | 140 to 175 | Long period ordered phase with ordered carbon atoms (A3) | Identified by superstructure spots in electron diffraction patterns |
| 100 to 200 | 212 to 390 | Precipitation of transition carbide as aligned 2 nm (0.08 μin.) diam particles (T1) | Recent work identifies carbides as η (orthorhombic, $Fe_2C$); earlier studies identified the carbides as $\epsilon$ (hexagonal, $Fe_{2.4}C$). |
| 200 to 350 | 390 to 660 | Transformation of retained austenite to ferrite and cementite (T2) | Associated with tempered-martensite embrittlement in low- and medium-carbon steels. |
| 250 to 700 | 480 to 1290 | Formation of ferrite and cementite; eventual development of well-spheroidized carbides in a matrix of equiaxed ferrite grains (T3) | This stage now appears to be initiated by χ-carbide formation in high-carbon Fe-C alloys. |
| 500 to 700 | 930 to 1290 | Formation of alloy carbides in chromium-, molybdenum-, vanadium- and tungsten-containing steels. The mix and composition of the carbides may change significantly with time (T4). | The alloy carbides produce secondary hardening and pronounced retardation of softening during tempering or long-time service exposure around 500 °C (930 °F). |
| 350 to 550 | 660 to 1020 | Segregation and cosegregation of impurity and substitutional alloying elements | Responsible for temper embrittlement |

**tempering** (metals). In heat treatment, reheating hardened steel to some temperature below the eutectoid temperature to decrease hardness and/or increase toughness.

**temper rolling**. Light cold rolling of sheet steel to improve flatness, to minimize the formation of *stretcher strains*, and to obtain a specified hardness or temper.

**temper time**. In resistance welding, that part of the postweld interval during which the current is suitable for tempering or heat treatment.

**template (templet)**. (1) A guide or a pattern in manufacturing items. (2) A gage or pattern made in a die department, usually from sheet steel; used to check dimensions on forgings and as an aid in sinking *die impressions* in order to correct dimensions. (3) A pattern used as a guide for cutting and laying plies of a *laminate*.

**temporary weld**. A weld made to attach a piece or pieces to a weldment for temporary use in handling, shipping, or working on the weldment.

**tenacity**. The term generally used in yarn manufacture and textile engineering to denote the strength of a yarn or of a filament of a given size. Numerically, it is the grams of breaking force per denier unit of yarn or filament size. Grams per denier is expressed as gpd. See also *denier*.

**tensile modulus**. See *Young's modulus*.

**tensile strength**. In tensile testing, the ratio of maximum load to original cross-sectional area. Also called *ultimate strength*. Compare with *yield strength*.

**tensile stress**. A stress that causes two parts of an elastic body, on either side of a typical stress plane, to pull apart. Contrast with *compressive stress*.

**tensile testing**. See *tension testing*.

**tension**. The force or load that produces elongation.

**tension set**. The condition in which a plastic material shows permanent deformation caused by a stress, after the stress is removed.

**tension testing**. A method of determining the behavior of materials subjected to uniaxial loading, which tends to stretch the material. A longitudinal specimen of known length and diameter is gripped at both ends and stretched at a slow, controlled rate until rupture occurs. Also known as tensile testing.

**tenth-scale vessel**. A filament wound material test vessel based on a one-tenth subscale of the prototype. See also *filament winding*.

**terminal erosion rate**. The final steady-state erosion rate that is reached (or appears to be approached asymptotically) after the erosion rate has declined from its maximum value. This occurs in some, but not all, cavitation and liquid impingement tests.

**terminal period**. In cavitation and liquid impingement erosion, a stage following the deceleration period, during which the erosion rate has leveled off and remains approximately constant (sometimes with superimposed fluctuations) at a value substantially lower than the maximum rate attained earlier.

**terminal phase**. A solid solution having a restricted range of compositions, one end of the range being a pure component of an alloy system.

**terminal solid solution**. In a multicomponent system, any solid phase of limited composition range that includes the composition of one of the components of the system. See also *solid solution*.

**ternary alloy**. An alloy that contains three principal elements.

**ternary system**. The complete series of compositions produced by mixing three components in all proportions.

**terne**. An alloy of lead containing 3 to 15% Sn, used as a *hot dip coating* for steel sheet or plate. The term long terne is used to describe terne-coated sheet, whereas short terne is used for terne-coated plate. Terne coatings, which are smooth and dull in appearance (terne means dull or tarnished in French), give the steel better corrosion resistance and enhance its ability to be formed, soldered, or painted.

**terpolymer**. A polymeric system that contains three monomeric units.

**terra cotta**. A term applied to ornamental units of fired clay used as facing material for buildings (Fig. 599).

**tertiary creep**. See *creep*.

**testing machine (load-measuring type)**. A mechanical device for applying a load (force) to a specimen.

**tetragonal**. Having three mutually perpendicular axes, two equal in length and unequal to the third.

**tex**. A unit for expressing linear density equal to the mass of weight in grams of 1000 meters of filament, fiber, yarn, or other textile strand.

**textile fibers**. Fibers or filaments used in reinforced plastics that can be processed into yarn or made into a fabric by interlacing in a variety of methods, including weaving, knitting, and braiding.

**textile oil**. (1) An oil used to lubricate thread or yarn to prevent breakage during spinning and weaving. (2) An oil acceptable for direct contact with fibers during textile production.

**texture**. In a polycrystalline aggregate, the state of distribution of crystal orientations. In the usual sense, it is synonymous with *preferred orientation*, in which the distribution is not random. Not to be confused with *surface texture*. See also *fiber* (metals) and *fiber texture* (metals).

FIG. 599

*Example of terra cotta exterior. Note the assortment of shapes above and beside the doorway.*

**TGA**. See *thermogravimetric analysis*.

**theoretical density**. The density of a material calculated from the number of atoms per unit cell and measurement of the lattice parameters.

**theoretical electrode force**. The force, neglecting friction and inertia, in making spot, seam, or projection welds by resistance welding, available at the electrodes of a resistance welding machine by virtue of the initial force application and the theoretical mechanical advantage of the system.

**theoretical stress-concentration factor**. See *stress-concentration factor*.

**theoretical throat**. See *throat of a fillet weld*.

**thermal aging**. Exposure of a material or component to a given thermal condition or a programmed series of conditions for prescribed periods of time.

**thermal alloying**. The act of uniting two different metals to make one common metal by the use of heat.

**thermal analysis**. A method for determining transformations in a metal by noting the temperatures at which thermal arrests occur. These arrests are manifested by changes in slope of the plotted or mechanically traced heating and cooling curves. When such data are secured under nearly equilibrium conditions of heating and cooling, the method is commonly used for determining certain critical temperatures required for the construction of *phase diagrams*.

**thermal conductivity**. (1) Ability of a material to conduct heat. (2) The rate of heat flow, under steady conditions, through unit area, per unit temperature gradient in the direction perpendicular to the area. Usually expressed in English units as Btu per square feet per degrees Fahrenheit (Btu/ft$^2 \cdot$ °F). It is given in SI units as watts per meter kelvin (W/m $\cdot$ K).

**thermal cutting**. A group of cutting processes which melts the metal (material) to be cut. See also *air carbon arc cutting, arc cutting, carbon arc cutting, electron beam cutting, laser beam cutting, metal powder cutting, oxyfuel gas cutting, oxygen arc cutting, oxygen cutting*, and *plasma arc cutting*.

**thermal cycling**. The cyclic change in thermal environment.

**thermal decomposition**. (1) The decomposition of a compound into its elemental species at elevated temperatures. (2) A process whereby fine solid particles can be produced from a gaseous compound. See also *carbonyl powder*.

**thermal electromotive force**. The *electromotive force* generated in a circuit containing two dissimilar metals when one junction is at a temperature different from that of the other. See also *thermocouple*.

**thermal embrittlement**. *Intergranular fracture* of maraging steels with decreased toughness resulting from improper processing after hot working. Thermal embrittlement occurs upon heating above 1095 °C (2000 °F) and then slow cooling through the temperature range of 980 to 815 °C

(1800 to 1500 °F), and has been attributed to precipitation of titanium carbides and titanium carbonitrides at austenite grain boundaries during cooling through the critical temperature range. See also *maraging steels*.

**thermal endurance**. The time required at a selected temperature for a material or system of materials to deteriorate to some predetermined level of electrical, mechanical, or chemical performance under prescribed test conditions.

**thermal etching**. Heating a specimen (usually a ceramic) in air, vacuum, or inert gases in order to delineate the grain structure (Fig. 600). Used primarily in high-temperature microscopy.

**thermal expansion**. The change in length of a material with change in temperature. See also *coefficient of thermal expansion*.

**thermal expansion molding**. In forming of plastics, a process in which elastomeric tooling details are constrained within a rigid frame to generate consolidation pressure by thermal expansion during the curing cycle of the autoclave molding process.

**thermal fatigue**. Fracture resulting from the presence of temperature gradients that vary with time in such a manner as to produce cyclic stresses in a structure. See also the figure accompanying the term *craze cracking*.

**thermal inspection**. A nondestructive test method in which heat-sensing devices are used to measure temperature variations in components, structures, systems, or physical processes. Thermal methods can be useful in the detection of subsurface flaws or voids, provided the depth of the flaw is not large compared to its diameter. Thermal inspection becomes less effective in the detection of subsurface flaws as the thickness of an object increases, because the possible depth of the defects increases.

FIG. 600

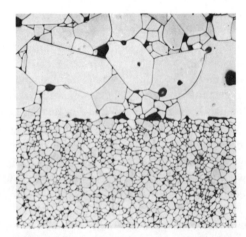

*Polished section of a diffusion-bonded joint between a coarse-grained and a fine-grained alumina ceramic (99.7% Al₂O₃) thermally etched in air at 1400 °C (2550 °F) for 1 h. 500×*

**thermally induced embrittlement**. See *embrittlement*.

**thermal-mechanical treatment**. See *thermomechanical working*.

**thermal noise**. The electrical noise produced in a resistor by thermally generated currents. These currents average zero, but produce electrical power having a non-zero average, which can affect instrument response. Also known as Johnson noise.

**thermal rating**. The maximum or minimum temperature at which a material or component will perform its function without undue degradation.

**thermal resistance**. A measure of a body's ability to prevent heat from flowing through it, equal to the difference between the temperatures of opposite faces of the body divided by the rate of heat flow. Also known as heat resistance.

**thermal shock**. The development of a steep temperature gradient and accompanying high stresses within a material or structure.

**thermal spraying**. A group of coating or welding processes in which finely divided metallic or nonmetallic materials are deposited in a molten or semimolten condition to form a coating. The coating material may be in the form of powder, ceramic rod, wire, or molten materials. See also the figures accompanying the terms *electric arc spraying*, *flame spraying*, *plasma spraying*, and *powder flame spraying*.

**thermal spraying gun**. A device for heating, feeding, and directing the flow of a thermal spraying material.

**thermal spray powder**. A metal, carbide, or ceramic powder mixture designed for use with *hardfacing* and *thermal spraying* operations.

**thermal stress cracking**. The crazing and cracking of some thermoplastic resins from overexposure to elevated temperatures. See also *stress-cracking failure*.

**thermal stresses**. Stresses in a material resulting from non-uniform temperature distribution.

**thermal taper**. See *thermal wedge*.

**thermal wear**. Removal of material due to softening, melting, or evaporation during sliding or rolling. Thermal shock and high-temperature erosion may be included in the general description of thermal wear. Wear by diffusion of separate atoms from one body to the other, at high temperatures, is also sometimes denoted as thermal wear.

**thermal wedge**. The increase in pressure due to the expansion of the lubricant, for example, in a parallel thrust bearing. Thermal distortion of the bearing surfaces may also form a wedge shape. This is referred to as thermal taper.

**thermionic cathode gun**. An *electron gun* that derives its electrons from a heated filament, which may also serve as the cathode. Also termed hot cathode gun.

**thermionic emission.** The ejection of a stream of electrons from a hot cathode, usually under the influence of an electrostatic field.

**thermit crucible.** The vessel in which *thermit reactions* take place.

**thermit mixture.** A mixture of metal oxide and finely divided aluminum with the addition of alloying metals as required.

**thermit mold.** In *thermit welding,* a mold formed around the parts to be welded to receive the molten metal.

**thermit reactions.** Strongly exothermic self-propagating reactions such as that where finely divided aluminum reacts with a metal oxide. A mixture of aluminum and iron oxide produces sufficient heat to weld steel, the filler metal being produced in the reaction. See also *thermit welding.*

**thermit welding.** A welding process which produces coalescence of metals by heating them with superheated liquid metal from a chemical reaction between a metal oxide and aluminum, with or without the application of pressure (Fig. 601). Filler metal, when used, is obtained from the liquid metal. The process is used primarily for welding railroad track.

**thermochemical machining.** Removal of workpiece material—usually only burrs and fins—by exposure to hot fuel gases which are formed by igniting an explosive, combustible mixture of natural gas and oxygen. Also known as the thermal energy method.

**thermochemical treatment.** Heat treatment for steels carried out in a medium suitably chosen to produce a change in the chemical composition of the object by exchange with the medium.

**thermocompression bonding.** See preferred term *hot pressure welding.*

**thermocouple.** A device for measuring temperatures, consisting of lengths of two dissimilar metals or alloys that are electrically joined at one end and connected to a voltage-measuring instrument at the other end. When one junction is hotter than the other, a *thermal electromotive force* is

FIG. 601

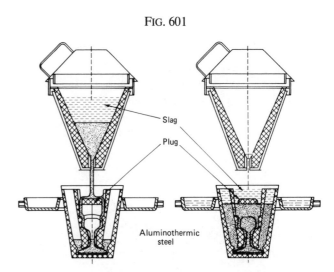

*Principles of the thermit welding process*

produced that is roughly proportional to the difference in temperature between the hot and cold junctions. Table 67 lists compositions and properties of standard thermocouple materials. Nonstandard materials include nickel-molybdenum, nickel-cobalt, iridium-rhodium, platinum-molybdenum, gold-palladium, palladium-platinum, and tungsten-rhenium alloys.

**thermoelastic instability.** In sliding contact, sharp variations in local surface temperatures with the passing of asperities leading to stationary or slowly moving hot spots of significant magnitude.

**thermoforming.** The process of forming a thermoplastic sheet into a three-dimensional shape after heating it to the point at which it is soft and flowable, and then applying differential pressure to make the sheet conform to the shape of a mold or die positioned below the frame. When the thermoplastic material has been reinforced, it should be heated to the point that it is soft enough to be formed without cracking or breaking the reinforcing fibers. There are three basic mold types: female (concave), male (con-

### Table 67  Properties of standard thermocouples

| Type | Thermoelements | Base composition | Melting point, °C | Resistivity, $n\Omega \cdot m$ | Recommended service | Max temperature °C | °F |
|---|---|---|---|---|---|---|---|
| J | JP | Fe | 1450 | 100 | Oxidizing or reducing | 760 | 1400 |
| | JN | 44Ni-55Cu | 1210 | 500 | | | |
| K | KP | 90Ni-9Cr | 1350 | 700 | Oxidizing | 1260 | 2300 |
| | KN | 94Ni-Al, Mn, Fe, Si, Co | 1400 | 320 | | | |
| N | NP | 84Ni-14Cr-1.4Si | 1410 | 930 | Oxidizing | 1260 | 2300 |
| | NN | 95Ni-4.4Si-0.15Mg | 1400 | 370 | | | |
| T | TP | OFHC Cu | 1083 | 17 | Oxidizing or reducing | 370 | 700 |
| | TN | 44Ni-55Cu | 1210 | 500 | | | |
| E | EP | 90Ni-9Cr | 1350 | 700 | Oxidizing | 870 | 1600 |
| | EN | 44Ni-55Cu | 1210 | 500 | | | |
| R | RP | 87Pt-13Rh | 1860 | 196 | Oxidizing or inert | 1480 | 2700 |
| | RN | Pt | 1769 | 104 | | | |
| S | SP | 90Pt-10Rh | 1850 | 189 | Oxidizing or inert | 1480 | 2700 |
| | SN | Pt | 1769 | 104 | | | |
| B | BP | 70Pt-30Rh | 1927 | 190 | Oxidizing, vacuum or | 1700 | 3100 |
| | BN | 94Pt-6Rh | 1826 | 175 | inert | | |

FIG. 602

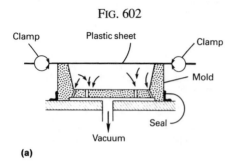

**(a)**

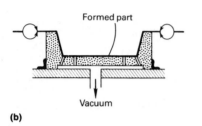

**(b)**

*Basic female mold thermoforming (straight vacuum forming). (a) Heated plastic sheet clamped over female mold cavity. (b) Vacuum pulls plastic sheet into mold, bringing it into contact with entire mold surface.*

FIG. 603

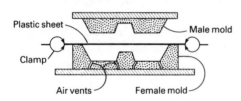

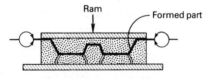

*Matched-mold thermoforming. (a) Heated plastic sheet between male and female molds. (b) Male and female molds forced together with sheet between them*

vex), and matched (a combination of the two). (1) In basic female forming, or straight vacuum forming, the heated sheet is positioned over the mold cavity and is pulled into the cavity by vacuum (Fig. 602). (2) In basic male forming, also called drape forming, the heated plastic is drawn over the mold. As soon as a seal is created, the vacuum is activated, and the plastic is forced directly against the surface of the mold. See also the figure accompanying the term *drape forming.* (3) In matched-mold thermoforming, the stamping force of matched male and female molds is used (Fig. 603). The male mold pushes the heated sheet

into the female cavity. See also the figures accompanying the terms (thermoforming variations) *air-slip thermoforming, plug-assist thermoforming, pressure bubble plug-assist thermoforming, trapped-sheet, contact heat, pressure thermoforming,* and *vacuum snapback thermoforming.*

**thermogalvanic corrosion.** Corrosion resulting from an *electrochemical cell* caused by a thermal gradient.

**thermogravimetric analysis (TGA).** The study of the change in mass of a material under various conditions of temperature and pressure.

**thermomechanical working.** A general term covering a variety of metal forming processes combining controlled thermal and deformation treatments to obtain synergistic effects, such as improvement in strength without loss of toughness. Same as thermal-mechanical treatment.

**thermoplastic.** Capable of being repeatedly softened by an increase in temperature and hardened by a decrease in temperature. Those polymeric materials that, when heated, undergo a substantially physical rather than chemical change and that in the softened stage can be shaped into articles by molding or extrusion.

**thermoplastic fluoropolymers.** See *fluoroplastics.*

**thermoplastic injection molding.** A process in which melted plastic is injected into a mold cavity, where it cools and takes the shape of the cavity. Bosses, screw threads, ribs, and other details can be integrated, which allows the molding operation to be accomplished in one step. The finished part usually does not require additional work before assembling. See also the figure accompanying the term *reciprocating-screw injection molding.*

**thermoplastic polyesters (TPES).** A class of thermoplastic polymers in which the repeating units are joined by ester groups. The two important types are *polyethylene terephthalate (PET),* which is widely used as film, fiber, and soda bottles; and *polybutylene terephthalate (PBT),* which is primarily a molding compound.

**thermoplastic polyimides (TPI).** Fully imidized, linear polymers with exceptionally good thermomechanical performance characteristics. Aromatic TPIs are generally produced by the polycondensation reaction of aromatic dianhydrides with aromatic diamines or aromatic diisocyanates in a suitable reaction medium.

**thermoplastic polyurethanes (TPUR).** Linear (segmented) block copolymers. Such a copolymer consists of repeating groups of diisocyanate and short-chain diol, or chain extender, for a rigid block, and repeating groups of diisocyanate and long-chain diol, or polyol, for a flexible block. TPURs are commonly injection molded, blow molded, or extruded. See also the figure accompanying the term *polyurethanes* (Technical Brief 32).

**thermoreactive deposition/diffusion process (TRD).** A method of coating steels with a hard, wear-resistant layer of carbides, nitrides, or carbonitrides. In the TRD process, the carbon and nitrogen in the steel substrate diffuse into a deposited layer with a carbide-forming or nitride-forming element such as vanadium, niobium, tantalum, chromium, molybdenum, or tungsten. The diffused carbon or nitrogen reacts with the carbide- and nitride-forming elements in the deposited coating so as to form a dense and metallurgically bonded carbide or nitride coating at the substrate surface.

**thermoset.** A resin that is cured, set, or hardened, usually by heating, into a permanent shape. The polymerization reaction is an irreversible reaction known as cross linking. Once set, a thermosetting plastic cannot be remelted, although most soften with the application of heat.

**thermoset injection molding.** A process in which thermoset material that has been heated to a liquid state is caused to flow into a cavity or several cavities and held at an elevated temperature for a specific time (Fig. 604). After cross linking is completed, the hardened part is removed from the open mold. The molds used are usually of through-hardened tool steel that has been chrome plated and highly polished.

FIG. 604

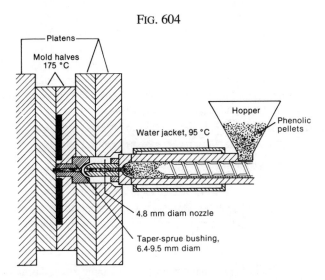

*Reciprocating-screw machine for thermoset injection molding*

**thermosetting.** Having the property of undergoing a chemical reaction by the action of heat, catalysts, ultraviolet light, and so on, leading to a relatively infusible state.

**thermosetting polyesters.** A class of resins produced by dissolving unsaturated, generally linear, alkyd resins in a vinyl-type active monomer such as styrene, methyl styrene, or diallyl phthalate. Cure is effected through vinyl polymerization using peroxide catalysts and promoters or heat to accelerate the reaction. One important commercial type is liquid resins that are cross linked with styrene and used either as impregnants for glass or carbon fiber rein-

forcements in laminates, filament-wound structures, and other built-up constructions, or as binders for chopped-fiber reinforcements in molding compounds, such as *sheet molding compound (SMC)* and *bulk molding compound (BMC)*. A second important type is liquid or solid resins that are cross linked with other esters in chopped-fiber and mineral-filled molding compounds, such as alkyd and diallyl phthalate. See also unsaturated polyesters.

**thermotropic liquid crystal.** A liquid crystalline polymer that can be processed using thermoforming techniques.

**thickener.** A solid material dispersed in a liquid lubricant to produce a grease. Silica, clays, and metallic soaps are widely used as thickeners.

**thick-film circuit.** A circuit that is fabricated by the deposition of materials having between 5 and 20 μm (0.2 and 0.8 mil) thickness, such as screen-printed cermet pastes on a ceramic substrate, which are fired in a kiln to create permanent conductive patterns. Compare with *thin-film circuit.*

**thick-film lubrication.** A condition of lubrication in which the film thickness of the lubricant is appreciably greater than that required to cover the surface asperities when subjected to the operating load, so that the effect of the surface asperities is not noticeable. Also known as full-film lubrication. See also *thin-film lubrication* and the figure accompanying the term *lubrication regimes.*

**thief.** A racking device or nonfunctional pattern area used in the electroplating process to provide a more uniform current density on plated parts. Thieves absorb the unevenly distributed current on irregularly shaped parts, thereby ensuring that the parts will receive an electroplated coating of uniform thickness. See also *robber.*

**thin-film circuit.** A circuit fabricated by the deposition of material several thousand angstroms thick (such as a circuit fabricated by vapor deposition). Compare with *thick-film circuit.*

**thin-film lubrication.** A condition of lubrication in which the film thickness of the lubricant is such that the friction and wear between the surfaces is determined by the properties of the surfaces as well as the viscosity of the lubricant. Under thin-film conditions, the coefficient of friction is often 10 to 100 times greater than under thick-film conditions and wear is no longer negligible. Compare with *thick-film lubrication* and see also the figure accompanying the term *lubrication regimes.*

**thin-layer chromatography (TLC).** A microtype of *chromatography* in which a thin layer of special absorbent is applied to a glass plate, a drop of a solution of the material being investigated is applied to an edge, and that side of the plate is then dipped in an appropriate solvent. The solvent travels up the thin layer of absorbent, which selectively separates the molecules present in the material being investigated.

**thinner**. A volatile liquid added to an adhesive to modify the consistency or other properties. See also *diluent* and *extenders*.

**thin-wall casting**. A term used to define a casting that has the minimum wall thickness to satisfy its service function.

**thixotropic**. Pertaining to the tendency of a fluid to decrease in viscosity as the time of exposure to a given shear rate increases. The shear stress versus shear strain rate curve of a thixotropic material should show a hysteresis loop. A purely pseudoplastic material will not give a hysteresis loop because this property is not time dependent.

**thixotropy**. A property of certain gels or adhesive systems to thin upon isothermal agitation (shearing) and to thicken upon subsequent rest.

**Thomas converter**. A Bessemer converter having a basic bottom and lining, usually dolomite, and employing a basic slag.

**thread**. See *fiber* (composites).

**thread count**. The number of yarns (threads) per inch in either the lengthwise (warp) or crosswise (fill, or weft) direction of woven fabrics used for reinforcing plastics.

**threading**. Producing external threads on a cylindrical surface.
1. Die threading: A process for cutting external threads on cylindrical or tapered surfaces by the use of solid or self-opening dies.
2. Single-point threading: Turning threads on a lathe.
3. Thread grinding: See definition under *grinding*.
4. Thread milling: A method of cutting screw threads with a milling cutter (Fig. 605).

**threading and knurling** (thermal spraying). A method of surface roughening in which spiral threads are prepared, followed by upsetting with a knurling tool.

**thread rolling**. The production of threads by rolling the piece between two grooved die plates, one of which is in motion

FIG. 605

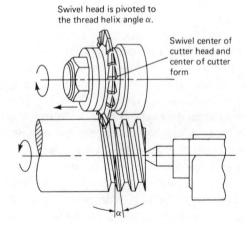

Swivel head is pivoted to the thread helix angle α.

Swivel center of cutter head and center of cutter form

*Thread milling operation*

FIG. 606

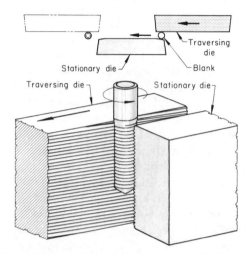

Traversing die

Stationary die

Blank

Traversing die

Stationary die

*Operating principle of flat traversing die thread rolling*

(Fig. 606), or between rotating grooved circular rolls. Also known as roll threading.

**three-quarters hard**. A *temper* of nonferrous alloys and some ferrous alloys characterized by tensile strength and hardness about midway between those of *half hard* and *full hard* tempers.

**three-point bending**. The bending of a piece of metal or a structural member in which the object is placed across two supports and force is applied between and in opposition to them (Fig. 607). See also *V-bend die*.

**threshold stress**. Threshold stress for *stress-corrosion cracking*. The critical gross section stress at the onset of stress-corrosion cracking under specified conditions.

**throat depth** (resistance welding). The distance from the centerline of the electrodes or platens to the nearest point of interference for flat sheets in a resistance welding machine. In the case of a resistance seam welding machine with a universal head, the throat depth is measured with the machine arranged for transverse welding.

**throat height** (resistance welding). The unobstructed dimension between arms throughout the throat depth in a resistance welding machine.

FIG. 607

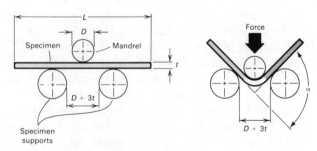

$L$

$D$

Specimen

Mandrel

$t$

$D + 3t$

Specimen supports

Force

$\alpha$

$D + 3t$

*Three-point bending setup*

**throat of a fillet weld**. A term that includes the theoretical throat, the actual throat, and the effective throat. (1) The theoretical throat is the distance from the beginning of the root of the joint perpendicular to the hypotenuse of the largest right triangle that can be inscribed within the fillet weld cross section. This dimension is based on the assumption that the root opening is equal to zero. (2) The actual throat is the shortest distance from the root of the weld to its face. (3) The effective throat is the minimum distance minus any reinforcement from the root of the weld to its face. See also the figures accompanying the terms *concave fillet weld* and *convex fillet weld*.

**throat of a groove weld**. See preferred term *size of weld*.

**throat opening**. See preferred term *horn spacing*.

**throughput**. Volume of charge passed in a time unit through a production sintering furnace.

**through weld**. A nonpreferred term sometimes used to indicate a weld of substantial length made by melting through one member of a lap or tee joint and into the other member.

**throw**. The distance from the centerline of the crankshaft or main shaft to the centerline of the crankpin or eccentric in crank or eccentric presses. Equal to one-half of the stroke. See also *crank press* and the figure accompanying the term *eccentric gear*.

**throwing power**. (1) The relationship between the *current density* at a point on a surface and its distance from the *counterelectrode*. The greater the ratio of the surface resistivity shown by the electrode reaction to the volume resistivity of the electrolyte, the better is the throwing power of the process. (2) The ability of a plating solution to produce a uniform metal distribution on an irregularly shaped *cathode*. Compare with *covering power*.

**thrust bearing**. A bearing in which the load acts in the direction of the axis of rotation.

**tide marks**. See *beach marks*.

**tie bar**. A bar-shaped connection added to a casting to prevent distortion caused by uneven contraction between two separated members of the casting.

**tiger stripes**. Continuous bright lines on sheet or strip in the rolling direction.

**tight fit**. A loosely defined fit of slight negative allowance the assembly of which requires a light press or driving force.

**TIG welding**. Tungsten inert-gas welding; see preferred term *gas tungsten-arc welding*.

**tile**. (1) A ceramic surfacing unit, usually relatively thin in relation to facial area, made from clay or a mixture of clay and other ceramic materials, having either a glazed or unglazed face and fired above red heat in the course of manufacture to a temperature sufficiently high to produce specific physical properties and characteristics.

**tilt**. In electron microscopy, the angle of the specimen relative to the axis of the electron beam; at zero tilt the specimen is perpendicular to the beam axis.

FIG. 608

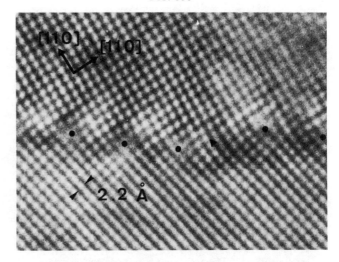

*High-resolution image of a tilt boundary in molybdenum. Foil orientation [001] for both grains*

**tilt boundary**. A subgrain boundary consisting of an array of edge *dislocations* (Fig. 608).

**tilt furnace**. A furnace for the infiltration of copper into porous sintered tungsten for heavy-duty contacts. The furnace is tilted to one position for the separate sintering of the tungsten and melting of the copper, then tilted to the opposite position to let the melt run to and contact or infiltrate, respectively, the tungsten pieces.

**tilting-pad bearing**. A pad bearing in which the pads are free to take up a position at an angle to the opposing surface according to the hydrodynamic pressure distribution over its surface.

**tilt mold**. A casting mold, usually a book (permanent) mold, that rotates from a horizontal to a vertical position during pouring, which reduces agitation and thus the formation and entrapment of oxides. See also the figure accompanying the term *permanent mold*.

**tilt mold ingot**. An ingot made in a *tilt mold*.

**time profile**. A plot of the modulus, damping, or both, of a material versus time.

**time quenching**. A term used to describe a quench in which the cooling rate of the part being quenched must be changed abruptly at some time during the cooling cycle.

**time-temperature curve**. A curve produced by plotting time against temperature.

**time-temperature-transformation (TTT) diagram**. See *isothermal transformation (IT) diagram*.

**timing marks**. Sharp lines produced by slightly changing the direction of the applied tension at regular time intervals (Fig. 609). Also known as *arrest marks*.

**tin and tin alloys**. See *Technical Brief 54*.

**tinning**. Coating metal with a very thin layer of molten solder or brazing filler metal.

## Technical Brief 54: Tin and Tin Alloys

TIN, which is a soft, brilliant white, low-melting-point material, was one of the first metals known to man. Throughout ancient history, various cultures recognized the virtues of tin in coatings, alloys, and compounds, and the use of the metal increased with advancing technology. Today, tin is an important metal in industry even though the annual tonnage used is relatively small compared with many other metals. One reason for this fact is that, in most applications, only very small amounts of tin are used at a time.

Tin is produced from both primary and secondary sources. Secondary tin is produced from recycled materials. Primary tin originates from the mineral cassiterite, a naturally occurring oxide of tin. These ores are smelted and refined to produce high-purity tin, which is cast into ingots weighing 12 to 25 kg (26 to 56 lb) or bars in weights of 1 kg (2.2 lb).

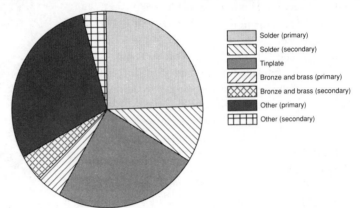

*Relative consumption of tin in the United States by application. 1988 data. Source: U.S. Bureau of Mines*

Because of its low strength (yield strength of only 11 MPa, or 1.6 ksi, at room temperature), the pure metal is not regarded as a structural material and is rarely used in monolithic form. Rather, tin is most frequently used as a coating for other metals. The largest single application of tin worldwide is in the manufacture of tinplate (steel sheet coated with tin), which accounts for about 40% of total world tin consumption. Since 1940, the traditional hot dip method of making tinplate has been largely replaced by electrodeposition of tin on continuous strips of rolled steel. Electrolytic tinplate can be produced with either equal or unequal amounts of tin on the two surfaces of the steel base metal. Coating thicknesses range from about 0.1 µm up to 0.60 mm (0.004 mil to 0.02 in.). More than 90% of tinplate is used for containers (tin cans). Tin coatings are also deposited on nonferrous alloys, primarily copper and copper-base alloys. These include tin-cadmium, tin-cobalt, tin-copper, tin-lead (see *terne*), tin-nickel, and tin-zinc alloys.

Tin also finds wide use in alloys, the most important of which are tin-base soft solders (see the table accompanying the term *solder*), bearing alloys (see the table accompanying the term *Babbitt metal*), and copper-base bronzes. Solders (primarily tin-lead solders) account for the largest use of tin in the United States. Tin-base solders are used to join food can seams, electronic and electrical components, and plumbing fixtures. Other applications for tin include jewelry and servingware (see *pewter* and *white metal*), organ pipes, *type metal*, and *fusible alloys*. Tin is also an alloying element in battery grid alloys, cast irons, dental amalgams, titanium alloys, and zirconium alloys.

### Recommended Reading

- W.B. Hampshire, Tin and Tin Alloys, *Metals Handbook*, 10th ed., Vol 2, ASM International, 1990, p 517-526
- D.J. Maykuth and W.B. Hampshire, Corrosion of Tin and Tin Alloys, *Metals Handbook*, 9th ed., Vol 13, 1987, p 770-783

**tin pest**. A polymorphic modification of tin that causes it to crumble into a powder known as gray tin (Fig. 610). It is generally accepted that the maximum rate of transformation occurs at about –40 °C (–40 °F), but transformation can occur at as high as about 13 °C (55 °F).

**tin sweat**. See *sweat*.

**tint etching**. Immersing metallographic specimens in specially formulated chemical etchants in order to produce a stable film on the specimen surface. When viewed under an optical microscope, these surface films produce colors which correspond to the various phases in the alloy. Also known as *color etching*.

**tinting**. See *heat tinting*.

**tin tossing**. Oxidizing impurities in molten tin by pouring it from one vessel to another in air, forming a dross that is mechanically separable.

**tip skid**. See preferred term *electrode skid*.

**TIR**. Abbreviation for *total indicator reading*.

FIG. 609

*Timing marks on the surface of a glass plate broken in nearly uniform tension. Crack propagated from left to right. 75×*

FIG. 610

*Tin pest resulting from the storage of a pure tin specimen at –20 °C (–4 °F)*

**titania.** A white water-insoluble powder of composition $TiO_2$ that is produced commercially from the minerals ilmenite and rutile. Used in paints and cosmetics and as an ingredient in porcelain enamels, ceramic whiteware, and ophthalmic glasses. Pure $TiO_2$ is also used as thin- or thick-film semiconductors.

**titanium and titanium alloys.** See *Technical Brief 55*.

**titanium carbide.** Very hard, heat-resistant ceramic materials of the composition TiC used in cermets and tungsten carbide cutting tools. Chemical vapor deposited TiC coatings are also used to extend the life of cemented carbide cutting tools. Table 68 compares the properties of TiC with other refractory carbides. See also *cemented carbides* (Technical Brief 9), *cermets* (Technical Brief 11), and the figure accompanying the term *composite coating*.

**titanium nitride.** A hard, high-melting-point ceramic (2950 °C, or 5342 °F) of the composition TiN that is used in cermets and as a coating material for cemented carbide cutting tools. See also *cermets* (Technical Brief 11) and the figure accompanying the term *carbide tools*.

**titration.** A method of determining the composition of a sample by adding known volumes of a solution of known concentration until a given reaction (color change, precipitation, or conductivity change) is produced. See also *volumetric analysis*.

**TIV.** Abbreviation for *total indicator variation*.

**T-joint.** See *tee joint*.

**TLC.** See *thin-layer chromatography*.

**toe crack.** A crack in the base metal occurring at the toe of a weld. See also the figure accompanying the term *weld crack*.

**toe of weld.** The junction between the face of a weld and the base metal. See also the figure accompanying the term *fillet weld*.

**toggle press.** A *mechanical press* in which the *slide* is actuated by one or more toggle links or mechanisms (Fig. 611).

**tolerance.** The specified permissible deviation from a specified nominal dimension, or the permissible variation in size or other quality characteristic of a part.

**Table 68  Properties of refractory metal carbides**

| Carbide | Hardness, HV (50 kg) | Crystal structure | Melting point °C | °F | Theoretical density, g/cm³ | Modulus of elasticity GPa | psi × 10⁶ | Coefficient of thermal expansion, µm/m · K |
|---|---|---|---|---|---|---|---|---|
| TiC | 3000 | Cubic | 3100 | 5600 | 4.94 | 451 | 65.4 | 7.7 |
| VC | 2900 | Cubic | 2700 | 4900 | 5.71 | 422 | 61.2 | 7.2 |
| HfC | 2600 | Cubic | 3900 | 7050 | 12.76 | 352 | 51.1 | 6.6 |
| ZrC | 2700 | Cubic | 3400 | 6150 | 6.56 | 348 | 50.5 | 6.7 |
| NbC | 2000 | Cubic | 3600 | 6500 | 7.80 | 338 | 49.0 | 6.7 |
| Cr₃C₂ | 1400 | Orthorhombic | 1800(a) | 3250 | 6.66 | 373 | 54.1 | 10.3 |
| WC | (0001) 2200 (10T0) 1300 | Hexagonal | ~2800(a) | 5050 | 15.7 | 696 | 101 | (0001) 5.2 (10T0) 7.3 |
| Mo₂C | 1500 | Hexagonal | 2500 | 4550 | 9.18 | 533 | 77.3 | 7.8 |
| TaC | 1800 | Cubic | 3800 | 6850 | 14.50 | 285 | 41.3 | 6.3 |

(a) Not congruently melting, dissociation temperature

## Technical Brief 55: Titanium and Titanium Alloys

TITANIUM is the fourth most abundant structural metal in the crust of the earth after aluminum, iron, and magnesium. The development of its alloys and processing technologies started only in the late 1940s; thus, titanium metallurgy just missed being a factor in World War II. The difficulty in extracting titanium from ores, its high reactivity in the molten state, its forging complexity, its machining difficulty, and its sensitivity to segregation and inclusions necessitated the development of special processing techniques. These techniques have contributed to the high cost of titanium raw materials, alloys, and final products. On the other hand, the low density of titanium alloys (about 60% of the density of steel) provides high structural efficiencies based on a wide range of mechanical properties, coupled with an excellent resistance to aggressive environments. These alloys have contributed to the quality and durability of military aircraft, helicopters, and turbofan jet engines as well as to the increased reliability of heat exchangers and surgical implants.

Titanium metal passes through three major steps during processing from ore to finished product: (1) reduction of titanium ore (rutile, or $TiO_2$) to a porous form of titanium metal called "sponge," (2) melting of sponge to form an ingot, and (3) remelting and casting into finished shape, or primary fabrication, in which ingots are converted into general mill products followed by secondary fabrication of finished shapes from mill products. Powder metallurgy processing is also commonly used.

Titanium exists in two crystallographic forms. At room temperature, unalloyed (commercially pure) titanium has a hexagonal close-packed (hcp) crystal structure referred to as alpha ($\alpha$) phase. At 883 °C (1621 °F), this transforms to a body-centered cubic (bcc) structure known as beta ($\beta$) phase. The manipulation of these crystallographic variations through alloying additions and thermomechanical processing is the basis for the development of a wide range of alloys and properties. These phases also provide a convenient way to categorize titanium alloys. Depending on their microstructure, titanium alloys fall into one of four classes: $\alpha$, near-$\alpha$, $\alpha$-$\beta$, or $\beta$. These classes denote the general type of microstructure after processing.

Aerospace applications—including use in both structural (airframe) components and jet engines—account for the largest share of titanium alloy use, because titanium saves weight in highly loaded components that operate at low to moderately elevated temperatures. Many titanium alloys have been custom designed to have optimum tensile, compressive, and/or creep strength at selected temperatures and, at the same time, to have sufficient workability to be fabricated into products suitable for specific applications. During the life of the titanium industry, alloy Ti-6Al-4V has been consistently responsible for about 45% of total industry applications. Titanium is also often used in applications that exploit its corrosion resistance. These applications include chemical processing, the pulp and paper industry, marine applications, energy production and storage, and biomedical applications that take advantage of titanium's inertness in the human body for use in surgical implants and prosthetic devices. Other materials of interest are titanium-niobium alloys used for superconductors (see *superconductors*, Technical Brief 53) and titanium-nickel alloys that exhibit a shape memory effect (see *shape memory alloys*, Technical Brief 42). Titanium-matrix composites and titanium-base intermetallics are also under development (see *metal-matrix composites*, Technical Brief 26, and *intermetallics*, Technical Brief 22).

**tolerance limits**. The extreme values (upper and lower) that define the range of permissible variation in size or other quality characteristic of a part. See also *quality characteristics*.

**tomographic plane**. A section of the part imaged by the tomographic process. Although in computed tomography the tomographic plane or slice is displayed as a two-dimensional image, the measurements are of the materials within a defined slice thickness associated with the plane. See also *slice* and the figure accompanying the term *computed tomography*.

**tomography**. From the Greek "to write a slice or section." The process of imaging a particular plane or slice through an object. See also *computed tomography*.

**tonghold**. The portion of a forging billet, usually on one end, that is gripped by the operator's tongs (Fig. 612). It is removed from the part at the end of the forging operation. Common to drop hammer and press-type forging.

**tooling**. A generic term applying to die assemblies and related items used for forming and forging metals.

**tooling marks**. Indications imparted to the surface of the forged part from dies containing surface imperfections or

## Summary of commercial and semicommercial grades and alloys of titanium

| Designation | Tensile strength (min) MPa | ksi | 0.2% yield strength (min) MPa | ksi | Impurity limits, wt % N (max) | C (max) | H (max) | Fe (max) | O (max) | Nominal composition, wt % Al | Sn | Zr | Mo | Others |
|---|---|---|---|---|---|---|---|---|---|---|---|---|---|---|
| **Unalloyed grades** | | | | | | | | | | | | | | |
| ASTM Grade 1 | 240 | 35 | 170 | 25 | 0.03 | 0.10 | 0.015 | 0.20 | 0.18 | ... | ... | ... | ... | ... |
| ASTM Grade 2 | 340 | 50 | 280 | 40 | 0.03 | 0.10 | 0.015 | 0.30 | 0.25 | ... | ... | ... | ... | ... |
| ASTM Grade 3 | 450 | 65 | 380 | 55 | 0.05 | 0.10 | 0.015 | 0.30 | 0.35 | ... | ... | ... | ... | ... |
| ASTM Grade 4 | 550 | 80 | 480 | 70 | 0.05 | 0.10 | 0.015 | 0.50 | 0.40 | ... | ... | ... | ... | ... |
| ASTM Grade 7 | 340 | 50 | 280 | 40 | 0.03 | 0.10 | 0.015 | 0.30 | 0.25 | ... | ... | ... | ... | 0.2 Pd |
| **Alpha and near-alpha alloys** | | | | | | | | | | | | | | |
| Ti-0.3Mo-0.8Ni | 480 | 70 | 380 | 55 | 0.03 | 0.10 | 0.015 | 0.30 | 0.25 | ... | ... | ... | 0.3 | 0.8 Ni |
| Ti-5Al-2.5Sn | 790 | 115 | 760 | 110 | 0.05 | 0.08 | 0.02 | 0.50 | 0.20 | 5 | 2.5 | ... | ... | ... |
| Ti-5Al-2.5Sn-ELI | 690 | 100 | 620 | 90 | 0.07 | 0.08 | 0.0125 | 0.25 | 0.12 | 5 | 2.5 | ... | ... | ... |
| Ti-8Al-1Mo-1V | 900 | 130 | 830 | 120 | 0.05 | 0.08 | 0.015 | 0.30 | 0.12 | 8 | ... | ... | 1 | 1 V |
| Ti-6Al-2Sn-4Zr-2Mo | 900 | 130 | 830 | 120 | 0.05 | 0.05 | 0.0125 | 0.25 | 0.15 | 6 | 2 | 4 | 2 | ... |
| Ti-6Al-2Nb-1Ta-0.8Mo | 790 | 115 | 690 | 100 | 0.02 | 0.03 | 0.0125 | 0.12 | 0.10 | 6 | ... | ... | 1 | 2 Nb, 1 Ta |
| Ti-2.25Al-11Sn-5Zr-1Mo | 1000 | 145 | 900 | 130 | 0.04 | 0.04 | 0.008 | 0.12 | 0.17 | 2.25 | 11.0 | 5.0 | 1.0 | 0.2 Si |
| Ti-5Al-5Sn-2Zr-2Mo(a) | 900 | 130 | 830 | 120 | 0.03 | 0.05 | 0.0125 | 0.15 | 0.13 | 5 | 5 | 2 | 2 | 0.25 Si |
| **Alpha-beta alloys** | | | | | | | | | | | | | | |
| Ti-6Al-4V(b) | 900 | 130 | 830 | 120 | 0.05 | 0.10 | 0.0125 | 0.30 | 0.20 | 6.0 | ... | ... | ... | 4.0 V |
| Ti-6Al-4V-ELI(b) | 830 | 120 | 760 | 110 | 0.05 | 0.08 | 0.0125 | 0.25 | 0.13 | 6.0 | ... | ... | ... | 4.0 V |
| Ti-6Al-6V-2Sn(b) | 1030 | 150 | 970 | 140 | 0.04 | 0.05 | 0.015 | 1.0 | 0.20 | 6.0 | 2.0 | ... | ... | 0.75 Cu, 6.0 V |
| Ti-8Mn(b) | 860 | 125 | 760 | 110 | 0.05 | 0.08 | 0.015 | 0.50 | 0.20 | ... | ... | ... | ... | 8.0 Mn |
| Ti-7Al-4Mo(b) | 1030 | 150 | 970 | 140 | 0.05 | 0.10 | 0.013 | 0.30 | 0.20 | 7.0 | ... | ... | 4.0 | ... |
| Ti-6Al-2Sn-4Zr-6Mo(c) | 1170 | 170 | 1100 | 160 | 0.04 | 0.04 | 0.0125 | 0.15 | 0.15 | 6.0 | 2.0 | 4.0 | 6.0 | ... |
| Ti-5Al-2Sn-2Zr-4Mo-4Cr(a)(c) | 1125 | 163 | 1055 | 153 | 0.04 | 0.05 | 0.0125 | 0.30 | 0.13 | 5.0 | 2.0 | 2.0 | 4.0 | 4.0 Cr |
| Ti-6Al-2Sn-2Zr-2Mo-2Cr(a)(b) | 1030 | 150 | 970 | 140 | 0.03 | 0.05 | 0.0125 | 0.25 | 0.14 | 5.7 | 2.0 | 2.0 | 2.0 | 2.0 Cr, 0.25 Si |
| Ti-10V-2Fe-3Al(a)(c) | 1170 | 170 | 1100 | 160 | 0.05 | 0.05 | 0.015 | 2.5 | 0.16 | 3.0 | ... | ... | ... | 10.0 V |
| Ti-3Al-2.5V(d) | 620 | 90 | 520 | 75 | 0.015 | 0.05 | 0.015 | 0.30 | 0.12 | 3.0 | ... | ... | ... | 2.5 V |
| **Beta alloys** | | | | | | | | | | | | | | |
| Ti-13V-11Cr-3Al(c) | 1170 | 170 | 1100 | 160 | 0.05 | 0.05 | 0.025 | 0.35 | 0.17 | 3.0 | ... | ... | ... | 11.0 Cr, 13.0 V |
| Ti-8Mo-8V-2Fe-3Al(a)(c) | 1170 | 170 | 1100 | 160 | 0.05 | 0.05 | 0.015 | 2.5 | 0.17 | 3.0 | ... | ... | 8.0 | 8.0 V |
| Ti-3Al-8V-6Cr-4Mo-4Zr(a)(b) | 900 | 130 | 830 | 120 | 0.03 | 0.05 | 0.020 | 0.25 | 0.12 | 3.0 | ... | 4.0 | 4.0 | 6.0 Cr, 8.0 V |
| Ti-11.5Mo-6Zr-4.5Sn(b) | 690 | 100 | 620 | 90 | 0.05 | 0.10 | 0.020 | 0.35 | 0.18 | ... | 4.5 | 6.0 | 11.5 | ... |

(a) Semicommercial alloy; mechanical properties and composition limits subject to negotiation with suppliers. (b) Mechanical properties given for annealed condition; may be solution treated and aged to increase strength. (c) Mechanical properties given for solution treated and aged condition; alloy not normally applied in annealed condition. Properties may be sensitive to section size and processing. (d) Primarily a tubing alloy; may be cold drawn to increase strength.

## Recommended Reading

- S. Lampman, Wrought Titanium and Titanium Alloys, *Metals Handbook*, 10th ed., Vol 2, ASM International, 1990, p 592-633
- D. Eylon, J.B. Newman, and J.K. Thorne, Titanium and Titanium Alloy Castings, *Metals Handbook*, 10th ed., Vol 2, ASM International, 1990, p 634-646
- D. Eylon and F.H. Froes, Titanium P/M Products, *Metals Handbook*, 10th ed., Vol 2, ASM International, 1990, p 647-660
- R.W. Schutz and D.E. Thomas, Corrosion of Titanium and Titanium Alloys, *Metals Handbook*, 9th ed., Vol 13, ASM International, 1987, p 669-706

dies on which some repair work has been done. These marks are usually slight rises or depressions in the metal.

**tooling resin.** Resins that have applications as tooling aids, coreboxes, prototypes, hammer forms, stretch forms, foundry patterns, and so forth. Epoxy and silicone are common examples.

**tool set.** See *die*.

**tool side.** The side of a plastic part that is cured against the tool (mold or mandrel).

**tool steels.** See *Technical Brief 56*.

**tooth.** (1) A projection on a multipoint tool (such as on a saw, milling cutter, or file) designed to produce cutting. (2) A projection on the periphery of a wheel or segment thereof—as on a gear (Fig. 613), spline, or sprocket, for example—designed to engage another mechanism and thereby transmit force or motion, or both. A similar projection on a flat member such as a rack.

**tooth point.** On a *face mill,* the chamfered cutting edge of the blade, to which a flat is sometimes added to produce a shaving effect and to improve finish.

**top-and-bottom process.** A process for separating copper and nickel, in which their molten sulfides are separated

FIG. 611

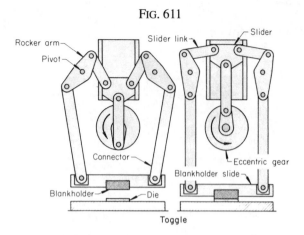

*Operating principles of a toggle drive system for mechanical presses*

FIG. 612

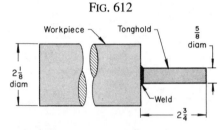

*Welded-on tonghold that substantially reduced forging crop-end loss. Dimensions given in inches.*

FIG. 613

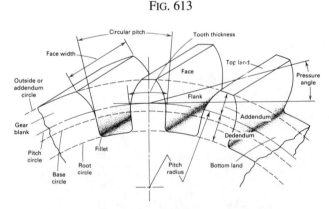

*Schematic of typical gear tooth nomenclature*

into two liquid layers by the addition of sodium sulfide. The lower layer holds most of the nickel.

**torch.** See preferred terms *cutting torch* and *welding torch*.

**torch brazing.** A brazing process in which the heat required is furnished by a fuel gas flame.

**torch soldering.** A soldering process in which the heat required is furnished by a fuel gas flame.

FIG. 614

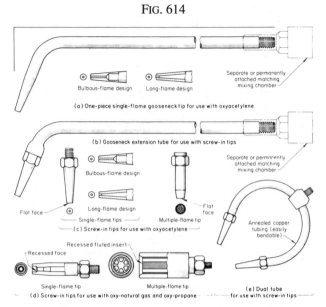

*Tip designs used in manual torch brazing*

**torch tip** (brazing). That part of an oxyfuel gas brazing torch from which the gases issue (Fig. 614). See also *cutting tip* and *welding tip*.

**toroid.** Doughnut-shaped piece of magnetic material, together with one or more coils of current-carrying wire wound about the doughnut, with the permeability of the magnetic material high enough so that the magnetic flux is almost completely confined within it. Also known as *toroidal coil* and *toroidal magnetic circuit*.

**torque-coil magnetometer.** A magnetometer that depends for its operation on the torque developed by a known current in a coil that can turn in the field to be measured. See also *magnetometer*.

**torsion.** (1) A twisting deformation of a solid or tubular body about an axis in which lines that were initially parallel to the axis become helices. (2) A twisting action resulting in shear stresses and strains.

**torsional moment.** In a body being twisted, the algebraic sum of the couples or the moments of the external forces about the axis of twist, or both.

**torsional pendulum.** A device for performing dynamic mechanical analysis of plastics, in which the sample is deformed torsionally and allowed to oscillate in free vibration. Modulus is determined by the frequency of the resultant oscillation, and damping is determined by the decreasing amplitude of the oscillation.

**torsional stress.** The shear stress on a transverse cross section caused by a twisting action.

**torsion test.** A test designed to provide data for the calculation of the *shear modulus*, *modulus of rupture in torsion*, and *yield strength* in shear.

**total carbon.** The sum of the *free carbon* and *combined carbon* (including carbon in solution) in a ferrous alloy.

**total cyanide.** Cyanide content of an electroplating bath (including both simple and complex ions).

**total elongation.** The total amount of permanent extension of a test piece broken in a tensile test usually expressed as a percentage over a fixed gage length. See also *elongation, percent.*

**total-extension-under-load yield strength.** See *yield strength.*

**total indicator reading.** See *total indicator variation.*

**total indicator variation.** The difference between the maximum and minimum indicator readings during a checking cycle.

**total transmittance.** The ratio of the radiant energy leaving one side of a region between two parallel planes to the radiant energy entering from the opposite side.

**toughness.** Ability of a material to absorb energy and deform plastically before fracturing. Toughness is proportional to the area under the *stress-strain curve* from the origin to the breaking point. In metals, toughness is usually measured by the energy absorbed in a notch impact test. See also *impact test.*

**tough pitch copper.** Copper containing from 0.02 to 0.04% oxygen, obtained by refining copper in a reverberatory furnace.

**tow.** An untwisted bundle of continuous filaments, usually referring to man-made fibers, particularly carbon and graphite, but also fiberglass and aramid. A tow designated as 140 K has 140,000 filaments. Table 69 lists fiber tow characteristics.

**T-peel strength.** The average load per unit width of adhesive bond line required to produce progressive separation of two bonded, flexible adherends, under standard test conditions.

**TPI.** See *thermoplastic polyimides.*

**TPUR.** See *thermoplastic polyurethanes.*

**tracer.** In composites fabrication, fiber, tow, or yarn added to a prepreg for verifying fiber alignment and, in the case of woven materials, for distinguishing warp fibers from fill fibers.

**tracer milling.** Duplication of a three-dimensional form by means of a cutter controlled by a tracer that is directed by a master form.

**track.** The mark made by a seal on the surface with which it mates.

**tracking pattern.** The path a seal ring makes when in rubbing contact with the mating ring or seal plate.

**traction.** In rolling contacts, the tangential stress transmitted across the interface. The traction will in general vary from point to point over the contact area. More generally, traction may denote the force per unit area of contact.

**tractive force.** The integral of the tangential surface stress over the area of contact.

**traditional ceramics.** See *Technical Brief 57.*

**traffic mark.** See *abrasion.*

**tramp alloys.** Residual alloying elements that are introduced into steel when unidentified alloy steel is present in the scrap charge to a steelmaking furnace.

**tramp element.** Contaminant in the components of a furnace charge, or in the molten metal or castings, whose presence is thought to be either unimportant or undesirable to the quality of the casting. Also called trace element.

**transcrystalline.** See *transgranular.*

**transcrystalline cracking.** Cracking or fracturing that occurs through or across a crystal. Also termed intracrystalline cracking.

**transfer.** In tribology, the process by which material from one sliding surface becomes attached to another surface, possibly as the result of interfacial adhesion (Fig. 615). Transfer is usually associated with adhesion, but the possibility of mechanical interlocking adherence, without adhesive bonding, exists in certain occurrences. Material may also back transfer to the surface from which it came. See also *selective transfer.*

**transference.** The movement of ions through the *electrolyte* associated with the passage of the electric current. Also called transport or migration.

**transference number.** The proportion of total electroplating current carried by ions of a given kind. Also called transport number.

**transfer ladle.** A ladle that can be supported on a monorail or carried in a shank and used to transfer metal from the melting furnace to the holding furnace or from the furnace to the pouring ladles.

**transfer molding.** A method of molding thermosetting materials in which the plastic is first softened by heat and pressure in a transfer chamber and then forced by high pressure through suitable sprues, runners, and gates into a closed mold for final shaping and curing. See also *resin transfer molding.*

### Table 69 Fiber tow characteristics before impregnation

| Material | Yield/tow | | Filament size | |
|---|---|---|---|---|
| | m/kg | yd/lb | μm | μin. |
| Graphite (1000-12 000 filaments/tow) | 300-1200 | 150-600 | 5-10 | 200-390 |
| Fiberglass (2450-12 240 filaments/tow) | 490-2400 | 245-1200 | 4-13 | 160-510 |
| Aramid (800-3200 filaments/tow) | 2000-7850 | 980-3900 | 12 | 470 |

## Technical Brief 56: Tool Steels

A TOOL STEEL is any steel used to make tools for cutting, forming, or otherwise shaping a material into a final part or component. These complex alloy steels, which contain relatively large amounts of tungsten, molybdenum, vanadium, manganese, and chromium, make it possible to meet increasingly severe service demands. In service, most tools are subjected to extremely high loads that are applied rapidly. The tools must withstand these loads a great number of times without breaking and without undergoing excessive wear or deformation. In many applications, tool steels must provide this capability under conditions that develop high temperatures in the tool. Most tool steel are wrought products, but precision castings can be used in some applications. The powder metallurgy process is also used in making tool steels. It provides, first, a more uniform carbide size and distribution in large sections and, second, special compositions that are difficult or impossible to produce in wrought or cast alloys.

Tool steels are classified according to their composition, application, or method of quenching. Each group is identified by a capital letter; individual tool steel types are assigned code numbers (see the accompanying table). High-speed steel are tool materials developed largely for use in high-speed metal cutting applications. There are two classifications of high-speed steels: molybdenum high-speed steels, or group M, which contain from 0.75 to 1.52% C and 4.50 to 11.00% Mo, and tungsten high-speed steels, or group T, which have similar carbon contents but high (11.75 to 21.00%) tungsten contents. Group M steels constitute more than 95% of all high-speed steel produced in the United States.

Hot-work steels (group H) have been developed to withstand the combinations of heat, pressure, and abrasion associated with punching, shearing, or forming of metals at high temperatures. Group H steels usually have medium carbon contents (0.35 to 0.45%) and chromium, tungsten, molybdenum, and vanadium contents of 6 to 25%. H steels are divided into chromium hot-work steels, tungsten hot-work steels, and molybdenum hot-work steels.

Cold-work tool steels are restricted in application to those uses that do not involve prolonged or repeated heating above 205 to 260 °C (400 to 500 °F). There are three categories of cold-work steels: air-hardening steels, or group A; high-carbon, high-chromium steels, or group D; and oil-hardening steels, or group O.

Shock-resisting, or group S, steels contain manganese, silicon, chromium, tungsten, and molybdenum, in various combinations; carbon content is about 1.50%. Group S steels are used primarily for chisels, rivet sets, punches, and other applications requiring high toughness and resistance to shock loading.

The low-alloy special-purpose, or group L, tool steels contain small amounts of chromium, vanadium, nickel, and molybdenum. Group L steels are generally used for machine parts and other special applications requiring good strength and toughness.

Mold steels, or group P, contain chromium and nickel as principal alloying elements. Because of their low resistance to softening at elevated temperatures, group P steels are used almost exclusively in low-temperature die casting dies and in molds for injection or compression molding of plastics.

Water-hardening, or group W, tool steels contain carbon as the principal alloying element (0.70 to 1.50% C). Group W steels, which also have low resistance to softening at elevated temperatures, are suitable for cold heading, coining, and embossing tools, woodworking tools, metal-cutting tools, and wear-resistant machine tool components.

## Recommended Reading

- A.M. Bayer, Wrought Tool Steels, *Metals Handbook*, 10th ed., Vol 1, ASM International, 1990, p 757-779
- K.E. Pinnow and W. Stasko, P/M Tool Steels, *Metals Handbook*, 10th ed., Vol 1, 1990, p 780-792
- G.A. Roberts and R.A. Gary, *Tool Steels*, 4th ed., American Society for Metals, 1980

## Composition limits of selected types of wrought tool steels

| Designation | | Composition(a), % | | | | | | | | |
|---|---|---|---|---|---|---|---|---|---|---|
| AISI | UNS | C | Mn | Si | Cr | Ni | Mo | W | V | Co |
| **Molybdenum high-speed steels** | | | | | | | | | | |
| M1 | T11301 | 0.78-0.88 | 0.15-0.40 | 0.20-0.50 | 3.50-4.00 | 0.30 max | 8.20-9.20 | 1.40-2.10 | 1.00-1.35 | ... |
| M2 | T11302 | 0.78-0.88; 0.095-1.05 | 0.15-0.40 | 0.20-0.45 | 3.75-4.50 | 0.30 max | 4.50-5.50 | 5.50-6.75 | 1.75-2.20 | ... |
| M4 | T11304 | 1.25-1.40 | 0.15-0.40 | 0.20-0.45 | 3.75-4.75 | 0.30 max | 4.25-5.50 | 5.25-6.50 | 3.75-4.50 | ... |
| M35 | T11335 | 0.82-0.88 | 0.15-0.40 | 0.20-0.45 | 3.75-4.50 | 0.30 max | 4.50-5.50 | 5.50-6.75 | 1.75-2.20 | 4.50-5.50 |
| M42 | T11342 | 1.05-1.15 | 0.15-0.40 | 0.15-0.65 | 3.50-4.25 | 0.30 max | 9.00-10.00 | 1.15-1.85 | 0.95-1.35 | 7.75-8.75 |
| M62 | T11362 | 1.25-1.35 | 0.15-0.40 | 0.15-0.40 | 3.50-4.00 | 0.30 max | 10.00-11.00 | 5.75-6.50 | 1.80-2.10 | |
| **Tungsten high-speed steels** | | | | | | | | | | |
| T1 | T12001 | 0.65-0.80 | 0.10-0.40 | 0.20-0.40 | 3.75-4.50 | 0.30 max | ... | 17.25-18.75 | 0.90-1.30 | ... |
| T15 | T12015 | 1.50-1.60 | 0.15-0.40 | 0.15-0.40 | 3.75-5.00 | 0.30 max | 1.00 max | 11.75-13.00 | 4.50-5.25 | 4.75-5.25 |
| **Chromium hot-work steels** | | | | | | | | | | |
| H11 | T20811 | 0.33-0.43 | 0.20-0.50 | 0.80-1.20 | 4.75-5.50 | 0.30 max | 1.10-1.60 | ... | 0.30-0.60 | ... |
| H19 | T20819 | 0.32-0.45 | 0.20-0.50 | 0.20-0.50 | 4.00-4.75 | 0.30 max | 0.30-0.55 | 3.75-4.50 | 1.75-2.20 | 4.00-4.50 |
| **Tungsten hot-work steels** | | | | | | | | | | |
| H21 | T20821 | 0.26-0.36 | 0.15-0.40 | 0.15-0.50 | 3.00-3.75 | 0.30 max | ... | 8.50-10.00 | 0.30-0.60 | ... |
| H23 | T20823 | 0.25-0.35 | 0.15-0.40 | 0.15-0.60 | 11.00-12.75 | 0.30 max | ... | 11.00-12.75 | 0.75-1.25 | ... |
| H26 | T20826 | 0.45-0.55(b) | 0.15-0.40 | 0.15-0.40 | 3.75-4.50 | 0.30 max | ... | 17.25-19.00 | 0.75-1.25 | ... |
| **Molybdenum hot-work steels** | | | | | | | | | | |
| H42 | T20842 | 0.55-0.70(b) | 0.15-0.40 | ... | 3.75-4.50 | 0.30 max | 4.50-5.50 | 5.50-6.75 | 1.75-2.20 | ... |
| **Air-hardening, medium-alloy, cold-work steels** | | | | | | | | | | |
| A2 | T30102 | 0.95-1.05 | 1.00 max | 0.50 max | 4.75-5.50 | 0.30 max | 0.90-1.40 | ... | 0.15-0.50 | ... |
| A6 | T30106 | 0.65-0.75 | 1.80-2.50 | 0.50 max | 0.90-1.20 | 0.30 max | 0.90-1.40 | ... | ... | ... |
| A10 | T30110 | 1.25-1.50(c) | 1.60-2.10 | 1.00-1.50 | ... | 1.55-2.05 | 1.25-1.75 | ... | ... | ... |
| **High-carbon, high-chromium, cold-work steels** | | | | | | | | | | |
| D2 | T30402 | 1.40-1.60 | 0.60 max | 0.60 max | 11.00-13.00 | 0.30 max | 0.70-1.20 | ... | 1.10 max | ... |
| D3 | T30403 | 2.00-2.35 | 0.60 max | 0.60 max | 11.00-13.50 | 0.30 max | ... | 1.00 max | 1.00 max | ... |
| **Oil-hardening cold-work steels** | | | | | | | | | | |
| O1 | T31501 | 0.85-1.00 | 1.00-1.40 | 0.50 max | 0.40-0.60 | 0.30 max | ... | 0.40-0.60 | 0.30 max | ... |
| O2 | T31502 | 0.85-0.95 | 1.40-1.80 | 0.50 max | 0.50 max | 0.30 max | 0.30 max | ... | 0.30 max | ... |
| O6 | T31506 | 1.25-1.55(c) | 0.30-1.10 | 0.55-1.50 | 0.30 max | 0.30 max | 0.20-0.30 | ... | ... | ... |
| **Shock-resisting steels** | | | | | | | | | | |
| S1 | T41901 | 0.40-0.55 | 0.10-0.40 | 0.15-1.20 | 1.00-1.80 | 0.30 max | 0.50 max | 1.50-3.00 | 0.15-0.30 | ... |
| S2 | T41902 | 0.40-0.55 | 0.30-0.50 | 0.90-1.20 | ... | 0.30 max | 0.30-0.60 | ... | 0.50 max | ... |
| S7 | T41907 | 0.45-0.55 | 0.20-0.90 | 0.20-1.00 | 3.00-3.50 | ... | 1.30-1.80 | ... | 0.20-0.30(d) | ... |
| **Low-alloy special-purpose tool steels** | | | | | | | | | | |
| L2 | T61202 | 0.45-1.00(b) | 0.10-0.90 | 0.50 max | 0.70-1.20 | ... | 0.25 max | ... | 0.10-0.30 | ... |
| L6 | T61206 | 0.65-0.75 | 0.25-0.80 | 0.50 max | 0.60-1.20 | 1.25-2.00 | 0.50 max | ... | 0.20-0.30(d) | ... |
| **Low-carbon mold steels** | | | | | | | | | | |
| P2 | T51602 | 0.10 max | 0.10-0.40 | 0.10-0.40 | 0.75-1.25 | 0.10-0.50 | 0.15-0.40 | ... | ... | ... |
| P5 | T51605 | 0.10 max | 0.20-0.60 | 0.40 max | 2.00-2.50 | 0.35 max | ... | ... | ... | ... |
| P20 | T51620 | 0.28-0.40 | 0.60-1.00 | 0.20-0.80 | 1.40-2.00 | ... | 0.30-0.55 | ... | ... | ... |
| **Water-hardening tool steels** | | | | | | | | | | |
| W1 | T72301 | 0.70-1.50(e) | 0.10-0.40 | 0.10-0.40 | 0.15 max | 0.20 max | 0.10 max | 0.15 max | 0.10 max | ... |
| W2 | T72302 | 0.85-1.50(e) | 0.10-0.40 | 0.10-0.40 | 0.15 max | 0.20 max | 0.10 max | 0.15 max | 0.15-0.35 | ... |

(a) All steels except group W contain 0.25 max Cu, 0.03 max P, and 0.03 max S; group W steels contain 0.20 max Cu, 0.025 max P, and 0.025 max S. Where specified, sulfur may be increased to 0.06 to 0.15% to improve machinability of group A, D, H, M, and T steels. (b) Available in several carbon ranges. (c) Contains free graphite in the microstructure. (d) Optional. (e) Specified carbon ranges are designated by suffix numbers.

## Technical Brief 57: Traditional Ceramics

TRADITIONAL CERAMICS are generally classified as those ceramic products that use clay or have a significant clay component in the batch. A clay-based ceramic body usually consists of one or more clays or clay minerals mixed with nonclay mineral powders such as fluxes (for example, feldspar) and fillers (for example, silica and alumina). Each of these constituents contributes to the plastic forming and fired characteristics of the body, with the clay acting as a plasticizer and binder for the other constituents.

Commercial clays are grouped as being kaoline (china) clay or ball clay. Kaolines consist primarily of ordered kaolinite, $Al_2Si_2O_5(OH)_4$, with some mica and free quartz. Ball clays,

### Characteristics of commercial clays found in the United States

| Properties | Coarse kaolin, sedimentary(a) | Fine kaolin, sedimentary(b) | Dark, fine ball clay(c) | Light, coarse ball clay(d) |
|---|---|---|---|---|
| Compound, wt% | | | | |
| $SiO_2$ | 45.7 | 46.7 | 50.5 | 60.4 |
| $Al_2O_3$ | 38.3 | 38.2 | 28.7 | 27.0 |
| $Fe_2O_3$ | 0.41 | 0.60 | 0.91 | 0.93 |
| $TiO_2$ | 1.55 | 1.42 | 1.48 | 1.62 |
| CaO | 0.08 | 0.12 | 0.40 | 0.28 |
| MgO | 0.06 | 0.20 | 0.30 | 0.26 |
| $K_2O$ | 0.06 | 0.15 | 0.89 | 1.70 |
| $Na_2O$ | 0.14 | 0.03 | 0.18 | 0.50 |
| Ignition loss | 13.65 | 13.79 | 16.58 | 7.59 |
| Minerals, wt% | | | | |
| Montmorillonite | nil | 3 | 8 | 7 |
| Kaolin group | 96 | 93 | 58 | 44 |
| Mica | 2 | 2 | 10 | 21 |
| Free quartz | trace | 1 | 14 | 26 |
| Organic | trace | trace | 8 | 0.5 |
| Particle size, % | | | | |
| <20 μm (<800 μin.) | 95 | 99 | 99 | 98 |
| <5 μm (<200 μin.) | 69 | 88 | 95 | 79 |
| <2 μm (<80 μin.) | 52 | 72 | 82 | 61 |
| <1 μm (<40 μin.) | 35 | 56 | 69 | 43 |
| <0.5 μm (<20 μin.) | 28 | 41 | 51 | 29 |
| Surface area, methylene blue index, meq/100 g | 1.6 | 10.5 | 12.1 | 5.6 |

(a) Washington County, Georgia. (b) Wilkinson County, Georgia. (c) Graves County, Kentucky. (d) Weakley County, Tennessee

which are dug out of the ground in "blocks" or "balls," tend to be very fine grained and are composed of ordered and disordered kaolinite and varying percentages of mica, illite, montmorillonite, free quartz, and organic matter.

Traditional ceramic bodies are formed into shapes using many different techniques. A general sequence of unit operations would include raw material preparation, batch preparation, forming, drying, prefire operations (glazing, decorating, etc.), firing, and postfire operations (glazing, decorating, machining, and/or cleaning). Foring techniques include hand molding and pottery wheels, extrusion, die pressing, and *slip casting*.

The five principal product areas for traditional ceramics are whitewares, glazes and porcelain enamels, structural clay products, cement, and refractories. Whiteware is the name given to a group ceramic products characterized by a white or light-colored body with a fine-grained structure. Most whiteware products are glazed or decorated

**transferred arc.** A plasma arc established between the electrode and the workpiece during plasma arc welding, cutting, and thermal spraying (Fig. 616). See also the figure accompanying the term *nontransferred arc*.

**transformation hardening.** Heat treatment of steels comprising austenitization followed by cooling under conditions such that the austenite transforms more or less completely into martensite and possibly into bainite.

**transformation-induced plasticity.** A phenomenon, occurring chiefly in certain highly alloyed steels that have been heat treated to produce metastable austenite or metastable austenite plus martensite, whereby, on subsequent deformation, part of the austenite undergoes strain-induced transformation to martensite. Steels capable of transform-

ing in this manner, commonly referred to as TRIP steels, are highly plastic after heat treatment, but exhibit a very high rate of strain hardening and thus have high tensile and yield strengths after plastic deformation at temperatures between about 20 and 500 °C (70 and 930 °F). Cooling to −195 °C (−320 °F) may or may not be required to complete the transformation to martensite. Tempering usually is done following transformation.

**transformation ranges.** Those ranges of temperature within which austenite forms during heating and transforms during cooling. The two ranges are distinct, sometimes overlapping but never coinciding. The limiting temperatures of the ranges depend on the composition of the alloy and on the rate of change of temperature, par-with patterns or designs. Examples of whiteware products are sanitaryware, tableware, artware, stoneware, and floor and wall tile.

A glaze is defined as a continuous adherent layer of glass on the surface of a ceramic body that is hard, nonabsorbent, and easily cleaned. A glaze is usually applied as a suspension of glaze-forming ingredients in water. After the glaze layer dries on the surface of the piece, it is fired, whereupon the ingredients melt to form a thin layer of glass. Porcelain enamels are very durable alkaliborosilicate glass coatings bonded by fusion to metal substrates at temperatures above 425 °C (800 °F). Porcelain enamels are applied primarily to steel sheet, cast iron, aluminum alloys (in sheet or cast form), and aluminum-coated steels. See also the composition tables accompanying the term *porcelain enamel*.

Structural clay products are ceramic materials used in construction. The raw materials are naturally occurring clays or shales. A distinguishing manufacturing characteristic of structural clays is their exposure to elevated firing temperatures in order to develop a bond between the particulate constituents and to develop the desired pore structure (pore quantity, pore size distribution, and pore connection) for the intended application. Structural clay products include facing materials for buildings, building brick, paving brick, roofing tile, sewer pipe, and drain tile.

Cements are inorganic powders consisting predominantly of calcium silicates, which, when mixed with water to form a paste, react slowly at ambient temperatures to produce a coherent, hardened mass with valuable engineering properties. The hardened powder product is porous and consists primarily of calcium silicate hydrate. Uses of cement include steel-reinforced pipe, panels, columns, and beams, highway pavements, foundations, canal linings, dams, bridge decks, and floor slabs. See also *cement* (Technical Brief 8).

Refractories are construction materials that can withstand high temperatures and maintain their physical properties. They are used extensively in structures associated with iron and steel production, copper and aluminum smelting, and glass and ceramic manufacturing. The primary types of clay refractories are fireclay and high alumina. Each type is used to produce bricks, as well as insulating refractories. See also *refractories* (Technical Brief 37).

## Recommended Reading

- R.A. Haber and P.A. Smith, Overview of Traditional Ceramics, *Engineered Materials Handbook*, Vol 4, ASM International, 1991, p 3-15
- G. Lewis, Ed., Applications for Traditional Ceramics, *Engineered Materials Handbook*, Vol 4, ASM International, 1991, p 893-959

ticularly during cooling. See also *transformation temperature*.

**transformation temperature.** The temperature at which a change in phase occurs. This term is sometimes used to denote the limiting temperature of a transformation range. The following symbols are used for irons and steels:

$Ac_{cm}$. In hypereutectoid steel, the temperature at which solution of cementite in austenite is completed during heating.

$Ac_1$. The temperature at which austenite begins to form during heating.

$Ac_3$. The temperature at which transformation of ferrite to austenite is completed during heating.

$Ac_4$. The temperature at which austenite transforms to delta ferrite during heating.

$Ae_{cm}$, $Ae_1$, $Ae_3$, $Ae_4$. The temperatures of phase changes at equilibrium.

$Ar_{cm}$. In hypereutectoid steel, the temperature at which precipitation of cementite starts during cooling.

$Ar_1$. The temperature at which transformation of austenite to ferrite or to ferrite plus cementite is completed during cooling.

$Ar_3$. The temperature at which austenite begins to transform to ferrite during cooling.

$Ar_4$. The temperature at which delta ferrite transforms to austenite during cooling.

FIG. 615

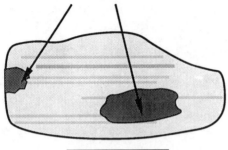

TOP VIEW OF THE
WORN SURFACE

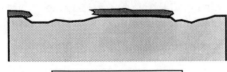

SIDE VIEW
OF THE WORN SURFACE

*Schematic of tribological transfer*

FIG. 616

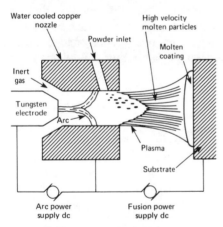

*Transferred arc plasma spraying*

**Ar′.** The temperature at which transformation of austenite to pearlite starts during cooling.

**M$_f$.** The temperature at which transformation of austenite to martensite is completed during cooling.

**M$_s$ (or Ar″).** The temperature at which transformation of austenite to martensite starts during cooling.

Note: All these changes, except formation of martensite, occur at lower temperatures during cooling than during heating, and depend on the rate of change of temperature.

**transformation toughened zirconia.** A generic term applied to stabilized zirconia systems in which the tetragonal symmetry is retained as the primary zirconia phase. The four most popular tetragonal phase stabilizers are ceria

FIG. 617

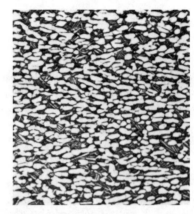

*Elongated "primary" $\alpha$ grains (light) in aged transformed $\beta$ matrix containing acicular $\alpha$ in a titanium alloy forging. 500×*

($CeO_2$), calcia (CaO), magnesia (MgO), and yttria ($Y_2O_3$). See also *zirconia* (Technical Brief 63).

**transformed beta.** A local or continuous structure in titanium alloys consisting of decomposition products arising by nucleation and growth processes during cooling from above the local or overall $\beta$ transus. Primary and regrowth $\alpha$ may be present. Transformed $\beta$ typically consists of $\alpha$ platelets that may or may not be separated by $\beta$ phase (Fig. 617).

**transgranular.** Through or across crystals or grains. Also called intracrystalline or transcrystalline.

**transgranular cracking.** Cracking or fracturing that occurs through or across a crystal or grain. Also called transcrystalline cracking. Contrast with *intergranular cracking*. See also the figure accompanying the term *step fracture* (metals) (1).

**transgranular fracture.** Fracture through or across the crystals or grains of a material. Also called transcrystalline fracture or intracrystalline fracture. Contrast with *intergranular fracture*.

**transient creep.** See *creep* and *primary creep*.

**transistor.** An active semiconductor device capable of providing power amplification and having three or more terminals.

**transitional fit.** A fit that may have either clearance or interference resulting from specified tolerances on hole and shaft.

**transition diagram.** In *tribology*, a plot of two or more experimental or operating variables that indicates the boundaries between various regimes of wear or surface damage. The *IRG transition diagram* is a plot of normal force (ordinate) versus sliding velocity (abscissa), and is used to identify three regions with differing lubrication effectiveness. Various plots have been called transition diagrams, and the context of usage must be established.

**transition lattice.** An unstable crystallographic configuration that forms as an intermediate step in a solid-state reaction such as precipitation from solid solution or eutectoid decomposition.

**transition metal.** A metal in which the available electron energy levels are occupied in such a way that the *d*-band contains less than its maximum number of ten electrons per atom, for example, iron, cobalt, nickel, and tungsten. The distinctive properties of the transition metals result from the incompletely filled *d*-levels.

**transition phase.** A nonequilibrium state that appears in a chemical system in the course of transformation between two equilibrium states.

**transition point.** At a stated pressure, the temperature (or at a stated temperature, the pressure) at which two solid phases exist in equilibrium—that is, an allotropic transformation temperature (or pressure).

**transition scarp.** A *rib mark* generated when a crack changes from one mode of growth to another, as when a wet crack accelerates abruptly from Region II (plateau) to Region III (dry) of a crack acceleration curve. See also *intersection scarp*.

**transition structure.** In precipitation from solid solution, a metastable precipitate that is coherent with the matrix.

**transition temperature.** The temperature at which the properties of a material change. Depending on the material, the transition change may or may not be reversible.

**transition temperature** (metals). (1) An arbitrarily defined temperature that lies within the temperature range in which metal fracture characteristics (as usually determined by tests of notched specimens) change rapidly, such as the ductile-to-brittle transition temperature (DBTT) (Fig. 618). The DBTT can be assessed in several ways, the most common being the temperature for 50% ductile and 50% brittle fracture (50% fracture appearance transition temperature, or FATT), or the lowest temperature at which the fracture is 100% ductile (100% fibrous criterion). The DBTT is commonly associated with *temper embrittlement* and *radiation damage* (neutron irradiation) of low-alloy steels. (2) Sometimes used to denote an arbitrarily defined temperature within a range in which the ductility changes rapidly with temperature.

**transmission electron microscope.** A microscope in which the image-forming rays pass through (are transmitted by) the specimen being observed (Fig. 619). Using the transmission electron microscope, microstructural features can be imaged at 1000 to 450,000×.

**transmission electron microscopy (TEM).** An analytical technique in which an image is formed on a cathode-ray tube whose raster is synchronized with the raster of an electron beam over an area of the sample surface. Image contrast is formed by the scattering of electrons out of the beam. TEM is used for very high magnification characterization of metals, ceramics, minerals, polymers, and biological materials.

**transmission grating.** In electron optics, a transparent diffraction grating through which light is transmitted. See also *concave grating*, *diffraction grating*, *plane grating*, and *reflection grating*.

**transmission method.** A method of x-ray or electron diffraction in which the recorded diffracted beams emerge on the

FIG. 618

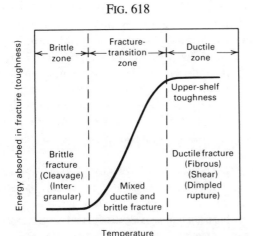

*Schematic of the effect of temperature on toughness of metals that exhibit a ductile-to-brittle fracture transition*

FIG. 619

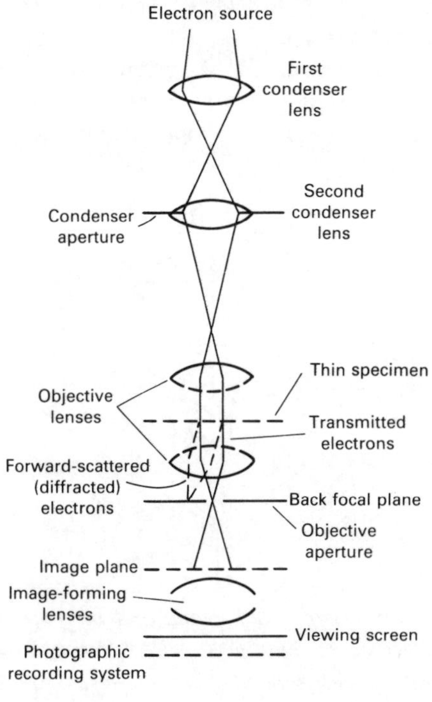

Electron source

First condenser lens

Second condenser lens

Condenser aperture

Objective lenses

Thin specimen

Transmitted electrons

Forward-scattered (diffracted) electrons

Back focal plane

Objective aperture

Image plane

Image-forming lenses

Viewing screen

Photographic recording system

*Electron optical column in a transmission electron microscope*

same side of the specimen as the transmitted primary beam.

**transmission oil.** (1) Oil used for transmission of hydraulic power. (2) Oil used to lubricate automobile transmission systems.

**transmittance.** The ratio of the light intensity transmitted by a material to the light intensity incident upon it. In emission spectrochemical analysis, the transmittance of a developed photographic emulsion, including its film or glass supporting base, is measured by a microphotometer. In absorption spectroscopy, the material is the sample. See also *diffuse transmittance*, *relative transmittance*, *specular transmittance*, and *total transmittance*.

**transpassive region.** The region of an *anodic polarization curve*, noble to and above the passive *potential* range, in which there is a significant increase in current density (increased metal dissolution) as the potential becomes more positive (noble).

**transpassive state.** (1) State of anodically passivated metal characterized by a considerable increase of the corrosion current, in the absence of pitting, when the *potential* is increased. (2) The noble region of potential where an electrode exhibits a higher than passive current density.

**transport.** See *transference*.

**transport number.** Same as *transference number*.

**trans stereoisomer.** A stereoisomer in which atoms or groups of atoms are arranged on opposite sides of a chain of atoms. See also *isotactic stereoisomerism*.

**transverse direction.** Literally, "across," usually signifying a direction or plane perpendicular to the direction of working (Fig. 620). In rolled plate or sheet, the direction across the width is often called long transverse; the direction through the thickness, short transverse.

**transversely isotropic.** (1) In reference to a material, exhibiting a special case of orthotropy in which properties are

Fig. 620

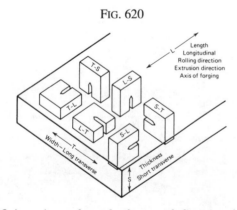

*Orientations of crack plane and direction for fracture specimens. First letter designates the direction perpendicular to the crack plane; second letter designates the direction parallel to the direction of crack growth.*

identical in two orthotropic dimensions but not the third. (2) Having identical properties in both transverse (short and long) but not in the longitudinal direction.

**transverse resistance seam welding.** The making of a resistance seam weld in a direction essentially at right angles to the throat depth of a resistance seam welding machine. See also the figure accompanying the term *resistance seam welding*. Contrast with *longitudinal resistance seam welding*.

**transverse rolling machine.** Equipment for producing complex preforms or finished forgings from round billets inserted transversely between two or three rolls that rotate in the same direction and drive the billet. The rolls, carrying replaceable die segments with appropriate impressions, make several revolutions for each rotation of the workpiece.

**transverse rupture strength.** The stress, as calculated from the flexure formula, required to break a sintered powder metallurgy specimen. The test for determining the transverse rupture strength involves applying the load at the center of a 31.8 by 12.7 by 6.4 mm (1.25 by 0.5 by 0.25 in.) beam which is supported near its ends.

**transverse strain.** The linear strain in a plane perpendicular to the loading axis of a specimen.

**trapped-sheet, contact heat, pressure thermoforming.** A *thermoforming* process for making plastic parts in which a hot, porous blow plate is used in both the heating and forming process (Fig. 621). The plastic sheet lies between the female mold cavity and the hot blow plate. Air forced through the plate and pressure from the female mold push the sheet onto the hot plate. When the sheet is sufficiently heated, air pressure forces it into the female mold.

**travel angle.** The angle that a welding electrode makes with a reference line perpendicular to the axis of the weld in the plane of the weld axis. This angle can be used to define the position of welding guns, welding torches, high-energy beams, welding rods, thermal cutting and thermal spraying torches, and thermal spraying guns. See also *drag angle* and *push angle* and the figures accompanying the terms *backhand welding* and *forehand welding*.

**travel angle** (pipe). The angle that a welding electrode makes with a reference line extending from the center of the pipe through the molten weld pool in the plane of the weld axis (Fig. 622).

**traverse speed.** The lineal velocity at which the torch is passed across the substrate during the thermal spraying operation.

**trees.** Visible projections of electrodeposited metal formed at sites of high current density.

**trepanning.** A machining process for producing a circular hole or groove in solid stock, or for producing a disk, cylinder, or tube from solid stock, by the action of a tool

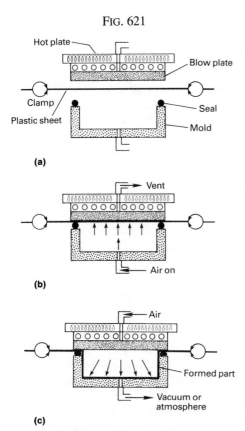

Fig. 621

**(a)**

**(b)**

**(c)**

*Trapped-sheet, contact heat, pressure thermo-forming. (a) Cold sheet trapped between heated blow plate and female mold. (b) Heated plates close on mold cavity. (c) Vacuum pulls heated sheet into mold cavity.*

Fig. 622

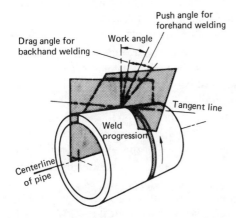

*Electrode orientation (travel angle) for a pipe weld*

containing one or more cutters (usually single-point) revolving around a center (Fig. 623).

**triaxiality**. In a *triaxial stress* state, the ratio of the smallest to the largest principal stress, all stresses being tensile.

Fig. 623

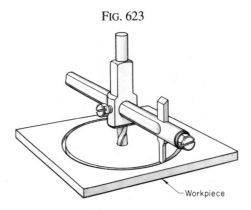

*Drill-mounted adjustable fly cutter used for trepanning various sizes of disks from flat stock, or grooves around centers*

**triaxial stress**. A state of stress in which none of the three principal stresses is zero. See also *principal stress (normal)*.

**tribo-**. A prefix indicating a relationship to interacting surfaces in relative motion.

**tribochemistry**. The part of chemistry dealing with interacting surfaces in relative motion. Tribochemistry broadly encompasses such areas as lubricant chemistry, changes in reactivity of surfaces due to mechanical contact, oxidative wear, and other phenomena.

**triboelement**. A solid body that is bounded by one or more *tribosurfaces* and that resides within a *tribosystem*. For example, in a pin-on-disk tribosystem, the pin is one triboelement and the disk is another. See also the figure accompanying the term *pin-on-disk machine*.

**tribology**. (1) The science and technology of interacting surfaces in relative motion and of the practices related thereto. (2) The science concerned with the design, friction, lubrication, and wear of contacting surfaces that move relative to each other (as in bearings, cams, or gears, for example).

**tribometer**. (1) An instrument or testing rig to measure normal and frictional forces of relatively moving surfaces. (2) Any device constructed for or capable of measuring the friction, lubrication, and wear behavior of materials or components.

**tribophysics**. That part of physics dealing with interacting surfaces in relative motion.

**triboscience**. The scientific discipline devoted to the systematic study of interacting surfaces in relative motion. Triboscience includes the scientific aspects of *tribochemistry*, *tribophysics*, contact mechanics, and materials and surface sciences as related to *tribology*.

**tribosurface**. Any solid surface whose intermittent, repeated, or continous contact with another surface or surfaces, in relative motion, results in friction, wear, and/or surface damage. The surface of a body subjected to a catastrophic collision would not generally be considered a tribosurface because significant damage to the entire body is involved.

**tribosystem.** Any functional combination of *triboelements*, including thermal and chemical surroundings.

**tribotechnology.** The aspect of *tribology* that involves the engineering application of *triboscience* and the design, development, analysis, and repair of components for tribological applications.

**triclinic.** Having three axes of any length, none of the included angles being equal to one another or equal to 90°.

**triggered capacitor discharge.** A high-voltage electrical discharge used in *emission spectroscopy* for vaporization and excitation of a sample material. The energy for the discharge is obtained from capacitors that are charged from an ac or dc electrical supply. Each discharge may be either oscillatory, critically damped, or overdamped. It is initiated by separate means and is extinguished when the voltage across the analytical gap falls to a value that no longer is sufficient to maintain it.

**trimetal bearing.** A bearing consisting of three layers. Trimetal bearings are often made of bronze with a white metal facing and a steel backing. See also the figure accompanying the term *sleeve bearing*.

**trimmer.** The dies used to remove the flash or excess stock from a forging.

**trimmer blade.** The portion of the trimmers through which a forging is pushed to shear off the flash.

**trimmer die.** The punch press die used for trimming flash from a forging.

**trimmer punch.** The upper portion of the trimmer that contacts the forging and pushes it through the trimmer blades; the lower end of the trimmer punch is generally shaped to fit the surface of the forging against which it pushes.

**trimmers.** The combination of *trimmer punch, trimmer blades*, and perhaps *trimming shoe* used to remove the flash from the forging.

**trimming.** (1) In forging, removing any parting-line flash or excess material from the part with a trimmer in a trim press; can be done hot or cold. (2) In drawing, shearing the irregular edge of the drawn part. (3) In casting, the removal of gates, risers, and fins.

**trimming press.** A power press suitable for trimming flash from forgings.

**trimming shoe.** The holder used to support *trimmers*. Sometimes called trimming chair.

**triple-action press.** A mechanical or hydraulic press having three slides with three motions properly synchronized for triple-action drawing, redrawing, and forming. Usually, two slides—the blankholder slide and the plunger—are located above and a lower slide is located within the bed of the press. See also *hydraulic press*, *mechanical press*, and *slide*.

**triple curve.** In a *P-T diagram*, a line representing the se-

quence of pressure and temperature values along which two conjugate phases occur in univariant equilibrium.

**triple point.** (1) A point on a phase diagram where three phases of a substance coexist in equilibrium. (2) The intersection of the boundaries of three adjoining grains, as observed in a metallographic section.

**tripoli.** Friable and dustlike silica used as an abrasive.

**TRIP steel.** A commercial steel product exhibiting *transformation-induced plasticity*.

**triton.** The nucleus of tritium ($^3$H), the triton is the only known radioactive nuclide belonging to hydrogen and β-decays to $^3$He with a half-life of 12.4 years.

**trommel.** A revolving cylindrical screen used in grading coarsely crushed ore.

**troostite** (obsolete). A previously unresolvable, rapidly etching, fine aggregate of carbide and ferrite produced either by tempering martensite at low temperature or by quenching a steel at a rate slower than the critical cooling rate. Preferred terminology for the first product is tempered martensite; for the latter, fine pearlite.

**Troy ounce.** A unit of weight for *precious metals* that is equal to 31.1034768 g (1.0971699 oz avoirdupois).

**true current density.** See preferred term *local current density*.

**true rake.** See preferred term *effective rake*.

**true strain.** (1) The ratio of the change in dimension, resulting from a given load increment, to the magnitude of the dimension immediately prior to applying the load increment. (2) In a body subjected to axial force, the natural logarithm of the ratio of the gage length at the moment of observation to the original gage length. Also known as natural strain.

**true stress.** The value obtained by dividing the load applied to a member at a given instant by the cross-sectional area over which it acts.

**truing.** The removal of the outside layer of abrasive grains on a grinding wheel for the purpose of restoring its face.

**trunnion bearing.** A bearing used as a pivot to swivel or turn an assembly.

**tryout.** In metal forming or forging, a preparatory run to check or test equipment, lubricant, stock, tools, or methods prior to a production run. Production tryout is run with tools previously approved; new die tryout is run with new tools not previously approved.

**tube furnace.** A furnace used for continuous or batch sintering powder metallurgy parts that utilizes a dense ceramic tube or a metallic retort to contain the controlled sintering atmosphere.

**tuberculation.** The formation of *localized corrosion* products scattered over the surface in the form of knoblike mounds called tubercles (Fig. 624). The formation of tubercles is usually associated with *biological corrosion*.

**tube reducing.** Reducing both the diameter and wall thick-

FIG. 624

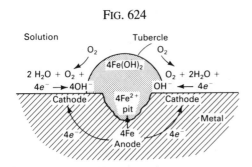

*Schematic of pit initiation and tubercle formation due to an oxygen concentration cell under a biological deposit*

ness of tubing with a mandrel and a pair of rolls. See also *spinning*.

**tube sinking**. Drawing tubing through a die or passing it through rolls without the use of an interior tool (such as a mandrel or plug) to control inside diameter; sinking generally produces a tube of increased wall thickness and length (Fig. 625).

**tube stock**. A semifinished tube suitable for subsequent reduction and finishing.

**tubular products, steel**. The general term used to cover all hollow carbon- and low-alloy steel products used as conveyors of fluids and as structural members. Although these products are usually produced in cylindrical form, they are often subsequently altered by various processing methods to produce square, oval, rectangular, and other symmetrical shapes. Table 70 provides a classification of steel tubular products.

**tumbling** (metals). Rotating workpieces, usually castings or forgings, in a barrel partly filled with metal slugs or abrasives, to remove sand, scale, or fins. It may be done dry, or with an aqueous solution added to the contents of the barrel. See also *barrel finishing*.

**tumbling** (plastics). Finishing operation for small plastic articles by which gates, flash, and fins are removed and/or surfaces are polished by rotating them in a barrel with wooden pegs, sawdust, and polishing compounds.

FIG. 625

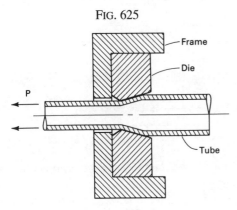

*Tube drawing without a mandrel (tube sinking)*

**tumble grinding**. Various surfacing operations ranging from deburring and polishing to honing and microfinishing metallic parts before and after plating.

**tungsten carbide**. See *cemented carbides* (Technical Brief 9).

**tungsten electrode**. A non-filler metal electrode used in arc welding or cutting, made principally of tungsten. Table 71 lists compositions of tungsten electrodes used for gas tungsten arc welding. See also the figure accompanying the term *welding torch* (arc).

**tungsten filaments**. Thin tungsten wire used principally in incandescent lamps (Fig. 626) and other filament applications that require resistance to creep at high temperatures.

FIG. 626

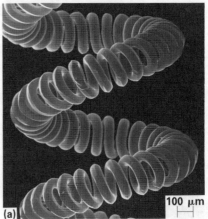

*Tungsten filament in a light bulb*

### Table 70 Classification of steel tubular products

| Product | Typical use | Production processes | Outside diameter(a), in. | Typical grades | Usual finished status |
|---|---|---|---|---|---|
| Oil country goods | | | | | |
| Casing ........ | To line oil and gas wells to prevent collapse of the hole | Seamless, electric resistance welding | 4.5-20 | H-40, J-55, K-55 C-75, L-80, N-80, C-90, G-95, P-110, Q-125 | As-rolled Normalize or quench and temper |
| Tubing ........ | To convey oil or gas from the producing strata to the earth's surface | Seamless, continuous welding, electric resistance welding | 1.050-4.5 | All others H-40, J-55, N-80, P-105 | Quench and temper As-rolled Normalize or quench and temper |
| Drill pipe ...... | Rotary stem for drill bits | Seamless | 2.375-6.625 | E | Normalize and temper, or quench and temper |
| | | | | X-95, G-105, S-135 | Quench and temper |
| Line pipe ........ | Conveys oil, gas, or water | Seamless, electric resistance welding, continuous welding, double submerged arc welding | 0.125(nom)-80 | All grades B, X42, X46, X52, X60, X65, X70 | As-rolled As-rolled Control rolled |
| Standard pipe .... | Plumbing, electrical conduit, low pressure conveyance of fluids, and nonstringent structural applications | Seamless, electric resistance welding, continuous welding, double submerged arc welding | 0.125(nom)-80 | All grades | As-rolled |
| Mechanical tubing ........ | Variety of round, hollow mechanical parts, such as automotive axles, bearing races, and hydraulic pistons | Seamless, electric resistance welding | 0.375-10.75 | Carbon and alloy | Hot rolled or cold drawn |
| Pressure tubing ........ | Boiler tubes, condenser tubes, heat exchanger tubes, and refrigeration tubes | Seamless, electric resistance welding | 0.5-10.75 | Carbon and alloy | Hot rolled or cold drawn |

Note: Because steel tubular products manufactured in the United States are customarily produced to standard inch and fractional inch sizes, tubular product sizes are given only in inches in this article. 1 in. = 25.4 mm or 2.54 cm.
(a) nom: nominal

### Table 71 American Welding Society classifications and composition limits for GTAW electrodes

| AWS classi-fication | Tungsten (min)(a), % | Thoria, % | Zirconia, % | Other (max)(b), % |
|---|---|---|---|---|
| EWP ..................... | 99.5 | ... | ... | 0.5 |
| EWTh-1 ..................... | 98.5 | 0.8-1.2 | ... | 0.5 |
| EWTh-2 ..................... | 97.5 | 1.7-2.2 | ... | 0.5 |
| EWTh-3(c) ................... | 98.95 | 0.35-0.55 | ... | 0.5 |
| EWZr ..................... | 99.2 | ... | 0.15-0.40 | 0.5 |

(a) By difference. (b) Total. (c) EWTh-3 is a tungsten electrode with an integral lateral segment throughout its length that contains 1.0 to 2.0% thoria; average thoria content of the electrode is as shown in this table.

**tungsten inert-gas welding.** See preferred term *gas tungsten arc welding.*

**tup impact test.** A falling-weight (tup) impact test developed specifically for plastic pipe and fittings.

**turbine oil.** An oil used to lubricate bearings in a steam or gas turbine.

**Turk's-head rolls.** Four undriven working rolls, arranged in a square or rectangular pattern, through which metal strip,

FIG. 627

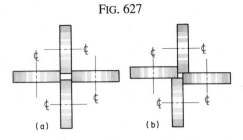

*Turk's-head rolls. (a) Positioned in line to form a rectangular cross section. (b) Offset to form a square section*

FIG. 628

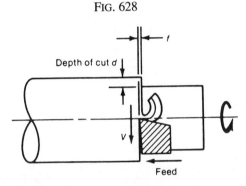

*Schematic of turning operation*

FIG. 629

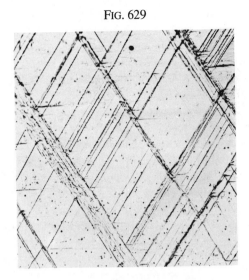

*Groups of extremely fine twin bands in polished and plastically deformed titanium. 200×*

wire, or tubing is drawn to form square or rectangular sections (Fig. 627).

**turning.** Removing material by forcing a single-point cutting tool against the surface of a rotating workpiece (Fig. 628). The tool may or may not be moved toward or along the axis of rotation while it cuts away material.

**turns per inch (tpi).** In composites, a measure of the amount of twist produced in a *yarn*, *tow*, or *roving* during its processing history. See also *twist*.

**tuyere.** An opening in a cupola, blast furnace, or converter for the introduction of air or inert gas. See also the figures accompanying the terms *blast furnace* and *cupola*.

**twill weave.** A basic fabric weave for reinforced plastics characterized by a diagonal rib, or twill line. Each end floats over at least two consecutive picks, allowing a greater number of yarns per unit area than in a plain weave, while not losing a great deal of fabric stability. See also *plain weave*.

**twin.** Two portions of a crystal with a definite orientation relationship; one may be regarded as the parent, the other as the twin. The orientation of the twin is a mirror image of the orientation of the parent across a twinning plane or an orientation that can be derived by rotating the twin portion about a twinning axis. See also *annealing twin* and *mechanical twin*.

**twin bands.** Bands across a crystal grain, observed on a polished and etched section, where crystallographic orien-

tations have a mirror-image relationship to the orientation of the matrix grain across a composition plane that is usually parallel to the sides of the band (Fig. 629).

**twin carbon arc brazing.** A brazing process which produces coalescence of metals by heating them with an electric arc between two carbon electrodes. The filler metal is distributed in the joint by capillary attraction.

**twin carbon arc welding.** A *carbon arc welding* process variation which produces coalescence of metals by heating them with an electric arc between two carbon electrodes. No shielding is used. Pressure and filler metal may or may not be used.

**twin-sheet thermoforming.** A technique for *thermoforming* hollow plastic objects by introducing high-pressure air between two sheets and blowing the sheets into the mold halves (vacuum is also applied).

**twist.** (1) In a yarn or other textile strand, the spiral turns about its axis per unit of length. Twist may be expressed as *turns per inch (tpi)*. Twist provides additional integrity to yarn before it is subjected to the weaving process, a typical twist consisting of up to one turn per inch. In many instances heavier yarns are needed for the weaving operation. This is normally accomplished by twisting together two or more single strands, followed by a plying operation. Plying essentially involves retwisting the twisted strands in the opposite direction from the original twist. The two types of twist normally used are known as S and Z, which indicate the direction in which the twisting is done. Usually, two or more strands twisted together with an S twist are plied with a Z twist in order to give a balanced yarn. Thus, the yarn properties, such as strength, bundle diameter, and yield, can be manipulated by the twisting and plying operations. (2) In pultruded parts, twist describes a condition of longitudinal, progressive rotation that can be easily detected for a noncircular cross section by placing the pultrusion on a plane surface, holding one end flat with the surface, and observing whether one edge or side of the other end does not lie parallel with that surface.

**twist boundary**. A subgrain boundary consisting of an array of screw *dislocations*.

**twist hackle** (ceramics and glasses). A *hackle* that separates portions of the crack surface, each of which has rotated from the original crack plane in response to a twist in the axis of principal tension. In a single crystal, a twist hackle separates portions of the crack surface, each of which follows the same cleavage plane, the normal to the cleavage plane being inclined to the principal tension. In a bicrystal or polycrystalline material, a hackle is initiated at a twist grain boundary.

**two-component adhesive**. An adhesive supplied in two parts that are mixed before application. Such adhesives usually cure at room temperature.

**two-high mill**. A type of rolling mill in which only two rolls, the working rolls, are contained in a single housing. Compare with *four-high mill* and *cluster mill*.

**T-X diagram**. A two-dimensional graph of the isobaric phase relationships in a binary system; the coordinates of the graph are temperature and concentration.

**type metal**. Any of a series of alloys containing lead (58.5 to 95%), antimony (2.5 to 25%), and tin (2.5 to 20%) used to make printing type. Small amounts of copper (1.5 to 2.0%) are added to increase hardness in some applications.

**typical-basis**. An average property value. No statistical assurance is associated with this basis.

# U

**U-bend die**. A die, commonly used in press-brake forming, that is machined horizontally with a square or rectangular cross-sectional opening that provides two edges over which metal is drawn into a channel shape (Fig. 630).

**Ugine-Sejournet process**. A direct extrusion process for metals that uses molten glass to insulate the hot billet and to act as a lubricant.

**UHMWPE**. See *ultrahigh molecular weight polyethylene*.

**ultimate elongation**. The elongation at rupture.

**ultimate strength**. The maximum stress (tensile, compressive, or shear) a material can sustain without fracture; determined by dividing maximum load by the original cross-sectional area of the specimen. Also known as nominal strength or maximum strength.

**ultimate tensile strength**. The ultimate or final (highest) stress sustained by a specimen in a tension test.

**ultrahard tool materials**. Very hard, wear-resistant materials—specifically, polycrystalline diamond and polycrystalline cubic boron nitride—that are fabricated into solid or layered cutting tool blanks for machining applications (Fig. 631). See also *superabrasives* (Technical Brief 51).

**ultrahigh molecular weight polyethylene (UHMWPE)**. Those polyethylene resins having weight-average molecular weights ranging from $3 \times 10^6$ to $6 \times 10^6$. These materials have both the highest abrasion resistance and

FIG. 630

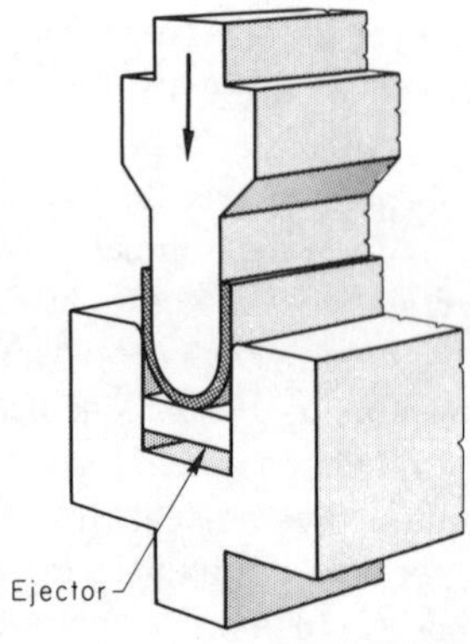

*U-bend press-brake forming*

FIG. 631

Tungsten carbide backed

Sandwich

Solid

Solid brazeable

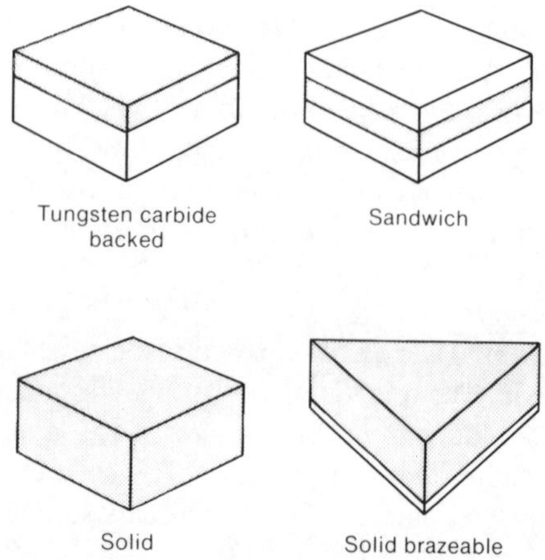

*Formats for polycrystalline diamond and polycrystalline cubic boron nitride tools*

**Table 72  Compositions of ultrahigh-strength steels**

| Designation or trade name | Composition, wt%(a) | | | | | | | |
|---|---|---|---|---|---|---|---|---|
| | C | Mn | Si | Cr | Ni | Mo | V | Co |
| **Medium-carbon low-alloy steels** | | | | | | | | |
| 4130 | 0.28–0.33 | 0.40–0.60 | 0.20–0.35 | 0.80–1.10 | . . . | 0.15–0.25 | . . . | . . . |
| 4140 | 0.38–0.43 | 0.75–1.00 | 0.20–0.35 | 0.80–1.10 | . . . | 0.15–0.25 | . . . | . . . |
| 4340 | 0.38–0.43 | 0.60–0.80 | 0.20–0.35 | 0.70–0.90 | 1.65–2.00 | 0.20–0.30 | . . . | . . . |
| AMS 6434 | 0.31–0.38 | 0.60–0.80 | 0.20–0.35 | 0.65–0.90 | 1.65–2.00 | 0.30–0.40 | 0.17–0.23 | . . . |
| 300M | 0.40–0.46 | 0.65–0.90 | 1.45–1.80 | 0.70–0.95 | 1.65–2.00 | 0.30–0.45 | 0.05 min | . . . |
| D-6a | 0.42–0.48 | 0.60–0.90 | 0.15–0.30 | 0.90–1.20 | 0.40–0.70 | 0.90–1.10 | 0.05–0.10 | . . . |
| 6150 | 0.48–0.53 | 0.70–0.90 | 0.20–0.35 | 0.80–1.10 | . . . | . . . | 0.15–0.25 | . . . |
| 8640 | 0.38–0.43 | 0.75–1.00 | 0.20–0.35 | 0.40–0.60 | 0.40–0.70 | 0.15–0.25 | . . . | . . . |
| **Medium-alloy air-hardening steels** | | | | | | | | |
| H11 mod | 0.37–0.43 | 0.20–0.40 | 0.80–1.00 | 4.75–5.25 | . . . | 1.20–1.40 | 0.40–0.60 | . . . |
| H13 | 0.32–0.45 | 0.20–0.50 | 0.80–1.20 | 4.75–5.50 | . . . | 1.10–1.75 | 0.80–1.20 | . . . |
| **High fracture toughness steels** | | | | | | | | |
| AF1410(b) | 0.13–0.17 | 0.10 max | 0.10 max | 1.80–2.20 | 9.50–10.50 | 0.90–1.10 | . . . | 13.50–14.50 |
| HP 9-4-30(c) | 0.29–0.34 | 0.10–0.35 | 0.20 max | 0.90–1.10 | 7.0–8.0 | 0.90–1.10 | 0.06–0.12 | 4.25–4.75 |

(a) P and S contents may vary with steelmaking practice. Usually, these steels contain no more than 0.035 P and 0.040 S. (b) AF1410 is specified to have 0.008P and 0.005S composition. Ranges utilized by some producers are narrower. (c) HP 9-4-30 is specified to have 0.10 max P and 0.10 max S. Ranges utilized by some producers are narrower.

highest impact strength of any plastic. See also *high-density polyethylenes* (Technical Brief 19).

**ultrahigh-strength steels.** Structural steels with minimum yield strengths of 1380 MPa (200 ksi). Such steels include medium-carbon low-alloy steels, medium-alloy air-hardening steels, and high fracture toughness ($K_{Ic}$ of 100 MPa$\sqrt{m}$, or 91 ksi$\sqrt{in}$.) steels. Table 72 lists compositions of commercial ultrahigh-strength steels.

**ultramicroscopic.** See *submicroscopic.*

**ultrasonic beam.** A beam of acoustical radiation with a frequency higher than the frequency range for audible sound —that is, above about 20 kHz.

**ultrasonic bonding.** A method of joining plastics using vibratory mechanical pressure at ultrasonic frequencies. Electrical energy is changed to ultrasonic vibrations by means of either a magnetostrictive or piezoelectric transducer. The ultrasonic vibrations generate frictional heat, melting the plastics and allowing them to join.

**ultrasonic cleaning.** Immersion cleaning aided by ultrasonic waves that cause microagitation.

**ultrasonic coupler.** In ultrasonic welding and soldering, the elements through which ultrasonic vibration is transmitted from the transducer to the tip (Fig. 632).

**ultrasonic C-scan inspection.** A method for displaying the relative attenuation of ultrasonic waves across the surface (plan view) of a structural component (Fig. 633). An ultrasonic transducer is used to scan the surface of a material mechanically in an *x-y* raster scan mode while generating and receiving waves. Either the material is immersed in a water bath or columns of water are provided between the transducer and the material as a medium for ultrasonic energy transmissions. The received wave signals are electronically conditioned and measured to determine relative energy losses of the wave as it progresses through the material at each particular location on the specimen. Ultrasonic C-scan has been used extensively to determine both the initial integrity of a manufactured part and the void content, and to follow the initiation and progression of damage resulting from environmental loading.

**ultrasonic frequency.** A frequency, associated with elastic waves, that is greater than the highest audible frequency, generally regarded as being higher than 20 kHz.

FIG. 632

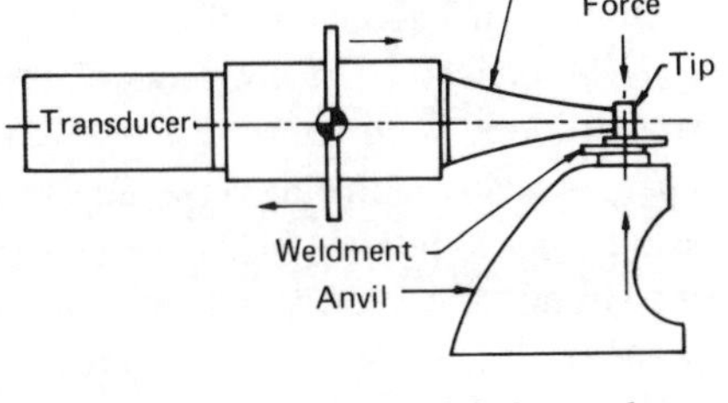

*Coupler placement in a lateral-drive ultrasonic spot and seam welding setup*

FIG. 633

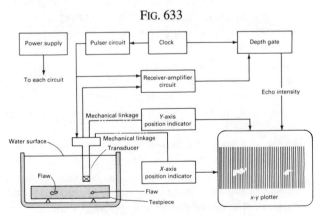

*Typical ultrasonic C-scan setup, including display, for basic pulse-echo ultrasonic immersion inspection*

FIG. 634

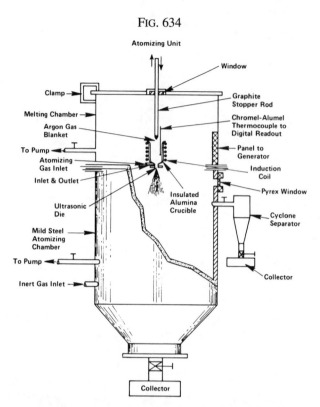

*Schematic of an ultrasonic gas atomization chamber*

FIG. 635

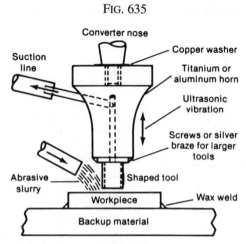

*Principal elements of ultrasonic impact grinding*

**ultrasonic gas atomization.** A variation of the gas atomization process which uses high-frequency gas pulses with velocities up to 4600 m/s (15,100 ft/s) to break up the molten metal stream (Fig. 634). This process can produce high yields of powder with particle diameters less than 20 μm.

**ultrasonic impact grinding.** Material removal by means of an abrasive slurry and the ultrasonic vibration of a non-rotating tool. The abrasive slurry flows through a gap between the workpiece and the vibrating tool (Fig. 635). Material removal occurs when the abrasive particles, suspended in the slurry, are struck on the downstroke of the vibrating tool. The velocity imparted to the abrasive particles causes microchipping and erosion as the particles impinge on the workpiece. See also *ultrasonic machining*.

**ultrasonic inspection.** A nondestructive method in which beams of high-frequency sound waves are introduced into materials for the detection of surface and subsurface flaws in the material. The sound waves travel through the material with some attendant loss of energy (attenuation) and are reflected at interfaces. The reflected beam is displayed and then analyzed to define the presence and location of flaws or discontinuities. Most ultrasonic inspection is done at frequencies between 0.1 and 25 MHz—well above the range of human hearing, which is about 20 Hz to 20 kHz.

**ultrasonic machining.** A process for machining of hard, brittle, nonmetallic materials that involves the ultrasonic vibration of a rotating diamond core drill or milling tool. Rotary ultrasonic machining is similar to the conventional drilling of glass and ceramic with diamond core drills, except that the rotating core drill is vibrated at an ultrasonic frequency of 20 kHz. Rotary ultrasonic machining does not involve the flow of an abrasive slurry through a gap between the workpiece and the tool. Instead, the tool contacts and cuts the workpiece, and a liquid coolant, usually water, is forced through the bore of the tube to cool and flush away the removed material. See also *ultrasonic impact grinding*.

**ultrasonic soldering.** A soldering process variation in which high-frequency vibratory energy is transmitted through molten solder to remove undesirable surface films and thereby promote wetting of the base metal. This operation is usually accomplished without a flux.

**ultrasonic testing.** See *ultrasonic inspection*.

**ultrasonic welding.** A solid-state process in which materials are welded by locally applying high-frequency vibratory energy to a joint held together under pressure (Fig. 636). Ultrasonic energy is produced through a transducer, which converts high-frequency electrical vibrations to mechanical vibrations at the same frequency, usually above 15 kHz (above the audible range). Mechanical vibrations are transmitted through a coupling system to the welding tip and into the workpieces. The tip vibrates laterally, essentially parallel to the weld interface, while static force is applied perpendicular to the interface. See also the figure accompanying the term *ultrasonic coupler*.

**ultraspeed welding.** See preferred term *commutator-controlled welding*.

**ultraviolet (UV).** Pertaining to the region of the electromagnetic spectrum from approximately 10 to 380 nm. The

FIG. 636

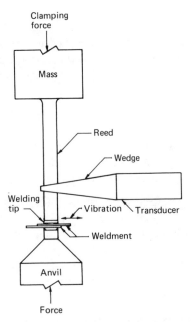

*High-power "wedge-reed" ultrasonic spot welding system*

term ultraviolet without further qualification usually refers to the region from 200 to 380 nm.

**ultraviolet degradation.** The degradation caused by long-term exposure of a material to sunlight or other ultraviolet rays containing radiation.

**ultraviolet/electron beam cured adhesives.** See *Technical Brief 58.*

**ultraviolet radiation.** Electromagnetic radiation in the wavelength range of 10 to 380 nm. See also *electromagnetic radiation.*

**ultraviolet stabilizer.** Any chemical compound that, when admixed with a resin, selectively absorbs ultraviolet rays.

**ultraviolet/visible (UV/VIS) absorption spectroscopy.** An analytical technique that measures the wavelength-dependent attenuation of ultraviolet, visible, and near-infrared light by an atomic or molecular species; used in the detection, identification, and quantification of numerous atomic and molecular species.

**unary system.** Composed of one component.

**unbond.** An area within an adhesively bonded interface between two adherends in which the intended bonding action failed to take place, or an area in which two layers of prepreg in a cured component do not adhere. Also used to denote specific areas deliberately prevented from bonding in order to simulate a defective bond, such as in the generation of quality standards specimens.

**uncertainty.** (1) An indication of the variability associated with a measured value that takes into account two major components of error: (a) *bias*, and (b) the random error attributed to the imprecision of the measurement process.

(2) The range of values within which the true value is estimated to lie. It is a best estimate of possible inaccuracy due to both random and systematic error.

**unctuous.** A general term expressing the slippery feel of a material, such as a lubricant, when rubbed with the fingers.

**underbead crack.** A crack in the heat-affected zone of a weld generally not extending to the surface of the base metal. See also the figure accompanying the term *weld crack.*

**undercoat.** A deposited coat of material which acts as a substrate for a subsequent thermal spray deposit. See also *bond coat.*

**undercooling.** Same as *supercooling.*

**undercure.** An undesirable condition of a molded plastic article resulting from the allowance of too little time and/or temperature or pressure for adequate hardening of the molding.

**undercut.** (1) In weldments, a groove melted into the base metal adjacent to the toe or root of a weld and left unfilled by weld metal (Fig. 637). (2) For castings or forgings, same as *back draft.*

**underdraft.** A condition wherein a metal curves downward on leaving a set of rolls because of higher speed in the upper roll.

**underfill.** (1) In weldments, a depression on the face of the weld or root surface extending below the surface of the adjacent base metal (Fig. 637). (2) A portion of a forging that has insufficient metal to give it the true shape of the impression.

**underfilm corrosion.** Corrosion that occurs under organic films in the form of randomly distributed threadlike filaments or spots. In many cases this is identical to *filiform corrosion.*

FIG. 637

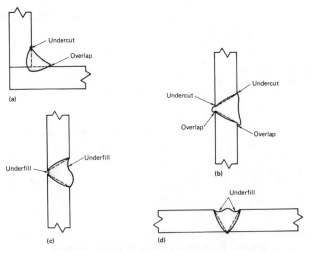

*Weld discontinuities affecting weld shape and contour. (a) Undercut and overlapping in a fillet weld. (b) Undercut and overlapping in a groove weld. (c) and (d) Underfill in groove welds*

## Technical Brief 58: Ultraviolet/Electron Beam Adhesives

RADIATION CURING involves the rapid conversion of specially formulated, 100% reactive liquids to solids. Potential energy sources include microwaves, visible infrared (IR), ultraviolet (UV), and electron beam (EB) sources, the latter two being the most commercially important. Radiation-cured materials are used as coatings, inks, adhesives, sealants, and potting compounds.

The UV curing process typically involves the exposure of a reactive liquid that contains a photoinitiator to UV radiation at a wavelength between 200 and 400 nm. The liquid is rapidly converted to a solid, usually in less than 60 s. In the EB curing process, electrons are artificially generated and accelerated to energies of less than 100 keV to greater than 1 billion keV. Generally, 50 to 350 keV electrons are used to cure adhesives that have bond lines with thicknesses of 25 to 38 μm. The reactive liquid in the EB process does not contain a photoinitiator.

Because the main advantage of UV/EB-curable adhesives is rapid curing at room temperature, they can be used to bond heat-sensitive substrates, such as polyvinyl chloride. In addition, the rapid cure often eliminates the need to fixture parts and greatly increases production rates. UV/EB-cured adhesives have been used to replace solvent-base adhesives because of the increasing cost of properly recovering and disposing of solvents. The cross-linked nature of UV/EB-cured adhesives results in good chemical, heat, and abrasion resistance, toughness, dimensional stability, and adhesion to many substrates. Unlike thermal curing, EB curing can be selective, and the depth of penetration can be controlled.

Most UV/EB adhesives are based on an addition polymerization curing mechanism. Materials consist of acrylic acid esters of various forms or combinations of acrylates with aliphatic or aromatic epoxies, urethanes, polyesters, or polyethers. Although the epoxy-base systems have higher tensile strengths, their elongations are less than those of the urethane-base systems. In addition, the urethane-base systems have better abrasion resistance.

Typical UV-curable adhesive applications include the electronics, automotive, medical, optics, and packaging markets, as well as tapes and labels. EB-curable adhesives are used in magnetic tapes and floppy disks, where magnetic particles are bonded to films, as well as in packaging, tapes, and labels.

### Comparison of selected UV-curable adhesives

| Property | Epoxy | Acrylic | Polyester | Methacrylates | | Acrylics | | Modified acrylates |
|---|---|---|---|---|---|---|---|---|
| Cure . . . . . . . . . . . . . . . . . . . . . UV | UV | UV | UV only | UV, anaerobic | UV, anaerobic | UV, anaerobic | UV, heat |
| Brookfield viscosity, Pa · s . . 2.5 | 0.55 | 6.5 | 0.105–6.0 | 0.500 | 2.25–20.0 | 0.300–2.0 | 7.0 |
| Service temperature range, | | | | | | | | |
| °C (°F) . . . . . . . . . . . . . . . −54 to 175 | −54 to 135 | −34 to 121 | −54 to 121 | −54 to 177 | −54 to 135 | −54 to 177 | −54 to 135 |
| (−65 to 347) | (−65 to 275) | (−30 to 250) | (−65 to 250) | (−65 to 350) | (−65 to 275) | (−65 to 350) | (−65 to 275) |
| Hardness· . . . . . . . . . . . . . . . Shore D65 | Barcol 70 | · · · | Shore D63–D82 | · · · | · · · | · · · | Barcol 75 |
| Bonded substrates(a) . . . . . . Glass | Glass, metals | Plastics, metals | Glass, metals, plastics | Glass, metals | Glass, metals | Metals | Glass, metals |

(a) For a UV cure, one substrate must be transparent.

### Recommended Reading

- M.M. Gauthier, Types of Adhesives, *Engineered Materials Handbook*, Vol 3, ASM International, 1990, p 74-93
- *Radiation Curing—An Introduction to Coatings, Varnishes, Adhesives and Inks*, Education Committee of the Radiation Curing Division, The Association for Finishing Processes, Society of Manufacturing Engineers, 1984, p 3-20

**undersize powder.** Powder particles smaller than the minimum permitted by a particle size specification.

**understressing.** Applying a cyclic stress lower than the *endurance limit*. This may improve fatigue life if the member is later cyclically stressed at levels above the endurance limit.

**uniaxial compacting.** Compacting of powder along one axis, either in one direction or in two opposing directions. Contrast with *isostatic pressing*.

**uniaxial load.** A condition in which a material or component is stressed in only one direction along its axis or centerline.

**uniaxial strain.** See *axial strain*.

**uniaxial stress.** A state of stress in which two of the three principal stresses are zero. See also *principal stress (normal)*.

**unidirectional compacting.** Compacting of powder in one direction.

**unidirectional laminate.** A reinforced plastic laminate in which substantially all of the fibers are oriented in the same direction. See also the figures accompanying the terms *laminate* and *quasi-isotropic laminate*.

**uniform corrosion.** (1) A type of corrosion attack (deterioration) uniformly distributed over a metal surface. (2) Corrosion that proceeds at approximately the same rate over a metal surface. Also called general corrosion.

**uniform elongation.** The elongation at maximum load and immediately preceding the onset of necking in a tensile test.

**uniformly distributed impact test.** See *distributed impact test*.

**uniform strain.** The strain occurring prior to the beginning of localization of strain (necking); the strain to maximum load in the tension test.

**unimeric.** Pertaining to a single molecule that is not monomeric, oligomeric, or polymeric, such as saturated hydrocarbons.

**unipolarity operation.** A resistance welding process variation in which succeeding welds are made with pulses of the same polarity.

**unit cell.** A parallelepiped element of crystal structure, containing a certain number of atoms, the repetition of which through space will build up the complete crystal (Fig. 638). See also the table accompanying the term *lattice*.

**unit power.** The net amount of power required during machining or grinding to remove a unit volume of material in unit time.

**univariant equilibrium.** A stable state among several phases equal to one more than the number of components, that is, having one degree of freedom.

**universal forging mill.** A combination of four hydraulic presses arranged in one plane equipped with billet manipulators and automatic controls, used for radial or draw forging. See also the figure accompanying the term *radial forging*.

**universal gas constant.** See *gas constant*.

**universal mill.** A rolling mill in which rolls with a vertical axis roll the edges of the metal stock between some of the passes through the horizontal rolls.

**unlubricated sliding.** Sliding without lubricant but not necessarily under completely dry conditions. Unlubricated sliding is often used to mean "not intentionally lubricated," but surface species such as naturally formed surface oxides and other interfacial contaminants may act in a *lubricious* manner in nominally unlubricated sliding.

FIG. 638

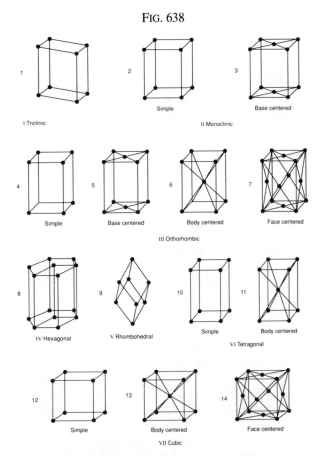

*Bravais lattices. Atoms in each crystalline structure (unit cell) are in the positions denoted by closed circles.*

**unsaturated compounds.** Any chemical compound having more than one bond between two adjacent atoms, usually carbon atoms, and being capable of adding other atoms at that point to reduce it to a single bond; for example, *olefins*.

**unsaturated polyesters.** See *Technical Brief 59*.

**unsymmetric laminate.** A *laminate* having an arbitrary stacking sequence without midplane symmetry.

**upper punch.** The member of a die assembly or tool set for forming powder metallurgy parts that closes the die and forms the top of the part being produced.

**upper ram.** The part of a pneumatic or hydraulic press for forming powder metallurgy parts that is moving in an upper cylinder and transmits pressure to the upper punch or set of upper punches.

**upset.** (1) The localized increase in cross-sectional area of a workpiece or weldment resulting from the application of pressure during mechanical fabrication or welding. (2) That portion of a welding cycle during which the cross-sectional area is increased by the application of pressure. (3) Bulk deformation resulting from the application of pressure in welding. The upset may be measured as a percent increase in interfacial area, a reduction

---

## Technical Brief 59: Unsaturated Polyesters

UNSATURATED POLYESTERS, commonly referred to as polyester resins, are prepared by the condensation polymerization of dibasic acids or anhydrides with dihydric alcohols, with the dibasic acid or anhydride being partially or completely composed of a 1,2-ethylenically unsaturated material, such as maleic anhydride or fumaric acid. Unlike thermoplastic polyesters (which do not contain any 1,2-ethylenically unsaturated material), unsaturated polyesters are processed only to a molecular weight in the range of 2000 to 20,000. The resultant polymer is then dissolved in a liquid reactive vinyl (1,2-ethylenically unsaturated) monomer, such as styrene, vinyl toluene, diallyl phthalate, or methyl methacrylate, to give a solution with a viscosity in the range of 0.200 to 2.00 Pa · s.

### Unsaturated polyester markets

| Market | Application |
|---|---|
| Construction | Pipe, building panels, portable buildings, swimming pools, floor grating, doors |
| Marine | Powerboats, sailboats, canoes, kayaks, gel coats, docks |
| Transportation | Automobile body panels, truck hoods, trailer panels, structural components, seating |
| Industrial | Corrosion control, tanks, process vessels, pipes, fittings, valves, fans, pollution control equipment, scrubbers, hoods, blowers, ducts, stacks, ladders, linings, chutes, breechings, sewer lines, waste water treatment equipment |
| Electrical | Appliance covers and housings, EDP equipment, circuit boards, insulators, switchgear, radomes |
| Sanitary ware | Bathtubs, shower stalls, hot tubs, spas, cultured marble, food-handling containers |
| Miscellaneous | Hobby castings, decorative art, buttons, bowling balls, skis, fishing rods, nonstructural furniture parts |

A wide variety of dibasic acids or anhydrides and dihydric alcohols can be combined with either maleic anhydride or fumaric acid to form unsaturated polyesters. For simplicity, however, unsaturated polyesters are commonly grouped into five classifications. Orthophthalic polyester resins are prepared from combinations of phthalic anhydride and either maleic anhydride or fumaric acid. Commonly referred to as general-purpose resins, orthophthalic resins are widely used. They are limited, however, with regard to thermal stability, chemical resistance, and processibility. A combination using isophthalic acid or terephthalic acid results in an isophthalic or terephthalic polyester resin. Although higher in cost than the equivalent orthophthalic resins, these materials are higher-quality resins with better thermal resistance, mechanical properties, and chemical resistance. Bisphenol A (BPA) fumarates are prepared by the reaction of propoxylated or ethoxylated BPA with fumaric acid. The aromatic BPA group imparts a higher degree of hardness and rigidity and improved thermal performance. Chlorendics are prepared using a combination of chlorendic anhydride or hexochlorocyclopentadiene acid with maleic anhydride or fumaric acid. These polymers show

---

in length, or a percent reduction in thickness (for lap joints).

**upset forging.** A forging obtained by *upset* of a suitable length of bar, billet, or bloom.

**upset pressing.** The pressing of a powder compact in several stages, which results in an increase in the cross-sectional area of the part prior to its ejection.

**upsetter.** A horizontal mechanical press used to make parts from bar stock or tubing by *upset forging*, piercing, bending, or otherwise forming in dies. Also known as a header.

**upsetting.** The working of metal so that the cross-sectional area of a portion or all of the stock is increased. See also *heading*.

**upsetting force.** In *upset welding*, the force exerted at the faying surfaces during upsetting. See also *upset* (3).

**upsetting time.** In *upset welding*, the time during upsetting. See also *upset* (3).

**upset weld.** A weld made by *upset welding*.

**upset welding.** A resistance welding process in which the weld is produced, simultaneously over the entire area of abutting surfaces or progressively along a joint, by applying mechanical force (pressure) to the joint, then causing electrical current to flow across the joint to heat the abutting surfaces. Pressure is maintained throughout the heating period. See also *open-gap upset welding*.

**upslope time.** In resistance welding, the time during which the welding current continuously increases from the beginning of welding current. See also *slope control*.

**uranium and uranium alloys.** See *Technical Brief 60*.

**uranyl.** The chemical name designating the $UO_2^{2+}$ group and compounds containing this group.

**urea-formaldehyde adhesive.** (1) An aqueous colloidal dispersion of urea-formaldehyde polymer which may contain modifiers and secondary binders to provide specific adhesive properties. (2) A type of adhesive based on a dry urea-formaldehyde polymer and water. A curing agent is commonly used with this type of adhesive.

**urethane hybrids.** Urethane acrylic polymers that are formed by the reaction of two liquid components, an acrylesterol and a modified diphenyl-methane-4,4'-diisocyanate (MDI) (Fig. 639). The acrylesterol is a hybrid of a urethane

excellent chemical resistance and, because of the presence of chlorine, some flame resistance. Dicyclopentadiene can also be incorporated into unsaturated polyesters. The resultant alicyclic groups enhance resistance to thermal oxidative decomposition at high temperatures. Because of this, dicyclopentadiene-containing unsaturated polyesters find wide use in high-temperature electrical applications.

### Mechanical properties of fiberglass-polyester resin composites

| Property | Orthophthalic | Isophthalic | BPA fumarate | Chlorendic | Dicyclopentadiene |
|---|---|---|---|---|---|
| Glass content, % | 40 | 40 | 40 | 40 | 34 |
| Barcol hardness | ... | 45 | 40 | 40 | ... |
| Tensile strength, MPa (ksi) | 152 (22) | 193 (28) | 124 (18) | 138 (20) | 96 (14) |
| Tensile modulus, GPa ($10^6$ psi) | 5.5 (0.8) | 11.7 (1.7) | 11.0 (1.6) | 9.7 (1.4) | 7.6 (1.1) |
| Tensile elongation, % | 1.7 | 2.0 | 1.2 | 1.4 | 1.9 |
| Flexural strength, MPa (ksi) | 220 (32) | 240 (35) | 160 (23) | 190 (28) | 160 (23) |
| Flexural modulus, GPa ($10^6$ psi) | 6.9 (1.0) | 7.6 (1.1) | 9.0 (1.3) | 9.7 (1.4) | 6.2 (0.9) |
| Compressive strength, MPa (ksi) | ... | 205 (30) | 180 (26) | 125 (18) | ... |
| Izod impact, J/m (ft · lbf/in.) | ... | 571 (10.7) | 640 (12) | 374 (7.0) | ... |

Unsaturated polyesters are used in a wide variety of markets, including construction, marine, transportation, industrial, electrical, and sanitary ware. In these applications, unsaturated polyesters resins are used as replacements for natural materials, such as wood, concrete, and marble, or for metals, such as steel and aluminum. Unsaturated polyester resins also compete with other plastic materials, most notably thermoplastics, in some end-use markets. Unsaturated polyester resins are commonly chosen because of their ease of fabrication, lower weight, higher strength, corrosion resistance, or lower cost. More than 75% of the unsaturated polyesters produced in the United States are used as fiberglass-reinforced composites. These materials are used for such applications as body panels on medium- and heavy-duty trucks and low-volume specialty automobiles, and as rear lifts on vans, station wagons, and sports cars.

### Recommended Reading

- C.D. Dudgeon, Unsaturated Polyesters, *Engineered Materials Handbook*, Vol 2, ASM International, 1988, p 246-251
- H.V. Boenig, *Unsaturated Polyesters: Structures and Properties*, Elsevier, 1964

FIG. 639

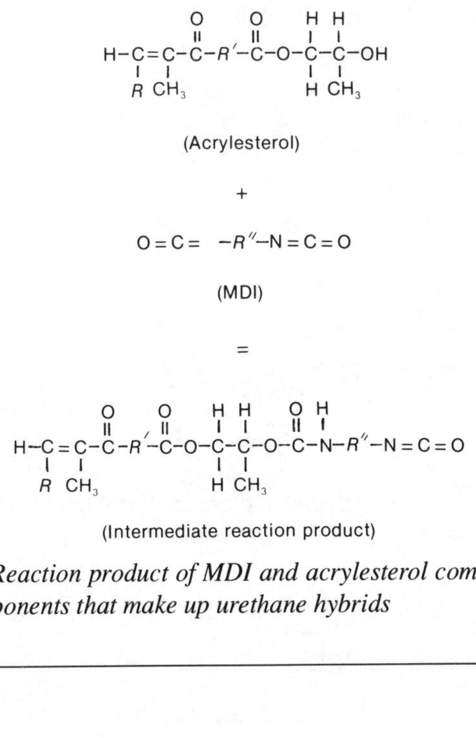

(Acrylesterol)

+

O=C= −R″−N=C=O

(MDI)

=

(Intermediate reaction product)

*Reaction product of MDI and acrylesterol components that make up urethane hybrids*

(monoalcohol) and an acrylic (unsaturated monoalcohol). The liquid-modified MDI contains two or more isocyanate groups that can react with the hydroxyl portion of the acrylesterol molecule. Acrylamate resin systems are reinforced with glass (30 to 40%) and are used in automotive applications and recreational products. When reinforced with carbon mat or metallized glass cloth, these materials can be used in communication equipment, such as electromagnetic interference/radio frequency interference (EMI/RFI) devices.

**urethane plastics.** Plastics based on resins made by the condensation of organic isocyanates with compounds or resins that contain hydroxyl groups. The resin is furnished as two component liquid monomers or prepolymers that are mixed in the field immediately before application. A great variety of materials are available, depending on the monomers used in the prepolymers and polyols and the type of diisocyanate employed. Extremely abrasion and impact resistant. See also *isocyanate plastics* and *polyurethanes* (Technical Brief 32).

**UV.** See *ultraviolet*.

## Technical Brief 60: Uranium and Uranium Alloys

URANIUM is a moderately strong and ductile metal that can be cast, formed, and welded by a variety of standard methods. It is used in nonnuclear applications primarily because of its very high density (19.1 g/cm$^3$; 68% greater than lead). Uranium is frequently selected over other very dense metals because it is easier to cast and/or fabricate than the refractory metal tungsten and much less costly than such precious metals as gold and platinum. Typical nonnuclear applications for uranium and uranium alloys include radiation shields, counterweights, and armor-piercing kinetic energy penetrators.

Natural uranium contains approximately 0.7% of the fissionable isotope U-235 and 99.3% U-238. Ore of this isotopic ratio is processed by mineral beneficiation and chemical procedures to produce uranium hexafluoride (UF$_6$). Isotopic separation is performed at this stage. This produces both enriched (radioactive) UF$_6$, which contains more than the natural isotopic abundance of U-235 and is subsequently processed and used for nuclear applications, and depleted UF$_6$, which typically contains 0.2% U-235. Access to enriched UF$_6$ is tightly controlled, but depleted material can be purchased for industrial applications. The UF$_6$ is reduced to uranium tetrafluoride (UF$_4$) by chemical reduction with hydrogen. The UF$_4$ is then reduced with magnesium or calcium in a closed vessel at elevated temperature, producing 150 to 500 kg (330 to 1100 lb) ingots of metallic uranium commonly referred to as derbies. These derbies are typically vacuum induction remelted and cast into shapes required for engineering components or for subsequent mechanical working at elevated temperatures.

Uranium is frequently alloyed to improve its corrosion resistance and mechanical properties. Alloying results in substantial decreases in density; therefore, it is desirable to obtain the necessary properties with small amounts of alloying additions. Principal alloying elements include titanium, niobium, molybdenum, and zirconium. Like unalloyed uranium, uranium alloys are produced by vacuum induction or vacuum arc melting and can be fabricated at elevated temperatures. A wide range of properties can be obtained by post-fabrication heat treatment of uranium alloys.

### Nominal compositions and properties of uranium alloys

| Alloy | Density, g/cm$^3$ | Hardness | Yield strength, MPa (ksi) | Tensile strength, MPa (ksi) | Elongation(a), % | Reduction of area(a), % |
|---|---|---|---|---|---|---|
| Unalloyed uranium(b) | 19.1 | 93 HRB | 295 (43) | 700 (101) | 22 | ... |
| | 19.1 | 94 HRB | 270 (39) | 720 (104) | 31 | ... |
| U-0.75Ti(b) | 18.6 | 36 HRC | 650 (94) | 1310 (190) | 31 | 52 |
| | 18.6 | 42 HRC | 965 (140) | 1565 (227) | 19 | 29 |
| | 18.6 | 52 HRC | 1215 (176) | 1660 (241) | <2 | <2 |
| U-2.0Mo | 18.5 | 34 HRC | 675 (98) | 1100 (160) | 23 | 25 |
| U-2.3Nb | 18.5 | 32 HRC | 545 (79) | 1060 (154) | 28 | 33 |
| U-4.5Nb | 17.9 | 42 HRC | 900 (130) | 1190 (173) | 10 | 8 |
| U-6.0Nb | 17.3 | 82 HRB | 160 (23) | 825 (120) | 31 | 34 |
| U-10Mo | 16.3 | 28 HRC | 900 (130) | 930 (134) | 9 | 30 |
| U-7.5Nb-2.5Zr | 16.4 | 20 HRC | 540 (78) | 850 (123) | 23 | 50 |

(a) All based on high-purity alloys with low hydrogen contents. (b) Property values may vary due to heat treating practice.

### Recommended Reading

- K.H. Eckelmeyer, Uranium and Uranium Alloys, *Metals Handbook*, 10th ed., Vol 2, ASM International, 1990, p 670-682
- G.M. Ludtka and E.L. Bird, Heat Treating of Uranium and Uranium Alloys, *ASM Handbook*, Vol 4, ASM International, 1991, p 928-938

# V

**vacancy**. A structural imperfection in which an individual atom site is temporarily unoccupied.

**vacuum annealing**. Annealing carried out at subatmospheric pressure.

**vacuum arc remelting (VAR)**. A consumable-electrode remelting process in which heat is generated by an electric arc between the electrode and the ingot. The process is performed inside a vacuum chamber. Exposure of the droplets of molten metal to the reduced pressure reduces the amount of dissolved gas in the metal. See also the figure accompanying the term *consumable-electrode remelting*.

**vacuum atomization**. A commercial batch powder production process based on the principle that, when a molten metal supersaturated with gas under pressure is suddenly exposed to vacuum, the gas expands, comes out of solution, and causes the liquid metal to be atomized (Fig. 640). Alloy powders based on nickel, copper, cobalt, iron, and aluminum can be vacuum atomized with hydrogen. Powders are spherical, clean, and of a high purity (Fig. 641).

**vacuum bag**. A flexible bag in which pressure may be applied to an assembly (inside the bag) by means of evacuation of the bag. See also *vacuum bag molding*.

**vacuum bag molding**. A process for manufacturing reinforced plastics in which a sheet of flexible, transparent material plus a bleeder cloth and release film are placed over the lay-up on the mold and sealed at the edges (Fig. 642). A vacuum is applied between the sheet and the lay-up. The entrapped air is mechanically worked out of the lay-up and removed by the vacuum, and the part is cured with temperature, pressure, and time. Also called *bag molding* or pressure bag molding. See also *lay-up*.

**vacuum brazing**. A nonpreferred term used to denote *furnace brazing* which takes place in a chamber or retort below atmospheric pressure.

**vacuum carburizing**. A high-temperature gas carburizing process using furnace pressures between 13 and 67 kPa (0.1 to 0.5 torr) during the carburizing portion of the cycle. Steels undergoing this treatment are austenitized in a rough vacuum, carburized in a partial pressure of hydrocarbon gas, diffused in a rough vacuum, and then quenched in either oil or gas. Both batch and continuous furnaces (Fig. 643) are used. See also the table accompanying the term *vacuum furnace*.

**vacuum casting**. A casting process in which metal is melted and poured under very low atmospheric pressure; a form of permanent mold casting in which the mold is inserted into liquid metal, vacuum is applied, and metal is drawn up into the cavity.

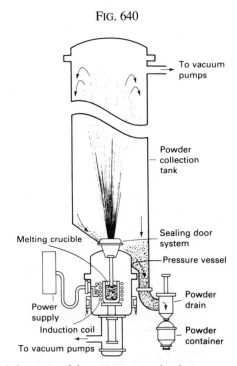

FIG. 640

*Schematic of the vacuum atomization process*

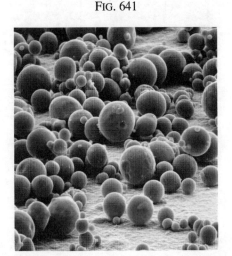

FIG. 641

*Scanning electron micrograph of a vacuum-atomized –100 mesh nickel-base powder. 650×*

**vacuum degassing**. The use of vacuum techniques to remove dissolved gases from molten alloys.

**vacuum deposition**. Deposition of a metal film onto a substrate in a vacuum by metal evaporation techniques.

FIG. 642

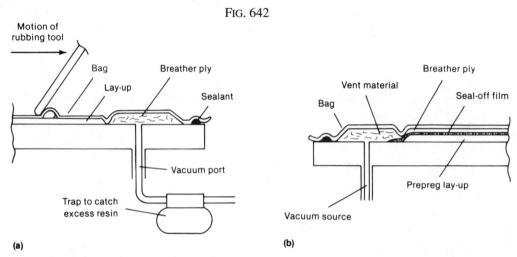

*Vacuum bag molding techniques. (a) Wipeout process for wet lay-ups. (b) Seal-off method for prepreg lay-ups*

FIG. 643

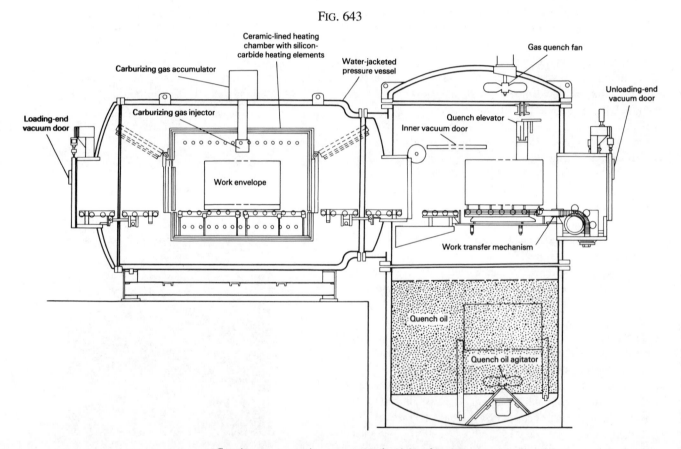

*Continuous ceramic vacuum carburizing furnace*

**vacuum deposition** (of interference films). A method of revealing the microstructures of metals and carbides with the aid of vacuum deposited interference layers (films). The phase shift in light reflected at the interference film/substrate interface contributes to the colors produced by a transparent film on a metallographic specimen. This phase shift depends on the optical properties of the film and substrate. Materials that have been found to produce phase contrast and color when vacuum deposited include titanium dioxide ($TiO_2$), silicon dioxide ($SiO_2$), zirconium

FIG. 644

*Typical arrangement for vacuum deposition of interference films. Arrow indicates the tungsten wire basket filled with material for evaporation.*

dioxide (ZrO$_2$), zinc sulfide (ZnS), tin oxide (SnO$_2$), and carbon. Of these, TiO$_2$ is the most commonly used. All these (except carbon) are supplied in powder or chips and must be contained in a tungsten wire basket, which is approximately 100 mm (4 in.) from the specimens (Fig. 644). The vacuum chamber is evacuated to $10^{-4}$ torr or lower. See also *reactive sputtering* (of interference films).

**vacuum forming.** A method of sheet forming in which the plastic sheet is clamped in a stationary frame, heated, and

FIG. 645

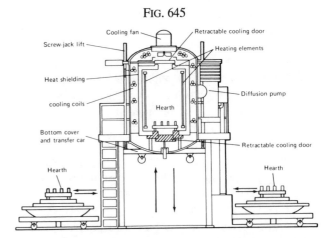

*Bottom-loading cold-wall vacuum furnace*

drawn down by a vacuum into a mold. In a broad sense, the term is sometimes used to refer to all sheet-forming techniques, including *drape forming* involving the use of vacuum and stationary molds. See also the figures accompanying the terms *thermoforming* and *vacuum snapback thermoforming*.

**vacuum furnace.** A furnace using low atmospheric pressures instead of a protective gas atmosphere like most heat-treating furnaces. Vacuum furnaces are categorized as hot wall or cold wall (Fig. 645), depending on the location of the heating and insulating components. Table 73 provides pressure ranges required for vacuum furnace operations.

**vacuum fusion.** An analytic technique for determining the amount of gases in metals; ordinarily used for hydrogen and oxygen, and sometimes for nitrogen. Applicable to many metals, but not to alkali or alkaline earth metals.

**vacuum hot pressing.** A method of processing materials

**Table 73  Pressure ranges required for selected vacuum furnace operations relative to standard atmospheric (0 gage) pressure**

| Gage pressure classification | Furnace application | Vacuum classification | Equivalent pressures | | | | | | | | |
| --- | --- | --- | --- | --- | --- | --- | --- | --- | --- | --- | --- |
| | | | Pa | torr | mm Hg(a) | μm Hg | in. Hg | psia(b) | psig | atm | bar |
| Pressure | Pressure quenching | | | | | | | | | | |
| | High gas | . . . | . . . | . . . | . . . | . . . | 177.17 | 87.02 | 72.32 | 5.92 | 6 |
| | | | . . . | . . . | . . . | . . . | 147.65 | 72.52 | 57.82 | 4.93 | 5 |
| | | | . . . | . . . | . . . | . . . | 118.12 | 58.02 | 43.32 | 3.95 | 4 |
| | | | . . . | . . . | . . . | . . . | 88.59 | 43.51 | 28.81 | 2.96 | 3 |
| | Gas | | . . . | . . . | . . . | . . . | 59.06 | 29.01 | 14.31 | 1.97 | 2 |
| Zero | . . . | . . . | $1.01\times10^5$ | 760 | 760 | $7.6\times10^5$ | 29.92 | 14.696 | 0 | 1 | 1.01 |
| Negative | Vacuum treatment | . . . | | | | | | | | | |
| | Normal backfill | Rough | $1.00\times10^5$ | 750 | 750 | $7.5\times10^5$ | 29.53 | 14.50 | . . . | 0.99 | 1 |
| | | | $1.3\times10^4$ | 100 | 100 | $10^5$ | . . . | . . . | . . . | . . . | . . . |
| | | | $1.3\times10^3$ | 10 | 10 | $10^4$ | . . . | . . . | . . . | . . . | . . . |
| | | | 130 | 1 | 1 | $10^3$ | . . . | . . . | . . . | . . . | . . . |
| | Normal range | Soft | 13 | 0.1 | 0.1 | 100 | . . . | . . . | . . . | . . . | . . . |
| | | | 1.3 | 0.01 | 0.01 | 10 | . . . | . . . | . . . | . . . | . . . |
| | | | 0.13 | $10^{-3}$ | $10^{-3}$ | 1 | . . . | . . . | . . . | . . . | . . . |
| | Maximum | Hard | 0.013 | $10^{-4}$ | $10^{-4}$ | 0.1 | . . . | . . . | . . . | . . . | . . . |
| | | | $1.3\times10^{-3}$ | $10^{-5}$ | $10^{-5}$ | 0.01 | . . . | . . . | . . . | . . . | . . . |
| | | | $1.3\times10^{-4}$ | $10^{-6}$ | $10^{-6}$ | $10^{-3}$ | . . . | . . . | . . . | . . . | . . . |
| | | | $1.3\times10^{-5}$ | $10^{-7}$ | $10^{-7}$ | $10^{-4}$ | . . . | . . . | . . . | . . . | . . . |
| | | | $1.3\times10^{-6}$ | $10^{-8}$ | $10^{-8}$ | $10^{-5}$ | . . . | . . . | . . . | . . . | . . . |

(a) Equal to 133.322387415 Pa, it differs from torr by one part in $7 \times 10^6$. (b) psia = psig + 14.7 psi.

(especially metal and ceramic powders) at elevated temperatures, consolidation pressures, and low atmospheric pressures.

**vacuum induction melting (VIM).** A process for remelting and refining metals in which the metal is melted inside a vacuum chamber by induction heating. The metal can be melted in a crucible and then poured into a mold.

**vacuum injection molding.** A molding process for fabricating reinforced plastics which utilizes both a male and female mold in which reinforcements are placed, a vacuum is applied, and a room temperature curing liquid resin is introduced to saturate the reinforcement.

**vacuum melting.** Melting in a vacuum to prevent contamination from air and to remove gases already dissolved in the metal; the solidification can also be carried out in a vacuum or at low pressure.

**vacuum metallizing.** A process in which surfaces are thinly coated by exposing them to a metal vapor under vacuum.

**vacuum molding.** See *V process.*

**vacuum nitrocarburizing.** A subatmospheric *nitrocarburizing* process using a basic atmosphere of 50% ammonia/50% methane, containing controlled oxygen additions of up to 2%.

**vacuum refining.** Melting in a vacuum to remove gaseous contaminants from the metal.

**vacuum residue.** The residue from vacuum distillation of crude oil.

**vacuum sintering.** Sintering of ceramics or metals at subatmospheric pressure.

**vacuum sintering furnace.** A furnace wherein sintering of ceramics or metals is conducted in a vacuum. The furnace may be of a design either for batch sintering or for continuous sintering. See also *vacuum sintering.*

**vacuum snapback thermoforming.** A *thermoforming* process for production of plastic items with external deep draws, such as auto parts and luggage. First, the sheet is clamped over the female cavity (Fig. 646). Air pressure is then introduced through the channel in the base plate, stretching the plastic. When the material has been sufficiently stretched, the pressure is turned off, and vacuum is turned on, pulling the plastic into the mold. There are many variations of this method, some of which employ plug assists. See also *plug-assist forming.*

**valence.** A positive number that characterizes the combining power of an element for other elements, as measured by the number of bonds to other atoms that one atom of the given element forms upon chemical combination; hydrogen is assigned valence 1, and the valence is the number of hydrogen atoms, or their equivalent, with which an atom of the given element combines.

**van der Waals bond.** A secondary bond arising from the fluctuating dipole nature of an atom with all occupied electron shells filled.

FIG. 646

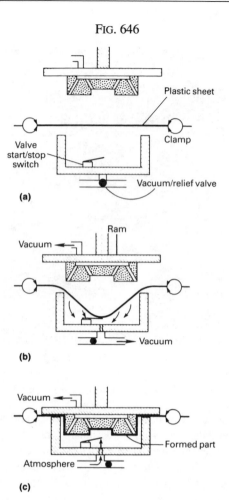

(a)

(b)

(c)

*Vacuum snapback thermoforming. (a) Heat-softened sheet positioned between male mold and casing cavity. (b) Sag stretches sheet and activates vacuum switch. (c) Vacuum snaps sheet back onto mold.*

**vapor.** The gaseous form of substances that are normally in the solid or liquid state, and that can be changed to these states either by increasing the pressure or decreasing the temperature.

**vapor blasting.** Same as *liquid honing.*

**vapor degreasing.** Degreasing of work in the vapor over a boiling liquid solvent, the vapor being considerably heavier than air. At least one constituent of the soil must be soluble in the solvent. Modifications of this cleaning process include vapor-spray-vapor, warm liquid-vapor, boiling liquid-warm liquid-vapor, and ultrasonic degreasing (Fig. 647).

**vapor-deposited replica.** A *replica* formed of a metal or a salt by the condensation of the vapors of the material onto the surface to be replicated.

**vapor deposition.** See *chemical vapor deposition, physical vapor deposition,* and *sputtering.*

F<small>IG</small>. 647

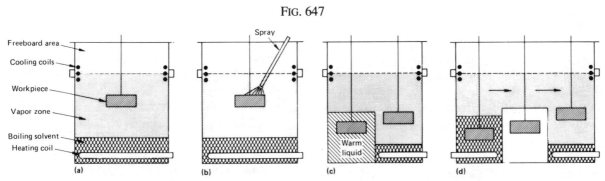

*Principal systems of vapor degreasing. (a) Vapor phase only. (b) Vapor-spray-vapor. (c) Warm liquid-vapor. (d) Boiling liquid-warm liquid-vapor*

**vapor-liquid-solid process**. A process that utilizes vapor feed gases and a liquid catalyst to produce solid crystalline whiskers, such as silicon carbide whiskers used in composite materials.

**vapor-phase lubrication**. A type of lubrication in which one or more gaseous reactants are supplied to the vicinity of the surface to be lubricated and which subsequently react to form a *lubricious* deposit on that surface.

**vapor plating**. Deposition of a metal or compound on a heated surface by reduction or decomposition of a volatile compound at a temperature below the melting points of the deposit and the base material. The reduction is usually accomplished by a gaseous reducing agent such as hydrogen. The decomposition process may involve thermal dissociation or reaction with the base material. Occasionally used to designate deposition on cold surfaces by vacuum evaporation. See also *vacuum deposition*.

**variability**. The number of degrees of freedom of a heterogeneous phase equilibrium.

**variance**. A measure of the squared dispersion of observed values or measurements expressed as a function of the sum of the squared deviations from the population mean or sample average.

**varistor**. A material, such as zinc oxide (ZnO), having an electrical resistance that is sensitive to changes in applied voltage.

**varnish**. (1) In lubrication, a deposit resulting from the oxidation and/or polymerization of fuels, lubricating oils, or organic constituents of bearing materials. Harder deposits are described as *lacquers*, softer deposits are described as *gums*. (2) A transparent surface coating which is applied as a liquid and then changes to a hard solid; all varnishes are solutions of resinous materials in a solvent.

**V-bend die**. A die commonly used in press-brake forming, usually machined with a triangular cross-sectional opening to provide two edges as fulcrums for accomplishing *three-point bending* (Fig. 648). See also the figure accompanying the term *press-brake forming*.

**V-cone blender**. A machine for blending metal powders that

F<small>IG</small>. 648

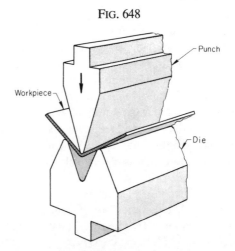

*V-bend press-brake forming*

has two cone-shaped containers arranged in a V and open to each other. See also *blending*.

**vector field**. Same as *resultant field*.

**Vegard's law**. The relationship that states that the lattice parameters of substitutional solid solutions vary linearly between the values for the components, with composition expressed in atomic percentage.

**veil**. An ultrathin mat for reinforcing plastics similar to a *surfacing mat*, often composed of organic fibers as well as glass fibers. See also *mat*.

**veining**. A sub-boundary structure in a metal that can be delineated because of the presence of a greater than average concentration of precipitate or solute atoms.

**Vello process**. A process for continuously drawing glass tubing (or cane) in which glass is fed downward to the draw through an annular orifice.

**vent** (metals). A small opening in a foundry mold for the escape of gases.

**vent** (plastics). A small hole or shallow channel in a mold that allows air or gas to exit as the plastic molding material enters.

**vent cloth**. A layer or layers of open-weave cloth used to provide a path for vacuum to "reach" the area over a *laminate* being cured, such that volatiles and air can be removed. Also causes the pressure differential that results in the application of pressure to the part being cured. Also called breather cloth.

**venting**. In autoclave curing of a composite part or assembly, turning off the vacuum source and venting the vacuum bag to the atmosphere. The pressure on the part is then the difference between pressure in the autoclave and atmospheric pressure. In injection molding, gases evolve from the melt and escape through vents machined in the barrel or mold.

**vent mark**. A small protrusion resulting from the entrance of metal into die vent holes.

**verification**. Checking or testing an instrument to ensure conformance with the specification.

**verified loading range**. In the case of testing machines, the range of indicated loads for which the testing machine gives results within the permissible variation specified.

**vermicular iron**. Same as *compacted graphite cast iron*.

**vermiculite**. A granular, clay mineral constituent that is used as a textural material in painting, as an aggregate in certain plaster formulations used in sculpture, or mixed with a resin to form a filler of relatively high compressive strength. See also *filler* (1).

**vernier**. A short auxiliary scale that slides along the main test instrument scale to permit more accurate fractional reading of the least main division of the main scale. See also *least count*.

**vertical illumination**. Light incident on an object from the objective side of an optical microscope so that smooth planes perpendicular to the optical axis of the objective appear bright. See also *objective* and *optical microscope*.

**vertical position**. The position of welding in which the axis of the weld is approximately vertical. See also the figure accompanying the term *welding position*.

**vertical position** (pipe welding). The position of a pipe joint in which welding is performed in the horizontal position and the pipe may or may not be rotated. See also the figures accompanying the terms *horizontal fixed position* (pipe welding) and *horizontal rolled position* (pipe welding).

**vibration density**. The apparent density of a powder mass when the volume receptacle is vibrated under specified conditions while being loaded. Similar to *tap density*. See also *apparent density* (1).

**vibratory cavitation**. Cavitation caused by the pressure fluctuations within a liquid, induced by the vibration of a solid surface immersed in the liquid.

**vibratory compaction**. A powder compacting process where vibration of the die assembly is used in addition to the usual pressure.

**vibratory finishing**. A process for deburring and surface finishing in which the product and an abrasive mixture are placed in a container and vibrated.

**vibratory mill**. A *ball mill* wherein comminution is aided by subjecting the balls or rods to a vibratory force. See also comminution.

**vibratory polishing**. A mechanical polishing process in which a metallographic specimen is made to move around the polishing cloth by imparting a suitable vibratory motion to the polishing system. See also *polishing* (4).

**Vicat softening point**. The temperature at which a flat-ended needle of 1 mm$^2$ (0.0015 in.$^2$) circular or square cross section will penetrate a thermoplastic specimen to a depth of 1 mm (0.040 in.) under a specified load, using a uniform rate of temperature rise.

**Vickers hardness number (HV)**. A number related to the applied load and the surface area of the permanent impression made by a square-based pyramidal diamond indenter having included face angles of 136°, computed from:

$$HV = 2P \sin \frac{\alpha/2}{d^2} = \frac{1.8544P}{d^2}$$

where $P$ is applied load (kgf), $d$ is mean diagonal of the impression (mm), and $\alpha$ is the face angle of the indenter (136°).

**Vickers hardness test**. A microindentation hardness test employing a 136° diamond pyramid indenter (Vickers) and variable loads (Fig. 649), enabling the use of one hardness scale for all ranges of hardness—from very soft lead to tungsten carbide. Also known as diamond pyramid hardness test. See also *microindentation* and *microindentation hardness number*.

FIG. 649

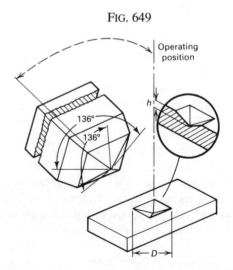

*Diamond pyramid indenter used for the Vickers test and resulting indentation in the workpiece. D is the mean diagonal of the indentation in millimeters.*

**vidicon.** A camera tube in which a charge-density pattern is formed by photoconduction and stored on a photoconductor surface that is scanned by an electron beam.

**VI improver.** An additive, usually a polymer, that reduces the variation of viscosity with temperature, thereby increasing the viscosity index of an oil.

**vinyl acetate plastics.** Plastics based on polymers of vinyl acetate or copolymers of vinyl acetate with other monomers, the vinyl acetate being the greatest amount by mass.

**vinyl chloride plastics.** Plastics based on polymers of vinyl chloride or copolymers of vinyl chloride with other monomers, the vinyl chloride being the greatest amount by mass.

**vinyl esters.** A class of thermosetting resins containing esters of acrylic and/or methacrylic acids, many of which have been made from epoxy resin. Cure is accomplished, as with unsaturated polyesters, by copolymerization with other vinyl monomers, such as styrene. Glass-reinforced vinyl esters are used in corrosion-resistant products, such as piping and storage tanks, used in the pulp and paper, chemical process, wastewater, and mining industries.

**vinylidene chloride plastics.** Plastics based on polymer resins made by the polymerization of vinylidene chloride or copolymerization of vinylidene chloride with other unsaturated compounds, the vinylidene chloride being the greatest amount by weight.

**virgin filament.** An individual *filament* that has not been in contact with any other fiber or any other hard material.

**virgin material.** A plastic material in the form of pellets, granules, powder, flock, or liquid that has not been subjected to use or processing other than that required for its initial manufacture.

**virgin metal.** Same as *primary metal.*

**viscoelasticity.** A property involving a combination of elastic and viscous behavior that makes deformation dependent upon both temperature and strain rate. A material having this property is considered to combine the features of a perfectly elastic solid and a perfect fluid.

**viscosity.** The bulk property of a fluid, semifluid, or semisolid substance that causes it to resist flow. Viscosity is defined by the equation:

$$\eta = \frac{\tau}{(dv/ds)}$$

where $\tau$ is the shear stress, $v$ is the velocity, and $s$ is the thickness of an element measured perpendicular to the direction of flow; $(dv/ds)$ is known as the rate of shear. Newtonian viscosity is often called dynamic viscosity, or absolute viscosity. Kinematic viscosity, or static viscosity $(v)$, is the ratio of dynamic viscosity $(\eta)$ to density $(\rho)$ at a specified temperature and pressure $(v = \eta/\rho)$. Recommended units of measure for dynamic viscosity are the pascal second (Pa · s) in SI units and poise (P) in English units. Recommended units of measure for kinematic vis-

cosity are square meters per second ($m^2$/s) in SI units and the stoke, or centistoke (cSt) in English units.

**viscosity coefficient.** The shearing stress tangentially applied that will induce a velocity gradient in a material.

**viscosity index (VI).** A commonly used measure of the change in viscosity of a fluid with temperature. The higher the viscosity index, the smaller the relative change in viscosity with temperature. Two different indices are used: The earlier usage applies to oils having a VI from 0 to 100. Extended VI applies to oils having a VI of at least 100. It compares the oil with a reference oil of VI 100.

**viscous.** Possessing viscosity. This term is frequently used to imply high viscosity.

**viscous deformation.** Any portion of the total deformation of a body that occurs as a function of time when load is applied but that remains permanently when the load is removed. Generally referred to as an *elastic deformation.*

**viscous friction.** See *fluid friction.*

**visible.** Pertaining to radiant energy in the electromagnetic spectral range visible to the normal human eye (~380 to 780 nm).

**visible-light-emitting diode.** An optoelectronic device containing a semiconductor junction that emits visible light when forward biased. Material is usually gallium phosphide or gallium arsenide phosphide. See also *gallium and gallium compounds* (Technical Brief 15).

**visible radiation.** Electromagnetic radiation in the spectral range visible to the human eye (~380 to 780 nm).

**visual examination.** The qualitative observation of physical characteristics, observed by using the unaided eye or perhaps aided by the use of a simple hand-held lens (up to 10×).

**vitreous.** Partially or completely comprised of a glass; often containing solid particles distributed therein.

**vitreous enamel.** See *porcelain enamel.*

**vitrification.** (1) The formation of a glassy or noncrystalline material. (2) That characteristic of a clay product resulting when the kiln temperature is sufficient to fuse grains and close the surface pores, forming an impervious mass. (3) The progressive reduction in porosity of a ceramic composition as a result of heat treatment, or the process involved.

**vitrify.** To render vitreous, generally by heating; usually, achieving enough glassy phase to render impermeable.

**V-mixer.** A machine for mixing metal powders that has two cylindrical containers arranged in the shape of a V and open to each other. See also *mixing.*

**void (composites).** Air or gas that has been trapped and cured into a laminate. Porosity is an aggregation of microvoids. Voids are essentially incapable of transmitting structural stresses or nonradiative energy fields.

**void (metals).** (1) A *shrinkage cavity* produced in castings or weldments during solidification. (2) A term generally applied to paints to describe *holidays*, holes, and skips in a film.

**void content.** Volume percentage of voids, usually less than 1% in a properly cured composite. The experimental determination is indirect, that is, it is calculated from the measured density of a cured laminate and the "theoretical" density of the starting material.

**volatile content.** The percent of *volatiles* that is driven off as a vapor from a plastic or an impregnated reinforcement.

**volatiles.** Materials, such as water and alcohol, in a sizing or resin formulation, that are capable of being driven off as a vapor at room temperature or at slightly elevated temperature. See also *size*.

**volatilization.** The conversion of a chemical substance from a liquid or solid state to a gaseous or vapor state by the application of heat, by reducing pressure, or by a combination of these processes. Also known as vaporization.

**volt.** The unit of potential difference or electromotive force in the meter-kilogram-second system, equal to the potential difference between two points for which 1 coulomb of electricity will do 1 joule of work in going from one point to the other. Symbolized V.

**voltage alignment.** A condition of alignment of an electron microscope so that the image expands or contracts symmetrically about the center of the viewing screen when the accelerating voltage is changed. See also *alignment*.

**voltage contrast.** In scanning electron microscopy, additional contrast in an image arising from increased emission of secondary electrons from negatively biased regions of a sample. This type of contrast is often used to advantage in the examination of microelectronic devices.

**voltage drop.** The amount of voltage loss from original input in a conductor of given size and length.

**voltage efficiency.** The ratio, usually expressed as a percentage, of the equilibrium-reaction potential in a given electrochemical process to the bath voltage.

**voltage regulator.** An automatic electrical control device for maintaining a constant voltage supply to the primary of a welding transformer.

**voltage stress.** That stress found within a material when subjected to an electrical charge.

**voltammetry.** An electrochemical technique in which the current between working (indicator) electrodes and counterelectrodes immersed in an electrolyte is measured as a function of the potential difference between the indicator electrode and a reference electrode.

**volume diffusion.** One of the primary diffusion mechanisms during sintering (Fig. 650). It is predominant for larger particles and higher temperatures, and its diffusion coefficient for the same conditions is smaller than that for grain boundary diffusion, and much smaller than that for surface diffusion. See also *grain boundary diffusion* and *surface diffusion*.

FIG. 650

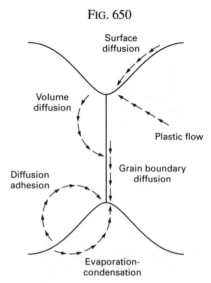

*Mass flow paths sketched with respect to the neck region between two sintering particles. Surface diffusion and evaporation-condensation are both surface transport mechanisms. Volume diffusion, diffusion adhesion, grain boundary diffusion, and plastic flow are all classified as bulk transport mechanisms.*

**volume filling.** Filling the volume of a die cavity or receptacle with loose powder, and striking off any excess amount.

**volume fraction.** Fraction of a constituent material, such as fibers in a composite material, based on its volume.

**volume ratio.** The volume percentage of solid in the total volume of a sintered body.

**volume resistance.** The ratio of the direct voltage applied to two electrodes in contact with or embedded in a specimen to that portion of the current between them that is distributed through the volume of the specimen.

**volume shrinkage.** The volumetric size reduction a powder compact undergoes during sintering. Contrast with *linear shrinkage*.

**volumetric analysis.** Quantitative analysis of solutions of known volume but unknown strength by adding reagents of known concentration until a reaction end point (color change or precipitation) is reached; the most common technique is by *titration*.

**volumetric modulus of elasticity.** See *bulk modulus of elasticity*.

**voxel.** Shortened term for volume element. In computed tomography, the volume within the object that corresponds to a single pixel element in the image. The box-shaped volume defined by the area of the pixel and the height of the slice thickness (Fig. 651). See also *pixel*.

**V process.** A molding (casting) process in which the sand is held in place in the mold by vacuum. The mold

FIG. 651

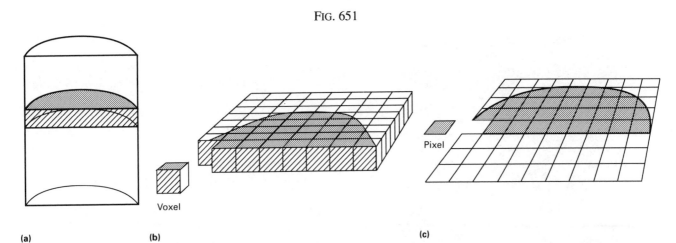

(a)                          (b)                                                (c)

*Components of a computed tomography (CT) image. (a) The x-ray transmission measurements and resultant CT image correspond to a defined slice through the object. (b) The cross-sectional slice can be considered to contain a matrix of volume elements, or voxels. (c) The reconstructed CT image consists of a matrix of picture elements, or pixels, each having a numeric value proportional to the measured x-ray attenuation characteristics of the material in the corresponding voxel.*

halves are covered with a thin sheet of plastic to retain the vacuum.

**V-ring seal**. A seal consisting of a ring or nested rings that have a V-shaped cross section and that are commonly made from elastomeric material. Spring loading is sometimes used to maintain contact between the seal and its mating surface. It is normally used to seal against axial motion.

**vulcanization**. A chemical reaction in which the physical properties of a rubber are changed in the direction of decreased plastic flow, less surface tackiness, and increased tensile strength by reacting it with sulfur or other suitable agents. See also *self-vulcanizing*.

**vulcanize**. To subject to *vulcanization*.

**V-X diagram**. A graph of the isothermal or isobaric phase relationships in a binary system, the coordinates of the graph being specific volume and concentration.

**Vycor**. See *glass* (Technical Brief 17).

# W

**wafer**. A slice of a semiconductor crystalline ingot. See also the figure accompanying the term *slices*.

**wake hackle**. A *hackle* line extending from a singularity at the crack front in the direction of cracking. Such markings are associated with inclusions in ceramics and glasses—pores, bubbles, solid particles—and are useful in determining the direction of crack propagation.

**walking-beam furnace**. A continuous-type heat treating or sintering furnace consisting of two sets of rails, one stationary and the other movable, that lift and advance parts inside the hearth (Fig. 652). With this system, the moving rails lift the work from the stationary rails, move it forward, and then lower it back onto the stationary rails. The moving rails then return to the starting position and repeat the process to advance the parts again.

FIG. 652

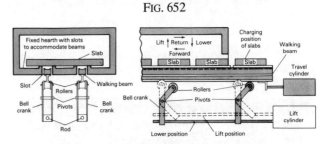

*Schematic of a walking-beam mechanism for advancing slabs through a furnace*

**Wallner line** (ceramics and glasses). A fracture surface marking, having a wavelike profile in the fracture surface (Fig. 653). Such marks frequently appear as a series of curved

FIG. 653

*Fracture surface of a piece of glass broken by striking it with a hammer. Origin is at the lower left; the wavelike lines are Wallner lines. 75×*

FIG. 654

1 μm

*Wallner lines (arrow) on the surface of a fractured WC-Co specimen. TEM formvar replica*

lines, indicating the direction of propagation of the fracture from the concave to the convex side of a given Wallner line. Also known as ripple mark.

**Wallner lines** (metals). A distinct pattern of intersecting sets of parallel lines (Fig. 654), sometimes producing a set of V-shaped lines, sometimes observed when viewing brittle fracture surfaces at high magnification in an electron microscope. Wallner lines are attributed to interaction between a shock wave and a brittle crack front propagating

at high velocity. Sometimes Wallner lines are misinterpreted as fatigue striations. See also the figure accompanying the term *fatigue striation* (metals).

**wandering sequence**. Same as *random sequence*.

**warm-setting adhesive**. Same as *intermediate-temperature-setting adhesive*.

**warm working**. Deformation of metals at elevated temperatures below the recrystallization temperature. The flow stress and rate of strain hardening are reduced with increasing temperature; therefore, lower forces are required than in cold working. See also *cold working* and *hot working*.

**warp**. A significant variation from the original true, or plane, surface or shape.

**warp** (composites). The yarn running lengthwise in a woven fabric. A group of yarns in long lengths and approximately parallel. Also, in laminates, a change in dimension of a cured laminate from its original molded shape.

**warpage** (metals). (1) Deformation other than contraction that develops in a casting between solidification and room temperature. (2) The distortion that occurs during annealing, stress relieving, and high-temperature service.

**warpage**. Dimensional distortion in a plastic object.

**wash**. (1) A coating applied to the face of a mold prior to casting. (2) An imperfection at a cast surface similar to a *cut (3)*.

**wash metal**. Molten metal used to wash out a furnace, ladle, or other container.

**wash primer**. A thin, inhibiting paint, usually chromate pigmented with a polyvinyl butyrate binder.

**water absorption**. The ratio of the weight of water absorbed by a material to the weight of the dry material.

**water atomization**. See *atomization* (powder metallurgy).

**water-base adhesives**. See *Technical Brief 61*.

**water break**. The appearance of discontinous film of water on a surface signifying nonuniform wetting and usually associated with a surface contamination.

**water break test**. A test to determine if a surface is chemically clean by the use of a drop of water, preferably distilled water. If the surface is clean, the water will break and spread; a contaminated surface will cause the water to bead.

**water-extended polyester**. A casting formulation in which water is suspended in the polyester resin.

**waterjet/abrasive waterjet machining**. A hydrodynamic machining process that uses a high-velocity stream of water as a cutting tool (Fig. 655). This process is limited to the cutting of nonmetallic materials when the jet stream consists solely of water. However, when fine abrasive particles are injected into the water stream, the process can be used to cut harder and denser materials. Abrasive waterjet machining has expanded the range of fluid jet machining to include the cutting of metals (Fig. 656), glass, ceramics,

## Technical Brief 61: Water-Base Adhesives

A WATER-BASE ADHESIVE formulation can be classified as a solution in which the polymer is totally soluble in water or alkaline water, or as a latex, which consists of a stable dispersion of polymer in an essentially aqueous medium. Solids content in water-base dispersions can be as high as 50% by volume. Many of the materials used in water-base adhesives are also used in organic-solvent-base adhesives. In recent years, exposure to organic solvents has been increasingly controlled by federal regulations. Therefore, it is advantageous to use water-base rather than organic-solvent-base adhesives. The use of water as a solvent results in lower cost, nonflammability, and lower toxicity.

Water-base adhesives have fairly good resistance to organic solvents. However, moisture resistance is usually poor, and adhesives are subject to freezing, which can affect properties. Although there are many advantages to using water-base adhesives, there is still a large and important market for solvent-base adhesives. Water-base adhesives are usually unsuitable for hydrophobic surfaces, such as plastics, because of poor wettability. In addition, water shrinks some substrates, such as paper, textiles, and cellulosics, and is corrosive to selected metals, such as copper. Organic-solvent-base adhesives are suitable for application on hydrophobic surfaces and are compatible with most metal surfaces.

Water-base adhesive solutions consist largely of natural adhesives. Materials that are soluble in water alone include animal glues, starch, dextrin, methylcellulose, and polyvinyl alcohol. Materials that are soluble in alkaline water include casein, rosin, carboxymethylcellulose, shellac, vinyl acetate, and acrylate copolymers containing carboxyl groups. A water-base adhesive latex consists of a stable dispersion in an aqueous medium. Latices can be classified as natural, synthetic, or artificial. A natural latex is formed from natural rubber. A synthetic latex is based on an aqueous dispersion of polymers obtained by emulsion polymerization. Adhesive families in this category include neoprene, styrene-butadiene rubber (SBR), nitrile rubber, polyvinyl acetate, polyacrylates, and polymethacrylates. An artificial latex is made simply by dispersing the solid polymer. Included in this category are natural rosin and its derivatives, synthetic butyl rubber, and reclaimed rubber.

Water-base adhesive families can be subdivided into materials used in making water-base adhesives only and materials used in making both water-base and organic-solvent-base adhesives. Included in the former classification are casein, dextrin, starch, animal glues, polyvinyl alcohol, sodium carboxymethylcellulose, and sodium silicate. Included in the latter group are amino resins (urea and melamine formaldehydes), phenolic resins (phenol and resorcinol formaldehydes), polacrylates, polyvinyl acetates, polyvinyl esters, neoprene, nitrile rubber, SBR, butyl rubber, natural rubber, and reclaimed rubber.

Water-base adhesive applications by demand include construction (15%), transportation (<1%), rigid bonding (<5%), packaging (60%), nonrigid bonding (<17%), consumer goods (<1%), and tapes (<3%). Rigid bonding applications for water-base adhesives include appliances, housewares, machinery, and electronics. Nonrigid bonding applications include fabrics, shoes, filters, and rugs.

## Typical water-base adhesive latex bonding substrates, applications, and other characteristics by chemical family

| Characteristic | Natural rubber | SBR | Nitrile | Polyvinyl acetate | Neoprene | Ethylene-vinyl acetate | Acrylic |
|---|---|---|---|---|---|---|---|
| Typical solids content, % | 60 | 55 | 64–65 | 50 | 50 | 34–47 | 37–65 |
| Color | White | Cream (wet), tan (dry) | White | White | Natural | White | Clear to white |
| Bonding substrates | Textiles, various others | Textiles, metals | Plastics | Wood, plastics | Plastics, wood, leather, plaster | Plastics, wood | Plastics, textiles, foams |
| Applications | Textiles, various others | Textiles, lightweight metals | PSAs, tapes, labels | Woodworking, general purpose | General industrial, construction | Packaging, film bonding | Packaging, tapes, labels, textiles, film bonding |

### Recommended Reading

- M.M. Gauthier, Types of Adhesives, *Engineered Materials Handbook*, Vol 3, ASM International, 1990, p 74-93
- J. Shields, *Adhesives Handbook*, 3rd ed., Butterworths, 1984, p 32-79

FIG. 655

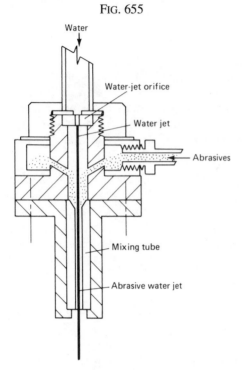

*Waterjet nozzle principle of operation*

FIG. 656

*Abrasive waterjet machining of an Inconel disk to produce turbine wheel blades*

and composite materials. Water pressures up to 410 MPa (60 ksi) are used. The coherent jet of water is propelled at speeds up to approximately 850 m/s (2800 ft/s).

**water quenching**. A quench in which water is the quenching medium. The major disadvantage of water quenching is its poor efficiency at the beginning or hot stage of the quenching process. See also *quenching*.

**water wash** (thermal spraying). The forcing of exhaust air and fumes from a spray booth through water so that the vented air is free of thermally sprayed particles or fumes.

**watt**. A unit of electrical power equal to 1 joule per second. Symbolized W.

**wavelength-dispersive spectroscopy (WDS)**. A method of x-ray analysis that employs a crystal spectrometer to discriminate characteristic x-ray wavelengths. Compare with *energy-dispersive spectroscopy*.

**wave number**. The number of waves per unit length. Commonly used in infrared and Raman spectroscopy, the wave number is expressed as the reciprocal of the wavelength. The usual unit of wave number is the reciprocal centimeter ($cm^{-1}$).

**wave soldering**. An automatic soldering process where work parts (usually printed circuit boards) are automatically passed through a wave of molten solder. See also *dip soldering*.

**waviness**. A wavelike variation from a perfect surface, generally much larger and wider than the roughness caused by tool or grinding marks. See also the figure accompanying the term *roughness*.

**wax**. (1) Any of a group of substances resembling beeswax in appearance and character, and in general distinguished by their composition of esters and higher alcohols, and by their freedom from fatty acids. (2) Preferred lubricant for pressing cemented carbide powder mixtures.

**wax pattern**. A precise duplicate, allowing for shrinkage, of the casting and required gates, usually formed by pouring or injecting molten wax into a die or mold. See also *investment casting*.

**wax pattern** (thermit welding). Wax molded around the parts to be welded to the form desired for the completed weld. See also the figure accompanying the term *thermit welding*.

**wear**. Damage to a solid surface, generally involving progressive loss of material, due to a relative motion between that surface and a contacting surface or substance. Compare with *surface damage*.

**wear debris**. Particles that become detached in a wear process.

**wear pad**. In forming, an expendable pad of rubber or rubber-like material of nominal thickness that is placed against the diaphragm to lessen the wear on it. See also *diaphragm* (3).

**wear rate**. The rate of material removal or dimensional change due to *wear* per unit of exposure parameter—for example, quantity of material removed (mass, volume, thickness) in unit distance of sliding or unit time.

**wear rate** (of seals). The amount of seal-surface wear, stated in terms of mils, worn in some designated time period. One commonly used unit is mils per hundred hours.

**wear resistance**. The resistance of a body to removal of

**Table 74 Representative compositions of weathering steels**

| Proprietary grade | Composition, wt%(a) | | | | | | | | |
|---|---|---|---|---|---|---|---|---|---|
| | C | Mn | P | S | Si | Cu | Ni | Cr | V |
| USS COR-TEN . . . | 0.19(a) | 0.80–1.25 | 0.04(a) | 0.05(a) | 0.30–0.65 | 0.25–0.40 | 0.40(a) | 0.40–0.65 | 0.02–0.10 |
| Mayari R . . . . . . . . | 0.20(a) | 0.75–1.35 | 0.04(a) | 0.05(a) | 0.15–0.50 | 0.20–0.40 | 0.50(a) | 0.40–0.70 | 0.01–0.10 |

(a) Maximum

material by wear processes, expressed as the reciprocal of the wear rate. Wear resistance is a function of the conditions under which the wear process takes place. These conditions should always be carefully specified.

**wear scar.** The portion of a solid surface that exhibits evidence that material has been removed from it due to the influence of one or more wear processes.

**wear transition.** Any change in the wear rate or in the dominant wear process occurring at a solid surface. Wear transitions can be produced by an external change in the applied conditions (for example, load, velocity, temperature, or gaseous environment) or by time-dependent changes (aging) of the materials and restraining fixtures in the *tribosystem*.

**weathering.** Exposure of materials to the outdoor environment.

**weathering steels.** Copper-bearing *high-strength low-alloy steels* that exhibit high resistance to atmospheric corrosion in the unpainted condition. Table 74 lists compositions of representative weathering steels.

**weave.** The particular manner in which a fabric is formed by interlacing yarns. Usually assigned a style number. See also the figures accompanying the terms *basketweave* (composites), *crowfoot satin*, *eight-harness satin*, *leno weave*, and *plain weave*.

**weave bead.** A weld bead made with oscillations transverse to the axis of the weld. Contrast with *stringer bead*.

**web (metals).** (1) A relatively flat, thin portion of a forging that effects an interconnection between ribs and bosses; a panel or wall that is generally parallel to the forging plane. See also *rib*. (2) For twist drills and reamers, the central portion of the tool body that joins the lands. (3) A plate or thin portion between stiffening ribs or flanges, as in an I-beam, H-beam or other similar section.

**web (plastics).** A thin plastic sheet in process in a machine. The molten web is that which issues from the die. The substrate web is the substrate being coated.

**webbing.** Filaments or threads that sometimes form when adhesively bonded surfaces are separated. See also *legging* and *stringiness*. Compare with *teeth*.

**wedge effect.** The establishment of a pressure wedge in a lubricant. See also *wedge formation* (2).

**wedge formation.** (1) In sliding metals, the formation of a wedge or wedges of plastically sheared metal in local regions of interaction between sliding surfaces. This type of wedge is also known as a prow. It is similar to a *built-up*

*edge*. (2) In hydrodynamic lubrication, the establishment of a pressure gradient in a fluid flowing into a converging channel. This is also known as *wedge effect*.

**weepage.** A minute amount of liquid leakage by a seal. It is commonly considered to be a leakage rate of less than one drop of liquid per minute.

**weft.** The transverse threads or fibers in a woven fabric. Those fibers running perpendicular to the warp. Also called *fill*, *filling yarn*, or *woof*.

**weight percent.** Percentage composition by weight. Contrast with *atomic percent*.

**weld.** A localized coalescence of metals or nonmetals produced either by heating the materials to suitable temperatures, with or without the application of pressure, or by the application of pressure alone and with or without the use of filler material.

**weldability.** A specific or relative measure of the ability of a material to be welded under a given set of conditions. Implicit in this definition is the ability of the completed weldment to fulfill all functions for which the part was designed.

**weld bead.** A deposit of filler metal from a single welding *pass* (3).

**weldbonding.** A joining method which combines resistance spot welding or resistance seam welding with adhesive bonding. The adhesive may be applied to a faying surface before welding or may be applied to the areas of sheet separation after welding.

**weld brazing.** A joining method which combines resistance welding with brazing.

**weld crack.** A crack in weld metal (Fig. 657). See also *crater crack*, *root crack*, *toe crack*, and *underbead crack*.

**weld cracking.** Cracking that occurs in the weld metal. See also *cold cracking*, *hot cracking*, *lamellar tearing*, and *stress-relief cracking*.

**weld decay.** *Intergranular corrosion*, usually of stainless steels or certain nickel-base alloys, that occurs as the result of sensitization in the *heat-affected zone* during the welding operation. See also the figure accompanying the term *sensitization*.

**weld-delay time** (resistance welding). The amount of time the beginning of welding current is delayed with respect to the initiation of the forge-delay timer in order to synchronize the forging force with welding current flow.

**weld gage.** A device for checking the shapes and sizes of welds.

FIG. 657

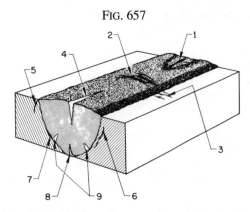

*Identification of cracks according to location in weld and base metal. 1, crater crack in weld metal; 2, transverse crack in weld metal; 3, transverse crack in heat-affected zone; 4, longitudinal crack in weld metal; 5, toe crack in base metal; 6, underbead crack in base metal; 7, fusion-line crack; 8, root crack in weld metal; 9, hat cracks in weld metal*

**weld-heat time** (resistance welding). The time from the beginning of welding current to the beginning of post-heat time.

**welding**. (1) Joining two or more pieces of material by applying heat or pressure, or both, with or without filler material, to produce a localized union through fusion or recrystallization across the interface. The thickness of the filler material is much greater than the capillary dimensions encountered in *brazing*. (2) May also be extended to include brazing and soldering. (3) In *tribology*, adhesion between solid surfaces in direct contact at any temperature.

**welding current** (automatic arc welding). The current in the welding circuit during the making of a weld, but excluding upslope, downslope, start, and crater fill current.

**welding current** (resistance welding). The current in the welding circuit during the making of a weld, but excluding preweld or postweld current.

**welding cycle**. The complete series of events involved in the making of a weld.

**welding electrode**. See preferred term *electrode* (welding).

**welding force**. See preferred terms *electrode force* and *platen force*.

**welding generator**. A generator used for supplying current for welding.

**welding ground**. Same as *work lead*.

**welding head**. The part of a welding machine or automatic welding equipment in which a welding gun or torch is incorporated.

**welding leads**. The electrical cables that serve as either *work lead* or *electrode lead* of an arc welding circuit.

**welding machine**. Equipment used to perform the welding operation. For example, spot welding machine, arc welding machine, seam welding machine, etc.

**welding position**. See definition of weld positions in Fig. 658. See also *flat position, horizontal position, horizontal fixed position, horizontal rolled position, inclined position, overhead position,* and *vertical position.*

**welding pressure**. The pressure exerted during the welding operation on the parts being welded. See also *electrode force* and *platen force.*

**welding procedure**. The detailed methods and practices, including joint preparation and welding procedures, involved in the production of a *weldment.*

**welding process**. A materials joining process which produces coalescence of materials by heating them to suitable temperatures, with or without the application of pressure or by the application of pressure alone, and with or without the use of filler metal.

**welding rectifier**. A device in a welding machine for converting alternating current to direct current.

FIG. 658

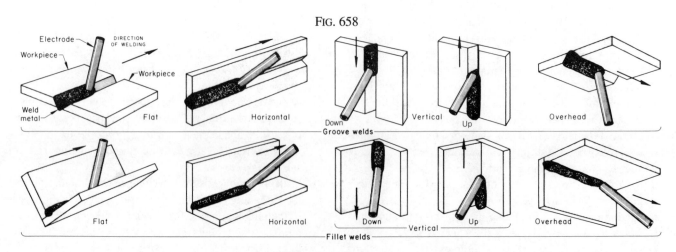

*Welding positions*

FIG. 659

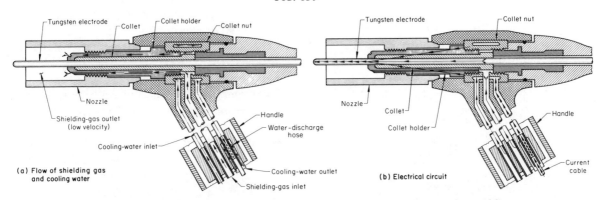

*Sectional views of a typical water-cooled torch for manual gas tungsten arc welding*

**welding rod**. A form of filler metal used for welding or brazing which does not conduct the electrical current, and which may be either fed into the weld pool or preplaced in the joint.

**welding sequence**. The order in which the various component parts of a weldment or structure are welded.

**welding stress**. *Residual stress* caused by localized heating and cooling during welding.

**welding tip**. A welding torch tip designed for welding. See also the figure accompanying the term *welding torch (oxyfuel gas)*.

**welding torch** (arc). A device used in the gas tungsten (Fig. 659) and plasma arc welding (Fig. 660) processes to control the position of the electrode, to transfer current to the arc, and to direct the flow of shielding and plasma gas. See also *gas tungsten arc welding* and *plasma arc welding*.

**welding torch** (oxyfuel gas). A device used in oxyfuel gas welding, torch brazing, and torch soldering for directing the heating flame produced by the controlled combustion of fuel gases (Fig. 661). See also *oxyfuel gas welding*.

FIG. 660

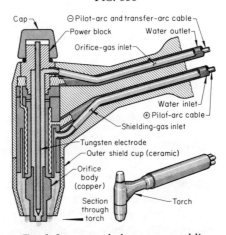

*Torch for manual plasma arc welding*

**welding transformer**. A transformer used for supplying current for welding.

**welding voltage**. See *arc voltage*.

**welding wire**. See preferred terms *electrode* and *welding rod*.

FIG. 661

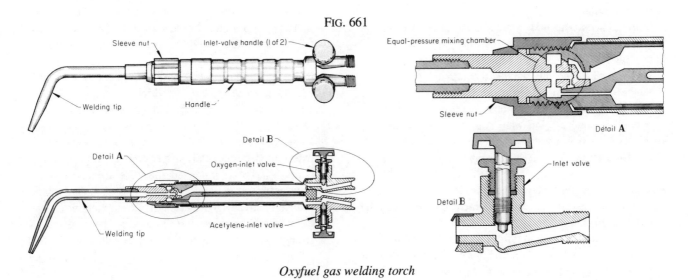

*Oxyfuel gas welding torch*

FIG. 662

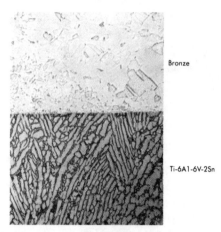

*Weld interface in a diffusion-bonded joint*

**weld interface**. The interface between weld metal and base metal in a fusion weld, between base metals in a solid-state weld without filler metal (Fig. 662), or between filler metal and base metal in a solid-state weld with a filler metal and in a braze.

**weld interval** (resistance welding). The total of all heat and cool time when making one multiple-impulse weld.

**weld-interval timer**. A device used in resistance welding to control heat and cool times and weld interval when making multiple-impulse welds singly or simultaneously.

**weld length**. See preferred term *effective length of weld*.

**weld line** (metals). See preferred term *weld interface*.

**weld line** (plastics). The mark visible on a finished plastic part made by the meeting of two flow fronts of plastic material during molding. Also called weld mark, flow line, knit line, or stria.

**weldment**. An assembly whose component parts are joined by welding.

**weld metal**. That portion of a weld which has been melted during welding.

**weld metal area**. The area of the *weld metal* as measured on the cross section of a weld.

**weld nugget**. The weld metal in spot, seam or projection welding. See also the figures accompanying the terms *nugget* and *resistance spot welding*.

**weld pass**. A single progression of a welding or surfacing operation along a joint, weld deposit, or substrate. The result of a pass is a weld bead, layer, or spray deposit.

**weld penetration**. See preferred terms *joint penetration* and *root penetration*.

**weld size**. See preferred term *size of weld*.

**weld tab**. Additional material on which the weld may be initiated or terminated.

**weld time** (automatic arc welding). The time interval from the end of start time or end of upslope to beginning of crater fill time or beginning of downslope.

**weld time** (resistance welding). The time that welding current is applied to the work in making a weld by single-impulse welding or flash welding.

**weld timer**. A device used in resistance welding to the weld time only.

**weld voltage**. See *arc voltage*.

**Wenstrom mill**. A rolling mill similar to a universal mill but where the edges and sides of a rolled section are acted on simultaneously.

**wet, wettability**. The property of a liquid, such as molten metal, to spread on a solid surface due to a low contact angle. This angle is a measure of the degree of wetting obtained in the solid-liquid system.

**wet-bag tooling**. A rubber or plastic sheet mold used in cold isostatic or hydrostatic pressing of powders. See also *cold isostatic pressing*, *hydrostatic mold*, and *hydrostatic pressing*.

**wet blasting**. A process for cleaning or finishing by means of a slurry of abrasive in water directed at high velocity against the workpieces. Many different kinds and sizes of abrasives can be used in wet blasting. Sizes range from 20-mesh (very coarse) to 5000-mesh (which is much finer than face powder). Among the types of abrasives used are: organic or agricultural materials such as walnut shells and peach pits; novaculite, which is a soft type (6 to 6.5 Mohs hardness) of silica (99.46% silica); silica, quartz, garnet, and aluminum oxide; other refractory abrasives; and glass beads.

**wet etching**. Development of microstructure in metals with liquids, such as acids, bases, neutral solutions, or mixtures of solutions.

**wet installation**. A bolted joint in which sealant is applied to the head and shank of the fastener such that after assembly a seal is provided between the fastener and the elements being joined. See also *sealant*.

**wet lay-up**. A method of making reinforced plastics by applying the resin system as a liquid when the reinforcement is put in place. Polyesters and vinyl esters are the most commonly used family of resins used for wet lay-up.

**wet-out**. In composites fabrication, the condition of an impregnated *roving* or *yarn* in which substantially all voids between the sized strands and filaments are filled with resin.

**wet strength** (adhesives). The strength of an adhesive joint determined immediately after removal from a liquid in which it has been immersed under specified conditions of time, temperature, and pressure. The term is commonly used alone to designate strength after immersion in water. In latex adhesives the term is used to describe the joint strength when the adherends are brought together with the adhesives still in the wet state. Compare with *dry strength* (adhesives).

FIG. 663

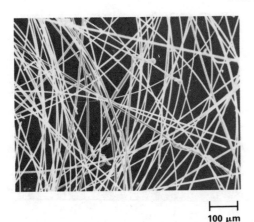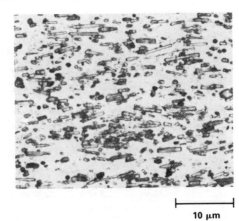

100 µm　　　　　　10 µm

*SiC whiskers produced by the vapor-liquid-solid process (left) and an SiC-whisker-reinforced aluminum alloy sheet (right) with the whiskers aligned in the direction of rolling*

**wet strength** (composites). The strength of an organic matrix composite when the matrix resin is saturated with absorbed moisture, or is at a defined percentage of absorbed moisture less than saturation (saturation is an equilibrium condition in which the net rate of absorption under prescribed conditions falls essentially to zero).

**wetting**. (1) The spreading, and sometimes absorption, of a fluid on or into a surface. (2) A condition in which the interface tension between a liquid and a solid is such that the contact angle if 0° to 90°. (3) The phenomenon whereby a liquid filler metal or flux spreads and adheres in a thin continuous layer on a solid base metal. (4) The formation of a relatively uniform, smooth, unbroken, and adherent film of solder to a basis metal.

**wetting agent**. (1) A substance that reduces the surface tension of a liquid, thereby causing it to spread more readily on a solid surface. (2) A surface-active agent that produces *wetting* by decreasing the *cohesion* within the liquid.

**wet winding**. The process of *filament winding* of composites in which strands are impregnated with resin before or during winding onto the mandrel. See also *dry winding*.

**wheelabrating**. Deflashing molded plastic parts by bombarding them with small particles at high velocity. See also *flash* (plastics).

**whirl** (oil). Instability of a rotating shaft associated with instability in the fluid film.

**whisker**. (1) A short single crystal fiber or filament used as a reinforcement in a matrix (Fig. 663). Whisker diameters range from 1 to 25 µm, with *aspect ratios* generally between 50 and 150. (2) Single-crystal growths resembling fine wire, which may extend to 0.64 mm (0.025 in.) high. They most frequently occur on printed circuit boards or electronic components that have been electroplated with tin (Fig. 664). (3) Metallic filamentary growths, often

FIG. 664

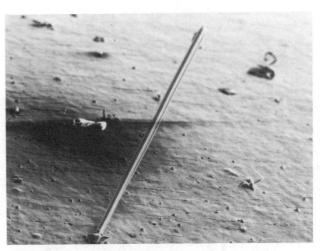

*SEM micrograph of tin whisker on the interior surface of a hybrid device lid. 245×*

microscopic, sometimes formed during electrodeposition and sometimes spontaneously during storage or service, after finishing.

**whiskers** (glass). See *striation* (glass).

**white-etching layer**. A surface layer in a steel that, as viewed in a section after etching, appears whiter than the base metal. The presence of the layer may be due to a number of causes, including plastic deformation induced by machining (Fig. 665), or surface rubbing, heating during a metallographic preparation stage to such an extent that the layer is austenitized and then hardened during cooling, and diffusion of extraneous elements into the surface.

**whiteheart malleable**. See *malleable cast iron*.

FIG. 665

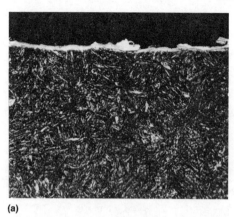 

(a)                                                        (b)

*White-etching layer in an electrical discharge machined steel specimen. (a) Recast metal and a thin white layer (0.0025 mm, or 0.0001 in.). (b) Higher-magnification view. 620×*

**white iron**. A *cast iron* that is essentially free of graphite, and most of the carbon content is present as separate grains of hard Fe₃C (Fig. 666). White iron exhibits a white, crystalline fracture surface because fracture occurs along the iron carbide platelets. White cast irons, which have alloy contents well above 4%, fall into three major groups:

1. Nickel-chromium white irons are low-chromium alloys containing 3 to 5% Ni and 1 to 4% Cr, with one alloy modification that contains 7 to 11% Cr. The nickel-chromium irons are also commonly identified as Ni-Hard types 1 to 4.

2. The high chromium irons contain 11 to 23% Cr, up to 3% Mo, and are often additionally alloyed with nickel or copper.

3. A third group comprises the 25% or 28% Cr white irons,

which may contain other alloying additions of molybdenum and/or nickel up to 1.5%.

The high-alloy white irons are primarily used for abrasion-resistant applications and are readily cast into parts needed in machinery for crushing, grinding, and handling of abrasive materials. The chromium content of high-alloy white irons also enhances their corrosion-resistant properties.

**white layer**. (1) Compound layer that forms in steels as a result of the *nitriding* process (Fig. 667). (2) In tribology, a *white-etching layer*, typically associated with ferrous alloys, that is visible in metallographic cross sections of bearing surfaces. See also *Beilby layer* and *highly deformed layer*.

**white liquor**. Cooking liquor from the kraft pulping process produced by recausticizing *green liquor* with lime. See also *kraft process*.

**white metal**. (1) A general term covering a group of white-

FIG. 666

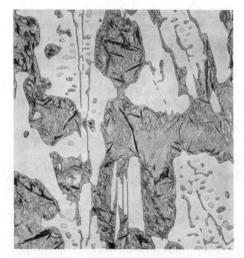

*The continuous network of M₃C (iron carbide) that forms in nickel-chromium white irons. 340×*

FIG. 667

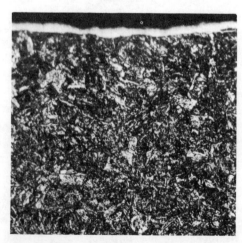

*White layer in a gas-nitrided 4140 steel specimen*

colored metals of relatively low melting points based on tin or lead. These materials are used for bearings and jewelry. (2) A copper matte of about 77% Cu obtained from smelting of sulfide copper ores.

**white radiation**. See *continuum*.

**white rust**. Zinc oxide; the powder product of corrosion of zinc or zinc-coated surfaces.

**whiteware**. A group of ceramic products characterized by a white or light colored body with a fine-grained structure which consist primarily of clay minerals, feldspars, and quartz. Most whiteware products are glazed—in whole or in part—and whiteware glazes may range from clear to completely opaque, white, or colored. Many whiteware products, such as tableware, are decorated with patterns or designs to enhance their beauty and appearance. Examples of whiteware products are sanitaryware, tableware, electrical porcelain, artware, stoneware, and tile.

**wicking**. (1) The flow of a liquid along a surface into a narrow space. This capillary action is caused by the attraction of the liquid molecules to each other and to the surface. (2) The flow of solder away from the desired area by *wetting* or *capillary action*.

**Widmanstätten structure**. A structure characterized by a geometrical pattern resulting from the formation of a new phase along certain crystallographic planes of the parent solid solution (Fig. 668). The orientation of the lattice in the new phase is related crystallographically to the orientation of the lattice in the parent phase. The structure was originally observed in meteorites, but is readily produced in many alloys, such as titanium, by appropriate heat treatment. See also the figure accompanying the term *basketweave* (titanium).

**width**. In the case of a beam, the shorter dimension perpendicular to the direction in which the load is applied.

**wildness**. A condition that exists when molten metal, during

FIG. 668

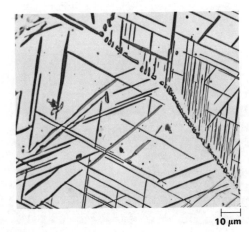

10 μm

*Widmanstätten precipitation in a Cu-3Ti alloy aged 10 h at 730 °C (1345 °F)*

cooling, evolves so much gas that it becomes violently agitated, forcibly ejecting metal from the mold or other container.

**Williams riser**. An *atmospheric riser*.

**wind angle**. In filament winding of composites, the angular measure in degrees between the direction parallel to the filaments and an established reference. In filament-wound structures, it is the convention to measure the wind angle with reference to the centerline through the polar bosses, that is, the axis of rotation. See also *filament winding*.

**winding pattern**. In filament winding of composites, the total number of individual circuits required for a winding path to begin repeating by laying down immediately adjacent to the initial circuit. A regularly recurring pattern of the filament path after a certain number of mandrel revolutions, leading eventually to the complete coverage of the mandrel. See also *filament winding*.

**winding tension**. In filament winding or tape wrapping of composites, the amount of tension on the reinforcement as it makes contact with the mandrel. See also *filament winding*.

**window**. A defect in thermoplastic film, sheet, or molding, caused by the incomplete plastication of a piece of material during processing. It appears as a globule in an otherwise blended mass. See also *fisheye* (plastics).

**winning**. Recovering a metal from an ore or chemical compound using any suitable hydrometallurgical, pyrometallurgical, or electrometallurgical method.

**wiped coat**. A hot dipped galvanized coating from which virtually all free zinc is removed by wiping prior to solidification, leaving only a thin zinc-iron alloy layer.

**wiped joint**. A joint made with solder having a wide melting range and with the heat supplied by the molten solder poured onto the joint. The solder is manipulated with a hand-held cloth or paddle so as to obtain the required size and contour.

**wipe etching**. See *swabbing*.

**wiper**. A pad of felt or other material used to supply lubricant or to remove debris.

**wiper forming, wiping**. Method of curving sheet metal sections or tubing over a form block or die in which this form block is moved relative to a wiper block or slide block (Fig. 669).

**wiping**. In tribology, the smearing or removal of material from one point, often followed by the redeposition of the material at another point, on the surface of two bodies in sliding contact. The smeared metal is usually softened or melted.

**wiping effect**. Activation of a metal surface by mechanical rubbing or wiping to enhance the formation of conversion coatings, such as phosphate coatings.

**wire**. (1) A thin, flexible, continuous length of metal, usually of circular cross section, and usually produced by drawing

FIG. 669

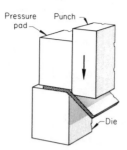

*Wiper forming setup*

through a die. The size limits for round wire sections range from approximately 0.13 mm (0.005 in.) to 25 mm (1 in.). Larger rounds are commonly referred to as *bars*. See also *flat wire*. (2) A length of single metallic electrical conductor, it may be of solid, stranded or tinsel construction, and may be either bare or insulated.

**wire bar.** A cast shape, particularly of tough pitch copper, that has a cross section approximately square with tapered ends, designed for hot rolling to rod for subsequent drawing into wire.

**wire coating.** The covering or coating of wire and cable in continuous length with insulating thermoplastics by extrusion (Fig. 670). See also the figure accompanying the term *extruder* (plastics).

**wire drawing.** Reducing the cross section of wire by pulling it through a die (Fig. 671; see facing page).

**wire feed speed.** The rate of speed in mm/s or in./min at which a welding filler metal is consumed in arc welding or thermal spraying.

**wire flame spray gun.** A flame spraying device utilizing an oxyacetylene flame to provide the heat, and the metallic material to be sprayed in wire or rod form. See also *wire flame spraying*.

**wire flame spraying.** A *thermal spraying* process variation

FIG. 670

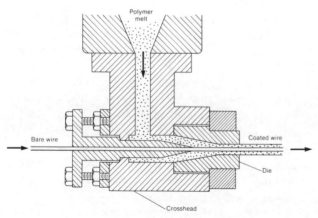

*Wire coating process*

FIG. 672

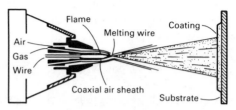

*Schematic of the wire flame spray process*

in which the material to be sprayed is in wire or rod form (Fig. 672). See also *flame spraying*.

**wire rod.** Hot-rolled coiled stock that is to be cold drawn into wire (Fig. 671; see facing page).

**wire straightener.** A device used for controlling the cast of coiled weld filler-metal wire to enable it to be easily fed into the welding gun.

**wiring.** Formation of a curl along the edge of a shell, tube, or sheet and insertion of a rod or wire within the curl for stiffening the edge. See also *curling*.

**wood failure.** The rupturing of wood fibers in strength tests on adhesively bonded specimens, usually expressed as the percentage of the total area involved which shows such failure.

**wood flour.** A pulverized wood product used in the foundry to furnish a reducing atmosphere in the mold, help overcome sand expansion, increase flowability, improve casting finish, and provide easier shakeout.

**wood laminate.** See *built-up laminated wood*, *glue-laminated wood*, and *plywood*.

**wood veneer.** A thin sheet of wood, generally within the thickness range from 0.3 to 6.3 mm (0.01 to 0.25 in.), to be used in a laminate.

**woody structure.** A macrostructure, found particularly in wrought iron and in extruded rods of aluminum alloys, that shows elongated surfaces of separation when fractured (Fig. 673).

FIG. 673

*"Woody" fracture appearance in a notched impact specimen of wrought iron*

Fig. 671

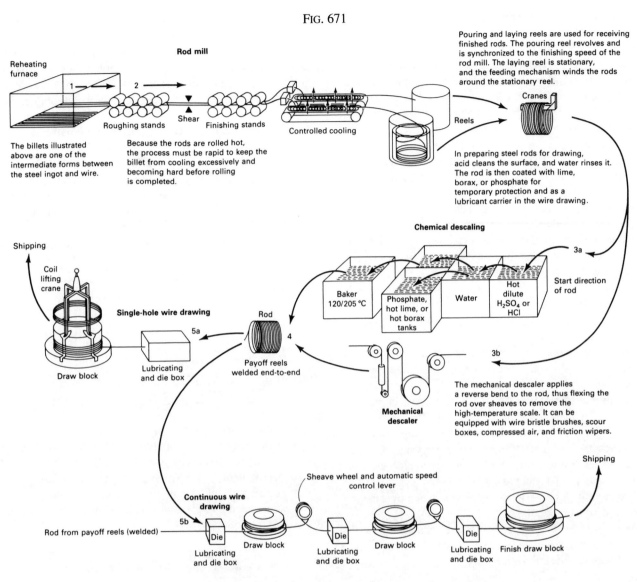

*Schematic illustrating how steel wire is drawn from rods*

**woof**. See *weft*.

**workability**. See *formability*.

**work angle**. The angle that the electrode makes with the referenced plane or surface of the base metal in a plane perpendicular to the axis of the weld (Fig. 674). See also *drag angle* and *push angle*.

**work angle** (pipe). The angle that the electrode makes with the referenced plane extending from the center of the pipe through the molten weld pool. See also the figure accompanying the term *travel angle* (pipe).

**work coil**. The inductor used when welding, brazing, or soldering with induction heating equipment. See also *induction work coil* and the figure accompanying the term *inductor*.

Fig. 674

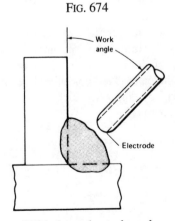

*Weld electrode work angle*

**work connection**. In welding, the connection of the work lead to the work.

**work factor**. A measure of the stability of a lubricant when subjected to an endurance test. The work factor is expressed as the average value of the ratio of three characteristics (viscosity, carbon residue, and neutralization number) as measured before the test to those same characteristics as measured after the test.

**work hardening**. Same as *strain hardening*.

**working distance**. The distance between the surface of the specimen being examined microscopically and the front surface of the objective lens.

**working electrode**. The test or specimen electrode in an *electrochemical cell*.

**working life**. The period of time during which a liquid resin or adhesive, after mixing with catalyst, solvent, or other compounding ingredients, remains usable. See also *gelation time* and *pot life*.

**work lead**. The electrical conductor connecting the source of arc welding current to the work. Also called work connection, welding ground, or ground lead.

**worm**. An exudation (sweat) of molten metal forced through the top crust of solidifying metal by gas evolution. See also *zinc worms*.

**woven fabric**. A material (usually a planar structure) constructed by interlacing yarns, fibers, or filaments to form such fabric patterns as plain, harness satin, and leno weaves. See also the figures accompanying the terms *basketweave* (composites), *crowfoot satin*, *eight-harness satin*, *leno weave*, and *plain weave*.

**woven roving**. A heavy glass fiber fabric made by weaving roving, or yarn bundles.

**wrap-around bend**. The bend obtained when a specimen is wrapped in a closed helix around a cylindrical mandrel.

**wrap forming**. See *stretch forming*.

**wrapped bush** (bearing). A thin-walled steel bush lined with a bearing alloy, or any other bearing bush made from strip.

**wrap seam**. A depression or step in the surface finish of a reinforced plastic caused by the lap of the flexible mold or carrier strip after it is removed from the cured pultrusion. See also *pultrusion*.

**wringing fit**. A fit of nominally zero allowance.

**wrinkle**. A surface imperfection in laminated plastics that has the appearance of a crease or fold in one or more outer sheets of the paper, fabric, or other base. Also occurs in vacuum bag molding when the bag is improperly placed, causing a crease.

**wrinkle depression**. An undulation or series of undulations or waves on the surface of a pultruded composite part.

**wrinkling**. A wavy condition obtained in deep drawing of sheet metal, in the area of the metal between the edge of the flange and the draw radius. Wrinkling may also occur in other forming operations when unbalanced compressive forces are set up.

**wrist pin bearing**. The bearing at the crankshaft end of an articulated connecting rod in a "V" engine.

**wrought and cast aluminum alloy designations**. Systems for designating wrought and cast aluminum alloys that have been devised by the Aluminum Association in the United States and the American National Standards Institute. For wrought alloys a four-digit system is used to produce a list of wrought compositions families as follows:

| | |
|---|---|
| Aluminum, ≥99.00% | 1xxx |
| Aluminum alloys grouped by major alloying element(s): | |
| Copper | 2xxx |
| Manganese | 3xxx |
| Silicon | 4xxx |
| Magnesium | 5xxx |
| Magnesium and silicon | 6xxx |
| Zinc | 7xxx |
| Other elements | 8xxx |
| Unused series | 9xxx |

Casting compositions are described by a three-digit system followed by a decimal value. The decimal .0 in all cases pertains to casting alloy limits. Decimals .1, and .2 concern ingot compositions, which after melting and processing should result in chemistries conforming to casting specification requirements. Alloy families for casting compositions are:

| | |
|---|---|
| Aluminum, ≥99.00% | 1xx.x |
| Aluminum alloys grouped by major alloying element(s): | |
| Copper | 2xx.x |
| Silicon, with added copper and/or magnesium | 3xx.x |
| Silicon | 4xx.x |
| Magnesium | 5xx.x |
| Zinc | 7xx.x |
| Tin | 8xx.x |
| Other elements | 9xx.x |
| Unused series | 6xx.x |

See also the table accompanying the term *aluminum and aluminum alloys* (Technical Brief 4).

**wrought iron**. A commercial iron consisting of slag (iron silicate) fibers entrained in a ferrite matrix.

**X**

*x*-axis. In reinforced plastic laminates, an axis in the plane of the laminate that is used as the 0° reference for designating the angle of a lamina. See also *laminate* and the figure accompanying the term *quasi-isotropic laminate.*

xenon. A rare gas (symbol Xe; atomic number 54) that is used in photographic flash lamps, luminescent tubes, and lasers.

x-ray. A penetrating electromagnetic radiation, usually generated by accelerating electrons to high velocity and suddenly stopping them by collision with a solid body. Wavelengths of x-rays range from about $10^{-1}$ to $10^{2}$ Å, the average wavelength used in research being about 1 Å. Also known as roentgen ray or x-radiation. See also *electromagnetic radiation.*

x-ray diffraction (XRD). An analytical technique in which measurements are made of the angles at which x-rays are preferentially scattered from a sample (as well as of the intensities scattered at various angles) in order to deduce information on the crystalline nature of the sample—its crystal structure, orientations, and so on.

x-ray diffraction residual stress techniques. Diffraction method in which the strain in the crystal lattice is measured, and the residual stress producing the strain is calculated assuming a linear elastic distortion of the crystal lattice. See also *macroscopic stress*, *microscopic stress*, and *residual stress.*

x-ray emission spectroscopy. Pertaining to *emission spectroscopy* in the x-ray wavelength region of the electromagnetic spectrum.

x-ray fluorescence. Emission by a substance of its characteristic x-ray line spectrum on exposure to x-rays.

x-ray map. An intensity map (usually corresponding to an image) in which the intensity in any area is proportional to the concentration of a specific element in that area.

x-ray photoelectron spectroscopy (XPS). An analytical technique that measures the energy spectra of electrons emitted from the surface of a material when exposed to monochromatic x-rays.

x-ray spectrograph. A photographic instrument for x-ray emission analysis. If the instrument for x-ray emission analysis does not employ photography, it is better described as an x-ray spectrometer.

x-ray spectrometry. Measurement of wavelengths of x-rays by observing their diffraction by crystals of known lattice spacing.

x-ray spectrum. The plot of the intensity or number of x-ray photons versus energy (or wavelength).

x-ray topography. A technique that comprises topography and x-ray diffraction. The term topography refers to a detailed description and mapping of physical (surface) features in a region. In the context of the x-ray diffraction, topographic methods are used to survey the lattice structure and imperfections in crystalline materials.

x-ray tube. A device for the production of x-rays by the impact of high-speed electrons on a metal target.

*xy*-plane. In reinforced plastic laminates, the reference plane parallel to the plane of the laminate.

**Y**

yarn. An assemblage of twisted filaments, fibers, or strands, either natural or manufactured, to form a continuous length that is suitable for use in weaving into textile materials used to reinforce plastics.

yarn bundle. See *bundle.*

*y*-axis. In composite laminates, the axis in the plane of the laminate that if perpendicular to the *x*-axis. See also *x-axis.*

Y-block. A single *keel block.*

yellow brass. A name sometimes used in reference to the 65Cu-35Zn type of *brass.*

yield. (1) Evidence of plastic deformation in structural materials. Also known as plastic flow or creep. See also *flow.* (2) The ratio of the number of acceptable items produced in a production run to the total number that were attempted to be produced. (3) Comparison of casting weight to the total weight of metal poured into the mold.

yield point. The first stress in a material, usually less than the maximum attainable stress, at which an increase in strain occurs without an increase in stress. Only certain materials—those which exhibit a localized, heterogeneous type of transition from elastic to plastic deformation—produce a yield point. If there is a decrease in stress after

FIG. 675

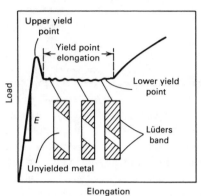

*Typical yield point behavior of low-carbon steel. The slope of the initial linear portion of the stress-strain curve, designated by E, is the modulus of elasticity.*

yielding, a distinction may be made between upper and lower yield points (Fig. 675). The load at which a sudden drop in the flow curve occurs is called the upper yield point. The constant load shown on the flow curve is the lower yield point.

**yield point elongation**. In materials that exhibit a yield point,

the difference between the elongation at the completion and at the start of discontinous yield (Fig. 675).

**yield strength**. The stress at which a material exhibits a specified deviation from proportionality of stress and strain. An offset of 0.2% is used for many materials, particularly metals. Compare with *tensile strength*.

**yield stress**. The stress level of highly ductile materials at which large strains take place without further increase in stress.

**yield value**. The stress (either normal or shear) at which a marked increase in deformation occurs without an increase in load.

**Young's modulus**. A term used synonymously with modulus of elasticity. The ratio of tensile or compressive stresses to the resulting strain. See also *modulus of elasticity*.

**yttria**. A rare earth oxide of composition $Y_2O_3$ that is added in small amounts (0.5 to 1.3 wt%) to nickel-base powder metallurgy superalloy compositions in order to improve their high-temperature creep resistance and oxidation resistance. Such alloys are referred to as oxide-dispersion-strengthened (ODS) alloys. Table 75 lists a number of commercial ODS alloys. Yttria is also an additive to ceramic compositions, most notably zirconia. See also *dispersion-strengthened material*, *mechanical alloying*, and *zirconia* (Technical Brief 63).

**Table 75  Oxide-dispersion-strengthened alloys**

| Designation | Alloy type | Rare earth oxide | Amount, wt% | Remarks |
|---|---|---|---|---|
| MA 956 | Fe superalloy | $Y_2O_3$ | 0.5 | . . . |
| Haynes 8077 | Ni superalloy | $Y_2O_3$ | 1.0 | Developmental alloy |
| IN 853 | Ni superalloy | $Y_2O_3$ | 1.2 | Corrosion resistance |
| MA 753 | Ni superalloy | $Y_2O_3$ | 1.3 | |
| MA 754 | Ni superalloy | $Y_2O_3$ | 0.6 | High-temperature alloy |
| MA 758 | Ni superalloy | $Y_2O_3$ | 0.6 | Resistant to molten glass |
| MA 953 | Ni superalloy | $La_2O_3$ | 0.9 | . . . |
| MA 957 | Ni superalloy | $Y_2O_3$ | 0.25 | Intermediate-temperature alloy |
| MA 6000 | Ni superalloy | $Y_2O_3$ | 1.1 | Creep resistance |

# Z

**ZAF corrections**. A quantitative x-ray program that corrects for atomic number ($Z$), absorption ($A$), and fluorescence ($F$) effects in a matrix.

**z-axis**. In laminates, the reference axis normal to the *xy* plane of the laminate.

**Zeeman effect**. A splitting of a degenerate electron energy level into states of slightly different energies in the presence of an external magnetic field. This effect is useful for background correction in *atomic absorption spectrometry*.

**zero bleed**. A laminate fabrication procedure that does not allow loss of resin during cure. Also describes prepreg made with the amount of resin desired in the final part, such that no resin has to be removed during cure. See also *laminate* and *prepreg*.

**zero time**. The time at which the given loading or constraint conditions are initially obtained in creep and stress-relaxation tests, respectively.

**zeta potential**. See *electrokinetic potential*.

## Technical Brief 62: Zinc and Zinc Alloys

ZINC, its alloys, and its chemical compounds represent the fourth most industrially utilized metal (behind iron, aluminum, and copper). Zinc is used in five principal areas of application: in coatings and anodes for corrosion protection of irons and steels, in zinc casting alloys, as an alloying element in copper, aluminum, magnesium, and other alloys, in wrought zinc alloys, and in zinc chemicals.

The use of zinc as a coating to protect iron and steel from corrosion is the largest single application for zinc worldwide. Metallic zinc coatings are applied to steels from a molten metal bath (hot dip galvanizing), by electrochemical means (electrogalvanizing), from a spray of molten metal (metallizing), and in the form of zinc powder by chemical/mechanical means (mechanical galvanizing). Zinc coatings

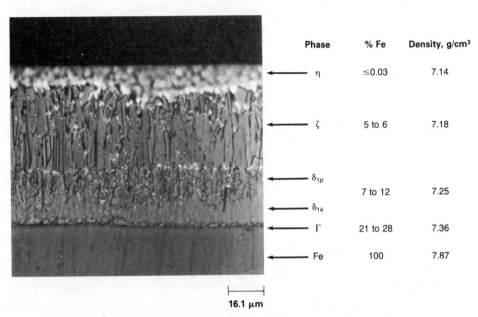

| Phase | % Fe | Density, g/cm³ |
|---|---|---|
| η | ≤0.03 | 7.14 |
| ζ | 5 to 6 | 7.18 |
| δ$_{1p}$ | 7 to 12 | 7.25 |
| δ$_{1k}$ | | |
| Γ | 21 to 28 | 7.36 |
| Fe | 100 | 7.87 |

16.1 μm

*Typical hot dip galvanized coating. Note the gradual transition from layer to layer, which results in a strong bond between base metal and coating.*

are applied to many different types of products, ranging in size from small fasteners to continuous strip to large structural shapes and assemblies. The hot dip galvanizing industry is currently the largest consumer of zinc in the coatings field.

Almost all of the zinc used in zinc casting alloys is employed in die casting compositions. Two alloy groups—hypoeutectic and hypereutectic—make up zinc alloy castings. The hypoeutectic alloys contain about 4% Al. Of these, Alloy 3 (Zn-4Al-0.4Mg) is the most commonly used. Alloy 7 (Zn-4Al-0.015Mg) is a modification of this alloy. Alloy 5 (Zn-4Al-1Cu-0.05Mg) is used when higher tensile strength and/or hardness is required. The hypereutectic alloys have higher aluminum contents (>5% Al) and are used for high-performance applications. These alloys include ZA-8 (Zn-8Al-1Cu-0.02Mg), ZA-12 (Zn-11Al-1Cu-0.025Mg), and ZA-27 (Zn-27Al-2Cu-0.015Mg). Zinc castings are used extensively in the transportation industry for parts such as carburetors, fuel pump bodies, wiper parts, speedometer frames, heater components, radio bodies, instrument panels, and body moldings. Zinc castings are also used extensively in electronic and electrical fittings of all kinds as well as for hardware used in the computer industry, in business machines, and in such items as recording machines and cameras.

Among zinc-containing alloys, copper-base alloys such as brasses are the largest zinc consumers. Rolled zinc is the principal form in which wrought products are supplied, although drawn zinc wire for metallizing is showing increasing usage. In the zinc-chemical category, zinc oxide is the major compound utilized.

## Recommended Reading

- R.J. Barnhurst, Zinc and Zinc Alloys, *Metals Handbook*, 10th ed., Vol 2, ASM International, 1990, p 527-542
- D.C.H. Nevison, Cast Zinc and Zinc Alloys, *Metals Handbook*, 9th ed., Vol 15, 1988, p 786-797
- D.C.H. Nevison, Corrosion of Zinc, *Metals Handbook*, 9th ed., Vol 13, 1987, p 755-769

## Technical Brief 63: Zirconia

ZIRCONIA ($ZrO_2$) is a heavy white powder that forms in nature as the mineral baddeleyite. Most ceramic-grade zirconium, however, is produced from zircon, which is the naturally occurring zirconium silicate sand ($ZrO_2 \cdot SiO_2$). Zirconia and zirconia-bearing oxides are utilized widely in the metallurgical and high-temperature chemical engineering industries because of their refractory nature and their relative inertness to many hostile environments. Zirconia-base solid solutions also exhibit oxygen ion conduction, and this property has been exploited in solid-state electrochemical oxygen monitors.

Zirconia exhibits three defined polymorphs— the monoclinic, tetragonal, and cubic phases. The monoclinic phase is stable up to about

### Typical properties of zirconia-base ceramics

| Material | Bulk density, $g/cm^3$ | Flexure strength MPa | Flexure strength ksi | Fracture toughness $MPa\sqrt{m}$ | Fracture toughness $ksi\sqrt{in.}$ | Hardness, GPa | Hardness, $10^6$ psi | Elastic modulus GPa | Elastic modulus $10^6$ psi |
|---|---|---|---|---|---|---|---|---|---|
| ZTA | 4.1–4.3 | 600–700 | 87–101 | 5–8 | 4.6–7.3 | 15–16 | 2–2.3 | 330–360 | 48–52 |
| Mg-PSZ | 5.7–5.8 | 600–700 | 87–101 | 11–14 | 10–13 | 12 | 1.7 | 210 | 30 |
| Y-TZP | 6.1 | 900–1200 | 130–174 | 8–9 | 7.3–8.2 | 12 | 1.7 | 210 | 30 |

1170 °C (2140 °F) where it transforms to the tetragonal phase, which is stable up to 2370 °C (4300 °F), while the cubic phase exists up to the melting point of 2680 °C (4855 °F). Of greatest significance is the tetragonal to monoclinic transformation, unusual in that on cooling through the transformation temperature (the monoclinic), it is associated with a large volume change (3 to 5%). This is sufficient to exceed elastic and fracture limits even in relatively small grains of $ZrO_2$ and can only be accommodated by cracking. Consequently, the fabrication of large components of pure $ZrO_2$ is not possible due to spontaneous failure upon cooling from the sintering temperature. Additives such as calcia (CaO), magnesia (MgO), yttria ($Y_2O_3$), or ceria ($CeO_2$) must be mixed with $ZrO_2$ to stabilize the material in either the tetragonal or cubic phase. Cubic-stabilized zirconia (CSZ) is generally used as a refractory and can be fabricated as ceramic bricks, foam, or wool.

While the CSZ materials have long been established, transformation-toughened zirconia (TTZ) is a relatively new material class. Transformation-toughened zirconia is a generic term applied to stabilized zirconia systems in which the tetragonal symmetry is retained as the primary zirconia phase. The two most popular tetragonal phase stabilizers are $Y_2O_3$ and MgO. Materials containing these additives are referred to as partially stabilized zirconia. MgO-stabilized $ZrO_2$ (Mg-PSZ) has enjoyed immense commercial success. Its combination of moderately high strength, high fracture toughness, and flaw tolerance enables the use of Mg-PSZ in the most demanding structural ceramic applications. Among the applications for this material are extrusion nozzles in steel production, wire-drawing cap stands, and compacting dies. Yttria-stabilized $ZrO_2$ (Y-TZP) is a fine-grain, high-strength material with moderate fracture toughness. Nearly 100% of the $ZrO_2$ is in the tetragonal symmetry, and the average grain size is about 0.6 to 0.8 µm. Among the applications for Y-TZP are ferrules for fiber-optic assemblies. The fine-grain microstructure and good mechanical properties lend Y-TZP as a candidate material for knife-edge applications, including scissors, slitter blades, and scalpels.

A third material of interest is zirconia-toughened alumina (ZTA), which is the generic term applied to alumina-zirconia systems where alumina is considered the primary or continuous (70 to 95%) phase. Zirconia particulate additions from 5 to 30% represent the second phase. ZTA is a material of interest primarily because it has a significantly higher strength and fracture toughness than alumina. ZTA compositions have been used in some cutting tool applications.

### Recommended Reading

- R. Stevens, Engineering Properties of Zirconia, *Engineered Materials Handbook*, Vol 4, ASM International, 1991, p 775-786
- G.L. DePoorter, T.K. Brog, and M.J. Readey, Structural Ceramics, *Metals Handbook*, 10th ed., Vol 2, ASM International, 1990, p 1019-1024

**Ziegler-Natta catalysts.** Initially, a catalyst consisting of an alkylaluminum compound with a compound of the titanium group of the periodic table, a typical combination being triethylaluminum and either titanium tetrachloride or titanium trichloride. Subsequently, an enormous variety of such mixtures is used in polymerization to provide stereospecificity (isotactic os syndiotactic). See also *isotactic stereoisomerism* and *syndiotactic stereoisomerism*.

## Technical Brief 64: Zirconium and Zirconium Alloys

ZIRCONIUM was discovered in 1789, but it was not isolated as a metal until 1824, when an impure zirconium metal powder was produced by the reduction of potassium fluorozirconate with potassium. In 1925, a purified metal was produced using the iodide decomposition process. This process is still used today to purify zirconium metal extracted from its ores. In 1947, the magnesium reduction method for extracting the metal from zirconium tetrachloride was developed at the U.S. Bureau of Mines by W.J. Kroll.

Although zirconium is sometimes described as an exotic or rare element, it is in fact plentiful. It is ranked 19th in abundance of the chemical elements occurring in the earth's crust, and it is more abundant than many common metals, such as nickel, chromium, and cobalt. The most important source for zirconium is zircon ($ZrSiO_4$), which occurs in several regions throughout the world in the form of beach sand.

Zirconium has excellent resistance to many corrosive media, including superheated water, and it is transparent to thermal energy neutrons. These properties prompted the U.S. Navy to use zirconium in water-cooled nuclear reactors as cladding for uranium fuel. In 1958, zirconium became available for industrial use and began to supplant stainless steel as a fuel cladding in commercial power station nuclear reactors. Also, the chemical processing industries began to use zirconium in several severe corrosion environments. Today, nuclear applications account for a large portion of all zirconium consumed. The excellent corrosion resistance of zirconium to strong acids and alkalies, salts, seawater, and other agents has attracted increasing attention for applications in chemical processing equipment. Zirconium is used as a getter in vacuum tubes and in the manufacture of such diverse items as surgical appliances, photoflash bulbs, and explosive primers. Along with niobium, zirconium is superconductive at low temperatures and is used to make superconductive magnets.

Zirconium forms intermetallic compounds with most metallic elements, and only a limited number of alloys have been developed. For nuclear service, it is desirable to have zirconium alloys with improved strength and corrosion resistance in high-temperature water or steam. The most common alloys—Zircaloy-2 and Zircaloy-4—contain tin, iron, chromium, and nickel. The other alloys of commercial importance are Zr-2.5Nb and Zr-1Nb. The nonnuclear alloys, which have mechanical properties similar to the nuclear grades, contain up to 4.5% Hf. Because hafnium is a neutron absorber, it is not added to zirconium alloys used for nuclear applications. Zirconium alloys are supplied in both wrought and cast forms.

### Compositions and tensile properties for nuclear grades of zirconium

| Grade | UNS No. | Sn | Fe | Cr | Ni | Nb | O(a) | Minimum tensile strength MPa | ksi | Minimum 0.2% yield strength MPa | ksi | Elongation in 50 mm (2 in.), % |
|---|---|---|---|---|---|---|---|---|---|---|---|---|
| Unalloyed reactor grade | R60001 | ... | ... | ... | ... | ... | 0.8 | 290 | 42 | 138 | 20 | 25 |
| Zircaloy-2 | R60802 | 1.4 | 0.1 | 0.1 | 0.05 | ... | 0.12 | 413 | 60 | 241 | 35 | 20 |
| Zircaloy-4 | R60804 | 1.4 | 0.2 | 0.1 | ... | ... | 0.12 | 413 | 60 | 241 | 35 | 20 |
| Zr-2.5Nb | R60901 | ... | ... | ... | ... | 2.6 | 0.14 | 448 | 65 | 310 | 45 | 20 |

(a) Typical content

### Recommended Reading

- R.T. Webster, Zirconium and Hafnium, *Metals Handbook*, 10th ed., Vol 2, ASM International, 1990, p 661-669
- T.L. Yau and R.T. Webster, Corrosion of Zirconium and Hafnium, *Metals Handbook*, 9th ed., Vol 13, ASM International, 1987, p 707-721

**zinc and zinc alloys.** See *Technical Brief 62*.

**zinc worms.** Surface imperfections, characteristic of high-zinc brass castings, that occur when zinc vapor condenses at the mold/metal interface, where it is oxidized and then becomes entrapped in the solidifying metals.

**zircon.** The mineral zircon silicate ($ZrSiO_4$), a very high melting point acid refractory material used as a molding sand, and in ceramic products such as *refractories* and *whiteware*.

**zirconia.** See *Technical Brief 63*.

**zirconium and zirconium alloys.** See *Technical Brief 64*.

**zincrometal.** A steel coil-coated product consisting of a mixed-oxide underlayer containing zinc particles and a zinc-rich organic (epoxy) topcoat. It is weldable, form-able, paintable, and compatible with commonly used adhesives. Zincrometal is used to protect outer body door panels in automobiles from corrosion.

**zone.** Any group of crystal planes that are all parallel to one line, which is called the zone axis.

**zone melting.** Highly localized melting, usually by induction heating, of a small volume of an otherwise solid metal piece, usually a metal rod. By moving the induction coil along the rod, the melted zone can be transferred from one end to the other. In a binary mixture where there is a large difference in composition on the liquidus and solidus lines, high purity can be attained by concentrating one of the constituents in the liquid as it moves along the rod.

**zone sintering.** Highly localized, progressive heating during sintering to produce a desired grain structure, such as grain orientation, and directional properties without subsequent working.

# Bibliography

- *ASM Handbook*, Vol 4, *Heat Treating*, ASM International, 1991
- *ASM Handbook*, Vol 18, *Friction, Lubrication, and Wear Technology*, ASM International, 1992
- C.M. Bower, Ed., *Glossary to the Science of Composites*, Monsanto Company and Washington State University, for Advanced Research Projects Agency, Department of Defense, 1965-1967
- T.G. Byrer, S.L. Semiatin, and D.C. Vollmer, Ed., *Forging Handbook*, Forging Industry Association and American Society for Metals, 1985
- *Compilation of ASTM Standard Definitions*, 7th ed., ASTM, 1990
- *Concise Encyclopedia of Science and Technology*, McGraw-Hill, 1984
- *Electronic Materials Handbook*, Vol 1, *Packaging*, ASM International, 1989
- *Engineered Materials Handbook*, Vol 1, *Composites*, ASM International, 1987
- *Engineered Materials Handbook*, Vol 2, *Engineering Plastics*, ASM International, 1988
- *Engineered Materials Handbook*, Vol 3, *Adhesives and Sealants*, ASM International, 1990
- *Engineered Materials Handbook*, Vol 4, *Ceramics and Glasses*, ASM International, 1991
- B. LeWark, Sr., *Composites Glossary*, Society of Manufacturing Engineers, 1985
- A.D. Merriman, *A Dictionary of Metallurgy*, Pitman Publishing, 1958
- *Metals Handbook*, 10th ed., Vol 1, *Properties and Selection: Irons, Steels, and High-Performance Alloys*, ASM International, 1990
- *Metals Handbook*, 10th ed., Vol 2, *Properties and Selection: Nonferrous Alloys and Special-Purpose Materials*, ASM International, 1990
- *Metals Handbook*, 9th ed., Vol 5, *Surface Cleaning, Finishing, and Coating*, American Society for Metals, 1982
- *Metals Handbook*, 9th ed., Vol 6, *Welding, Brazing, and Soldering*, American Society for Metals, 1983
- *Metals Handbook*, 9th ed., Vol 7, *Powder Metallurgy*, American Society for Metals, 1984
- *Metals Handbook*, 9th ed., Vol 8, *Mechanical Testing*, American Society for Metals, 1985
- *Metals Handbook*, 9th ed., Vol 9, *Metallography and Microstructures*, American Society for Metals, 1985
- *Metals Handbook*, 9th ed., Vol 10, *Materials Characterization*, American Society for Metals, 1986
- *Metals Handbook*, 9th ed., Vol 11, *Failure Analysis and Prevention*, American Society for Metals, 1986
- *Metals Handbook*, 9th ed., Vol 12, *Fractography*, ASM International, 1987
- *Metals Handbook*, 9th ed., Vol 13, *Corrosion*, ASM International, 1987
- *Metals Handbook*, 9th ed., Vol 14, *Forming and Forging*, ASM International, 1988
- *Metals Handbook*, 9th ed., Vol 15, *Casting*, ASM International, 1988
- *Metals Handbook*, 9th ed., Vol 16, *Machining*, ASM International, 1989
- *Metals Handbook*, 9th ed., Vol 17, *Nondestructive Evaluation and Quality Control*, ASM International, 1989
- *Metals Handbook*, 8th ed., Vol 1, *Properties and Selection of Metals*, American Society for Metals, 1961
- *Metals Handbook, Desk Edition*, American Society for Metals, 1985
- *NACE Glossary of Corrosion Related Terms*, National Association of Corrosion Engineers, 1985
- S.P. Parker, Ed., *McGraw-Hill Dictionary of Scientific and Technical Terms*, 4th ed., 1989
- W.W. Perkins, Ed., *Ceramic Glossary*, The American Ceramic Society, 1984
- *Standard Welding Terms and Definitions*, ANSI/AWS A30-89, American Welding Society, 1989
- L.R. Whittington, *Whittington's Dictionary of Plastics*, 2nd ed., Technomic, 1978

# Appendix

## SI prefixes—names and symbols

| Exponential expression | Multiplication factor | Prefix | Symbol |
|---|---|---|---|
| $10^{18}$ | 1 000 000 000 000 000 000 | exa | E |
| $10^{15}$ | 1 000 000 000 000 000 | peta | P |
| $10^{12}$ | 1 000 000 000 000 | tera | T |
| $10^{9}$ | 1 000 000 000 | giga | G |
| $10^{6}$ | 1 000 000 | mega | M |
| $10^{3}$ | 1 000 | kilo | K |
| $10^{2}$ | 100 | hecto(a) | h |
| $10^{1}$ | 10 | deka(a) | da |
| $10^{0}$ | 1 | BASE UNIT | |
| $10^{-1}$ | 0.1 | deci(a) | d |
| $10^{-2}$ | 0.01 | centi(a) | c |
| $10^{-3}$ | 0.001 | milli | m |
| $10^{-6}$ | 0.000 001 | micro | μ |
| $10^{-9}$ | 0.000 000 001 | nano | n |
| $10^{-12}$ | 0.000 000 000 001 | pico | p |
| $10^{-15}$ | 0.000 000 000 000 001 | femto | f |
| $10^{-18}$ | 0.000 000 000 000 000 001 | atto | a |

(a) Nonpreferred. Prefixes should be selected in steps of $10^3$ so that the resultant number before the prefix is between 0.1 and 1000. These prefixes should not be used for units of linear measurement, but may be used for higher order units. For example, the linear measurement, decimeter, is nonpreferred, but square decimeter is acceptable.

## Base, supplementary, and derived SI units

| Measure | Unit | Symbol |
|---|---|---|
| **Base units** | | |
| Amount of substance | mole | mol |
| Electric current | ampere | A |
| Length | meter | m |
| Luminous intensity | candela | cd |
| Mass | kilogram | kg |
| Thermodynamic temperature | kelvin | K |
| Time | second | s |
| **Supplementary units** | | |
| Plane angle | radian | rad |
| Solid angle | steradian | sr |
| **Derived units** | | |
| Absorbed dose | gray | Gy |
| Acceleration | meter per second squared | $m/s^2$ |
| Activity (of radionuclides) | becquerel | Bq |
| Angular acceleration | radian per second squared | $rad/s^2$ |
| Angular velocity | radian per second | rad/s |
| Area | square meter | $m^2$ |
| Capacitance | farad | F |
| Concentration (of amount of substance) | mole per cubic meter | $mol/m^3$ |
| Conductance | siemens | S |
| Current density | ampere per square meter | $A/m^2$ |
| Density, mass | kilogram per cubic meter | $kg/m^3$ |
| Electric charge density | coulomb per cubic meter | $C/m^3$ |
| Electric field strength | volt per meter | V/m |
| Electric flux density | coulomb per square meter | $C/m^2$ |
| Electric potential, potential difference, electromotive force | volt | V |
| Electric resistance | ohm | Ω |
| Energy, work, quantity of heat | joule | J |
| Energy density | joule per cubic meter | $J/m^3$ |

| Measure | Unit | Symbol |
|---|---|---|
| Entropy | joule per kelvin | J/K |
| Force | newton | N |
| Frequency | hertz | Hz |
| Heat capacity | joule per kelvin | J/K |
| Heat flux density | watt per square meter | $W/m^2$ |
| Illuminance | lux | lx |
| Inductance | henry | H |
| Irradiance | watt per square meter | $W/m^2$ |
| Luminance | candela per square meter | $cd/m^2$ |
| Luminous flux | lumen | lm |
| Magnetic field strength | ampere per meter | A/m |
| Magnetic flux | weber | Wb |
| Magnetic flux density | tesla | T |
| Molar energy | joule per mole | J/mol |
| Molar entropy | joule per mole kelvin | J/mol · K |
| Molar heat capacity | joule per mole kelvin | J/mol · K |
| Moment of force | newton meter | N · m |
| Permeability | henry per meter | H/m |
| Permittivity | farad per meter | F/m |
| Power, radiant flux | watt | W |
| Pressure, stress | pascal | Pa |
| Quantity of electricity, electric charge | coulomb | C |
| Radiance | watt per square meter steradian | $W/m^2$ · sr |
| Radiant intensity | watt per steradian | W/sr |
| Specific heat capacity | joule per kilogram kelvin | J/kg · K |
| Specific energy | joule per kilogram | J/kg |
| Specific entropy | joule per kilogram kelvin | J/kg · K |
| Specific volume | cubic meter per kilogram | $m^3/kg$ |
| Surface tension | newton per meter | N/m |
| Thermal conductivity | watt per meter kelvin | W/m · K |
| Velocity | meter per second | m/s |
| Viscosity, dynamic | pascal second | Pa · s |
| Viscosity, kinematic | square meter per second | $m^2/s$ |
| Volume | cubic meter | $m^3$ |
| Wavenumber | 1 per meter | l/m |

# Appendix

## Conversion factors

| To convert from | to | multiply by |
|---|---|---|
| **Angle** | | |
| degree | rad | 1.745 329 E − 02 |
| **Area** | | |
| in.$^2$ | mm$^2$ | 6.451 600 E + 02 |
| in.$^2$ | cm$^2$ | 6.451 600 E + 00 |
| in.$^2$ | m$^2$ | 6.451 600 E − 04 |
| ft$^2$ | m$^2$ | 9.290 304 E − 02 |
| **Bending moment or torque** | | |
| lbf · in. | N · m | 1.129 848 E − 01 |
| lbf · ft | N · m | 1.355 818 E + 00 |
| kgf · m | N · m | 9.806 650 E + 00 |
| ozf · in. | N · m | 7.061 552 E − 03 |
| **Bending moment or torque per unit length** | | |
| lbf · in./in. | N · m/m | 4.448 222 E + 00 |
| lbf · ft/in. | N · m/m | 5.337 866 E + 01 |
| **Current density** | | |
| A/in.$^2$ | A/cm$^2$ | 1.550 003 E − 01 |
| A/in.$^2$ | A/mm$^2$ | 1.550 003 E − 03 |
| A/ft$^2$ | A/m$^2$ | 1.076 400 E + 01 |
| **Electricity and magnetism** | | |
| gauss | T | 1.000 000 E − 04 |
| maxwell | μWb | 1.000 000 E − 02 |
| mho | S | 1.000 000 E + 00 |
| Oersted | A/m | 7.957 700 E + 01 |
| Ω · cm | Ω · m | 1.000 000 E − 02 |
| Ω circular-mil/ft | μΩ · m | 1.662 426 E − 03 |
| **Energy (impact, other)** | | |
| ft · lbf | J | 1.355 818 E + 00 |
| Btu (thermochemical) | J | 1.054 350 E + 03 |
| cal (thermochemical) | J | 4.184 000 E + 00 |
| kW · h | J | 3.600 000 E + 06 |
| W · h | J | 3.600 000 E + 03 |
| **Flow rate** | | |
| ft$^3$/h | L/min | 4.719 475 E − 01 |
| ft$^3$/min | L/min | 2.831 000 E + 01 |
| gal/h | L/min | 6.309 020 E − 02 |
| gal/min | L/min | 3.785 412 E + 00 |
| **Force** | | |
| lbf | N | 4.448 222 E + 00 |
| kip (1000 lbf) | N | 4.448 222 E + 03 |
| tonf | kN | 8.896 443 E + 00 |
| kgf | N | 9.806 650 E + 00 |
| **Force per unit length** | | |
| lbf/ft | N/m | 1.459 390 E + 01 |
| lbf/in. | N/m | 1.751 268 E + 02 |
| **Fracture toughness** | | |
| ksi $\sqrt{\text{in.}}$ | MPa$\sqrt{\text{m}}$ | 1.098 800 E + 00 |
| **Heat content** | | |
| Btu/lb | kJ/kg | 2.326 000 E + 00 |
| cal/g | kJ/kg | 4.186 800 E + 00 |

| To convert from | to | multiply by |
|---|---|---|
| **Heat input** | | |
| J/in. | J/m | 3.937 008 E + 01 |
| kJ/in. | kJ/m | 3.937 008 E + 01 |
| **Length** | | |
| Å | nm | 1.000 000 E − 01 |
| μin. | μm | 2.540 000 E − 02 |
| mil | μm | 2.540 000 E + 01 |
| in. | mm | 2.540 000 E + 01 |
| in. | cm | 2.540 000 E + 00 |
| ft | m | 3.048 000 E − 01 |
| yd | m | 9.144 000 E − 01 |
| mile | km | 1.609 300 E + 00 |
| **Mass** | | |
| oz | kg | 2.834 952 E − 02 |
| lb | kg | 4.535 924 E − 01 |
| ton (short, 2000 lb) | kg | 9.071 847 E + 02 |
| ton (short, 2000 lb) | kg × 10$^3$(a) | 9.071 847 E − 01 |
| ton (long, 2240 lb) | kg | 1.016 047 E + 03 |
| **Mass per unit area** | | |
| oz/in.$^2$ | kg/m$^2$ | 4.395 000 E + 01 |
| oz/ft$^2$ | kg/m$^2$ | 3.051 517 E − 01 |
| oz/yd$^2$ | kg/m$^2$ | 3.390 575 E − 02 |
| lb/ft$^2$ | kg/m$^2$ | 4.882 428 E + 00 |
| **Mass per unit length** | | |
| lb/ft | kg/m | 1.488 164 E + 00 |
| lb/in. | kg/m | 1.785 797 E + 01 |
| **Mass per unit time** | | |
| lb/h | kg/s | 1.259 979 E − 04 |
| lb/min | kg/s | 7.559 873 E − 03 |
| lb/s | kg/s | 4.535 924 E − 01 |
| **Mass per unit volume (includes density)** | | |
| g/cm$^3$ | kg/m$^3$ | 1.000 000 E + 03 |
| lb/ft$^3$ | g/cm$^3$ | 1.601 846 E − 02 |
| lb/ft$^3$ | kg/m$^3$ | 1.601 846 E + 01 |
| lb/in.$^3$ | g/cm$^3$ | 2.767 990 E + 01 |
| lb/in.$^3$ | kg/m$^3$ | 2.767 990 E + 04 |
| **Power** | | |
| Btu/s | kW | 1.055 056 E + 00 |
| Btu/min | kW | 1.758 426 E − 02 |
| Btu/h | W | 2.928 751 E − 01 |
| erg/s | W | 1.000 000 E − 07 |
| ft · lbf/s | W | 1.355 818 E + 00 |
| ft · lbf/min | W | 2.259 697 E − 02 |
| ft · lbf/h | W | 3.766 161 E − 04 |
| hp (550 ft · lbf/s) | kW | 7.456 999 E − 01 |
| hp (electric) | kW | 7.460 000 E − 01 |
| **Power density** | | |
| W/in.$^2$ | W/m$^2$ | 1.550 003 E + 03 |
| **Pressure (fluid)** | | |
| atm (standard) | Pa | 1.013 250 E + 05 |
| bar | Pa | 1.000 000 E + 05 |
| in. Hg (32 °F) | Pa | 3.386 380 E + 03 |
| in. Hg (60 °F) | Pa | 3.376 850 E + 03 |
| lbf/in.$^2$ (psi) | Pa | 6.894 757 E + 03 |
| torr (mm Hg, 0 °C) | Pa | 1.333 220 E + 02 |

| To convert from | to | multiply by |
|---|---|---|
| **Specific heat** | | |
| Btu/lb · °F | J/kg · K | 4.186 800 E + 03 |
| cal/g · °C | J/kg · K | 4.186 800 E + 03 |
| **Stress (force per unit area)** | | |
| tonf/in.$^2$(tsi) | MPa | 1.378 951 E + 01 |
| kgf/mm$^2$ | MPa | 9.806 650 E + 00 |
| ksi | MPa | 6.894 757 E + 00 |
| lbf/in.$^2$ (psi) | MPa | 6.894 757 E − 03 |
| MN/m$^2$ | MPa | 1.000 000 E + 00 |
| **Temperature** | | |
| °F | °C | 5/9 · (°F − 32) |
| °R | °K | 5/9 |
| **Temperature interval** | | |
| °F | °C | 5/9 |
| **Thermal conductivity** | | |
| Btu · in./s · ft$^2$ · °F | W/m · K | 5.192 204 E + 02 |
| Btu/ft · h · °F | W/m · K | 1.730 735 E + 00 |
| Btu · in./h · ft$^2$ · °F | W/m · K | 1.442 279 E − 01 |
| cal/cm · s · °C | W/m · K | 4.184 000 E + 02 |
| **Thermal expansion** | | |
| in./in. · °C | m/m · K | 1.000 000 E + 00 |
| in./in. · °F | m/m · K | 1.800 000 E + 00 |
| **Velocity** | | |
| ft/h | m/s | 8.466 667 E − 05 |
| ft/min | m/s | 5.080 000 E − 03 |
| ft/s | m/s | 3.048 000 E − 01 |
| in./s | m/s | 2.540 000 E − 02 |
| km/h | m/s | 2.777 778 E − 01 |
| mph | km/h | 1.609 344 E + 00 |
| **Velocity of rotation** | | |
| rev/min (rpm) | rad/s | 1.047 164 E − 01 |
| rev/s | rad/s | 6.283 185 E + 00 |
| **Viscosity** | | |
| poise | Pa · s | 1.000 000 E + 01 |
| strokes | m$^2$/s | 1.000 000 E − 04 |
| ft$^2$/s | m$^2$/s | 9.290 304 E − 02 |
| in.$^2$/s | mm$^2$/s | 6.451 600 E + 02 |
| **Volume** | | |
| in.$^3$ | m$^3$ | 1.638 706 E − 05 |
| ft$^3$ | m$^3$ | 2.831 685 E − 02 |
| fluid oz | m$^3$ | 2.957 353 E − 05 |
| gal (U.S. liquid) | m$^3$ | 3.785 412 E − 03 |
| **Volume per unit time** | | |
| ft$^3$/min | m$^3$/s | 4.719 474 E − 04 |
| ft$^3$/s | m$^3$/s | 2.831 685 E − 02 |
| in.$^3$/min | m$^3$/s | 2.731 177 E − 07 |
| **Wavelength** | | |
| Å | nm | 1.000 000 E − 01 |

(a) kg × 10$^3$ = 1 metric ton

## Mathematical signs and symbols

| Symbol | Definition | Symbol | Definition | Symbol | Definition |
|---|---|---|---|---|---|
| + | plus (sign of addition); positive | $\equiv$ | identical with | $\pi$ | (pi) = 3.14159+ |
| − | minus (sign of subtraction); negative | ~ | similar to; approximately | ° | degrees |
| ± (∓) | plus or minus (minus or plus) | $\approx$ | approximately equals | ′ | minutes |
| × | times, by (multiplication sign); diameters (magnification) | $\cong$ | approximately equals; congruent | ″ | seconds |
| · | multiplied by | $\leq$ | equal to or less than | $\angle$ | angle |
| ÷ | sign of division | $\geq$ | equal to or greater than | $dx$ | differential of $x$ |
| / | divided by; per | $\neq$ | not equal to | $\Delta$ | (delta) difference |
| : | ratio sign; divided by; is to | $\rightarrow \doteq$ | approaches | $\partial$ | partial derivative |
| :: | equals; as (proportion) | $\propto$ | varies as; is proportional to | $\int$ | integral of |
| < | less than | $\infty$ | infinity | $\oint$ | line integral around a closed path |
| > | greater than | $\sqrt{}$ | square root of | $\Sigma$ | (sigma) summation of |
| << | much less than | $\|$ | parallel to | $\nabla$ | del or nabla; vector differential operator |
| >> | much greater than | ( ) [ ] { } | parentheses, brackets, and braces; quantities enclosed by them to be taken together in multiplying, dividing, etc. | $|x|$ | absolute value of $x$ |
| = | equals | | | | |

## Greek alphabet

| Upper and lower cases | Name | Upper and lower cases | Name | Upper and lower cases | Name | Upper and lower cases | Name |
|---|---|---|---|---|---|---|---|
| A α | Alpha | H η | Eta | N ν | Nu | T τ | Tau |
| B β | Beta | Θ θ ϑ | Theta | Ξ ξ | Xi | Υ υ | Upsilon |
| Γ γ | Gamma | I ι | Iota | O o | Omicron | Φ φ φ | Phi |
| Δ δ ∂ | Delta | K κ | Kappa | Π π | Pi | X χ | Chi |
| E ε | Epsilon | Λ λ | Lambda | P ρ | Rho | Ψ ψ | Psi |
| Z ζ | Zeta | M μ | Mu | Σ σ ς | Sigma | Ω ω | Omega |

## Symbols and atomic numbers for the chemical elements(a)

| Name | Symbol | Atomic No. | Name | Symbol | Atomic No. | Name | Symbol | Atomic No. | Name | Symbol | Atomic No. |
|---|---|---|---|---|---|---|---|---|---|---|---|
| Actinium | Ac | 89 | Erbium | Er | 68 | Mendelevium | Md | 101 | Ruthenium | Ru | 44 |
| Aluminum | Al | 13 | Europium | Eu | 63 | Mercury | Hg | 80 | Rutherfordium | Rf | 104 |
| Americium | Am | 95 | Fermium | Fm | 100 | Molybdenum | Mo | 42 | Samarium | Sm | 62 |
| Antimony | Sb | 51 | Fluorine | F | 9 | Neodymium | Nd | 60 | Scandium | Sc | 21 |
| Argon | Ar | 18 | Francium | Fr | 87 | Neon | Ne | 10 | Selenium | Se | 34 |
| Arsenic | As | 33 | Gadolinium | Gd | 64 | Neptunium | Np | 93 | Silicon | Si | 14 |
| Astatine | At | 85 | Gallium | Ga | 31 | Nickel | Ni | 28 | Silver | Ag | 47 |
| Barium | Ba | 56 | Germanium | Ge | 32 | Niobium | Nb | 41 | Sodium | Na | 11 |
| Berkelium | Bk | 97 | Gold | Au | 79 | Nitrogen | N | 7 | Strontium | Sr | 38 |
| Beryllium | Be | 4 | Hafnium | Hf | 72 | Nobelium | No | 102 | Sulfur | S | 16 |
| Bismuth | Bi | 83 | Hahnium | Ha | 105 | Osmium | Os | 76 | Tantalum | Ta | 73 |
| Boron | B | 5 | Helium | He | 2 | Oxygen | O | 8 | Technetium | Tc | 43 |
| Bromine | Br | 35 | Holmium | Ho | 67 | Palladium | Pd | 46 | Tellurium | Te | 52 |
| Cadmium | Cd | 48 | Hydrogen | H | 1 | Phosphorus | P | 15 | Terbium | Tb | 65 |
| Calcium | Ca | 20 | Indium | In | 49 | Platinum | Pt | 78 | Thallium | Tl | 81 |
| Californium | Cf | 98 | Iodine | I | 53 | Plutonium | Pu | 94 | Thorium | Th | 90 |
| Carbon | C | 6 | Iridium | Ir | 77 | Polonium | Po | 84 | Thulium | Tm | 69 |
| Cerium | Ce | 58 | Iron | Fe | 26 | Potassium | K | 19 | Tin | Sn | 50 |
| Cesium | Cs | 55 | Krypton | Kr | 36 | Praseodymium | Pr | 59 | Titanium | Ti | 22 |
| Chlorine | Cl | 17 | Lanthanum | La | 57 | Promethium | Pm | 61 | Tungsten | W | 74 |
| Chromium | Cr | 24 | Lawrencium | Lr (Lw) | 103 | Protactinium | Pa | 91 | Uranium | U | 92 |
| Cobalt | Co | 27 | Lead | Pb | 82 | Radium | Ra | 88 | Vanadium | V | 23 |
| Copper | Cu | 29 | Lithium | Li | 3 | Radon | Rn | 86 | Xenon | Xe | 54 |
| Curium | Cm | 96 | Lutetium | Lu | 71 | Rhenium | Re | 75 | Ytterbium | Yb | 70 |
| Dysprosium | Dy | 66 | Magnesium | Mg | 12 | Rhodium | Rh | 45 | Yttrium | Y | 39 |
| Einsteinium | Es | 99 | Manganese | Mn | 25 | Rubidium | Rb | 37 | Zinc | Zn | 30 |
| | | | | | | | | | Zirconium | Zr | 40 |

(a) Elements 106, 107, and 109 have been reported, but no official names or symbols have yet been assigned.

# Appendix

## Periodic table of the elements

Metals ← → | Nonmetals ← →

Key to chart:
- Atomic number → **50** +2 ← Oxidation states
- Symbol → **Sn** +4
- Atomic weight → 118.69
- -18-18-4 ← Electron configuration

| Ia | IIa | IIb | IVb | Vb | VIb | VIIb | VIII | | | Ib | IIb | IIIa | IVa | Va | VIa | VIIa | 0 | Orbit |
|---|---|---|---|---|---|---|---|---|---|---|---|---|---|---|---|---|---|---|
| **1** +1<br>**H** −1<br><br>1.0079<br>1 | | | | | | | | | | | | | | | | | **2** 0<br>**He**<br><br>4.00260<br>2 | K |
| **3** +1<br>**Li**<br><br>6.939<br>2-1 | **4** +2<br>**Be**<br><br>9.0122<br>2-2 | | | | | | | | | | | **5** +3<br>**B**<br><br>10.81<br>2-3 | **6** +2<br>**C** +4<br>−4<br>12.011<br>2-4 | **7** +1<br>**N** +2<br>+3<br>+4<br>+5<br>−1<br>−2<br>−3<br>14.0067<br>2-5 | **8** −2<br>**O**<br><br>15.9994<br>2-6 | **9** −1<br>**F**<br><br>18.998403<br>2-7 | **10** 0<br>**Ne**<br><br>10.17₉<br>2-8 | K-L |
| **11** +1<br>**Na**<br><br>22.9898<br>2-8-1 | **12** +2<br>**Mg**<br><br>24.312<br>2-8-2 | Transition elements | | | | | | | | | | **13** +3<br>**Al**<br><br>26.98154<br>2-8-3 | **14** +2<br>**Si** +4<br>−4<br>28.08<br>2-8-4 | **15** +3<br>**P** +5<br>−3<br>30.97376<br>2-8-5 | **16** +4<br>**S** +6<br>−2<br>32.06<br>2-8-6 | **17** +1<br>**Cl** +5<br>+7<br>−1<br>35.453<br>2-8-7 | **18** 0<br>**Ar**<br><br>39.984<br>2-8-8 | K-L-M |
| **19** +1<br>**K**<br><br>39.09<br>-8-8-1 | **20** +2<br>**Ca**<br><br>40.08<br>-8-8-2 | **21** +3<br>**Sc**<br><br>44.9559<br>-8-9-2 | **22** +2<br>**Ti** +3<br>+4<br>47.9<br>-8-10-2 | **23** +2<br>**V** +3<br>+4<br>+5<br>50.941<br>-8-11-2 | **24** +2<br>**Cr** +3<br>+6<br>51.996<br>-8-13-1 | **25** +2<br>**Mn** +3<br>+4<br>+7<br>54.9380<br>-8-13-2 | **26** +2<br>**Fe** +3<br><br>55.847<br>-8-14-2 | **27** +2<br>**Co** +3<br><br>58.9332<br>-8-15-2 | **28** +2<br>**Ni** +3<br><br>58.71<br>-8-16-2 | **29** +1<br>**Cu** +2<br><br>63.54<br>-8-18-1 | **30** +2<br>**Zn**<br><br>65.38<br>-8-18-2 | **31** +3<br>**Ga**<br><br>69.72<br>-8-18-3 | **32** +2<br>**Ge** +4<br><br>72.59<br>-8-18-4 | **33** +3<br>**As** +5<br>−3<br>74.9216<br>-8-18-5 | **34** +4<br>**Se** +6<br>−2<br>78.96<br>-8-18-5 | **35** +1<br>**Br** +5<br>−1<br>79.904<br>-8-18-7 | **36** 0<br>**Kr**<br><br>83.80<br>-8-18-8 | -L-M-N |
| **37** +1<br>**Rb**<br><br>85.467<br>-18-8-1 | **38** +2<br>**Sr**<br><br>87.62<br>-18-8-2 | **39** +3<br>**Y**<br><br>88.9059<br>-18-9-2 | **40** +4<br>**Zr**<br><br>91.22<br>-18-10-2 | **41** +3<br>**Nb** +5<br><br>92.9064<br>-18-12-1 | **42** +6<br>**Mo**<br><br>95.94<br>-18-13-1 | **43** +4<br>**Tc** +6<br>+7<br>98.9062<br>-18-13-2 | **44** +3<br>**Ru**<br><br>101.07<br>-18-15-1 | **45** +3<br>**Rh**<br><br>102.905<br>-18-16-1 | **46** +2<br>**Pd** +4<br><br>106.4<br>-18-18-0 | **47** +1<br>**Ag**<br><br>107.868<br>-18-18-1 | **48** +2<br>**Cd**<br><br>112.40<br>-18-18-2 | **49** +3<br>**In**<br><br>114.82<br>-18-18-3 | **50** +2<br>**Sn** +4<br><br>118.69<br>-18-18-4 | **51** +3<br>**Sb** +5<br>−1<br>121.75<br>-18-18-5 | **52** +4<br>**Te** +6<br>−2<br>127.60<br>-18-18-6 | **53** +1<br>**I** +5<br>+7<br>−3<br>126.9045<br>-18-18-7 | **54** 0<br>**Xe**<br><br>131.30<br>-18-18-8 | -M-N-O |
| **55** +1<br>**Cs**<br><br>132.9054<br>-18-8-1 | **56** +2<br>**Ba**<br><br>137.3<br>-18-8-2 | **57*** +3<br>**La**<br><br>138.9055<br>-18-9-2 | **72** +4<br>**Hf**<br><br>178.49<br>-32-10-2 | **73** +5<br>**Ta**<br><br>180.948<br>-32-11-2 | **74** +6<br>**W**<br><br>183.85<br>-32-12-2 | **75** +4<br>**Re** +6<br>+7<br>186.207<br>-32-13-2 | **76** +3<br>**Os** +4<br><br>190.2<br>-32-14-2 | **77** +3<br>**Ir** +4<br><br>192.9<br>-32-15-2 | **78** +2<br>**Pt** +4<br><br>195.09<br>-32-16-2 | **79** +1<br>**Au** +3<br><br>196.9665<br>-32-18-1 | **80** +1<br>**Hg** +2<br><br>200.59<br>-32-18-2 | **81** +1<br>**Tl** +3<br><br>204.37<br>-32-19-3 | **82** +2<br>**Pb** +4<br><br>207.19<br>-32-18-4 | **83** +3<br>**Bi** +5<br><br>208.980<br>-32-18-5 | **84** +2<br>**Po** +4<br><br>(209)<br>-32-18-6 | **85**<br>**At**<br><br>(210)<br>-32-18-7 | **86** 0<br>**Rn**<br><br>(222)<br>-32-18-8 | -N-O-P |
| **87** +1<br>**Fr**<br><br>(223)<br>-18-8-1 | **88** +2<br>**Ra**<br><br>226.0254<br>-18-8-2 | **89** ** +3<br>**Ac**<br><br>(227)<br>-18-9-2 | **104** +4<br>**Rf**<br><br>(261)<br>-32-10-2 | **105**<br>**Ha**<br><br>(262)<br>-32-11-2 | **106**<br><br>(263)<br>-32-12-2 | | | | | | | | | | | | | -O-P-Q |

| *Lanthanides | **58** +3<br>**Ce** +4<br>140.12<br>-20-8-2 | **59** +3<br>**Pr**<br>140.9077<br>-21-8-2 | **60** +3<br>**Nd**<br>144.24<br>-22-8-2 | **61** +3<br>**Pm**<br>147<br>-23-8-2 | **62** +2<br>**Sm** +3<br>150.4<br>-24-8-2 | **63** +2<br>**Eu** +3<br>151.96<br>-25-8-2 | **64** +3<br>**Gd**<br>157.25<br>-25-9-2 | **65** +3<br>**Tb**<br>158.925<br>-27-8-2 | **66** +3<br>**Dy**<br>162.50<br>-28-8-2 | **67** +3<br>**Ho**<br>164.9304<br>-29-8-2 | **68** +3<br>**Er**<br>167.26<br>-30-8-2 | **69** +3<br>**Tm**<br>168.9342<br>-31-8-2 | **70** +2<br>**Yb** +3<br>173.04<br>-32-8-2 | **71** +3<br>**Lu**<br>174.967<br>-32-9-2 | -N-O-P |
|---|---|---|---|---|---|---|---|---|---|---|---|---|---|---|---|
| **Actinides | **90** +4<br>**Th**<br>232.038<br>-18-10-2 | **91** +5<br>**Pa** +4<br>231.0359<br>-20-9-2 | **92** +3<br>**U** +4<br>+5<br>+6<br>238.029<br>21-9-2 | **93** +3<br>**Np** +4<br>+5<br>+6<br>237.0482<br>-22-9-2 | **94** +3<br>**Pu** +4<br>+5<br>+6<br>239.052<br>-24-8-2 | **95** +3<br>**Am** +4<br>+5<br>+6<br>(243)<br>-25-8-2 | **96** +3<br>**Cm**<br>(247)<br>-25-9-2 | **97** +3<br>**Bk** +4<br>(247)<br>-27-8-2 | **98** +3<br>**Cf**<br>(251)<br>-28-8-2 | **99** +3<br>**Es**<br>(254)<br>-29-8-2 | **100** +3<br>**Fm**<br>(257)<br>-30-8-2 | **101** +2<br>**Md** +3<br>(258)<br>-31-8-2 | **102** +2<br>**No** +3<br>(259)<br>-32-8-2 | **103** +3<br>**Lr**<br>(260)<br>-32-9-2 | -O-P-Q |

Numbers in parentheses are mass numbers of most stable isotope of that element.

# Abbreviations and Symbols

| | |
|---|---|
| $a$ | crack length; crystal lattice length along the $a$ axis; wheel depth of cut in grinding |
| $a_f$ | final cross-sectional area |
| $a_o$ | original cross-sectional area |
| A | amp; ampere; area |
| $A$ | amplitude; area; ratio of the alternating stress amplitude to the mean stress; absorbance |
| Å | angstrom |
| AA | Aluminum Association |
| AAC | air carbon arc cutting |
| AAR | Association of American Railroads |
| AAS | atomic absorption spectroscopy/spectrometry |
| AASHTO | American Association of State Highway Transportation Officials |
| AAW | air acetylene welding |
| AB | arc brazing |
| ABS | acrylonitrile-butadiene-styrene; American Bureau of Shipping |
| ABST | alpha-beta solution treatment |
| ac | alternating current |
| $Ac_1$ | temperature at which austenite begins to form on heating |
| $Ac_3$ | temperature at which transformation of ferrite to austenite is completed on heating |
| $Ac_{cm}$ | in hypereutectoid steel, temperature at which cementite completes solution in austenite |
| AC | acetal; air cooled; adaptive control |
| ACD | annealed cold drawn |
| ACGIH | American Conference of Governmental Industrial Hygienists |
| ACI | Alloy Casting Institute |
| A/D | analog-to-digital converter |
| ADC | analog-to-digital converter |
| ADCI | American Die Casting Institute |
| ADI | austempered ductile iron |
| ADTT | average daily truck traffic |
| $Ae_{cm}$, $Ae_1$, $Ae_3$ | equilibrium transformation temperatures in steel |
| AE | acoustic emission |
| AECL | Atomic Energy of Canada Limited |
| AECMA | Association Européenne des Constructeurs de Matériel Aérospatial |
| AEM | analytical electron microscopy |
| AES | Auger electron spectroscopy; atomic emission spectrometry |
| AFD | automated forging design |
| AFM | abrasive flow machining |
| AFNOR | Association Francaise de Normalisation |
| AFS | American Foundrymen's Society; atomic fluorescence spectrometry |
| AFWAL | Air Force Wright Aeronautical Laboratories |
| AGA | American Gas Association |
| AGMA | American Gear Manufacturers Association |
| AGV | automatic guided vehicle |
| AHW | atomic hydrogen welding |
| AI | artificial intelligence |
| AIA | Aerospace Industries Association |
| AIME | American Institute of Mining, Metallurgical and Petroleum Engineers |
| AIP | American Institute of Physics |
| AISC | American Institute of Steel Construction |
| AISI | American Iron and Steel Institute |
| AJM | abrasive jet machining |
| AKDQ | aluminum-killed drawing quality |
| AKS | aluminum-potassium-silicon |
| ALPID | Analysis of Large Plastic Incremental Deformation (bulk deformation modeling software) |
| AM | analytical modeling |
| AMMRC | Army Materials and Mechanics Research Center |
| AMS | Aerospace Material Specification |
| amu | atomic mass unit |
| ANSI | American National Standards Institute |
| AOAC | Association of Official Analytical Chemists |
| AOC | oxygen arc cutting |
| AOCS | American Oil Chemists' Society |
| AOD | argon oxygen decarburization |
| AP | armor piercing; atom probe |
| APA | 3-aminophenylacetylene |
| APB | 1,3-bis(3-aminophenoxy) benzene; antiphase boundary |
| API | American Petroleum Institute |
| APM | atom probe microanalysis |
| APMI | American Powder Metallurgy Institute |
| APT | ammonium paratungstate |
| APU | auxiliary power unit (space shuttle) |
| AQ | as quenched |
| AQL | acceptable quality levels |
| $Ar_1$ | temperature at which transformation to ferrite or to ferrite plus cementite is completed on cooling |
| $Ar_3$ | temperature at which transformation of austenite to ferrite begins on cooling |

| | | | | |
|---|---|---|---|---|
| $Ar_{cm}$ | temperature at which cementite begins to precipitate from austenite on cooling | | **BWG** | Birmingham wire gage |
| **AREA** | American Railway Engineering Association | | **BWR** | boiling water reactor(s) |
| **ASA** | acrylonitrile-styrene-acrylate terpolymer | | $c$ | crystal lattice length along the $c$ axis; velocity (speed of light) |
| **ASCE** | American Society of Civil Engineers | | | |
| **ASIP** | Aircraft Structural Integrity Program | | **C** | coulomb; cementite |
| **ASM** | American Society for Metals | | $C$ | capacitance; heat capacity |
| **ASME** | American Society of Mechanical Engineers | | **CAB** | calcium argon blowing |
| **ASP** | antisegregation process | | **CAC** | carbon arc cutting |
| **AS/RS** | automatic storage and retrieval systems | | **CAD/CAM** | computer-aided design/computer-aided manufacturing |
| **ASTM** | American Society for Testing and Materials | | | |
| **at.%** | atomic percent | | **CAE** | computer-aided engineering |
| **ATEM** | analytical transmission electron microscope/microscopy | | **cal** | calorie |
| | | | **CAM** | computer-aided-manufacturing |
| **atm** | atmospheres (pressure) | | **CANDU** | Canadian deuterium uranium (reactor) |
| **at.ppm** | atomic parts per million | | **CAP** | consolidation by atmospheric pressure |
| **ATR** | attenuated total reflectance | | **CARES** | Ceramic Analysis and Reliability Evaluation of Structures |
| **AWG** | American wire gage | | | |
| **AWJ** | abrasive waterjet | | **CARP** | Committee for Acoustic Emission in Reinforced Plastics |
| **AWM** | abrasive waterjet machining | | | |
| **AWS** | American Welding Society | | **CASS** | copper-accelerated acetic acid-salt spray (test) |
| **b** | barn; Burgers vector | | | |
| $b$ | crystal lattice length along the $b$ axis | | **CAT** | computer-aided tomography |
| $B$ | magnetic induction | | **CAW** | carbon arc welding |
| **bal** | balance; remainder | | **CBED** | convergent-beam electron diffraction |
| **bcc** | body-centered cubic | | **CBEDP** | convergent-beam electron diffraction pattern |
| **BCIRA** | British Cast Iron Research Association | | **CBN** | cubic boron nitride |
| **bct** | body-centered tetragonal | | **CC** | combined carbon |
| **BDT** | brittle-ductile transition | | **C-C** | carbon-carbon |
| **Bé** | Baumé (specific-gravity scale) | | **CCD** | charge-coupled device |
| **BE** | backscattered electron | | **CCI** | crevice corrosion index |
| **BET** | Brunauer-Emmett-Teller | | **CCM** | Crucible ceramic mold |
| **BF** | bright-field (illumination) | | **CCPA** | Cemented Carbide Producers Association |
| **BHP** | Broken Hill Proprietary | | **CCR** | conventional controlled rolling |
| $BI$ | basicity index | | **CCT** | critical crevice temperature; center-cracked tension (specimen); continuous cooling transformation |
| **BMA** | butadiene-co-maleic anhydride | | | |
| **BMAW** | bare metal arc welding | | | |
| **BMC** | bulk molding compound | | **cd** | candela |
| **BMI** | bismaleimide (resin) | | **CDA** | Copper Development Association |
| **BNI** | bisnadimide | | **CE** | carbon equivalent |
| **BOP** | basic oxygen process | | **CEBAF** | continuous electron beam accelerator facility |
| **BOS** | basic oxygen steelmaking | | **CEN** | Comité Européen de Normalisation (European Committee for Standardization) |
| **BPO** | benzoyl peroxide | | | |
| **Bq** | becquerel | | **CERT** | constant extension rate test |
| **BS** | British Standard | | **CET** | columnar-equiaxed transition |
| **BSCCO** | Bi-Sr-Ca-Cu-O | | **CFC** | corrosion-fatigue cracking |
| **BST** | beta solution treatment | | **CFRP** | carbon fiber reinforced plastic |
| **BTDE** | benzophenone tetracarboxylic acid dimethyl ester | | **CFTA** | Committee of Foundry Technical Associations |
| | | | **CG** | compacted graphite |
| **Btu** | British thermal unit | | **CGA** | Compressed Gas Association |
| **BUE** | built-up edge | | **CHIP** | cold and hot isostatic pressing |
| **BUS** | broken-up structure | | **CHR** | conventional hot rolling |

| | |
|---|---|
| Ci | curie |
| CIE | Commission Internationale de l'Eclairage (International Commission on Illumination) |
| CIM | computer-integrated manufacturing |
| CINDAS | Center for Information and Numerical Data Analysis and Synthesis |
| CIP | cold isostatic pressing |
| CIRCLE | cylindrical internal reflection cell |
| CL | confidence limits; cathodoluminescence |
| CLA | counter-gravity low-pressure casting of air-melted alloys |
| CLAS | counter-gravity low-pressure air-melted sand casting |
| CLV | counter-gravity low-pressure casting of vacuum-melted alloys |
| cm | centimeter |
| CM | chemical machining (milling) |
| cmc | critical micelle concentration |
| CMC | ceramic matrix composite |
| CMM | coordinate measuring machine |
| CMOS | complementary metal oxide semiconductor |
| CN | cyanogen |
| CNB | Chevron notch bend |
| CNC | computer numerical control |
| COD | crack opening displacement |
| cos | cosine |
| cot | cotangent |
| CP | commercially pure |
| cpm | cycles per minute |
| CPM | Crucible Particle Metallurgy |
| cps | cycles per second; counts per second |
| CPVC | chlorinated polyvinyl chloride |
| CPU | central processing unit |
| CQ | commercial quality |
| CR | cold rolled |
| CRE | carbon removal efficiency |
| CRO | cathode-ray oscilloscope |
| CRR | carbon removal rate |
| CRT | cathode ray tube |
| CS | ceramic shell |
| CSA | Canadian Standards Association |
| C-SAM | C-mode scanning acoustic microscopy |
| CSD | controlled spray deposition |
| CSP | compact strip production |
| cSt | centiStokes |
| CT | computed tomography; compact type (specimen); compact tension (test specimen); continuous transformation |
| CTBN | carboxyl-terminated butadiene acrylonitrile |
| CTE | coefficient of thermal expansion |
| CTFE | polychlorotrifluoroethylene |
| CTOA | crack tip opening angle |
| CTOD | crack tip opening displacement |
| CVD | chemical vapor disposition |
| CVI | chemical vapor infiltration (impregnation) |
| CVN | Charpy V-notch (impact test or specimen) |
| cw | continuous-wave (spectrometer) |
| d | day |
| $d$ | depth; diameter; lattice spacing of crystal planes; an operator used in mathematical expressions involving a derivative (denotes rate of change) |
| $D$ | diameter; distance |
| DAC | digital-to-analog converter |
| $da/dN$ | fatigue crack growth rate |
| $da/dt$ | crack growth rate per unit time |
| DADPS | diaminodiphenylsulfone |
| DAIP | diallyl isophthalate |
| DAP | diallyl phthalate |
| DARPA | Defense Advanced Research Projects Agency |
| DAS | dendrite arm spacing |
| dB | decibel |
| DBMS | data base management system |
| DBTT | ductile-brittle transition temperature |
| dc | direct current |
| DCB | double-cantilever beam |
| DCEN | direct current electrode negative |
| DCEP | direct current electrode positive |
| DCP | direct-current plasma |
| DCRF | Die Casting Research Foundation |
| DDS | diaminodiphenyl sulfone |
| deca | decahydronaphthalene |
| DESY | Deutsche Electronen Syncrotron |
| DETA | diethylene triamine |
| d.f. | degrees of freedom |
| DF | dark-field (illumination) |
| $Df$ | dilution factor |
| DFB | diffusion brazing |
| DFW | diffusion welding |
| dhcp | double hexagonal close-packed |
| Di | didymium (mixture of the rare earth elements praseodymium and neodymium) |
| diam | diameter |
| DIB | diiodobutane |
| DIC | differential interference contrast (illumination) |
| DIN | Deutsche Industrie-Normen (German Industrial Standards) |
| DIP | dual-in-line package (electronic component) |
| DIS | Draft International Standard; Ductile Iron Society |
| dm | decimeter |
| DMAC | dimethyl acetamide |

| | |
|---|---|
| **DMF** | dimethyl formamide |
| **DMSO** | dimethyl sulfoxide |
| **DMW** | dissimilar-metal weld |
| **DNC** | direct numerical control |
| **DOA** | dead on arrival |
| **DOC** | Department of Commerce; depth of cut |
| **DoD** | Department of Defense |
| **DOF** | device operating failure |
| **DOT** | Department of Transportation |
| **DP** | dual phase |
| **DPH** | diamond pyramid hardness (Vickers hardness) |
| **DQ** | drawing quality |
| **DQSK** | drawing quality special killed |
| **DR** | digital radiography; digital radiograph |
| **DRAM** | dynamic random access memory |
| **DRCR** | dynamic recrystallization |
| **DRI** | direct reduced iron |
| **DRS** | diffuse reflectance spectroscopy |
| **DRX** | dynamic recrystallization |
| **DS** | directional solidification; dip soldering |
| **DSA** | dispersion-strengthened alloy |
| **DSC** | differential scanning calorimetry |
| **DT** | drop tower; dynamic tear (test) |
| **DTA** | differential thermal analysis |
| **DU** | depleted uranium |
| **DWT** | drop-weight test |
| **DWTT** | drop-weight tear test |
| | |
| $e$ | natural log base, 2.71828; charge of an electron; engineering strain (see also $\varepsilon$) |
| $e_1$ | major engineering strain |
| $e_2$ | minor engineering strain |
| $E$ | energy; modulus of elasticity (Young's modulus) |
| $E_{cell}$ | measured cell potential |
| $E_{corr}$ | corrosion potential |
| **EAA** | poly(ethylene-co-acrylic acid) |
| **EAF** | electric arc furnace |
| **EB** | electron beam |
| **EBHT** | electron beam hardening treatment |
| **EBIC** | electron beam induced current |
| **EBM** | electron beam machining |
| **EBW** | electron beam welding |
| **EBW-HV** | electron beam welding—high vacuum |
| **EBW-MV** | electron beam welding—medium vacuum |
| **EBW-NV** | electron beam welding—nonvacuum |
| **EC** | eddy current |
| **ECAP** | energy-compensated atom probe |
| **ECEA** | end cutting edge angle |
| **ECG** | electrochemical grinding |
| **ECM** | electrochemical machining |
| **EDM** | electrical discharge machining |

| | |
|---|---|
| **EDS** | energy-dispersive x-ray spectrometry; energy-dispersive spectroscopy |
| **EDTA** | ethylenediamine tetraacetic acid |
| **EDXA** | electron dispersive analysis by x-ray |
| **EEC** | European Economic Community |
| **EELS** | electron energy loss spectroscopy |
| **EGW** | electrogas welding |
| **EIA** | Electronics Industries Association |
| **ELCl** | extra-low chlorine powder |
| **ELI** | extra-low interstitial |
| **ELP** | electropolishing |
| **emf** | electromotive force |
| **EMF** | electromagnetic fields |
| **EMI** | electromagnetic iron; electromagnetic interference |
| **EMSA** | Electron Microscopy Society of America |
| **ENAA** | epithermal neutron activation |
| **ENSIP** | Engine Structural Integrity Program |
| **EP** | epoxy; extreme pressure |
| **EPA** | Environmental Protection Agency |
| **EPC** | evaporative pattern casting |
| **EPDM** | ethylene-propylene-diene monomer |
| **EPMA** | electron probe x-ray microanalysis; electron probe microanalysis |
| **EPR** | ethylene propylene rubber |
| **EPRI** | Electric Power Research Institute |
| **EPS** | expanded polystyrene pattern |
| **Eq** | equation |
| **ESC** | environmental stress cracking |
| **ESCA** | electron spectroscopy for chemical analysis |
| **ESD** | electrostatic discharge |
| **ESR** | electron spin resonance; electroslag remelting |
| **ESW** | electroslag welding |
| *et al.* | and others |
| **ETP** | electrolytic tough pitch (copper) |
| **eV** | electron volt |
| **EXAFS** | extended x-ray absorption fine structure |
| **exp** | exponent, exponential |
| **EXW** | explosion welding |
| | |
| **f** | fiber |
| $f$ | frequency; transfer function; precipitate volume fraction |
| **F** | farad; ferrite; fluorescence |
| $F$ | force; Faraday constant (96,486 C/mol) |
| **FAC** | forced-air cool |
| **FATT** | fracture-appearance transition temperature |
| **FC** | furnace cool |
| **FCAW** | flux cored arc welding |
| **fcc** | face-centered cubic |
| **FCC** | Federal Communications Commission |
| **fct** | face-centered tetragonal |

| | |
|---|---|
| FDA | Food and Drug Administration |
| FDM | finite-difference method |
| FEA | finite-element analysis |
| FEG | field emission gun |
| FEM | finite element modeling/method |
| FEP | fluorinated ethylene propylene |
| FET | field-effect transistor |
| FFT | fast Fourier transform |
| FG | flake graphic |
| FGD | flue gas desulfurization |
| FGHAZ | fine grain heat-affected zone |
| FHWA | Federal Highway Administration |
| FIA | flow injection analysis |
| FIFO | first in, first out |
| Fig. | figure |
| FIM | field ion microscopy |
| FIOR | fluid iron ore reduction |
| FLD | forming limit diagram |
| fm | femtometer |
| FM | frequency modulation; full mold; ferromagnet |
| FM (process) | *fonte mince* (thin iron) |
| FMR | ferromagnetic resonance |
| FMS | flexible manufacturing system |
| FN | ferrite number |
| FNAA | fast neutron activation analysis |
| FOC | chemical flux cutting |
| FOLZ | first-order Laue zone |
| FOW | forge welding |
| FP | polycrystalline alumina fiber |
| FRC | free radical cure |
| FRM | fiber-reinforced metals |
| FRP | fiber-reinforced plastic |
| FRS | fiber-reinforced superalloys |
| FRTP | fire-refined tough pitch (copper) |
| FRW | friction welding |
| FSS | fatigue striation spacing |
| ft | foot |
| ftc | footcandle |
| FTIR | Fourier transform infrared (spectroscopy) |
| ft-L | footlambert |
| FTS | Fourier transform spectrometer |
| FW | flash welding |
| FWHM | full width at half maximum |
| FZ | fusion zone |
| g | gram |
| *g* | acceleration due to gravity |
| G | gauss; graphite |
| *G* | shear modulus; modulus of rigidity; thermal gradient; Gibbs free energy |
| GA | gas atomization |
| gal | gallon |

| | |
|---|---|
| GAR | grain aspect ratio |
| GB | grain boundary |
| GBq | gigabecquerel |
| GC | grain-coarsened; gas chromatography |
| GCHAZ | grain-coarsened heat-affected zone |
| GC-IR | gas chromatography-infrared (spectroscopy) |
| GC/MS | gas chromatograph/mass spectrometer/spectrometry |
| gcp | geometrically close-packed |
| GdIG | gadolinium iron garnet |
| gf | gram-force |
| GFAAS | graphite furnace atomic absorption spectrometry |
| GFN | grain fineness number |
| GGG | gallium gadolinium garnet |
| GHz | gigahertz |
| GJ | gigajoule |
| GMAW | gas metal arc welding |
| GOR | gas-oil ratio (in petroleum production) |
| GP | Guinier-Preston (zone) |
| GPa | gigapascal |
| GPC | gel permeation chromatography |
| gpd | grams per denier |
| gr | grain |
| Gr | graphite |
| GS | grain size |
| GSGG | gallium scandium gadolinium garnet |
| GTAW | gas tungsten arc welding |
| GTO | gate turnoff |
| Gy | gray (unit of absorbed radiation) |
| h | hour |
| *h* | Planck's constant |
| H | Henry |
| *H* | enthalpy; Grossmann number; height; magnetic field; magnetic field strength |
| $H_a$ | applied magnetic field |
| $H_c$ | coercive force; critical magnetic field; thermodynamic critical field |
| HAD | high aluminum defect |
| HAZ | heat-affected zone |
| HB | Brinell hardness |
| HBN | hexagonal boron nitride |
| HCF | high-cycle fatigue |
| HCL | hollow cathode lamp |
| hcp | hexagonal close-packed |
| HDI | high-density inclusion |
| HDPE | high-density polyethylene |
| HEC | hydrogen embrittlement cracking |
| HEM | hydrogen embrittlement |
| HEP | high-energy physics |
| HERF | high-energy-rate forging |
| hexa | hexamethylene tetramine |

| | |
|---|---|
| **HF** | hardenability factor |
| **HFP** | hexafluoropropylene |
| **HFRSc** | Scleroscope hardness (Model C) |
| **HFRSd** | Scleroscope hardness (Model D) |
| **HFRW** | high frequency resistance welding |
| **HIC** | hydrogen-induced cracking |
| **HID** | high interstitial defect |
| **HIP** | hot isostatic pressing |
| **HIPS** | high-impact polystyrene |
| **HK** | Knoop hardness |
| **HLW** | high-level waste |
| **HM** | high modulus |
| **HOLZ** | higher order Laue zone |
| **hp** | horsepower |
| **HPLC** | high-pressure liquid chromatography |
| **HPSN** | hot-pressed silicon nitride |
| **HPW** | hot pressure welding |
| **HR** | Rockwell hardness (requires scale designation, such as HRC for Rockwell C hardness) |
| **HRA** | Rockwell "A" hardness |
| **HRB** | Rockwell "B" hardness |
| **HRE** | Rockwell "E" hardness |
| **HRF** | Rockwell "F" hardness |
| **HRH** | Rockwell "H" hardness |
| **HSLA** | high-strength, low-alloy (steel) |
| **HSS** | high-speed steel |
| **HTLA** | heat-treatable low-alloy (steel) |
| **HV** | Vickers hardness (diamond pyramid hardness) |
| **HVC** | hydrovac process |
| **HWR** | heavy water reactor(s) |
| **Hz** | hertz |
| | |
| **i** | current (measure of number of electrons) |
| *i* | current density |
| *I* | bias current; current; electrical current; emergent intensity; intensity |
| $i_0$ | exchange current |
| $i_{corr}$ | corrosion current |
| $i_{crit}$ | critical current for passivation |
| $i_{pass}$ | passive current |
| **IACS** | International Annealed Copper Standard |
| **IARC** | International Agency for Research on Cancer |
| **IASCC** | irradiation-assisted stress-corrosion cracking |
| **IC** | integrated circuit; ion chromatography |
| **ICBM** | Intercontinental Ballistic Missile |
| **ICFTA** | International Committee of Foundry Technical Associations |
| **ICHAZ** | intercritical heat-affected zone |
| **ICP** | inductively coupled plasma |

| | |
|---|---|
| **ICP-AES** | inductively coupled plasma atomic emission spectroscopy |
| **ICPMS** | inductively coupled plasma mass spectrometry |
| **ICR** | intensified controlled rolling |
| **ID** | inside diameter |
| **IEEE** | Institute of Electrical and Electronics Engineers |
| **IF** | interstitial free |
| **IFI** | Industrial Fasteners Institute |
| **IGA** | intergranular attack; inert-gas atomization |
| **IGP** | intergranular penetration |
| **IGSCC** | intergranular stress-corrosion cracking |
| **IIR** | isoprene-isobutylene rubber |
| **IIW** | International Institute of Welding |
| **ILZRO** | International Lead Zinc Research Organization |
| **I/M** | ingot metallurgy |
| **in.** | inch |
| **INCRA** | International Copper Research Association |
| **INPO** | Institute for Nuclear Power Operations |
| **IPC** | Institute for Interconnecting and Packaging Electronic Circuits |
| **ipm** | inches per minute |
| **ips** | inches per second |
| **ipt** | inch per tooth |
| **IPTS** | International Practical Temperature Scale |
| **IR** | infrared (radiation); infrared (spectroscopy); injection refining |
| **IRAS** | infrared astronomy satellite |
| **IRB** | infrared brazing |
| **IRGCHAZ** | intercritically reheated grain-coarsened heat-affected zone |
| **IRRAS** | infrared reflection absorption spectroscopy |
| **IRS** | infrared soldering |
| **IS** | induction soldering |
| **ISA** | Instrument Society of America |
| **ISCC** | intergranular stress-corrosion cracking |
| **ISO** | International Organization for Standardization |
| **ISRI** | Institute of Scrap Recycling Industries |
| **IT** | isothermal transformation |
| **ITER** | international thermonuclear experimental reactor |
| **ITS** | International Temperature Scale |
| **IVD** | ion vapor deposited |
| **IW** | induction welding |
| | |
| **J** | joule |
| *J* | crack growth energy release rate (fracture mechanics) |
| $J_c$ | critical current density |
| $J_{ec}$ | Jominy equivalent cooling |

| | |
|---|---|
| $J_{eh}$ | Jominy equivalent hardness |
| **JCPDS** | Joint Committee on Powder Diffraction Standards |
| **JIC** | Joint Industry Conference |
| **JIS** | Japanese Industrial Standard |
| **JIT** | just in time (manufacturing) |
| | |
| **k** | karat |
| $k$ | Boltzmann constant; notch sensitivity factor; thermal conductivity |
| **K** | Kelvin |
| $K$ | coefficient of thermal conductivity; bulk modulus of elasticity; stress-intensity factor in linear elastic fracture mechanics |
| $\Delta K$ | stress-intensity factor range |
| $K_0$ | crack initiation toughness |
| $K_a$ | crack arrest toughness |
| $K_c$ | plane-stress fracture toughness |
| $K_f$ | fatigue notch factor |
| $K_I$ | stress-intensity factor |
| $K_{Ia}$ | plane-strain crack arrest toughness |
| $K_{Ic}$ | plane-strain fracture toughness |
| $K_{Id}$ | dynamic fracture toughness |
| $K_{IHE}$ | threshold stress intensity for hydrogen embrittlement |
| $K_{ISCC}$ | threshold stress intensity to produce stress-corrosion cracking |
| $K_t$ | theoretical stress-concentration factor |
| $K_{th}$ | threshold stress-intensity factor |
| **KB** | kilobyte |
| **kbar** | kilobar (pressure) |
| **KBES** | knowledge-based expert system (computer software) |
| **K-BOP** | Kawasaki basic oxygen process |
| **KEK** | Japanese atomic energy facility |
| **keV** | kiloelectron volt |
| **kg** | kilogram |
| **kgf** | kilogram force |
| **kHz** | kilohertz |
| **km** | kilometer |
| **KMS** | Kloeckner metallurgy scrap |
| **kN** | kilonewton |
| **kPa** | kilopascal |
| **ksi** | kips (1000 lbf) per square inch |
| **kV** | kilovolt |
| **kW** | kilowatt |
| | |
| $l$ | length |
| **L** | liter; longitudinal |
| $L$ | length |
| $L_0$ | initial length |
| **LAMMA** | laser microscope mass analysis |
| **LAST** | lowest anticipated service temperature |

| | |
|---|---|
| **lb** | pound |
| **LBE** | lance bubble equilibrium |
| **lbf** | pound force |
| **LBM** | laser beam machining |
| **LBW** | laser beam welding |
| **LBZ** | local brittle zone |
| **LCD** | liquid crystal display |
| **LCF** | low cycle fatigue |
| **LCP** | large coil program; liquid crystal polymer |
| $L/D$ | length-to-diameter ratio |
| **LDH** | limiting dome height |
| **LDR** | limiting draw ratio |
| **LEC** | liquid-encapsulated Czochralski |
| **LED** | light-emitting diode |
| **LEED** | low-energy electron diffraction |
| **LEFM** | linear elastic fracture mechanics |
| **LEISS** | low-energy ion-scattering spectroscopy |
| **LF** | ladle furnace |
| **LF/VD** | ladle furnace vacuum degassing |
| **LF/VD-VAD** | ladle furnace vacuum degassing and vacuum arc degassing |
| **LIM** | liquid injection molding |
| **LiMCA** | liquid metal cleanness analyzer |
| **LIMS** | laser ionization mass spectroscopy |
| **lm** | lumen |
| **LME** | liquid metal embrittlement |
| **LMP** | Larsen-Miller parameter |
| **LMR** | liquid metal refining |
| **ln** | natural logarithm (base $e$) |
| **LNG** | liquified natural gas |
| **log** | common logarithm (base 10) |
| **LOI** | limiting oxygen index |
| **LOR** | loss on reduction |
| **LOX** | liquid oxygen |
| **LPCVD** | low pressure chemical vapor deposition |
| **LPE** | liquid-phase epitaxy |
| **LPG** | liquefied petroleum gas |
| **LRO** | long-range order |
| **LSG** | low-stress grinding |
| **LT** | long transverse (direction) |
| **LVDT** | linear variable differential transformer |
| **LWR** | light water reactor(s) |
| **lx** | lux |
| | |
| **m** | meter |
| $m$ | strain rate sensitivity factor; mass; molar (solution) |
| **M** | martensite |
| $M$ | magnetization; magnification; molar solution; molecular weight |
| $M_f$ | temperature at which martensite formation finishes during cooling |

| | | | |
|---|---|---|---|
| $M_s$ | temperature at which martensite starts to form from austenite on cooling | mpg | miles per gallon |
| mA | milliampere | mph | miles per hour |
| MA | microalloyed | MPIF | Metal Powder Industries Federation |
| MAA | methacrylic acid | MPS | main propulsion system (space shuttle) |
| MAE | microalloying elements | MQG | melt-quench technique |
| MAPP | | mrem | millirem |
| or MPS | methyl-acetylene-propadiene (gas) | MRI | magnetic resonance imaging |
| max | maximum | MRR | material removal rate |
| MB | megabyte | ms | millisecond |
| MBE | molecular beam epitaxy | MS | megasiemens; mass spectrometry |
| MC | metal carbide | mSv | millisievert |
| MCR | minimum creep rate | mT | millitesla |
| MDA | methylene dianiline | MTF | modulation transfer function |
| MDI | diphenylmethane-4,4'-diisocyanate | MTI | Materials Technology Institute (of the Chemical Process Industry) |
| MDRX | metadynamic recrystallization | | |
| ME | Mössbauer effect | mV | millivolt |
| MEA | monoethanolamine | MV | megavolt |
| MEK | methyl ethyl ketone | MVEMA | methyl vinyl ether-co-maleic |
| MEKP | methyl ethyl ketone peroxide | MW | molecular weight |
| MeV | megaelectronvolt | MWG | music wire gage |
| MFTF | mirror fusion test facility | | |
| mg | milligram | n | neutrons |
| Mg | megagram (metric tonne, or kg $\times$ 10$^3$) | $n$ | strain-hardening exponent; refractive index |
| MHD | magnetohydrodynamic (casting) | N | Newton |
| MHz | megahertz | $N$ | fatigue life (number of cycles); normal solution; speed of rotation (rev/min) |
| MIBK | methyl isobutyl ketone | | |
| MIE | metal-induced embrittlement | $N_f$ | number of cycles to failure |
| MIG | metal inert gas (welding) | NA | numerical aperture |
| min | minute; minimum | NAA | neutron activation analysis |
| MINT | metal in-line treatment | NACA | National Advisory Committee for Aeronautics |
| MIL | military | | |
| MIL-STD | military standard | NACE | National Association of Corrosion Engineers |
| MIM | metal injection molding | NASA | National Aeronautics and Space Administration |
| mips | million intrusions per second | | |
| MIT | Massachusetts Institute of Technology | NBR | acrylonitrile-butadiene rubber |
| MJ | megajoule | NBS | National Bureau of Standards |
| MJR | modified jelly roll | nC | nanocoulomb |
| mL | milliliter | NC | numerical control |
| mm | millimeter | ND | normal direction (of a sheet) |
| MM | misch metal | NDE | nondestructive evaluation |
| MMA | methyl methacrylate | NDI | nondestructive inspection |
| MMC | metal-matrix composite | NDT | nil-ductility transition; nondestructive testing |
| MMIC | monolithic microwave integrated | | |
| mo | month | NDTT | nil ductility transition temperatures |
| MO or MeO | metal oxide | Nd:YAG | neodymium: yttrium-aluminum-garnet (laser) |
| MOCVD | metallo-organic chemical vapor deposition | | |
| mol% | mole percent | NEMA | National Electrical Manufacturers Association |
| MOLE | molecular optical laser examiner | | |
| MOR | modulus of resilience; modulus of rupture | NFPA | National Fire Prevention Association |
| mPa | millipascal | NG | nuclear grade |
| MPa | megapascal | NGPA | Natural Gas Producers Association |
| | | NGR | nuclear gamma-ray resonance |
| MPC | Metal Properties Council | NIR | near infrared |

| | | | |
|---|---|---|---|
| NIST | National Institute of Standards and Technology | p | page |
| nm | nanometer | *p* | pressure |
| NMR | nuclear magnetic resonance | P | pearlite |
| NMTP | nonmartensitic transformation product | *P* | power; pressure; applied load |
| No. | number | *p*O₂ | partial pressure of oxygen |
| NPS | American National Standard Straight Pipe Thread | Pa | pascal |
| | | PA | prealloyed; polyamide |
| NPSC | American National Standard Straight Pipe Thread for Couplings | PAC | plasma arc cutting |
| | | PACVD | plasma-assisted chemical vapor deposition |
| NPT | American National Standard Taper Pipe Thread | PAEK | polyaryletherketone |
| | | PAI | polyamide-imide |
| NQR | nuclear quadruple resonance | PAN | polyacrylonitrile |
| NR | natural rubber; neutron radiography | PAR | polyarylate |
| NRC | Nuclear Regulatory Commission | PAS | polyaryl sulfone; photoacoustic spectroscopy |
| NRL | Naval Research Laboratories | PAW | plasma arc welding |
| ns | nanoseconds | PBI | polybenzimidazole |
| NSR | notch strength ratio | PBT | polybutylene terephthalate |
| NT | normalized and tempered | p-c | plastic-carbon (replica) |
| NTHM | net tonne of hot metal | PC | programmable controller; personal computer |
| NTSB | National Transportation and Safety Board | PCB | printed circuit board |
| NWTI | National Wood Tank Institute | PCBN | polycrystalline CBN (cubic boron nitride) |
| | | PCD | polycrystalline diamond |
| | | PCM | photochemical machining |
| O | oil | PD | preferred direction |
| OAW | oxyacetylene welding | PDCP | polydicyclopentadiene |
| OBM | oxygen blow method | PDF | probability density function |
| OD | outside diameter | PDMS | polydimethylsiloxane |
| ODMR | optical double magnetic resonance | PE | polyethylene |
| ODS | oxide dispersion strengthened | PEA | polyethylene adipate |
| Oe | oersted | PEEK | polyetheretherketone |
| OECD | Organization for Economic Cooperation and Development | PEI | polyether-imide |
| | | PEK | polyetherketone |
| OES | optical emission spectroscopy | PEL | permissible exposure limits |
| OF | oxygen-free | PESV | polyether sulfone |
| OFC | oxyfuel gas cutting | PET | polyethylene terephthalate |
| OFHC | oxygen-free high conductivity (copper) | PEW | percussion welding |
| OFW | oxyfuel gas welding | PFC | planar-flow casting |
| OHW | oxyhydrogen welding | PGAA | prompt gamma-ray activation analysis |
| OM | optical micrograph | PGM | platinum-group metals |
| OMS | orbital maneuvering system (space shuttle) | pH | negative logarithm of hydrogen-ion activity |
| ONERA | Office Nationale d'Etudes et de Recherches Aerospatiales | PH | precipitation-hardenable |
| | | PI | polyimide |
| ONIA | Office National Industrial d'Azote | PIB | polyisobutylene |
| OQ | oil quench | PIC | pressure-impregnation-carbonization |
| OQ & T | oil quenched and tempered | PIXE | proton-induced x-ray emission; particle-induced x-ray emission |
| ORNL | Oak Ridge National Laboratory | | |
| OSHA | Occupational Safety and Health Administration | pixel | picture element |
| | | PLAP | pulsed-laser atom probe |
| OSTE | one-step temper embrittlement | PLASTEC | Plastics Technical Evaluation Center |
| OTB | oxygen top blown | PLC | programmable logic controller |
| OTSG | once-through steam generator | PLZT | lead lanthanum zirconate titanate |
| oz | ounce | pm | picometer |

| | |
|---|---|
| **PM** | precious metal; paramagnet |
| **P/M** | powder metallurgy |
| **PMMA** | polymethyl methacrylate |
| **PMN** | lead magnesium niobate |
| **PMS** | Process Management System; lead-molybdenum-sulfide ($PbMo_6S_8$) |
| **PMT** | photomultiplier tube |
| **PNN** | lead nickel niobate |
| **POC** | metal powder cutting; products of combustion |
| **POM** | polyoxymethylene |
| **PP** | polypropylene |
| **ppb** | parts per billion |
| **PPB** | prior particle boundary |
| **ppba** | parts per billion atomic |
| **PPE** | polyphenylene ether |
| **ppm** | parts per million |
| **PPO** | polyphenylene oxide |
| **PPRIC** | Pulp and Paper Research Institute of Canada |
| **PPS** | polyphenylene sulfide; property prediction system |
| **ppt** | parts per trillion |
| **PPTA** | poly-*p*-phenylene terephthalamide |
| **PQ** | physical quality |
| *Pr* | Prandtl number |
| **PREP** | plasma rotating electrode process |
| **PROM** | programmable read-only memory |
| **PS** | polystyrene |
| **PSA** | pressure-sensitive adhesive |
| **psi** | pounds per square inch |
| **psia** | pounds per square inch (absolute) |
| **psid** | pounds per square inch (differential) |
| **psig** | gage pressure (pressure relative to ambient pressure) in pounds per square inch |
| **PSP** | plasma spraying |
| **PSU** | polysulfone |
| **PSZ** | partially stabilized zirconia |
| **PTFE** | polytetrafluoroethylene |
| **PTH** | plated through-holes |
| **PUCB** | phenolic urethane cold box |
| **PUN** | phenolic urethane no-bake |
| **PUR** | polyurethane |
| **PVA** | polyvinyl alcohol |
| **PVAC** | polyvinyl acetate |
| **PVAL** | polyvinyl alcohol |
| **PVB** | polyvinyl butyral |
| **PVC** | polyvinyl chloride |
| **PVD** | physical vapor deposition |
| **PVDF** | polyvinylidene fluoride |
| **PVF** | polyvinyl formal |
| **PVP** | polyvinylpyrrolidone |
| **PVRC** | Pressure Vessel Research Committee |
| **PWHT** | postweld heat treatment |
| **PWR** | pressurized water reactor(s) |
| **PZN** | lead zinc niobate |
| **PZT** | lead zirconium titanate |
| *q* | fatigue notch sensitivity factor |
| *Q* | quench factor; heat removal rate |
| **Q-BOP** | quick-quiet basic oxygen process |
| **QT** | quenched and tempered |
| *r* | radius |
| **R** | roentgen; Rankine; rare earth |
| *R* | radius; universal gas constant; ratio of the minimum stress to the maximum stress; bulk resistance; reluctance (reciprocal of permeance); rolling reduction ratio; resultant force |
| $R_a$ | surface roughness in terms of arithmetic average |
| $R_{max}$ | maximum peak-to-valley roughness height |
| $R_q$, $R_{rms}$ | root-mean-square roughness average |
| $R_t$ | total roughness peak-to-valley |
| $R_y$ | maximum peak-to-valley roughness height |
| $R_z$ | ten-point height (roughness average) |
| **RA** | recrystallization annealed; reduction of/in area; rosin activated |
| **rad** | radiation absorbed dose; radian |
| **RAD** | ratio-analysis diagram |
| **RAM** | random access memory |
| **RBS** | Rutherford backscattering spectroscopy/spectrometry |
| **RBSC** | reaction-bonded silicon carbide |
| **RBSN** | reaction-bonded silicon nitride |
| **RCF** | rolling contact fatigue |
| **RCR** | recrystallization controlled rolling |
| **RD** | rolling direction (of a sheet) |
| **RDF** | radial distribution function (analysis) |
| *Re* | Reynolds number |
| **RE** | rare earth |
| **Ref** | reference |
| **rem** | remainder or balance; roentgen equivalent man |
| **REP** | rotating-electrode process |
| **rf, RF** | radio frequency |
| **RFEC** | remote-field eddy current |
| **RGA** | residual gas analyzer |
| **RGB** | red, green, blue |
| **RH** | relative humidity; refrigeration hardened |
| **RHEED** | reflection high-energy electron diffraction |
| **RIM** | reaction injection molding |
| **rms** | root mean square |
| **ROC** | rapid omnidirectional compaction/consolidation |
| **rpm** | revolutions per minute |

| | |
|---|---|
| **RRIM** | reinforced reaction injection molding |
| **RS** | Raman spectroscopy; rapid solidification |
| **RSD** | relative standard deviation |
| **RSEW** | resistance seam welding |
| **RSR** | rapid solidification rate |
| **RSW** | resistance spot welding |
| **RT** | room temperature |
| **RTE** | reversible temper embrittlement |
| **RTM** | resin transfer molding |
| **RTR** | real-time radiography |
| **RTV** | room-temperature vulcanizing (rubber) |
| **RWMA** | Resistance Welder Manufacturers Association |
| | |
| **s** | second |
| *s* | sample standard deviation |
| **S** | siemens |
| *S* | nominal engineering stress or normal engineering stress |
| $S_a$ | alternating stress amplitude |
| $S_C$ | carbon saturation |
| $S_m$ | mean stress |
| $S_{max}$ | maximum stress |
| $S_{min}$ | minimum stress |
| $S_r$ | range of stress |
| **SACD** | spheroidized annealed cold drawn |
| **SACP** | selected-area channeling pattern |
| **SAD** | selected-area diffraction |
| **SADP** | selected-area diffraction pattern |
| **SAE** | Society of Automotive Engineers |
| **SAM** | scanning Auger microscopy; scanning acoustic microscopy |
| **SAN** | styrene-acrylonitrile |
| **SANS** | small-angle neutron scattering |
| **SATT** | shear-area transition temperature |
| **SAW** | submerged arc welding |
| **SAXS** | small-angle x-ray scattering |
| **SBC** | steel-bonded (titanium) carbide |
| **SBR** | styrene-butadiene rubber |
| **SBS** | styrene-butadiene-styrene |
| **Sc** | Schmidt number |
| **SC** | single-crystal |
| **SCC** | stress-corrosion cracking |
| **SCE** | saturated calomel electrode |
| **SCEA** | side cutting edge angle |
| **SCFH** | standard cubic feet per hour |
| **scfm** | standard cubic foot per minute |
| **SCFM** | subcritical fracture mechanics |
| **SCHAZ** | subcritical heat-affected zone |
| **SCR** | silicon controlled rectifier |
| **SCRATA** | Steel Castings Research and Trade Association |
| **SDI** | strategic defense initiative |

| | |
|---|---|
| **SE** | secondary electron |
| **SEM** | scanning electron microscope/microscopy |
| **SEN** | single-edge notched |
| **SENB** | single-edge notch beam |
| **SERS** | surface-enhanced Raman spectroscopy |
| **SF** | slow-fast (wave form) |
| **SFC** | supercritical fluid chromatography |
| **sfm** | surface feet per minute |
| **SG** | standard grade; spheroidal graphite; spin glass |
| **SGA** | soluble-gas atomization |
| **Sh** | Sherwood number |
| **SHE** | standard hydrogen electrode |
| **SHT** | Sumitomo high toughness |
| **SI** | Système International d'Unités; silicone |
| **SIC** | standard industry codes |
| **SIMA** | strain induced melt activated |
| **SIMS** | secondary ion mass spectroscopy |
| **sin** | sine |
| **SIS** | superconductor/insulating/superconductor |
| **SLAM** | scanning laser acoustic microscopy |
| **SLR** | single-lens reflex (camera) |
| **SMA** | shape memory alloy |
| **SMAW** | shielded metal arc welding |
| **SMC** | sheet molding compound |
| **SME** | Society of Manufacturing Engineers; shape memory effect; solid-metal embrittlement |
| **SMIE** | solid metal induced embrittlement |
| **SMS** | tin-molybdenum-sulfide ($SnMo_6S_8$) |
| **SMYS** | specified minimum yield stress |
| *S-N* | stress-number of cycles (fatigue) |
| **S/N** | signal-to-noise (ratio) |
| **SNECMA** | Societe Nationale d'Etude et de Construction de Moteurs |
| **SPC** | statistical process control |
| **SPE** | Society of Plastics Engineers |
| **spf** | seconds per foot |
| **SPF** | superplastic forming |
| **SPF/DB** | superplastic forming/diffusion bonding |
| **sp gr** | specific gravity |
| **SPI** | Society of the Plastics Industry |
| **SPS** | service propulsion system (space shuttle) |
| **SQ** | structural quality |
| **SQUID** | superconducting quantum interference device |
| **SRB** | sulfate reducing bacteria; solid rocket booster (space shuttle) |
| **SRGHAZ** | subcritically reheated grain-coarsened heat-affected zone |
| **SRIM** | structural reaction injection molding |
| **SRM** | Standard Reference Material(s) |
| **SRO** | short-range order |
| **SRX** | static recrystallization |

| | |
|---|---|
| **SS** | slow-slow (wave form); Swedish Standard |
| **SSC** | superconducting supercollider |
| **SSI** | small-scale integration |
| **SSMS** | spark source mass spectrometry |
| **SSPC** | Steel Structures Painting Council |
| **SSRT** | slow strain rate testing |
| **SSVOD** | strong stirred vacuum oxygen decarburization |
| **SSW** | solid-state welding |
| **ST** | short transverse (direction) |
| **STA** | solution treated and aged |
| **STB** | Sumitomo top and bottom blowing (process) |
| **std** | standard |
| **STEL** | short term exposure limit |
| **STEM** | scanning transmission electron microscope/ microscopy |
| **STOL** | short time overload |
| **STP** | standard temperature and pressure |
| **STQ** | solution treated and quenched |
| **SUS** | Saybolt universal second (measure of viscosity) |
| **Sv** | sievert |
| **SW** | stud arc welding |
| **SWG** | steel wire gage |
| | |
| **t** | metric tonne |
| *t* | thickness; time |
| **T** | tesla |
| *T* | temperature |
| $T_b$ | boiling temperature |
| $T_c$ | transition temperature from normal to superconducting state; critical ordering temperature; critical transition temperature; Curie temperature |
| $T_g$ | glass transition temperature |
| $T_L$ | liquidus temperature |
| $T_m, T_M$ | melt/melting temperature |
| $T_s$ | solidus temperature |
| **tan** | tangent |
| **TAPPI** | Technical Association of the Pulp and Paper Industry |
| **TB** | torch brazing |
| **TBCCO** | Tl-Ba-Ca-Cu-O |
| **TC** | total carbon |
| **TCA** | trichloroethane |
| **TCAB** | twin carbon arc brazing |
| **TCE** | trichloroethylene |
| **tcp** | topologically close-packed |
| **TCP** | thermochemical processing |
| **TCT** | thermochemical treatment |
| **TD** | thoria dispersed; thorium dioxide dispersion strengthened; transverse direction (of a sheet) |

| | |
|---|---|
| **TDI** | toluene diisocyanate |
| **TEA** | triethanolamine |
| **TEG** | triethylene glycol |
| **TEM** | transmission electron microscopy; thermal energy method |
| **TFE** | tetrafluoroethylene |
| **TG** | thermogravimetry |
| **TGA** | thermogravimetric analysis |
| **THSP** | thermal spraying |
| **TIG** | tungsten inert gas (welding) |
| **TIR** | total indicator reading |
| **TMA** | thermomechanical analysis |
| **TMCP** | thermomechanical controlled processing |
| **TME** | tempered martensite embrittlement |
| **TMP** | thermomechanical processing |
| **TNAA** | transmission electron microscopy |
| **TNT** | 2,4,6-trinitrotoluene |
| **TOF** | time of flight |
| **tonf** | tons of force |
| **TP** | thermoplastic |
| **TPI** | thermoplastic polyimide; turns per inch |
| **TPUR** | thermoplastic polyurethane |
| **TRIP** | transformation induced plasticity |
| **TRS** | transverse rupture strength |
| **tsi** | tons per square inch |
| **TTS** | tearing topography surface |
| **TTT** | time-temperature-transformation |
| **TTU** | through-transmission ultrasonics |
| **TTZ** | transformation-toughened zirconia |
| | |
| **UBC** | used beverage can/container |
| **UCL** | upper control limit |
| **UHC** | ultrahigh carbon |
| **UHF** | ultra-high frequency |
| **UHMWPE** | ultrahigh molecular weight polyethylene |
| **UHV** | ultrahigh vacuum |
| **UKAEA** | United Kingdom Atomic Energy Authority |
| **UL** | Underwriter's Laboratories |
| **ULCB** | ultralow-carbon bainitic |
| **UNI** | Ente Nazionale Italiano di Unificazione |
| **UNS** | Unified Numbering System (ASTM-SAE) |
| **UPS** | ultraviolet photoelectron spectroscopy |
| **USAF** | United States Air Force |
| **USBM** | United States Bureau of Mines |
| **USDA** | United States Department of Agriculture |
| **USM** | ultrasonic machining |
| **USSWG** | United States steel wire gage |
| **UST** | underground storage tank |
| **USW** | ultrasonic welding |
| **UT** | ultrasonic testing |
| **UTS** | ultimate tensile strength |
| **UV** | ultraviolet |
| **UV/VIS** | ultraviolet/visible (absorption spectroscopy) |

| | | | | |
|---|---|---|---|---|
| $v$ | velocity | | **XRPD** | x-ray powder diffraction |
| **V** | volt | | **XRF** | x-ray fluorescence |
| $V$ | volume; velocity | | **XRS** | x-ray spectroscopy |
| **VAC-ESR** | electroslag remelting under reduced pressure | | | |
| **VAD** | vacuum arc degassing | | **YAG** | yttrium-aluminum-garnet |
| **VADER** | vacuum arc double-electrode remelting | | **YBCO** | Y-Ba-Cu-O |
| **VAR** | vacuum arc remelted/remelting | | **YIG** | yttrium iron garnet |
| **V-D** | vacuum degassing | | **yr** | year |
| **VHF** | very high frequency | | **YS** | yield strength |
| **VHP** | vacuum hot pressing | | | |
| **VHS** | very high speed | | $z$ | ion change |
| **VHSIC** | very high speed integrated circuit | | **Z** | impedance; atomic number; standard normal distribution |
| **VID** | vacuum induction degassing | | | |
| **VIDP** | vacuum induction degassing and pouring | | **ZGS** | zirconia grain stabilized |
| **VIM** | vacuum induction melting | | **ZOLZ** | zero-order Laue zone |
| **VIM/VID** | vacuum induction melting and degassing | | **ZR** | zone refined |
| **VIS** | visible | | **ZTA** | zirconia-toughened-alumina |
| **VLSI** | very large scale integration | | | |
| **VM** | vacuum melted | | ° | angular measure; degree |
| **VOD** | vacuum oxygen decarburization (ladle metallurgy) | | °C | degree Celsius (centigrade) |
| | | | °F | degree Fahrenheit |
| **VODC** | vacuum oxygen decarburization (converter metallurgy) | | $\rightleftarrows$ | direction of reaction |
| | | | % | percent |
| **VOID** | vacuum oxygen induction decarburization | | $\alpha$ | angle of incidence; coefficient of thermal expansion |
| **vol** | volume | | $\gamma$ | surface energy; surface tension |
| **vol%** | volume percent | | $\Delta$ | change in quantity; an increment; a range |
| **voxel** | volume element | | $\Delta G°$ | standard free energy of formation change |
| | | | $\Delta H$ | change in enthalpy |
| **w** | whisker | | $\Delta T$ | temperature difference |
| **W** | watt | | $\varepsilon$ | strain |
| $W$ | width; weight | | $\dot{\varepsilon}$ | strain rate |
| **Wb** | weber | | $\eta$ | viscosity |
| **WDS** | wavelength dispersive spectroscopy | | $\theta$ | angle |
| **WJM** | waterjet machining | | $\mu$ | friction coefficient; magnetic permeability; x-ray absorption coefficient |
| **wk** | week | | | |
| **WOL** | wedge-opening load | | $\mu$B | Bohr magneton |
| **WORM** | write once read many | | ℄ | centerline |
| **WQ** | water quench | | $\mu$in. | microinch |
| **WQT** | water quenched and tempered | | $\mu$m | micrometer (micron) |
| **WRC** | Welding Research Council | | $\mu$s | microsecond |
| **WS** | wave soldering | | ℗ | parting line |
| **wt%** | weight percent | | $\nu$ | Poisson's ratio |
| **WXRFS** | wavelength-dispersion x-ray fluorescence spectroscopy | | $\rho$ | density |
| | | | $\sigma$ | stress |
| **XES** | x-ray emission spectroscopy | | $\tau$ | shear stress |
| **XLPE** | cross-linked polyethylene | | $\Omega$ | ohm |
| **XPS** | x-ray photoelectron spectroscopy | | $\Phi$ | angle of refraction |
| **XRD** | x-ray powder diffraction | | $\sqrt{}$ | surface roughness |
| **XRFS** | x-ray fluorescence spectroscopy | | | |